Lecture Notes in Computer Science 7514

Michela Milano (Ed.)

Principles and Practice of Constraint Programming

18th International Conference, CP 2012
Québec City, QC, Canada, October 8-12, 2012
Proceedings

 Springer

Volume Editor

Michela Milano
DEIS Universitá di Bologna
Via le Risorgimento, 2
40136 Bologna, Italy
E-mail: michela.milano@unibo.it

ISSN 0302-9743
ISBN 978-3-642-33557-0
DOI 10.1007/978-3-642-33558-7

e-ISSN 1611-3349
e-ISBN 978-3-642-33558-7

Springer Heidelberg Dordrecht London New York

Library of Congress Control Number: 2012947317

CR Subject Classification (1998): F.4.1, G.1.6, F.2.2, F.3, G.2.2, D.3.2, F.1, E.1, I.2.8

LNCS Sublibrary: SL 2 – Programming and Software Engineering

Typesetting: Camera-ready by author, data conversion by Scientific Publishing Services, Chennai, India

Printed on acid-free paper

Springer is part of Springer Science+Business Media (www.springer.com)

Dedication to David L. Waltz

Dave Waltz will always be remembered in the constraint programming community for his seminal work on arc consistency, a concept that lies at the heart of constraint processing. Moreover, one of the remarkable things about Dave is that constraint programming is only one of the fields in which his research served as a catalyst.

Dave also performed great service to the computer science community, for example as President of the Association for the Advancement of Artificial Intelligence, which recently presented him with its Distinguished Service Award. In recent years, he returned to the constraint programming community to serve generously on the Advisory Board of the Cork Constraint Computation Centre.

The research that produced Dave's pioneering contribution to constraint programming is a sterling illustration of several principles basic to scientific research:

- Opportunity from Challenge
- Simplicity from Complexity
- Globality from Locality

Dave's PhD thesis extended work in early machine vision on three-dimensional interpretation of two-dimensional line drawings. This involved assigning labels, which indicated three-dimensional edge interpetations, like "convex" or "concave", to the lines around junctions in the line drawing, and then finding a consistent assignment across the entire scene. In the "blocks world" being studied, there was a simple, powerful "constraint" requiring lines between junctions to have the same label at both ends, e.g. an edge could not be convex at one end and concave at the other. In seeking to provide the ability to interpret more complex scenes, Dave added many more junction labelings. However, the increased number of options to sift through presented a challenge, especially in the context of the modest computing power available in those days.

Out of this challenge arose the prototypical arc consistency filtering algorithm that became known as the "Waltz Algorithm". The additional richness of understanding that his new junctions added actually aided this process, helping the filtering algorithm to eliminate possibilities and disambiguate scenes. Finally, the filtering algorithm achieved a global effect simply by repeated, albeit cleverly implemented, propagation of an "atomic", local process.

In Dave's own words, from his Ph.D. thesis:

> There is one lesson which I think is important, perhaps more important than any other in terms of the ways it might aid future research. For a long time after I had found the ways of describing region illuminations and edge decompositions, I tried to find a clever way to collapse the large set of line labels these distinctions implied into a smaller and more manageable set which would retain all the "essential" distinctions, whatever they were. Frustrated in this attempt for quite a while, I finally decided to go ahead and include every possible labeling in the program, even though this promised to involve a good deal of typing. I hoped that when I ran the program certain regularities would appear, i.e. that when the program found a particular labeling for a junction it would always find another as well, so that the two labelings could be collapsed into one new one with no loss of information. Of course, as it turned out, it was the fact that I had made such precise distinctions that allowed the program to find unique labelings. The moral of this is that one should not be afraid of semi-infinities; a large number of simple facts may be needed to represent what can be deduced by computation using a few general ideas.

The willingness to face "semi-infinities" fearlessly and turn them to advantage may be seen as a thread running through Dave's contributions to computer science and society, including work that foreshadowed today's world of internet search engines and "big data".

Dave embraced life with great enthusiasm. If he had attended the conference this year, he would have been keen to hear about the latest directions in which his seminal work has led. The proceedings are dedicated to this inspiring man.

July 2012   Eugene C. Freuder

Preface

This volume contains the proceedings of the Eighteenth International Conference on Principles and Practice of Constraint Programming (CP 2012) that was held in Quebec City, Canada, October 8–12, 2012.

The CP conference is the annual international conference on constraint programming. It is concerned with all aspects of computing with constraints, including theory, algorithms, environments, languages, models, systems, and applications such as decision making, resource allocation, and agreement technologies.

Beside the technical program, CP 2012 featured two special tracks. The former was the traditional *application track*, which focused on industrial and academic uses of constraint technology and its comparison and integration with other optimization techniques (MIP, local search, SAT, etc.). The second track, featured for the first time in 2012, concentrated on *multidisciplinary papers: cross-cutting methodology and challenging applications*, collecting papers that link CP technology with other techniques like machine learning, data mining, game theory, simulation, knowledge compilation, visualization, control theory, and robotics. In addition the track focused on challenging application fields with a high social impact such as CP for life sciences, sustainability, energy efficiency, web, social sciences, finance, and verification.

The interest of the research community in this conference was witnessed by the record number of submissions received this year. We received 186 (long and short) papers as follows: 139 papers submitted to the main track, 28 to the multi-disciplinary track, and 18 to the application track.

The reviewing process was headed for the first time by a two-level program committee, consisting of senior and program committee members. Senior PC members were responsible for managing a set of papers in their respective areas of expertise. The main reason for splitting the program committee was that senior PC members physically met in Bologna for a two-day meeting on June 2–3, 2012. Each paper was extensively discussed and additional last minute reviews were added when needed. At the end of the reviewing process we accepted 48 papers for the main CP track, 8 papers for the application track, and 12 for the multi-disciplinary track.

Amongst the accepted papers we selected a best paper: "A Generic Method for Identifying and Exploiting Dominance Relations" by Geoffrey Chu and Peter Stuckey. Also, we selected a best application paper: "Scheduling Scientific Experiments on the Rosetta/Philae Mission" by Gilles Simonin, Christian Artigues, Emmanuel Hebrard, and Pierre Lopez. Finally, we awarded two honorable mentions to Mohamed Siala, Emmanuel Hebrard, and Marie-José Huguet for the paper "An Optimal Arc Consistency Algorithm for a Chain of Atmost Constraints with Cardinality" and to Hannes Uppman for the paper "Max-Sur-CSP on Two Elements".

This year, authors of all accepted papers were invited to make both an oral presentation and a poster. This volume includes the papers accepted for the three tracks and also contains extended abstracts of three invited talks. The invited speakers who honored the conference with their presentations were Laurent Michel (Connecticut University), with a talk on "Constraint Programming and a Usability Quest", Miguel F. Anjos (GERAD and École Polytechnique de Montréal), who gave a talk on "Optimization Challenges in Smart Grid Operations", and Barry O'Sullivan (Cork Constraint Computation Centre), who gave a talk entitled "Where Are the Interesting Problems?".

A tradition since CP 2001, the conference included a Doctoral Program, which allowed PhD students to come to the conference, present a poster on their PhD topic, and meet senior researchers working on constraint programming. I am very grateful to Michele Lombardi and Stanislav Živný, who did a wonderful job in organizing the Doctoral Program.

CP 2012 also included three tutorials on Monte-Carlo tree search (Michel Sebag), on Optimization for Disaster Management (Pascal Van Hentenryck), and on Constraint Programming with Decision Diagrams (Willem-Jan van Hoeve). In addition the Doctoral Program tutorial was given by Warren Powell on Stochastic Optimization. Many thanks to all the tutorialists for their great job.

I would like to thank the whole program committee for the time spent in reviewing papers and in (sometimes very long) discussions. A special thanks goes to all the senior program committee members for their hard work in driving the discussions and writing meta-reviews and for coming to the physical meeting in Bologna. I appreciated the friendly and constructive atmosphere of the meeting and the positive feedback that I received after the meeting. I personally think the process could always be improved, but we all did our best to have a fair and in-depth reviewing process.

I would like to thank Mark Wallace who coped with my conflicts of interest regarding papers from my colleagues, Barry O'Sullivan and Helmut Simonis for their great job in preparing and organizing the special tracks, and Pascal Van Hentenryck who enthusiastically accepted to organize a panel and a call for position papers on the "Future of Constraint Programming".

The conference would not have been possible without the great job done by the Conference Chairs Gilles Pesant and Claude-Guy Quimper. They took care of the local organization and patiently tried to go along with my requests and wishes.

For conference publicity I warmly thank Ian Miguel and Thierry Moisan, who did a great job in advertising the conference and the workshops and in preparing and continuously updating the conference web site. I am very grateful to Meinolf Sellmann who acted as Workshop Chair and put together an exciting program with five half-day workshops.

Louis-Martin Rousseau was the sponsorship chair and he did a wonderful job in collecting funds from a number of sponsors. I would like to thank the sponsors who make it possible to organize this conference: the Journal of Artificial Intelligence, Cercle des Ambassadeurs de Québec, IBM, Institute of Computational

Sustainability, NICTA, SICS, Cork Constraint Computation Centre, and CIR-RELT. My gratitude goes also to the Easychair support team who assisted me in the use of the system for managing submissions.

I would like to thank my family. I spent evenings and nights working on the CP program (submissions, reviews, proceedings) and replying to thousands of emails. They constantly supported me and shared my enthusiasm.

Last but not least, I want to thank the ACP Executive Committee, who honoured me with the invitation to serve as Program Chair of CP 2012. This has been a great opportunity for me to live through an intense, exciting, and informative experience. I hope I met at least some of their expectations for CP 2012.

October 2012   Michela Milano

Conference Organization

Conference Co-chairs

Gilles Pesant — École Polytechnique de Montréal
Claude-Guy Quimper — Université Laval

Program Chair

Michela Milano — Università di Bologna

Application Track Chair

Helmut Simonis — Cork Constraint Computation Centre

Multi-disciplinary Track Chair

Barry O'Sullivan — Cork Constraint Computation Centre

Doctoral Program Co-chairs

Michele Lombardi — Università di Bologna
Stanislav Živný — University of Oxford

Workshop Chair

Meinolf Sellmann — IBM Thomas J. Watson Research Center

Publicity Co-chairs

Ian Miguel — University of St Andrews
Thierry Moisan — Université Laval

Sponsorship Chair

Louis-Martin Rousseau — École Polytechnique de Montréal

Senior Program Committee

Christian Bessiere	CNRS and University of Montpellier
David Cohen	Royal Holloway, University of London
Pierre Flener	Uppsala University
Alan Frisch	University of York
Enrico Giunchiglia	Università di Genova
Carla Gomes	Cornell University
John Hooker	Carnegie Mellon
Amnon Meisels	Ben Gurion University of the Negev
Laurent Michel	University of Connecticut
Jean-Charles Régin	University of Nice-Sophia Antipolis
Thomas Schiex	INRA, Toulouse
Peter Stuckey	University of Melbourne
Pascal Van Hentenryck	NICTA and University of Melbourne
Mark Wallace	Monash University
Toby Walsh	NICTA and UNSW

CP Main Track Program Committee

Fahiem Bacchus	University of Toronto
Roman Barták	Charles University
Chris Beck	University of Toronto
Ramon Bejar	University of Lleida
Nicolas Beldiceanu	EMN
Stefano Bistarelli	Università di Perugia
Mats Carlsson	SICS
Hubie Chen	Universitat Pompeu Fabra
Martin Cooper	IRIT - University of Toulouse
Ivan Dotu	Boston College
Thibaut Feydy	NICTA and The University of Melbourne
Marco Gavanelli	Università di Ferrara
Carmen Gervet	German University in Cairo (GUC)
Emmanuel Hebrard	LAAS CNRS
Holger Hoos	University of British Columbia
Christopher Jefferson	University of St Andrews
Narendra Jussien	École des Mines de Nantes
George Katsirelos	INRA
Christophe Lecoutre	CRIL
Michele Lombardi	Università di Bologna
Pedro Meseguer	IIIA-CSIC
Ian Miguel	University of St Andrews
Nina Narodytska	UNSW and NICTA
Justin Pearson	Uppsala University
Laurent Perron	Google France

Gilles Pesant	École Polytechnique de Montréal
Karen Petrie	University of Dundee
Steve Prestwich	Cork Constraint Computation Centre
Claude-Guy Quimper	Université Laval
Louis-Martin Rousseau	École Polytechnique de Montréal
Michel Rueher	University of Nice - Sophia Antipolis
Ashish Sabharwal	IBM Research
Christian Schulte	KTH Royal Institute of Technology and SICS
Meinolf Sellmann	IBM Thomas J. Watson Research Center
Paul Shaw	IBM
Kostas Stergiou	University of Western Macedonia
Stefan Szeider	Vienna University of Technology
Guido Tack	Monash University
Willem Jan van Hoeve	Carnegie Mellon University
Brent Venable	Università di Padova
William Yeoh	Singapore Management University
Makoto Yokoo	Kyushu University
Tallys Yunes	University of Miami
Roie Zivan	Ben Gurion University of the Negev
Stanislav Živný	University of Oxford

Application Track Program Committee

Roman Barták	Charles University
Hadrien Cambazard	G-SCOP, University of Grenoble
Sophie Demassey	TASC, EMN, INRIA/LINA
Pierre Flener	Uppsala University
Angelo Oddi	ISTC-CNR, Italian National Research Council
Laurent Michel	University of Connecticut
Barry O'Sullivan	Cork Constraint Computation Centre
Laurent Perron	Google France
Michel Rueher	University of Nice - Sophia Antipolis
Paul Shaw	IBM
Michael Trick	Carnegie Mellon
Mark Wallace	Monash University
Armin Wolf	Fraunhofer FOKUS

Multi-disciplinary Track Program Committee

Rolf Backofen	Albert-Ludwigs-University Freiburg
Ken Brown	University College Cork
Simon De Givry	INRA - UBIA
Johan de Kleer	Palo Alto Research Center, Inc.
Agostino Dovier	Università di Udine
Alexander Felfernig	Graz University of Technology

Maria Fox	King's College London
Alan Holland	University College Cork
Christophe Jermann	LINA / Université de Nantes
Inês Lynce	IST/INESC-ID, Technical University of Lisbon
Pierre Marquis	CRIL-CNRS and Université d'Artois
Amnon Meisels	Ben Gurion University of the Negev
Dino Pedreschi	Università di Pisa
Claude-Guy Quimper	Université Laval
Martin Sachenbacher	Technische Universität München
Helmut Simonis	Cork Constraint Computation Centre
Stefan Szeider	Vienna University of Technology
Pascal Van Hentenryck	NICTA
Peter Van Roy	Université catholique de Louvain
Gérard Verfaillie	ONERA

Additional Reviewers

Akgun, Ozgur	Eveillard, Damien
Ansotegui, Carlos	Fernandez, Cesar
Argelich, Josep	Fichte, Johannes Klaus
Artigues, Christian	Formisano, Andrea
Audemard, Gilles	Frank, Barbara
Baioletti, Marco	Gao, Sicun
Bergman, David	Gent, Ian
Bonfietti, Alessio	Gogate, Vibhav
Bouveret, Sylvain	Goldsztejn, Alexandre
Brockbank, Simon	Gosti, Giorgio
Carman, Mark	Goultiaeva, Alexandra
Castañeda Lozano, Roberto	Grimes, Diarmuid
Cattafi, Massimiliano	Grinshpoun, Tal
Ceberio, Martine	Grossi, Valerio
Cerquides, Jesus	Grubshtein, Alon
Chu, Geoffrey	Gu, Hanyu
Cire, Andre Augusto	Gualandi, Stefano
Collavizza, Hélène	Gutierrez, Patricia
Cymer, Radosław	Harabor, Daniel
Davies, Jessica	Hashemi Doulabi, Seyed Hossein
de Givry, Simon	Heinz, Stefan
De Silva, Daswin	Hirayama, Katsutoshi
Dilkina, Bistra	Hjort Blindell, Gabriel
Downing, Nicholas	Hutter, Frank
Drescher, Conrad	Ishii, Daisuke
Dvorak, Wolfgang	Iizuka, Yasuki
Eickmeyer, Kord	Janota, Mikolas
Ermon, Stefano	Jeavons, Peter

Kell, Brian
Kelsey, Tom
Kilby, Philip
Kotthoff, Lars
Kroc, Lukas
Kullmann, Oliver
Lauria, Massimo
Le Bras, Ronan
Levitt, Vadim
Lhomme, Olivier
Li, Chu-Min
Liebman, Ariel
Lorca, Xavier
Magnin, Morgan
Malapert, Arnaud
Manquinho, Vasco
Marques-Silva, João
Martin, Bruno
Martins, Ruben
Mateu, Carles
Matsui, Toshihiro
Mehta, Deepak
Michel, Claude
Misra, Neeldhara
Moisan, Thierry
Morin, Michael
Morris, Robert
Mueller, Moritz
Nanni, Mirco
Neveu, Bertrand
Nguyen, Duc Thien
Nightingale, Peter
O'Mahony, Eoin
Okimoto, Tenda
Ordyniak, Sebastian
Papadopoulos, Alexandre
Paparrizou, Anastasia
Peano, Andrea
Peri, Or
Petit, Thierry
Petke, Justyna
Piazza, Carla
Piette, Cédric

Pini, Maria Silvia
Prud'homme, Charles
Pujol-Gonzalez, Marc
Quesada, Luis
Rasconi, Riccardo
Razgon, Igor
Refalo, Philippe
Robinson, Nathan
Rollon, Emma
Rossi, Francesca
Roussel, Olivier
Roy, Pierre
Ruggieri, Salvatore
Ruiz de Angulo, Vicente
Rümmele, Stefan
Sakti, Abdelilah
Salvagnin, Domenico
Samulowitz, Horst
Santini, Francesco
Schaus, Pierre
Schimpf, Joachim
Schwind, Nicolas
Scott, Joseph
Seidl, Martina
Simon, Laurent
Slivovsky, Friedrich
Soliman, Sylvain
Somogyi, Zoltan
Tesson, Pascal
Thiele, Sven
Thorstensen, Evgenij
Trilling, Laurent
Truszczyński, Mirosław
Vilím, Petr
Vinyals, Meritxell
Waldispuhl, Jerome
Will, Sebastian
Xu, Hu
Xu, Lin
Xue, Yexiang
Yap, Roland
Yoshida, Yuichi
Zhou, Yuan

Table of Contents

Application Track

Multi-disciplinary Track

Constraint Programming and a Usability Quest

Laurent D. Michel

University of Connecticut
Storrs, CT 06269-2155
`ldm@engr.uconn.edu`

In 2004, Jean-Francois Puget presented [2] an analysis of the "simplicity of Use" of Constraint Programming from which he articulated a series of challenges to make Constraint Programming systems accessible and easier to use. The core of the argument was a contrast between mathematical programming and constraint programming tools. Mathematical programming adopts a *model and run* paradigm, rely on a simple vocabulary to model problems (i.e., linear constraints), support standard formats for sharing models and benefit from extensive documentation on how to model [5]. Constraint programming features a *model and search* paradigm, rich modeling languages with combinatorial objects and has a distinctive flavor of programming. While it can be construed as CP's Achilles' heel, it is also its most potent strength and is supported by modeling aids [3,4]. The very existence of sophisticated parameter tuning solutions for SAT solvers and Math Programming solvers to determine ideal parameters (e.g., ParamILS [1]) certainly cast a major shadow on the potency of the *model and run* mantra that is evolving into *model and search for the right parameters*.

Accessibility to CP technology is a legit concern and the appeal of turnkey solutions cannot be underestimated. CP tools are extremely pliable and uniquely adapted to classes of problems where all else fails. Retaining CP's flexibility while delivering *model and run* solutions suitable for a large number of situations is the position adopted here. This talk explores developments and solutions to the apparent quandary. Specifically, it explores automatic search for Constraint-Based Local Search, Scheduling, and finite-domain systems, generic black-box search procedures, automatic parallelization and assisted hybridization.

References

1. Hutter, F., Hoos, H.H., Leyton-Brown, K., Stützle, T.: Paramils: an automatic algorithm configuration framework. J. Artif. Int. Res. 36(1), 267–306 (2009)
2. Puget, J.-F.: Constraint Programming Next Challenge: Simplicity of Use. In: Wallace, M. (ed.) CP 2004. LNCS, vol. 3258, pp. 5–8. Springer, Heidelberg (2004)
3. Van Hentenryck, P.: The OPL Optimization Programming Language. The MIT Press, Cambridge (1999)
4. Van Hentenryck, P., Michel, L.: Constraint-Based Local Search. The MIT Press, Cambridge (2005)
5. Wolsey, L.A., Nemhauser, G.L.: Integer and Combinatorial Optimization. Wiley-Interscience (1999)

M. Milano (Ed.): CP 2012, LNCS 7514, p. 1, 2012.

Optimization Challenges
in Smart Grid Operations

Miguel F. Anjos

Canada Research Chair in Discrete Nonlinear Optimization in Engineering,
GERAD & École Polytechnique de Montréal, Montreal, QC, Canada H3C 3A7
anjos@stanfordalumni.org
http://www.gerad.ca/files/Sites/Anjos

Abstract. A smart grid is the combination of a traditional power distribution system with two-way communication between suppliers and consumers. While this communication is expected to deliver energy savings, cost reductions and increased reliability and security, smart grids bring up challenges in the management of the power system, such as integrating renewable energy sources and incorporating demand-management approaches. We discuss how optimization-based techniques are being used to overcome some of these challenges and speculate on ways in which constraint programming could play a greater role in this area.

Keywords: optimization, smart grid, unit commitment, demand-reponse, load curtailment.

Electricity is a critical source of energy for society. With increasing demand for electricity and various constraints on new generation capacity, it is imperative to increase the overall efficiency of the power system. This need has motivated the concept of *smart grids*. A smart grid is the combination of a traditional power distribution system with two-way communication between suppliers and consumers. For example, a smart grid will in principle support two-way interaction with individual homes to achieve system-wide objectives [1]. Smart grids introduce important challenges in the management of the resulting system. These include integrating renewable energy sources such as wind and solar energy, managing bidirectional flows of power, and incorporating demand-response. We present two areas in which optimization-based techniques are being used to overcome some of these challenges. We also aim to encourage the constraint programming community to play a greater role in this area.

1 The Unit Commitment Problem

Given a set of power generators and a set of electricity demands, the unit commitment (UC) problem wants to minimize the cost to produce the required amount of electricity while ensuring that all generators operate within their physical limits [5]. The solution to the problem consists of an optimal production schedule

M. Milano (Ed.): CP 2012, LNCS 7514, pp. 2–3, 2012.

for each generator. Because of the operating constraints of the generator and the behaviour of the power network, the UC problem is a nonlinear mixed-integer optimization problem. Real-life instances of UC are typically large-scale and difficult to solve. Traditionally, the problem has been solved using Lagrangian relaxation. It can also be addressed using mixed-integer linear programming (MILP) and this approach is now commonly used in practice. We present some of our recent contributions to solving these MILP problems, including a tightened representation of the polytope of feasible power generation schedules [4] and the exploitation of symmetry in the presence of many generators with identical characteristics [3]. Important issues requiring further research will also be discussed.

2 Optimal Load Curtailment

Electricity demand changes continuously and these fluctuations are often considerable. With increasing demand for electricity and various constraints on new generation capacity, an important objective is to decrease peaks in order to increase the utilization of existing capacity and hence the overall efficiency of the system. Demand-response aims to induce changes in the consumption patterns of electricity consumers so as to reduce the peaks in power demand. Studies have shown that schemes such as dynamic pricing have limited success, because most consumers have limited price sensitivity [2]. Load curtailment refers to a customer's reduction of energy consumption upon request and in exchange for financial compensation. Load curtailment is one of the demand-response initiatives proposed in Europe and America during the last decade [6]. The problem of delivering load curtailments at minimum cost over a given time horizon leads to challenging optimization models with a strong combinatorial structure and a large search space. We will present some of our ongoing research in this area in collaboration with a major load-curtailment service provider.

References

1. Costanzo, G.T., Zhu, G., Anjos, M.F., Savard, G.: A system architecture for autonomous demand side load management in the smart grid. Cahier du GERAD G-2011-68, GERAD, Montreal, QC, Canada (2011)
2. Kim, J.-H., Shcherbakova, A.: Common failures of demand response. Energy 36(2), 873–880 (2011)
3. Ostrowski, J., Anjos, M.F., Vannelli, A.: Symmetry in scheduling problems. Cahier du GERAD G-2010-69, GERAD, Montreal, QC, Canada (2010)
4. Ostrowski, J., Anjos, M.F., Vannelli, A.: Tight mixed integer linear programming formulations for the unit commitment problem. IEEE Transactions on Power Systems 27(1), 39–46 (2012)
5. Padhy, N.P.: Unit commitment-a bibliographical survey. IEEE Transactions on Power Systems 19(2), 1196–1205 (2004)
6. Torriti, J., Hassan, M.G., Leach, M.: Demand response experience in Europe: policies, programmes and implementation. Energy 35(4), 1575–1583 (2010)

Where Are the Interesting Problems?[*]

Barry O'Sullivan

Cork Constraint Computation Centre
Department of Computer Science, University College Cork, Ireland
`b.osullivan@cs.ucc.ie`

Abstract. Constraint programming has become an important technology for solving hard combinatorial problems in a diverse range of application domains. It has its roots in artificial intelligence, mathematical programming, operations research, and programming languages. In this talk we will discuss a number of challenging application domains for constraint programming, and the technical challenges that these present to the research community.

1 Introduction

Constraint programming (CP) is a technology for solving combinatorial optimisation problems [3]. A major generic challenge that faces CP is *scalability* [2], largely because the problems to which it is usually applied are computationally intractable (NP-Hard). While CP has been successfully applied in domains such as scheduling, timetabling, planning, inventory management and configuration, many instances of these problems are extremely challenging for traditional CP methods due to their hardness.

However, an emerging dimension of scale relates to problem size, and the volume of data available that is relevant to solving a particular instance, e.g. extremely large domain sizes, or very large extensionally defined constraints of high arity. In 2009 information on the web was doubling every 18 months. It is now believed that this occurs in less than 12 months. This exponential growth in data, often referred to as the "big data" challenge, presents us with major opportunities. For example, McKinsey Global Institute estimates that European government administrations could benefit from over € 250 billion in operational efficiencies by properly exploiting "big data".

Another major challenge in real-world application domains is *uncertainty* [4,5]. For example, in scheduling, the duration of a task might be uncertain, while in inventory management there might be uncertainty related to customer demand. Surprising, even predicting the future does not imply that we can make better decisions. The interactions between the choices that face us are usually interlinked in complex ways. Being able to react appropriately to risk is more important than knowing about the risk or even modelling it. The traditional "get data – model – implement"–cycle is no longer sufficient in most domains. We often

---

[*] This work was supported by Science Foundation Ireland Grant 10/IN.1/3032.

M. Milano (Ed.): CP 2012, LNCS 7514, pp. 4–5, 2012.

need to deal with large amounts of rapidly changing data whereby adaptation becomes key. The study of managing complex sources of data upon which we must make complex, risky, economic or environmentally important, decisions provides a compelling context for constraint programming [1].

2 Challenge Domains

A number of important application domains are emerging due to the growth in large-scale and complex computing infrastructure, a desire to improve business efficiency, the need for improved quality in public services, and demands for sustainable environmental planning. In this talk we will discuss several examples including: data centre optimisation, e.g. energy management through workload consolidation; innovative enterprise and public service delivery, e.g. optimising access to diagnostic health services; optimised human mobility and smart cities, e.g. integrating mobility mining and constraint programming; and, natural resource management, sometimes referred to as computational sustainability.

3 Technical Challenges

In this talk a set of specific technical challenges are presented, motivated by the complexities of decision making in data-rich domains such as those highlighted above. There is the general challenge of integrating CP with other technical disciplines to provide a holistic solution to specific classes of problems, or to address the requirements of particular application domains. Therefore, CP must integrate with a variety of other technical domains in order to meet these challenges such as: machine learning; data mining; game theory; simulation; knowledge compilation; visualization; control theory; engineering; medicine and health; bioscience; and mathematics. Domain-specific integrations must also emerge in areas such as: life sciences, sustainability, energy efficiency, the web, social sciences, and finance. A sample of such opportunities are discussed in this talk.

References

1. O'Sullivan, B.: Opportunities and Challenges for Constraint Programming. In: Proceedings of AAAI (2012)
2. Perron, L., Shaw, P., Furnon, V.: Propagation Guided Large Neighborhood Search. In: Wallace, M. (ed.) CP 2004. LNCS, vol. 3258, pp. 468–481. Springer, Heidelberg (2004)
3. Rossi, F., van Beek, P., Walsh, T.: Handbook of Constraint Programming. Foundations of Artificial Intelligence. Elsevier, New York (2006)
4. Van Hentenryck, P., Bent, R.: Online Stochastic Combinatorial Optimization. MIT Press (2006)
5. Verfaillie, G., Jussien, N.: Constraint solving in uncertain and dynamic environments: A survey. Constraints 10(3), 253–281 (2005)

A Generic Method for Identifying and Exploiting Dominance Relations

Geoffrey Chu and Peter J. Stuckey

National ICT Australia, Victoria Laboratory,
Department of Computer Science and Software Engineering,
University of Melbourne, Australia
{gchu,pjs}@csse.unimelb.edu.au

Abstract. Many constraint problems exhibit *dominance relations* which can be exploited for dramatic reductions in search space. Dominance relations are a generalization of symmetry and conditional symmetry. However, unlike symmetry breaking which is relatively well studied, dominance breaking techniques are not very well understood and are not commonly applied. In this paper, we present formal definitions of dominance breaking, and a generic method for identifying and exploiting dominance relations via *dominance breaking constraints*. We also give a generic proof of the correctness and compatibility of symmetry breaking constraints, conditional symmetry breaking constraints and dominance breaking constraints.

1   Introduction

In a constraint satisfaction or optimization problem, dominance relations describe pairs of assignments where one is known to be at least as good as the other with respect to satisfiability or the objective function. When such dominance relations are known, we can often prune off many of the solutions without changing the satisfiability or the optimal value of the problem. Many constraint problems exhibit dominance relations which can be exploited for significant speedups (e.g., [16,25,3,18,24,22,6,12]).

Dominance relations are a generalization of symmetry and conditional symmetry and offer similar or greater potential for reductions in search space. Unlike symmetries however, dominance relations are not very widely exploited. Dominance relations can be hard to identify, and there are few standard methods for exploiting them. It it also often hard to prove that a particular method is correct, especially when multiple dominance relations are being exploited simultaneously. These issues have been overcome in the case of symmetry, which is why symmetry breaking is now standard and widely used. Dominance relations have been successfully applied in a number of problems, but their treatment is often very problem specific and yields little insight as to how they can be generalized. In this paper, we seek to advance the usage of dominance relations by making the following contributions:

- We describe a generic method for identifying and exploiting a large class of dominance relations using *dominance breaking constraints*.

M. Milano (Ed.): CP 2012, LNCS 7514, pp. 6–22, 2012.

- We show that our method naturally produces symmetry breaking and conditional symmetry breaking constraints as well (since they are simply special cases of dominance breaking).
- We give a generic theorem proving the correctness and compatibility of all symmetry breaking, conditional symmetry breaking and dominance breaking constraints generated by our method.

The layout of the paper is as follows. In Section 2, we give our definitions. In Section 3, we describe our method of identifying and exploiting dominance relations using dominance breaking constraints. In Section 4, we describe how our method can be extended to generate symmetry and conditional symmetry breaking constraints as well. In Section 5, we discuss related work. In Section 6, we provide experimental results. In Section 7, we discuss future work. In Section 8, we conclude.

2 Definitions

To facilitate rigorous proofs in the later sections, we will give our own definitions of variables, domains, constraints, constraint problems and dominance relations. These are slightly different from the standard definitions but are equivalent to them in practice.

Let $\equiv$ denote syntactical identity, $\Rightarrow$ denote logical implication and $\Leftrightarrow$ denote logical equivalence. We define variables and constraints in a problem independent way. A variable v is a mathematical quantity capable of assuming any value from a set of values called the *default domain* of v. Each variable is typed, e.g., Boolean or Integer, and its type determines its default domain, e.g., $\{0, 1\}$ for Boolean variables and $\mathbb{Z}$ for Integer variables. Given a set of variables V, let Θ_V denote the set of valuations over V where each variable in V is assigned to a value in its default domain. A constraint c over a set of variables V is defined by a set of valuations $solns(c) \subseteq \Theta_V$. Given a valuation θ over $V' \supseteq V$, we say θ satisfies c if the restriction of θ onto V is in $solns(c)$. Otherwise, we say that θ violates c. A domain D over variables V is a set of *unary constraints*, one for each variable in V. In an abuse of notation, if a symbol A refers to a set of constraints $\{c_1, \ldots, c_n\}$, we will often also use the symbol A to refer to the constraint $c_1 \wedge \ldots \wedge c_n$. This allows us to avoid repetitive use of conjunction symbols.

A *Constraint Satisfaction Problem* (CSP) is a tuple $P \equiv (V, D, C)$, where V is a set of variables, D is a domain over V, and C is a set of n-ary constraints. A valuation θ over V is a *solution* of P if it satisfies every constraint in D and C. The aim of a CSP is to find a solution or to prove that none exist. In a *Constraint Optimization Problem* (COP) $P \equiv (V, D, C, f)$, we also have an objective function f mapping Θ_V to an ordered set, e.g., $\mathbb{Z}$ or $\mathbb{R}$, and we wish to minimize or maximize f over the solutions of P. In this paper, we deal with finite domain problems only, i.e., where the initial domain D constrains each variable to take values from a finite set of values.

We define dominance relations over full valuations. We assume that all objective functions are to be minimized, and consider constraint satisfaction problems as constraint optimization problems with $f(\theta) = 0$ for all valuations θ.

Definition 1. *A* dominance relation $\prec$ *for COP* $P \equiv (V, D, C, f)$ *is a* transitive *and* irreflexive *binary relation on* Θ_V *such that if* $\theta_1 \prec \theta_2$, *then either: 1)* θ_1 *is a solution and* θ_2 *is a non-solution, or 2) they are both solutions or both non-solutions and* $f(\theta_1) \leq f(\theta_2)$.

If $\theta_1 \prec \theta_2$, we say that θ_1 *dominates* θ_2. Note that we require our dominance relations to be irreflexive. This means that no loops can exist in the dominance relation, and makes it much easier to ensure the correctness of the method. The following theorem states that it is correct to prune all dominated assignments.

Theorem 1. *Given a finite domain COP* $P \equiv (V, D, C, f)$, *and a dominance relation* $\prec$ *for* P, *we can prune all assignments* θ *such that* $\exists \theta'$ *s.t.* $\theta' \prec \theta$, *without changing the satisfiability or optimal value of* P.

Proof. Let θ_0 be an optimal solution. If θ_0 is pruned, then there exists some solution θ_1 s.t. $\theta_1 \prec \theta_0$. Then θ_1 must be a solution with $f(\theta_1) \leq f(\theta_0)$, so θ_1 is also an optimal solution. In general, if θ_i is pruned, then there must exist some θ_{i+1} s.t. $\theta_{i+1} \prec \theta_i$ and θ_{i+1} is also an optimal solution. Since $\prec$ is transitive and irreflexive, it is impossible for the sequence $\theta_0, \theta_1, \ldots$ to repeat. Then since there are finitely many solutions, the sequence must terminate in some θ_k which is an optimal solution and which is not pruned. $\qquad\square$

We can extend $\prec$ to relate search nodes in the obvious way.

Definition 2. *Let* D_1 *and* D_2 *be the domains from two different search nodes. If* $\forall \theta_2 \in solns(D_2), \exists \theta_1 \in solns(D_1)$ *s.t.* $\theta_1 \prec \theta_2$, *then we define* $D_1 \prec D_2$.

Clearly if $D_1 \prec D_2$, Theorem 1 tells us that we can safely prune the search node with D_2. We call the pruning allowed by Theorem 1 *dominance breaking* in keeping with *symmetry breaking* for symmetries.

Dominance relations can be derived either statically before search or dynamically during search in order to prune the search space. It is easy to see that static symmetry breaking (e.g., [7,10]) are a special case of static dominance breaking. For example, consider the lex-leader method of symmetry breaking. Suppose S is a symmetry group of problem P. Suppose $lex(\theta)$ is the lexicographical function being used in the lex-leader method. We can define a dominance relation: $\forall \sigma \in S, \forall \theta, \sigma(\theta) \prec \theta$ if $lex(\sigma(\theta)) < lex(\theta)$. Then applying Theorem 1 to $\prec$ gives the lex-leader symmetry breaking constraint (i.e., prune all solutions which are not the lex-leader in their equivalence class). Similarly, dynamic symmetry breaking techniques such as Symmetry Breaking During Search [15] and Symmetry Breaking by Dominance Detection [8,11] are special cases of dynamic dominance breaking. Nogood learning techniques such as Lazy Clause Generation [23,9] and Automatic Caching via Constraint Projection [5] are also examples of dynamic dominance breaking. We will discuss these two methods in more detail in Section 5.

Just as in the case of symmetry breaking, it is generally *incorrect* to simultaneously post dominance breaking constraints for multiple dominance relations. This is because dominance relations only ensure that one assignment is at least as good as the other (not strictly better than), thus when we have multiple dominance relations, we could have loops such as $\theta_1 \prec_1 \theta_2$ and $\theta_2 \prec_2 \theta_1$, and posting the dominance breaking constraint for both $\prec_1$ and $\prec_2$ would be wrong. We have to take care when breaking symmetries, conditional symmetries and dominances that all the pruning we perform are compatible with each other. As we shall show below, one of the advantages of our method is that all the symmetry breaking, conditional symmetry breaking and dominance breaking constraints generated by our method are provably compatible.

Dominance breaking constraints can be particularly useful in optimization problems, because they provide a completely different and complementary kind of pruning to the branch and bound paradigm. In the branch and bound paradigm, the only way to show that a partial assignment is suboptimal is to prove a sufficiently strong bound on its objective value. Proving such bounds can be very expensive, especially if the model does not propagate strong bounds on the objective. In the worst case, further search is required, which can take an exponential amount of time. On the other hand, dominance breaking can prune a partial assignment without having to prove any bounds on its objective value at all, since it only needs to know that the partial assignment is suboptimal. Once dominance relations expressing conditions for suboptimality are found and proved, the only cost in the search is to check whether a partial assignment is dominated, which can often be much lower than the cost required to prove a sufficiently strong bound to prune the partial assignment.

3   Identifying and Exploiting Dominance Relations

3.1   Overview of Method

We now describe a generic method for identifying and exploiting a fairly large class of dominance relations using dominance breaking constraints. The idea is to use mappings σ from valuations to valuations to construct dominance relations. Given a mapping σ, we ask: under what conditions does σ map a solution to a better solution? If we can find these conditions, then we can build a dominance relation using these conditions and exploit it by posting a dominance breaking constraint. More formally:

Step 1 Find mappings $\sigma : \Theta_V \to \Theta_V$ which are likely to map solutions to better solutions.

Step 2 For each σ, find a constraint $scond(\sigma)$ s.t. if $\theta \in solns(C \wedge D \wedge scond(\sigma))$, then $\sigma(\theta) \in solns(C \wedge D)$.

Step 3 For each σ, find a constraint $ocond(\sigma)$ s.t. if $\theta \in solns(C \wedge D \wedge ocond(\sigma))$, then $f(\sigma(\theta)) < f(\theta)$.

Step 4 For each σ, post the dominance breaking constraint $db(\sigma) \equiv \neg(scond(\sigma) \wedge ocond(\sigma))$.

The following theorem proves the correctness of this method.

Theorem 2. *Given a finite domain COP $P \equiv (V, D, C, f)$, a set of mappings S, and for each mapping $\sigma \in S$ constraints $scond(\sigma)$ and $ocond(\sigma)$ satisfying: $\forall \sigma \in S$, if $\theta \in solns(C \wedge D \wedge scond(\sigma))$, then $\sigma(\theta) \in solns(C \wedge D)$, and: $\forall \sigma \in S$, if $\theta \in solns(C \wedge D \wedge ocond(\sigma))$, then $f(\sigma(\theta)) < f(\theta)$, we can add all of the dominance breaking constraints $db(\sigma) \equiv \neg(scond(\sigma) \wedge ocond(\sigma))$ to P without changing its satisfiability or optimal value.*

Proof. Construct a binary relation $\prec$ as follows. For each σ, for each $\theta \in solns(C \wedge D \wedge scond(\sigma) \wedge ocond(\sigma))$, define $\sigma(\theta) \prec \theta$. Now, take the transitive closure of $\prec$. We claim that $\prec$ is a dominance relation. It is transitive by construction. Also, by construction, $\theta \in solns(C \wedge D \wedge scond(\sigma) \wedge ocond(\sigma))$ guarantees that $\sigma(\theta)$ is a solution and that $f(\sigma(\theta)) < f(\theta)$. Thus $\forall \theta_1, \theta_2, \theta_1 \prec \theta_2$ implies that θ_1 and θ_2 are solutions, and that $f(\theta_1) < f(\theta_2)$. This means that $\prec$ is irreflexive and satisfies all the properties of a dominance relation, thus by Theorem 1, we can prune any $\theta \in solns(C \wedge D \wedge scond(\sigma) \wedge ocond(\sigma))$ for any σ without changing the satisfiability or optimality of P. Thus it is correct to add $db(\sigma)$ for any σ to P. $\qquad\square$

Note that there are no restrictions on σ. It does not have to be injective or surjective. The $db(\sigma)$ are guaranteed to be compatible because they all obey the same strict ordering imposed by the objective function f, i.e., they prune a solution only if a solution with strictly better f value exists. We illustrate the method with two simple examples before we go into more details.

Example 1. Consider the Photo problem. A group of people wants to take a group photo where they stand in one line. Each person has preferences regarding who they want to stand next to. We want to find the arrangement which satisfies the most preferences.

We can model this as follows. Let $x_i \in \{1, \ldots, n\}$ for $i = 1, \ldots, n$ be variables where x_i represent the person in the ith place. Let p be a 2d integer array where $p[i][j] = p[j][i] = 2$ if person i and j both want to stand next to each other, $p[i][j] = p[j][i] = 1$ if only one of them wants to stand next to the other, and $p[i][j] = p[j][i] = 0$ if neither want to stand next to each other. The only constraint is: $alldiff(x_1, \ldots, x_n)$. The objective function to be minimized is given by: $f = -\sum_{i=1}^{n-1} p[x_i][x_{i+1}]$.

Step 1. Since this is a sequence type problem, mappings which permute the sequence in some way are likely to map solutions to solutions. For simplicity, consider the set of mappings which flip a subsequence of the sequence, i.e., $\forall i < j, \sigma_{i,j}$ maps x_i to x_j, x_{i+1} to x_{j-1}, $\ldots$, x_j to x_i.

Step 2. We want to find the conditions under which σ maps solutions to solutions. Since all of these σ are symmetries of $C \wedge D$, we do not need any conditions and it is sufficient to set $scond(\sigma_{i,j}) \equiv true$.

Step 3. We want to find the conditions under which $f(\sigma_{i,j}(\theta)) < f(\theta)$. If we compare the LHS and RHS, it is clear that the only difference is the terms $p[x_{i-1}][x_j]$, $p[x_i][x_{j+1}]$ on the LHS and the terms $p[x_{i-1}][x_i]$, $p[x_j][x_{j+1}]$ on the RHS. So it is sufficient to set $ocond(\sigma_{i,j}) \equiv p[x_{i-1}][x_j] + p[x_i][x_{j+1}] > p[x_{i-1}][x_i] + p[x_j][x_{j+1}]$.

Step 4. For each $\sigma_{i,j}$, we can post the dominance breaking constraint: $\neg(p[x_{i-1}][x_j] + p[x_i][x_{j+1}] > p[x_{i-1}][x_i] + p[x_j][x_{j+1}])$.
These dominance breaking constraints ensure that if some subsequence of the assignment can be flipped to improve the objective, then the assignment is pruned. □

Example 2. Consider the 0-1 knapsack problem where x_i are 0-1 variables, we have constraint $\sum w_i x_i \leq W$ and we have objective $f = -\sum v_i x_i$, where w_i and v_i are constants.

Step 1. Consider mappings which swap the values of two variables, i.e., $\forall i < j, \sigma_{i,j}$ swaps x_i and x_j.

Step 2. A sufficient condition for $\sigma_{i,j}$ to map the current solution to another solution is: $scond(\sigma_{i,j}) \equiv w_i x_j + w_j x_i \leq w_i x_i + w_j x_j$. Rearranging, we get: $(w_i - w_j)(x_i - x_j) \geq 0$.

Step 3. A sufficient condition for $\sigma_{i,j}$ to map the current solution to an assignment with a better objective function is: $ocond(\sigma_{i,j}) \equiv v_i x_j + v_j x_i > v_i x_i + v_j x_j$. Rearranging, we get: $(v_i - v_j)(x_i - x_j) < 0$.

Step 4. For each $\sigma_{i,j}$, we can post the dominance breaking constraint: $db(\sigma_{i,j}) \equiv \neg(scond(\sigma_{i,j}) \wedge ocond(\sigma_{i,j}))$. After simplifying, we have $db(\sigma_{i,j}) \equiv x_i \leq x_j$ if $w_i \geq w_j$ and $v_i < v_j$, $db(\sigma_{i,j}) \equiv x_i \geq x_j$ if $w_i \leq w_j$ and $v_i > v_j$, and $db(\sigma_{i,j}) \equiv$ *true* for all other cases.

These dominance breaking constraints ensure that if one item has worse value and greater or equal weight to another, then it cannot be chosen without choosing the other also. □

3.2   Step 1: Finding Appropriate Mappings σ

In general, we want to find σ's such that $scond(\sigma)$ and $ocond(\sigma)$ are as small and simple as possible, as this will lead to dominance breaking constraints that are easier to propagate and prune more. So we want σ such that it often maps a solution to a better solution. σ's which are symmetries or almost symmetries of the problem make good candidates, since their $scond(\sigma)$ will be simple, and all else being equal, there is around a 50% chance that it will map the solution to one with a better objective value. In general, we can try all the common candidates for symmetries such as swapping two variables, swapping two values, swapping two rows/columns in matrix type problems, flipping/moving a subsequence in a sequence type problem, etc. Mappings which are likely to map an assignment to one with better objective value are also good candidates, since their $ocond(\sigma)$ will be simple. For example, in scheduling problems minimizing makespan, we can try shifting items forwards in the schedule. There may also be problem specific σ's that we can try.

3.3 Step 2: Finding $scond(\sigma)$

We can calculate $scond(\sigma)$ straightforwardly with the help of the following definition.

Definition 3. *Given a mapping $\sigma : \Theta_V \to \Theta_V$, we can extend σ to map constraints to constraints as follows. Given a constraint c, $\sigma(c)$ is defined as a constraint over V such that θ satisfies $\sigma(c)$ iff $\sigma(\theta)$ satisfies c.*

For example, if $c \equiv x_1 + 2x_2 + 3x_3 \geq 10$, and σ swaps x_1 and x_3, then $\sigma(c) \equiv x_3 + 2x_2 + 3x_1 \geq 10$. Or if $c \equiv (x_1, x_2) \in \{(1,1), (2,3), (3,1)\}$, and σ permutes the values (1, 2, 3) to (2, 3, 1) on x_1 and x_2, then $\sigma(c) \equiv (x_1, x_2) \in \{(3,3), (1,2), (2,3)\}$.

It is easy to define $\sigma(c)$, however, $\sigma(c)$ may or may not be a simple logical expression. For example, if $c \equiv x_1 + 2x_2 \geq 5$ and σ swaps the values 1 and 2, then $\sigma(c) \equiv (x_1 = 1 \wedge x_2 = 1) \vee (x_1 = 2 \wedge x_2 = 1) \vee (x_1 \neq 1 \wedge x_1 \neq 2 \wedge x_2 \neq 1 \wedge x_2 \neq 2 \wedge x_1 + 2x_2 \geq 5)$ which does not simplify at all.

For each σ, we want to find a sufficient condition $scond(\sigma)$ so that if a solution satisfied $scond(\sigma)$, then σ maps it to another solution. A necessary and sufficient condition is: $C \wedge D \wedge scond(\sigma) \Rightarrow \sigma(C \wedge D)$, i.e., C and D together with $scond(\sigma)$ must imply the mapped versions of every constraint in C and D.

We can construct $scond(\sigma)$ as follows. We calculate $\sigma(c)$ for each $c \in C \cup D$. If it is not implied by $C \wedge D$, then we add a constraint c' to $scond(\sigma)$ such that $C \wedge D \wedge c' \to \sigma(c)$. For example, in the knapsack problem in Example 2, $\sigma_{i,j}(C) \equiv w_1 x_1 + \ldots + w_i x_j + \ldots + w_j x_i + \ldots w_n x_n \leq W$. It is easy to see that $\sum w_i x_i \leq W \wedge w_i x_j + w_j x_i \leq w_i x_i + w_j x_j \Rightarrow \sigma_{i,j}(C)$, hence we could set $scond(\sigma_{i,j}) \equiv w_i x_j + w_j x_i \leq w_i x_i + w_j x_j$.

If σ is a symmetry of $C \wedge D$ then $scond(\sigma) \equiv true$. If σ is almost a symmetry of $C \wedge D$, then $scond(\sigma)$ is usually fairly small and simple, because most of the $\sigma(c)$ are already implied by $C \wedge D$.

3.4 Step 3: Finding $ocond(\sigma)$

We assume that the objective function $f(\theta)$ is defined over all assignments (not just solutions). We first give a few definitions.

Definition 4. *Given a function σ mapping assignments to assignments, we extend σ to map functions to functions as follows: $\forall \theta$, $\sigma(f)(\theta) = f(\sigma(\theta))$.*

Definition 5. *Given two functions mapping assignments to the reals f and g, we use $f < g$ to denote a constraint such that: θ satisfies $f < g$ iff $f(\theta) < g(\theta)$.*

For each σ, we want to find a sufficient condition $ocond(\sigma)$ so that if a solution satisfied $ocond(\sigma)$, then σ maps it to an assignment with a strictly better objective value. A necessary and sufficient condition is: $C \wedge ocond(\sigma) \Rightarrow \sigma(f) < f$. We can typically just set $ocond(\sigma) \equiv \sigma(f) < f$. For example, in both the Photo and Knapsack examples above, we simply calculated $\sigma(f) < f$, eliminated equal terms from each side, and used that as $ocond(\sigma)$.

3.5   Step 4: Posting the Dominance Breaking Constraint

Once we have found $scond(\sigma)$ and $ocond(\sigma)$, we can construct the dominance breaking constraint $db(\sigma) \equiv \neg(scond(\sigma) \wedge ocond(\sigma))$ and simplify it as much as possible. If it is simple enough to implement efficiently, we can add it to the problem. If not, we can simply ignore it, as it is not required for the correctness of the method. It is quite common that the dominance breaking constraint for different σ's will have common subexpressions. We can take advantage of this to make the implementation of the dominance breaking constraints more efficient.

4   Generating Symmetry and Conditional Symmetry Breaking Constraints

The method described so far only finds dominance breaking constraints which prune a solution when its objective value is *strictly* worse than another. We can do better than this, as there are often pairs of solutions which have equally good objective value and we may be able to prune many of them. Exploiting such sets of equally good pairs of solution is called symmetry breaking and conditional symmetry breaking. We show that with a slight alteration, our method will generate dominance breaking constraints that will also break symmetries and conditional symmetries.

We modify the method as follows. We add in a Step 0, and alter Step 3 slightly.

Step 0   Choose a refinement of the objective function f' with the property that $\forall \theta_1, \theta_2, f(\theta_1) < f(\theta_2)$ implies $f'(\theta_1) < f'(\theta_2)$.

Step 3*   For each σ, find a constraint $ocond(\sigma)$ s.t. if $\theta \in solns(C \wedge D \wedge ocond(\sigma))$, then $f'(\sigma(\theta)) < f'(\theta)$.

We have the following theorem concerning the correctness of the altered method.

Theorem 3. *Given a finite domain COP $P \equiv (V, D, C, f)$, a refinement of the objective function f' satisfying $\forall \theta_1, \theta_2, f(\theta_1) < f(\theta_2)$ implies $f'(\theta_1) < f'(\theta_2)$, a set of mappings S, and for each mapping $\sigma \in S$ constraints $scond(\sigma)$ and $ocond(\sigma)$ satisfying: $\forall \sigma \in S$, if $\theta \in solns(C \wedge D \wedge scond(\sigma))$, then $\sigma(\theta) \in solns(C \wedge D)$, and: $\forall \sigma \in S$, if $\theta \in solns(C \wedge D \wedge ocond(\sigma))$, then $f'(\sigma(\theta)) < f'(\theta)$, we can add all of the dominance breaking constraints $db(\sigma) \equiv \neg(scond(\sigma) \wedge ocond(\sigma))$ to P without changing its satisfiability or optimal value.*

Proof. The proof is analogous to that of Theorem 2.

The $db(\sigma)$ are guaranteed to be compatible because they all obey the same strict ordering imposed by the refined objective function f', i.e., they prune a solution only if a solution with strictly better f' value exists. Theorem 3 is a very useful result as it is generally quite difficult to tell whether different symmetry, conditional symmetry or dominance breaking constraints are compatible. There are lots of examples in literature where individual dominance breaking constraints are proved correct, but no rigorous proof is given that they are correct when used together (e.g., [12,6,14]). The symmetry, conditional symmetry or dominance breaking constraints generated by our method are guaranteed to be compatible by Theorem 3, thus the user of the method does not need to prove anything themselves. We now show with some examples how the altered method can generate symmetry and conditional symmetry breaking constraints.

Example 3. Consider the Photo problem from Example 1. Suppose that in Step 0, instead of setting $f' = f$, we set $f' = lex(f, x_1, \ldots, x_n)$, the lexicographic least vector $(f, x_1, \ldots, x_n)$. That is, we order the solutions by their objective value, and then tie break by the value of x_1, then by x_2, etc. Clearly, $f(\theta_1) < f(\theta_2)$ implies $f'(\theta_1) < f'(\theta_2)$ so it is a refinement. Now, consider what happens in Step 3. In general, we have $\sigma(lex(f_1, \ldots, f_n)) < lex(f_1, \ldots, f_n) \Leftrightarrow lex(\sigma(f_1), \ldots, \sigma(f_n)) < lex(f_1, \ldots, f_n) \Leftrightarrow (\sigma(f_1) < f_1) \vee (\sigma(f_1) = f_1 \wedge \sigma(f_2) < f_2) \vee \ldots \vee (\sigma(f_1) = f_1 \wedge \ldots \wedge \sigma(f_{n-1}) = f_{n-1} \wedge \sigma(f_n) < f_n)$.

In this problem, we have: $\forall i < j, ocond(\sigma_{i,j}) \equiv \sigma(f') < f' \equiv (p[x_{i-1}][x_j] + p[x_i][x_{j+1}] > p[x_{i-1}][x_i] + p[x_j][x_{j+1}]) \vee (p[x_{i-1}][x_j] + p[x_i][x_{j+1}] = p[x_{i-1}][x_i] + p[x_j][x_{j+1}] \wedge x[j] < x[i])$. There is an additional term in $ocond(\sigma_{i,j})$ which says that we can also prune the current assignment if the flipped version has *equal* objective value but a better lexicographical value for $\{x_1, \ldots, x_n\}$. Thus $db(\sigma_{i,j})$ not only breaks dominances but also includes a conditional symmetry breaking constraint. Similarly, consider $\sigma_{1,n}$. Because it is a boundary case, the terms in $\sigma(f)$ and f all cancel and we have $ocond(\sigma_{1,n}) \equiv x[n] < x[1]$, so $db(\sigma_{1,n}) \equiv x[1] \leq x[n]$ which is simply a symmetry breaking constraint. $\square$

Example 4. Consider the Knapsack problem from Example 2. In Step 0, we can tie break solutions with equal objective value by the weight used, and then lexicographically, i.e., $f' = lex(f, \sum w_i x_i, x_1, \ldots, x_n)$. In Step 3, we have: $\forall i < j, ocond(\sigma_{i,j}) \equiv \sigma(f') < f' \equiv ((v_i - v_j)(x_i - x_j) < 0) \vee ((v_i - v_j)(x_i - x_j) = 0 \wedge (w_i - w_j)(x_i - x_j) > 0) \vee ((v_i - v_j)(x_i - x_j) = 0 \wedge (w_i - w_j)(x_i - x_j) = 0 \wedge x_j < x_i)$. In Step 4, after simplifying, in addition to the dominance breaking constraints we had before, we would also have: $db(\sigma_{i,j}) \equiv x_i \leq x_j$ if $w_i > w_j$ and $v_i = v_j$, $db(\sigma_{i,j}) \equiv x_i \geq x_j$ if $w_i < w_j$ and $v_i = v_j$, and $db(\sigma_{i,j}) \equiv x_i \leq x_j$ if $w_i = w_j$ and $v_i = v_j$ which is a symmetry breaking constraint. $\square$

We can also apply the altered method to satisfaction problems to generate symmetry and conditional symmetry breaking constraints.

Example 5. The Black Hole Problem [14] seeks to find a solution to the Black Hole patience game. In this game the 52 cards of a standard deck are laid out in 17 piles of 3, with the Ace of spades starting in a "black hole". Each turn, a card at the top of one of the piles can be played into the black hole if it is ± 1 from the card that was played previously, with king wrapping back around to ace. The aim is to play all 52 cards. We can model the problem as follows. Let the suits be numbered from 1 to 4 in the order spades, hearts, clubs, diamonds. Let the cards be numbered from 1 to 52 so that card i has suit $(i-1)/13+1$ and number $(i-1)\%13+1$, where 11 is jack, 12 is queen and 13 is king. Let $l_{i,j}$ be the jth card in the ith pile in the initial layout. Let x_i be the turn in which card i was played. Let y_i be the card which was played in turn i. We have:

$$x_1 = 1 \tag{1}$$

$$inverse(x, y) \tag{2}$$

$$x_{l_{i,j}} < x_{l_{i,j+1}} \qquad \forall 1 \le i \le 17, 1 \le j \le 2 \tag{3}$$

$$(y_{i+1} - y_i)\%13 \in \{-1, 1\} \qquad \forall 1 \le i \le 51 \tag{4}$$

We now apply our method. Since cards which are nearer to the top of the piles are much more likely to be played early on, we choose a lexicographical ordering which reflects this. We define $f' = lex(x_{l_{1,1}}, \ldots, x_{l_{17,1}}, \ldots, x_{l_{1,3}}, \ldots, x_{l_{17,3}})$. An obvious set of mappings that are likely to map solutions to solutions is to swap cards of the same number in the sequence of cards to be played. Consider $\sigma_{i,j}$ for $i - j\%13 = 0, i \ne 1, j \ne 1$ where $\sigma_{i,j}$ swaps x_i and x_j, and swaps the values of i and j among $\{y_1, \ldots, y_{52}\}$.

Now we construct $scond(\sigma_{i,j})$. For each constraint c in the problem, we need to find a c' such that $C \wedge D \wedge c' \Rightarrow \sigma_{i,j}(c)$ and add it to $scond(\sigma_{i,j})$. Clearly, the domain constraints and the constraints in (1), (2) and (4) are all symmetric in $\sigma_{i,j}$, so we do not need to add anything for them. However, there will be some constraints in (3) which are not symmetric in $\sigma_{i,j}$. For example, suppose we wished to swap $3\spadesuit$ and $3\heartsuit$, and they were in piles: $(2\spadesuit, 3\spadesuit, 5\clubsuit)$ and $(1\diamondsuit, 3\heartsuit, 6\diamondsuit)$, where $3\spadesuit$ is in lexicographically earlier pile than $3\heartsuit$. The constraints in 3 which are not symmetric in $\sigma_{i,j}$ are those involving $3\spadesuit$ or $3\heartsuit$, i.e., $2\spadesuit < 3\spadesuit, 3\spadesuit < 5\clubsuit, 1\diamondsuit < 3\heartsuit$ and $3\heartsuit < 6\diamondsuit$. Their symmetric versions are $2\spadesuit < 3\heartsuit, 3\heartsuit < 5\clubsuit, 1\diamondsuit < 3\spadesuit$ and $3\spadesuit < 6\diamondsuit$ respectively, so we can set $scond(\sigma_{2,15}) \equiv 2\spadesuit < 3\heartsuit \wedge 3\heartsuit < 5\clubsuit \wedge 1\diamondsuit < 3\spadesuit \wedge 3\spadesuit < 6\diamondsuit$. To construct $ocond(\sigma_{i,j})$, we can set $ocond(\sigma_{i,j}) \equiv \sigma_{i,j}(f') < f'$. For this example, we have $ocond(\sigma_{2,15}) \equiv 3\heartsuit < 3\spadesuit$. Combining, we have $db(\sigma_{2,15}) \equiv \neg(2\spadesuit < 3\heartsuit \wedge 3\heartsuit < 5\clubsuit \wedge 1\diamondsuit < 3\spadesuit \wedge 3\spadesuit < 6\diamondsuit \wedge 3\heartsuit < 3\spadesuit)$. We can use the constraints in the original problem to simplify this further. Since $3\spadesuit < 5\clubsuit$ is an original constraint and $3\heartsuit < 3\spadesuit \wedge 3\spadesuit < 5\clubsuit \Rightarrow 3\heartsuit < 5\clubsuit$, we can eliminate the second term in $db(\sigma_{2,15})$. Since $1\diamondsuit < 3\heartsuit$ is an original constraint and $1\diamondsuit < 3\heartsuit \wedge 3\heartsuit < 3\spadesuit \Rightarrow 1\diamondsuit < 3\spadesuit$, we can eliminate the third term in $db(\sigma_{2,15})$. The result is $db(\sigma_{2,15}) \equiv \neg(2\spadesuit < 3\heartsuit \wedge 3\spadesuit < 6\diamondsuit \wedge 3\heartsuit < 3\spadesuit)$. The other cases are similar. $\qquad\square$

Although the conditional symmetry breaking constraints derived in Example 5 are identical to those derived in [14], our method is much more generic and can be applied to other problems as well. Also, no rigorous proof of correctness is given in [14], whereas Theorem 3 shows that these conditional symmetry breaking constraints are compatible. In this problem it is quite possible to derive multiple incompatible conditional symmetry breaking constraints which are individually correct. For example, suppose in addition to $(2\spadesuit, 3\spadesuit, 5\clubsuit)$ and $(1\diamondsuit, 3\heartsuit, 6\diamondsuit)$, we had a third pile $(2\heartsuit, 3\diamondsuit, 7\spadesuit)$, then the following conditional symmetry breaking constraints are all individually correct: $\neg(2\spadesuit < 3\heartsuit \wedge 3\spadesuit < 6\diamondsuit \wedge 3\heartsuit < 3\spadesuit)$, $\neg(2\heartsuit < 3\spadesuit \wedge 3\diamondsuit < 5\clubsuit \wedge 3\spadesuit < 3\diamondsuit)$, $\neg(1\diamondsuit < 3\diamondsuit \wedge 3\heartsuit < 7\spadesuit \wedge 3\diamondsuit < 3\heartsuit)$, but they are incompatible. For example, no matter which permutation of $3\spadesuit$, $3\heartsuit$, and $3\diamondsuit$ is applied, the partial solution $1\spadesuit, 2\heartsuit, 1\diamondsuit, 2\heartsuit, 3\spadesuit, 4\spadesuit, 3\heartsuit, 4\diamondsuit, 3\diamondsuit$ is pruned by one of the three conditional symmetry breaking constraints. Our method will never produce such incompatible sets of dominance breaking constraints.

Example 6. The Nurse Scheduling Problem (NSP) is to schedule a set of nurses over a time period such that work and hospital regulations are all met, and as many of the nurses' preferences are satisfied. There are many variants of this problem in the literature (e.g., [21,2]). We pick a simple variant to illustrate our method. Each day has three shifts: day, evening, and overnight. On each day, each nurse should be scheduled into one of the three shifts or scheduled a day off. For simplicity, we can consider a day off to be a shift as well. We number the shifts as day: 1, evening: 2, over-night: 3, day-off: 4. Each shift besides day-off requires a minimum number r_i of nurses to be rostered. Nurses cannot work for more than 6 days in a row, and must work at least 10 shifts per 14 days. Each nurse i has a preference $p_{i,j}$ for which of the four shift they wish to take on day j. The objective is to maximize the number of satisfied preferences. Let n be the number of nurses and m be the number of days. Let $x_{i,j}$ be the shift that nurse i is assigned to on day j. Then the problem can be stated as follows:

$$\text{Maximize} \quad \sum_{i=1}^{n}\sum_{j=1}^{m}(x_{i,j} = p_{i,j})$$

Subject to

$$among(r_i, \infty, [x_{k,j} \mid 1 \le k \le n], i) \qquad \forall 1 \le i \le 3, 1 \le j \le m \qquad (5)$$

$$among(1, \infty, [x_{i,j} \mid k \le j < k+7], 4) \qquad \forall 1 \le i \le n, 1 \le k \le n-6 \qquad (6)$$

$$among(-\infty, 4, [x_{i,j} \mid k \le j < k+14], 4) \qquad \forall 1 \le i \le n, 1 \le k \le n-13 \qquad (7)$$

Where $among(l, u, [x_1, \ldots, x_n], v)$ means that there are at least l and at most u variables from among $[x_1, \ldots, x_n]$ which take the value v. We now apply our dominance breaking method. Firstly, we can potentially get some symmetry or conditional symmetry breaking in by refining the objective function to $f' = lex(f, x_{1,1}, x_{2,1}, \ldots, x_{n,m})$. Let us consider mappings which are likely to map solutions to solutions. An obvious set of candidates are mappings which swap the shifts of two nurses on the same day, i.e., mappings $\sigma_{i_1,i_2,j}$ which swap $x_{i_1,j}$ and $x_{i_2,j}$.

We wish to calculate $scond(\sigma_{i_1,i_2,j})$. For each $c \in C \cup D$, we need to find c' such that $C \wedge D \wedge c' \Rightarrow \sigma_{i_1,i_2,j}(c)$. It is easy to see that the constraints in (5) are all symmetric in $\sigma_{i_1,i_2,j}$, so we do not need to add anything to $scond(\sigma_{i_1,i_2,j})$. The constraints $among(1, \infty, [x_{i,j} \mid k \le j < k+7], 4)$ in (6) will be satisfied by

$\sigma(\theta)$ iff: $x_{i_1,j} \neq 4 \vee x_{i_2,j} = 4 \vee among(1, \infty, [x_{i,j} \mid k \leq j < k + 7, j \neq i_1], 4)$. Similarly, the constraints $among(-\infty, 4, [x_{i,j} \mid k \leq j < k + 14], 4)$ in (7) will be satisfied by $\sigma(\theta)$ iff: $x_{i_1,j} = 4 \vee x_{i_2,j} \neq 4 \vee among(-\infty, 3, [x_{i,j} \mid k \leq j < k + 14, j \neq i_1], 4)$. Since the *among* conditions are probably too expensive to check, we can simply throw them away. We lose some potential pruning, but it is still correct, since we had a disjunction of conditions. So we add $x_{i_1,j} \neq 4 \vee x_{i_2,j} = 4$ and $x_{i_1,j} = 4 \vee x_{i_2,j} \neq 4$ to $scond(\sigma_{i_1,i_2,j})$. Calculating $\sigma(f') > f'$ is straight forward. We simply try each pair of values for $x_{i_1,j}$ and $x_{i_2,j}$ and see if swapping them improves the refined objective function. For example, suppose $i_1 < i_2$, $p_{i_1,j} = 4$, $p_{i_2,j} = 2$, then $ocond(\sigma_{i_1,i_2,j}) \equiv \sigma(f') < f' \equiv (x_{i_1,j}, x_{i_2,j}) \in \{(1,4), (2,1), (2,3), (2,4), (3,1), (3,4)\}$. Then $db(\sigma_{i_1,i_2,j}) \equiv \neg(scond(\sigma_{i_1,i_2,j}) \wedge ocond(\sigma_{i_1,i_2,j})) \equiv (x_{i_1,j}, x_{i_2,j}) \notin \{(2,1), (2,3), (3,1)\}$, which is a table constraint encapsulating both dominance breaking and symmetry breaking constraints. $\qquad\square$

5   Related Work

There have been many works on problem specific applications of dominance relations, e.g., the template design problem [25], online scheduling problems [16], the Maximum Density Still Life problem, Steel Mill Design problem and Peaceable Armies of Queens problem [24], the Minimization of Open Stacks problem [6], and the Talent Scheduling Problem [12]. However, the methods used are typically very problem specific and offer little insight as to how they can be generalized and applied to other problems. The implementations of these methods are also often quite ad-hoc (e.g., pruning values from domains even though they do not explicitly violate any constraint), and it is not clear whether they can be correctly combined with other constraint programming techniques. In contrast, our new method rests on a much stronger theoretical foundation and is completely rigorous. Since our method simply adds constraints to the problem, the modified problem is a perfectly normal constraint problem and it is correct to use any other constraint programming technique on it. Another important advantage of our method is that we are able to use any search strategy we want on the modified problem. This is not the case with many of the problem specific dominance breaking methods as they rely on specific labeling strategies.

There are a small number of works on generic methods for detecting and exploiting dominance relations. Machine learning techniques have been proposed as a method for finding candidate dominance relations [27]. This method works by encoding problems and candidate dominance relations into forms amenable to machine learning. Machine learning techniques such as experimentation, deduction and analogy are then used to identify potential dominance relations. This method was able to identify dominance relations for the 0/1 knapsack problem and a number of scheduling problems. However, the main weakness of this method is that it only generates candidate dominance relations and does not prove their correctness. Each candidates has to be analyzed to see if they are in fact a dominance relation. Then the dominance relation has to be manually proved and exploited.

Recently, several generic and automatic methods have been developed for exploiting certain classes of dominance relations. These include nogood learning techniques such as Lazy Clause Generation [23,9] and Automatic Caching via Constraint Projection [5]. Both of these can be thought of as dynamic dominance breaking, where after some domain D_1 is found to fail, a nogood (constraint) n is found which guarantees that if D_2 violates n, then D_2 is dominated by D_1 and must also fail. The nogood n is posted as an additional redundant constraint to the problem. Lazy Clause Generation derives this n by resolving together clauses which explain the inferences which led to the failure. Automatic Caching via Constraint Projection derives n by finding conditions such that projection of the subproblem onto the subset of unfixed variables yield a more constrained problem. These methods are to a large extent complementary to the method presented in this paper. None of these methods exhausts all possible dominances occurring in a problem, and there are dominances which can be exploited by one method but not another. Thus we can often use them simultaneously to gain an even greater reduction in search space.

6   Experimental Results

We now give some experimental results for our method on a variety of problems. We have already discussed how our method applies to the Photo Problem, Knapsack Problem, Black Hole Problem, and Nurse Scheduling. For these problems, we generate random instances of several different sizes, with 10 instances of each size. We also give experimental results for four further problems:

RCPSP. The resource constrained project scheduling problem (RCPSP) [4] schedules n tasks using m renewable resources so that ordering constraints among tasks hold and resource usage limits are respected. A standard dominance rule for this problem, used in search, is that each task must start at time 0 or when another task ends, since any schedule not following this rule is dominated by one constructed by shifting tasks earlier until the rule holds. We use the instances from the standard J60 benchmark set [1] which are non-trivial (not solved by root propagation) and solvable by at least one of the methods.

Talent Scheduling Problem. In the Talent Scheduling Problem [12], we have a set of scenes and a set of actors. Each actor appears in a number of scenes and is paid a certain amount per day they are on location. They must stay on location from the first scene they are in till the last scene they are in. The aim is to find the schedule of scenes $x_1, \ldots, x_n$ which minimize the cost of the actors. We set $f' = lex(f, x_1, \ldots, x_n)$. We consider mappings which take one scene and move it to another position in the sequence. We generate 10 random instances of size 14, 16, and 18.

Steel Mill Problem. In the Steel Mill Problem [13], we have a set of orders to be fulfilled and the aim is to minimize the amount of wasted steel. Each order i has a size and a color (representing which path it takes in the mill) and is to be assigned to a slab x_i. Each slab can only be used for orders of two different colors. Depending on the sum of the sizes of the orders on each slab, a certain amount of steel will be wasted. We set $f' = lex(f, x_1, \ldots, x_n)$ and try mappings where we take all orders of a certain color from one slab, and all orders of a certain color from another slab, and swap the slabs they are assigned to. We generate 10 random instances of size 40 and 50.

PC Board Problem. In the PC Board Problem [19], we have $n \times m$ components of various types which need to be assigned to m machines. Each machine must be assigned exactly n components and there are restrictions on the sets of components that can go on the same machine. Each type of component gains a certain utility depending on which machine it is assigned to and the goal is to maximize the overall utility. We set $f' = lex(f, x_{1,1}, x_{1,2}, \ldots, x_{n,m})$ where $x_{i,j}$ is the type of component assigned to the jth spot on the ith machine. We consider mappings which swap two components on different machines. We generate 20 random instances of size 6×8.

The experiments were performed on Xeon Pro 2.4GHz processors using the CP solver CHUFFED. For each set of benchmarks, we report the geometric mean of time taken in seconds and the number of failed nodes for: the original problem with no dominance breaking of any form (base), with dominance breaking constraints generated by our method (db), with Lazy Clause Generation (lcg), and with dominance breaking constraints and Lazy Clause Generation (db+lcg). A timeout of 900 seconds was used. Fastest times and lowest node counts are shown in bold.

As Table 1 shows, adding dominance breaking constraints can significantly reduce the search space on a variety of problems, leading to large speedups which tend to grow exponentially with problem size. Our method can also often be combined with Lazy Clause Generation for additional speedup (e.g., Photo, Steel Mill, Talent Scheduling, Nurse Scheduling, PC Board). In some cases (e.g., Knapsack, Black Hole), even though adding LCG on top of our method can reduce the node count further, the extra overhead of LCG swamps out any benefit. In other cases (e.g., RCPSP), adding our dominance breaking constraints on top of LCG actually increases the run time and node count. In this problem, the dynamically derived dominances from LCG are stronger than the static ones that our method derives. Adding the dominance breaking constraints interferes with and reduces the benefit of LCG. In general however, our method appears to provide significant speedups over a wide range of problems for both non-learning and nogood learning solvers.

7   Future Work

Although we have developed this method in the context of Constraint Programming, the dominance relations we find can be applied to other kinds of search as well. For example, MIP solvers, which use branch and bound, can also benefit from the power of dominance relations, as they can encounter suboptimal partial assignments which nevertheless do not produce an LP bound strong enough to

Table 1. Comparison of the original model and the model augmented with dominance breaking constraints

Problem	base		db		lcg		db+lcg	
	Time	Nodes	Time	Nodes	Time	Nodes	Time	Nodes
Photo-14	1.09	57773	0.90	10967	0.30	5791	**0.25**	**1962**
Photo-16	8.38	441574	4.00	43373	6.49	44325	**1.40**	**8960**
Photo-18	60.68	2828622	22.09	206507	19.73	138926	**6.25**	**24523**
Knapsack-20	**0.01**	215	**0.01**	9	**0.01**	212	**0.01**	**7**
Knapsack-30	0.17	46422	**0.01**	91	0.85	45733	**0.01**	**65**
Knapsack-50	602	1×10^8	**0.01**	684	900	1×10^7	**0.01**	**507**
Knapsack-100	900	1×10^8	**0.40**	54705	900	1×10^7	1.05	**37571**
Black-hole	5.18	77542	**0.08**	607	0.97	2767	0.09	**347**
Nurse-15-7	900	9×10^7	900	8×10^7	1.72	55217	**0.91**	**24258**
Nurse-15-14	900	8×10^7	900	8×10^7	483.29	7×10^6	**140.95**	**1×10^6**
RCPSP	358.95	2779652	279.74	781399	**4.07**	**7890**	32.84	32770
Talent-Sched-14	1.66	39479	0.42	10122	0.45	4983	**0.27**	**3189**
Talent-Sched-16	16.08	349704	2.33	51993	3.71	27186	**1.28**	**12336**
Talent-Sched-18	252.05	5557959	13.88	299043	26.25	128810	**4.28**	**31829**
Steel-Mill-40	60.64	1×10^6	22.00	451636	16.31	75293	**4.53**	**27225**
Steel-Mill-50	379.21	7×10^6	231.95	3×10^6	249.39	714451	**32.24**	**129788**
PC-board	547.93	4×10^7	412.29	1×10^7	20.28	156933	**7.51**	**64320**

prune the subproblem. Simple dominance rules such as fixing a variable to its upper/lower bound if it is only constrained from below/above [17] are already in use in MIP, but our method can produce much more generic dominance rules. Similarly, local search can benefit tremendously from dominance relations, as they can show when a solution is suboptimal and map it to another solution which is better. Exploring how our method could be adapted for use in other kinds of search is an interesting avenue of future work.

It may also be possible to automate many or all of the steps involved in our method. Such automation would provide a great benefit for system users as they will be able to feed in a relatively "dumb" model and have the system automatically identify and exploit the dominances. Step 0 typically requires augmenting the objective function with an appropriate lexicographical ordering of the variables. For Step 1, there already exist automated methods for detecting symmetries in problem instances [20,26]. Such methods can be adapted to look for good candidates for σ. Step 2 and 3 involves algebraic manipulations which are not difficult for a computer to do. The difficulty lies in Step 4, where we need to simplify the logical expressions and determine whether the dominance breaking constraint is sufficiently simple, efficient and powerful that it is worth adding to to problem. Automating our method is another interesting avenue of future work.

8   Conclusion

We have described a generic method for identifying and exploiting dominance relations in constraint problems. The method generates a set of dominance breaking constraints which are provably correct and compatible with each other. The method also generates symmetry and conditional symmetry breaking constraints as a special case, thus it unifies symmetry breaking, conditional symmetry breaking and dominance breaking under one method. Experimental results show that the dominance breaking constraints so generated can lead to significant reductions in search space and run time on a variety of problems, and that they can be effectively combined with other dominance breaking techniques such as Lazy Clause Generation.

Acknowledgments. NICTA is funded by the Australian Government as represented by the Department of Broadband, Communications and the Digital Economy and the Australian Research Council. This work was partially supported by Asian Office of Aerospace Research and Development (AOARD) Grant FA2386-12-1-4056.

References

1. PSPLib - project scheduling problem library, `http://129.187.106.231/psplib/`, (accessed on March 1, 2012)
2. Abdennadher, S., Schlenker, H.: Nurse scheduling using constraint logic programming. In: AAAI/IAAI, pp. 838–843 (1999)
3. Aldowaisan, T.A.: A new heuristic and dominance relations for no-wait flowshops with setups. Computers & OR 28(6), 563–584 (2001)
4. Brucker, P., Drexl, A., Möhring, R.H., Neumann, K., Pesch, E.: Resource-constrained project scheduling: Notation, classification, models, and methods. European Journal of Operational Research 112(1), 3–41 (1999)
5. Chu, G., de la Banda, M.G., Stuckey, P.J.: Automatically Exploiting Subproblem Equivalence in Constraint Programming. In: Lodi, A., Milano, M., Toth, P. (eds.) CPAIOR 2010. LNCS, vol. 6140, pp. 71–86. Springer, Heidelberg (2010)
6. Chu, G., Stuckey, P.J.: Minimizing the Maximum Number of Open Stacks by Customer Search. In: Gent, I.P. (ed.) CP 2009. LNCS, vol. 5732, pp. 242–257. Springer, Heidelberg (2009)
7. Crawford, J.M., Ginsberg, M.L., Luks, E.M., Roy, A.: Symmetry-breaking predicates for search problems. In: Proceedings of the 5th International Conference on Principles of Knowledge Representation and Reasoning, pp. 148–159. Morgan Kaufmann (1996)
8. Fahle, T., Schamberger, S., Sellmann, M.: Symmetry Breaking. In: Walsh, T. (ed.) CP 2001. LNCS, vol. 2239, pp. 93–107. Springer, Heidelberg (2001)
9. Feydy, T., Stuckey, P.J.: Lazy Clause Generation Reengineered. In: Gent, I.P. (ed.) CP 2009. LNCS, vol. 5732, pp. 352–366. Springer, Heidelberg (2009)
10. Flener, P., Pearson, J., Sellmann, M., Van Hentenryck, P.: Static and Dynamic Structural Symmetry Breaking. In: Benhamou, F. (ed.) CP 2006. LNCS, vol. 4204, pp. 695–699. Springer, Heidelberg (2006)
11. Focacci, F., Milano, M.: Global Cut Framework for Removing Symmetries. In: Walsh, T. (ed.) CP 2001. LNCS, vol. 2239, pp. 77–92. Springer, Heidelberg (2001)
12. de la Banda, M.G., Stuckey, P.J., Chu, G.: Solving talent scheduling with dynamic programming. INFORMS Journal on Computing 23(1), 120–137 (2011)

13. Gargani, A., Refalo, P.: An Efficient Model and Strategy for the Steel Mill Slab Design Problem. In: Bessière, C. (ed.) CP 2007. LNCS, vol. 4741, pp. 77–89. Springer, Heidelberg (2007)

14. Gent, I.P., Kelsey, T., Linton, S.A., McDonald, I., Miguel, I., Smith, B.M.: Conditional Symmetry Breaking. In: van Beek, P. (ed.) CP 2005. LNCS, vol. 3709, pp. 256–270. Springer, Heidelberg (2005)

15. Gent, I.P., Smith, B.M.: Symmetry breaking in constraint programming. In: Horn, W. (ed.) Proceedings of the 14th European Conference on Artificial Intelligence, pp. 599–603. IOS Press (2000)

16. Getoor, L., Ottosson, G., Fromherz, M.P.J., Carlson, B.: Effective redundant constraints for online scheduling. In: AAAI/IAAI, pp. 302–307 (1997)

17. Hoffman, K.L., Padberg, M.: Improving LP-representations of zero-one linear programs for branch-and-cut. INFORMS Journal on Computing 3(2), 121–134 (1991)

18. Korf, R.E.: Optimal rectangle packing: New results. In: Proceedings of the Fourteenth International Conference on Automated Planning and Scheduling (ICAPS 2004), pp. 142–149 (2004)

19. Martin, R.: The Challenge of Exploiting Weak Symmetries. In: Hnich, B., Carlsson, M., Fages, F., Rossi, F. (eds.) CSCLP 2005. LNCS (LNAI), vol. 3978, pp. 149–163. Springer, Heidelberg (2006)

20. Mears, C., de la Banda, M.G., Wallace, M.: On implementing symmetry detection. Constraints 14(4), 443–477 (2009)

21. Miller, H.E., Pierskalla, W.P., Rath, G.J.: Nurse scheduling using mathematical programming. Operations Research, 857–870 (1976)

22. Monette, J.N., Schaus, P., Zampelli, S., Deville, Y., Dupont, P.: A CP approach to the balanced academic curriculum problem. In: Seventh International Workshop on Symmetry and Constraint Satisfaction Problems, vol. 7 (2007), http://www.info.ucl.ac.be/~yde/Papers/SymCon2007_bacp.pdf

23. Ohrimenko, O., Stuckey, P.J., Codish, M.: Propagation via lazy clause generation. Constraints 14(3), 357–391 (2009)

24. Prestwich, S., Beck, J.C.: Exploiting dominance in three symmetric problems. In: Fourth International Workshop on Symmetry and Constraint Satisfaction Problems, pp. 63–70 (2004), http://zeynep.web.cs.unibo.it/SymCon04/SymCon04.pdf

25. Proll, L.G., Smith, B.: Integer linear programming and constraint programming approaches to a template design problem. INFORMS Journal on Computing 10(3), 265–275 (1998)

26. Puget, J.-F.: Automatic Detection of Variable and Value Symmetries. In: van Beek, P. (ed.) CP 2005. LNCS, vol. 3709, pp. 475–489. Springer, Heidelberg (2005)

27. Yu, C.F., Wah, B.W.: Learning dominance relations in combinatorial search problems. IEEE Trans. Software Eng. 14(8), 1155–1175 (1988)

Scheduling Scientific Experiments
on the Rosetta/Philae Mission

Gilles Simonin, Christian Artigues, Emmanuel Hebrard, and Pierre Lopez

CNRS, LAAS, 7 avenue du colonel Roche, F-31400 Toulouse, France
Univ de Toulouse, LAAS, F-31400 Toulouse, France
`{gsimonin,artigues,hebrard,lopez}@laas.fr`

Abstract. The Rosetta/Philae mission was launched in 2004 by the European Space Agency (ESA). It is scheduled to reach the comet 67P/Churyumov-Gerasimenko in 2014 after traveling more than six billion kilometers. The Philae module will then be separated from the orbiter (Rosetta) to attempt the first ever landing on the surface of a comet. If it succeeds, it will engage a sequence of scientific exploratory experiments on the comet.

In this paper we describe a constraint programming model for scheduling the different experiments of the mission. A feasible plan must satisfy a number of constraints induced by energetic resources, precedence relations on activities, or incompatibility between instruments. Moreover, a very important aspect is related to the transfer (to the orbiter then to the Earth) of all the data produced by the instruments. The capacity of inboard memories and the limitation of transfers within visibility windows between lander and orbiter, make the transfer policy implemented on the lander's CPU prone to data loss. We introduce a global constraint to handle data transfers. The goal of this constraint is to ensure that data-producing activities are scheduled in such a way that no data is lost.

Thanks to this constraint and to the filtering rules we propose, mission control is now able to compute feasible plans in a few seconds for scenarios where minutes were previously often required. Moreover, in many cases, data transfers are now much more accurately simulated, thus increasing the reliability of the plans.

1 Introduction

The international Rosetta/Philae project is an European consortium mission approved in 1993, and is under the leadership of the German Aerospace Research Institute (DLR) and ESA[1]. The spacecraft was launched in 2004 by Ariane 5, and is set to travel more than six billion kilometers to finally reach and land on the comet 67P/Churyumov-Gerasimenko in 2014 in order to analyze the comet structure. It will follow a complex trajectory which includes four gravity assist maneuvers (3 x Earth, 1 x Mars) before finally reaching the comet and enter its orbit. Then, the lander Philae will be deployed and will land on the surface of the comet. Philae features ten instruments, each developed by a European laboratory, to accomplish a given scientific experiment when approaching, or once landed on the comet. For instance, *CIVA* and *ROLIS* are two imaging instruments, used to take panoramic pictures of the comet and microscopic images.

---

[1] European Space Agency.

M. Milano (Ed.): CP 2012, LNCS 7514, pp. 23–37, 2012.

The Alpha Proton X-ray Spectrometer (*APXS*) analyses the chemical composition of the landing site and its alteration during the comet's approach of the Sun. This data will be used to characterize the surface of the comet, to determine the chemical composition of the dust component, and to compare the dust with known meteorite types.

The exploratory mission will have three phases. The SDL (*Separation-Descent-Landing*) will run for 30 minutes during which many experiments will be done. The FSS (*First Science Sequence*) will last 5 days. This phase is critical because the execution of the most energetically greedy experiments requires battery power. The quality of this schedule conditions the longevity of the batteries and is therefore key to the success of the mission. Finally, during the LTS phase (*Long Term Science*), scientific activities will be resumed at a much slower pace, using the lander's own solar panels to partially reload the batteries. This phase will continue for months until the probe is destroyed due to the extreme temperatures of the Sun.

This project is a collaboration with CNES[2] in Toulouse (France). The goal of the *Scientific Operation and Navigation Centre* (SONC) is to plan the sequence of experiments and maneuvers to be done in each of these phases while making the best use of the available resources. This project has many similarities with the (interrupted) Net-Lander program [7]. A first software (called MOST) has been developed on top of the Ilog-Scheduler/Solver library by an industrial subcontractor. Every instrument, subsystem and experiment has been modeled precisely in this framework, and it is therefore possible to check solutions with a high degree of confidence on their feasibility.

The main scientific experiments need to be scheduled to satisfy a number of constraints involving the concurrent use of energy (batteries), and of the main CPUs as well as each instrument's memory. Moreover, each experiment produces data that must be transferred to the Earth. Each experiment has its own memory, collecting data as it is produced. This data is then transferred to a central mass-memory, then sent to Rosetta (the orbiter) when it is in *visibility*, i.e., above the horizon of the comet with respect to Philae. All transfers from the experiments to the mass memory, and from the mass memory to the orbiter are executed (that is, computed onboard) by the *Command and Data Management System* (CDMS). The transfer policy of the CDMS may lead to data loss when an experiment produces more data than its memory can store and its priority is not high enough to allow a transfer to the mass-memory. This is modeled within MOST using RESERVOIR constraints [6]. Data-producing activities fill the reservoir, while multiple pre-defined data transfer tasks of variable duration empty it. There are numerous problems related to data transfers in spatial applications [5,2], however the problem at hand is significantly different since plans are computed on the ground, and the data transfer policy is beyond our control.

This modeling choice has several drawbacks and it quickly became apparent that it was the critical aspect of the problem to tackle in order to find better solutions faster. The first problem with this model is that data transfers are not accurately represented. For each experiment, a sequence of tasks standing for data transfers are pre-defined. Their duration is constrained so that the experiment with the highest priority is allowed to transfer as much as possible, and no overlap is allowed among transfers. In the current implementation there is a transfer task every 120 seconds over the horizon, with a

---

[2] Centre National d'Etudes Spatiales.

maximum duration of 120 seconds. This is too few to accurately represent the policy of the CDMS, however, this is already too much for Ilog-Scheduler to handle (the planning horizon may be up to one day, i.e., about 700 transfer tasks for each experiment).

Instead we propose to encapsulate data transfers into a global constraint. The decision variables are start times of data-producing activities (data-producing rate and duration are known in advance) and the priority permutation. This allows us to very quickly check the satisfiability of a schedule with respect to data transfer. Moreover, we can compute bounds allowing to filter out the domain of the variables standing for start time of the data-producing activities. Unfortunately enforcing arc consistency or even bounds consistency on this constraint is NP-hard, so we do not give a complete filtering algorithm. However, our approach reduces the solving time dramatically: from hours in some cases to seconds in all scenarios currently considered by the SONC. Moreover, the result is much more accurate, to the point that some scenarios for which MOST could not show that transfers were feasible can now be solved efficiently.

In Section 2 we briefly outline the energetic aspect of the problem and more formally define the data transfer aspect. Then, in Section 3 we introduce our approach to modeling data transfers. In particular we give an efficient satisfiability checking procedure and two filtering rules for the introduced global constraint. Last in Section 5, we report experimental results and compare results between old and new models.

2   Problem Description

Each experiment and subsystem can be seen as a list of activities to be scheduled. Notwithstanding data transfers, the problem can be seen as a scheduling problem over a set of experiments with relatively standard constraints.

Precedences: Activities within the same experiment might have precedence constraints; for instance, the Lander also carries a Sampling Drilling and Distribution device (*SD2*), that drills 20 cm into the surface, collect samples and drop them in an oven. The oven then rotates to a position whereby it can be connected to the inlet of the gas management system of another instrument: Ptolemy. At this point, the oven must be heated so that volatiles are released and analysed by Ptolemy.

Cumulative Resources: Activities concurrently use the energy from batteries and solar panels. The energy needed to run each task is supplied by an auxiliary power line. Each auxiliary line is linked to a converter, and each converter is linked to the main power line. At each level, the total instant power delivered cannot exceed a given threshold. For each auxiliary power line, all the activities supplied by this line are constrained by a CUMULATIVE constraint [1] with capacity equal to this threshold. Similarly another CUMULATIVE constraint is associated to each converter, and a last one is associated to the main power line, involving all activities of the problem.

State Resources: Each instrument can have multiple states along the schedule. Some activities can trigger the modification of the state of an instrument, and the processing of certain activities might be subject to some instruments being a given state. This is modeled using predefined state resources constraints in Ilog-Scheduler.

Data Transfer and Memory Constraints: Every experiment has its own memory. Some activities produce data, temporarily stored on the experiment's memory. Then this data will be transferred onto the mass-memory and subsequently to the orbiter.

The CDMS controls all data transfers, from the experiments to the mass memory, and from the mass memory to the orbiter. Within a plan, experiments are ordered according to a priority function. Apart from this ordering, the CDMS is completely autonomous. It transfers data from the experiment with highest priority among those with transferable data. Moreover, it transfers data from the mass-memory to the orbiter whenever possible, that is, when there is some visibility. However, it does not ensure that all produced data will eventually be transferred back to the Earth. When too much data is produced simultaneously and not enough can be transferred on the mass-memory, or when there is no visibility with the orbiter and therefore the mass-memory cannot be emptied, the capacity of an experiment's memory may be overloaded and data is lost.

The Mars-Express mission, launched in 2003 and still in operation, also featured a data transfer planning problem, similar to Rosetta's in many respects. In both cases, data-producing activities are to be scheduled, data is kept into a number of memory storage devices on board and periodically transmitted to the Earth during visibility windows. However, a critical difference is that in Mars-Express, the transfers are actually decisions to be made at the planning level. A flow model was proposed to address the so-called Memory Dumping Problem in [4,8] and further improved in [10]. In our case, however, the CDMS policy is a given. As a consequence, data loss can only be controlled through the schedule of data-producing activities.

In other words, we shall consider data transfers as a global constraint on the start times of activities ensuring that no data will be lost with respect to the CDMS policy.

3   A Global Constraint for Data Transfer

Except for the data transfer aspect, all the constraints above can be modeled using the standard methods and algorithms [9] all available in Ilog-Scheduler. Hence, we focus on data transfers and propose a global constraint to reason about this aspect of the problem.

From now on, we consider a set $\{E_1, \ldots, E_m\}$ of m experiments. An *experiment* $E_k = \{t_{k1}, \ldots, t_{kn}\}$ is a set of data-producing tasks[3], and is associated with a memory of capacity M_k. A task t_{ki} produces data for a duration p_{ki} at a rate π_{ki} in the experiment's memory. The lander possesses a mass memory of capacity M_0, where data can be transferred from experiments.

The CDMS is given as input a priority ordering on experiments. For $i \in [1, \ldots, m]$, we denote by $P(i)$ the experiment at rank i in this ordering and its dual $R(k)$ standing for the rank of experiment E_k in the priority ordering ($P(i) = k \Leftrightarrow R(k) = i$). We shall say that experiment E_k has higher priority than experiment E_j iff $R(k) < R(j)$.

Data can only be transferred out to the orbiter when it is in *visibility*, that is in the line of sight of the lander over the horizon of the comet. Visibility is represented as a set of intervals $\{[a_1, b_1], \ldots, [a_v, b_v]\}$ in the scheduling horizon which lengths

---

[3] To simplify the notations, we assume that all experiments have the same number of activities. This is of course not the case, however it does not affect the methods we introduce.

and frequencies depend on the chosen orbit. We shall use $V(t)$ as a Boolean function which equals true iff time t is included in one of the visibility intervals. Moreover, data is transferred in and out memories by *block* units of 256 bytes.

We consider the following decision variables: $s_{11}, \ldots, s_{mn}$, with domain in $[0, \ldots, H]$, standing for the start times of data-producing tasks $\{t_{11}, \ldots, t_{mn}\}$, respectively. Here we will assume that the priority permutation is fixed. Indeed there are often few experiments and exploring all permutations would be easy.

The fact that data loss should be avoided can be seen as a relation (i.e., a constraint) between the decision variables above. It is relatively easy to understand this relation procedurally since the CDMS policy is deterministic. Given a priority ordering and a fixed schedule of the data-producing activities, one can unroll the rules outlined above to check whether the CDMS policy will lead to data loss or not.

In this section, we formally define the constraint DATATRANSFER$([s_{11}, \ldots, s_{mn}])$ ensuring that the schedule of tasks $\{t_{11}, \ldots, t_{mn}\}$ is such that no data is lost.

First, in Section 3.1 we discuss an "exact" definition based on following the transfer of each block of data individually. However, this formulation is not practical, so we propose an alternative model in Section 3.2. The basic idea is to represent all the data produced by an activity as a continuous quantity. With this viewpoint, tasks that do not produce data as fast as it can be transferred are challenging to model. Indeed, the CDMS actually waits for a block of data to be completed before transferring it, and can therefore use this waiting time to transfer data from other experiments with lower priority. We approximate this "vertical" partition of the data bus, by a "horizontal" partition, i.e., we consider that the bandwidth can be divided over parallel transfers. This model allows us to represent memory usage very precisely, with a computational complexity independent on the time horizon and on the amount of data produced.

3.1 CDMS Policy

In this section we detail the CDMS policy, then we define a constraint modeling the relation between the start time of data-producing activities induced by this policy.

The CDMS transfers data by blocks of 256 bytes. Its policy is relatively simple and can be described using a simple automated earliest transfer scheduling algorithm (AETS). AETS runs the two following processes in parallel:

- Repeat: Scan experiments by order of priority until one with at least one block of data on its memory is found. In that case, transfer one block from this experiment to the mass memory unless the mass memory is full.
- Repeat: If the orbiter is visible, and there is at least one block of data on the mass memory, then dump one block (transfer from the mass memory to the orbiter).

So we can define the DATATRANSFER constraint as the relation allowing only assignments of start times and priorities such that given the CDMS policy (AETS), no block of data is produced while the memory of the experiment is full.

In order to specify this constraint as precisely as possible we would need to consider each block of data, and its associated transfer task, individually. More precisely, we need $\pi_{ki} p_{ki}$ transfer tasks for each data-producing activity t_{ki}. The release time of the jth

block's transfer task is $s_{ki} + j(1/\pi_{ki})$, where s_{ki} is the start time of t_{ki}. Moreover, the start times and durations of these transfer tasks are functionally dependent on the start times of data-producing activities and experiment priorities. This dependence relation is a consequence of the AETS procedure.

The DATATRANSFER constraint is NP-complete to satisfy, hence NP-hard to filter. Indeed, consider the particular case where memory capacities are all of exactly one block of data, and the mass-memory is unlimited. In this case, each activity must transfer a block to the mass-memory as soon as it is produced. In other words, we can see a transfer task as non-interruptible. However, since there is a single transfer channel to the mass-memory, no overlap is possible between these tasks. Since we have time windows on the variables s_{ki}, this particular case is therefore equivalent to a disjunctive unary resource, i.e., it is strongly NP-hard.

3.2 Approximated Definition

It is difficult to capture very precisely the behavior of the CDMS. Moreover, it is not practical since it involves manipulating a very large number of transfer tasks. When we consider a data-producing activity in isolation, the number of transfer tasks and their frequency is easy to compute. However, when we consider several data-producing activities with different priorities and unknown start times, this viewpoint becomes impractical. We therefore propose an alternative model that approximate very closely the amount of transferred data with a reasonable time and space complexity.

The basic idea is straightforward. Consider a task t_{ki} that produces more data than it can transfer $\tau_{r(k,t)} \leq \pi_{ki}$, with $\tau_{r(k,t)}$ the transfer rate at time t from an experiment E_k to mass-memory. Suppose first that there is no task with higher priority. The transfer can be seen as a continuous task of duration $\frac{\pi_{ki}p_{ki}}{\tau_{r(k,t)}}$. However there are three difficulties.

First, blocks are transferred from experiments memories to the mass memory at a constant rate. However, when seeking which experiment to transfer from, the length of the scanning process depends on the number of *active* experiments and on the priority of the experiment eventually selected. An experiment E_k is active between the start of its first activity and the end time of its last activity, or if the experiment memory is not empty. The transfer rate is thus larger in practice for the higher priority experiments as they are scanned first. To emulate this, we use variable transfer rates. The potential transfer rate $\tau_{r(k,t)}$ depends on the number x of active experiments and on its relative priority y among them. The actual value is read in a table which entries were measured experimentally. The rate above applies to non-visibility periods. According to the same principles, another table gives us the transfer rate $\tau'_{r(k,t)}$ while in visibility (it is lower due to the parallelism with memory dumps). Transfers between mass-memory and the orbiter have a constant rate denoted τ^{mm}.

Second, transfer tasks can be interrupted, however, they are different from classic preemptive tasks in that we do not decide when the interruption occurs. When an experiment with higher priority starts producing data, it preempts any current transfer of lower priority. This is easy to model since this is the unique context where an interruption can happen. If there is no experiment with higher priority to interrupt the transfer, the usage of the experiment's memory increases at rate $\pi_{ki} - \tau_{r(k,t)}$ during p_{ki} seconds. Similarly, during $\frac{\pi_{ki}p_{ki}}{\tau_{r(k,t)}}$ seconds the usage of the mass memory increases at rate $\tau_{r(k,t)}$.

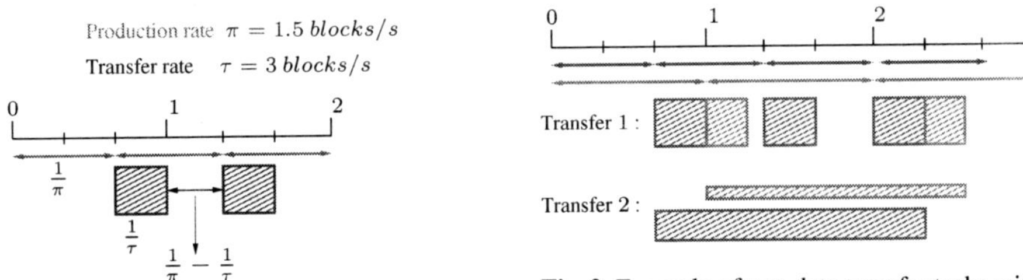

Fig. 1. Example of exact data transfer

Fig. 2. Example of two data transfer tasks with both model

The third difficulty concerns tasks producing data at a lower rate than the possible transfer rate (i.e., $\tau_{r(k,t)} > \pi_{ki}$). In this case, data is transferred one block at a time, with a lag between each transfer to wait for the next block to be produced (see Fig. 1).

Other tasks of lower priority with non-empty memory can use these gaps to begin the transfer of a block of data. In other words, the duration of the transfer of highest priority is still very close to $\frac{\pi_{ki}p_{ki}}{\tau_{r(k,t)}}$ seconds, however other transfer can be squeezed in that same period. In order to simulate this, we consider that the data bus has a capacity (bandwidth) normalized to 1. The demand of a task t_{ki} at time t is $\min(1, \frac{\pi_{ki}}{\tau_{r(k,t)}})$. Bandwidth is allocated recursively, according to priority (see Figure 2).

Let m_k^t stand for the quantity of data in the memory of experiment E_k at time t (with $t \in \mathbb{R}$) and m_0^t be the quantity of data in the mass-memory. Moreover, let π_k^t stand for the data-producing rate on experiment E_k at time t, let $p(\tau_k^t)$ stand for the potential transfer rate of experiment E_k at time t if it was of highest priority and let τ_k^t stand for the actual transfer rate from experiment E_k to the mass memory at time t.

Definition 1

$$\text{DATATRANSFER}(s_{11}, \ldots, s_{mn}) \Leftrightarrow$$

$$\forall t, k, \qquad \pi_k^t = \begin{cases} \pi_{ki} & \text{if } \exists i\ s.t.\ s_{ki} \leq t \leq s_{ki} + p_{ki} \\ 0 & \text{otherwise} \end{cases} \tag{1}$$

$$\forall t, k,\ \ p(\tau_k^t) = \begin{cases} 0 & \text{if } m_0^t = M_0 \vee (m_k^t = \pi_k^t = 0) \\ \tau_{r(k,t)} & \text{if } m_0^t < M_0 \wedge m_k^t > 0 \\ \min(\pi_k^t, \tau_{r(k,t)}) & \text{otherwise} \end{cases} \tag{2}$$

$$\forall t, k, \qquad \tau_k^t = \min\left(p(\tau_k^t), \tau_{r(k,t)}\left(1 - \sum_{i=1}^{R(k)-1} \frac{\pi_{P(i)}^t}{\tau_{r(P(i),t)}}\right)\right) \tag{3}$$

$$\forall t, k, \qquad m_k^t = \int_0^t (\pi_k^t - \tau_k^t)\mathrm{d}t \tag{4}$$

$$\forall t, \qquad m_0^t = \int_0^t \left(\sum_{k=1}^m \tau_k^t - V(t)\tau^{mm}\right)\mathrm{d}t \tag{5}$$

$$\forall t, k, \qquad m_k^t \leq M_k \tag{6}$$

Equation 1 simply states that if a data-producing activity t_{ki} is running at time t, then the data-producing rate π_k^t of an experiment k at that time is equal to the data-producing rate of t_{ki}, and it is zero otherwise.

Fig. 3. Comparison of the two representations: two data-producing activities t_1 and t_2 (bottom); The "exact" view of the corresponding transfers, sharing the transfer bus because of gaps due to the low data-producing rate (middle); The alternative reformulation, where this is modeled as sharing the bandwidth (top)

Equation 2 defines the *expected transfer rate* $p(\tau_k^t)$ of experiment E_k at time t if it is not trumped by other experiments of higher priority. If there is no data on the memory and no data being produced, or if the mass memory is full, this rate is zero. Otherwise, if there is some data on memory, it can be transferred at the maximum available rate ($\tau_{r(k,t)}$). If the memory is empty, but data is being produced, we assume that it cannot be transferred at a higher rate than it is produced.

Equation 3 gives the *real* transfer rate, i.e., taking into account experiments with higher priorities. The experiment with highest priority uses the bandwidth proportionally to the ratio between its expected transfer rate $p(\tau_k^t)$ and the maximum transfer rate $\tau_{r(k,t)}$. Then the residual bandwidth is attributed using recursively the same rule.

Finally, equations 4 and 5 link the usage of the different memories to the sum of the in and out transfer rates (π_k and τ_k are used here as functions of t) while equations 6 ensure that memory capacity is never exceeded.

In Figure 3 we show the difference between the two models.

4   Checking and Filtering Algorithms

In this section we introduce a filtering procedure for the DATATRANSFER constraint. We first introduce an efficient $O(nm \log(nm))$ procedure for computing transfers and memory usage of a given schedule. This procedure execute a *sweep* of the horizon similar to that described in [3]. Besides checking whether the constraint is violated we shall also use this algorithm to compute lower bounds in order to filter the domains.

4.1   Data Transfer Verification

Given a complete schedule of the data-producing activities, and a priority ordering, we now describe an algorithm that computes the effective transfer rate (in the sense of Definition 1) and the memory usage for each experiment over the whole horizon in time $O(nm \log(nm))$. Notice that both are step functions, moreover we will see that there are at most $O(nm)$ breaking points, so they can be stored on $O(nm)$ bits. This algorithm can be used to verify whether an assignment is consistent by simply checking that the usage of all experiments remains within the memory's capacity. We shall also use it to compute bounds on the memory usage of extreme scenarios (e.g., all tasks set to their earliest start time). It sweeps the time horizon chronologically, computing variations of various parameters only when certain *events* occur.

First, we build the list of events. Each event is time tagged and there are six types (for $O(nm)$ events in total): *Start/end of visibility*; *Start/end of a data-producing activity*; *Start/end of experiments*. Then, we sort them in chronological order and explore them in that order. For each time point t where at least an event occurs, we go through all events occurring at t and update the following arrays accordingly:

- *visibility* stands for whether there is a visibility line at time t. It is flipped whenever encountering a "Start of visibility" or "End of visibility" event;
- *production*(k) stands for the data-producing rate of experiment E_k at time t. It is increased (resp. decreased) by the data-producing rate of the activity whenever encountering a "Start of production" (resp. "End of production") event;
- *active* stands for the number of active experiments at time t. It is increased (resp. decreased) by one whenever encountering a "Start of experiment" (resp. "End of experiment") event.

At each step of the loop, we therefore know the complete state (data-producing rate on each experiment, whether we are in visibility or not, and how many experiments are active). Moreover, we also keep track of the memory usage with another array: *memory*. We then compute what are the current transfers, and partition the bandwidth between them. For each experiment E_k (visited by order of priority), if it has data on memory, or if it is currently producing data, and if the bandwidth is not zero, we create a transfer. We first compute its potential transfer rate $\tau_{r(k,t)}$ according to the rules described above. If it has some data on memory, all of the bandwidth is attributed to this transfer. Otherwise, its actual transfer rate is equal to the minimum between the nominal transfer rate and the current data-producing rate: $\tau = \min(production(k), \tau_{r(k,t)})$. The ratio $\tau/\tau_{r(k,t)}$ of the bandwidth is allocated to this transfer.

Then, for each experiment currently in transfer, we compute a theoretical deadline, i.e., the date at which it will be emptied at this rate of transfer if nothing changes. Notice that it can be *never*. Similarly, we compute a theoretical deadline for filling the mass-memory. If the earliest of all these deadlines happens earlier than the next scheduled event, we add it to the list of events. This type of events will do nothing on its own, however, it will allow the algorithm to recompute the transfers according to the new situation (the mass-memory being filled, or an experiment's memory being empty).

Finally, the usage of each memory at time t is updated according to the transfers.

This algorithm has a worst case time complexity of $O(nm \log(nm))$. The list of events has initially $O(nm)$ elements. There are two for each data-producing task, two for each visibility window, and two for each experiment (we assume that the number of visibility windows is less than nm). Sorting them can therefore be done in $O(nm \log(nm))$ time. In the main loop, events are processed only once, and this takes at most $O(m)$ time. Moreover, in some cases, "deadline" events can be added during the exploration of the event list. However, at most one such event can be added for each event initially in the list. Indeed, consider a deadline event. It is created only if no other event yet to process has an earlier date. In other words, transfer and data-producing rates as well as visibility do not change. The experiment memory that was emptied will therefore stay empty at least until the next standard event. The same is true for deadline event triggered by filled mass-memory: it will stay full at least until the next visibility event. Therefore, the worst case time complexity of the main loop is $O(nm)$.

4.2 Filtering Rules

In this section we introduce two propagation rules for the DATATRANSFER constraint.

Minimal transfer span: The first rule tries to guess a lower bound on the total span of a subset of activities of the same experiment E_k. The intuition is that if data is produced at a higher rate than it can be transferred out, the capacity of a memory could be reached and data will be lost. In other words, given a set $\Omega \subseteq E_k$ of data-producing activities of an experiment E_k, the total amount of data produced by these activities, minus what can be stored on the memory of E_k, need to be transferred out. The duration of this transfer is a lower bound on the span of this set of activities, i.e., the duration between the minimum start time and maximum end time of any activity in this set.

The total amount of data produced by activities in Ω is equal to $\sum_{t_{ki} \in \Omega} \pi_{ki} p_{ki}$. At most M_k can be stored on the experiment's own memory, hence at least $\sum_{t_{ki} \in \Omega} \pi_{ki} p_{ki} - M_k$ has to be transferred out *before* the end of the last data-producing activity. Let τ be the highest possible transfer rate for data out of the experiment's own memory. We can use this rate to derive a lower bound on the total duration of Ω:

$$\left(\max_{t_{ki} \in \Omega} (e_{ki}) - \min_{t_{ki} \in \Omega} (s_{ki}) \right) \geq \frac{\sum_{t_{ki} \in \Omega} \pi_{ki} p_{ki} - M_k}{\tau} \tag{7}$$

In real scenarios, data-producing activities of a given experiment cannot overlap, and in many cases the order is known a priori. Assuming that the activities in Ω are ordered, with t_{kf} being the first task and t_{kl} being the last task in Ω, we can often induce the simpler constraint: $e_{kl} - s_{kf} \geq \frac{\sum_{t_{ki} \in \Omega} \pi_{ki} p_{ki} - M_k}{\tau}$.

Example 1. Figure 4 depicts the application of this rule. We have two activities t_1, t_2, the former producing $\pi_1 = 5$ blocks/sec and the latter $\pi_2 = 4$ blocks/sec, both for 70 seconds. Therefore, $\pi_1 p_1 + \pi_2 p_2 = 630$ blocks are produced. Assume that the memory of this experiment has a capacity of 250 blocks. Consequently, 380 blocks need to be transferred out in order to avoid data loss. Since the maximum transfer rate is 2 blocks/seconds, this transfer will take at least 190 seconds. We can conclude that the end of t_2 is at least 190 seconds after the start of t_1. The grey scale gives the evolution of the memory for t_2 finishing exactly 190 seconds after the start of t_1.

Moreover, we can take into account the data produced by activities of experiments of higher priority, since their data transfers will preempt those of lower priority.

Consider an interval of time $[a, b]$. Any data produced by experiments of higher priority during this period must be transferred out before E_k can be allowed to transfer.

Let $\min(|t_{ki} \cap [a, b]|)$ be the minimum size of a common interval between $[a, b]$ and $[s_{ki}, s_{ki} + p_{ki}]$ for any value of s_{ki}. If $|[a, b] \cap [c, d]|$ stands for the size of the intersection

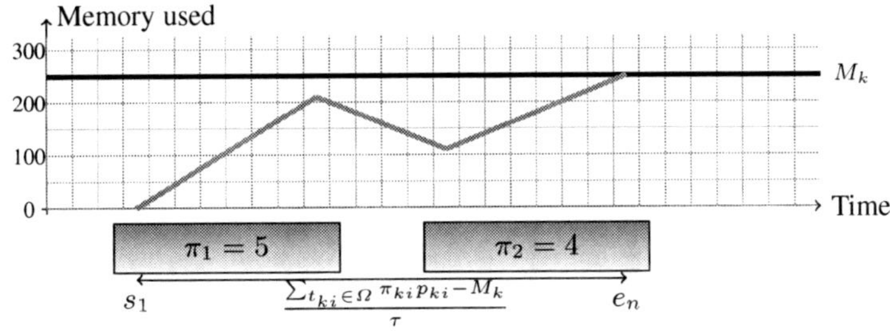

Fig. 4. Example of minimal span constraint

of intervals $[a, b]$ and $[c, d]$, then :

$$\min(|t_{ki} \cap [a, b]|) =$$
$$\min(|[a, b] \cap [\min(s_{ki}), \min(s_{ki}) + p_{ki}]|, |[a, b] \cap [\max(s_{ki}), \max(s_{ki}) + p_{ki}]|)$$

We can compute a lower bound $T_k(a, b)$ on the time required to transfer the data produced by experiments of higher priority than k on the interval $[a, b]$ as a lower bound on the data produced, divided by the maximum transfer rate:

$$T_k(a, b) = \left(\sum_{j=1}^{j<R(k)} \sum_{i=1}^{n} |t_{P(j)i} \cap [a, b]| * \pi_{P(j)i} \right)/\tau$$

Given a subset $\Omega \subseteq E_k$ of experiment E_k, consider the time interval $[a, b]$ between the latest start time of any task in Ω ($a = \min_{t_{ki}\in\Omega}(\max(s_{ki}))$) and the earliest end time of any task in Ω ($b = \max_{t_{ki}\in\Omega}(\min(s_{ki}) + p_{ki})$). The lower bound on the span given above assumes continuity of the transfer, and by definition this duration must include the interval $[a, b]$. Therefore, any interruption of the transfer during this period induces the same delay on the minimal span of Ω. In other words, any time taken to transfer data of experiments with higher priority during $[a, b]$ ($T_k(a, b)$) can be simply added to the lower bound above.

Hence we can tighten the constraint 7 as follows (with $a = \max_{t_{ki}\in\Omega}(\min(s_{ki}))$ and $b = \min_{t_{ki}\in\Omega}(\max(s_{ki}) + p_{ki})$):

$$\left(\max_{t_{ki}\in\Omega}(e_{ki}) - \min_{t_{ki}\in\Omega}(s_{ki}) \right) \geq \frac{\sum_{t_{ki}\in\Omega}\pi_{ki}p_{ki} - M_k}{\tau} + T_k(a, b) \qquad (8)$$

We apply this rule for every set of consecutive activities (with respect to their earliest start times) of every experiment. There are n^2m such sets, and computing the lower bound takes at most $O(nm)$ time. The whole procedure hence has a worst case time complexity of $O(n^3m^2)$.

Mass memory saturation: Since transfers from the lander to the orbiter are possible only during visibility, the data can only accumulate on the mass-memory while not in visibility. As a consequence, the period that precedes a visibility window is critical since the mass memory can be saturated hence blocking all transfers. When this happens, data produced by an experiment remains on its memory at least until the next visibility window, and it is possible to lose data when the experiment's memory itself is saturated.

We use this observation to deduce that data-producing activities that would generate too much data to hold on the mass memory and on their own memory should be either advanced or postponed. Suppose that we know that at time $\bar{t}$, the mass-memory will necessarily be filled. It will remain so until the next visibility. Now, if an activity t_{ki} produces more data in the interval between $\bar{t}$ and the next visibility window than its own memory can hold, it will be lost. Indeed, no data can be transferred onto the mass-memory as long as it is full, and it will start to be emptied only when the visibility allows it. We can thus deduce that the activity t_{ki} must start either early enough to produce before $\bar{t}$ or late enough so that the data in excess will be produced during the visibility period (in order to have a chance to be transferred).

First, we show how to compute an upper bound $\bar{t}$ on the time when the mass-memory will reach its maximum before a given visibility window. We consider a single visibility cycle $V = (a, v, b)$, where $a < v < b$ denote, respectively, the end of the previous cycle, the start of a visibility window, and the end of that visibility window. Let $\Omega(V)$ be the set of data-producing activities that are necessarily scheduled within the interval $[a, b]$.

Proposition 1. *Scheduling all data-producing activities in $\Omega(V)$ to their latest start time minimizes the memory usage of the mass-memory (m_0^t) for all $t \in [a, b]$.*

Proof (sketch). Clearly if we consider a data-producing activity t_{ki} in isolation, setting its start time to the latest possible time point $(\max(s_{ki}))$ delays the transfer onto the mass memory hence its memory usage for any time point in $[a, b]$.

When multiple data-producing activities can run in parallel, experiments of high priority can preempt transfer intervals of experiment of lower priority. Therefore, one could advance a data-producing activity t_{jl} in time in order to use the resource and therefore delay the transfer of some of the data produced by t_{ki}. However, since the transfer rate increases with the priority, for any time interval where the transfer of the data produced by t_{jl} preempts that produced by t_{ki}, data is being transferred to the mass memory at a higher rate. Thus, advancing a data-producing activity t_{jl} of higher priority never helps minimizing the mass memory usage. $\qquad\square$

Given a visibility cycle $V = (a, v, b)$, we can therefore get a lower bound on m_0^t on the usage of the mass memory for any t in the interval $[a, b]$ using the sweep algorithm. For every task in $\Omega(V)$, we tentatively fix it to its latest start time and execute the sweep algorithm. Hence, we can easily compute $\bar{t}$, the smallest value of t for which $m_0^t = m_0^v$.

Given an experiment E_k. We can bound the amount of data that can be produced by any task of this experiment in the period $[\bar{t}, v]$ *and* stored without loss. There are $m_0^{\bar{t}}$ blocks of data already on the mass-memory, so $M_0 - m_0^{\bar{t}}$ is free. Moreover, up to M_k can be stored on the experiment's own memory, for a total of $\delta_k = M_0 + M_k - m_0^{\bar{t}}$ blocks. Above this threshold, data produced by activities of experiment E_k between $\bar{t}$ and v will be lost. If $|t_{ki} \cap [\bar{t}, v]|$ stands for the length of the overlap between an activity t_{ki} and the interval $[\bar{t}, v]$, an activity t_{ki} produces $|t_{ki} \cap [\bar{t}, v]| * \pi_{ki}$ blocks of data in the interval $[\bar{t}, v]$. Therefore, the following relation must hold: $\sum_{i=1}^{n}(\min(|t_{ki} \cap [\bar{t}, v]|) * \pi_{ki}) \leq \delta_k$ from which we can deduce the following implied constraint:

$$|t_{ki} \cap [\bar{t}, v]| \leq (\delta_k - \sum_{j \neq i \in [1,n]} (\min(|t_{kj} \cap [\bar{t}, v]|) * \pi_{kj}))/\pi_{ki} \qquad (9)$$

We run the sweep algorithm once to obtain the value of $\bar{t}$. Then, for each experiment, we can compute δ_k and in time $O(n)$ the values of $\min(|t_{kj} \cap [\bar{t}, v]|$ for each activity t_{ki}. Finally we compute the implied constraint also in time $O(n)$ (it takes constant time for each activity, once $\min(|t_{kj} \cap [\bar{t}, v]|$ is known). Finally we apply it only when it collapses to a simple lower or upper bound on the start time s_{ki} of an activity t_{ki}. The total time complexity of this filtering rule is thus $O(nm \log(nm) + nm) = O(nm \log(nm))$.

5 Experimental Results

All the previous algorithms and filtering rules have been implemented on the latest version of MOST. We ran experimentations on different scenarios provided by the group SONC of CNES. Each scenario consists in one, two or three experiments which must be scheduled on a time window between 10 hours and 1 day. For each subset of experiments, several variations are tested in order to assess uncertain parameters. For instance, the visibility cycle depends on the exact mass and shape of the comet, the orbit selected by Rosetta, and the landing site chosen for Philae, all of which are unknown. Some scenarios have continuous visibility, while other have different periods for the visibility cycles. The hardware onboard the probe will have travelled in extreme temperatures for ten years, so the exact charge and efficiency of the batteries is also uncertain. Moreover, engineers of SONC test a range of variations on other parameters such as the memory capacity simply to stress-test the system (MOST).

5.1 Search Effort

We ran 8 scenarios and compared the results of the current version of MOST against the ad-hoc propagator introduced in this paper. Both were run on quad-core Sun $T5120$ running Solaris 2.10 with $8GB$ of RAM. The current version of MOST (denoted MOST+ILCRESERVOIR) models data transfers using Ilog-Scheduler ILCRESERVOIR constraints. In our version (denoted MOST+DATATRANSFER) we use only the first filtering rule described in Section 4.2.[4]

We report the results in Table 1. We present for each scenario the set of experiments involved, the memory capacities, and whether the visibility is continuous or not. Then we give the number of fails calculated by Ilog-Scheduler during search, the initialization time and finally the solving time.

We observe first that using our approach, solutions can be obtained without any fail, whereas the previous model explored a much larger search tree. The reformulation using ILCRESERVOIR constraints was indeed very loose, and did not allow to detect inconsistencies early. Moreover, to overcome this weakness, the scenarios produced by the group SONC are overly constrained in order to cut possibilities and allow the solver to converge more easily. Moreover, our propagator is relatively light and therefore more

---

[4] The second filtering rule was not implemented when the experiments were run.

Table 1. Old vs. new version of MOST on 8 standard scenarios

SCENARIO	PARAMETERS			MOST+ILC-RESERVOIR			MOST +DATATRANSFER	
	M_k	M_0	Visi.	Fail	Init. time (s)	Search time (s)	Init. time (s)	Search time (s)
Consert	500	17456	Periodic	295	4.06	20.07	0.88	0.08
Consert/Romap	500/250	17456	Periodic	7112	11.13	Time out	1.17	0.1
Consert/Romap	500/250	37456	Periodic	7051	11.03	Time out	1.17	0.1
SD2/Ptolemy	64/2000	17456	Periodic	234	26.71	41.72	3.37	0.09
SD2/Ptolemy	64/2000	17456	Continuous	211	32.78	79.48	3.25	0.08
SD2/Cosac/Civa	64/24000/4000	37456	Periodic	407	50.20	181.91	2.75	0.14
SD2/Cosac/Civa	64/24000/4000	17456	Periodic	413	50.84	179.19	2.95	0.15
SD2/Cosac/Civa	64/24000/4000	17456	Continuous	390	25.12	91.08	1.82	0.10

time effective, compared to the model using a large amount of transfer tasks throughout the horizon for each reservoir constraint.

In fact, the model was so large that the initialization time is very high. The few seconds of initialization time in our approach correspond to the rest of the model (cumulative and unary resources) which is common to both implementations.

In two cases, no solution was found by MOST+ILCRESERVOIR within the 600 seconds time cutoff. However, this is not explained (only) by performance issues. In fact, these two scenarios do not have a valid solution under the old model, whereas they are feasible.

5.2   A More Accurate Modeling

In MOST+ILCRESERVOIR, activities with very low data-producing rate are treated differently because of rounding issues: It is assumed that the data is produced all at once at the end of the activity. Therefore in these scenarios, transfers can be delayed by a substantial amount compared to the real behavior of the CDMS.

Moreover, since transfer tasks have a frequency of 120 seconds, they cannot accurately model situations where the CDMS frequently switches between different transfers. The scenario Consert/Romap highlights this problem on SONC's version of MOST. Both experiments have small data-producing activities and small memory capacities. Therefore, switches between transfers from these two experiments are extremely frequent. However, with MOST+ILCRESERVOIR it is not possible to switch frequently enough since data-transfer tasks are preallocated every 120 seconds on the time line. As a result, the model using the ILCRESERVOIR constraint has no solution, whereas transfers are actually possible.

6   Conclusion

In this paper we have presented an application of constraint programming for the international spatial mission ROSETTA/PHILAE. We have identified that the main problem is the management of data transfers and in particular, data loss. We have shown that the previous constraint programming approach was not well-adapted to this problem and we introduced a global constraint to forbid data-loss. In particular we proposed an efficient sweep algorithm which checks and computes the feasibility of data transfers.

We also have presented two propagation rules for the data transfer constraint. Overall, our approach greatly improves the results both for computing times, and accuracy of the solutions.

Acknowledgements. This work has been supported by CNES. We would like to thank members of the Science Operations and Navigation Centre (SONC) in Toulouse for their invaluable input.

References

1. Aggoun, A., Beldiceanu, N.: Extending CHIP in order to solve complex scheduling and placement problems. Mathematical and Computer Modelling 17(7), 57–73 (1993)
2. Beaumet, G., Verfaillie, G., Charmeau, M.-C.: Feasibility of autonomous decision making on board an agile earth-observing satellite. Computational Intelligence 27(1), 123–139 (2011)
3. Beldiceanu, N., Carlsson, M.: Sweep as a Generic Pruning Technique Applied to the Non-overlapping Rectangles Constraint. In: Walsh, T. (ed.) CP 2001. LNCS, vol. 2239, pp. 377–391. Springer, Heidelberg (2001)
4. Cesta, A., Cortellessa, G., Denis, M., Donati, A., Fratini, S., Oddi, A., Policella, N., Rabenau, E., Schulster, J.: Mexar2: AI solves mission planner problems. IEEE Intelligent Systems 22, 12–19 (2007)
5. Morris, R., Frank, J., Jónsson, A., Smith, D.E.: Planning and scheduling for fleets of earth observing satellites. In: 6th i-SAIRAS, pp. 18–22 (2001)
6. Laborie, P.: Algorithms for propagating resource constraints in AI planning and scheduling: existing approaches and new results. Artificial Intelligence 143(2), 151–188 (2003)
7. Mancel, C., Lopez, P.: Complex optimization problems in space systems. In: 13th International Conference on Automated Planning & Scheduling, ICAPS 2003 (2003)
8. Oddi, A., Policella, N.: Improving robustness of spacecraft downlink schedules. IEEE Transactions on Systems, Man and Cybernetics 37(5), 887–896 (2007)
9. Nuijten, W., Baptiste, P., Le Pape, C.: Constraint Based Scheduling. Springer (2001)
10. Righini, G., Tresoldi, E.: A mathematical programming solution to the Mars Express memory dumping problem. IEEE Transactions on Systems, Man and Cybernetics 40(3), 268–277 (2010)

Max-Sur-CSP on Two Elements

Hannes Uppman*

Department of Computer and Information Science,
Linköping University, SE-581 83 Linköping, Sweden
hannes.uppman@liu.se

Abstract. Max-Sur-CSP is the following optimisation problem: given a set of constraints, find a surjective mapping of the variables to domain values that satisfies as many of the constraints as possible. Many natural problems, e.g. Minimum k-Cut (which has many different applications in a variety of fields) and Minimum Distance (which is an important problem in coding theory), can be expressed as Max-Sur-CSPs. We study Max-Sur-CSP on the two-element domain and determine the computational complexity for all constraint languages (families of allowed constraints). Our results show that the problem is solvable in polynomial time if the constraint language belongs to one of three classes, and NP-hard otherwise. An important part of our proof is a polynomial-time algorithm for enumerating all near-optimal solutions to a generalised minimum cut problem. This algorithm may be of independent interest.

1   Introduction

In the *constraint satisfaction problem* (CSP) one is given a set of variables, a set of domain values and a collection of constraints, and is asked to determine if there is an assignment of values to the variables that satisfies all of the constraints. This general framework unifies a huge class of computational problems. However, some natural problems resist being formulated as CSPs and has motivated the introduction of several variants of CSP, e.g. the *maximum constraint satisfaction problem* (Max-CSP), where one tries to find an assignment that satisfies as many of the constraints as possible, and the *surjective constraint satisfaction problem* (Sur-CSP), where one asks if there is a surjective assignment that satisfies the constraints. Max-CSP has been intensively studied and a wealth of results can be found in the literature; it is understood on small domains [8,11,15,16,18], over some restricted classes of languages [19], and its approximability is understood under the assumption of the unique games conjecture [23]. Also Sur-CSP has been thoroughly investigated, mostly from a graph-theoretical perspective. Deciding the complexity, even for small special cases, of this problem has however turned out to be challenging, see [2,13,20,22].

Recently Bach and Zhou [1] introduced the *maximum surjective constraint satisfaction problem* (Max-Sur-CSP) which in a natural way combines Max-CSP and Sur-CSP. An example of a problem that is readily expressed as a Max-Sur-CSP, but not captured in a good way as a Max-CSP or a Sur-CSP, is Minimum k-Cut. This is a problem

---

* Partially supported by the National Graduate School in Computer Science (CUGS), Sweden.

M. Milano (Ed.): CP 2012, LNCS 7514, pp. 38–54, 2012.

from graph theory that has applications in a broad variety of fields both in science and engineering. Many other variations of cut problems are also nicely described as Max-Sur-CSPs. One such example is the *minimum asymmetric cut problem* [1]. Yet another well studied example that is easily expressed as a Max-Sur-CSP is the Minimum Distance problem, a central problem in coding theory. The complexity of this problem was unknown, but conjectured to be NP-complete, for nearly 20 years before Vardy [26] turned the conjecture into a theorem. It might seem peculiar that the examples above all deal with minimisation whereas Max-Sur-CSP is a maximisation problem. However, switching between the two forms is easy. Consider e.g. the Minimum Cut problem; we are given a graph and want to partition its vertices into two blocks so that the number of edges crossing the block-boundaries is minimised. Clearly we may think of this also as the maximisation problem in which we search for a partition that gives us as many block-internal edges as possible.[1]

In this paper we focus on Max-Sur-CSP over the two-element domain and determine the computational complexity for all constraint languages (classes of allowed constraints). We show that the problem either can be solved in polynomial time or is NP-hard. The tractable languages are described. Furthermore, we prove that if there is a polynomial-time algorithm that finds one optimal solution, then there is also a polynomial-delay algorithm that enumerates all of them.

It is known that α-minimum cuts (cuts of weight no more than α times the weight of a minimum cut) in a graph can be enumerated efficiently and that the number of such cuts is bounded by a polynomial in the size of the graph [17,21,27]. We show that similar results carry over to a (slightly) generalised version of the minimum asymmetric cut problem. This gives us a polynomial-time algorithm that plays an important role in our classification. In fact, we show that the languages handled by this algorithm together with languages in a previously known tractable class (the 2-monotone languages) make up the only tractable fragments of Max-Sur-CSP. Hardness results emerge from connections to the related problems Max-CSP and Sur-CSP, and the problem Minimum Distance.

The organisation of the paper is as follows. Section 1.1 covers preliminaries and related work, Sect. 2 contains hardness results and the classification, and Sect. 3 contains our algorithmic results.

1.1 Preliminaries and Related Work

We use the notation $\mathbb{N} = \mathbb{Z}_{\geq 0}$, $\mathbb{B} = \{0, 1\}$, $[n] = \{1, \ldots, n\}$, $\mathbf{0} = (0, \ldots, 0)$ and $\mathbf{1} = (1, \ldots, 1)$. By $\subset$ we mean proper subset. We use $\leq$ to denote the partial order over $\mathbb{B}^n$ where $s \leq t$ if and only if $s_i \leq t_i$ for every $i \in [n]$. By $\mathrm{proj}_s(x)$ we denote a (generalised) projection of x described by the tuple s, e.g. $\mathrm{proj}_{(3,2)}(a, b, c) = (c, b)$ and $\mathrm{proj}_{(1,1,1)}(a, b) = (a, a, a)$. The indicator function for a set A is denoted by χ^A, i.e. $\chi^A(x) = 1$ if $x \in A$ and $\chi^A(x) = 0$ otherwise. The arity of a tuple t (relation R) is written $\mathrm{ar}(t)$ ($\mathrm{ar}(R)$).

Let D (the domain) be a finite set and Γ (the constraint language) be a finite collection of finitary relations over D. We define the problems $\mathrm{CSP}(D, \Gamma)$, $\mathrm{Sur\text{-}CSP}(D, \Gamma)$,

---

[1] Note that this kind of construction does not necessarily preserve approximability properties.

Max-CSP(D, Γ) and Max-Sur-CSP(D, Γ) as follows. An *instance*, to any of the problems, is a function $f : D^n \to \mathbb{N}$ defined as

$$f(x) = \sum_{i=1}^{m} w_i \chi^{R_i}(\mathrm{proj}_{s^i}(x))$$

through a collection of weights $w_i \in \mathbb{N}$, constraint relations $R_i \in \Gamma$, and constraint scopes $s^i \in [n]^{\mathrm{ar}(R_i)}$. The problems CSP$(D, \Gamma)$ and Sur-CSP(D, Γ) are decision problems and asks for an *answer* to the questions: is there $x \in D^n$, respectively is there $x \in \{y \in D^n : \{y_1, \ldots, y_n\} = D\}$, such that $f(x) = \sum_{i=1}^{m} w_i$? The problems Max-CSP$(D, \Gamma)$ and Max-Sur-CSP(D, Γ) are optimisation problems where a *solution* is an element in D^n, respectively in $\{x \in D^n : \{x_1, \ldots, x_n\} = D\}$, and an *optimal solution* is a solution that maximises f.

Since we consider only the two-element domain, we will throughout the paper

- use Γ to refer to a finite set of finitary relations on $\mathbb{B}$, and
- write CSP(Γ) to mean CSP$(\mathbb{B}, \Gamma)$, Max-CSP(Γ) to mean Max-CSP$(\mathbb{B}, \Gamma)$, etc.

We begin by setting up some notation. We let maj denote the *majority operation* on $\mathbb{B}$, i.e. the unique ternary operation that satisfies $x = \mathrm{maj}(x, x, y) = \mathrm{maj}(x, y, x) = \mathrm{maj}(y, x, x)$ for every $x, y \in \mathbb{B}$, we let neg denote the unary operation $\{0 \mapsto 1, 1 \mapsto 0\}$, and we let aff denote the operation $(x, y, z) \mapsto x + y + z \pmod 2$. We lift the finitary operations on $\mathbb{B}$ to powers of $\mathbb{B}$ by coordinate-wise application, e.g. $\max(x, y) = (\max(x_1, y_1), \ldots, \max(x_n, y_n))$ if $x, y \in \mathbb{B}^n$. For any unary operation φ on $\mathbb{B}$ we take this further and lift φ to any relation R on $\mathbb{B}$ by $\varphi(R) = \{\varphi(t) : t \in R\}$, and to any set Γ of relations on $\mathbb{B}$ by $\varphi(\Gamma) = \{\varphi(R) : R \in \Gamma\}$. A relation R is said to be φ-*closed*, for some finitary operation φ on $\mathbb{B}$, if $\varphi(t^1, \ldots, t^k) \in R$ for every $t^1, \ldots, t^k \in R$. Finally, if φ and σ are binary operations on $\mathbb{B}$, we say that a relation $R \subseteq \mathbb{B}^m$ admits the (φ, σ) *multimorphism* if $\chi^R(x) + \chi^R(y) \le \chi^R(\varphi(x, y)) + \chi^R(\sigma(x, y))$ for every $x, y \in \mathbb{B}^m$ [7].

A relation R over $\mathbb{B}$ is said to be *0-valid* if $\mathbf{0} \in R$, *1-valid* if $\mathbf{1} \in R$, *Horn* if R is min-closed, *dual Horn* if R is max-closed [14], *bijunctive* if R is maj-closed [24], *affine* if R is aff-closed [10,24] and *2-monotone* if R admits the $(\min, \max)$ multimorphism [6]. We say that a constraint language Γ is of a certain class if every relation in Γ is of this class, e.g. Γ is Horn if every $R \in \Gamma$ is Horn.

We are now ready to state some complexity results on the two-element domain.

Theorem 1 (Schaefer [24]). *The problem CSP(Γ) is*

- *in P if Γ is 0-valid, 1-valid, Horn, dual Horn, affine or bijunctive,*
- *and NP-complete otherwise.*

Theorem 2 (Creignou [8] (see also [18])). *The problem Max-CSP(Γ) is*

- *in PO if Γ is 0-valid, 1-valid or 2-monotone,*
- *and APX-complete otherwise.*

Theorem 3 (Creignou and Hébrard [9]). *The problem Sur-CSP(Γ) is*

- *in P if Γ is Horn, dual Horn, affine or bijunctive,*
- *and NP-complete otherwise.*

For CSP a complete classification is known also on the three-element domain [3], and for Max-CSP the results has been extended to domains with up to four elements [15,16].

Some relations will show up from time to time, we let $=_\mathbb{B} = \{x \in \mathbb{B}^2 : x_1 = x_2\}$, $\leq_\mathbb{B} = \{x \in \mathbb{B}^2 : x_1 \leq x_2\}$, $\text{NAND}_k = \mathbb{B}^k \setminus \{\mathbf{1}\}$, $C_0 = \{0\}$, $C_1 = \{1\}$ and $A_k = \{x \in \mathbb{B}^k : \sum_{i=1}^{k} x_i = 0 \,(\text{mod } 2)\}$.

Finally, we introduce two new classes of relations.

Definition 4. *A relation $R \subseteq \mathbb{B}^m$ is called* almost-min-min *if there is $I \subseteq \bigcup_{k=1}^{m} [m]^k$ and relations $\{P_s : s \in I\} \subseteq \{=_\mathbb{B}, NAND_1, \ldots, NAND_k\}$ so that $R = \{x \in \mathbb{B}^m : \bigwedge_{s \in I} \text{proj}_s(x) \in P_s\}$. A relation R is called* dual almost-min-min *if $\text{neg}(R)$ is almost-min-min.*

Note that a relation is almost-min-min if it (interpreted as a predicate) can be expressed as a finite conjunction over (the predicates) $=_\mathbb{B}$ and NAND_k.

Example 5. The relation $R = \{(0,0,0), (0,1,1), (1,0,0)\}$ is almost-min-min since $R = \{x \in \mathbb{B}^3 : (x_1, x_2) \in \text{NAND}_2 \wedge (x_2, x_3) \in =_\mathbb{B}\}$.

2   Complexity Dichotomy

Our main result is the following dichotomy theorem.

Theorem 6. *The problem Max-Sur-CSP(Γ) is*
- *in PO if Γ is almost-min-min, dual almost-min-min or 2-monotone,*
- *and NP-hard otherwise.*

Bach and Zhou [1] showed that, for any finite domain D and any finite set Δ of finitary relations over D, the following holds.

Theorem 7. *If Max-CSP(D, Δ) is in PO, then Max-Sur-CSP(D, Δ) is in PTAS, and if Max-CSP(D, Δ) is APX-hard, then Max-Sur-CSP(D, Δ) is APX-hard.*

Combining Theorems 6, 7 and 2 gives the following more nuanced classification.

Theorem 8. *The problem Max-Sur-CSP(Γ) is*
- *in PO if Γ is almost-min-min, dual almost-min-min or 2-monotone,*
- *otherwise it is NP-hard to solve exactly, but*
  - *in PTAS if Γ is 0-valid or 1-valid,*
  - *and APX-hard otherwise.*

We can also show that if we can find one optimal solution, then we can enumerate all (with polynomial delay). Polynomial-delay enumeration has in several settings related to CSP been studied before, see e.g. [4,5,9,12,25,27].

Proposition 9. *If Max-Sur-CSP(Γ) is in PO, then there is a polynomial-delay algorithm that enumerates every optimal solution to any instance of Max-Sur-CSP(Γ).*

In the rest of this section we will, with some help of algorithmic results from Sect. 3, prove Theorem 6 and Proposition 9.

We will need to do reductions among problems. A key building block of many of these reductions will be the following restricted form of (strict) implementations [11].

Definition 10. *Let $\{c_1, \ldots, c_p\} \subseteq \mathbb{B}$, $S \subseteq \mathbb{B}^q$ and $R \subseteq \mathbb{B}^r$. We say that the relation S implements the relation R using constants $c_1, \ldots, c_p$ if there is $s \in [r + p]^q$ such that $R = \{x \in \mathbb{B}^r : \mathrm{proj}_s(x_1, \ldots, x_r, c_1, \ldots, c_p) \in S\}$. In this case we write $S \xrightarrow{c_1, \ldots, c_p} R$.*

Example 11. Define $R = \left\{ \begin{array}{l} (0, 0, 0, 1), \\ (0, 0, 1, 1), \\ (0, 1, 0, 1) \end{array} \right\}$. Note that $R \xrightarrow{1} C_0$ since $C_0 = \{x \in \mathbb{B} : \mathrm{proj}_{(1,1,1,2)}(x, 1) \in R\}$, $R \to R$ since $R = \{x \in \mathbb{B}^m : \mathrm{proj}_{(1,2,3,4)}(x) \in R\}$ and $R \xrightarrow{0,1} \mathrm{NAND}_2$ since $\mathrm{NAND}_2 = \{x \in \mathbb{B}^2 : \mathrm{proj}_{(3,1,2,4)}(x; 0, 1) \in R\}$.

We use implementations to construct polynomial-time reductions as follows.

Lemma 12. *If $\{S, C_{c_1}, \ldots, C_{c_p}\} \subseteq \Gamma$ and $S \xrightarrow{c_1, \ldots, c_p} R$, then Max-Sur-CSP$(\Gamma \cup \{R\})$ $\leq_p$ Max-Sur-CSP(Γ).*

Proof. We prove the lemma for the case when $(c_1, \ldots, c_p) = (0, 1)$, the other cases are handled analogously. Let $f : \mathbb{B}^n \to \mathbb{N}$ be a Max-Sur-CSP$(\Gamma \cup \{R\})$ instance $f(x) = f_1(x) + \sum_{i=1}^{m} w_i \chi^R(\mathrm{proj}_{s^i}(x))$. Let s be such that $z \in R$ if and only if $\mathrm{proj}_s(z; 0, 1) \in S$. We know that such a tuple exists since $S \xrightarrow{0,1} R$. For every $p, q \in [n]$ we create the instance

$$f_{p,q}(x) = f_1(x) + \sum_{i=1}^{m} w_i \chi^S(\mathrm{proj}_s(\mathrm{proj}_{s^i}(x); x_p, x_q)) + M\chi^{C_0}(x_p) + M\chi^{C_1}(x_q),$$

where the weight M is chosen large enough (e.g. as twice the sum of all weights in the instance f) so that any optimal solution x to this instance satisfies $x_p = 0$ and $x_q = 1$, assuming $p \neq q$. This construction yields an instance of Max-Sur-CSP(Γ) and can be carried out in polynomial time. Note that for at least one pair $(p, q) \in [n]^2$ any optimal solution to $f_{p,q}$ is also an optimal solution to f. $\square$

Note that implementations are transitive in the following sense: if R, S and T are relations over $\mathbb{B}$ such that $R \xrightarrow{c_1, \ldots, c_u} S \xrightarrow{d_1, \ldots, d_v} T$, then there is $\{e_1, \ldots, e_w\} \subseteq \{c_1, \ldots, c_u, d_1, \ldots, d_v\}$ such that $R \xrightarrow{e_1, \ldots, e_w} T$.

Another observation that will be useful to us several times is the following.

Lemma 13. *Max-Sur-CSP$(\Gamma) \leq_p$ Max-Sur-CSP$(\mathrm{neg}(\Gamma))$*

Proof. Let $f : \mathbb{B}^n \to \mathbb{N}$ be an instance $f(x) = \sum_{i=1}^{m} w_i \chi^{R_i}(\mathrm{proj}_{s^i}(x))$ of Max-Sur-CSP(Γ) and let g the corresponding instance of Max-Sur-CSP$(\mathrm{neg}(\Gamma))$, i.e. $g(x) = \sum_{i=1}^{m} w_i \chi^{\mathrm{neg}(R_i)}(\mathrm{proj}_{s^i}(x))$. It is not hard to verify that $x \in \mathbb{B}^n$ is an optimal solution to g if and only if $\mathrm{neg}(x)$ is an optimal solution to f. $\square$

Affine Relations. We will here show hardness for some affine languages by establishing a connection to the Minimum Distance problem. An instance of this NP-complete [26] decision problem is a matrix $H \in \mathbb{B}^{m \times n}$ and an integer $w > 0$. The question is if there is $x \in \mathbb{B}^n$ such that $0 < \sum_{i=1}^{n} x_i \leq w$ and $Hx = 0 \,(\mathrm{mod}\ 2)$. We begin by connecting Minimum Distance to Max-Sur-CSP for one of the most simple affine relations.

Lemma 14. *Minimum Distance $\leq_p$ Max-Sur-CSP($\{A_3\}$).*

Proof. From an instance $H \in \mathbb{B}^{m \times n}, w \in \mathbb{N}$ of the Minimum Distance problem, we can find $s_{i,j}$ and a_i such that the system $Hx = 0 \,(\mathrm{mod}\ 2)$ can be written $\sum_{j=1}^{a_i} x_{s_{i,j}} = 0 \,(\mathrm{mod}\ 2)$, $i \in [m]$. Define the function $f_i : \mathbb{B}^{n+1+a_i} \to \mathbb{N}$ as

$$f_i(x_1, \ldots, x_n, y_0, \ldots, y_{a_i}) = \chi^{C_0}(y_0) + \sum_{j=1}^{a_i} \chi^{A_3}(y_{j-1}, x_{s_{i,j}}, y_j) + \chi^{C_0}(y_{a_i}).$$

Note that $(x_1, \ldots, x_n, y_0, \ldots, y_{a_i}) \in \mathbb{B}^{n+1+a_i}$ maximises f_i if and only if $y_0 = 0$, $y_j = \sum_{k=1}^{j} x_{s_{i,k}} \,(\mathrm{mod}\ 2)$, and $y_{a_i} = \sum_{j=1}^{a_i} x_{s_{i,j}} = 0 \,(\mathrm{mod}\ 2)$. To solve the Minimum Distance instance we can therefore create an instance $f : \mathbb{B}^{n+(1+a_1)+\cdots+(1+a_m)} \to \mathbb{N}$ of Max-Sur-CSP($\{A_3, C_0\}$) as

$$f(x; y^1; \ldots; y^m) = n \sum_{i=1}^{m} f_i(x; y^i) + \sum_{i=1}^{n} \chi^{C_0}(x_i).$$

It is not hard to see (assuming the system $Hx = 0 \,(\mathrm{mod}\ 2)$ has a non-zero solution) that if $(x; y^1; \ldots; y^m)$ is any optimal solution to f, then x is a non-zero solution to $Hx = 0 \,(\mathrm{mod}\ 2)$ of minimum weight, $\sum_{i=1}^{n} x_i$. Comparing this weight with w then answers the Minimum Distance instance. So, since $A_3 \to C_0$, we are done. $\square$

From the following simple reduction we see that also the relation A_4 is hard.

Lemma 15. *Max-Sur-CSP($\{A_3\}$) $\leq_p$ Max-Sur-CSP($\{A_4\}$)*

Proof. Given an instance $f : \mathbb{B}^n \to \mathbb{N}$ of Max-Sur-CSP($\{A_3\}$) defined by $f(x) = \sum_{i=1}^{m} w_i \chi^{A_3}(\mathrm{proj}_{s_i}(x))$, we construct the Max-Sur-CSP($\{A_4\}$) instance $g : \mathbb{B}^{n+1} \to \mathbb{N}$ as $g(x) = \sum_{i=1}^{m} w_i \chi^{A_4}(\mathrm{proj}_{(s_i;n+1)}(x))$. Note that $g(\mathbf{0}; 1) = g(\mathbf{1}; 0) = 0$, $g(x, 0) = f(x)$ and $g(x, 1) = f(\mathrm{neg}(x))$ for every $x \in \mathbb{B}^n$. So if $(x_1, \ldots, x_n, y)$ is an optimal solution to g, then either x or $\mathrm{neg}(x)$ is an optimal solution to f. $\square$

Using the two previous lemmas we can prove hardness for many affine relations.

Lemma 16. *If $R \subseteq \mathbb{B}^m$ is affine and (0-valid or 1-valid), then R is bijunctive or Max-Sur-CSP($\{R\}$) is NP-hard.*

Proof. Consider first the case when R is 0-valid and not bijunctive. Since R is not bijunctive there are $t^1, t^2, t^3 \in R$ such that $\mathrm{maj}(t^1, t^2, t^3) \notin R$. Let $s^1 = \mathrm{aff}(\mathbf{0}, t^2, t^3)$, $s^2 = \mathrm{aff}(t^1, \mathbf{0}, t^3)$ and $s^3 = \mathrm{aff}(t^1, t^2, \mathbf{0})$. R is affine, so $s^1, s^2, s^3 \in R$. We claim $\mathrm{maj}(s^1, s^2, s^3) \notin R$. To see this note that $\mathrm{maj}(s_i^1, s_i^2, s_i^3) = 1$ if $|\{t_i^1, t_i^2, t_i^3\}| = 2$ and $\mathrm{maj}(s_i^1, s_i^2, s_i^3) = 0$ if $|\{t_i^1, t_i^2, t_i^3\}| = 1$. So if $\mathrm{maj}(s^1, s^2, s^3) \in R$, then $R \not\ni \mathrm{maj}(t^1, t^2, t^3) = \mathrm{aff}(\mathrm{aff}(t^1, t^2, t^3), \mathrm{maj}(s^1, s^2, s^3), \mathbf{0}) \in R$, which clearly is a contradiction. Let $\{p_1, \ldots, p_k\}$ be a partition of $[m]$ such that $i, j \in p_\ell$ if and only if $(s_i^1, s_i^2, s_i^3) = (s_j^1, s_j^2, s_j^3)$ and let $\varphi : [m] \to [k]$ be a function such that $\varphi(i) = j$ if $i \in p_j$. Now $R \to R'$ where $R' = \{x \in \mathbb{B}^k : \mathrm{proj}_{(\varphi(1), \ldots, \varphi(m))}(x) \in R\}$. By construction $(s_i^1, s_i^2, s_i^3) \in \{\mathbf{0}, (0, 1, 1), (1, 0, 1), (1, 1, 0)\}$, so $\mathrm{ar}(R') \leq 4$. Note that R' is affine, 0-valid and not bijunctive. Consider the following cases.

- If $\mathrm{ar}(R') \le 2$, then R' is bijunctive, so this case never occurs.
- If $\mathrm{ar}(R') = 3$, then $A_3 \subseteq R'$ or R' is bijunctive. Since $\mathrm{maj}(s^1, s^2, s^3) \notin R$ we have $\mathbf{1} \notin R'$. Also $(1,0,0) \notin R'$ since otherwise $\mathbf{1} = \mathrm{aff}((1,0,0),(0,1,1),\mathbf{0})) \in R'$, symmetrically $(0,1,0),(0,0,1) \notin R'$, hence $R' = A_3$.
- If $\mathrm{ar}(R') = 4$, then wlog we can assume $\{(0,0,1,1),(0,1,0,1),(0,1,1,0)\} \subseteq R'$.
  - If $\mathbf{1} \in R'$, then R' is neg-closed since $\mathrm{neg}(x) = \mathrm{aff}(\mathbf{1}, \mathbf{0}, x)$, so $A_4 \subseteq R'$. It is not hard to see (as in the previous case) that in this case $R' = A_4$.
  - If $\mathbf{1} \notin R'$, then $R' \to C_0$ (since R' is 0-valid and not 1-valid) and $R' \xrightarrow{0} R''$ where $R'' = \{x \in \mathbb{B}^3 : \mathrm{proj}_{(4,1,2,3)}(x; 0) \in R'\}$. By the same arguments as in the case $\mathrm{ar}(R') = 3$ we can conclude that $R'' = A_3$.

This means that Max-Sur-CSP($\{R\}$) is NP-hard since by Lemmas 14 and 15, Minimum Distance $\le_p$ Max-Sur-CSP($\{A_3\}$) $\le_p$ Max-Sur-CSP($\{A_4\}$).

What remains to be handled is the case when R is affine, 1-valid, not bijunctive and not 0-valid. In this case $\mathrm{neg}(R)$ is affine, 0-valid and not bijunctive, so by the result above and by Lemma 13 we are done. $\qquad\square$

Horn, Dual-Horn and Bijunctive Relations. Many constraint languages of the type studied here are strong enough to express the constant-relations C_0 and C_1. This will often make it easy for us to prove hardness results. A central lemma is the following.

Lemma 17. *If $\le_{\mathbb{B}} \in \Gamma$, then Max-Sur-CSP($\Gamma \cup \{C_0, C_1\}$) $\le_p$ Max-Sur-CSP(Γ).*

Proof. Let $f : \mathbb{B}^n \to \mathbb{N}$ be a Max-Sur-CSP($\Gamma \cup \{C_0, C_1\}$) instance

$$f(x) = f_1(x) + \sum_{i=1}^{m_0} w_i^0 \chi^{C_0}(x_{u_i}) + \sum_{i=1}^{m_1} w_i^1 \chi^{C_1}(x_{v_i}).$$

For every $p, q \in [n]$ we create the instance

$$f_{p,q}(x) = f_1(x) + \sum_{i=1}^{m_0} w_i^0 \chi^{\le_{\mathbb{B}}}(x_{u_i}, x_p) + \sum_{i=1}^{m_1} w_i^1 \chi^{\le_{\mathbb{B}}}(x_q, x_{v_i})$$
$$+ M \sum_{i=1}^{n} \chi^{\le_{\mathbb{B}}}(x_p, x_i) + M \sum_{i=1}^{n} \chi^{\le_{\mathbb{B}}}(x_i, x_q),$$

where the weight M is chosen large enough (e.g. as twice the sum of all weights in the instance f) so that any optimal solution x to this instance satisfies $x_p = 0$ and $x_q = 1$, assuming $p \ne q$. This construction yields an instance of Max-Sur-CSP(Γ) and can be carried out in polynomial time. Note that for at least one pair $(p, q) \in [n]^2$ any optimal solution to $f_{p,q}$ is also an optimal solution to f. $\qquad\square$

We will see that $\le_{\mathbb{B}}$ can be implemented from many min-closed relations, the following lemma handles one case.

Lemma 18. *If $R \subseteq \mathbb{B}^m$ is min-closed, 1-valid and not max-closed, then $R \to \le_{\mathbb{B}}$ or $R \to C_1$ and $R \xrightarrow{1} \le_{\mathbb{B}}$.*

Proof. Consider first the case when R is 0-valid. Note that R is not neg-closed since $\max(x, y) = \mathrm{neg}(\min(\mathrm{neg}(x), \mathrm{neg}(y)))$, so there is $t \in R$ such that $\mathrm{neg}(t) \notin R$. Hence, $R \to \leq_{\mathbb{B}}$ since $\leq_{\mathbb{B}} = \{x \in \mathbb{B}^2 : \mathrm{proj}_{(1+t_1,\ldots,1+t_m)}(x) \in R\}$. We now turn to the case when R is not 0-valid. In this case clearly $R \to C_1$. Note that R contains a $\leq$-minimum tuple t since if t^1 and t^2 are two $\leq$-minimal tuples in R, then $\min(t^1, t^2)$ is another $\leq$-minimal tuple, hence $t^1 = t^2$. Assume t contains k zeroes and define $\varphi : [m] \to [k + 1]$ as $\varphi(i) = k + 1$ if $t_i = 1$ and $\varphi(i) = |\{j \in [i] : t_j = 0\}|$ if $t_i = 0$. Now $R \xrightarrow{1} R'$ where $R' = \{x \in \mathbb{B}^k : \mathrm{proj}_{(\varphi(1),\ldots,\varphi(m))}(x; 1) \in R\}$. Clearly R' is min-closed, 0-valid, 1-valid and not max-closed, so by the case above and the transitivity of $\to$ we are done. $\qquad\square$

For convenience we introduce the operator, Ξ, that can be thought of as "removing duplicated columns" from relations.

Definition 19. *If $R \subseteq \mathbb{B}^m$, we let $\Xi(R) = \mathrm{proj}_{(s_1,\ldots,s_k)}(R)$, where $s_1 < s_2 < \cdots < s_k$ and $\{s_1, \ldots, s_k\} = \{i \in [m] : \text{if } j \in [m] \text{ and } t_j = t_i \text{ for all } t \in R, \text{ then } j \geq i\}$.*

Example 20. If $R = \left\{ \begin{array}{l} (1,\ 1,\ 1,\ 1), \\ (1,\ 0,\ 1,\ 0) \end{array} \right\}$ then $\Xi(R) = \left\{ \begin{array}{l} (1,\ 1), \\ (1,\ 0) \end{array} \right\}$.

We can now show that another collection of min-closed relations also implements $\leq_{\mathbb{B}}$.

Lemma 21. *Let R be a relation over $\mathbb{B}$. If R is 0-valid and min-closed, then either $\Xi(R)$ admits the $(\min, \min)$ multimorphism, or $R \to \leq_{\mathbb{B}}$ or $R \to C_0$ and $R \xrightarrow{0} \leq_{\mathbb{B}}$.*

Proof. The result holds for unary relations. Assume that it also holds for all relations of arity strictly less than m. Let R be of arity m and consider the following cases.

- $R \neq \Xi(R)$

  Note that $\Xi(R)$ is 0-valid and min-closed. Since $R \to \Xi(R)$ this means that the result follows from the inductive hypothesis.

- $R = \Xi(R)$

  - R is 1-valid

    * If R is not neg-closed, then there is $t \in R$ such that $\mathrm{neg}(t) \notin R$. In this case $R \to \leq_{\mathbb{B}}$ since $\leq_{\mathbb{B}} = \{x \in \mathbb{B}^2 : \mathrm{proj}_{(1+t_1,\ldots,1+t_m)}(x) \in R\}$.

    * If R is neg-closed, then $R = \mathbb{B}^m$. To prove this we pick any $s \in \mathbb{B}^m$ and show that $s \in R$. For $i \in [m]$, let s^i be the tuple where $s^i_i = s_i$ and $s^i_j = 1$ for all $j \neq i$. We claim $s^i \in R$. Note that if $s^i_i = 1$, then $s^i = 1 \in R$. If $s^i_i = 0$ we pick a set of tuples $\{y^{i,j}\}_{j \in [m]} \subseteq R$ such that $y^{i,j}_i = 0$ and $y^{i,j}_j = 1$ for $j \neq i$. Such a set exists; we know that there is at least one tuple $q \in R$ such that $q_i \neq q_j$ and that R is neg-closed, so we may set $y^{i,j}$ equal to either q or $\mathrm{neg}(q)$. This means, since R is max-closed as $\max(x, y) = \mathrm{neg}(\min(\mathrm{neg}(x), \mathrm{neg}(y)))$, that $s^i = \max(y^{i,1}, \max(\ldots \max(y^{i,m-1}, y^{i,m}) \ldots)) \in R$. Finally we have that $s = \min(s^1, \min(\ldots \min(s^{m-1}, s^m) \ldots)) \in R$ since R is min-closed. Since $R = \mathbb{B}^m$, clearly R admits the $(\min, \min)$ multimorphism.

  - R is not 1-valid

    Assume that R does not admit the $(\min, \min)$ multimorphism. Then, since R is min-closed, there is a pair $t^1, t^2 \in \mathbb{B}^m$ such that $t^1 \in R$, $t^2 \notin R$ and

$\min(t^1, t^2) \notin R$. Note that $R \to C_0$ since R is 0-valid and not 1-valid. Let R' be the projection of $\{x \in R : x \leq t^1\}$ onto the coordinates $\{i : t_i^1 = 1\}$. Clearly $R \xrightarrow{0} R'$. Note that R' is 0-valid, min-closed, has arity strictly less than m and does not admit the $(\min, \min)$ multimorphism.

Note also that $R' = \Xi(R')$. Assume this is false, then there are i, j such that $t_i = t_j$ for all $t \in R'$. Since $R = \Xi(R)$ there is $s \in R$ such that $s_{i'} \neq s_{j'}$ where i', j' are the coordinates in R that corresponds to the coordinates i, j in R'. Since R is min-closed $t^1 \geq \min(t^1, s) \in R$, so $\min(t^1, s)$ is projected to a tuple $s' \in R'$ satisfying $s_i' \neq s_j'$, which is a contradiction. Hence, by the inductive hypothesis, $R' \xrightarrow{0} \leq_{\mathbb{B}}$. $\qquad\square$

Relations that admit the $(\min, \min)$ multimorphism have a simple structure.

Lemma 22. *If $R \subseteq \mathbb{B}^m$ admits the* $(\min, \min)$ *multimorphism, then there exists $I \subseteq \bigcup_{k=1}^m [m]^k$ so that $R = \{x \in \mathbb{B}^m : \bigwedge_{s \in I} \mathrm{proj}_s(x) \in NAND_{|s|}\}$.*

Proof. The result holds for the unary relations. Assume it also holds for all relations of arity strictly less than m, and let R be of arity m.

First note that if R is 1-valid, then $R = \mathbb{B}^m$. We can therefore assume that R is not 1-valid. For $i \in [m]$, the relation $\mathrm{proj}_{(1,\ldots,i-1,i+1,\ldots,m)}(R)$ admits the $(\min, \min)$ multimorphism. Let $R' = \{x \in \mathbb{B}^m : x \in \mathrm{NAND}_m \wedge \bigwedge_{i=1}^m \mathrm{proj}_{(1,\ldots,i-1,i+1,\ldots,m)}(x) \in \mathrm{proj}_{(1,\ldots,i-1,i+1,\ldots,m)}(R)\}$. By the inductive hypothesis there is $I' \subseteq \bigcup_{k=1}^m [m]^k$ such that $R' = \{x \in \mathbb{B}^m : \bigwedge_{s \in I'} \mathrm{proj}_s(x) \in \mathrm{NAND}_{|s|}\}$. Clearly $R \subseteq R'$. Assume $R' \setminus R \neq \varnothing$. Let t be a $\leq$-maximal element in $R' \setminus R$ and pick any i such that $t_i = 0$. Since $t \in R'$ we have $\mathrm{proj}_{(1,\ldots,i-1,i+1,\ldots,m)}(t) \in \mathrm{proj}_{(1,\ldots,i-1,i+1,\ldots,m)}(R)$, so since $t \notin R$ we must have[2] $t_{i \leftarrow 1} \in R$. But R admits the $(\min, \min)$ multimorphism and $t \leq t_{i \leftarrow 1}$, so $t \in R$. This is a contradiction, hence $R' = R$. $\qquad\square$

Immediately from Lemma 22 we get the following result.

Corollary 23. *If $R \subseteq \mathbb{B}^m$ and $\Xi(R)$ admits the* $(\min, \min)$ *multimorphism, then R is almost-min-min.*

We now summarise the hardness results for Horn, dual Horn and bijunctive relations.

Lemma 24. *Let R be a finitary relation on $\mathbb{B}$. If R is* min-*closed or* max-*closed, then either R is 2-monotone, almost-min-min, dual almost-min-min, or Max-Sur-CSP($\{R\}$) is NP-hard.*

Proof. Note that Max-CSP($\{R\}$) $\leq_p$ Max-Sur-CSP($\{R\}$). Assume R is min-closed and consider the following cases.

- If R is 0-valid, then by Lemmas 17 and 21, Corollary 23 and Theorem 2 it follows that R is 2-monotone or almost-min-min, or Max-Sur-CSP($\{R\}$) is NP-hard.
- If R is not 0-valid, then
  - if R is not 1-valid, then R is 2-monotone or Max-Sur-CSP($\{R\}$) is NP-hard by Theorem 2, and

---

[2] By $t_{i \leftarrow x}$ we mean $(t_1, \ldots, t_{i-1}, x, t_{i+1}, \ldots, t_{\mathrm{ar}(t)})$.

- if R is 1-valid, then
    * if R is not max-closed, then R is not 2-monotone and Max-Sur-CSP($\{R\}$) is NP-hard by Lemmas 17 and 18, and Theorem 2, and
    * if R is max-closed, then $\mathrm{neg}(R)$ is min-closed, max-closed, 0-valid and not 1-valid, so (by a previous case) $\mathrm{neg}(R)$ is 2-monotone or almost-min-min, or Max-Sur-CSP($\{\mathrm{neg}(R)\}$) is NP-hard. By Lemma 13 this means that R is 2-monotone or dual almost-min-min, or Max-Sur-CSP($\{R\}$) is NP-hard.

If R is max-closed, then $\mathrm{neg}(R)$ is min-closed, so $\mathrm{neg}(R)$ is either 2-monotone, almost-min-min, dual almost-min-min, or Max-Sur-CSP($\{\mathrm{neg}(R)\}$) is NP-hard. This means, according to Lemma 13, that R is either 2-monotone, almost-min-min, dual almost-min-min, or Max-Sur-CSP($\{R\}$) is NP-hard. $\qquad\square$

Dichotomy. We are now ready to put the pieces from the last two subsections together.

Lemma 25. *Γ is 2-monotone, almost-min-min or dual almost-min-min, or Max-Sur-CSP(Γ) is NP-hard.*

Proof. Note that Sur-CSP(Γ) $\leq_p$ Max-Sur-CSP(Γ) and Max-CSP(Γ) $\leq_p$ Max-Sur-CSP(Γ). So, by Theorem 2 we know that Γ is 2-monotone, 0-valid or 1-valid, or Max-Sur-CSP(Γ) is NP-hard.

Consider first the case, when Γ is 0-valid and not 2-monotone. Assume Max-Sur-CSP(Γ) is not NP-hard. By Theorem 3 we know that Γ is Horn, dual Horn, bijunctive or affine. By Lemmas 16 and 24 we then get that every relation $R \in \Gamma$ is either 2-monotone, almost-min-min or dual almost-min-min. This means that every $R \in \Gamma$ is 2-monotone or almost-min-min. To see this, note that if R is dual almost-min-min, then $\Xi(R)$ is both 0-valid and admits the $(\max, \max)$ multimorphism, so $\Xi(R) = \mathbb{B}^m$ for some m. Hence, $\Xi(R)$ admits the $(\min, \min)$ multimorphism and R is almost-min-min. If Γ is not almost-min-min, then there exists $R \in \Gamma$ that is not almost-min-min (and not dual almost-min-min) and must therefore be 2-monotone and hence min-closed. By Corollary 23 and Lemmas 21 and 17 we can conclude that Max-Sur-CSP($\Gamma \cup \{C_0, C_1\}$) $\leq_p$ Max-Sur-CSP(Γ). Note that $\Gamma \cup \{C_0, C_1\}$ is not 0-valid, not 1-valid and not 2-monotone, so by Theorem 2, Max-Sur-CSP($\Gamma \cup \{C_0, C_1\}$), and therefore also Max-Sur-CSP(Γ), is NP-hard.

If Γ is 1-valid and not 2-monotone, then $\mathrm{neg}(\Gamma)$ is 0-valid and not 2-monotone, so we know that $\mathrm{neg}(\Gamma)$ is almost-min-min or dual almost-min-min, or Max-Sur-CSP($\mathrm{neg}(\Gamma)$) is NP-hard. Hence, by Lemma 13, Γ is almost-min-min or dual almost-min-min, or Max-Sur-CSP(Γ) is NP-hard. $\qquad\square$

Using results from Sect. 3 we can now prove Theorem 6 and Proposition 9.

Proof (of Theorem 6). Max-Sur-CSP(Γ) is known to be in PO if Γ is 2-monotone [1]. In the case when Γ is almost-min-min the result follow from Theorem 36 (via the definition of Min-Sur-CSP(Γ) in Sect. 3), and in the case when Γ is dual almost-min-min it follows by Lemma 13. Hardness follows from Lemma 25. $\qquad\square$

Proof (of Proposition 9). We know by Theorem 6 that Max-Sur-CSP(Γ) is in PO only if Γ is almost-min-min, dual almost-min-min or 2-monotone. If Γ is 2-monotone, then so is $\Gamma \cup \{C_0, C_1\}$, and in this case we may enumerate every optimal solution with a simple recursive algorithm (see e.g. [5,9]). If Γ is almost-min-min and we are given an instance $f : \mathbb{B}^n \to \mathbb{N}$ of Max-Sur-CSP(Γ), then either there is $x \in \mathbb{B}^n$ that satisfies every constraint, in which case (since Γ is Horn) the result follows from [9], or Theorem 36 is applicable. The case when Γ is dual almost-min-min is completely symmetric to the almost-min-min case. $\qquad\square$

3   A Generalised Minimum Cut Problem

In this section we focus on the problem Min-Sur-CSP($\{\mathrm{NAND}_k, =_\mathbb{B}\}$) (defined below) which is a slight generalisation of the minimum asymmetric cut problem [1]. We will show that we can enumerate *all* α-optimal solutions (solutions of cost no more than α times the cost of an optimal solution) to an instance of this problem in polynomial time. At the end of the section we also show how this generalises even further to languages that are almost-min-min.

Let D be a finite set and Δ be a finite family of finitary relations over D, and let $\bar\chi^R = 1 - \chi^R$. We define the problem Min-Sur-CSP(D, Δ) as follows. An *instance* is a function $f : D^n \to \mathbb{N}$ defined as

$$f(x) = \sum_{i=1}^{m} w_i \bar\chi^{R_i}(\mathrm{proj}_{s^i}(x))$$

through a collection of weights $w_i \in \mathbb{N}$, constraint relations $R_i \in \Delta$, and constraint scopes $s^i \in [n]^{\mathrm{ar}(R_i)}$. A *solution* is an element in $\{x \in D^n : \{x_1, \ldots, x_n\} = D\}$, and an *optimal solution* is a solution that minimises f. This problem is clearly very similar to Max-Sur-CSP(D, Δ); instead of maximising the number of satisfied constraints, we are minimising the number of unsatisfied ones. An optimal solution to an instance of one of the problems is certainly also an optimal solution to the corresponding instance of the other problem. However, the problems do differ when it comes to approximability and near-optimal solutions – and this is something we will make use of.

Let $f : \mathbb{B}^n \to \mathbb{N}$ be an instance of Min-Sur-CSP($\{\mathrm{NAND}_k, =_\mathbb{B}\}$). We can wlog assume that $f = f_1 + f_2$ where f_1 is an instance of Min-Sur-CSP($\{\mathrm{NAND}_k\}$) and f_2 is an instance of Min-Sur-CSP($\{=_\mathbb{B}\}$). We will throughout the section reserve f, f_1 and f_2 to refer to these particular instances.

Note that we can always express f_2 as $f_2(x) = \sum_{\{i,j\} \in E} w_{\{i,j\}} \bar\chi^{=_\mathbb{B}}(x_i, x_j)$. So finding a optimal solution to f_2 is equivalent to solving the Minimum Cut problem on the undirected graph $([n], E)$ with edge weights $w_{\{i,j\}}$.

For notational convenience, we will treat f, f_1 and f_2 as functions $2^{[n]} \to \mathbb{N}$, i.e. we write $f(X)$ for $f(\chi^X(1), \ldots, \chi^X(n))$.

Definition 26. *For $g : 2^V \to \mathbb{N}$, $U \in 2^V$ and $\alpha \in \mathbb{Q}_{\geq 0}$, let*

$$\lambda(g, U) = \min\{g(Y) : \varnothing \subset Y \subset U\} \text{ and}$$
$$\mathrm{Sol}(g, U, \alpha) = \{X : \varnothing \subset X \subset U \text{ and } g(X) \leq \alpha\lambda(g, U)\}.$$

We define $V = [n]$ and $\overline{X} = V \setminus X$. We will assume that $\lambda(f_2, V) > 0$ and show that in this case $\mathrm{Sol}(f, V, \alpha)$ can be computed in polynomial time for any fixed $\alpha \in \mathbb{N}$.

Note that f_1 and f_2 are functions $2^V \to \mathbb{N}$ satisfying

$$f_2(X) + f_2(Y) \geq f_2(X \cap Y) + f_2(X \cup Y), \tag{1}$$

$$f_2(X) = f_2(\overline{X}), \tag{2}$$

$$f_2(X) + f_2(Y) \geq f_2(X \setminus Y) + f_2(Y \setminus X), \tag{3}$$

$$f_2(Z) > 0 \quad \text{and} \tag{4}$$

$$f_1(X) \geq f_1(X \cap Y) + f_1(X \cap \overline{Y}) \tag{5}$$

for every $X, Y \in 2^V$ and every $Z \in 2^V$ such that $\varnothing \subset Z \subset V$. Here (3) is implied by (1) and (2) since $f_2(X) + f_2(Y) = f_2(\overline{X}) + f_2(Y) \geq f_2(\overline{X} \cap Y) + f_2(\overline{X} \cup Y) = f_2(\overline{X \cap Y}) + f_2(\overline{X \cup Y}) = f_2(Y \setminus X) + f_2(X \setminus Y)$. We will in this section frequently, often without mention, make use of these properties.

Our first step towards computing $\mathrm{Sol}(f, V, \alpha)$ will be an algorithm for computing $\mathrm{Sol}(f, V, 1)$. But, before stating this algorithm we will establish some properties of the elements in $\mathrm{Sol}(f, U, 1)$, $U \subseteq V$.

Lemma 27. *If $U \subseteq V$, $Y \in \mathrm{Sol}(f_2, U, 1)$, $X \in \mathrm{Sol}(f, U, 1)$ and no Z exists such that $Z \subset Y$ and $Z \in \mathrm{Sol}(f_2, U, 1)$, then $Y \cap X \in \{\varnothing, X, Y\}$.*

Proof. Assume that $Y \cap X \notin \{\varnothing, X, Y\}$. We have $f_2(X) + f_2(Y) \geq f_2(X \setminus Y) + f_2(Y \setminus X) > f_2(X \setminus Y) + f_2(Y)$ since $\varnothing \subset Y \setminus X \subset Y \subset U$ and Y is an inclusion-minimal element in $\mathrm{Sol}(f_2, U, 1)$. This means that $f(X) = f_1(X) + f_2(X) \geq f_1(X \setminus Y) + f_2(X) > f_1(X \setminus Y) + f_2(X \setminus Y) = f(X \setminus Y)$ and, since $\varnothing \subset X \setminus Y \subset U$, that $X \notin \mathrm{Sol}(f, U, 1)$. We have thus reached a contradiction. $\square$

Lemma 28. *If $U \subseteq V$, $X \in \mathrm{Sol}(f, U, 1)$ and $X \supseteq Y$ for some $Y \in \mathrm{Sol}(f_2, U, 1)$, then $X \in \mathrm{Sol}(f_2, U, 1)$.*

Proof. Note that $f_2(X) \geq f_2(Y) = f(Y) - f_1(Y) \geq f(Y) - f_1(X) \geq f(X) - f_1(X) = f_2(X)$, so $f_2(Y) = f_2(X)$. $\square$

Lemma 29. $\mathrm{Sol}(f_2, U, \alpha)$ *can, for any $U \subseteq V$ and any fixed $\alpha \in \mathbb{N}$, be computed in polynomial time.*

Proof. We have already remarked that $\mathrm{Sol}(f_2, V, \alpha)$ is the set of α-minimum cuts (cuts of weight no more than α times the weight of a minimum cut) in a graph $\mathcal{G}$ with integer edge-weights. This graph can be created in polynomial time, and it is known that the set of all α-minimum cuts in any graph can be computed in polynomial time [27].

In the case when $U \subset V$, create for every $u \in U$ the graph $\mathcal{G}_u$ from $\mathcal{G}$ by contracting all vertices in $(V \setminus U) \cup \{u\}$ into the single vertex u. Note that the set of α-minimum cuts in $\mathcal{G}_u$ is a superset of $\{S \in \mathrm{Sol}(f_2, U, \alpha) : u \notin S\}$. Computing the set of α-minimum cuts in $\mathcal{G}_u$ for each $u \in U$ therefore allows us to compute $\mathrm{Sol}(f_2, U, \alpha)$. $\square$

It was shown by Karger [17] that the number of α-minimum cuts in a graph (V, E) is less than $|V|^{2\alpha}$. This together with the proof of Lemma 29 gives us the following result.

Lemma 30. *If $U \subseteq V$ and $\alpha \in \mathbb{N}$, then $|\operatorname{Sol}(f_2, U, \alpha)| \leq |U|^{2\alpha+1}$.*

By Lemmas 27 and 28, we know that if $U \subseteq V$, $X \in \operatorname{Sol}(f, U, 1)$ and Y is an inclusion minimal element in $\operatorname{Sol}(f_2, U, 1)$, then either $X \subseteq Y$, $X \subseteq U \setminus Y$ or $X \in \operatorname{Sol}(f_2, U, 1)$. Using these properties we can compute $\operatorname{Sol}(f, U, 1)$ as follows.

$\textsc{Solutions}(U)$

```
1   if |U| = 1
2       return ∅
3   else
4       A := Sol(f₂, U, 1)
5       Y := arbitrary inclusion minimal element in A
6       C := Solutions(Y)
7       D := Solutions(U \ Y)
8       E := A ∪ C ∪ D ∪ {Y, U \ Y}
9       λ(f, U) := min{f(X) : X ∈ E}
10      return {X ∈ E : f(X) = λ(f, U)}
```

The correctness of the algorithm follows from earlier remarks, and from the algorithm we get a simple upper bound on the number of elements in $\operatorname{Sol}(f, U, 1)$.

Lemma 31. *If $U \subseteq V$, then $|\operatorname{Sol}(f, U, 1)| \leq |U|^4$.*

By using this bound, it is now easy to argue that the algorithm has a polynomial running time.

Lemma 32. *The algorithm* $\textsc{Solutions}$ *computes* $\operatorname{Sol}(f, U, 1)$ *in polynomial time for any* $U \subseteq V$.

Proof. The recursion tree generated by $\textsc{Solutions}$ has height less than or equal to $|U|$ and no more than $|U|$ leafs, and therefore contains at most $|U|^2$ vertices. By Lemmas 29, 30 and 31 the work needed in each vertex can be done in polynomial time. $\qquad\square$

Lemmas 27 and 28 describe some properties of elements in $\operatorname{Sol}(f, U, 1)$. The following lemma gives us a somewhat similar result for the elements in $\operatorname{Sol}(f, U, \alpha)$, $U \subseteq V$.

Lemma 33. *If $p > 3$, $U \subseteq V$, $X \in \operatorname{Sol}(f, U, \frac{p}{3})$, $Y \in \operatorname{Sol}(f_2, U, 1)$ and $\varnothing \notin \{X \setminus Y, Y \cap X\}$, then either $X \in \operatorname{Sol}(f_2, U, p+2)$ or $\{X \setminus Y, X \cap Y\} \subseteq \operatorname{Sol}(f, U, \frac{p-1}{3})$.*

Proof. Let $\gamma = \frac{p}{p+2}$. If $X \notin \operatorname{Sol}(f_2, U, p+2)$, i.e. $f_2(X) > (p+2)f_2(Y)$, then

$$
\begin{aligned}
f(X) &= f_1(X) + \gamma f_2(X) + \tfrac{2}{p+2} f_2(X) \\
&> f_1(X) + \gamma f_2(X) + 2 f_2(Y) \\
&\geq f_1(X) + \gamma(f_2(X \setminus Y) + f_2(Y \setminus X)) + f_2(Y) \\
&> f_1(X) + \gamma f_2(X \setminus Y) + \gamma(f_2((Y \setminus X) \setminus Y) + f_2(Y \setminus (Y \setminus X))) \\
&\geq f_1(X) + \gamma f_2(X \setminus Y) + \gamma f_2(X \cap Y) \\
&\geq f_1(X \setminus Y) + f_1(X \cap Y) + \gamma f_2(X \setminus Y) + \gamma f_2(X \cap Y) \\
&\geq \gamma f(X \setminus Y) + \gamma f(X \cap Y).
\end{aligned}
$$

Hence $f(X \cap Y) < \frac{p+2}{p} f(X) - f(X \setminus Y) \leq \frac{p+2}{p} \frac{p}{3} \lambda(f, U) - \lambda(f, U) = \frac{p-1}{3} \lambda(f, U)$, so $X \cap Y \in \operatorname{Sol}(f, U, \frac{p-1}{3})$. Symmetrically $X \setminus Y \in \operatorname{Sol}(f, U, \frac{p-1}{3})$. $\qquad\square$

By Lemma 33 we know that if $U \subseteq V$, $X \in \mathrm{Sol}(f, U, \frac{p}{3})$ and Y is an element in $\mathrm{Sol}(f_2, U, 1)$, then either $X \subseteq Y$, $X \subseteq U \setminus Y$, $X \in \mathrm{Sol}(f_2, U, p+2)$ or X is the union of two slightly "better" sets – sets that are both in $\mathrm{Sol}(f, U, \frac{p-1}{3})$. Using these properties we can compute $\mathrm{Sol}(f, U, \frac{p}{3})$. Note that the algorithm does not just compute $\mathrm{Sol}(f, U, \frac{p}{3})$, but the entire sequence $\mathrm{Sol}(f, U, \frac{3}{3}), \mathrm{Sol}(f, U, \frac{4}{3}), \ldots, \mathrm{Sol}(f, U, \frac{p}{3})$.

NEAR-SOLUTIONS(U, α)

  1 **if** $|U| = 1$
  2   **return** $(\varnothing, \ldots, \varnothing)$
  3 **else**
  4      $A := \mathrm{Sol}(f_2, U, 1)$
  5      $Y :=$ any element from A
  6      $(B^3, \ldots, B^{3\alpha}) :=$ NEAR-SOLUTIONS(Y, α)
  7      $(C^3, \ldots, C^{3\alpha}) :=$ NEAR-SOLUTIONS$(U \setminus Y, \alpha)$
  8      $S^3 :=$ SOLUTIONS(U)
  9      $\lambda(f, U) := f(X)$ for some $X \in S^3$
 10   **for** $p := 4$ **to** 3α
 11        $D^p := \mathrm{Sol}(f_2, U, p+2)$
 12        $E := \{X \cup Y : X \in S^{p-1}, Y \in S^{p-1}, \varnothing \subset X \cup Y \subset U\}$
 13        $F := B^p \cup C^p \cup D^p \cup E \cup \{Y, U \setminus Y\}$
 14        $S^p := \{X \in F : f(X) \leq \frac{p}{3}\lambda(f, U)\}$
 15   **return** $(S^3, \ldots, S^{3\alpha})$

The correctness of the algorithm follows from earlier remarks. From the algorithm we get a simple upper bound on the number of elements in $\mathrm{Sol}(f, U, \alpha)$.

Lemma 34. *If* $U \subseteq V$ *and* $p \in \mathbb{N}$*, then* $|\mathrm{Sol}(f, U, \frac{p}{3})| \leq |U|^{3^p}$.

Proof. By Lemma 30 we know that $|\mathrm{Sol}(f_2, U, p+2)| \leq |U|^{2p+5}$. If $p + |U| = 0$, then clearly the result holds. Assume the lemma is true when $p + |U| < N$ and consider the case when $p + |U| = N$. If $|U| = m \geq 2$ and $p \geq 3$, then

$$|\mathrm{Sol}(f, U, \tfrac{p}{3})| \leq \max_{\varnothing \subset Y \subset U} |\mathrm{Sol}(f, Y, \tfrac{p}{3})| + |\mathrm{Sol}(f, U \setminus Y, \tfrac{p}{3})| + 2 + |\mathrm{Sol}(f_2, U, p+2)|$$

$$+ |\mathrm{Sol}(f, U, \tfrac{p-1}{3})|^2 \leq \max_{0 < k < m} k^{3^p} + (m-k)^{3^p} + 2 + m^{2p+5} + (m^{3^{p-1}})^2$$

$$\leq 1 + (m-1)^{3^p} + 2 + m^{2p+5} + m^{\frac{2}{3}3^p} \leq (m-1)^{3^p} + 2m^{\frac{2}{3}3^p}$$

$$= m^{3^p}\left((1 - \tfrac{1}{m})^{3^p} + 2m^{-3^{p-1}}\right) \leq m^{3^p}(1 - \tfrac{1}{m} + \tfrac{1}{m}) = m^{3^p} = |U|^{3^p},$$

and if $m < 2$ or $p < 3$ we clearly (since $\lambda(f, V) > 0$) have $|\mathrm{Sol}(f, U, \frac{p}{3})| \leq |U|^{3^p}$. $\qquad\square$

By using this bound, it is now easy to argue that the algorithm has a polynomial running time.

Lemma 35. *The algorithm* NEAR-SOLUTIONS *computes* $\mathrm{Sol}(f, U, \alpha)$ *in polynomial time for any* $U \subseteq V$ *and any fixed* $\alpha \in \mathbb{N}$.

Finally, we show that the ability to compute all α-optimal solutions to instances of Min-Sur-CSP$(\{\mathrm{NAND}_k, =_\mathbb{B}\})$ allows us to compute also the α-optimal solutions to instances of Min-Sur-CSP(Γ), for any Γ that is almost-min-min.

Theorem 36. *Let Γ be almost-min-min and $\alpha \in \mathbb{N}$. If g is any instance of Min-Sur-CSP(Γ), then either there is a solution x to g such that $g(x) = 0$, or all α-optimal solutions to g can be enumerated in polynomial time.*

Proof. Let $g : \mathbb{B}^n \to \mathbb{N}$ be any instance of Min-Sur-CSP(Γ) for which there is no $x \in \mathbb{B}^n \setminus \{\mathbf{0}, \mathbf{1}\}$ such that $g(x) = 0$. For any function $h : \mathbb{B}^n \to \mathbb{N}$, let $\lambda(h) = \min\{h(x) : \mathbf{0} < x < \mathbf{1}\}$ and $\mathrm{Sol}(h, \alpha) = \{x \in \mathbb{B}^n : \mathbf{0} < x < \mathbf{1} \text{ and } h(x) \le \alpha\lambda(h)\}$. We will show that we, in polynomial time, can compute $\mathrm{Sol}(g, \alpha)$, for any fixed $\alpha \in \mathbb{N}$.

Recall that g is defined as $g(x) = \sum_{i=1}^{m} w_i \bar{\chi}^{R_i}(\mathrm{proj}_{s^i}(x))$, where $R_i = \{x \in \mathbb{B}^{a_i} : \bigwedge_{j=1}^{k_i} \mathrm{proj}_{p^{i,j}}(x) \in P_{i,j}\}$ for some collection of relations $\{P_{i,j} : i \in [m], j \in [k_i]\} \subseteq \{\mathrm{NAND}_1, \ldots, \mathrm{NAND}_a, =_{\mathbb{B}}\}$, where $a = \max_{i \in [m]} a_i$. We may wlog assume that $k_1 = \cdots = k_m$ and let $k = k_1$. For every $j \in [k]$, define $h_j : \mathbb{B}^n \to \mathbb{N}$ as $h_j(x) = \sum_{i=1}^{m} w_i \bar{\chi}^{P_{i,j}}(\mathrm{proj}_{p^{i,j}}(\mathrm{proj}_{s^i}(x)))$. Set $h(x) = \sum_{j=1}^{k} h_j(x)$ and note that $h(x) \le kg(x) \le kh(x)$ for every $x \in \mathbb{B}^n$. So if $g(x) \le \alpha\lambda(g)$, then $h(x) \le kg(x) \le k\alpha\lambda(g) \le k\alpha\lambda(h)$ which means that $\mathrm{Sol}(g, \alpha) \subseteq \mathrm{Sol}(h, k\alpha)$. To compute $\mathrm{Sol}(g, \alpha)$ we can therefore compute the larger set $\mathrm{Sol}(h, k\alpha)$ and then throw the unwanted elements out. Note that $\mathrm{NAND}_u \to \mathrm{NAND}_v$ if $u \ge v$, so we can wlog assume that $\{P_{i,j} : i \in [m], j \in [k]\} \subseteq \{\mathrm{NAND}_a, =_{\mathbb{B}}\}$. This means that we can represent h as the sum of a Min-Sur-CSP($\{\mathrm{NAND}_a\}$) instance h_1 and a Min-Sur-CSP($\{=_{\mathbb{B}}\}$) instance h_2. We are therefore close to being able to compute $\mathrm{Sol}(h, k\alpha)$ by the algorithms in Sect. 3. However, we might have $\lambda(h_2) = 0$, which prevents us from directly using Lemma 35. To overcome this issue, let $M = n^2$ and $h' = h'_1 + h'_2$, where $h'_1(x) = Mh_1(x)$ and $h'_2(x) = Mh_2(x) + \sum_{(i,j) \in [n]^2} \bar{\chi}^{=_{\mathbb{B}}}(x_i, x_j)$. Clearly $\lambda(h'_2) > 0$. Note that if $h(x) \le k\alpha\lambda(h)$, then $h'(x) \le 2Mh(x) \le 2Mk\alpha\lambda(h) \le 2k\alpha\lambda(h')$ since (because $h(x) \ge g(x) > 0$ for every $\mathbb{B}^n \setminus \{\mathbf{0}, \mathbf{1}\}$) we have $Mh(x) \le h'(x) \le 2Mh(x)$ for every $x \in \mathbb{B}^n \setminus \{\mathbf{0}, \mathbf{1}\}$. Hence, $\mathrm{Sol}(h, k\alpha) \subseteq \mathrm{Sol}(h', 2k\alpha)$ and, by Lemma 35, we know that $\mathrm{Sol}(h', 2k\alpha) \supseteq \mathrm{Sol}(g, \alpha)$ can be computed in polynomial time, for any fixed α. $\square$

4   Remarks and Open Ends

The ultimate goal in this direction of research would naturally be a complexity classification for Max-Sur-CSP on any finite domain. Preferably both for computation of exact and approximate solutions. This will likely not be easy, but we hope that our results together with [1] may act as a base to build from. However, there are also plenty of smaller questions. One immediate question is if Proposition 9 holds also for Max-Sur-CSP on larger domains. Another area to investigate is the bound in Lemma 34 of the number of α-optimal solutions of an instance of Min-Sur-CSP($\{\mathrm{NAND}_k, =_{\mathbb{B}}\}$). A tighter bound would certainly be interesting.

Acknowledgements. I am greatly indebted to Peter Jonsson for stimulating discussions and for suggesting that hardness results from coding theory could be relevant (which indeed turned out to be the case). I am also thankful to Hang Zhao who shared with me her knowledge of the problem and commented on an early draft, and to the anonymous reviewers for their helpful remarks.

References

1. Bach, W., Zhou, H.: Approximation for maximum surjective constraint satisfaction problems. arXiv:1110.2953v1 [cs.CC] (2011)
2. Bodirsky, M., Kára, J., Martin, B.: The complexity of surjective homomorphism problems – a survey. Discrete Applied Mathematics 160(12), 1680–1690 (2012)
3. Bulatov, A.A.: A dichotomy theorem for constraint satisfaction problems on a 3-element set. J. ACM 53, 66–120 (2006)
4. Bulatov, A.A., Dalmau, V., Grohe, M., Marx, D.: Enumerating homomorphisms. J. Comput. Syst. Sci. 78(2), 638–650 (2012)
5. Cohen, D.A.: Tractable decision for a constraint language implies tractable search. Constraints 9, 219–229 (2004)
6. Cohen, D., Cooper, M., Jeavons, P., Krokhin, A.: Supermodular functions and the complexity of max CSP. Discrete Applied Mathematics 149(1-3), 53–72 (2005)
7. Cohen, D.A., Cooper, M.C., Jeavons, P.G., Krokhin, A.A.: The complexity of soft constraint satisfaction. Artif. Intell. 170, 983–1016 (2006)
8. Creignou, N.: A dichotomy theorem for maximum generalized satisfiability problems. J. Comput. Syst. Sci. 51, 511–522 (1995)
9. Creignou, N., Hébrard, J.J.: On generating all solutions of generalized satisfiability problems. Informatique Théorique et Applications 31(6), 499–511 (1997)
10. Creignou, N., Hermann, M.: Complexity of generalized satisfiability counting problems. Information and Computation 125(1), 1–12 (1996)
11. Creignou, N., Khanna, S., Sudan, M.: Complexity classifications of boolean constraint satisfaction problems. Society for Industrial and Applied Mathematics, Philadelphia (2001)
12. Dechter, R., Itai, A.: Finding all solutions if you can find one. In: AAAI 1992 Workshop on Tractable Reasoning, pp. 35–39 (1992)
13. Golovach, P.A., Paulusma, D., Song, J.: Computing Vertex-Surjective Homomorphisms to Partially Reflexive Trees. In: Kulikov, A., Vereshchagin, N. (eds.) CSR 2011. LNCS, vol. 6651, pp. 261–274. Springer, Heidelberg (2011)
14. Horn, A.: On sentences which are true of direct unions of algebras. The Journal of Symbolic Logic 16(1), 14–21 (1951)
15. Jonsson, P., Klasson, M., Krokhin, A.: The approximability of three-valued max CSP. SIAM J. Comput. 35, 1329–1349 (2006)
16. Jonsson, P., Kuivinen, F., Thapper, J.: Min CSP on Four Elements: Moving beyond Submodularity. In: Lee, J. (ed.) CP 2011. LNCS, vol. 6876, pp. 438–453. Springer, Heidelberg (2011)
17. Karger, D.R.: Global min-cuts in RNC, and other ramifications of a simple min-out algorithm. In: Proceedings of the Fourth Annual ACM-SIAM Symposium on Discrete Algorithms, SODA 1993, pp. 21–30. Society for Industrial and Applied Mathematics, Philadelphia (1993)
18. Khanna, S., Sudan, M., Trevisan, L., Williamson, D.P.: The approximability of constraint satisfaction problems. SIAM J. Comput. 30, 1863–1920 (2001)
19. Kolmogorov, V., Živný, S.: The complexity of conservative valued CSPs. In: Proceedings of the Twenty-Third Annual ACM-SIAM Symposium on Discrete Algorithms, SODA 2012, pp. 750–759. SIAM (2012)
20. Martin, B., Paulusma, D.: The Computational Complexity of Disconnected Cut and $2K_2$-Partition. In: Lee, J. (ed.) CP 2011. LNCS, vol. 6876, pp. 561–575. Springer, Heidelberg (2011)
21. Nagamochi, H., Nishimura, K., Ibaraki, T.: Computing all small cuts in an undirected network. SIAM J. Discret. Math. 10, 469–481 (1997)

22. Golovach, P.A., Lidický, B., Martin, B., Paulusma, D.: Finding vertex-surjective graph homomorphisms. arXiv:1204.2124v1 [cs.DM] (2012)
23. Raghavendra, P.: Optimal algorithms and inapproximability results for every CSP? In: Proceedings of the 40th Annual ACM Symposium on Theory of Computing, STOC 2008, pp. 245–254. ACM, New York (2008)
24. Schaefer, T.J.: The complexity of satisfiability problems. In: Proceedings of the Tenth Annual ACM Symposium on Theory of Computing, STOC 1978, pp. 216–226. ACM, New York (1978)
25. Valiant, L.G.: The complexity of enumeration and reliability problems. SIAM Journal on Computing 8(3), 410–421 (1979)
26. Vardy, A.: Algorithmic complexity in coding theory and the minimum distance problem. In: Proceedings of the Twenty-Ninth Annual ACM Symposium on Theory of Computing, STOC 1997, pp. 92–109. ACM, New York (1997)
27. Vazirani, V., Yannakakis, M.: Suboptimal Cuts: Their Enumeration, Weight and Number. In: Kuich, W. (ed.) ICALP 1992. LNCS, vol. 623, pp. 366–377. Springer, Heidelberg (1992)

An Optimal Arc Consistency Algorithm
for a Chain of Atmost Constraints with Cardinality

Mohamed Siala[1,2], Emmanuel Hebrard[1,3], and Marie-José Huguet[1,2]

[1] CNRS, LAAS, 7 avenue du colonel Roche, F-31400 Toulouse, France
[2] Univ de Toulouse, INSA, LAAS, F-31400 Toulouse, France
[3] Univ de Toulouse, LAAS, F-31400 Toulouse, France
`{siala,hebrard,huguet}@laas.fr`

Abstract. The ATMOSTSEQCARD constraint is the conjunction of a cardinality constraint on a sequence of n variables and of $n - q + 1$ constraints ATMOST u on each subsequence of size q.

This constraint is useful in car-sequencing and crew-rostering problems. In [18], two algorithms designed for the AMONGSEQ constraint were adapted to this constraint with a $O(2^q n)$ and $O(n^3)$ worst case time complexity, respectively. In [10], another algorithm similarly adaptable to filter the ATMOSTSEQCARD constraint with a time complexity of $O(n^2)$ was proposed.

In this paper, we introduce an algorithm for achieving Arc Consistency on the ATMOSTSEQCARD constraint with a $O(n)$ (hence optimal) worst case time complexity. We then empirically study the efficiency of our propagator on instances of the car-sequencing and crew-rostering problems.

1 Introduction

In many applications there are restrictions on the successions of decisions that can be taken. Some sequences are allowed or preferred while other are forbidden. For instance, in crew-rostering applications, it is often not recommended to have an employee work on an evening or a night shift and then again on the morning shift of the next day.

Several constraints have been proposed to deal with this type of problems. The REGULAR [12] and COST-REGULAR constraints [7] make it possible to restrict sequences in an arbitrary way. However, there might often exist more efficient algorithm for the particular case at hand. For instance, filtering algorithms have been proposed for the AMONGSEQ constraint in [6,10,19,18]. This constraint ensures that all subsequences of size q have at least l but no more than u values in a set v. This constraint is often applied to car-sequencing and crew-rostering problems. However, the constraints in these two benchmarks do not correspond exactly to this definition. Indeed, there are often no lower bound restriction on the number of values ($l = 0$). Instead, the number of values in the set v is often constrained by an overall demand.

In this paper we consider the constraint ATMOSTSEQCARD. This constraint, posted on n variables $x_1, \ldots, x_n$, ensures that, in every subsequence of length q, no more than u are set to a value in a set v, and that over all the sequence, exactly d are set to values in v. In car-sequencing, this constraint allows to state that given an option, no sub-sequence of length q can involve more than u classes of cars requiring this option,

M. Milano (Ed.): CP 2012, LNCS 7514, pp. 55–69, 2012.

and that exactly d cars require it overall. In crew-rostering problems, one can state, for instance, that a worker must have at least a 16h break between two 8h shifts, or that a week should not involve more than 40h of working time, while enforcing a total number of worked hours over the scheduling period.

The rest of the paper is organized as follows: In Section 2 we give a brief state of the art of the sequence constraints. Then in Section 3, we give a linear time (hence optimal) algorithm for filtering this constraint. Last, in Section 4 we evaluate our new propagator on car-sequencing benchmarks, before concluding in Section 5.

2   CSP and SEQUENCE Constraints

A constraint satisfaction problem (CSP) is a triplet $P = (\mathcal{X}, \mathcal{D}, \mathcal{C})$ where $\mathcal{X}$ is a set of variables, $\mathcal{D}$ is a set of domains and $\mathcal{C}$ is a set of constraints that specify allowed combinations of values for subsets of variables. We denote by $\min(x)$ and $\max(x)$ the minimum and maximum values in $\mathcal{D}(x)$, respectively. An assignment of a set of variables $\mathcal{X}$ is a tuple w, where $w[i]$ denotes the value assigned to the variable x_i. A constraint $C \in \mathcal{C}$ on a set of variables $\mathcal{X}$ defines a relation on the domains of variables in $\mathcal{X}$. An assignment w is consistent for a constraint C iff it belongs to the corresponding relation. A constraint C is *arc consistent* (AC) iff, for every value j of every variable x_i in its scope there exists a consistent assignment w such that $w[i] = j$, i.e., a *support*.

There are several variants of the SEQUENCE constraints, we first review them and then motivate the need for the variant proposed in this paper: ATMOSTSEQCARD. In the following definitions, v is a set of integers and l, u, q are integers. Sequence constraints are conjunctions of AMONG constraints, constraining the number of occurrences of a set of values in a set of variables.

Definition 1. $\text{AMONG}(l, u, [x_1, \ldots, x_q], v) \Leftrightarrow l \leq |\{i \mid x_i \in v\}| \leq u$

Chains of AMONG Constraints: The AMONGSEQ constraint, first introduced in [2], is a chain of AMONG constraints of width q slid along a vector of n variables.

Definition 2. $\text{AMONGSEQ}(l, u, q, [x_1, \ldots, x_n], v) \quad \Leftrightarrow \quad \bigwedge_{i=0}^{n-q} \text{AMONG}(l, u, [x_{i+1}, \ldots, x_{i+q}], v)$

The first (incomplete) algorithm for filtering this constraint was proposed in 2001 [1]. Then in [18,19], two complete algorithms for filtering the AMONGSEQ constraint were introduced. First, a reformulation using the REGULAR constraint using 2^{q-1} states and hence achieving AC in $O(2^q n)$ time. Second, an algorithm achieving AC with a worst case time complexity of $O(n^3)$. Moreover, this last algorithm is able to handle arbitrary sets of AMONG constraints on consecutive variables (denoted GEN-SEQUENCE), however in $O(n^4)$. Last, two flow-based algorithms were introduced in [10]. The first achieves AC on AMONGSEQ in $O(n^{3/2} \log n \log u)$, while the second achieves AC on GEN-SEQUENCE in $O(n^3)$ in the worst case. These two algorithms have an amortised complexity down a branch of the search tree of $O(n^2)$ and $O(n^3)$, respectively.

Chain of ATMOST **Constraints:** Although useful in both applications, the AMONGSEQ constraint does not model exactly the type of sequences useful in car-sequencing and crew-rostering applications.

First, there is often no lower bound for the cardinality of the sub-sequences, i.e., $l = 0$. Therefore AMONGSEQ is unnecessarily general in that respect. Moreover, the capacity constraint on subsequences is often paired with a cardinality requirement.

For instance, in car-sequencing, classes of car requiring a given option cannot be clustered together because a working station can only handle a fraction of the cars passing on the line (*at most* u times in any sequence of length q). The total number of occurrences of these classes is a requirement, and translates as an overall cardinality constraint rather than lower bounds on each sub-sequence.

In crew-rostering, allowed shift patterns can be complex, hence the REGULAR constraint is often used to model them. However, working in at most u shifts out of q is a useful particular case. If days are divided into three 8h shifts, ATMOSTSEQ with $u = 1$ and $q = 3$ makes sure that no employee work more than one shift per day *and* that there must be a 24h break when changing shifts. Moreover, similarly to car-sequencing, the lower bound on the number of worked shifts is global (monthly, for instance).

In other words, we often have a chain of ATMOST constraints, defined as follows:

Definition 3. $\text{ATMOST}(u, [x_1, \ldots, x_q], v) \Leftrightarrow \text{AMONG}(0, u, [x_1, \ldots, x_q], v)$

Definition 4. $\text{ATMOSTSEQ}(u, q, [x_1, \ldots, x_n], v) \qquad \Leftrightarrow \qquad \bigwedge_{i=0}^{n-q} \text{ATMOST}(u, [x_{i+1}, \ldots, x_{i+q}], v)$

However, it is easy to maintain AC on this constraint. Indeed, the ATMOST constraint is monotone, i.e., the set of supports for value 0 is a super-set of the set of supports for value 1. Hence a ATMOSTSEQ constraint is AC iff each ATMOST is AC [4].

A good tradeoff between filtering power and complexity can be achieved by reasoning about the total number of occurrences of values from the set v together with the chain of ATMOST constraints.[1] We therefore introduce the ATMOSTSEQCARD constraint, defined as the conjunction of an ATMOSTSEQ with a cardinality constraint on the total number of occurrences of values in v:

Definition 5. $\text{ATMOSTSEQCARD}(u, q, d, [x_1, \ldots, x_n], v) \Leftrightarrow$

$$\text{ATMOSTSEQ}(u, q, [x_1, \ldots, x_n], v) \wedge |\{i \mid x_i \in v\}| = d$$

The two AC algorithms introduced in [19] were adapted in [18] to achieve AC on the ATMOSTSEQCARD constraint. First, in the same way that AMONGSEQ can be encoded with a REGULAR constraint, ATMOSTSEQCARD can be encoded with a COST-REGULAR constraint, where the cost stands for the overall demand, and it is increased on transitions labeled with the value 1. This procedure has the same worst case time complexity, i.e., $O(2^q n)$. Second, the more general version of the polynomial algorithm (GEN-SEQUENCE) is used, to filter the following decomposition of the ATMOSTSEQCARD constraint into a conjunction of AMONG:

---

[1] This modeling choice is used in [18] on car-sequencing.

$$\textsc{AtMostSeqCard}(u, q, d, [x_1, \ldots, x_n], v) \Leftrightarrow$$

$$\bigwedge_{i=0}^{n-q} \textsc{Among}(0, u, [x_{i+1}, \ldots, x_{i+q}], v) \wedge \textsc{Among}(d, d, [x_1, \ldots, x_n], v)$$

Since the number of $\textsc{Among}$ constraints is linear, the algorithm of van Hoeve et al. runs in $O(n^3)$ on this decomposition. Similarly, the algorithm of Maher et al. runs in $O(n^2)$ on $\textsc{AtMostSeqCard}$, which is the best known complexity for this problem.

Global Sequencing Constraint: Finally, in the particular case of car-sequencing, not only we have an overall cardinality for the values in v, but each value corresponds to a class of car and has a required number of occurrences. Therefore, Puget and Régin proposed to consider the conjunction of a $\textsc{AmongSeq}$ and a $\textsc{Gcc}$ constraint. Let c_l and c_u be two mapping on integers such that $c_l(j) \leq c_u(j) \ \forall j$, and let $\mathcal{D} = \bigcup_{i=1}^{n} \mathcal{D}(x_i)$. The Global Cardinality Constraint ($\textsc{Gcc}$) is defined as follows:

Definition 6. $\textsc{Gcc}(c_l, c_u, [x_1, \ldots, x_n]) \Leftrightarrow \bigwedge_{j \in \mathcal{D}} c_l(j) \leq |\{i \mid x_i = j\}| \leq c_u(j)$

The Global Sequencing Constraint is defined as follows:

Definition 7. $\textsc{Gsc}(l, u, q, c_l, c_u, [x_1, \ldots, x_n], v) \Leftrightarrow$

$$\textsc{AmongSeq}(l, u, q, [x_1, .., x_n], v) \wedge \textsc{Gcc}(c_l, c_u, [x_1, .., x_n])$$

The mappings c_l and c_u are defined so that for a value v, both $c_l(v)$ and $c_u(v)$ map to the number of occurrences of the corresponding class of car. A reformulation of this constraint into a set of $\textsc{Gcc}$ constraints was introduced in [15]. However, achieving AC on this constraint is NP-hard [3]. In fact, so is achieving bounds consistency.[2]

3   The $\textsc{AtMostSeqCard}$ Constraint

In this section we introduce a linear filtering algorithm for the $\textsc{AtMostSeqCard}$ constraint. We first give a simple greedy algorithm for finding a support with a $O(nq)$ time complexity. Then, we show that one can use two calls to this procedure to enforce AC. Last, we show that its worst case time complexity can be reduced to $O(n)$.

It was observed in [18] and [10] that we can consider Boolean variables and $v = \{1\}$, since the following decomposition of $\textsc{Among}$ (or $\textsc{AtMost}$) does not hinder propagation as it is Berge-acyclic [6]:

$$\textsc{Among}(l, u, [x_1, \ldots, x_q], v) \Leftrightarrow \bigwedge_{i=1}^{q} (x_i \in v \leftrightarrow x'_i = 1) \wedge l \leq \sum_{i=1}^{q} x'_i \leq u$$

Therefore, throughout the paper, we shall consider the following restriction of the $\textsc{AtMostSeqCard}$ constraint, defined on Boolean variables, and with $v = \{1\}$:

Definition 8.

$$\textsc{AtMostSeqCard}(u, q, d, [x_1, \ldots, x_n]) \Leftrightarrow \bigwedge_{i=0}^{n-q} (\sum_{l=1}^{q} x_{i+l} \leq u) \wedge (\sum_{i=1}^{n} x_i = d)$$

---

[2] Comics note, Nina Narodytska.

3.1 Finding a Support

Let w be an n-tuple in $\{0,1\}^n$, $w[i]$ denotes the ith element of w, $|w| = \sum_{i=1}^{n} w[i]$ its cardinality, and $w[i : j]$ the $(|j - i| + 1)$-tuple equal to w on the sub-sequence $[x_i, \ldots, x_j]$.

We first show that one can find a support by greedily assigning variables in a lexicographical order to the value 1 whenever possible, that is, while taking care of not violating the ATMOSTSEQ constraint. More precisely, doing so leads to an assignment of maximal cardinality, which may easily be transformed into an assignment of cardinality d. This greedy rule, computing an assignment w maximizing the cardinality of the sequence $(x_1, \ldots, x_n)$ subject to a ATMOSTSEQ constraint (with parameters u and q), is shown in Algorithm 1.

First, the tuple w is initialized to the minimum value in the domain of each variable in Line 1. Then, at each step $i \in [1, \ldots, n]$ of the main loop, $c(j)$ is the cardinality of the jth subsequence involving the variable x_i, i.e. at step i, $c(j) = \sum_{l=\max(1,i-q+j)}^{\min(n,i+j-1)} w[l]$. According to the greedy rule sketched above, we set $w[i]$ to 1 iff it is not yet assigned $(\mathcal{D}(x_i) = \{0,1\})$ and if this does not violate the capacity constraints, that is, there is no subsequence involving x_i of cardinality u or more. This is done by checking the maximum value of $c(j)$ for $j \in [1, \ldots, q]$ (Condition 2). In that case, the cardinality of every subsequence involving x_i is incremented (Line 3). Finally when moving to the next variable, the values of $c(j)$ are shifted (Line 4), and the value of $c(q)$ is obtained by adding the value of $w[i + q]$ and subtracting $w[i]$ to its previous value (Line 5). From now on, we shall denote $\overrightarrow{w}$ the assignment found by leftmost on the sequence $x_1, \ldots, x_n$. Moreover, we shall denote $\overleftarrow{w}$ the assignment found by the same algorithm, however on the sequence $x_n, \ldots, x_1$, that is, right to left. Notice that, in order to simplify the notations, $\overleftarrow{w}[i]$ shall denote the value assigned by leftmost to the variable x_i, and not x_{n-i+1} as it would really be if we gave the reversed sequence as input.

Lemma 1. leftmost $\quad$ maximizes $\quad \sum_{i=1}^{n} x_i \quad$ subject $\quad$ to $\quad$ ATMOSTSEQ$(u, q, [x_1, \ldots, x_n])$.

Proof. Let $\overrightarrow{w}$ be the assignment found by leftmost, and suppose that there exists w another assignment (consistent for ATMOSTSEQ$(u, q, [x_1, \ldots, x_n])$) such that $|w| > |\overrightarrow{w}|$. Let i be the smallest index such that $\overrightarrow{w}[i] \neq w[i]$. By definition of $\overrightarrow{w}$, we know that $\overrightarrow{w}[i] = 1$ hence $w[i] = 0$. Now let j be the smallest index such that $\overrightarrow{w}[j] < w[j]$ (it must exists since $|w| > |\overrightarrow{w}|$).

We first argue that the assignment w' equal to w except that $w'[i] = 1$ and $w'[j] = 0$ (as in $\overrightarrow{w}$) is consistent for ATMOSTSEQ. Clearly, its cardinality is not affected by this swap, hence $|w'| = |w|$. Now, we consider all the sum constraints whose scopes are changed by this swap, that is, the sums $\sum_{l=a}^{a+q-1} w'[l]$ on intervals $[a, a + q - 1]$ such that $a \leq i < a + q$ or $a \leq j < a + q$. There are three cases:

1. Suppose first that $a \leq i < j < a + q$: in this case, the value of the sum is the same in w and w', therefore it is lower than or equal to u.
2. Suppose now that $i < a \leq j < a + q$: in this case, the value of the sum is decreased by 1 from w to w', therefore it is lower than or equal to u.

Algorithm 1. `leftmost`
$count \leftarrow 0$;
1 **foreach** $i \in [1, \ldots, n]$ **do** $w[i] \leftarrow \min(x_i)$;
foreach $i \in [1, \ldots, q]$ **do** $w[n + i] \leftarrow 0$;
$c(1) \leftarrow w[1]$;
foreach $j \in [2, \ldots, q]$ **do** $c(j) \leftarrow c(j - 1) + w[j]$;
foreach $i \in [1, \ldots, n]$ **do**
2   **if** $\|\mathcal{D}(x_i)\| > 1$ & $\max_{j \in [1,q]}(c(j)) < u$ **then**
$w[i] \leftarrow 1$;
$count \leftarrow count + 1$;
3   **foreach** $j \in [1, \ldots, q]$ **do** $c(j) \leftarrow c(j) + 1$;
4   **foreach** $j \in [2, \ldots, q]$ **do** $c(j - 1) \leftarrow c(j)$;
5    $c(q) \leftarrow c(q - 1) + w[i + q] - w[i]$;
return w;

x_i	w	c 1	2	3	4	max
.	**1**	**0**	**0**	**0**	**1**	**1**
0	0	1	1	2	1	2
.	0	1	2	1	1	2
1	1	2	1	1	1	2
.	**1**	**1**	**1**	**1**	**0**	**1**
.	0	2	2	1	0	2
.	0	2	1	0	0	2
0	0	1	0	0	1	1
.	**1**	**0**	**0**	**1**	**1**	**1**
0	0	0	2	2	1	2
1	1	2	2	1	2	2
.	0	2	1	2	1	2
.	0	1	2	1	1	2
1	1	2	1	1	1	2
.	**1**	**1**	**1**	**1**	**0**	**1**
.	0	2	2	1	0	2

Fig. 1. An algorithm to compute an assignment satisfying $\text{ATMOSTSEQ}(2, 4, [x_1, \ldots, x_n])$ with maximal cardinality (left), and an example of its execution (right). Dots in the first column stand for unassigned variables. The second column shows the computed assignment w, and the next columns show the state of the variables $c(1), c(2), c(3)$ and $c(4)$ at the start of each iteration. The last column stands for the maximum value among $c(1), c(2), c(3)$ and $c(4)$.

3. Last, suppose that $a \leq i < a + q \leq j$: in this case, for any $l \in [a, a + q - 1]$, we have $w'[l] \leq \overrightarrow{w}[l]$ since j is the smallest integer such that $\overrightarrow{w}[j] < w[j]$, hence the sum is lower than or equal to u.

Therefore, given a sequence w of maximum cardinality and that differs with $\overrightarrow{w}$ at rank i, we can build one of equal cardinality and that does not differs from $\overrightarrow{w}$ until rank $i + 1$. By iteratively applying this argument, we can obtain a sequence identical to $\overrightarrow{w}$, albeit with cardinality $|w|$, therefore contradicting our hypothesis that $|w| > |\overrightarrow{w}|$. $\qquad\square$

Corollary 1. *Let $\overrightarrow{w}$ be the assignment returned by* `leftmost`. *There exists a solution of* $\text{ATMOSTSEQCARD}(u, q, d, [x_1, \ldots, x_n])$ *iff the three following propositions hold:*
(1) $\text{ATMOSTSEQ}(u, q, [x_1, \ldots, x_n])$ is satisfiable
(2) $\sum_{i=1}^{n} min(x_i) \leq d$
(3) $|\overrightarrow{w}| \geq d$.

Proof. It is easy to see that these conditions are all necessary: (1) and (2) come from the definition, and (3) is a direct application of Lemma 1. Now we prove that they are sufficient by showing that if these properties hold, then a solution exists. Since $\text{ATMOSTSEQ}(u, q, [x_1, \ldots, x_n])$ is satisfiable, $\overrightarrow{w}$ does not violate the chain of ATMOST constraints as the value 1 is assigned to x_i only if all subsequences involving x_i have cardinality $u - 1$ or less. Moreover, since $\sum_{i=1}^{n} min(x_i) \leq d$ there are at least $|\overrightarrow{w}| - d$ variables such that $min(x_i) = 0$ and $\overrightarrow{w}[i] = 1$. Assigning them to 0 does not violate the ATMOSTSEQ constraint, hence there exists a support. $\qquad\square$

Lemma 1 and Corollary 1 give us a polynomial support-seeking procedure for ATMOSTSEQCARD. Indeed, the worst case time complexity of Algorithm 1 is in $O(nq)$. There are n steps and on each step, Lines 2, 3 and 4 involve $O(q)$ operations. Therefore, for each variable x_i, a support for $x_i = 0$ or $x_i = 1$ can be found in $O(nq)$.

Consequently we have a naive AC procedure running in $O(n^2 q)$ time.

3.2   Filtering the Domains

In this section, we show that we can filter out all the values inconsistent with respect to the ATMOSTSEQCARD constraint within the same time complexity as Algorithm 1.

First, we show that there can be inconsistent values only in the case where the cardinality $|\overrightarrow{w}|$ of the assignment returned by `leftmost` is exactly d: in any other case, the constraint is either violated (when $|\overrightarrow{w}| < d$) or AC, (when $|\overrightarrow{w}| > d$). The following proposition formalises this:

Proposition 1. *The constraint* ATMOSTSEQCARD$(u, q, d, [x_1, \ldots, x_n])$ *is* AC *if the three following propositions hold:*

(1) ATMOSTSEQ$(u, q, [x_1, \ldots, x_n])$ *is* AC

(2) $\sum_{i=1}^{n} min(x_i) \leq d$

(3) $|\overrightarrow{w}| > d$

Proof. By Corollary 1 we know that ATMOSTSEQCARD$(u, q, d + 1, [x_1, \ldots, x_n])$ is satisfiable. Let w be a satisfying assignment, and consider without loss of generality a variable x_i such that $|\mathcal{D}(x_i)| > 1$. Assume first that $w[i] = 1$. The solution w' equal to w except that $w'[i] = 0$ satisfies ATMOSTSEQCARD$(u, q, d, [x_1, \ldots, x_n])$. Indeed, $|w'| = |w| - 1 = d$ and since ATMOSTSEQ$(u, q, [x_1, \ldots, x_n])$ was satisfied by w it must be satisfied by w'. Hence, for every variable x_i s.t. $|\mathcal{D}(x_i)| > 1$, there exists a support for $x_i = 0$.

Suppose that $w[i] = 0$, and let $a < i$ (resp. $b > i$) be the largest (resp. smallest) index such that $w[a] = 1$ and $\mathcal{D}(x_a) = \{0, 1\}$ (resp. $w[b] = 1$ and $\mathcal{D}(x_b) = \{0, 1\}$). Let w' be the assignment such that $w'[i] = 1$, $w'[a] = 0$, $w'[b] = 0$, and $w = w'$ otherwise. We have $|w'| = d$, and we show that it satisfies ATMOSTSEQ$(u, q, [x_1, \ldots, x_n])$. Consider a subsequence $x_j, \ldots, x_{j+q-1}$. If $j + q \leq i$ or $j > i$ then $\sum_{l=j}^{j+q-1} w'[l] \leq \sum_{l=j}^{j+q-1} w[l] \leq u$, so only indices j s.t. $j \leq i < j + q$ matter. There are two cases:

1. Either a or b or both are in the sub-sequence ($j \leq a < j + q$ or $j \leq b < j + q$). In that case $\sum_{l=j}^{i+q-1} w'[l] \leq \sum_{l=j}^{i+q-1} w[l]$.
2. Neither a nor b are in the sub-sequence ($a < j$ and $j + q \leq b$). In that case, since $\mathcal{D}(x_i) = \{0, 1\}$ and since condition (1) holds, we know that $\sum_{l=j}^{i+q-1} min(x_l) < u$. Moreover, since $a < j$ and $j + q \leq b$, there is no variable x_l in that sub-sequence such that $w[l] = 1$ and $0 \in \mathcal{D}(x_l)$. Therefore, we have $\sum_{l=j}^{i+q-1} w[l] < u$, hence $\sum_{l=j}^{i+q-1} w'[l] \leq u$.

In both cases w satisfies all capacity constraints. $\square$

Remember that achieving AC on ATMOSTSEQ is trivial since AMONG is monotone. Therefore we focus of the case where ATMOSTSEQ is AC, and $|\overrightarrow{w}| = d$. In particular, Propositions 2, 3, 4 and 5 only apply in that case. The equality $|\overrightarrow{w}| = d$ is therefore implicitly assumed in all of them.

Proposition 2. *If* $|\overrightarrow{w}[1 : i - 1]| + |\overleftarrow{w}[i + 1 : n]| < d$ *then* $x_i = 0$ *is not* AC.

Proof. Suppose that we have $|\overrightarrow{w}[1 : i - 1]| + |\overleftarrow{w}[i + 1 : n]| < d$ and suppose that there exists an assignment w such that $w[i] = 0$ and $|w| = d$.

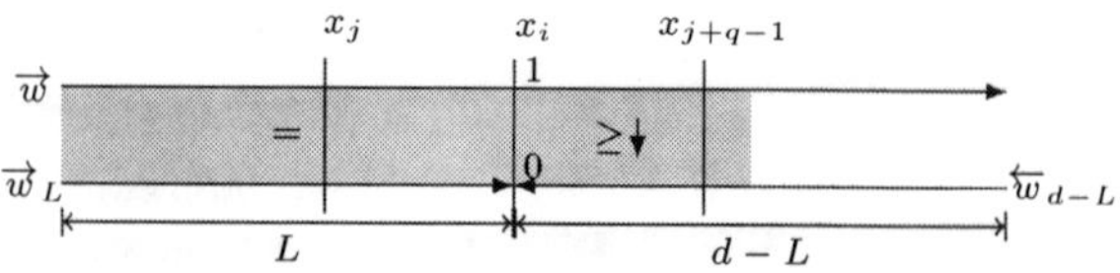

Fig. 2. Illustration of Proposition 4's proof. Horizontal arrows represent assignments.

By Lemma 1 on the sequence $x_1, \ldots, x_{i-1}$ we know that $\sum_{l=1}^{i-1} w[l] \leq |\vec{w}[1 : i - 1]|$.

By Lemma 1 on the sequence $x_n, \ldots, x_{i+1}$ we know that $\sum_{l=i+1}^{n} w[l] \leq |\overleftarrow{w}[i + 1 : n]|$.

Therefore, since $w[i] = 0$, we have $|w| = \sum_{l=1}^{n} w[l] < d$, thus contradicting the hypothesis that $|w| = d$. Hence, there is no support for $x_i = 0$. □

Proposition 3. *If* $|\vec{w}[1 : i]| + |\overleftarrow{w}[i : n]| \leq d$ *then* $x_i = 1$ *is not* AC.

Proof. Suppose that we have $|\vec{w}[1 : i]| + |\overleftarrow{w}[i : n]| \leq d$ and suppose that there exists an assignment w' such that $w'[i] = 1$ and $|w'| = d$.

By Lemma 1 on the sequence $x_1, \ldots, x_i$ we know that $\sum_{l=1}^{i} w'[l] \leq |\vec{w}[1 : i]|$.

By Lemma 1 on the sequence $x_n, \ldots, x_i$ we know that $\sum_{l=i}^{n} w'[l] \leq |\overleftarrow{w}[i : n]|$.

Therefore, since $w'[i] = 1$, we have $|w'| = \sum_{l=1}^{i} w'[l] + \sum_{l=i}^{n} w'[l] - 1 < d$, thus contradicting the hypothesis that $|w'| = d$. Hence there is no support for $x_i = 1$. □

Propositions 2 and 3 entail a pruning rule. In a first pass, from left to right, one can use an algorithm similar to `leftmost` to compute and store the values of $|\vec{w}[1 : i]|$ for all $i \in [1, \ldots, n]$. In a second pass, the values of $|\overleftarrow{w}[i : n]|$ for all $i \in [1, \ldots, n]$ are similarly computed by simply running the same procedure on the same sequence of variables, however *reversed*, i.e., from right to left. Using these values, one can then apply Proposition 2 and Proposition 3 to filter out the value 0 and 1, respectively. We detail this procedure in the next section.

We first show that these two rules are complete, that is, if ATMOSTSEQ is AC, and the overall cardinality constraint is AC then an assignment $x_i = 0$ (resp. $x_i = 1$) is inconsistent iff Proposition 2 (resp. Proposition 3) applies. The following Lemma shows that the greedy rule maximises the density of 1s on any subsequence starting on x_1, and therefore minimises it on any subsequence finishing on x_n. Let `leftmost`(k) denote the algorithm corresponding to applying `leftmost`, however stopping whenever the cardinality of the assignment reaches a given value k.

Lemma 2. *Let* w *be a satisfying assignment of* ATMOSTSEQ$(u, q, [x_1, \ldots, x_n])$. *If* $k \leq |w|$ *then the assignment* $\vec{w}_k$ *computed by* `leftmost`(k) *is such that, for any* $1 \leq i \leq n$: $\sum_{l=i}^{n} \vec{w}_k[l] \leq \sum_{l=i}^{n} w[l]$.

Proof. Let m be the index at which `leftmost`(k) stops. We distinguish two cases. If $i > m$, for any value l in $[m + 1, \ldots, n]$, $\vec{w}_k[l] \leq w[l]$ (since $\vec{w}_k[l] = \min(x_l)$), hence $\sum_{l=i}^{n} \vec{w}_k[l] \leq \sum_{l=i}^{n} w[l]$. When $i \leq m$, clearly for $i = 1$, $\sum_{l=i}^{n} \vec{w}_k[l] \leq \sum_{l=i}^{n} w[l]$ since $|\vec{w}_k| \leq |w|$. Now consider the case of $i \neq 1$. Since $|\vec{w}_k| \leq |w|$, then

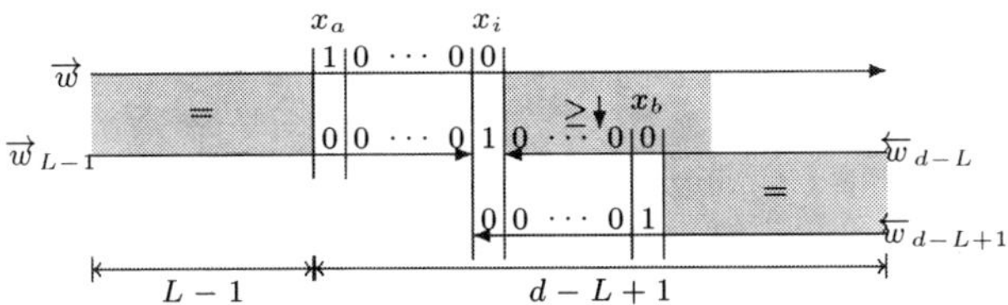

Fig. 3. Illustration of Proposition 5's proof. Horizontal arrows represent assignments.

$\sum_{l=i}^{n} \vec{w}_k[l] \leq |w| - \sum_{l=1}^{i-1} \vec{w}_k[l]$. Thus, $\sum_{l=i}^{n} \vec{w}_k[l] \leq \sum_{l=i}^{n} w[l] + (\sum_{l=i}^{i-1} w[l] - \sum_{l=1}^{i-1} \vec{w}_k[l])$. Moreover, by applying Lemma 1, we show that $\sum_{l=1}^{i-1} \vec{w}_k[l]$ is maximum, hence larger than or equal to $\sum_{l=1}^{i-1} w[l]$. Therefore, $\sum_{l=i}^{n} \vec{w}_k[l] \leq \sum_{l=i}^{n} w[l]$.
$\square$

Proposition 4. *If* ATMOSTSEQ$(u, q, [x_1, \ldots, x_n])$ *is* AC, *and* $|\vec{w}[1 : i-1]| + |\overleftarrow{w}[i+1 : n]| \geq d$ *then* $x_i = 0$ *has a support.*

Proof. Let $\vec{w}$ be the assignment found by `leftmost`. We consider, without loss of generality, a variable x_i such that $\mathcal{D}(x_i) = \{0, 1\}$ and $|\vec{w}[1 : i-1]| + |\overleftarrow{w}[i+1 : n]| \geq d$ and show that one can build a support for $x_i = 0$. If $\vec{w}[i] = 0$ or $\overleftarrow{w}[i] = 0$ then there exists a support for $x_i = 0$, hence we only need to consider the case where $\vec{w}[i] = \overleftarrow{w}[i] = 1$.

Let $L = |\vec{w}[1 : i-1]|$, and let $\overleftarrow{w}_{d-L}$ be the result of `leftmost`$(d-L)$ on the subsequence $x_n, \ldots, x_i$. We will show that w, defined as the concatenation of $\vec{w}[1 : i-1]$ and $\overleftarrow{w}_{d-L}[i : n]$ is a support for $x_i = 0$.

First, we show that $w[i] = 0$, that is $\overleftarrow{w}_{d-L}[i] = 0$. By hypothesis, since $|\vec{w}[1 : i-1]| + |\overleftarrow{w}[i+1 : n]| \geq d$, we have $|\overleftarrow{w}[i+1 : n]| \geq d - L$. Now consider the sequence $x_i, \ldots, x_n$, and let w' be the assignment such that $w'[i] = 0$, and $w' = \overleftarrow{w}[i+1 : n]$ otherwise. Since $|w'| = |\overleftarrow{w}[i+1 : n]| \geq d - L$, by Lemma 2, we know that w' has a higher cardinality than $\overleftarrow{w}_{d-L}$ on any subsequence starting in i, hence $w[i] = \overleftarrow{w}_{d-L}[i] = w'[i] = 0$.

Now we show that w does not violate the ATMOSTSEQ constraint. Obviously, since it is the concatenation of two consistent assignments, it can violate the constraint only on the junction, i.e., on a sub-sequence $x_j, \ldots, x_{j+q-1}$ such that $j \leq i$ and $i < j + q$.

We show that the sum of any such subsequence is less or equal to u by comparing with $\vec{w}$, as illustrated in Figure 2. We have $\sum_{l=j}^{j+q-1} \vec{w}[l] \leq u$, and $\sum_{l=j}^{i-1} \vec{w}[l] = \sum_{l=j}^{i-1} w[l]$. Moreover, by Lemma 2, since $|\vec{w}[i : n]| = |\overleftarrow{w}_{d-L}| = d - L$ we have $\sum_{l=i}^{j+q-1} \overleftarrow{w}_{d-L}[l] \leq \sum_{l=i}^{j+q-1} \vec{w}[l]$ hence $\sum_{l=i}^{j+q-1} w[l] \leq \sum_{l=i}^{j+q-1} \vec{w}[l]$. Therefore, we can conclude that $\sum_{l=j}^{j+q-1} w[l] \leq u$.
$\square$

Proposition 5. *If* ATMOSTSEQ$(u, q, [x_1, \ldots, x_n])$ *is* AC, $|w[1 : i]| + |w[n : i]| > d$ *then* $x_i = 1$ *has a support.*

Proof. Let $\vec{w}$ and $\overleftarrow{w}$ be the assignments found by `leftmost`, on respectively $x_1, \ldots, x_n$ and $x_n, \ldots, x_1$. We consider, without loss of generality, a variable x_i such that $\mathcal{D}(x_i) = \{0, 1\}$ and $|\vec{w}[1 : i]| + |\overleftarrow{w}[i : n]| > d$ and show that one can build a support for $x_i = 1$. If $\overrightarrow{w}[i] = 1$ or $\overleftarrow{w}[i] = 1$ then there exists a support for $x_i = 1$, hence we only need to consider the case where $\overrightarrow{w}[i] = \overleftarrow{w}[i] = 0$.

Let $L = |\overrightarrow{w}[1:i]| = |\overrightarrow{w}[1:i-1]|$ (this equality holds since $\overrightarrow{w}[i] = 0$). Let $\overrightarrow{w}_{L-1}$ be the assignment obtained by using $\mathtt{leftmost}(L-1)$ on the subsequence $x_1, \ldots, x_{i-1}$, and let $\overleftarrow{w}_{d-L}$ be the assignment returned by $\mathtt{leftmost}(d-L)$ on the subsequence $x_n, \ldots, x_{i+1}$.

We show that w such that $w[i] = 1$, equal to $\overrightarrow{w}_{L-1}$ on $x_1, \ldots, x_{i-1}$ and to $\overleftarrow{w}_{d-L}$ on $x_{i+1}, \ldots, x_n$, is a support.

Clearly $|w| = d$, therefore we only have to make sure that all capacity constraints are satisfied. Moreover, since it is the concatenation of two consistent assignments, it can violate the constraint only on the junction, i.e., on a sub-sequence $x_j, \ldots, x_{j+q-1}$ such that $j \leq i$ and $i < j + q$.

We show that the sum of any such subsequence is less or equal to u by comparing with $\overrightarrow{w}$ and $\overleftarrow{w}_{d-L}$ (see Figure 3). First, observe that on the subsequence $x_1, \ldots, x_{i-1}$, $\overrightarrow{w}_{L-1} = \overrightarrow{w}$, except for the largest index a such that $\overrightarrow{w}[a] = 1$ and $w[a] = 0$. Similarly on $x_n, \ldots, x_{i+1}$, we have $\overleftarrow{w}_{d-L} = \overleftarrow{w}_{d-L+1}$, except for the smallest b such that $\overleftarrow{w}_{d-L+1}[b] = 1$. There are two cases:

Suppose first that $j > a$. In that case, $\sum_{l=j}^{j+q-1} w[l] = \sum_{l=i}^{j+q-1} \overleftarrow{w}_{d-L+1}[l]$ if $j + q - 1 \geq b$, and otherwise it is equal to 1. It is therefore alway less than or equal to u since $i \geq j$ (and we assume $u \geq 1$).

Now suppose that $j \leq a$. In that case, consider first the subsequence $x_j, \ldots, x_i$. On this interval, the cardinality of w is the same as that of $\overrightarrow{w}$, i.e., $\sum_{l=j}^{i} w[l] = \sum_{l=j}^{i-1} \overrightarrow{w}_{L-1}[l] + 1 = \sum_{l=j}^{i} \overrightarrow{w}[l]$. On the subsequence $x_{i+1}, \ldots, x_{j+q-1}$, observe that $|w[i+1:n]| = |\overrightarrow{w}[i+1:n]| = d - L$, hence by Lemma 2, we have $\sum_{l=i+1}^{j+q-1} w[l] = \sum_{l=i+1}^{j+q-1} \overleftarrow{w}_{d-L}[l] \leq \sum_{l=i+1}^{j+q-1} \overrightarrow{w}[l]$. Therefore $\sum_{l=j}^{j+q-1} w[l] \leq \sum_{l=j}^{j+q-1} \overrightarrow{w}[l] \leq u$. $\square$

3.3 Algorithmic Complexity

Using Propositions 2, 3, 4 and 5 one can design a filtering algorithm with the same worst case time complexity as $\mathtt{leftmost}$. In this section, we introduce a linear time implementation of $\mathtt{leftmost}$, denoted $\mathtt{leftmost\_count}$, that returns the values of $\overrightarrow{w}[1:i]$ for all values of i (Algorithm 2).

It is easy to see that $\mathtt{leftmost\_count}$ has a $O(n)$ worst case time complexity. In order to prove its correctness, we will show that the assignment computed by $\mathtt{leftmost\_count}$ is the same as that computed by $\mathtt{leftmost}$.

Proof (sketch). We only sketch the proof here. The following three invariants are true at the beginning of each step i of the main loop:

- The cardinality of the sub-sequence j is given by $c[(i+j-2) \bmod q] + count[i-1]$.
- The number of sub-sequences of cardinality k is given by $occ[n - count[i-1] + k]$.
- The cardinality maximum of any sub-sequence is given by max_c.

1st *invariant:* The value of $c[j]$ is updated in two ways in $\mathtt{leftmost}$. First, at each step of the loop the values in $c[1]$ through to $c[q]$ are shifted to the left. Therefore, there is only one really new value. By using the modulo operation, we can update only one

---

Algorithm 2. `leftmost_count`

Data: $u, q, [x_1, \dots, x_n]$
Result: $count : [0, \dots, n] \mapsto [0, \dots, n]$
foreach $i \in [1, \dots, n]$ **do**
 $w[i] \leftarrow \min(x_i)$;
 $occ[i] = 0$;

foreach $i \in [0, \dots, n]$ **do** $count[i] \leftarrow 0$;
$c[0] \leftarrow w[1]$;
foreach $i \in [1, \dots, u]$ **do** $occ[n + i] = 0$;
foreach $i \in [1, \dots, q]$ **do**
 $w[n + i] \leftarrow 0$;
 if $i < q$ **then** $c[i] \leftarrow c[i - 1] + w[i + 1]$
 $occ[n + c[i - 1]] \leftarrow occ[n + c[i - 1]] + 1$;

$max\_c \leftarrow \max(\{c[i] \mid i \in [0, \dots, q - 1]\})$;
foreach $i \in [1, \dots, n]$ **do**
 if $max\_c < u$ & $|\mathcal{D}(x_i)| > 1$ **then**
  $max\_c \leftarrow max\_c + 1$;
  $count[i] \leftarrow count[i - 1] + 1$;
  $w[i] \leftarrow 1$;

 else $count[i] \leftarrow count[i - 1]$;
 $prev \leftarrow c[(i - 1) \bmod q]$;
 $next \leftarrow c[(i + q - 2) \bmod q] + w[i + q] - w[i]$;
 $c[(i - 1) \bmod q] \leftarrow next$;
 if $prev \neq next$ **then** $occ[n + next] \leftarrow occ[n + next] + 1$;
 if $next + count[i] > max\_c$ **then** $max\_c \leftarrow max\_c + 1$;
 $occ[n + prev] \leftarrow occ[n + prev] - 1$;
 if $occ[n + prev] = 0$ & $prev + count[i] = max\_c$ **then**
  $max\_c \leftarrow max\_c - 1$;

return $count$;

$\mathcal{D}(x_i)$	$\vec{w}[i]$	$\overleftarrow{w}[i]$	$L[i]$	$R[n-i+1]$	$L[i]+R[n-i+1]$	$L[i-1]+R[n-i]$	$\mathrm{AC}(\mathcal{D}(x_i))$
			0				
.	1	1	1	10	11	**9**	1
0	0	0	1	9	10	10	0
.	1	0	2	9	11	10	.
.	1	1	3	9	12	10	.
.	1	1	4	8	12	10	.
.	0	1	4	7	11	10	.
.	0	0	4	6	**10**	10	**0**
.	0	0	4	6	**10**	10	**0**
0	0	0	4	6	10	10	0
1	1	1	4	6	10	10	1
0	0	0	4	6	10	10	0
.	1	1	5	6	11	**9**	1
.	1	1	6	5	11	**9**	1
.	1	1	7	4	11	**9**	1
.	0	0	7	3	**10**	10	**0**
.	0	0	7	3	**10**	10	**0**
.	0	0	7	3	**10**	10	**0**
.	1	0	8	3	11	10	.
.	0	1	8	3	11	10	.
.	1	1	9	2	11	**9**	1
.	1	1	10	1	11	**9**	1
1	1	1	10	0	10	10	1
			0				

Fig. 4. Example of the execution of Algorithm 3 for $u = 4$, $q = 8$, $d = 12$. The first line stands for current domains, dots are unassigned variables (hence $ub = 10$). The two next lines give the assignments $\vec{w}$ and $\overleftarrow{w}$ obtained by running `leftmost_count` from left to right and from right to left, respectively. The third and fourth lines stand for the values of $L[i] = |\vec{w}[1 : i]|$ and $R[n - i + 1] = |\overleftarrow{w}[i : n]|$. The fifth and sixth lines correspond to the application of, respectively, Proposition 2 and 3. Last, the seventh line give the arc consistent closure of the domains.

of these values. Second, when $w[i]$ takes the value 1, we increment $c[1]$ up to $c[q]$. Since this happened exactly $count[i - 1]$ times at the start of step i, we can simply add $count[i - 1]$ to obtain the same value as in `leftmost`.

2nd *invariant:* The data structure occ is a table, storing at index $n + k$ the number of subsequences involving x_i with cardinality k for the current assignment w. Therefore, by decrementing the pointer to the first element of the table we in effect shift the entire table. Here again the value of $count[i - 1]$, or rather the expression $(n - count[i - 1])$ points exactly to the required starting point of the table.

3rd *invariant:* The maximum (or minimum) cardinality (of subsequences involving x_i) can change by a unit at the most from one step to the next. Therefore, when the variable max_c needs to change, it can only be incremented (when $occ[n - count[i - 1] + max\_c + 1]$ goes from the value 0 to 1) or decremented (when $occ[n - count[i - 1] + max\_c]$ goes from the value 1 to 0). $\qquad\square$

Algorithm 3 computes the AC closure of a constraint ATMOSTSEQCARD$(u, q, d, [x_1, \dots, x_n])$. In the first line, the AC closure of ATMOSTSEQ$(u, q, [x_1, \dots, x_n])$

Algorithm 3. AC(ATMOSTSEQCARD)

Data: $[x_1, \ldots, x_n], u, q, d$
Result: AC on ATMOSTSEQCARD$(u, q, d, [x_1, .., x_n])$
AC(ATMOSTSEQ)$(u, q, [x_1, \ldots, x_n])$;
$ub \leftarrow d - \sum_{i=1}^n \min(x_i)$;
$L \leftarrow$ leftmost_count$([x_1, \ldots, x_n], u, q, d)$;
if $L[n] = ub$ **then**
    $R \leftarrow$ leftmost_count$([x_n, \ldots, x_1], u, q, d)$;
    foreach $i \in [1, \ldots, n]$ *such that* $\mathcal{D}(x_i) = \{0, 1\}$ **do**
        if $L[i] + R[n - i + 1] \le ub$ **then** $\mathcal{D}(x_i) \leftarrow \{0\}$;
        if $L[i - 1] + R[n - i] < ub$ **then** $\mathcal{D}(x_i) \leftarrow \{1\}$;

else if $L[n] < ub$ **then** Fail

---

is computed. It ensures that the filtering rules introduced in this paper hold. For lack of space, we do not give a pseudo-code for achieving AC on ATMOSTSEQ. However, it can be done in linear time using a procedure similar to leftmost_count. We want to compute, for the assignment corresponding to the lower bound of each domain, if an unassigned variable is covered by a subsequence of cardinality u for the lower bounds of the domains. We do it using a truncated version of leftmost_count: the values of $w[i]$ are never updated, i.e., they are set to the minimum value in the domain and we never enter the if-then-else block starting at condition 1 in Algorithm 2. Moreover, we store the value of max_c for each value of i in a table that we can subsequently use to achieve AC on ATMOSTSEQ, by going through it and assigning 0 to any unassigned variable covered by at least one subsequence of cardinality u.

The remainder is a straight application of Propositions 2, 3, 4 and 5. We give an example of its execution in Figure 4. The worst case time complexity of Algorithm 3 is therefore $O(n)$, hence optimal.

4   Experimental Results

We tested our filtering algorithm on two benchmarks: car-sequencing and crew-rostering. All experiments ran on Intel Xeon CPUs 2.67GHz under Linux. For each instance, we launched 5 randomized runs with a 20 minutes time cutoff.[3] All models are implemented using Ilog-Solver.

Since we compare propagators, we averaged the results across several branching heuristics to reduce the bias that these can have on the outcome. For each considered data set, we report the total number of successful runs (*#sol*). Then, we report the CPU time (*time*) in seconds and number of backtracks (*avg bts*) both averaged over all successful runs, instances and branching heuristics. Moreover, we report the maximum number of backtracks (*max bts*) in the same scope. We emphasize the statistics of the best method (w.r.t. *#sol*) for each data set using bold face fonts.

4.1   Car-Sequencing

In the car-sequencing problem [8,17], n vehicles have to be produced on an assembly line, subject to capacity and demand constraints.

---

[3] For a total of approximately 200 days of CPU time.

Table 1. Evaluation of the filtering methods (averaged over all heuristics)

Method	set1 ($70 \times 34 \times 5$)			set2 ($4 \times 34 \times 5$)			set3 ($5 \times 34 \times 5$)			set4 ($7 \times 34 \times 5$)		
	#sol	avg bts	time	#sol	avg bts	time	#sol	avg bts	time	#sol	avg bts	time
sum	8480	231.2K	13.93	95	1.4M	76.60	0	-	> 1200	64	543.3K	43.81
gsc	11218	1.7K	3.60	325	131.7K	110.99	31	55.3K	276.06	140	25.2K	56.61
amsc	10702	39.1K	4.43	**360**	**690.8K**	**72.00**	16	40.3K	8.62	**153**	**201.4K**	**33.56**
amsc+gsc	**11243**	**1.2K**	**3.43**	339	118.4K	106.53	**32**	**57.7K**	**285.43**	147	23.8K	66.45

We use a standard model, implemented in Ilog-Solver. We have n integer variables standing for the class of vehicles in each slot of the assembly line and nm boolean variables y_i^j standing for whether the vehicle in the ith slot requires option j. The demand for each class is enforced with a GCC [14]. We compare four models for the capacity constraints coupled with the demand on each option (derived from the demand on classes). In the first model (*sum*) a simple decomposition into a chain of sum constraints plus an extra sum for the demand is used. In the second (*gsc*), we use one GSC constraint per option. In the third, (*amsc*), we use the AC procedure introduced in this paper for the ATMOSTSEQCARD constraint. Finally, in the fourth (*amsc+gsc*) we combine the GSC constraint with our filtering algorithm.

We use 34 different heuristics, obtained by combining different ways of *exploring* the assembly line either in lexicographic order or from the middle to the sides); of *branching* on affectation of a class to a slot or of an option to a slot; of *selecting* the best class or option among a number of natural criteria (such as maximum demand, minimum u/q ratio, as well as other criteria described in or derived from [5,16]).

We use benchmarks available in the CSPLib [9] divided into four sets of respectively 70, 4, 5 and 7 instances ranging from 100 to 400 cars. Instances of the third set are all unsatisfiable all other are satisfiable.

The results are given in Table 1. In all cases, the best number of solved problems is obtained either by *amsc+gsc* (for small or unsatisfiable instances), or by *amsc* alone (for larger instances of set2 and set4). Overall, we observe that the GSC constraint allows to prune much more values than ATMOSTSEQCARD. However, it slows down the search by a substantial amount (we observed a factor 12.5 on the number of nodes explored per second). Moreover, the amounts of filtering obtained by these two methods are incomparable. Therefore combining them is always better than using GSC alone.

In [18] the authors applied their method to set1, set2 and set3 only. For their experiments, they considered the best result provided by 2 heuristics (σ and min domain). When using COST-REGULAR or GEN-SEQUENCE filtering alone, 50.7% of problems are solved and when combining either COST-REGULAR or GEN-SEQUENCE with GSC, 65.2% of problems are solved (with a time out of 1 hour). In our experiments, in average over the 34 heuristics and the 5 re-starts, ATMOSTSEQCARD and GSC solve respectively 82.5% and 86.11% of instances and combining ATMOSTSEQCARD with GSC solves 86.36% instances in a time out of 20 minutes.

4.2 Crew-Rostering

In this problem, working shifts have to be attributed to employees over a period, so that the required service is met at any time and working regulations are respected. The latter

Table 2. Evaluation of the filtering methods (averaged over all heuristics)

Benchmarks	underconstrained ($5 \times 2 \times 126$)				hard ($5 \times 2 \times 111$)				overconstrained ($5 \times 2 \times 44$)			
	#sol	avg bts	max bts	time	#sol	avg bts	max bts	time	#sol	avg bts	max bts	time
sum	1229	110.5K	10.1M	12.72	574	370.7K	13.4M	38.45	272	52.1K	5.7M	5.56
gsc	1210	6.2K	297.2K	29.19	579	23.5K	433.5K	77.78	276	7.7K	378.9	24.14
amsc	**1237**	**34.2K**	**7.5M**	**5.82**	**670**	**213.4K**	**7.5M**	**31.01**	**284**	**51.3K**	**7.5M**	**6.22**

condition can entail a wide variety of constraints. Previous works [11] [13] used allowed (or forbidden) patterns to express successive shift constraints. For example, with 3 shifts of 8 hours per day: D (day), E (evening) and N (night), ND can be forbidden since employees need some rest after night shifts. In this paper we consider a simple case involving 20 employees with 3 shifts of 8 hours per days where no employee can work more than one 8h shift per day, no more than 5 days a week, and the break between two worked shifts must be at least 16h. The planning horizon is of 28 days, and each employee must work 34 hours per week in average (17 shifts over the 4 weeks period).

We use a model with one Boolean variable e_{ij} for each of the m employees and each of the n shifts stating if employee i works on shift j. The demand d_j^s on each shift j is enforced through a sum constraint $\sum_{i=1}^{m} e_{ij} = d_j^s$. The other constraints are stated using two ATMOSTSEQCARD constraints per employee, one with ratio $u/q = 1/3$, another with ratio $5/21$, and both with the same demand $d = 17$ corresponding to 34 hours of work per week. We compare three models. In the first (*sum*), we use a decomposition in a chain of sum constraints. In the second (*gsc*), we use the GSC constraint to encode it. Observe that in this case, since the domains are Boolean, the GCC within the GSC constraint is in fact nothing more than a sum. Therefore, it cannot prune more than ATMOSTSEQCARD (however it is stronger than the simple sum decomposition). In the third model (*amsc*) we use the algorithm presented in this paper.

281 instances were generated, with employee unavailability ranging from from 18% to 46% by increment of 0.1. We partition the instances into three sets, with in the first 126 instances with lowest unavailability (all satisfiable), in the third, the 44 instances with highest unavailability (mostly unsatisfiable), and the rest in the second group.

We used two branching heuristic. In the first we chose the employee with minimum slack, and assign it to its possible shift of maximum demand. In the second we use the same criteria, but select the shift first and then the employee.

We report the results in Table 2. The AC algorithm achieves more filtering than the sum decomposition and the GSC decomposition. However, depending on the heuristic choices and the random seed, the size of the search tree is not always smaller. We observe that our propagator is only marginally slower in terms of nodes explored per second than the sum decomposition and much faster (by a factor 20.4 overall) than GSC. It is able to solve 15.7% and 16.7% more problems within the 20 minutes cutoff in the "hard" set than the GSC and sum decompositions, respectively.

5   Conclusion

We introduced a linear algorithm for achieving arc consistency on the ATMOSTSEQCARD constraint. Previously, the best AC algorithm had a $O(n^2)$ time complexity [10]. However, it ran in $O(n^2 \log n)$ time down a branch since subsequents calls cost $O(n \log n)$, whilst our algorithm is not incremental hence requires up to $O(n^2)$ steps down a branch.

The empirical evaluation on car-sequencing and crew-rostering benchmarks shows that this propagator is useful on these applications.

Acknowledgments. We would like to thank Nina Narodytska for her precious help and comments.

References

1. Beldiceanu, N., Carlsson, M.: Revisiting the *Cardinality* Operator and Introducing the *Cardinality-Path* Constraint Family. In: Codognet, P. (ed.) ICLP 2001. LNCS, vol. 2237, pp. 59–73. Springer, Heidelberg (2001)
2. Beldiceanu, N., Contejean, E.: Introducing Global Constraints in CHIP. Mathematical Computation Modelling 20(12), 97–123 (1994)
3. Bessiere, C., Hebrard, E., Hnich, B., Kiziltan, Z., Walsh, T.: The Slide Meta-Constraint. In: CPAI Workshop, held alongside CP (2006)
4. Bessiere, C., Hebrard, E., Hnich, B., Kiziltan, Z., Walsh, T.: Slide: A Useful Special Case of the Cardpath Constraint. In: ECAI, pp. 475–479 (2008)
5. Boivin, S., Gravel, M., Krajecki, M., Gagné, C.: Résolution du Problème de Car-sequencing à l'Aide d'une Approche de Type FC. In: JFPC (2005)
6. Brand, S., Narodytska, N., Quimper, C.-G., Stuckey, P.J., Walsh, T.: Encodings of the SEQUENCE Constraint. In: Bessière, C. (ed.) CP 2007. LNCS, vol. 4741, pp. 210–224. Springer, Heidelberg (2007)
7. Demassey, S., Pesant, G., Rousseau, L.-M.: A Cost-Regular Based Hybrid Column Generation Approach. Constraints 11(4), 315–333 (2006)
8. Dincbas, M., Simonis, H., Van Hentenryck, P.: Solving the Car-Sequencing Problem in Constraint Logic Programming. In: ECAI, pp. 290–295 (1988)
9. Gent, I.P., Walsh, T.: Csplib: a benchmark library for constraints (1999)
10. Maher, M.J., Narodytska, N., Quimper, C.-G., Walsh, T.: Flow-Based Propagators for the SEQUENCE and Related Global Constraints. In: Stuckey, P.J. (ed.) CP 2008. LNCS, vol. 5202, pp. 159–174. Springer, Heidelberg (2008)
11. Menana, J., Demassey, S.: Sequencing and Counting with the multicost-regular Constraint. In: van Hoeve, W.-J., Hooker, J.N. (eds.) CPAIOR 2009. LNCS, vol. 5547, pp. 178–192. Springer, Heidelberg (2009)
12. Pesant, G.: A Regular Language Membership Constraint for Finite Sequences of Variables. In: Wallace, M. (ed.) CP 2004. LNCS, vol. 3258, pp. 482–495. Springer, Heidelberg (2004)
13. Pesant, G.: Constraint-Based Rostering. In: PATAT (2008)
14. Régin, J.-C.: Generalized Arc Consistency for Global Cardinality Constraint. In: AAAI, pp. 209–215(2) (1996)
15. Régin, J.-C., Puget, J.-F.: A Filtering Algorithm for Global Sequencing Constraints. In: Smolka, G. (ed.) CP 1997. LNCS, vol. 1330, pp. 32–46. Springer, Heidelberg (1997)
16. Smith, B.M.: Succeed-first or Fail-first: A Case Study in Variable and Value Ordering (1996)
17. Solnon, C., Cung, V., Nguyen, A., Artigues, C.: The car sequencing problem: Overview of state-of-the-art methods and industrial case-study of the ROADEF'2005 challenge problem. EJOR 191, 912–927 (2008)
18. van Hoeve, W.J., Pesant, G., Rousseau, L.-M., Sabharwal, A.: New Filtering Algorithms for Combinations of Among Constraints. Constraints 14(2), 273–292 (2009)
19. van Hoeve, W.-J., Pesant, G., Rousseau, L.-M., Sabharwal, A.: Revisiting the Sequence Constraint. In: Benhamou, F. (ed.) CP 2006. LNCS, vol. 4204, pp. 620–634. Springer, Heidelberg (2006)

Conflict Directed Lazy Decomposition

Ignasi Abío[1] and Peter J. Stuckey[2]

[1] Technical University of Catalonia (UPC), Barcelona, Spain
[2] Department of Computing and Information Systems, and
NICTA Victoria Laboratory, The University of Melbourne, Australia

Abstract. Two competing approaches to handling complex constraints in satisfaction and optimization problems using SAT and LCG/SMT technology are: *decompose* the complex constraint into a set of clauses; or *(theory) propagate* the complex constraint using a standalone algorithm and explain the propagation. Each approach has its benefits. The decomposition approach is prone to an explosion in size to represent the problem, while the propagation approach may require exponentially more search since it does not have access to intermediate literals for explanation. In this paper we show how we can obtain the best of both worlds by *lazily decomposing* a complex constraint propagator using conflicts to direct it. If intermediate literals are not helpful for conflicts then it will act like the propagation approach, but if they are helpful it will act like the decomposition approach. Experimental results show that it is never much worse than the better of the decomposition and propagation approaches, and sometimes better than both.

1 Introduction

Compared with other systematic constraint solving techniques, SAT solvers have many advantages for non-expert users. They are extremely efficient off-the-shelf black boxes that require no tuning regarding variable (or value) selection heuristics. However, propositional logic cannot directly deal with complex constraints: we need either *to enrich the language* in which the problems are defined, or *to reduce the complex constraints* to propositional logic.

Lazy clause generation (LCG) or *SAT Modulo Theories* (SMT) approaches correspond to an enrichment of the language: the problem can be expressed in first-order logic instead of propositional logic. A specific theory solver for that (kind of) constraint, called a *propagator*, takes care of the non-propositional part of the problem, propagating and explaining the propagations, whereas the SAT Solver deals with the propositional part. On the other hand, reducing the constraints to propositional logic corresponds to *encoding* or *decomposing* the constraints into SAT: the complex constraints are replaced by an equivalent set of auxiliary variables and clauses.

The advantages of the propagator approach is that the size of the propagator and its data structures are typically quite small (in the size of the constraint) compared to the size of a decomposition, and we can make use of specific global algorithms for efficient propagation. The advantages of the decomposition approach are that the resulting propagation uses efficient SAT data structures and

M. Milano (Ed.): CP 2012, LNCS 7514, pp. 70–85, 2012.
© Springer-Verlag Berlin Heidelberg 2012

are inherently incremental, and more importantly, the auxiliary variables give the solver more scope for learning appropriate reusable nogoods.

In this paper we examine how to get the best of each approach, and illustrate our method on two fundamental constraints: cardinality and pseudo-Boolean constraints.

An important class of constraints are the so-called *cardinality constraints*, that is, constraints of the form $x_1 + \cdots + x_n \ \# \ K$, where the K is an integer, the x_i are Boolean (0/1) variables, and the relation operator $\#$ belongs to $\{\leqslant, \geqslant, =\}$. Cardinality constraints are omnipresent in practical SAT applications such as timetabling [1] and scheduling constraint solving [2]. Some optimization problems, such as MaxSAT or close-solutions problems (see [3]), can be reduced to a set of problems with a single cardinality constraint (see Section 4.1).

The two different approaches for solving complex constraints have both been studied for cardinality constraints. In the literature one can find different decompositions using adders [4], binary trees [5] or sorting networks [6], among others. The best decomposition, to our knowledge, is the cardinality network-based encoding [7]. On the other hand, we can use a propagator for deal with these constraints, and using either an SMT Solver [8] or LCG Solver [9].

Another important class of constraints are the *pseudo-Boolean (PB) constraints*, that is, constraints of the form $a_1x_1 + \cdots + a_nx_n \ \# \ K$, where K and a_i are integers, the x_i are Boolean (0/1) variables, and the relation operator $\#$ belongs to $\{\leqslant, \geqslant, =\}$. These constraints are very important and appear frequently in application areas such as cumulative scheduling [10], logic synthesis [11] or verification [12].

In the literature one can find different decompositions of PB constraints using adders [4,6], BDDs or similar tree-like structures [6,13,14] or sorting networks [6]. As before, LCG and SMT approaches are also possible.

To see why both approaches, both propagator and decomposition, have advantages consider the following two scenarios:

- Consider a problem with hundreds of large cardinality constraints where all but 1 never cause failure during search. Decomposing each of these constraints will cause a huge burden on the SAT solver, adding many new variables and clauses, all of which are actually useless. The propagation approach will propagate much faster, and indeed just the decomposition step could overload the SAT solver.
- Consider the problem with the cardinality constraint $x_1 + \cdots + x_n \leqslant K$ and some propositional clauses implying $x_1 + \cdots + x_n \geqslant K + 1$. The problem is obviously unsatisfiable, but if we use a propagator for the cardinality constraint, it will need to generate all the ^nC_k explanations possible in order to prove the unsatisfiability. However with a decomposition approach the problem can be solved in polynomial time due to the auxiliary variables.

In conclusion it seems likely that in every problem there are *some* auxiliary variables that will produce more general nogoods and will help the SAT solver, and *some* other variables that will only increase the search space size, making the problem more difficult. The intuitive idea of Lazy Decomposition is to try to generate only the *useful* auxiliary variables. The solver initially behaves as a basic LCG solver. If it observes that an auxiliary variable would appear in many nogoods, the solver generates it.

While there is plenty of research on combining SAT and propagation-based methods, for example all of SAT modulo theories and lazy clause generation, we are unaware of any previous work where a complex constraint is partially decomposed. There is some recent work [15] where the authors implement an incremental method for solving pseudo-Boolean constraints with SAT, by decomposing the pseudo-Booleans one by one. However, they do not use propagators for dealing with the non-decomposed constraints, and the decomposition is done in one step for a single constraint.

The remainder of the paper is organized as follows. In the next section we give SAT and LCG/SMT solving as well as decompositions and propagator definitions for both cardinality and psuedo-Boolean constraints. In Section 3 we define a framework for lazy decomposition propagators, and instantiate it for cardinality and psuedo-Boolean constraints. In Section 4 we show results of experiments, and in Section 5 we conclude.

2 Preliminaries

2.1 SAT Solving

Let $\mathcal{X} = \{x_1, x_2, \ldots\}$ be a fixed set of propositional *variables*. If $x \in \mathcal{X}$ then x and $\overline{x}$ are *positive and negative literals*, respectively. The *negation* of a literal l, written $\overline{l}$, denotes $\overline{x}$ if l is x, and x if l is $\overline{x}$. A *clause* is a disjunction of literals $\overline{x}_1 \vee \ldots \vee \overline{x}_p \vee x_{p+1} \vee \ldots \vee x_n$, sometimes written as $x_1 \wedge \ldots \wedge x_p \rightarrow x_{p+1} \vee \ldots \vee x_n$. A *CNF formula* is a conjunction of clauses.

A (partial) *assignment* A is a set of literals such that $\{x, \overline{x}\} \nsubseteq A$ for any x, i.e., no contradictory literals appear. A literal l is *true* in A if $l \in A$, is *false* in A if $\overline{l} \in A$, and is *undefined* in A otherwise. True, false or undefined is the *polarity* of the literal l. A clause C is true in A if at least one of its literals is true in A. A formula F is true in A if all its clauses are true in A. In that case, A is a *model* of F. Systems that decide whether a formula F has any model are called SAT-solvers, and the main inference rule they implement is *unit propagation*: given a CNF F and an assignment A, find a clause in F such that all its literals are false in A except one, say l, which is undefined, add l to A and repeat the process until reaching a fix-point.

Clauses are not the only constraints that can be defined over the propositional variables. Sometimes some clauses of the formula are expressed more compactly as a single complex constraint.

2.2 SMT/LCG Solver

An SMT solver or a LCG solver[1] is a system for finding models of a formula F and a set of complex constraints $\{c_i\}$. It is composed of two parts: a SAT

---

[1] In this paper we do not distinguish between SMT solvers and LCG solvers. The two techniques are very similar and both of fit the sketch presented here, although arguably the propagator centric view is more like LCG.

solver engine and a *propagator* for every constraint c_i. The SAT solver searches a model of the formula and the propagators infer consequences of the assignment and the set of constraints (this is, *propagate*), and, on demand of the SAT solver, provide the reason of some of the propagated literals (called the *explanation*).

2.3   Cardinality Constraints

A cardinality constraint takes the form $x_1 + \cdots + x_n \,\#\, K$, where the K is an integer, the x_i are literals, and the relation operator $\#$ belongs to $\{\leqslant, \geqslant, =\}$.

Propagation. For a $\leqslant$ constraint, the propagator keeps a count of the number of literals of the constraint which are *true* in the current assignment. The propagator increments this value every time the SAT solver assigns *true* a literal of the constraint. The count is decremented when the SAT solver unassigns one of these literals. When this value is equal to K, no other literal can be *true*: the propagator sets to *false* all the remaining literals. The explanation for setting a literal x_j to *false* can be built by searching for the K literals $\{x_{i_1}, \ldots, x_{i_K}\}$ of the constraint which are *true* to give the explanation $x_{i_1} \wedge x_{i_2} \wedge \cdots \wedge x_{i_K} \rightarrow \overline{x_j}$.

Similarly, in a $\geqslant$ constraint the propagator keeps a count of the literals which are *false* in the current assignment. When this value is equal to $n - K$, the propagator sets to *true* the non-propagated literals. A propagator for an equality constraint keeps track of both values.

Cardinality Network Decomposition. A *k-cardinality network* of size n is a logical circuit with n inputs and n outputs satisfying two properties:

1. The number of *true* outputs of the network equals the number of *true* inputs.
2. For every i with $1 \leqslant i \leqslant k$, the i-th output of the network is *true* if and only if there were at least i *true* inputs.

An example of cardinality networks are sorting networks e.g. [6], with size $O(n \log^2 n)$. An example is shown in Figure 1(b). The smallest decomposition for a k-cardinality network is $O(n \log^2 k)$ [7].

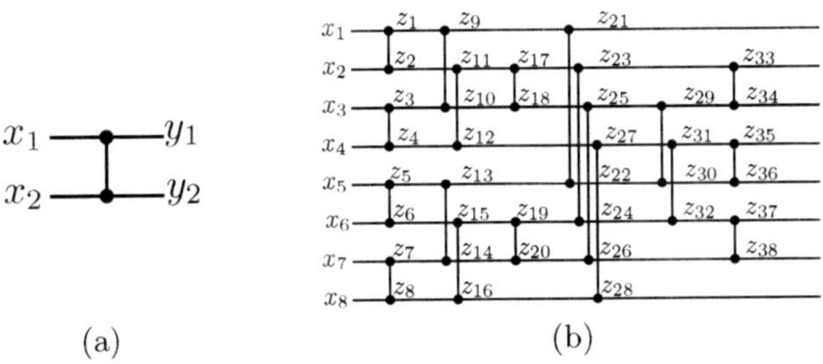

(a)   (b)

Fig. 1. (A) A 2-comparator $2\mathsf{comp}(x_1, x_2, y_1, y_2)$ is shown as a vertical bar joining two lines. (b) An odd-even merge sorting network for $n = 8$. Each line segment (broken at nodes) represents a Boolean variable.

Cardinality networks are composed of 2-comparators. A *2-comparator* is a circuit $2\mathsf{comp}(x_1, x_2, y_1, y_2)$ with inputs x_1, x_2 and outputs $y_1 = x_1 \vee x_2$ and $y_2 = x_1 \wedge x_2$ illustrated in Figure 1(a). 2-comparators can be easily encoded into SAT through the Tseitin transformations [16] using the clauses: $x_1 \rightarrow y_1$, $x_2 \rightarrow y_1$, $x_1 \wedge x_2 \rightarrow y_2$, $\overline{x_1} \rightarrow \overline{y_2}$, $\overline{x_2} \rightarrow \overline{y_2}$ and $\overline{x_1} \wedge \overline{x_2} \rightarrow \overline{y_1}$. A cardinality network can be decomposed into SAT by encoding all its 2-comparators.

Cardinality constraints can be decomposed into SAT through cardinality networks. For instance, a constraint $x_1 + \cdots + x_n \leqslant K$ can be decomposed in two steps: firstly, we build a $K + 1$-cardinality network and encode it into SAT. Secondly, we add the clause $\overline{y_{K+1}}$, where y_{K+1} is the $K + 1$-th output of the network. Notice that this implies that no $K + 1$ inputs $(x_1, x_2, \ldots, x_n)$ are *true*, since the $K + 1$-th output of a $K + 1$-cardinality network is *true* if and only if there are at least $K + 1$ *true* inputs.

Similarly, the constraint $x_1 + \cdots + x_n \geqslant K$ can be decomposed into a K-cardinality network by adding the clause y_K, and $x_1 + \cdots + x_n = K$ can be decomposed with a $K+1$-Cardinality Network adding the clauses y_K and $\overline{y_{K+1}}$.[2]

Example 1. Figure 1 shows an 8-cardinality network. Constraint $x_1 + \cdots + x_8 \leqslant 3$ can be decomposed into SAT by adding the auxiliary variables $z_1, z_2, \ldots, z_{38}$; the definition clauses $x_1 \rightarrow z_1, x_2 \rightarrow z_1, x_1 \wedge x_2 \rightarrow z_2, x_3 \rightarrow z_3, \ldots$; and the unit clause $\overline{z_{35}}$.

2.4   Pseudo-Boolean Constraints

PB constraints are another kind of complex constraint. They take the form $a_1 x_1 + \cdots + a_n x_n \# K$, where K and a_i are integers, the x_i are literals, and the relation operator $\#$ belongs to $\{\leqslant, \geqslant, =\}$. In this paper we assume that the operator $\#$ is $\leqslant$ and the coefficients a_i and K are positive. Other cases can be easily reduced to this one (see [6]).

PB Propagator. The propagator must keep the current sum s during the search, defined as the sum of all coefficients a_i for which x_i is true. This value can be easily incrementally computed: every time the SAT solver sets a literal x_i of the constraint to true, the propagator adds a_i to s, and when the literal is unassigned by the SAT solver it subtracts a_i. For each $i \in \{1, \ldots, n\}$ such that x_i is unassigned and $K - s < a_i$, the propagator sets x_i to false. The propagator can produce explanations in the same way as in the cardinality case: if it has propagated x_j to false, $x_{i_1} \wedge \cdots \wedge x_{i_r} \rightarrow \overline{x_j}$, is returned as the explanation, where $x_{i_1}, \cdots, x_{i_r}$ are all the literals of the constraint with true polarity.

PB Decomposition. PB decomposition into SAT can be made in two steps: first, we build the reduced ordered binary decision diagram (BDD) [17] of the PB constraint; second, we decompose the BDD into SAT.

---

[2] Actually, cardinality networks for $\leqslant$ constraints can be encoded into SAT only adding the first 3 clauses of every 2-comparator. Similarly, in $\geqslant$ constraints we only need the last 3 clauses of 2-comparators [7].

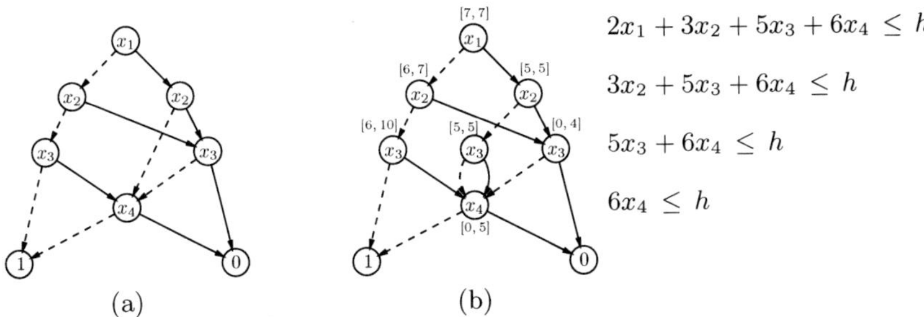

Fig. 2. (a) The BDD of PB constraint $2x_1+3x_2+5x_3+6x_4 \leqslant 7$, and (b) the BDD with long arcs to internal nodes replaced using intermediate nodes and with the intervals given for each node. Each node represents the constraint shown to the right of the BDD for the values of h given by the range.

A BDD of a PB constraint $a_1x_1 + \cdots + a_nx_n \leqslant K$ is a decision diagram that represents this constraint: it has a root node with *selector variable* x_1, the BDD of $a_2x_2 + \cdots + a_nx_n \leqslant K - a_1$ as a *true child* and the BDD of $a_2x_2 + \cdots + a_nx_n \leqslant K$ as *false child*. Moreover, it is *reduced*, i.e., there is no node with identical true and false child, and there are no isomorphic subtrees.

Example 2. Let us consider the PB constraint $c \equiv 2x_1 + 3x_2 + 5x_3 + 6x_4 \leqslant 7$. The BDD of that constraint is shown in Figure 2(a). False (true) children are indicated with dashed (solid) arrows. Terminal node 0 (1) represents the BDDs of Boolean false (true) function.

This BDD represents the constraint in the following sense: assume x_1 and x_4 are true and x_2 is false. The constraint is false no matter the polarity of x_3, since $2x_1 + 3x_2 + 5x_3 + 6x_4 \geqslant 2x_1 + 6x_4 = 8 > 7$. In the BDD of the figure 2(a), if we follow the first solid arrow (since x_1 is true), then the dashed arrow (x_2 is false) and finally the solid one (x_4 is true), we arrive to the false terminal node: that is, the assignment does not satisfy the constraint.

For lazy decomposition we will decompose the BDD one layer at a time from the bottom-up. To simplify this process we create a (non-reduced) BDD which does not have any arcs that skip a level unless they go direct to a terminal node, by introducing artificial nodes. Figure 2(b) shows the resulting BDD for c.

Given a node ν with selector variable x_i, we define *the interval* of ν as the set of integers h such that $a_ix_i + \cdots + a_nx_n \leqslant h$ is represented by the BDD rooted at node ν. This set is always an interval (see [14]). Figure 2(b) shows the intervals of constraint $2x_1 + 3x_2 + 5x_3 + 6x_4 \leqslant 7$. The BDDs for a PB constraint can be efficiently built, as shown in [18]. The algorithm, moreover, returns the interval of every BDD's node.

We follow the encoding proposed in [14]: for every node, we introduce a fresh variable. Let ν be a node with selector variable x_j and true and false children t and f. We add the clauses $\nu \rightarrow f$ and $\nu \wedge x_j \rightarrow t$. We also add a unit clause for setting the root of the BDD to true, and unit clauses setting the true and false terminal nodes to true and false respectively.

This encoding has the following property. Let A be a partial assignment of the variables $x_1, x_2, \ldots, x_n$, and let ν be the node of the BDD of the PB constraint $c \equiv a_1 x_1 + \ldots + a_n x_n \leqslant K$, with selector variable x_i and interval $[\alpha, \beta]$. Then, the unit propagation of the partial assignment A and the encoding of the constraint c produces:

- ν if and only if $c \wedge A \models (a_i x_i + \cdots + a_n x_n \leq \beta)$.
- $\overline{\nu}$ if and only if $c \wedge A \models (a_i x_i + \cdots + a_n x_n > \beta)$.

In other words, if $a_1 x_1 + \cdots + a_{i-1} x_{i-1} \geqslant K - \beta$ in a partial assignment, unit propagation sets ν to true. If ν is false, unit propagation assures that $a_1 x_1 + \cdots + a_{i-1} x_{i-1} \leqslant K - \beta - 1$.

The way of ordering the constraint before constructing the BDD has a big impact on the BDD size. Computing the optimal ordering with respect the BDD size is a NP-hard problem [19], but experimentally the increasing order ($a_1 \leq a_2 \leq \cdots \leq a_n$) is shown to be a good choice. In this paper we use this order.

3   Lazy Decomposition

The idea of lazy decomposition is quite simple: a Lazy Decomposition (LD) solver is, in some sense, a combination of a Lazy Clause Generation solver and an eager decomposition. LD solvers, as LCG solvers, are composed of a SAT solver engine (that deals with the propositional part of the problem) and propagators, each one in charge of a complex constraint. The difference between LCG and LD solvers lies in the role of the propagators: LCG propagators only propagate and give explanations. LD propagators, in addition, detect which variables of the decomposition would be helpful. These variables and the clauses from the eager decomposition involving them are added to the SAT solver engine.

LD is not specific to a few complex constraints, but a general methodology. Given a complex constraint type and an eager decomposition method for it, an LD propagator must be able to perform the following actions:

- **Identify (dynamically) which parts of the decomposition would be helpful to learning:** LD can be seen as a combined methodology that aims to take advantage of the most profitable aspects of LCG and eager decomposition. This point assures that the solver moves to the decomposition when it is the best option.
- **Propagate the constraint when any subset of the decomposition has been added:** The propagator must work either without decomposition or with a part of it.
- **Avoid propagation for the constraint which is handled by the current decomposition:** auxiliary variables from the eager decomposition have their own meanings. The propagator must use these meanings in order to efficiently propagate the constraint when it is partially decomposed. For example, if the entire decomposition is added, we want the propagator to do no work at all.

In this paper we present two examples of LD propagators: the first one, for cardinality constraints, is based on the eager decomposition of Cardinality Networks [7]. The second one, a propagator for pseudo-Boolean constraints, is based on a BDD decomposition [14].

3.1   Lazy Decomposition Propagator for Cardinality Constraints

In this section we describe the LD propagator for a cardinality constraint of the form $x_1 + x_2 + \ldots + x_n \leqslant K$. LD propagators for $\geqslant$ or $=$ cardinality constraints can be defined in a similar way.

According to Section 2.3, the decomposition of a cardinality constraint based on cardinality networks consists in the encoding of 2-comparators into SAT. A key property of the 2-comparator $\mathsf{2comp}(x_1, x_2, y_1, y_2)$ of Figure 1(a) is that $x_1 + x_2 = y_1 + y_2$. This holds since $y_1 = x_1 \vee x_2$ and $y_2 = x_1 \wedge x_2$. Thus we can define a *2-comparator decomposition step* for 2-comparator $\mathsf{2comp}(x_1, x_2, y_1, y_2)$ as replacing the current cardinality constraint $x_1 + x_2 + x_3 + \ldots + x_n \leqslant K$ by $y_1 + y_2 + x_3 + \ldots + x_n \leqslant K$ and adding a SAT decomposition for the 2-comparator. The resulting constraint system is clearly equivalent. The decomposition introduces the new variables y_1 and y_2 to the SAT Solver engine.

The propagation of the LD propagator works just as in the LCG case. As decomposition occurs the cardinality constraint that is being propagated changes by substituting newly defined decomposition variables for older variables.

Example 3. Figure 1 shows an 8-cardinality network. A LD propagator for the constraint $x_1 + \ldots + x_8 \leqslant 3$ initially behaves as an LCG propagator for that constraint. When variables $z_1, z_2, \ldots, z_{12}$ are introduced by decomposing the corresponding six 2-comparators, the substitutions result in the cardinality constraint $z_5 + z_6 + z_7 + z_8 + z_9 + z_{10} + z_{11} + z_{12} \leqslant 3$.

An LD propagator must determine parts of the decomposition that should be added to the SAT solver. For efficiency, our LD solver adds variables only when it performs a restart: restarts occurs often enough for generating the important variables *not too late*, but occasionally enough to not significantly affect solver performance. Moreover, it is much easier to add variables and clauses to the solver at the root of search.

The propagator assigns a natural number act_i, the *activity*, to every literal x_i of the constraint. Every time a nogood is constructed, the activity of the literals belonging to the nogood is incremented by one. Each time the solver restarts the propagator checks if the activities of the literals of the constraint are greater than λN, where N is the number of conflicts since the last restart and λ is a parameter of the LD solver.

If $act_i \leqslant \lambda N$ then $act_i := act_i/2$. This is done in order to focus on the recent activity. If $act_i > \lambda N$, there are three possibilities:

- If x_i is not the input of a 2-comparator (i.e. an output of the cardinality network) nothing is done.

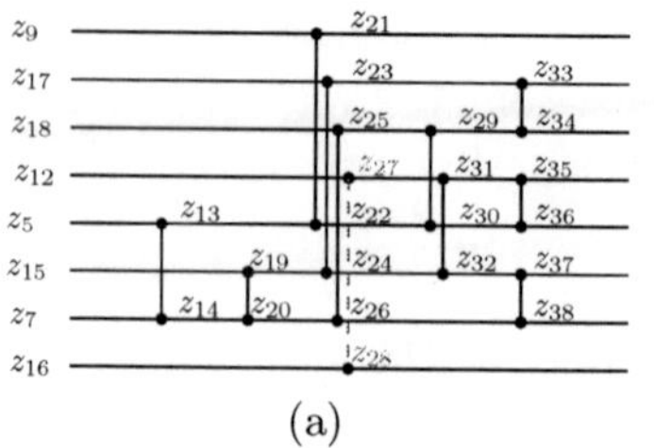
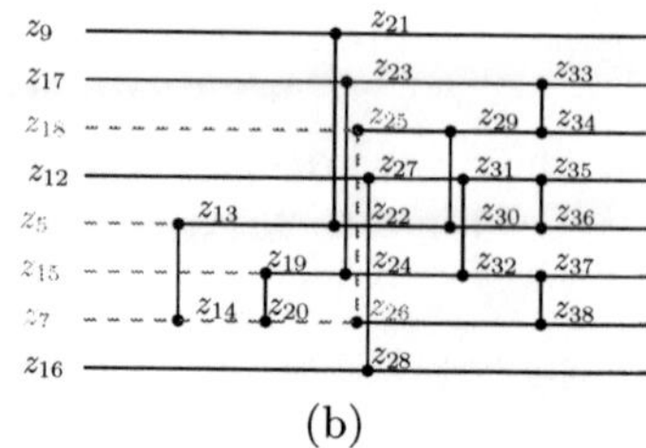

(a)   (b)

Fig. 3. The remaining undecomposed sorting network after decomposing some 2-comparators with (a) $2\mathrm{comp}(z_{12}, z_{16}, z_{27}, z_{28})$ shown dotted, and (b) inputs leading to $2\mathrm{comp}(z_{18}, z_{20}, z_{25}, z_{26})$ shown dotted.

– If x_i is an input of a 2-comparator $2\mathrm{comp}(x_i, x_j, y_1, y_2)$, and its other input x_j has been already generated by the decomposition, we perform a decomposition step on the comparator.
– If x_i is an input of a 2-comparator $2\mathrm{comp}(x_i, x_j, y_1, y_2)$, and its other input x_j has not been generated by the decomposition yet, we proceed as follows: let $S = \{x_{k_1}, x_{k_2}, \ldots, x_{k_s}\}$ be the literals in the current constraint that, after some decomposition steps, can reach x_j. We perform a decomposition step on all the comparators whose inputs both appear in S. Thus x_j is "closer" to being generated by decomposition.

Example 4. Assume the LD propagator for the constraint $x_1 + \ldots + x_8 \leqslant 3$ has generated some variables, so the current constraint is $z_9 + z_{17} + z_{18} + z_{12} + z_5 + z_{15} + z_7 + z_{16} \leqslant 3$. The remaining undecomposed cardinality network is shown in Figure 3.1(a).

In a restart, if the activity of z_{12} is greater than λN we decompose the comparator $2\mathrm{comp}(z_{12}, z_{16}, z_{27}, z_{28})$ generating new literals z_{27} and z_{28} and using them to replace z_{12} and z_{16} in the constraint.

However, if the activity of z_{18} is greater than λN, we cannot decompose $2\mathrm{comp}(z_{18}, z_{20}, z_{25}, z_{26})$ since z_{20} has not been generated yet. The literals reaching z_{20} are z_5, z_{15} and z_7 (see Figure 3.1(b)). Since z_5 and z_7 are the inputs of a 2-comparator $2\mathrm{comp}(z_5, z_7, z_{13}, z_{14})$, this comparator is encoded: z_{13} and z_{14} are introduced and they replace z_5 and z_7 in the constraint.

3.2 Lazy Decomposition Propagator for PB Constraints

In this section we describe the LD propagator for a PB constraint of the form $c \equiv a_1 x_1 + \cdots + a_n x_n \leqslant K$ with $a_i > 0$, since other PB constraints can be reduced to this one.

Suppose $\mathcal{B}$ is the BDD for the PB constraint c. The decomposition of the constraint works as follows: if ν is a node with selector variable x_i and interval $[\alpha, \beta]$, ν is set to true if $a_1 x_1 + \cdots + \ldots + a_{i-1} x_{i-1} \geqslant K - \beta$. If ν is set to false, the encoding assures that $a_1 x_1 + \cdots + a_{i-1} x_{i-1} \leqslant K - \beta + 1$. The LD propagator must maintain this property for nodes ν which have been created as a literal via decomposition.

In our LD propagator, the BDD is lazily encoded from bottom to top: all nodes with the same selector variable are encoded together, thus removing a layer from the bottom of the BDD. Therefore, the LD propagator must deal with the nodes ν at some level i which all represent expressions of the form $a_i x_i + \cdots + a_n x_n \leqslant \beta_\nu$ or equivalently $a_1 x_1 + \cdots + a_{i-1} x_{i-1} \geqslant K - \beta_\nu$. Suppose ν' is the node at level i with highest β_ν where ν' is currently false. The decomposed part of the original PB constraint thus requires that $a_1 x_1 + \cdots + a_{i-1} x_{i-1} \leqslant K - \beta_{\nu'} - 1$. Define $K_i = K - \beta_{\nu'} - 1$, and $node_i = \nu'$

The LD propagator works as follows. The propagator maintains the current sum (lower bound) of the expression $s = a_1 x_1 + \cdots + a_{i-1} x_{i-1}$, just as in the LCG case. If this value is greater than $K - \beta_\nu$ for some leaf node ν with selector variable x_i and interval $[\alpha_\nu, \beta_\nu]$, this node variable ν is set to true. If some leaf node ν (with selector variable x_i and interval $[\alpha_\nu, \beta_\nu]$) is set to false, we set $K_i = K = \beta_\nu - 1$ and $node_i = \nu$. If, at some moment, $s + a_j$ for some $1 \leqslant j < i$ where x_j is undefined is greater than K_i, the propagator sets x_j to false. The explanation is the literals in $x_1, \ldots, x_{i-1}$ that are true and $node_i$.

The policy for lazy decomposition is as follows. Every time a nogood is generated that requires explanation from the PB constraint c, an activity act_c for the constraint c is incremented. If at restart $act_c \geqslant \mu N$ where N is the number of nogoods since last restart we decompose the bottom layer of c and set $act_c = 0$. Otherwise $act_c := act_c / 2$.

Note that the fact that the coefficients a_i in c are in increasing order is important. Big coefficients are more important to the constraint and hence their corresponding variables are likely to be the most valuable for decomposition.

4   Experimental Results

The goals of this section are, first, to check that Lazy Decompositions solvers do in fact significantly reduce the number of auxiliary variables generated and, secondly, to compare them to the LCG and eager decomposition solving approaches. For some problems we include other related solving approaches to illustrate we are not optimizing a very slow system.

All the methods are programmed in the Barcelogic SAT solver [20]. All experiments were performed on a 2Ghz Linux Quad-Core AMD. All the experiments used a value of $\lambda = 0.3$ and $\mu = 0.1$. We experimented with different values and found values for λ between 0.1–0.5 give similar performance, while values for μ between 0.05–0.5 also give similar performance. While there is more to investigate here, it is clear that no problem specific tuning of these parameters is required.

4.1   Cardinality Optimization Problems

Many of the benchmarks on which we have experimented are pure SAT problems with an optimal cardinality function (i.e., an objective function $x_1 + \cdots + x_n$) to minimize.

These problems can be solved by branch and bound: first, we search for an initial solution solving the SAT problem. Let O be the value of $x_1 + \cdots + x_n$ in this solution. Then, we include the cardinality constraint $x_1 + \cdots + x_n \leqslant O - 1$. We repeatedly solve replacing the cardinality constraint by $x_1 + \cdots + x_n \leqslant O - 1$, where O is the last solution found. The process finishes when the last problem is unsatisfiable, which means that O is the optimal solution.

Notice that this process can be used for all approaches considered. In the cardinality network decomposition approach, the encoding is not re-generated every time a new solution is found: we just have to add a unit clause setting the O-th output variable of the network to false. LCG and LD solvers can also easily be adapted as branch and bound solvers, by modifying the bound on the constraint.

For all the benchmarks of this section we have compared the LCG solver for cardinality constraints (LCG), the eager cardinality constraint decomposition approach (DEC), our Lazy Decomposition solver for cardinality constraints (LD), and the three best solvers for industrial partial MaxSAT problems in the past Partial MaxSAT Evaluation 2011: versions 1.1 (QMaxSAT1.1) and 4.0 (QMaxSAT4.0) of QMaxSAT [21] and Pwbo solver, version 1.2 (Pwbo) [22].

Partial MaxSAT. The first set of benchmarks we used were obtained from the MaxSAT Evaluation 2011 (http://maxsat.ia.udl.cat/introduction/), industrial partial MaxSAT category. The benchmarks are encodings of different problems: filter design, logic synthesis, minimum-size test pattern generation, haplotype inference or maximum-quartet consistency.

We can easily transform these problems into SAT problems by introducing one fresh variable to any soft clause. The objective function is the sum of all these new variables. Time limit was set to 1800 seconds per benchmark as in the Evaluation. Table 1(a) shows the number of problems (up to 497) solved by the different methods after, respectively, 15 seconds, 1 minute, etc.

In these problems the eager decomposition approach is much better than the LCG solver. Our LD approach has a similar behavior to the decomposition approach, but LD is faster in the easiest problems. Notice that with these results we would be the best solver in the evaluation, even though our method for solving these problems (adding a fresh variable per soft clause) is a very naive one!

Table 1. Number of instances solved of (a) 497 partial MaxSAT benchmarks and (b) 600 DES benchmarks

Method	15s	1m	5m	15m	30m
DEC	211	296	367	**382**	386
LCG	144	209	265	275	279
LD	**252**	**319**	**375**	381	**386**
QMaxSAT4.0	191	274	352	370	377
Pwbo	141	185	260	325	354
QMaxSAT1.1	185	278	356	373	383

(a)

Method	15s	1m	5m	15m
DEC	409	490	530	541
LCG	151	186	206	228
LD	370	482	528	539
QMaxSAT4.0	275	421	534	**557**
Pwbo	265	361	423	446
QMaxSAT1.1	378	488	**537**	556
SARA-09	**411**	**501**	**537**	549

(b)

Discrete-Event System Diagnosis Suite. The next benchmarks we used are for discrete-event system (DES) diagnosis as presented in [23]. In these problems, we consider a plant modeled by a finite automaton. Its transitions are labeled by the events that occur when the transition is triggered. A sequence of states and transitions on the DES is called a trajectory; it models a behavior of the plant. Some events are observable, that is, an observation is emitted when they occur. The goal of the problem is, knowing that there is a set of faulty events in the DES, find a trajectory consistent with the observations that minimizes the number of faults. As all the problems in this subsection, it is modeled by a set of SAT clauses and a cardinality function to minimize.

In addition to the previously mentioned methods, we have also compared the best SAT encoding proposed in [23] (denoted by SARA-09). It is a specific encoding for these problems. Table 1(b) shows the number of benchmarks solved by the different methods after 15 seconds, 1 minute, etc.

The best method is that described in [23]. However, DEC and LD methods are not far from it. This is a strong argument for these methods, since SARA-09 is a specific method for these problems while eager and lazy decomposition are general methods. On the other hand, LCG does not perform well in these problems, and LD performs more or less as DEC. Both versions of QMaxSAT also performs very well on these problems.

Close Solution Problems. Another type of optimization problems is suggested in [3]. In these problems, we have a set of SAT clauses and a model, and we want to find the most similar solution (w.r.t the Hamming distance) to the given model if we add some few extra clauses. Table 2(a) contains the number of solved instances of the original paper after different times.

For the original problems LD is slightly better than eager decomposition (DEC) and much better than the other approaches.

Since the number of instances of the original paper was small, we created more instances. We selected the 55 satisfiable instances from SAT Competition 2011, industrial division, that we could solve in 10 minutes. For each of these 55 problems, we generated 10 close-solution benchmarks adding a single randomly generated new clause (with at most 5 literals) that falsified the previous model. 100 of the 550 benchmarks were unsatisfiable, so we removed them (searching

Table 2. Number of instances solved of the (a) 40 original close-solution problems and (b) 450 new close-solution problems

Method	15s	1m	5m	15m	60m
DEC	18	24	**31**	**34**	34
LCG	16	18	24	27	30
LD	**19**	**26**	31	**34**	**36**
QMaxSAT4.0	9	14	18	20	22
Pwbo	5	6	7	7	7
QMaxSAT1.1	6	11	16	17	19

(a)

Method	15s	1m	5m	15m	60m
DEC	143	168	208	226	243
LCG	181	223	242	255	268
LD	**187**	**230**	**252**	**262**	**279**
QMaxSAT4.0	55	55	63	69	80
Pwbo	102	144	179	204	215
QMaxSAT1.1	54	55	57	57	64

(b)

the closest solution does not make sense in an unsatisfiable problems). Table 2(b) shows the results on the remaining 450 instances.

For the new problems LCG and LD are the best methods with similar behaviour. Notice that for these problems the cardinality constraint size involves all the variables of the problem, so it can be huge. In a few cases, the encoding approach runs out of memory since the encoding needed more than 2^{25} variables. We considered these cases as a timeout.

4.2   MSU4

Another type of cardinality benchmarks also comes from the MaxSAT Evaluation 2008. In this case we solved them using the *msu4* algorithm [24], which transforms a partial MaxSAT problem into a set of SAT problems with multiple cardinality constraints.[3]

We have grouped all the problems that came from the same partial MaxSAT problem, and we set a timeout of 900 seconds for solving all the family of problems. We had 1883 families of problems (i.e., there were originally 1883 partial MaxSAT problems), but in many cases all the problems of the family could be solved by any method in less than 5 seconds, so we removed them. Table 3(a) contains the results on the remaining 479 benchmarks.

Table 3. (a) Number of families solved from 479 non-trivial MSU4 problems, and (b) number of instances solved from 669 problems PB Competition-2011

Method	15s	1m	5m	15m
DEC	190	282	352	411
LCG	123	168	212	241
LD	**263**	**336**	**410**	**435**

(a)

Method	15s	1m	5m	15m	60m
DEC	318	354	390	407	427
LCG	**372**	387	400	415	433
LD	369	382	401	423	439
borg	280	**406**	**438**	**445**	**467**

(b)

In these problems the LD approach is clearly the best, particularly in the first minute. The reason is that for most of the problems, DEC is faster than LCG, and LD performs similarly to DEC. But there are some problems where LCG is much faster than DEC: in these cases, LD is also faster than LCG, so in total it beats both other methods. Moreover, in some problems there are some *important* constraints which should be decomposed and some other which shouldn't. The LD approach can do this, while DEC and LCG methods either decompose all the constraints or none.

4.3   PB Competition Problems

To compare pseudo-Boolean propagation approaches we used benchmarks from the pseudo-Boolean Competition 2011 (http://www.cril.univ-artois.fr/PB11/),

---

[3] We thanks Albert Oliveras and Carlos Ansótegui for his assistance with these benchmarks.

category *DEC-SMALLINT-LIN* (no optimisation, small integers, linear constraints). In this problems we have compared the LCG, DEC and LD approaches for PB constraints and the winner of the pseudo-Boolean Competition 2011, the solver borg (**borg**) [25] version pb-dec-11.04.03. Table 3(b) contains the number of solved instances (up to 669) after 15 seconds, 1 minute, etc.

In this case, LCG approach is better than DEC, while LD is slightly better than LCG and much better than DEC since presumably it is worth decomposing some of the PB constraints to improve learning, but not all of them. The borg solver is clearly the best, but again it is a tuned portfolio solver specific for pseudo-Boolean problems and makes use of techniques (as in linear programming solvers) which treat all PB constraints simultaneously.

4.4   Variables Generated

One of the goals of Lazy Decomposition is to reduce the search space of the problem. In this section we examine the "raw" search space size in terms of the number of Boolean variables in the model.

Table 4 shows the results of all the problem classes. DEC gives the multiplication factor of Boolean variables created by eager decomposition. For example if the original problem has 100 Boolean variables and the decomposition adds 150 auxiliary variables, we have 250 Boolean variables in total and the multiplication factor will be 2.5. LD gives the multiplication factor of Boolean variables resulting from lazy decomposition. Finally *aux. %* gives the percentage of auxiliary decomposition variables actually created using lazy decomposition. The values in the table are the average over all the problems in that class.

In the optimization problems, there is just one cardinality constraint and most of the time is devoted to proving the optimality of the best solution. Therefore, the cardinality constraint appears in most nogoods since we require many explanations to prove the optimality of the solution. For these classes, the number of auxiliary variables we need is high 35-60 %. Still this reduction is significant.

In the MSU4 and PB problems, on the other hand, there are lots of complex constraints. Most of them have little impact in the problem (i.e., during the search they cause few propagations and conflicts). These constraints are not

Table 4. The average variable multiplication factor for (DEC) eager decomposition and (LD) lazy decomposition, and the average percentage of auxiliary decomposition variables created by lazy decomposition

Class of problems	DEC	LD	*aux. %*
Partial MaxSAT	7.46	5.41	61.72
DES	1.55	1.16	26.62
Original close-solution	12.21	7.48	45.33
New close-solution	24.55	12.38	35.88
MSU4	1.77	1.01	2.18
PB Competition	44.21	17.52	3.24

decomposed in the lazy approach. The LD solver only decomposes part of the most active constraints, so, the number of auxiliary variables generated in these problems is highly reduced.

5   Conclusions and Future Work

We have introduced a new general approach for dealing with complex constraints in complete methods for combinatorial optimization, that combines the advantages of decomposition and global constraint propagation. We illustrate this approach on two different constraints: cardinality and pseudo-Boolean constraints. The results show that, in both cases, our new approach is nearly as good as the best of the eager decomposition and global propagation approaches, and often better. Note that the strongest results for lazy decomposition arise when we have many complex constraints, since many of them will not be important for solving the problem, and hence decomposition is completely wasteful. But we can see that for the important constraints it is worthwhile to decompose.

There are many directions for future work. First we can clearly improve our policies for when and what part of a constraint to decompose. We will also investigate how to decide the right form of decomposition for a constraint during execution rather than fixing on a decomposition prior to search. We also plan to create lazy decomposition propagators for other complex constraints such as linear integer constraints, regular, lex, and incorporate the technology into a full lazy clause generation solver.

Acknowledgments. The first author is partially supported by Spanish Min. of Educ. and Science through the LogicTools-2 project (TIN2007-68093-C02-01) and by an FPU grant. NICTA is funded by the Australian Government as represented by the Department of Broadband, Communications and the Digital Economy and the Australian Research Council through the ICT Centre of Excellence program.

References

1. Asín Achá, R., Nieuwenhuis, R.: Curriculum-based course timetabling with SAT and MaxSAT. Annals of Operations Research, 1–21 (February 2012)
2. Mcaloon, K., Tretkoff, C., Wetzel, G.: Sports league scheduling. In: Proceedings of the 3th Ilog International Users Meeting (1997)
3. Abío, I., Deters, M., Nieuwenhuis, R., Stuckey, P.J.: Reducing Chaos in SAT-Like Search: Finding Solutions Close to a Given One. In: Sakallah, K.A., Simon, L. (eds.) SAT 2011. LNCS, vol. 6695, pp. 273–286. Springer, Heidelberg (2011)
4. Warners, J.P.: A Linear-Time Transformation of Linear Inequalities into Conjunctive Normal Form. Information Processing Letters 68(2), 63–69 (1998)
5. Bailleux, O., Boufkhad, Y.: Efficient CNF Encoding of Boolean Cardinality Constraints. In: Rossi, F. (ed.) CP 2003. LNCS, vol. 2833, pp. 108–122. Springer, Heidelberg (2003)
6. Eén, N., Sörensson, N.: Translating Pseudo-Boolean Constraints into SAT. Journal on Satisfiability, Boolean Modeling and Computation 2, 1–26 (2006)

7. Asín, R., Nieuwenhuis, R., Oliveras, A., Rodríguez-Carbonell, E.: Cardinality networks: a theoretical and empirical study. Constraints 16(2), 195–221 (2011)
8. Nieuwenhuis, R., Oliveras, A., Tinelli, C.: Solving SAT and SAT Modulo Theories: From an abstract Davis–Putnam–Logemann–Loveland procedure to DPLL(T). Journal of the ACM, JACM 53(6), 937–977 (2006)
9. Ohrimenko, O., Stuckey, P., Codish, M.: Propagation via lazy clause generation. Constraints 14(3), 357–391 (2009)
10. Schutt, A., Feydy, T., Stuckey, P.J., Wallace, M.G.: Why Cumulative Decomposition Is Not as Bad as It Sounds. In: Gent, I.P. (ed.) CP 2009. LNCS, vol. 5732, pp. 746–761. Springer, Heidelberg (2009)
11. Aloul, F.A., Ramani, A., Markov, I.L., Sakallah, K.A.: Generic ILP versus specialized 0-1 ILP: an update. In: Pileggi, L.T., Kuehlmann, A. (eds.) 2002 International Conference on Computer-aided Design, ICCAD 2002, pp. 450–457. ACM (2002)
12. Bryant, R.E., Lahiri, S.K., Seshia, S.A.: Deciding CLU Logic Formulas via Boolean and Pseudo-Boolean Encodings. In: Proc. Intl. Workshop on Constraints in Formal Verification (September 2002); Associated with Intl. Conf. on Principles and Practice of Constraint Programming (CP 2002)
13. Bailleux, O., Boufkhad, Y., Roussel, O.: A Translation of Pseudo Boolean Constraints to SAT. Journal on Satisfiability, Boolean Modeling and Computation, JSAT 2(1-4), 191–200 (2006)
14. Abío, I., Nieuwenhuis, R., Oliveras, A., Rodríguez-Carbonell, E.: BDDs for Pseudo-Boolean Constraints – Revisited. In: Sakallah, K.A., Simon, L. (eds.) SAT 2011. LNCS, vol. 6695, pp. 61–75. Springer, Heidelberg (2011)
15. Manolios, P., Papavasileiou, V.: Pseudo-boolean solving by incremental translation to SAT. In: Formal Methods in Computer-Aided Design, FMCAD (2011)
16. Tseitin, G.S.: On the Complexity of Derivation in the Propositional Calculus. Zapiski Nauchnykh Seminarov LOMI 8, 234–259 (1968)
17. Bryant, R.E.: Graph-Based Algorithms for Boolean Function Manipulation. IEEE Trans. Computers 35(8), 677–691 (1986)
18. Mayer-Eichberger, V.: Towards Solving a System of Pseudo Boolean Constraints with Binary Decision Diagrams. Master's thesis, Lisbon (2008)
19. Tani, S., Hamaguchi, K., Yajima, S.: The Complexity of the Optimal Variable Ordering Problems of Shared Binary Decision Diagrams. In: Ng, K.W., Balasubramanian, N.V., Raghavan, P., Chin, F.Y.L. (eds.) ISAAC 1993. LNCS, vol. 762, pp. 389–398. Springer, Heidelberg (1993)
20. Bofill, M., Nieuwenhuis, R., Oliveras, A., Rodríguez-Carbonell, E., Rubio, A.: The Barcelogic SMT Solver. In: Gupta, A., Malik, S. (eds.) CAV 2008. LNCS, vol. 5123, pp. 294–298. Springer, Heidelberg (2008)
21. Koshimura, M., Zhang, T., Fujita, H., Hasegawa, R.: Qmaxsat: A partial max-sat solver. JSAT 8(1/2), 95–100 (2012)
22. Martins, R., Manquinho, V., Lynce, I.: Parallel Search for Boolean Optimization. In: RCRA International Workshop on Experimental Evaluation of Algorithms for Solving Problems with Combinatorial Explosion (2011)
23. Anbulagan, Grastien, A.: Importance of Variables Semantic in CNF Encoding of Cardinality Constraints. In: Bulitko, V., Beck, J.C. (eds.) Eighth Symposium on Abstraction, Reformulation, and Approximation, SARA 2009. AAAI (2009)
24. Manquinho, V., Marques-Silva, J., Planes, J.: Algorithms for Weighted Boolean Optimization. In: Kullmann, O. (ed.) SAT 2009. LNCS, vol. 5584, pp. 495–508. Springer, Heidelberg (2009)
25. Silverthorn, B., Miikkulainen, R.: Latent class models for algorithm portfolio methods. In: Proceedings of the Twenty-Fourth AAAI Conference on Artificial Intelligence (2010)

Improving SAT-Based Weighted MaxSAT Solvers*

Carlos Ansótegui[1], Maria Luisa Bonet[2], Joel Gabàs[1], and Jordi Levy[3]

[1] DIEI, Univ. de Lleida
{carlos,joel.gabas}@diei.udl.cat
[2] LSI, UPC
bonet@lsi.upc.edu
[3] IIIA-CSIC
levy@iiia.csic.es

Abstract. In the last few years, there has been a significant effort in designing and developing efficient Weighted MaxSAT solvers. We study in detail the WPM1 algorithm identifying some weaknesses and proposing solutions to mitigate them. Basically, WPM1 is based on iteratively calling a SAT solver and adding blocking variables and cardinality constraints to relax the unsatisfiable cores returned by the SAT solver. We firstly identify and study how to break the symmetries introduced by the blocking variables and cardinality constraints. Secondly, we study how to prioritize the discovery of higher quality cores. We present an extensive experimental investigation comparing the new algorithm with state-of-the-art solvers showing that our approach makes WPM1 much more competitive.

1 Introduction

Many combinatorial optimization problems can be modelled as Weighted Partial MaxSAT formulas. Therefore, Weighted Partial MaxSAT solvers can be used in several domains as: combinatorial auctions, scheduling and timetabling problems, FPGA routing, software package installation, etc.

The Maximum Satisfiability (MaxSAT) problem is the optimization version of the satisfiability (SAT) problem. The goal is to maximize the number of satisfied clauses in a SAT formula, in other words, to minimize the number of falsified clauses. The clauses can be divided into hard and soft clauses, depending on whether they must be satisfied (hard) or they may or may not be satisfied (soft). If our formula only contains soft clauses it is a MaxSAT formula, and if it contains both, hard and soft clauses, it is a Partial MaxSAT formula. The Partial MaxSAT problem can be further generalized to the Weighted Partial MaxSAT problem. The idea is that not all soft clauses are equally important.

---

* This research has been partially founded by the CICYT research projects TASSAT (TIN2010-20967-C04-01/03/04) and ARINF (TIN2009-14704-C03-01).

M. Milano (Ed.): CP 2012, LNCS 7514, pp. 86–101, 2012.

The addition of weights to soft clauses makes the formula Weighted, and lets us introduce preferences between them. The weights indicate the penalty for falsifying a clause. Given a Weighted Partial MaxSAT problem, our goal is to find an assignment that satisfies all the hard clauses, and the sum of the weights of the falsified clauses is minimal. Such an assignment will be optimal in this context.

SAT technology has evolved to a mature state in the last decade. SAT solvers are really successful at solving industrial decision problems. The next challenge is to use this technology to solve more efficiently industrial optimization problems. Although there has been important work in this direction, we have not reached the success of SAT solvers yet. The present work is one more step in MaxSAT technology to achieve full industrial applicability.

Originally, MaxSAT solvers such as WMaxSatz [12], MiniMaxSat [10], IncWMaxSatz [13] and akmaxsat where depth-first branch and bound based. Recently, there has been a development of SAT based approaches which essentially iteratively call a SAT solver: SAT4J [5], WBO and MSUNCORE [14], WPM1 [1], WPM2 [2], *BINC* and *BINCD* [11] and maxHS [8]. While branch and bound based solvers are competitive for random and crafted instances, SAT based solvers are better for industrial instances.

The WPM1, WBO and MSUNCORE solvers implement weighted versions of the Fu and Malik's algorithm [9]. Essentially, they perform a sequence of calls to a SAT solver, and if the SAT solver returns an unsatisfiable core, they reformulate the problem by introducing new auxiliary variables and cardinality constraints which relax the clauses in the core. Further details are given in section 3 and 5. In this work, we analyze in more detail the WPM1 algorithm to identify and mitigate some weaknesses. The first weakness we have observed is that the addition of the auxiliary variables naturally introduce symmetries which should be broken to achieve better performance. The second weakness has to do with the quality of the cores returned by the SAT solver. Since the SAT solver is used as a black box, we need to come up with new strategies to lead the solver to find better quality cores.

We have conducted an extensive experimental investigation with the best solvers at the last MaxSAT evaluation and other solvers that did not take part in the evaluation, but have been reported to show very good performance. We can see that our current approach can boost radically the performance of the WPM1 becoming the most robust approach.

This paper proceeds as follows: Section 2 introduces some preliminary concepts; Section 3 presents the Fu and Malik's algorithm; Section 4 describes the problem of symmetries and shows how to break them; Section 5 presents the WPM1 algorithm and describes the problem of the quality of the cores; Section 6 introduces an stratified approach to come up with higher quality cores; Section 7 presents some previous concepts needed to describe a general stratified approach discussed in Section 8 and finally Section 9 presents the experimental evaluation.

2 Preliminaries

We consider an infinite countable set of *boolean variables* $\mathcal{X}$. A *literal* l is either a variable $x_i \in \mathcal{X}$ or its negation $\overline{x_i}$. A *clause* C is a finite set of literals, denoted as $C = l_1 \vee \cdots \vee l_r$, or as $\square$ for the empty clause. A *SAT formula* φ is a finite set of clauses, denoted as $\varphi = C_1 \wedge \cdots \wedge C_m$.

A *weighted clause* is a pair (C, w), where C is a clause and w is a natural number or infinity, indicating the penalty for falsifying C. A clause is called *hard* if the corresponding weight is infinity, otherwise the clause is called *soft*.

A *(Weighted Partial) MaxSAT formula* is a multiset of weighted clauses

$$\varphi = \{(C_1, w_1), \ldots, (C_m, w_m), (C_{m+1}, \infty), \ldots, (C_{m+m'}, \infty)\}$$

where the first m clauses are soft and the last m' clauses are hard. The set of variables occurring in a formula φ is noted as $\mathrm{var}(\varphi)$.

A *total truth assignment* for a formula φ is a function $I : \mathrm{var}(\varphi) \to \{0, 1\}$, that can be extended to literals, clauses, SAT formulas and MaxSAT formulas, the following way:

$$I(\overline{x}_i) = 1 - I(x_i)$$
$$I(l_1 \vee \ldots \vee l_r) = \max\{I(l_1), \ldots, I(l_r)\}$$
$$I(\{C_1, \ldots, C_m\}) = \min\{I(C_1), \ldots, I(C_m)\}$$
$$I(\{(C_1, w_1), \ldots, (C_m, w_m)\}) = w_1 \cdot (1 - I(C_1)) + \ldots + w_m \cdot (1 - I(C_m))$$

We define the *optimal cost* of a MaxSAT formula as

$$\mathrm{cost}(\varphi) = \min\{I(\varphi) \mid I : \mathrm{var}(\varphi) \to \{0, 1\}\}$$

and an *optimal assignment* as an assignment I such that $I(\varphi) = \mathrm{cost}(\varphi)$.

We also define *partial truth assignments* for φ as a partial function $I : \mathrm{var}(\varphi) \to \{0, 1\}$ where instantiated falsified literals are removed and the formula is simplified accordingly.

Example 1. Given $\varphi = \{(\overline{y}, 6), (x \vee y, 2), (x \vee z, 3), (y \vee z, 2)\}$ and $I : \{y, z\} \to \{0, 1\}$ such that $I(y) = 0$ and $I(z) = 0$, we have $I(\varphi) = \{(x, 5), (\square, 2)\}$. We also have $\mathrm{cost}(I(\varphi)) = 2$ and $\mathrm{cost}(\varphi) = 0$.

Notice that, for any MaxSAT formula φ and partial truth assignment I, we have $\mathrm{cost}(\varphi) \leq \mathrm{cost}(I(\varphi))$. Notice also that when w is finite, the pair (C, w) is equivalent to having w copies of the clause $(C, 1)$ in our multiset.

We say that a truth assignment I *satisfies* a literal, clause or a SAT formula if it assigns 1 to it, and *falsifies* it if it assigns 0. A SAT formula is *satisfiable* if there exists a truth assignment that satisfies it. Otherwise, it is *unsatisfiable*. Given an unsatisfiable SAT formula φ, an *unsatisfiable core* φ_c is a subset of clauses $\varphi_c \subseteq \varphi$ that is also unsatisfiable. A *minimal unsatisfiable core* is an unsatisfiable core such that any proper subset of it is satisfiable.

The *Weighted Partial MaxSAT problem* for a weighted partial MaxSAT formula φ is the problem of finding an *optimal assignment*. If the optimal cost is infinity, then the subset of hard clauses of the formula is unsatisfiable, and we say that the formula is *unsatisfiable*. The *Weighted MaxSAT problem* is the Weighted Partial MaxSAT problem when there are no hard clauses. The *Partial MaxSAT problem* is the Weighted Partial MaxSAT problem when the weights of soft clauses are all equal. The *MaxSAT problem* is the Partial MaxSAT problem when there are no hard clauses. Notice that the *SAT problem* is equivalent to the Partial MaxSAT problem when there are no soft clauses.

3   The Fu and Malik's Algorithm

The first SAT-based algorithm for Partial MaxSAT algorithm was the Fu and Malik's algorithm described in [9]. It was implemented in the MaxSAT solver msu1.2 [17,18], and its correctness was proved in [1].

The algorithm consists in iteratively calling a SAT solver on a working formula φ. This corresponds to the line $(st, \varphi_c) := SAT(\{C \mid (C_i, w_i) \in \varphi\})$. The SAT solver will say whether the formula is satisfiable or not (variable st), and in case the formula is unsatisfiable, it will give an unsatisfiable core (φ_c). At this point the algorithm will produce new variables, blocking variables (BV in the code), one for each soft clause in the core. The new working formula φ will consist in adding the new variables to the soft clauses of the core, adding a cardinality constraint saying that exactly one of the new variables should be true ($CNF(\sum_{b \in BV} b = 1)$ in the code), and adding one to the counter of falsified clauses. This procedure is applied until the SAT solver returns SAT.

For completeness, we reproduce the code of the Fu and Malik's algorithm in Algorithm 1.

Next we present an example of execution that will be used in the next section.

Example 2. Consider the pigeon-hole formula PHP_1^5 with 5 pigeons and one hole where the clauses saying that no two pigeons can go to the same hole are hard, while the clauses saying that each pigeon goes to a hole are soft:

$$\varphi = \{(x_1, 1), (x_2, 1), (x_3, 1), (x_4, 1), (x_5, 1), (\overline{x}_1 \vee \overline{x}_2, \infty), \ldots, (\overline{x}_4 \vee \overline{x}_5, \infty)\}$$

In what follows, the new b variables will have a super-index indicating the number of the unsatisfiable core, and a subindex indicating the index of the original soft clause.

Suppose that applying the FuMalik algorithm, the SAT solver computes the (minimal) unsatisfiable core $C_1 = \{1, 2\}$. Here we represent the core by the set of indexes of the soft clauses contained in the core. The new formula will be as shown on the right. At this point, the variable *cost* takes value 1.

$$\varphi_1 = \{ (x_1 \vee b_1^1, 1),$$
$$(x_2 \vee b_2^1, 1),$$
$$(x_3, 1),$$
$$(x_4, 1),$$
$$(x_5, 1) \} \cup$$
$$\{(\overline{x}_i \vee \overline{x}_j, \infty) \mid i \neq j\} \cup$$
$$CNF(b_1^1 + b_2^1 = 1, \infty)$$

Algorithm 1. The pseudo-code of the FuMalik algorithm (with a minor correction).

Input: $\varphi = \{(C_1, 1), \ldots, (C_m, 1), (C_{m+1}, \infty), \ldots, (C_{m+m'}, \infty)\}$

1: **if** $SAT(\{C_i \mid w_i = \infty\}) = (\text{UNSAT}, \_)$ **then return** $(\infty, \emptyset)$
　　　　　　　　　　　　　　　　　　　　　　▷Hard clauses are unsatisfiable
2: $cost := 0$ 　　　　　　　　　　　　　　　　　　　　　　▷Optimal
3: **while** *true* **do**
4: 　　$(st, \varphi_c) := SAT(\{C_i \mid (C_i, w_i) \in \varphi\})$ 　▷Call to the SAT solver without weights
5: 　　**if** $st = \text{SAT}$ **then return** $(cost, \varphi)$
6: 　　$BV := \emptyset$ 　　　　　　　　　　　　　　　　▷Set of blocking variables
7: 　　**foreach** $C_i \in \varphi_c$ **do**
8: 　　　　**if** $w_i \neq \infty$ **then** 　　　　　　　　　　▷If the clause is soft
9: 　　　　　　$b := \text{new\_variable}(\ )$
10: 　　　　　$\varphi := \varphi \setminus \{(C_i, 1)\} \cup \{(C_i \vee b, 1)\}$ 　　　　▷Add blocking variable
11: 　　　　　$BV := BV \cup \{b\}$
12: 　　$\varphi := \varphi \cup \{(C, \infty) \mid C \in CNF(\sum_{b \in BV} b = 1)\}$
　　　　　　　　　　　　　　　　　▷Add cardinality constraint as hard clauses
13: 　　$cost := cost + 1$

If the next unsatisfiable cores found by the SAT solver are $C_2 = \{3, 4\}$ and $C_3 = \{1, 2, 3, 4\}$, then the new formula will be:

$$\varphi_2 = \{ \ (x_1 \vee b_1^1 \qquad , 1),$$
$$(x_2 \vee b_2^1 \qquad , 1),$$
$$(x_3 \vee \quad b_3^2 , 1),$$
$$(x_4 \vee \quad b_4^2 , 1),$$
$$(x_5 \qquad , 1) \ \} \ \cup$$
$$\{(\overline{x}_i \vee \overline{x}_j, \infty) \mid i \neq j\} \ \cup$$
$$CNF(b_1^1 + b_2^1 = 1, \infty) \ \cup$$
$$CNF(b_3^2 + b_4^2 = 1, \infty)$$

$$\varphi_3 = \{ \ (x_1 \vee b_1^1 \vee \qquad b_1^3, 1),$$
$$(x_2 \vee b_2^1 \vee \qquad b_2^3, 1),$$
$$(x_3 \vee \qquad b_3^2 \vee b_3^3, 1),$$
$$(x_4 \vee \qquad b_4^2 \vee b_4^3, 1),$$
$$(x_5 \qquad , 1) \ \} \ \cup$$
$$\{(\overline{x}_i \vee \overline{x}_j, \infty) \mid i \neq j\} \ \cup$$
$$CNF(b_1^1 + b_2^1 = 1, \infty) \ \cup$$
$$CNF(b_3^2 + b_4^2 = 1, \infty) \ \cup$$
$$CNF(b_1^3 + b_2^3 + b_3^3 + b_4^3 = 1, \infty)$$

After the third iteration, the variable *cost* has value 3. Finally, after finding the core $C_4 = \{1, 2, 3, 4, 5\}$ we get the following satisfiable MaxSAT formula:

$$\varphi_4 = \{ \ (x_1 \vee b_1^1 \vee \qquad b_1^3 \vee b_1^4 , 1),$$
$$(x_2 \vee b_2^1 \vee \qquad b_2^3 \vee b_2^4 , 1),$$
$$(x_3 \vee \qquad b_3^2 \vee b_3^3 \vee b_3^4 , 1),$$
$$(x_4 \vee \qquad b_4^2 \vee b_4^3 \vee b_4^4 , 1),$$
$$(x_5 \vee \qquad b_5^4 , 1) \ \} \ \cup$$
$$\{(\overline{x}_i \vee \overline{x}_j, \infty) \mid i \neq j\} \ \cup$$
$$CNF(b_1^1 + b_2^1 = 1, \infty) \ \cup$$
$$CNF(b_3^2 + b_4^2 = 1, \infty) \ \cup$$
$$CNF(b_1^3 + b_2^3 + b_3^3 + b_4^3 = 1, \infty) \ \cup$$
$$CNF(b_1^4 + b_2^4 + b_3^4 + b_4^4 + b_5^4 = 1, \infty)$$

At this point *cost* is 4. The algorithm will now call the SAT solver on φ_4, and the solver will return the answer "satisfiable". The algorithm returns $cost = 4$.

4 Breaking Symmetries

It is well known that formulas that contain a great deal of symmetries cause SAT solvers to explore many redundant truth assignments. Adding symmetry breaking clauses to a formula has the effect of removing the symmetries, while keeping satisfiability the same. Therefore it is a way to speed up solvers by pruning the search space.

In the case of executions of the FuMalik algorithm, symmetries can appear in two ways. On one hand, there are formulas that naturally contain many symmetries. For instance, in the case of the pigeon-hole principle we can permute the pigeons or the holes, leaving the formula intact. On the other hand, in each iteration of the FuMalik algorithm, we modify the formula adding new variables and hard constraints. In this process we can also introduce symmetries. In the present paper, we are no concerned with eliminating natural symmetries of a MaxSAT formula as in [16], since that might be costly, and it is not the aim of the present work. Instead we will eliminate the symmetries that appear in the process of performing the algorithm. In this case, it is very efficient to extract the symmetries given our implementation of the algorithm.

Before we formally describe the process of eliminating the symmetries, we will see an example.

Example 3. Consider again the pigeon-hole formula PHP_1^5 of Example 2. The working formula φ_3 from the previous section is still unsatisfiable, this is the reason to find a fourth core C_4. However, if we do not consider the clause x_5 the formula is satisfiable, and has 8 distinct models (two for each variable among $\{x_1, \ldots, x_4\}$ set to true). Here, we show 2 of the models, marking the literals set to true (we do not include the clauses $\overline{x}_i \vee \overline{x}_j$, for $i \neq j$ and put the true literals in boxes):

$$x_1 \vee \boxed{b_1^1} \vee \qquad b_1^3 \qquad\qquad x_1 \vee b_1^1 \vee \qquad \boxed{b_1^3}$$
$$x_2 \vee b_2^1 \vee \qquad \boxed{b_2^3} \qquad\qquad x_2 \vee \boxed{b_2^1} \vee \qquad b_2^3$$
$$x_3 \vee \qquad \boxed{b_3^2} \vee b_3^3 \qquad\qquad x_3 \vee \qquad \boxed{b_3^2} \vee b_3^3$$
$$\boxed{x_4} \vee \qquad b_4^2 \vee b_4^3 \qquad\qquad \boxed{x_4} \vee \qquad b_4^2 \vee b_4^3$$
$$\boxed{b_1^1} + b_2^1 = 1 \qquad\qquad b_1^1 + \boxed{b_2^1} = 1$$
$$\boxed{b_3^2} + b_4^2 = 1 \qquad\qquad \boxed{b_3^2} + b_4^2 = 1$$
$$b_1^3 + \boxed{b_2^3} + b_3^3 + b_4^3 = 1 \qquad\qquad \boxed{b_1^3} + b_2^3 + b_3^3 + b_4^3 = 1$$

The previous two models are related by the permutation $b_1^1 \leftrightarrow b_2^1, b_1^3 \leftrightarrow b_2^3$. The two ways of assigning values to the b variables are equivalent. The existence of so many *partial* models makes the task of showing unsatisfiability of the formula (including x_5) much harder.

The mechanism to eliminate the symmetries caused by the extra variables is as follows: suppose we are in the s iteration of the FuMalik algorithm, and we have obtained the set of cores $\{\varphi_1, \ldots, \varphi_s\}$. We assume that the clauses in the cores follow a total order. For clarity we will name the new variables of core φ_l for l such that $1 \leq l \leq s$ as b_i^l, where i is an index in φ_l. Now, we add the clauses:

$$b_i^s \to \overline{b}_j^l \qquad \text{for } l = 1, \ldots, s-1 \text{ and } i, j \in \varphi_l \cap \varphi_s \text{ and } j > i$$

This clauses implies that in Example 3 we choose the model on the left rather than the one on the right.

Example 4. For the Example 3, after finding the third unsatisfiable core C_3, we would add the following clauses to break symmetries (written in form of implications):

$$b_1^3 \to \overline{b}_2^1$$
$$b_3^3 \to \overline{b}_4^2$$

Adding these clauses, instead of the 8 partial models, we only have 4, one for each possible assignment of x_i to true.

After finding the fourth core C_4, we also add (written in compact form):

$$b_1^4 \to (\overline{b}_2^1 \wedge \overline{b}_2^3 \wedge \overline{b}_3^3 \wedge \overline{b}_4^3)$$
$$b_2^4 \to (\overline{b}_3^3 \wedge \overline{b}_4^3)$$
$$b_3^4 \to (\overline{b}_4^2 \wedge \overline{b}_4^3)$$

5   The WPM1 Algorithm

Algorithm 2 is the weighted version of the FuMalik algorithm described in section 3 [1,14] In this algorithm, we iteratively call a SAT solver with a weighted working formula, but excluding the weights. When the SAT solver returns an unsatisfiable core, we calculate the minimum weight of the clauses of the core (w_{min} in the algorithm.). Then, we transform the working formula in the following way: we duplicate the core having on one of the copies, the clauses with weight the original minus the minimum weight, and on the other copy we put the blocking variables and we give it the minimum weight. Finally we add the cardinality constraint on the blocking variables, and we add w_{min} to the *cost*.

The process of doubling the clauses might imply to end up converting clauses with weight say w into w copies of the clause of weight 1. When this happens, the process becomes very inefficient. In the following we show a (tiny) example that reflects this situation.

Example 5. Consider the formula $\varphi = \{(x_1, 1), (x_2, m), (\overline{x}_2, \infty)\}$.

Assume that the SAT solver always includes the first soft clause in the returned unsatisfiable core, even if this makes the core not minimal. After one iteration, the new formula would be:

$$\varphi_1 = \{(x_1 \vee b_1^1, 1), (x_2 \vee b_2^1, 1), (x_2, m - 1), (\overline{x}_2, \infty), (b_1^1 + b_2^1 = 1, \infty)\}$$

Algorithm 2. The pseudo-code of the WPM1 algorithm.

Input: $\varphi = \{(C_1, w_1), \ldots, (C_m, w_m), (C_{m+1}, \infty), \ldots, (C_{m+m'}, \infty)\}$

1: **if** $SAT(\{C_i \mid w_i = \infty\}) = (\text{UNSAT}, \_)$ **then return** $(\infty, \emptyset)$ ▷Hard clauses are unsatisfiable

2: $cost := 0$ ▷Optimal

3: **while** $true$ **do**

4:    $(st, \varphi_c) := SAT(\{C_i \mid (C_i, w_i) \in \varphi\})$ ▷Call to the SAT solver without weights

5:    **if** $st = \text{SAT}$ **then return** $(cost, \varphi)$

6:    $BV := \emptyset$ ▷Blocking variables of the core

7:    $w_{min} := \min\{w_i \mid C_i \in \varphi_c \wedge w_i \neq \infty\}$ ▷Minimum weight

8:    **foreach** $C_i \in \varphi_c$ **do**

9:      **if** $w_i \neq \infty$ **then**

10:        $b := \text{new\_variable}()$

11:        $\varphi := \varphi \setminus \{(C_i, w_i)\} \cup \{(C_i, w_i - w_{min}), (C_i \vee b, w_{min})\}$
                         ▷Duplicate soft clauses of the core

12:        $BV := BV \cup \{b\}$

13:    $\varphi := \varphi \cup \{(C, \infty) \mid C \in CNF(\sum_{b \in BV} b = 1)\}$
                         ▷Add cardinality constraint as hard clauses

14:    $cost := cost + w_{min}$

If from now on, at each iteration i, the SAT solver includes the first clause along with $\{(x_2, m - i + 1), (\overline{x}_2, \infty)\}$ in the unsatisfiable core, then at iteration i, the formula would be:

$$\varphi_i = \{ (x_1 \vee b_1^1 \vee \cdots \vee b_1^i, 1), (x_2 \vee b_2^1, 1), \ldots, (x_2 \vee b_2^i, 1), (x_2, m - i), (\overline{x}_2, \infty),$$
$$(b_1^1 + b_2^1 = 1, \infty), \ldots, (b_1^i + b_2^i = 1, \infty)\}$$

The WPM1 algorithm would need m iterations to solve the problem.

Obviously, a reasonable good SAT solver would return a better quality core than in previous example. However, unless it can guarantee that it is minimal, a similar example (but more complicated) could be constructed.

6 A Stratified Approach for WPM1

In Algorithm 3 we present a modification of the WPM1 algorithm that tries to prevent the situation described in Example 5 by carrying out a stratified approach. The main idea is to restrict the set of clauses sent to the SAT solver to force it to concentrate on those with higher weights. As a result, the SAT solver returns unsatisfiable cores with clauses with higher weights. These are better quality cores and contribute to increase the cost faster. When the SAT solver returns SAT, then we allow it to use clauses with lower weights.

In Algorithm 3 we use a variable w_{max}, and we only send to the SAT solver the clauses with weight greater or equal than it. As in Algorithm 2, we start by checking that hard clauses are satisfiable. Then, we initialize w_{max} to the

---

Algorithm 3. The pseudo-code of the stratified approach for WPM1 algorithm.

Input: $\varphi = \{(C_1, w_1), \ldots, (C_m, w_m), (C_{m+1}, \infty), \ldots, (C_{m+m'}, \infty)\}$

1: **if** $SAT(\{C_i \mid w_i = \infty\}) = (\text{UNSAT}, \_)$ **then return** $(\infty, \emptyset)$

2: $cost := 0$ ▷Optimal

3: $w_{max} := \max\{w_i \mid (C_i, w_i) \in \varphi \wedge w_i < w_{max}\}$

4: **while** *true* **do**

5:    $(st, \varphi_c) := SAT(\{C_i \mid (C_i, w_i) \in \varphi \wedge w_i \geq w_{max}\})$ ▷Call without weights

6:    **if** $st = \text{SAT}$ **and** $w_{max} = 0$ **then return** $(cost, \varphi)$

7:    **else**

8:       **if** $st = \text{SAT}$ **then** $w_{max} := \max\{w_i \mid (C_i, w_i) \in \varphi \wedge w_i < w_{max}\}$

9:       **else**

10:         $BV := \emptyset$ ▷Blocking variables of the core

11:         $w_{min} := \min\{w_i \mid C_i \in \varphi_c \wedge w_i \neq \infty\}$ ▷Minimum weight

12:         **foreach** $C_i \in \varphi_c$ **do**

13:            **if** $w_i \neq \infty$ **then**

14:               $b := \text{new\_variable}()$

15:               $\varphi := \varphi \setminus \{(C_i, w_i)\} \cup \{(C_i, w_i - w_{min}), (C_i \vee b, w_{min})\}$
              ▷Duplicate soft clauses of the core

16:               $BV := BV \cup \{b\}$

17:         $\varphi := \varphi \cup \{(C, \infty) \mid C \in CNF(\sum_{b \in BV} b = 1)\}$
             ▷Add cardinality constraint as hard clauses

18:         $cost := cost + w_{min}$

---

highest weight smaller than infinite. If the SAT solver returns SAT, there are two possibilities. Either w_{max} is zero (it means that we have already sent all clauses to the SAT solver) and we finish; or it is not yet zero, and we decrease w_{max} to the highest weight smaller than w_{max}, allowing the SAT solver to use clauses with smaller weights. If the SAT solver returns UNSAT, we proceed like in Algorithm 2. This algorithm was submitted to the MaxSAT evaluation 2011 as WPM1 (version 2011). It was the best performing solver for the weighted partial industrial category. The description of the solver was never published in a paper before.

We can use better strategies to decrease the value of w_{max}. Notice that, in the worst case, we could need more executions of the SAT solver than Algorithm 2, because the calls that return SAT but $w_{max} > 0$ do not contribute to increase the computed cost. Therefore, we need to find a balance between the number of those unproductive SAT calls, and the minimum weight of the cores. For example, one of the possible strategies is to decrease w_{max} until the following condition is satisfied

$$\frac{|C_i \mid (C_i, w_i) \in \varphi \wedge w_i < w_{max}\}|}{|\{w_i \mid (C_i, w_i) \in \varphi \wedge w_i < w_{max}\}|} > \alpha$$

or $w_{max} = 0$. This strategy tends to send more new clauses to the SAT solver when they have bigger diversity of weights. In our implementation of WPM1 submitted to the MaxSAT evaluation 2012, we use this strategy, called *diversity heuristic*, with $\alpha = 1.25$.

The proof of the correctness of this algorithm is like the proof for WPM1. The only additional point is that the new algorithm is forcing the SAT solver to find some cores before others. In the proof of correctness of WPM1 there is no assumption on what cores the SAT solver finds first.

7   MaxSAT Reducibility

Our algorithms solve a MaxSAT formula by successively transforming it until we get a satisfiable formula. To prove the soundness of the algorithms it suffices to prove that these transformations preserve the cost of the formula. However, apart from this notion of *cost-preserving transformation*, we can define other (stronger) notions of formula transformation, like *MaxSAT equivalence* and *MaxSAT reducibility*.

Definition 1.
*We say that φ_1 and φ_2 are **cost-equivalent** if* $\mathrm{cost}(\varphi_1) = \mathrm{cost}(\varphi_2)$.
*We say that φ_1 and φ_2 are **MaxSAT equivalent** if, for any assignment $I : var(\varphi_1) \cup var(\varphi_2) \rightarrow \{0,1\}$, we have $\mathrm{cost}(I(\varphi_1)) = \mathrm{cost}(I(\varphi_2))$.*
*We say that φ_1 is **MaxSAT reducible** to φ_2 if, for any assignment $I : var(\varphi_1) \rightarrow \{0,1\}$, we have $\mathrm{cost}(I(\varphi_1)) = \mathrm{cost}(I(\varphi_2))$.*

Notice that the distinction between MaxSAT equivalence and MaxSAT reduction is the domain on the partial assignment. In one case it is $var(\varphi_1) \cup var(\varphi_2)$, and in the other $var(\varphi_1)$.

The notion of cost-preserving transformation is the weakest of all three notions, and suffices to prove the soundness of the algorithms. However, it does not allow us to replace *sub*-formulas by cost-equivalent *sub*-formulas, in other words $\mathrm{cost}(\varphi_1) = \mathrm{cost}(\varphi_2)$ does not imply $\mathrm{cost}(\varphi_1 \cup \varphi_3) = \mathrm{cost}(\varphi_2 \cup \varphi_3)$. On the other hand, the notion of MaxSAT equivalence is the strongest of all three notions, but too strong for our purposes, because the formula transformations we use does not satisfy this notion. When φ_2 has variables not occurring in φ_1, it is convenient to use the notion of MaxSAT reducibility, that, in these cases, is weaker than the notion of MaxSAT equivalence.

In the following we show some examples of the notions of Definition 1.

Example 6. The following example shows a formula transformation that preserves the cost, but not MaxSAT reducibility. Consider $\varphi_1 = \{(x, 2), (\overline{x}, 1)\}$ and $\varphi_2 = \{(\Box, 1)\}$. We have $\mathrm{cost}(\varphi_1) = \mathrm{cost}(\varphi_2) = 1$, hence the transformation of φ_1 into φ_2 is cost-preserving. However, φ_1 is not MaxSAT reducible to φ_2, because the assignment $I : \{x\} \rightarrow \{0,1\}$ with $I(x) = 0$, makes $\mathrm{cost}(I(\varphi_1)) = 2 \neq 1 = \mathrm{cost}(I(\varphi_2))$.

On the contrary, φ_2 is MaxSAT reducible to φ_1, because there is a unique assignment $I : \emptyset \to \{0,1\}$, and it satisfies $\mathrm{cost}(I(\varphi_1)) = \mathrm{cost}(I(\varphi_2))$. Hence, MaxSAT reducibility is not a symmetric relation.

The following example shows that MaxSAT reducibility does not imply MaxSAT equivalence. Consider $\varphi_1 = \{(x,2),(\overline{x},1)\}$ and $\varphi_3 = \{(\square,1),(x,1),(x \vee y,1),(\overline{x} \vee z,1),(\overline{y} \vee \overline{z},\infty)\}$. We have that φ_1 is MaxSAT reducible to φ_3. To prove this, we must consider two interpretations I_1 and I_2, defined by $I_1(x) = 0$ and $I_2(x) = 1$. In the first case, we obtain $I_1(\varphi_1) = \{(\square,2)\}$ and $I_1(\varphi_3) = \{(\square,2),(y,1),(\overline{y} \vee \overline{z},\infty)\}$ that have the same cost 2. In the second case, we obtain $I_2(\varphi_1) = \{(\square,1)\}$ and $I_2(\varphi_3) = \{(\square,1),(z,1),(\overline{y} \vee \overline{z},\infty)\}$ that have also the same cost 1. However, φ_1 and φ_3 are not MaxSAT equivalent because for $I : \{x,y,z\} \to \{0,1\}$ defined by $I(x) = I(y) = I(z) = 1$ we have $\mathrm{cost}(I(\varphi_1)) = 1 \neq \infty = \mathrm{cost}(I(\varphi_3))$.

Finally, φ_1 is MaxSAT equivalent to $\varphi_4 = \{(\square,1),(x,1)\}$.

The notion of MaxSAT equivalence was implicitly defined in [7]. In this paper a MaxSAT resolution rule that preserves MaxSAT equivalence is defined, and proved complete for MaxSAT.

For lack of space we state without proof:

Lemma 1. (1) *If φ_1 is MaxSAT-reducible to φ_2 and $var(\varphi_2) \cap var(\varphi_3) \subseteq var(\varphi_1)$, then $\varphi_1 \cup \varphi_3$ is MaxSAT-reducible to $\varphi_2 \cup \varphi_3$.*
(2) *MaxSAT-reducibility is transitive: if φ_1 is MaxSAT-reducible to φ_2, φ_2 is MaxSAT-reducible to φ_3, and $var(\varphi_1) \cap var(\varphi_3) \subseteq var(\varphi_2)$, then φ_1 is MaxSAT-reducible to φ_3.*

Example 7. Notice that the side condition of Lemma 1 (1) is necessary. For instance, if we take $\varphi_1 = \{(\square,1)\}$, $\varphi_2 = \{(x,1),(\overline{x},\infty)\}$ and $\varphi_3 = \{(x,1)\}$, where the side condition $var(\varphi_2) \cap var(\varphi_3) = \{x\} \not\subseteq \emptyset = var(\varphi_1)$ is violated, we have that φ_1 is MaxSAT reducible to φ_2, but $\varphi_1 \cup \varphi_3$ is not MaxSAT reducible to $\varphi_2 \cup \varphi_3$.

Similarly, the side condition in Lemma 1 (2) is also necessary. For instance, if we take $\varphi_1 = \{(x,1),(\overline{x},1)\}$, $\varphi_2 = \{(\square,1)\}$ and $\varphi_3 = \{(x,1),(\overline{x},\infty)\}$, where the side condition $var(\varphi_1)) \cap var(\varphi_3) = \{x\} \not\subseteq \emptyset = var(\varphi_2)$ is also violated, we have that φ_1 is MaxSAT reducible to φ_2 and this to φ_3. However, φ_1 is not MaxSAT reducible to φ_3.

There are two side conditions in Lemma 1 (1) and (2) (see Example 7) that restrict the set of variables that can occur in the MaxSAT problems. However, if we ensure that problem transformations only *introduce fresh* variables, i.e. when φ_1 is MaxSAT reduced to φ_2, all new variables introduced in φ_2 do not occur elsewhere, then these conditions are trivially satisfied. In our algorithms, all formula transformations satisfy this restriction.

8   Generic Stratified Approach

In Algorithm 4 we show how the stratified approach can be applied to any *generic* weighted MaxSAT solver WPM. In the rest of the section we will describe what

---

Algorithm 4. The pseudo-code of a generic MaxSAT algorithm that follows a stratified approach heuristics.

Input: $\varphi = \{(C_1, w_1), \ldots, (C_m, w_m)\}$

1: $cost := 0$

2: $w_{max} = \infty$

3: **while** *true* **do**

4:     $\varphi_{w_{max}} := \{(C_i, w_i) \in \varphi \mid w_i \geq w_{max}\}$

5:     $(cost', \varphi_{sat}, \varphi_{res}) = \mathrm{WPM}(\varphi_{w_{max}})$

6:     $cost = cost + cost'$

7:     **if** $cost = \infty$ **or** $w_{max} = 0$ **then return** $(cost, \varphi_{sat})$

8:     $W = \sum \{w_i \mid (C_i, w_i) \in \varphi \setminus \varphi_{w_{max}} \cup \varphi_{res}\}$

9:     $\varphi_{sat} = \{(C_i, \mathrm{harden}(w_i, W)) \mid (C_i, w_i) \in \varphi_{sat}\}$

10:     $\varphi = (\varphi \setminus \varphi_{w_{max}}) \cup \varphi_{sat} \cup \varphi_{res}$

11:     $w_{max} = \mathrm{decrease}(w_{max})$

12: **return** $(cost, \varphi)$

13: **function** harden(w,W)

14: **begin**

15:     **if** $w > W$ **then return** ∞

16:     **else return** w

---

properties the generic algorithm WPM has to satisfy in order to ensure the correctness of this approach.

We assume that, given a weighted MaxSAT formula φ, $WPM(\varphi)$ returns a triplet $(cost, \varphi_{sat}, \varphi_{res})$ such that φ is MaxSAT reducible to $\{(\square, cost)\} \cup \varphi_{sat} \cup \varphi_{res}$, φ_{sat} is satisfiable (has cost zero), and clauses of φ_{res} have cost strictly smaller than w_{max}. Given φ, WPM1 return a pair $(cost, \varphi')$ where φ is MaxSAT reducible to $\{(\square, cost)\} \cup \varphi'$ and φ is satisfiable, hence satisfies the requirements taking $\varphi_{res} = \emptyset$. Moreover, we can also think of WPM as an algorithm that *partially* solves the formula, and returns a lower bound cost, a satisfiable part of the formula φ_{sat}, and an unsolved residual φ_{res}.

The algorithm uses a variable w_{max} to restrict the clauses sent to the MaxSAT solver. The first time $w_{max} = \infty$, and we run WPM only on the hard clauses. Then, in each iteration we send clauses with weight w_{max} or bigger to WPM. We add the return cost to the current cost, and decrease w_{max}, until w_{max} is zero.

Algorithm 3 is an instance of this generic schema where WPM is a partial execution of WPM1 where clauses generated during duplication with weight smaller than w_{max} are put apart in φ_{res}.

Lines 8 and 9 are optional and can be removed from the algorithm without affecting to its correctness. They are inspired in [15]. The idea is to harden all soft clauses with weight bigger than the sum of the weights of the clauses not sent to the WPM plus the clauses returned in φ_{res}. The proof of the correctness of these lines is based in the following lemma (not proved for lack of space).

Lemma 2. *Let* $\varphi_1 = \{(C_1, w_1), \ldots, (C_m, w_m), (C_{m+1}, \infty), \ldots, (C_{m+m'}, \infty)\}$ *be a satisfiable MaxSAT formula,* $\varphi_2 = \{(C'_1, w'_1), \ldots, (C'_r, w'_r)\}$ *be a MaxSAT formula without hard clauses and* $W = \sum_{j=1}^{r} w'_j$. *Let*

$$harden(w) = \begin{cases} w & \text{if } w \leq W \\ \infty & \text{if } w > W \end{cases}$$

and $\varphi'_1 = \{(C_i, harden(w_i)) \mid (C_i, w_i) \in \varphi_1\}$. *Then* $\text{cost}(\varphi_1 \cup \varphi_2) = \text{cost}(\varphi'_1 \cup \varphi_2)$.

Notice that we check the applicability of this lemma dynamically, recomputing the value W in every iteration in line 8 of Algorithm 4.

Theorem 1. *Assuming that WPM, given a formula* φ*, returns a triplet* $(cost, \varphi_{sat}, \varphi_{res})$ *such that* φ *is MaxSAT reducible to* $\{(\Box, cost)\} \cup \varphi_{sat} \cup \varphi_{res}$*,* φ_{sat} *is satisfiable, and* φ_{res} *only contain clauses with weight strictly smaller than* w_{max}*, Algorithm 4 is a correct algorithm for Weighted Partial MaxSAT. Moreover, when for a formula* φ*, the algorithm returns* (c, φ')*, then* $c = \text{cost}(\varphi)$ *and any assignment satisfying* φ' *is an optimal assignment of* φ*.*

9   Experimental Results

We conducted our experimentation on the same environment as the MaxSAT evaluation [4] (processor 2 GHz). We increased the timeout from half hour to two hours, and the memory limit from 0.5G to 1G. The solvers that implement our Weighted Partial MaxSAT algorithms are built on top of the SAT solver picosat (v.924) [6]. The solver *wpm1* implements the original WPM1 algorithm [1]. The cardinality constraints introduced by WPM1 are translated into SAT through the regular encoding [3]. This encoding assures a linear complexity on the size of the cardinality constraint. This is particularly important for the last queries where the size of the cores can be potentially close to the number of soft clauses. We use the subscript b to indicate that we break symmetries as described in section 4, s to indicate we apply the stratified approach and d to indicate that we apply the diversity heuristic to compute the next w_{max}, both described in section 6. *wpm1* was the solver submitted to the 2009 and 2010 MaxSAT evaluations, and *wmp1$_s$* the one submitted to the 2011 evaluation. The hardening soft clauses (lines 8 and 9 in Algorithm 4) had not impact in our implementations' performance.

In the following we present results for the benchmarks of the Weighted Partial MaxSAT categories of the MaxSAT 2011 evaluation. We compare our solvers with the best three solvers of the evaluation, and other solvers which did not compete but have been reported to exhibit good performance, such as, *binc* and *bincd* [11], maxhs [8] and the Weighted CSP solver toulbar2 [19].

We present the experimental results following the same classification criteria as in the MaxSAT evaluation. For each solver and set of instances, we present the number of solved instances in parenthesis and the mean time required to solve them. Solvers are ordered from left to right according to the total number

Table 1. Experimental results

set	#	wpm1$_{bsd}$	wpm1$_{bs}$	wpm1$_s$	wpm1$_b$	wbo1.6	wpm1	wpm2	maxhs	bincd	maxhs	sat4j	binc	toulbar2
haplotyping	100	**423(95)**	425(95)	378(88)	376(79)	93.4(72)	390(65)	350(42)	1043(34)	196(21)	1247(35)	108(20)	408(27)	383(5)
timetabling	26	685.60(9)	**683(9)**	585(8)	992(9)	776(4)	1002(7)	707(9)	1238(4)	766(6)	2388(5)	0(0)	1350(4)	0(0)
upgradeability	100	35.8(100)	37.2(100)	36.5(100)	36.9(100)	63.3(100)	114(100)	311(100)	**29(100)**	637(78)	28.4(50)	844(30)	0(0)	0(0)
Total	226	**204**	204	196	188	176	172	151	138	105	90	50	31	5

(a) Weighted Partial - Industrial

set	#	wpm1$_{bsd}$	incw maxsatz	ak maxsat	toulbar2	wpm1$_{bs}$	wmax satz09z	maxhs	sat4j	wpm1$_s$	bincd	binc	wpm1	wbo1.6	wpm1$_b$
auc-paths	86	274(53)	7.59(86)	**4.6(86)**	28.8(86)	332(53)	570(80)	72.2(86)	994(44)	22(33)	1828(2)	0(0)	0(0)	0(0)	0(0)
auc-sched	84	**11.9(84)**	220(84)	123(84)	133(84)	12.2(84)	92(84)	1125(69)	716(80)	7.6(80)	130(50)	103(45)	0(0)	0(0)	0(0)
planning	56	27.9(52)	92.2(38)	354(40)	149(41)	**26.3(56)**	220(50)	306(29)	3.27(55)	12.5(54)	59.5(47)	51.1(46)	1.46(28)	1.33(30)	3.63(29)
warehouses	18	44(14)	1184(18)	37(2)	0.03(1)	571(3)	0.32(1)	0.37(1)	1.34(1)	1644(3)	7.03(1)	8.67(1)	**4.23(18)**	0.51(4)	0.88(12)
miplib	12	1187(4)	**1419(5)**	0.47(2)	63.2(3)	1165(4)	266(3)	0.07(1)	693(4)	34(3)	618(3)	699(3)	0.21(1)	0(0)	1507(1)
random-net	74	**241(39)**	1177(1)	1570(2)	0(0)	0(0)	0(0)	2790(6)	0(0)	0(0)	0(0)	0(0)	615(27)	63(37)	439(12)
spot5dir	21	257(10)	1127(5)	1106(5)	217(5)	383(10)	11.5(2)	199(6)	1.95(2)	1.41(5)	**66.7(11)**	51.7(6)	1.03(4)	2.60(5)	12.8(6)
spot5log	21	**532(14)**	0.63(4)	200(5)	170(5)	574(14)	15.6(2)	710(6)	6.04(3)	44.4(6)	124(11)	79.3(7)	21.5(6)	25.5(6)	131(7)
Total	372	**270**	241	226	225	224	222	204	189	184	125	108	84	82	67

(b) Weighted Partial - Crafted

Instance set	#	maxhs	wbo1.6	wpm1$_b$	wpm1$_{bsd}$	wpm1	wpm1$_{bs}$	wpm1$_s$	bincd	sat4j	binc	toulbar2
Table4 [8]	13	**4.41(13)**	5.03(13)	5.54(13)	8.84(13)	18.92(13)	534.13(13)	559.23(13)	56.69(11)	3485.52(6)	19.50(1)	0.00(0)
Total	13	**13**	13	13	13	13	13	13	11	6	1	0

(c) Table 4 from [8]. Linux upgradibility family forcing diversity of weights

of instances they solved. We present in bold the results for the best performing solver in each set. '# ' stands for number of instances of the given set.

Table 1(a) presents the results for the industrial instances of the Weighted Partial MaxSAT category. As we can see, our original solver *wpm1* would have ranked as the second best solver after *wbo1.6*. By breaking symmetries ($wpm1_b$) we solve 12 more instances than *wbo1.6*, and 20 more if we apply the stratified approach. Combining both, we solve 28 more instances. The addition of the diversity heuristic to the stratified approach has no impact for the instances of this category. We do not present any result on branch and bound based solvers since they typically do not perform well on industrial instances.

Table 1(b) presents the results for the crafted instances of the Weighted Partial MaxSAT category. The best ranked solvers in this category for the MaxSAT 2011 evaluation were: *incwmaxsatz*, *akmaxsat* and *wmaxsatz09*, in this order. All are branch and bound based solvers, which typically dominate the crafted and random categories. We can see that our solver *wpm1* shows a poor performance in this category. However, by applying the stratified approach ($wpm1_s$) we jump from 84 solved instances to 184. If we also break symmetries ($wpm1_{bs}$) we solve 224 instances, ranking as the third best solver respect to the participants of the MaxSAT 2011 evaluation, very close to *akmaxsat*. If we compare carefully the results of *wpm1* and $wpm1_{bs}$, we notice that there are two sets where *wpm1* behaves much better (warehouses and random-net). This suggests that we must make our stratified approach more flexible, for example, by incorporating the diversity heuristic ($wpm1_{bsd}$). Using $wpm1_{bsd}$ we solve up to 270 instances, outperforming all the branch and bound solvers.

In [8] it is pointed out that instances with a great diversity of weights can be a bottleneck for some Weighted MaxSAT solvers. To test this hypothesis they generate 13 instances from the Linux upgradibility set in the Weighted Partial MaxSAT industrial category preserving the underlying CNF but modifying the weights to force a greater diversity. We have reproduced that experiment in Table 1(c). As we can see, *wpm1* compares well to the best performing solvers, and by breaking symmetries ($wpm1_b$) we reach the performance of *maxhs* and *wbo1.6*. On the other hand, the stratified approach impacts negatively ($wpm1_s$ or $wpm1_{bs}$), but the diversity heuristic fixes this problem.

Taking into consideration the experimental results obtained in the different categories, we can see that our approach $wpm1_{bsd}$ is the most robust solver for Weighted Partial MaxSAT instances. We also checked the effectiveness of breaking symmetries for Unweighted Partial MaxSAT instances. For industrial instances we improve from 181 to 262 solved instances, and for crafted from 55 to 115.

References

1. Ansótegui, C., Bonet, M.L., Levy, J.: Solving (Weighted) Partial MaxSAT through Satisfiability Testing. In: Kullmann, O. (ed.) SAT 2009. LNCS, vol. 5584, pp. 427–440. Springer, Heidelberg (2009)
2. Ansótegui, C., Bonet, M.L., Levy, J.: A new algorithm for weighted partial MaxSAT. In: AAAI 2010 (2010)

3. Ansótegui, C., Manyà, F.: Mapping Problems with Finite-Domain Variables to Problems with Boolean Variables. In: Hoos, H.H., Mitchell, D.G. (eds.) SAT 2004. LNCS, vol. 3542, pp. 1–15. Springer, Heidelberg (2005)
4. Argelich, J., Li, C.M., Manyà, F., Planes, J.: MaxSAT evaluations (2006, 2007, 2008, 2009, 2010, 2011), http://www.maxsat.udl.cat
5. Berre, D.L.: SAT4J, a satisfiability library for java (2006), http://www.sat4j.org
6. Biere, A.: PicoSAT essentials. Journal on Satisfiability 4, 75–97 (2008)
7. Bonet, M.L., Levy, J., Manyà, F.: Resolution for Max-SAT. Artif. Intell. 171(8-9), 606–618 (2007)
8. Davies, J., Bacchus, F.: Solving MAXSAT by Solving a Sequence of Simpler SAT Instances. In: Lee, J. (ed.) CP 2011. LNCS, vol. 6876, pp. 225–239. Springer, Heidelberg (2011)
9. Fu, Z., Malik, S.: On Solving the Partial MAX-SAT Problem. In: Biere, A., Gomes, C.P. (eds.) SAT 2006. LNCS, vol. 4121, pp. 252–265. Springer, Heidelberg (2006)
10. Heras, F., Larrosa, J., Oliveras, A.: MiniMaxSat: A New Weighted Max-SAT Solver. In: Marques-Silva, J., Sakallah, K.A. (eds.) SAT 2007. LNCS, vol. 4501, pp. 41–55. Springer, Heidelberg (2007)
11. Heras, F., Morgado, A., Marques-Silva, J.: Core-guided binary search algorithms for maximum satisfiability. In: AAAI 2011 (2011)
12. Li, C.M., Manyà, F., Mohamedou, N., Planes, J.: Exploiting Cycle Structures in Max-SAT. In: Kullmann, O. (ed.) SAT 2009. LNCS, vol. 5584, pp. 467–480. Springer, Heidelberg (2009)
13. Lin, H., Su, K., Li, C.M.: Within-problem learning for efficient lower bound computation in Max-SAT solving. In: AAAI 2008, pp. 351–356 (2008)
14. Manquinho, V., Marques-Silva, J., Planes, J.: Algorithms for Weighted Boolean Optimization. In: Kullmann, O. (ed.) SAT 2009. LNCS, vol. 5584, pp. 495–508. Springer, Heidelberg (2009)
15. Marques-Silva, J., Argelich, J., Graça, A., Lynce, I.: Boolean lexicographic optimization: algorithms & applications. Ann. Math. Artif. Intell. 62(3-4), 317–343 (2011)
16. Marques-Silva, J., Lynce, I., Manquinho, V.: Symmetry Breaking for Maximum Satisfiability. In: Cervesato, I., Veith, H., Voronkov, A. (eds.) LPAR 2008. LNCS (LNAI), vol. 5330, pp. 1–15. Springer, Heidelberg (2008)
17. Marques-Silva, J., Manquinho, V.: Towards More Effective Unsatisfiability-Based Maximum Satisfiability Algorithms. In: Kleine Büning, H., Zhao, X. (eds.) SAT 2008. LNCS, vol. 4996, pp. 225–230. Springer, Heidelberg (2008)
18. Marques-Silva, J., Planes, J.: On using unsatisfiability for solving maximum satisfiability. CoRR abs/0712.1097 (2007)
19. Sanchez, M., Bouveret, S., Givry, S.D., Heras, F., Jégou, P., Larrosa, J., Ndiaye, S., Rollon, E., Schiex, T., Terrioux, C., Verfaillie, G., Zytnicki, M.: Max-CSP competition 2008: toulbar2 solver description (2008)

Distributed Tree Decomposition with Privacy

Vincent Armant, Laurent Simon, and Philippe Dague

Laboratoire de Recherche en Informatique, LRI, Univ. Paris-Sud & CNRS, Orsay, F-91405

Abstract. Tree Decomposition of Graphical Models is a well known method for mapping a graph into a tree, that is commonly used to speed up solving many problems. However, in a distributed case, one may have to respect the privacy rules (a subset of variables may have to be kept secret in a peer), and the initial network architecture (no link can be dynamically added). In this context, we propose a new distributed method, based on token passing and local elections, that shows performances (in the jointree quality) close to the state of the art Bucket Elimination in a centralized case (*i.e.* when used without these two restrictions). Until now, the state of the art in a distributed context was using a Depth-First traversal with a clever heuristic. It is outperformed by our method on two families of problems sharing the small-world property.

1 Introduction

Tree decomposition was introduced in [21]. It aims at mapping the graphical representation of a problem into a tree such that all linked variables in the initial graph stay linked (directly, or indirectly) in any node (also called *cluster* of the new tree). Once decomposed, the solving time of the initial problem can be bounded for a large class of problems. This nice property explains why tree decomposition has been widely studied, in many centralized applications (graph theory [21], constraints optimization [16,5,15], planning [13], databases [11,6], knowledge representation [14,22,10]), but also in distributed ones (distributed constraints optimization [19,3,7,17], ...). Intuitively, the decomposition guides the reasoning mechanism through the network, and can bound the number of messages. When dealing with distributed systems, it is indeed essential to bound the maximal number of messages, for instance to ensure some quality of services.

When the problem is tree-decomposed, its complexity can be bounded by an exponent of its *width*, which is the size of the largest cluster in the tree (minus 1). Many applications rely on good tree decompositions, and many polynomial classes are based on the existence of a good decomposition. Because of its exponential impact on the bound, even a small improvement in the quality of the decomposition may lead to large improvements in practice.

However, when the system is intrinsically distributed, or subject to privacy settings, no peer can have a global view of the system, and new algorithms must be explored. For instance, adding a link between two peers may not be feasible (only already existing links can be allowed). In all previous approaches, the initial distributed system was supposed to fulfil an additional strong characteristic: two peers that share a common variable must be directly linked by the network.

In this paper, we propose to explore the general case, *i.e.* when links between peers are not forced to follow the above characteristic. In this more general case, we propose

M. Milano (Ed.): CP 2012, LNCS 7514, pp. 102–117, 2012.

a new distributed algorithm for distributed tree decomposition, based on local elections with a token. Using a token was necessary to prevent concurrent decisions, which was identified as one of the main issues for good jointree construction. We conclude this paper by an experimental analysis of our algorithm on families of small-world networks, and show that the produced jointrees are significantly better than state of the art distributed algorithms, while allowing more general networks.

2   Distributed Tree Decomposition

Introduced By Robertson and Seymour [21], the tree decomposition was applied in many problems. It was noticeably used in circuit diagnosis in [9], in the centralized case. In this work, Dechter and El Fattah considered a graphical model of the set of equations describing the circuit. Each node of the graph is labelled by a variable, and each edge models a semantic dependency between variables in the same equation.

Figure 1 shows a problem described by a conjunction of f_i formulae signatures (e.g. constraints, properties, theories). On the right we show the corresponding interaction graph s.t. each node is labelled by a variable and there exists a link between two nodes iff they appear in the scope of a same function. For example, in light grey, we surround parts of the interaction graph modeling f_1, f_2 and f_3 of the initial problem.

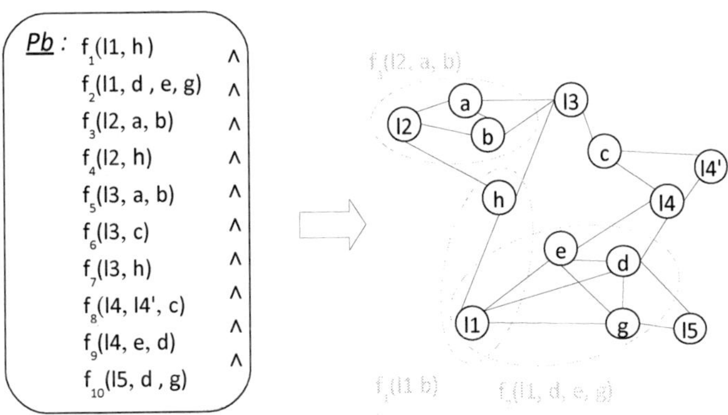

Fig. 1. Left: Centralized problem description. Right: The corresponding problem interaction graph.

The tree decomposition of an interaction graph is a tree of clusters of variables having the running intersection property and preserving the initial dependency schema of variables. Here we recall the definition of [21], keeping in mind that a node corresponds to a variable.

Definition 1 (tree decomposition [21]). *A tree decomposition of a graph G is a pair (χ, T) in which $T = (CT, F)$ is a tree where the nodes $ct_i \in CT$ are clusters names and the edges F model the inter-dependencies between clusters, and $\chi = \{\chi_{ct_i} : ct_i \in CT\}$ is a set of subsets of vertices(G) s.t. each subset χ_{ct_i} is the set of vertices of the cluster ct_i. The tree decomposition fulfils the following properties:*

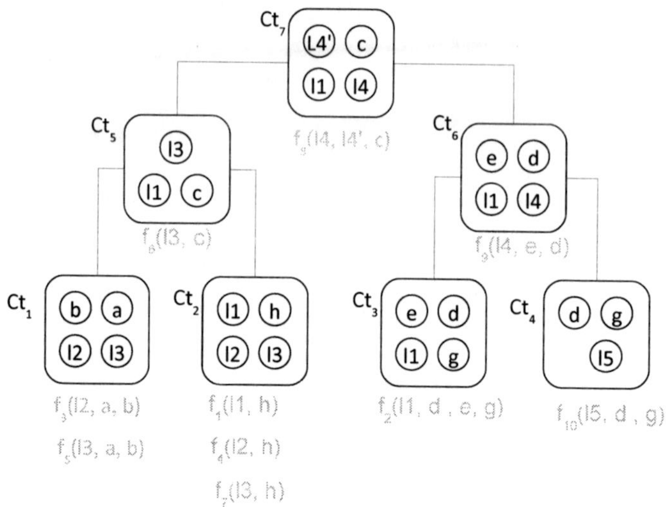

Fig. 2. Tree decomposition of the interaction graph of Figure 1

1. $\bigcup_{ct_i \in CT} \chi_{ct_i} = vertices(G),$
    *All nodes in the initial graph are at least in a cluster, and the tree contains no new variables (**vertices compliance**).*
2. $\forall \{x, y\} \in Edges(G), \exists ct_i \in CT \text{ with } \{x, y\} \in \chi_{ct_i}$
    *Each pair of variables connected by an edge in the initial graph must be found together in a cluster (**dependencies compliance**).*
3. $\forall ct_i, ct_j, ct_k \in CT$, *if* ct_k *is on the path from* ct_i *to* ct_j *in* T, *then* $\chi_{ct_i} \cap \chi_{ct_j} \subseteq \chi_{ct_k}$,
    *Two clusters that contain the same node are connected by clusters that also contain this node (**running intersection**).*

The result of a tree decomposition is also called a jointree. Figure 2 represents the tree decomposition of the interaction graph of Figure 1. In this figure, one can check that each node labelled by a variable in the interaction graph belongs to at least one cluster of the tree-decomposition. In addition, we can notice that clusters $ct_2, ct_5, ct_7, ct_6, ct_3$ that contain $l1$ are connected in the tree. Then, the running intersection property is satisfied for $l1$. It is easy to check that the running intersection is satisfied for all variables.

2.1   Distributed Theories and Privacy

Let us point out that, in the interaction graph, each variable is directly connected to all other variables that appear with it in at least one formula. Now, in a distributed context, the framework is made up of peers, where each peer knows a subset of formulae and can interact with its neighbourhood by messages in order to solve a global problem. In that case we need to consider at the same time the network interactions between peers defined by peers' acquaintances and the semantic interactions between variables defined

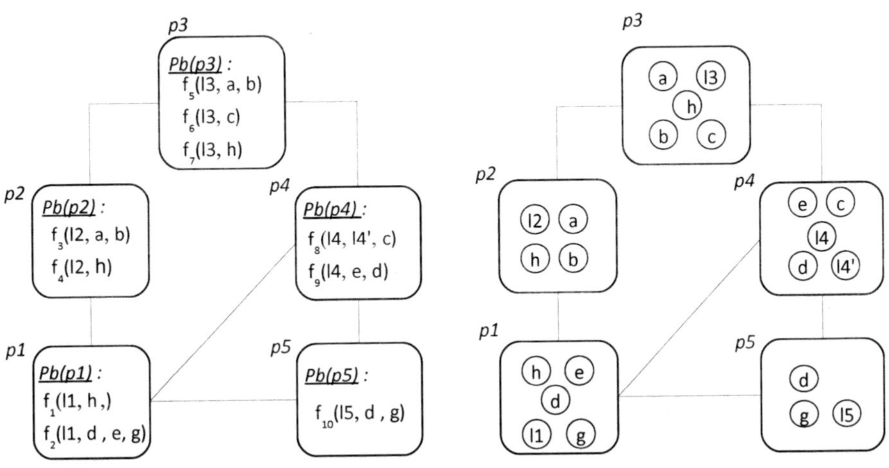

Fig. 3. Left: distributed problem of figure 1. Right: Its acquaintance graph.

by formulae (from a global point of view). In our approach, we model a distributed setting of *peers* and formulae by the notion of acquaintance graph.

Definition 2 (acquaintance graph). *Given a distributed problem setting defined on variables $Vars$, an acquaintance graph $G((P, V), ACQ)$ is a graph defined by:*

1. *$P = \{P_i\}_{i=1..n}$ is the set of peers,*
2. *$V : P \to 2^{Vars}$ is the node labelling function where $V(p_i)$ represents the vocabulary used by p_i to describe its problem.*
3. *$ACQ \subseteq P \times P$ defines the peers' acquaintances i.e. the neighbourhood with which each peer can exchange messages.*

This definition of an acquaintance graph differs from [1], which represents a multigraph where edges are labelled by shared variables.

Figure 3 (right) shows an example of an acquaintance graph of a distributed theory (left). We can emphasize that the acquaintance graph differs from the interaction graph. p_1 and p_3 have in common the variable h, but they do not share it in the network via a direct link like it would be the case in an interaction graph. Let us also point out that, without loss of generality, we can assume that two peers sharing an acquaintance link share also one or several variables. In addition, **we will not allow the creation of new acquaintance links** once the acquaintance graph is given. This is a first restriction for building a tree decomposition. One may also notice, figure 3, that a particular set of variables (written l_i), are only in one peer. We will consider these variables "local", and the other ones "shared". The privacy rule of our framework ensures that **local variables stay local**, *i.e.* no local variable is sent to any other peer in the network. This is our second restriction for building a tree decomposition. We can notice that the tree decomposition shown in figure 4 does respect privacy while the one in figure 2 does not.

2.2   Distributed Tree Decomposition

Because an acquaintance link between two peers may not follow the interaction graph of variables, we need to adapt Definition 1 for the distributed context.

Definition 3 (Distributed Tree Decomposition (DTD)).

Let $G((P, V), ACQ)$ be the acquaintance graph of a distributed system. A tree of clusters $T((CT, \chi), F)$ s.t. each cluster $ct \in CT$ is labelled by a set of variables $\chi(ct)$, is a distributed tree decomposition of G iff:

1. *$\bigcup_{p \in P} V(p) = \bigcup_{ct \in CT} \chi(ct)$ **(compliance of the vocabulary)***
2. *$\forall p \in P, \exists ct \in CT$ s.t. $V(p) \subseteq \chi(ct)$ **(compliance of variables dependency)***
3. *$\forall ct, ct', ct" \in CT$, if ct' is on the path from ct to $ct"$ in T, then $\chi(ct) \cap \chi(ct") \subseteq \chi(ct')$ **(running intersection)***
4. *There exists a function $\gamma : CT \to P$ that represents the cluster provenance s.t. $\forall \{ct, ct'\} \in F$:*
   - *$\gamma(ct) = \gamma(ct')$ or*
   - *$\{\gamma(ct), \gamma(ct')\} \in ACQ$*

   *Each cluster is hosted by one peer. Neighbouring clusters are hosted by the same peer or by neighbour peers in the acquaintance graph. **(compliance of acquaintances)***
5. *$\forall v \in V(p)$ s.t. $\forall p' \neq p, v \notin V(p'), \forall ct$ s.t. $v \in \chi(ct)$: $\gamma(ct) = p$*
   *Local variables belonging to p can only appear in a cluster hosted by p. **(privacy rule)***

In the above definition, the first three properties are similar to the classical tree decomposition. If the later preserves the initial set of nodes and their dependencies, in our definition, the distributed tree decomposition of an acquaintance graph preserves the set of peer variables and their dependencies. We also need to add the fourth one, to adapt our DTD definition to the distributed case. The compliance of acquaintances also forces DTDs to have the following property: a cluster is hosted by one (and only one) peer, and a peer hosts at least one cluster (if its formula is not empty), and possibly many. In addition, all interactions between clusters are following the initial peer acquaintance. Our privacy rule, in the context of DTD is expressed by the fifth property. All clusters hosted by one peer p can contain any shared variable from any peer, but only local variables of p itself. If we look at figure 4, we see that all variables l_i stay at their initial peer. At the opposite, shared variables h and g are sent to p_4 for ensuring the running intersection.

2.3   Tree Decomposition: Choose Your Father

Tree decomposition aims at directing the reasoning during search, or fixing a bound on the reasoning task. Then, the complexity of most classical reasoning tasks is bounded by the size of the largest obtained cluster. There is a strong link between the problem of finding a good decomposition, a good forgetting order, and the graph triangulation problem. However, in a distributed approach, only orders based on depth-first (DFS) and breadth-first (BFS) searches have been proposed so far. Building a distributed decomposition tree by BFS was first proposed by Elzahir et al. [7] for distributed CSPs. GrunBach and Wu [12] proposed to apply the tree decomposition for software verification, based on a BFS. However, we experimentally checked (not reported here) that the quality of the decompositions by BFS order was significantly worse than those based on DFS. So, we limit this section to the presentation of DFS-based orders and Bucket-Elimination (BE) orders.

Distributed, Depth-First, Tree Decomposition. In distributed constraints optimization, Distributed DFS (DDFS) is the state of the art [19,8]. Initially, a peer is chosen as the root of the tree (this is a global decision). It chooses one of its neighbours and sends it a special message. When a peer receives such a message for the first time, it records the sender as its father. Then, it forwards this message to one of its neighbours that have not yet been chosen by it. If all its neighbours have been visited, then it sends the return message to its father. The DDFS algorithm sends messages through all edges of the network of peers. Its complexity (in the number of non concurrent messages) is $2 * |E|$, where E is the set of edges.

Distributed Bucket Elimination. Bucket Elimination is a general method that can be used for decomposition. It builds a tree of clusters from the leaves to the root of the tree. When applied with the DFS order, BE does not have to build the tree, it follows the DFS one: DFS induces a total order on viewed peers s.t., for each peer, each of its neighbours is an ancestor or a descendant. Initially, an empty cluster χ_{ct_p} is associated to each peer; then, for each peer p_i, if it is a descendant of some peer p_j then the vocabulary common to the two peers p_i, p_j is added to the cluster $\chi_{ct_{p_i}}$. Then, the algorithm typically proceeds bottom-up in two stages: (1) a peer p_i sends to its father p_j a projection $pjct_{p_i}$ of the variables contained in $\chi_{ct_{p_i}}$ and belonging to one of its ancestor; (2) when p_j receives $pjct_{p_i}$ from p_i, it adds $pjct_{p_i}$ in its cluster $\chi_{ct_{p_i}}$. When it has received a projection from all its children it projects to his father the variables of $\chi_{ct_{p_j}}$ shared with one of its ancestors. For more details, the piece of work [3] assumes a DTD based on Cluster Tree Elimination [4] applied to a decentralized context.

We show the jointree built by Distributed BE with DDFS and the Maximum Cardinality Set (MSC) heuristic in figure 4. In the latter, privacy is preserved. One may notice here that variables are shared among peers if needed. The notion of projection ($pjct_{p_i}$ in the above algorithm) will also appear in our algorithm. This notion is essential for ensuring the running intersection property in the final jointree.

2.4   Good Properties of Centralized Methods

Centralized methods often rely on equivalent problems for tree decomposition, such as graph triangulation. In this problem, one wants to add as few edges as possible to a given graph in order to obtain a chordal graph (all cycles of length four or more have a chord, which is an edge joining two non-adjacent nodes in the cycle). The triangulation algorithm iteratively eliminates all nodes. At each iteration, (1) a variable is chosen; (2) edges are added between pairs of nodes from its neighbours that are not already connected (clique-fill); (3) the node is removed. The quality of a triangulation can be

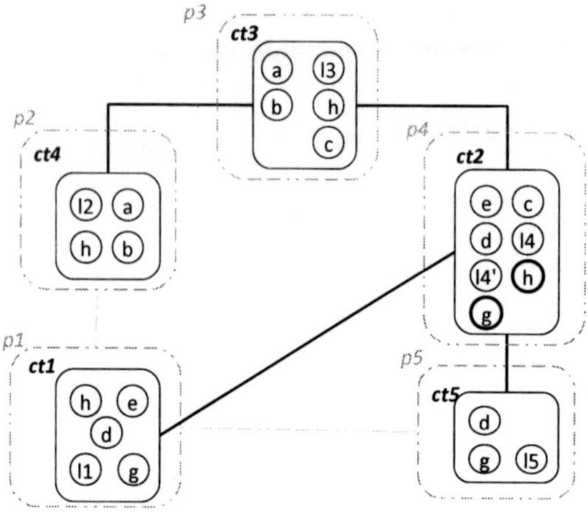

Fig. 4. Tree Decomposition given by BE and driven by DFS and MCS heuristic

trivially measured by the number of new edges. The node chosen at step (1) must add as few edges as possible, in a short and in a long-term point of view. For instance, a node in a clique with its entire neighbours is a very good candidate. A node with a poorly connected neighbourhood is a bad choice. The heuristic *Min-Fill-In* is based on this very simple observation. Each node is evaluated w.r.t. the number of new edges one may have to add to fill its neighbourhood as a clique.

In order to get a global elimination order from the above algorithm, we must add a fourth step: a cluster containing all the neighbourhood of the removed variable is built (and memorized) at the new step. In this context, the Min-Fill-In algorithm can be viewed as trying to reduce the size of the clusters. However, when adapting this idea to the distributed case, it may not be possible to add links to any pair of peers. In our formulation, the acquaintance graph is given, and no network link can be added "on the fly".

About concurrency, it is trivial to say that centralized methods are not concurrent. However, we think that the non-concurrency property is essential to obtain good join-trees. Let us consider a simple path $v_1 - ... - v_n$. One can check that the sequential elimination of one or two variables (i.e. (v_1, v_n)) at ends of the path will lead to the creation of clusters with at most two variables. Unfortunately, the concurrent eliminations of at least three variables at the same time will lead to the creation of a cluster of at least three variables. This example can easily be generalized for tree structured graph or cyclic graph, and clearly illustrates the fact that even few concurrent decisions can get any algorithm away from the optimal decomposition. In order to prevent those bad decisions to be taken, we propose to use a token to limit concurrent choices.

3   Distributed Token Elimination

The idea of the algorithm is the following. Each peer votes for its best neighbour (including itself). It has to decide, locally, which peer should be removed first. A peer that

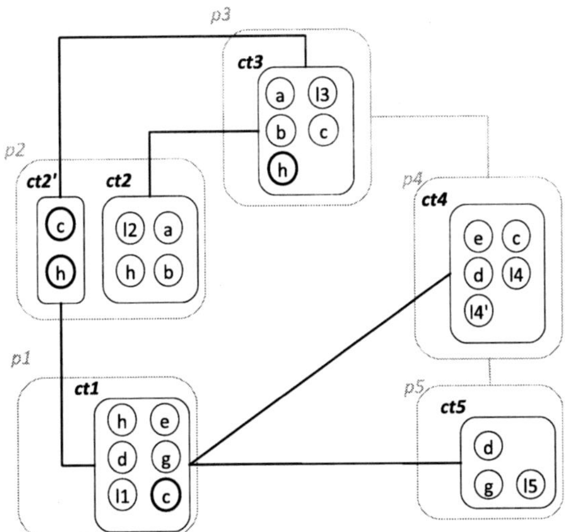

Fig. 5. Tree Decomposition obtained by the Token Elimination on the running example

has been chosen by all its neighbours can be removed (it is a sink node in the graph of
"best" choices). When removed, a peer builds a cluster of variables for the distributed
jointree, and memorizes it locally. This peer will be the root of a sub-jointree in the
final jointree. All the variables that are not local to this sub-tree are in its *projection*
(exactly in the same sense as it is used in BE, see 2.3). A removed peer remains active,
and participates to the elections until all its neighbourhood has also been removed (after
that, it will only be active for routing token messages).

In order to build a distributed jointree (DJT) at the end of the algorithm, we consider
the first removed peers as the leaves of the DJT. The final DJT is built bottom-up to
the root. This is done when all the peers have been eliminated. The algorithm has to
connect all the sub-jointrees together. Because we do not want to allow any new link in
the network, while ensuring the running intersection, this stage is a little bit tricky. Each
peer asks the network for its son by flooding its name and its projection, and when the
son answers, all peers on the path between them create new "connection" clusters with
all the variables needed for ensuring the running intersection property in the final DJT.
So, we have the following properties: First, during the whole algorithm, each cluster
is only in one peer. Second, before this last reconnection stage, each peer can only
contains one cluster (called the "main" cluster). Third, at the end of the algorithm, each
peer contains exactly one "main" cluster, and an unbounded number of "connection"
clusters.

An illustration of the final solution of our algorithm is shown in figure 5. One can especially notice that privacy is preserved, and that clusters, thanks to connection clusters,
are connected on the top of the initial peers network (see the $P2$ clusters).

Additionally, as reported section 2.4, we need to carefully handle concurrent decisions. For this, a single token is circulating on the network, following the current votes.
Because the directed graph of "best" choices is not guaranteed to be strongly connected,

we must add an escape strategy. For this, we suppose a cycle that goes through all nodes exactly once. In our approach, we build this cycle by a DFS traversal of the peers.

3.1 Data Structures and Notations

The *Score* of a peer is the estimated size of its main cluster (the algorithm tries to choose the next peer with minimal score). In addition, each peer maintains an array *score*[] of the "cluster" score of its neighbours. This "cluster" score takes into account the current projections of orphan sub-jointrees. More generally, this is where the heuristics take place.

The *Token* circulating in the network is also decorated with precious information. It contains, intuitively, the set of current orphan sub-jointrees that are being built, rooted in removed peers so far. To allow any peer in the network to smartly choose which sub-jointree to attach as a son, each sub-jointree also contains its projection, *i.e.* all the non local variables of the sub-jointree.

neighbours∗ is the set of all the neighbours of a peer, including removed peers, and itself. *neighbours*+ is the set of all the neighbours of a peer, including itself, but not including removed peers. *best* is the "best" neighbour of a peer p, *i.e.* the peer having the smallest value in the *score*[] array, including p. *Children* is the set of direct children a peer has registered so far.

3.2 The Distributed Token Elimination Algorithm

DTE is given as Algorithm 1. Initially, the token is sent to one peer.

Local Elections. Each time a peer p_{token} receives (or initially has) the token (line 2), it organizes local elections (line 4) and waits for the votes of all its neighbours (including removed ones). The reception of this message is treated in line 31. The computed score of p takes into account the size of the set of variables made up from current projections in the token (set of non local variables of sub-jointrees built so far), local variables and shared variables with ancestors. The underlying idea of this score is to estimate the size of the cluster if p, the peer that received the $ELEC$ message (line 31), would be removed at this point. The computed score of the peer p is sent back to p_{token} but also to all its other neighbours, line 33, and received line 37. So, if we get back to the initial peer p_{token} that organized the local election, it has updated its own *score*[] values for all its neighbours, taking into account the current open projections in the token, and the local election is closed.

One may notice here that *score*[] arrays have been updated only in the neighbourhood of p_{token}, and thus the votes of the neighbours of p_{token} are based only on partial information: if p is a neighbour of p_{token}, p only updates the scores of its neighbours that are also connected to p_{token}. All other peers are ranked according to an old value of *score*[], that can be based on an old set of orphans sub-jointrees. At some point, we must accept this, because this is just a heuristic, and trying to have a better estimation will cost too much. However, one may also notice here that if no peer is elected, the token remains the same, and follows the "best" links or the global cycle, eventually touching all the peers at the end of the cycle. The algorithm will end because once a peer has updated its own score according to the token, if the token does not change (no peer is removed), then its score will remain the same. If the token touches all the peers, all $score[]$ values will have been updated according to the same orphan sub-jointrees (this remark is important for termination, we have the guarantee to find a node to remove after at most one cycle, and in any case after visiting at most two times each edge).

At this step, after line 5, the peer p_{token} has now two possibilities, removing itself (see section 3.2), or giving the token to its best neighbour/cycle (see section 3.2).

Elected and Removed. When p_{token} is a local minimum, *i.e.* all its neighbourhood, including itself, has elected it, then it has two steps to perform before being removed from the network. First, it needs to compute the new projection, and register itself as a new potential orphan sub-jointree with this projection.

This is done from lines 9 to line 17 in the algorithm. Pj_p is the projection of p over its alive neighbourhood. $Children$ is the set of peers that p has chosen as sons: all the peers that are roots in the current orphan sub-jointrees with which p shares at least one variable (these allow one to preserve the running intersection). New_χ is the set of variables in the projections of orphan sub-jointrees that p has to take into account when it will be removed, in addition to its own projection variables. So, for the new sub-jointree rooted in p, its projection χ_p is computed in line 12. Now, p has to update the set of sub-jointrees attached to the token. First, it removes its sons, then it adds itself as a new sub-jointree. At last, it is important that p tells its neighbourhood it has been removed (sending and acknowledging $ELIM$ messages). During this stage, it is also important to notice that all neighbours of p update their $best$ values (including p itself).

We postpone the explanation of lines regarding the reconnection of sub-jointrees to the end of the section.

Sending the Token. The idea of lines 24 to 29 is to let the token following the "best" links in the graph, and rely on the static DFS order for escaping. So, if p is not its best choice (not in a local minimum), then, if p still has alive neighbours, it sends the token to the best of its alive neighbours. Otherwise, because $neighbours^+$ initially contains p itself, p has been removed with all its neighbours. In this case, we rely on the static cycle built on the top of DFS to send the token and escape.

If p was its own best choice, we are in a local minimum, and we also need to escape. Observing the fact that at least one of the alive neighbours is not considering p as the best choice (if p was the best choice for all its neighbourhood, then it has just been eliminated, and has sent the $ELIM$ messages to its neighbourhood forcing its neighbours to vote for another peer), we know that there is a peer with a better local score than p if we follow any $best$ link from any neighbour of p that has not voted for it. So, in our solution, we propose to use this heuristic, as shown line 29.

Running Example. An example (limited to two steps) of the algorithm is given in figure 6. First, suppose that the token was in p_1. According to the votes, the token is sent to p_5. Then, as shown in figure 6 (left), p_5 is selected for elimination (see the scores in the figure). It adds its own projection set $\{d, g\}$ (all its variable except local ones) to the token and removes itself. The token then goes through p_1 and p_2 (following elections results), and then (Figure 6 right) p_2 is removed and updates the token accordingly. In

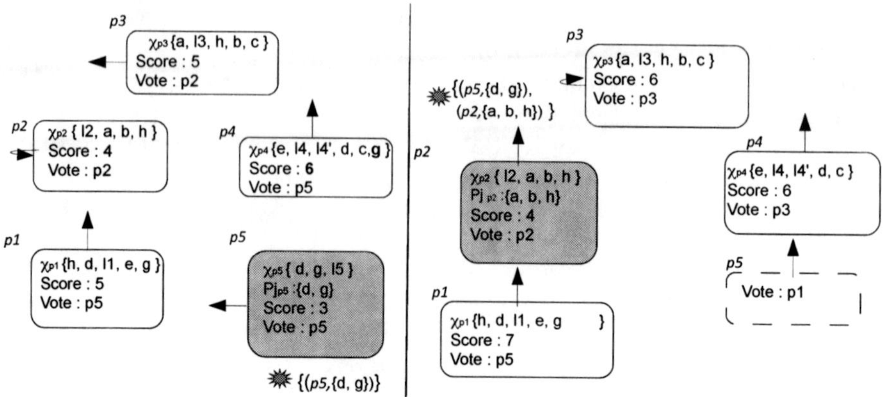

Fig. 6. Two iterations of Distributed Token Elimination on our example

the next step (not shown on the figure), the token will be sent to p_3 which will remove the couple $(p_2, \{a, b, h\})$ from the token (p_3 shares at least one variable in this set) and will add the couple $(p_3, \{h, c\})$ in the token (a and b are not anymore shared).

Final Stage: Re-connecting All Sub-jointrees When the above algorithm finishes, it ends with a forest of sub-jointrees, distributed on the network. Each node has its own main cluster, fixed when the node was removed, but all the sub-jointrees have to be merged in one jointree, by reconnecting them while respecting the running intersection.

The first peer that detects the termination broadcasts the *Reconnection* message in the network (line 20). Then, (not shown here), each non-leaf peer p that receives this message initiates the flooding algorithm 2: each p broadcasts a new message $ReqCON$ to the network with the set of its children. When receiving a message $ReqCON$ for the first time, a peer p records the sender p' as a father link and broadcasts the message to all its neighbourhood (excluding p'). Of course, due to the flooding mechanism, the message $ReqCON$ visits all links of the network (lines 4 to 6).

Now, when a peer has received the $ReqCON$ message from all its neighbours (line 10), it initiates the answer by sending back the message $AnsCON$ with the list of children of the initial peer p (noted $Wanted$ in the algorithm, lines 12 to 14). The message then goes bottom-up in the tree, following all the fathers links created above. Now, each peer that recognizes itself in the list attached to $AnsCON$ when the message goes up adds its own shared variables to the list of new shared variables (noted $ProjChild$ in the algorithm, lines 25 to 26). Those variables need to be added along the path to ensure the running intersection property. They are shared between the peer and its direct neighbourhood. In addition, each time a peer receives $AnsCON$ with a non empty list $projChild$, it creates a new "connection" cluster containing all variables of $projChild$ (named $conClusters(Wanted)$ in the algorithm for simplicity, lines 21 to 22).

Complexity Analysis. At the end of the Distributed Token Elimination n peers have been eliminated. For each elimination the token will potentially visit all edges in both directions (in the worst case) making a local election at each step. Afterwards, during

Algorithm 1. Distributed Token Elimination

(2) p **receives** $TOKEN(SubT)$ **from** p'

(3)   *// local election*

(4) $\forall p' \in neighbours^*$, send $ELEC(SubT)$ to p'

(5)   wait $VOTE(b_{p'})$ from all $p' \in neighbours^*$

(6)

(7)   **if** (p has received $VOTE(\top)$ from all $p' \in neighbours^*$)

(8)   *// Eliminate p (see section 3.2)*

(9) $Pj_p \leftarrow \bigcup_{p' \in neighbours^+} V_p \cap V_{p'}$

(10) $Children \leftarrow \bigcup_{(p',V') \in SubT} p' \text{ s.t. } V' \cap V_p \neq \emptyset$

(11) $New_\chi \leftarrow \bigcup_{(p',V') \in SubT} V' \text{ s.t. } p' \in Children$

(12) $\chi_p \leftarrow Pj_p \cup New_\chi \setminus \bigcup_{v \in V_p} \text{ s.t. } v \notin Pj_p$

(13)   *// Update Orphan Sub-Jointrees*

(14) $SubT \leftarrow SubT \setminus \bigcup_{(p',V') \in SubT} (p',V') \text{ s.t. } p' \in Children$

(15) $SubT \leftarrow SubT \cup (p, \chi_p)$

(16)   send $ELIM$ to all $p' \in neighbours^*$

(17)   wait $AckELIM$ from all $p' \in neighbours^*$

(18)   **if** $(Pjct = \{\{\}\})$

(19)   *// re-connecting sub-jointrees (see section 3.2)*

(20)   broadcasts the $Reconnection$ message in the network.

(21)

(22)   *// sending the token(see section 3.2)*

(23)   **if** $(best \neq p)$

(24)   **if** $(neighbours^+ \neq \emptyset)$ *// token passing*

(25)   send $TOKEN(SubT)$ to $best$

(26)   **else**

(27)   send $TOKEN(SubT)$ to $succ_{DFS}(p')$

(28)   **else**

(29)   send $TOKEN(SubT)$ to p'' s.t. p as received $VOTE(\bot)$ from p''

(31) p **receives** $ELEC(SubT)$ **from** p'

(32) $score[p] \leftarrow eval(SubT)$

(33)   send $SCORE(score[p])$ to its neighbours and wait $AckSCORE$

(34) $best \leftarrow min(score[])$

(35)   send $VOTE(best == p')$ to p'

(37) p **receives** $SCORE(i)$ **from** p'

(38) $score[p'] \leftarrow i$

(39)   send $AckSCORE$ to p'

(41) p **receives** $ELIM$ **from** p'

(42)   remove p' from $score$ and $neighbours^+$

(43) $best \leftarrow min(score[])$

(44)   send $AckELIM$ to p'

Algorithm 2. Sub-jointrees reconnection

(2) p **receives** $ReqCON(Wanted)$ **from** p'
(3)   // *top-down flooding*
(4)   **if** $(father(Wanted) \neq \emptyset)$
(5) $father(Wanted) \leftarrow p'$
(6)   send $ReqCON(Wanted)$ to $neighbours - p'$
(7)
(8)   // *bottom up connections*
(9) $nbMsg(Wanted) + +$
(10)   **if** $(nbMsg(Wanted) == |neighbours|)$
(11)   **if** $(p \in Children)$
(12)   send $AnsCON(Wanted, \{Pj_p\})$ to $father(Wanted)$
(13)   **else**
(14)   send $AnsCON(Wanted, \{\})$ to $father(Wanted)$

(16) p **receives** $AnsCON(Wanted, projChild)$ **from** p'
(17)
(18)   **if** $(Wanted == Children)$
(19) $sonLinks \leftarrow sonLinks \cup p'$ //*for future use*
(20)   **else**
(21)   **foreach** $V' \in projChild$
(22) $conClusters(Wanted) \leftarrow conClusters(Wanted) \cup V'$
(23) $nbMsg(Wanted) + +$
(24)   **if** $(nbMsg(Wanted) == |neighbours|)$
(25)   **if** $(p \in Wanted)$
(26)   send $AnsCON(Wanted, conClusters(Wanted) \cup Pj_p)$ to $father(Wanted)$
(27)   **else**
(28)   send $AnsCON(Wanted, conClusters(Wanted))$ to $father(Wanted)$

the sub jointree reconnection phase, each peer will be required. Since the local election requires a constant number of non concurrent messages and the time complexity of the sub-jointrees connection is bound by the diameter of the graph, the time complexity of the whole algorithm is in $O(N * E)$. In addition, the algorithm is complete and correct as soon as the initial graph respects the running intersection. The correctness and the completeness of our algorithm can be ensured by the following remark: the execution of our algorithm can be easily reduced to a particular execution of a centralized, classical, decomposition tree algorithm.

4   Experimental Evaluation

We implemented a java-based simulator of a distributed system for analysing the performance of our Distributed Token Elimination algorithm. We used a generator of small-world networks, following the model generation of Barabassi and Albert [2] (called BA Graphs) and Watts and Strogatz [24] (called WS Graphs). Let us point out that BA Graphs are extremely heterogeneous. The degree of nodes follows a power law, a strong characteristic of many real world problems (World Wide Web, email exchanges, scientific citations, ...) [18]. WS Graphs are more homogeneous and have been used for

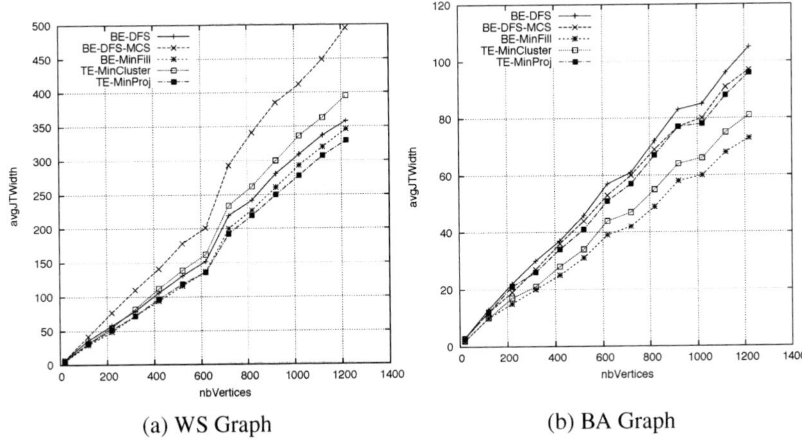

(a) WS Graph   (b) BA Graph

Fig. 7. Width of obtained jointrees, average values over 10 experiments per point

modelling diagnosis circuits [20] or various CSP instances [23]. On this kind of networks, we were able to scale up to more than thousands of variables. Before analyzing the results, let us also point out that we do not consider the privacy characteristics of our algorithms in this experiment: all variables are shared; and we consider that each peer has exactly one and only one variable for simplicity. However, as shown in the algorithm, line 12, private variables are treated apart. They are automatically removed from any projection the peer can send away. Thus, our experiment does not necessarily have to take privacy into account.

We report in figure 7 the quality of the jointrees produced by our algorithm, Distributed Token Elimination, with a set of selected state of the art algorithms. Bucket Elimination with the MinFill heuristic (BE-MinFill) is well known and represents the current state of the art for centralized methods. We also report two DFS variants (Maximum Cardinality Set and no heuristic). The first one is certainly the most used in distributed systems. We report the Token Elimination algorithm with two variants. The first one is the MinCluster (the one previously reported), and the second one MinProjection (instead of scoring the peers with their potential cluster size, we score them according to the size of the set of projection variables).

The reported results show a significant improvement of Token Elimination over DFS. The obtained results, with the MinCluster variant, are now very close to BE on BA Graph. Surprisingly, after 700 nodes TE-Minproj gives better decompositions than BE on WS-Graph. Let us recall here that the jointrees built by Token Elimination are forced to follow the original links in the initial graph, which is not the case of BE.

The above results must be nuanced by the cost of the distributed compilation itself. As shown in figure 8, if DFS does not produce very good jointrees, it can produce them very quickly. BE needs more time, but produces the best jointrees. The total time needed by Token Elimination is clearly the worst on BA Graph figure, and quickly increasing. However, let us point out that for WS Graph TE is faster than BE.

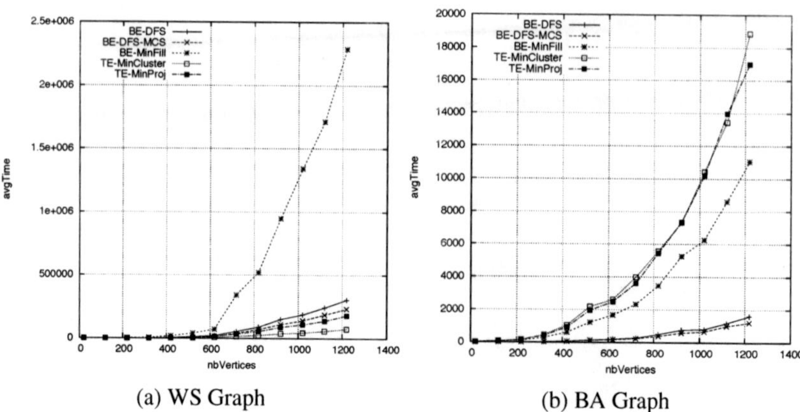

(a) WS Graph   (b) BA Graph

Fig. 8. Average CPU time for compilation of jointrees over 10 experiments per point

Given the improvement in the quality of the obtained jointrees, we strongly believe Token Elimination is a good solution, particularly suited to a distributed compilation approach of the network.

5   Conclusion

In this paper, we proposed a new distributed method for building distributed jointrees. Our method can handle privacy rules by keeping secrets secret (local variables stay local), and allow the jointree to be built only on the top of existing links between nodes. We show that this method can handle large structured instances and, even if the compilation cost is clearly above the simple Distributed DFS algorithm, the quality of the obtained jointree clearly outperforms DFS, even with a clever heuristic, and can even surpass the quality of Bucket Elimination in its centralized version. We believe that our algorithm can improved previous results in many distributed applications.

Acknowledgments. We want to thank the reviewers for the very useful comments they provided during the reviewing process and Philippe Jégou for his fruitful remarks.

References

1. Adjiman, P., Chatalic, P., Goasdoué, F., Rousset, M.-C., Simon, L.: Distributed reasoning in a peer-to-peer setting: Application to the semantic web. Journal of Artificial Intelligence Research (JAIR) 25, 269–314 (2006)
2. Barabási, A.-L., Albert, R.: Emergence of scaling in random networks. Science 286(5439), 509–512 (1999)
3. Brito, I., Meseguer, P.: Cluster tree elimination for distributed constraint optimization with quality guarantees. Fundam. Inform. 102(3-4), 263–286 (2010)
4. Dechter, R.: Constraint processing. Elsevier Morgan Kaufmann (2003)
5. Dechter, R.: Tractable structures for constraint satisfaction problems. In: Handbook of Constraint Programming, part I, ch. 7, pp. 209–244. Elsevier (2006)

6. Dermaku, A., Ganzow, T., Gottlob, G., McMahan, B., Musliu, N., Samer, M.: Heuristic Methods for Hypertree Decomposition. In: Gelbukh, A., Morales, E.F. (eds.) MICAI 2008. LNCS (LNAI), vol. 5317, pp. 1–11. Springer, Heidelberg (2008)
7. Ezzahir, R., Bessiere, C., Benelallam, I., Belaissaoui, M.: Asynchronous breadth first search dcop algorithm. Applied Mathematical Sciences 2(37-40), 1837–1854 (2008)
8. Faltings, B., Léauté, T., Petcu, A.: Privacy guarantees through distributed constraint satisfaction. In: IAT, pp. 350–358 (2008)
9. Fattah, Y.E., Dechter, R.: Diagnosing tree-decomposable circuits. In: IJCAI, pp. 1742–1749 (1995)
10. Favier, A., de Givry, S., Jégou, P.: Exploiting Problem Structure for Solution Counting. In: Gent, I.P. (ed.) CP 2009. LNCS, vol. 5732, pp. 335–343. Springer, Heidelberg (2009)
11. Gottlob, G., Leone, N., Scarcello, F.: A comparison of structural CSP decomposition methods. Artificial Intelligence 124, 2000 (2000)
12. Grumbach, S., Wu, Z.: Distributed tree decomposition of graphs and applications to verification. In: IPDPS Workshops, pp. 1–8 (2010)
13. Guestrin, C., Gordon, G.J.: Distributed planning in hierarchical factored MDPs. In: UAI, pp. 197–206 (2002)
14. Huang, J., Darwiche, A.: The language of search. J. Artif. Intell. Res. (JAIR) 29, 191–219 (2007)
15. Jégou, P., Ndiaye, S., Terrioux, C.: Dynamic heuristics for backtrack search on tree-decomposition of CSPs. In: IJCAI, pp. 112–117 (2007)
16. Kask, K., Dechter, R., Larrosa, J., Dechter, A.: Unifying tree decompositions for reasoning in graphical models. Artif. Intell. 166(1-2), 165–193 (2005)
17. Léauté, T., Faltings, B.: Coordinating logistics operations with privacy guarantees. In: IJCAI, pp. 2482–2487 (2011)
18. Newman, M.E.J.: The structure and function of complex networks. SIAM Review 45(2), 167–256 (2003)
19. Petcu, A., Faltings, B.: A scalable method for multiagent constraint optimization. In: IJCAI, pp. 266–271 (2005)
20. Provan, G., Wuang, J.: Automated benchmark model generators for model-based diagnostic inference. In: IJCAI, pp. 513–518 (2007)
21. Robertson, N., Seymour, P.D.: Graph minors. ii. algorithmic aspects of tree-width. Journal of Algorithms 7(3), 309–322 (1986)
22. Subbarayan, S., Bordeaux, L., Hamadi, Y.: Knowledge compilation properties of tree-of-BDDs. In: AAAI, pp. 502–507 (2007)
23. Walsh, T.: Search in a small world. In: IJCAI, pp. 1172–1177 (1999)
24. Watts, D.J., Strogatz, S.H.: Collective dynamics of ṡmall-worldńetworks. Nature 393(6684), 440–442 (1998)

Refining Restarts Strategies for SAT and UNSAT

Gilles Audemard[1,*] and Laurent Simon[2]

[1] Univ Lille-Nord de France CRIL / CNRS UMR8188,
Lens, F-62307
`audemard@cril.fr`
[2] Univ Paris-Sud, LRI / CNRS UMR8623
Orsay, F-91405
`simon@lri.fr`

Abstract. So-called Modern SAT solvers are built upon a few – but essential – ingredients: branching, learning, restarting and clause database cleaning. Most of them have been greatly improved since their first introduction, more than ten years ago. In many cases, the initial reasons that lead to their introduction do not explain anymore their current usage (for instance: very rapid restarts, aggressive clause database cleaning). Modern SAT solvers themselves share fewer and fewer properties with their ancestor, the classical backtrack search DPLL procedure.

In this paper, we explore restart strategies in the light of a new vision of SAT solvers. Following the successful results of GLUCOSE, we consider CDCL solvers as resolution-based producers of clauses. We show that this vision is particularly salient for targeting UNSAT formulae. In a second part, we show how detecting sudden increases in the number of variable assignments can help the solver to target SAT instances too. By varying our restart strategy, we show an important improvement over GLUCOSE 2.0, the winner of the 2011 SAT Competition, category Application SAT+UNSAT formulae. Finally we would like to point out that this new version of GLUCOSE was the winner of the SAT Challenge 2012.

1 Introduction

Despite the important progress constantly observed in the practical solving of SAT/-UNSAT problems, most of these components are commonly viewed as improvements over the DPLL-62 [4] procedure. In many works, including recent ones, the vision of CDCL (Conflict Driven Clause Learning, also called "Modern SAT") solvers [10,13] as backtrack-search engines prevails. In this paper, we discuss this point of view and focus our work on an essential ingredient of Modern SAT solvers: restarts. Restart strategies used in CDCL solvers are also often understood as heirs of the DPLL framework. If the search seems to not be close to finding a model, then a restart often means "restarts tree-search elsewhere".

Restarting was first proposed to prevent the search tree from making mistakes in its top decisions. This idea was formally observed by the *Heavy Tailed* distribution of CPU time of DPLL-like solvers. However, initially, restarting was thought of as a witness of

---

* This work has been partially supported by CNRS and OSEO, under the ISI project "Pajero".

M. Milano (Ed.): CP 2012, LNCS 7514, pp. 118–126, 2012.

branching heuristics failure. Thus, restarts were not (*i.e.* after thousands of explored branches) and, to ensure completeness of approaches, restarts were triggered according to a geometric series. More recently, restarts strategies were following Luby series, a linear increasing series.

In this work, we take an alternate vision of CDCL solvers. Our goal is to view them as resolution-based "clauses producers" instead of "search-space explorers". This vision is illustrated by our solver GLUCOSE. It was already embedded in GLUCOSE 1.0 [1], thanks to a good measure over learnt clauses, called LBD (Literal Block Distance), however, some details of the restart strategy used in the version of GLUCOSE that participated to the SAT Competition 2009 [1] [8] were never published. Additionally, in this paper, we describe a new restart strategy, used in GLUCOSE 2.1, that shows significant improvements over GLUCOSE 2.0.[2] This new restart strategy is twice as aggressive and still does not have any guarantee that long runs will be allowed. This property, associated with the fact that GLUCOSE 2.0 deleted 93% of the learnt clauses during the SAT competition 2011,[3] shows that GLUCOSE can hardly be considered as a systematic search solver, and is increasingly distant from the DPLL vision of Modern SAT Solvers. We also show in this paper that such an aggressive strategy is especially good for UNSAT problems. In order to allow an improvement also on SAT instances, we propose to add a "blocking" parameter to our strategy, which is able to postpone one or many aggressive restarts when the solver seems to suddenly reach a global assignment. Intuitively, our work tries to reconcile the two opposing forces working in CDCL solvers. One (the UNSAT force) is trying to decrease the number of decision levels, by learning good clauses. The other (the SAT force) is trying to increase the number of decision levels, by assigning more and more variables.

2   Background

For lack of space, we assume the reader is familiar with Satisfiability notions (variables x_i, literal x_i or $\neg x_i$, clause, unit clause and so on). We just recall the global schema of CDCL solvers [5,10]: a typical branch can be seen as a sequence of decisions (usually with the VSIDS heuristic) followed by propagations, repeated until a conflict is reached. Each decision literal is assigned at its own level shared with all propagated literals assigned at the same level (a block is defined as all literals which are assigned at the same decision level). Each time a conflict is reached, a *nogood* is extracted. The learnt clause is then added to the clause database and a *backjumping* level is computed from it. Solvers also incorporate other components such as restarts (see section 3) and a learnt clause database reduction policy.

This last component is clearly crucial to solver performance. Indeed, keeping too many learnt clauses will slow down the unit propagation process, while deleting too many of them will break the overall learning benefit. Then, more or less frequently,

---

[1] GLUCOSE 1.0 won the first prize in the category Application UNSAT, SAT Competition 2009.

[2] GLUCOSE 2.0 won the first prize in the category Application SAT+UNSAT, 2011.

[3] If we study all GLUCOSE's traces of the 2011 competition, phase 2, in the categories Applications and Crafted, GLUCOSE 2.0 learnt 973,468,489 clauses (sum over all traces) but removed 909,123,525 of them during computation, i.e. more than 93end.

half of the learnt clauses are removed from the database. Consequently, identifying good learnt clauses - relevant to the proof derivation - is clearly an important challenge. The first proposed quality measure follows the success of the activity based VSIDS heuristic. More precisely, a learnt clause is considered relevant to the proof, if it is involved more often in recent conflicts, *i.e.* usually used to derive asserting clauses. Clearly, this deletion strategy is based on the assumption that a useful clause in the past could be useful in the future. More recently, a more accurate measure called LBD (*Literal Block Distance*) was proposed to estimate the quality of a learnt clause. This measure is defined as follow [1]:

Definition 1 (Literal Block Distance (LBD)). *Given a clause c, and a partition of its literals into n subsets according to the current assignment, s.t. literals are partitioned w.r.t their level. The LBD of C is exactly n.*

Extensive experiments demonstrated that clauses with small LBD values are used more often (in propagation and conflict) than those with higher LBD values [1]. Intuitively, the LBD score measures the number of decisions that a given clause links together. It was shown that "glue clauses", *i.e.*, clauses of LBD 2, are very important. Then, this new measure can be combined with a very agressive cleaning strategy: bad clauses are often deleted.

Another important component for our purposes is the literal polarity to be chosen when the next decision variable is selected, by the VSIDS heuristic. Usually, a default polarity (e.g. false) is defined and used each time a decision literal is assigned. Based on the observation that restarting and backjumping might lead to repetitive solving the same subformulas, Pipatsrisawat and Darwiche [12] proposed to dynamically save for each variable the last seen polarity. This literal polarity based heuristic, called phase saving, prevents the solver from solving the same satisfiable subformulas several times. As we will see in the next section, this component is crucial for rapid restarts.

3   State of the (Rest)Art

In CDCL solvers, restarting is not restarting. The solver maintains all its heuristic values between them and restarts must be seen as dynamic rearrangements of variable dependencies. Additionally, hidden restarts are triggered in all solvers: when a unit clause is learnt (which is often the case in many problems, during the first conflicts), the solver usually cancels all its decisions without trying to recover them, in order to immediately consider this new fact. Thus, on problems with a lot of learnt unit clauses, the solver will restart more often, but silently.

There has been substantial previous work in this area. We only review here the most significant ones for our own purposes. In 2000 [6], the heavy-tailed distribution of back-tracking (DPLL) procedures was identified. To avoid the high variance of the runtime and improve the average performance of SAT solvers, it was already proposed to add restarts to DPLL searches. As we will see, this explanation does not hold anymore for explaining the performance of recent solvers.

In MINISAT 1.4 [5], a scheduled restart policy was incorporated, following a geometric series (common ratio of 1.5, starting after 100 conflicts). In [7] the Luby series [9] was demonstrated to be very efficient. This series follows a slow-but-exponentially increasing law like 1 1 2 1 1 2 4 1 1 2 4 8 1 1 2 1 ... Of course, in practice, this series is multiplied by a factor (in general between 64 and 256). This series is interesting because it was proven to be optimal for search algorithms in which individual runs are independent, i.e., share no information. We will discuss the problems of this restart strategy later.

It was proposed in [3] to nest two series, a geometric one and a Luby. The idea was to ensure that restarts are guaranteed to increase (allowing the search to, theoretically, be able to terminate) and that fast restarts occur more often than in the geometric series. In [2], it was proposed to postpone any scheduled restart by observing the "agility" of the solver. This measure is based on the polarity of the phase saving mechanism (see section 2). If most of the variables are forced against their saved polarity, then the restart is postponed: the solver might find a refutation soon. If polarities are stalling, the scheduled restart is triggered.

We should probably note here that the phase saving scheme [12] is crucial in the case of fast restarts. Indeed, it allows the solver to recover most of the decisions (but not necessarily in the same order) before and after the restart.

3.1   A Note on the Luby Restart Strategy

Luby restarts were shown to be very efficient even for very small values (a factor of 6 was shown to be optimal [7] in terms of number of conflicts). However, in most of the CDCL solvers using Luby restarts, the CPU time needs to be taken into account and a typical Luby factor is often between 32 and 256.

To have an intuition of the Luby behavior, if we take a constant ratio of 50 (to compare it with GLUCOSE 2.1 that will be presented in the following section), after 1,228,800 conflicts, the CDCL solver will have triggered 4095 restarts, with an average of 300 conflicts per restart and one long run of 102,400 conflicts (and two runs of 51,200 conflicts, four runs of 25,600 conflicts, ...). Thus, if a few top-level decisions of the longest run were wrong, 8% of the total effort will be lost. Considering the fact that this longest run is triggered without any special "stronger" heuristic (one can branch on very recent variables only for instance, leading this long run to a very local sub-problem), this is a risky strategy. Clearly, there is room for improvements.

4   Faster and More Reactive Restarts for UNSAT

In the first version of GLUCOSE we experimented with a restart policy based on the decrease of decision levels. The aim was to encourage the solver to minimize the number of decisions before reaching conflicts. We showed that, on many instances, the faster the decreasing is, the better the performance (this observation was the initial motivation for GLUCOSE). However, even if this strategy was used in [1], additional experiments quickly pointed out the importance of good clauses (having a low LBD score). The overall architecture of GLUCOSE was quickly fully-oriented to focus on LBD scores only, so it was natural to rely on the production of good clauses instead of the production of fewer decision levels as possible (this last measure was shown to be stalling after a while, or even constantly increasing on some particular benchmarks. Taking

Table 1. Number of solved instances with different values of X (size of the `queueLBD`). Results are refined by category of benchmarks (SAT, UNSAT) and the margin ratio (0.7 or 0.8). Tests were (here) conducted on SAT 2011 Application benchmarks only. The CPU CutOff was 900s.

					X			
K	SAT?	50	75	100	200	300	400	500
0.7	N	**89**	86	86	81	85	75	80
0.8	N	**93**	**93**	90	86	85	87	82
0.7	Y	78	81	78	77	81	77	**83**
0.8	Y	78	77	76	**79**	76	78	74

LBD scores only into account was shown to be more informative). Thus, we decided to change the restart policy of the version of GLUCOSE that participated to the SAT 2009 competition.

The idea behind our restart strategy is the following: since we want good clauses (w.r.t. LBD), we perform a restart if the last produced ones have high LBDs. To do that, if K times ($0 < K < 1$) the average of LBD scores of the last X conflicts was above the average of all LBD scores of produced clauses so far, then a restart is triggered. To be able to compute the moving average of the X last LBDs we use a bounded queue (called `queueLBD`) of size X. Of course, whenever the bounded queue is not full, we can not compute the moving average, then at least X conflicts are performed before making a restart. We produce here a sketch of the algorithm that triggers restarts.

```
// In case of conflict
compute learnt clause c;
sumLBD+= c.lbd(); conflicts++;// Used for global average
queueLBD.push(c.lbd());
if(queueLBD.isFull() && queueLBD.avg()*K>sumLBD/conflicts) {
    queueLBD.clear();
    restart();
}
```

The magic constant K, called the *margin ratio*, provides different behaviors. The larger K is, the fewer restarts are performed. In the first version of GLUCOSE, we use $X = 100$ and $K = 0.7$. These values were experimentally fixed to give good results on both SAT and UNSAT problems. This strategy was not changed in GLUCOSE 2.0.

We show now how we can improve the value of X by observing the performance of GLUCOSE on SAT/UNSAT problems. We must also say, as a preliminary, that improving SAT solvers is often a cruel world. To give an idea, improving a solver by solving at least ten more instances (on a fixed set of benchmarks of a competition) is generally showing a critical new feature. In general, the winner of a competition is decided based on a couple of additional solved benchmarks.

Table 1 provides for two margin ratios (0.7 and 0.8, chosen for their good performance), the number of solved instances from the SAT 2011 competition when changing the size of the bounded queue (X). We also need to clarify that 0.7 was used in GLUCOSE 1.0 and 2.0, while 0.8 was, also, already proposed in GLUEMINISAT [11]. As we can see, the size X of the bounded queue has a major impact on UNSAT formulae, and not much effect on SAT instances: for both values of K, the number of solved UNSAT instances are clearly decreasing, while the SAT ones are stalling. Thus, bounded queue size of 50 seems to be a very good value for UNSAT problems. In order to understand the impact of changing the bounded queue size, let us notice here that fixing X has not only the effect of restarting more or less frequently (this is a strict minimal interval length for restarts: no restart is triggered while the bounded queue is still not full), it has also an important side effect: the longer the queue is, the slower it can react to important but short variations. Large average windows can "absorb" short but high variances. Thus, a small value for X will have a more reactive behavior, leading to even more restarts.

As a short conclusion for this section, we see that, by assigning more reactive values to the restart strategy of GLUCOSE 1.0, GLUEMINISAT used very good parameters for UNSAT problems. This observation certainly explain why GLUEMINISAT won a first prize in the UNSAT category in 2011. Next, we show how we can keep these parameters while also allowing more SAT instances to be solved.

5   Blocking a Fast Restart When Approaching a Full Assignment

As mentioned above, learning good clauses should allow the solver to make fewer and fewer decisions before reaching conflicts. However, we have clearly shown that reducing the bounded queue size has no outstanding effect on SAT instances. In this section, we show how we can add a new blocking strategy for postponing restarts, when the solver is approaching a global solution in order to improve our solver on SAT instances.

The problem we have is the following. GLUCOSE is firing aggressive clause deletion (see section 2) and fast restarts. On some instances, restarts are really triggered every 50 conflicts. So, if the solver is trying to reach a global assignment, it has now only a few tries before reaching it. Additionally, the aggressive clause deletion strategy may have deleted a few clauses that are needed to reach the global assignment directly. The idea we present here is to simply delay for one turn the next restart (the bounded queue is emptied and the restart possibility will be tested only when it is full again) each time the number of total assignments are significantly above the average measured during a window of last conflicts (we chose a rather large window of 5,000 conflicts).

The total number of assignments (called trail size) is the current number of decisions plus the number of propagated literals. Let us now formalize the notion of "significantly above" expressed above. We need for this an additional margin value, that we will call R. Our idea is to empty the bounded queue of LBD (see section 4), thus postponing a possible restart, each time a conflict is reached with a trail size greater than R times the moving average of the last 5,000 conflicts trail sizes (computed using an other bounded queue called `trailQueue`). Of course, this can occur many times during an interval of restarts, and thus the bounded queue can be emptied before it is again full, or can be emptied many times during the interval. We produce here the sketch of the algorithm that blocks restarts.

```
// In case of conflict
queueTrail.push(trail.size());
if(queueLBD.isFull() && queueTrail.isFull() &&trail.size()>R*queueTrail.avg()) {
    queueLBD.clear();
}
```

Table 2. Number of solved instances with different values of R with, in parenthesis, the average number of conflicts between two restarts ("No" means no blocking). Tests were conducted on SAT 2011 Application benchmarks only. The CPU CutOff was 900s.

(X, K)	SAT?	No	1.2	1.3	1.4	1.5
				R		
$(100, 0.7)$	N	**86** (4500)	85 (4700)	**86** (3600)	83 (4600)	**86** (4400)
$(50, 0.8)$	N	93 (318)	89 (364)	93 (375)	**94** (369)	93 (400)
$(100, 0.7)$	Y	78 (8200)	77 (7700)	79 (7500)	76 (8600)	**80** (12000)
$(50, 0.8)$	Y	78 (433)	80 (710)	78 (750)	**87** (520)	83 (561)

Table 2 helps us determine the right value for R, according to the different choices of (X,K): $((100, 0.7)$ and $(50, 0.8))$ Here are a few conclusions that can be drawn from this Table:

- For UNSAT problems, we observe that the R value does not play any significant role. If we take a look on the average number of conflicts between each restart (average over all solved instances) we can notice that the size of the bounded queue increases a lot the number of conflicts between each restart. Here, however, the blocking strategy has no real impact.
- For SAT problems, we observe a few interesting things. First, varying the value of the R parameter does not impact performance when $(X, K) = (100, 0.7)$ (we also observed this on larger values than 100). Second, for the couple $(50, 0.8)$, blocking restarts seems very promising. Restarts are very aggressive and the blocking strategy it may be crucial to postpone a restart. Setting R to 1.4 is clearly the best choice. The results improve the number of solved SAT instances by solving 9 additional SAT benchmarks. We can also notice that, (1) the different values of R have an impact on the average number of conflicts between two restarts and, (2) in case of SAT instances the average number of conflicts is always greater than for UNSAT ones.

As a conclusion here, we experimentally show that playing on the R value has a great impact on SAT instances, and almost no impact on UNSAT ones. This can be observed by looking at the average number of conflicts between restarts. $R = 1.4$ allows larger windows for SAT while maintaining small ones for UNSAT.

6   Experimental Evaluation of GLUCOSE 2.1

Progresses made in the practical solving of SAT instances are constantly observed and important, even in the last few years. We illustrate here the progress made since our first release of GLUCOSE 1.0, 3 years ago. To understand the progress made, we must also notice that GLUCOSE 2.0 is based on MINISAT 2.2 (instead of MINISAT 2).

Our methodology was the following. We worked on GLUCOSE 2.1 on 2011 benchmarks only, then tested our final ideas on all benchmarks (2009+2011 benchmarks). Our aim was somehow to counterbalance the possibility that GLUCOSE 2.1 would have been specialized for 2011 benchmarks (the 2.0 version had no access to them). As we will see, we also measured an important improvement on benchmarks from the 2009 competition.

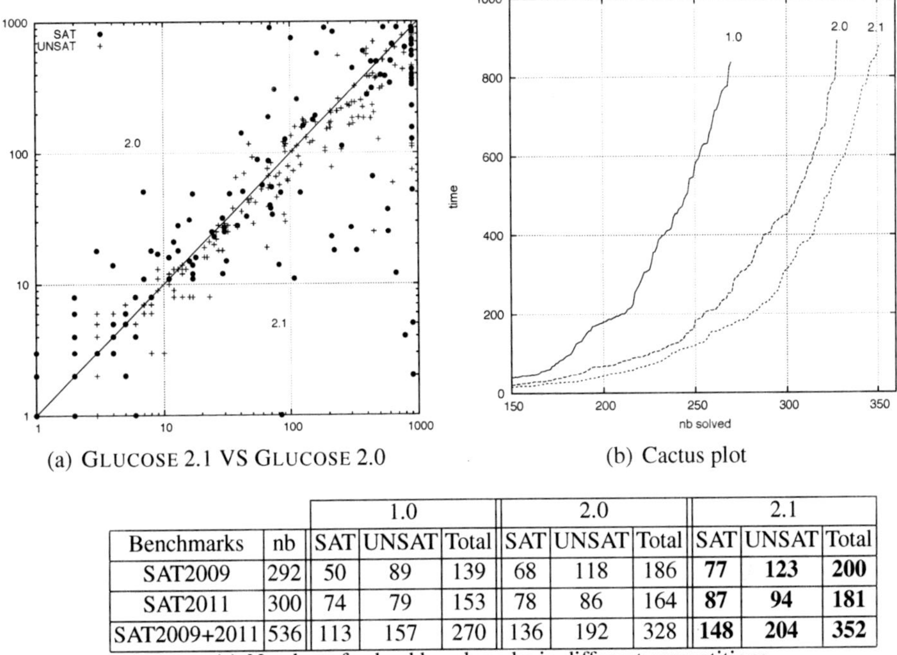

(a) GLUCOSE 2.1 VS GLUCOSE 2.0   (b) Cactus plot

Benchmarks	nb	1.0			2.0			2.1		
		SAT	UNSAT	Total	SAT	UNSAT	Total	SAT	UNSAT	Total
SAT2009	292	50	89	139	68	118	186	**77**	**123**	**200**
SAT2011	300	74	79	153	78	86	164	**87**	**94**	**181**
SAT2009+2011	536	113	157	270	136	192	328	**148**	**204**	**352**

(c) Number of solved benchmarks in different competitions

Fig. 1. a. Scatter plot: Each dot represents an instances, x-axis (resp. y-axis) the time needed by version 2.0 (resp. 2.1) to solve it. Dots below the diagonal represent instances solved faster with version 2.1 (log scale).

b. Cactus plot of 3 versions: The x-axis gives the number of solved instances and the y-axis the time needed to solve them.

c. For each competition, the number of benchmarks (nb) is provided with, for each versions of GLUCOSE, the number of solved instances.

First of all, the scatter plot in Figure 1(a) provides a graphical comparison of version 2.0 and 2.1 on all instances. Since most dots are below the diagonal, it is clear that version 2.1 is faster than 2.0. We also see on the traditional cactus plot shown in Figure 1(b) the breakthrough improvements made in the last 3 years. The novelty of GLUCOSE 2.1 vs GLUCOSE 2.0 is less important, but still significant and relies only on the improvements presented in this paper.

Finally, Figure 1(c) details the obtained results on the 2 sets of benchmarks, with the same CPU CutOff of 900 (of course benchmarks common to both sets are reported only one time in last line). Clearly, the new version of GLUCOSE, based on more aggressive restarts presented in this paper, obtained the best results.

7   Conclusion

In this paper we show how, by refining the dynamic strategy of GLUCOSE for UNSAT problems, and adding a new and simple blocking strategy to it, that is specialized for SAT problems, we are able to solve significantly more problems, more quickly. The overall idea of this paper is also to push a new vision of CDCL SAT solvers. We think they may now be closer to resolution-based producers of good clauses rather than back-track search engines.

References

1. Audemard, G., Simon, L.: Predicting learnt clauses quality in modern SAT solvers. In: Proceedings of IJCAI, pp. 399–404 (2009)
2. Biere, A.: Adaptive Restart Strategies for Conflict Driven SAT Solvers. In: Kleine Büning, H., Zhao, X. (eds.) SAT 2008. LNCS, vol. 4996, pp. 28–33. Springer, Heidelberg (2008)
3. Biere, A.: Picosat essentials. JSAT 4, 75–97 (2008)
4. Davis, M., Logemann, G., Loveland, D.: A machine program for theorem-proving. Communication of ACM 5(7), 394–397 (1962)
5. Eén, N., Sörensson, N.: An Extensible SAT-solver. In: Giunchiglia, E., Tacchella, A. (eds.) SAT 2003. LNCS, vol. 2919, pp. 502–518. Springer, Heidelberg (2004)
6. Gomes, C., Selman, B., Crato, N., Kautz, H.: Heavy-tailed phenomena in satisfiability and constraint satisfaction problems. Journal of Automated Reasoning 24, 67–100 (2000)
7. Huang, J.: The effect of restarts on the efficiency of clause learning. In: Proceedings of IJCAI, pp. 2318–2323 (2007)
8. Le Berre, D., Jarvisalo, M., Roussel, O., Simon, L.: SAT competitions (2002-2011), http://www.satcompetition.org/
9. Luby, M., Sinclair, A., Zuckerman, D.: Optimal speedup of las vegas algorithms. In: Israel Symposium on Theory of Computing Systems, pp. 128–133 (1993)
10. Moskewicz, M., Madigan, C., Zhao, Y., Zhang, L., Malik, S.: Chaff: Engineering an efficient SAT solver. In: Proceedings of DAC, pp. 530–535 (2001)
11. Nabeshima, H., Iwanuma, K., Inoue, K.: Glueminisat2.2.5. System Desciption, http://glueminisat.nabelab.org
12. Pipatsrisawat, K., Darwiche, A.: A Lightweight Component Caching Scheme for Satisfiability Solvers. In: Marques-Silva, J., Sakallah, K.A. (eds.) SAT 2007. LNCS, vol. 4501, pp. 294–299. Springer, Heidelberg (2007)
13. Silva, J.M., Sakallah, K.: Grasp - a new search algorithm for satisfiability. In: ICCAD, pp. 220–227 (1996)

Boosting Local Consistency Algorithms over Floating-Point Numbers*

Mohammed Said Belaid, Claude Michel, and Michel Rueher

I3S (UNS/CNRS)
2000, route des Lucioles - Les Algorithmes - bât. Euclide B - BP 121
06903 Sophia Antipolis Cedex, France
{MSBelaid,Claude.Michel}@i3s.unice.fr, Michel.Rueher@gmail.com

Abstract. Solving constraints over floating-point numbers is a critical issue in numerous applications notably in program verification. Capabilities of filtering algorithms over the floating-point numbers ($\mathcal{F}$) have been so far limited to 2b-consistency and its derivatives. Though safe, such filtering techniques suffer from the well known pathological problems of local consistencies, e.g., inability to efficiently handle multiple occurrences of the variables. These limitations also have their origins in the strongly restricted floating-point arithmetic. To circumvent the poor properties of floating-point arithmetic, we propose in this paper a new filtering algorithm, called FPLP, which relies on various relaxations over the real numbers of the problem over $\mathcal{F}$. Safe bounds of the domains are computed with a mixed integer linear programming solver (MILP) on safe linearizations of these relaxations. Preliminary experiments on a relevant set of benchmarks are promising and show that this approach can be effective for boosting local consistency algorithms over $\mathcal{F}$.

1   Introduction

Critical systems are more and more relying on floating-point (FP) computations. For instance, embedded systems are typically controlled by software that store measurements and environment data as floating-point number ($\mathcal{F}$). The initial values and the results of all operations must therefore be rounded to some nearby float. This rounding process can lead to significant changes, and, for example, can modify the control flow of the program. Thus, the verification of programs performing FP computations is a key issue in the development of critical systems.

Methods for verifying programs performing FP computations are mainly derived from standard program verification methods. Bounded model checking (BMC) techniques have been widely used for finding bugs in hardware design [3] and software [11]. SMT solvers are now used in most of the state-of-the-art BMC tools to directly work on high level formula (see [2,9,11]). The bounded model checker CBMC encodes each FP operation of the program with a set of logic

---

* This work was partially supported by ANR VACSIM (ANR-11-INSE-0004), ANR AEOLUS (ANR-10-SEGI-0013) and OSEO ISI PAJERO projects.

M. Milano (Ed.): CP 2012, LNCS 7514, pp. 127–140, 2012.

functions on bit-vectors which requires thousands of additional variables and becomes quickly intractable [6]. Tools based on abstract interpretation [10,22] can show the absence of run-time errors (e.g., division by zero) on program working with FP numbers. Tools based on abstract interpretation are safe since they over-approximate FP computations. However, over-approximations may be very large and these tools may generate many false alerts, and thus reject many valid programs. For instance, Chen's polyhedral abstract domains [7] rely on coarse approximations of floating-point operations that do not take advantage of the rounding mode. Constraint programming (CP) has also been used for program testing [13,14] and verification [8]. CP offers many benefits like the capability to deduce information from partially instantiated problems or to exhibit counterexamples. The CP framework is very flexible and simplifies the integration of new solvers for handling a specific domain, for instance FP solvers. However, it is important to understand that solvers over real numbers ($\mathcal{R}$) cannot correctly handle FP arithmetic. Dedicated constraint solvers are required in safe CP-based framework and BMC-SMT tools for testing or verifying numerical software[1].

Techniques to solve FP constraints are based on adaptations of classical consistencies (e.g. box-consistency and 2B-consistency) over $\mathcal{R}$ [21], [20,5]. However FP solvers based on these techniques do not really scale up to large constraint systems. That is why we introduce here a new method to handle constraints over the FP numbers by taking advantage of solvers over $\mathcal{R}$. The basic tenet is to build correct but tight relaxations over $\mathcal{R}$ of the FP operations. To ensure the tightness of the result, each FP operation is approximated according to its rounding mode. For example, assume that x and y are positive normalized FP numbers[2], then the FP product $x \otimes y$ with a rounding mode set to $-\infty$, is bounded by $\alpha \times (x \times y) < x \otimes y \leq x \times y$ where $\alpha = 1/(1 + 2^{-p+1})$ and p is the size of the significand. Approximations for special cases have also been refined, e.g., for the addition with a rounding mode set to zero, or for the multiplication by a constant.

Using these relaxations, a problem over the FP numbers is first translated into a set of nonlinear constraints over $\mathcal{R}$. A linearization of the nonlinear constraints is then applied to obtain a mixed integer linear problem (MILP) over $\mathcal{R}$. In this process, binary variables are used to handle concave domains to prevent too loose over-approximations. This last set of constraints can directly be solved by available MILP solvers over $\mathcal{R}$ which are relieved from the drawbacks of FP arithmetic. Efficient MILP solvers rely on FP computations and thus, might miss some solutions. In order to ensure a safe behavior of our algorithm, correct rounding directions are applied to the relaxation coefficients [19,4] and a

---

[1] See FPSE (`http://www.irisa.fr/celtique/carlier/fpse.html`), a solver for FP constraints coming from C programs.

[2] A FP number is a triple (s, e, m) where s is the sign, e the exponent and m the significand. Its value is given by $(-1)^s \times 1.m \times 2^e$. r and p are the size of the exponent and the significand. The IEEE standard 754 defines the single format with $(r, p) = (8, 23)$ and the double format with $(r, p) = (11, 52)$. A normalized FP number's significand has no non-zero digits to the left of the decimal point and a non-zero digit just to the right of the decimal point.

procedure [23] to compute a safe minimizer from the unsafe result of the MILP solver is also applied. Preliminary experiments are promising and this new filtering technique should really help to scale up all verifications tools that uses a FP solver.

Our method relies on a high level representation of the FP operations and, thus, does not suffer from the same drawbacks than bit vector encoding. The bit vector encoding used in CBMC generates thousands of additional binary variables for each FP operation of the program. For example, an addition of two 32 bits floats requires 2554 binary variables [6]. The mixed approximations proposed in [6] reduce the number of additional binary variables significantly but the resulting system remains expensive in memory consumption. For instance, a single addition with only 5 bits of precision still requires 1035 additional variables. Our method does also generate additional variables: temporary variables are used to decompose complex expressions into elementary operations over the FP numbers and some binary variables are used to handle the different cases of our relaxations. However, the number of generated variables is negligible compared to the ones required by a bit vector encoding.

1.1   An Illustrative Example

Before going into the details, let us illustrate our approach on a very simple example. Consider the simple constraint

$$z = x \oplus y \ominus x \tag{1}$$

where x, y and z are 32 bits FP variables, and $\oplus$ and $\ominus$ are the addition and the subtraction over $\mathcal{F}$, respectively. Over the real numbers, such an expression can be simplified to $z = y$. However, this is not true with FP numbers. For example, over $\mathcal{F}$ and with a rounding mode set to the nearest, $10.0 \oplus 10.0^{-8} \ominus 10.0$ is not equal to 10.0^{-8} but to 0. This absorption phenomenon illustrates why expressions over the FP numbers cannot be simplified in the same way than expressions over the real numbers.

Now, let us assume that $x \in [0.0, 10.0]$, $y \in [0.0, 10.0]$ and $z \in [0.0, 10.0^8]$. FP2B, a 2B-consistency [16] algorithm adapted to FP constraints [20], first performs forward propagation of the domains of x and y on the domain of z using an interval arithmetic where interval bounds are computed with a rounding mode set to the nearest. Backward propagation being of no help here, the filtering process yields:

$$x \in [0.0, 10.0], \ y \in [0.0, 10.0], \ z \in [0.0, 20.0]$$

This poor filtering is due to the fact that 2B-consistency algorithms cannot handle efficiently constraints with multiple occurrences of the variables. A stronger consistency like 3B-consistency [16] will reduce the domain of z to the interval $[0.0, 10.01835250854492188]$. However, 3B-consistency will fail to reduce the domain of z when x and y occur more than two times, like in $z = x \oplus y \ominus x \ominus y \oplus x \oplus y \ominus x$.

Algorithm FPLP, introduced in this paper, first builds safe nonlinear relaxations over $\mathcal{R}$ of the constraints over $\mathcal{F}$ derived from the program. Of course, these relaxations are computed according to the rounding mode. Applied to constraint (1), it yields the following relaxations over $\mathcal{R}$:

$$\begin{cases} (1 - \frac{2^{-p}}{(1-2^{-p})})(x + y) \leq tmp1 \\ tmp1 \leq (1 + \frac{2^{-p}}{(1+2^{-p})})(x + y) \\ (1 - \frac{2^{-p}}{(1-2^{-p})})(tmp1 - x) \leq tmp2 \\ tmp2 \leq (1 + \frac{2^{-p}}{(1+2^{-p})})(tmp1 - x) \\ z = tmp2 \end{cases}$$

where p is the size of the significand of the FP variables. $tmp1$ approximates the result of the operation $x \oplus y$ by means of two planes over $\mathcal{R}$ which encompass all the results of this addition over $\mathcal{F}$. $tmp2$ does the same for the subtraction. Some relaxations, like the one of the product, include nonlinear terms. In such a case, a linearization process is applied to get a MILP. Once the problem is fully linear, a MILP solver is used to shrink the domain of each variable, respectively, minimizing and maximizing it.

FPLP, which stands for Floating-Point Linear Program, implements the algorithm previously sketched. A call to FPLP on constraint (1) immediately yields:

$$x \in [0, 10], \ y \in [0, 10], \ z \in [0, 10.0000023841859]$$

which is a much tighter result than the one computed by FP2B. Contrary to 3B-consistency, FPLP still gives the same result with FPLP provides the same result for constraint $z = x \oplus y \ominus x \ominus y \oplus x \oplus y \ominus x$ whereas 3B-consistency cannot reduce the upper bound of z on the latter constraint.

1.2   Outline of the Paper

The rest of this paper is organized as follows: the next section introduces the nonlinear relaxations over $\mathcal{R}$ of the constraints over $\mathcal{F}$. The following section shows how the nonlinear terms of the relaxations are linearized. Then, the filtering algorithm is detailed and the results of our experiments are given before concluding the paper.

2   Relaxations of FP Constraints

This section introduces nonlinear relaxations over $\mathcal{R}$ of the FP constraints from the initial problem. These relaxations are the cornerstone of the filtering process described in this paper. They must be *correct*, i.e., they must preserve the whole set of solutions of the initial problem, and *tight*, i.e., they should enclose the smallest amount of non FP solutions.

These relaxations are built using two techniques: the *relative error* and the *correctly rounded* operations. The former is a technique frequently used to analyze the precision of the computation. The latter property is ensured by any

IEEE 754 compliant implementation of the FP arithmetic: a correctly rounded operation is an operation whose result over $\mathcal{F}$ is equal to the rounding of the result of the equivalent operation over $\mathcal{R}$. In other word, let x and y be two FP numbers, $\odot$ and $\cdot$, respectively, an operation over $\mathcal{F}$ and its equivalent over $\mathcal{R}$, if $\odot$ is correctly rounded then, $x \odot y = round(x \cdot y)$.

In the rest of this section, we first detail how to build these relaxations for a specific case before defining the relaxations in the general cases. Then, we will show how the different cases can be simplified.

2.1   A Specific Case

In order to explain how these relaxations are built, let us consider the case where an operation is computed with a rounding mode set to $-\infty$ and the result of this operation is a positive and normalized FP number. Such an operation, denoted $\odot$, could be any of the four basic binary operations from the FP arithmetic. The operands are all supposed to have the same FP type, i.e., either float, double or long double. Then, the following property holds:

Proposition 1. *Let x and y be two FP numbers whose significand is represented by p bits. Assume that the rounding mode is set to $-\infty$ and that the result of $x \odot y$ is a normalized positive FP numbers smaller than max_f, the biggest FP number, then the following property holds:*

$$\frac{1}{1 + 2^{-p+1}}(x \cdot y) < x \odot y \leq (x \cdot y)$$

where $\odot$ is a basic operation over the FP numbers and, $\cdot$ is the equivalent operation over the real numbers.

Proof. Since IEEE 754 basic operations are correctly rounded and the rounding mode is set to $-\infty$, we have:

$$x \odot y \leq x \cdot y < (x \odot y)^+ \tag{2}$$

$(x \odot y)^+$, the successor of $(x \odot y)$ within the set of FP numbers, can be computed by

$$(x \odot y)^+ = (x \odot y) + ulp(x \odot y)$$

as, ulp, which stands for unit in the last place, is defined by $ulp(x) = x^+ - x$. Thus, it results from (2) that

$$x \odot y \leq x \cdot y < (x \odot y) + ulp(x \odot y)$$

From the second inequality, we have

$$\frac{1}{x \odot y + ulp(x \odot y)} < \frac{1}{x \cdot y}$$

By multiplying each side of the inequality by $x \odot y$ – which is a positive number – we get

$$\frac{x \odot y}{x \odot y + ulp(x \odot y)} < \frac{x \odot y}{x \cdot y}$$

By multiplying each side of the above inequality by -1 and by adding one to each side, we obtain

$$1 - \frac{x \odot y}{x \cdot y} < 1 - \frac{x \odot y}{x \odot y + ulp(x \odot y)} = \frac{ulp(x \odot y)}{x \odot y + ulp(x \odot y)} \tag{3}$$

Now, consider ϵ, the relative error defined by

$$\epsilon = \left| \frac{real\_value - float\_value}{real\_value} \right|$$

ϵ is the absolute value of the difference between the result over $\mathcal{R}$ and the result over $\mathcal{F}$ divided by the result over $\mathcal{R}$. In the considered case, the result of $x \odot y$ being a positive normalized floating-point number and $x \cdot y \geq x \odot y$, the relative error is given by

$$0 \leq \epsilon = \frac{x \cdot y - x \odot y}{x \cdot y} = 1 - \frac{x \odot y}{x \cdot y}$$

Thus, thanks to (3), we have

$$0 \leq \epsilon < \frac{ulp(x \odot y)}{x \odot y + ulp(x \odot y)}$$

z, the result of the operation $x \odot y$, is a binary positive and normalized FP number that can be written $z = 1.m_z 2^{e_z}$, where m_z has p bits. Moreover, $ulp(z) = 2^{-p+1} 2^{e_z}$. Therefore,

$$0 \leq \epsilon < \frac{2^{-p+1} 2^{e_z}}{m_z 2^{e_z} + 2^{-p+1} 2^{e_z}} = \frac{2^{-p+1}}{m_z + 2^{-p+1}}$$

The value of the significand of a normalized FP number belongs to the interval $[1.0, 2.0[$. An upper bound of the relative error ϵ is given by the minimum of $m_z + 2^{-p}$ which is reached when $m_z = 1$. Thus

$$0 \leq \epsilon < \frac{2^{-p+1}}{1 + 2^{-p+1}}$$

Since we have

$$\epsilon = \frac{x \cdot y - x \odot y}{x \cdot y}$$

we have

$$0 \leq \frac{x \cdot y - x \odot y}{x \cdot y} < \frac{2^{-p+1}}{1 + 2^{-p+1}}$$

and

$$0 \leq x \cdot y - x \odot y < (x \cdot y) \frac{2^{-p+1}}{1 + 2^{-p+1}}$$

By multiplying each side of the inequality by -1 and adding $x \cdot y$ to each side, we finally obtain

$$\frac{1}{1 + 2^{-p+1}} (x \cdot y) < x \odot y \leq x \cdot y \qquad \square$$

Table 1. Relaxations of $x \odot y$ for each rounding mode where $z_r = x \cdot y$

Rounding mode	Negative normalized	Negative denormalized	Positive denormalized	Positive normalized
to $-\infty$	$[(1 + 2^{-p+1})z_r, z_r]$	$[z_r - min_f, z_r]$	$[z_r - min_f, z_r]$	$[\frac{1}{(1+2^{-p+1})}z_r, z_r]$
to $+\infty$	$[z_r, \frac{1}{(1+2^{-p+1})}z_r]$	$[z_r, z_r + min_f]$	$[z_r, z_r + min_f]$	$[z_r, (1 + 2^{-p+1})z_r]$
to 0	$[z_r, \frac{1}{(1+2^{-p+1})}z_r]$	$[z_r - min_f, z_r]$	$[z_r, z_r + min_f]$	$[\frac{1}{(1+2^{-p+1})}z_r, z_r]$
to nearest	$[(1 + \frac{2^{-p}}{(1+2^{-p})})z_r,$ $(1 - \frac{2^{-p}}{(1-2^{-p})})z_r]$	$[z_r - \frac{min_f}{2},$ $z_r + \frac{min_f}{2}]$	$[z_r - \frac{min_f}{2}$ $z_r + \frac{min_f}{2}]$	$[(1 - \frac{2^{-p}}{(1-2^{-p})})z_r,$ $(1 + \frac{2^{-p}}{(1+2^{-p})})z_r]$

2.2 Generalization

Table 1 summarizes the relaxations for each rounding mode in the different cases, i.e., positive or negative FP numbers, as well as, normalized and denormalized FP numbers. Each case has a dedicated correct and tight approximation built in a way similar to the one of the case detailed in the previous subsection.

Note that tighter approximations for specific cases could also be computed. For example, the approximation of an addition with a rounding mode sets to $\pm\infty$ could be slightly improved. In a similar way, the structure of the problem is another source of improvements of the approximations. For example, $2 \otimes x$ being exactly computed[3], it can directly be evaluated over $\mathcal{R}$.

2.3 Simplified Relaxations

The main issue with the previous relaxations is that the solving process will have to handle the different cases. As a result, for n basic operations, the solver has to deal with 4^n potential combinations of the relaxations. To decrease substantially this complexity, we provide here a combination of the four cases of each rounding mode into a single case.

Let us first consider the case where the rounding mode is set to $-\infty$:

Proposition 2. *Let x and y be two FP numbers whose significand size is p and, assume that the rounding mode is set to $-\infty$ and, that $-max_f < x \odot y < max_f$, then,*

$$z_r - 2^{-p+1}|z_r| - min_f \leq x \odot y \leq z_r$$

where min_f is the smallest positive FP number, $\odot$ and $\cdot$ are respectively a basic binary operation over $\mathcal{F}$ and its equivalent over $\mathcal{R}$, and $z_r = x \cdot y$.

Proof. In a first step, the normalized and denormalized approximations are combined. If $z_r > 0$ then $\frac{1}{1+2^{-p+1}}z_r < z_r$. Thus,

$$\frac{1}{1 + 2^{-p+1}}z_r - min_f < z_r - min_f$$

---
[3] Provided that no overflow occurs.

134   M.S. Belaid, C. Michel, and M. Rueher

Table 2. Simplified relaxations of $x \odot y$ for each rounding mode (with $z_r = x \cdot y$)

Rounding mode	The approximation of $x \odot y$
to $-\infty$	$[z_r - 2^{-p+1}|z_r| - min_f, z_r]$
to $+\infty$	$[z_r, z_r + 2^{-p+1}|z_r| + min_f]$
to 0	$[z_r - 2^{-p+1}|z_r| - min_f,$ $z_r + 2^{-p+1}|z_r| + min_f]$
to the nearest	$[z_r - \frac{2^{-p}}{(1-2^{-p})}|z_r| - \frac{min_f}{2},$ $z_r + \frac{2^{-p}}{(1-2^{-p})}|z_r| + \frac{min_f}{2}]$

and

$$\frac{1}{1+2^{-p+1}} z_r - min_f < \frac{1}{1+2^{-p+1}} z_r$$

Therefore,

$$\frac{1}{1+2^{-p+1}} z_r - min_f < x \odot y \le z_r, \; z_r \ge 0$$

When $z_r \le 0$, we get

$$(1 + 2^{-p+1}) z_r - min_f < x \odot y \le z_r, \; z_r \le 0$$

These two approximations can be rewritten as follows,

$$\begin{cases} z_r - \frac{2^{-p+1}}{1+2^{-p+1}} z_r - min_f < x \odot y \le z_r, z_r \ge 0 \\ z_r + 2^{-p+1} z_r - min_f < x \odot y \le z_r, \quad z_r \le 0 \end{cases}$$

To combine the negative and positive approximations together we can use the absolute value:

$$\begin{cases} z_r - \frac{2^{-p+1}}{1+2^{-p+1}} |z_r| - min_f < x \odot y \le z_r, z_r \ge 0 \\ z_r - 2^{-p+1} |z_r| - min_f < x \odot y \le z_r, \quad z_r \le 0 \end{cases}$$

As $max\{\frac{2^{-p+1}}{1+2^{-p+1}}, 2^{-p+1}\} = 2^{-p+1}$, we get

$$z_r - 2^{-p+1} |z_r| - min_f \le x \odot y \le z_r \qquad \square$$

The same reasoning holds for other rounding modes. Table 2 summarizes the simplified relaxations for each rounding mode. Note that these approximations define concave sets.

3   Linearization of the Relaxations

The relaxations introduced in the previous section contain nonlinear terms that cannot be directly handled by a MILP solver. In this section, we describe how these terms are approximated by sets of linear constraints.

3.1   Absolute Value Linearization

Simplified relaxations that allow to handle all numerical FP values with a single set of two inequalities require absolute values. Absolute values can either be loosely approximated by three linear inequalities or by a tighter decomposition based on big M rewriting method:

$$\begin{cases} z = z_p - z_n \\ |z| = z_p + z_n \\ 0 \leq z_p \leq M \times b \\ 0 \leq z_n \leq M \times (1-b) \end{cases}$$

where b is a boolean variable, z_p and z_n are real positive variables and, M is a FP number such that $M \geq max\{|\underline{z}|, |\overline{z}|\}$. The method separates z_p, the positive values of z, from z_n, its negative values. When $b = 1$, z gets its positive values and we have $z = z_p = |z|$. If $b = 0$, z gets its negative values and we have $z = -z_n$ and $|z| = z_n$.

If the underlying MILP solver allows indicator constraints, the two last set of inequalities can be replaced by:

$$\begin{cases} b = 0 \rightarrow z_p = 0 \\ b = 1 \rightarrow z_n = 0 \end{cases}$$

3.2   Linearization of Nonlinear Operations

Bilinear terms, square terms, and quotient linearizations are based on standard techniques used by Sahinidis et al [24]. They have been also used in the Quad system [15] designed to solve constraints over the real numbers. $x \times y$ is linearized according to Mc Cormick [18]:

Let $x \in [\underline{x}, \overline{x}]$ and $y \in [\underline{y}, \overline{y}]$, then

$$\begin{cases} z - \underline{x}y - \underline{y}x + \underline{x}\underline{y} \geq 0 \\ -z + \underline{x}y + \overline{y}x - \underline{x}\overline{y} \geq 0 \\ -z + \overline{x}y + \underline{y}x - \overline{x}\underline{y} \geq 0 \\ z - \overline{x}y - \overline{y}x + \overline{x}\overline{y} \geq 0 \end{cases}$$

These linearizations have been proved to be optimal by Al-Khayyal and Falk [1].

Each time $x = y$, i.e., in case of $z = x \otimes x$, the linearization can be improved. x^2 convex hull is underestimated by all the tangents at x^2 curve between $\underline{x}$ and $\overline{x}$ and overestimated by the line that join $(\underline{x}, \underline{x}^2)$ to $(\overline{x}, \overline{x}^2)$. A good balance is obtained with the two tangents at the bounds of x. Thus, x^2 linearization yields:

$$\begin{cases} z + \underline{x}^2 - 2\underline{x}x \geq 0 \\ z + \overline{x}^2 - 2\overline{x}x \geq 0 \\ (\underline{x} + \overline{x})x - z - \underline{x}\overline{x} \geq 0 \\ z \geq 0 \end{cases}$$

The division takes advantage of the properties of real arithmetic: the essential observation is that $z = x/y$ is equivalent to $x = z \times y$. Therefore, Mc Cormick [18] linearizations can be used here. These linearizations need the bounds of z which can directly be computed by interval arithmetic:

$$[\underline{z}, \overline{z}] = [\nabla(min(\underline{x}/\underline{y}, \underline{x}/\overline{y}, \overline{x}/\underline{y}, \overline{x}/\overline{y})),$$
$$\Delta(max(\underline{x}/\underline{y}, \underline{x}/\overline{y}, \overline{x}/\underline{y}, \overline{x}/\overline{y}))]$$

where ∇ and Δ are respectively the rounding modes towards $-\infty$ and $+\infty$.

4   Filtering Algorithm

The proposed filtering algorithm relies on the linearizations of the relaxations over $\mathcal{R}$ of the initial problem to attempt to shrink the domain of the variables by means of a MILP solver. Algorithm 1 details the steps of this filtering process.

First, function **Approximate** relaxes initial FP constraints to nonlinear constraints over $\mathcal{R}$. Then, function **Linearize** linearizes the nonlinear terms of these relaxations to get a MILP.

The filtering loop starts with a call to FP2B, a filtering process relying on an adaptation of 2B-consistency to FP constraints that attempts to reduce the bounds of the variables. FP2B propagates bound values to intermediate variables. The cost of this filtering process is quite light: it stops as soon as domain size reduction between two iterations is less than 10%. Thanks to function **UpdateLinearizations**, newly computed bounds are used to tighten the MILP. Note that this function updates variable domains as well as linearization coefficients.

After that, MILP is used to compute a lower bound and an upper bound of the domain of each variable by means of function **safeMin**. This function computes a safe global minimizer of the MILP.

This process is repeated until the percentage of reduction of the domains of the variables is lower than a given ϵ.

4.1   Getting a Safe Minimizer

Using an efficient MILP solver like CPLEX to filter the domains of the variables raises two important issues related to FP computations.

First, linearization coefficients are computed with FP arithmetic and are subject to rounding errors. Therefore, to avoid the loss of solutions, special attention must be paid to the rounding directions. Correct linearizations rely on FP computations done using the right rounding directions. For instance, consider the linearization of x^2 where $\underline{x} \geq 0$ and $\overline{x} \geq 0$:

$$\begin{cases} y + \Delta(\underline{x}^2) - \Delta(2\underline{x})x \geq 0 \\ y + \Delta(\overline{x}^2) - \Delta(2\overline{x})x \geq 0 \\ \Delta(\underline{x} + \overline{x})x - y - \nabla(\underline{x}\overline{x}) \geq 0 \\ y \geq 0 \end{cases}$$

Algorithm 1. FPLP

```
 1: Function FPLP(V, D, C, ε)
 2: % V: FP variables
 3: % D: Domains of the variables
 4: % C: Constraints over FP numbers
 5: % ε: Minimal reduction between two iterations
 6: C' ← Approximate (C);
 7: C'' ← Linearize (C', D);
 8: boxSize ← ∑ₓ∈V (x̄_D − x_D);
 9: repeat
10:     D' ← FP2B(V, D, C, ε);
11:     if ∅ ∈ D' then
12:        return ∅;
13:     end if
14:     C'' ← UpdateLinearizations(C'', D');
15:     for all x ∈ V do
16:        [x_D', x̄_D'] ← [safeMin(x, C''), −safeMin(−x, C'')];
17:        if [x_D', x̄_D'] = ∅ then
18:           return ∅;
19:        end if
20:     end for
21:     oldBoxSize ← boxSize;
22:     boxSize ← ∑ₓ∈V (x̄_D' − x_D');
23:     D ← D'
24: until boxSize ≥ oldBoxSize * (1 − ε);
25: return D;
```

This process that ensures that all the linearizations are safe is called within the **Linearize** and **UpdateLinearizations** functions. For more details on how to compute safe coefficients see [19,4].

Second, efficient MILP solvers use FP arithmetic. Thus, the computed minimizer might be wrong. The unsafe MILP solver is made safe thanks to the correction procedure introduced in [23]. It consists in computing a safe lower bound of the global minimizer. The **safeMin** function implements these corrections and return a safe minimizer of the MILP.

5   Experiments

This section compares the results of different filtering techniques for FP constraints with the method introduced in this paper. Experiments have been done on a laptop with an Intel Duo Core at 2.8Ghz and 4Gb of memory running under Linux.

Our experiments are based on the following set of benchmarks:

- `Absorb 1` detects if, in a simple addition, x absorbs y while `Absorb 2` checks if y absorbs x.

Table 3. Experiments

Program	n	n_T	n_B	2B $t(ms)$	3B $t(ms)$	%(2B)	FPLP (without 2B) $t(ms)$	%(2B)	FPLP $t(ms)$	%(2B)
Absorb1	2	1	1	TO	TO	-	3	98.91	5	98.91
Absorb2	2	1	1	1	24	0.00	3	100.00	4	100.00
Fluctuat1	3	12	2	4	156	99.00	264	99.00	172	99.00
Fluctuat2	3	10	2	1	4	0.00	29	0.00	21	0.00
MeanValue	4	28	6	3	82	97.45	530	97.46	78	97.46
Cosine	5	33	7	5	153	33.60	104	33.61	43	33.61
SqrtV1	11	140	29	9	27198	99.63	1924	100.00	1187	100.00
SqrtV2	21	80	17	7	TO	-	2337	100.00	1321	100.00
SqrtV3	5	46	8	5	573	53.80	185	54.83	82	54.83
Sine taylor	6	44	9	5	452	63.29	313	63.29	227	63.29
Sine iter	16	109	21	8	4503	39.20	5885	39.31	165	39.31
Qurt	6	21	3	4	26	43.56	163	43.56	38	43.56
Poly	6	51	9	5	1569	49.17	765	76.66	309	76.66
Newton	7	69	14	5	1542	45.16	479	45.16	195	45.16

- Fluctuat1 and Fluctuat2 are program pathes that come from a presentation of the Fluctuat tool in [12].
- MeanValue returns true if an interval contains a mean value and false otherwise.
- Cosine is a program that computes the function $cos()$ with a Taylor formula.
- SqrtV1 computes sqrt in $[0.5, 2.5]$ using a two variable iterative method.
- SqrtV2 computes sqrt with a Taylor formula.
- SqrtV3 computes the square root of $(x + 1)$ using a Taylor formula. This program comes from CDFPL benchmarks[4].
- Sine taylor computes the function sine using a Taylor formula.
- Sine iter computes the function sine with an iterative method and comes from the SNU real time library[5].
- Qurt computes the real and imaginary roots of a quadratic equation and also comes from the SNU library.
- Poly tries to compare two different writings of a polynomial. This program is available on Eric Goubault web page[6]
- Newton computes one or two iterations of a Newton on the polynomial $x - x^3/6 + x^5/120x^7/5040$ and comes from CDFPL benchmarks.

Table 3 summarizes experiment results for the following filtering methods: FP2B, an adaptation of 2B-consistency to FP constraints that takes advantage of the

---

[4] See http://www.cprover.org/cdfpl/.
[5] See http://archi.snu.ac.kr/realtime/
[6] See http://www.lix.polytechnique.fr/~goubault/.

property described in [17] to avoid some slow convergences, FP3B, an adaptation of 3B-consistency to FP constraints, FPLP(without FP2B), an implementation of algorithm 1 without the call to FP2B and, FPLP, an implementation of algorithm 1. First column of table 3 gives program's names, column 2 gives the number of variables of the initial problem and column 3 gives the amount of temporary variables used to decompose complex expressions in elementary operations. Column 4 gives the number of binary variables used by FPLP. For each filtering algorithm, table 3 gives the amount of milliseconds required to filter the constraints (columns $t(ms)$). For all filtering algorithm but FP2B, table 3 gives also the percentage of reduction compared to the reduction obtained by FP2B (columns %(FP2B)). The time out (TO) was set to 2 minutes.

The results from table 3 show that FPLP achieves much better domain reductions than 2B-consistency and 3B-consistency filtering algorithms. FPLP requires more times than FP2B but the latter achieves a very weaker pruning on theses benchmarks. This is exemplified by the two `Absorb1` and `SqrtV1` benches. Here, FP2B suffers from the multiple occurrences of the variables. FPLP also consistently outperforms FP3B : it almost always provides much smaller domains and it requires much less time.

A comparison of FPLP with and without a call to FP2B shows that a cooperation between these two filtering methods can significantly decrease the computation time but does not change the filtering capabilities.

6   Conclusion

In this paper, we have introduced a new filtering algorithm for handling constraints over FP numbers. This algorithm benefits from the linearizations of the relaxations over $\mathcal{R}$ of the initial constraints over $\mathcal{F}$ to reduce the domains of the variables with a MILP solver. Experiments show that FPLP drastically improves the filtering process, especially when combined with a FP2B filtering process. MILP benefits from a more global view of the constraint system than local consistencies, and thus provides an effective way to handle multiple occurrences of variables.

Additional experiments are required to better understand the interactions between the two algorithms and to improve their performances.

References

1. Al-Khayyal, F.A., Falk, J.E.: Jointly constrained biconvex programming. Mathematics of Operations Research 8(2), 273–286 (1983)
2. Armando, A., Mantovani, J., Platania, L.: Bounded model checking of software using SMT solvers instead of SAT solvers. Int. J. Softw. Tools Technol. Transf. 11, 69–83 (2009)
3. Biere, A., Cimatti, A., Clarke, E., Zhu, Y.: Symbolic Model Checking without BDDs. In: Cleaveland, W.R. (ed.) TACAS 1999. LNCS, vol. 1579, pp. 193–207. Springer, Heidelberg (1999)
4. Borradaile, G., Van Hentenryck, P.: Safe and tight linear estimators for global optimization. Mathematical Programming (2005)

5. Botella, B., Gotlieb, A., Michel, C.: Symbolic execution of floating-point computations. Softw. Test., Verif. Reliab. 16(2), 97–121 (2006)
6. Brillout, A., Kroening, D., Wahl, T.: Mixed abstractions for floating-point arithmetic. In: Proceedings of FMCAD 2009, pp. 69–76. IEEE (2009)
7. Chen, L., Miné, A., Cousot, P.: A Sound Floating-Point Polyhedra Abstract Domain. In: Ramalingam, G. (ed.) APLAS 2008. LNCS, vol. 5356, pp. 3–18. Springer, Heidelberg (2008)
8. Collavizza, H., Rueher, M., Hentenryck, P.: CPBPV: a constraint-programming framework for bounded program verification. Constraints 15(2), 238–264 (2010)
9. Cordeiro, L., Fischer, B., Marques-Silva, J.: SMT-based bounded model checking for embedded ANSI-C software. IEEE Transactions on Software Engineering (May 2011)
10. Cousot, P., Cousot, R., Feret, J., Miné, A., Mauborgne, L., Monniaux, D., Rival, X.: Varieties of static analyzers: A comparison with astree. In: TASE 2007, pp. 3–20. IEEE (2007)
11. Ganai, M.K., Gupta, A.: Accelerating high-level bounded model checking. In: Proceedings of the 2006 IEEE/ACM International Conference on Computer-Aided Design, ICCAD 2006, pp. 794–801. ACM, New York (2006)
12. Ghorbal, K., Goubault, E., Putot, S.: A Logical Product Approach to Zonotope Intersection. In: Touili, T., Cook, B., Jackson, P. (eds.) CAV 2010. LNCS, vol. 6174, pp. 212–226. Springer, Heidelberg (2010)
13. Gotlieb, A., Botella, B., Rueher, M.: Automatic test data generation using constraint solving techniques. In: ISSTA, pp. 53–62 (1998)
14. Gotlieb, A., Botella, B., Rueher, M.: A CLP Framework for Computing Structural Test Data. In: Lloyd, J., Dahl, V., Furbach, U., Kerber, M., Lau, K.-K., Palamidessi, C., Pereira, L.M., Sagiv, Y., Stuckey, P.J. (eds.) CL 2000. LNCS (LNAI), vol. 1861, pp. 399–413. Springer, Heidelberg (2000)
15. Lebbah, Y., Michel, C., Rueher, M., Daney, D., Merlet, J.-P.: Efficient and safe global constraints for handling numerical constraint systems. SIAM J. Numer. Anal. 42, 2076–2097 (2005)
16. Lhomme, O.: Consistency techniques for numeric CSPs. In: IJCAI, pp. 232–238 (1993)
17. Marre, B., Michel, C.: Improving the Floating Point Addition and Subtraction Constraints. In: Cohen, D. (ed.) CP 2010. LNCS, vol. 6308, pp. 360–367. Springer, Heidelberg (2010)
18. McCormick, G.P.: Computability of global solutions to factorable nonconvex programs – part i – convex underestimating problems. Mathematical Programming 10, 147–175 (1976)
19. Michel, C., Lebbah, Y., Rueher, M.: Safe embedding of the simplex algorithm in a CSP framework. In: Proc. of CPAIOR 2003, CRT, Université de Montréal, pp. 210–220 (2003)
20. Michel, C.: Exact projection functions for floating point number constraints. In: AMAI (2002)
21. Michel, C., Rueher, M., Lebbah, Y.: Solving Constraints over Floating-Point Numbers. In: Walsh, T. (ed.) CP 2001. LNCS, vol. 2239, pp. 524–538. Springer, Heidelberg (2001)
22. Miné, A.: Weakly Relational Numerical Abstract Domains. PhD thesis, École Polytechnique, Palaiseau, France (December 2004)
23. Neumaier, A., Shcherbina, O.: Safe bounds in linear and mixed-integer programming. Math. Programming A 99, 283–296 (2004)
24. Ryoo, H.S., Sahinidis, N.V.: A branch-and-reduce approach to global optimization. Journal of Global Optimization, 107–138 (1996)

A Model Seeker: Extracting Global Constraint Models from Positive Examples

Nicolas Beldiceanu[1] and Helmut Simonis[2,*]

[1] TASC team (INRIA/CNRS), Mines de Nantes, France
`Nicolas.Beldiceanu@mines-nantes.fr`
[2] Cork Constraint Computation Centre
Department of Computer Science, University College Cork, Ireland
`h.simonis@4c.ucc.ie`

Abstract. We describe a system which generates finite domain constraint models from positive example solutions, for highly structured problems. The system is based on the global constraint catalog, providing the library of constraints that can be used in modeling, and the Constraint Seeker tool, which finds a ranked list of matching constraints given one or more sample call patterns.

We have tested the modeler with 230 examples, ranging from 4 to 6,500 variables, using between 1 and 7,000 samples. These examples come from a variety of domains, including puzzles, sports-scheduling, packing & placement, and design theory. When comparing against manually specified "canonical" models for the examples, we achieve a hit rate of 50%, processing the complete benchmark set in less than one hour on a laptop. Surprisingly, in many cases the system finds usable candidate lists even when working with a single, positive example.

1 Introduction

In this paper we present the *Model Seeker* system which generates constraint models from example solutions. We focus on problems with a regular structure (this encompasses *matrix models* [14]), whose models can be compactly represented as a small set of conjunctions of identical constraints. We exploit this structure in our learning algorithm to focus the search for the strongest (i.e. most restrictive) possible model.

In our system we use global constraints from the global constraint catalog [2] mainly as modeling constructs, and not as a source of filtering algorithms. The global constraints are the primitives from which our models are created, each capturing some particular aspect of the overall problem. Using existing work on global constraints for mixed integer programming [20] or constraint based local search [16], our results are not only applicable for finite domain constraint programming, but can potentially reach a wider audience.

The input format we have chosen consists of a flat vector of integer values, allowing for different representations of the same problem. We do not force the user to adapt his input to any particular technology, but rather aim to be able to handle examples taken from a variety of existing sources.

---

* Supported by EU FET grant ICON (project number 284715).

M. Milano (Ed.): CP 2012, LNCS 7514, pp. 141–157, 2012.

In our method we extensively use meta-data about the constraints in the catalog, which describe their properties and their connection. We have added a number of new, useful information classes during our work, which prove to be instrumental in recognizing the structure of different models.

The main contribution of this paper is the presentation of the implemented Model Seeker tool, which can deal with a variety of problem types at a practical scale. The examples we have studied use up to 6,500 variables, and deal with up to 7,000 samples, even though the majority of the problems are restricted to few, and often unique solution samples. We currently only work with positive examples, which seems to provide enough information to achieve quite accurate models of problems. As a side-effect of our work we also have strengthened the constraint description in the constraint catalog with new categories of meta-data, in particular to show implications between different constraints.

Our paper is structured in the following way: We first introduce a running example, that we will use to explain the core features of our system. In Section 2, we describe the basic workflow in our system, also detailing the types of meta-data that are used in its different components. We present an overview of our evaluation in Section 3, which is followed by a discussion of related work (Section 4), before finishing with limitations and possible future work in Section 5. For space reasons we can only give an overview of the learning algorithm and the obtained results. A full description can be found in a companion technical report at `http://4c.ucc.ie/~hsimonis/modelling/report.pdf`.

1.1 A Running Example

As a running example we use the 2010/2011 season schedule of the Bundesliga, the German soccer championship. We take the data given in `http://www.weltfussball.de/alle_spiele/bundesliga-2010-2011/`, replacing team names with numbers from 1 to 18. The schedule is given as a set of games on each day of the season. Table 1 shows days 1, 2, 3, 18 and 19 of the schedule. Each line shows all games of one day; on the first day, team 1 (at home) is playing against team 2 (away), team 3 (at home) plays team 4, etc. The second half of the season (days 18-34) repeats the games of the first half, exchanging the home and away teams, on day 18, for example, team 18 (at home) plays team 17, team 2 (at home) plays team 1, and so on. Overall, each team plays each other twice, once at home, and once away in a double round-robin scheme.

Table 1. Bundesliga Running Example: Input Data

1	2	3	4	5	6	7	8	9	10	11	12	13	14	15	16	17	18
8	1	14	11	4	7	2	15	12	13	6	9	10	3	18	5	16	17
3	14	17	2	13	6	5	12	9	16	11	18	1	4	15	8	7	10

. . .

18	17	2	1	4	3	6	5	10	9	16	15	14	13	12	11	8	7
13	12	11	14	17	16	15	2	9	6	1	8	7	4	5	18	3	10

. . .

As input data we receive the flat vector of numbers, we will reconstruct the matrix as part of our analysis. Note that for most sports scheduling problems we will have access to only one example solution, the published schedule for a given year, schedules from different years encode different teams and constraints, and are thus incomparable.

2   Workflow

We will now describe how we proceed from the given positive examples to a candidate list of constraints modeling the problem. The workflow is described in Figure 1. Data are shown in green, software modules in blue/bold outline, and specific global constraint catalog meta-data are shown in yellow/italics. We first give a brief overview of the modules, and then discuss each step in more detail.

Transformation. In a first step, we try to convert the input samples to other, more appropriate representations. This might involve replacing a 0/1 format with finite domain values, or converting different graph representations into the successor variable form used by the global constraints in the catalog. For some transformations, we keep both the original and the transformed representation for further analysis, for others we replace the original sample with the transformed data.

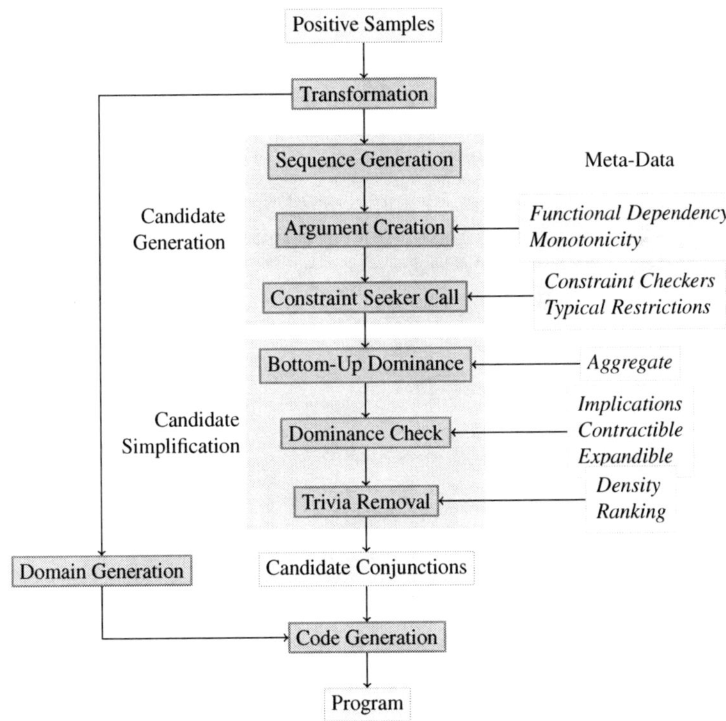

Fig. 1. Workflow in the Model Seeker

Candidate Generation. The second step (Sequence Generation) tries to group the variables of the sample into regular subsets, for example interpreting the input vector as a matrix and creating subsequences for all rows and all columns. In the Argument Creation step, we create call patterns for constraints from the subsequences. We can try each subsequence on its own, or combine pairs of subsequences or use all subsequences together in a single collection. We also try to add additional arguments based on functional dependencies and monotonic arguments of constraints, described as meta-data in the global constraint catalog. For each of these generated call patterns, we then call the Constraint Seeker to find matching constraints, which satisfy all subsequences of all samples. For this we enforce the *Typical Restrictions*, meta-data in the catalog, which describe how a constraint is typically used in a constraint program. Only the highest ranking candidates are retained for further analysis.

Candidate Simplification. After the seeker calls, we potentially have a very large list of possible candidate conjunctions (up to 2,000 conjunctions in our examples), we now have to reduce this set as much as possible. We first apply a Dominance Check to remove all conjunctions of constraints that we can show are implied by other conjunctions of constraints in our candidate list. Instead of showing the implication from first principles, we rely on additional meta-data in the catalog, which describe implications between constraints, but we also use conditional implications which only hold if certain argument restrictions apply, and expandible and contractible [22] properties, which state that a constraint still holds if we add or remove some of its decision variables. The dominance check is the core of our modeling system, helping to remove irrelevant constraints from the candidate list. In the last step before our final candidate list output the system removes trivial constraints and simplifies some constraint pattern. This also performs a ranking of the candidates based on the constraint and sequence generator used, trying to place the most relevant conjunction of constraints at the top of the candidate list.

Code Generation. As a side effect of the initial transformation, we also create potential domains for the variables of our problem. In the default case, we just use the range of all values occurring in the samples, but for some graph-based transformations a more refined domain definition is used. Given the candidate list and domains for the variables, we can easily generate code for a model using the constraints. At the moment, we produce SICStus Prolog code using the call format of the catalog. The generated code can then be used to validate the samples given, or to find solutions for the model that has been found.

After this brief overview, we will now discuss the different process steps in more detail.

2.1   Transformation

In finite domain programming, there are implicit conventions on how to express models leading to effective solution generation. In our system, we can not assume that the user is aware of these conventions, nor that the sample solutions are already provided in the "correct" form. We therefore try to apply a series of (currently 12) input transformations, that convert between different possible representations of problems, and that retain the form which matches the format expected by the constraint catalog.

Table 2. Bundesliga Running Example: Transformed Problem

2	−1	4	−3	6	−5	8	−7	10	−9	12	−11	14	−13	16	−15	18	−17
−8	15	−10	7	−18	9	−4	1	−6	3	−14	13	−12	11	−2	17	−16	5
4	−17	14	−1	12	−13	10	−15	16	−7	18	−5	6	−3	8	−9	2	−11

. . .

−2	1	−4	3	−6	5	−8	7	−10	9	−12	11	−14	13	−16	15	−18	17
8	−15	10	−7	18	−9	4	−1	6	−3	14	−13	12	−11	2	−17	16	−5

. . .

In each case, some pre- and post-conditions must be satisfied for the transformation to be considered valid. We now give some examples.

Converting 0/1 Samples. If a solution is given using only 0/1 values, there may be a way of re-interpreting the sample as a permutation with finite domain variables. If we consider the 0/1 values as an $n \times n$ matrix (x_{ij}) where each row and each column contains a single one, we can transform this into a vector v_i of n finite domain values based on the equivalence

$$\forall_{1 \leq i \leq n} \forall_{1 \leq j \leq n} : \quad x_{ij} = 1 \Rightarrow v_i = j$$

This transformation is the equivalent of a channeling constraint between 0/1 and finite domain variables, described for example in [19].

Using Successor Notation for Graph Constraints. Most graph constraints in the catalog use a *successor* notation to describe circuits, cycles or paths, i.e. the value of a node indicates the next node in the final assignment. But this is not the only way of describing graphs. In the original Knight's Tour formulation [31], the value of a node is the position in the path, i.e. the first node has value one, the second value two, and so on. We have defined transformations which map between these (and several other) formats, while checking that the resulting graph can be compactly described.

Using Schreuder Tables. Another transformation is linked to sports scheduling problems. In many cases, users like to give the schedule as a list of fixtures, listing which games will be played on each day. The first team is the home team, the second the away team. For constraint models, the format of Schreuder tables [28], as shown in Table 2 for our running example, can lead to more compact models [24,30,17,18]. For each time-point t over q rounds and for an even number of teams n, they can be obtained from the fixture representation as follows:

$$\forall_{1 \leq t \leq q(n-1)} \forall_{1 \leq i \leq \lfloor n/2 \rfloor} : \quad x_{2i-1,t} = k, x_{2i,t} = l \Rightarrow v_{k,t} = l, v_{l,t} = -k$$

2.2   Sequence Generator

After the input transformation, we have to consider possible, regular substructures which group the samples into subsequences. For space reasons again, we only give some examples of the sequence generators used in our running example, the full list (containing 21 generators) with their formal definition can be found in the technical report, some were already described in [4].

vector(n). This is the most basic sequence generator of treating all elements of the sample as a single sequence of size n.

scheme(n, r, c, a, b). By far the most common sequence generator treats a sample of length n as an $r \times c$ matrix, and creates non-overlapping blocks of size $a \times b$, creating n/ab sequences of size ab. The number of such partitions depends on the number of factors of n, as $n = rc$. For our running example (Section 1.1) with 612 values, we have to consider the matrices 2×306, 3×204, 4×153, 6×102, 9×68, 17×36, 18×34, 34×18, 36×17, 68×9, 102×6, 153×4, 204×3 and 306×2. Some of the blocks created from these matrices lead to the same partition of the variables, only one representative is kept.

repart(n, r, c, a, b). This sequence generator also treats the sample of size n as a $r \times c$ matrix, and considers blocks of size $a \times b$. But it groups elements in the same position from each block, creating $a \times b$ sequences of size $n/(ab)$.

For the running example, a total of 68 subsequence collections are generated. Note that the subsequences often, but not always, have the same size. We also provide an API where the user can provide his own sequence generators, this can be helpful to deal with known, but irregular structure in the problem.

2.3 Argument Creation

In the next step of the operation, we convert the generated subsequences into call patterns for the Constraint Seeker [3]. In order to consider more of the constraints in the catalog, we have to provide different argument signatures by organizing the subsequences in different ways, and by adding arguments.

Single, Pairs and Collection. In the first part we decide how we want to use the subsequences. Consider we have k subsequences, each of length m. If we use each subsequence on its own, we create k call patterns with a single argument, each a collection of m variables. This corresponds to the argument pattern used by `alldifferent` constraints, for example. We can also consider pairs of subsequences, creating a call patterns with two arguments, for $k - 1$ calls to a predicate like `lex_greater`. Finally we can use all subsequences as a single collection of collections, which creates one call with a collection of k collections of m elements each. This would match a constraint like `lex_alldifferent`. We generate all these potential calls in parallel, and perform the steps described in the following two paragraphs.

Value Projection. For some problems (like our transformed, running example), a projection from the original domain to a smaller domain can lead to a more compact model. If, for example, some of the values in the sample are positive, and others are negative, we can try a projection using the absolute value or the sign of the numbers, in addition to the original values.

Adding Arguments. Many global constraints use additional arguments besides the main decision variables. If we do not generate these arguments in the call pattern, we can not find such constraints with the Constraint Seeker. But just enumerating all possible values for these additional arguments would lead to a combinatorial explosion. Fortunately, we can compute values for these arguments in case of functional dependencies and monotonic arguments. This is similar to the argument generation discussed for the `gcc` constraint discussed in [8].

2.4   Constraint Seeker

The Constraint Seeker [3] will find a ranked list of global constraints that satisfy a collection of positive and negative sample calls, using the available constraint checkers of the catalog. We use this seeker as a black-box for all call patterns with all additional argument values and value projections defined in the previous section.

Using Multiple, Positive Samples. The seeker first checks that the call signature matches the constraint, then tries to evaluate the constraint on the samples. In our case, these are the call patterns prepared in the previous step for all subsequences of all positive examples given. In our modeling system we currently do not consider negative examples. They would require a slightly different treatment, as a negative example can be rejected by just one constraint, while all positive examples must be accepted by all constraints found.

Typical Restrictions. In addition to the restrictions that must hold for the constraint to be applied, in our modeling tool we also check for the *typical* restrictions that are specified in the catalog. The `alldifferent` constraint for example can be called with an empty collection, but a typical use would have more than two variables in the collection. The typical constraints are expressed using the same language as the formal restrictions of the catalog, checking their validity thus does not require any additional code.

Selecting Top-Ranked Elements. The Constraint Seeker returns a ranked list of candidates, this ranking is a combination of structural properties (functional dependencies or monotonic arguments), implications between constraints, estimated solution density and estimated popularity of the constraint described in [3]. In our system we only use the top ranked element that satisfies all subsequences of all samples. This reduces the number of candidates to be considered, while at the same time it does not seem to exclude constraints that are required for the highly structured problems considered.

For our running example, we perform 1,099 calls to the Constraint Seeker, which performs 82,458 constraint checks, and which results in 589 possible candidate conjunctions. We now face the task of reducing this candidate list as much as possible, keeping only interesting conjunctions.

2.5   Bottom-up Dominance

Some constraints like `sum` or `gcc` have the *aggregate* property, one can combine multiple such constraints over disjoint variable sets by adding the right hand sides or summing the counter values. As an example, we can combine

$$x_1 + x_2 = 5 \land x_3 + x_4 = 2 \Rightarrow x_1 + x_2 + x_3 + x_4 = 7$$

We want to remove aggregated constraints of this type, as they are implied by conjunctions of smaller constraints. We perform a bottom-up saturation of combining constraints with the aggregate property up to a limited size, and remove any candidate conjunctions where all constraints are dominated.

2.6   Dominance Check

The dominance check compares all conjunction candidates against each other (worst case quadratic number of comparisons), and marks dominated entries. Note that dominated entries may be used to dominate other entries, and thus can not be removed immediately. We use a number of meta-data fields to check for dominance.

Implications. In our final candidate list, we are interested in only the strongest, most restrictive constraints, all constraints that are implied by other candidate constraints can be excluded. Note that this will sometimes lead to overly restrictive solutions, especially if only a few samples are given.

Checking if some conjunction is implied by some other conjunction for a particular set of input values is a complex problem, a general solution would require sophisticated theorem proving tools like those used in [11] for a restricted problem domain. We do not attempt such a generic solution, but instead rely on meta-data in the catalog linking the constraints. That meta-data is useful also for understanding the relations between constraints, and thus serves multiple purposes. This syntactic implication check is easy to implement, but only can be used if both constraints have the same arguments.

Conditional Implications. For some constraints additional implications exist, but only if certain restrictions apply. The `cycle` constraint for example implies the `circuit` constraint, but only if the NCYCLE argument is equal to one. For conditional implications the arguments do not have to be the same, but the main decision variables used must match.

Contractibility and Expandability. Other useful properties are contractibility [22] and expandibility. A constraint (like `alldifferent`) is *contractible* if it still holds if we remove some of its decision variables. This allows us to dominate large conjunctions of constraints with few variables with small conjunctions of constraints with many variables. Due to the way we systematically generate all subsequence collections, this is often possible. In a similar way, some constraints like `atleast` are *expandible*, they still hold if we add decision variables. We can again use this property to dominate some conjunctions of constraints. Details and possible extensions have been described in [4].

Hand-Coded Domination Rules. Some dominance rules are currently hand-crafted in the program, if the required meta-data have not yet been formalized in the catalog description. Such examples can be an important source of requirements for the catalog itself, enhancing the expressive power of the constraint descriptions.

2.7   Trivia Removal

Even after the dominance check, we can still have candidate explanations which are valid and not dominated, but which are not useful for modeling. In the trivia removal section, we eliminate or replace most of these based on sets of rules.

Functional Dependencies on Single Samples. In Section 2.3 we have described how we can add some arguments to a call pattern for functional dependencies. In the case of pure functional dependencies [1], we have to worry about pattern consisting of a single subsequence with a single sample. In that case, the constraint does not filter any pattern, as for each pattern the correct value can be selected. We therefore remove such candidates.

Constraint Simplification. At this point we can also try to simplify some constraints that have particular structure. A typical example are `lex_chain` constraints on a subsequence, where already the first entries of the collections are ordered in strictly increasing order. We can therefore replace the `lex_chain` constraint on the subsequences with a `strictly_increasing` constraint on the first elements of the collections, using a special `first` sequence generator. These constraints often occur as symmetry-breaking restrictions in models, which we find if all the samples given respect the symmetry breaking.

Uninteresting Constraints. Even with the typical restrictions in the Constraint Seeker, we often find candidates (like `not_all_equal`) which are not very interesting for defining models. As a final safe-guard, we use a black-list to remove some combinations of constraints and sequence generators that should not be included in our models.

2.8   Candidate List for Bundesliga Schedule

Table 3 shows the list of the candidate conjunctions generated for our transformed example problem. Entries in green match a manually defined model, ten other candidates are also proposed. The arguments of constraints in the Constraint Conjunction column indicate any additional parameters, the $*n$ indicates how many constraints form the conjunction. The value projections `absolute_value` and `sign` convert each element of the input data, `id` denotes the identity projection.

   Some of the constraints mentioned are perhaps unfamiliar, we provide a short definition. The constraint `symmetric_alldifferent`$([x_1, .., x_n])$ [2, page 1854] in line 4 states that

$$\forall_{1 \leq i \leq n} : \quad x_i \in [1, n]; x_i = j \iff x_j = i$$

It expresses the constraint that if team A plays team B on some day, then team B will play team A. The constraints `twin`$([\langle x_1, y_1 \rangle, ..., \langle x_n, y_n \rangle])$ [2, page 1896] in lines 7, 19 and 20 state that

$$\forall_{1 \leq i \leq n} : \quad (x_i = u \wedge y_i = v) \Rightarrow (\forall_{1 \leq j \leq n} : x_j = u \iff y_j = v)$$

These constraints express the fact that the tournament is played in two symmetric half-seasons, with home and away games swapped. Note that constraints 8, 21 and 23 also express this condition, but using an `elements` constraint, pairing positive and negative numbers. The `alldifferent` constraint in line 1 expresses that no repeat games occur in the season, while that of line 5 states that all teams play on each day. The `strictly_increasing` constraint in line 9 results from the simplification of a symmetry breaking `lex_chain` constraint. The `gcc` in line 14 states that each team plays 17 home (positive value) and 17 away (negative value) games. Finally, the `among_seq` constraint in line 22 states that no team has more than two consecutive away games.

Table 3. Constraint Conjunctions for Problem Bundesliga

-	Sequence Generator	Projection	Constraint Conjunction
1	scheme(612,34,18,34,1)	id	alldifferent*18
2	scheme(612,34,18,2,2)	id	alldifferent*153
3	scheme(612,34,18,1,18)	id	alldifferent*34
4	scheme(612,34,18,1,18)	absolute_value	symmetric_alldifferent([1..18])*34
5	scheme(612,34,18,17,1)	absolute_value	alldifferent*36
6	repart(612,34,18,34,9)	id	sum_ctr(0)*306
7	repart(612,34,18,34,9)	id	twin*1
8	repart(612,34,18,34,9)	id	elements([i,-i])*1
9	first(9,[1,3,5,7,9,11,13,15,17])	id	strictly_increasing*1
10	vector(612)	id	global_cardinality([-18.. -1-17,0-0,1..18-17])*1
11	repart(612,34,18,34,9)	id	sum_powers5_ctr(0)*306
12	repart(612,34,18,34,9)	id	sum_cubes_ctr(0)*306
13	repart(612,34,18,34,3)	sign	global_cardinality([-1-3,0-0,1-3])*102
14	scheme(612,34,18,34,1)	sign	global_cardinality([-1-17,0-0,1-17])*18
15	repart(612,34,18,17,9)	sign	global_cardinality([-1-2,0-0,1-2])*153
16	repart(612,34,18,2,9)	sign	global_cardinality([-1-17,0-0,1-17])*18
17	scheme(612,34,18,1,18)	sign	global_cardinality([-1-9,0-0,1-9])*34
18	repart(612,34,18,34,9)	sign	sum_ctr(0)*306
19	repart(612,34,18,34,9)	sign	twin*1
20	repart(612,34,18,34,9)	absolute_value	twin*1
21	repart(612,34,18,34,9)	sign	elements([i,-i])*1
22	scheme(612,34,18,34,1)	sign	among_seq(3,[-1])*18
23	repart(612,34,18,34,9)	absolute_value	elements([i,i])*1
24	first(9,[1,3,5,7,9,11,13,15,17])	absolute_value	strictly_increasing*1
25	first(6,[1,4,7,10,13,16])	absolute_value	strictly_increasing*1
26	scheme(612,34,18,34,1)	absolute_value	nvalue(17)*18

2.9   Domain Creation

By default, the domains of the variables in our generated models are the interval between the smallest and largest value occurring in the samples. Based on the transformation used, we can use more restricted domains for graph models like graph partitioning and domination [15], where the domain of each variable/node specifies the initial graph.

2.10   Code Generation

The code generation builds flat models for the given instances. The programs consist of five parts, we first define all variables with their domains, then state all restrictions due to fixed values as assignments, state any projections used to simplify the variables, then build the constraints in the catalog syntax, and finally call a generic value assignment routine to search for a solution. We can use the generated model as a test to check if it accepts the given samples, or to generate new solutions for the problem. Many puzzles have a unique solution, we can count solutions of our program to see if the generated model is restrictive enough to capture this property.

It would be straightforward to generate the code for other systems than SICStus Prolog, provided that the catalog constraints are supported. A version generating FlatZinc[23] or XCSP [27] would be especially attractive to benefit from the variety of backend solvers which support these formats.

3   Evaluation

Table 4 shows summary results for selected problems of our evaluation set. The problems range from sports scheduling (ACC Basketball Scheduling, csplib11; Bundesliga; DEL2011 (German ice hockey league); Scotish Premier League (soccer); Rabodirect Pro 12 (rugby)), to scheduling (Job-shop 10x10 [10]) and placement (Duijvestijn, csplib9; Conway 5x5x5 [5]; Costas Array [12]), design theory (BIBD, csplib28; Kirkman [13]; Orthogonal Latin Squares [9]), event scheduling (Social Golfer, csplib10; Progressive Party, csplib13) and puzzles. Details of these problems can be found in the technical report mentioned before. Smaller problems are solved within seconds, even the largest require less than 5 minutes on a single core of a MacBook Pro (2.2GHz) with 8Gb of memory.

The columns denote: *Transformation Id*: the number of the transformation applied (if any), *Instance Size*: the number of values in the solution, i.e. the number of variables in the model, *Nr Samples*: the number of solutions given as input, *Nr Sequences*: the number of sequence sets generated, *Nr Seeker Calls*: the number of times the Constraint Seeker is called, *Constraint Checks*: the number of calls to constraint checkers inside the seeker, *Nr Relevant*: the number of initial candidate conjunctions found by the Constraint Seeker, *Nr Non Dom*: the number of non-dominated candidates remaining after the dominance checkers, *Nr Specified*: the number of conjunctions specified in the manual, "canonical" model, *Nr Models*: the number of conjunctions given as output of the Model Seeker, *Nr Missing*: how many of the manually defined conjunctions were not found by the system, *Hit Rate*: the percentage rate of Nr Specified to Nr Models, a value of 100% indicates that exactly the candidates of the canonical model were found, and *Time*: the execution time in seconds.

For two of the problems, we only find part of the complete model. The Progressive party problem [29] requires a bin-packing constraint that we currently do not recognize, as it relies on additional data for the boat sizes, while the ACC basketball problem contains several constraints which apply only for specific parts of the schedule, and which can not be learned from a single solution. Also note that for the De Jaenisch problem [26], we show results with and without a transformation. This problem combines a "near" magic square, found without transformation, with an Euler Knight tour, using transformation 7.

For our full evaluation, we have used 230 examples from various sources. For 10 of the examples no reasonable model was generated, either because we did not have the right sequence generator, or we are currently missing the global constraint required to express the problem. For a further 37 problems, only part of the model was found. This is typically caused by some constraint requiring additional data, not currently given as input, or by an over-specialization of the output, where the Model Seeker finds a more restrictive constraint than the one specified manually. Overall, we considered 73 constraints in the Constraint Seeker, and selected 53 different global constraints as potential solutions. This is only a fraction of the 380 constraint in the catalog, many of the missing constraints have more complex argument signatures or use finite sets, which are currently not available in SICStus Prolog.

Figure 2 shows the number of candidates found for all examples studied as a function of the instance size, split between single samples and multiple samples. Note that the plot uses a log-log scale. The results indicate that even with a single sample, the number of candidate conjunctions found is quite limited, this drops further if multiple samples are used.

Another view of all the results is shown in Figure 3. It shows the relationship between number of variables and execution time, again grouped by problems with a single sample and problems for which multiple samples were provided. While no formal complexity analysis has been attempted, as several subproblems are expressed as constraint problems, results seem to indicate a low-polynomial link between problem size and execution time. The non-linear least square fit for the single sample problems is $8.5e^{-2}x^{0.90}$, and for multiple samples $6.1e^{-3}x^{1.45}$.

Table 4. Selected Example Results

Name	Transformation Id	Instance Size	Nr Samples	Nr Sequences	Nr Seeker Calls	Constraint Checks	Nr Relevant	Nr Non Dom	NrSpecified	Nr Models	Nr Missing	Hit Rate	Time [s]
12 Amazons All	-	12	156	36	127	35596	55	9	5	8	0	62.50	2,64
8 Queens All	-	8	92	20	100	12077	34	7	3	5	0	60.00	0.83
ACC Basketball Schedule	-	162	1	109	1786	29117	772	263	23	36	7	n/a	9.17
BIBD (8,14,7,4,3)	-	112	92	151	626	92461	232	112	4	15	0	26.67	26.01
Bundesliga	18	612	1	68	1099	52933	589	169	16	26	0	61.54	51.44
Conway 5x5x5 Packing	-	102	1	60	184	2619	78	35	1	2	0	50.00	0.42
Costas Array 12	-	12	48	36	121	14820	42	2	2	2	0	100.00	1.01
DEL 2011	-	728	1	235	1334	66999	555	173	3	8	0	37.50	54.23
De Jaenisch Tour	-	64	1	83	568	10130	283	58	2	13	0	15.38	1.66
De Jaenisch Tour	7	64	1	36	219	12952	113	67	1	1	0	100.00	0.46
Dominating Knights 8	9	64	2	36	141	11021	51	42	1	1	0	100.00	0.31
Duijvestijn 21	-	84	1	111	504	11625	240	102	1	12	0	8.33	1.77
Euler Knight Cube 4x4x4	7	64	1	36	208	12759	97	58	1	1	0	100.00	0.42
Job Shop 10x10 (10 sol)	-	400	10	326	1521	80589	582	130	2	2	0	100.00	40.27
Kirkman Wikipedia	-	105	1	40	179	2634	89	39	3	5	0	60.00	1.21
Leaper Tour 18x18	7	324	1	60	298	105955	140	61	1	1	0	100.00	2.18
Magic Square All	-	16	7040	33	176	1068574	57	5	4	4	0	100.00	115.07
Magic Square Duerer	-	16	1	33	212	2074	115	44	9	15	0	60.00	0.25
Orthogonal Latin Squares 10	-	200	1	147	910	15441	443	118	3	8	0	37.50	6.04
Progressive Party	-	174	1	45	171	4279	61	31	4	3	1	n/a	0.70
Rabodirect Pro12	18	264	1	66	1041	46898	539	155	8	13	0	61.54	7.91
Scotish Premier League	18	396	1	68	992	58959	459	157	9	12	0	75.00	14.18
Social Golfer	-	288	1	528	2813	69681	1221	256	5	36	0	13.89	61.93
Sudoku 81x81	-	6561	1	91	657	101075	334	68	3	5	0	60.00	244.16

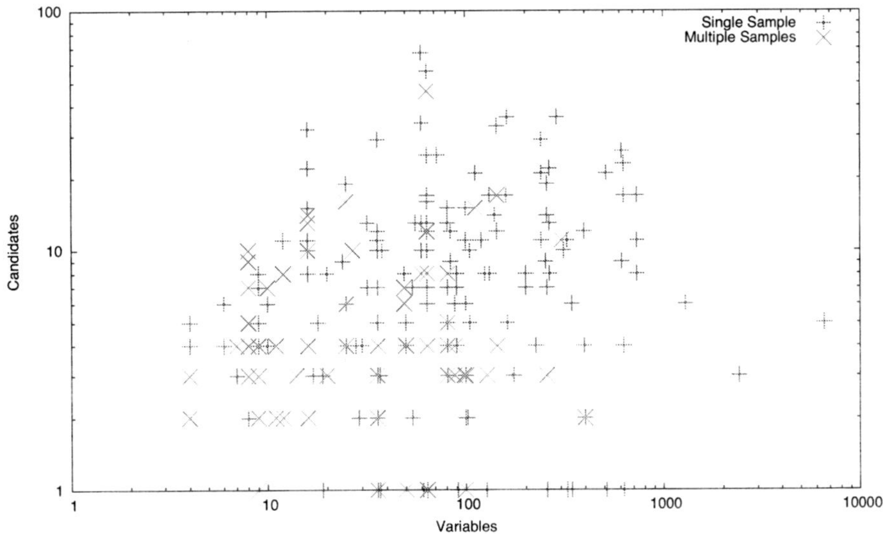

Fig. 2. Candidates as a Function of Problem Size (Variables)

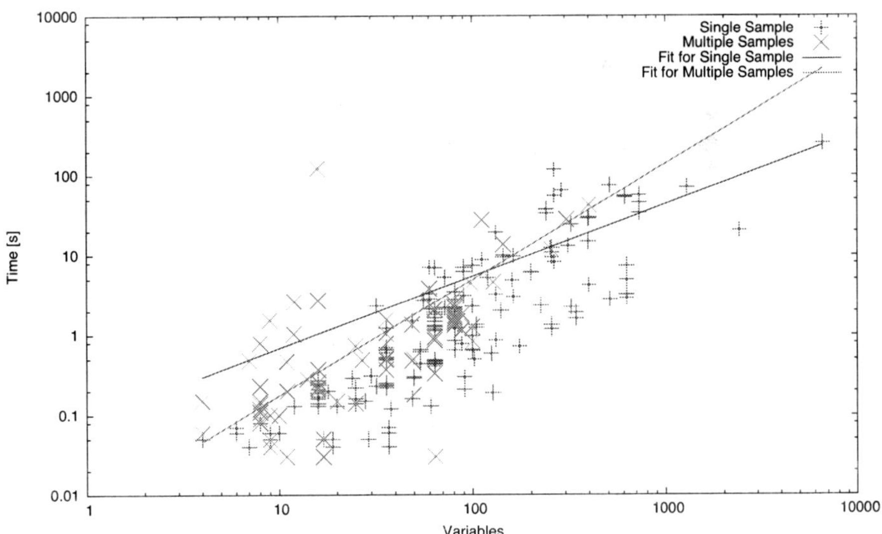

Fig. 3. Execution Time as a Function of Problem Size (Variables)

Table 5. Lines of Code / Run Time per Module over all 230 Examples

Module	Lines	Time [s]	% of Total
Transformation	1,500	1	0.03
Sequence Generation	1,000	53	2.81
Argument Creation	1,000	150	7.84
Constraint Seeker Call	300	464	24.22
Bottom-Up Check	200	506	26.42
Dominance Check	800	739	38.61
Trivia Removal	500	1	0.03
Glue / IO / Test	2,000	-	-

Table 5 shows the number of lines required for the different components of the system, as well as accumulated execution times over all 230 examples measured for these components. The programming effort is fairly evenly split amongst the different components, while the two dominance checkers require nearly two-thirds of the total execution time, with the constraint seeker using another quarter of the time. The system is written in SICStus Prolog 4.2, and uses the Constraint Seeker [3] with an additional 6,500 lines of code, and the global constraint catalog meta-data description of 60,000 lines of Prolog.

4   Related Work

Our approach of searching for conjunctions of identical constraints generalizes the idea of *matrix models* [14], which are an often-used pattern in constraint modeling.

The method proposed is a special, restricted case of *constraint acquisition* [25], which is the process of finding a constraint network from a training set of positive and negative examples. The learning process operates on a library of allowed constraints, and a resulting solution is a conjunction of constraints from that library, each constraint ranging over a subset of the variables.

The most successful of these systems is the CONACQ system [6], which proposes the use of version space learning to generate models interactively with the user, relying on an underlying SAT model to perform the learning. This is shown to work for binary constraints, but the method breaks down for global constraints over an unlimited number of variables.

In [7], the authors study the problem of determining argument values for global constraints like the `gcc` from example solutions, in the context of timetabling problems. This is similar to the argument creation we describe in Section 2.3.

The more recent work of [21] considers the use of inductive logic programming for finding models for problems given as a set of logic formulas. This can be powerful to find generic, size-independent models for a problem, but again, it is unclear how to deal with a library of given global constraints, which may not have a simple description as logic formulas.

Our dominance check based on meta-data is related to the work described in [11], where they use a theorem prover to find certain implications between constraints for a restricted domain. This does not rely on meta-data provided in the system, but instead would require a very powerful theorem prover to work for a collection of constraints for problems of the size considered here.

Common to all these results is that they have not been evaluated on a large variety of problems, that they consider only a limited number of potential constraints, and that problem sizes have been quite small.

5   Limitations, Future Work and Conclusions

We are currently only considering some 70 constraints in the global constraint catalog in our seeker calls. Many of the missing constraints require additional information (cost matrix, lookup tables) which have to be provided as additional input data to the system. For some problems, such additional data, like a precedence graph, may also express implicit, less regular sequence generators, which define for which variables a constraint should be stated. Extending our input format to allow for such data would drastically increase both the number of constraints that can be considered, as well as the range of application problems that can be modelled.

Most other constraint acquisition systems use both positive and negative examples. The negative examples are used to interactively differentiate between competing models of the system. We currently only use positive examples, but given recent results on global constraint reification [1], we could extend our system to include this functionality.

If we want to provide the functionality we have presented here to end-users, we will have to consider issues of usability and interactivity, allowing the user to filter and change constraint candidates, as well as being able to suggest custom sequence generators tailored to a specific problem.

Ultimately, we are looking for a modeling tool which can analyze samples of different sizes, and generate a generic, size independent model. Building on top of our existing framework, this would require to express both the sequence generator parameters and any additional arguments for constraints in terms of a variable problem size, to produce more compact, iterative code instead of the flat models currently generated.

Exploiting the idea that many highly structured combinatorial problems can be described by a small set of conjunctions of identical global constraints, this paper proposes a Model Seeker for extracting global constraint models from positive sample solutions. It relies on a detailed description of the constraints in terms of meta-data in the global constraint catalog. The system provides promising results on a variety of problems even when working from a limited number of examples.

Acknowledgement. The help of Hakan Kjellerstrand in finding example problems is gratefully acknowledged.

References

1. Beldiceanu, N., Carlsson, M., Flener, P., Pearson, J.: On the reification of global constraints. Technical Report T2012:02, SICS (2012)
2. Beldiceanu, N., Carlsson, M., Rampon, J.: Global constraint catalog, 2nd edn. (revision a). Technical Report T2012:03, SICS (2012)

3. Beldiceanu, N., Simonis, H.: A Constraint Seeker: Finding and Ranking Global Constraints from Examples. In: Lee, J. (ed.) CP 2011. LNCS, vol. 6876, pp. 12–26. Springer, Heidelberg (2011)

4. Beldiceanu, N., Simonis, H.: Using the global constraint seeker for learning structured constraint models: a first attempt. In: The 10th International Workshop on Constraint Modelling and Reformulation (ModRef 2011), Perugia, Italy, pp. 20–34 (September 2011)

5. Berlekamp, E.R., Conway, J.H., Guy, R.K.: Winning ways for your mathematical plays, 2nd edn., vol. 4. A K Peters/CRC Press (2004)

6. Bessière, C., Coletta, R., Freuder, E.C., O'Sullivan, B.: Leveraging the Learning Power of Examples in Automated Constraint Acquisition. In: Wallace, M. (ed.) CP 2004. LNCS, vol. 3258, pp. 123–137. Springer, Heidelberg (2004)

7. Bessière, C., Coletta, R., Koriche, F., O'Sullivan, B.: A SAT-Based Version Space Algorithm for Acquiring Constraint Satisfaction Problems. In: Gama, J., Camacho, R., Brazdil, P.B., Jorge, A.M., Torgo, L. (eds.) ECML 2005. LNCS (LNAI), vol. 3720, pp. 23–34. Springer, Heidelberg (2005)

8. Bessière, C., Coletta, R., Petit, T.: Acquiring Parameters of Implied Global Constraints. In: van Beek, P. (ed.) CP 2005. LNCS, vol. 3709, pp. 747–751. Springer, Heidelberg (2005)

9. Bose, R.C., Shrikhande, S.S., Parker, E.T.: Further results on the construction of mutually orthogonal latin squares and the falsity of Euler's conjecture. Canadian Journal of Mathematics 12, 189–203 (1960)

10. Carlier, J., Pinson, E.: An algorithm for solving the job shop problem. Management Science 35, 164–176 (1989)

11. Charnley, J., Colton, S., Miguel, I.: Automatic generation of implied constraints. In: Brewka, G., Coradeschi, S., Perini, A., Traverso, P. (eds.) ECAI. Frontiers in Artificial Intelligence and Applications, vol. 141, pp. 73–77. IOS Press (2006)

12. Drakakis, K.: A review of Costas arrays. Journal of Applied Mathematics, 1–32 (2006)

13. Dudeney, H.E.: Amusements in Mathematics. Dover, New York (1917)

14. Flener, P., Frisch, A., Hnich, B., Kiziltan, Z., Miguel, I., Walsh, T.: Matrix modelling. Technical Report 2001-023, Department of Information Technology, Uppsala University (September 2001)

15. Haynes, T.W., Hedetniemi, S.T., Slater, P.J.: Fundamentals of Domination in Graphs. Monographs and Textbooks in Pure and Applied Mathematics. Marcel Dekker (1998)

16. Van Hentenryck, P., Michel, L.: Constraint-Based Local Search. MIT Press, Boston (2005)

17. Henz, M.: Scheduling a major college basketball conference - revisited. Operations Research 49, 163–168 (2001)

18. Henz, M., Müller, T., Thiel, S.: Global constraints for round robin tournament scheduling. European Journal of Operational Research 153(1), 92–101 (2004)

19. Hernández, B.M.: The Systematic Generation of Channelled Models in Constraint Satisfaction. PhD thesis, University of York, York, YO10 5DD, UK, Department of Computer Science (2007)

20. Hooker, J.N.: Integrated Methods for Optimization. Springer Science + Business Media, LLC, New York (2007)

21. Lallouet, A., Lopez, M., Martin, L., Vrain, C.: On learning constraint problems. In: ICTAI (1), pp. 45–52. IEEE Computer Society (2010)

22. Maher, M.J.: Open Constraints in a Boundable World. In: van Hoeve, W.-J., Hooker, J.N. (eds.) CPAIOR 2009. LNCS, vol. 5547, pp. 163–177. Springer, Heidelberg (2009)

23. Marriott, K., Nethercote, N., Rafeh, R., Stuckey, P.J., de la Banda, M.G., Wallace, M.: The design of the Zinc modelling language. Constraints 13(3), 229–267 (2008)

24. Nemhauser, G., Trick, M.: Scheduling a major college basketball conference. Operations Research 46, 1–8 (1998)

25. O'Sullivan, B.: Automated modelling and solving in constraint programming. In: Fox, M., Poole, D. (eds.) AAAI, pp. 1493–1497. AAAI Press (2010)
26. Petkovic, M.S.: Famous Puzzles of Great Mathematicians. American Mathematical Society, Providence (2009)
27. Roussel, O., Lecoutre, C.: XML representation of constraint networks format XCSP 2.1. Technical Report arXiv:0902.2362v1, Universite Lille-Nord de France, Artois (2009)
28. Schreuder, J.A.M.: Combinatorial aspects of construction of competition Dutch professional football leagues. Discrete Applied Mathematics 35(3), 301–312 (1992)
29. Smith, B.M., Brailsford, S.C., Hubbard, P.M., Paul Williams, H.: The progressive party problem: Integer linear programming and constraint programming compared. Constraints 1(1/2), 119–138 (1996)
30. Walser, J.P.: Domain-Independent Local Search for Linear Integer Optimization. PhD thesis, Technical Faculty of the University des Saarlandes, Saarbruecken, Germany (October 1998)
31. Watkins, J.J.: Across the Board: The Mathematics of Chessboard Problems. Princeton University Press, Princeton (2004)

On Computing Minimal Equivalent Subformulas

Anton Belov[1], Mikoláš Janota[2], Inês Lynce[2], and Joao Marques-Silva[1,2]

[1] CASL, University College Dublin, Ireland
[2] IST/INESC-ID, Technical University of Lisbon, Portugal

Abstract. A propositional formula in Conjunctive Normal Form (CNF) may contain redundant clauses — clauses whose removal from the formula does not affect the set of its models. Identification of redundant clauses is important because redundancy often leads to unnecessary computation, wasted storage, and may obscure the structure of the problem. A formula obtained by the removal of all redundant clauses from a given CNF formula $\mathcal{F}$ is called a Minimal Equivalent Subformula (MES) of $\mathcal{F}$. This paper proposes a number of efficient algorithms and optimization techniques for the computation of MESes. Previous work on MES computation proposes a simple algorithm based on iterative application of the definition of a redundant clause, similar to the well-known deletion-based approach for the computation of Minimal Unsatisfiable Subformulas (MUSes). This paper observes that, in fact, most of the existing algorithms for the computation of MUSes can be adapted to the computation of MESes. However, some of the optimization techniques that are crucial for the performance of the state-of-the-art MUS extractors cannot be applied in the context of MES computation, and thus the resulting algorithms are often not efficient in practice. To address the problem of efficient computation of MESes, the paper develops a new class of algorithms that are based on the iterative analysis of subsets of clauses. The experimental results, obtained on representative problem instances, confirm the effectiveness of the proposed algorithms. The experimental results also reveal that many CNF instances obtained from the practical applications of SAT exhibit a large degree of redundancy.

1 Introduction

A propositional formula in Conjunctive Normal Form (CNF) is redundant if some of its clauses can be removed without changing the set of models of the formula. Formula redundancy is often desirable. For example, modern Conflict-Driven Clause Learning (CDCL) Boolean Satisfiability (SAT) solvers learn redundant clauses [33,1], which are often essential for solving practical instances of SAT. However, redundancy can also be undesirable. For example, in knowledge bases formula redundancy leads to the use of unnecessary storage and computational resources [25]. Another example is the undesirable redundant clauses in the CNF representation of belief states in a conformant planner [37,36]. In the context of probabilistic reasoning systems, concise representation of conditional

M. Milano (Ed.): CP 2012, LNCS 7514, pp. 158–174, 2012.

independence information can be computed by removing redundant clauses from certain propositional encodings [30]. More generally, given the wide range of applications of SAT, one can pose the question: does a given problem domain encoder introduce redundancy, and if so, how significant is the percentage of redundant clauses? Removal of redundancies can also find application in solving problems from different complexity classes, for example Quantified Boolean Formulas (QBF). Besides propositional logic formulas, the problem of the identification of redundant constraints is relevant in other domains. Concrete examples include Constraint Satisfaction Problems (CSP) [12,10,9], Satisfiability Modulo Theories (SMT) [35], and Ontologies [19].

This paper addresses the problem of computing an irredundant subformula $\mathcal{E}$ of a redundant CNF formula $\mathcal{F}$, such that $\mathcal{F}$ and $\mathcal{E}$ have the same set of models. Such subformula $\mathcal{E}$ will be referred to as a *Minimal Equivalent Subformula (MES)* of $\mathcal{F}$. Previous work on removing redundant clauses from CNF formulas proposes a direct approach [7], which iteratively checks the definition of redundant clause and removes the clauses that are found to be redundant. While the direct approach is similar to the well-known deletion-based approach for the computation of Minimal Unsatisfiable Subformulas (MUSes), most of the techniques developed for the extraction of MUSes (e.g. see [18,14,27]) have not been extended to the computation of MESes.

The paper has five main contributions, summarized as follows. First, the paper shows that many of the existing MUS extraction algorithms can be extended to the computation of MESes. Since efficient MUS extraction uses a number of key techniques for reducing the total number of SAT solver calls — namely, clause set refinement [13,29,28] and model rotation [28,4] — this paper analyzes these techniques in the context of MES extraction. The second contribution of the paper is, then, to show that model rotation can be integrated, and, in fact, improved, in MES extraction, however clause set refinement *cannot* be used in MES algorithms derived from the existing MUS algorithms. Third, the paper proposes a reduction from MES computation problem to group-MUS computation problem [26,29]; this reduction enables the use of *both* model rotation and clause set refinement for MES extraction. Fourth, given that the approach of reduction to group-MUS can result in hard instances of SAT, the paper proposes an incremental reduction of MES to group-MUS extraction, that involves the separate analysis of subsets of clauses. Fifth, and finally, the paper develops solutions for checking that computed MESes are correct. These solutions find application in settings where independent certification is required.

Experimental results, obtained on representative satisfiable instances from past SAT competitions, show that the new algorithms for MES computation achieve significant performance gains over the basic algorithms, and allow targeting redundancy removal for reasonably sized formulas. In addition, the experimental results show that *many* CNF formulas, from a wide range of application domains, contain a significant percentage of redundant clauses, in some cases exceeding 90% of the original clauses.

2   Preliminaries

Standard definitions for propositional logic are assumed. Propositional formulas are defined over a set of propositional variables $X = \{x_1, \ldots, x_n\}$. A CNF formula $\mathcal{F}$ is a conjunction of disjunctions of literals (*clauses*), where a literal is a variable or its complement. Unless otherwise stated, a CNF formula will be referred to as a formula. Formulas are represented by letters with calligraphic fonts, e.g. $\mathcal{F}$, $\mathcal{E}$, $\mathcal{S}$, $\mathcal{W}$, etc. When necessary, subscripts are used. Clauses are represented by c or c_i, $i = 1, \ldots, m$. A CNF formula can also be viewed as a multiset of non-tautologous clauses where a clause is a (multi)set of literals. The two representations are used interchangeably, and are clear from the context. A *truth assignment* μ is a mapping from X to $\{0, 1\}$, $\mu : X \to \{0, 1\}$. The truth value of a clause c or formula $\mathcal{F}$, given a truth assignment μ, is represented as $c[\mu]$ and $\mathcal{F}[\mu]$, respectively. A truth assignment is a *model* if it satisfies all clauses in $\mathcal{F}$, i.e. $\mathcal{F}[\mu] = 1$. Two formulas $\mathcal{F}_1$ and $\mathcal{F}_2$ are *equivalent*, $\mathcal{F}_1 \equiv \mathcal{F}_2$, if they have the same set of models. The negation of a clause c, denoted by $\neg c$ represents a conjunction of unit clauses, one for each literal in c. It will also be necessary to negate CNF formulas, e.g. $\neg \mathcal{F}$. The following CNF encoding for $\neg \mathcal{F}$ is used [34]. An auxiliary variable u_i is associated with each $c_i \in \mathcal{F}$, defining a set of variables U. For each $l_{i,j} \in c_i$, create binary clauses $(\neg l_{i,j} \vee \neg u_i)$. Finally, create a clause $(\vee_{u_i \in U} u_i)$. This CNF encoding is represented as $\mathrm{CNF}(\cdot)$. For example, $\mathrm{CNF}(\neg c)$ and $\mathrm{CNF}(\neg \mathcal{F})$ denote, respectively, the CNF encoding of $\neg c$ and $\neg \mathcal{F}$ described above; both $\mathrm{CNF}(\neg c)$ and $\mathrm{CNF}(\neg \mathcal{F})$ can be used in set context to denote sets of clauses. Calls to a SAT solver are represented with $\mathrm{SAT}(\cdot)$.

2.1   MUSes and MESes

The following definition of Minimal Unsatisfiable Subformulas (MUSes) is used [6].

Definition 1 (MUS). $\mathcal{M} \subseteq \mathcal{F}$ *is a* Minimal Unsatisfiable Subformula *(MUS) iff* $\mathcal{M}$ *is unsatisfiable and* $\forall_{\mathcal{S} \subsetneq \mathcal{M}}, \mathcal{S}$ *is satisfiable.*

MUS extraction algorithms can be broadly characterized as deletion-based [8,3] or insertion-based. Moreover, insertion-based algorithms can be characterized as linear search [11] or dichotomic search [23,21]. Recent work proposed insertion-based MUS extraction with relaxation variables and AtMost1 constraints [28]. In practice, the most efficient MUS extraction algorithms are organized as deletion-based but operate as insertion-based to allow the integration of essential pruning techniques. These algorithms are referred to as *hybrid* [28]. Examples of pruning techniques used to reduce the number of SAT solver calls include *clause set trimming* (during preprocessing), *clause set refinement* and *model rotation* [13,29,28,4]. Clause set refinement [13,28] exploits the SAT solver *false* (or *unsatisfiable*) outcomes to reduce the set of clauses that need to be analyzed. This consists of removing clauses that are *not* included in the MUS being constructed, e.g. by restricting the target set of clauses to the unsatisfiable subset computed by the

SAT solver. In contrast, model rotation exploits the SAT solver *true* (or *satisfiable*) outcomes to also reduce the set of clauses that need to be analyzed. In this case, models are used to identify clauses that *must* be included in the MUS being constructed. Recent experimental data [28] indicates that these two techniques are *essential* for MUS extraction on large application problem instances.

Motivated by several applications, MUSes and related concepts have been extended to CNF formulas where clauses are partitioned into disjoint sets called *groups* [26,29].

Definition 2 (Group-Oriented MUS). *Given an explicitly partitioned unsatisfiable CNF formula $\mathcal{F} = \mathcal{D} \cup \mathcal{G}_1 \cdots \cup \mathcal{G}_n$, a group oriented MUS (or, group-MUS) of $\mathcal{F}$ is a subset $\mathcal{F}' = \mathcal{D} \cup \mathcal{G}_{i_1} \cup \cdots \cup \mathcal{G}_{i_k}$ of $\mathcal{F}$ such that $\mathcal{F}'$ is unsatisfiable and, for every $1 \leq j \leq k$, $\mathcal{F}' \setminus \mathcal{G}_{i_j}$ is satisfiable.*

The group $\mathcal{D}$ in the above definition is called a *don't care* group, and the explicitly partitioned CNF formulas as above are referred to as *group-CNF* formulas.

MUSes are a special case (for unsatisfiable formulas) of irredundant subformulas [25]. The following definitions will be used throughout.

Definition 3 (Redundant/Irredundant Clause/Formula). *A clause $c \in \mathcal{F}$ is said to be* redundant *in $\mathcal{F}$ if $\mathcal{F} \setminus \{c\} \models c$ or, equivalently, $\mathcal{F} \setminus \{c\} \cup \mathrm{CNF}(\neg c) \models \bot$. Otherwise, c is said to be* irredundant *in $\mathcal{F}$. A formula $\mathcal{F}$ is redundant if it has at least one redundant clause; otherwise it is irredundant.*

Irredundant subformulas of (redundant) formulas are referred to as *irredundant equivalent subsets* [25] and as *irredundant cores* [24]. In this paper, irredundant subformulas are referred to as *Minimal Equivalent Subformulas* (MESes), by analogy with MUSes.

Definition 4 (MES). $\mathcal{E} \subseteq \mathcal{F}$ *is a Minimal Equivalent Subformula (MES) iff $\mathcal{E} \equiv \mathcal{F}$ and $\forall_{\mathcal{Q} \subsetneq \mathcal{E}}, \mathcal{Q} \not\equiv \mathcal{F}$.*

Clearly, an MES is irredundant. Moreover, deciding whether a CNF formula is an MES is D^P-complete [25]. In the case of group-CNF formulas, the concept of *group-oriented MES (group-MES)* can be defined analogously to Definition 2.

2.2   Related Work

MUSes find a wide range of practical applications, and have been extensively studied (see [18,14,27] for recent overviews, and [23,21,16,17] for connections with CSP). The problem of computing minimal (or irredundant) representations of CNF formulas (and related problems) has been the subject of extensive research (e.g. [2,20,7,25,24]). Complexity characterizations of redundancy problems in logic can be found in [25]. An algorithm for computing an MES based on the direct application of the definition of clause redundancy is studied in [7]. More recently, properties of MESes are studied in [24]. Approximate solutions for redundancy removal based on unit propagation are proposed in [15,31]. Applications (of restricted forms) of redundancy removal can

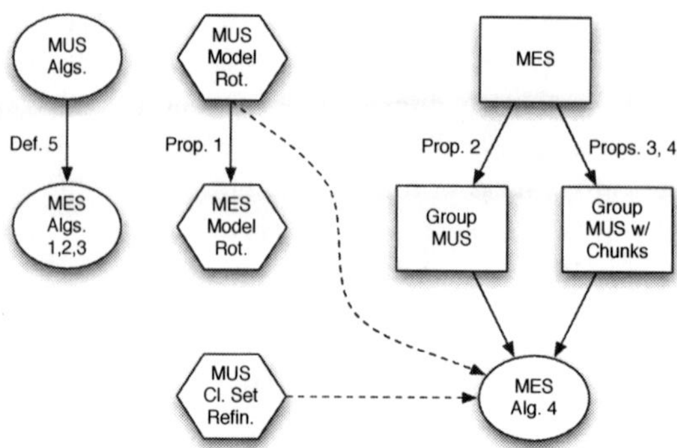

Fig. 1. Approaches to MES extraction

be found in [22,37,36,15,25,30]. The importance of redundant clauses in CDCL SAT solvers is addressed in [33,1]. Redundancy problems have been studied in many other settings, e.g. [12,10,9,35,19].

3   MES Extraction Algorithms

This section develops several new approaches for computing one MES of a CNF formula $\mathcal{F}$. The first solution consists of adapting *any* MUS extraction algorithm based on identification of so-called transition clauses, for MES extraction. Afterwards, key techniques used in MUS extraction are studied. Model rotation [28,4] is applied to MES extraction, and it is argued that clause set refinement [13,29,28] cannot be directly applied to MES extraction algorithms resulting from adapting existing MUS extraction algorithms. Next, a reduction from MES to group-MUS formulation [26,29] is developed, which enables the use of both model rotation and clause set refinement. Although the reduction of MES to group-MUS extraction enables the integration of key techniques, it is also the case that the resulting instances of SAT are hard to solve. This section concludes by developing an incremental reduction from MES to group-MUS extraction, which produces much easier instances of SAT. Figure 1 summarizes the approaches to MES extraction described in the remainder of this section.

3.1   From MUS Extraction to MES Extraction

A key definition in MUS extraction algorithms is that of *transition clause* [18]. A transition clause c is such that, if added to a satisfiable subformula $\mathcal{R}$, the resulting subformula is unsatisfiable. This definition can be generalized for MES extraction as follows. Let $\mathcal{R} \subsetneq \mathcal{F}$ denote a (reference) subformula of $\mathcal{F}$ which is to be extended to be equivalent to $\mathcal{F}$. Observe that $\mathcal{F} \vDash \mathcal{R}$, and so our goal is to extend $\mathcal{R}$ with a clause c, such that $\mathcal{R} \cup \{c\} \vDash \mathcal{F}$, and so $\mathcal{R} \cup \{c\} \equiv \mathcal{F}$.

Definition 5 (Witness of Equivalence). *Let $\mathcal{S}$ denote a subformula of $\mathcal{F}$, $\mathcal{S} \subsetneq \mathcal{F}$, with $\mathcal{S} \nvDash \mathcal{F} \setminus \mathcal{S}$, and let $c \in \mathcal{F} \setminus \mathcal{S}$. If $\mathcal{S} \cup \{c\} \vDash \mathcal{F} \setminus (\mathcal{S} \cup \{c\})$, then c is a witness of $\mathcal{S} \cup \{c\} \equiv \mathcal{F}$.*

Algorithm 1: Plain deletion-based MES extraction

> **Input** : Formula $\mathcal{F}$
> **Output**: MES $\mathcal{E}$
> 1 **begin**
> 2 $\mathcal{S} \leftarrow \mathcal{F}$
> 3 $\mathcal{E} \leftarrow \emptyset$ // MES under-approximation
> 4   **while** $\mathcal{S} \neq \emptyset$ **do**
> 5 $c \leftarrow$ SelectRemoveClause($\mathcal{S}$)   // Get a clause from $\mathcal{S}$
> 6 $\mathcal{W} \leftarrow \{c\}$
> 7   **if** SAT($\mathcal{E} \cup \mathcal{S} \cup$ CNF($\neg\mathcal{W}$)) **then**
> 8 $\mathcal{E} \leftarrow \mathcal{E} \cup \{c\}$ // Add clause c to $\mathcal{E}$ if $\mathcal{E} \cup \mathcal{S} \nvDash c$
> 9   **return** $\mathcal{E}$ // Final $\mathcal{E}$ is MES
> 10 **end**

Algorithm 2: Plain insertion-based MES extraction

> **Input** : Formula $\mathcal{F} = \{c_1, \ldots, c_m\}$
> **Output**: MES $\mathcal{E}$
> 1 **begin**
> 2 $\mathcal{W} \leftarrow \mathcal{F}$
> 3 $\mathcal{E} \leftarrow \emptyset$ // MES under-approximation
> 4   **while** $\mathcal{W} \neq \emptyset$ **do**
> 5 $(\mathcal{S}, c_r) \leftarrow (\emptyset, \emptyset)$
> 6   **while** SAT($\mathcal{E} \cup \mathcal{S} \cup$ CNF($\neg\mathcal{W}$)) **do**
> 7 $c_r \leftarrow$ SelectRemoveClause($\mathcal{W}$)   // Extend $\mathcal{S}$ while $\mathcal{E} \cup \mathcal{S} \nvDash \mathcal{W}$
> 8 $\mathcal{S} \leftarrow \mathcal{S} \cup \{c_r\}$
> 9 $\mathcal{E} \leftarrow \mathcal{E} \cup \{c_r\}$ // c_r is in MES
> 10 $\mathcal{W} \leftarrow \mathcal{S} \setminus \{c_r\}$
> 11   **return** $\mathcal{E}$ // Final $\mathcal{E}$ is an MES
> 12 **end**

The above definition can now be used to adapt *any* MUS extraction algorithm for MES extraction. The remainder of this section illustrates how this can be achieved. Let $\mathcal{F}$ be a CNF formula partitioned as follows, $\mathcal{F} = \mathcal{E} \cup \mathcal{R} \cup \mathcal{S}$. A sub-formula $\mathcal{R}$ is redundant in $\mathcal{F}$ iff $\mathcal{E} \cup \mathcal{S} \vDash \mathcal{R}$. Given an under-approximation $\mathcal{E}$ of an MES of $\mathcal{F}$, and a working subformula $\mathcal{S} \subsetneq \mathcal{F}$, the objective is to decide whether $\mathcal{E}$ and $\mathcal{S}$ entail all the other clauses of $\mathcal{F}$. MUS extraction algorithms can be organized as (see Section 2.1): (i) deletion-based [8,3], (ii) insertion-based [11,38], (iii) insertion-based with dichotomic search [23,21], and (iv) insertion-based with relaxation variables [28].

The pseudo-code for deletion-based MES extraction is shown in Algorithm 1. At each step, the algorithm uses an MES under-approximation $\mathcal{E}$, a set of (remaining) clauses $\mathcal{S}$, and a target clause c, to check whether $\mathcal{E} \cup \mathcal{S} \vDash c$. If this is not the case, i.e. if $\mathcal{E} \cup \mathcal{S} \nvDash c$, then c is witness of $\mathcal{E} \cup \mathcal{S} \cup \{c\} \equiv \mathcal{F}$, and so c is added to $\mathcal{E}$; otherwise, c is discarded. Observe that the deletion-based MES extraction algorithm corresponds to the direct implementation of the definition of redundant clause (e.g. see [7]).

Algorithm 3: MES extraction with dichotomic search

Input : Formula $\mathcal{F} = \{c_1, \ldots, c_m\}$
Output: MES $\mathcal{E}$

```
 1 begin
 2 │   W ← F
 3 │   E ← ∅                                    // MES under-approximation
 4 │   while W ≠ ∅ do
 5 │   │   (min, mid, max) ← (0, 0, |W|)
 6 │   │   repeat
 7 │   │   │   S ← {c₁, …, c_mid}               // Extract sub-sequence of W
 8 │   │   │   if SAT(E ∪ S ∪ CNF(¬(W \ S))) then
 9 │   │   │   │   min ← mid + 1                // Extend S if E ∪ S ⊭ W \ S
10 │   │   │   else
11 │   │   │   │   max ← mid                    // Reduce S if E ∪ S ⊨ W \ S
12 │   │   │   mid ← ⌊(min + max)/2⌋
13 │   │   until min = max
14 │   │   if min > 0 then
15 │   │   │   E ← E ∪ {c_min}
16 │   │   W ← {cᵢ | i < min}
17 │   return E                                 // Final E is an MES
18 end
```

The pseudo-code for insertion-based linear and dichotomic search MES extraction are shown in Algorithms 2 and 3, respectively. Both algorithms iteratively add clauses to set $\mathcal{S}$ while $\mathcal{E} \cup \mathcal{S} \nvDash \mathcal{W}$. The last clause included in $\mathcal{S}$ such that $\mathcal{E} \cup \mathcal{S} \vDash \mathcal{W}$ is the witness of equivalence. The main difference between algorithms 2 and 3 is how the witness of equivalence is searched for.

Recent experimental data indicates that the most efficient MUS extraction algorithms are deletion-based (or variants) [28]. Since insertion-based MES algorithms require SAT solver calls with (possibly large) complemented formulas, deletion-based MES extraction algorithms are expected to outperform the insertion-based ones. This is confirmed by the results in Section 5.

Efficient MUS extraction algorithms [13,29,28] *must* use a number of additional techniques for reducing the number of SAT solver calls. These techniques include *clause set refinement* [13,28] and *model rotation* [28,4]. The next section shows how model rotation can be used in MES extraction. In contrast, clause set refinement *cannot* be applied in the algorithms described above.

Example 1. Let $\mathcal{F} = (x_1) \wedge (x_1 \vee x_3) \wedge (x_2)$, and consider the execution of Algorithm 1. First, clause (x_1) is removed and the resulting formula is satisfiable with $x_1 = 0$, $x_2 = x_3 = 1$; hence (x_1) is irredundant. Second, clause $(x_1 \vee x_3)$ is removed and the resulting formula is unsatisfiable; hence $(x_1 \vee x_3)$ is redundant. Moreover, the computed unsatisfiable core is $\{(x_1), (\neg x_1)\}$, where $(\neg x_1)$ is taken from $\neg(x_1 \vee x_3)$. However, (x_2) is not in the unsatisfiable core, but it cannot be removed.

As illustrated by the previous example, since MES extraction algorithms require adding the negation of a subformula of $\mathcal{F}$, the computed unsatisfiable cores depend on this negation, which changes as the algorithm executes. Thus, a computed unsatisfiable core provides no information about which clauses need not be considered further. Despite this negative result, Sections 3.3 and 3.4 develop solutions for MES extraction that enable clause set refinement.

3.2   Using Model Rotation in MES Extraction

The definition of irredundant clause (see Definition 3) can be associated with specific truth assignments, that can serve as *witnesses* of irredundancy.

Proposition 1. *A clause $c \in \mathcal{F}$ is irredundant in $\mathcal{F}$ if and only if there exists a truth assignment μ, such that $(\mathcal{F} \setminus \{c\})[\mu] = 1$ and $c[\mu] = 0$.*

Proof. Follows directly from the definition of an irredundant clause and the semantics of $\mathcal{F} \setminus \{c\} \not\vdash c$. $\square$

In the context of MUS extraction, Proposition 1 is used as a basis of *model rotation* [28,4] — a powerful optimization technique that in practice allows to significantly reduce the number of calls to SAT solver, and results in multiple orders of magnitude reduction in run-times of hybrid MUS extraction algorithms on industrial problem instances (cf. [4]). Proposition 1 has also been used in the context of local search for MUS extraction [32].

In the context of MES extraction, model rotation can be instrumented as follows. If a SAT solver call returns satisfiable (see, for example, line 7 in Algorithm 1), then we can use the model computed by the solver to look for other irredundant clauses using Proposition 1. Note that the clauses of the negation of the working formula (e.g. $\mathrm{CNF}(\neg\mathcal{W})$ in Algorithm 1) can be disregarded, and the objective is to modify the returned model such that *exactly* one other clause of the working formula becomes falsified. Iteratively, each literal from the currently falsified clause is flipped, and one checks whether the formula has another *single* falsified clause, which can then be declared irredundant according to Proposition 1, *without* a SAT solver call. This process is continued recursively from the newly detected irredundant clause.

Model rotation for MES extraction can be improved further. Since the formula is satisfiable, model rotation may reach assignments where *all* clauses are satisfied (note that this situation is not possible in MUS extraction). In this case, rather than terminating the process, clauses with the smallest number of satisfied literals are selected, the satisfied literals are flipped, and again one checks whether the formula has a single unsatisfied clause. As demonstrated in Section 5, this *improved model rotation* is very effective for redundancy removal.

3.3   A Reduction of MES to Group-MUS

As argued in Section 3.1, although MUS extraction algorithms can be modified for MES extraction, clause set refinement [13,29,28] cannot be used. In the context of MUS computation, this technique is *paramount* for reducing the number of SAT solver calls in instances with many redundant clauses. This section develops a reduction of MES computation problem to group-MUS computation problem [26,29] — recall Definition 2. A key advantage of this reduction is that it enables clause set refinement.

Proposition 2 (MES to Group-MUS Reduction). *Given a CNF formula $\mathcal{F}$, and any $\mathcal{E} \subseteq \mathcal{F}$, let $R_{\mathcal{F}}(\mathcal{E})$ be the group-CNF formula $\mathcal{D} \cup \bigcup_{c \in \mathcal{E}} \mathcal{G}_c$, where $\mathcal{D} = \mathrm{CNF}(\neg\mathcal{F})$ is the don't care group, and $\mathcal{G}_c = \{c\}$. Then, $\mathcal{E}$ is an MES of $\mathcal{F}$ if and only if $R_{\mathcal{F}}(\mathcal{E})$ is a group-MUS of $R_{\mathcal{F}}(\mathcal{F})$.*

Proof. We prove both directions simultaneously. Since $R_{\mathcal{F}}(\mathcal{E}) = \mathrm{CNF}(\neg\mathcal{F}) \cup \mathcal{E}$, we have that $\mathcal{E} \vDash \mathcal{F}$ (and so $\mathcal{E} \equiv \mathcal{F}$), if and only if $R_{\mathcal{F}}(\mathcal{E})$ is unsatisfiable. Let c be any clause of $\mathcal{E}$. Then, by Proposition 1, c is irredundant in $\mathcal{E}$ if and only if there exists an assignment μ, such that $(\mathcal{E} \setminus \{c\})[\mu] = 1$ and $c[\mu] = 0$. Equivalently, there is an extension μ' of μ such that $\mathrm{CNF}(\neg\mathcal{F})[\mu'] = 1$ and $(\mathcal{E} \setminus \{c\})[\mu'] = 1$, i.e. $R_{\mathcal{F}}(\mathcal{E} \setminus \{c\})$ is satisfiable. $\square$

Expressing the MES computation problem as a group-MUS computation problem enables the use of optimization techniques for (group-)MUS extraction in computing irredundant CNF formulas. Most importantly, the clause-set refinement becomes usable and effective again. We demonstrate this with the following example.

Example 2. Consider the CNF formula $\mathcal{F} = (x_1) \wedge (x_1 \vee y_i \vee y_j)$, with $1 \leq i < j \leq k$, and $k \geq 2$. All clauses with a literal in the y variables are redundant, for a total of $k(k-1)/2$ redundant clauses. The reduction in Proposition 2 produces the group-CNF formula $R_{\mathcal{F}}(\mathcal{F})$ with the don't care group $\mathcal{D} = \mathrm{CNF}(\neg\mathcal{F})$, and a singleton group for each clause in $\mathcal{F}$, i.e. $\mathcal{G}_{11} = \{(x_1)\}$, $\mathcal{G}_{ij} = \{(x_1 \vee y_i \vee y_j)\}$, with $1 \leq i < j \leq k$. For this example, let us assume that the group-MUS of $R_{\mathcal{F}}(\mathcal{F})$ is computed using a deletion-based algorithm. Let $\mathcal{G}_{11}$ be the first group to be analyzed. This is done by removing the clause in $\mathcal{G}_{11}$ from the formula. The resulting formula has a model μ, with $\mu(x_1) = 0$ and $\mu(y_i) = \mu(y_j) = 1$, with $1 \leq i < j \leq k$. As a result, (x_1) is declared irredundant, and added back to the formula. Afterwards, pick one of the other groups, e.g. $\mathcal{G}_{12}$. If the corresponding clause is removed from the formula, the resulting formula is unsatisfiable, and so the clause is declared redundant. More importantly, a CDCL SAT solver will produce an unsatisfiable core $\mathcal{U} \subseteq \{x_1\} \cup \mathrm{CNF}(\neg\mathcal{F})$. Hence, clause set refinement serves to eliminate *all* of the remaining groups of clauses, and so the MES is computed with *two* SAT solver calls. As noted earlier, MES algorithms based on adapting existing MUS algorithms are unable to implement clause set refinement, and so cannot drop $k(k-1)/2$ clauses after the second SAT solver call.

Observe that the *group-MUS approach* to the MES extraction problem, i.e. using the reduction in Proposition 2, is *independent* of the actual group-MUS extraction algorithm used. Hence, any existing group-MUS extraction algorithm can be used for the MES extraction problem. In practice, given the significant performance difference between deletion-based and insertion-based MUS extraction algorithms [28], our implementation is based on deletion-based group-MUS extraction and its most recent instantiation, i.e. the *hybrid* approach in [28].

3.4   Incremental Reduction of MES to Group-MUS

A major drawback of the group-MUS approach is that for large input formulas the resulting instances of SAT can be hard. This is due to the CNF encoding of $\neg\mathcal{F}$ that produces a large disjunction of auxiliary variables. A solution to this issue is based on an *incremental* reduction of MES extraction to group-MUS extraction. Let $\mathcal{T}$ be any subset of clauses of $\mathcal{F}$ — we refer to clauses of $\mathcal{T}$ as *target* clauses, and to the set $\mathcal{T}$ itself as a *chunk* of $\mathcal{F}$. The incremental reduction is based on the observation that in group-MUS approach the redundancy of any target clause $c \in \mathcal{T}$ can be established by analysing c with respect to $\mathrm{CNF}(\neg\mathcal{T})$ rather than $\mathrm{CNF}(\neg\mathcal{F})$. This observation is stated precisely below.

Proposition 3. *Let $\mathcal{T} \subseteq \mathcal{F}$ be a set of target clauses. For any $\mathcal{E} \subseteq \mathcal{T}$, let $R_\mathcal{T}(\mathcal{E})$ be the group-CNF formula $\mathcal{D} \cup \bigcup_{c \in \mathcal{E}} \mathcal{G}_c$, where $\mathcal{D} = \mathcal{F} \setminus \mathcal{T} \cup \mathrm{CNF}(\neg\mathcal{T})$ is the don't care group, and $\mathcal{G}_c = \{c\}$. Then, $\mathcal{E}$ is irredundant in $\mathcal{F}$ and $\mathcal{F} \setminus \mathcal{T} \cup \mathcal{E} \equiv \mathcal{F}$ if and only if $R_\mathcal{T}(\mathcal{E})$ is a group-MUS of $R_\mathcal{T}(\mathcal{T})$.*

Note that $R_\mathcal{T}(\mathcal{T})$ is unsatisfiable. Also, in the case when the chunk $\mathcal{T}$ is taken to be the whole formula $\mathcal{F}$, the group-CNF formula $R_\mathcal{T}(\mathcal{E})$ is exactly the formula $R_\mathcal{F}(\mathcal{E})$ from Proposition 2, and so the claim of Proposition 2 is a special case of the claim of Proposition 3. We omit the proof of Proposition 3 as it essentially repeats the steps of the proof of Proposition 2, using the definition of $R_\mathcal{T}(\mathcal{E})$ instead of $R_\mathcal{F}(\mathcal{E})$.

For the general case, consider a partition $\mathcal{F}_1, \ldots, \mathcal{F}_k$ of $\mathcal{F}$ into chunks. Then, an MES of $\mathcal{F}$ can be computed by applying the group-MUS approach of Proposition 3 to each chunk. Proposition 3 is applied in order, with already computed irredundant subformulas replacing the original (redundant) subformulas. Explicitly, for the iteration j, $1 \leq j \leq k$, the input group-CNF formula in Proposition 3 is defined as follows:

$$\mathcal{D} \cup \bigcup_{c \in \mathcal{F}_j} \mathcal{G}_c, \text{ with } \mathcal{D} = \mathcal{E}_1 \cup \ldots \cup \mathcal{E}_{j-1} \cup \mathcal{F}_{j+1} \cup \ldots \mathcal{F}_k \cup \mathrm{CNF}(\neg\mathcal{F}_j) \qquad (1)$$

as a don't care group, where $\mathcal{E}_i$, $1 \leq i < j$, is the computed irredundant set of clauses in $\mathcal{F}_i$, and, as before, $\mathcal{G}_c = \{c\}$.

Proposition 4. *Let $\mathcal{F}_1, \ldots, \mathcal{F}_k$ be a partition of $\mathcal{F}$ into chunks. Let $\mathcal{E}_j$ be the set of clauses obtained from applying group-MUS extraction to formula in (1), and by considering each $\mathcal{F}_j$ as the set of target clauses. Let $\mathcal{E} = \mathcal{E}_1 \cup \mathcal{E}_2 \cup \ldots \cup \mathcal{E}_k$. Then $\mathcal{E}$ is an MES of $\mathcal{F}$ [1].*

Proof sketch. The proof uses induction on the sets $\mathcal{F}_j$ and Proposition 3 to prove the following inductive invariant: for $1 \leq j \leq k$, $\left(\bigcup_{r \leq j} \mathcal{E}_r \cup \bigcup_{r > j} \mathcal{F}_r \right) \equiv \mathcal{F}$, and the formula $\bigcup_{r \leq j} \mathcal{E}_r$ is irredundant in $\mathcal{F}$.
Base case: Take $\mathcal{T} = \mathcal{F}_1$, and apply Proposition 3 to formula $\mathcal{F}$.
Inductive step: For $1 < j \leq k$, take $\mathcal{T} = \mathcal{F}_j$, and apply Proposition 3 to the formula $\bigcup_{r < j} \mathcal{E}_r \cup \bigcup_{r \geq j} \mathcal{F}_r$. Then, using the inductive hypothesis, establish the required invariant. $\qquad\square$

---

[1] The proposition is stated slightly informally as to avoid additional notation.

Algorithm 4: Deletion-based group-MUS extraction of MES with chunks

```
    Input   : Formula F = {F_1, ..., F_k} with k chunks
    Output: MES E
 1  begin
 2  │   for j ← 1 to k do                          // Analyze each of the k chunks
 3  │   │   D = E_1 ∪ ... ∪ E_{j-1} ∪ F_{j+1} ∪ ... ∪ F_k ∪ CNF(¬F_j) // Don't care group
 4  │   │   W = F_j                                 // Target group of clauses F_j
 5  │   │   E_j ← ∅                                 // Irredundant clauses in chunk j
 6  │   │   while W ≠ ∅ do
 7  │   │   │   c ← SelectRemoveClause(W)
 8  │   │   │   (st, ν, U) = SAT(D ∪ E_j ∪ W)
 9  │   │   │   if st = true then                   // If SAT, c is irredundant in F
10  │   │   │   │   E_j ← E_j ∪ {c}
11  │   │   │   │   (W, E_j) ← Rotate(W, E_j, ν)     // Apply model rotation
12  │   │   │   else
13  │   │   │   │   W ← U ∩ W                        // Clause-set refinement
14  │   │   E = E_1 ∪ ... ∪ E_k
15  │   │   return E                                 // E is an MES
16  end
```

While Proposition 4 can be used to compute an MES of an input formula $\mathcal{F}$ by iteratively calling a group-MUS extractor, it can also be integrated into a unified algorithm to enable certain optimizations (e.g. incremental SAT solving). Algorithm 4 shows the pseudo-code for deletion-based group-MUS approach to MES extraction using chunks. Different chunk sizes can be considered. Concrete examples include chunks of size 1 or a single chunk aggregating all clauses (i.e. the reduction to group-MUS defined in Proposition 2). For chunks of size great than 1, the group-MUS approach has the ability to prove several clauses redundant by using clause-set refinement. Generally, the chunk size in the group-MUS approach controls the *trade-off* between the potential power of clause-set refinement and the difficulty of the instances of SAT given to the SAT solver.

Nevertheless, an essential issue with Algorithm 4 is the selection of the size of the chunks. The current implementation of the algorithm uses chunks of fixed size, independently of the size of the formula. Alternative solutions include using chunk sizes dependent on the size of the formula, and also *adaptive* chunk sizes.

4   Certification of Correctness

In some applications, it is paramount to guarantee that the computed subformula is indeed irredundant and, possibly more importantly, that each computed irredundant subformula $\mathcal{E}$ is equivalent to the original CNF formula $\mathcal{F}$. Given a computed CNF formula $\mathcal{E}$, it is simple to validate whether it is irredundant. Essentially, one can run one of the algorithms outlined in earlier sections. The problem of validating whether $\mathcal{E} \equiv \mathcal{F}$ looks more challenging.

To check whether $\mathcal{E} \equiv \mathcal{F}$, it suffices to exhibit a truth assignment that is a model of one formula and not of the other. This condition corresponds to testing the satisfiability of the following CNF formula:

$$(\mathcal{E} \wedge \mathrm{CNF}(\neg\mathcal{F})) \vee (\mathcal{F} \wedge \mathrm{CNF}(\neg\mathcal{E})) \tag{2}$$

Clearly, the resulting instances of SAT are expected to be hard to solve, given the disjunction and the negation of formulas in (2). Nevertheless, it is also the case that $\mathcal{E} \subseteq \mathcal{F}$ and so, $\mathcal{F} \vDash \mathcal{E}$. Hence it is only necessary to check whether $\mathcal{E} \vDash \mathcal{F}$. Since $\mathcal{F} = \mathcal{E} \wedge \mathcal{R}$, $\mathcal{E} \vDash \mathcal{F}$ can be represented as $\mathcal{E} \vDash \mathcal{E} \wedge \mathcal{R}$, or equivalently, $\mathcal{E} \wedge (\neg\mathcal{E} \vee \neg\mathcal{R}) \vDash \bot$, which simplifies to $\mathcal{E} \wedge \neg\mathcal{R} \vDash \bot$. This condition corresponds to testing the satisfiability of the following CNF formula:

$$\mathcal{E} \cup \mathrm{CNF}(\neg\mathcal{R}) \tag{3}$$

If $\mathcal{E} \cup \mathrm{CNF}(\neg\mathcal{R})$ has a model, then one can satisfy $\mathcal{E}$ while unsatisfying one or more clauses in $\mathcal{R}$. Hence, the two formulas would not be equivalent.

Although easier than (2), (3) can still result in hard instances of SAT (similarly to what happens with the reduction of MES to group-MUS extraction). A technique to reduce the complexity of the resulting instances of SAT is to partition $\mathcal{R}$ into chunks, and check each chunk separately for equivalence. The objective is to check whether $\mathcal{E} \vDash \mathcal{R}$. If this condition holds, then $\mathcal{E} \equiv \mathcal{F}$; otherwise $\mathcal{E} \not\equiv \mathcal{F}$. If $\mathcal{R}$ is partitioned into a number of subformulas (or chunks), we get $\mathcal{R} = \mathcal{R}_1 \wedge \ldots \wedge \mathcal{R}_k$, and so the condition becomes, $\mathcal{E} \vDash \mathcal{R}_1 \wedge \ldots \wedge \mathcal{R}_k$, that can also be represented as $\mathcal{E} \wedge (\neg\mathcal{R}_1 \vee \ldots \vee \neg\mathcal{R}_k) \vDash \bot$. This condition holds if and only if, $\forall_{1 \le j \le k}$, $\mathcal{E} \wedge \neg\mathcal{R}_j \vDash \bot$. Hence, the use of chunks allows splitting a potentially hard (and believed unsatisfiable) instance of SAT, into k (likely) easier (and also believed unsatisfiable) instances of SAT.

5   Experimental Results

The algorithms described in the previous sections were implemented within the MUS extraction framework of a state-of-the-art MUS extractor `MUSer2` [2]; the framework was configured to use SAT solver `picosat-935` [5] in the incremental mode. The experiments were performed on an HPC cluster, where each node is dual quad-core Intel Xeon E5450 3 GHz with 32 GB of memory. Each algorithm was run with a timeout of 1800 seconds and a memory limit of 4 GB per input instance. To evaluate the algorithms, we selected 300 problem instances from practical application domains of SAT used in past SAT competitions [3]. The instances were selected using the following criteria: the instance is solvable within 1 second by `picosat-935`, and the number of clauses in the instance is less than 100,000. These criteria were derived from the worst-case analysis of deletion-based MES algorithms (whereby the number of SAT calls is linear in the size of the input formula) and our previous experience with the effects various optimization techniques in the context of MUS extraction.

---

[2] `http://logos.ucd.ie/wiki/doku.php?id=muser`.
[3] `http://www.satcompetition.org/`.

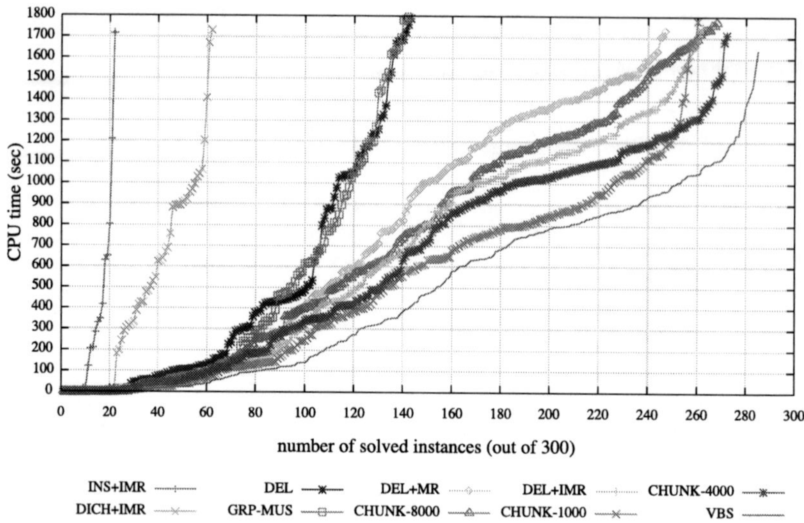

Fig. 2. Cactus plot with the run times of all algorithms

The cactus plot[4] in Fig. 2 provides an overview of the results of our experimental study. The legend in this, and the subsequent plots, is as follows: DEL represents the implementation of the deletion-based algorithm (Algorithm 1); INS (resp. DICH) is the implementation of insertion-based (resp. dichotomic) algorithms from Section 3.1; +MR indicates the addition of model rotation, while +IMR indicates the addition of the improved version of model rotation (cf. Section 3.2); GRP-MUS is the implementation of the reduction of MES to group-MUS from Section 3.3; CHUNK-x is the implementation of the deletion-based chunked group-oriented MUS algorithm from Section 3.3 with chunk size x; VBS refers to a *virtual best solver* [5] — we elaborate on its composition shortly.

A number of conclusions can be drawn from the plot in Fig. 2. First, we note that the improvements to the plain deletion-based algorithm (DEL) suggested in Section 3.1, namely the addition of model rotation (DEL+MR) and the improved model rotation (DEL+IMR), have a very significant positive effect on the performance of the algorithm; also observe that the improved model rotation provides a notable boost over model rotation. Further, it is clear that the the insertion-based and the dichotomic algorithms, even with the addition of IMR, do not scale. As a side note, the gap between the performance of these and the deletion-based algorithm in the MES setting is *significantly* larger than that in the MUS setting (see for example [28]) — this is due to the addition of the negation of the working formula, which is unavoidable in the MES setting. We

---

⁴ Cactus plots show the sorted run times of algorithms over all instances and are commonly used in SAT competitions to compare performance of multiple solvers.

⁵ VBS represents a solver obtained from running a number of algorithms in parallel. It can be seen as a naive portfolio solver — we note that portfolio solvers are the current tour de force in SAT solving.

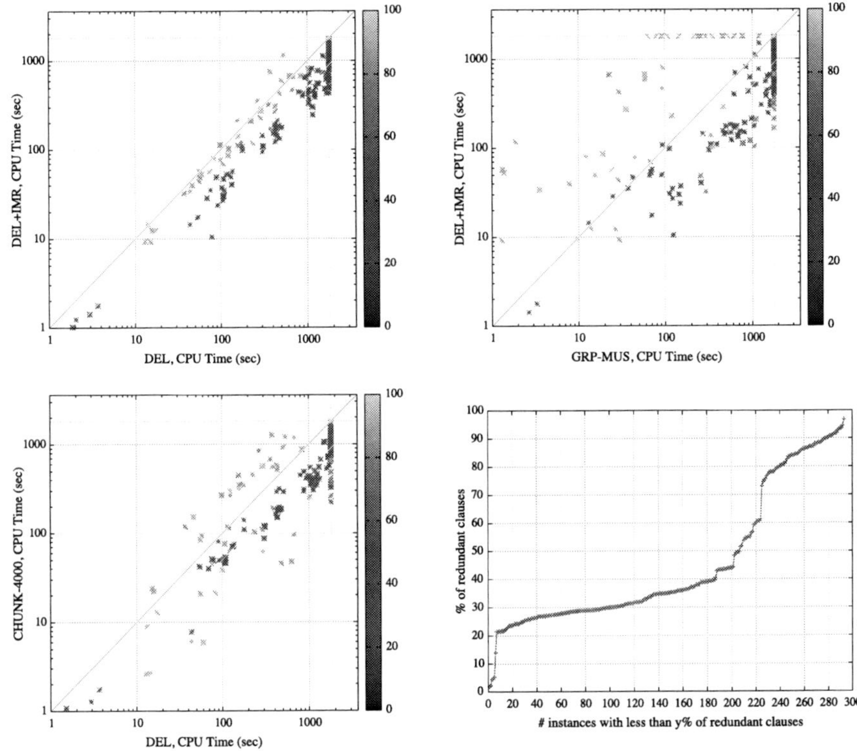

Fig. 3. Top and bottom-left: selected scatter plots (timeout 1800 sec.); color range represents % of redundant clauses in an instance. Bottom-right: cumulative histogram of the % of redundant clauses in the set of instances; note that the histogram is rotated 90° to be consistent with the cactus plot in Fig.2.

also observe the weak performance of the GRP-MUS approach — this is not surprising, since for large formulas, GRP-MUS can produce hard instances of SAT; this deficiency, in fact, was the motivation for the chunked approach. While DEL+IMR is among the best performing algorithms, on most of the instances it is significantly outperformed by the chunked group-oriented MUS algorithm with chunk size 1000. However, CHUNK-1000 loses to DEL+IMR on some of the hard instances — increasing the chunk size to 4000 pushes the performance of the algorithm ahead, however a further increase of chunk size to 8000 begins to affect the performance negatively. The VBS in Fig. 2 is constructed from DEL+IMR, GRP-MUS, CHUNK-1000 and CHUNK-4000. The former two are taken because they represent the extremes of the chunked approach — chunks of size 1 for DEL+IMR and a single chunk of the size of the input formula for GRP-MUS. The fact that the results for the VBS configuration (285 solved instances) are clearly superior to any of the individual algorithms indicates that the proposed algorithms are highly complementary, and are suitable for a multi-core/portfolio implementation. We emphasize that the plain deletion-based DEL is, to our knowledge, the current published state-of-the-art in MES computation, and so the algorithms proposed in this paper constitute a significant advancement, with the best algorithms solving more than twice the number of instances within the timeout.

The scatter plots in Fig. 3 provide additional insights into the performance of some of the algorithms. The color code indicates the amount of redundant clauses in the formulas: yellow (lighter) indicates close to 100% redundant clauses, whereas blue (darker) indicates less than 20% redundant clauses. The top-left plot (DEL vs. DEL+IMR) demonstrates the impact of improved model rotation (IMR) on the performance of the deletion-based approach. We note that while IMR allows to solve significantly more instances (263 vs 144 – see Fig. 2), the technique has little impact on many highly redundant instances (yellow). This behaviour is expected as IMR can only help to detect irredundant clauses, and it was the motivation for the development of the group-MUS approach. The top-right plot of Fig. 3 (GRP-MUS vs. DEL+IMR) demonstrates the effectiveness of the group-MUS approach on the highly redundant instances. The plot clearly shows that the direct and the group-MUS approaches are complementary – the former excels on mostly irredundant instances, while the latter is best on highly redundant ones. This observation provides additional, empirical, justification for the development of the chunked group-oriented MUS algorithm. The bottom-left plot (DEL vs. CHUNK-4000) confirms the effectiveness of the chunked approach — we observe significant performance improvements on instances with diverse degrees of redundancy.

The bottom-right plot in Fig. 3 presents the cumulative histogram of the degree of redundancy in the problem instances used in our experiments. Note that the instances were selected from the SAT competition benchmark sets *prior* to the experiments, and so the selection was not biased by the degree of redundancy. Nevertheless, approximately 2/3 of the instances have between 20% and 50% redundant clauses, the remaining instances have over 50% redundant clauses, and close to 5% of the instances have in excess of 90% redundant clauses. We conclude that problem instances from practical applications may exhibit very significant levels of redundancy.

6   Conclusions

This paper proposes novel algorithms for computing MESes. The main contributions of the paper can be summarized as follows: (i) Adapting existing MUS extraction algorithms for MES extraction; (ii) Development of model rotation for MES extraction, and analysis of why clause set refinement cannot be applied; (iii) Reduction of MES to group-MUS extraction, which enables both model rotation and clause set refinement to be applied; (iv) Use of chunks for incremental reduction of MES to group-MUS extraction, aiming at reducing the hardness of instances of SAT when the formula includes the negation of a CNF formula; and (v) Development of a solution for the independent certification of the correctness of computed MESes. The experimental results indicate that the algorithms proposed in this paper improve the direct approach [7] significantly, more than doubling the number of instances that can be solved, and with significant performance gains, which can exceed one order of magnitude.

The experimental evaluation carried out in the paper demonstrates that the developed algorithms are relevant for the practical applications of SAT. Indeed, our results show that many real-world SAT instances have significant percentages of redundant clauses. These findings clearly motivate revisiting the CNF encoding techniques used for generating these instances.

Although in this paper we do not address group-MES computation problem explicitly, the algorithms developed in the paper can be extended without difficulty to this setting. This could enable the identification and the removal of redundant constraints in other domains, such as CSP [12,10,9], SMT [35], and Ontologies [19]. This is the subject of future work.

References

1. Atserias, A., Fichte, J.K., Thurley, M.: Clause-learning algorithms with many restarts and bounded-width resolution. J. Artif. Intell. Res. 40, 353–373 (2011)
2. Ausiello, G., D'Atri, A., Saccà, D.: Minimal representation of directed hypergraphs. SIAM J. Comput. 15(2), 418–431 (1986)
3. Bakker, R.R., Dikker, F., Tempelman, F., Wognum, P.M.: Diagnosing and solving over-determined constraint satisfaction problems. In: IJCAI, pp. 276–281 (1993)
4. Belov, A., Marques-Silva, J.: Accelerating MUS extraction with recursive model rotation. In: FMCAD, pp. 37–40 (2011)
5. Biere, A.: Picosat essentials. Journal on Satisfiability, Boolean Modeling and Computation 4, 75–97 (2008)
6. Biere, A., Heule, M., van Maaren, H., Walsh, T. (eds.): Handbook of Satisfiability. Frontiers in Artificial Intelligence and Applications, vol. 185. IOS Press (2009)
7. Boufkhad, Y., Roussel, O.: Redundancy in random SAT formulas. In: AAAI, pp. 273–278 (2000)
8. Chinneck, J.W., Dravnieks, E.W.: Locating minimal infeasible constraint sets in linear programs. INFORMS Journal on Computing 3(2), 157–168 (1991)
9. Chmeiss, A., Krawczyk, V., Sais, L.: Redundancy in CSPs. In: ECAI, pp. 907–908 (2008)
10. Choi, C.W., Lee, J.H.-M., Stuckey, P.J.: Removing propagation redundant constraints in redundant modeling. ACM Trans. Comput. Log. 8(4) (2007)
11. de Siqueira, N.J.L., Puget, J.-F.: Explanation-based generalisation of failures. In: ECAI, pp. 339–344 (1988)
12. Dechter, A., Dechter, R.: Removing redundancies in constraint networks. In: AAAI, pp. 105–109 (1987)
13. Dershowitz, N., Hanna, Z., Nadel, A.: A Scalable Algorithm for Minimal Unsatisfiable Core Extraction. In: Biere, A., Gomes, C.P. (eds.) SAT 2006. LNCS, vol. 4121, pp. 36–41. Springer, Heidelberg (2006)
14. Desrosiers, C., Galinier, P., Hertz, A., Paroz, S.: Using heuristics to find minimal unsatisfiable subformulas in satisfiability problems. J. Comb. Optim. 18(2), 124–150 (2009)
15. Fourdrinoy, O., Grégoire, É., Mazure, B., Saïs, L.: Eliminating Redundant Clauses in SAT Instances. In: Van Hentenryck, P., Wolsey, L.A. (eds.) CPAIOR 2007. LNCS, vol. 4510, pp. 71–83. Springer, Heidelberg (2007)
16. González, S.M., Meseguer, P.: Boosting MUS Extraction. In: Miguel, I., Ruml, W. (eds.) SARA 2007. LNCS (LNAI), vol. 4612, pp. 285–299. Springer, Heidelberg (2007)

17. Grégoire, É., Mazure, B., Piette, C.: MUST: Provide a Finer-Grained Explanation of Unsatisfiability. In: Bessière, C. (ed.) CP 2007. LNCS, vol. 4741, pp. 317–331. Springer, Heidelberg (2007)

18. Grégoire, É., Mazure, B., Piette, C.: On approaches to explaining infeasibility of sets of Boolean clauses. In: ICTAI, pp. 74–83 (November 2008)

19. Grimm, S., Wissmann, J.: Elimination of Redundancy in Ontologies. In: Antoniou, G., Grobelnik, M., Simperl, E., Parsia, B., Plexousakis, D., De Leenheer, P., Pan, J. (eds.) ESWC 2011, Part I. LNCS, vol. 6643, pp. 260–274. Springer, Heidelberg (2011)

20. Hammer, P.L., Kogan, A.: Optimal compression of propositional horn knowledge bases: Complexity and approximation. Artif. Intell. 64(1), 131–145 (1993)

21. Hemery, F., Lecoutre, C., Sais, L., Boussemart, F.: Extracting MUCs from constraint networks. In: ECAI, pp. 113–117 (2006)

22. Järvisalo, M., Heule, M.J.H., Biere, A.: Inprocessing Rules. In: Gramlich, B., Miller, D., Sattler, U. (eds.) IJCAR 2012. LNCS, vol. 7364, pp. 355–370. Springer, Heidelberg (2012)

23. Junker, U.: QUICKXPLAIN: Preferred explanations and relaxations for over-constrained problems. In: AAAI, pp. 167–172 (2004)

24. Kullmann, O.: Constraint satisfaction problems in clausal form II: Minimal unsatisfiability and conflict structure. Fundam. Inform. 109(1), 83–119 (2011)

25. Liberatore, P.: Redundancy in logic I: CNF propositional formulae. Artif. Intell. 163(2), 203–232 (2005)

26. Liffiton, M.H., Sakallah, K.A.: Algorithms for computing minimal unsatisfiable subsets of constraints. J. Autom. Reasoning 40(1), 1–33 (2008)

27. Marques-Silva, J.: Minimal unsatisfiability: Models, algorithms and applications. In: ISMVL, pp. 9–14 (2010)

28. Marques-Silva, J., Lynce, I.: On Improving MUS Extraction Algorithms. In: Sakallah, K.A., Simon, L. (eds.) SAT 2011. LNCS, vol. 6695, pp. 159–173. Springer, Heidelberg (2011)

29. Nadel, A.: Boosting minimal unsatisfiable core extraction. In: FMCAD, pp. 121–128 (October 2010)

30. Niepert, M., Gucht, D.V., Gyssens, M.: Logical and algorithmic properties of stable conditional independence. Int. J. Approx. Reasoning 51(5), 531–543 (2010)

31. Piette, C.: Let the solver deal with redundancy. In: ICTAI, pp. 67–73 (2008)

32. Piette, C., Hamadi, Y., Saïs, L.: Efficient Combination of Decision Procedures for MUS Computation. In: Ghilardi, S., Sebastiani, R. (eds.) FroCoS 2009. LNCS, vol. 5749, pp. 335–349. Springer, Heidelberg (2009)

33. Pipatsrisawat, K., Darwiche, A.: On the power of clause-learning SAT solvers as resolution engines. Artif. Intell. 175(2), 512–525 (2011)

34. Plaisted, D.A., Greenbaum, S.: A structure-preserving clause form translation. Journal of Symbolic Computation 2(3), 293–304 (1986)

35. Scholl, C., Disch, S., Pigorsch, F., Kupferschmid, S.: Computing Optimized Representations for Non-convex Polyhedra by Detection and Removal of Redundant Linear Constraints. In: Kowalewski, S., Philippou, A. (eds.) TACAS 2009. LNCS, vol. 5505, pp. 383–397. Springer, Heidelberg (2009)

36. To, S.T., Son, T.C., Pontelli, E.: On the use of prime implicates in conformant planning. In: AAAI (2010)

37. To, S.T., Son, T.C., Pontelli, E.: Conjunctive representations in contingent planning: Prime implicates versus minimal CNF formula. In: AAAI (2011)

38. van Maaren, H., Wieringa, S.: Finding Guaranteed MUSes Fast. In: Kleine Büning, H., Zhao, X. (eds.) SAT 2008. LNCS, vol. 4996, pp. 291–304. Springer, Heidelberg (2008)

Including Soft Global Constraints in DCOPs[*]

Christian Bessiere[1], Patricia Gutierrez[2], and Pedro Meseguer[2]

[1] University of Montpellier, France
bessiere@lirmm.fr
[2] IIIA - CSIC, Universitat Autònoma de Barcelona
08193 Bellaterra, Spain
{patricia,pedro}@iiia.csic.es

Abstract. In the centralized context, global constraints have been essential for the advancement of constraint reasoning. In this paper we propose to include soft global constraints in distributed constraint optimization problems (DCOPs). Looking for efficiency, we study possible decompositions of global constraints, including the use of extra variables. We extend the distributed search algorithm BnB-ADOPT$^+$ to support these representations of global constraints. In addition, we explore the relation of global constraints with soft local consistency in DCOPs, in particular for the generalized soft arc consistency (GAC) level. We include specific propagators for some well-known soft global constraints. Finally, we provide empirical results on several benchmarks.

1 Introduction

Distributed Constraint Optimization Problems (DCOPs) are commonly used for modeling many multi-agent coordination problems. DCOPs are formalized in terms of agents, variables with finite domains and cost functions (a particular case of soft constraints [12]). Cost functions are used to evaluate the cost of variable assignments. Agents should find a complete value assignment with minimum sum of costs. It is usually assumed that each agent handles a single variable, and it also knows about the domain and cost functions associated with that variable.

In the centralized context, global constraints have been essential for the advancement of constraint reasoning. The well-known *alldifferent(T)* global constraint means that all the variables in the set T must assign a different value (independently of the cardinality of T). Soft global constraints are associated with a violation measure that defines the costs of value assignments. For example, the *soft-alldifferent(T)* is associated with the violation measure μ_{var} (the number of variables in T that have to change their value to satisfy that all are different), or with μ_{dec} (the number of pairs of variables in T with the same value [14]).

In the distributed context, global constraints have been studied in the satisfaction case [2]. However, to the best of our knowledge, no relation between DCOPs and soft

---

[*] Christian Bessiere is partially supported by the FP7-FET ICON project 284715 and by the "Agence Nationale de la Recherche" project ANR-10-BLA-0214. Patricia Gutierrez and Pedro Meseguer are partially supported by the projects TIN2009-13591-C02-02 and Generalitat de Catalunya 2009-SGR-1434. Patricia Gutierrez has an FPI scholarship BES-2008-006653.

M. Milano (Ed.): CP 2012, LNCS 7514, pp. 175–190, 2012.

$$D_{x_1} = \{a\} \quad D_{x_2} = \{a,b\} \quad D_{x_3} = \{a,b\}$$

x_1	x_2	x_3	μ_{var}
a	a	a	2
a	a	b	1
a	b	a	1
a	b	b	1

x_1	x_2	μ_{var}
a	a	1
a	b	0

x_1	x_3	μ_{var}
a	a	1
a	b	0

x_2	x_3	μ_{var}
a	a	1
a	b	0
b	a	0
b	b	1

Fig. 1. (*Left*) *soft-alldifferent* global constraint with μ_{var} violation measure; (*right*) a decomposition in binary constraints. However, *soft-alldifferent* is not binary decomposable with μ_{var}, because $\mu_{var}^{x_1,x_2,x_3}(a,a,a) = 2 \neq \mu_{var}^{x_1,x_2}(a,a) + \mu_{var}^{x_1,x_3}(a,a) + \mu_{var}^{x_2,x_3}(a,a) = 3$.

global constraints have been established. In this paper, we advocate for the inclusion of soft global constraints in DCOPs. We assume that a soft constraint instance can be expressed as a cost function in the weighted model [12], so these terms are used interchangeably in the paper. In DCOPs it is a common assumption that cost functions are binary, that is, defined over two variables. However, not every cost function relation can be decomposed into an equivalent set of binary ones. For example, consider the *soft-alldifferent* constraint with the violation measure μ_{var} (represented by the cost function that appears in Figure 1 left). Observe that the tuple $(x_1 = a, x_2 = a, x_3 = a)$ has a different cost in the global formulation –involving all variables– and in the binary formulation (three cost functions in Figure 1 right). Hence, this soft constraint is not binary decomposable with violation measure μ_{var}.[1] In general, most soft global constraints are not binary decomposable, so working with their original formulations is crucial for their effective inclusion in DCOPs.

Our proposal enhances DCOP expressivity since not every cost function can be expressed as a set of binary cost functions. We also investigate several decompositions of soft global constraints, including decompositions with extra variables, looking for the one that provides the best performance in DCOP solving. We extend the distributed search algorithm BnB-ADOPT$^+$ to support different decompositions of soft global constraints. In addition, we explore the relation of global constraints with soft local consistency in DCOPs, in particular with the generalized soft arc consistency (GAC) level. On the one hand, the quality of the bounds obtained as result of applying local consistency is often better when the problem contains global constraints than when it contains an equivalent binary formulation. On the other hand, enforcing GAC on global constraints can be expensive using generic propagators. In the worst case, this is exponential in the number of variables. However, efficient propagators have been proposed for some global constraints that exploit constraint semantics, reaching the consistency level with lower complexity (usually polynomial) than with generic propagators.

The paper is structured as follows. In Section 2 we define some concepts needed for the rest of the paper. We analyze the decomposition of global constraints in a polynomial number of fixed arity constraints in Section 3. We explain how to include global constraints in DCOPs in Section 4, jointly with the extension of the distributed search algorithm BnB-ADOPT$^+$ to handle them (without and with GAC enforcement). We provide an empirical evaluation of the proposed techniques in Section 5. Finally, we conclude the paper in Section 6.

---

[1] However *soft-alldifferent* is binary decomposable with violation measure μ_{dec} [14].

2   Preliminaries

Concepts such as constraint optimization, soft global constraints and soft arc consistency have been defined using a centralized point of view, but they can easily be generalized to the distributed context. Here we recall the original definitions, the DCOP generalization and a short description of the BnB-ADOPT$^+$ algorithm.

COP. A *Constraint Optimization Problem* (COP) is defined by $(\mathcal{X}, \mathcal{D}, \mathcal{C})$ where: $\mathcal{X} = \{x_1, \ldots, x_n\}$ is a set of variables. $\mathcal{D} = \{D_{x_1}, \ldots, D_{x_n}\}$ is a set of finite domains such that D_{x_i} is the value set for x_i. $\mathcal{C}$ is a finite set of cost functions, where every cost function $C(T) : \prod_{x_i \in T} D_{x_i} \mapsto N \cup \{0, \infty\}$ on the ordered subset of variables $T = (x_1, \ldots, x_r)$ specifies the costs of every combination of values on T. It is worth noting that hard constraints can be modelled in this formalism associating $0/\infty$ costs with permitted/forbidden tuples, respectively. When a cost function $C(T)$ is evaluated on a value tuple t we follow the notation: $C_T(t)$. The cost of a tuple t is calculated adding all individual cost functions evaluated on t. If $\top$ is the lowest unacceptable cost, a *solution* is a tuple t containing a complete variable assignment with cost lower than $\top$. An *optimal solution* is a solution with minimum cost.

These problems are in many cases solved using a branch-and-bound schema, usually enhanced with sophisticated methods to improve lower bound computation (maintaining some forms of local consistency at each node). This facilitates pruning of the current branch and removal of future values, which improves performance.

Soft Global Constraints. A soft global constraint C is a class of soft constraints whose arity is not fixed. Constraints with different arities can be defined by the same class. For instance, *soft-alldifferent*(x_1, x_2, x_3) and *soft-alldifferent*(x_1, x_4, x_5, x_6) are two instances of the *soft-alldifferent* global constraint. The cost of global constraints is evaluated using a violation measure μ. A soft global constraint C with violation measure μ is *contractible* iff μ is a non-decreasing function [10].[2] A soft global constraint C with violation measure μ admits a *binary decomposition without extra variables* iff for any instance $C(x_1, \ldots, x_p)$ of C, there exists a set S of binary soft constraints involving only variables $x_1, \ldots, x_p$ such that for any value tuple t on $x_1, \ldots, x_p$, $\sum_{C(x_i, x_j) \in S} C_{x_i, x_j}(t[x_i, x_j]) = \mu(t)$. We also say that C is semantically decomposable in S.

Soft Arc Consistency. We consider a COP: (i, a) means x_i taking value a, $\top$ is the lowest unacceptable cost, $C(x_i)$ is the unary cost function on x_i values, C_ϕ is a zero arity cost function that represents a lower bound of the cost of any solution. As [8,9], we consider the following local consistencies:

- *Node Consistency**: (i, a) is node consistent* (NC*) if $C_\phi + C_{x_i}(a) < \top$; x_i is NC* if all its values are NC* and there is $a \in D_{x_i}$ s.t. $C_{x_i}(a) = 0$; a problem is NC* if every variable is NC*.

---

[2] Function f on a sequence is non-decreasing if $f(\mathbf{a}) \leq f(\mathbf{b})$, for every sequence $\mathbf{a}$ and $\mathbf{b}$ such that $\mathbf{a}$ is a prefix of $\mathbf{b}$ [10]. The intuition behind is as follows: C with μ is contractible when $\mu(a, b, c) \leq \mu(a, b, c, d) \leq \mu(a, b, c, d, e)....$, so shortening by the right the sequence of variables on which C is defined gives a valid lower bound to the cost of C. This is in relation with the nested representation, defined in Section 4.

- *Generalized Arc Consistency**: (i, a) is generalized arc consistent (GAC) wrt. a non-unary cost function $C(T)$, if there exist a value tuple t on T such that $(i, a) \in t$ and $C_T(t) = 0$; x_i is GAC if all its values are GAC wrt. every cost function involving x_i; a problem is GAC* if every variable is GAC and NC*.

In the following we refer to NC* and GAC* as NC and GAC, without asterisk. GAC can be reached by shifting costs from the problem and deleting values not NC. Cost are shifted with *equivalent preserving transformations* in the following way: first projecting the minimum cost from non-unary cost functions to unary costs functions, and then projecting the minimum cost from unary cost functions into C_ϕ. After projection, node inconsistent values are deleted. When a value is deleted in x_i, GAC is rechecked on every variable that x_i is constrained with, so a deleted value might cause further deletions. The GAC check must be performed until no further values are deleted. The systematic application of these operations (projection and deletion of node inconsistent values) does not change the optimum (for details on projections and optimality, see [8]).

DCOP. A *Distributed Constraint Optimization Problem* (DCOP) [13] is defined by $(\mathcal{X}, \mathcal{D}, \mathcal{C}, \mathcal{A}, \alpha)$ where $\mathcal{X}, \mathcal{D}$ and $\mathcal{C}$ define a COP and: $\mathcal{A} = \{a_1, \ldots, a_p\}$ is a set of agents. $\alpha : \mathcal{X} \rightarrow \mathcal{A}$ maps each variable to one agent. Solving a DCOP is an NP-hard task. Agents communicate and coordinate while looking for the optimal solution through messages. In this paper, it is assumed that: messages are never lost; messages sent from one agent to another are delivered in the same order they were sent; α maps a single variable to each agent, so we use the terms variable and agent interchangeably.

BnB-ADOPT$^+$. BnB-ADOPT [16] is a reference algorithm for optimal DCOP solving. Agents are arranged in a depth-first search (DFS) pseudo-tree and asynchronously perform a depth-first-branch-and-bound search until an optimal solution is found. Agents may have a parent, children (connected by tree edges of the pseudo-tree), pseudoparents and pseudochildren (connected by back-edges of the pseudo-tree) [13]. Each agent *self* holds a *context* that is updated with message exchange. The *context* holds a set of assignments involving *self* ancestors. Agents exchange the following messages:

- VALUE(i, j, val, th): agent i informs child or pseudochild j that it has taken value *val* with threshold th;
- COST$(k, j, context, lb, ub)$: agent k informs parent j that with *context* its bounds are lb and ub;
- TERMINATE(i, j): agent i informs child j that agent i terminates.

A BnB-ADOPT agent executes the following loop: it reads and processes all incoming messages and assigns a value. Then, it sends a VALUE to each child or pseudochild and a COST to its parent. When BnB-ADOPT terminates, each agent has assigned the optimum value for its variable. We use the BnB-ADOPT$^+$ version [7], which saves redundant messages. For more details, see [16,7].

3   Soft Global Constraint Decompositions in DCOP

In this Section we analyze the different forms of decomposing a soft global constraint into a polynomial number of smaller constraints of fixed arity, in the DCOP context (with the standard assumption that each agent owns a single variable).

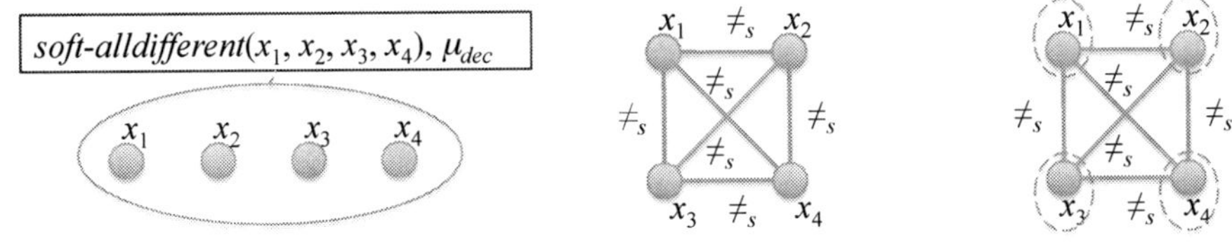

Fig. 2. *Left*: The *soft-alldifferent*(x_1, x_2, x_3, x_4) global constraint with the μ_{dec} violation measure. *Center*: Its binary decomposition, $\neq_s$ stands for soft binary inequality. *Right*: Binary decomposition in DCOP; agents are represented with discontinuous lines.

3.1 Decompositions without Extra Variables

As previously mentioned, some global constraints are semantically decomposable in a set of binary constraints on the variables of the global constraint. For example, in the hard case *alldifferent* is semantically decomposable in a clique of binary inequality constraints between the variables involved in the global constraint. Passing to the soft case, the *soft-alldifferent* global constraint with the violation measure μ_{dec} is semantically equivalent to a clique of soft binary inequalities. A soft binary inequality has 0 cost if the involved variables have different values and a cost of 1 if they have the same value. Including the binary decomposition of a soft global constraint does not cause extra difficulties in most DCOP solving algorithms (you are simply adding some extra soft binary constraints that are treated as any other soft constraint). Figure 2 shows the decomposition of *soft-alldifferent* into a clique of soft binary inequalities.

3.2 Decompositions with Extra Variables

In the hard case, there are global constraints that are not binary decomposable but they can be decomposed in a polynomial number of smaller, fixed arity constraints [4,3], if we allow a polynomial number of *extra variables*. For example, the hard $atmost[k, v](y_1, ..., y_p)$ global constraint establishes that value v cannot appear more than k times in $\{y_1, ..., y_p\}$. Allowing $p+1$ extra variables $\{z_0, z_1, ..., z_p\}$ with domains $D_{z_j} = \{0, 1, ..., j\}$, p new ternary constraints:

$$\text{if } y_i = v \text{ then } z_i = z_{i-1} + 1 \text{ else } z_i = z_{i-1} \qquad i : 1, ..., p$$

and one unary constraint:

$$z_p \leq k$$

It is easy to see that the original constraint is semantically equivalent to this set of new constraints. Variables $\{z_0, z_1, ..., z_p\}$ are acting as counters: z_i contains the number of times value v appears in the original variables $y_1, ..., y_i$. Variables $\{z_0, z_1, ..., z_p\}$ are called extra variables because they are not present in the original problem definition. However, they are treated as any other problem variable.

Passing to the soft case, the *soft-atmost*$[k, v](y_1, ..., y_p)$ has the following meaning: if value v appears less than or k times in the set $\{y_1, ..., y_p\}$ that assignment costs

0, otherwise it costs the number of times v appears minus k. This soft constraint can be decomposed with extra variables as follows. We keep the same extra variables as in the hard decomposition $\{z_0, z_1, ..., z_p\}$ with the same domains $D_{z_j} = \{0, 1, ..., j\}$. Previous p ternary constraints remain as hard constraints modelled in the soft formalism (permitted/forbidden tuples $0/\infty$ cost) plus the unary constraint that becomes the soft one:

$$\text{if } z_p \leq k \text{ then cost} = 0 \text{ else cost} = z_p - k$$

In tabular form with μ as cost, each ternary constraint generates a table similar to the table on the left, while the unary constraint generates a table similar to the table on the right:

z_{i-1}	y_i	z_i	μ
c	v	$c+1$	0
c	$\neq v$	c	0
	otherwise		∞

z_p	μ
$\leq k$	0
$> k$	$z_p - k$

The proposed decomposition appears in Figure 3.[3]

Allowing extra variables in DCOP, a question naturally follows: which agent owns these extra variables, which have no real existence? To solve this issue we propose to add a number of *virtual agents*, to own these extra variables. While this approach allows to keep the assumption that each agent owns a single variable, a new issue appears on the existence and activity of virtual agents with respect to real agents. Previous approaches have used the idea of virtual agents to accommodate modifications or extensions that deviate from original problem structure [13]. In addition, all variables are treated in the same way, one variable per agent, so no preference is given to a particular subset of variables in front of others. Implementation maintains uniformity for all variables.

Virtual agents can be simulated by real agents. If some real agents have substantial computational/communication resources, they can host some virtual agents. The precise allocation of virtual agents depends on the nature of the particular application to solve.

4 Adding Soft Global Constraints in DCOP

We consider several ways to model the inclusion of a soft global constraint in DCOPs, looking for the one that gives the best performance. The user chooses one of the three representations and the solving is done on that representation. In the rest of this Section, the term "constraint" always mean "soft constraint" (either global or not).

We assume that agents are ordered. The evaluation of a global constraint $C(T)$ by every agent depends on the selected model. We analyze the three following representations:

- Direct representation. C is treated as a generic constraint of arity $|T|$. Only one agent involved in the constraint evaluates it: the one that appears last in pseudo-tree ordering.

---

[3] Observe that local consistencies on decompositions of global cost functions has recently been explored in [1].

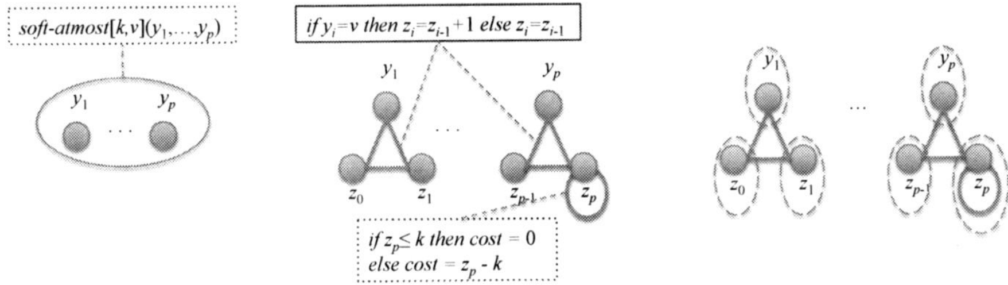

Fig. 3. *Left*: The *soft-atmost*$[k, v](y_1, ..., y_p)$ soft global constraint. *Center*: Its decomposition in p ternary and one unary constraint. *Right*: This decomposition in the distributed context; agents are represented with discontinuous lines.

- Nested representation. If C is contractible, then C allows nested representation. The nested representation of $C(T)$ with $T = \{x_{i_1}, \ldots, x_{i_p}\}$ is the set of constraints $\{C(x_{i_1}, \ldots, x_{i_j})$ with $j \in 2 \ldots p\}$. For instance, the nested representation of *soft-alldifferent*(x_1, x_2, x_3, x_4) is the set $S = \{$*soft-alldifferent*(x_1, x_2), *soft-alldifferent*(x_1, x_2, x_3), *soft-alldifferent* $(x_1, x_2, x_3, x_4)\}$. The nested representation has the following benefit. Since x_2, x_3 and x_4 are the last agent of a constraint in S, any of them is able to evaluate that particular constraint. When assignments are made following the order x_1, x_2, x_3, x_4, every intermediate agent is able to aggregate costs and calculate a lower bound of the current partial solution. Since C is contractible, this bound increases monotonically on every agent. By this, it is possible to calculate updated lower bounds during search and backtrack earlier if the current solution has unacceptable cost.
- Bounded arity representation. If C is binary decomposable without extra variables, agent *self* includes all constraints of the binary decomposition of C that involve x_{self} in their scope. Otherwise, if C is decomposable with extra variables, agent *self* includes all constraints of the decomposition of C that involve x_{self} in their scope. On the contrary with previous representations (direct and nested), constraints included by *self* are non-global.

Since the nested representation allows to calculate updated bounds and performs efficient backtracking, it is expected to be more efficient than the direct representation. However not all global constraints are contractible, so the direct representation has to be analyzed. In the bounded arity representation every intermediate agent is an evaluator, as in the nested representation.

4.1   Search with BnB-ADOPT$^+$

From now on, we differentiate between two types of constraint instances: global constraints (as a result of the direct and nested representations) and non-global constraints (coming from bounded arity representation of global constraints, as well as particular constraints that may exist in the considered problem).

BnB-ADOPT$^+$ can be generalized to handle constraints of any arity. It is required that each constraint is evaluated by the last of the agents involved in the constraint in the partial ordering of the pseudo-tree, while other agents have to send their values to the evaluator. This simple strategy is mentioned in [16,13]. We assume that our version of BnB-ADOPT$^+$ includes this generalization.

We have extended the distributed search algorithm BnB-ADOPT$^+$ [7] to support global constraints. The following modifications are needed:

1. (*self* denotes a generic agent) *self* keeps a set of global constraints, separated from the set of non-global constraints it is involved. Agent *self* knows about (and stores) a constraint C iff *self* is involved in C. Every constraint $C(T)$ implicitly contains the agents involved in T (neighbors of *self*). For some global constraints, additional information can be stored. For example, for the *soft-atmost*$[k, v]$ constraint, parameters k (number of repetitions) and v (value) are stored.
2. During the search process, every time *self* needs to evaluate the cost of a given value v, all local costs are aggregated. Non-global constraints are evaluated as usual, and global constraints are evaluated according to their violation measure.
3. VALUE messages are sent to agents, depending on the constraint type:
   - For a non-global constraint, VALUE messages are sent to all the children and the last pseudochild in the ordering (the deepest agent in the DFS tree involved in the constraint evaluates it; VALUE to children are needed because they include a threshold required in BnB-ADOPT; observe that for binary constraints this is the original BnB-ADOPT behavior).
   - For a global constraint, there are two options:
     - For the direct representation, VALUE messages are sent to all the children and the last pseudochild in the ordering (the deepest agent in the DFS tree involved in the constraint evaluates it; VALUE to children are needed because they include a threshold required in BnB-ADOPT). [4]
     - For the nested representation, VALUE messages are sent to all children and all pseudochildren (any child or pseudochild is able to evaluate a constraint of the nested representation).
4. COST messages include a list of all the agents that have evaluated a global constraint. This is done to prevent duplication of costs when using the nested representation and it is explained in the next paragraphs.

Figure 4 shows the pseudocode for cost aggregation in BnB-ADOPT$^+$ (lines [1-4]). Costs coming from non-global constraints are calculated as usual, aggregating all non-global constraint costs evaluated on *self* value and the assignments of the current context (lines [5-13]). Costs coming from global constraints are calculated in lines [14-27]. Although there is no need to separate non-global from global cost aggregation, we have presented them in separate procedures for a better understanding of the new modifications.

For every global constraint of the set (*globalConstraintSet*) *self* creates a tuple with the assignments in its current context (*assignments*, lines [17-19]). If *self* is the deepest agent in the DFS tree (taking into account the variables involved in the global constraint) then *self* evaluates the constraint (lines [20-21]). If *self* is an intermediate

---

[4] In distributed search, a global constraint in the direct representation has the same treatment as a non-global one. However, when GAC is enforced, global and non-global constraints are treated differently (Section 4.2).

```
(1)  procedure CalculateCost(value)
(2)     cost = cost + NonGlobalCostWithValue(value);
(3)     cost = cost + GlobalCostWithValue(value);
(4)     return cost;

(5)  function NonGlobalCostWithValue(value)
(6)     cost = 0;
(7)     for each nonGlobal ∈ nonGlobalConstraintSet do
(8)        assignments = new list(); assignments.add(self, value);
(9)        for each (xi, di) ∈ context do
(10)          if xi ∈ nonGlobal.vars then assignments.add(xi, di);
(11)       if assignments.size == nonGlobal.vars.size then          //self is the last evaluator
(12)          cost = cost + nonGlobal.Evaluate(assignments);
(13)    return cost;

(14) function GlobalCostWithValue(value)
(15)     cost = 0;
(16)    for each global ∈ globalConstraintSet do
(17)       assignments = new list(); assignments.add(self, value);
(18)       for each (xi, di) ∈ context do
(19)          if xi ∈ global.vars then assignments.add(xi, di);
(20)       if assignments.size == global.vars.size then             //self is the last evaluator
(21)          cost = cost + global.μ.Evaluate(assignments);
(22)       else                                        //self is an intermediate agent in the restriction
(23)          if NESTED representation then
(24)             for each xi ∈ global.vars do
(25)                if lowerGlobalEvaluators.contain(xi) then cost = cost + 0;
(26)                else cost = cost + global.μ.Evaluate(assignments);
(27)    return cost;
```

Fig. 4. Aggregating costs of binary and global cost functions

agent, it does the following. If representation is direct, $self$ cannot evaluate the global constraint: it does nothing and cost remains unchanged. If representation is nested, it requires some care. A nested global constraint is evaluated more than once by intermediate agents and if these costs are simply aggregated duplication of costs may occur. To prevent this, COST messages include the set of agents that have evaluated global constraints (*lowerGlobalEvaluators*). When a COST message arrives, $self$ knows which agents have evaluated its global constraints and contributed to the lower bound. If some of them appear in the scope of C, then $self$ does not evaluate C (lines [25-26]). By doing this, the deepest agent in the DFS tree evaluating the global constraint precludes any other agent in the same branch to evaluate the constraint, avoiding cost duplication. Preference is given to the deepest agent because it is the one that receives more value assignments and can perform a more informed evaluation. When bounds coming from a branch of the DFS are reinitialized (this happens under certain conditions in BnB-ADOPT, for details see [16]), the agents in the set *lowerGlobalEvaluators* lying on that branch are removed.

4.2   Propagation with BnB-ADOPT$^+$

Specific propagators exploiting the semantics of global constraints have been proposed in the centralized case [9]. These propagators allow to achieve generalized arc consistency in polynomial time whereas a generic propagator is exponential in the number of variables in the scope of the constraint.

Soft local consistency is based on *equivalent preserving transformations* where costs are shifted from non-unary cost functions to unary cost functions. The same technique can be applied in distributed. We project costs from non-global/global cost functions to

BnB-ADOPT$^+$-UGAC messages:
VALUE($sender, destination, value, threshold, \top, C_\phi$)
COST($sender, destination, context[], lb, ub, \underline{subtreeContr}, \underline{lowerGlobalEvaluators}$)
STOP($sender, destination, emptydomain$)
$\underline{\text{DEL}(sender, destination, value)}$

Fig. 5. Messages of BnB-ADOPT$^+$-UGAC. New parts wrt. BnB-ADOPT$^+$ are underlined

unary cost functions and finally project unary costs to C_ϕ. After projections are made agents check their domains searching for inconsistent values. For this, some modifications are needed:

- The domain of neighboring agents (agents connected with *self* by soft constraints) are represented in *self*.
- A new DEL message is added to notify value deletions.
- COST and VALUE messages include extra information.

Following the technique proposed in [6], we maintain GAC during search performing only unconditional deletions, so we call it unconditional generalized arc consistency (UGAC). [5] An agent *self* deletes a value v *unconditionally* if this value is assured to be sub-optimal and does not need to be restored again during the search process. If *self* contains a value v not NC ($C_{self}(v) + C_\phi > \top$) then v can be deleted unconditionally because the cost of a solution containing the assignment *self* $= v$ is necessarily greater than $\top$. We also detect unconditional deletions in the following way. Let us consider agent *self* executing BnB-ADOPT$^+$. Suppose *self* assigns value v and sends the corresponding VALUE messages. As response, COST messages arrive. We consider those COST messages whose context is simply $(self, v)$. This means that the bounds informed in these COST messages only depend on *self* assignment (observe that the *root* agent always receives such COST messages). If the sum of the lower bounds contained in those COST messages exceeds $\top$, v can be deleted unconditionally because the cost of a solution containing the assignment *self* $= v$ is necessarily greater than $\top$.

As in [6], messages include information required to perform deletions, namely $\top$ (the lowest unacceptable cost), C_ϕ (the minimum cost of any complete assignment), the subtree contribution to C_ϕ (each node k computes the contribution to the C_ϕ of the subtree rooted at k), and the set of agents lower than the current one that are evaluators of global constraints in which the current is involved. These four elements travel in existing BnB-ADOPT$^+$ messages (the first two in VALUE messages, the last two in COST messages). In addition, a new message DEL($self, k, v$) is added, to notify agent k that *self* deletes value v. The structure of these new messages appears in Figure 5. When *self* receives a VALUE message, *self* updates its local copies of $\top$ and C_ϕ if the values contained in the received message are better (lower $\top$ or higher C_ϕ). When *self* receives a COST message from a child c, *self* records c subtree contribution to C_ϕ and the list of lower agent global evaluators. When *self* receives a DEL message, *self* removes the deleted value from its domain copy of the sender agent and performs

---

[5] Previous results [5] indicate that conditional deletions do not always pay off in terms of communication cost. Because of that, we concentrate here on unconditional deletions.

```
(1)   procedure ProjectFromAllDiffToUnary(global, v)
(2)   graph = graphsSet.get(global);              //the graph associated with global is fetched
(3)   minCost = minCost + getMinCostFlow(graph);
(4)   for each x_i ∈ global.vars do
(5)     minCost = minCost + CostWithFlow(graph, x_i, v);
(6)     if minCost > 0 then
(7)       graph.getArc(x_i, v).cost = cost − minCost;
(8)       if x_i = self then C_self(v) = C_self(v) + minCost;

(9)   function getMinCostFlow(graph)
(10)    graph.SuccesiveShortestPath();
(11)    minCostFlow = 0;
(12)    for each arc ∈ graph.arcs do
(13)      minCostFlow = minCostFlow + (arc.flow * arc.cost);
(14)    return minCostFlow;

(15)  function CostWithFlow(graph, x_i, v)
(17)    if graph.arc(x_i, v).flow = 0 then return 0;
(19)    path = graph.residualGraph.FindShortestPath(x_i, v);//shortest path from v to x_i in the residual graph
(20)    flow = ∞; cost = 0;                        //calculate flow as the minimum capacity in this path
(21)    for each arc ∈ path do
(22)      if arc.capacity < flow then flow = arc.capacity;
(24)    for each arc ∈ path do
(25)      cost = flow * arc.cost;
(26)    return cost;
```

Fig. 6. Projection with *soft-alldifferent* global constraint

projections from the soft constraints involving the sender agent to its unary costs and to C_ϕ. When $\top$ or C_ϕ change, D_{self} is tested for possible deletions.

This mechanism described to detect and propagate unconditional deletions is similar to the one proposed in [6]. However, to reach the GAC level agents need to project costs not only from binary cost functions, but from global cost functions as well. In the following, we describe how to project binary and global costs specifically from the *soft-alldifferent* and *soft-atmost* global constraints.

Projecting Cost with Bounded Arity Constraints. The projection of costs from cost function $C(T)$ to the unary cost function $C_{x_i}(a)$, where T is a fixed set of variables, $x_i \in T$ and $a \in D_{x_i}$ is a flow of costs defined as follows. Let α_v be the minimum cost in the set of tuples of $C(T)$ where $x_i = a$ (namely $\alpha_a = min_{t\in tuples\ s.t.\ x_i=a}C_T(t)$). The projection consists in adding α_a to $C_{x_i}(a)$ (namely, $C_{x_i}(a) = C_{x_i}(a) + \alpha_a, \forall a \in D_{x_i}$) and subtracting α_a from $C_T(t)$ (namely, $C_T(t) = C_T(t) - \alpha_a, \forall t \in tuples\ s.t.\ x_i = a, \forall a \in D_{x_i}$). Every agent in T performs projections following a fixed order (projections are done first over higher agents in the pseudo tree). As a result, cost functions are updated the same way in all agents.

Projecting Costs with *soft-alldifferent*. We follow the approach described in [9] for the centralized case, where GAC is enforced on the *soft-alldifferent* constraint in polynomial time, whereas it is exponential when a generic algorithm is used.

A graph for every *soft-alldifferent* constraint is constructed following [15]. This graph is stored by the agent and updated during execution. Every time a projection operation is required, instead of exhaustively looking at all tuples of the global constraint, the minimum cost that can be projected is computed as the flow of minimum cost of the graph associated with the constraint [9]. Minimum flow cost computation is based on the successive shortest path algorithm, which searches shortest paths in the graph until no more flows can be added to the graph. Pseudocode appears in Figure 6.

```
(1)    procedure ProjectFromAtMostToUnary(global, v)
(2)    if global.v = v and D_self.contains(v) then
(3)      singletonCounter = 0; cost = 0;
(4)      for each x_i ∈ global.vars do
(5)        if D_{x_i}.contains(v) and D_{x_i}.size() = 1 then
(6)          singletonCounter = singletonCounter + 1;
(7)      if singletonCounter > global.k then
(8)        cost = singletonCounter − global.k;
(9)        if cost > global.projectedCost then
(10)         cost = temp;
(11)         cost = cost − global.projectedCost;
(12)         global.projectedCost = temp;
(13)         if global.vars[0] = self then C_self(v) = C_self(v) + cost;
```

Fig. 7. Projection with *soft-atmost*$[k, v]$ global constraint

Evaluation of these propagators in the distributed context is an extra issue because it is not based on table look-ups. In the centralized case, they are usually evaluated by their CPU time. An evaluation proposal appears in Section 5.

Projecting costs with *soft-atmost*. For the *soft-atmost* global constraint we propose the following technique to project costs from the global constraint *soft-atmost*$[k, v](T)$ to the unary cost functions $C_{x_i}(v)$. Agent x_i counts how many agents in T have a singleton domain $\{v\}$. If the number of singleton domains $\{v\}$ is greater than k, a minimum cost equal to the number of singleton domains $\{v\}$ minus k can be added to the unary cost $C_{x_i}(v)$ in one of the agents of the global constraint. We always project on the first agent of the constraint (we choose the first agent because in case of value deletion the search space reduction is larger). To maintain equivalence, the *soft-atmost* constraint stores this cost, that will be decremented from any future projection performed. Pseudocode appears in Figure 7.

5   Experimental Results

To evaluate the impact of including soft global constraints, we tested on several random DCOPs sets including *soft-alldifferent* and *soft-atmost* global constraints. The first set of experiments considers binary random DCOPs with 10 variables and domain size of 5. The number of binary cost functions is $n(n-1)/2 * p_1$, where n is the number of variables and p_1 varies in the range $[0.2, 0.9]$ in steps of 0.1. Binary costs are selected from an uniform cost distribution. Two types of binary cost functions are used, cheap and expensive. Cheap cost functions extract costs from the set $\{0, ..., 10\}$ while expensive ones extract costs from the set $\{0, ..., 1000\}$. The proportion of expensive cost functions is 1/4 of the total number of binary cost functions (this is done to introduce some variability among binary tuple costs [6]). In addition to binary constraints, global constraints are included. The first set of experiments includes 2 *soft-alldifferent*(T) global constraints in every instance, where T is a set of 5 randomly chosen variables. The violation measure is μ_{dec}. The second set of experiments includes 2 *soft-atmost*$[k,v](T)$ global constraints in every instance, where T is a set of 5 randomly chosen variables, k (number of repetitions) is randomly chose from the set $\{0, ..., 3\}$ and v (value) is randomly selected from the variable domain. The violation measure used is μ_{var}. To

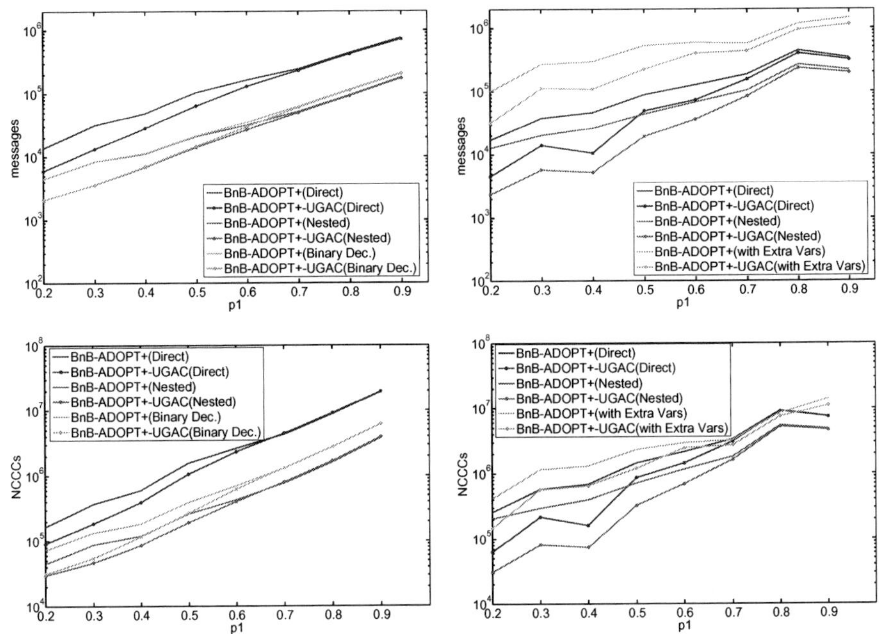

Fig. 8. Experimental results of random DCOPs including (*left*) *soft-alldifferent* global constraints with the violation measure μ_{dec}; (*right*) *soft-atmost* global constraints with the violation measure μ_{var}

balance binary and global costs, the cost of the *soft-alldifferent* and *soft-atmost* constraints is calculated as the amount of the violation measure multiplied by 1000.

We tested the extended versions of BnB-ADOPT$^+$ and BnB-ADOPT$^+$-UGAC able to handle global constraints using a discrete event simulator. Computational effort is evaluated in terms of non-concurrent constraint checks (NCCCs) [11]. Network load is evaluated in terms of the number of messages exchanged. Execution considers the different models to incorporate soft global constraints. For *soft-atmost* with μ_{var} we tested direct and nested representation, and the decomposition with extra variables. For *soft-alldifferent* with μ_{dec} we tested the direct and nested representations, jointly with its binary decomposition.

Specifically for UGAC enforcement, computational effort is measured as follows. For the sets including *soft-alldifferent* global constraints, we use a special propagator proposed in [9]. Every time a projection operation is required, instead of exhaustively looking at all tuples of the global constraint (which would increment the NCCC counter for every tuple), we compute the minimum flow of this graph. Minimum flow cost computation is based on the successive shortest path algorithm. We can think of every shortest path computation as a variable assignment of the global constraint.

We assess the computational effort of computing the shortest path algorithm as the number of nodes of the graph where this algorithm is executed (looks reasonable for small graphs, which is the case here). Each time the successive shortest path algorithm is executed, we add this number to the NCCC counter of the agent.

For the sets including the *soft-atmost* global constraints, every time the cost of the violation measure is computed as the number of singleton domains $\{v\}$ minus k, the NCCC counter is incremented.

Figure 8 (left) contains the results of the first experiment including *soft-alldifferent* with violation measure μ_{dec}. Figure 8 (right) contains the results of the second experiment including *soft-atmost* with violation measure μ_{var}.

From these results, we observe the following facts. First, for BnB-ADOPT$^+$, the nested representation offers the best performance both in terms of communication cost and computation effort (number of messages and NCCCs, observe the logarithmic scale), at substantial distance. In the direct representation, VALUE messages are sent to all children and the last pseudochild in the global constraint, whereas in the nested representation VALUE messages are sent to all children and all pseudochildren in the global constraint. However the early detection of dead-ends compensates by far the extra number of messages that should be sent in the nested representation.

Second, for BnB-ADOPT$^+$, the nested representation is substantially better (in terms of messages and NCCCs) than the binary decomposition (in *soft-alldifferent*) and than the decomposition with extra variables (in *soft-atmost*). If an agent changes value, it will send the same number of VALUE messages in the nested representation than in the clique of binary constraints (assuming that the binary decomposition is a clique, as happens with *soft-alldifferent*). However, in the nested representation receivers will evaluate larger constraints (with arity greater than 2), so they are more effective and as global effect this representation requires less messages than the binary decomposition. The decomposition with extra variables includes many new variables in the problem, causing many extra messages. These messages lead to more computational effort (more NCCCs).

Third, for all the representations and for most of the problems considered, UGAC maintenance always pays off (in terms of messages and NCCCs). In other words, BnB-ADOPT$^+$-UGAC consistently uses less computational and communication resources than BnB-ADOPT$^+$, no matter the used representation. Savings are substantial, specially for low and medium connectivities. Although UGAC causes to do more work each time a message is exchanged, the reduction in messages is so drastic that the overall effect is less computation (NCCC curves have the same shape as number of messages). This important fact indicates the impact of this limited form of soft GAC maintenance in distributed constraint optimization.

A closer look to the *soft-alldifferent* results of nested representation and binary decomposition indicate that as connectivity increases, messages/NCCCs required by the binary decomposition grow higher than those of the nested representation. The number of value deletions using the nested representation is higher than using the binary decomposition (because pruning using global constraints is more powerful than using the binary decomposition); as consequence, the search space is slightly smaller when using the nested representation, and due to this, less messages are required for its complete exploration. Processing less messages causes to decrease the computational effort measured in NCCCs.

From these results, we conclude that the nested decomposition offers the best performance, at a substantial distance of the other considered representations. Decomposition with extra variables using virtual agents and the direct representation are models to avoid when representing contractible global constraints in distributed constraint optimization. Enforcing UGAC pays off, causing nice savings in all representations.

6   Conclusions

In this paper we have introduced the use of soft global constraints in distributed constraint optimization. We have proposed several ways to represent soft global constraints in a distributed constraint network, depending on soft global constraint properties. We extended the distributed search algorithm BnB-ADOPT$^+$ to support the inclusion of global constraints. We evaluated its performance with and without the UGAC consistency level (generalized arc consistency with unconditional deletions).

From this work, we can extract the following conclusions:

- the use of global constraints is necessary in distributed constraint optimization to extend DCOP expressivity,
- considering two global constraints as a proof of concept, we show,
    - if the added global constraint is contractible, the nested representation is the one that, at substantial distance from others, offers the best performance both in terms of communication cost (number of messages) and computational effort (NCCCs),
    - UGAC maintenance always pays off in terms of number of messages, causing also less NCCCs in a very substantial portion of the experiments.

As future work, we plan to consider the extension of this work to other DCOP solving algorithms, as well as extending the empirical evaluation to other global constraints.

References

1. Allouche, D., Bessiere, C., Boizumault, P., de Givry, S., Gutierrez, P., Loudni, S., Meétivier, J.P., Schiex, T.: Filtering decomposable global cost functions. In: Proc. AAAI 2012 (2012)
2. Bessiere, C., Brito, I., Gutierrez, P., Meseguer, P.: Global constraints in distributed constraint satisfaction. In: Proc. AAMAS 2012 (2012)
3. Bessiere, C., Hebrard, E., Hnich, B., Kiziltan, Z., Walsh, T.: Slide: A useful special case of the cardpath constraint. In: Proc. ECAI 2008, pp. 475–479 (2008)
4. Bessière, C., Van Hentenryck, P.: To Be or Not to Be ... a Global Constraint. In: Rossi, F. (ed.) CP 2003. LNCS, vol. 2833, pp. 789–794. Springer, Heidelberg (2003)
5. Brito, I., Meseguer, P.: Connecting ABT with Arc Consistency. In: Stuckey, P.J. (ed.) CP 2008. LNCS, vol. 5202, pp. 387–401. Springer, Heidelberg (2008)
6. Gutierrez, P., Meseguer, P.: BnB-ADOPT$^+$ with several soft AC levels. In: Proc. ECAI 2010, pp. 67–72 (2010)
7. Gutierrez, P., Meseguer, P.: Saving redundant messages in BnB-ADOPT. In: Proc. AAAI 2010, pp. 1259–1260 (2010)
8. Larrosa, J., Schiex, T.: In the quest of the best form of local consistency for weighted CSP. In: Proc. IJCAI 2003, pp. 239–244 (2003)
9. Lee, J.H.M., Leung, K.L.: Towards efficient consistency enforcement for global constraints in weighted constraint satisfaction. In: Proc. IJCAI 2009, pp. 559–565 (2009)

10. Maher, M.J.: SOGgy Constraints: Soft Open Global Constraints. In: Gent, I.P. (ed.) CP 2009. LNCS, vol. 5732, pp. 584–591. Springer, Heidelberg (2009)
11. Meisels, A., Kaplansky, E., Razgon, I., Zivan, R.: Comparing Performance of Distributed Constraint Processing Algorithms, pp. 86–93 (2002)
12. Meseguer, P., Rossi, F., Schiex, T.: Soft Constraints. In: Handbook of Constraint Programming, ch. 9, pp. 281–328. Elsevier (2006)
13. Modi, P.J., Shen, W.M., Tambe, M., Yokoo, M.: Adopt: asynchronous distributed constraint optimization with quality guarantees. Artificial Intelligence 161(1-2), 149–180 (2005)
14. Petit, T., Régin, J.-C., Bessière, C.: Specific Filtering Algorithms for Over-Constrained Problems. In: Walsh, T. (ed.) CP 2001. LNCS, vol. 2239, pp. 451–463. Springer, Heidelberg (2001)
15. van Hoeve, W.J., Pesant, G., Rousseau, L.M.: On global warming: flow-based soft global constraints. Journal of Heuristics 12, 347–373 (2006)
16. Yeoh, W., Felner, A., Koenig, S.: BnB-ADOPT: An asynchronous branch-and-bound DCOP algorithm. JAIR 38, 85–133 (2010)

The Weighted Average Constraint

Alessio Bonfietti and Michele Lombardi

DEIS, University of Bologna
{alessio.bonfietti,michele.lombardi2}@unibo.it

Abstract. Weighted average expressions frequently appear in the context of allocation problems with balancing based constraints. In combinatorial optimization they are typically avoided by exploiting problems specificities or by operating on the search process. This approach fails to apply when the weights are decision variables and when the average value is part of a more complex expression. In this paper, we introduce a novel **average** constraint to provide a convenient model and efficient propagation for weighted average expressions appearing in a combinatorial model. This result is especially useful for Empirical Models extracted via Machine Learning (see [2]), which frequently count average expressions among their inputs. We provide basic and incremental filtering algorithms. The approach is tested on classical benchmarks from the OR literature and on a workload dispatching problem featuring an Empirical Model. In our experimentation the novel constraint, in particular with incremental filtering, proved to be even more efficient than traditional techniques to tackle weighted average expressions.

1 Introduction

A weighted average expression is a function in the form:

$$avg(\overline{v}, \overline{w}) = \frac{\sum_{i=0}^{n-1} v_i \cdot w_i}{\sum_{i=0}^{n-1} w_i} \tag{1}$$

where v_i and w_i respectively are the *value* and the *weight* of each term in the average and the notation $\overline{v}$, $\overline{w}$ refers to the whole vector of values/weights and n is the number of terms. We assume all weights are non-negative. Moreover, we assume $avg(\overline{v}, \overline{w}) = 0$ if all weights are 0. When appearing in a combinatorial model, both the term values and weights of the expression may be variables, say V_i and W_i. Weighted average expressions are found in a number of application settings, such as balancing the workload of assembly lines, fair scheduling of work shifts, uniform distribution of travel distances in routing problems.

Average expressions also frequently appear as inputs of Empirical Models. Those are Machine Learning models (such as Neural Networks, Support Vector Machines or Decision Trees), obtained via training (over a real world system or a simulator) and encoded using some combinatorial optimization technology. Empirical Models have been recently proposed as a mean to enable decision making over complex systems [2]. The approach has in principle broad applicability, but

M. Milano (Ed.): CP 2012, LNCS 7514, pp. 191–206, 2012.

the extracted models often feature average values as part of their input (e.g. [1]). Hence the ability to deal with weighted average expressions is essential for the applicability of such methods.

Research efforts so far have been focused on balancing issues (see the works on the **spread** [7,9] and **deviation** constraint [10,11]). Balancing involves keeping the average close to a target value and avoiding outliers. A general formulation for a **balancing** constraint has been given in terms of p-norm in [10] and [8], for the case where all weights are constant and equal to 1. In this situation the average expression has fixed denominator and is therefore linear and easy to handle. Hence, the **spread** and **deviation** constraint focus on providing more powerful filtering by respectively taking into account restrictions on the variance and on the total deviation.

There are however cases where non-constant weights are definitely handy. Assume we want to balance the average per-depot travel time in a Vehicle Routing Problem with n vehicles and multiple depots. The assignment of a vehicle i to depot j can be conveniently modeled as a $\{0,1\}$ variable $W_{i,j}$, leading to:

$$Y_j = \frac{\sum_{i=0}^{n-1} V_i \cdot W_i}{\sum_{i=0}^{n-1} W_i} \qquad \text{or} \qquad \sum_{i=0}^{n-1} V_i \cdot W_i = Y_j \cdot \sum_{i=0}^{n-1} W_i \qquad (2)$$

where V_i is the travel distance for vehicle i and Y_j is the average travel distance for depot j. The left-most formulation requires the denominator to be strictly positive. Unless the number of vehicles per depot is a-priori known, the sum of $W_{i,j}$ variables is non-fixed, the expression non linear and the resulting propagation weak, thus making non-uniform weights very difficult to deal with. Many allocation problems with balancing restrictions raise similar issues.

In this paper, we introduce a novel **average** global constraint to provide a convenient model and effective filtering for average expressions with non-uniform weights. Unlike **spread** or **deviation**, we do not explicitly try to limit outliers, since filtering for the average value is sufficiently challenging by itself. The paper is structured as follows: in Section 2 we introduce the **average** constraint and provide filtering algorithms with the assumption that all term *values* are fixed. Several improvements to the basic algorithms are discussed in Section 3. In Section 4 the efficiency and effectiveness of our filtering is investigated; moreover, our approach is compared to an alternative method requiring no ad-hoc constraint and to the decomposed constraint from Equation (2). Finally, we show how to deal with variable term values in Section 5 and we provide concluding remarks in Section 6.

2   The Average Constraint

The **average** global constraint provides a convenient model and effective propagation for weighted average expressions. We will initially assume that the term *weights* are decision variables, while the *values* are constant: this setting captures

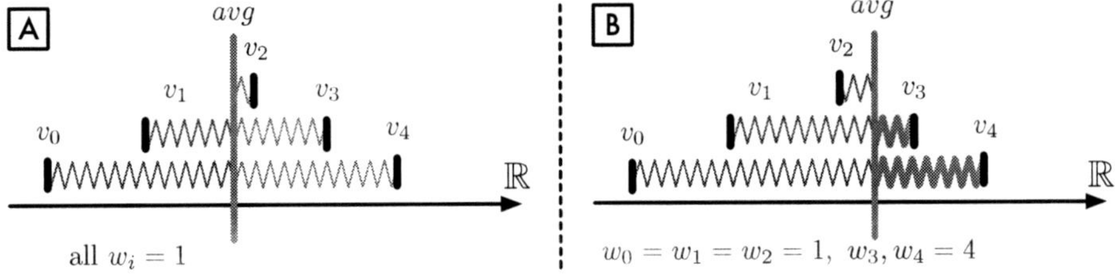

Fig. 1. A: Initial weight configuration ($w_i \in \{1..4\}\forall i$); B: Upper bound configuration

most of the difficulties found by CP methods with weighted average expressions. In the considered case, the constraint signature is:

$$\texttt{average}(\overline{v}_i, \overline{\mathtt{W}}, \mathtt{Y}) \tag{3}$$

where $\overline{v}$ is a vector containing the integer value of each term and $\overline{\mathtt{W}}$ is a vector of integer variables representing the weights. Weights must be non-negative, i.e. for each variable we have $\min(\mathtt{W_i}) \geq 0$ (where $\min(\mathtt{W_i})$ is the minimum value in the domain of $\mathtt{W_i}$). The integer variable $\mathtt{Y}$ stores the weighted average value. The constraint maintains bound consistency on the expression:

$$\mathtt{Y} = round\left(\frac{\sum_{i=0}^{n-1} v_i \cdot \mathtt{W_i}}{\sum_{i=0}^{n-1} \mathtt{W_i}} \right) \tag{4}$$

where n is the number of terms and $round(\cdot)$ is the usual rounding operator. The use of integer values and weights is a design choice, motivated by the lack of support for real-valued variables in many current CP solvers. In the following, we provide filtering algorithms for the constraint as from signature (3), while the case of variable term values is considered in Section 5.

2.1   Computing Bounds on Variable Y

We focus on computing an upper bound for $\mathtt{Y}$ (a lower bound can be obtained analogously). On this purpose, observe that all value necessarily appear in the expression at least with minimum weight. Hence, the formula:

$$\frac{\sum_{i=0}^{n-1} v_i \cdot \min(\mathtt{W_i})}{\sum_{i=0}^{n-1} \min(\mathtt{W_i})} \tag{5}$$

provides a first estimate of the weighted average, which can be increased or lowered by including some terms with non-minimal weight. It is convenient to think of the expression as a system of physical springs with fixed anchored points, all pulling a single bar (see Figure 1). Anchor points represents the term values and the spring strengths are the weights. The initial estimate corresponds to a configuration where each spring has minimum strength (see Figure 1A). Intuitively, increasing the spring strength of one of the right-most anchor points will increase the average value. Formally, we have:

Statement 1. *Let $\overline{w}^*$ be a fixed weight assignment. Then, increasing a weight w_j^* by an amount $\delta > 0$ leads to an increased weighted average value iff:*

$$v_j > \frac{\sum_{i=0}^{n-1} v_i \cdot w_i^*}{\sum_{i=0}^{n-1} w_i^*} \tag{6}$$

where we assume the right most part of the inequality is 0 if all weights are 0.

The statement can be proved by algebraical manipulation of Equation (1). Note that the effect of increasing a weight does not depend on δ: if increasing w_i makes the current estimate larger, then the maximum increment leads to the largest increase. Hence we can improve the initial estimate by repeatedly choosing a v_i that satisfies the condition from Statement 1 and then maximizing w_i.

The process may however stop at a local maximum if the v_i are not properly chosen. For example, in Figure 1A we may decide to maximize v_2, since it is higher than the current average estimate. However, this decision makes it impossible to reach the configuration from Figure 1B, corresponding to the actual maximum of the average expression. Luckily, there exists a sequence of weight increments guaranteed to stop at a *global* maximum.

Statement 2. *The maximum value of the weighted average can be obtained starting from Expression (5), by repeatedly maximizing the weight of the term j with maximum v_j, as long as the condition from Theorem 1 is satisfied.*

Assume we have $v_j \geq v_i$: then one can check that maximizing both w_i and w_j cannot lead to a lower average than maximizing w_j alone. Therefore, considering w_j before w_i is safer than the opposite order. By induction we obtain that the term with highest value should alway be considered first. In the example from Figure 1, this kind of reasoning leads to the configuration in Figure 1B, corresponding to the actual maximum.

A filtering rule to propagate changes of $\mathtt{W_i}$ domains to $\mathtt{Y}$ is given by the pseudo-code in Algorithm 1. The algorithm computes the numerator and denominator of Expression 5 (line 2), which are then updated by repeatedly maximizing weights (lines 4-6) as long as the condition from Theorem 1 holds (line 3). The rounding operation at line 6 is needed since $\mathtt{Y}$ is an integer variable. The worst case complexity at search time is $O(n)$, since the initial sorting operation (line 1) can be done once for all at model loading time.

Algorithm 1. Filtering on $\max(\mathtt{Y})$

Require: a change in the bound of some $\overline{\mathtt{W}}_\mathtt{i}$ domain
1: sort terms by decreasing v_i
2: compute $v_{ub} = \sum_{i=0}^{n-1} v_i \cdot \min(\mathtt{W_i})$ and $w_{ub} = \sum_{i=0}^{n-1} \min(\mathtt{W_i})$
3: **for** $i = 0$ to $n-1$, as long as $v_i > {}^{v_{ub}}/_{w_{ub}}$ **do**
4:     $\delta = \max(\overline{\mathtt{W}}_\mathtt{i}) - \min(\overline{\mathtt{W}}_\mathtt{i})$
5:     $v_{ub} = v_{ub} + v_i \cdot \delta$
6:     $w_{ub} = w_{ub} + \delta$
7: $\max(\mathtt{Y}) \leftarrow round({}^{v_{ub}}/_{w_{ub}})$

2.2 Computing Bounds on Variables W_i

Lower and upper bounds on the W_i variables can be deduced based on the domain of Y. In this section we focus on the deductions performed from the domain max, i.e. $\max(Y)$ (filtering with the minimum is analogous). In first place, note that if $\max(Y)$ is equal to its maximum computed as from Section 2.1, no pruning on the W_i domains is possible. If this is not the case, we may be able to do some filtering. In particular we have:

$$\frac{\sum_{i=0}^{n-1} v_i \cdot W_i}{\sum_{i=0}^{n-1} W_i} \leq \max(Y) \tag{7}$$

Let us assume initially that there is at least a non-zero weight. Then we get:

$$\sum_{i=0}^{n-1} W_i \cdot rv_i(Y) \leq 0 \qquad \text{with} \ \ rv_i(Y) = v_i - \max(Y) \tag{8}$$

where we refer to $rv_i(Y)$ as *reduced value* or term i. The *maximum slack* is obtained by minimizing W_i if the reduced value is positive and maximizing the variable if the reduced value is negative. Formally:

$$ms(\overline{W}, Y) = \sum_{i=0}^{n-1} rv_i(Y) \cdot \begin{cases} \min(W_i) & \text{if } rv_i(Y) > 0 \\ \max(W_i) & \text{otherwise} \end{cases} \tag{9}$$

By keeping a single variable free, we get:

$$rv_j(Y) \cdot W_j \leq - \sum_{i=0,i\neq j}^{n-1} rv_i(Y) \cdot \begin{cases} \min(W_i) & \text{if } rv_i(Y) > 0 \\ \max(W_i) & \text{otherwise} \end{cases} \tag{10}$$

The right-most member of the Equation (10) is constant and referred to as θ_j in the followings. Then, depending on the sign of the reduced value we get either a lower or an upper bound:

$$W_j \leq {}^{\theta_j}\!/_{rv_j(Y)} \ \text{ if } rv_j(Y) > 0, \qquad\qquad W_j \geq {}^{\theta_j}\!/_{rv_j(Y)} \ \text{if } rv_j(Y) < 0 \tag{11}$$

and no propagation in case $rv_j(Y) = 0$. If all weights in θ_j are 0, then it may be the case that all weights are zero and Equation (7) turns into:

$$\frac{v_j \cdot W_j}{W_j} \leq \max(Y) \tag{12}$$

From which we can deduce that: 1) If $\max(Y) < 0$, then W_i must be non-zero since a weighted average is 0 if all weights are 0. 2) If $v_j > \max(Y)$, then W_i cannot be higher than 0. Note that if both the rules are triggered we have a fail.

The reasoning leads to Algorithm 2. Line 1 performs a de-round operation (needed since all variables are integer). As a side effect, the expression $v_j - y_{max}$

Algorithm 2. Filtering from max(Y) to W_i variables

Require: a change in max(Y), or in the domain bounds of some W_i

1: Let y_{max} be the largest value v such that $round(v) = \max(Y)$. If y_{max} is greater or equal to the maximum from Section 2.1, then immediately return.

2: Compute the maximum slack $ms(\overline{W}, Y)$ according to Equation (10), using y_{max} instead of max(Y). Additionally, let w_{ms} be the corresponding sum of weights.

3: **for** $j = 0$ to $n - 1$ **do**

4:     **if** $v_j - y_{max} > 0$ **then**

5:         $\theta_j = -(ms(\overline{W}, Y) - (v_j - y_{max}) \cdot \min(W_j))$

6:         $w_{\theta_j} = w_{ms} - \min(W_j)$

7:     **else**

8:         $\theta_j = -(ms(\overline{W}, Y) - (v_j - y_{max}) \cdot \max(W_j))$

9:         $w_{\theta_j} = w_{ms} - \max(W_j)$

10:     **if** $w_{\theta_i} = 0$ **then**

11:         **if** $y_{max} < 0$ **then** $\min(W_j) \leftarrow 1$

12:         **if** $v_j > y_{max}$ **then** $\max(W_j) \leftarrow 0$

13:     **else**

14:         **if** $v_j - y_{max} > 0$ **then** $\max(W_j) \leftarrow \left\lfloor \theta_j / (v_j - y_{max}) \right\rfloor$

15:         **if** $v_j - y_{max} < 0$ **then** $\min(W_j) \leftarrow \left\lceil \theta_j / (v_j - y_{max}) \right\rceil$

---

is used in the algorithm in place of the reduced value. The maximum slack is computed at line 2. Then, for each term we compute the θ_i value at lines 5 and 8, by subtracting the contribution of term j from $ms(\overline{W}, Y)$; we also compute the sum w_{θ_i} of weights in θ_i (lines 6 and 9). If $w_{\theta_i} = 0$ we apply the filtering rules from Equation (12) at lines 11-12, otherwise, we use the regular filtering rules from Equation (11) at lines 14-15. The floor/ceil operations at lines 14-15 are needed since all variables are integer. The overall time complexity is $O(n)$.

3   Optimizations

Incremental Upper Bound and Maximum Slack Computation: It is possible to improve the efficiency of the filtering algorithm via incremental computation of the upper bound from Section 2.1 (let this be $ub(\overline{W})$) and the maximum slack $ms(\overline{W}, Y)$ from Section 2.2. The approach requires all terms to be sorted by decreasing v_j (this can be done once for all at model loading time).

Specifically, let v_{ub}, w_{ub} be the current numerator and denominator of $ub(\overline{W})$. Moreover, let j_{ub} be the index of the last term whose weight was maximized by Algorithm 1. In case the domain minimum/maximum of some W_j variable changes, new values for v_{ub}, w_{ub} and j_{ub} can be computed in two steps. In first place, we have to perform the updates:

$$v_{ub} = v_{ub} + v_j \cdot \begin{cases} \max(W_j)_{new} - \max(W_j)_{old} & \text{if } j \leq j_{ub} \\ \min(W_j)_{new} - \min(W_j)_{old} & \text{otherwise} \end{cases} \tag{13}$$

Algorithm 3. Update j_{ub}

---

while $v_{j_{ub}} >^{v_{ub}}/w_{ub}$ **do**
$\quad \delta = \max(\overline{W}_{j_{ub}}) - \min(\overline{W}_{j_{ub}})$
$\quad v_{ub} = v_{ub} + v_{j_{ub}} \cdot \delta$
$\quad w_{ub} = w_{ub} + \delta$
$\quad j_{ub} = j_{ub} + 1$

Algorithm 4. Update j_{ms}

---

while $v_{j_{ms}} > y_{max}$ **do**
$\quad \delta = \min(\overline{W}_{j_{ms}}) - \max(\overline{W}_{j_{ms}})$
$\quad v_{ms} = v_{ms} + v_{j_{ms}} \cdot \delta$
$\quad w_{ms} = w_{ms} + \delta$
$\quad j_{ms} = j_{ms} + 1$

$$w_{ub} = w_{ub} + \begin{cases} \max(W_j)_{new} - \max(W_j)_{old} & \text{if } j \leq j_{ub} \\ \min(W_j)_{new} - \min(W_j)_{old} & \text{otherwise} \end{cases} \tag{14}$$

Due to the resulting changes of $ub(\overline{W})$, it may become necessary to maximize *one or more* terms previously not satisfying the condition from Statement 1. This can be done by means of Algorithm 3, which is analogous to lines 3-6 of Algorithm 1, but makes use of the stored j_{ub}. Note that in case the domain of more than a single W_i variable changes, then *all* the updates from Equation (13) and (14) should be performed before running Algorithm 3. This behavior is easy to implement if the solver provides a delayed constraint queue.

The incremental computation has constant space complexity, since only three values need to be stored (v_{ub}, w_{ub} and j_{ub}). Updating v_{ub} and w_{ub} using Equation (13) and (14) has constant time complexity and the number of possible updates from the root to a leaf of the search tree is $O(n \times \max \text{ domain size})$. In the important case of $\{0, 1\}$ weights, no more than $2 \times n$ updates can occur. Since the j_{ub} index cannot advance more than n times, the overall complexity is $O(n)$ along a full branch of the search tree.

Incremental computation for the maximum slack from Section 2.2 can be achieved in a similar fashion. In first place, observe that Equation (9) can be decomposed as $ms(\overline{W}, Y) = v_{ms} - \max(Y) \cdot w_{ms}$, where the dependency on $\overline{W}$ is omitted for sake of simplicity and:

$$v_{ms} = \sum_{i=0}^{n-1} v_i \cdot \begin{cases} \min(W_i) & \text{if if } rv_i(Y) > 0 \\ \max(W_i) & \text{otherwise} \end{cases} \tag{15}$$

$$w_{ms} = \sum_{i=0}^{n-1} \begin{cases} \min(W_i) & \text{if if } rv_i(Y) > 0 \\ \max(W_i) & \text{otherwise} \end{cases} \tag{16}$$

The values v_{ms} and w_{ms} can be updated incrementally similarly to v_{ub} and w_{ub}. We then need to update the index j_{ms} of the last minimized element in the maximum slack configuration, which can be done by means of Algorithm 4.

Reducing the Complexity of Term Weight Filtering: The techniques described in the previous section make the computation of $ub(\overline{W})$ and $ms(\overline{W}, Y)$ very efficient. However, pruning the W_i variables as described in Algorithm 2 still takes linear time and acts a bottleneck for the propagator complexity. It may be possible to achieve higher efficiency by exploiting ideas in [4]: here however we apply a simpler technique, described in the case of positive reduced values (i.e. terms j with $j < j_{ms}$ from Section 3). The corresponding θ_j value is:

$$\theta_j = -ms(\overline{W}, Y) + rv_j(Y) \cdot \min(W_j) \tag{17}$$

with $ms(\overline{W}, Y) \leq 0$. The upper bound computed by Algorithm 2 on W_j is ${\theta_j}/{rv_j(Y)}$ and can be decomposed as:

$$\frac{\theta_j}{rv_j(Y)} = -\frac{ms(\overline{W}, Y)}{rv_j(Y)} + \min(W_j) \tag{18}$$

We are interested in finding a sufficient condition to *avoid the bound computation for some W_j variables*. Now, observe that for every $k > j$, we have:

$$-\frac{ms(\overline{W}, Y)}{rv_j(Y)} + \min_i\{\min(W_i)\} \leq -\frac{ms(\overline{W}, Y)}{rv_k(Y)} - \min(W_k) \tag{19}$$

since $v_k \leq v_j$ (due to the initial sorting step). In other words, the left-most quantity in Equation (19) bounds the upper-bound value of all terms k with $k > j$. Therefore, if at line 14 of Algorithm 1 we realize that:

$$-\frac{ms(\overline{W}, Y)}{rv_j(Y)} + \min_i\{\min(W_i)\} \geq \max_i\{\max(W_i)\} \tag{20}$$

we can immediately stop filtering all $j < j_{ms}$. In order to simplify the computation, we statically compute the minimum and maximum among all domains (i.e. $\min_i\{\min(W_i)\}$ and $\max_i\{\max(W_i)\}$) at model loading time. As a result, checking this stop condition has constant time complexity.

Avoiding Useless Activations of Term Weight Filtering: Finally, we can rely on the stored j_{ms} value to reduce the number of calls to Algorithm 2. On this purpose, observe that the bound computation at lines 14-15 is based on the maximum slack. Note that in the $ms(\overline{W}, Y)$ expression (see Equation (9)) the terms with positive $rv(Y)$ appear with weight minimum, while the terms with non-positive reduced value appear with maximum weight.

As a consequence, changes in $\max(W_i)$ with $i > j_{ms}$ or changes in $\min(W_i)$ with $i \leq j_{ms}$ do not change the value of the maximum slack and do no require to run Algorithm 1. Hence, immediately after the execution of Algorithm 4 we can check which W_i boundaries have changed after the last update, and avoid performing additional filtering if this is not needed. The same does not hold for changes of $\max(Y)$, that always require to re-execute Algorithm 2.

4   Experimental Results

The described approach was implemented in Google OR-tools [6] and tested on two different benchmarks. In first place, we tackled a variant of the Single

Source Capacitated Facility Location Problem (SSCFLP) [12] with a balancing objective. The problem structure was suitable for different solution approaches, respectively: a model with our average constraint, a model using the decomposed formulation from Equation 2 and a method that employs an alternative search strategy to avoid the use of an explicit average constraint. We performed a comparison to show that our method has is competitive with alternative approaches.

In a second experimentation we use the average constraint to provide input to an Empirical Model, namely a Neural Network capturing the thermal behavior of a multi-core CPU. The network is employed within a large-size workload dispatching problem with a maximal efficiency objective. On this problem, we compare only our constraint and the decomposed formulations, since the the third method cannot be applied. Moreover, we take advantage of the large scale of the instances to compare the performance of our incremental filtering algorithm w.r.t its non-incremental version.

4.1 Balancing the Travel Cost in the SSCFLP

Problem Formulation. The Capacitated Facility Location Problems represent a major class of location problems, well known in the OR-literature [12]. The SSCFLP deals with opening facilities j at a set of candidate locations, to serve at minimum cost a set of customers i distributed in a discrete space. Different facility sites have distinct (fixed) opening costs f_j. Each customer must be supplied from a single facility and has a fixed demand d_i. The sum of the demands for a single facility should not exceed its capacity s_j. Assigning a customer to a facility incurs a transportation charge $c_{i,j}$. The usual goal of the problem is to decide which facilities to open and to assign customers so as to minimize the overall cost. Here, we consider instead a variant where the objective is to minimize the worst-case average transportation cost per facility. Additionally, we require that the total opening cost does not exceed the given threshold. We adopt the following model (with n customers and m facilities):

$$\min Z = \max_{j=0..m-1} A_j \tag{21}$$

$$\text{s.t.:} \quad X_{i,j} = 1 \Leftrightarrow (W_i = j) \quad \forall j = 0..m-1, \; i = 0..n-1 \tag{22}$$

$$0 \leq X_{i,j} \leq Y_j \leq 1 \quad \forall j = 0..m-1, \; i = 0..n-1 \tag{23}$$

$$\sum_{i=0}^{n-1} d_i \cdot X_{i,j} \leq s_j \cdot Y_j \quad \forall j = 0..m-1 \tag{24}$$

$$\sum_{j=0}^{m-1} f_j \cdot Y_j \leq F_{max} \tag{25}$$

$$\texttt{average}(\bar{c}_j, \bar{X}_j, A_j) \quad \forall j = 0..m-1 \tag{26}$$

The decision variable are:

$$W_i \in \{0..m-1\} \qquad W_i = j \text{ iff customer } i \text{ is assigned to facility } j$$

$$X_{i,j} \in \{0,1\} \qquad X_{i,j} = 1 \text{ iff customer } i \text{ is assigned to facility } j$$

$$Y_j \in \{0,1\} \qquad\qquad Y_j = 1 \text{ iff facility } j \text{ is open}$$
$$A_j \in \{0..\inf\} \qquad\qquad \text{average transportation cost for each facility } j$$

where $\overline{X}_j$ and $\overline{c}_j$ are the subsets of $X_{i,j}$ and $c_{i,j}$ that refer to facility j. F_{max} is the opening cost limit which was obtained in a preliminary experimentation.

Solution Methods. We solved the problem with three different approaches:

1. The model reported in the previous paragraph (referred to as AVG), using the incremental filtering for the average constraint.
2. A model (referred to as DEC) where the **average** constraint is replaced by its decomposed formulation from Equation 2.
3. A method (referred to as SWEEP) where the optimization problem is slit into a sequence of satisfaction problems. Each subproblem is obtained by 1) removing the A_j variables; 2) removing the cost function; 3) replacing the average constraints with the following linear inequalities:

$$\sum_{i=0}^{n-1} c_{i,j} \cdot X_{i,j} < z_k^* \cdot \sum_{i=0}^{n-1} X_{i,j} \qquad\qquad \forall j = 0..m-1 \qquad (27)$$

where z_k^* for the first subproblem is a safe upper bound and z_k^* for the k-th problem in the sequence is equal to the cost of the solution the $k-1$-problem.

The AVG and SWEEP approaches are equivalent in term of propagation, but SWEEP has to explore a larger space due to the decoupling. The DEC approach results instead in weaker (but more efficient) propagation. In all cases, we use tree search with random variable and value selection, performing restarts when a fixed fail limit is reached (20000). This approach was chosen since the considered instance size makes a heuristic approach more viable and because restarts make the performance of all methods less erratic. We run the experiments on a subset of the benchmarks by Beasley in the OR-lib [3], in particular the $cap6X$, $cap7X$, $cap9X$, $cap10X$, $cap12X$, and $cap13X$ groups, having different configuration number of facilities, customers and costs. All the experiments are run with a time limit of 60 seconds on a 2.8 GHz Intel i7 and repeated 10 times per instance.

Results. Table 1 shows the average solution gaps over time (%) between the AVG, the DEC and the SWEEP approach[1]. In particular, we measured the gap between the best solution found by both approaches at some pre-defined sampling points (one per column in the table).

Note that the performance of the AVG approach is considerably better than DEC, in spite of the fact that the heavy use of restarts favors approaches based on weaker (and faster) propagation. This is likely due to the fact that for this

---

[1] The gap for a single instance is $\frac{z_{\text{SWEEP}}}{z_{\text{AVG}}} - 1$, resp. $\frac{z_{\text{DEC}}}{z_{\text{AVG}}} - 1$. The reported value is the average gap for all instances in the group.

Table 1. Average Solution Gap (%) for each benchmarks subset

Instance set	vs.	\multicolumn{6}{c}{Solution Gap (%)}					
		1s	3s	5s	10s	30s	60s
cap6X	SWEEP	18.9 %	16.5 %	18.55 %	19.7 %	11.45 %	11.6 %
	DEC	37.2 %	47.5 %	49.225 %	17.8 %	6.725 %	7.9 %
cap7X	SWEEP	7 %	7.7 %	7.7 %	7.8 %	10.3 %	9.7 %
	DEC.	22.275 %	14.1 %	11.05 %	8.6 %	10.7 %	11.075 %
cap9X	SWEEP	25.8 %	31.025 %	30.775 %	28.525 %	23.95 %	22.075 %
	DEC.	62.1 %	66.125 %	65.025 %	36.75 %	12 %	12.75 %
cap10X	SWEEP	4.9 %	7.3 %	5.8 %	5.45 %	5.175 %	4.6%
	DEC.	8.75 %	8.6 %	6.7 %	5.775 %	6 %	5.9 %
cap12X	SWEEP	29 %	29 %	35.55 %	28.575 %	23.2 %	20 %
	DEC.	15.1 %	23.275 %	11.625 %	12.225 %	14.9 %	12.5 %
cap13X	SWEEP	4.95 %	4.475 %	4.075 %	4.95 %	4.4 %	3.625 %
	DEC.	10.9 %	6.95 %	7.15 %	6.7 %	5.7 %	4.6 %

problem the weighted average expressions appear straight in the cost function. More surprisingly, AVG works better than SWEEP, which is equivalent in term of propagation effectiveness and faster in terms of propagation efficiency: apparently, the overhead for the repeated replacement of constraints (27) has a negative impact on the solver performance, which overcomes that of the more complex filtering of the average constraint.

4.2   Empirical Models and Incremental Filtering

Problem Formulation. We consider a workload dispatching problem for a 48-core CPU by Intel [5]. The platform features some temperature control policies, which may slow down the execution to avoid overheating. We want to map a set of n tasks i on the platform cores j so as to incur the smallest possible slowdown due to the thermal controllers. As in many realistic settings, we assume tasks are preemptive and have unknown duration. Each task must be assigned to a platform core and is characterized by a parameter named CPI (Clock Per Instruction): the smaller the CPI_i value, the more computation intensive the task and the higher the generated heat.

The thermal behavior of a core is also affected by the workload of neighboring cores. The resulting correlation between workload and platform efficiency is strongly non-linear (due to the thermal controllers and thermal effects) and very tough to model via conventional means. In [1] we proposed an approach were such a complex function is captured via a Neural Network, trained over a system simulator. Here, we consider a slightly simplified version of the problem, corresponding to the following model (with n tasks and $m = 48$ cores):

$$\max Z = \min_{j=0..m-1} E_j \tag{28}$$

$$\text{s.t.:} \quad \mathtt{E_j} = \mathtt{ANN_j}(temp, \{\mathtt{A_k} \mid \text{core } k \text{ neighbor of } j\}) \quad \forall j = 0..m-1 \tag{29}$$

$$\mathtt{average}(\overline{CPI}, \overline{\mathtt{X}}_\mathtt{j}, \mathtt{A_j}) \quad\quad\quad \forall j = 0..m-1 \tag{30}$$

$$\sum_{i=0}^{n-1} \mathtt{X_{i,j}} \geq 1 \quad\quad\quad \forall j = 0..m-1 \tag{31}$$

$$\mathtt{X_{i,j}} = 1 \Leftrightarrow (\mathtt{W_i} = j) \quad \forall i = 0..n-1, j = 0..m-1 \tag{32}$$

The decision variable are:

$\mathtt{W_i} \in \{0..m-1\}$ $\quad\quad\quad$ $\mathtt{W_i} = j$ iff task i is assigned to core j

$\mathtt{X_{i,j}} \in \{0,1\}$ $\quad\quad\quad$ $\mathtt{X_{i,j}} = 1$ iff task i is assigned to core j

$\mathtt{A_j} \in \{0..\infty\}$ $\quad\quad\quad$ average CPI of tasks assigned to core j

The $\mathtt{E_j}$ variables represent the efficiency of core j. The notation $\mathtt{ANN_j}(\cdot)$ stands for the set of Neuron Constraints [2] corresponding the Neural Network we trained for core j. In principle $\mathtt{E_j}$ should take real values in $]0,1]$, in practice we use a fixed precision integer representation.

Solution Method. We solved the problem using tree search with random variable and value selection, performing restarts when a fixed fail limit is reached (6000 fails). We compared our approach with a model making use of the decomposed average formulation, referred to as DEC. Moreover, we used this problem to investigate the speed-up of the incremental **average** filtering (referred to as INC) over the non incremental one (referred to as NO INC).

For the experimentation we used the benchmarks from [1], consisting of 3 groups, each counting 2 worksets of 10 instances[2]. Groups differ by the number of tasks (120, 240, or 480). Worksets within each group correspond to different task compositions: all instances in Workset *I* contain 75% low-CPI and 25% high-CPI tasks, while the mix is instead 85%/15% for Workset *II*. Each instance was solved 10 times with each method, on an 2.8 GHz Intel i7.

Results. Table 2 shows the results of the experimentation. The benchmarks are divided according to the size of the instances (120, 240, and 480 tasks). For each workset (I or II), each column in the table refers to a different sampling time and shows the solution quality gap between AVG and NO INC, and between AVG and DEC (the gap is computed analogously to the previous experimentation). The table also reports the search speed, measured in terms of average number of branches per second. As expected, the incremental approach is considerably better then the non incremental one. Interestingly, AVG has approximatively the same branching speed on the 240 and 480 task instances, suggesting that the propagator is processing a roughly constant number of weight variables (thanks to the stopping condition from Section 3). Interestingly, AVG often works better than DEC: the presence of a the network of Neuron Constraints makes in principle the propagation on the average far less critical, and yet employing the global **average** constraint seems to definitely pay off.

---

[2] The benchmarks are available for download at `http://ai.unibo.it/data/TAWD`

Table 2. Incremental Speed-up: solution efficiency and gaps

Inst. size	Workset	Algorithm	Average Solution Efficiency (%)					
			1 sec	3 sec	5 sec	10 sec	30 sec	60 sec
120	Set I	NO INC	28.66 %	9.03 %	12.68 %	8.08 %	4.62 %	4.21 %
		DEC	0.74 %	-5.71 %	-5.64 %	-4.67 %	-3.86 %	-2.25 %
	Set II	NO INC	18.14 %	11.06 %	11.01 %	4.41 %	6.07 %	3.83 %
		DEC	1.39 %	-1.07 %	2.78 %	3.07 %	4.11 %	4.82 %
Average speed: (branch/secs)		AVG	4043.33		NO INC	2571.88	DEC	9332.96
240	Set I	NO INC	7.58 %	12.14 %	9.19 %	5.85 %	4.23 %	3.97 %
		DEC	0.02 %	1.98 %	4.47 %	2.56 %	1.65 %	2.27 %
	Set II	NO INC	5.79 %	10.14 %	8.62 %	6.11 %	4.18 %	3.14 %
		DEC	3.98 %	1.47 %	2.49 %	2.78 %	3.36 %	2.35 %
Average speed: (branch/secs)		AVG	1845.00		NO INC	949.31	DEC	2904.68
480	Set I	NO INC	*inf*	24.01 %	25.38 %	12.65 %	4.06 %	2.51 %
		DEC	7.82 %	1.41 %	2.73 %	1.72 %	0.86 %	0.63 %
	Set II	NO INC	*inf*	22.79 %	26.10 %	16.24 %	8.65 %	5.71 %
		DEC	6.58 %	2.60 %	3.65 %	5.13 %	4.26 %	3.61 %
Average speed: (branch/secs)		AVG	2119.86		NO INC	635.75	DEC	2797.21

5   Dealing with Variable Term Values

It is possible to extend the average constraint to deal with variable term values. The resulting signature is:

$$\mathtt{average}(\overline{\mathtt{V}}, \overline{\mathtt{W}}, \mathtt{Y}) \tag{33}$$

where $\overline{V}$ is a vector of integer variables. Variable term values allow to tackle a broader class of problems, such as the routing example mentioned in Section 1. Filtering for the new constraint signature requires a few modifications to the procedures described in Section 2.1 and 2.2.

Computing Bounds on Variable Y: Obtaining an upper bound for Y is relatively easy, due to the following theorem:

Theorem 1. *Let $\overline{v}^*$, $\overline{w}^*$ be fixed assignments for $\overline{V}$ and $\overline{W}$. Then, increasing a value v_j^* by an amount $\delta > 0$ always leads to an increased weighted average value.*

Proof. Directly follows from:

$$\frac{\sum_{i=0}^{n-1} v_i^* \cdot w_i^* + \delta \cdot w_j^*}{\sum_{i=0}^{n-1} w_i^*} \geq \frac{\sum_{i=0}^{n-1} v_i^* \cdot w_i^*}{\sum_{i=0}^{n-1} w_i^*} \tag{34}$$

since $\delta > 0$ and $w_j^* \geq 0$. $\qquad\square$

In other words, higher term values are always to be preferred. Hence, an upper bound on Y can be computed by means of Algorithm 1 by replacing v_i with

$\max(V_i)$. Unlike for the fixed value case, however, the sorting operation at line 1 cannot be performed once for all at model loading time, so that the resulting complexity is $O(n \log n)$.

Computing Bounds on Variables W_i: Bounds for the weight variables in the fixed value case are based on Equation (8), from which the maximum slack value is obtained. The equivalent formula when term values are non constant is:

$$\sum_{i=0}^{n-1} W_i \cdot (V_i - \max(Y)) \leq 0 \tag{35}$$

The inequality is valid if $\sum_{i=0}^{n-1} W_i$ can be > 0. The maximum slack corresponds to an assignment of V_i variables, besides W_i. The following theorem holds:

Theorem 2. *Let $\overline{v}^*$, $\overline{w}^*$ be fixed assignment for $\overline{V}$ and $\overline{W}$. Then, decreasing a value v_j^* by an amount $\delta > 0$ always leads to an increased slack.*

The proof is analogous to that of Theorem 5. As a consequence, choosing the minimum possible value for each V_i leads to maximum slack, i.e.:

$$ms = \sum_{i=0}^{n-1} (\min(V_i) - \max(Y)) \cdot \begin{cases} \min(W_i) & \text{if } \min(V_i) - \max(Y) > 0 \\ \max(W_i) & \text{otherwise} \end{cases} \tag{36}$$

The full notation for the slack expression is $ms(\overline{V}, \overline{W}, Y)$. By keeping a single term free and the remaining ones in the maximum slack configuration, we obtain:

$$(V_j - \max(Y)) \cdot W_j \leq \theta_i \tag{37}$$

which corresponds to Equation (10) in Section 2.2. The θ_i value is obtained analogously to the case of fixed values. Moreover, we also use w_{θ_i} to denote the sum of the weights employed in the computation of θ. Equation (37) can be used to obtain bounds for W_i and V_i.

In particular, the domain of variable V_i can be pruned as described in Algorithm 5 (that should be integrated in Algorithm 2). Lines 2 and 5 are needed to handle the $W_i = 0$ case. Lines 3 and 6 update the domain max of V_i according to Equation (37). Note that the W_i assignment leading to the correct bound depends on the sign of θ_i. The rounding operations at lines 3 and 6 are necessary since V_i is an integer variable. Finally, y_{max} is defined as from Algorithm 2.

Bounds on the weight variable W_i can be obtained via Algorithm 6, which should replace lines 14-15 in Algorithm 2. The maximum slack is obtained in this case by assigning V_i to the minimum possible value. Then, depending on the sign of the resulting reduced value and on the sign of θ_i we obtain either a lower bound or an upper bound on W_i. As a particular case, line 3 corresponds to a negative upper bound (hence to a fail) and line 5 to a negative lower bound (hence to no pruning).

Algorithm 5. Filtering on V_i
1: **if** $\theta_i < 0$ **then**
2:     **if** $\max(W_i) = 0$ **then** fail
3:     **else** $\max(V_i) \leftarrow \left\lfloor \frac{\theta_i}{\min(W_i)} + y_{max} \right\rfloor$
4: **else**
5:     **if** $\max(W_i) = 0$ **then** do nothing
6:     **else** $\max(V_i) \leftarrow \left\lfloor \frac{\theta_i}{\max(W_i)} + y_{max} \right\rfloor$

Algorithm 6. Filtering on W_i
1: **if** $\theta_i < 0$ **then**
2:     **if** $\min(V_i) - y_{max} < 0$ **then**
3:         $\min(W_i) \leftarrow \left\lceil \frac{\theta}{\min(V_i) - y_{max}} \right\rceil$
4:     **else** fail
5: **else**
6:     **if** $\min(V_i) - y_{max} \leq 0$ **then**
7:         do nothing
8:     **else** $\max(W_i) \leftarrow \left\lfloor \frac{\theta}{\min(V_i) - y_{max}} \right\rfloor$

Similarly to Section 2.2, the case of $w_\theta = 0$ should be handle separately, since it may compromise the assumption that $\sum_{i=0}^{n-1} W_i > 0$. In this case, we have:

$$\frac{V_i \cdot W_i}{W_i} \leq \max(Y) \tag{38}$$

If $\max(Y) < 0$, then we have $V_i \leq \max(Y)$ and the minimum allowed value for W_i is 1. Conversely, if $\max(Y) \geq 0$ then we have $V_i \leq \max(Y)$ if $\min(W_i) > 0$, and no propagation if $\min(W_i) = 0$.

6   Concluding Remarks

We have introduced a novel global constraint to provide a convenient model and effective filtering for weighted average expressions. The **average** constraint can be fruitfully applied to allocation problems with balancing constraints or objectives. Furthermore, **average** is a key enabler for optimization approaches making use of a complex system models, extracted via Machine Learning. We provided an efficient incremental filtering algorithm for the case with fixed term values. This was used to successfully tackle a large scale combinatorial problem featuring a Neural Network model learned from a system simulator. Finally, we discussed how to perform non-incremental filtering when both term values and weights are variable. Further developments of the constraint are strongly connected to ongoing research on the hybridization of CP, Machine Learning and Simulation.

Acknowledgement. This project if funded by a *Google Focused Grant Program on Mathematical Optimization and Combinatorial Optimization in Europe*, with title: "Model Learning in Combinatorial Optimization: a case study on thermal aware dispatching".

References

1. Bartolini, A., Lombardi, M., Milano, M., Benini, L.: Optimization and Controlled Systems: A Case Study on Thermal Aware Workload Dispatching. Accepted for Publication at AAAI 2012 (2012)

2. Bartolini, A., Lombardi, M., Milano, M., Benini, L.: Neuron Constraints to Model Complex Real-World Problems. In: Lee, J. (ed.) CP 2011. LNCS, vol. 6876, pp. 115–129. Springer, Heidelberg (2011)
3. Beasley, J.E.: OR-Library: Distributing Test Problems by Electronic Mail. Journal of the Operational Research Society 41(11), 1069–1072 (1990)
4. Harvey, W., Schimpf, J.: Bounds consistency techniques for long linear constraints. In: Proc. of the TRICS Workshop, at CP (2002)
5. Howard, J., Dighe, S., et al.: A 48-Core IA-32 message-passing processor with DVFS in 45nm CMOS. In: Proc. of ISSCC, pp. 108–109 (February 2010)
6. Perron, L.: Operations Research and Constraint Programming at Google. In: Lee, J. (ed.) CP 2011. LNCS, vol. 6876, p. 2. Springer, Heidelberg (2011)
7. Pesant, G., Régin, J.-C.: SPREAD: A Balancing Constraint Based on Statistics. In: van Beek, P. (ed.) CP 2005. LNCS, vol. 3709, pp. 460–474. Springer, Heidelberg (2005)
8. Régin, J.C.: Global Constraints: a survey. In: Hybrid Optimization, pp. 63–134. Springer, New York (2011)
9. Schaus, P., Deville, Y., Dupont, P., Régin, J.C.: Simplification and extension of the spread constraint. In: Third Workshop on Constraint Propagation and Implementation, in CP 2006, pp. 72–92 (2006)
10. Schaus, P., Deville, Y., Dupont, P., Régin, J.-C.: The Deviation Constraint. In: Van Hentenryck, P., Wolsey, L.A. (eds.) CPAIOR 2007. LNCS, vol. 4510, pp. 260–274. Springer, Heidelberg (2007)
11. Schaus, P., Deville, Y., Dupont, P.: Bound-Consistent Deviation Constraint. In: Bessière, C. (ed.) CP 2007. LNCS, vol. 4741, pp. 620–634. Springer, Heidelberg (2007)
12. Sridharan, R.: The capacitated plant location problem. European Journal of Operational Research 87(2), 203–213 (1995)

Weibull-Based Benchmarks for Bin Packing

Ignacio Castiñeiras[1], Milan De Cauwer[2], and Barry O'Sullivan[3]

[1] Dpto. de Sistemas Informáticos y Computación
Universidad Complutense de Madrid, Spain
`ncasti@fdi.ucm.es`
[2] Département Informatique, Université de Nantes, France
`milan.de-cauwer@etu.univ-nantes.fr`
[3] Cork Constraint Computation Centre, University College Cork, Ireland
`b.osullivan@cs.ucc.ie`

Abstract. Bin packing is a ubiquitous problem that arises in many practical applications. The motivation for the work presented here comes from the domain of data centre optimisation. In this paper we present a parameterisable benchmark generator for bin packing instances based on the well-known Weibull distribution. Using the shape and scale parameters of this distribution we can generate benchmarks that contain a variety of item size distributions. We show that real-world bin packing benchmarks can be modelled extremely well using our approach. We also study both systematic and heuristic bin packing methods under a variety of Weibull settings. We observe that for all bin capacities, the number of bins required in an optimal solution increases as the Weibull shape parameter increases. However, for each bin capacity, there is a range of Weibull shape settings, corresponding to different item size distributions, for which bin packing is hard for a CP-based method.

1 Introduction

The one-dimensional bin packing problem is ubiquitous in operations research. It is defined as follows. Given a set $S = \{s_1, \ldots, s_n\}$ of n indivisible items, each of a known positive size s_i, and m bins each of capacity C, the challenge is to decide whether we can pack all n items into the m bins such that the sum of sizes of the items in each bin does not exceed C. The one-dimensional bin packing problem is NP-Complete. Amongst the many applications of this problem are timetabling, scheduling, stock cutting, television commercial break scheduling, and container packing [3,5]. Bin packing is closely related to a variety of other problems such as rectangle packing [30,29,33,6]. Recent work has focused on geometric generalisations of bin packing [4].

Typical bin packing methods rely on either heuristics [1], genetic algorithms [7], operations research methods [3], satisfiability techniques [11], or constraint programming [23,28]. There are many known bounds on the optimal number of bins which can be used in most of the techniques mentioned above [5,14,18].

M. Milano (Ed.): CP 2012, LNCS 7514, pp. 207–222, 2012.
© Springer-Verlag Berlin Heidelberg 2012

The motivation for the work presented here comes from the domain of data centre optimisation. Workload consolidation involves ensuring that the total amount of resource required by the set of jobs assigned to a server does not exceed the capacity of that resource. The application of constraint programming to this domain has only very recently attracted attention [12,21]. While there are many benchmark suites for bin packing in the literature [7,13,24,26,27,35], these are all artificial and lacking a practical basis. Typically, as for example in the benchmarks by Scholl and Klein [24][1], item sizes are generated using either uniform or normal distributions. As Gent has pointed out, current benchmark suites in this area are often unrealistic and trivial to solve [9]. More recently, Regin et al. [21] have called for more realistic benchmark suites for use in studying large-scale data centre problems. It is this requirement that this paper fulfills.

Section 2 presents a parameterisable benchmark generator for bin packing instances based on the well-known Weibull distribution [36]. Using the shape and scale parameters of this distribution a variety of item size distributions can be generated. In Section 3 we show that a number of real-world bin packing benchmarks can be modelled extremely well using this approach, for example the instances from the 2012 ROADEF/EURO Challenge which focuses on a data centre problem provided by Google, as well as the bin packing components of a several real world examination timetabling problems. We study the behaviour of both systematic (Section 4) and heuristic (Section 5) bin packing methods under a variety of Weibull settings. We show that our Weibull-based framework allows for very controlled experiments in a bin packing setting in which the distribution of item sizes can be precisely controlled. We discuss how the difficulty of bin packing is affected by the item size distribution and by bin capacity. Specifically, we observed that for all bin capacities, the number of bins required in an optimal solution increases as the Weibull shape parameter increases. However, for each bin capacity, it seems that there is a range of Weibull shape settings, corresponding to different item size distributions, for which bin packing is hard for a CP-based method.

2   The Weibull Distribution

In probability theory and statistics, the Weibull distribution is a continuous probability distribution. It is named after Waloddi Weibull, who presented the distribution in a seminal paper in 1951 [36]. The Weibull distribution is defined by a shape parameter, $k > 0$, and a scale parameter, $\lambda > 0$. The probability density function, $f(x; \lambda, k)$, of a random variable x distributed according to a Weibull distribution is defined as follows:

$$f(x; \lambda, k) = \begin{cases} \frac{k}{\lambda} \cdot \left(\frac{x}{\lambda}\right)^{k-1} \cdot e^{-(x/\lambda)^k} & x \geq 0, \\ 0, & \text{otherwise} \end{cases}$$

---

[1] `http://www.wiwi.uni-jena.de/Entscheidung/binpp/`

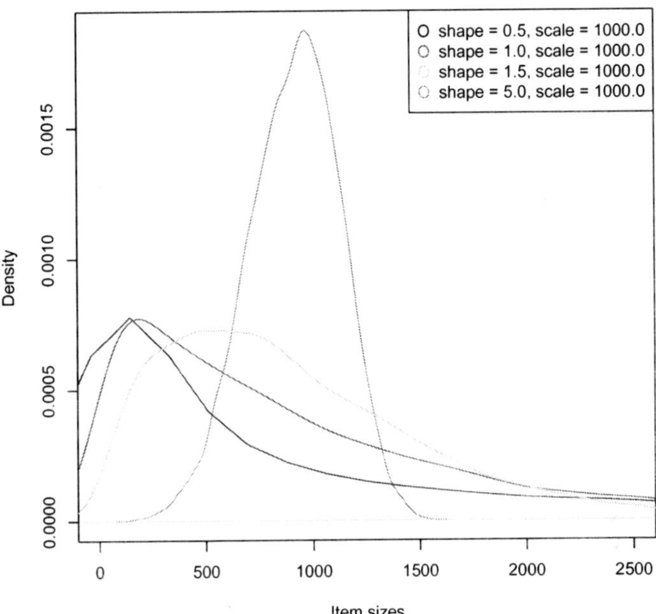

Fig. 1. Weibull distributions with small shape parameters and fixed scale parameter

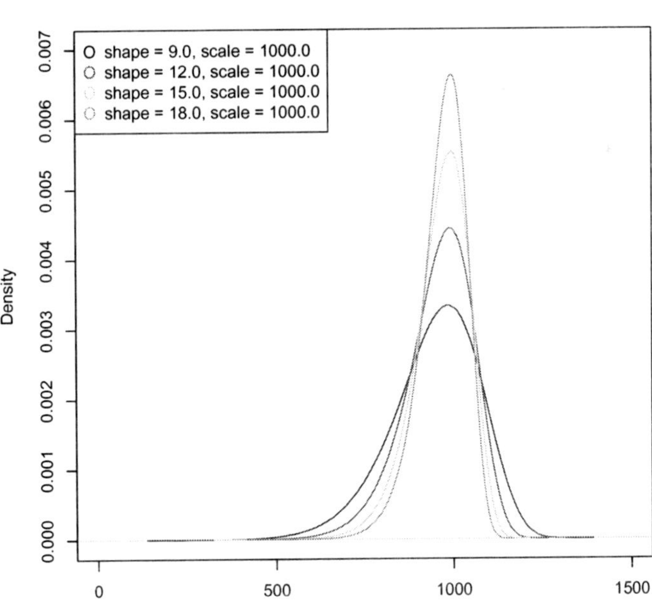

Fig. 2. Weibull distributions with larger shape parameters and fixed scale parameter

The Weibull can model many distributions that naturally occur in a variety of problem domains involving distributions of time horizons, time slots or lot sizes [36].

Figures 1 and 2 present several examples of different distributions that can be obtained by instantiating the Weibull distribution. Figure 1 presents four different distributions for small values of the shape parameter, k. Clearly very different regimes are possible, some exhibiting extremely high skew around the value specified by the distribution's scale parameter, λ. In Figure 2, larger values of the shape parameter are considered. In these examples we can see that the distribution exhibits lower variation as shape increases.

In this paper we propose using the Weibull distribution as the basis for a parameterisable benchmark generator for bin packing instances in which the *item sizes* are generated according to a Weibull distribution parameterised by specific values of k and λ. Using these parameters a variety of item size distributions can be generated. In Section 3 we show that some real-world bin packing benchmarks can be modelled extremely well using a Weibull distribution.

3   Fitting Weibull Distributions to Real-world Instances

In this section we demonstrate the flexibility of the Weibull distribution in fitting to a variety of bin packing problems coming from real-world applications. In Section 3.1 we show a specific example of how well the Weibull distribution can fit to a problem instance arising from the 2012 ROADEF/EURO Challenge. This example will show, visually, the quality of the fit that can be obtained. However, in Section 3.2 we present a more rigorous analysis of the goodness-of-fit that can be achieved through the use of two standard statistical tests.

3.1   An Example Problem in Data Centre Management

The 2012 ROADEF/EURO Challenge[2] is concerned with the problem of machine reassignment, with data and sponsorship coming from Google. The subject of the challenge is to find a best-cost mapping of processes, which have specific resource requirements, onto machines, such that a variety of constraints are satisfied. A core element of the problem are bin packing constraints stating that the total amount of a given resource required by the processes assigned to a machine does not exceed the amount available. An important element of this challenge is the mapping of processes to machines such that the availability of each resource on the machine is not exceeded by the requirements of the set of services assigned to it. This subproblem is a multi-capacity bin packing problem: each machine is a bin with many elements defined by the set of resources available, and each process corresponds to an item that consumes different amounts of each resource.

Figure 3 presents an example probability distribution for Resource 10 from instance a2(5) of the benchmarks available for the ROADEF/EURO Challenge.

---

[2] http://challenge.roadef.org/2012/en/index.php

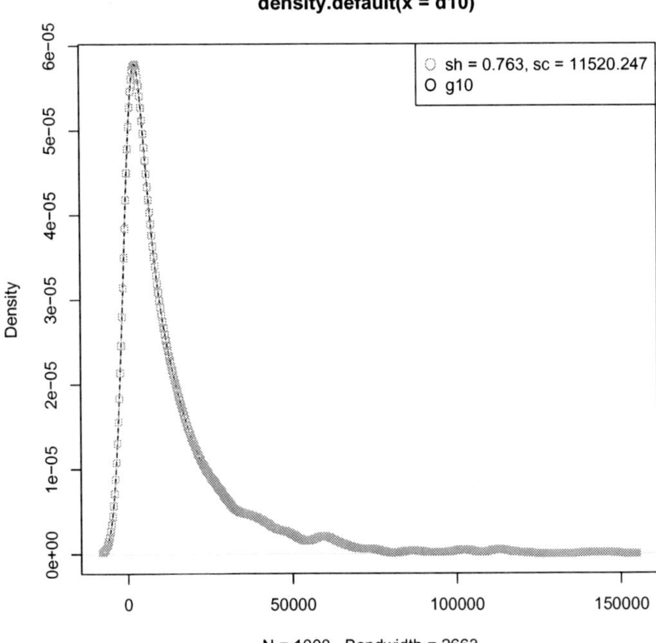

Fig. 3. An example of the quality of fit one can achieve when using a Weibull distribution for a real-world bin packing problem. Here we present the data and Weibull fit associated with Resource 10 of instance **a2(5)** from the 2012 ROADEF/EURO Challenge sponsored by Google.

The probability density function that corresponds to the actual data is plotted as a line. We can see clearly that this distribution is extremely skewed, with the majority of the probability mass coming from smaller items. Another characteristic of the data is the spread along the x-axis, showing that the range of likely item sizes spans several orders-of-magnitude, and there is a very small possibility of encountering extremely large items.

We used R, the open-source statistical computing platform [32], to fit a Weibull distribution to this data, using maximum likelihood fitting [16]. Specifically, we have used the R Weibull Distribution Maximum Likelihood Fitting implementation by Wessa, which is available as an online service [37]. The resulting Weibull is presented in Figure 3 as the circles imposed on the density function from the data. By observation we can see that the fit is extremely good. In the next section we will study the quality of fit more rigorously, demonstrating that it is statistically significant.

3.2   Verifying the Goodness of Fit

We study a variety of benchmark bin packing problems. As mentioned above, the 2012 ROADEF/EURO Challenge provides a publicly available set of problem instances that contains many bin packing instances. In addition to those, we consider real-world examination timetabling benchmarks. The bin packing component of these problems involves scheduling examinations (items) involving specified numbers of students (item sizes), into rooms of specified capacity (bin capacities) within time-slots (number of bins). We consider the data sets available from universities in Toronto, Melbourne and Nottingham [20]. These are available from the OR library.[3]

We used two goodness-of-fit tests to evaluate whether or not the Weibull distribution is capable of modeling the distribution of item sizes in these data sets. We discuss each of these tests in the following sections.

The Kolmogorov-Smirnov Test. The two-sided Kolmogorov-Smirnov (KS) test is a non-parametric test for the equality of continuous, one-dimensional, probability distributions.[4] As implemented in R, this test requires two sample sets: one representing the observed data, and the other representing a sample from the hypothesised distribution. In our setting, the observed data is represented by the item sizes from the benchmark we wish to model, while the second set is a vector of items generated according to the best-fit Weibull distribution with parameters (shape and scale) estimated from the observed data [22]. The null hypothesis of this statistical test is that *the two data sets come from the same underlying distribution*. For a 95% level of confidence, if the p-value from the test is at least 0.05, then we cannot reject the null hypothesis.

The KS test was performed on all instances from our exam timetabling (ETT) and ROADEF/EURO benchmark suites; the details of a randomly selected subset are presented in Table 1. We can see that most of the ETT item size distributions can be accurately modeled by a Weibull distribution since the corresponding p-values are above 5% (highlighted in bold). However, the KS test clearly rejects the null hypothesis for the ROADEF/EURO instances, most likely due to a both the size of the data sets and the presence of outliers in the tail of the distribution. It is known that when dealing with large data sets with a small number of large outliers, this test tends to underestimate the p-value. This means that even if the null hypothesis is rejected the candidate distribution might still characterise the data set [17]. For this reason we use a second test, the χ^2, to further validate the results.

The χ^2 Test. As a complementary approach, we used the χ^2 goodness-of-fit test which is less sensitive to outliers in the sample data.[5] The null hypothesis is that the observed and expected distributions are not statistically different. The procedure requires grouping items into γ categories according to their size. Based on these categories, we can compute the expected number of values in each category, assuming that the item sizes are drawn from a Weibull distribution with shape and scale parameters estimated from the data set. The χ^2 statistic is then computed as:

---

[3] http://people.brunel.ac.uk/~mastjjb/jeb/info.html
[4] http://mathworld.wolfram.com/Kolmogorov-SmirnovTest.html
[5] http://mathworld.wolfram.com/Chi-SquaredTest.html

$$\chi^2 = \sum_{i=1}^{\gamma} (O_i - E_i)^2 / E_i,$$

from which we can obtain the corresponding p-value, where O_i and E_i are the observed and expected frequencies of each category i, respectively. We model the tail of the distribution, in the standard way, by building a wider category that counts all items in the tail of the distribution. The other $\gamma - 1$ categories are equally sized.

As shown in Table 1, the null hypothesis cannot be rejected for any of the benchmarks that are presented. Therefore, the conclusion is that the Weibull distribution provides a good fit for the item size distributions in the benchmark instances we considered. We conjecture that it will also do so in very many other cases encountered in practice.

4 Systematic Search for Bin Packing

We consider the performance of a systematic constraint-based bin packing method on a wide number of classes of Weibull-based bin packing benchmarks. Our experiment involved varying the parameters of the Weibull distribution so that item sets for bin packing instances could be generated. A range of bin capacities were studied. The details of the experimental setup are described in Section 4.1.

4.1 Bin Packing Instances and Solver

Bin Packing Instances. We considered problems instances involving 100 items. We fixed the scale parameter, λ, of the Weibull to 1000. As experimental parameters we varied both the capacity of the bins and the shape of the Weibull distribution, generating 100 instances for each combination of parameters. The capacities we considered were $c \times max(I)$, where $c \in [1.0, 1.1, \ldots, 1.9, 2.0]$ and $max(I)$ is the the maximum item size encountered in the instance. Therefore, the capacity of the bins considered were at least equal to the largest item, or at most twice that size.

For the shape parameter of the Weibull we considered a very large range: $[0.1, 0.2, \ldots, 19.9]$, giving 199 settings of this parameter. By fixing the scale parameter to 1000 we considered item sizes that could span over three orders-of-magnitude. To build our problem generator we used the Boost library [2]. This is a C++ API that includes type definitions for random number generators and a Weibull distribution, which is parameterized by the random number generator, the shape and the scale. Iteration capabilities for traversing the distribution of generated values are also provided. We generated 100 instances for each combination of shape and scale, giving 199 classes of item sets, providing 19,900 item sets. For each of these sets we generated bin packing instances by taking each set and associating it with a bin capacity in the range described above. In this way we could be sure that as we changed bin capacity, the specific sets of items to be considered was controlled.

Table 1. The parameters of the best-fit Weibull distributions obtained for randomly selected instances of a number of real-world examination timetabling benchmarks and instances from the 2012 ROADEF/EURO Challenge sets. We present *p-values* for both the KS and χ^2 goodness-of-fit tests, highlighting in bold the results that show a statistically significant fit. #(cat) and lbTail are the number of categories and the lower bound of the category representing the tail for the χ^2 test.

| | | Weibull Best-fit | | KS test | χ^2 test | | |
	Set Instance	shape	scale	*p*-value	#(cat)	lbTail	*p*-value
ETT	Nott	1.044	43.270	**0.7864**	7	100	**0.059**
	MelA	0.946	109.214	**0.091**	10	427	**0.073**
	MelB	0.951	117.158	**0.079**	5	47	**0.051**
	Cars	1.052	85.438	0.037	18	53	**0.109**
	hec	1.139	138.362	**0.436**	10	293	**0.204**
	yor	1.421	37.049	**0.062**	7	117	**0.068**
RAODEF	$a1_3^2$	0.447	104,346.70	0.005	30	163,000	**0.105**
	$a1_3^3$	0.549	88,267.85	0.001	15	54,800	**0.068**
	$a2_1^5$	0.562	67,029.83	0.000	30	470,000	**0.768**
	$a2_4^4$	0.334	103,228.30	0.001	30	500,000	**0.051**
	b_6^3	0.725	40,469.74	0.000	20	185,000	**0.060**
	b_3^5	0.454	91,563.28	0.000	30	140,000	**0.088**

Hardware. Our experiments were run on an Intel Dual Core 2.4Ghz processor with 4GB RAM memory, running Windows XP(SP3). Microsoft Visual Studio 2008 tools have been used for compiling and linking the C++ code.

Constraint-Based Bin Packing Model. For our experiments we have used Gecode 3.7.0 [8]. The bin packing model used is the most efficient one included in the Gecode distribution for finding the minimum number of bins for a given bin packing instance [25]. This model employs the L_1 lower bound on the minimum number of bins by Martello and Toth [19]. It uses an upper bound based on the first-fit bin packing heuristic which packs each item into the first bin with sufficient capacity.

The model uses the following variables: one variable to represent the number of bins used to pack the items; one variable per item representing which bin the item is assigned to; and a variable per bin representing its load. The main constraint included in the model is the global bin packing constraint proposed by Paul Shaw [28], enforcing that the packing of items into bins corresponds to the load variables. Those items whose size is greater than half of the bin capacity are directly placed into different bins. If a solution uses a number of bins smaller than the upper bound, then the load associated with unused bins is set to 0, and symmetry breaking constraints ensure that this reasoning applies to the lexicographically last variables first. Additional symmetry breaking constraints ensure that search avoids different solutions involving permutations of items with equal size.

The search strategy used is as follows. The variable representing the number of bins used in the solution is labelled first, and in increasing order, thus ensuring that the first solution found is optimal. The variables representing the item assignments to bins, and the load on each bin, are then labelled using the *complete decreasing best fit* strategy proposed by Gent and Walsh [10], which tries to place the items into those bins with sufficient but least free space. In our experiments a timeout of 10 seconds is used to ensure that our experiments take a reasonable amount of time. We verified that increasing this to five minutes does not significantly increase the proportion of solved instances. However, of course, for some classes a large number of time-outs were observed, so further empirical study is needed in those cases.

4.2   Scenario 1: Small Weibull Shape Parameter Values

In this section we explore the behaviour of a systematic search on bin packing instances generated using our Weibull-based approach when considering small values of the distribution's shape parameter, specifically values ranging from 0.5 to 5.0 in steps of 0.1. Figure 4 presents the results – Figure 4(a) and Figure 4(b) present the average time required to those instances solved within the timeout, and the proportion of instances involved, respectively. In these plots we only consider capacity factors 1.0, 1.5, and 2.0.

It is clear that the shape factor, which defines the spread of item sizes, has a dramatic impact on the average time taken to find the optimal solution to a bin packing instance. By referring back to Figure 1 one can observe how the distribution of item sizes is changing. The lower values of the shape parameters correspond to distributions that have greater skew towards smaller items. As the shape parameter increases, consider value 1.5, there is a much greater range of possible item sizes. Once we get to higher shape values, consider value 5.0, the distribution of item sizes becomes more symmetric.

This shift in item size distribution impacts the difficulty of bin packing earlier when the capacity of the bin is smaller. Consider the effort required when the bin capacity is equal to the largest item, i.e. capacity factor 1.0, in Figure 4(a). The range of shapes over which these problems are hard is quite narrow, and we shall see in the next section, that this is influenced by the bin capacity associated with the problem instance. This difficulty arises from the interaction between item size distribution and bin capacity whereby finding the best combinations of items to place in the same bin becomes challenging. As the shape parameter increases, the range of item sizes again decreases which, given the small bin capacity, makes the instance easy once more. For a bin capacity equal to the largest item size the hard region corresponds to values of Weibull shape between 1.5 and 3.0. Increasing the capacity of the bins dramatically increases the computational challenge of the problems, since again, search effort is invested in finding a good combination of items to fit into each bin. Clearly, from Figure 4(a), we can see that problem difficulty increases very dramatically as bin capacity increases.

Using our proposed Weibull-based model for generating bin packing instances we claim that not only can one model many real-world bin packing settings, as shown earlier in this paper, but it is possible to carry out very controlled experiments on the behaviour of bin packing methods, studying the effect of the

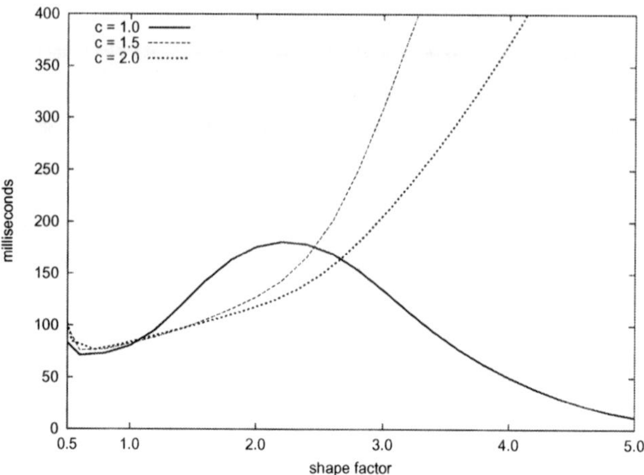

(a) Average running time for instances that did not timeout.

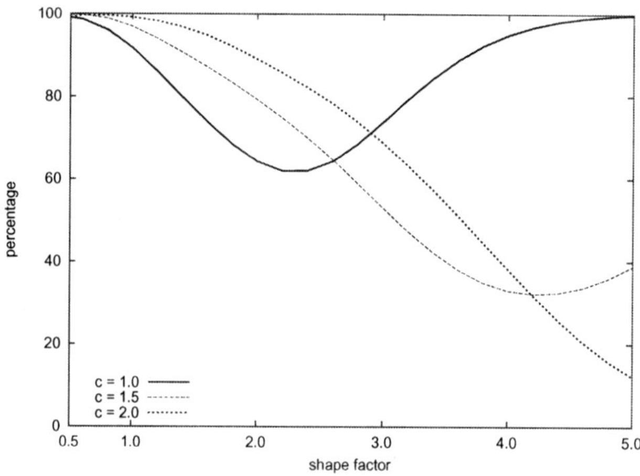

(b) Percentage of instances solved within the timeout.

Fig. 4. Average runtime and percentage of solved instances for values of the shape parameter in the 0.5,. . .,5.0 range

various aspects of the problem, such as bin capacity and item size distribution in isolation, or together.

4.3 Scenario 2: Full Range of Shape Parameters

We have also performed a more wide ranging study of the interaction between the shape of the Weibull distribution, bin capacity, and the hardness of bin packing for a systematic method. In this section we will briefly present a set of experiments that exhibit the various behaviours discussed above. We consider all values of the shape parameter in our data set, $0.1 \leq k \leq 19.9$. Figure 2 shows how

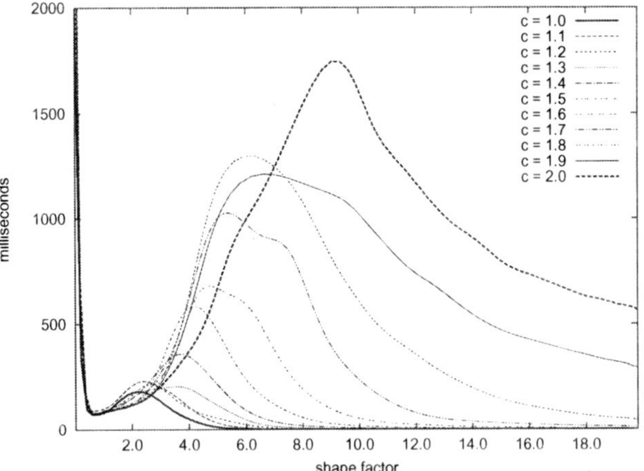

(a) Average running time for instances that did not time-out.

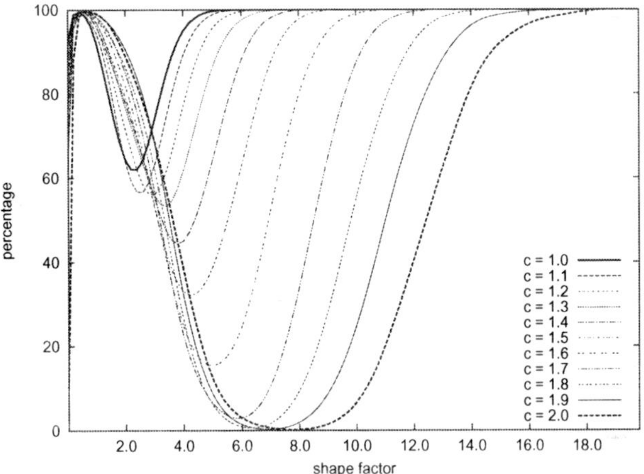

(b) Percentage of instances solved within the timeout.

Fig. 5. Average runtime and percentage of solved instances for values of the shape parameter for range of Weibull shapes that is sufficiently wide to exhibit the easy-hard-easy behaviour in search effort

the distributions with larger shape parameter values differ from those with the smaller values studied above. Essentially, these distributions have lower spread shown by successively taller density functions centering towards the value of the scale parameter.

Figure 5 presents both the average running time (Figure 5(a)) of the instances solved within the timeout and the percentage of instances that this corresponds to (Figure 5(b)). The average number of bins associated with these instances is presented in Figure 6.

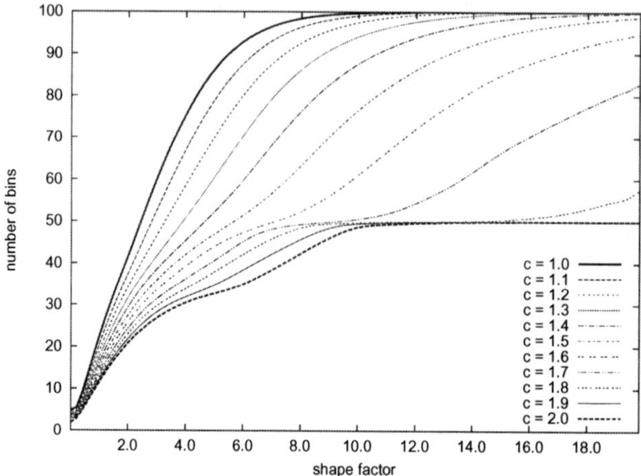

Fig. 6. Average number of bins associated with the optimal solutions to the instances presented in Figure 5.

Again, in these plots we can see that, as before, problem difficulty peaks at a specific value of Weibull shape for different values of capacity (Figure 5). From Figure 6, which presents the average number of bins in an optimal solution, we can extract the average number of items per bin for each class, since all of our instances have 100 items. As before, the range of shapes over which search efforts are hard, correspond to specific ranges of numbers of bins (or average number of items per bin). Therefore, there is an obvious interrelationship between bin capacity, item size distribution, and both problem hardness and numbers of items per bin.

5   Bin Packing Heuristics

Our earlier experiments considered the performance of a systematic search method for bin packing. In this section, for completeness, we use the same set of instances to present bin packing performance when using some well-known heuristics: MaxRest, FirstFit, BestFit and NextFit. Briefly these heuristics operate as follows. MaxRest places the next item into the bin with maximum remaining space capacity; FirstFit places the next item into the first bin that can accommodate it; BestFit places the next item into the bin that will have the least remaining capacity once the item has been accommodated by it; finally, NextFit keeps the last bin open and creates a new bin if the next item cannot be accommodated in the current bin, which it will then close. For our experiments we used a publicly available implementation of these heuristics by Rieck.[6] Because our benchmark generator produces instances that have items sorted in decreasing order of size, the difference in performance between

---

[6] http://bastian.rieck.ru/uni/bin_packing/

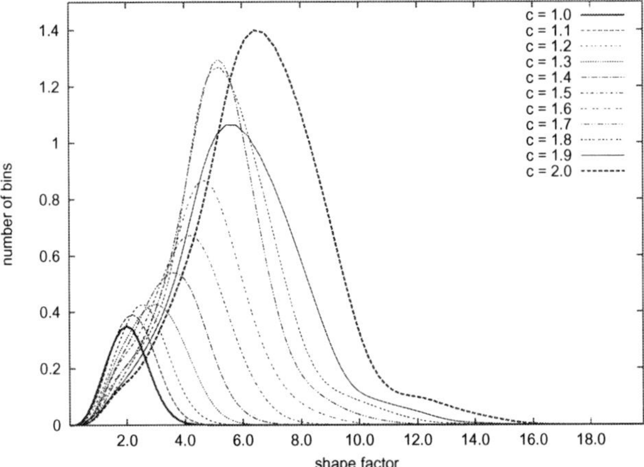

(a) The difference in the average number of bins in solutions found using heuristics like MAXREST, FIRSTFIT, or BESTFIT. Because these perform similarly we only present results for MAXREST.

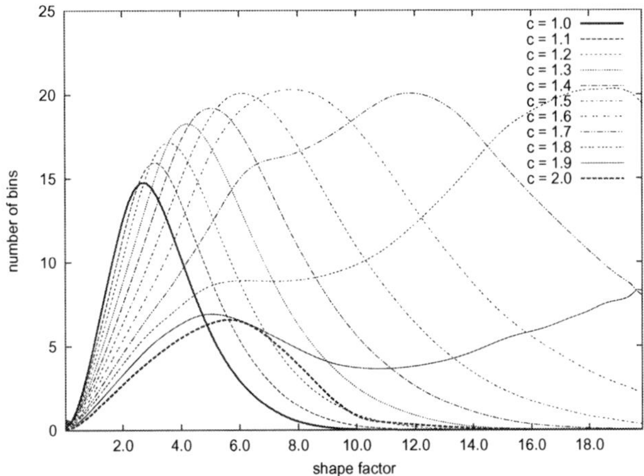

(b) The difference in the average number of bins in solutions found using NEXTFIT.

Fig. 7. The difference in the average number of bins required by each of the heuristics and the optimal solutions - if a heuristic finds the optimal solution the difference is 0

MAXREST, FIRSTFIT, and BESTFIT is very small, so we will only present results for MAXREST, while NEXTFIT will be presented separately.

Figure 7 presents the results, in each case showing the difference in the average number of bins as compared with the optimal value found by the systematic search method used earlier. A representative set of results for MAXREST, FIRSTFIT, and BESTFIT are presented in Figure 7(a) using MAXREST as the example, while those for NEXTFIT are presented in Figure 7(b). Interestingly the quality of the solutions found using the MAXREST, FIRSTFIT, and BESTFIT heuristics closely follows the difficulty of the problem when using a systematic solver. This makes intuitive sense, since for these problems, finding a good combination of items to give a good quality solution is difficult.

This performance contrasts starkly with that of the NEXTFIT heuristic (Figure 7(b)) which does significantly worse than optimal across almost all values of Weibull shape. While, the greediness of this heuristic does not pay off, the more considered reasoning used by the other heuristic does, except when the problem is even challenging for systematic search.

6   Conclusions

In this paper we have presented a parameterisable benchmark generator for bin packing instances based on the well-known Weibull distribution. The motivation for our work in this area comes from the domain of data centre optimisation and, in particular, workload consolidation which can be viewed as multi-capacity bin packing. We have demonstrated how our approach can very accurately model real-world bin packing problems, e.g. those from the ROADEF/EURO Challenge, and from real-world examination timetabling problems. We also presented an empirical analysis of both systematic search and heuristic methods for bin packing based on a large benchmark suite generated using our approach, showing a variety of interesting behaviours that are otherwise difficult to observe systematically. We observed that for all bin capacities, the number of bins required in an optimal solution increases as the Weibull shape parameter increases. However, for each bin capacity, there is a range of Weibull shape settings, corresponding to different item size distributions, for which bin packing is hard for a CP-based method.

In future work we will gather a large set of bin packing instances from real world applications and compute best-fit Weibull distributions to them. In a completely orthogonal direction, our approach provides a two parameter model for item size distributions, giving us a three parameter bin packing model: bin capacity, and Weibull shape and scale parameters. These parameters provide a basis for tuning bin packing methods and generating portfolio-based bin packing solvers that rely on these parameters for learning their best configuration.

The model we have presented here can, of course, be trivially extended to produce benchmark generators for a variety of other important problems, such as knapsacks, multi-processor scheduling, job shop scheduling, timetabling, to name but a few. We have begun to develop these generators, with the ultimate plan being a fully parameterised benchmark generator which we will integrate into the Numberjack combinatorial optimisation platform,[7] amongst others.

Acknowledgements. This work was funded by Science Foundation Ireland under Grant 10/IN.1/I3032, and was undertaken during research visits by Ignacio Castiñeiras and Milan De Cauwer to the Cork Constraint Computation Centre.

---

[7] `http://numberjack.ucc.ie`

Ignacio Castiñeiras is also partially supported by the Spanish Ministry of Education and Science (Grant TIN2008-06622-C03-01), Universidad Complutense de Madrid and Banco Santander (Grant UCM-BSCH-GR58/08-910502), and the Madrid regional government (Grant S2009TIC-1465). We would like to thank Arnaud Malapert, Jean-Charles Regin, and Mohamed Rezgui for sharing copies of the standard benchmarks from the literature with us.

References

1. Alvim, A.C.F., Ribeiro, C.C., Glover, F., Aloise, D.J.: A hybrid improvement heuristic for the one-dimensional bin packing problem. Journal of Heuristics 10(2), 205–229 (2004)
2. Boost: free peer-reviewed portable C++ source libraries. Version 1.47.0, http://www.boost.org/
3. Cambazard, H., O'Sullivan, B.: Propagating the Bin Packing Constraint Using Linear Programming. In: Cohen, D. (ed.) CP 2010. LNCS, vol. 6308, pp. 129–136. Springer, Heidelberg (2010)
4. Carlsson, M., Beldiceanu, N., Martin, J.: A geometric constraint over k-dimensional objects and shapes subject to business rules. In: Stuckey (ed.) [31], pp. 220–234
5. de Carvalho, J.V.: Exact solution of cutting stock problems using column generation and branch-and-bound. International Transactions in Operational Research 5(1), 35–44 (1998)
6. Dupuis, J., Schaus, P., Deville, Y.: Consistency check for the bin packing constraint revisited. In: Lodi, et al. (eds.) [15], pp. 117–122
7. Falkenauer, E.: A hybrid grouping genetic algorithm for bin packing. Journal of Heuristics 2(1), 5–30 (1996)
8. Gecode: generic constraint development environment. Version 3.7.0, http://www.gecode.org/
9. Gent, I.P.: Heuristic solution of open bin packing problems. Journal of Heuristics 3(4), 299–304 (1998)
10. Gent, I.P., Walsh, T.: From approximate to optimal solutions: constructing pruning and propagation rules. In: International Joint Conference on Artifical Intelligence, pp. 1396–1401 (1997)
11. Grandcolas, S., Pinto, C.: A SAT encoding for multi-dimensional packing problems. In: Lodi, et al. (eds.) [15], pp. 141–146
12. Hermenier, F., Demassey, S., Lorca, X.: Bin Repacking Scheduling in Virtualized Datacenters. In: Lee, J. (ed.) CP 2011. LNCS, vol. 6876, pp. 27–41. Springer, Heidelberg (2011)
13. Korf, R.E.: An improved algorithm for optimal bin packing. In: Proceedings of the 18th International Joint Conference on Artificial Intelligence, pp. 1252–1258 (2003)
14. Labbé, M., Laporte, G., Martello, S.: Upper bounds and algorithms for the maximum cardinality bin packing problem. European Journal of Operational Research 149(3), 490–498 (2003)
15. Lodi, A., Milano, M., Toth, P. (eds.): CPAIOR 2010. LNCS, vol. 6140. Springer, Heidelberg (2010)
16. Steenberger, M.R.: Maximum likelihood programming in R, http://www.unc.edu/~monogan/computing/r/MLE_in_R.pdf
17. Denis, J.-B., Delignette-Muller, M.L., Pouillot, R., Dutang, C.: Use of the package fitdistrplus to specify a distribution from non-censored or censored data (April 2011)

18. Martello, S., Toth, P.: Knapsack Problems: Algorithms and Computer Implementations. John Wiley & Sons, Inc. (1990)
19. Martello, S., Toth, P.: Lower bounds and reduction procedures for the bin packing problem. Discrete Appl. Math. 28, 59–70 (1990)
20. Qu, R., Burke, E., McCollum, B., Merlot, L.T., Lee, S.Y.: A survey of search methodologies and automated approaches for examination timetabling (2006)
21. Régin, J.-C., Rezgui, M.: Discussion about constraint programming bin packing models. In: Proceedings of the AIDC Workshop (2011)
22. Rici, V.: Fitting distribution with R (2005)
23. Schaus, P.: Solving Balancing and Bin-Packing Problems with Constraint Programming. PhD thesis, Université Catholique de Louvain-la-Neuve (2009)
24. Scholl, A., Klein, R., Jürgens, C.: Bison: A fast hybrid procedure for exactly solving the one-dimensional bin packing problem. Computers & Operations Research 24(7), 627–645 (1997)
25. Schulte, C., Tack, G., Lagerkvist, M.Z.: Modeling and Programming with Gecode (2012), http://www.gecode.org/doc-latest/MPG.pdf
26. Schwerin, P., Wäscher, G.: The bin-packing problem: A problem generator and some numerical experiments with FFD packing and MTP. International Transactions in Operational Research 4(5-6), 377–389 (1997)
27. Schwerin, P., Wäscher, G.: A New Lower Bound for the Bin-Packing Problem and its Integration Into MTP. Martin-Luther-Univ. (1998)
28. Shaw, P.: A constraint for bin packing. In: Wallace (ed.) [34], pp. 648–662
29. Simonis, H., O'Sullivan, B.: Search strategies for rectangle packing. In: Stuckey (ed.) [31], pp. 52–66
30. Simonis, H., O'Sullivan, B.: Almost Square Packing. In: Achterberg, T., Beck, J.C. (eds.) CPAIOR 2011. LNCS, vol. 6697, pp. 196–209. Springer, Heidelberg (2011)
31. Stuckey, P.J. (ed.): CP 2008. LNCS, vol. 5202. Springer, Heidelberg (2008)
32. The R project for statistical computing. Version 1.47.0, http://www.r-project.org/
33. Vidotto, A.: Online constraint solving and rectangle packing. In: Wallace (ed.) [34], p. 807
34. Wallace, M. (ed.): CP 2004. LNCS, vol. 3258. Springer, Heidelberg (2004)
35. Wäscher, G., Gau, T.: Heuristics for the integer one-dimensional cutting stock problem: A computational study. OR Spectrum 18(3), 131–144 (1996)
36. Weibull, W.: A statistical distribution function of wide applicability. Journal of Appl. Mech.-Transactions 18(3), 293–297 (1951)
37. Wessa, P.: Free Statistics Software, Office for Research Development and Education, version 1.1.23-r7 (2011), http://www.wessa.net/

Space-Time Tradeoffs for the Regular Constraint

Kenil C.K. Cheng, Wei Xia, and Roland H.C. Yap

School of Computing
National University of Singapore
{xiawei,ryap}@comp.nus.edu.sg,
kenilc@gmail.com

Abstract. Many global constraints can be described by a regular expression or a DFA. Originally, the regular constraint, uses a DFA to describe the constraint, however, it can also be used to express a table constraint. Thus, there are many representations for an equivalent constraint. In this paper, we investigate the effect of different representations on natural extensions of the regular constraint focusing on the space-time tradeoffs. We propose a variant of the regular constraint, $nfac(X)$, where X can be an NFA, DFA or MDD. Our nfac algorithm enforces GAC directly on any of the input representations, thus, generalizing the mddc algorithm. We also give an algorithm to directly convert an NFA representation to an MDD for nfac(MDD) or mddc. Our experiments show that the NFA representation not only saves space but also time. When the ratio of the NFA over the MDD or DFA size is small enough, nfac(NFA) is faster. When the size ratio is larger, nfac(NFA) still saves space and is a bit slower. In some problems, the initialization cost of MDD or DFA can also be a significant overhead, unlike NFA which has low initialization cost. We also show that the effect of the early-cutoff optimization is significant in all the representations.

1   Introduction

Global constraints based on grammars can be used to express arbitrary constraints including other global constraints. The most well known is perhaps the regular constraint [1] which can also express other global constraints, e.g. contiguity, among, lex, precedence, stretch, etc. The regular constraint (originally) takes a deterministic finite automata (DFA) as the constraint definition but a non-deterministic finite automata (NFA) is also possible. On the other hand, the table constraint defines an arbitrary constraint explicitly as a set of solutions. Two efficient generalised arc consistency (GAC) algorithms for non-binary table constraints are mddc [2] and str2 [3]. In mddc, the constraint is specified as a multi-valued decision diagram (MDD)[1] while str2 works with the explicit table.

We see then that there can be many input representations[2] defining the same set of solutions (tuples) to a constraint. However, it is well known that transforming an NFA to a DFA can lead to an exponentially larger automata. The MDDs

---

[1] An algorithm for constructing MDD from an explicit table constraint is given in [2].

[2] These representations may be regarded as different constraint forms with equivalent semantics.

M. Milano (Ed.): CP 2012, LNCS 7514, pp. 223–237, 2012.

corresponding to an NFA can also be large. Thus, there are different constraint representation choices with different space requirements for the associated GAC algorithms. Experiments in [3] also show that the representation size can determine solver runtime, mddc is faster than str2 when the MDD representation is more compact and vice versa.[3]

In this paper, we investigate the space-time tradeoffs of the NFA, DFA and MDD representations of the same equivalent constraint. We extend mddc to nfac which can enforce GAC on a constraint defined as an NFA, DFA and MDD. An algorithm which converts a constraint in NFA or DFA form directly to an MDD is also given. We extend nfac to grammarc, which works on a grammar constraint expressed in Greibach normal form.

We experiment with hard CSPs whose size differences in their NFA, DFA and MDD representations are large. We show that the early-cutoff optimization is a significant optimization for NFA, DFA and MDD in nfac. There are two space-time tradeoffs. Firstly, the NFA representation is more efficient for the nfac algorithm once the corresponding DFA or MDD becomes big enough. If memory itself is an issue, e.g. the CSP has many large constraints, it may be worthwhile to use NFA even when the DFA or MDD is smaller. Secondly, when the size of the DFA or MDD is large enough, once the total time including construction and initialization times are taken into account, the NFA can be much faster then DFA or MDD. We also compare the nfac solver with regular and find nfac to be faster.

1.1 Related Work

The regular constraint was introduced by Pesant [1]. He proposed a GAC algorithm based on a layered graph data structure expanded from the DFA input into the constraint. Furthermore, the layered graph algorithm can also be made to process NFAs [5]. The regular constraint is generalized from automata as the specification to context-free grammars using parsing algorithms [6,7]. GAC algorithms based on the CYK parser, which requires the CFG in Chomsky normal from (CNF), and the Earley parser, which is not restricted to CNF, are proposed. [6] also proposed an NFA constraint propagator by encoding the NFA transitions into a sequence of ternary constraints. The grammar constraint can be converted into an automaton which is proposed by [8]. We investigate the space-time tradeoffs once the constraint is in automata form to show when such conversions are beneficial.

Arbitrary constraints can also be represented using multi-valued decision diagrams (MDDs) [9]. The bddc [10], mddc [2] and case [11] propagators achieve GAC on various forms of an MDD constraint by traversing the graph in a depth-first manner. The regular constraint can also be used as the propagator for MDD constraints [2]. However, the question of space-time tradeoffs of MDD versus the most naturally related representations, NFA and DFA, has not been experimentally investigated in detail.

---

[3] See also the results for str3 with table size [4].

2   Background

A *constraint satisfaction problem* (CSP) $\mathcal{P} = (X, \mathcal{C})$ consists of a finite set X of variables and a finite set $\mathcal{C}$ of constraints. Every *variable* $x_i \in X$ can only take values from a finite *domain* $dom(x_i)$. An *assignment* (x_i, a) denotes $x_i = a$. An *r-ary constraint* $C \in \mathcal{C}$ on an ordered set of r distinct variables $x_1, \ldots, x_r$ is a subset of the Cartesian product $\prod_{i=1}^{r} dom(x_i)$ that restricts the values of the variables in C can take simultaneously. The *scope* is the set of variables and denoted by $var(C)$. The *arity* of C is the size of $var(C)$, and will usually be denoted by r. Sometimes we write $C(x_1, \ldots, x_r)$ to make the scope explicit. A set of assignments $\theta = \{(x_1, a_1), \ldots, (x_r, a_r)\}$ *satisfies* C, and is a *solution* of C, iff $(a_1, \ldots, a_r) \in C$. Solving a CSP requires finding an assignment for each variable from its domain so that all constraints are satisfied. Two constraints C_1 and C_2 are *equivalent*, written as $C_1 \equiv C_2$, iff $\theta \in C_1 \iff \theta \in C_2$.

Consider a CSP $\mathcal{P} = (X, \mathcal{C})$. An assignment (x_i, a) is *generalized arc consistent* (GAC) [12] iff for every constraint $C \in \mathcal{C}$ such that $x_i \in var(C)$, there is a solution θ of C where $(x_i, a) \in \theta$ and $a \in dom(x_i)$ for every $(x_i, a) \in \theta$. This solution is called a *support* for (x_i, a) in C. A variable $x_i \in X$ is GAC iff (x_i, a) is GAC for every $a \in dom(x_i)$. A constraint is GAC iff every variable in its scope is GAC. A CSP is GAC iff every constraint in $\mathcal{C}$ is GAC.

A *non-deterministic finite state automaton* (NFA) $G = \langle Q, \Sigma, \delta, q_0, F \rangle$ consists of a set Q of *states*, a set Σ of *symbols*,[4] a *transition function* $\delta : Q \times \Sigma \mapsto 2^Q$, an *initial state* $q_0 \in Q$, and a set $F \subseteq Q$ of *final states*. G is known as a *deterministic finite state automaton* (DFA) iff $|\delta(q, a)| = 1$ for each $q \in Q$ and $a \in \Sigma$. If $q' \in \delta(q, a)$, there is a *transition* or a *directed edge* from q to q' via a. $|\delta| = |\{(q, q') | q \in Q, q' \in \delta(q, a) \text{ for some } a \in \Sigma\}|$ is used to denote the number of transitions in an NFA. Note that our definition does not allow for ϵ-transitions in the NFA or DFA.

Let $G = \langle Q, \Sigma, \delta, q_0, F \rangle$ be an NFA. An *r-ary NFA constraint* represented by G is $\Phi_G^r(\{q_0\}, 1)$, which is recursively defined as

$$
\Phi_G^r(S, i) \equiv
\begin{cases}
false & : S = \emptyset \\
S \cap F \neq \emptyset & : i = r + 1 \\
\bigvee_{a \in \Sigma} \left(x_i = a \wedge \Phi_G^r(\delta'(S, a), i + 1) \right) & : \text{otherwise}
\end{cases}
$$

where $\delta'(S, a) = \bigcup_{q \in S} \delta(q, a)$ and $1 \leq i \leq r + 1$. We use $\mathtt{tt}$ to denote *true* and $\mathtt{ff}$ to denote *false*.

A *reduced multi-valued decision diagram* (MDD) [9] is an acyclic DFA $G = \langle Q, \Sigma, \delta, q_0, \{\mathtt{tt}\} \rangle$, such that for any states $q, q' \in Q$, $\Phi_G^r(\{q\}, i) \equiv \Phi_G^r(\{q'\}, i)$ implies $q = q'$, where $1 \leq i \leq r + 1$. In particular, a *binary decision diagram* (BDD) [13] is an MDD where $\Sigma = \{0, 1\}$.

In the rest of this paper, an MDD or NFA constraint refers to a constraint represented by an MDD or NFA respectively. A DFA constraint is a special case of an NFA constraint where the automaton is deterministic. Fig. 1 gives a graphical representation of equivalent NFA, DFA and MDD constraints.

---

[4] W.L.O.G., we assume Σ be a set of integers in this paper.

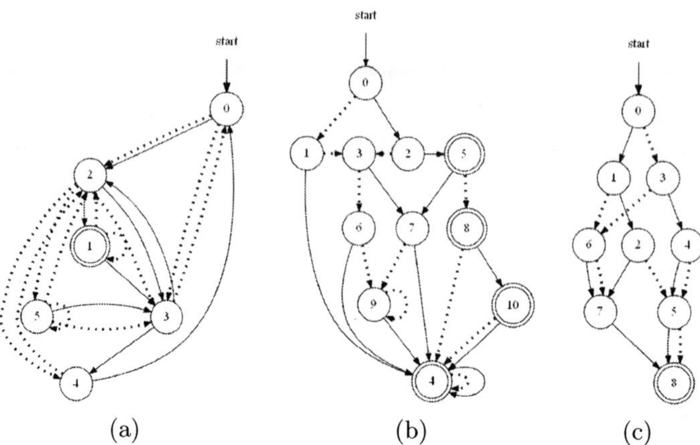

Fig. 1. (a) an NFA and (b) an equivalent DFA (non-minimized). Each (double) circle with label i represents a (final) state q_i. There is a directed edge from i to j iff $q_j \in \delta(q_i, a)$, where the edge is solid if $a = 1$ and dotted if $a = 0$. (c) an equivalent MDD constraint for a 4-ary regular constraint represented by the NFA in (a) or the DFA in (b).

3   NFA-to-MDD Conversion

We first present an algorithm to convert an r-ary NFA constraint into an equivalent MDD constraint. One approach could be to first construct a DFA from the NFA, e.g. by using the subset construction algorithm. The DFA can then be traversed to generate solutions lexicographically which are then inserted into an MDD using the mddify construction algorithm [2]. We propose instead to convert directly from the NFA into the corresponding MDD constraint. More precisely, given an NFA $G = \langle Q, \Sigma, \delta, q_0, F \rangle$, the algorithm nfa2mdd creates an MDD $G' = \langle Q', \Sigma, \delta, q'_0, \{\mathtt{tt}\} \rangle$ such the r-ary constraints represented by G and G' are equivalent. The pseudo-code of the algorithm nfa2mdd is in Fig. 2. The idea is to simulate the NFA [14] in a depth-first fashion and build the MDD from bottom to up. In other words, it combines NFA-to-DFA conversion (or subset construction) and trie-to-MDD construction.

The execution of the main procedure nfa2mdd-recur is as follows: If the current subset S of states of the NFA G is at depth $i = r + 1$ (line 1), the unique final state $\mathtt{tt}$ of the MDD G' is returned iff S contains a final state of G. This is because, by definition, any path from the start state to a final state corresponds to a solution of Φ_G^r iff its length is r. Otherwise ($i \leq r$), the traversal continues by visiting collective successors for every $a \in \Sigma$, the initial domain of the variable (line 3). The recursive call returns the start state q'_a of the sub-MDD. If q'_a is not the failure state, the pair (a, q'_a) is recorded (line 5), so the MDD only contains paths leading to $\mathtt{tt}$ (true). Now, if all successors of the states in S lead to a dead-end, the failure state $\mathtt{ff}$ will be returned (line 6), which means the current sub-MDD has no solution. Otherwise, we need to create a (new) start state for this sub-MDD. By definition, two equivalent (sub-)MDD constraints

```
nfa2mdd(G, r)
// input: NFA G = ⟨Q, Σ, δ, q₀, F⟩ and arity r
begin
    Q' := ∅
    initialize cache[], unique[] // dictionaries
    δ' := ∅
    q'₀ := nfa2mdd-recur({q₀}, 1, r)
    return ⟨Q', Σ, δ', q'₀, {tt}⟩ // output: MDD G'

nfa2mdd-recur(S, i, r)
begin
1    if i = r + 1 then
         if S ∩ F = ∅ then
         |   return ff // failure state
         else
             return tt // final state (unique)

2    q' := cache[⟨S, i⟩] // dictionary lookup
     if q' ≠ Null then return q'

3    E := ∅
     for a ∈ Σ do
4        T := ⋃_{q∈S} δ(q, a)
         if T ≠ ∅ then
             q'ₐ := nfa2mdd-recur(T, i + 1, r)
5            if q'ₐ ≠ ff then E := E ∪ {⟨a, q'ₐ⟩}

6    if E = ∅ then
     |   q' := ff
     else
7        q' := unique[E] // dictionary lookup
         if q' = Null then
8            make a new state q'
             Q' := Q' ∪ {q'} // insert new state
             for ⟨a, q'ₐ⟩ ∈ E do
9                δ'(q', a) := {q'ₐ} // insert new transitions
             unique[E] := q'

10   cache[⟨S, i⟩] := q'
     return q'
```

Fig. 2. NFA to MDD Construction

should be represented by the same sub-MDD, so we use *unique*, implemented as a dictionary, to store all created states in the MDD. With our bottom-up construction, we can identify equivalent sub-MDDs easily by using E as a key (line 7). When we cannot find an existing state, we generate a new one (line 8) and the corresponding transitions (line 9).

Proposition 1. nfa2mdd-recur *constructs reduced MDDs.*

Proof. According to the algorithm, one state q will be generated only when there is no other state q' that $\Phi^r_G(\{q\}, i) \equiv \Phi^r_G(\{q'\}, i')$. This guarantees that no two different states q, q' will be generated if $\Phi^r_G(\{q\}, i) \equiv \Phi^r_G(\{q'\}, i')$ and the MDD being constructed is reduced.

The depth-first traversal will implicitly enumerate an exploration tree of size $|\Sigma|^r$. To reduce the size of the traversal, we make use of caching (lines 2 and 10) to ensure that any sub-MDD is only be expanded once. In the worst case, nfa2mdd-recur visits $O(2^{|Q|}r)$ NFA states. This is also the maximum number of slots in *cache*. However, the cache size can be smaller, thus trading time for space. Note that if we convert an NFA into a DFA using subset construction the time complexity will be $O(2^{2|Q|})$ since the DFA will have at most $O(2^{|Q|})$ states.

Given an NFA $G = \langle Q, \Sigma, \delta, q_0, F \rangle$ and arity r, the runtime and space complexities of nfa2mdd are given as follows:

Proposition 2. *The worst-case time complexity of* nfa2mdd *is* $O(|Q|^2 \cdot |\Sigma| \cdot \min\{2^{|Q|} \cdot r, |\Sigma|^{r+1}\})$.

Proof. In each call of nfa2mdd-recur, the caching operations on *cache* and *unique* takes $O(|Q|)$ and $O(|\Sigma|)$ respectively. Every set union at line 4 can take $O(|Q|^2)$ time. The last term accounts for the maximum number of recursive calls: (1) each subset of Q is at most visited r times, and (2) the generation tree has $O(|\Sigma|^r)$ non-leaf nodes.

In practice, the use of caching may significantly reduce the runtime.

Proposition 3. *The worst-case space complexity of* nfa2mdd *is* $O(\max\{2^{|Q|} \cdot |Q| \cdot r, |\Sigma|^r \cdot |\Sigma|\})$.

Proof. (1) for *cache*, the maximum number of slots is $O(2^{|Q|}r)$ and for each slot, the key S takes $O(|Q|)$ space in the dictionary entry, and (2) for *unique*, the MDD size is in $O(|\Sigma|^r)$ and same as *cache*, each slot takes $O(|\Sigma|)$ space.

To reduce this high memory requirement, we can trade time for space and restrict the number of entries in *cache* and/or *unique*. In the latter case, the resulting MDD may not be reduced, i.e., it may contain equivalent sub-MDDs.

Proposition 4. *The output MDD G' is in the worst case r times larger than the DFA equivalent to the input NFA G, but it is exponentially smaller than the DFA in the best case.*

Proof. For the worst case, suppose nfa2mdd is run on a DFA equivalent to G. Since nfa2mdd-recur visits every DFA state at most r times, there will be at most r MDD states created for each DFA state. For the best case, consider an NFA that represents the regular expression $\{0,1\}^*0\{0,1\}^n$. The equivalent DFA has 2^n nodes; whereas, for any fixed $r > n$, the corresponding MDD has $r+1$ states, which represents the constraint $(x_{r-n} = 0) \wedge \bigwedge_{i=1}^{r} x_i \in \{0,1\}$.

$\mathsf{nfac}(G)$
// enforce GAC on the constraint $\Phi_G^r(x_1, \ldots, x_r)$
// represented by the NFA $G = \langle Q, \Sigma, \delta, q_0, F \rangle$
begin

11    $\mathsf{clear}(cache)$

   for $i := 1$ **to** r **do**
     $E_i := dom(x_i)$ *// no value has support yet*

12    $\Delta := r + 1$

   $\mathsf{nfac\text{-}recur}(q_0, 1, r)$ *// update $E_1, \ldots, E_r$*

   for $i := 1$ **to** $\Delta - 1$ **do**

13      $dom(x_i) := dom(x_i) \setminus E_i$ *// $E_j = \emptyset$ for all $j \geq \Delta$*

$\mathsf{nfac\text{-}recur}(q, i, r)$
begin

14   **if** $i = r + 1$ **then**
     if $q \in F$ **then**
       | **return** *Yes*
     else
       return *No*

15    $st := cache[\langle q, i \rangle]$
   if $st \neq Null$ **then return** st

   $st := No$

16   **for** $a \in dom(x_i)$ **do**

17   **for** $q' \in \delta(q, a)$ **do**
       if $\mathsf{nfac\text{-}recur}(q', i + 1, r) = Yes$ **then**
         $st := Yes$
         $E_i := E_i \setminus \{a\}$

18   **if** $i + 1 = \Delta$ **and** $E_i = \emptyset$ **then**
           $\Delta := i$
           go to *line 20*

19   **if** $i \geq \Delta$ **then**
           go to *line 20*

20    $cache[\langle q, i \rangle] := st$
   return st

Fig. 3. nfac and nfac-recur

4   Maintaining GAC on an NFA Constraint

There are a number of algorithms to enforce GAC on an NFA constraint. In this paper, we propose an algorithm which generalizes the mddc algorithm [2] for MDD constraints to enforce GAC on NFA or DFA constraints. Fig. 3 shows the algorithm nfac, which enforces GAC on an r-ary NFA constraint $\Phi_G^r(x_1, \ldots, x_r)$ represented by an NFA $G = \langle Q, \Sigma, \delta, q_0, F \rangle$. The nfac algorithm can be applied to a constraint represented in NFA, DFA or MDD form as a DFA or MDD constraint is a special case of an NFA.

Throughout the execution, for each x_i, nfac keeps a set E_i of values in the domain of x_i which have no support (found yet). The function nfac-recur traverses G recursively and updates $E_1, \ldots, E_r$ on the fly. Line 13 then removes for each variable all values that have no support from its domain.

The function nfac-recur works as follows: If the current state q is at depth $i = r + 1$ (line 14), *Yes* is returned iff q is a final state of G. This is because, by definition, any path from the start state to a final state corresponds to a solution of Φ_G^r iff its length is r. Otherwise ($i \leq r$), the traversal continues for every $a \in dom(x_i)$ (line 16) and every $q' \in \delta(q, a)$ (line 17). In the case of a DFA or MDD, the set $\delta(q, a)$ only contains one state for a DFA so the loop at line 17 will only be executed once.

Then if nfac-recur returns *Yes*, we will remove a from E_i because (x_i, a) has at least one support. Line 18 and 19 terminate the (outermost) iteration as soon as every value in every domain of $x_i, x_{i+1}, \ldots, x_r$ has a support. This is called the *early-cutoff* optimization and it avoids traversing parts of the MDDs.[5] At last the function returns *st*, which is *Yes* if the current sub-NFA constraint is satisfiable and *No* otherwise.

Proposition 5. *When* nfac-recur *terminates, the value* $a \in E_i$ *iff the assignment* (x_i, a) *has no support.*

The use of caching at lines 15 and 20 guarantees nfac-recur traverses each sub-NFA at most r times. The entries in *cache* are emptied by the procedure clear (line 11). To make nfac incremental during search, we can implement *cache* with the sparse set data structure used in mddc [2].

Given an NFA $G = \langle Q, \Sigma, \delta, q_0, F \rangle$ and arity r, we have the following results on the runtime and space complexity of nfac.

Proposition 6. *The time complexity of* nfac *is* $O(|\delta| \cdot r)$.

Proof. With caching, each state in G is visited $O(r)$ times, and the two for-loops at line 16 traverse each outgoing edge of the state at most once (operations on *cache* takes $O(1)$ time [2]).

Proposition 7. *The space complexity of* nfac *is* $O(|\delta| + |Q| \cdot r)$ *if the cache is maximal.*

Proof. The first term corresponds to the number of transitions in G and the second term gives the space requirement for *cache* (each entry in *cache* takes $O(1)$ space [2]).

However, if the cache has a fixed size, the space complexity becomes $O(|\delta|)$.

---

[5] A form of early-cutoff is also applied in the str2 algorithm [3].

```
    grammarc(G)
21  // same as nfac() but calls grammarc-recur(S, 1, r)

    grammarc-recur(α, i, r)
    begin
22  |   if i = r + 1 then
    |   |   if α is empty then
    |   |   |   return Yes
    |   |   else
    |   |   |   return No
    |
23  |   if |α| > r − i + 1 then
    |   |   return No
    |
24  |   st := cache[⟨α, i⟩]
    |   if st ≠ Null then return st
    |
    |   st := No
25  |   for a ∈ dom(xᵢ) do
26  |   |   for head(α) → aβ do
    |   |   |   if grammarc-recur(β++tail(α), i + 1, r) = Yes then
    |   |   |   |   st := Yes
    |   |   |   |   Eᵢ := Eᵢ \ {a}
27  |   |   |   |   if i + 1 = Δ and Eᵢ = ∅ then
    |   |   |   |   |   Δ := i
    |   |   |   |   |   go to line 29
    |   |   |   |   if i ≥ Δ then
28  |   |   |   |   |   go to line 29
    |
29  |   cache[⟨α, i⟩] := st
    |   return st
```

Fig. 4. grammarc-recur

5   Extension of nfac to Grammar Constraints

The nfac algorithm can also be extended to the global grammar constraints. Our algorithm grammarc, which enforces GAC on a constraint defined as a context-free grammar (CFG) is shown in Fig 4. The algorithm requires the grammar to be in Greibach normal form (GNF), namely, all productions are of the form $A \to a\alpha$ where A is a non-terminal, a is a terminal and α is a (possibly empty) sequence of non-terminals.[6]

The main difference with nfac is that it uses a sequence of non-terminals, i.e. α and β are two sequences of non-terminals. We use the following notation: $|\alpha|$ is the number of symbols in α; head(α) and tail(α) are the first symbol and the sub-sequence of α following the first symbol respectively; and ++ is the concatenation of two sequences. The grammarc calls grammarc-recur by passing

---

[6] Any CFG G can be converted into an equivalent G' in GNF. The size of G' is $O(|G|^4)$ in general, or $O(|G|^3)$ if G has no chain productions [15].

Table 1. Sizes of NFA, DFA and MDD constraints in the benchmarks. The number of CSP instances in each group is given by #instances.

Benchmark	#instances	NFA size	DFA det size	DFA min size	MDD size
nfa-34-40-3-18	14	25	1005	843	4976
nfa-50-30-5-11	14	30	15287	15031	11617
nfa-36-30-5-15	9	40	48908	38479	123473
nfa-60-30-7-10	14	30	40865	40290	27766
nfa-57-30-7-12	12	30	39915	39790	81400
nfa-54-30-7-15	11	30	41018	40113	195927

the start symbol S of CFG. Once the grammarc-recur reaches level $r + 1$, Yes will be returned if α is an empty string. At line 23, due to the arity of the constraint, if the non-terminals left in current string are more than the terminals needed, No will be returned (as the grammar is in GNF and every non-terminal will generate at least one terminal). grammarc can also work with an ambiguous grammar (line 26) and uses the early-cutoff optimization (line 27 and 28).

grammarc is an alternative algorithm to enforce GAC on grammar constraints. We have not experimented with grammar constraints, and have focused on investigating time-space tradeoffs from automata representations to MDDs.[7] We expect that the use of cache and early-cutoff is still helpful in grammarc.

6   Experimental Results

Our implementations are in C++. The MDD construction with nfa2mdd is implemented with the STL library for convenience. As STL is not so efficient, we can expect that the runtime can be further improved but still the trends from the experiments are clear. We implement nfac in Gecode 3.5.0 as it has an efficient implementation of regular[8] based on the layer graph updating algorithm in [1]. The *cache* implementation in nfac follows [2] – using a sparse set (for fast incrementality). Experiments were run on a Intel i7/960 @3.20GHz with 12G RAM on 64-bit Linux. We use FSA Utilities 6.276[9] to do the NFA generation and dk.brics.automaton[10] to do DFA determinization and minimization.

We want to investigate time-space tradeoffs, so the benchmarks need to have: (i) sufficient variation in size between NFA, DFA and MDD representations ranging from small to large; and (ii) be sufficiently hard with the chosen search heuristic so that we are able to adequately exercise the GAC algorithms. Existing benchmarks do not meet our goals. We thus chose to generate hard random

---

[7] This is because it is unclear how to generate hard grammar constraints which will further exemplify the space-time tradeoffs above the ones in automata.

[8] We also tested Gecode 3.7.3 and found similar runtimes to Gecode 3.5.0 in our tests.

[9] It is implemented in Prolog, http://www.let.rug.nl/~vannoord/Fsa/fsa.html

[10] It is implemented in Java, http://www.brics.dk/automaton/

Table 2. (a) Average construction time of DFA and MDD CSPs ; and (b) Average solving time of NFA, DFA, MDD and Gecode Regular CSPs (timeout is 3600s)

(a) Construction time (seconds)

	DFA		MDD		
Benchmarks	det time	min time	n2m time	det2m time	min2m time
nfa-34-40-3-18	0.6	0.6	1.0	0.4	0.4
nfa-50-30-5-11	24.0	72.0	16.8	6.8	6.4
nfa-36-30-5-15	66.0	534.7	83.9	58.4	56.5
nfa-60-30-7-10	115.9	556.9	69.6	34.2	33.0
nfa-57-30-7-12	108.4	522.1	113.3	60.6	60.0
nfa-54-30-7-15	103.8	517.7	182.6	119.5	115.3

(b) Solving time (seconds)

	nfac-init			$nfac^c$			$nfac^{rc}$			regular	
Benchmarks	NFA	DFA	MDD	NFA	DFA	MDD	NFA	DFA	MDD	init	DFA
nfa-34-30-3-18	0	0	0.1	7.1	5.4	5.6	15.5	28.9	23.0	0.1	19.6
nfa-50-30-5-11	0	1.3	1.0	11.5	10.6	8.0	26.9	116.0	32.3	1.8	73.0
nfa-36-30-5-15	0	2.7	6.7	54.7	66.5	54.6	132.9	1445.8	577.7	5.1	643.4
nfa-60-30-7-10	0	6.3	4.6	40.3	44.9	32.0	103.5	638.4	169.9	8.1	445.8
nfa-57-30-7-12	0	6.0	13.1	75.0	90.9	72.9	192.7	1524.7	576.2	7.9	855.6
nfa-54-30-7-15	0	5.8	31.0	88.7	123.4	112.5	220.6	(5 out)	(4 out)	8.5	1249.2

CSPs for these goals using NFA constraints. We generate NFAs using FSA Utilities having deterministic density[11] around 1.1 or 1.2 which gives a constraint tightness following model RB and we found these to take sufficient runtimes to exercise the GAC algorithms. Half of the states in Q are chosen as final states to prevent the DFA/MDD from being small.

We generated 74 CSP instances in 6 different groups. The groups are named using the notation nfa-c-v-d-r where c is the number of constraints, v is the number of variables, d is the size of domain and r is the constraint arity. The sizes of NFA, DFA, MDD are made differently, so as to avoid any bias to the different algorithms considered. The average size of a single NFA and DFA constraint (in number of states) and MDD constraint (in number of nodes) is given in Table 1. The "det size" column gives the size of the DFA constructed from the NFA and the "min size" column gives the size after minimizing the DFA. To illustrate the size tradeoffs, one CSP instance in nfa-54-30-7-15 has has 1620 NFA states and 13K transitions in the NFA CSP; while the DFA CSP has 2.2M DFA states and 15.9M transitions; and the MDD CSP has 11M nodes and 76M edges.

First, we investigate the performance of our nfa2mdd algorithm. Table 2a gives the average construction time of DFA and MDD CSPs from the respective NFA

---

[11] Deterministic density of an NFA is defined as $\frac{|\delta|}{|Q||\Sigma|}$. If the deterministic density is 2, the equivalent DFA is expected to be exponentially larger [16]. However, the resulting constraints are loose and the random CSPs easy to solve.

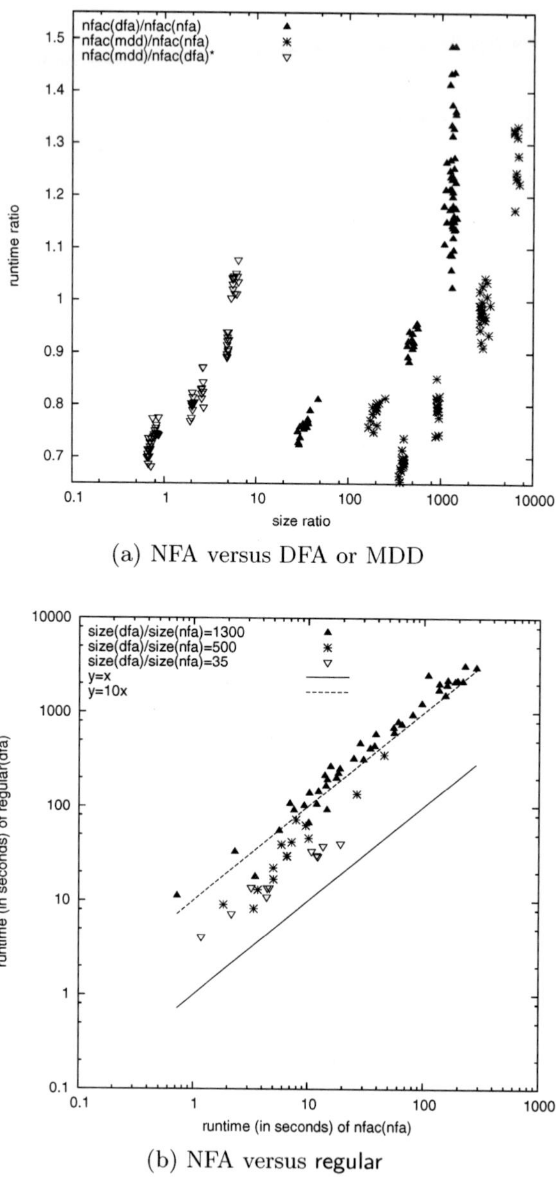

(a) NFA versus DFA or MDD

(b) NFA versus regular

Fig. 5. (a) Solver time compared with NFA or DFA versus input DFA or MDD size; and (b) Regular compared with NFA versus input DFA size

CSP instance. The time to construct the NFA CSP instance itself is small because the number of states is at most 40 and can be effectively ignored. A DFA CSP can be built from the NFA CSP, e.g. using subset construction, the "det time" column. The DFA constraints in the DFA CSP can also be minimized, the "min time" column. There are several ways of building an MDD: (i) from the NFA

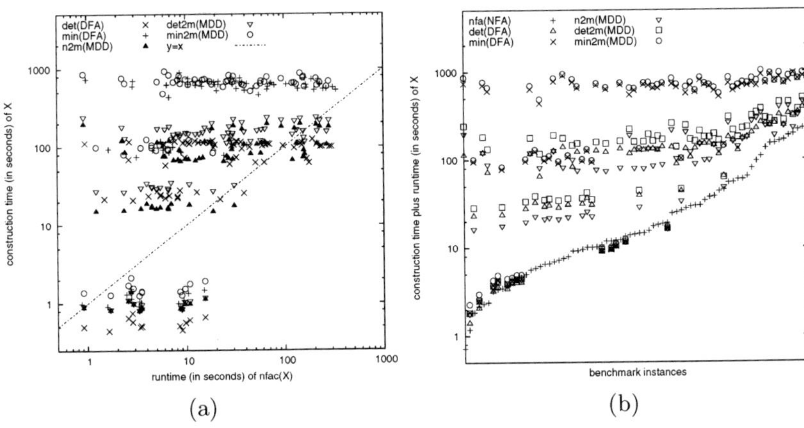

Fig. 6. (a) shows the construction time over runtime for different original input representations but ignoring initialization time; and (b) shows the total time ignoring initialization across input representations across all benchmark instances. For an NFA, the construction time is effectively 0, and the total time is nfac (nfa).

CSP to an MDD CSP using nfa2mdd on the NFA constraints, the n2m column; (ii) from the non-minimized DFA CSP to an MDD CSP using nfa2mdd on the DFA constraints, the det2m column; and (iii) from the minimized DFA CSP to an MDD CSP, the min2m column. The total construction time for an MDD CSP from each way is: (i) just n2m; (ii) "det time"+det2m; and (iii) "det time"+"min time"+min2m. In our benchmarks, directly constructing the MDD from the NFA is the fastest in most cases. Even constructing the DFA from NFA takes significant time. Minimizing the DFA in these benchmarks is too costly except for small instances. Later we compare construction time with solving time to better understand the tradeoffs.

Secondly, we evaluate the efficiency of nfac to understand the space-time tradeoffs for enforcing GAC on the well known NFA, DFA and MDD representations. We use a static variable ordering, max-deg, and lexicographic value ordering in solving all the CSP instances to ensure the search space remains the same. Table 2b gives the solving runtime averaged over the CSP instances in a group. The nfac-init column shows the time to initialize and load a previously constructed CSP instance. The $nfac^c$ column gives the time to solve the CSP using the nfac algorithm with the early-cutoff optimization while $nfac^{rc}$ is without the optimization. Both sets of timings do not include the initialization time which is already given.

Fig 5a shows the detailed space-time tradeoffs for each problem instance. The notation nfac(X) denotes the nfac algorithm with early-cutoff on representation X. The Y-axis shows the speedup for the indicated label, a ratio > 1 means that nfac(nfa) is faster than nfac(dfa) or nfac(mdd). The X-axis shows the ratio in the size of the input representation compared to the NFA. We see that when the size ratio increases (for dfa/nfa > 1000 and for mdd/nfa > 3000); enforcing

GAC using nfac on the NFA is faster than on the DFA or MDD. This shows that nfac(nfa) will be more efficient than nfac(dfa) and nfac(mdd) (and also mddc) when the size of DFAs and MDDs are large. We also show a similar comparison between nfac(mdd) and nfac(dfa) (the $\triangledown$ label, asterixed as the denominator is different). It also shows a similar trend, nfac(dfa) can be faster than nfac(mdd) as long as the size of the DFA is relatively small.

Fig 5b does a similar comparison of nfac(nfa) with the Gecode regular constraint. We see that regular is slower than nfac(nfa). Once the size of the DFA becomes large, the size ratio > 1300, regular is about one order of magnitude slower. In addition, if we compare $nfac^{rc}(mdd)$ with regular, even without the early-cutoff optimization, it is still faster than regular except on the benchmark group nfa-34-30-3-18, which has the smallest DFA size of all the benchmarks.

Since construction time is significant, we also compare the total time = construction time + solving time, but not counting initialization time.[12] Fig. 6a shows that for the DFA and MDD representations, the construction time is not only significant but in most cases exceeds the solving time (above the y=x diagonal). Note that nfac(nfa) is not in the diagram since the other representations are constructed from the NFA form. Fig. 6b shows the total solving time for each benchmark and choice of construction and input representation. It adds nfac(nfa) back into the comparison. We see that now under the total time metric, nfac with NFA outperforms almost all the other choices. The runner-up for most instances is mddc with nfa2mdd(nfa).

Thirdly, we evaluate the effect of early-cutoff, which helps to avoid traversing parts of the graphs in Table 2b. We see that the early cutoff optimization is effective in reducing the GAC time significantly across the board between NFA, DFA and MDD. The optimization is more effective for DFAs and MDDs. regular is also slower than nfac(nfa) and nfac(mdd) even without the early-cutoff optimization.

7   Conclusion

The regular constraint is an expressive and useful global constraint. There are many input representations to define a regular constraint, the most well known and natural ones being NFA, DFA and MDD. In this paper, we propose a new GAC algorithm, nfac for the regular constraint, which can enforce GAC on a constraint given in NFA, DFA and MDD directly. We also give a direct MDD construction algorithm to convert from NFA constraints to MDD constraints. We also extend nfac to deal with grammarc constraints.

The motivation for investigating different representations is to understand better the time and space tradeoffs when the CSP is large in one representation but smaller in another. We investigate the tradeoffs for CSPs with variations in size ratio between all three representations. We show that the direct conversion to an MDD constraint from an NFA one can be done more efficiently than through the DFA. Although it is common to minimize the DFA, it may be

---

[12] We have not included initialization time simply to avoid I/O from the CSP loading, though, it might be even worse for large sizes.

too costly when the DFA is large and the savings may not be sufficient. It turns out that using an NFA representation not only saves space (which may be significant) but also the total runtime (factoring in construction, initialization and solving time). Thus, nfac(nfa) can save both space and time. However, when the problem size is smaller, nfac(mdd) (or the mddc propagator) is faster but at the cost of increased space over an NFA. We also show the importance of the early-cutoff optimization originally shown in the mddc and str2 [3]. We also found that the regular propagator using layered graphs was slower than nfac with all input representations and we conjecture it is due to the use of the early-cutoff and cache optimizations.

References

1. Pesant, G.: A Regular Language Membership Constraint for Finite Sequences of Variables. In: Wallace, M. (ed.) CP 2004. LNCS, vol. 3258, pp. 482–495. Springer, Heidelberg (2004)
2. Cheng, K.C.K., Yap, R.H.C.: An mdd-based generalized arc consistency algorithm for positive and negative table constraints and some global constraints. Constraints 15, 265–304 (2010)
3. Lecoutre, C.: Str2: optimized simple tabular reduction for table constraints. Constraints 16, 341–371 (2011)
4. Lecoutre, C., Likitvivatanavong, C., Yap, R.H.C.: A path-optimal gac algorithm for table constraints. In: ECAI (2012)
5. Lagerkvist, M.Z.: Techniques for Efficient Constraint Propagation. Ph.D. thesis, Royal Institute of Technology (2008)
6. Quimper, C.-G., Walsh, T.: Global Grammar Constraints. In: Benhamou, F. (ed.) CP 2006. LNCS, vol. 4204, pp. 751–755. Springer, Heidelberg (2006)
7. Sellmann, M.: The Theory of Grammar Constraints. In: Benhamou, F. (ed.) CP 2006. LNCS, vol. 4204, pp. 530–544. Springer, Heidelberg (2006)
8. Katsirelos, G., Narodytska, N., Walsh, T.: Reformulating Global Grammar Constraints. In: van Hoeve, W.-J., Hooker, J.N. (eds.) CPAIOR 2009. LNCS, vol. 5547, pp. 132–147. Springer, Heidelberg (2009)
9. Srinivasan, A., Kam, T., Malik, S., Brayton, R.: Algorithms for discrete function manipulation. In: Computer Aided Design, pp. 92–95 (1990)
10. Cheng, K.C.K., Yap, R.H.C.: Maintaining generalized arc consistency on ad-hoc n-ary boolean constraints. In: ECAI, pp. 78–82 (2006)
11. Carlsson, M.: Filtering for the case constraint. Talk given at Advanced School on Global Constraints (2006)
12. Mackworth, A.K.: On reading sketch maps. In: IJCAI, pp. 598–606 (1977)
13. Bryant, R.E.: Graph-based algorithms for Boolean function manipulation. IEEE Trans. on Comp. 35(8), 667–691 (1986)
14. Thompson, K.: Regular expression search algorithm. Comm. of the ACM 11(6), 419–422 (1968)
15. Blum, N., Koch, R.: Greibach normal form transformation, revisited. Information and Computation 150, 47–54 (1997)
16. Leslie, T.: Efficient approaches to subset construction. Master's thesis, University of Waterloo (1995)

Inter-instance Nogood Learning
in Constraint Programming

Geoffrey Chu and Peter J. Stuckey

National ICT Australia, Victoria Laboratory,
Department of Computer Science and Software Engineering,
University of Melbourne, Australia
{gchu,pjs}@csse.unimelb.edu.au

Abstract. Lazy Clause Generation is a powerful approach to reducing search in Constraint Programming. This is achieved by recording sets of domain restrictions that previously led to failure as new clausal propagators called nogoods. This dramatically reduces the search and provides orders of magnitude speedups on a wide range of problems. Current implementations of Lazy Clause Generation only allows solvers to learn and utilize nogoods *within* an individual problem. This means that everything the solver learns will be forgotten as soon as the current problem is finished. In this paper, we show how Lazy Clause Generation can be extended so that nogoods learned from one problem can be retained and used to significantly speed up the solution of other, similar problems.

1   Introduction

Lazy Clause Generation (LCG) [8,5] is a powerful approach to reducing search in Constraint Programming (CP). Finite domain propagation is instrumented to record an explanation for each inference. This creates an implication graph like that built by a SAT solver [7], which may be used to derive nogoods that explain the reason for the failure. These nogoods can be propagated efficiently using SAT unit propagation technology, and can lead to exponential reductions in search space on many problems. Lazy clause generation provides state of the art solutions to a number of combinatorial optimization problems such as resource constrained project scheduling [9] and carpet cutting [10].

Current implementations of Lazy Clause Generation derive nogoods which are only valid within the problem in which they were derived. This means the nogoods cannot be correctly applied to other problems and everything that is learned has to be thrown away after the problem is finished. There are many real life situations where a user might want to solve a series of similar problems, e.g., when problem parameters such as customer demands, tasks, costs or availability of resources change. Clearly, it would be beneficial if the nogoods learned in one problem can be used to speedup the solution of other similar problems.

There are many methods that attempt to reuse information learned from solving previous problems in subsequent problems. However, the information learned by such methods differ significantly from those discussed in this paper.

M. Milano (Ed.): CP 2012, LNCS 7514, pp. 238–247, 2012.

Portfolio based methods (e.g., [11]) use machine learning or similar techniques to try to learn which solver among a portfolio of solvers will perform best on a new problem given characteristics such as the problem size or the properties of its constraint graph. Methods such as [6] attempt to learn effective search heuristics for specific problem domains by using reinforcement learning or similar techniques. In the case where we wish to find the solution to a modified problem which is as similar to the old solution as possible, methods such as [1], which keep track of the value of each variable in the old solution and reuses it as a search heuristic, can be effective. In the special case where we have a series of *satisfaction* problems where each problem is strictly more constrained than the previous one, e.g. satisfaction based optimization, all nogoods can trivially be carried on and reused in the subsequent problems. In this paper, we are interested in the more general case where subsequent problems can be more constrained, less constrained, or simply different because the parameters in some constraints have changed. We show how to generalize Lazy Clause Generation to produce *parameterized nogoods* which can be carried from instance to instance and used for an entire problem class.

2   Background

Let $\equiv$ denote syntactic identity, $\Rightarrow$ denote logical implication and $\Leftrightarrow$ denote logical equivalence. A *constraint optimization problem* is a tuple $P \equiv (V, D, C, f)$, where V is a set of variables, D is a set of domain constraints $v \in D_v, v \in V$, C is a set of constraints, and f is an objective function to minimize (we can write $-f$ to maximize). An assignment θ assigns each $v \in V$ to an element $\theta(v) \in D_v$. It is a *solution* if it satisfies all constraints in C. An assignment θ is an *optimal solution* if for all solutions θ', $\theta(f) \leq \theta'(f)$. In an abuse of notation, if a symbol C refers to a set of constraints $\{c_1, \ldots, c_n\}$, we will often also use the symbol C to refer to the conjunction $c_1 \wedge \ldots \wedge c_n$.

CP solvers solve CSP's by interleaving search with inference. We begin with the original problem at the root of the search tree. At each node in the search tree, we propagate the constraints to try to infer variable/value pairs which cannot be taken in any solution to the problem. Such pairs are removed from the current domain. If some variable's domain becomes empty, then the subproblem has no solution and the solver backtracks. If all the variables are assigned and no constraint is violated, then a solution has been found and the solver can terminate. If inference is unable to detect either of the above two cases, the solver further divides the problem into a number of more constrained subproblems and searches each of those in turn.

In an LCG solver, each propagator is instrumented to explain each of its inferences with a clause called the *explanation*. Each clause consists of *literals* of the form $x = v, x \neq v, x \geq v$ or $x \leq v$ where x is a variable and v is a value.

Definition 1. *Given current domain D, suppose the propagator for constraint c makes an unary inference m, i.e., $c \wedge D \Rightarrow m$. An explanation for this inference is a clause: $expl(m) \equiv l_1 \wedge \ldots \wedge l_k \rightarrow m$ s.t. $c \Rightarrow expl(m)$ and $D \Rightarrow l_1 \wedge \ldots \wedge l_k$.*

The explanation $expl(m)$ explains why m has to hold given c and the current domain D. We can consider $expl(m)$ as the fragment of the constraint c from which we inferred that m has to hold. For example, given constraint $x \leq y$ and current domain $x \in \{3, 4, 5\}$, the propagator may infer that $y \geq 3$, with the explanation $x \geq 3 \rightarrow y \geq 3$.

These explanations form an acyclic implication graph. Whenever a conflict is found by an LCG solver, the implication graph can be analyzed in order to derive a set of sufficient conditions for the conflict. This is done by repeatedly resolving the conflicting clause (the clause explaining the conflict) with explanation clauses, resulting in a new nogood for the problem.

Example 1. Let $C_1 \equiv \{x_2 < x_3, x_1 + 2x_2 + 3x_3 \leq 13\}$. Suppose we tried $x_1 = 1$ and $x_2 = 2$. We would infer $x_3 \geq 3$ from the first constraint with explanation: $x_2 \geq 2 \rightarrow x_3 \geq 3$. The second constraint would then fail with explanation: $x_1 \geq 1 \wedge x_2 \geq 2 \wedge x_3 \geq 3 \rightarrow false$. Resolving the conflicting clause with the explanation clause gives the nogood: $x_1 \geq 1 \wedge x_2 \geq 2 \rightarrow false$.

Let $expls(n)$ be the set of explanations from which a nogood n was derived. Then $expls(n) \Rightarrow n$.

3   Parameterized Nogoods

Suppose we want to solve several similar problem instances from the same problem class. Currently, Lazy Clause Generation produces nogoods which are only correct within the instance in which it was derived, thus we cannot carry such nogoods from one instance to the next and reuse them. In this section, we show how we can generalize the nogoods produced by Lazy Clause Generation to *parameterized* nogoods which are valid for a whole problem class.

Each nogood derived by Lazy Clause Generation represents a resolution proof that a certain subtree in the problem contains no solutions. For us to correctly reuse this nogood in a different problem, we have to show that the resolution proof is valid in the other problem. We have the following result:

Theorem 1. *Let $P_1 \equiv (V, D, C_1, f)$ and $P_2 \equiv (V, D, C_2, f)$ be two constraint optimization problems. Let n be a nogood derived while solving P_1. If $C_2 \Rightarrow expls(n)$, then n is also a valid nogood for P_2.*

Proof. $C_2 \Rightarrow expls(n) \Rightarrow n$.

Theorem 1 tells us that if every explanation used to derive a nogood n is also implied by a second problem P_2, then n is also a valid nogood in P_2.

Example 2. Let $C_1 \equiv \{x_1 < x_2, x_1 + 2x_2 \leq 9\}$ and $C_2 \equiv \{x_1 + 1 < x_2, x_1 + 2x_2 \leq 10\}$. Suppose we tried $x_1 = 3$ in the first problem. We would infer $x_1 \geq 3 \rightarrow x_2 \geq 4$ from the first constraint, and the second constraint would then fail with $x_1 \geq 3 \wedge x_2 \geq 4 \rightarrow false$. The nogood would simply be $x_1 \geq 3 \rightarrow false$. Now, this nogood is also valid in the second problem because: $x_1 + 1 < x_2 \Rightarrow x_1 \geq 3 \rightarrow x_2 \geq 4$ and $x_1 + 2x_2 \leq 10 \Rightarrow x_1 \geq 3 \wedge x_2 \geq 4 \rightarrow false$, so both explanations used to derive the nogood are implied in the second problem.

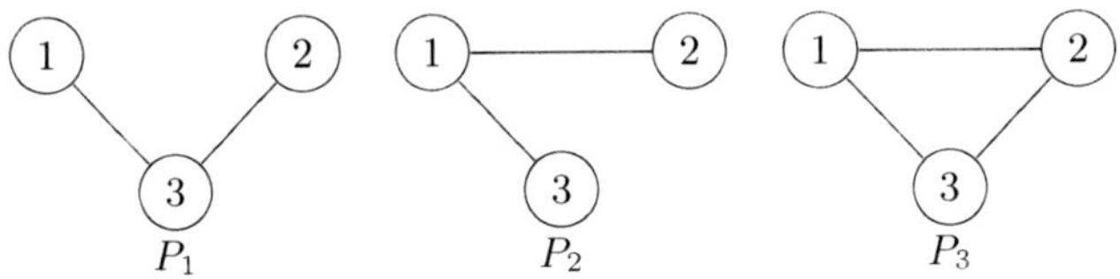

Fig. 1. 3 different graphs for colouring

Note that the constraints in the second problem do not have to be the same as the first. They can be stronger (like the first constraint in Example 2) or weaker (like the second constraint in Example 2), and there can be more or fewer constraints, as long as the constraints are strong enough to imply all the explanations used to derive the nogood in the first problem.

To determine whether a nogood can be reused in a different problem, we need an efficient way to keep track of whether all the explanations used to derive a nogood are implied in the new problem. We can alter the problem models in order to achieve this. Instead of modeling each instance as an individual constraint problem, we create a generic *problem class model* which is then parameterized to produce the individual instances. That is, we create a problem class model $P_{class} \equiv (V \cup Q, D, C, f)$, where Q are *parameter variables*. Individual instances P_i are then created by fixing the variables in Q to instance specific values R_i.

Example 3. Consider the graph coloring problem with n nodes. Let $Q \equiv \{a_{i,j} \mid i, j = 1, \ldots, n\}$ be a set of Boolean parameter variables representing whether there is an edge between node i and j. We can define $V \equiv \{v_1, \ldots, v_n\}$, $D \equiv \{v_i \in \{1, \ldots, n\}$, $C \equiv \{a_{i,j} \rightarrow v_i \neq v_j\}$ and $f = max(v_i)$. Each instance would then be created by setting the variables in Q to certain values to represent the adjacency matrix for that instance.

In the traditional way of modeling, parameters are considered as constants and we have separate problems for each problem instance. In our approach however, parameters are variables and there is only a single problem for the entire problem class. Since the parameters are now variables instead of constants, when LCG is used on an instance, the explanations generated by the propagators will include literals on the parameter variables. These additional literals describe sufficient conditions on the parameter values to make the inferences valid. When such parameterized explanations are resolved together to form a nogood, the nogood will also have literals describing sufficient conditions on the parameter values to make the nogood valid. Thus each nogood becomes a *parameterized nogood* that is valid across the whole problem class. On instances where all the conditions on the parameters are satisfied, the nogood will be *active* and will be able to prune things as per normal. On instances where any of the conditions on the parameters are not satisfied, the nogood will be *inactive* (trivially satisfied) and will not prune anything.

Example 4. Consider three graph coloring instances: P_1, where $a_{1,2} = false$, $a_{1,3} = true, a_{2,3} = true$, P_2, where $a_{1,2} = true, a_{1,3} = true, a_{2,3} = false$, and

P_3, where $a_{1,2} = true, a_{1,3} = true, a_{2,3} = true$, illustrated in Figure 1. Suppose in P_1, we are looking for solutions with $f \leq 2$ and we made search decisions $v_1 = 1, v_2 = 2$. In the unparameterized model, we would use explanations: $f \leq 2 \rightarrow v_3 \leq 2$, $v_1 = 1 \rightarrow v_3 \neq 1$, $v_2 = 2 \rightarrow v_3 \neq 2$, $v_3 \neq 1 \wedge v_3 \neq 2 \wedge v_3 \leq 2 \rightarrow false$ and would derive a nogood $v_1 = 1 \wedge v_2 = 2 \wedge f \leq 2 \rightarrow false$. Now, it is incorrect to apply this nogood in P_2, because it is simply not true. For example $v_1 = 1, v_2 = 2, v_3 = 2$ is a perfectly valid solution for P_2. In the parameterized model, we would use the explanations: $f \leq 2 \rightarrow v_3 \leq 2$, $v_1 = 1 \wedge a_{1,3} \rightarrow v_3 \neq 1$, $v_2 = 2 \wedge a_{2,3} \rightarrow v_3 \neq 2$, $v_3 \neq 1 \wedge v_3 \neq 2 \wedge v_3 \leq 2 \rightarrow false$ and would derive a nogood $a_{1,3} \wedge a_{2,3} \wedge v_1 = 1 \wedge v_2 = 2 \wedge f \leq 2 \rightarrow false$, which correctly encapsulates the fact that the nogood is only valid if the graph has edges between node 1 and 3 and node 2 and 3. The parameterized nogood can be correctly applied to any instance of the graph coloring problem. It is inactive in P_2 because $a_{2,3} = false$ in P_2, but it might prune something in P_3 because $a_{1,3} = a_{2,3} = true$ in P_3.

4   Implementation

Naively, we could implement the parameter variables as actual variables with constraints over them. Then simply running the normal LCG solver on this model will generate parameterized nogoods. However, such an implementation is less efficient than using an instance specific model where the parameters are considered as constants. For example, a linear constraint: $\sum a_i x_i$ where a_i are parameters would be a simple linear constraint in an instance model, but would be a quadratic constraint if we consider parameter variables as actual variables. We can improve the implementation by taking advantage of the fact that the parameters will always be fixed when we are solving an instance. To do this we use normal propagators which treat parameters as constants, but alter their explanations so that they include literals representing sufficient conditions on the parameters to make the inference true.

For example, given a linear constraint $a_1 x_1 + a_2 x_2 + a_3 x_3 \leq 10$, $a_1 = 1, a_2 = 2, a_3 = 3$ and current bounds $x_1 \geq 1$, $x_2 \geq 2$, we can infer $x_3 \leq 1$, and we would explain it using: $a_1 \geq 1 \wedge a_2 \geq 2 \wedge a_3 \geq 3 \wedge x_1 \geq 1 \wedge x_2 \geq 2 \rightarrow x_3 \leq 1$. Given a constraint c which may be added to or removed from an instance depending on a Boolean parameter b, we would modify the explanations for c's inferences by adding the literal b to each explanation. Given a *cumulative* constraint where task durations, resource usage and the capacity of the machine are parameters, we would add lower bound literals on the duration and resource usage of each task involved in the inference, and an upper bound literal on the machine capacity to each explanation. The modifications to the explanations of other parameterized constraints are similarly straightforward and we do not describe them all.

A LCG solver can generate an enormous number of parameterized nogoods during search, most of which are not particularly useful. Clearly, it would be inefficient to retain all of these nogoods. We take advantage of the inbuilt capabilities of LCG solvers for deleting useless nogoods. The LCG solver CHUFFED maintains an activity score for each nogood based on how often it is used. When the number of nogoods in the constraint store reaches 100000, the least active half are deleted. We only reuse the, at most 100000, parameterized nogoods which survive till the end of the solve. At the beginning of each new instance, we check each parameterized nogood to see if the conditions on the parameters are satisfied. If not, we ignore the parameterized nogood, as it cannot prune anything in this particular instance. If the conditions are satisfied, we add it to the constraint store and handle it in the same way as nogoods learned during search, i.e., we periodically remove inactive ones. This ensures that if the parameterized nogoods learned from previous instances are useless, they will quickly be removed and will no longer produce any overhead.

4.1   Strengthening Explanations for Inter-instance Reuse

An important optimization in LCG is to *strengthen* explanations so that the nogoods derived from it are more reusable. For example, consider a constraint: $x_1 + 2x_2 + 3x_3 \leq 13$, and current domains: $x_1 \geq 4, x_2 \geq 3$. Clearly, we can infer that $x_3 \leq 1$. Naively, we might explain this using the current bounds as: $x_1 \geq 4 \wedge x_2 \geq 3 \rightarrow x_3 \leq 1$. This explanation is valid, but it is not the strongest possible explanation. For example, $x_1 \geq 2 \wedge x_2 \geq 3 \rightarrow x_3 \leq 1$ is also a valid explanation and is strictly stronger logically. Using these stronger explanations result in stronger nogoods which may prune more of the search space. It is often the case that there are multiple ways to strengthen an explanation and it is not clear which one is best. For example, $x_1 \geq 4 \wedge x_2 \geq 2 \rightarrow x_3 \leq 1$ is another way to strengthen the explanation for the above inference.

In the context of inter-instance learning, there is an obvious choice of which strengthening to pick. We can preferentially strengthen the explanations so that they are more reusable across different problems. We do this by weakening the bounds on parameter variables in preference to those on normal variables. For example, if x_1 was a parameter variable and x_2 was a normal variable, we would prefer the first strengthening above rather than the second, as that explanation places weaker conditions on the parameter variables and will allow the nogood to prune things in more instances of the problem class.

4.2   Hiding Parameter Literals

The linear constraint is particularly difficult for our approach as its explanations often involve very specific conditions on the parameters, and these conditions might not be repeated in other instances of the problem we are interested in. For example, suppose: $\sum_{i=1}^{k} a_i x_i \leq m$ where a_i are positive parameter variables and x_i are normal variables. Suppose each a_i is fixed to a value of r_i, and each x_i is fixed to a value of b_i and we have a failure. A naive explanation of this inference would be of the form: $\wedge_{i=1}^{k}(a_i \geq r_i \wedge x_i \geq b_i) \rightarrow false$. This places a lower bound condition on all of the a_i involved in the linear constraint, which may be hard to meet in any other instance of the problem. We can improve the situation by decomposing long linear constraints into ternary linear constraints involving partial sum variables. This serves to "hide" some of the parameters away and makes the explanation more reusable.

Example 5. Suppose we had constraints: $x_3 < x_4$ and $a_1x_1 + a_2x_2 + a_3x_3 + a_4x_4 \leq 29$, and in this instance, the parameters are set to $a_1 = 1, a_2 = 2, a_3 = 3, a_4 = 4$. Suppose we tried $x_1 = 1$, $x_2 = 2$, $x_3 = 3$. We can infer that $x_4 \geq 4$ with explanation $x_3 \geq 3 \rightarrow x_4 \geq 4$. The second constraint then fails with explanation: $(a_1 \geq 1 \wedge a_2 \geq 2 \wedge a_3 \geq 3 \wedge a_4 \geq 4) \wedge x_1 \geq 1 \wedge x_2 \geq 2 \wedge x_3 \geq 3 \wedge x_4 \geq 4 \rightarrow \textit{false}$, leading to nogood: $(a_1 \geq 1 \wedge a_2 \geq 2 \wedge a_3 \geq 3 \wedge a_4 \geq 4) \wedge x_1 \geq 1 \wedge x_2 \geq 2 \wedge x_3 \geq 3 \rightarrow \textit{false}$. The condition on the parameters: $(a_1 \geq 1 \wedge a_2 \geq 2 \wedge a_3 \geq 3 \wedge a_4 \geq 4)$ is difficult to satisfy. However, suppose we decomposed the linear constraint into: $a_1x_1 + a_2x_2 \leq s_2, s_2 + a_3x_3 \leq s_3, s_3 + a_4x_4 \leq 29$. Now, after inferring $x_4 \geq 4$ from the first constraint, we would have a chain of inferences and explanations: $(a_1 \geq 1 \wedge a_2 \geq 2) \wedge x_1 \geq 1 \wedge x_2 \geq 2 \rightarrow s_2 \geq 5$, $(a_3 \geq 3) \wedge s_2 \geq 5 \wedge x_3 \geq 3 \rightarrow s_3 \geq 14$, $(a_4 \geq 4) \wedge s_3 \geq 14 \wedge x_4 \geq 4 \rightarrow \textit{false}$. The nogood would be derived by resolving the last two clause with $x_3 \geq 3 \rightarrow x_4 \geq 4$, which gives the nogood: $(a_3 \geq 3 \wedge a_4 \geq 4) \wedge s_2 \geq 5 \wedge x_3 \geq 3 \rightarrow \textit{false}$. This nogood only has conditions on a_3 and a_4 and is strictly stronger logically. By introducing the partial sum variables, the conditions on a_1 and a_2 have been "hidden" into the bound literal on s_2 instead, producing a more reusable nogood.

5   Experiments

We evaluate our method on four problems. We briefly describe each problem, their parameters, and situations where we may wish to solve several similar instances of the problem.

Radiation Therapy Problem. In the Radiation Therapy Problem [2], a doctor develops a treatment plan for a patient consisting of an intensity matrix describing the amount of radiation to be delivered to each part of the treatment area. The aim is to find the configuration of collators and beam intensities which minimizes the total amount of radiation delivered and the treatment time. The intensities are parameters. The doctor may alter the treatment plan (change some of the intensities) and we may wish to re-optimize. We use instances of with 15 rows, 12 columns and a max intensity of 10.

Minimization of Open Stacks Problem. In the Minimization of Open Stacks Problem (MOSP) [4], we have a set of customers each of which requires a subset of the products. The products are produced one after another. Each customer has a stack which must be opened from the time when the first product they require is produced till the last product they require is produced. The aim is to find the production order which minimizes the number of stacks which are open at any time. The parameters are whether a customer requires a certain product. Customers may change their orders and we may wish to re-optimize the schedule. We use instances with 35 customers and a shared product density of 0.2.

Table 1. Comparison of solving (a) Radiation Therapy instances, (b) MOSP instances, (c) Graph Coloring instances, and (d) Knapsack instances: from scratch (**scratch**) and solving them making use of parameterized nogoods (**para**) from a similar instance

(a) Radiation Therapy

diff	scratch		para		reuse	speedup
1%	7.10	23000	0.08	32	97%	88.8
2%	7.25	23219	0.31	433	93%	23.8
5%	7.03	22556	1.17	2496	81%	6.02
10%	7.57	23628	2.52	6198	74%	3.00
20%	7.68	23326	4.35	11507	55%	1.77

(b) MOSP

diff	scratch		para		reuse	speedup
1%	38.67	111644	1.03	921	93%	37.5
2%	41.65	114358	4.57	6700	88%	9.11
5%	42.80	111745	31.02	75600	71%	1.38
10%	47.80	95521	40.77	88909	57%	1.17
20%	27.37	86231	33.26	88186	45%	0.82

(c) Graph Coloring

diff	scratch		para		reuse	speedup
1%	15.12	54854	7.72	17026	74%	1.96
2%	18.78	61630	16.17	35994	61%	1.16
5%	23.65	70710	24.96	52026	35%	0.94
10%	45.14	96668	45.54	79763	20%	0.99
20%	48.12	101668	45.96	87431	9%	1.04

(d) Knapsack

diff	scratch		para		reuse	speedup
1%	18.21	48714	6.74	11339	100%	2.70
2%	18.47	48640	7.32	13042	100%	2.52
5%	18.80	49154	16.91	37786	100%	1.11
10%	19.38	49298	21.12	44952	100%	0.92
20%	20.83	50007	24.15	49387	100%	0.86

Graph Coloring. The existence of an edge between each pair of nodes is a parameter. Edges may be added or removed and we may wish to re-optimize. We use instances with 55 nodes and an edge density of 0.33.

Knapsack. In the 0-1 Knapsack Problem, the value, weight and availability of items are parameters. New items might become available, or old ones might become unavailable and we may wish to re-optimize. We use instances with 100 items.

For each of these problems, we generate 100 random instances. From each of these base instances, we generate modified versions of the instance where 1%, 2%, 5%, 10% or 20% of the parameters have been randomly changed. The instances are available online [3]. We solve these instances using the Lazy Clause Generation solver CHUFFED running on 2.8 GHz Xeon Quad Core E5462 processors. As a baseline, we solve every instance from scratch (**scratch**). To compare with our method, we first solve each base instance and learn parameterized nogoods from it. We then solve the corresponding modified versions while making use of these parameterized nogoods (**para**). The geometric mean of the run times in seconds and the nodes required to solve each set of 100 instances is shown in Table 1. We also show the geometric mean of the percentage of parameterized nogoods which are active in the second instance of each pair of instances (**reuse**), and the speedup (**speedup**).

As can be seen from the results, parameterized nogoods can provide significant reductions in node count and run times on a variety of problems. The speedups vary between problem classes and are dependent on how similar the instance is to one that has been solved before. Dramatic speedups are possible for Radiation and MOSP when the instances are similar enough, whereas the speedups are smaller for Knapsack and Graph Coloring. The more similar an instance is to one that has been solved before, the greater the number of parameterized nogoods which are active in this instance and the greater the speedup tends to be. When the instance is too dissimilar, parameterized nogoods provide little to no benefit.

The percentage of parameterized nogoods which are active in the second instance is highly dependent on the problem class. This is because depending on the structure of the problem, each parameterized nogood can involve a small or large number of the instance parameters. The fewer the parameters involved, the fewer the conditions on the parameters and the more likely it is that the nogood will be reusable in another instance. The "hiding parameter literals" optimization described in Section 4.2 is clearly beneficial for the Knapsack Problem, raising the reusability to 100%. Without it, few of the parameterized nogoods are active in the second instance and there is no speedup (not shown in table). While parameterized nogoods must be active in order to provide any pruning, there is no guarantee that an active nogood will actually provide "useful" pruning. This can be seen in the results for Knapsack, where despite the fact that all the parameterized nogoods can potentially prune something, they do not prune anything useful when the second instance is too different from the first.

6   Conclusion

We have generalized the concept of nogoods, which are valid only for an instance, to *parameterized nogoods* which are valid for an entire problem class. We have described the modifications to a Lazy Clause Generation solver required to generate such parameterized nogoods. We evaluated the technique experimentally and found that parameterized nogoods can provide significant speedups on a range of problems when several similar instances of the same problem need to be solved. The more similar the instances are, the greater the speedup from using parameterized nogoods.

Acknowledgments. NICTA is funded by the Australian Government as represented by the Department of Broadband, Communications and the Digital Economy and the Australian Research Council. This work was partially supported by Asian Office of Aerospace Research and Development (AOARD) Grant FA2386-12-1-4056.

References

1. Abío, I., Deters, M., Nieuwenhuis, R., Stuckey, P.J.: Reducing Chaos in SAT-Like Search: Finding Solutions Close to a Given One. In: Sakallah, K.A., Simon, L. (eds.) SAT 2011. LNCS, vol. 6695, pp. 273–286. Springer, Heidelberg (2011)
2. Baatar, D., Boland, N., Brand, S., Stuckey, P.J.: Minimum Cardinality Matrix Decomposition into Consecutive-Ones Matrices: CP and IP Approaches. In: Van Hentenryck, P., Wolsey, L.A. (eds.) CPAIOR 2007. LNCS, vol. 4510, pp. 1–15. Springer, Heidelberg (2007)
3. Chu, G.: Interproblem nogood instances,
   `www.cis.unimelb.edu.au/~pjs/interprob/`

4. Chu, G., Stuckey, P.J.: Minimizing the Maximum Number of Open Stacks by Customer Search. In: Gent, I.P. (ed.) CP 2009. LNCS, vol. 5732, pp. 242–257. Springer, Heidelberg (2009)
5. Feydy, T., Stuckey, P.J.: Lazy Clause Generation Reengineered. In: Gent, I.P. (ed.) CP 2009. LNCS, vol. 5732, pp. 352–366. Springer, Heidelberg (2009)
6. Langley, P.: Learning effective search heuristics. In: IJCAI, pp. 419–421 (1983)
7. Moskewicz, M.W., Madigan, C.F., Zhao, Y., Zhang, L., Malik, S.: Chaff: Engineering an Efficient SAT Solver. In: Proceedings of the 38th Design Automation Conference, pp. 530–535. ACM (2001)
8. Ohrimenko, O., Stuckey, P.J., Codish, M.: Propagation = Lazy Clause Generation. In: Bessière, C. (ed.) CP 2007. LNCS, vol. 4741, pp. 544–558. Springer, Heidelberg (2007)
9. Schutt, A., Feydy, T., Stuckey, P.J., Wallace, M.G.: Why Cumulative Decomposition Is Not as Bad as It Sounds. In: Gent, I.P. (ed.) CP 2009. LNCS, vol. 5732, pp. 746–761. Springer, Heidelberg (2009)
10. Schutt, A., Feydy, T., Stuckey, P.J., Wallace, M.G.: Explaining the cumulative propagator. Constraints 16(3), 250–282 (2011)
11. Xu, L., Hutter, F., Hoos, H.H., Leyton-Brown, K.: Satzilla: Portfolio-based algorithm selection for sat. J. Artif. Intell. Res. (JAIR) 32, 565–606 (2008)

Solving Temporal Problems Using SMT:
Strong Controllability

Alessandro Cimatti, Andrea Micheli, and Marco Roveri

Fondazione Bruno Kessler — Irst
`{cimatti,amicheli,roveri}@fbk.eu`

Abstract. Many applications, such as scheduling and temporal planning, require the solution of Temporal Problems (TP's) representing constraints over the timing of activities. A TP with uncertainty (TPU) is characterized by activities with uncontrollable duration. Depending on the Boolean structure of the constraints, we have simple (STPU), constraint satisfaction (TCSPU), and disjunctive (DTPU) temporal problems with uncertainty.

In this work we tackle the problem of strong controllability, i.e. finding an assignment to all the controllable time points, such that the constraints are fulfilled under any possible assignment of uncontrollable time points. We work in the framework of Satisfiability Modulo Theory (SMT), where uncertainty is expressed by means of universal quantifiers. We obtain the first practical and comprehensive solution for strong controllability: the use of quantifier elimination techniques leads to quantifier-free encodings, which are in turn solved with efficient SMT solvers.

We provide a detailed experimental evaluation of our approach over a large set of benchmarks. The results clearly demonstrate that the proposed approach is feasible, and outperforms the best state-of-the-art competitors, when available.

1   Introduction

Many applications require the scheduling of a set of activities over time, subject to constraints of various nature. Scheduling is often expressed as a *Temporal Problem* (TP), where each activity is associated with two time points, representing the start time and the end time, and with a duration, all subject to constraints. Several kinds of temporal problems have been identified, depending on the nature and structure of the constraints. If the constraints are expressible as a simple conjunction of constraints over distances of time points, then we have the so-called *Simple Temporal Problem* (STP). A more complex class is *Temporal Constraint Satisfaction Problem* (TCSP), where a distance between time points can be constrained to a list of disjoint intervals. Constraints in TCSP's can be seen as a restricted form of Boolean combinations. When arbitrary Boolean combinations are allowed, we have a *Disjunctive Temporal Problem* (DTP). A temporal problem is said to be *consistent* if there exists an assignment for the

M. Milano (Ed.): CP 2012, LNCS 7514, pp. 248–264, 2012.
© Springer-Verlag Berlin Heidelberg 2012

time points, such that all the constraints are satisfied [1]. Such an assignment is called a *schedule*, and it corresponds to sequential time-triggered programs, that are often used in control of satellites and rovers.

In many practical cases, however, the duration of activities is uncontrollable. TP are thus extended with *uncertainty* in the duration of activities, thus obtaining the classes of STPU, TCSPU and DTPU. As in the case of consistency, we look for a schedule. However, the schedule only determines the start of the activities, and must satisfy the constraints *for all* the uncontrollable durations of the activities. If such a schedule exists, the problem is said to be *strongly controllable* [2].

In this paper, we propose a comprehensive and effective approach to strong controllability of TPU. The approach relies on the Satisfiability Modulo Theory (SMT) framework [3]. This framework provides representation capabilities, based on fragments of first order formulae. Reasoning is carried out within a decidable fragments of first order logic, where interpretations are constrained to satisfy a specific theory of interest (i.e. linear arithmetic). Modern SMT solvers are a tight integration of a Boolean SAT solver, that is highly optimized for the *case split* required by the Boolean combination of constraints, with dedicated constraint solvers for the theories of interest. Moreover, some SMT solvers provide embedded efficient primitives to handle *quantifiers*, and dedicated techniques for quantifier-elimination are available. Several effective SMT solvers are available (e.g. MathSAT [4,5], Z3 [6], Yices [7], OpenSMT [8]). SMT solving has had increasing applications in many areas including Answer Set Programming (ASP) [9], formal verification [10], and test case generation [11].

We tackle the strong controllability problem of TPUs by reduction to SMT problems, that are then fed into efficient SMT solvers. First, we show how to encode a TPU into the theory of quantified linear Real arithmetic (LRA) and, by leveraging the specific nature of the problem, we optimize the encoding by reducing the scope of quantifiers. The resulting formula can be fed to any SMT solver for (quantified) LRA. Second, we present a general reduction procedure from strong controllability to consistency, based on the application of quantifier elimination techniques upfront. The resulting formulae can be directly fed into any SMT solver for the quantifier-free LRA. This gives the first general comprehensive solver for strong controllability of TPUs. Third, we generalize the results by Vidal and Fargier [2], originally stated for STPU, to the class of TCSPU. In this way, we avoid the use of expensive general purpose quantifier elimination techniques, with significant performance improvements.

The proposed approach has been implemented in a solver based on state-of-the-art SMT techniques. To the best of our knowledge, this is the first solver for strong controllability of TPUs. We carried out a thorough experimental evaluation, over a large set of benchmarks. We analyze the merits of the various encodings, and demonstrate the overall feasibility of the approach. We also compare the proposed approaches with state-of-the-art algorithms on consistency problems. SMT solvers are competitive with, and often outperform, the best known dedicated solving techniques.

Structure of the Paper. In sections 2 and 3 we present some technical preliminaries and background about SMT. In section 4 we formally define temporal problems, while in section 5 we present several SMT encodings for consistency. Encodings for strong controllability are described in section 6 and an overview of the related work is given is section 7. We report the results of the performed experimental evaluation in section 8, and in section 9 we draw some conclusions and outline directions for future work.

2   Technical Preliminaries

Our setting is standard first order logic. The first-order signature is composed of constants, variables, function symbols, Boolean variables, and predicate symbols. A term is either a constant, a variable, or the application of a function symbol of arity n to n terms. A theory constraint (also called a theory atom) is the application of a predicate symbol of arity n to n terms. An atom is either a theory constraint or a Boolean variable. A literal is either an atom or its negation. A clause is a finite disjunction of literals. A formula is either true ($\top$), false ($\bot$), a Boolean variable, a theory constraint, the application of a propositional connective of arity n to n formulae, or the application of a quantifier to an individual variable and a formula. We use $x, y, v, \ldots$ for variables, and $\vec{x}, \vec{y}, \vec{v}, \ldots$ for vectors of individual or Boolean variables. Terms and formulae are referred to as expressions, denoted with $\phi, \psi, \ldots$ We write $\phi(x)$ to highlight the fact that x occurs in ϕ, and $\phi(\vec{x})$ to highlight the fact that each x_i occurs in ϕ.

Substitution is defined in the standard way (see for instance [12]). We write $\phi[t/s]$ for the substitution of every occurrence of term t in ϕ with term s. Let $\vec{t}$ and $\vec{s}$ be vectors of terms, we write $\phi[\vec{t}/\vec{s}]$ for the parallel substitution of every occurrence of t_i (the i-th element of $\vec{t}$) in ϕ with s_i.

We use the standard semantic notion of interpretation and satisfiability. We call *satisfying assignment* or *model* of a formula $\phi(\vec{x})$ a total function μ that assigns to each x_i an element of its domain such that the formula $\phi[\vec{x}/\mu(\vec{x})]$ evaluates to $\top$. A formula $\phi(\vec{x})$ is *satisfiable* if and only if it has a satisfying assignment.

Checking the satisfiability (SAT) of a formula consists in finding a satisfying assignment for the formula. This problem is approached in propositional logic with enhancements of the DPLL algorithm: the formula is converted into an equi-satisfiable one in Conjunctive Normal Form (CNF); then, a satisfying assignment is incrementally built, until either all the clauses are satisfied, or a conflict is found, in which case back-jumping takes place (i.e. certain assignments are undone). Keys to efficiency are heuristics for the variable selection, and learning of conflicts (see e.g. [13]).

3   Satisfiability Modulo Theories

Given a first-order formula ψ in a decidable background theory T, *Satisfiability Modulo Theory* (SMT) [3] is the problem of deciding whether there exists a satisfying assignment to the free variables in ψ. For example, consider the formula $(x \leq y) \wedge (x + 3 = z) \vee (z \geq y)$ in the theory of real arithmetic ($x, y, z \in \mathbb{R}$). The formula is satisfiable and a satisfying assignment is $\{x := 5,\ y := 6,\ z := 8\}$. The theory of real arithmetic interprets 3 as a real number and $+, =, <, >, \leq, \geq$ as the corresponding operations and relations over $\mathbb{R}$.

In this work we concentrate on the theory of Linear Arithmetic over the Real numbers (LRA). A formula in LRA is an arbitrary Boolean combination, or universal ($\forall$) and existential ($\exists$) quantification, of atoms in the form $\sum_i a_i x_i \bowtie c$ where $\bowtie \in \{>, <, \leq, \geq, \neq, =\}$, every x_i is a real variable and every a_i and c is a real constant. Given two real constants l, u such that $l \leq u$, we denote with $t \in [l, u]$ the formula $l \leq t \wedge t \leq u$. Difference logic (RDL) is the subset of LRA such that atoms have the form $x_i - x_j \bowtie c$. We denote with QF_LRA and QF_RDL the quantifier-free fragments of LRA and RDL, respectively.

An SMT solver [3] is a decision procedure which solves the satisfiability problem for a formula expressed in a decidable subset of First-Order Logic. The most efficient implementations of SMT solvers use the so-called "lazy approach", where a SAT solver is tightly integrated with a T-solver. The role of the SAT solver is to enumerate the truth assignments to the Boolean abstraction of the first-order formula. The Boolean abstraction has the same Boolean structure of the first-order formula, but "replaces" the predicates which contain T information with fresh Boolean variables. The Boolean abstraction of $(x \leq y) \wedge (x + 3 = z) \vee (z \geq y)$ is $a \wedge (b \vee c)$, where a, b, c are fresh Boolean variables. The T-solver is invoked when the SAT solver finds a satisfying assignment for the Boolean abstraction: the satisfying assignment to Boolean abstraction maps directly to a conjunction of T atoms, which the T-solver can handle. If the conjunction is satisfiable also the original formula is satisfiable. Otherwise the T-solver returns a conflict set which identifies a reason for the unsatisfiability. Then, the negation of the conflict set is learned by the SAT solver in order to prune the search. Examples of solvers based on the "lazy approach" are MathSAT [4] and Z3 [6].

In order to deal with quantifiers in LRA many techniques have been developed and implemented in SMT solvers. Some solvers, like e.g. Z3 [6] natively support quantifiers. However, many SMT solvers cannot deal with them. Several techniques have been developed for removing quantifiers from an LRA formula (e.g. Fourier-Motzkin [14], Loos-Weispfenning [15,16]): they transform an LRA formula into an *equivalent* QF_LRA formula. These techniques enable for the use of solvers with no native support for quantifiers at a cost that is doubly exponential in time and space in the original formula size [14,16,15].

4   Temporal Problems

A Temporal Problem (TP) is a formalism that is used to represent temporal constraints over time-valued variables representing time points. This formalism is expressive enough to express Allen's interval algebra [17] and also quantitative constraints over intervals and time points. Two families of TP's have been presented in literature over the years: TP without uncertainty, in which all the time points can be freely assigned by the solver [1,18]; TP with uncertainty (TPU), in which only some of the time points can be assigned by the solver, while the others are intended to be assigned by an adversary. As such, TPU's can be seen as a form of game between the solver and an adversarial environment [2,19].

Definition 1. *A TPU is a tuple* (X_c, X_u, C_c, C_f), *where* $X_c \doteq \{b_1, ..., b_n\}$ *is the set of* controllable *time points,* $X_u \doteq \{e_1, ..., e_m\}$ *is the set of* uncontrollable *time points,* $C_c \doteq \{cc_1, ..., cc_m\}$ *is the set of* contingent *constraints, and* $C_f \doteq \{cf_1, ..., cf_h\}$ *is the set of* free *constraints.*

$$cc_i \doteq (e_i - b_{j_i}) \in [l_i, u_i] \qquad cf_i \doteq \bigvee_{j=1}^{D_i} (x_{i,j} - y_{i,j}) \in [l_{i,j}, u_{i,j}]$$

such that: $j_i \in [1 \dots n]$, $l_i, u_i \in \mathbb{R}$, $l_i \leq u_i$, $l_{i,j}, u_{i,j} \in \mathbb{R} \cup \{+\infty, -\infty\}$, $l_{i,j} \leq u_{i,j}$, D_i *is the number of disjuncts for the i-th constraint,* $x_{i,j}, y_{i,j} \in X_c \cup X_u$

Intuitively, time points belonging to X_c are time decisions that can be controlled by the solver, while time points in X_u are under the control of the environment. A similar subdivision is imposed on the constraints: free constraints C_f are constraints that the solver is required to fulfill, while contingent constraints (C_c) are the assumptions that the environment will fulfill. As in [2] we consider only contingent constraints that start with a controllable time point. Thus, each uncontrollable time point is linked by exactly one contingent constraint to a controllable time point. We remark that this assumption does not affect the generality of the formalism, as for each contingent constraint $(e_i - e_j) \in [l, u]$ we can add an artificial new controllable time point b, and add $(b - e_j) \in [0, 0]$ to the free constraints and $(e_i - b)$ to the contingent constraints.

A *TP without uncertainty* is a TPU $(X_c, \emptyset, \emptyset, C_f)$, i.e. the set of uncontrollable time points is empty (from which it also follows that the set of contingent constraints is empty).

Depending on the generality of the constraints in C_c and C_f, three classes of TPU's are identified [19]. Definition 1 in its general form identifies *Disjunctive Temporal Problem with Uncertainty* (DTPU). If each constraint contains at most two time points, the resulting problem is a *Temporal Constraint Satisfaction Problem with Uncertainty* (TCSPU). If each constraint has exactly one disjunct (i.e. $D_i = 1$ for all i), we obtain a *Simple Temporal Problem with Uncertainty* (STPU). Similarly, we can define the corresponding TP without uncertainty (DTP [18], TCSP, and STP [1]).

We define an assignment to the time points as a total function from time points to real values. Given a TP without uncertainty, checking *consistency* corresponds to deciding the existence of an assignment that fulfills all the constraints of the problem. We call such an assignment a *consistent schedule*, and we say that the TP is *consistent*. Checking the consistency of a TPU (X_c, X_u, C_c, C_f) is defined as checking the consistency of the TP without uncertainty $(X_c \cup X_u, \emptyset, \emptyset, C_c \cup C_f)$.

Intuitively, when checking consistency of a TPU, the behavior of the environment is assumed to be "cooperative" with the solver. In this paper, we focus on Strong Controllability (SC) [2] for a TPU, where the environment is adversarial. SC consists in deciding the existence of a *strong schedule*, i.e. an assignment to controllable time points that fulfills the free constraints under any assignment of uncontrollable time points that satisfies the contingent constraints. A TPU for which there exists a strong schedule is said to be *strongly controllable*.

If a TPU is strongly controllable, it is also consistent. However, the converse does not hold in general. Consider for example the STPU such that $X_c = \{A, B\}$, $X_u = \{C\}$, $C_c = \{(B - A) \in [1, 10]\}$, and $C_f = \{(C - A) \in [0, 12], (C - B) \in [1, 5]\}$. The problem is consistent, and a consistent schedule is $\{A = 0, B = 3, C = 5\}$. However, the problem is not strongly controllable because if the duration of the interval $B - A = 1$, the window of opportunity for scheduling C is $[2, 6]$, that is disjoint from the window of opportunity when the duration is equal to 10, that is $[11, 12]$. Since $B - A$ is decided by the adversarial environment, there is no strong schedule that allows the solver to win.

5   Encoding of Consistency Problems in SMT

We first focus on the consistency problem, i.e. the case in which there is no uncontrollability. The consistency problem can be reduced to checking the satisfiability of a quantifier-free formula modulo the LRA theory. The temporal problem is consistent if and only if the corresponding SMT formula is satisfiable, and any satisfying assignment for the formula corresponds to a consistent schedule for the problem.

The use of SMT to check the consistency of TP without uncertainty has been investigated in the past (e.g. [20]). Here, consistency checking plays the role of backend for strong controllability. In the following, we present several SMT encodings, that turn out to have different performance in the solvers, depending on the nature of the constraints.

In the following, we assume that a TP $(X_c, \emptyset, \emptyset, C_f)$ is given. The first encoding in SMT of the consistency problem can be directly obtained as follows: for every time point in X_c we introduce a real variable, and we denote with $\vec{X}_c$ the vector of such variables; each constraint in C_f is directly mapped on the corresponding SMT formula; the encoding is the SMT formula shown in Equation 1.

$$\bigwedge_{i=1}^{|C_f|} \bigvee_{j=1}^{D_i} (((x_{i,j} - y_{i,j}) \geq l_{i,j}) \wedge ((x_{i,j} - y_{i,j}) \leq u_{i,j})) \tag{1}$$

This encoding is linear in the size of the original TP, but does not exploit any knowledge on the structure of the problem, and is thus referred to as *naïve encoding*. In particular, we notice that the resulting SMT formula is not in CNF.

In the rest of this section we introduce three optimizations: the switch encoding (applicable to any TP), the switch encoding with mutual exclusion and the hole encoding (both for TCSPs only). The *switch encoding* performs a CNF conversion of the formula in Equation 1 by means of a polarity-based CNF labeling conversion [21]. To this extent, we introduce $\sum_{i=0}^{|C_f|} D_i$ Boolean "switch" variables $s_{i,j}$, and the resulting encoding is the one in Equation 2.

$$\bigwedge_{i=1}^{|C_f|} ((\bigwedge_{j=1}^{D_i} ((\neg s_{i,j} \vee ((x_{i,j} - y_{i,j}) \geq l_{i,j})) \wedge \\ (\neg s_{i,j} \vee ((x_{i,j} - y_{i,j}) \leq u_{i,j})))) \wedge (\bigvee_{j=1}^{D_i} s_{i,j})) \tag{2}$$

This encoding is also linear in the size of the original TP, and it directly produces a CNF formula. We notice that the clauses involving theory atoms are binary; furthermore, if a switch variable is assigned to false, the corresponding clauses are satisfied without any theory reasoning. These factors have a positive impact on the performance of the SMT solver.

If we focus on the TCSP class, we can exploit the problem structure to further improve our encodings. In TCSP each constraint is composed of disjuncts of the form $t \in [l_j, u_j]$, where t is the difference of two variables, and for all j, $l_j \leq u_j$ and $u_j < l_{j+1}$. Clearly, the disjuncts are mutually exclusive. However, with the previous encoding it is left to the solver to discover this property. We strengthen the switch encoding by statically adding mutual exclusion constraints of the form $(\neg s_h \vee \neg s_k)$, with $h \neq k$. Adding this information to the encoding is a form of static learning, and it can guide the Boolean search by pruning branches that are unsatisfiable in the theory. The *switch encoding with mutual exclusion* is presented in Equation 3.

$$
\begin{aligned}
\bigwedge_{i=1}^{|C_f|} (\bigwedge_{j=1}^{D_i} ((\neg s_{i,j} \vee ((x_{i,j} - y_{i,j}) \geq l_{i,j})) \wedge \\
(\neg s_{i,j} \vee ((x_{i,j} - y_{i,j}) \leq u_{i,j}))) \wedge \\
(\bigvee_{j=1}^{D_i} s_{i,j}) \wedge (\bigwedge_{j=1}^{D_i} \bigwedge_{k=j+1}^{D_i} (\neg s_{i,j} \vee \neg s_k)))
\end{aligned}
\tag{3}
$$

This encoding is in CNF, but its size is quadratic in the size of the TP— or, more specifically, in $(\max_i D_i)$, i.e. the size of the longest clause.

A different encoding for the TCSP problem class is obtained as follows. For each constraint, we constrain $t \in [l_1, u_D]$, and we exclude the "holes" between intervals, a hole being an open interval (u_j, l_{j+1}). The result is the *hole encoding* reported in Equation 4.

$$
\begin{aligned}
\bigwedge_{i=1}^{|C_f|} (((x_i - y_i) \geq l_{i,1}) \wedge ((x_i - y_i) \leq u_{i,D_i}) \wedge \\
(\bigwedge_{j=1}^{D_i-1} ((x_i - y_i) \leq u_{i,j}) \vee ((x_i - y_i) \geq l_{i,(j+1)})))
\end{aligned}
\tag{4}
$$

This encoding is linear in the size of the original TP, does not introduce any additional variable, and, most importantly, results in a 2-CNF formula. These properties are noteworthy and will be exploited in the following sections.

Finally, we notice that Equation 4 is logically equivalent to Equation 1 (in the applicable case of TCSP), while Equations 2 and 3 are only are equi-satisfiable to it, because of the added switch variables. The solution to the temporal problem is still obtained directly from any satisfying assignment, gathering the values for the variables in $\vec{X}_c$.

6   Encoding of SC Problems in SMT

We now consider the SC problem, in which some time points are not schedulable by the solver, and are considered uncontrollable when looking for a schedule for the controllable time points. We describe the reduction of the SC problem to SMT. We developed a number of encodings that are satisfiable if and only if the temporal problem is strongly controllable, and such that a model of each encoding yields a solution for the original problem.

In the following, we assume that a TPU (X_c, X_u, C_c, C_f) is given.

6.1 Encodings into Quantified LRA

As in the previous section, each time point is associated with an SMT variable. The encoding in Equation 5 is a direct logical mapping of the notion of strong controllability; we call this encoding *direct encoding*.

$$\forall \vec{X}_u.(C_c(\vec{X}_c, \vec{X}_u) \rightarrow C_f(\vec{X}_c, \vec{X}_u)) \tag{5}$$

Equation 5 is satisfiable if and only if there exists an assignment to the controllable variables X_c such that, for all assignments to the uncontrollable variables X_u satisfying the contingent constraints C_c, the free constraints C_f are also satisfied. In the above formula, the controllable variables are implicitly existentially quantified. In case of satisfiability, the SMT solver returns a satisfying assignment to the controllable variables that is exactly a strong schedule.

In order to enable further simplifications, we notice that contingent constraints depend both on controllable and uncontrollable time points, and we re-code the problem as follows. We rewrite each uncontrollable time point e_i in terms of the time difference with its starting time point b_{j_i} by means of an uncontrollable offset variable y_i. For every contingent constraint $cc_i = e_i - b_{j_i} \in [l_i, u_i]$, let $y_i \in \mathbb{R}$ be the uncontrollable offset variable associated to e_i such that: $0 \leq y_i \leq u_i - l_i$ and $e_i = b_{j_i} + u_i - y_i$. Intuitively, y_i represents the offset w.r.t. maximum duration, and can be used to rewrite all the constraints involving e_i in terms of b_{j_i} and y_i only. We formalize this rewriting as a function ρ such that $\rho(e_i) \doteq b_{j_i} + u_i - y_i$. With a small abuse of notation, we denote with $\rho(\vec{X}_u)$ the vector of formulae obtained by the application of ρ to all the elements of X_u. To simplify the notation, we also introduce the vector $\vec{Y}_u$ that is the vector of uncontrollable offset variables $(y_1, ..., y_m)$. Thanks to the redefinition of each e_i in terms of y_i, the rewriting of the contingent constraints depends on $\vec{Y}_u$ only.

Let $\Gamma(\vec{Y}_u)$ be the formula representing the conjunction of all the contingent constraints after the recoding, and $\Psi(\vec{X}_c, \vec{Y}_u)$ be the conjunction of all the free constraints rewritten in terms of $\vec{X}_c$ and $\vec{Y}_u$.

$$\Gamma(\vec{Y}_u) \doteq \bigwedge_{k=1}^{m}(y_k \in [0, (u_k - l_k)]) \qquad \Psi(\vec{X}_c, \vec{Y}_u) \doteq \bigwedge_{c \in C_f} c[\vec{X}_u / \rho(\vec{X}_u)](\vec{X}_c, \vec{Y}_u)$$

In this setting, the SC consists in finding a value for $\vec{X}_c$ that satisfies the free constraints $\Psi(\vec{X}_c, \vec{Y}_u)$ under any possible value of $\vec{Y}_u$ that satisfies $\Gamma(\vec{Y}_u)$.

The SC encoding in Equation 5 can be recoded as an LRA formula in the free variables $\vec{X}_c$ as follows.

$$\forall \vec{Y}_u.(\Gamma(\vec{Y}_u) \rightarrow \Psi(\vec{X}_c, \vec{Y}_u)) \tag{6}$$

We call this encoding *offset encoding*. This formulation corresponds to a quantified SMT problem in LRA, and still requires a solver that supports quantified formulae, but the part of the encoding representing the contingent constraint is now dependent on $\vec{Y}_u$ only.

The main problem in the previous encodings is the scope of the universal quantifier. Since the computational cost of quantification is very high, we can rewrite the offset encoding in Equation 6 in order to obtain a possibly more efficient encoding. Let us assume that $\Psi(\vec{X}_c, \vec{Y}_u)$ is written as a conjunction of h clauses $\psi_h(\vec{X}_{c_h}, \vec{Y}_{u_h})$, where $X_{c_h} \subseteq X_c$ and $Y_{u_h} \subseteq Y_u$ are the variables used in the clause ψ_h. This assumption can be easily satisfied by converting $\Psi(\vec{X}_c, \vec{Y}_u)$ in CNF. We can rewrite $\neg\Gamma(\vec{Y}_u)$ as $\bar{\Gamma}(\vec{Y}_u) \doteq \bigvee_{k=1}^{m}((y_k < 0) \vee (y_k > (u_k - l_k)))$. Let $\bar{\Gamma}(\vec{Y}_u)|_{Y_{u_k}} \doteq \bigvee_{y_k \in Y_{u_k}}((y_k < 0) \vee (y_k > (u_k - l_k)))$.

Assuming the temporal problem is consistent, we have that $\bigwedge_h \forall\vec{Y}_u.(\bar{\Gamma}(\vec{Y}_u) \vee \psi_h(\vec{X}_{c_h}, \vec{Y}_{u_h}))$ if and only if $\bigwedge_h \forall\vec{Y}_{u_h}.(\bar{\Gamma}(\vec{Y}_u)|_{Y_{u_h}} \vee \psi_h(\vec{X}_{c_h}, \vec{Y}_{u_h}))$, and we obtain the *distributed encoding* of Equation 7.

$$\bigwedge_h \forall\vec{Y}_{u_h}.(\bar{\Gamma}(\vec{Y}_u)|_{Y_{u_h}} \vee \psi_h(\vec{X}_{c_h}, \vec{Y}_{u_h})) \tag{7}$$

The size of the produced (quantified) formula is linear with respect to the original TPU. This encoding still requires a solver that supports quantified formulae, and contains as many quantifiers as clauses. However, each quantification is now restricted to the offset variables $Y_{u_h} \subseteq Y_u$ occurring in each clause ψ_h. This encoding also limits the scope of the universal quantifiers, which turns out to be beneficial in practice. Intuitively, this is related to the fact that a number of quantifier eliminations in LRA on smaller formulae may be much cheaper than a single, monolithic quantifier elimination over a large formula.

6.2   Encodings into Quantifier-Free LRA

In order to exploit solvers that do not support quantifiers, we propose an encoding of strong controllability into a quantifier-free SMT(LRA) formula. This is obtained by resorting to an external procedure for quantifier elimination.

We rewrite Equation 7 as $\bigwedge_h \neg(\exists\vec{Y}_{u_h}.(\neg\bar{\Gamma}(\vec{Y}_u)|_{Y_{u_h}} \wedge \neg\psi_h(\vec{X}_{c_h}, \vec{Y}_{u_h})))$, in order to apply a procedure for the elimination of existential quantifiers from a conjunction of literals (e.g. Fourier-Motzkin [14]). Notice that both $\bar{\Gamma}(\vec{Y}_u)|_{Y_{u_h}}$ and $\psi_h(\vec{X}_{c_h}, \vec{Y}_{u_h})$ are clauses, and thus their negations are both conjunctions of literals. The result of each quantifier elimination is again a conjunction of literals, which, once negated, yields a clause, in the following referred to as $\psi_h^{\Gamma}(\vec{X}_{c_h})$. The resulting encoding, reported in Equation 8, is called *eager for-all elimination encoding*.

$$\bigwedge_h \psi_h^{\Gamma}(\vec{X}_{c_h}) \tag{8}$$

For the TCSPU class, it is not necessary to apply a general purpose quantifier elimination procedure. Given the specific nature of the constraints, only few cases are possible, and for each of them we use a pattern-based encoding, that in essence precomputes the result of quantifier elimination. This result can be thought of as generalizing to TCSPU the result proposed in [2] for the case of STPU. We start from the distributed encoding of Equation 7, where each (sub)clause ψ_h is generated by the hole encoding. We treat each clause as a separate existential quantification problem, and provide static results for each case. The final result is logically equivalent to the corresponding $\psi_h^\Gamma(\vec{X}_{c_h})$ in Equation 8.

Each clause under analysis results from the encoding of a free constraint in the TCSPU over variables v and w, with D intervals. Let t be $v - w$. The encoding results in two unit clauses ($t \geq l_1$ and $t \leq u_D$), and in $D - 1$ binary clauses in the form $(t \leq u_i) \vee (t \geq l_{i+1})$.

The static elimination procedure must deal with four possible cases, depending on v and w being controllable or uncontrollable[1]. For the two unit clauses, we proceed as in [2]. Here we show the more complex cases of binary clauses. Let the binary clause have the form $(v - w \leq u) \vee (v - w \geq l)$ (notice that $u < l$ because u is the upper bound of the "lower" interval). When v is uncontrollable, we write x_v for its starting point, y_v for its offset, and L_v and U_v for the lower and upper bound of the contingent constraint relative to v[2]; similarly for w.

1. $v \in X_c$ and $w \in X_c$. The clause does not contain quantified variables, and therefore the quantifier can be simply removed.

2. $v \in X_c$ and $w \in X_u$. The formula $\neg \bar{\Gamma}(\vec{Y}_u)|_{\{y_w\}} \wedge \neg \psi(v, x_w, y_w)$ can be represented by:

$$(0 \leq y_w) \wedge (y_w \leq U_w - L_w) \wedge$$
$$(y_w < l - v + x_w + U_w) \wedge (x_w - v + U_w + u < y_w).$$

Using quantifier elimination over $\exists y_w.(\neg \bar{\Gamma}(\vec{Y}_u)|_{\{y_w\}} \wedge \neg \psi(v, x_w, y_w))$, we obtain the following formula (given that $(l - u > 0)$ and $(U_w - L_w > 0)$):

$$((l - v + x_w + U_w > 0) \wedge (l - v + x_w + L_w \leq 0)) \vee$$
$$((l - v + x_w + L_w \geq 0) \wedge (v - x_w - u - L_w > 0)).$$

Since in eager for-all elimination encoding we need the negation of the existential quantification we can rewrite the formula as follows:

$$((l - v + x_w + U_w \leq 0) \vee (l - v + x_w + L_w > 0)) \wedge$$
$$((l - v + x_w + L_w < 0) \vee (v - x_w - u - L_w \leq 0)).$$

3. The case when $v \in X_u$ and $w \in X_c$ is dual:

$$((x_v + U_v - w - u \leq 0) \vee (x_v - w - u + L_v > 0)) \wedge$$
$$((x_v - l - w + L_v \geq 0) \vee (x_v - w - u + L_v < 0)).$$

4. $v \in X_u$ and $w \in X_u$. The formula $\neg \bar{\Gamma}(\vec{Y}_u)|_{\{y_v, y_w\}} \wedge \neg \psi(x_v, x_w, y_v, y_w)$ is thus $\neg \bar{\Gamma}(\vec{Y}_u)|_{\{y_v, y_w\}} \wedge (v - w < l) \wedge (v - w > u)$ which in turn can be rewritten as

$$(x_v + U_v - x_w - U_w + y_w - l < y_v) \wedge$$
$$(y_v < x_v + U_v - x_w - U_w + y_w - u) \wedge$$
$$(0 \leq y_v) \wedge (y_v \leq U_v - L_v) \wedge (-y_w \leq 0) \wedge (y_w \leq U_w - L_w).$$

---

[1] The possible cases are actually eight but $v - w \geq k$ can be rewritten as $w - v \leq -k$.

[2] We assume $L_v < U_v$; in the other cases the problem is not interesting.

Using the assumptions detailed above and negating the quantification result, we obtain the following formula:

$$\begin{aligned}
&((x_v + U_v - x_w - U_w - u > 0) \vee (x_v + U_v - x_w - u - L_w \le 0)) \wedge \\
&((x_v + U_v - x_w - U_w - u < 0) \vee (x_v - x_w - U_w - u + L_v \ge 0)) \wedge \\
&((x_v - x_w - U_w - l + L_v \ge 0) \vee (x_v - x_w - l + L_v - L_w < 0)) \wedge \\
&((x_v - x_w - l + L_v - L_w > 0) \vee (x_v - x_w - u + L_v - L_j \le 0)).
\end{aligned}$$

The construction described above can be used in Equation 8. This specialized quantification technique results in a 2-CNF formula that has size linear in the original TCSPU. This is because the size of the hole encoding is linear, and for each clause, we statically resolve the quantification by creating at most four new binary clauses. As far as encoding time is concerned, for a TCSPU with m free constraints, the encoding can be generated in $O(m * max(D_i)log(max(D_i)))$ time, because of the sorting time needed in the hole encoding. This encoding spares the computational cost of quantifier elimination and produces a highly optimized QF_LRA formula.

7   Related Work

The seminal work on strong controllability is [2]. The problem is tackled for the limited case of STPU problem class. Vidal and Fargier identify a clever, constant time quantification technique for SC reasoning, which is at the core of their procedure. Compared to [2], we propose a comprehensive solution and an implementation for the cases of TCSPU and DTPU. Furthermore, we generalize to the case of TCSPU the specialized quantifier elimination techniques proposed in [2] for STPU.

Strong controllability for the cases beyond STPU have been tackled in [19], where specialized algorithms based on meta-CSP are proposed. The work in [19] tackles the same problem addressed here; however, it is purely theoretical, and to the best of our knowledge no implementation exists. Furthermore, the approach is based on the use of explicit CSP search to deal with case splits, while we rely on the symbolic expressive power of the SMT framework.

The use of SMT techniques to solve temporal problems is not new. The most advanced work is presented in [20], where the $TSAT++$ tool is presented. TSAT++ can be seen as a specialized SMT solver for DTP problems. The work does not deal with strong controllability, and is limited to consistency for temporal problems. The performance of TSAT++ relative to more modern SMT solvers is analyzed in the next section, on temporal consistency problems.

As far as the consistency problem of STP is concerned, the work in [22] represents the state-of-the-art. Planken, de Weerdt and van der Krogt presented an efficient algorithm for computing all-pairs shortest paths in a directed graph with possibly negative Real weights. As pointed out by the authors, the proposed algorithm can be used to solve STP (but not TCSP or DTP). We used their tool in our experimental comparison for STP consistency. We remark that the focus of our work is on the strong controllability (and not consistency) problem.

We also mention two other forms of controllability for TPUs: weak controllability (WC) and dynamic controllability (DC). A TPU is said to be WC if, for every possible evolution of the uncontrollable environment, there exists an allocation to the controllable time points that fulfills the free constraints of the problem. This notion is much weaker than SC, because the allocation strategy for the controllable time points is allowed to depend on the allocation of the uncontrollable time points. In this setting, the solver is assumed to be "clairvoyant" and is able to decide its moves based on the past and also the future moves of the opponent. In their seminal paper, Vidal and Fargier [2] address the WC problem for the STPU class. Algorithms for deciding WC for TCSPU and DTPU are provided in [23]. The use of SMT techniques to deal with weak controllability has been recently investigated in [24], addressing both the decision and the strategy extraction problems (i.e. the problem of checking if a TPU is WC, and the problem of building a strategy for the solver). The work presented in this paper, compared to [24], tackles a radically different problem. An important difference between SC and WC is the shape of the solution: while in SC a solution is a static assignment to controllable time points, in WC the strategy requires conditional structures to be expressed. Thus, the use of SMT techniques in [24] is also substantially different from what is done here.

DC is similar to WC, but the choices of the scheduler can be based on past environment decisions only. As pointed out in [2], if a problem is SC then it is also DC and if it is DC then it is also WC, but the implication chain is not reversible. In [25] the authors focus on deciding DC for the STPU problem class, while in [23] the result is extended for TCSPUs. However, no effective solutions to DC exists for the DTPU problem class.

8   Experimental Evaluation

8.1   Implementation

We developed a tool that automatically encodes the various classes of temporal problems in SMT problems. The tool can deal with consistency problems by generating SMT (QF_LRA) encodings. As for strong controllability problems, the tool implements the two reductions to SMT (LRA) (with quantifiers), and can obtain SMT (QF_LRA) by applying quantifier elimination techniques. The quantifier elimination step in the eager for-all elimination encoding is carried out by calling the formula simplifier provide by Z3 [6], and a quantifier elimination functionality built on top of MathSAT5 [5].

The tool is currently connected to three different SMT solvers: namely *MathSAT4* [4], *MathSAT5* [5] and *Z3* [6]. Given that the encodings are written in SMT-LIB2 (and also in SMT-LIB1 format), it would be straightforward to connected it to any SMT solver that is able to parse the SMT-LIB language. We remark however that our purpose is to compare the performance of the encodings we propose, and not to compare the various SMT solvers. Z3 can be seen as a representative for solvers that support quantified theories, and MathSAT as representative for quantifier-free solvers. We expect other solvers (e.g. Yices [7], OpenSMT [8]) to exhibit a similar behavior (see [26]).

8.2   Experimental Set-Up

In order to experimentally assess the performance of the techniques presented in this paper, we used a set of randomly-generated benchmarks. Consistency problems were generated using the random instance generator presented in [20]; the same generator was extended to introduce random uncertainty, and to generate strong controllability problems. The benchmark set contains 2108 instances for each problem class in TP without uncertainty (STP, TCSP and DTP), and 1054 instances for each TPU class (STPU, TCSPU and DTPU). We used random instance generators because they are typically used in literature (e.g. [20]), and because they can be easily scaled to stress the solvers.

We executed all our experiments on a machine running Scientific Linux 6.0, equipped with two quad-core Xeon processors @ 2.70GHz. We considered a memory limit of 2GB and a time-out of 300 seconds. The benchmarks and the results are available from `https://es.fbk.eu/people/roveri/tests/cp2012`.

For consistency problems, we analyzed the performance of the various solvers on the various encodings. We also compared our encodings with all the available solver for TP without uncertainty (i.e. Snowball for the case of STP, and TSAT++).

For strong controllability problems, to the best of our knowledge there are no solvers available. Thus, we only evaluated the different approaches, to highlight the difference in performance and the respective merits.

8.3   Results for Consistency

The results for consistency problems are reported in Figure 1 (left). We plotted in logarithmic scale the cumulative time in seconds to solve the considered set of benchmarks. For STP problems, we compared the naïve encoding with various algorithms available in the *SnowBall3* [22] tool, and with TSAT++ [20]. (In the case of STP, the other encodings coincide with the naïve encoding.) In TCSP and DTP, we tested all the applicable encodings with all the SMT solvers under analysis and with TSAT++. The plots show that the SMT approach is competitive with dedicated techniques. MathSAT4 implements a dedicated algorithm for the theory of difference logic [27], and is thus faster than MathSAT5, that uses a general purpose algorithm for LRA [28]. Both solvers greatly benefit from the hole encoding, compared to the switch encoding with mutual exclusion (switch me) and the plain switch encoding. This encoding produces a formula that has just one real variable for every time point and has at most two literals per clause: this greatly simplifies the SMT search procedure by augmenting the number of unit propagations and by reducing the size of the search space.

Z3 is extremely efficient, and the attempts to improve the encodings may result in performance degradation. The TSAT++ solver is outperformed by state-of-the-art SMT solvers, but again notice that the hole encoding yields substantial improvements in performance.

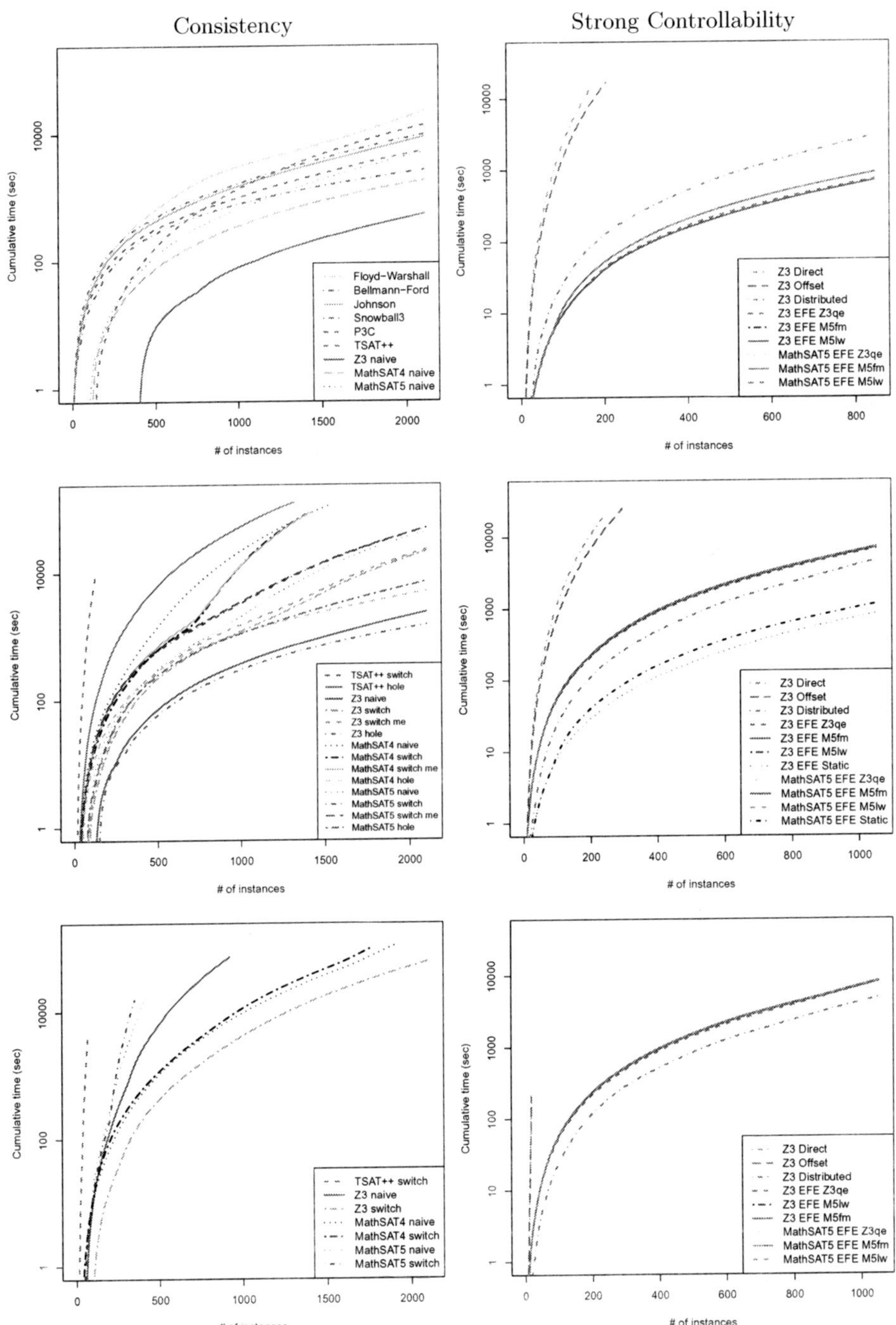

Fig. 1. Results for experimental evaluation: consistency of STP (left-top), TCSP (left-center), and DTP (left-bottom); strong controllability of STPU (right-top), TCSPU (right-center) and DTPU (right-bottom)

8.4   Results for Strong Controllability

The results for strong controllability are reported in Figure 1 (right). We plotted in logarithmic scale the cumulative time in seconds to solve the considered set of benchmarks. Differently from the consistency case, the time here considers also the encoding time which dominates solving time for the case of quantifier-free encodings. The quantified encodings (Direct, Offset and Distributed) are solved with Z3. The quantifier-free encodings resulting from eager for-all elimination are obtained by the application of three quantifier elimination procedures: the internal simplifier of Z3 (EFE Z3qe), and two implementations in MathSAT5 of the Fourier-Motzkin (EFE M5fm) and the Loos-Weispfenning (EFE M5lw) elimination procedures. The resulting encodings are solved using Z3 and MathSAT5 on the quantifier-free theory of Reals.

The plots clearly show that both the offset and direct encodings quickly reach the resource limits, and are unable to solve all the instances. The behavior of the distributed encoding is slightly better than the eager for-all elimination approach. The difference can be explained in purely technological terms: the quantifier elimination modules are called via pipe in our implementation; on the other hand, when Z3 solves the distributed encoding, it can perform quantifier elimination "in-memory".

We notice the impact of the static quantification techniques (EFE Static), when applicable (i.e. for TCSPU). The substantial difference in performance resides only in the quantification technique, that all produce the same quantifier-free formula.

9   Conclusions

In this paper, we presented a comprehensive approach to strong controllability for temporal problems with uncertainty. The formalization is based on the SMT framework, and encompasses constraints with arbitrary disjunctions. We deal with uncertainty by means of logic-based quantifier elimination techniques. The experiments demonstrate the scalability of the approach, based on the use of efficient SMT solvers.

In the future, we will investigate the problem of searching schedules that optimize a given cost function, and the addition of constraints over resources associated to activities. Finally, within the SMT-based framework, we will investigate the case of dynamic controllability.

References

1. Dechter, R., Meiri, I., Pearl, J.: Temporal constraint networks. Artif. Intell. 49, 61–95 (1991)
2. Vidal, T., Fargier, H.: Handling contingency in temporal constraint networks: from consistency to controllabilities. Journal of Experimental Theoretical Artificial Intelligence 11, 23–45 (1999)

3. Barrett, C.W., Sebastiani, R., Seshia, S.A., Tinelli, C.: Satisfiability modulo theories. In: Handbook of Satisfiability, pp. 825–885. IOS Press (2009)

4. Bruttomesso, R., Cimatti, A., Franzén, A., Griggio, A., Sebastiani, R.: The MATH-SAT 4 SMT Solver. In: Gupta, A., Malik, S. (eds.) CAV 2008. LNCS, vol. 5123, pp. 299–303. Springer, Heidelberg (2008)

5. Cimatti, A., Griggio, A., Sebastiani, R., Schaafsma, B.: The MathSAT5 SMT solver (2011), http://mathsat.fbk.eu

6. de Moura, L., Bjørner, N.: Z3: An Efficient SMT Solver. In: Ramakrishnan, C.R., Rehof, J. (eds.) TACAS 2008. LNCS, vol. 4963, pp. 337–340. Springer, Heidelberg (2008)

7. Dutertre, B., de Moura, L.: The Yices SMT solver (2006), Tool paper at http://yices.csl.sri.com/tool-paper.pdf

8. Bruttomesso, R., Pek, E., Sharygina, N., Tsitovich, A.: The OpenSMT Solver. In: Esparza, J., Majumdar, R. (eds.) TACAS 2010. LNCS, vol. 6015, pp. 150–153. Springer, Heidelberg (2010)

9. Niemelä, I.: Integrating Answer Set Programming and Satisfiability Modulo Theories. In: Erdem, E., Lin, F., Schaub, T. (eds.) LPNMR 2009. LNCS, vol. 5753, p. 3. Springer, Heidelberg (2009)

10. Franzén, A., Cimatti, A., Nadel, A., Sebastiani, R., Shalev, J.: Applying SMT in symbolic execution of microcode. In: Bloem, R., Sharygina, N. (eds.) Formal Methods in Computer-Aided Design - FMCAD, pp. 121–128. IEEE (2010)

11. Godefroid, P., Levin, M.Y., Molnar, D.A.: Automated whitebox fuzz testing. In: Network and Distributed System Security Symposium - NDSS. The Internet Society (2008)

12. Kleene, S.: Mathematical Logic. J. Wiley & Sons (1967)

13. Moskewicz, M.W., Madigan, C.F., Zhao, Y., Zhang, L., Malik, S.: Chaff: Engineering an Efficient SAT Solver. In: Design Automation Conference - DAC, pp. 530–535. ACM Press, New York (2001)

14. Schrijver, A.: Theory of Linear and Integer Programming. J. Wiley & Sons (1998)

15. Loos, R., Weispfenning, V.: Applying linear quantifier elimination. Computer Journal 36, 450–462 (1993)

16. Monniaux, D.: A Quantifier Elimination Algorithm for Linear Real Arithmetic. In: Cervesato, I., Veith, H., Voronkov, A. (eds.) LPAR 2008. LNCS (LNAI), vol. 5330, pp. 243–257. Springer, Heidelberg (2008)

17. Allen, J.F.: Maintaining knowledge about temporal intervals. Communication of the ACM 26, 832–843 (1983)

18. Tsamardinos, I., Pollack, M.E.: Efficient solution techniques for disjunctive temporal reasoning problems. Artificial Intelligence 151, 43–89 (2003)

19. Peintner, B., Venable, K.B., Yorke-Smith, N.: Strong Controllability of Disjunctive Temporal Problems with Uncertainty. In: Bessière, C. (ed.) CP 2007. LNCS, vol. 4741, pp. 856–863. Springer, Heidelberg (2007)

20. Armando, A., Castellini, C., Giunchiglia, E.: SAT-Based Procedures for Temporal Reasoning. In: Biundo, S., Fox, M. (eds.) ECP 1999. LNCS, vol. 1809, pp. 97–108. Springer, Heidelberg (2000)

21. de la Tour, T.B.: Minimizing the Number of Clauses by Renaming. In: Stickel, M.E. (ed.) CADE 1990. LNCS, vol. 449, pp. 558–572. Springer, Heidelberg (1990)

22. Planken, L., de Weerdt, M., van der Krogt, R.: Computing all-pairs shortest paths by leveraging low treewidth. Journal of Artificial Intelligence Research (JAIR) 43, 353–388 (2012)

23. Venable, K.B., Volpato, M., Peintner, B., Yorke-Smith, N.: Weak and dynamic controllability of temporal problems with disjunctions and uncertainty. In: Workshop on Constraint Satisfaction Techniques for Planning & Scheduling, pp. 50–59 (2010)
24. Cimatti, A., Micheli, A., Roveri, M.: Solving Temporal Problems using SMT: Weak Controllability. In: Hoffmann, J., Selman, B. (eds.) American Association for Artificial Intelligence - AAAI. AAAI Press (2012)
25. Morris, P.H., Muscettola, N., Vidal, T.: Dynamic control of plans with temporal uncertainty. In: Nebel, B. (ed.) International Joint Conference on Artificial Intelligence - IJCAI, pp. 494–502. Morgan Kaufmann (2001)
26. Barrett, C., Deters, M., de Moura, L., Oliveras, A., Stump, A.: 6 Years of SMT-COMP. Journal of Automated Reasoning (to appear, 2012)
27. Cotton, S., Maler, O.: Fast and Flexible Difference Constraint Propagation for DPLL(T). In: Biere, A., Gomes, C.P. (eds.) SAT 2006. LNCS, vol. 4121, pp. 170–183. Springer, Heidelberg (2006)
28. Dutertre, B., de Moura, L.: A Fast Linear-Arithmetic Solver for DPLL(T). In: Ball, T., Jones, R.B. (eds.) CAV 2006. LNCS, vol. 4144, pp. 81–94. Springer, Heidelberg (2006)

A Characterisation of the Complexity
of Forbidding Subproblems in Binary Max-CSP[*]

Martin C. Cooper[1], Guillaume Escamocher[1], and Stanislav Živný[2]

[1] IRIT, University of Toulouse III, 31062 Toulouse, France
{cooper,escamocher}@irit.fr
[2] Department of Computer Science, University of Oxford, OX1 3QD Oxford, UK
standa.zivny@cs.ox.ac.uk

Abstract. Tractable classes of binary CSP and binary Max-CSP have recently been discovered by studying classes of instances defined by excluding subproblems. In this paper we characterise the complexity of all classes of binary Max-CSP instances defined by forbidding a single subproblem. In the resulting dichotomy, the only non-trivial tractable class defined by a forbidden subproblem corresponds to the set of instances satisfying the so-called joint-winner property.

1   Introduction

Max-CSP is a generic combinatorial optimization problem which consists in finding an assignment to the variables which satisfies the maximum number of a set of constraints. Max-CSP is NP-hard, but much research effort has been devoted to the identification of classes of instances that can be solved in polynomial time.

One classic approach consists in identifying tractable constraint languages, i.e. restrictions on the constraint relations which imply tractability. For example, if all constraints are supermodular, then Max-CSP is solvable in polynomial time, since the maximization of a supermodular function (or equivalently the minimization of a submodular function) is a well-known tractable problem in Operations Research [16]. Over two-element domains [7], three-element domains [12], and fixed-valued languages [10], a dichotomy has been given: supermodularity is the only basic reason for tractability. However, over four-element domains it has been shown that other tractable constraint languages exist [13]. Another classic approach consists in identifying structural reasons for tractability, i.e. restrictions on the graph of constraint scopes (known as the constraint graph) which imply the existence of a polynomial-time algorithm. In the case of binary CSP the only class of constraint graphs which ensures tractability (subject to certain complexity theory assumptions) are essentially graphs of bounded treewidth [8,11]. It is well known that structural reasons for tractability generalise to optimisation versions of the CSP [1,9].

---

[*] Martin Cooper is supported by Projects ANR-10-BLAN-0210 and 0214. Stanislav Živný is supported by a Junior Research Fellowship at University College, Oxford.

M. Milano (Ed.): CP 2012, LNCS 7514, pp. 265–273, 2012.
© Springer-Verlag Berlin Heidelberg 2012

Recently, a new avenue of research has led to the discovery of tractable classes of CSP or Max-CSP instances defined by forbidding a specific (set of) subproblem(s). Novel tractable classes have been discovered by forbidding simple 3-variable subproblems [3,6]. In the present paper we consider all classes of binary Max-CSP instances defined by forbidding occurrences of a single subproblem. The dichotomy that we give can be seen as an important first step towards a complete characterisation of the complexity of classes of binary Max-CSP instances defined by forbidding *sets* of subproblems.

To relate this to similar work on the characterisation of the complexity of forbidden patterns [2], we should point out that a pattern can represent a set of subproblems by leaving the compatibility of some pairs of assignments undefined. Another difference between subproblems and patterns is that in a subproblem all variable-value assignments are assumed distinct, whereas in a pattern two assignments may represent the same assignment in an instance [2].

The complexity of classes of binary Max-CSP instances defined by local properties of (in)compatibilities have previously been characterised, but only for properties on exactly 3 assignments to 3 distinct variables [4]. In the present paper we consider classes defined by forbidding subproblems of any size and possibly involving several assignments to the same variable, thus allowing more refinement in the definition of classes of Max-CSP instances.

2 Definitions and Basic Results

A *subproblem* P is simply a binary Max-CSP instance: variables are distinct, each variable has its own domain composed of distinct values, and a cost of 0 or 1 is associated with each pair of assignments to two distinct variables. In a Max-CSP instance defined in this way, the goal of maximising the number of satisfied constraints is clearly equivalent to minimising the total cost. We consider that in any subproblem or instance a constraint is given for each pair of distinct variables (even if the constraint corresponds to a constant-0 cost function). We only consider subproblems with all binary constraints but no unary constraints. As we will show later, our results are independent of the presence of unary constraints. It will sometimes be more convenient to consider an instance as a set of variable-value assignments together with a function $cost$, such that $cost(p, q) \in \{0, 1\}$ denotes the cost of simultaneously making the pair of assignments p, q, together with a function var such that $var(p)$ indicates the variable associated with assignment p.

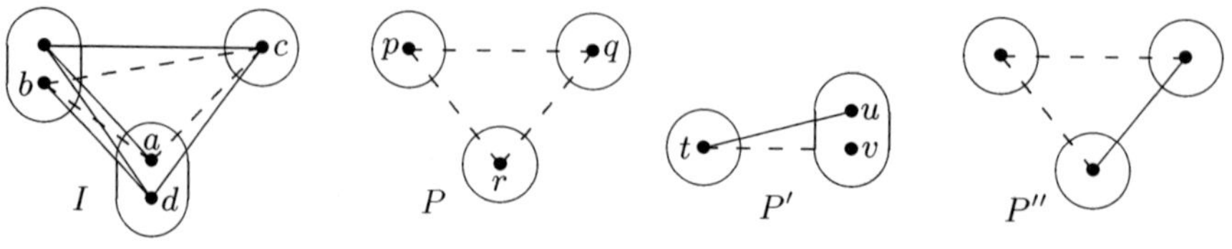

Fig. 1. The instance I contains P and P' as subproblems but not P''

A subproblem P *occurs* in a binary Max-CSP instance (or, equivalently, another subproblem) I if P is isomorphic to some sub-instance of I obtained by taking a subset U of the variables of I and subsets of each of the domains of the variables in U. We also say that I *contains* P as a subproblem. To illustrate this notion, consider the instance I and the three subproblems P, P', P'' shown in Fig. 1. A bullet point represents a variable-value assignment, assignments to the same variable are grouped together in the same oval, a dashed line between points a and b means $cost(a, b) = 1$ and a solid line means $cost(a, b) = 0$. In this example, subproblem P occurs in I with the corresponding isomorphism $p \mapsto a$, $q \mapsto b$, $r \mapsto c$. Similarly, P' occurs in I with the corresponding isomorphism $t \mapsto c$, $u \mapsto d$, $v \mapsto a$. On the other hand, P'' does not occur in I.

In this paper we denote by $\mathcal{F}(P)$ the set of Max-CSP instances in which the subproblem P is forbidden, i.e. does not occur. Thus if I, P' and P'' are as shown in Fig. 1, $I \in \mathcal{F}(P'')$ but $I \notin \mathcal{F}(P')$. If $\Sigma = \{P_1, \ldots, P_s\}$ is a set of subproblems, then we use $\mathcal{F}(\Sigma)$ or $\mathcal{F}(P_1, \ldots, P_s)$ to denote the set of Max-CSP instances in which no subproblem $P_i \in \Sigma$ occurs. The following lemma follows from the above definitions, by transitivity of the occurrence relation.

Lemma 1. *If $\forall P \in \Sigma_1$, $\exists Q \in \Sigma_2$ such that Q occurs in P, then $\mathcal{F}(\Sigma_2) \subseteq \mathcal{F}(\Sigma_1)$.*

We say that $\mathcal{F}(\Sigma)$ is *tractable* if there is a polynomial-time algorithm to solve it. We say that $\mathcal{F}(\Sigma)$ is *intractable* if it is NP-hard. We assume throughout this paper that $P \neq NP$. Suppose that $\mathcal{F}(\Sigma_1) \subseteq \mathcal{F}(\Sigma_2)$. Clearly, $\mathcal{F}(\Sigma_1)$ is tractable if $\mathcal{F}(\Sigma_2)$ is tractable and $\mathcal{F}(\Sigma_2)$ is intractable if $\mathcal{F}(\Sigma_1)$ is intractable. Our aim is to characterise the tractability of $\mathcal{F}(P)$ for all subproblems P. We first show that we only need to consider subproblems with domains of size at most 2.

Lemma 2. *Let P be a subproblem with three or more values in the domain of some variable and let $\mathcal{F}(P)$ be the set of Max-CSP instances in which the subproblem P is forbidden. Then $\mathcal{F}(P)$ is intractable.*

Proof. Max-Cut is intractable and can be reduced to Max-CSP on Boolean domains [7]. Thus $\mathcal{F}(P)$ is intractable since it includes all instances of Max-CSP on Boolean domains. $\qquad\square$

3   Subproblems on Two Variables

We now consider the subproblems on just two variables shown in Fig. 2. Modulo independent permutations of the variables and of the two domains, these are the only possible subproblems with domains of size at most 2.

Lemma 3. *If Q_1 is the subproblem shown in Fig. 2, then $\mathcal{F}(Q_1)$ is intractable.*

Proof. Let I be an instance in $\mathcal{F}(Q_1)$. It is easy to see that all binary cost functions between any pair of variables in I must be constant. Hence I is equivalent to a trivial Max-CSP instance with no binary cost functions. $\qquad\square$

Lemma 4. *If Q_0 and U are as shown in Fig. 2, then $\mathcal{F}(\{Q_0, U\})$ is intractable.*

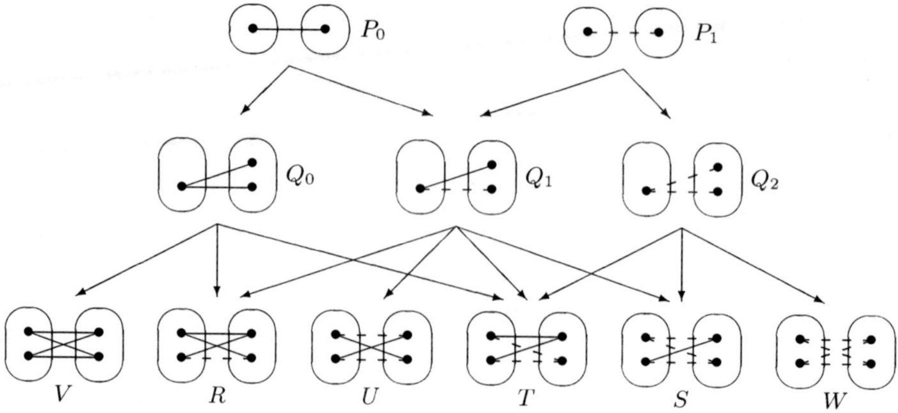

Fig. 2. Subproblems on two variables (showing inclusions between subproblems)

Proof. Max-Cut can be coded as Max-CSP over Boolean domains in which all constraints are of the form $X_i \neq X_j$. We can replace each constraint $X_i \neq X_j$ by an equivalent gadget G with two extra variables Y_{ij}, Z_{ij}, where G is given by $\neg X_i \wedge Y_{ij}$, $\neg Y_{ij} \wedge \neg X_j$, $X_i \wedge \neg Z_{ij}$, $Z_{ij} \wedge X_j$. It is easily verified that placing the gadget G on variables X_i, X_j is equivalent to imposing the constraint $X_i \neq X_j$; when $X_i = X_j$ at most one of these constraints can be satisfied and when $X_i \neq X_j$ at most two constraints can be satisfied.

For each pair of variables X, X' in the resulting instance of Max-CSP such that there is no constraint between X and X', we place a binary constraint on X, X' of constant cost 1. In the resulting Max-CSP instance, there are no two zero costs in the same binary cost function. Thus, this polynomial reduction from Max-Cut produces an instance in $\mathcal{F}(\{Q_0, U\})$. Intractability of $\mathcal{F}(\{Q_0, U\})$ then follows from the NP-hardness of Max-Cut. $\qquad\square$

Lemma 5. *If Q_2 and U are as shown in Fig. 2, then $\mathcal{F}(\{Q_2, U\})$ is intractable.*

Proof. As in the proof of Lemma 4, the proof is again by a polynomial reduction from Max-Cut. This time each constraint $X_i \neq X_j$ is replaced by the gadget G' where G' is $\neg X_i \vee Y_{ij}$, $\neg Y_{ij} \vee \neg X_j$, $X_i \vee \neg Z_{ij}$, $Z_{ij} \vee X_j$. When $X_i = X_j$ at most three of these constraints can be satisfied and when $X_i \neq X_j$ all four constraints can be satisfied.

For each pair of variables X, X' in the resulting instance of Max-CSP such that there is no constraint between X and X', we place a binary constraint on X, X' of constant cost 0. The resulting instance is in $\mathcal{F}(\{Q_2, U\})$. $\qquad\square$

This provides us with a dichotomy for subproblems on just two variables.

Theorem 1. *If P is a 2-variable binary Max-CSP subproblem, then $\mathcal{F}(P)$ is tractable if and only if P occurs in Q_1 (shown in Fig. 2).*

Proof. By Lemma 2, we only need to consider subproblems in which each domain is of size at most two.

Since each of P_0 and P_1 occur in Q_1, it follows from Lemma 3 and Lemma 1 that $\mathcal{F}(P_0)$ and $\mathcal{F}(P_1)$ are also tractable. Since Q_0 occurs in R, T, V and Q_2 occurs in S, W, it follows from Lemmas 4, 5 and Lemma 1 that $\mathcal{F}(Q_0)$, $\mathcal{F}(Q_2)$, $\mathcal{F}(R)$, $\mathcal{F}(S)$, $\mathcal{F}(T)$, $\mathcal{F}(U)$, $\mathcal{F}(V)$, $\mathcal{F}(W)$ are all intractable. This covers all the possible subproblems with domains of size at most 2 as shown in Fig. 2.   □

4   Subproblems on Three Variables

We recall the following result which follows directly from Theorem 5 of [4].

Lemma 6. *A class of binary Max-CSP instances defined by forbidding a single subproblem comprised of a triangle of three assignments to three distinct variables is tractable if and only if the three binary costs are 0,1,1.*

Binary Max-CSP instances in which the triple of binary costs 0,1,1 does not occur in any triangle satisfy the so-called joint-winner property [6]. This class has recently been generalised to the hierarchically-nested convex class which is a tractable class of Valued CSP instances involving cost functions of arbitrary arity [5]. The following corollary is just a translation of Lemma 6 into the notation of forbidden subproblems.

Corollary 1. *Let A, B, C, D be the subproblems shown in Fig. 3. Then $\mathcal{F}(C)$ is tractable, but $\mathcal{F}(A)$, $\mathcal{F}(B)$, and $\mathcal{F}(D)$ are intractable.*

Lemma 7. *Given the subproblem E shown in Fig. 3 and the set $\mathcal{F}(E)$ of Max-CSP instances in which the subproblem E is forbidden, then $\mathcal{F}(E)$ is intractable.*

Proof. The constraint graph of a Max-CSP instance is the graph $\langle V, E \rangle$ where V is the set of variables and $\{X_i, X_j\} \in E$ if there is a pair of assignments p, q with $var(p) = X_i$, $var(q) = X_j$ and such that $cost(p, q) = 1$. Clearly the constraint graph of any instance in which E occurs contains a triangle. Max-Cut is NP-hard even on triangle-free graphs [15]. Any such instance of Max-Cut coded as an instance I of binary Max-CSP does not contain E as a subproblem since the constraint graph of I is triangle-free. Hence $\mathcal{F}(E)$ is intractable.   □

Lemma 8. *The only 3-variable subproblem P for which the set $\mathcal{F}(P)$ is tractable is the subproblem C shown in Fig. 3.*

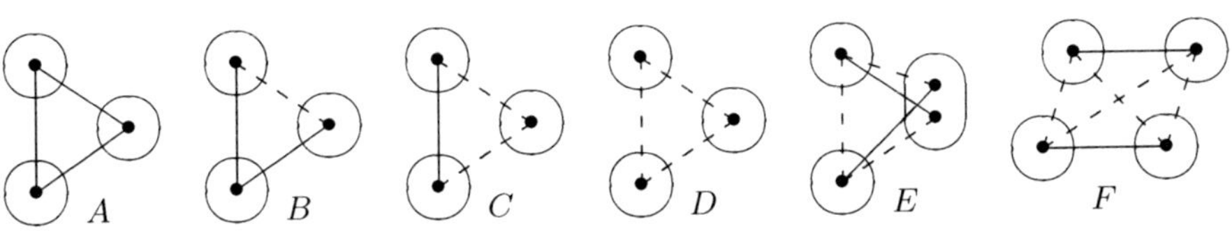

Fig. 3. Subproblems on three or four variables

Proof. Let P be a 3-variable subproblem. For $\mathcal{F}(P)$ to be tractable, P must not have as a subproblem any of Q_0, Q_2, A, B, D, E which have all been shown to define intractable classes (Lemmas 4, 5, 7 and Corollary 1). The only 3-variable subproblem which does not contain any of Q_0, Q_2, A, B, D, E is C. The result then follows from Lemma 1. $\qquad\square$

5 Subproblems on More than Three Variables

It turns out that the tractable classes we have already identified, defined by forbidden subproblems on two or three variables, are the only possible tractable classes defined by forbidding a single subproblem. To complete our dichotomy, we require one final lemma.

Lemma 9. *If F is the subproblem shown in Fig. 3, then $\mathcal{F}(F)$ is intractable.*

Proof. It is known that Max-Cut on C_4-free graphs is NP-hard [14]. To see this, let G be a graph and G' a version of G in which each edge is replaced by a path composed of three edges. Clearly, G' is C_4-free and the maximum cut of G' is of the same size as the maximum cut of G.

When a Max-Cut instance on a C_4-free graph is coded as a Max-CSP instance I, the subproblem F cannot occur since there can be no length-4 cycles of non-trivial constraints in I. Hence $\mathcal{F}(F)$ is intractable. $\qquad\square$

By looking at all possible combinations of edges in a subproblem, it is possible to show that F is the only subproblem on four variables in which neither A, B, D nor E shown in Fig. 3 occur. Since $\mathcal{F}(A)$, $\mathcal{F}(B)$, $\mathcal{F}(D)$ and $\mathcal{F}(E)$ are intractable, then from Lemma 9 the classes of Binary Max-CSP instances defined by forbidding a single subproblem on four or more variables are all intractable and we can now state our dichotomy.

Theorem 2. *If P is a binary Max-CSP subproblem, then $\mathcal{F}(P)$ is tractable if and only if P occurs either in Q_1 (shown in Fig. 2) or in C (shown in Fig. 3).*

It follows that $\mathcal{F}(P)$ is tractable only for $P = P_0$, P_1, Q_1 (shown in Fig. 2) or C (shown in Fig. 3). It follows that the only non-trivial tractable class defined by a forbidden subproblem corresponds to the set of instances satisfying the so-called joint-winner property. The joint-winner property encompasses, among other things, codings of non-intersecting graph-based or variable-based SoftAllD-iff constraints together with arbitrary unary constraints [6]. It is worth pointing out that Theorem 2 is independent of the presence of unary cost functions, in the sense that tractable classes remain tractable when arbitrary unary costs are allowed and NP-hardness results are valid even if no unary costs are allowed.

6 Forbidding Sets of Subproblems

Certain known tractable classes can be defined by forbidding more than one subproblem. For example, in [4] it was shown that $\mathcal{F}(\{A, B\})$, $\mathcal{F}(\{B, D\})$ and

$\mathcal{F}(\{A, D\})$ are all tractable (where A, B, C, D are the subproblems given in Fig. 3). The most interesting of these three tractable classes is $\mathcal{F}(\{A, B\})$ which is equivalent to maximum matching in graphs.

In this section we give a necessary condition for a forbidden set of subproblems to define a tractable class of binary Max-CSP instances.

Definition 1. *A subproblem (or an instance) P is* Boolean *if the size of the domain of each variable in P is at most two.*

A negative edge pair *is a set of variable-value assignments p, q, r, s such that $var(p) = var(r) \neq var(q) = var(s)$, $cost(p, q) = cost(r, s) = 1$ and $p \neq r$. A* positive edge pair *is a set of variable-value assignments p, q, r, s such that $var(p) = var(r) \neq var(q) = var(s)$, $cost(p, q) = cost(r, s) = 0$ and $p \neq r$.*

A negative cycle *is a set of variable-value assignments $p_1, \ldots, p_m$, with $m > 2$, such that the variables $var(p_i)$ $(i = 1, \ldots, m)$ are all distinct, $cost(p_i, p_{i+1}) = 1$ $(i = 1, \ldots, m)$ and $cost(p_m, p_1) = 1$. A* positive cycle *is a set of assignments $p_1, \ldots, p_m$ $(m > 2)$, such that the variables $var(p_i)$ $(i = 1, \ldots, m)$ are all distinct, $cost(p_i, p_{i+1}) = 0$ $(i = 1, \ldots, m)$ and $cost(p_m, p_1) = 0$.*

A negative pivot point *is a variable-value assignment p such that there are two assignments q, r with $var(p)$, $var(q)$, $var(s)$ all distinct and $cost(p, q) = cost(p, r) = 1$. A* positive pivot point *is an assignment p such that there are two assignments q, r with $var(p)$, $var(q)$, $var(s)$ all distinct and $cost(p, q) = cost(p, r) = 0$.*

Proposition 1. *If Σ is a finite set of subproblems, then $\mathcal{F}(\Sigma)$ is tractable only if*

1. *There is a Boolean subproblem $P \in \Sigma$ such that P contains no negative edge pair, no negative cycle and at most one negative pivot point, and*
2. *There is a Boolean subproblem $Q \in \Sigma$ such that Q contains no positive edge pair, no positive cycle and at most one positive pivot point, and*
3. *There is a Boolean subproblem $B \in \Sigma$ such that B contains neither Q_0 nor Q_2 (as shown in Fig. 2).*

Proof. Suppose that condition (1) is not satisfied. We will show that $\mathcal{F}(\Sigma)$ is NP-hard. Let t be an odd integer strictly greater than the number of variables in any subproblem in Σ. As in Lemma 5 the proof is by a polynomial reduction from Max-Cut. This time each constraint $X_i \neq X_j$ is replaced by the gadget G_t where G_t is $\neg X_i \vee Y_1$, $\neg Y_k \vee Y_{k+1}$ $(k = 1, \ldots, t - 1)$, $\neg Y_t \vee \neg X_j$, and $X_i \vee \neg Z_1$, $Z_k \vee \neg Z_{k+1}$ $(k = 1, \ldots, t - 1)$, $Z_t \vee X_j$. The gadget G_t is equivalent to $X_i \neq X_j$ since when $X_i = X_j$ one of its constraints must be violated, but when $X_i \neq X_j$ all of its constraints can be satisfied. For each pair of variables X, X' in the resulting instance of Max-CSP such that there is no constraint between X and X', we place a binary constraint on X, X' of constant cost 0.

The resulting instance I has no domain of size greater than two, and contains no negative edge pair, no negative cycle of length at most t and no two negative pivot points at a distance at most t. Let $P \in \Sigma$. Since (1) is not satisfied, and by definition of t, either P has a domain of size more than two, or contains a negative edge pair, a negative cycle of length at most t or two negative pivot points at a distance at most t. It follows that P cannot occur in I. Thus, we have demonstrated a polynomial reduction from Max-Cut to $\mathcal{F}(\Sigma)$.

The proof for condition (2) is similar. This time each constraint $X_i \neq X_j$ is replaced by the gadget G'_t given by $\neg X_i \wedge Y_1$, $\neg Y_k \wedge Y_{k+1}$ ($k = 1, \ldots, t-1$), $\neg Y_t \wedge \neg X_j$, and $X_i \wedge \neg Z_1$, $Z_k \wedge \neg Z_{k+1}$ ($k = 1, \ldots, t$), $Z_t \wedge X_j$. The gadget G'_t is equivalent to the constraint $X_i \neq X_j$; when $X_i = X_j$ at most t of its constraints can be satisfied and when $X_i \neq X_j$ at most $t+1$ of its constraints can be satisfied. For each pair of variables X, X' in the resulting instance of Max-CSP such that there is no constraint between X and X', we place a binary constraint on X, X' of constant cost 1.

The resulting instance I has no domain of size greater than two, and contains no positive edge pair, no positive cycle of length at most t and no two positive pivot points at a distance at most t. Let $P \in \Sigma$. If condition (2) is not satisfied, no $P \in \Sigma$ can occur in I. Thus, this polynomial reduction is from Max-Cut to $\mathcal{F}(\Sigma)$.

If condition (3) is not satisfied, then, by Lemma 1, $\mathcal{F}(\Sigma)$ contains all Boolean instances in $\mathcal{F}(Q_0, Q_2)$. But $\mathcal{F}(Q_0, Q_2)$ is equivalent to the set of Boolean instances of Max-CSP in which for each pair of variables X_i, X_j there is a constraint between X_i and X_j with this constraint being either $X_i = X_j$ or $X_i \neq X_j$. This set of Max-CSP instances is equivalent to the 2-Cluster Editing problem whose decision version is known to be NP-complete [17][1]. Hence $\mathcal{F}(\Sigma)$ is NP-hard if condition (3) is not satisfied.   □

7   Conclusion

We have given a dichotomy concerning the tractability of classes of binary Max-CSP instances defined by forbidding a single subproblem. We have also given a necessary condition for the tractability of classes defined by forbidding sets of subproblems.

Classes defined by forbidding (sets of) subproblems are closed under permutations of the set of variables and independent permutations of each variable domain. An interesting avenue of future research is to place structure, such as an ordering, on the set of variables or on domain elements within the forbidden subproblems with the aim to uncover novel tractable classes.

References

1. Bertelé, U., Brioshi, F.: Nonserial dynamic programming. Academic Press (1972)
2. Cooper, M.C., Escamocher, G.: A Dichotomy for 2-Constraint Forbidden CSP Patterns. In: AAAI 2012 (2012)
3. Cooper, M.C., Jeavons, P.G., Salamon, A.Z.: Generalizing constraint satisfaction on trees: hybrid tractability and variable elimination. Artificial Intelligence 174(9-10), 570–584 (2010)

---

[1] We are grateful to Peter Jeavons for pointing out this equivalence.

4. Cooper, M.C., Živný, S.: Tractable Triangles. In: Lee, J. (ed.) CP 2011. LNCS, vol. 6876, pp. 195–209. Springer, Heidelberg (2011)
5. Cooper, M.C., Živný, S.: Hierarchically Nested Convex VCSP. In: Lee, J. (ed.) CP 2011. LNCS, vol. 6876, pp. 187–194. Springer, Heidelberg (2011)
6. Cooper, M.C., Živný, S.: Hybrid tractability of valued constraint problems. Artificial Intelligence 175(9-10), 1555–1569 (2011)
7. Creignou, N., Khanna, S., Sudan, M.: Complexity classification of Boolean constraint satisfaction problems. SIAM Monographs on Discrete Mathematics and Applications, vol. 7 (2001)
8. Dalmau, V., Kolaitis, P.G., Vardi, M.Y.: Constraint Satisfaction, Bounded Treewidth, and Finite-Variable Logics. In: Van Hentenryck, P. (ed.) CP 2002. LNCS, vol. 2470, pp. 310–326. Springer, Heidelberg (2002)
9. Dechter, R.: Constraint Processing. Morgan Kaufmann (2003)
10. Deineko, V., Jonsson, P., Klasson, M., Krokhin, A.: The approximability of Max CSP with fixed-value constraints. Journal of the ACM 55(4) (2008)
11. Grohe, M.: The complexity of homomorphism and constraint satisfaction problems seen from the other side. Journal of the ACM 54(1), 1–24 (2007)
12. Jonsson, P., Klasson, M., Krokhin, A.: The approximability of three-valued MAX CSP. SIAM J. Comput. 35(6), 1329–1349 (2006)
13. Jonsson, P., Kuivinen, F., Thapper, J.: Min CSP on Four Elements: Moving beyond Submodularity. In: Lee, J. (ed.) CP 2011. LNCS, vol. 6876, pp. 438–453. Springer, Heidelberg (2011)
14. Kamiński, M.: MAX-CUT and Containment Relations in Graphs. In: Thilikos, D.M. (ed.) WG 2010. LNCS, vol. 6410, pp. 15–26. Springer, Heidelberg (2010)
15. Lewis, J.M., Yannakakis, M.: The node-deletion problem for hereditary properties is NP-complete. Journal of Computer System Sciences 20(2), 219–230 (1980)
16. Orlin, J.B.: A faster strongly polynomial time algorithm for submodular function minimization. Mathematical Programming Ser. A 118(2), 237–251 (2009)
17. Shamir, R., Sharan, R., Tsur, D.: Cluster graph modification problems. Discrete Applied Mathematics 144, 173–182 (2004)

Optimisation Modelling for Software Developers

Kathryn Francis, Sebastian Brand, and Peter J. Stuckey

National ICT Australia, Victoria Research Laboratory,
The University of Melbourne, Victoria 3010, Australia
{kathryn.francis,sebastian.brand,peter.stuckey}@nicta.com.au

Abstract. Software developers are an ideal channel for the distribution of Constraint Programming (CP) technology. Unfortunately, including even basic optimisation functionality in an application currently requires the use of an entirely separate paradigm with which most software developers are not familiar.

We suggest an alternative interface to CP designed to overcome this barrier, and describe a prototype implementation for Java. The interface allows an optimisation problem to be defined in terms of procedures rather than decision variables and constraints. Optimisation is seamlessly integrated into a wider application through automatic conversion between this definition and a conventional model solved by an external solver.

This work is inspired by the language CoJava, in which a simulation is automatically translated into an optimal simulation. We extend this idea to support a general interface where optimisation is triggered on-demand. Our implementation also supports much more advanced code, such as object variables, variable-sized collections, and complex decisions.

1 Introduction

This paper is concerned with the usability of Constraint Programming (CP) tools, specifically for general software developers. The reason we have chosen to target general software developers is because they have the potential to pass on the benefits of CP technology to many more end users by incorporating optimisation functionality into application-specific software.

Much of the work on usability for CP is aimed at expert users involved in research or the development of large scale industrial applications. For example, modelling languages such as OPL [12] and MiniZinc [10] are designed to address the requirement of these advanced users to easily experiment with different models and solving strategies, with a fine level of control.

Software developers aiming to incorporate basic optimisation functionality into an application have no such requirement for experimentation and control, as they do not have the expertise to benefit from this. Furthermore, it is highly inconvenient to have to learn and use a separate paradigm in order to implement a single application feature.

We propose an alternative interface to CP technology which hides from the user all reference to CP specific concepts. The problem is defined within the

M. Milano (Ed.): CP 2012, LNCS 7514, pp. 274–289, 2012.
© Springer-Verlag Berlin Heidelberg 2012

procedural paradigm, by specifying the procedure used to combine individual decisions and evaluate the outcome. Automatic translation from this definition to a conventional CP model allows the problem to be solved using an external constraint solver. The solution giving optimal values for decision variables is then translated back into a structured representation of the optimal outcome.

This sort of interface would greatly improve the accessibility of CP for general software developers. Procedural programmers are well practiced at defining procedures, so most software developers will find this form of problem definition easy to create and understand. Integration of optimisation functionality into a wider application is also straightforward, as the programmer is freed from the burden of writing tedious and error-prone code to manually translate between two different representations of the same problem (one using types and structures appropriate for the application, and another using primitive decision variables).

This project is inspired by the language CoJava [5] and uses the same core technique to transform procedural code into declarative constraints. We extend significantly beyond CoJava in a number of ways.

2   Background and Related Work

Constraint Programming. In order to apply CP techniques to a given satisfaction or optimisation problem, it must first be modeled as a set of constrained decision variables. The model can be defined using a special purpose modelling language, or a CP library embedded within a host programming language; see e.g. [11,8,9]. Such libraries typically provide data types to represent decision variables, constraints, the model, and the solver. Although one does not have to use a foreign language, it is necessary to understand CP modelling concepts and to work directly with these to build a declarative model.

Simulation. An alternative approach to optimisation is to model the situation using a simulation, and then experiment with parameters, searching for a good selection. This technique, called *simulation optimisation*, treats the simulation as a black box. It is inherently heuristic and thus unable to find provably optimal solutions. For a survey of simulation optimisation, see e.g. [6,7].

Most research into simulation optimisation assumes that the true objective function is not known: the simulation incorporates some non-determinism and gives a noisy estimate of this function. If the simulation is actually deterministic given a choice of parameters, then it is possible to convert the simulation code into an equivalent constraint model and use this to find a provably optimal set of parameters. Performing this conversion automatically was the goal of CoJava.

CoJava. The system described in this paper is inspired by and builds on techniques developed for the language CoJava, introduced in [4] and described further in [5]. CoJava is an extension to the Java language, intended to allow programmers to use optimisation technology to solve a problem modelled as a simulation.

The simulation is written in Java using library functions to choose random numbers, assert conditions which must hold, and nominate a program variable as the optimisation objective. The CoJava compiler transforms the simulation code into a constraint model whose solution gives a number to return for each random choice so that all assertions are satisfied and the objective variable is assigned the best possible value. The original code is then recompiled with the random choices replaced by assignments to these optimal values. The result is a program which executes the optimal execution path of the original simulation.

CoJava was later extended to SC-CoJava and CoReJava [2,1,3], introducing simpler semantics, but also narrowing in focus to supply chain applications.

3   Contributions

Architecture. Despite using the same basic technique to translate procedural code into a declarative constraint model, the purpose and architecture of our system are fundamentally different to CoJava.

- Optimisation is triggered explicitly as required during program execution, and problem instance data is not required at compile time. In CoJava, optimisation is implicit and takes place at compile time.
- Performing optimisation at run time allows us to support interactive applications, while CoJava is restricted to simulation-like programs. This is an important distinction, as interactive applications can enable non-technical end users to access optimisation technology independently.
- The code written by the programmer is pure Java and does not take on any new semantics. CoJava on the other hand is an extension of the Java language, with semantics which depend on the mode of compilation.

Supported Code. We dramatically expand on the type of code supported, allowing a much more natural coding style.

- We extend support for non-determinism from arithmetic numeric and Boolean types to arbitrary object types.
- We allow variable collections of objects such as sets, lists and maps to be constructed and manipulated.
- We introduce generic higher level decisions to aid modelling, in contrast to the approach taken in SC-CoJava of supplying application-specific library components.

4   A Prototype for Java: The Programmer's Perspective

This paper proposes an alternative interface to CP technology which allows an optimisation problem to be specified through code which constructs a solution using the results of pre-supplied decision making procedures, and code which evaluates that solution, determining whether or not it is valid and calculating a measure of its quality. This idea is applicable to any host programming language, although the precise design would obviously depend on the features of the language. We describe here a proof-of-concept implementation for Java, consisting of a Java library and a plug-in for the Eclipse IDE.

The library includes only three public classes/interfaces. To include optimisation capability in an application, the programmer implements the Solution interface, defining two methods: build and evaluate. The build method is passed a ChoiceMaker object which provides decision making procedures, and uses these to build a solution. The evaluate method calculates and returns the value of the current solution, or throws an exception if the current solution is invalid.

Optimisation is triggered using the buildMinimal and buildMaximal methods provided by the Optimiser class. These take a Solution object and apparently call its build method with a special ChoiceMaker able to make optimal decisions, so that a subsequent call to evaluate will return the minimal or maximal value without encountering any exceptions. Additional versions of these methods with an extra parameter allow the programmer to request the best solution found within a given time limit.

The Eclipse plug-in performs compile time program manipulation to support run time conversion into a conventional constraint model. This step is performed transparently during program launch, or as an independent operation.

As an illustration of the system we consider a simple project planning application.[1] Input to the program is a list of tasks, each of which may have dependencies indicating other tasks which must be scheduled at least a given number of days earlier. The application chooses a day to schedule each task so that all dependencies are satisfied and the project finishes as early as possible. The generated project plan is displayed to the user, who is then allowed to repeatedly reschedule the project after adjusting the tasks and dependencies. Note that this sort of interactive application cannot be achieved using the CoJava architecture.

The Solution interface is implemented by a class ProjectPlan, whose build and evaluate methods are shown below. A ProjectPlan is initialised with a reference to a list of tasks alltasks. The build method chooses a day for each task and passes this to the task to be recorded. The evaluate method first checks that all task dependencies are satisfied, and then calculates and returns the latest scheduled day as the value of the solution.

```java
void build(ChoiceMaker chooser) {           Integer evaluate() throws Exception {
   for(Task task : alltasks) {                 for(Task task : alltasks)
      int day = chooser.chooseInt(1, max);        if(!task.dependenciesSatisfied())
      task.setScheduledDay(day);                     throw new Exception();
   }                                            return getFinishDay();
}                                            }
```

An optimal project plan is obtained by passing a ProjectPlan to the buildMinimal method of Optimiser (as we wish to find the solution with smallest evaluation

---

[1] Example code is available online: `www.csse.unimelb.edu.au/~pjs/optmodel/`

result). After this method is called the tasks currently recorded in tasklist will have scheduled days corresponding to an optimal project plan. The application displays the plan using the Task objects, and is then free to make changes to the task list and dependencies (based on user input) before calling buildMinimal again, which will update the scheduled days to once again represent an optimal solution given the new tasks and dependencies.

5   A Prototype for Java: Implementation

To solve the optimisation problem specified by the build and evaluate methods of a Solution object, an equivalent constraint model must be constructed and sent to a solver. As the complete problem specification depends on the program state when optimisation is requested, the constraint model cannot be constructed at compile time. Our approach is to generate at compile time a transformed version of the original code which can be used at run time to create a complete model based on the current state.

5.1   Compile Time

The transformed version of the original code works with symbolic expressions rather than concrete values. It builds an expression for the new value of each possibly affected field (thus capturing all possible state changes), and an expression for the value of the solution.

The compilation process is illustrated in Figure 1. Fortunately it is not necessary to create a transformed version of the entire program: only code used within the build and evaluate methods is relevant to optimisation. Transformed versions of the relevant methods are added to the project, and then this generated code is compiled along with the original source code using a regular Java compiler.

The code transformation is based on that described for CoReJava [3]. The logic of the original code is captured by executing all reachable statements, while ensuring that assignments are predicated on the conditions under which normal execution would reach the assignment. A variable assigned a new value is constrained to equal the assigned value if the *path constraint* holds and the old value otherwise. Pseudo-code showing the code generated for an assignment statement is shown below, where P is the current path constraint, and C and V are respectively the constraints and solver variables collected so far.

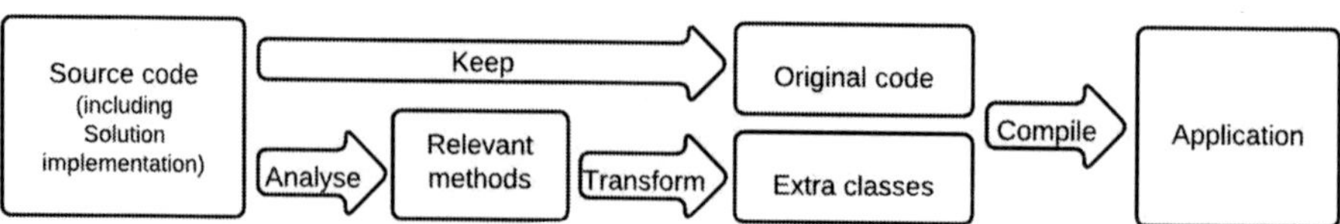

Fig. 1. The compile time operation

$$x = e; \qquad \begin{aligned} &x' := \text{newVar}(); \; V := V \cup \{x'\} \\ &C := C \wedge (x' = \textbf{if } P \textbf{ then } e \textbf{ else } x) \\ &x := x' \end{aligned}$$

Where the original code contains a branch whose condition depends on the outcome of some decision, the transformed code executes both branches.

```
if(condition){
    <then-block>
}
else{
    <else-block>
}
```

$$\begin{aligned} &P_0 := P \\ &P := P_0 \wedge condition \\ &\text{<translation of then-block>} \\ &P := P_0 \wedge \neg condition \\ &\text{<translation of else-block>} \\ &P := P_0 \end{aligned}$$

A throw statement introduces a constraint that the current path constraint must evaluate to false (as normal execution is not allowed to reach this point).

We have found it helpful to introduce a separate type to represent variables (local variables, method parameters, and fields), rather than using the expression type directly. The Variable type provides a method to look up an expression for the current value of the variable, and an **assign** method to replace assignment statements. The types of affected local variables and method parameters are changed to this Variable type, so that assignments can be handled correctly as discussed above. The types of fields are not changed, instead when a field is accessed for the first time a Variable is created and recorded against the object and field name, and future uses of the same field refer to this Variable.

One complication not discussed in the CoJava work is return statements. A return statement provides an expression for the return value of a method, but it also indicates that the remainder of the method should not be executed. To correctly handle methods with multiple return statements, at each return a condition is added to the path constraint for the remainder of the method ensuring that all further assignments are predicated on not reaching this return statement. The condition added is the negation of the part of the current path constraint falling within the scope of the current method. The loop exit statements break and continue are supported using the same technique. Return handling is illustrated by the translation shown below (where a is a field of the current object):

```
int exampleMethod() {
    if(a > 5) {
        return a;
    }
    a = a + 1;

    return a;
}
```

$$\begin{aligned} &P_0 := P \\ &P := P_0 \wedge (a > 5); \\ &r := a \; \textit{// introduce variable r for return value} \\ &P := P_0 \wedge \neg(a > 5) \; \textit{// add return constraint} \\ &a' := \text{newVar}(); \; V := V \cup \{a'\} \\ &C := C \cup (a' = \textbf{if } P \textbf{ then } (a+1) \textbf{ else } a); \; a := a' \\ &r' := \text{newVar}(); \; V := V \cup \{r'\} \\ &C := C \cup (r' = \textbf{if } P \textbf{ then } a \textbf{ else } r); \; r := r' \\ &P := P_0; \; \textbf{return } r \end{aligned}$$

The transformed versions of affected methods are added to new classes, leaving the original classes unchanged. The transformed methods are made static but are given an extra argument for the object on which the method is called. A further object of type ModelBuilder is threaded through all method calls to keep track of state information. It maintains the path constraint, and also records variables introduced by decision procedures and as intermediate variables, and constraints caused by throw statements or introduced to constrain intermediate variables.

5.2   Run Time

At run time, optimisation is triggered by a call to the Optimiser method buildMinimal or buildMaximal, with an object implementing the Solution interface passed as an argument. The Optimiser first uses reflection techniques to find the transformed build and evaluate methods corresponding to the type of the received Solution object, and then executes these methods.

The variables and constraints recorded in the ModelBuilder during execution of the transformed methods, along with the objective expression returned by the evaluate method, are used to generate a constraint problem which is then sent to an external solver. We chose to express the problem in MiniZinc [10] because it gives us an easy choice of constraint problem solvers.

The ModelBuilder also records a Variable object for each potentially updated field. When a solution is obtained, the optimal decisions are substituted into the expression giving the final value for each of these Variable objects, and reflection techniques are used to update the corresponding fields accordingly. The run time process is illustrated in Figure 2.

Returning to the project planning example, consider the method getFinishDay (Figure 3), which is called by the evaluate method of ProjectPlan. In the transformed version of this method, when getScheduledDay is called on a task the result is an integer expression which is not constant (as the task's scheduled day is assigned in the build method to the result of chooseInt). The greater-than operator is therefore replaced with a method which creates a comparison expression. This means that the if statement has a non-constant condition. As discussed above, the body of the if statement is executed, but the condition (the comparison expression) is added to the current path constraint at the beginning of the then block, and removed at the end.

This means that when the assignment of finishDay is reached, the path constraint will not be constant, so finishDay is given a Variable type, and the assign

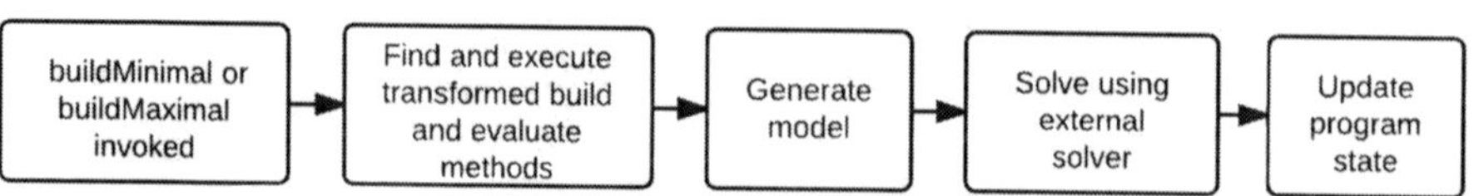

Fig. 2. The optimisation process, triggered on demand during application execution

```
int getFinishDay() {
    int finishDay = 1;
    for(Task task : alltasks)
        if(task.getScheduledDay() > finishDay)
            finishDay = task.getScheduledDay();
    return finishDay;
}
```

Fig. 3. ProjectPlan getFinishDay method, called from within evaluate

method is used in place of the assignment. The assign method creates a new intermediate variable for the current value of finishDay, and records a constraint that this new expression is equal to the assigned expression (this task's scheduled day) if the path constraint (the greater-than comparison) evaluates to true, and otherwise it is equal to the previous value. The expression finally returned by getLatestDay is an intermediate variable constrained via a series of these varUpdate constraints to equal the latest scheduled day.

The definition of varUpdate, and an example MiniZinc model for a project with 3 tasks is shown in Figure 4. The three day variables in this model are core decision variables introduced by calls to chooseInt. The finishDay variables are intermediate variables used to represent the value stored in the finishDay variable in the getFinishDay method discussed above. The varUpdate constraints also originate from here, while the two constraints enforcing task dependencies originate from the evaluate method, each being the negation of the path constraint when the throw statement was reached.

```
var 1..6: day0;
var 1..6: day1;
var 1..6: day2;
var {1,2,3,4,5,6}: finishDay31;
var {1,2,3,4,5,6}: finishDay33;
var {1,2,3,4,5,6}: finishDay35;
constraint (not (day1 < day0));
constraint (not (day2 < (1 + day1))) /\ (not (day2 < (2 + day0)));
constraint varUpdate(finishDay31,(day0 > 1),day0,1);
constraint varUpdate(finishDay33,(day1 > finishDay31),day1,finishDay31);
constraint varUpdate(finishDay35,(day2 > finishDay33),day2,finishDay33);
solve :: int_search([day0, day1, day2], input_order, indomain_split, complete)
            minimize finishDay35;

predicate varUpdate(var int: out, var bool: update, var int: new, var int: old) =
    (out = [old,new][1+bool2int(update)]);
```

Fig. 4. MiniZinc model for project planning example

5.3 Refining the Translation

Many values computed within the build and evaluate methods are unaffected by the results of decisions. Measures to take this into account can reduce unnecessary complexity in the model. At compile time, translation is only required for a method if at some time it is passed a non-constant argument, or if the code within the method uses decision procedures provided by the ChoiceMaker, changes the state of some object, or reads a field which is updated elsewhere in translated code. Within a method, a variable that is never assigned to a value that depends on the outcome of decisions, and is never assigned conditionally depending on the outcome of a decision, does not need its type changed.

At run time, if the condition for a branching statement is constant, only the corresponding branch is executed. Also, if the current path constraint is constant, an assignment can overwrite the previous value of a variable unconditionally. Actually, it is only required that the part of the path constraint which falls within the variable's scope is constant. If the part of the path constraint outside this scope is not satisfied, then normal execution would never reach the declaration of the variable, making its value irrelevant.

An example of this is illustrated in the translation below:

```
if(X) {                              P₀ := P
                                     P := P₀ ∧ X
    int sum = 0;                     sum := 0
    for(int item : list) {           for item in list do
        sum = sum + item;                sum' := newVar(); V := V ∪ {sum'}
    }                                    C := C ∪ (sum' = sum + item); sum := sum'
    a = sum;                         a' := newVar(); V := V ∪ {a'}
                                     C := C ∪ (a' = if P then sum else a); a := a'
}                                    P := P₀
```

The Boolean expression X is added to the path constraint for the duration of the then block. The assignment to a (which is declared outside the if statement) is conditional on X, but all assignments to sum are unconditional.

6 Object Variables

For real-world problems, natural code will almost always involve decisions at the object level. We describe here our support for this, including a procedure to choose one object from a collection. Note that this is a major extension over CoJava, which restricts non-determinism to primitive typed variables only.

Let us consider a new version of the project planning application. This time, resources are no longer infinite. Instead only a given number of hours are available on each day. We introduce a Day class, with fields recording the day number and the maximum number of hours available, as well as the number of hours currently assigned. Each Task now also has a duration, and we need to ensure that the total duration of all tasks assigned to a day does not exceed the hours available.

The new build method for ProjectPlan is shown below. For each task a Day object is chosen, and the task is assigned to this Day using the addTask method. This method updates the hours assigned to the day, returning false if the total is now greater than the available hours. If the return value is false an exception is thrown to indicate that the solution is not acceptable.

```
public void build(ChoiceMaker chooser) throws Exception {
    for(Task task : alltasks) {
        Day chosenDay = chooser.chooseFrom(allDays);
        if(!chosenDay.addTask(task))
            throw new Exception("Failed to add task");
    }
}
```

The object represented by chosenDay depends on the outcome of the choose-From decision. This means that in the transformed version of the code we need to be able to call addTask without knowing which Day object is the target.

In order to support this sort of code, we first need to be able to represent the choice of an object using primitive solver variables. This is achieved by assigning an integer key to each distinct object. Each expression with a non-primitive type has a domain of possible objects, which can be translated into a corresponding integer domain. This representation allows straightforward equality comparisons between object expressions. The chooseFrom method simply creates a new expression whose domain is given by the provided collection, with a corresponding integer decision variable.

The next consideration is field accesses. As with single objects, each accessed field of an object expression is assigned a Variable object. This Variable must be able to produce an expression for the current value of the field, and to accept an expression for a newly assigned value. For object expressions, a special type of Variable is used which has a reference to the Variable for the corresponding field of each object possibly represented by the expression, as well as an integer expression for the index of the chosen object.

When an expression for the current value is requested, an intermediate solver variable is created and constrained to equal the current field expression for the object at the chosen index. This is a simple element constraint.

An assignment must update the corresponding field for every object in the expression's domain. A new intermediate variable is created for each, and a user-defined predicate fieldUpdate is used to ensure that all except the one at the correct index are equal to the previous values, while the one at the correct index is equal to the assigned expression if the current path constraint holds, or the old value otherwise.

We also need to handle the calling of methods on object expressions. Fortunately, the only way the target object affects the outcome of a method invocation is through the values stored in its fields. This means that all uncertainty can be pushed down to the field level. Translated methods are already converted to be static with an argument indicating the object called on, so it is straightforward to allow this argument to be non-constant. Then, for each field access the Variable retrieved is of the special type discussed above.

7   Variable Collections

Combinatorial problems commonly involve collections such as sets or lists. It is therefore valuable to allow the use of these in the Solution code. It should be possible not only to store variable objects in collections, but also to make arbitrary changes to the collection in variable contexts, so that the resulting size and composition of the collection depends on the outcome of decisions. Furthermore, it should be possible to iterate over these variable collections.

Returning to the project planning example, imagine we are now allowed to hire an external contractor to perform some tasks, paying an hourly rate plus a callout fee for each day the contractor's services are required. Instead of finishing as early as possible we wish to minimise cost while meeting a deadline. Extracts from the revised build and evaluate methods for ProjectPlan are shown in Figure 5.

```
Day day = chooser.chooseFrom(days);          int cost = dayFee * contractorDays.size();
if(chooser.chooseBool()) {                   for(Task t : contractedTasks) {
    contractedTasks.add(task);                   cost += hourlyRate * t.duration();
    contractorDays.add(day);                 }
}                                            return cost;
```

Fig. 5. Extracts from revised build and evaluate methods making use of collections

Note first that the chosen day for the task, a decision variable, is added to the contractorDays set. Second, this set and the list of contracted tasks are both updated conditionally depending on whether or not this task is to be contracted out (as decided by the chooseBool method). The crucial aspect here is that the for loop in the evaluate code iterates over a collection whose size depends on the values of decision variables.

We have implemented a special purpose translation for the Set, List and Map interfaces in order to support this kind of code. The VariableSet, VariableList and VariableMap classes provide specialised transformed versions of (almost) all methods included in these collection interfaces, using a special-purpose representation for the state of the collection.

7.1   Sets, Lists and Maps

The VariableSet class represents a set using a list of possible members of the set, and a corresponding list of Boolean expressions indicating whether or not each item is actually in the set. Each possible item is also an expression which may represent a choice between several actual objects. When a VariableSet is initialised, all items are constant expressions, and the Boolean conditions are true. It is only through operations on the set that variability is introduced.

As an example of an operation we consider add. This method is required to add the given item to the set if it is not already present, and return true if the set has changed. The pseudo-code below defines the effect of calling add on a set s having current state $vs = \langle n, x_{1..n}, c_{1..n} \rangle$, where n is the number of possible members of the set, x_i is the possible member at index i, and c_i is the Boolean condition indicating whether or not x_i is actually in the set.

$$
\begin{aligned}
&a := \bigwedge_{i=1..n} \neg(x_i = y \wedge c_i) \\
&b := P \wedge a \ \ // \ P: \textit{path constraint} \\
&vs := \langle n + 1, \langle x_{1..n}, y \rangle, \langle c_{1..n}, b \rangle \rangle \\
&result := a
\end{aligned}
$$

boolean $result = s.\mathrm{add}(y);$

The Boolean expression a represents the condition that y was not already in the set. This expression is also used as the return value. It is possible for the set to remain unchanged even if y was not already present, as the current path constraint may not be satisfied. However, in this case execution would not reach this method call, so the return value is irrelevant.

Clearly it is important to handle constants well, as otherwise the model becomes unnecessarily complex. For example, the add method does not add a new possible item if one of the existing possible items is identical to the added item. In this case the existing item's Boolean condition is updated instead, and if this is already constant and true, no change is required.

The VariableList class maintains a list of possible members in the same fashion as VariableSet, but instead of a Boolean expression indicating whether or not the item is present, an integer expression for each possible item indicates its (0-based) index in the list, with all items not actually in the list having indices greater than the length of the list. A separate integer expression is maintained for the current length of the list.

The VariableMap class is implemented as an extension of the VariableSet class, with an added expression for each possible key giving its currently assigned value.

7.2   Iteration

Iteration in Java is performed using the Iterator interface, with two methods: hasNext to check whether there are remaining items, and next to retrieve the next item. We support iteration over variable collections using a VariableIterator class which implements the transformed versions of these operations. That is, the hasNext method returns a Boolean expression which evaluates to true if at least one of the remaining items is actually in the collection, while the next method returns an expression for the next item.

Enhanced for loops (for example that on the right of Figure 5) are converted into the equivalent while loop using an explicit iterator. At the beginning of each loop iteration the hasNext method is called, and the resulting Boolean expression is added to the path constraint, to be removed at the end of the loop body. When the hasNext method returns an expression which is constant and false (as it does when it runs out of possible members of the collection), the loop is terminated.

Although the VariableSet state includes a list of expressions for possible members of the set, the iterator cannot simply return these in order. To correctly reflect loop exit logic all items which are actually in the set must be returned before any which are not. For this reason, each item returned by the VariableIterator is actually a new expression which may represent any of the possible members of the set. An integer variable is created for each returned item, giving the corresponding index into the list of possible members. These indices are constrained to ensure that an item which is not in the set is never returned before an item which is in the set, and further (to avoid symmetry) that within these two groups items are returned in order of index.

Iteration over lists is implemented similarly except that the order in which items are returned is determined by the indices stored as part of the VariableList state. For both lists and sets, the Boolean expression returned by hasNext is simply a comparison between the number of items already returned and an expression for the size of the collection.

Obviously constant detection is very important to avoid excessive complexity. We have implemented some simplifications, such as returning all items which are definitely in the set first, and excluding entirely any which are definitely not in the set, but have not yet investigated all possible simplifications.

8   Complex Decisions

With support for variable collections, it becomes possible to provide more complex decision procedures allowing decisions to be specified at a higher level. As an illustration, let us return to the project planning example. Imagine that instead of choosing a single worker for each task, we assign a team of workers. For each task there are a set of allowed team sizes, and the task duration varies according to the size chosen. Furthermore, tasks may be performed across multiple days, as long as an integral number of hours is assigned to each day and the total number of hours matches the task duration.

Below is a build method appropriate for this situation, making use of complex decision procedures.

```java
public void build(ChoiceMaker chooser) {
    for(Task task : allTasks) {
        int teamSize = chooser.chooseFrom(task.allowedTeamSizes());
        int taskDuration = task.getDuration(teamSize);
        Set<Worker> team = chooser.chooseSubset(allWorkers, teamSize);
        Map<Day,Integer> chosenDays = chooser.allocate(taskDuration, allDays);
        for(Worker worker : team)
            worker.assignTask(task, chosenDays);
    }
}
```

The procedure first chooses a team size and a corresponding team. Then the total task duration is allocated to days using the allocate method. This method decides how much of the given quantity (in this case the task duration) should be allocated to each object in the given collection (in this case the list of days). Finally, the workers' schedules are updated appropriately.

Not only do complex decision procedures greatly simplify the build method, they also provide opportunities to make use of global constraints. For example, the allocate method can make use of a global sum constraint to ensure that the total quantity allocated is correct.

9   Experimental Results and Future Work

Having developed a working system, our next concern is performance. We present in Table 1 preliminary experimental results demonstrating the relative performance of the system compared with equivalent hand-written models. It is unrealistic to expect that automatically generated models will be able to compete with models produced by an expert. Our aim is to be able to handle problems arising for small businesses or individuals. The results show that we still have some work to do to achieve this goal. Note that compilation time is not shown as this was insignificant: the entire suite compiles in 20 seconds (with around 15 seconds spent performing code transformation).

For most problems the vast majority of the total time is spent solving the model (rather than generating it). This suggests the potential to greatly decrease the running time by improving the model. Our initial analysis of the automatically generated models has led to the identification of two easily detected programming patterns for which stronger constraints are available. The table below shows the original Java code for each of these patterns, the constraints which would be generated using the standard transformation, and the alternative stronger constraints.

Java code	Standard translation	Specialised translation
if(c) x++;	var x' = [x, x+1][bool2int(c)+1];	var x' = x + bool2int(c);
if(a > x) x = a;	var x' = [x, a][bool2int(a > x)+1];	var x' = max(x, a);

After adding a step during the model generation phase to automatically detect just these two programming patterns and replace the constraints, we have observed a significant improvement in performance on several benchmarks. We anticipate further gains can be made by identifying other patterns for which straightforward model refinements are beneficial. Some patterns may involve larger portions of the code, for example a loop which sums a collection of values or counts the number of objects satisfying some property.

A complementary approach is to attempt to exploit the structure of the generated models, which tend to be quite unidirectional. We may be able to take advantage of this property for efficient propagation scheduling or through the development of a specialised search strategy.

Other future work could design global constraints to improve the treatment of collection operations as well as variable and field updates. There is also room for further exploration of alternative representations which could be used for the state of variable collections. The current implementations are correct but certainly not as efficient as possible.

Table 1. Experimental results for project planning example and various well-known problems. In order the figures give the total time (secs) for the optimisation step, the percentage of this total used by the solver, the new total time using basic model refinement with percentage improvement in brackets, the solving time for an equivalent hand-written model, and the number of times faster this hand-written model is compared with the improved total time. Timing figures are the average over 30 instances.

Problem	Size	Total	Solving	Improved	Hand	Factor
project planning 1	250 tasks	1.13	78.1%	0.69 (39%)	0.16	4.4
project planning 2	18 tasks	16.22	97.6%	13.31 (18%)	1.18	11.3
project planning 3	14 tasks	32.38	96.6%	26.38 (19%)	0.27	96.9
bin packing	8 items	8.67	98.2%	5.81 (33%)	0.34	17.2
golomb ruler	7 ticks	3.45	33.6%	3.41 (0.9%)	2.03	1.7
knapsack (0-1)	30 items	1.97	80.4%	1.95 (0.9%)	0.15	13.2
knapsack (bounded)	30 items	6.40	95.2%	6.36 (0.6%)	1.10	5.8
routing (pickup-del)	8 stops	6.48	96.5%	6.45 (0.6%)	0.33	19.7
social golfers	9 golfers	2.65	75.0%	2.60 (1.9%)	0.14	18.3
talent scheduling	8 scenes	39.47	99.7%	20.87 (47%)	2.20	9.5

10   Conclusion

If software developers with no specialised CP expertise could easily incorporate CP technology into their applications, this would greatly increase its impact.

We have designed an alternative interface to CP which aims to be more intuitive and convenient for software developers. The necessary translation between paradigms is automated, allowing the programmer to work exclusively with a procedure based definition of the optimisation problem. This significantly reduces the burden on the programmer and allows straightforward integration of optimisation functionality within a wider application.

A natural coding style is allowed with support for object variables, variable collections, and high level decision procedures, building on and significantly extending techniques used to implement the language CoJava.

Preliminary experiments show that further work is required to achieve our goal of performance sufficient for small business and personal applications, but that there are gains to be made using very simple local adjustments to the model. We have also identified several other avenues of investigation which may lead to further improvements in performance.

Acknowledgments. NICTA is funded by the Australian Government as represented by the Department of Broadband, Communications and the Digital Economy and the Australian Research Council.

References

1. Al-Nory, M., Brodsky, A.: Unifying simulation and optimization of strategic sourcing and transportation. In: Winter Simulation Conference (WSC), pp. 2616–2624 (2008)

2. Brodsky, A., Al-Nory, M., Nash, H.: Service composition language to unify simulation and optimization of supply chains. In: Hawaii International Conference on System Sciences, p. 74. IEEE Computer Society, Los Alamitos (2008)
3. Brodsky, A., Luo, J., Nash, H.: CoReJava: learning functions expressed as Object-Oriented programs. In: Machine Learning and Applications, pp. 368–375. IEEE Computer Society, Los Alamitos (2008)
4. Brodsky, A., Nash, H.: CoJava: a unified language for simulation and optimization. In: Conference on Object-oriented Programming, Systems, Languages, and Applications (OOPSLA), pp. 194–195. ACM, New York (2005)
5. Brodsky, A., Nash, H.: CoJava: Optimization Modeling by Nondeterministic Simulation. In: Benhamou, F. (ed.) CP 2006. LNCS, vol. 4204, pp. 91–106. Springer, Heidelberg (2006)
6. Carson, Y., Maria, A.: Simulation optimization: methods and applications. In: Winter Simulation Conference (WSC), pp. 118–126. IEEE Computer Society, Atlanta (1997)
7. Fu, M.C., Glover, F.W., April, J.: Simulation optimization: a review, new developments, and applications. In: Winter Simulation Conference, WSC (2005)
8. Hebrard, E., O'Mahony, E., O'Sullivan, B.: Constraint Programming and Combinatorial Optimisation in Numberjack. In: Lodi, A., Milano, M., Toth, P. (eds.) CPAIOR 2010. LNCS, vol. 6140, pp. 181–185. Springer, Heidelberg (2010)
9. Jussien, N., Rochart, G., Lorca, X.: The CHOCO constraint programming solver. In: CPAIOR 2008 Workshop on OpenSource Software for Integer and Constraint Programming OSSICP 2008 (2008)
10. Nethercote, N., Stuckey, P.J., Becket, R., Brand, S., Duck, G.J., Tack, G.: MiniZinc: Towards a Standard CP Modelling Language. In: Bessière, C. (ed.) CP 2007. LNCS, vol. 4741, pp. 529–543. Springer, Heidelberg (2007)
11. Schulte, C., Lagerkvist, M., Tack, G.: GECODE – an open, free, efficient constraint solving toolkit, http://www.gecode.org/
12. Van Hentenryck, P., Michel, L., Perron, L., Régin, J.-C.: Constraint Programming in OPL. In: Nadathur, G. (ed.) PPDP 1999. LNCS, vol. 1702, pp. 98–116. Springer, Heidelberg (1999)

Adaptive Bisection of Numerical CSPs

Laurent Granvilliers

LINA, CNRS, Université de Nantes, France
2 rue de la Houssinière, BP 92208, 44322 Nantes Cedex 3
laurent.granvilliers@univ-nantes.fr

Abstract. Bisection is a search algorithm for numerical CSPs. The main principle is to select one variable at every node of the search tree and to bisect its interval domain. In this paper, we introduce a new adaptive variable selection strategy following an intensification diversification approach. Intensification is implemented by the maximum smear heuristic. Diversification is obtained by a round-robin ordering on the variables. The balance is automatically adapted during the search according to the solving state. Experimental results from a set of standard benchmarks show that this new strategy is more robust. Moreover, it is particularly efficient for solving the well-known Transistor problem, illustrating the benefits of an adaptive search.

1 Introduction

A Numerical Constraint Satisfaction Problem (NCSP) is a set of numerical constraints over continuous variables having interval domains. For the sake of simplicity, we consider from now on square systems of equations $f(x) = 0$ where $f : [x] \to \mathbb{R}^n$ is defined on a n-dimensional interval box $[x]$. Solving such a problem with interval methods generally consists in paving its solution set $\Sigma(f, [x])$ with interval boxes at a given precision. To this end, interval branch-and-prune algorithms alternate contraction steps and branching steps until reaching small enough boxes [7,6]. The contraction component typically combines the interval Newton method and consistency techniques in order to prune the variable domains. The branching step usually generates two sub-problems by bisecting the domain of one variable. The variable selection strategy used in the branching step is called here the bisection strategy.

Several general bisection strategies have been proposed [8,2,6,11]. MaxDom selects the variable having the largest domain. The goal is to minimize the width of intervals resulting from the evaluation of f using its natural interval extension. MaxSmear is a greedy strategy that follows the same idea applied to the mean value extension of f. To this end, the variable having the maximum *smear value* is chosen. RR is a fair strategy using a round-robin ordering that leads to regularly select every variable. Even if MaxSmear is known to be good in average, determining the good strategy for a given problem may not be easy. Moreover, we have observed that MaxSmear is more useful when boxes are small enough.

We introduce in this paper a new adaptive bisection strategy implementing a balance between MaxSmear and RR. This strategy can be more RR-oriented

M. Milano (Ed.): CP 2012, LNCS 7514, pp. 290–298, 2012.

in the early stages of the solving process to allow a diversification on the set of all variables. A MaxSmear behavior can be strengthened during the search according to the contraction of boxes. Our hybridization scheme is inspired from greedy randomized adaptive search procedures [3]. It consists in making a fair choice in a subset of the variables having the greatest smear values. The balance is determined by the size of the subset containing the most interesting variables. The benefits of this adaptation scheme are illustrated by the experimentations on a large set of classical problems.

The bisection scheme and interval methods are presented in Section 2. The new adaptive strategy is introduced in Section 3. The different strategies are analyzed and compared using a set of classical benchmarks and the experimental results are reported in Section 4. A conclusion follows.

2   Bisection Scheme

2.1   Interval Arithmetic

We consider (closed) intervals $[u] \doteq [\underline{u}, \overline{u}] \doteq \{u \in \mathbb{R} : \underline{u} \leqslant u \leqslant \overline{u}\}$. The set of intervals is denoted by $\mathbb{IR}$. A n-dimensional interval box $[u]$ is a vector of $\mathbb{IR}^n$ that can be seen as a Cartesian product $[u_1] \times \cdots \times [u_n]$. The width of an interval $[u]$ is the difference $w([u]) \doteq \overline{u} - \underline{u}$. The width of a box is the maximum width taken componentwise. The hull of a set $S \subseteq \mathbb{R}$ is the interval $\square S \doteq [\inf S, \sup S]$.

Interval operations are set extensions of real operations. An *inclusion form* of a function $g : \mathbb{R}^n \to \mathbb{R}$ is an interval function $[g] : \mathbb{IR}^n \to \mathbb{IR}$ such that for all $[u] \in \mathbb{IR}^n$, $u \in [u]$ implies $g(u) \in [g]([u])$. An inclusion form can then be used to enclose the range of a real function over an interval domain. The *natural extension* of g is an inclusion form extending an explicit expression of g by replacing every real operation by the corresponding interval operation and every real number by its hull. The *mean value extension* of g is an inclusion form obtained from a first-order approximation of g in a given interval domain.

2.2   Bisection Algorithm

A bisection algorithm calculates a set of boxes enclosing the solution set $\Sigma(f, [x])$ of a NCSP at a given precision $\epsilon > 0$. This algorithm maintains a list of boxes L until it becomes empty. At the beginning, $[x]$ is added in L. At each step, one box $[u]$ is extracted from L and a contracted box $[v] \subseteq [u]$ is computed. Then:

- $[v]$ is discarded if it is empty;
- $[v]$ is labelled as a solution box if its width is smaller than ϵ;
- otherwise $[v]$ is bisected in two boxes $[v']$ and $[v'']$, which are added in L.

Bisecting $[v]$ consists in choosing one variable x_k such that the width of $[v_k]$ is greater than ϵ. Let c be the midpoint of $[v_k]$. Then two sub-boxes are created:

$$[v'] = ([v_1], \ldots, [v_{k-1}], [\underline{v_k}, c], [v_{k+1}], \ldots, [v_n])$$
$$[v''] = ([v_1], \ldots, [v_{k-1}], [c, \overline{v_k}], [v_{k+1}], \ldots, [v_n])$$

The contraction procedure may combine consistency techniques and interval methods. For instance 2B consistency [9] and box consistency [1] are able to early reduce large boxes during the search at a moderate cost. The Newton operator acts as a global constraint. In favorable cases, this operator is able to contract small boxes enclosing the solutions with a quadratic convergence.

The efficiency of the bisection algorithm strongly depends on the coupling of the bisection strategy and the contraction procedure. In fact, the bisection strategy must be tuned to improve future contractions, hence reducing the size of the search tree.

2.3   Bisection Strategies

The MaxDom heuristic selects the variable having the largest domain in the current box $[v]$. The idea is to minimize the width of the evaluation of the natural extension of f.

The MaxSmear heuristic aims at selecting the variable with respect to which f varies most rapidly. Let $[A] = ([a_{ij}])$ be an inclusion form of the Jacobian matrix of f evaluated at $[v]$. The smear value of x_i in $[v]$ is defined as:

$$s(x_i, [v]) \doteq \max_{j=1,\ldots,n} \left\{ \max(|\underline{a_{ji}}|, |\overline{a_{ji}}|) \right\} \times w([v_i])$$

The selected variable is the one having the maximum smear value. This choice tends to minimize the width of the evaluation of the mean value extension of f.

The RR heuristic manages the variables in turn according to a given ordering. Given the ordering $x_1 < x_2 < \cdots < x_n$, variable x_1 is selected first, and then x_2, and so on, such that only the domains larger than ϵ are considered. This is a fair strategy that leads to forget no variable.

It is admitted that the maximum smear heuristic is often the best one. However, the other heuristics can be more efficient for some problems. These remarks argue for the design of an hybrid strategy. These techniques will be discussed in the next section and a new adaptive strategy will be introduced.

3   Adaptive Bisection Strategy

3.1   Motivation

The smear criterion has been designed to improve a bisection algorithm that used the interval Newton operator as a contraction procedure [8]. It turns out that this operator solves an interval linear system derived from the mean value extension. By choosing variables with the maximum smear values, the largest interval coefficients of such linear systems will become narrower, hence increasing the contraction power of the algorithm. However, the smear criteria may not always be useful to distinguish the good variables. This is the case when the intervals in the Jacobian matrix are large, which generally happens when the domains are large or when the derivatives are overestimated.

When solving difficult problems, it is common not to contract boxes in the early stages of the bisection process. At this phase, the goal of bisection may be to allow local consistencies to contract boxes. Given an equation $f_j(x_1, \ldots, x_n) = 0$, the domain of x_i in a box $[u]$ can be reduced by considering an equivalent functional form of the equation $x_i = g(x_1, \ldots, x_{i-1}, x_{i+1}, \ldots, x_n)$ and by intersecting $[u_i]$ with the evaluation of the natural extension of g at $[u]$. The key point is that the interval evaluation of $g(x_1, \ldots, x_{i-1}, x_{i+1}, \ldots, x_n)$ may be tight enough only after bisecting at least once the $(n-1)$ variables. For this purpose the round-robin strategy is very useful.

Our conclusion is that the strategy must be adaptive. If the smear criteria are not exploitable then the round-robin strategy has to be used. The problem is to determine a balance between the two heuristics.

3.2   New Adaptive Strategy

We introduce a new adaptive bisection scheme composed of three consecutive steps described hereafter. Let $[v]$ be the current box to be bisected and let ϵ be the desired precision.

First step: Selection of the candidate variables. The list of candidate variables L is the set of variables whose domains in $[v]$ are larger than ϵ. In practice, due to the machine precision, the variables whose domains cannot be bisected must also be excluded from L.

Second step: Generation of the RCL. The restricted candidate list (RCL) is a subset of L defining the variables that will be eligible for bisection. Let s be the vector of smear values $s(x_i, [v])$ for all $x_i \in L$. Let s_l be the lowest value in s and let s_u be the greatest value in s. Given a threshold $S \in [s_l, s_u]$, let RCL be the set of all the variables $x_i \in L$ such that $s(x_i, [v]) \geqslant S$. We propose to compute the threshold as

$$S = s_l + \alpha(s_u - s_l)$$

where $\alpha \in [0, 1]$ is a parameter controlling the size of the RCL. Fixing $\alpha = 1$ clearly leads to the maximum smear heuristic. Fixing $\alpha = 0$ implies $RCL = L$.

We assume that the smear values are interesting when the mean value extensions are tight enough. As a consequence, the value of α must increase when the smear values of the variables in L decrease. We then define

$$\alpha = \phi(\psi(s))$$

where $\phi : \mathbb{R}^+ \mapsto [0, 1]$ is continuous and decreasing and $\psi(s)$ is a scalar of the same order of the smear values in s. The ψ function can be defined as the mean or the standard deviation. The ϕ function can be defined as

$$\phi_1(y) = \frac{1}{1 + \beta y}$$

where $\beta > 0$ is a parameter determining the curvature of the function.

Third step: Choice of the variable. The selected variable is the variable from the *RCL* having the least number of bisections in the current branch of the search tree. This approach mimics a round-robin strategy on a subset of the best variables. As a consequence, the adaptive strategy is more round-robin-oriented when α has a low value, and it behaves like the maximum smear heuristics when α gets closer to 1.

4   Experimental Results

4.1   Problems and Solvers

We consider a set of 23 standard problems from the COCONUT benchmarking suite. All the experiments were made on a Core I7/2.67GHz using the interval solver Realpaver [4] based on the interval arithmetic library Gaol[1]. The contraction procedure combines 2B consistency, box consistency, and the interval Newton operator. The precision is assigned to $\epsilon = 10^{-8}$. The number of bisection steps is used as a measure to compare the different strategies. Solving times are not given here since they vary in proportion to the numbers of bisection steps.

4.2   Results

The ψ function used to calculate a scalar from the smear vector s is the standard deviation, which seems better than the mean. The ϕ function that transforms $\psi(s)$ to α is ϕ_1 with $\beta = 0.5$, which gives good average behaviors.

General results. The experimental results are reported in Table 1. In each row but the last two the problem name and its number of variables n are given. MaxDom (maximum domain), RR (round-robin), Smear, and ABS (adaptive bisection) are the number of bisection steps required to solve the problem with the corresponding strategy. The last two rows are the sums per columns: Sum is the sum of all values while TrimmedSum is equal to Sum minus the two largest values and the two smallest values in the given column.

The problems are ordered according to the best strategy for solving them. Problems from the first set are better solved with MaxDom. However, the other heuristics compare very well. It seems that the variables are equally important in these problems.

RR is the best for the second set of problems. MaxDom is clearly inefficient for Neuro and the smear criteria must be used instead of the domain widths. However, the smear-based strategy is not good for Combustion and the result of ABS shows that it is useful to diversify the choice of variables. Here, it is interesting to examine the evolution of α. In fact, α is close to 0 at the beginning of the solving process; the smear values are huge and a round-robin choice is preferred. The value of α increases according to the reduction of boxes and the final values are close to 1, leading to a smear-based behavior. For Trigo1 the

---

[1] http://sourceforge.net/projects/gaol/

Table 1. Comparison of bisection strategies (number of bisections)

Problem	n	MaxDom	RR	Smear	ABS
Chemistry	5	**65**	140	87	91
Kinematics2	8	**2 692**	3 476	2 795	2 812
PotraRheinboldt	20	**154**	708	159	184
Trigo2	10	**1 115**	1 747	1 414	1 178
Trinks	6	**63**	92	77	67
Yamamura	7	**45**	336	46	46
Celestial	3	3 361	**2 599**	4 258	3 074
Combustion	10	559	**463**	1 303	509
Neuro	6	1 288 779	**6 367**	7 223	7 215
Trigo1	10	1 258	**1 244**	2 086	2 071
Geineg	6	8 546	8 116	**1 707**	2 400
I3	20	10 034	9 178	**4 760**	5 314
Kapur	5	2 791	1 651	**231**	323
Nbody	8	1 976	2 049	**1 532**	1 765
BroydenTri	20	24	723	**23**	**23**
Brown	10	12 443	5 850	6 827	**4 996**
CzaporGeddes2	4	17 165	5 880	6 101	**5 569**
DiscreteBoundary	40	**6**	126	**6**	**6**
Economics	6	1 108	1 530	1 063	**1 023**
Kinematics1	12	226	221	215	**206**
Nauheim	8	1 204	722	816	**708**
Solotarev	4	2 668	267	325	**259**
Transistor	9	112 795	121 765	83 711	**42 085**
Sum		1 469 077	175 250	126 765	81 924
TrimmedSum		67 473	44 089	35 802	32 595

good balance is not reached. This suggests that the value of β must be increased for this problem in order to allow a more RR-oriented strategy.

The third set of problems illustrate the capabilities of the maximum smear heuristics. MaxDom and RR are clearly outperformed. The loss of performances of ABS is limited. The adaptation of α seems rather good, though this could be improved.

The new strategy ABS is the best one for the last set of problems. The most striking result concerns `Transistor`, which is known as a difficult problem with exponential functions. The smear criteria are clearly useless for large boxes, which is due to the evaluation of the exponential terms resulting in huge intervals. At the end of the solving process, α is approximatively equal to 0.5 and the strategy is balanced between RR and Smear.

The sum of values in the penultimate line shows that ABS is more robust, allowing to significantly decrease the total number of bisection steps. There is no problem for which ABS is much worse than the other strategies. Moreover, ABS remains the best strategy if the extremal cases are not considered (see TrimmedSum). The gains with respect to the other strategies are respectively equal to 9% (MaxSmear), 26% (RR), and 52% (MaxDom). It is worth noticing that ABS competes well even with default parameter settings.

Filtering algorithms. It is interesting to test whether ABS remains efficient when different filtering techniques are used. It turns out that very similar results (as in Table 1) have been obtained with a modified algorithm including probing operators [12]. In this case, the trimmed sums are equal to 19455 (MaxDom), 17409 (RR), 13142 (MaxSmear), and 12139 (ABS).

Growth of problem dimension. Three generic problems parameterized by the dimension n are considered now. The results reported in Table 2 show that the behaviour of ABS does not depend on the dimension for these problems. In particular, ABS is the best strategy for solving Brown and Economics. However, it seems that MaxDom can be useful, especially for Yamamura.

Table 2. Study of heuristics according to the dimension of problems

Problem	n	MaxDom	RR	Smear	ABS
Yamamura	7	**45**	336	46	46
	8	**148**	1 774	241	260
	9	**409**	6 216	833	852
	10	**1 097**	24 637	7 093	7 462
Brown	9	4 903	3 977	**2 494**	2 995
	10	12 443	5 850	6 827	**4 996**
	11	18 611	22 044	17 073	**14 455**
	12	155 052	53 563	30 529	**27 886**
Economics	5	**216**	226	218	258
	6	1 108	1 530	1 063	**1 023**
	7	7 117	9 042	6 449	**5 875**
	8	40 737	53 507	34 552	**29 048**

4.3   Parameter Tuning

The behavior of the ABS strategy depends on the ϕ function. Now we fix $\phi = \phi_1$ and different values of β are tested. Profiles of the function are illustrated below.

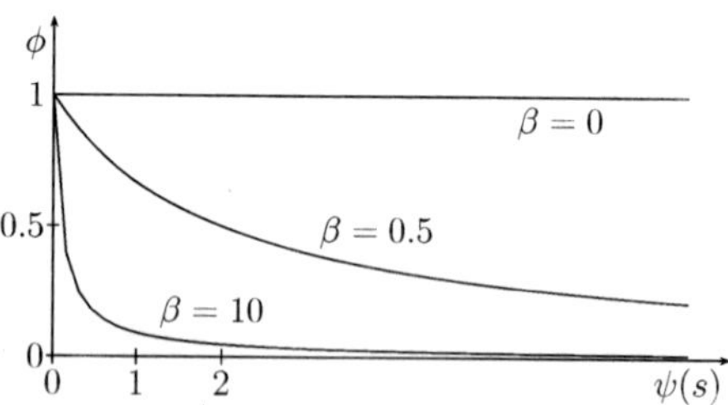

The results are reported in Table 3. Problems from the first set are better solved by the RR strategy and we clearly see that the performances decrease with β. On the contrary, the second set of problems is favorable for the MaxSmear strategy and the number of bisections decreases with β.

Table 3. Behavior of ABS according to β in ϕ_1 (number of bisections)

Problem	n	$\beta = 100$	$\beta = 10$	$\beta = 1$	$\beta = 0.5$	$\beta = 0$
Celestial	3	3042	3048	3069	3074	4258
Combustion	10	495	507	505	509	1406
Neuro	6	7059	7051	7151	7215	7273
Trigo1	10	1211	1616	2023	2071	2086
Geineg	6	6543	6485	3310	2400	1655
I3	20	9919	9407	6544	5314	4745
Kapur	5	323	323	323	323	255
Nbody	8	1809	1799	1794	1765	1541
Sum		30 401	30 236	24 719	22 671	23 219

The sums per columns are reported in the last line. The least number of bisections is obtained for $\beta = 0.5$. As a consequence, this specific value has been chosen as the default value. This value could probably be refined. Moreover, dynamically tuning this parameter could be useful.

5   Conclusion

The adaptive bisection strategy introduced in this paper aims at selecting variables according to observations of the solving state. The smear values are used to distinguish the good variables but more confidence is given when their magnitude is small. Balancing the use of the MaxSmear and RR strategies leads to a robust strategy that is also able to solve more efficiently some difficult problems.

The algorithm presented here is a general scheme and it can be parameterized to implement many concrete algorithms. The computation of α could be refined with other ψ and ϕ functions, possibly including automatic parameter tuning strategies [5]. The relative smear criterion [11] could be more efficient than the smear criterion. It could also be useful to learn the impact of variables [10]. Finally, more random strategies can be easily implemented, in particular for calculating the value of α or for choosing the variable in the restricted candidate list. In fact, we have tested such strategies but they are not robust in general. However, we have observed interesting behaviors and very good performances for some problems and this approach could be used as an experimental tool.

Acknowledgements. The author thanks Xavier Gandibleux and Christophe Jermann for interesting discussions and valuable suggestions on early drafts of this paper. The author is also grateful to the three anonymous reviewers whose remarks have been taken into account to improve the presentation of this work.

References

1. Benhamou, F., McAllester, D., Van Hentenryck, P.: CLP(Intervals) Revisited. In: Proc. ILPS, pp. 124–138 (1994)
2. Csendes, T., Ratz, D.: Subdivision Direction Selection in Interval Methods for Global Optimization. SIAM J. Numerical Analysis 34, 922–938 (1997)

3. Feo, T.A., Resende, M.G.C.: Greedy Randomized Adaptive Search Procedures. Journal of Global Optimization 6, 109–133 (1995)
4. Granvilliers, L., Benhamou, F.: Algorithm 852: Realpaver: an Interval Solver using Constraint Satisfaction Techniques. ACM Trans. Mathematical Software 32(1), 138–156 (2006)
5. Hamadi, Y., Monfroy, E., Saubion, F. (eds.): Autonomous Search. Springer (2012)
6. Van Hentenryck, P., Mcallester, D., Kapur, D.: Solving Polynomial Systems Using a Branch and Prune Approach. SIAM J. Numerical Analysis 34, 797–827 (1997)
7. Kearfott, R.B.: Some Tests of Generalized Bisection. ACM Trans. Mathematical Software 13(3), 197–220 (1987)
8. Kearfott, R.B., Novoa, M.: Algorithm 681: INTBIS, a portable interval Newton/bisection package. ACM Trans. Mathematical Software 16(2), 152–157 (1990)
9. Lhomme, O.: Consistency Techniques for Numeric CSPs. In: Proc. IJCAI, pp. 232–238 (1993)
10. Refalo, P.: Impact-Based Search Strategies for Constraint Programming. In: Wallace, M. (ed.) CP 2004. LNCS, vol. 3258, pp. 557–571. Springer, Heidelberg (2004)
11. Trombettoni, G., Araya, I., Neveu, B., Chabert, G.: Inner Regions and Interval Linearizations for Global Optimization. In: Proc. AAAI, pp. 99–104 (2011)
12. Trombettoni, G., Chabert, G.: Constructive Interval Disjunction. In: Bessière, C. (ed.) CP 2007. LNCS, vol. 4741, pp. 635–650. Springer, Heidelberg (2007)

Resource Constrained Shortest Paths with a Super Additive Objective Function

Stefano Gualandi[1] and Federico Malucelli[2]

[1] Università di Pavia, Dipartimento di Matematica
[2] Politecnico di Milano, Dipartimento di Elettronica e Informazione
`stefano.gualandi@unipv.it`, `malucell@elet.polimi.it`

Abstract. We present an exact solution approach to the constrained shortest path problem with a super additive objective function. This problem generalizes the resource constrained shortest path problem by considering a cost function $c(\cdot)$ such that, given two consecutive paths P_1 and P_2, $c(P_1 \cup P_2) \geq c(P_1) + c(P_2)$. Since super additivity invalidates the Bellman optimality conditions, known resource constrained shortest path algorithms must be revisited. Our exact solution algorithm is based on a two stage approach: first, the size of the input graph is reduced as much as possible using resource, cost, and Lagrangian reduced-cost filtering algorithms that account for the super additive cost function. Then, since the Lagrangian relaxation provides a tight lower bound, the optimal solution is computed using a near-shortest path enumerative algorithm that exploits the lower bound. The behavior of the different filtering procedures are compared, in terms of computation time, reduction of the input graph, and solution quality, considering two classes of graphs deriving from real applications.

1   Introduction

Consider a directed graph $G = (N, A)$, a set of resources K, a length l_e and a resource consumption r_e^k for each arc e in A and each resource k in K, the problem of finding the shortest path between a pair of given vertices s and t in N such that the consumption of each resource k does not exceed a given upper limit U^k is known in the literature as the *Resource Constrained Shortest Path problem* (RCSP). In this problem, the resource consumption of a path is given as an additive (linear) function. Thus, representing a path P as list of arcs, $r^k(P) = \sum_{e \in P} r_e^k$. Similarly, the cost of P is additive in the arc lengths, that is, $c(P) = \sum_{e \in P} l_e$. The *additivity* property plays an important role in the solution approaches since, given two consecutive paths P_1 and P_2, the cost and the resource consumption of the union of the two paths are $c(P_1 \cup P_2) = c(P_1) + c(P_2)$ and $r^k(P_1 \cup P_2) = r^k(P_1) + r^k(P_2)$. Note that RCSP is weakly NP-hard even in the case of a single resource, and approximate algorithms perform poorly in practice.

In this paper, we study a generalization of RCSP that considers super additive cost functions, that is such that $c(P_1 \cup P_2) \geq c(P_1) + c(P_2)$. This is motivated by

M. Milano (Ed.): CP 2012, LNCS 7514, pp. 299–315, 2012.

several applications. For instance, consider the allowances paid for extra hours of works in a crew scheduling problem. Usually, allowances are not due for the regular working hours, but a fixed allowance is paid for each hour exceeding the regular working time. Indeed, allowances bring in the cost function a step-wise component that makes the cost function super additive. Another application is in the transportation context, where the passenger perceived travel times have to be minimized. The travel time is perceived differently if it is below or above a certain threshold, thus also in this case the sum of the perceived travel times of two consecutive paths is smaller than the perceived travel time of the path obtained by joining the two paths.

The additive RCSP has been widely studied in the literature, and for a recent survey refer to [9]. Most of the existing algorithms follow a three stage approach:

1. Compute a lower bound (LB), usually via Lagrangian relaxations, and an upper bound (UB) with some heuristic.
2. Exploit such bounds to remove nodes and arcs that cannot provably lead to feasible or optimum paths.
3. Close the duality gap between the lower and the upper bound with a search strategy.

In order to use this approach for RCSP with super additive cost functions, we need to reconsider all the three stages, and this is the main contribution of our paper. As shown in [20,7], in Constraint Programming terminology, phase two is related to constraint filtering and propagation, and phase three to branch-and-bound search. We review next the main literature on RCSP.

The seminal work in [8] is the first that combines a Lagrangian relaxation to compute a lower bound, with k-shortest path enumeration to close the duality gap. In [1], reduction techniques based on the resource consumption are introduced to shrink the graph. In [3], RCSP is formulated as an integer linear program, and a Lagrangian relaxation, solved via subgradient optimization, is used to compute lower bounds and to perform additional problem reductions exploiting the near-optimal Lagrangian multipliers. To compute optimal solutions, the authors use a branch-and-bound depth-first search, running their reduction algorithm at every node of the search tree. In [15,25], the hull approach is introduced to solve the Lagrangian relaxation in the case of a single resource, and, in order to close the gap, paths are ranked by reduced costs and enumerated with a k-shortest path algorithm. A near-shortest path algorithm, introduced in [14], based on the reduced cost lengths is used instead of k-shortest path in [4]. In [16], Lagrangian relaxations and preprocessing techniques are pushed to their limit, since a Lagrangian relaxation is solved for every arc, while simultaneously performing problem reductions. The authors show that, for difficult instances, the computational overhead is paid-off by the graph reduction achieved. The computational experiments in [3] are among the few that consider up to 10 resources, since in most of the cases, only one resource is considered in the experiments.

The works in [8,1,3,15,25,4,16] are all based on an enumerative search algorithm to close the gap between the lower and upper bounds found in phase one. Since the problem is weakly NP-hard, pseudo-polynomial labeling algorithms are efficiently used to solve RCSP in practice. Recently in [5], a systematic computational comparison of different algorithms has shown that a labeling algorithm that fully exploits all the information gathered during phases one and two is the state-of-the-art algorithm for several types of graphs. Label-setting algorithms strongly rely on the existence of *dominance rules* that permit to fathom every dominated subpath. Since in many application there exist additional side-constraints that can weaken or even vanish the strength of such dominance rules, we have focused on enumerative search algorithms. A similar argument has motivated the use of CP in a recent route finder application developed by IBM [13]. Other interesting approaches to RCSP are the use of constrained-based local search as proposed in [17], and the branch-and-cut algorithm proposed in [10]. The use of dominators could lead to more efficient implementation of the reduction techniques [18].

The main applications of RCSP are in the context of column generation for vehicle routing problems, where the pricing subproblem is formulated as an RCSP on an auxiliary network. Similarly, also the Quality of Service routing in telecommunications [12,11] adopts the same approach. In column generation, the RCSP is often defined on a directed acyclic graph (e.g., see [24]), for which generalized arc consistency on the resource consumption can be achieved in linear time in the size of the graph [22]. For these reasons, our preliminary computational test are focused on directed acyclic graph, though the proposed approach can be directly applied to general graphs with non-negative arc lengths by using a Dijkstra-like algorithm to compute additive shortest paths instead of using dynamic programming.

Shortest path problems with non additive cost functions (but without resource constraints) emerged as a core subproblem in finding traffic equilibria [6]. The case of a single attribute shortest path is considered in [23], where the super additive component is a composite (differentiable) function of a linear summation on the arcs. The authors called their problem the Nonadditive Shortest Path Problem (NASP), and they use the hull approach introduced in [15] to compute lower bounds and a labeling algorithm to close the duality gap. The only other work about non additive shortest path problems we are aware of is presented in [19], where the authors solve multi-objective and multi-constrained non additive shortest path problems via labeling algorithms that exploit so called *gradient dominance* rules.

The outline of the paper is as follows. Section 2 introduces the notation and the formal statement of the problem. Section 3 presents the two Lagrangian relaxations used to compute lower bounds. Section 4 details the implementation of our constrained path solver for super additive cost functions. Section 5 presents computational results on random grid graphs used in the literature, and a set of graphs extracted by a column generation algorithm for a crew scheduling problem arising from real data. Finally, in Section 6 we discuss further research topics.

2   Problem Stament and Notation

Let $G = (N, A)$ be a simple weighted directed graph, and let s and t be the designated source and target node, respectively. For each arch $e = (i, j) \in A$, let l_e denote its length and let r_e^k be the quantity of the resource k consumed on the arc e. Let K be a set of resources. The resource consumption is assumed to be additive on the arcs along the path. For each resource $k \in K$ a lower and an upper bound L^k and U^k are given.

Definition 1. *A path P in G is resource feasible if and only if*

$$L^k \leq \sum_{e \in P} r_e^k \leq U^k, \quad \forall k \in K. \tag{1}$$

Since we consider simple paths only, a path P is represented by a set of arcs. With a slight abuse of notation, we denote by $P_1 \cup P_2$ the union of the set of arcs in P_1 and P_2, even if such union does not correspond to a path. Given a path P_{st} from s to t and two subpaths P_{si} and P_{it}, we denote by $P_{si} \uplus P_{it}$ the concatenation of the two subpaths that yields the path P_{st}.

Considering two arc disjoint paths P_1 and P_2 we can introduce the following definitions:

Definition 2. *A cost function $f(\cdot)$ is* **additive** *if*

$$f(P_1 \cup P_2) = f(P_1) + f(P_2). \tag{2}$$

Definition 3. *A cost function $f(\cdot)$ is* **super additive** *if*

$$f(P_1 \cup P_2) \geq f(P_1) + f(P_2). \tag{3}$$

In our problem, the cost of a path P is the sum of an additive length function $l(\cdot)$ and a super additive cost function $f(\cdot)$:

$$c(P) = l(P) + f(P) = \sum_{e \in P} l_e + f(P). \tag{4}$$

Clearly, cost function $c(\cdot)$ is super additive.

The problem of finding the shortest path P_{st}^* in G from s to t such that constraints (1) are satisfied and the objective function (4) is minimized is called the *Resource Constrained Superadditive Shortest Path Problem (RCSP-S)*.

Note that super additivity invalidates the optimality conditions based on the Bellmann's equation, and therefore known algorithms for the resource constrained shortest path problem need to be revisited. This motivates our work.

In order to simplify the exposition, and at the same time to get closer to a real-life application, we assume that the super additive function $f(\cdot)$ is a composite non-decreasing function of a linear function in one of the resource, i.e., $f(P) = f\left(\sum_{e \in P} r_e^1\right)$. As an example, consider the allowances of extra working time: they are usually a step-wise linear function of the overall time, that is you get paid an allowance for each extra hour of work. In addition, to easy the notation and without loss of generality, we only consider upper bounds on the resource consumption, i.e. $L^k = 0$. Further on we will denote by $r^k(P) = \sum_{e \in P} r_e^k$.

3   Filtering Based on Costs and Reduced Costs

3.1   Resource-Based Filtering

Constrained shortest path algorithms usually exploit resource additivity to filter out nodes and arcs of the graph that cannot lead to a feasible path. Considering resource k, let F^k denote the shortest path tree rooted at source node s (hereafter called the *forward* tree), and let B^k be the shortest path tree in the reversed graph of G, denoted by $\overleftarrow{G}$, from the target node t (hereafter called the *backward* tree). Note that the two trees F^k and B^k are obtained with two shortest path computations starting from s and using the forward arcs, and starting from t and using the backward arcs, where the arc lengths are given by the resource k consumption r_{ij}^k. As the notation, $r_i^k(F^k)$ and $r_i^k(B^k)$ are the minimum resource consumptions from node s to i and from i to t, respectively.

Resource-based filtering is based on the following relations and a direct application of resource additivity [1,3]. A node i can be removed from N whenever $r_i^k(F^k) + r_i^k(B^k) > U^k$. Similarly an arc (i,j) can be removed from A whenever $r_i^k(F^k) + r_{ij}^k + r_j^k(B^k) > U^k$. For instance, the first relation is valid because $r^k(P_{si}) + r^k(P_{it}) = r^k(P_{si} \uplus P_{it})$ and a path is resource feasible if and only if $r^k(P_{si} \uplus P_{it}) \leq U^k$ for each $k \in K$.

3.2   Cost-Based Filtering

For additive cost functions, we apply the same procedure to remove nodes and arcs. First, we compute the forward and backward shortest path trees. Again, this is done with two shortest path computations using the original arc lengths. Then, for every node (arc) we check whether a feasible path passing by the node (arc) improves the current best path. If it is not the case we filter out the node (arc).

For super additive cost functions, we need to revise the two relations that allow us to remove nodes and arcs. Due to (3), given a node i, the path obtained concatenating the shortest paths from s to i and from i to t can have a cost higher than the shortest path from s to t, as shown in the following example.

Example 1. Consider the cost function $c(P) = \sum_{e \in P} l_e + \left(\sum_{e \in P} r_e^1 \right)^2$. Note that in this cost function the super additive component depends on the consumption of resource 1. This type of function is the same used in [23]. Consider the graph in Figure 1, and suppose that we have found a path of total cost 32 (not shown in the figure). The shortest path P_{si} from s to i is $\{(s,a),(a,i)\}$ and has cost $5 + 5 + (1+1)^2 = 14$, and the shortest path P_{it} from i to t is $\{(i,b),(b,t)\}$ with cost 14. The path that we obtain as $P_{si} \uplus P_{it}$ has cost $(10+10) + (2+2)^2 = 36$. Since the cost is greater than 32, it is tempting to remove node i. However, the path $\{(s,a),(a,i),(i,d),(d,t)\}$, that is not obtained as the concatenation of the two shortest subpaths, has cost $25 + 2^2 = 29 < 36$.

Let us consider a super additive non negative function $f(\cdot)$ that linearly depends on the consumption of one resource. W.l.o.g let r^1 be such a resource (for simplicity, further on we will omit apex 1 in the function definition). In order to

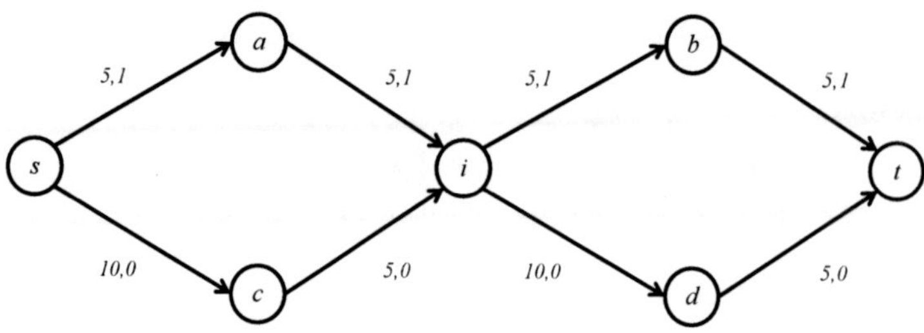

Fig. 1.

apply a filtering on the path costs, we need to compute four shortest path trees per node, two in the original graph G and two in the reversed graph $\overleftarrow{G}$. Let F^l and F^r be the forward shortest path trees, the first with respect to arc lengths l_e and the second with respect to the arc resource consumption r_e. Let B^l and B^r be the backward shortest path trees. Then we can filter out nodes and arcs as follows.

Proposition 1. *Given F^l, F^r, B^l, B^r, and an upper bound of value UB, any node i that satisfies the following relation:*

$$l_i(F^l) + l_i(B^l) + f\left(r_i(F^r) + r_i(B^r)\right) \geq UB \tag{5}$$

can be removed from N. Any arc $e = (i,j)$ that satisfies:

$$l_i(F^l) + l_e + l_j(B^l) + f\left(r_i(F^r) + r_e + r_i(B^r)\right) \geq UB \tag{6}$$

can be removed from A.

Proof. The cost of the shortest path $P_{st} = P_{si} \uplus P_{it}$ passing by a node i is:

$$\begin{aligned}
c(P_{si} \uplus P_{it}) &= l(P_{si} \uplus P_{it}) + f(P_{si} \uplus P_{it}) \\
&= l(P_{si}) + l(P_{it}) + f(P_{si} \uplus P_{it}) \\
&\geq l_i(F^l) + l_i(B^l) + f(P_{si} \uplus P_{it}) \\
&\geq l_i(F^l) + l_i(B^l) + f(r_i(F^r) + r_i(B^r)),
\end{aligned}$$

since $f(r_i(F^r) + r_i(B^r))$ is the least cost due to the super additive function. Thus $l_i(F^l) + l_i(B^l) + f(r_i(F^r) + r_i(B^r))$ is a valid lower bound for any path passing by node i, and if its value is greater than or equal to the upper bound, we can safely remove node i from N. Similarly, we can prove relation (6).   □

Note that if we keep track of the path yielding the best upper bound, whenever the target node t is no longer reachable from s, then the current best path is proved to be optimum.

3.3   Lagrangian Relaxation and Cost-Based Filtering

Let us introduce the arc-flow formulation of RCSP-S that we use to derive a first standard Lagrangian relaxation. Let x_{ij} be a 0-1 variable that denotes if arc (i,j) is part of the shortest path. Let r^1 be the resource contributing to the super additive cost function. The non-linear arc flow formulation of RCSP-S is as follows:

$$\min \quad \sum_{(i,j)\in A} l_{ij}x_{ij} + f\left(\sum_{(i,j)\in A} r^1_{ij}x_{ij}\right) \tag{7}$$

$$\text{s.t.} \quad \sum_{(i,j)\in A} x_{ij} + \sum_{(j,i)\in A} x_{ji} = \begin{cases} -1 \text{ if } i = s \\ 0 \text{ if } i \neq s, i \neq t \\ +1 \text{ if } i = t \end{cases}, \qquad \forall i \in N, \tag{8}$$

$$\sum_{(i,j)\in A} r^k_{ij}x_{ij} \leq U^k, \qquad \forall k \in K, \tag{9}$$

$$x_{ij} \in \{0,1\}, \qquad \forall(i,j) \in A. \tag{10}$$

Constraints (8) impose the flow balance, and constraints (9) force the upper limits on the resource consumption.

Since the domain of the super additive function $f(\cdot)$ linearly depends on r^1, the linearization proposed in [23] for the NASP can be applied. Let us introduce a new variable z that equals the consumption of resource r^1 on the path, the problem can be reformulated as follows:

$$\min \quad \sum_{ij\in A} l_{ij}x_{ij} + f(z) \tag{11}$$

$$\text{s.t.} \quad \sum_{(i,j)\in A} x_{ij} + \sum_{ji\in A} x_{ji} = \begin{cases} -1 \text{ if } i = s \\ 0 \text{ if } i \neq s, i \neq t \\ +1 \text{ if } i = t \end{cases}, \qquad \forall i \in N, \tag{12}$$

$$\sum_{(i,j)\in A} r^k_{ij}x_{ij} \leq U^k, \qquad \forall k \in K, \tag{13}$$

$$\sum_{(i,j)\in A} r^1_{ij}x_{ij} = z, \tag{14}$$

$$x_{ij} \in \{0,1\}, \qquad \forall(i,j) \in A. \tag{15}$$

Since we are minimizing and $f(\cdot)$ is non-negative, the equality in constraint (14) is replaced by inequality '$\leq$'.

The Lagrangian relaxation obtained by penalizing in the objective function constraints (13) and (14) gives rise to two separable subproblems that can be solved easly. Let $\lambda_k, k \in K$ be the dual multipliers of constraints (13) and let β be the dual multiplier of constraint (14). Both are restricted in sign to be non-positive, i.e., $\lambda_k \leq 0$ and $\beta \leq 0$. The Lagrangian relaxation of problem (11)–(15) is:

$$\Phi(\lambda, \beta) = \min \sum_{(i,j)\in A} \left(l_{ij} + \sum_{k\in K} \lambda_k r_{ij}^k + \beta r_{ij}^1 \right) x_{ij} - \sum_{k\in K} \lambda_k U^k + f(z) - \beta z$$

$$(16)$$

$$\text{s.t.} \sum_{(i,j)\in A} x_{ij} + \sum_{(j,i)\in A} x_{ji} = \begin{cases} -1 & \text{if } i = s \\ 0 & \text{if } i \neq s, i \neq t \\ +1 & \text{if } i = t \end{cases}, \quad \forall i \in N, \quad (17)$$

$$x_{ij} \in \{0, 1\}, \quad \forall (i,j) \in A. \tag{18}$$

Note that variable z does not appear in any constraint, thus the problem decomposes into two independent subproblems. One subproblem involves variables x_{ij} and is a simple shortest path problem without additional constraints and having arc lengths defined by the reduced costs $\bar{c}_{ij}(\lambda, \beta) = l_{ij} + \sum_{k\in K} \lambda_k r_{ij}^k + \beta r_{ij}^1$. The other problem involves only variable z and is a unidimensional unconstrained optimization problem.

The Lagrangian dual is:

$$\max_{\lambda, \beta}\{\Phi(\lambda, \beta)\} = \max_{\lambda, \beta} \left\{ -\sum_{k\in K} \lambda_k U^k + \min \sum_{(i,j)\in A} \bar{c}(\lambda, \beta)_{ij} x_{ij} + f(z) - \beta z \right\}$$

$$(19)$$

and its optimal solution can be obtained by standard mathematical programming methods. Let $(\bar{\lambda}, \bar{\beta})$ be the optimal dual multipliers, and the values $\bar{x}_{ij}$ and $\bar{z}$ the corresponding primal solutions, though not necessarily feasible for the original problem. Let $F^{(\bar{\lambda},\bar{\beta})}$ and $B^{(\bar{\lambda},\bar{\beta})}$ be the forward and backward additive shortest path trees according to reduced costs $\bar{c}_{ij}(\lambda, \beta)$.

A simple lower bound to the Lagrangian dual is obtained by neglecting the super additive component of the cost function thus we consider only the effects of the resource constraints on the additive cost function. Even in a weaker way, this evaluation can still filter out some nodes and arcs.

Proposition 2. *Given $F^{\bar{\lambda}}, B^{\bar{\lambda}}, F^r$ and B^r, and an upper bound of value UB, any node i that satisfies the following relation:*

$$\bar{c}_i(F^{\bar{\lambda}}) + \bar{c}_i(B^{\bar{\lambda}}) + f(r_i(F^r) + r_i(B^r)) \geq UB + \sum_{k\in K} \bar{\lambda}_k U^k \tag{20}$$

can be removed from N. Any arc $e = (i, j)$ that satisfies:

$$\bar{c}_i(F^{\bar{\lambda}}) + c_e^{\bar{\lambda}} + \bar{c}_j(B^{\bar{\lambda}}) + f(r_i(F^r) + r_e + r_j(B^r)) \geq UB + \sum_{k\in K} \bar{\lambda}_k U^k \tag{21}$$

can be removed from A.

Proof. Note that $\bar{c}_i(F^{\bar{\lambda}}) + \bar{c}_j(B^{\bar{\lambda}}) - \sum_{k\in K} \bar{\lambda}_k U^k$ is the Lagrangian lower bound obtained using the optimal multipliers $\bar{\lambda}$ and by setting $\beta = 0$. As in Proposition (1), $f(r_i(F^r) + r_i(B^r))$ is a valid lower bound for shortest path with the super additive function. $\qquad\square$

Proposition 3. *Given $F^{(\bar\lambda,\bar\beta)}$ and $B^{(\bar\lambda,\bar\beta)}$, and an upper bound of value UB, any node i that satisfies the following relation:*

$$\bar c_i(F^{(\bar\lambda,\bar\beta)}) + \bar c_i(B^{(\bar\lambda,\bar\beta)}) \geq UB + \sum_{k\in K} \bar\lambda_k U^k - f(\bar z) + \bar\beta\bar z \tag{22}$$

can be removed from N. Any arc $e = (i,j)$ that satisfies:

$$\bar c_i F^{(\bar\lambda,\bar\beta)} + \bar c_e(\bar\lambda,\bar\beta) + \bar c_j(B^{(\bar\lambda,\bar\beta)}) \geq UB + \sum_{k\in K} \bar\lambda_k U^k - f(\bar z) + \bar\beta\bar z \tag{23}$$

can be removed from A.

Proof. (Sketch) In the Lagrangian dual we have the quantity $(-\sum_{k\in K} \lambda_k U^k + f(z) - \beta z)$ that does not depend on the path. Therefore, once we have solved to optimality the Lagrangian dual and obtained the optimal values $(\bar\lambda, \bar\beta, \bar z)$ the fixed quantity $(-\sum_{k\in K} \bar\lambda_k U^k + f(\bar z) - \bar\beta\bar z)$ can be added to reduced cost of any shortest path passing by i to obtain a valid lower bound. $\qquad\square$

3.4   An Alternative Lagrangian Relaxation

The RCSP-S problem can be formulated using path variables instead of arc-flow variables. This type of formulation is used in [15] for RCSP and in [23] for NASP.

The path-based formulation has a 0-1 variable y_p for each path p in G. Let $\mathcal{P}$ denote the collection of all paths of G. Given a path p, its super additive cost is c_p and, for each resource k, its resource consumption is r_p^k. The path-based formulation is:

$$\min \quad \sum_{p\in\mathcal{P}} c_p y_p \tag{24}$$

$$\text{s.t.} \quad \sum_{p\in\mathcal{P}} y_p = 1 \tag{25}$$

$$\sum_{p\in\mathcal{P}} r_p^k y_p \leq U^k \qquad \forall k \in k \tag{26}$$

$$y_p \in \{0,1\} \qquad \forall p \in \mathcal{P}. \tag{27}$$

Note that this formulation is linear, differently from (11)–(15).

If we relax the integrality constraint (27) into $y_p \geq 0$, and consider the dual of the so obtained LP relaxation, we get the following LP problem:

$$\max \quad u + \sum_{k\in K} U^k u^k \tag{28}$$

$$\text{s.t.} \quad u + \sum_{k\in K} r_p^k u^k \leq c_p \qquad \forall p \in \mathcal{P} \tag{29}$$

$$u^k \leq 0 \qquad \forall k \in K. \tag{30}$$

The difference of this formulation in comparison with those in [15,23] resides in the fact that path c_p are here defined by the super additive cost function, thus requiring ad hoc methods for solving the problem. When solving this problem with a cutting plane algorithm, the separation problem, amounts to solve a NASP with the reduced costs equal to $\bar{c}_e = c_e + \sum_{k \in K} r_e^k \bar{u}^k$. In [15], the separation problem is an additive shortest problem, solvable in polynomial time, while in our case the separation is a weakly NP-hard problem.

4   The Constrained Path Solver

Instead of implementing a global constraint in a general purpose constraint programming solver, we have implemented our constrained path solver from scratch using a simplified three stage approach, where the first two phases are joint and executed at same time. First, the solver computes the Lagrangian relaxation and filters out (i.e. it removes from G) every possible node and/or arc using resource, cost, and reduced-cost filtering as presented in the previous section. Then, the solver performs an exhaustive search, by enumerating every near-shortest path.

As originally proposed in [5], our filtering algorithm looks for improving feasible solutions *while* filtering, since better UBs make cost filtering more effective in practice. That is why we called our algorithm FILTERANDDIVE. The shorter path constraint discussed in [7,22] does not have the dive component, but it additionaly detects those arcs that must be part of a feasible path. In our implementation, reduced-cost filtering is done with both optimal and suboptimal Lagrangian multipliers [21].

4.1   Constraint Filtering and Bounding

The constrained path solver executes a general filtering framework using different arc-length functions. Hereafter, we denote by $g(\cdot)$ a generic arc-length function. The arc-length functions that we consider for filtering are (i) the consumption of each k-th resource to perform resource-based filtering, (ii) the super additive cost function $c(\cdot)$ to check Proposition (1), and (iii) the Lagrangian reduced-cost function parametric in $(\bar{\lambda}, \bar{\beta}, \bar{z})$ to check Propositions (2) and (3).

Before applying the filtering for any function, we need to compute the forward and backward shortest path trees. The current implementation deals with directed acyclic graphs and uses dynamic programming to find shortest path trees that has complexity linear in the size of the graph. We use a modified shortest path algorithm that, besides the usual node distance and predecessor labels, it stores also a tuple of labels $\langle$cost, reduced-cost, length, resources$\rangle$, that is the distance labels for the other objective functions. The output of our SHORTESTPATH procedure is an "augmented" shortest path tree. The additional information stored at each node is used to *dive* for improving paths. Indeed, given a pair of forward and backward shortest path trees F^g and B^g and an arc (i,j), we use the tuple of labels stored at the nodes of the two trees, F_i^g and B_j^g along with the properties of arc (i,j) to check if $P_{si} \uplus \{i,j\} \uplus P_{jt}$ is feasible. In the presence of a feasible path that improves also the cost of the incumbent, we update the UB. In case $LB \geq UB$ we can also prove the optimality.

Algorithm 1 shows our FILTERANDDIVE algorithm that filters out nodes and arcs while diving for new incumbent solutions. It takes as input a directed graph $G = (N, A)$, a pair (LB, UB), a pair of forward and backward shortest path trees (F^g, B^g) and an upper limit U^g that depends on the arc-length function $g(\cdot)$. The filtering algorithm uses three sub-procedures:

- PATHCOST(F_i^g, e, B_j^g): using the tuple of labels stored at each node, it computes the cost of $P_{si} \uplus \{i, j\} \uplus P_{jt}$ without building the path.
- PATHFEASIBLE(F_i^g, e, B_j^g): using the tuple of labels stored at each node, it checks if $P_{si} \uplus \{i, j\} \uplus P_{jt}$ is feasible without building the path.
- MAKEPATH(F_i^g, e, B_j^g): using the predecessor labels stored in F_i^g and B_j^g it finds and stores the path $P_{si} \uplus \{i, j\} \uplus P_{jt}$.

While PATHCOST and PATHFEASIBLE require constant time, the time complexity of filtering is linear in the graph size, to which we must add the complexity of computing the shortest path trees, that amounts to another linear time complexity in the graph size if we deal with acyclic graphs.

Algorithm 1. FILTERANDDIVE(G, LB, UB, F^g, B^g, U^g)

Input: $G = (N, A)$ directed graph and distance function $g(\cdot)$
Input: (LB, UB) lower and upper bounds on the optimal path
Input: F^g, B^g forward and backward shortest path tree as function of $g(\cdot)$
Input: U^g upper bound on the path length as function of $g(\cdot)$
Output: An optimum path, or updated UB, or a reduced graph

1   **foreach** $i \in N$ **do**
2   **if** $F_i^g + B_i^g > U^g$ **then**
3      $N \leftarrow N \setminus \{i\}$
4   **else**
5   **foreach** $e = (i, j) \in A$ **do**
6   **if** $F_i^g + g(e) + B_j^g > U^g$ **then**
7        $A \leftarrow A \setminus \{e\}$
8   **else**
9   **if** PATHCOST(F_i^g, e, B_j^g) $< UB \wedge$ PATHFEASIBLE(F_i^g, e, B_j^g) **then**
10         $P_{st}^* \leftarrow$ MAKEPATH(F_i^g, e, B_j^g);
11   Update UB and store P_{st}^*;
12   **if** $LB \geq UB$ **then**
13   **return** P_{st}^* (that is an optimum path)
14   **else**
15         $A \leftarrow A \setminus \{e\}$

Note that our constraint solver can handle more complex rules than just checking resource consumptions. However, the PATHFEASIBLE procedure should run in constant time or at most linear time, since it must be executed several times.

Algorithm 2 details the algorithm that performs the Lagrangian reduced-cost filtering. The first step consists of computing the Lagrangian dual using a subgradient procedure as in [3], while paying attention to solve $\min\{f(z) - \beta z\}$. Then the optimal Lagrangian multipliers $(\bar{\lambda}, \bar{\beta})$ are used to define $g(\cdot)$ as the reduced costs $\bar{c}_{ij}(\bar{\lambda}, \bar{\beta}) = l_{ij} + \sum_{k \in K} \bar{\lambda}_k r_{ij}^k + \bar{\beta} r_{ij}^1$. Together with the optimal value $\bar{z}$, they are also used to define $U^{(\bar{\lambda}, \bar{\beta}, \bar{z})} = UB + \sum_{k \in K} \lambda_k U^k - f(\bar{z}) + \bar{\beta}\bar{z}$. If the shortest reduced-cost path is feasible, then it is an optimum path. Otherwise, if the shortest reduced-cost path has a cost higher than UB, then the path giving UB is the optimum path. If the shortest reduced-cost is higher than the current lower bound, then we can update the lower bound. In any of these three conditions holds, we can compute the shortest forward and backward reduced-cost trees and call FILTERANDDIVE with $F^{(\bar{\lambda}, \bar{\beta})}$, $B^{(\bar{\lambda}, \bar{\beta})}$, and $U^{(\bar{\lambda}, \bar{\beta}, \bar{z})}$.

Algorithm 2. Lagrangian Reduced Cost-based Filtering

Input: $G = (N, A)$ directed graph
Input: (LB, UB) lower and upper bounds on the optimal path
Output: An optimum path, or updated (LB, UB), or a reduced graph
1 $(\bar{\lambda}, \bar{\beta}), \bar{z} \leftarrow$ SOLVELAGRANGIANRELAXATION;
2 $g(\cdot) \leftarrow$ REDUCEDCOSTFUNCTION$(\bar{\lambda}, \bar{\beta})$;
3 $U^g \leftarrow UB + \sum_{k \in K} \lambda_k U^k - f(\bar{z}) + \bar{\beta}\bar{z} + \epsilon$;
4 $F^g \leftarrow$ SHORTESTPATH$(G, s, g(\cdot))$;
5 **if** $c_t(F^g) \geq UB$ **then**
6 $\quad$ **return** the path giving UB is optimum

7 **if** $\bar{c}_t(F^g) > LB$ **then**
8 $\quad$ $LB \leftarrow \bar{c}_t(F^g)$;

9 **if** PATHFEASIBLE(F_t^g) **then**
10 $\quad$ $P_{st}^* \leftarrow$ MAKEPATH(F_t^g);
11 $\quad$ **return** P_{st}^* (that is an optimum path)

12 $B^g \leftarrow$ SHORTESTPATH$(\overleftarrow{G}, t, g(\cdot))$;
13 FILTERANDDIVE$(G, LB, UB, F^g, B^g, U^g)$;

Constraint propagation is quite simple: every time a call to the filtering algorithm removes a node/arc or update UB, the filtering algorithm is called again for each length function $g(\cdot)$.

4.2 Gap Closing: Near-Shortest Path Enumeration

The duality gap between the lower bound obtained with the Lagrangian relaxation and the value of the best feasible path obtained during the execution of FILTERANDDIVE is closed with a near-shortest path enumerative algorithm [14], as in [4,16]. This algorithm is similar to a search algorithm in any traditional

Constraint Programming solver: it uses a depth-first branch-and-bound. The filtering algorithms, however, are only executed at the root node of the search tree until propagation reaches a fix point, and this happens when any call to the filtering algorithm (with any of the considered length function $g(\cdot)$) has neither removed an arc/node nor updated UB.

Algorithm 3 sketches our near-shortest path algorithms used to close the duality gap. The depth-first search is implemented directly with a stack. For each node we maintain a tuple of labels $\langle$cost, reduced-cost, length, resources$\rangle$. The search algorithm computes once the backward search trees B^l and B^k for each resource $k \in K$. The backward search trees are used to get heuristic distances to the destination. Every time a node i is extracted from the stack, we check for every outgoing arc $e = (i, j)$ the following:

$$l(P_{si}) + l_e + B_j^l + f\left(r(P_{si}) + r_{ij} + B_j^r\right) \geq UB, \tag{31}$$

where $l(P_{si})$ and $r(P_{si})$ are the cumulated length and resource consumption from s to i. Whenever the above relation holds or the path form s to j is infeasible, the algorithm do not extend the search to node j. Similarly, the Lagrangian reduced-costs are used to prune the vertices of the search tree that yield to paths whose lower bounds are higher than UB (cumulated reduced-costs appear in the tuple of labels stored L_i at each node i).

5   Computational Results

The graphs we consider are both randomly generated and real-life instances. The randomly generated instances are directed acyclic grid graphs with negatively correlated arc lengths and resource consumptions. These type of graphs proved to be harder to solve than real-life structured instances [11], where usually lengths and resource consumptions are (weakly) correlated. As super additive cost function, we define $c(P) = l(P) + \left(r^1(P)\right)^2$, the same function used in the computational experiments in [23]. We use up to 10 resources. In addition, we consider a set of real instances coming from a bus driver scheduling solver of an Italian transportation company.

The path solver is implemented in C++ using the Boost Library and the QSopt linear programming library [2] to solve by cutting plane problem (28)–(30). The code is compiled with the gcc-4.3 compiler. All the tests were executed on a MacBookPro dual core 2.4Ghz with 4GB of RAM.

Random grid graphs. Table 1 shows the effects of resource, cost, and reduced-cost based filtering in terms of computation time in seconds, percentage of removed arcs, and a "pessimistic" duality gap computed as $\frac{UB-LB}{LB} \times 100$. We do not report the gap for the resource-based filtering, because without considering costs we do not have any significant lower bound. Each row gives the averages for 20 random graphs of the same size, but different costs and resource consumptions. All the instances have at least a feasible solution, the limits on the resource

---

Algorithm 3. Near-Shortest Path Enumeration

Input: $G = (N, A)$ directed graph
Input: (LB, UB) lower and upper bounds on the optimal path
Data: S stack of visited vertices, L tuple of labels
Output: An optimum path P_{st}^*

1 Compute backward shortest reduced-cost tree $B^{(\bar{\lambda}, \bar{\beta})}$;
2 Initialize tuple of labels L_s;
3 $S \leftarrow s$;
4 **forall the** $v \in N$ **do**
5 $\quad$ $\delta^+(v) \leftarrow$ forward star of v

6 **while** S *is not empty* **do**
7 $\quad$ $v \leftarrow$ front(S);
8 $\quad$ **if** $\exists e = (i, j) \in \delta^+(i)$ **then**
9 $\quad\quad$ $\delta^+(i) \leftarrow \delta^+(i) \setminus (i, j)$;
10 $\quad\quad$ **if** $j \notin S$ **and** $\textsc{EstimateCost}(L_i, e, B_j^{(\bar{\lambda}, \bar{\beta})}) < UB$ **then**
11 $\quad\quad\quad$ **if** $\textsc{PathFeasible}$ **then**
12 $\quad\quad\quad\quad$ **if** $j = t$ **then**
13 $\quad\quad\quad\quad\quad$ $P_{st}^* \leftarrow \textsc{MakePath};\quad UB \leftarrow c(P_{st}^*)$;
14 $\quad\quad\quad\quad\quad$ **if** $LB \geq UB$ **then**
15 $\quad\quad\quad\quad\quad\quad$ **return** P_{st}^* is an optimum path

16 $\quad\quad\quad\quad$ **else**
17 $\quad\quad\quad\quad\quad$ $S \leftarrow j;\quad L_j \leftarrow L_i \uplus e$;

18 $\quad$ **else**
19 $\quad\quad$ $S \leftarrow S \setminus \{i\},\quad \delta^+(i) \leftarrow$ forward star of i;

---

consumption are not tight taken singularly, but their combination makes the problem difficult.

Regarding the computation time, all the three types of filtering are efficient and scale gracefully with the size of the graph. As the percentage of arcs removed, it is clear that the reduced-cost filtering outperforms the other two. This is due to the Lagrangian relaxation that takes into account at the same time the costs and all the resource constraints. On the contrary, the resource-based filtering acts on a single resource constraint at a time. The little effect of cost-based filtering is due to the lack of tight upper bounds. Another important consequence of having a smaller graph due to reduced-cost filtering is the ability to find increasingly better lower and upper bounds that give very small duality gaps. We do not report here computational results, but without the diving component none of the cost filtering were so effective.

Bus driver scheduling. Table 2 shows a similar comparison but for real life instances. We collected the instances by running a commercial column generation algorithm, where the pricing subproblem is a RCSP with a (super additive) step-wise function that models extra working time allowances. We selected 42

Table 1. Comparison of filtering algorithms for directed acyclic graphs with n vertices, more than $2n$ arcs, and 10 resources. Each row gives the averages over 20 instances. Time is in seconds; $\Delta = \frac{m}{m_0} \times 100$ is the percentage of removed arcs ($m_0 = |A|$ is the number of arcs in G, m is after filtering), for resources, cost, and reduced cost filtering, respectively; gap is $\frac{UB-LB}{LB} \times 100$.

	RESOURCE		COST			REDUCED COST		
n	Time	Δ	Time	Δ	Gap	Time	Δ	Gap
1600	0.07	7.3%	0.09	9.7%	81.6%	0.10	92.3%	2.6%
6400	0.58	3.7%	0.81	7.4%	49.9%	0.68	94.8%	0.6%
25600	2.71	2.2%	3.59	4.2%	63.6%	2.76	95.3%	0.2%

Table 2. Comparison of filtering algorithms of real life instances. Here the gap is computed on the optimum value Opt, i.e. gap is $\frac{UB-Opt}{Opt} \times 100$.

Graphs		RESOURCE		COST			REDUCED COST			EXACT
n	m	Time	Δ	Time	Δ	Gap	Time	Δ	Gap	Time
4137	135506	0.77	22.5%	4.54	23.2%	71.3%	3.12	30.2%	0.0%	75.1
2835	132468	0.59	40.3%	4.40	41.0%	64.8%	2.35	45.4%	0.0%	30.6
3792	134701	0.92	30.2%	4.58	30.8%	88.7%	2.87	37.4%	0.0%	69.3

challenging instances. Each instance has 7 resources used to model complex regulation constraints. The instances refer to 3 different graphs, and each instance has different additive arc length l_e. Since the duality gap is large, we report also the computation time of the exact near-shortest path algorithm.

In this case, the resource-based filtering has more impact in reducing the number of arcs. The cost based filtering takes longer, since it removes a few arcs at the time, and therefore, it continues to trigger propagation, but with a small effect in the quality of the solution found and on the numbers of removed arcs. Apparently, the impact on the arc filtering due to reduced-cost filtering is less impressive. However, note that the duality gap computed with respect to the optimal solution is equal to zero, and the computation time is much less than considering only the costs, in order to test the approach on a variety of applications.

6 Conclusions

We have presented cost and reduced-cost based filtering algorithms for the RCSP problem with a super additive objective function, where the super additivity component is a composite function of a linear function. Reduced-cost filtering exploits Lagrangian relaxations and is very effective for reducing the graph size and to dive for improving solutions. The current implementation deals with directed acyclic graphs, but we are currently extending the solver to cyclic graphs with non-negative arc lengths.

References

1. Aneja, Y.P., Aggarwal, V., Nair, K.P.K.: Shortest chain subject to side constraints. Networks 13(2), 295–302 (1983)
2. Applegate, D., Cook, W., Dash, S., Mevenkamp, M.: QSopt linear programming solver, http://www.isye.gatech.edu/~wcook/qsopt/ (last visited, April 2012)
3. Beasley, J.E., Christofides, N.: An algorithm for the resource constrained shortest path problem. Networks 19(4), 379–394 (1989)
4. Carlyle, W.M., Royset, J.O., Wood, R.K.: Lagrangian relaxation and enumeration for solving constrained shortest-path problems. Networks 52(4), 256–270 (2008)
5. Dumitrescu, I., Boland, N.: Improved preprocessing, labeling and scaling algorithms for the weight-constrained shortest path problema. Networks 42(3), 135–153 (2003)
6. Gabriel, S.A., Bernstein, D.: Nonadditive shortest paths: subproblems in multi-agent competitive network models. Computational & Mathematical Organization Theory 6(1), 29–45 (2000)
7. Gellermann, T., Sellmann, M., Wright, R.: Shorter Path Constraints for the Resource Constrained Shortest Path Problem. In: Barták, R., Milano, M. (eds.) CPAIOR 2005. LNCS, vol. 3524, pp. 201–216. Springer, Heidelberg (2005)
8. Handler, G.Y., Zang, I.: A dual algorithm for the constrained shortest path problem. Networks 10(4), 293–309 (1980)
9. Irnich, S., Desaulniers, G.: Shortest Path Problems with Resource Constraints. In: Desaulniers, G., Desrosiers, J., Solomon, M. (eds.) Column Generation, pp. 33–65. Springer (2005)
10. Jepsen, M.K., Petersen, B., Spoorendonk, S.: A branch-and-cut algorithm for the elementary shortest path problem with a capacity constraint. Technical Report 08/01, Dept. of Computer Science. University of Copenhagen, Copenhagen (2008)
11. Kuipers, F., Korkmaz, T., Krunz, M., Van Mieghem, P.: Performance evaluation of constraint-based path selection algorithms. IEEE Network 18(5), 16–23 (2004)
12. Kuipers, F., Van Mieghem, P., Korkmaz, T., Krunz, M.: An overview of constraint-based path selection algorithms for qos routing. IEEE Communications Magazine 40(12), 50–55 (2002)
13. Lefebvre, M.P., Puget, J.-F., Vilím, P.: Route Finder: Efficiently Finding k Shortest Paths Using Constraint Programming. In: Lee, J. (ed.) CP 2011. LNCS, vol. 6876, pp. 42–53. Springer, Heidelberg (2011)
14. Matthew Carlyle, W., Kevin Wood, R.: Near-shortest and k-shortest simple paths. Networks 46(2), 98–109 (2005)
15. Mehlhorn, K., Ziegelmann, M.: Resource Constrained Shortest Paths. In: Paterson, M. (ed.) ESA 2000. LNCS, vol. 1879, pp. 326–337. Springer, Heidelberg (2000)
16. Muhandiramge, R., Boland, N.: Simultaneous solution of lagrangean dual problems interleaved with preprocessing for the weight constrained shortest path problem. Networks 53(4), 358–381 (2009)
17. Pham, Q.D., Deville, Y., Van Hentenryck, P.: Constraint-Based Local Search for Constrained Optimum Paths Problems. In: Lodi, A., Milano, M., Toth, P. (eds.) CPAIOR 2010. LNCS, vol. 6140, pp. 267–281. Springer, Heidelberg (2010)
18. Quesada, L., Van Roy, P., Deville, Y., Collet, R.: Using Dominators for Solving Constrained Path Problems. In: Van Hentenryck, P. (ed.) PADL 2006. LNCS, vol. 3819, pp. 73–87. Springer, Heidelberg (2006)
19. Reinhardt, L.B., Pisinger, D.: Multi-objective and multi-constrained non-additive shortest path problems. Computers & Operations Research 38(3), 605–616 (2011)

20. Sellmann, M.: Reduction Techniques in Constraint Programming and Combinatorial Optimization. PhD thesis, University of Paderborn (2003)
21. Sellmann, M.: Theoretical Foundations of CP-Based Lagrangian Relaxation. In: Wallace, M. (ed.) CP 2004. LNCS, vol. 3258, pp. 634–647. Springer, Heidelberg (2004)
22. Sellmann, M., Gellermann, T., Wright, R.: Cost-based filtering for shorter path constraints. Constraints 12, 207–238 (2007)
23. Tsaggouris, G., Zaroliagis, C.: Non-additive Shortest Paths. In: Albers, S., Radzik, T. (eds.) ESA 2004. LNCS, vol. 3221, pp. 822–834. Springer, Heidelberg (2004)
24. Zhu, X., Wilhelm, W.E.: A three-stage approach for the resource-constrained shortest path as a sub-problem in column generation. Computers & Operations Research 39(2), 164–178 (2012)
25. Ziegelmann, M.: Constrained shortest paths and related problems. PhD thesis, Universität des Saarlandes, Germany (2001)

Relating Proof Complexity Measures
and Practical Hardness of SAT

Matti Järvisalo[1], Arie Matsliah[2], Jakob Nordström[3], and Stanislav Živný[4]

[1] Department of Computer Science & HIIT, University of Helsinki, Finland
[2] IBM Research and Technion, Haifa, Israel
[3] KTH Royal Institute of Technology, Stockholm, Sweden
[4] University of Oxford, United Kingdom

Abstract. Boolean satisfiability (SAT) solvers have improved enormously in performance over the last 10–15 years and are today an indispensable tool for solving a wide range of computational problems. However, our understanding of what makes SAT instances hard or easy in practice is still quite limited. A recent line of research in proof complexity has studied theoretical complexity measures such as *length*, *width*, and *space* in resolution, which is a proof system closely related to state-of-the-art conflict-driven clause learning (CDCL) SAT solvers. Although it seems like a natural question whether these complexity measures could be relevant for understanding the practical hardness of SAT instances, to date there has been very limited research on such possible connections. This paper sets out on a systematic study of the interconnections between theoretical complexity and practical SAT solver performance. Our main focus is on space complexity in resolution, and we report results from extensive experiments aimed at understanding to what extent this measure is correlated with hardness in practice. Our conclusion from the empirical data is that the resolution space complexity of a formula would seem to be a more fine-grained indicator of whether the formula is hard or easy than the length or width needed in a resolution proof. On the theory side, we prove a separation of general and tree-like resolution space, where the latter has been proposed before as a measure of practical hardness, and also show connections between resolution space and backdoor sets.

1 Introduction

In the last 10–15 years, SAT solvers have become a standard tool for solving a wide variety of real-world computational problems [1]. Although all known SAT solvers have exponential running time in the worst case, dramatic improvements in performance have led to modern SAT solvers that can handle formulas with millions of variables. At the same time, very small formulas with just a few hundred variables are known which are completely beyond the reach of even the very best solvers. Understanding what makes a SAT instance hard or easy for state-of-the-art SAT solvers is therefore a fundamental problem. In particular, a natural, but not at all well-understood, question is whether one can find a good *measure on the practical hardness* of SAT instances.

The current work addresses this question from the viewpoint of *conflict-driven clause learning* (CDCL) solvers [2, 3, 4], which—applying efficient data structures, clause

M. Milano (Ed.): CP 2012, LNCS 7514, pp. 316–331, 2012.
© Springer-Verlag Berlin Heidelberg 2012

learning, forgetting, restarting, phase saving, and other important schemes—form the most prominent SAT solver paradigm today. Our goal is to explore possible connections between *practical hardness*—as witnessed by running times of CDCL solvers on SAT instances—and *proof complexity measures* employed in the formal study of the resolution proof systems that can be seen to underlie CDCL SAT solvers.

The main bottleneck for CDCL solvers—apart from the obvious exponential worst case behavior—is the amount of memory used. In practice, it is completely infeasible to store all clauses learned during a CDCL run, and one therefore needs to design a highly selective and efficient clause caching scheme that learns and keeps the clauses needed for the CDCL solver to finish fast. Thus, understanding time and memory requirements for clause learning algorithms, and how these resources are related to each other, is a question of great practical importance.

Proof complexity provides a possible approach for analyzing the potential and limitations of SAT solvers by studying the formal systems of reasoning which the solvers use to generate proofs. A lower bound for a proof system tells us that any algorithm, even an optimal (non-deterministic) one making all the right choices, must necessarily use at least the amount of a certain resource specified by this bound. In the other direction, theoretical upper bounds on a proof complexity measure give hope that SAT solvers can perform well with respect to the measure if an efficient search algorithm can be designed. Whereas the plain Davis–Putnam–Logemann–Loveland procedure (DPLL) [5, 6] is known to correspond to *tree-like* resolution, by recent theoretical accounts (including [7, 8]) CDCL solvers can be understood as deterministic instantiations of the general resolution proof system. In this context, the proof complexity measures of *length* (a.k.a. size) and *space* are interesting since they in some sense model the running time and memory consumption of (optimal) CDCL solvers, while *width* is another measure that seems relevant for practical performance in view of e.g. [9]. An informal description of these measures is as follows (see Section 2 for the formal definitions):

Length: The number of clauses in a resolution proof.
Width: The size of a largest clause in a resolution proof.
Space: The number of clauses needed "on the board" in a self-contained presentation of a proof, where inferences can only be made from what is currently on the board.

The length, width, or space of proving a formula is defined in the natural way by taking the minimum over all possible proofs with respect to the measure in question. As will be discussed in more detail below, these measures have been proven to form a strict hierarchy in the sense that for any formula we have the (informally stated) relations

$$\text{space} \geq \text{width} \geq \log \text{length} \tag{1}$$

(scaling length so that all measures have the same magnitude—length can be exponential while space and width are always at most linear), and these inequalities can all be asymptotically strict.

When discussing whether these theoretical complexity measures have any bearing on the practical hardness of formulas, perhaps the most obvious candidate to start with is *length*. Indeed, if the shortest resolution proof is very long then clearly no CDCL solver can do well, since a resolution proof can be extracted from the run of such a solver.

Thus a strong lower bound on length is a sure indicator of hardness. But in the opposite direction, the fact that there exists a short resolution proof does not mean that this proof is easy to find. On the contrary, [10] indicates that it might be computationally infeasible to find any proof in reasonable length even given the guarantee that a short proof exists. Therefore, length seems too "optimistic" as a measure of hardness in practice.

Turning next to *width*, if the shortest proof for a formula is very long, then the width of a "narrowest" proof must by necessity also be large (simply by counting the number of possible clauses). However, the fact that a resolution proof needs to be wide, i.e., has to contain some large clause, does *not* necessarily imply that the minimum length is large [11]. Width is thus a stricter hardness measure than length. Recently, is was shown in [9] (under some theoretical assumptions) that if the minimum width of any resolution proof for a formula is w, then a CDCL solver (that does not at all care about width per se) will with high probability decide the formula in time something like n^w. This seems to indicate that resolution width could be a measure that correlates well with practical hardness for CDCL solvers. On the other hand, some of the technical assumptions needed to establish this result seem somewhat idealized compared to how CDCL solvers work in practice.

Since, as already discussed, memory consumption is a major concern in practice, the related theoretical measure of *space* clearly seems interesting to study in this context. Indeed, this was arguably the main reason why research into proof space complexity was initiated in the late 1990s. It was shown in [12] that if there is no narrow proof for a formula, then there is no small-space proof either. However, large space does *not* imply anything about the width [13], and hence space is an even stricter hardness measure than width. Intuitively, one could argue that for a formula with high space complexity, CDCL solvers would need to learn many clauses and keep them in memory, and therefore such formulas should be hard in practice. A stronger conjecture would be that it also holds that if a formula has low space complexity—which also guarantees that there are both short and narrow proofs—then a CDCL solver should (at least in principle) be able to learn the few clauses needed to quickly produce a short proof.

The purpose of our work is to provide an empirical evaluation of these proof complexity measures and their relevance for hardness in practice, focusing in particular on space. Our work can be seen to implement a program outlined previously in [14]. It should be noted, however, that [14] suggested *tree-like space*, i.e., space measured in the subsystem of resolution producing only tree-like proofs, as a measure of practical hardness. We will return to the question of general versus tree-like space later, but let us just remark here that in tree-like resolution space is tightly correlated with length in the sense that there is a short tree-like proof if and only if there is a space-efficient one [15] (which is provably not true in general resolution by [13]). If tree-like space were a good indicator of hardness in practice, this would mean that CDCL solvers could decide a formula efficiently if and only if it had a short *tree-like* proof, which in turn would seem to imply that in practice CDCL solvers cannot provide any significant improvements in performance over plain DPLL solvers. This does not seem consistent with observations that the former can vastly outperform the latter; observations that have been corroborated by theoretical papers arguing that CDCL proof search is closely related to *general* resolution [7, 8] instead of mere *tree-like* resolution.

1.1 Contributions of This Paper

The main contribution of this work is to provide a first extensive empirical evaluation of the connections between resolution space complexity and practical hardness for CDCL solvers. We remark that although [14] suggested this, as far as we are aware no such evaluation has previously been carried out (and, indeed, has only been made possible by recent advances in proof complexity providing formulas with the necessary theoretical guarantees). We investigate to what extent resolution space correlates with running times for state-of-the-art CDCL solvers. In the experiments, we employ highly structured SAT instances—so-called *pebbling formulas*—allowing us to make controlled observations on the relation between space complexity and SAT solver performance. Pebbling formulas are guaranteed to be very easy with respect to length (the short proof consists of just listing the clauses of the formula in the right order and doing a very small constant number of intermediate derivation steps in between each such clause) and also with respect to width (the proofs just sketched will have width equal to the width of the formula, which in all cases is a one-digit constant) but can be varied with respect to their space complexity from constant all the way up to almost linear. This makes it possible to study which level of granularity (if any) is the best one in the hierarchy of measures in (1) when we are looking for indicators of hardness in practice. If length or width were the most relevant measure, then CDCL solvers should behave in essentially the same way for all formulas in our experiments. If, however, the more fine-grained measure of space is a more precise indicator of hardness, then we would expect to see a clear correlation between running times and theoretical space complexity. As we will argue below, the conclusion from our experiments is that the latter case seems to hold.

Complementing the empirical evaluation, we also present some theoretical results related to resolution space. In particular, we prove the first non-trivial separation of general and tree-like resolution space. Previously, only a constant-factor separation was known [16], meaning that it could not be ruled out that the two measures were essentially the same except for small multiplicative constants. We improve this to a logarithmic separation, which, while still leaving room for further improvements, shows that the measures are fundamentally different. This in turn motivates our focus on general space as a measure of practical hardness for CDCL solvers. Furthermore, elaborating on and extending related results in [14], we also address the relation between resolution space and *backdoor sets* [17], a somewhat more practically motivated measure previously proposed as a proxy for practical hardness.

2 Proof Complexity Preliminaries

We now give a brief overview of the relevant proof complexity background; for more details, see e.g. [18]. We assume familiarity with CNF formulas, which are conjunctions of clauses, where a clause is a disjunction of literals (unnegated or negated variables, with negation denoted by overbar). It is convenient to view clauses as sets, so that there is no repetition of literals and order is irrelevant. A k-CNF formula has all clauses of size at most k, which is implicitly assumed to be some (small, say one-digit) constant throughout this paper. Below we will focus on k-CNF formulas to get cleaner statements of the theoretical results (analogous results hold in general but are not as simple to state).

A *resolution refutation* $\pi\colon F \vdash \perp$ of an unsatisfiable CNF formula F, also known as a *resolution proof* for F, is an ordered sequence of clauses $\pi = (D_1, \ldots, D_\tau)$ such that $D_\tau = \perp$ is the empty clause containing no literals, and each line D_i, $1 \leq i \leq \tau$, is either one of the clauses in F (*axioms*) or is derived from clauses D_j, D_k in π with $j, k < i$ by the *resolution rule* $\frac{B \vee x \quad C \vee \bar{x}}{B \vee C}$ (where the clause $B \vee C$ is the *resolvent* of the clauses $B \vee x$ and $C \vee \bar{x}$ on x). With every resolution proof π we can associate a graph G_π by having a sequence of vertices v_i labelled by the clauses D_i on a line in order of increasing i, and with edges from v_j and v_k to v_i if D_i was derived by resolution from D_j and D_k. Note that there might be several occurrences of a clause D in the proof π, and if so each occurrence gets its own vertex in G_π.

The *length* $L(\pi)$ of a resolution proof π is the number of clauses in it (counted with repetitions). The *width* $W(C)$ of a clause C is $|C|$, i.e., the number of literals, and the width $W(\pi)$ of a proof π is the size of a largest clause in π. The *space* (sometimes referred to as *clause space*) of a proof at step i is the number of clauses C_j, $j < i$, with edges to clauses C_k, $k \geq i$, plus 1 for the clause C_i derived at this step. That is, intuitively space measures the number of clauses we need to keep in memory at step i, since they were derived before step i but will be used to infer new clauses after step i (or possibly at step i). The space of a proof is the maximum space over all steps in the proof. *Tree-like (clause) space* is defined in exactly the same way except that the graph G_π representing π is constrained to be a (binary) tree.

We next briefly review what is known about these measures. As shown in [19, 20] and many later papers, the length of refuting a CNF formula F can be exponential in the size of F (measured as the total number of literals counted with repetitions), and it is easy to show that the worst case is at most exponential. For width, clearly the size of the formula (and, in particular, the number of distinct variables in it) is an upper bound, and there are matching lower bounds up to constant factors [21]. If a formula has a narrow proof then this proof must also be short (simply by counting the total number of distinct clauses). The opposite does not necessarily hold as proven in [11] (although very strong lower bounds on width do imply strong lower bounds on length by [21]).

Just as for width, although somewhat less obviously, space is also at most linear (even for tree-like space) as shown in [15], and again there are matching lower bounds, e.g., in [22, 23]. In [12] it was shown that if a formula can be refuted in small space, this implies there is also a small-width proof (although in general this will not be the same proof). The converse of this is false in the strongest sense possible—there are formulas with constant-width proofs that require almost linear (i.e., worst-case) space [13]. Since space upper-bounds width, and also width upper-bounds length as discussed above, it follows that upper bounds on space imply upper bounds on length. Conversely, in [15] it was shown that small length implies small space for the restricted case of tree-like resolution. In general resolution, however, the fact that a formula is refutable in small length says essentially nothing about the space complexity [13].

3 Pebbling Formulas

To study the proof complexity measures of length, width and space, and to relate them to the practical hardness of CNF formulas, we focus on so-called *pebbling formulas* (also known as *pebbling contradictions*). Our main motivation for using pebbling formulas is that, as explained below, recent theoretical advances allow us to construct such formulas with varying (and fully specified) space complexity properties while keeping the length and width complexity fixed.

Pebbling formulas are so called since they encode instances of *pebble games* played on directed acyclic graphs (DAGs). Wee refer to the survey [24] for more information about such games. The pebbling formula over G associates one variable with each vertex, postulates the source vertices (with no incoming edges) to be true and the (unique) sink vertex (with no outgoing edges) to be false, and then specifies that truth propagates from the sources to the sink. More formally, as defined in [21] the pebbling formula Peb_G over a DAG G consists of:

- for all source vertices s in G, a unit clause s (*source axioms*),
- for all non-sources v with incoming edges from the vertex set $pred(v)$ of immediate predecessors of v, the clause $\bigvee_{u \in pred(v)} \overline{u} \vee v$ (*pebbling axioms*),
- for the (unique) sink z of G, the unit clause $\overline{z}$ (*sink axiom*).

If G has n vertices and max fan-in ℓ, then Peb_G is an unsatisfiable $(1+\ell)$-CNF formula with $n + 1$ clauses over n variables. For all graphs used in this paper we have $\ell = 2$.

Pebbling formulas are not of much use to us as such—they are very easy with respect to all proof complexity measure we have discussed, and are easily seen to be solvable simply by unit propagation. However, they can be transformed into much more interesting formulas by substituting Boolean functions for the variables as follows.

Given any CNF formula F, we can fix a Boolean function $f \colon \{0,1\}^d \mapsto \{0,1\}$ and substitute every variable x in F by $f(x_1, \ldots, x_d)$, where $x_1, \ldots, x_d$ are new variables that do not appear anywhere else. Then we expand this out to get an equivalent CNF formula over this new set of variables. For a small example, if we let $\oplus$ denote binary exclusive or, then the clause $\overline{x} \vee y$ after substitution becomes $\neg(x_1 \oplus x_2) \vee (y_1 \oplus y_2)$ which is expanded out to the set of clauses

$$\{x_1 \vee \overline{x}_2 \vee y_1 \vee y_2, \; x_1 \vee \overline{x}_2 \vee \overline{y}_1 \vee \overline{y}_2, \; \overline{x}_1 \vee x_2 \vee y_1 \vee y_2, \; \overline{x}_1 \vee x_2 \vee \overline{y}_1 \vee \overline{y}_2\} \; . \quad (2)$$

(For general f there might be some choices in exactly how to do this expansion, but such implementation details do not affect this discussion so we ignore them for brevity.)

The *pebbling price* of a DAG G measures how much space is needed to pebble G. As shown in [13, 25], making substitutions in pebbling formulas using *robust* functions f, meaning that the truth value of f can never be fixed by just assigning to one variable, yields substituted CNF formulas for which the *space complexity of the formula coincides with the pebbling price of G.*[1] The canonical example of a robust function is exclusive or. A non-robust function is ordinary or, but [26, 27] show that even for this function the same connection holds at least for certain fairly general families of graphs. This means that if we pick the right graphs, we can generate CNF formulas with known

---

[1] Actually, this is an oversimplification and formally speaking not correct—the space will be somewhere in between the deterministic black and the (smaller) non-deterministic black-white pebbling price, but these two measures are within a small constant factor for all graphs considered in this paper so this is immaterial. We emphasize that all constants involved are explicitly known and are very small, with the single exception of the gtb graphs discussed below.

Table 1. DAG families and properties of the resulting CNF formula families

Name	Description	Space cplx
pyr$\langle h\rangle$seq	Sequence of pyramid graphs of (constant) height h	$\Theta(h)$
width$\langle w\rangle$chain	Chain graph of (constant) width w	$\Theta(w)$
bintree	Complete binary tree	$\Theta(\log n)$
pyrofpyr	Pyramid of height $\sqrt[4]{n}$ with each node expanded to pyramid	$\Theta(\sqrt[4]{n})$
pyrseqsqrt	Sequence of pyramids of (growing) height $\sqrt[4]{n}$	$\Theta(\sqrt[4]{n})$
pyramid	Pyramid graph of (growing) height $\sqrt{n}$	$\Theta(\sqrt{n})$
gtb	DAGs from [28] with butterfly graphs as superconcentrators	$\Theta(n/\log^2 n)$

space complexity. In addition, it is easy to show that any pebbling formula, even after substitution, can be refuted in (small) constant width and (small) linear length.[2] Thus, in this way we can get formulas that are uniformly *very* easy with respect to length and width, but for which the space complexity varies.

Such formulas would seem like excellent benchmarks for testing the correlation between theoretical complexity measures and hardness in practice, and in particular for investigating at which level of granularity theoretical hardness should be measured given the hierarchy in (1). If the minimum width, or length, of a proof for a formula F were good indicators of whether F is hard or easy, then we would expect to get similar running times for pebbling formulas over all graphs of the same size (when fixing the substitution function). If the more fine-grained measure of space is a more precise indicator, however, we would expect running time to correlate with space complexity. Carrying out large-scale experiments along these lines and analyzing the results is the main practical contribution of this paper. When designing such experiments, one needs to choose (a) graphs from which to generate the benchmarks, and (b) substitution functions to apply. We discuss this next.

An overview of our choice of graph families and their space complexities is given in Table 1. Let us first explain two important building blocks. A *pyramid* of height h is a layered DAG with $h + 1$ layers, where there is one vertex in the highest layer, two vertices in the next layer, et cetera, down to $h + 1$ vertices in the lowest layer, and where the ith vertex at layer L has incoming edges from the ith and $(i + 1)$st vertices at layer $L - 1$. A *chain* of width w is a layered graph with w vertices at each layer, and with vertices i and $i - 1$ at layer $L - 1$ having edges to vertex i in layer L $(\bmod\, w)$.

To obtain two different types of graphs of constant space complexity, we consider sequences of pyramids of constant height h with the sink of each pyramid connected to the leftmost source of next pyramid (pyr$\langle h\rangle$seq) and chains of constant width w (width$\langle w\rangle$chain). Another graph family that should yield easy formulas are complete binary trees (bintree), the space complexity of which is equal to the height of the tree.

To get "medium-hard" DAGs, we use pyramids in two different ways. In pyramid-of-pyramids graphs (pyrofpyr) we take a pyramid of height h and expand each of its

---

[2] We will not elaborate on exact constants due to space constraints, but all k-CNF formulas considered have $4 \le k \le 9$ and the refutation width coincides with the formula width. As to length, a pebbling formula generated with binary XOR substitution and having L clauses is refutable in length $2.25 \cdot L$, and the blow-up for other substitution functions is similar.

Table 2. Substitution functions

Name	Description	Output	CNF encoding (for $d = 3$ variables)
or_d	OR of d vars	$x_1 \vee \cdots \vee x_d$	$x_1 \vee x_2 \vee x_3$
xor_d	parity of d vars	$x_1 \oplus \cdots \oplus x_d$	$x_1 \vee x_2 \vee x_3,\ x_i \bigvee_{j \neq i} \overline{x}_j,\ i = 1, 2, 3$
maj_d	majority of d vars	$2(x_1 + \cdots + x_d) > d$	$x_1 \vee x_2,\ x_1 \vee x_3,\ x_2 \vee x_3$
eq_d	all d vars equal	$x_1 = \cdots = x_d$	$x_1 \vee \overline{x}_2,\ \overline{x}_1 \vee x_2,\ x_1 \vee \overline{x}_3,\ \overline{x}_1 \vee x_3$
el_d	exactly one of d	$x_1 + \cdots + x_d = 1$	$x_1 \vee x_2 \vee x_3,\ \overline{x}_1 \vee \overline{x}_2,\ \overline{x}_1 \vee \overline{x}_3,\ \overline{x}_2 \vee \overline{x}_3$
s_id	if-then-else	$x_1 ? x_2 : x_3$	$\overline{x}_1 \vee x_2,\ x_1 \vee x_3$

vertices v to pyramid of same height h with sink z_v. Every incoming edge to v is drawn to all sources of the pyramid, and all outgoing vertices from v are drawn from z_v. It is an easy argument that the space complexity is roughly $2h$ and the size of the graph is roughly h^4, so the space complexity grows like $\sqrt[4]{n}$ for graphs of size n. Another way of getting graphs of the same space complexity is to use the same construction as in pyr$\langle h \rangle$seq above but employ graphs of height $\sqrt[4]{n}$. These are our pyrseqsqrt graphs.

Finally to get "really hard" graphs we consider two well-known graph families. The first one is simply pyramids of height (and hence space complexity) $\sqrt{n}$. The second is based on DAGs with maximal space complexity $\Theta(n/\log n)$ [28]. These graphs cannot be used as-is, however. The bound in [28] is asymptotic, with the smallest instances of huge size (due to the need for so-called *superconcentrators* of linear size). Therefore, we modify the construction to use much simpler, but asymptotically worse, superconcentrators made from butterfly graphs. It is not hard to verify that the proofs in [28] still go through, and we get much smaller graphs (gtb) that we can actually use to generate CNF formulas, at the price of paying a log factor in the space complexity.

When choosing the substitution functions to apply for pebbling formulas generated from graphs in these families, we want to achieve two objectives. On the one hand, we would like the functions to be robust (as explained above). On the other hand, however, we do not want too large a blow-up in formula size when substituting functions for variables. Again due to space constraints, we cannot go into too much details, but Table 2 presents our choice of substitution functions, which seem to provide a good trade-off between the two goals, and describes their CNF encodings. Note that we also use the (non-robust) standard non-exclusive or functions, which nevertheless provably preserves the space complexity (albeit with worse guarantees for the hidden constants) for all graph families considered here except the gtb family.

4   Improved Separation of General and Tree-Like Resolution Space

Before reporting on the empirical part of this work, we return to the question motivated by [14] of whether tree-like space or general space is likely to be the most relevant space measure when it comes to hardness. We already explained in the introduction the reasons for our skepticism regarding tree-like space as a good hardness measure for CDCL. In this section, we complement this with a more theoretical argument, showing that the tree-like and general resolution space measures are different. Our logarithmic separation, stated next, improves on the constant-factor separation in [16] which is all that was known previously.

Theorem 1. *There are families of 4-CNF formulas $\{F_n\}_{n=1}^{\infty}$ of size $\Theta(n)$ such that their space complexity in general resolution is $O(1)$ whereas the tree-like space complexity grows like $\Theta(\log n)$.*

Proof. To prove Theorem 1, we apply the equivalence of the tree-like space of a CNF formula F and Prover-Delayer game on F as described in [16]. The Prover asks about variable assignments, and the delayer answers true, false or $*$ to each query. If the answer is $*$, Prover picks an assignment adversarially but Delayer scores a point. The game ends when Prover has forced a partial truth value assignment that falsifies some clause of F. If Delayer can score exactly p points with an optimal strategy, then the tree-like space complexity is $p + 2$, and the opposite also holds.

Consider a graph that is just a line $(v_1, v_2, \ldots, v_n)$ of length n with edges from each vertex v_i to the next vertex v_{i+1} on the right. Let F_n be the pebbling formula over this graph with substitution by (binary) XOR $\oplus$ as described in Section 3.

It is immediate that the general resolution space complexity is constant. Just start with the leftmost node v_1, for which the XOR of the two associated variables holds in view of the source axioms. Then derive step by step, using pebbling axioms, that if the XOR holds at one vertex v_i, then this implies it also holds at the next vertex v_{i+1}. Finally we reach the rightmost vertex v_n, where the sink axioms say that XOR does not hold. Contradiction.

Now we give a Delayer strategy that scores $\log_2 n$ points. For every vertex, the first time Prover asks about any of the two variables associated to the vertex Delayer answers $*$ and scores. The first time Prover asks about a second variable, Delayer looks whether the vertex v_i is in the leftmost or rightmost half of (the remaining part of) the graph. In the former case, Delayer answers so that the XOR of the two variables is satisfied, which gives a problem instance of the same type of at least half the size over $(v_{i+1}, \ldots, v_n)$. In the latter case, Delayer makes sure the XOR is false. Then Prover has to continue playing in $(v_1, \ldots, v_{i-1})$ to falsify the formula since the rest is now satisfiable. Since Prover can only halve the size of the graph for each second question, and Delayer scores a point for each first question, the tree-like space complexity is $\Omega(\log n)$, and since Prover can use precisely this strategy, the space bound is tight. $\quad\square$

As a final remark, let us note that this proof works equally well for CNF formulas generated from the pyr$\langle h \rangle$seq and width$\langle w \rangle$chain graphs in Section 3.

5   Experimental Evaluation

This section summarizes results of our experiments running state-of-the-art CDCL SAT solvers on pebbling formulas with varying resolution space.[3] As benchmarks, we used CNF formulas that we generated for *all* combinations of the graphs mentioned in Table 1 (we used pyr$\langle h \rangle$seq with $h \in \{1, 3, 5, 10\}$ and width$\langle w \rangle$chain with $h \in \{2, 5, 10\}$; overall 12 graph families) and the substitution functions mentioned in Table 2 (we

---

[3] The only experiments previously reported for CDCL solvers on pebbling formulas we are aware of were on "grid formulas", i.e., formulas over pyramids with or_2 using zChaff, with a somewhat different motivation [29].

used el_3, maj_3, or_2, or_3, or_4, s_id, xor_2, eq_3; overall 8 functions) yielding a total of 96 families of CNF formulas. Due to the massive amount of data produced, here we only provide a snapshot of the results. Complete data for all experiments as well as a more detailed description of the formula instances used can be found at `http://www.csc.kth.se/~jakobn/publications/cp12/`.

5.1 Experiment Setup

For the experiments, we used the CDCL solvers Minisat 2.2.0 [30] and Lingeling[4] [31]. We ran the solvers on all CNF families in two modes: "as-is", and with all preprocessing (and all inprocessing for Lingeling) disabled. The experiments were run under Linux on a Intel Core i5-2500 3.3-GHz quad-core CPU with 8 GB of memory. We limited the run-time of each solver to 1 hour per instance.

5.2 Results

First recall that the only parameter that varies among the different formula families is their resolution space complexity. Namely, for each of the 96 families, every formula with n variables has $O(1)$ resolution width and $O(n)$ resolution length, with small hidden constants depending on the substitution function only; only the resolution space varies from $O(1)$ to $\Omega(n/\log^2 n)$ as explained in Section 3.

Overall, the results show a notable difference in running times between different families. In particular, we observed that the hardest families with respect to space complexity are also hardest in practice. In other words, *for these families we observed a clear correlation between space complexity and practical hardness.* While there are some observed exceptions (mainly in the lower-end spectrum of the space complexity), there is a positive correlation between run-times and resolution space for almost all of the families.

An example of the results for both of the solvers with preprocessing turned off is given in Figure 1. For clarity, we only include the following 5 families in each plot (listed in non-increasing order of space complexity, cf. Table 1): pyr1seq, bintree, pyrseqsqrt, pyramid, and gtb.

One reason to run experiments over all combinations of graphs and substitution function is to distill the dependence on space and filter out other factors if possible. For instance, different substitution function can (and will) have different properties in practice although their theoretical guarantees are the same. By aggregating results over all substitution functions instead of just considering one or two functions, we get an overall picture. This is summarised in Table 3, which gives average run-times per instance, normalized by the number of variables and calculated over all considered substitution functions with and without preprocessing. In particular, for each family, we take run-times/size (where size is the number of variables in the formula) for each instance, and sum all these numbers for the chosen family and then take an arithmetic mean; that is, $(\text{run-time}_1/\text{size}_1 + \ldots + \text{run-time}_k/\text{size}_k)/k$. The families in Table 3 are *listed in non-increasing order of space complexity* (cf. Table 1). The numbers, which are multiplied by 10^4, show that the average run-times (without preprocessing, first column) correlate

---

[4] Version 774, with an option to disable pre- and inprocessing, was provided by Armin Biere.

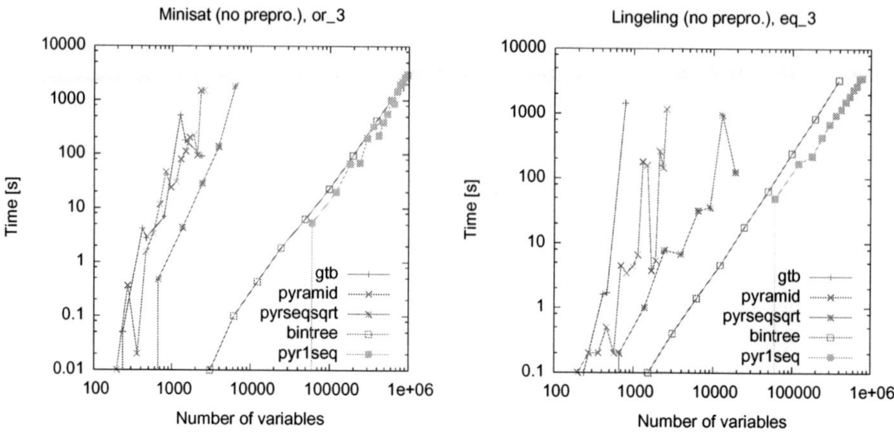

Fig. 1. Results without preprocessing

strongly with space complexity, the bintree family being the only significant exception. However, bintree graphs have an exceptionally large number (half) of source nodes, which get translated into simpler clauses compared to clauses from non-source nodes. (For a majority of the substitution functions, namely or_2, xor_2, maj_3, e1_d, and s_id, most or all of the resulting clauses are binary.)

Table 3. Average run-times for all families

Time [seconds] $\times$ 10e4		
formula family	**no preprocessing**	**with preprocessing**
gtb	637.65	9.36
pyramid	569.13	1.14
pyrofpyr	234.17	9.45
pyrseqsqrt	122.77	1.64
bintree	12.27	0.27
width10chain	59.87	5.79
pyr10seq	44.12	2.70
width5chain	56.77	5.97
pyr5seq	31.60	2.09
pyr3seq	33.35	1.24
width2chain	45.24	3.25
pyr1seq	39.43	0.36

5.3 Effects of Preprocessing

While a clear correlation between space complexity and solver running times was observed on a variety (but not all) of the formula families, we observed that preprocessing

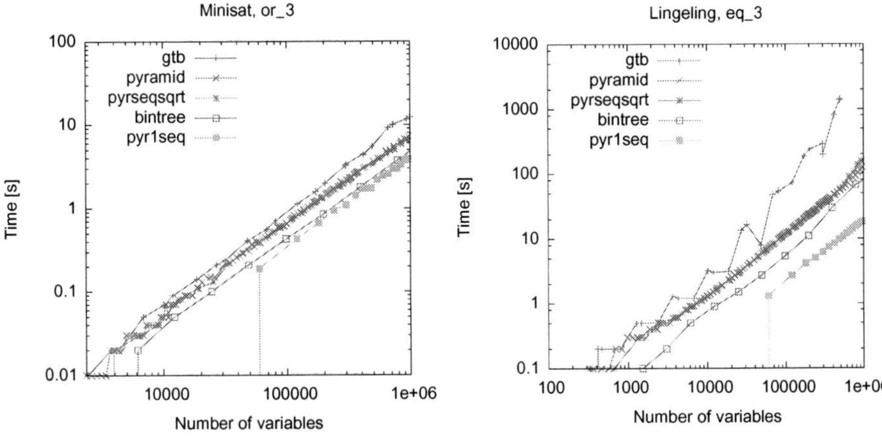

Fig. 2. Results with preprocessing

and inprocessing resulted in rather different results. In particular, as exemplified in Figure 2, although smaller correlations can still be observed (especially for Lingeling), preprocessing appears to even out many of the observed differences in the data for the solvers without preprocessing. While this is an observation that clearly sets apart the preprocessing techniques from the behavior of the core CDCL techniques, we do not find it too surprising. Simply, the theoretical space measure that we study can be expected to correlate more or less well with what is going on during clause learning. When preprocessing is applied to a formula F, however, what is fed to the clause learning solver is another formula F' for which we have no theoretical guarantees as to the space complexity (it can a priori be both lower and higher). Furthermore, the fact that our benchmarks have been chosen specifically to be very easy with respect to length and width also means that they are likely to be amenable to the kind of heuristics used in preprocessing. We see much scope for future work here, including ideas how to modify formulas so that they have the same theoretical guarantees but so that these guarantees are more likely to "survive" the preprocessing stage.

5.4   Considerations

We remark that one reasonable objection is that since our benchmarks are pebbling formulas generated from graph it is not clear that we are measuring space complexity per se—maybe we are measuring some other, unrelated graph property. And indeed, some graph properties, such as the number of source nodes, translate into properties of the formulas (many small clauses) that are not captured by the space complexity.

This problem is hard to get around, however. Resolution space complexity seems likely to be **PSPACE**-complete, and it is an easy argument that is is **NP**-hard to approximate in any meaningful way, so we cannot expect to be able to determine the space complexity of arbitrary formulas. Instead, we have to pick special instances where we know the space complexity for other reasons, which is the case for the pebbling formula families considered in our experiments.

6   Relating Resolution Space and Backdoors

Compared to the proof complexity hardness measures of resolution length, width, and space, a more practically motivated well-known hardness measure is the size of *(strong) backdoor sets*. Backdoors sets were first studied in [17], and this and subsequent works have shown that real-world SAT instances often have small backdoors, which might offer an explanation why modern SAT solvers perform notably well on such instances.

The definition of backdoor sets is made with respect to a polynomial-time subsolver A. Given a subsolver A and a unsatisfiable formula F, a (strong) backdoor set S is a subset of the variables in F such that for every truth assignment ρ over S, the subsolver A determines unsatisfiability of $F\restriction_\rho$.[5] The algorithm A might be, for instance, unit propagation, polynomial-time restricted DPLL, or a 2-SAT algorithm.

In [14], it was shown that given a subsolver that only accepts formulas with tree-like resolution space k, a CNF formula F, and a strong backdoor set S of F, the tree-like resolution space of F is bounded from above by $|S| + k$. In fact, when restricting to such abstract subsolvers that only accept formulas with tree-like resolution space k, the minimum backdoor set size is a *proper* upper bound for tree-like resolution space. However, [14] does not seem to elaborate too much on what such abstract subsolvers might be. Our following theorem states the relationship between resolution space and backdoor sets in a more concrete way.

Theorem 2. *The following claims hold for any CNF formula F over n variables.*

1. *If F has a backdoor set B of size b with respect to 2-CNF, then the space complexity of F is at most $b + \mathrm{O}(1)$.*
2. *If F has a backdoor set B of size b with respect to unit propagation, then the space complexity of F is at most $b + \mathrm{O}(1)$.*
3. *If F has a backdoor set B of size b with respect to DPLL running in time* $\mathrm{poly}(n)$, *then the space complexity of F is at most $b + \mathrm{O}(\log n)$.*

Proof. Suppose that ρ is a (partial) truth value assignment to the b variables in the backdoor set B, and that C_ρ is the unique minimal clause falsified by ρ. Then if π_ρ is a resolution refutation of $F\restriction_\rho$ in clause space s, by plugging in F instead of $F\restriction_\rho$ we get from π_ρ a resolution derivation of the clause C_ρ in the same space s. Also, it is easy to show that the set of clauses $\{C_\rho \mid \rho$ all total assignments to $B\}$ can be refuted by a (tree-like) resolution refutation π_B in simultaneous space $b + \mathrm{O}(1)$ and length 2^{b+1}. Thus, if each clause C_ρ is derivable in clause space at most s, then we can combine these derivations π_ρ with the refutation π_B of the set of clauses $\{C_\rho \mid \rho$ all total assignments to $B\}$ to get a refutation of F in space $b + s + \mathrm{O}(1)$, simply by running π_B, and whenever a clause C_ρ is needed (which will only be once per clause) call on π_ρ as a subroutine.

What remains is to upper bound the space complexity of $F\restriction_\rho$.

1. Any 2-CNF formula can be refuted by resolution in clause space at most 4 [15].
2. Unit propagation can be seen as a resolution proof DAG that is a long chain with every chain vertex having a unique predecessor except for its predecessor on the chain. Such a proof is in (tree-like) clause space 3.

---

[5] $F\restriction_\rho$ denotes F restricted by ρ, i.e., with variables set according to ρ after which the formula is simplified by removing satisfied clauses and unsatisfied literals from clauses.

3. If DPLL runs in time n^c, then it produces a tree-like resolution proof in size n^c. According to [15], tree-like resolution length L implies that one can do length L and space $\log L + O(1)$ simultaneously. Hence we have space $b + O(\log n)$. $\square$

Recently, the concept of *learning-sensitive backdoors* [32] was proposed as a concept more tightly connected with CDCL solvers than the original definition of backdoors. It was shown that strong learning-sensitive backdoors can be exponentially smaller than traditional backdoors [32]. It remains an interesting open question whether general resolution space is bounded from above by the size of learning-sensitive backdoors, i.e., whether small learning-sensitive backdoors imply low (general) resolution space complexity.

7   Concluding Remarks

This paper advances and expands on the program outlined first in [14], namely, to shed light on possible connections between theoretical complexity measures and practical hardness of SAT, and in particular on whether space complexity is a good indicator of hardness. We provide an extensive empirical evaluation on the correlation between resolution space and practical hardness, running state-of-the-art CDCL SAT solvers on benchmark formulas with theoretically proven properties based on recent results in proof complexity. To the best of our knowledge, no such experiments have previously been done, and we consider this a conceptually important step towards the more general goal of relating complexity of SAT solving in theory and practice. Furthermore, complementing the empirical evaluation, we prove new theoretical results related to resolution space, in particular separating general and tree-like resolution space and thus showing that the two measures are indeed different. We also address the relation between resolution space and backdoor sets.

Regarding the empirical work, while the results presented here do not provide conclusive evidence for resolution space being the "ultimate right measure" of practical hardness (it may be safe to assume that theory and practice most often do not behave exactly identically), the important observation is that we do see nontrivial correlations. We therefore argue that our results are consistent with the hypothesis that resolution space complexity should be a relevant measure of hardness in practice for CNF formulas. In particular, space might be a more precise indicator of practical hardness than length or width, in the sense that the latter two measures give too optimistic estimates for formulas which have very low length or width complexity but which might nevertheless be hard for state-of-the-art CDCL SAT solvers to solve in practice.

We have already discussed at the end of Section 5 why the experiments by necessity had to be run on designed combinatorial benchmark formulas rather than on real-world instances. Another possible issue is that to truly understand the relation between practical hardness on one hand, and length, width and space complexity on the other, one would need experiments that vary all three parameters. However, while this a priori seems like a very reasonable request, the hierarchy between these measures in (1) means that such experiments unfortunately are provably impossible to perform. As soon as the length or width complexity increases, the space complexity will increase as well.

Thus, all we can hope to study is whether space provides a more precise indication of hardness than is given by width or length.

We view our work as only a first step in an interesting line of research, and see many important questions to investigate further. One such question closely related to the current paper would be to study the recent theoretical results on trade-offs between proof length and proof space for resolution in [25, 33], and perform experiments on whether these results translate into trade-offs between running time and memory consumption for CDCL solvers in practice.

Acknowledgements. We thank Armin Biere for providing a modified version of Lingeling with pre- and inprocessing switched off, and Niklas Eén and Niklas Sörensson for useful discussions and advice regarding MiniSAT. We also thankfully acknowledge technical assistance at KTH Royal Institute of Technology from Torbjörn Granlund and Mikael Goldmann. The 3rd author gratefully acknowledges the discussions with Noga Zewi that led to the separation of tree-like and general resolution.

The 1st author was supported by the Academy of Finland (grants 132812 and 251170). The 3rd author was supported by Swedish Research Council grant 621-2010-4797 and by the European Research Council under the European Union's Seventh Framework Programme (FP7/2007–2013) / ERC grant agreement no 279611. The 4th author was supported by a Junior Research Fellowship at University College, Oxford. Part of this work was done during the visits of the 1st, 2nd and 4th authors to KTH supported in part by the foundations *Johan och Jakob Söderbergs stiftelse*, *Magnus Bergvalls Stiftelse*, *Stiftelsen Längmanska kulturfonden*, and *Helge Ax:son Johnsons stiftelse*.

References

[1] Biere, A., Heule, M.J.H., van Maaren, H., Walsh, T. (eds.): Handbook of Satisfiability. Frontiers in Artificial Intelligence and Applications, vol. 185. IOS Press (2009)

[2] Marques-Silva, J.P., Sakallah, K.A.: GRASP: a search algorithm for propositional satisfiability. IEEE Trans. Computers 48(5), 506–521 (1999)

[3] Moskewicz, M.W., Madigan, C.F., Zhao, Y., Zhang, L., Malik, S.: Chaff: engineering an efficient SAT solver. In: Proc. DAC, pp. 530–535. ACM (2001)

[4] Marques-Silva, J.P., Lynce, I., Malik, S.: Conflict-driven clause learning SAT solvers. In: [1], pp. 131–153

[5] Davis, M., Logemann, G., Loveland, D.: A machine program for theorem proving. Communications of the ACM 5(7), 394–397 (1962)

[6] Davis, M., Putnam, H.: A computing procedure for quantification theory. Journal of the ACM 7(3), 201–215 (1960)

[7] Buss, S.R., Hoffmann, J., Johannsen, J.: Resolution trees with lemmas: Resolution refinements that characterize DLL-algorithms with clause learning. Logical Methods in Computer Science 4(4:13) (2008)

[8] Pipatsrisawat, K., Darwiche, A.: On the power of clause-learning SAT solvers as resolution engines. Artificial Intelligence 175, 512–525 (2011)

[9] Atserias, A., Fichte, J.K., Thurley, M.: Clause-learning algorithms with many restarts and bounded-width resolution. Journal of Artificial Intelligence Research 40, 353–373 (2011)

[10] Alekhnovich, M., Razborov, A.A.: Resolution is not automatizable unless W[P] is tractable. In: Proc. FOCS, pp. 210–219. IEEE (2001)

[11] Bonet, M.L., Galesi, N.: Optimality of size-width tradeoffs for resolution. Computational Complexity 10(4), 261–276 (2001)

[12] Atserias, A., Dalmau, V.: A combinatorial characterization of resolution width. Journal of Computer and System Sciences 74(3), 323–334 (2008)

[13] Ben-Sasson, E., Nordström, J.: Short proofs may be spacious: An optimal separation of space and length in resolution. In: Proc. FOCS, pp. 709–718. IEEE (2008)

[14] Ansótegui, C., Bonet, M.L., Levy, J., Manyà, F.: Measuring the hardness of SAT instances. In: Proc. AAAI, pp. 222–228. AAAI Press (2008)

[15] Esteban, J.L., Torán, J.: Space bounds for resolution. Inf. Comput. 171(1), 84–97 (2001)

[16] Esteban, J.L., Torán, J.: A combinatorial characterization of treelike resolution space. Information Processing Letters 87(6), 295–300 (2003)

[17] Williams, R., Gomes, C.P., Selman, B.: Backdoors to typical case complexity. In: Proc. IJCAI, pp. 1173–1178. Morgan Kaufmann (2003)

[18] Nordström, J.: Pebble games, proof complexity and time-space trade-offs. Logical Methods in Computer Science (to appear, 2012),
`http://www.csc.kth.se/~jakobn/research`

[19] Haken, A.: The intractability of resolution. Theoret. Comp. Sci. 39(2-3), 297–308 (1985)

[20] Urquhart, A.: Hard examples for resolution. Journal of the ACM 34(1), 209–219 (1987)

[21] Ben-Sasson, E., Wigderson, A.: Short proofs are narrow—resolution made simple. Journal of the ACM 48(2), 149–169 (2001)

[22] Alekhnovich, M., Ben-Sasson, E., Razborov, A.A., Wigderson, A.: Space complexity in propositional calculus. SIAM Journal on Computing 31(4), 1184–1211 (2002)

[23] Ben-Sasson, E., Galesi, N.: Space complexity of random formulae in resolution. Random Structures and Algorithms 23(1), 92–109 (2003)

[24] Nordström, J.: New wine into old wineskins: A survey of some pebbling classics with supplemental results. Foundations and Trends in Theoretical Computer Science (to appear, 2012), Draft version available at
`http://www.csc.kth.se/~jakobn/research/`

[25] Ben-Sasson, E., Nordström, J.: Understanding space in proof complexity: Separations and trade-offs via substitutions. In: Proc. ICS, pp. 401–416 (2011)

[26] Nordström, J.: Narrow proofs may be spacious: Separating space and width in resolution. SIAM Journal on Computing 39(1), 59–121 (2009)

[27] Nordström, J., Håstad, J.: Towards an optimal separation of space and length in resolution (Extended abstract). In: Proc. STOC, pp. 701–710 (2008)

[28] Gilbert, J.R., Tarjan, R.E.: Variations of a pebble game on graphs. Technical Report STAN-CS-78-661, Stanford University (1978)

[29] Beame, P., Kautz, H., Sabharwal, A.: Towards understanding and harnessing the potential of clause learning. Journal of Artificial Intelligence Research 22, 319–351 (2004)

[30] Eén, N., Sörensson, N.: An Extensible SAT-solver. In: Giunchiglia, E., Tacchella, A. (eds.) SAT 2003. LNCS, vol. 2919, pp. 502–518. Springer, Heidelberg (2004)

[31] Biere, A.: Lingeling, Plingeling, PicoSAT and PrecoSAT at SAT Race 2010. FMV Tech. Report 10/1, Johannes Kepler University (2010)

[32] Dilkina, B., Gomes, C.P., Sabharwal, A.: Backdoors in the Context of Learning. In: Kullmann, O. (ed.) SAT 2009. LNCS, vol. 5584, pp. 73–79. Springer, Heidelberg (2009)

[33] Beame, P., Beck, C., Impagliazzo, R.: Time-space tradeoffs in resolution: Superpolynomial lower bounds for superlinear space. In: Proc. STOC, pp. 213–232. ACM (2012)

The SEQBIN Constraint Revisited*

George Katsirelos[1], Nina Narodytska[2], and Toby Walsh[2]

[1] UBIA, INRA, Toulouse, France
`george.katsirelos@toulouse.inra.fr`
[2] NICTA and UNSW, Sydney, Australia
`{nina.narodytska,toby.walsh}@nicta.com.au`

Abstract. We revisit the SEQBIN constraint [1]. This meta-constraint subsumes a number of important global constraints like CHANGE [2], SMOOTH [3] and INCREASINGNVALUE [4]. We show that the previously proposed filtering algorithm for SEQBIN has two drawbacks even under strong restrictions: it does not detect bounds disentailment and it is not idempotent. We identify the cause for these problems, and propose a new propagator that overcomes both issues. Our algorithm is based on a connection to the problem of finding a path of a given cost in a restricted n-partite graph. Our propagator enforces domain consistency in $O(nd^2)$ and, for special cases of SEQBIN that include CHANGE, SMOOTH and INCREASINGNVALUE in $O(nd)$ time.

1 Introduction

Global constraints are some of the jewels in the crown of constraint programming. They identify common structures such as permutations, and exploit powerful mathematical concepts like matching theory, and computational techniques like flow algorithms to deliver strong pruning of the search space efficiently. Particularly eye-catching amongst these jewels are the meta-constraints: global constraints that combine together other constraints. For example, the CARDPATH meta-constraint [3] counts how many times a constraint holds down a sequence of variables. The SEQBIN meta-constraint was recently introduced in [1] to generalize several different global constraints used in time-tabling, scheduling, rostering and resource allocation. It also generalizes the CARDPATH constraint where the constraint being counted is binary. Our aim is to revisit the SEQBIN meta-constraint and give a new and efficient propagation algorithm.

2 Background

We write $D(X)$ for the domain of possible values for X, $lb(X)$ for the smallest value in $D(X)$, $ub(X)$ for the greatest. We will assume values range over 0 to d. A constraint is *domain consistent* (*DC*) if and only if when a variable is assigned any of the values in its domain, there exist compatible values in the domains of all the other variables of

* NICTA is funded by the Australian Government's Department of Broadband, Communications, and the Digital Economy and the Australian Research Council. This work was partially funded by the "Agence nationale de la Recherche", reference ANR-10-BLA-0214.

M. Milano (Ed.): CP 2012, LNCS 7514, pp. 332–347, 2012.

the constraint. Such an assignment is called a *support*. A constraint is *bound consistent* (*BC*) if and only if when a variable is assigned the lower or upper bound in its domain, there exist compatible values between the lower and upper bounds for all the other variables. Such an assignment is called a *bound support*. A constraint is *bounds disentailed* when there exists no solution such that each variable takes value between its lower and upper bounds. A constraint is *monotone* if and only if there exists a total ordering $\prec$ of the domain values such that for any two values v, w if $v \prec w$ then v can be replaced by w in any support [5]. We define $\pi = (\pi^{bottom} := 0 \prec \ldots \prec d =: \pi^{top})$. A binary constraint is *row-convex* if, in each row of the matrix representation of the constraint, all supported values are consecutive (i.e., no two values with support are separated by a value in the same row without support) [6]. We use $x_{i,j}$ to represent the variable-value pair $X_i = j$. Let C be a binary constraint. We write $(j, k) \in C$ if C allows the tuple (j, k). Consider a soft binary constraint C. We denote the cost of the tuple $c(j, k)$. If $(j, k) \in C$ then $c(j, k) = 0$ and $c(j, k) = 1$ otherwise. Given two sets of integers S and R, we denote $S \uplus R = \{s + r \mid s \in S, r \in R\}$. Given a constant c, we write $S \uplus c = \{s + c \mid s \in S\}$. We denote $I[X]$ an instantiation of the variable sequence $X = [X_1, \ldots, X_n]$.

3   The SEQBIN Constraint

The SEQBIN meta-constraint ensures that a binary constraint B holds down a sequence of variables, and counts how many times another binary constraint C is violated.

Definition 1. *Given an instantiation $I[N, X_1, \ldots, X_n]$ and binary constraints B and C, the meta-constraint* SEQBIN(N, X, C, B) *is satisfied if and only if for any $i \in [1, n-1]$, $(I[X_i], I[X_{i+1}]) \in B$ holds, and $I[N]$ is equal to the number of violations of the constraint C, $(I[X_i], I[X_{i+1}]) \notin C$, in $I[X]$ plus 1.*

Note that we add 1 for consistency with the definition of SEQBIN in [1].

Example 1. Consider the SEQBIN$(N, [X_1, \ldots, X_7], C, B)$ constraint where $N = \{3\}$, B is TRUE and $C(X_i, X_j)$ is a monotone constraint with one satisfying tuple $(1, 1) \in C$, $D(X_1) = D(X_3) = D(X_5) = D(X_7) = 1$ and $D(X_2) = D(X_4) = D(X_6) = \{0, 1\}$. Consider an instantiation $I[N = 3, X_1 = 1, X_2 = 0, X_3 = 1, \ldots, X_7 = 1]$. The constraint C is violated twice: $(X_1 = 1, X_2 = 0)$ and $(X_2 = 0, X_3 = 1)$. Hence, the cost of the assignment is $N = 2 + 1 = 3$. $\square$

A number of global constraints can be used to propagate SEQBIN including REGULAR [7,8], cost REGULAR [9], CARDPATH [3] and SLIDE [5]. However, all are more expensive than the propagator proposed here. A thorough analysis of related work is presented in [1]. We will assume that, as a preprocessing step, all binary constraints B are made DC which takes just $O(nd)$ time for monotone B. We say that an instantiation $I[X]$ is B-coherent iff $(I[X_i], I[X_{i+1}]) \in B$, $i = 1, \ldots, n$. A value $v \in D(X_i)$ is B-coherent iff there exists a B-coherent instantiation $I[X]$ with $I[X_i] = v$.

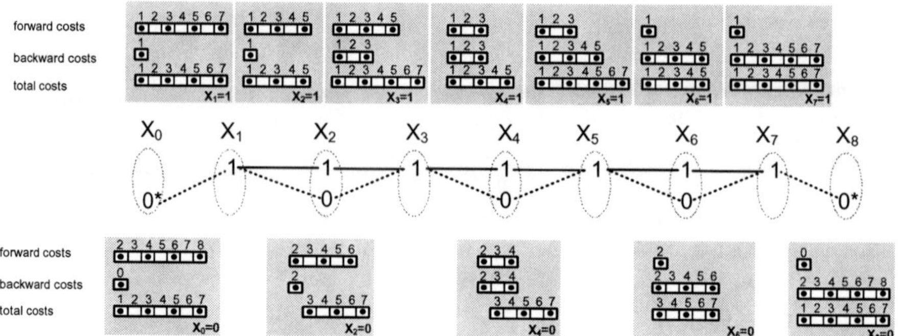

Fig. 1. A 9-partite graph that corresponds to the SEQBIN constraint from Example 1. Dashed edges have cost one and solid edges have zero cost.

3.1   A Graph Representation of SEQBIN

We present a connection between finding a solution of the SEQBIN constraint and the problem of finding a path of a given cost in a special n-partite graph where the cost of an edge is either 0 or 1. We start with a description of the graph $G(V, E)$. For each variable-value pair $x_{i,j}$ we introduce a vertex in the graph that we label $x_{i,j}$, $V = \{x_{i,j} | i = 1, \ldots, n, j \in D(X_i)\}$. For each pair $(x_{i,j}, x_{i+1,v})$ we introduce an edge iff the tuple $(j, v) \in B$, hence, $E = \{(x_{i,j}, x_{i+1,v}) | i = 1, \ldots, n - 1, j \in D(X_i), v \in D(X_{i+1}) \land (j, v) \in B\}$. An edge (j, v) is labeled with $c(j, v)$. Note that vertices $x_{i,j}$, $j \in D(X_i)$, form the ith partition as they do not have edges between them. Moreover, there are edges only between neighbor partitions i and $i + 1$, $i = 1, \ldots, n - 1$. Hence, the resulting graph is a special type of n-partite graph that we call *layered*. To keep the presentation clear, we introduce dummy variables X_0 and X_{n+1} with a single vertex 0^*, and edges from $x_{0,0^*}$ to all vertices (values) of X_1 with cost 1, and from all vertices of X_n to $x_{n+1,0^*}$ with cost 1. To simplify notation, we label a vertex $x_{i,j}$ that is at the ith layer simply as j in all figures. We also use solid lines for edges of cost zero and dashed lines for edges of cost one. As variables correspond to layers in the graph we refer to layers and variables interchangeably. Similarly, as variable-value pairs correspond to vertices in the graph we refer to vertices at the ith layer and values in $D(X_i)$ interchangeably. Given two values j at the ith layer and v at the $(i + 1)$th layer we say that j/v is a support value for v/j iff there exists an edge (j, v) in the graph.

Example 2. Consider the SEQBIN$(N, [X_1, \ldots, X_7], C, B)$ constraint from Example 1. Figure 1 shows the corresponding graph representation of the constraint.   □

We now describe an algorithm, PATHDP to find a path of a given cost in a layered graph. PATHDP is a special case of the dynamic programming algorithm for the knapsack problem where all items have unit costs. Both the existing propagator for SEQBIN and our new one are specializations of PATHDP. Another specialization of PATHDP is the propagator for cost REGULAR [9]. We denote by $c(X)$ the set of all possible numbers of violations achieved by an assignment to X: $c(X) = \{k \mid I \text{ is } B\text{-coherent} \land c(I) = k\}$ and similarly $c(x_{i,j}) = \{k \mid I \text{ is } B\text{-coherent} \land$

$I[X_i] = j \wedge c(I) = k\}$. We denote the *forward cost* from the variable X_i to X_n by $c_f(x_{i,j}) = \{k \mid I[X_i, \ldots, X_n]$ is B-coherent $\wedge I[X_i] = j \wedge c(I) = k\}$. This set contains all the distinct costs that are achievable by paths from the vertex $x_{i,j}$ to the vertex $x_{n+1,0^*}$. We write $lb_f(x_{i,j}) = \min(c_f(x_{i,j}))$ and $ub_f(x_{i,j}) = \max(c_f(x_{i,j}))$. Similarly, we denote the *backward cost* from the variable X_1 to X_i by $c_b(x_{i,j}) = \{k \mid I[X_1, \ldots, X_i]$ is B-coherent $\wedge I[X_i] = j \wedge c(I) = k\}$. It contains all the distinct costs that are achievable by paths from the vertex $x_{0,0^*}$ to the vertex $x_{i,j}$. We denote by $lb_b(x_{i,j}) = \min(c_b(x_{i,j}))$ and $ub_b(x_{i,j}) = \max(c_b(x_{i,j}))$.

Algorithm 1. The pseudocode code for the PATHDP algorithm

```
 1: procedure PATHDP ( G(V, E))
 2:     for i = n → 0; j ∈ D(Xᵢ) do                        ▷ Compute the forward cost
 3:         c_f(x_{i,j}) = ∅
 4:         for k ∈ D(X_{i+1}), (j, k) ∈ B do
 5:             c_f(x_{i,j}) = c_f(x_{i,j}) ⋃ (c_f(x_{i+1,k}) ⊎ c(j, k))
 6:     for i = 1 → n + 1; j ∈ D(Xᵢ) do                    ▷ Compute the backward cost
 7:         c_b(x_{i,j}) = ∅
 8:         for k ∈ D(X_{i-1}), (k, j) ∈ B do
 9:             c_b(x_{i,j}) = c_b(x_{i,j}) ⋃ (c_b(x_{i-1,k}) ⊎ c(k, j))
10:     for i = 0 → n + 1; j ∈ D(Xᵢ) do                    ▷ Compute the total cost
11:         c(x_{i,j}) = c_f(x_{i,j}) ⊎ c_b(x_{i,j})
```

PATHDP performs two scans of the layered graph, one from X_n to X_1 to compute forward costs, and one from X_1 to X_n to compute backward costs. The backward pass processes one layer at a time and computes the set $c_f(x_{i,j})$ for each variable X_i and value $j \in D(X_i)$ (lines 2–5). Dually, the forward pass computes for each variable X_i and value $j \in D(X_i)$, the backward cost $c_b(x_{i,j})$(lines 6–9). Finally, for each vertex the set of costs achievable on paths from $x_{0,0^*}$ to $x_{n+1,0^*}$ that pass through $x_{i,j}$ is $c_f(x_{i,j}) \uplus c_b(x_{i,j})$. To match the semantics of SEQBIN, we compute $c_f(x_{i,j}) \uplus c_b(x_{i,j}) \uplus (-1)$ for each vertex.

The time complexity for SEQBIN using PATHDP is $O(n^2 d^2)$: the number of distinct costs is at most n, so getting the union of two cost sets takes $O(n)$ time. Each vertex has at most d outgoing edges, so the set $c_f(x_{i,j})$ can be computed in $O(nd)$ time for each $x_{i,j}$. There are $O(nd)$ vertices in total, giving the stated complexity of $O(n^2 d^2)$.

Example 3. Consider the SEQBIN$(N, [X_1, \ldots, X_7], C, B)$ constraint from Example 1. Figure 1 shows the forward cost $c_f(x_{i,j})$, the backward cost $c_b(x_{i,j})$ and the total cost $c_f(x_{i,j}) \uplus c_b(x_{i,j}) \uplus (-1)$, $j \in D(X_i)$, $i = 0, \ldots, 8$ in gray rectangles. We have one rectangle for each variable-value pair $X_i = j$. Consider, for example, the vertex '1' at layer X_5. We compute the forward cost $c_f(x_{5,1}) = (c_f(x_{6,0}) \uplus c(1,0)) \cup (c_f(x_{6,1}) \uplus c(1,1)) = \{3\} \cup \{1\} = \{1, 3\}$ and the backward cost $c_b(x_{5,1}) = (c_b(x_{4,0}) \uplus c(0,1)) \cup (c_b(x_{4,1}) \uplus c(1,1)) = \{3, 5\} \cup \{1, 3\} = \{1, 3, 5\}$. Then $c_f(x_{5,1}) \uplus c_b(x_{5,1} \uplus (-1) = \{1, 3\} \uplus \{1, 3, 5\} \uplus (-1) = \{1, 3, 5, 7\}$. □

Lemma 1. *Let $G(V, E)$ be a layered graph constructed from the SEQBIN(N, X, C, B) constraint as described above. There exists a bijection between B-coherent assignments $I[X]$ of cost s and paths in the graph $G(V, E)$ of cost $s + 1$.*

3.2 Revisiting SEQBIN

A domain consistency algorithm for the $\text{SEQBIN}(N, X, C, B)$ constraint, SEQBINALG was proposed in [1] under the restriction that B is a monotone constraint. In this section we identify two drawbacks of this algorithm that make it incomplete. We show that SEQBINALG does not detect bounds disentailment and it is not idempotent even if B is a monotone constraint. It was observed idependently in [10] that SEQBINALG does not enforce DC. However, the authors do not explicitly explain the source of the problems of SEQBINALG and only identify a very restricted class of SEQBIN instances where SEQBINALG does enforce DC.

We will identify the main reason that SEQBINALG fails to enforce DC. This is important to develop a new algorithm that does enforce DC in $O(nd^2)$ time when B is monotone. SEQBINALG uses Algorithm 1 to compute only the lower and upper bounds of the forward and backward cost (Lemma 1 and 2 in [1]). Namely, using the notations in [1], we compute $\underline{s}(x_{i,j}) = lb(c_f(x_{i,j}))$, $\overline{s}(x_{i,j}) = ub(c_f(x_{i,j}))$, $\underline{p}(x_{i,j}) = lb(c_b(x_{i,j}))$ and $\overline{p}(x_{i,j}) = ub(c_b(x_{i,j}))$ in $O(nd^2)$. SEQBINALG is based on these values and runs in 4 steps [1]:

Phase 1 Remove all non B-coherent values in $D(X)$.
Phase 2 For all values in $D(X)$, compute $\underline{s}(x_{i,j}), \overline{s}(x_{i,j}), \underline{p}(x_{i,j})$ and $\overline{p}(x_{i,j})$.
Phase 3 Adjust the min and max value of N with respect to $\underline{s}(X)$ and $\overline{s}(X)$.
Phase 4 Using the result of Phase 3 and Proposition 4 [1], prune the remaining B-coherent values.

The correctness of SEQBINALG relies on Proposition 3. Unfortunately, this proposition is not correct, and the algorithm is consequently incomplete.

Proposition 3 (in [1]). *Given an instance of* $\text{SEQBIN}(N, X, C, B)$ *with monotone* B, $\text{SEQBIN}(N, X, C, B)$ *has a solution iff* $[\underline{s}(X), \overline{s}(X)] \cap N \neq \emptyset$ *where* $\underline{s}(X) = \min_{j \in D(X_1)} \underline{s}(x_{1,j})$ *and* $\overline{s}(X) = \max_{j \in D(X_1)} \overline{s}(x_{1,j})$.

Issue 1. Bounds disentailment.

Lemma 2. *The algorithm* SEQBINALG *for* $\text{SEQBIN}(N, X, C, B)$ *with monotone* B *does not detect bounds disentailment.*

Proof. Consider the $\text{SEQBIN}(N, [X_1, \ldots, X_7], C, \text{TRUE})$ constraint in Example 1. The constraint TRUE is monotone. Consider $N = 4$. Then $\underline{s}(X) = 1$ and $\overline{s}(X) = 7$. Hence, $[1, 7] \cap \{4\} \neq \emptyset$. However, there is no solution with cost 4. The problem with the proof of Proposition 3 in [1] is the last sentence which claims that there is a solution for each value $k \in [\underline{s}(x_{1,v}), \overline{s}(x_{1,v})]$ for some v. This is not true as Example 1 demonstrates. Note also that $[\underline{s}(x_{i,j}) + \underline{p}(x_{i,j}), \overline{s}(x_{i,j}) + \overline{p}(x_{i,j})] \cap \{4\} \neq \emptyset$. Hence, according to Proposition 4 [1] each variable-value pair is DC which is also incorrect. $\qquad\square$

Issue 2. Idempotency. As a consequence of not detecting bounds disentailment, SEQBINALG is also not idempotent.

Lemma 3. *The filtering algorithm* SEQBINALG *for* $\text{SEQBIN}(N, X, C, B)$ *with monotone* B *is not idempotent.*

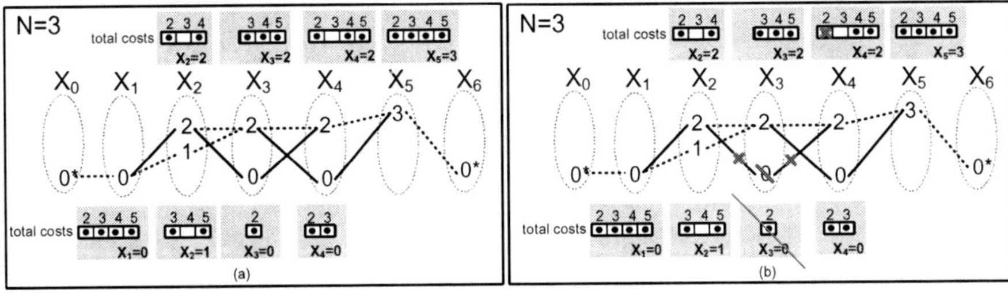

Fig. 2. A 7-partite graph that corresponds to the SEQBIN constraint from the proof of Lemma 3. Dashed edges have cost one and solid edges have cost zero. (a) shows initial costs; (b) shows costs after $X_3 = 0$ is pruned.

Proof. Consider the $\text{SEQBIN}(N, [X_1, \ldots, X_7], C, B)$ where $N = \{3\}$, $B = \{(j, k) | j, k \in [0, 3], (j, k) \notin (0, 0), (1, 0)\}$. $C(X_i, X_j)$ is a monotone constraint with three satisfying tuples $(2, 0), (0, 2), (0, 3) \in C$. Finally, $D(X_1) = \{0\}, D(X_2) = \{1, 2\}, D(X_3) = D(X_4) = \{0, 2\}$ and $D(X_5) = \{3\}$. Figure 2(a) shows the graph representation of the example. Note that $c(x_{3,0}) \cap N = \{2\} \cap \{3\} = \emptyset$. Hence, the value 0 is pruned from $D(X_3)$. Therefore, the value $X_4 = 2$ loses its support with cost 2 (Figure 2(b)). The new cost of $x_{4,2}$ is $\{4, 5\} \cap N = \emptyset$ and the value 2 is pruned from $D(X_4)$. Note that the removal of $X_4 = 2$ triggers further propagation as $X_2 = 2$ loses its support of cost 5, and 2 is removed from $D(X_2)$ at the next step. □

We note that if B is not monotone, SEQBINALG may need $O(n)$ iterations to reach its fixpoint and Proposition 2 in [1] only works if B is monotone.

Remedy for SEQBINALG. As seen in Lemmas 2–3, the main cause of incompleteness in SEQBINALG is that the set of costs for each vertex is a set rather than an interval even when B is monotone. One way to overcome this problem is to restrict $\text{SEQBIN}(N, X, C, B)$ to those instances where it is an interval. This approach was taken in [10] where $\text{SEQBIN}(N, X, C, B)$ was restricted to *counting-continuous* constraints.

Definition 2. *The constraint* $\text{SEQBIN}(N, X, C, B)$ *is* counting-continuous *if and only if for any instantiation $I[X]$ with k stretches in which C holds, for any variable $X_i \in X$, changing the value of X_i in $I[X]$ leads to k, $k + 1$, or $k - 1$ violations.*

This restriction ensures that the structure of the cost for each variable-value pair is an interval and, indeed, the filtering algorithm SEQBINALG enforces DC. However, this approach has a number of drawbacks. First, restricting $\text{SEQBIN}(N, X, C, B)$ to counting-continuous with monotone B excludes useful combinations of B and C. Example 1 shows that $\text{SEQBIN}(N, X, C$ is monotone, B is TRUE) does not satisfy this property. Secondly, many practically interesting examples [1] that can be propagated in $O(nd)$ time do not satisfy these conditions. As was observed in [10], constraints $\text{CHANGE}^{\{=, \neq\}} = \text{SEQBIN}(N, X, C \in \{=, \neq\}, \text{TRUE})$ and SMOOTH are not counting-continuous. The INCREASINGNVALUE constraint which is $\text{SEQBIN}(N, X, =, \leq)$ violates the condition that B is monotone. The only remaining constraint that satisfies these restrictions on B and C is $\text{CHANGE}^{\{<,\le\}} = \text{SEQBIN}(N, X, C \in \{<,\le\}, \text{TRUE})$. Unfortunately, the proof relies on the claim that C is monotone, which is false for $C \in \{<,\le\}$. Thirdly, we do not currently have a test to check if $\text{SEQBIN}(N, X, C, B)$ is counting-continuous. Despite the problems pointed out above, the filtering algorithm SEQBINALG enforces DC on INCREASINGNVALUE and $\text{CHANGE}(C \in \{<,\le\})$ in $O(nd)$ as the counting-continuous property together with the row and column convexity of C are sufficient to achieve this complexity.

In this work we take a different approach. We focus on an extension of the algorithm to handle non-interval cost sets. The challenge is to perform this extension in $O(nd^2)$ as the generic dynamic programming algorithm PATHDP that handles sets natively runs in $O(n^2d^2)$ time. Note that if the cost structure is an unrestricted set of values then the time complexity of PATHDP is going to be hard to improve as it is a specialization of a well-studied dynamic programming algorithm for the knapsack problem where all items have unit cost. Hence, we show that the structure of the costs for a variable-value pair is restricted if B is monotone. This allows us to perform union operations on sets in $O(1)$ time rather than $O(n)$.

3.3 Cost Structure

We show that the structure of the cost for each variable-value pair is restricted. First, we introduce definitions to formalize the structure of forward and backward costs.

Definition 3. *A set S is a zipper set if it can be obtained from an interval $[a, b]$ by removing all odd or all even values. We denote a zipper set as $[a \sim b]$.*

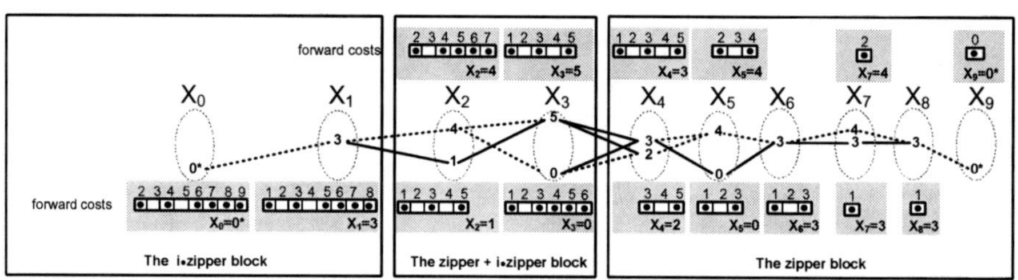

Fig. 3. A 10-partite graph that corresponds to the SEQBIN constraint from Examples 4–6. Dashed edges have cost one and solid edges have zero cost.

Note that in a zipper set $[a \sim b]$, a and b have the same parity. If both are odd, $[a \sim b]$ is an *odd zipper set*, while if both are even, $[a \sim b]$ is an *even zipper set*.

Definition 4. *A set S is an i·zipper set if it can be written as $[a \sim b] \cup [b, c] \cup [c \sim d]$, $a \le b < c \le d$. We denote an i·zipper set as $[a \sim b - c \sim d]$. If $a = b$, we write the set as $[b - c \sim d]$ and if $c = d$ we write it as $[a \sim b - c]$.*

Given an i·zipper set $[a \sim b] \cup [b, c] \cup [c \sim d]$, we denote the left part $[a \sim b]$ as $l \cdot zip$, the middle part $[b - c]$ as $i \cdot val$ and the right part $[c \sim d]$ as $r \cdot zip$.

Example 4. Consider the $\text{SEQBIN}(N, [X_1, \ldots, X_8], C, B)$ constraint that Figure 3 presents. We only show the forward cost sets. For example, the forward cost set $c_f(x_{3,5})$ is a zipper $[1 \sim 5]$, $c_f(x_{5,4}) = [2 \sim 4]$ is an even zipper and $c_f(x_{3,5}) = [1 \sim 5]$ is an odd zipper. An example of an i·zipper set is $c_f(x_{1,3}) = [1 \sim 5 - 8]$. $\square$

Our filtering algorithm is based on the following theorem.

Theorem 1. *Consider a* $\text{SEQBIN}(N, X, C, B)$ *constraint with monotone B and arbitrary C. Let* $[b, c]$, $c > b$ *be the maximal interval such that* $[b, c] \subseteq c_f(x_{i,v})$, $i = 1, \ldots, n, v \in D(X_i)$. *If such an interval does not exist we define* $[b, c] = \emptyset$. *Then the following holds for any value* $j, k, \{j, k\} \in D(X_i)$ *and* $i = 1, \ldots, n$:

1. ***Uniqueness.*** *The set* $c_f(x_{i,v})$ *is either a zipper or i·zipper set.*
2. ***Overlapping.*** *If* $c_f(x_{i,j})$ *and* $c_f(x_{i,k})$ *are i·zipper sets,* $c_f(x_{i,j}) = [a \sim b - c \sim d]$ *and* $c_f(x_{i,k}) = [s \sim r - q \sim t]$, *then* $[b, c] \cap [r, q] \neq \emptyset$.
3. ***Structure.***
   - *Bounded holes. If* $c_f(x_{i,j})$ *is an i·zipper set,* $[a \sim b - c \sim d]$ *then* $b - a \leq 4$ *and* $d - c \leq 4$
   - *Closeness.* $|lb_f(x_{i,j}) - lb_f(x_{i,k})| \leq 2$ *and* $|ub_f(x_{i,j}) - ub_f(x_{i,k})| \leq 2$.

Theorem 1 shows that the structure of $c_f(x_{i,j})$, $j \in D(X_i)$ is limited to few distinct structures of sets: a zipper and an i·zipper. This allows us to deal with such restricted sets efficiently. We give an overview of the proof. We identify two key properties of the problem. The first property is that for all but at most two layers the cost structure is homogeneous. All costs $c_f(x_{i,j})$ are either zippers or i·zippers. Moreover, layers that only contain zippers (i·zippers) are consecutive. The layers $[n_2, \ldots, n]$ only contain zippers for some n_2. The layers $[1, \ldots, n_1]$ only contain i·zippers for some $n_1 < n_2$. There are at most two heterogeneous layers between these sequences.

Example 5. Consider Figure 3. We only show the forward cost $c_f(x_{i,j})$ for each variable-value pair in a gray rectangle. The homogeneous consecutive layers $[n_2 = 4, \ldots, n = 8]$ only contain zippers. The two heterogeneous consecutive layers $[2, 3]$ contain zippers and i·zippers. The homogeneous consecutive layers $[0, n_1 = 1]$ only contain i·zippers. $\square$

The second property is that if we consider all cost sets at one layer then their lower(upper) bounds are at most distance two from each other. This is stated as the closeness property of the structure in Theorem 1. Section 3.4 proves the first property and Section 3.5 proves the second property. The rest of the proof of Theorem 1 uses induction on the number of layers, taking these properties into account. Due to lack of space, these proofs are in the Appendices in [11]. Appendix C.1 proves Theorem 1 for the sequence of layers that only contain cost sets that are zippers. Moreover, it imposes an additional property on the structure of zippers. Appendix C.2 proves Theorem 1 for the two heterogeneous layers. This is the most tedious part of the proof using enumeration of all possible distinct structures of the forward(backward) cost. This enumeration is feasible because of the properties of the cost structure in the first sequence. Appendix C.3 proves Theorem 1 for the last sequence that only contains i·zippers. We show that no new cost structures may appear in this sequence. Overall, we prove that there are a bounded number of cost structures at each layer.

3.4   Partitioning of Layers

The proof of Theorem 1 is based on the following lemma that partitions variables $X_1, \ldots, X_n$ into three groups based on the structure of the forward costs (the backward costs are similar, but the partition may be different).

Lemma 4. *Consider a* $\textsc{SeqBin}(N, X, C, B)$ *with monotone B and arbitrary C. Let $X_t = j$ be the first variable in the reverse order of variables such that there exists a value j and an interval $[a, b]$, $a < b$ such that $[a, b] \subseteq c_f(x_{t,j})$, i.e., for all $t' \in [t+1, n]$, $j \in D(X_{t'})$, there does **not** exist $[a', b']$, $a' < b'$, such that $[a', b'] \subseteq c_f(x_{t',j})$. Then for all $c_f(x_{s,j})$, $j \in D(X_s)$, $s \in [1, t-2]$, there exists an interval $[a_{s,j}, b_{s,j}]$ such that $[a_{s,j}, b_{s,j}] \subseteq c_f(x_{s,j})$.*

Proof. Consider the pair of variables X_t and X_{t-1}. We recall that we consider variables in the reverse order form n to 0. Let v be the maximum value in the total order π such that $v \in D(X_{t-1})$. By the monotonicity of B and the fact that $B(X_{t-1}, X_t)$ is DC, we conclude that $(v, j) \in B$. Otherwise, if $(v, j) \notin B$, the value j had to be pruned from $D(X_i)$ by enforcing DC on $B(X_{t-1}, X_t)$ as v is the top value in the ordering in $D(X_{t-1})$. Therefore, there exists an interval $[c, d] \in \{[a, b], [a+1, b+1]\}$, $c < d$ such that $[c, d] \subseteq c_f(x_{t-1,v})$.

Consider the pair of variables X_{t-1} and X_{t-2}. Due to monotonicity of B we know that $(k, v) \in B$, $q \in D(X_{t-2})$ as v is the top value in π such that $v \in D(X_{t-1})$. Hence, v is a support for all k and $c_f(x_{t-2,k})$ must contain an interval as $c_f(x_{t-2,k}) = \bigcup_{w \in D(X_{t-1})} (c_f(x_{t-1,w}) \uplus c(j, w))$ and $[c, d] \subseteq c_f(x_{t-1,v})$, $v \in D(X_{t-1})$. Hence, there exists an interval $[c', d']$, $c' < d'$ such that $[c', d'] \subseteq c_f(x_{t-2,k})$ for all $k \in D(X_{t-2})$ including the top value in the ordering π, k', such that $k' \in D(X_{t-2})$. We repeat the argument for layers s, $s \in [1, \ldots, t-3]$. $\square$

Corollary 1. *Consider a* $\textsc{SeqBin}(N, X, C, B)$ *with monotone B and arbitrary C. Then there are three blocks of consecutive variables $[X_1, X_{n_1}] \cup [X_{n_1+1}, X_{n_2-1}] \cup [X_{n_2}, X_n]$ with $n_1 < n_2 \leq n_1 + 3$, i.e., the size of the partition $[X_{n_1+1}, X_{n_2-1}]$ is at most 2, and:*

Zipper block. *For all i, j, $i \in [n_2, n]$, $j \in D(X_i)$, there does* not *exist an interval $[a, b] \subseteq [1, \pi^{top}]$, $a < b$, such that $[a, b] \subseteq c_f(x_{i,j})$.*

Zipper + i·Zipper block. *There exist i, j, $i \in [n_1 + 1, n_2 - 1]$, $j \in D(X_i)$ and an interval $[a, b] \subseteq [1, \pi^{top}]$, $a < b$, such that $[a, b] \subseteq c_f(x_{i,j})$.*

i·Zipper block. *For all i, j, $i \in [1, n_1]$, $j \in D(X_i)$ there exists an interval $[a, b] \subseteq [1, \pi^{top}]$, $a < b$, such that $[a, b] \subseteq c_f(x_{i,j})$.*

Example 6. Consider Figure 3. The zipper block includes $[X_4, \ldots, X_8]$. The zipper + i·zipper block includes variables X_2 and X_3. The i·zipper block contains X_1. $\square$

3.5   Closeness of Costs

We show that if B is a monotone constraint then the forward cost of the values of a variable cannot deviate too much from each other. Hence, we prove the closeness property of the cost structure in Theorem 1.

Lemma 5. *Consider a* SEQBIN(N, X, C, B) *with monotone B and arbitrary C. Consider a variable X_i, $i = [1, \ldots n]$. Then for any two values $j, k \in D(X_i)$, $j \prec k$, either* $ub_f(x_{i,j}) \in [ub_f(x_{i,k}), ub_f(x_{i,k}) + 1]$ *or* $ub_f(x_{i,k}) \in [ub_f(x_{i,j}), ub_f(x_{i,j}) + 2]$.

Proof. By induction on the distance from n. The base case is trivial, as $ub_f(x_{n,i}) = lb_f(x_{n,i}) = 1$ for all i. Suppose this holds for all $X_{t+1}, \ldots, X_n$. We show that it holds for X_t. Let v be a value such that $ub_f(x_{t,k}) = ub_f(x_{t+1,v}) + c(k, v)$ and w be a value such that $ub_f(x_{t,j}) = ub_f(x_{t+1,w}) + c(j, w)$.

Property 1. If $w = v$ or $ub_f(x_{t+1,v}) = ub_f(x_{t+1,w})$ then lemma holds. Proof: This follows from the assumption that all costs in C are zero or one. Hence, $|ub_f(x_{t,j}) - ub_f(x_{t,k})| \le 1$.

Property 2. The tuple $(k, w) \in B$. Proof: This follows from monotonicity of B and the assumption that $j \prec k$ and from $(j, w) \in B$.

Property 3. If $w \prec v$ then $(j, v) \in B$. Proof: This follows from monotonicity of B and the assumptions $w \prec v$ and $(j, w) \in B$.

Property 4. If $(j, v) \in B$ then $w = v$. Proof: In this case the bipartite subgraph over four vertices k, j, v, w is complete (Figure 4(a)). Hence, $v' = argmax_{w,v}(ub_f(x_{t+1,v}), ub_f(x_{t+1,w}))$ is a potential support for both $ub_f(x_{t,j})$ and $ub_f(x_{t,k})$ and w and v coincide.

From Properties 1–4 we know that we only have to prove Lemma in the following case: $v \prec w$, $v \ne w$, $ub_f(x_{t+1,v}) \ne ub_f(x_{t+1,w})$ and $(j, v) \notin B$.

By the induction hypothesis, there exist two cases: $ub_f(x_{t+1,v}) \in [ub_f(x_{t+1,w}), ub_f(x_{t+1,w}) + 1]$ or $ub_f(x_{t+1,w}) \in [ub_f(x_{t+1,v}), ub_f(x_{t+1,v}) + 2]$.

Case 1. We assume $ub_f(x_{t+1,v}) \in [ub_f(x_{t+1,w}), ub_f(x_{t+1,w}) + 1]$. As $ub_f(x_{t+1,v}) \ne ub_f(x_{t+1,w})$ we know that $ub_f(x_{t+1,v}) = ub_f(x_{t+1,w}) + 1$. We denote $p = ub_f(x_{t+1,w})$. Figure 4(b) shows this case. Note as costs of the edges are zero or one, $ub_f(x_{t,k}) \in \{p + 1, p + 2\}$. On the other hand, $ub_f(x_{t,j}) \in \{p, p + 1\}$. Hence, $ub_f(x_{t,k}) \in [ub_f(x_{t,j}), ub_f(x_{t,j}) + 2]$ as required.

Case 2. We assume $ub_f(x_{t+1,w}) \in [ub_f(x_{t+1,v}), ub_f(x_{t+1,v}) + 2]$ (Figure 4 (c)) and since $ub_f(x_{t+1,v}) \ne ub_f(x_{t+1,w})$, $ub_f(x_{t+1,w}) > ub_f(x_{t+1,v})$. As $v \prec w$ the value w is a support value for both j and k. Hence, either $v = w$ or $ub_f(x_{t+1,v}) = ub_f(x_{t+1,w})$. This contradicts $v \ne w$ and $ub_f(x_{t+1,v}) \ne ub_f(x_{t+1,w})$. $\qquad\square$

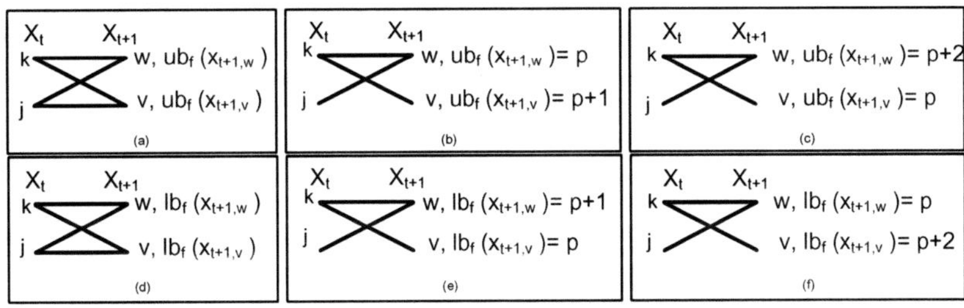

Fig. 4. Computation of the forward cost upper bound, $ub_f(x_{i,j})$, (a)–(c) and the forward cost lower bound, $lb_f(x_{i,j})$, (d)–(f). Note that we do not distinguish between 0 and 1 cost edges.

Lemma 6. *Consider a* SEQBIN(N, X, C, B) *with monotone B and arbitrary C. Then, either $lb_f(x_{i,j}) \in [lb_f(x_{i,k}), lb_f(x_{i,k}) + 2]$ or $lb_f(x_{i,k}) \in [lb_f(x_{i,j}), lb_f(x_{i,j}) + 1]$ for all variables X_i and values $j \prec k$.*

Proof. Analogous to Lemma 5 (Figure 4, (d)–(f)).   □

We omit the rest of the proof here due to space limitation (see Appendix B–C). We only mention Appendix C.1, Lemma 15 that refines Theorem 1 for layers in the zipper block as we use this result in Section 4. Lemma 15 shows that at the ith layer in the zipper block, $i \in [n_1 + 3, n]$, there are at most 4 possible distinct sets $c_f(x_{i,j})$, $j \in D(X_i)$.

3.6   Total Cost

Lemma 7. *Consider a* SEQBIN(N, X, C, B) *constraint with monotone B and arbitrary C. The set $c(x_{i,j}) = c_f(x_{i,j}) \uplus c_b(x_{i,j}) \uplus (-1)$, $j \in D(X_i)$, $i = 1, \ldots, n$ is either a zipper or an i·zipper set. For any i·zipper set $c(x_{i,j}) = [a \sim b - c \sim d]$ it holds $b - a \leq 4$ and $d - c \leq 4$. Moreover, $c(x_{i,j})$ can be computed in $O(1)$ time.*

Proof. It is sufficient to consider $c(x_{i,j}) = c_f(x_{i,j}) \uplus c_b(x_{i,j})$ as a shift by a constant does not change the structure of the set. As $c_f(x_{i,j})$ and $c_b(x_{i,j})$ satisfy Theorem 1, they are either zipper or i·zipper sets. We consider 3 cases.

  Case 1. Both $c_f(x_{i,j})$ and $c_b(x_{i,j})$ are zipper sets. Consider zipper sets $c_f(x_{i,j}) = [a \sim b] = \{a, a+2, \ldots, b\}$ and $c_b(x_{i,j}) = [c \sim d] = \{c, c+2, \ldots, d\}$. Then $c(x_{i,j}) = \{a+c, a+2+c, \ldots, b+c, b+c+2, \ldots, \ldots b+d\} = [(a+c) \sim (b+d)]$.

  Case 2. Both $c_f(x_{i,j})$ and $c_b(x_{i,j})$, are i·zipper sets. Consider $c_f(x_{i,j}) = [a \sim b - r \sim q]$ and $c_b(x_{i,j}) = [c \sim d - f \sim e]$. We consider the most general case where $a < b, r < q, c < d$ and $f < e$.

  We perform the operation $\uplus$ in three steps, $c(x_{i,j}) = c^1 \cup c^2 \cup c^3$ where $c^1 = c_f(x_{i,j}) \uplus [d, f] = [a + d, q + f]$, $c^2 = c_f(x_{i,j}) \uplus [c \sim d] = [(a + c) \sim (b + d)] \cup [b + c, r + d] \cup [(r + c) \sim (b + q)]$, and $c^3 = c_f(x_{i,j}) \uplus [f \sim e] = [(a + f) \sim (b + e)] \cup [b + f, r + e] \cup [(r + f) \sim (e + q)]$. As $(b + c) \leq (b + d)$ and $(r + c) \leq (r + d)$ we get $c^2 = [(a + c) \sim (b + c) - (r + d) \sim (b + q)]$. Similarly, we get $c^3 = [(a + f) \sim (b + f) - (r + e) \sim (e + q)]$.

  Finally, $c(x_{i,j}) = c^1 \cup c^2 \cup c^3 = [(a+c) \sim (\min((b+c), (a+d)) - \max((r+e), (q+f)) \sim (e+q)]$. Consider the value $\min((b+c), (a+d) - (a+c))$. If $b+c \leq a+d$ then we have $(b+c) - (a+c) = b - a \leq 4$. If $a+d \leq b+c$ then we have $(a+d) - (a+c) = d - c \leq 4$. Similarly, we prove the result $(e + q) - \max((r + e), (q + f)) \leq 4$. Hence, the statement of the lemma holds.

  Case 3. Exactly one of $\{c_f(x_{i,j}), c_b(x_{i,j})\}$ is a zipper set. Similar to Case 2.

  Complexity. In all three cases above, the proof is constructive and we give an analytic expression to compute $c(x_{i,j})$. Hence, this can be done in $O(1)$ time.   □

Example 7. Suppose $c_f(x_{i,j}) = [2 \sim 6 - 8 \sim 12]$ and $c_b(x_{i,j}) = [10 \sim 16 - 20 \sim 22]$. Both $c_f(x_{i,j})$ and $c_b(x_{i,j})$ are i·zipper sets. Hence, to compute $c(x_{i,j}) = (c_f(x_{i,j}) \uplus c_b(x_{i,j}) \uplus (-1))$ we use the expression $[(2+10) \sim (\min((6+10), (2+16)) - \max((8+22), (12 + 20)) \sim (12 + 22)] \uplus (-1) = [11 \sim 15 - 31 \sim 33]$.   □

4   Domain Consistency Algorithm

In this section we present SEQBINALGNEW, a domain consistency algorithm for SEQBIN(N, X, C, B) with monotone B. It has the same structure as SEQBINALG:

Phase 1 Remove all non B-coherent values in the domains of X.
Phase 2 For all values in the domains of X, compute $c_f(x_{i,j})$ and $c_b(x_{i,j})$.
Phase 3 Prune the domain of N with respect to $c_f(x_{0,0^*})$.
Phase 4 Prune the remaining B-coherent values.

The main complexity bottleneck is Phase 2 and Phase 4. If we do not put any restrictions on B and C then it takes $O(n^2 d^2)$ in total to compute these sets. We show that the complexity of SEQBINALGNEW decreases as we put restrictions on constraints B and C. With respect to phase 3, we note that the cardinality of both $D(N)$ and $c_f(x_{0,0^*})$ is at most n, so their intersection can be computed in time $O(n)$.

4.1   Domain Consistency Algorithm in $O(nd^2)$ with Monotone B

Phase 2 of SEQBINALGNEW. We exploit the structure of the costs established by Theorem 1 to improve PATHDP (Phase 2). We show that lines 4–5 and 8–9 can be done in $O(d)$ time if B is monotone.

Lemma 8. *Consider a* SEQBIN(N, X, C, B) *constraint such that B is monotone. For all $j \in D(X_i)$, $i \in [1, \ldots, n]$, $c_f(x_{i,j}) = \bigcup_{v \in D(X_{i+1})} (c_f(x_{i+1,v}) \uplus c(j,v))$ can be computed in $O(d)$ time.*

Proof. We partition all supports v into two groups based on the value of $c(j, v)$. The first group S_0 contains values such that $c(j, v) = 0$ and the second group S_1 contains values such that $c(j, v) = 1$. We find $c^1 = \bigcup_{v \in S_0} c_f(x_{i+1,v})$ and $c^2 = \bigcup_{v \in S_1} c_f(x_{i+1,v})$. Then we find $c_f(x_{i,j}) = c^1 \cup (c^2 \uplus 1)$. We prove the lemma for c^1 (c^2 is analogous.)

Compute c^1. We assume that p is the smallest lower bound among the forward cost sets of the values in S_0 and $q + 2$ is the greatest upper bound: $p = \min_{v \in S_0} lb_f(x_{i+1,v})$ and $q + 2 = \max_{v \in S_0} ub_f(x_{i+1,v})$. We refer to $l \cdot zip$ of $c_f(x_{i+1,v})$ as $l \cdot zip(x_{i+1,v})$ to simplify notation (similarly, for the other two parts $i \cdot val$ and $r \cdot zip$). By Theorem 1 we know that $lb(l \cdot zip(x_{i+1,v})) \in [p, p+2]$, $ub(r \cdot zip(x_{i+1,v})) \in [q, q+2]$, $lb(i \cdot val(x_{i+1,v})) \in [p, p+6]$ and $ub(i \cdot val(x_{i+1,v})) \in [q-4, q+2]$. Hence, we compute the 20 *indicator values* $J_y^{l \cdot zip}(v)$, $y \in [p, p+2]$, $J_y^{r \cdot zip}(v)$, $y \in [q, q+2]$, $J_y^{i \cdot val_{lb}}(v)$, $y \in [p, p+6]$, and $J_y^{i \cdot val_{ub}}(v)$, $y \in [q-4, q+2]$, $v \in S_0$. For example, we define $J_y^{l \cdot zip}(v) = 1$, iff $lb(l \cdot zip(x_{i+1,v})) = y$ and $J_y^{l \cdot zip}(S_0) = \max_{v \in S_0} J_y^{l \cdot zip}(v)$, $y \in [p, p+2]$. Similarly, we compute the other 19 indicators. This can be done in $O(d)$ time with a linear scan over $c_f(x_{i+1,v})$, $v \in S_0$. Then we can compute $\bigcup_{v \in S_0} c_f(x_{i+1,v}) = [a^* \sim b^* - c^* \sim d^*]$ in 4 steps, each of which takes $O(1)$ time.

Union of $i \cdot val$. Theorem 1 shows that all $i \cdot val$ sets must overlap. Hence, the union of $i \cdot val(x_{i+1,j})$ forms an interval. We find the minimum value y, $y \in \{p, \ldots, p+6\}$ such that $J_y^{i \cdot val_{lb}}(S_0) = 1$. If such a value y exists then we set $b^* = y$. Then we find the largest value $y' \in \{q-4, \ldots, q+2\}$ such that $J_{y'}^{i \cdot val_{ub}}(S_0) = 1$ and set $c^* = y'$. Note that if y exists then y' exists. If y does not exist we know that all $c_f(x_{i+1,v})$, $v \in S_0$ are zipper sets and we set $b^* = c^* = \emptyset$.

Union of $l \cdot zip$. Suppose $b^* \neq \emptyset$. We find indicators $J_y^{l \cdot zip}(S_0)$, $y \in [p, p+2]$, that are set to one. Set p' to the minimum among $[p, p+2]$, for which there exists $J_{p'}^{l \cdot zip}(S_0) = 1$. If $J_{p+1}^{l \cdot zip}(S_0) = 1$ and $J_p^{l \cdot zip}(S_0) = 1$ or $J_{p+2}^{l \cdot zip}(S_0) = 1$ or $b^* \in \{p, p+2\}$ then set a^* and reset b^*, so that $a^* = b^* = \min(p+1, p')$ otherwise set $a^* = p'$ and leave b^* unchanged. Union of $r \cdot zip$ is similar to union of $l \cdot zip$.

Union of zippers. Suppose $b^* = \emptyset$. Then we determine which of 4 distinct sets (Appendix C.1, Lemma 15) are present among $c_f^1(x_{i+1,v})$, $v \in S_0$. As there are at most 4 such that are zippers we can union them in $O(1)$ time and identify the values a^*, b^*, c^* and d^*.

We can compute $c^1 \cup (c^2 \uplus 1)$ in $O(1)$. We omit the proof here due to space considerations (see Appendix D, Lemma 19).

Complexity. For each $j \in D(X_i)$, $i \in [1, \ldots, n]$, the forward cost set $c_f(x_{i,j})$, can be computed in $O(d)$. As we have $O(nd)$ such sets, the total time complexity is $O(nd^2)$. One way to reduce this complexity is to compute $c_f(x_{i,j})$ in $O(1)$. $\square$

Corollary 2. *Phase 2 of the algorithm* SEQBINALGNEW *runs in* $O(nd^2)$ *time.*

Phase 4 of SEQBINALGNEW. We present the final phase of SEQBINALGNEW.

Lemma 9. *Consider a* SEQBIN(N, X, C, B) *constraint such that B is monotone. For each $i \in [1, \ldots, n]$, the **total** time complexity to compute $c(x_{i,j}) \cap D(N) \neq \emptyset$, $j \in D(X_i)$, is $O(d)$. The total time complexity of Phase 4 is $O(nd)$.*

Proof. Preprocessing of $D(N)$. We use a preprocessing step to compute cumulatively sums s_v^{odd} and s_v^{even} to collect information about the presence of odd and even values in $D(N)$. Hence, $s_0^{odd} = 0$, $s_{j+1}^{odd} = s_j^{odd} + (j \in D(N) \wedge j \text{ is odd })$, $j \in [1, \ldots, \pi^{top}]$. Similarly, we compute s_j^{even}. This can be done in $O(d)$. Then the value $s_{j_1}^{odd} - s_{j_2-1}^{odd}$ shows how many odd values of $D(N)$ are in the interval $[j_2, j_1]$.

Performing the check. By Lemma 7 we know that $c(x_{i,j})$ is either zipper or i·zipper. If $c(x_{i,j})$ is an even zipper set $[a \sim b]$ we check if $s_b^{even} - s_{a-1}^{even} \neq 0$. If so the variable-value pair $X_i = j$ is supported. Similarly, if $c(x_{i,j})$ is an odd zipper set. Suppose $c(x_{i,j})$ is an i·zipper set $[a \sim b - c \sim d]$. Then, we can check separately whether each of three parts $[a \sim b] \cup [b - c] \cup [c \sim d]$ has an intersection with $D(N)$ using the cumulative sum values. Hence, the check can be done in $O(1)$ time. There are $O(d)$ sets $c(x_{i,j})$, $j \in D(X_i)$. Hence, the total time complexity of one layer is $O(d)$.

Complexity. The graph has $O(n)$ layers. So, the total time complexity is $O(nd)$. $\square$

4.2   DC Algorithm with Monotone B and Row and Column Convex C

Finally, we show that if C is row and column convex then SEQBINALGNEW runs in $O(nd)$ time. The only remaining bottleneck is Phase 2.

Lemma 10. *Consider a* SEQBIN(N, X, C, B) *constraint such that B is monotone and C is row and column convex under that same ordering π that gives monotonicity. The sets $c_f(x_{i,j})$ and $c_b(x_{i,j})$, $j \in D(X_i)$, $i \in [1, \ldots, n]$, can be computed in time $O(d)$.*

Proof. We give an algorithm to compute $c_f(x_{i,j})$. Computing $c_b(x_{i,j})$ is similar. Recall that in PATHDP (lines 4–5), $c_f(x_{i,j}) = \bigcup_{v \in D(X_{i+1}), (j,v) \in B} (c_f(x_{i+1,v}) \uplus c(j,v))$. Since B is monotone, the set of B supports of $X_i = j$, $Supports(x_{i,j}) = \{v | (j,v) \in B \wedge v \in D(X_{i+1})\}$, forms the *interval* $[a, v^t]$, $v^t \leq \pi^{top}$ for some a such that v^t is that maximum value in $D(X_{i+1})$.

As C is row convex, the interval $[a, v^t]$ is partitioned into 3 subintervals $[a, v^t] = [a, b] \cup [b, c] \cup [c, v^t]$ such that $c(j, v) = 1, v \in [a, b] \cup [c, v^t]$ and $c(j, v) = 0, v \in [b, c]$ and we can write $c_f(x_{i,j}) = c^1 \cup c^2 \cup c^3$ where $c^1 = \bigcup_{v \in [a,b] \cap D(X_{i+1})} c_f(x_{i+1,v}) \uplus 1$, $c^2 = \bigcup_{v \in [b,c] \cap D(X_{i+1})} c_f(x_{i+1,v})$ and $c^3 = \bigcup_{v \in [c,d] \cap D(X_{i+1})} c_f(x_{i+1,v}) \uplus 1$. We exploit the fact that c^1, c^2, c^3 are computed over intervals to avoid recomputation of the indicator values for each $c(x_{i,j})$, as was necessary in Lemma 8. We do this with an $O(d)$ time preprocessing step that allows us to then compute each $c(x_{i,j})$ in $O(1)$. This reduces the complexity of lines 2–5 from $O(n^2 d^2)$ to $O(nd)$.

The preprocessing step consists in computing cumulative sums over the indicator values in an interval. For each indicator value J_y^z, $z \in \{l \cdot zip, r \cdot zip, i \cdot val\}$, we compute the array $cs_y^z(i)$, which counts the number of values in $[1, i]$ for which the indicator value is 1. For example, $cs_y^{l \cdot zip}(0) = 0$, $cs_y^{l \cdot zip}(v) = cs_y^{l \cdot zip}(v-1) + J_y^{l \cdot zip}(v)$, $y \in [p, p+2]$, $v \in D(X_{i+1})$. To compute the cumulative sums we do a linear scan over $c_f(x_{i+1,v})$, $v \in D(X_{i+1})$. Given these sums we can compute whether, for example, $lb(l \cdot zip(x_{i+1,v})) = y, v \in [a', b']$ in constant time by checking whether $cs_y^{l \cdot zip}(b') - cs_y^{l \cdot zip}(a' - 1) > 0$.

The rest of the proof is identical to Lemma 8 (subsection 'Compute c^1'.). It takes $O(1)$ time to compute c^1 given the cumulative sums. We do this for d sets $c_f(x_{i,j})$ and the preprocessing step takes $O(d)$, so the total time complexity is $O(d)$. $\square$

Corollary 3. *Lemma 10 holds if the negation of C is row and column convex under the same ordering π that gives monotonicity.*

Proof. The only difference from the proof of Lemma 10 is that the interval $[a, v^t]$ of supports of each value $j \in D(X_i)$ is partitioned into $[a, v^t] = [a, b] \cup [b, c] \cup [c, v^t]$, such that $c(j, v) = 0, v \in [a, b] \cup [c, v^t]$ and $c(j, v) = 1, v \in [b, c]$. $\square$

Example 8. Suppose $c_f(x_{i+1,v})$, $v \in [1, 2, 3]$ contain the following forward costs: $c_f(x_{i+1,1}) = [1 \sim 5 - 8 \sim 12]$, $c_f(x_{i+1,2}) = [3 \sim 5 - 6 \sim 10]$ and $c_f(x_{i+1,3}) = [2 \sim 6 - 8 \sim 10]$. The min value p is 1 and the max value $q + 2$ is 12. First we compute cumulative sums. The table below shows the non-zero vectors of cumulative sums.

		v				v				v				v
Cumulative sums	0	1	2	3	Cumulative sums	0	1	2	3	Cumulative sums	0	1	2	3	Cumulative sums	0	1	2	3
$cs_1^{l \cdot zip}(v)$		[0	1	1	1]	$cs_{10}^{r \cdot zip}(v)$	[0	0	1	2]	$cs_5^{i \cdot val_{lb}}(v)$	[0	1	2	2]	$cs_6^{i \cdot val_{ub}}(v)$	[0	0	1	1]
$cs_2^{l \cdot zip}(v)$		[0	0	0	1]	$cs_{12}^{r \cdot zip}(v)$	[0	1	1	1]	$cs_6^{i \cdot val_{lb}}(v)$	[0	0	0	1]	$cs_8^{i \cdot val_{ub}}(v)$	[0	1	1	2]
$cs_3^{l \cdot zip}(v)$		[0	0	1	1]															

Suppose that the values $1, 2$ and 3 are supports for $c_f(x_{i,1}) = [a^* \sim b^* - c^* \sim d^*]$. Using Lemma 10, we find $\cup_{j=[1,3]} i \cdot val(x_{i+1,j}) = [5-8] \cup [5-6] \cup [6-8] = [5-8]$. So $b^* = 5$ and $c^* = 8$ Then, we check if there exists $lb(l \cdot zip(x_{i+1,j})) = y, y \in \{1, 2, 3\}$ using cumulative sums. For $y \in \{1, 2, 3\}$ we get that $cs_y^{l \cdot zip}(3) - cs_y^{l \cdot zip}(0) > 0$.

So we set a^* and reset b^* so that $a^* = b^* = \min(2,1) = 1$. Finally, we check if there exists $ub(r \cdot zip(x_{i+1,j})) = y$, $y \in \{10, 11, 12\}$. For $y \in \{10, 12\}$ we get that $cs_y^{r \cdot zip}(3) - cs_y^{l \cdot zip}(0) > 0$. Moreover, the value $q + 1 = 11$ does not occur among $ub(r \cdot zip(x_{i+1,j}))$. Hence, we set $d^* = 12$. This gives $c_f(x_{i,1}) = [1 - 8 \sim 12]$ □

Corollary 4. *The filtering algorithm* SEQBINALGNEW *enforces domain consistency on* CHANGE *and* SMOOTH *in* $O(nd)$ *time.*

Proof. CHANGE is SEQBIN$(N, X, C \in \{=, \neq, <, \leq, >, \geq\}, \text{TRUE})$. This satisfies Lemma 10 as $\{=, <, \leq, >, \geq\}$ are row/column convex as is the negation of $\{\neq\}$. SMOOTH is SEQBIN$(N, X, C$ is $\{|X_i - X_{i+1}| > cst\}, \text{TRUE})$, $cst \in N$, is the negation of row/column convex constraint $\{|X_i - X_{i+1}| \leq cst\}$.   □

Corollary 5. *The filtering algorithm* SEQBINALGNEW *enforces domain consistency on* INCREASINGNVALUE *in* $O(nd)$ *time.*

Proof. INCREASINGNVALUE(X, N) is SEQBIN$(X, N, \leq, =)$ [1]. This version of the SEQBIN constraint is counting-continuous and therefore $c(x_{i,j})$ is an interval. Hence, all costs c_f, c_b are intervals. Moreover, $=$ is row and column convex, so SEQBINALGNEW reduces to SEQBINALG and enforces GAC in $O(nd)$.   □

Finally, we note that we can slightly generalize SEQBIN so that it does not require the same B and C for every pair of variables as the proof of Theorem 1 does not rely on the property that B and C are the same for each pair of consecutive variables.

5   Conclusions

The SEQBIN meta-constraint subsumes a number of important global constraints like CHANGE, SMOOTH and INCREASINGNVALUE. We have shown that the filtering algorithm for SEQBIN proposed in [1] has two drawbacks even under strong restrictions: it does not detect bounds disentailment and it is not idempotent. We identified the cause for these problems, and proposed a new propagator that overcomes both issues. Our algorithm is based on a connection to the problem of finding a path of a given cost in a restricted n-partite graph. Our propagator enforces domain consistency in $O(nd^2)$ and, for special cases of SEQBIN that include CHANGE, SMOOTH, and INCREASINGNVALUE, in $O(nd)$ time.

References

1. Petit, T., Beldiceanu, N., Lorca, X.: A generalized arc-consistency algorithm for a class of counting constraints. In: Proceedings of the 22nd International Joint Conference on Artificial Intelligence (IJCAI 2011), pp. 643–648 (2011)
2. Cosytec: CHIP Reference manual (1997)
3. Beldiceanu, N., Carlsson, M., Rampon, J.X.: Global constraint catalog. Technical report 2005-08, Swedish Institute of Computer Science (2010)
4. Beldiceanu, N., Hermenier, F., Lorca, X., Petit, T.: The Increasing Nvalue Constraint. In: Lodi, A., Milano, M., Toth, P. (eds.) CPAIOR 2010. LNCS, vol. 6140, pp. 25–39. Springer, Heidelberg (2010)

5. Bessiere, C., Hebrard, E., Hnich, B., Kiziltan, Z., Walsh, T.: Slide: A useful special case of the cardpath constraint. In: Proceedings of the 18th European Conference on Artificial Intelligence, pp. 475–479 (2008)
6. Dechter, R., van Beek, P.: Local and Global Relational Consistency. In: Montanari, U., Rossi, F. (eds.) CP 1995. LNCS, vol. 976, pp. 240–257. Springer, Heidelberg (1995)
7. Pesant, G.: A Regular Language Membership Constraint for Finite Sequences of Variables. In: Wallace, M. (ed.) CP 2004. LNCS, vol. 3258, pp. 482–495. Springer, Heidelberg (2004)
8. Beldiceanu, N., Carlsson, M., Petit, T.: Deriving Filtering Algorithms from Constraint Checkers. In: Wallace, M. (ed.) CP 2004. LNCS, vol. 3258, pp. 107–122. Springer, Heidelberg (2004)
9. Demassey, S., Pesant, G., Rousseau, L.M.: A cost-regular based hybrid column generation approach. Constraints 11(4), 315–333 (2006)
10. Petit, T., Beldiceanu, N., Lorca, X.: A generalized arc-consistency algorithm for a class of counting constraints: Revised edition that incorporates one correction. CoRR abs/1110.4719 (2011)
11. Katsirelos, G., Narodytska, N., Walsh, T.: The SeqBin Constraint Revisited (2012), Available from CoRR http://arxiv.org/abs/1207.1811

Eigenvector Centrality in Industrial SAT Instances[*]

George Katsirelos[1] and Laurent Simon[2]

[1] UBIA, INRA, Toulouse, France
`george.katsirelos@inra-toulouse.fr`
[2] Univ Paris-Sud, LRI / CNRS UMR8623
Orsay, F-91405
`simon@lri.fr`

Abstract. Despite the success of modern SAT solvers on industrial instances, most of the progress relies on intensive experimental testing of improvements or new ideas. In most cases, the behavior of CDCL solvers cannot be predicted and even small changes may have a dramatic positive or negative effect. In this paper, we do not try to improve the performance of SAT solvers, but rather try to improve our understanding of their behavior. More precisely, we identify an essential structural property of industrial instances, based on the Eigenvector centrality of a graphical representation of the formula. We show how this static value, computed only once over the initial formula casts new light on the behavior of CDCL solvers.

We also advocate for a better partitionning of industrial problems. Our experiments clearly suggest deep discrepancies among the families of benchmarks used in the last SAT competitions.

1 Introduction

Despite the impressive progress made in the practical solving of SAT problems in recent years, little work has been done to experimentally study the behavior of those so-called "Modern SAT Solvers". Those solvers are based on a variant of the backtrack search DPLL [4] procedure with learning [7]. While the lookahead architecture of DPLL solvers was relatively easy to understand (all the power of solvers were based on efficient pruning heuristics), the behavior of CDCL (Conflict Driven Clause Learning algorithms, *i.e.* the "Modern SAT solvers") is highly unpredictable, due to its lookback architecture and dynamic branching heuristics.

Nowadays the picture is quite complex: we know the ingredients to build an efficient SAT solver (highly reactive heuristic, resolution-based learning, frequent restarts, frequent clause database reduction), and we are constantly improving them. However, we can hardly explain why those ingredients are so efficient on "real-world" problems. It is claimed that those instances have a particular structure, well suited to the CDCL mechanisms. But what exactly is this particular structure? It is understood that real world instances are different from uniform random instances, but only recently has

[*] This work was partially funded by the "Agence nationale de la Recherche", reference ANR-10-BLA-0214.

M. Milano (Ed.): CP 2012, LNCS 7514, pp. 348–356, 2012.

some progress been made toward characterizing this structure by finding that these instances exhibit modularity [1,2]. It is also known [8] that many real-world instances share the small-world property of graphs.

In this paper, we use a new approach to identify hidden structure in "application" instances. First, we base our approach on a directed graph representation of the formula that separates positive and negative literals, which is close to the factor graph representation. Because some industrial instances are really huge, we compute the eigenvector centralities of the vertices by an efficient iterative algorithm inspired by the Google PageRank algorithm. It provides a very good approximation of the centrality of a clause or a literal in the formula. Intuitively, this measure gives us the frequency with which an infinite random walk in the graph will traverse this vertex. Second, we relate this static measure, computed only once on a preprocessed instance, to measures computed during the search of a CDCL solver. Our experiments are performed with a typical CDCL solver (GLUCOSE). Finally, we show how some structural properties of the initial formula may guide the CDCL search. Moreover, we clearly show that the granularity of SAT problems in the SAT competitions is now too coarse: some families of benchmarks exhibit distinct behavior.

2 Eigenvector Centrality on SAT Directed Graphical Models

Due to lack of space, we assume the reader is familiar with CDCL solvers.

2.1 The SAT Directed Graphical Model

In previous work, SAT instances were studied via the Variable Incidence Graph (VIG) or the Clause-Variable Incidence Graph (CVIG) [2]. Those graphs are undirected. In the VIG representation, each variable of the initial formula is encoded by exactly one vertex in the graphical model. An edge is added between the vertices of two variables if they both occur in at least one clause in the initial formula. In the CVIG, the graph is bipartite (each clause and each variable correspond to exactly one vertex). An edge links a clause vertex and a variable vertex iff the variable occurs in the clause.

Variations are possible. Weights can be added, for instance (links induced by shorter clauses being stronger). In the factor graph representation [3], based on CVIG, edges are additionally labelled by the polarity of the variables in the CVIG graph.

Our graphical representation is contructed such that a random walk on it mimics a random walk of a repair algorithm that would try to satisfy as many clauses as possible. The graph is bipartite and vertices are labeled by clauses and literals (not variables). There is an edge from a clause C to a literal l iff l occurs positively in the clause. There is also an edge from a literal l' to a clause C' iff the literal l' occurs negatively in the clause. A random walk on this graph mimics the operation of a naive algorithm which repairs a global assignment by randomly selecting a clause, randomly flipping a variable that does not satisfy it and then moving on to a clause that is not satisfied by the new value of that variable, repeating the process until it finds a satisfying assignment.

2.2   The Pagerank Algorithm

The PageRank algorithm [5] is an efficient iterative algorithm approximating the stationary distribution of a random walk on a graph. It computes an approximation of eigenvector centrality and is particularly well suited to large graphs, such as the web. To ensure and accelerate convergence, it uses a *damping* factor, that allows the random walk to jump, with a small probability, to any vertex, uniformly at random.

The centrality of a literal C_l is exactly the approximation of the eigenvector centrality of the vertex returned by the above algorithm. We extend it to the centrality of a variable by taking the geometric average over the two literals, $Cv_x = \sqrt{C_x{}^2 + C_{\neg x}{}^2}$. In the rest of this paper, we use both literal and variable centrality.

Once we have computed the centrality of all vertices in the graph, it is possible to infer additional information. For instance, a vertex being the most central of its neighborhood is likely to be on the fringe [6] between communities.

2.3   Pagerank on SAT Problems

We used a damping factor of 0.95 (Google used 0.85) and an accepted total error of $1e-9$ (sum of the changes of all vertices during one iteration). In most cases, the algorithm converged after a few hundred iterations.

We chose to use the Satelite preprocessor on all instances before computing the centrality and running the CDCL solver, for two reasons. First, the real instances passed to CDCL solvers are, in most of the cases, filtered by preprocessing, so it makes sense to work with the same input as the CDCL solver. Second, preprocessing get rid of noise like unit clauses which may complicate computation (the Markov chain of a graph that represents unit clauses is not ergodic) and produce measures that have a worse fit.

In our experiments, we tested all 658 benchmarks from SatRace 2008, SatCompetition 2009 and 2011, in the Application category. We fixed a cutoff of 5 Million conflicts, but placed no bound on CPU time.

3   Observations and Analysis of CDCL Behavior

3.1   Centrality of Variables for Decisions / Propagations

In this first part, we focus only on variable centrality (not literal). We study the centrality of variables picked by the branching heuristic. In figure 1(left), we compare the average centrality of picked variables against the average centrality of all variables. In this figure, as in the rest of the figures in this paper, each point representing one instance. This figure clearly shows that picked variables are the most central variables in the formula. They are almost always above the average centrality of the formula. This may be one factor of the efficiency of the VSIDS heuristics: by branching on central variables, it encourages decomposition of the formula.

The comparison with the figure 1 (right) is also very interesting. Propagated variables are, in almost all the cases, more central than average. However, closer examination shows that the points on the left cluster below those on the right. This raises the question of whether decision variables are more central than propagated variables. We answer

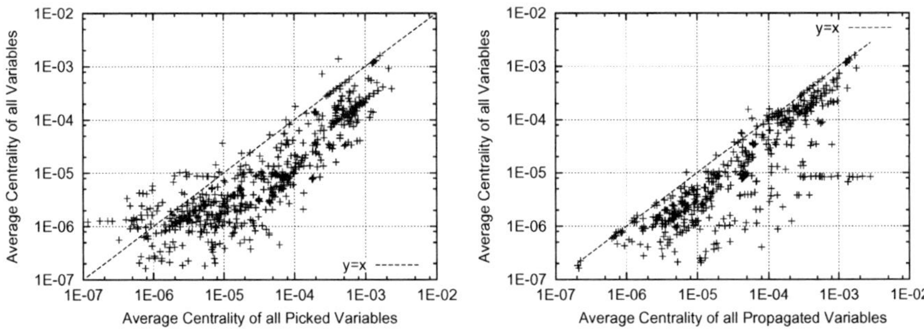

Fig. 1. (**Left**) Average Centrality of Picked Variables (x-axis) against Average Centrality of all Variables (y-axis). (**Right**) Average Centrality of Propagated Variables (x-axis) against Average Centrality of all Variables (y-axis).

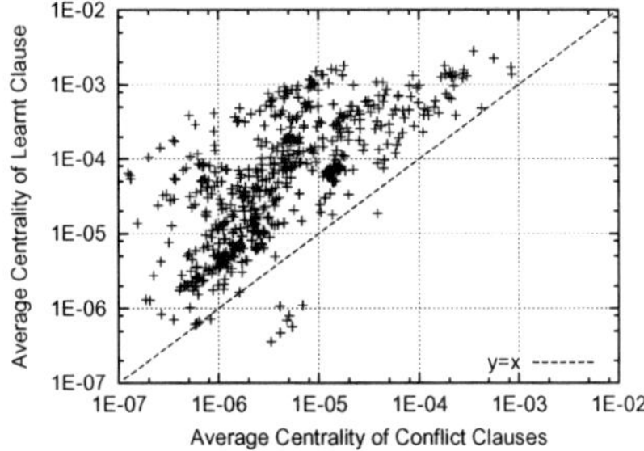

Fig. 2. Average Centrality of Variables (x-axis) occurring in conflict clauses against Average Centrality of all Variables (y-axis) occurring in learnt clauses

this question (positively) in section 3.4, by first refining the benchmarks into families: a few families of benchmarks display opposite behaviors and no clear global picture can be drawn without splitting the set of benchmarks into families.

3.2   Centrality of Variables Occurring in Conflicts and Learnt Clauses

In this experiment, we compare the centrality of conflict clauses (clauses found to be empty during search) against learnt clauses. Because it was impractical to recompute the centrality of the entire formula at each conflict, we measure, for a clause, its centrality as the average of the centrality of its variables (this is again not based on literal centrality).

The results, shown in figure 2, clearly state that learnt clauses contain more central variables than conflict clauses. This is a surprising result. If learnt clauses link together central variables, it intuitively follows that conflict clauses should also be central, but this is not the case. One explanation may come from the notion of fringe (see section 2.2). Learnt clauses may be built upon a few variables from fringes. Conflicts could be detected inside a cluster, thus having no fringe variables in it. This is clearly worth further investigation.

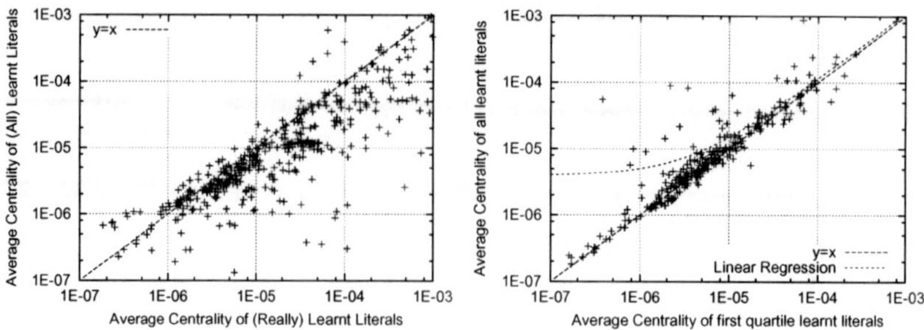

Fig. 3. (**Left**) Average Centrality of Learnt Literals during conflict analysis (x axis) against the Average Centrality of all learnt literals. (**Right**) Average Centrality of Learnt Literals during the first quartile of the computation (x axis) against the Average Centrality of all learnt literals.

3.3   Centrality of Learnt Unit Clauses

Unit clauses learnt during search are particularly important. They clearly simplify the problem, and may be considered as witnesses of which parts of the search space the solver has explored. Typically, in an UNSAT instance, the solver will learn unit clauses at a rate that is relatively high in the beginning of the search and slows as the search goes on, suggesting that each new unit clause is harder and harder to prove. In this experiment, we explore whether there is a relationship between the "hardness" of a literal (when it was learnt) and its centrality. We also note here that this behavior is also observed on SAT instances, but is not general. On some problems, the solver learns no unit clauses until the very end.

There are in fact two kinds of unit clauses that can be learnt. When conflict analysis produces a unit clause, the solver immediately adds this literal as a new fact in the formula and, in many cases, a few other unit clauses are immediately propagated. We distinguish this first literal from the propagated literals, even though they were propagated at the root of the search tree, and call it a "really" learnt literal.

From now on, we consider literal centrality instead of variable centrality. First, figure 3 (left) shows the different kind of unit clauses learnt. This figure clearly shows that the "really" learnt literal is the most central one. Intuitively, other literals simply follow from that assignment. This is compatible with our hypothesis of CDCL solvers working on fringes. This can also be explained by the fact that the branching heuristic favors more central variables (see figure 1). Thus, central literals are the most likely to be learnt first.

We now answer the initial question of this subsection. The answer is given in figure 3 (right). We compare the average centrality of the first quartile of all learnt literals against the centrality of the learnt literals during the whole computation (until the formula is solved or the cutoff is reached). Even if it is not clearly above the $y = x$ line, there is a general tendency showing that centrality of learnt literals tends to increase as the search progresses (the regression line is clearly above the line). This result is interesting for several reasons. First, it explains in part how CDCL solvers focus on particular parts of the search space. Second, it casts new light on parallelization of SAT solvers. In order to efficiently parallelize a CDCL solver, we need to understand which parts of the search

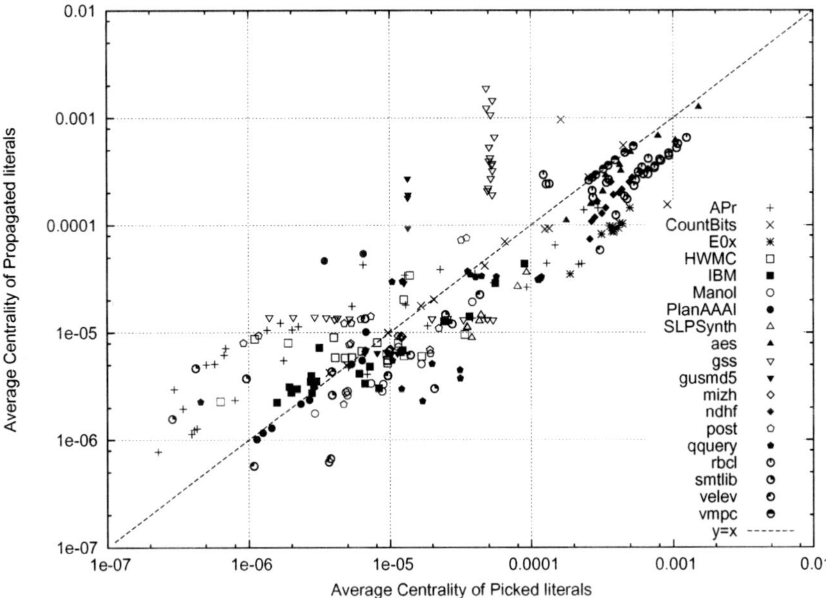

Fig. 4. Average Centrality of Picked Literals against Average Centrality of Propagated Literals, refined by series of benchmarks

space the solver explores, in order to distribute the work evenly. More precisely, by knowing which parts of the search space are relevant to the current proof, it is possible to distribute the effort by giving each process a relevant and precise part of the proof to build, instead of only ensuring orthogonal searches. Moreover, it shows that, intuitively, each unit clause may have a price, in terms of proof length. That means that trying to prove it by orthogonal search may not be the best choice (each search will have to pay more or less the same price).

3.4   Using Families of Benchmarks for Refining Results

We now refine our study by selecting subfamilies among the 653 benchmarks we tested. It is indeed hard to identify general tendencies, because of the discrepancy between different families. We start with figure 4. It shows that, in general, CDCL solvers branch on central literals and propagate less central ones. Most of the instances are below the $y = x$ line except a few families. For instance, the crypto problems (gss, gusmd5) do not follow this trend. This result suggests that these families of instances might benefit from specialized CDCL solvers.

The first UIP literal [9], during conflict analysis, is also an essential ingredient of CDCL solvers. Figure 5 shows that, except for a few families (smtlib, velev, vmpc), the FUIP literal tends to be one of the most central literal in learnt clauses. Despite the discrepancy between decision literals and propagated literals, even in crypto problems, first UIP literals tend to be very central (see the very small cloud of the gss problems for instance).

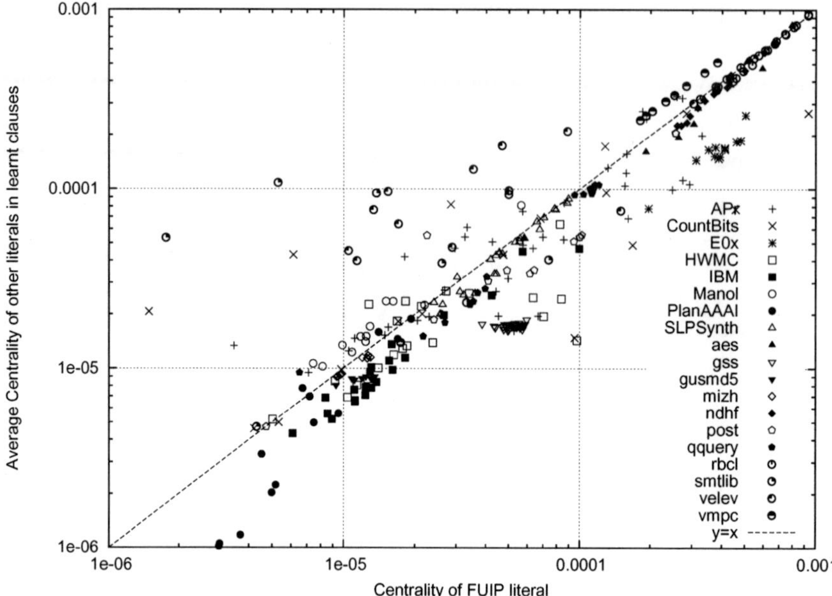

Fig. 5. Average Centrality of FUIP Literals against Average Centrality of other literals in learnt clauses. Most of the instances are below the line $y = x$ except a few families.

3.5   Influence of the Biases of Variables on the Search

Once the centrality of literals has been computed, one may define the biases of a variable as the tendency of the random walk to visit more often the variable positively or negatively. In this last experiment, we explore the relationship between this tendency and the actual CDCL search. The biases of a variable x is defined as $bi(x) = (Cx_{pos} - Cx_{neg})/(Cx_{pos} + Cx_{neg})$, and takes values between -1 and 1. We also defined the "observed Biases" as $ob(x)(Nx_{pos} - Nx_{neg})/(Nx_{pos} + Nx_{neg})$. In the above notations, Cx_{pos} (resp. Cx_{neg}) is the centrality of literal x (resp. $\neg x$). Nx_{pos} (resp. Nx_{neg}) is the number of times the literal x (resp. $\neg x$) is propagated during the CDCL search (until the solution is found, or the bound of 5M conflicts is reached).

In order to study the possible relationships between $bi(x)$ and $ob(x)$, we computed the disagreement between them, $d(x) = |bi(x) - ob(x)|$ (a value between 0 and 2). A value of 1, for instance, means that the CDCL solver did not follow the prediction given by $bi(x)$. A value of 0 means that it strictly follows it, and a value of 2 means that it systematically take the opposite. We also refined the predictive power of $bi(x)$ by considering that only variables having $|bi(x)| > 0.5$ are "biased", meaning that the bias value is significant. We do the same for $ob(x)$ by considering it as meaningful if the measure for the variable x is based on more than 10 assignments during search. The x-axis of figure 6 takes all variables into account (even variables with biases of 0, or variables that were never assigned during search), while the y-axis takes only biased and meaningful variables. Note that the existence of biased variables is not

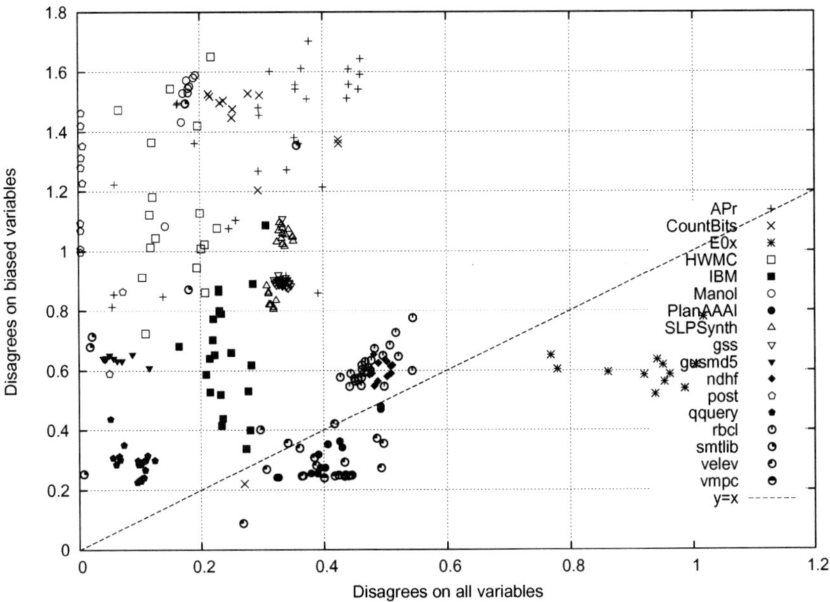

Fig. 6. Predicting phase of variables according to their biases

guaranteed. On some instances, there was only a few biased variables. However, as shown figure 6, some instances show strong correlation between the two values, especially when restricting the cases to biased variables. Clearly, no general rule can be drawn here. However, benchmarks from the same families generally cluster around a common disagreement value.

4   Conclusion

We have shown in this paper that the the centrality of literals and variables in industrial SAT instances is correlated with various aspects of the behavior of CDCL solvers during search. Let us recall that the centrality is computed only at the beginning. The values we compute do not change during search. Despite this approximation, the results we obtained clearly show that centrality plays an important role. This is also one of the first experimental studies of CDCL solvers, that link such a static measure to their actual behavior. We also showed that, even in the application category, families of benchmarks show a large discrepancy in their behavior. This clearly suggests that a finer granularity of categories is needed, in order to specialize solvers and continue to improve them.

However, we were not yet able to turn these observations into predictions and guess, for instance, which parts of the search space the solve is likely to explore or which literal are likely to be first UIP literals or unit clauses. Our quest to explain and understand the CDCL solvers needs more experimental study of those complex systems we built.

References

1. Ansótegui, C., Giráldez-Cru, J., Levy, J.: The Community Structure of SAT Formulas. In: Cimatti, A., Sebastiani, R. (eds.) SAT 2012. LNCS, vol. 7317, pp. 410–423. Springer, Heidelberg (2012)
2. Ansótegui, C., Levy, J.: On the modularity of industrial SAT instances. In: CCIA (2011)
3. Braunstein, A., Mézard, M., Zecchina, R.: Survey propagation: An algorithm for satisfiability. Random Structures & Algorithms 27(2), 201–226 (2005)
4. Davis, M., Logemann, G., Loveland, D.: A machine program for theorem-proving. Communication of ACM 5(7), 394–397 (1962)
5. Langville, A., Meyer, C.: Google's PageRank and Beyond: The Science of Search Engine Rankings. Princeton Unviersity Press (2006)
6. Shah, D., Zaman, T.: Community detection in networks: The leader-follower algorithm. arXiv:1011.0774v1 (2010)
7. Silva, J.M., Sakallah, K.: Grasp - a new search algorithm for satisfiability. In: ICCAD, pp. 220–227 (1996)
8. Walsh, T.: Search in a small world. In: IJCAI 1999: Proceedings of the Sixteenth International Joint Conference on Artificial Intelligence, pp. 1172–1177. Morgan Kaufmann Publishers Inc., San Francisco (1999)
9. Zhang, L., Madigan, C., Moskewicz, M., Malik, S.: Efficient conflict driven learning in a boolean satisfiability solver. In: Proceedings of IEEE/ACM International Conference on Computer Design (ICCAD), pp. 279–285 (2001)

Classifying and Propagating Parity Constraints

Tero Laitinen, Tommi Junttila, and Ilkka Niemelä

Aalto University
Department of Information and Computer Science
PO Box 15400, FI-00076 Aalto, Finland
`{Tero.Laitinen,Tommi.Junttila,Ilkka.Niemela}@aalto.fi`

Abstract. Parity constraints, common in application domains such as circuit verification, bounded model checking, and logical cryptanalysis, are not necessarily most efficiently solved if translated into conjunctive normal form. Thus, specialized parity reasoning techniques have been developed in the past for propagating parity constraints. This paper studies the questions of deciding whether unit propagation or equivalence reasoning is enough to achieve full propagation in a given parity constraint set. Efficient approximating tests for answering these questions are developed. It is also shown that equivalence reasoning can be simulated by unit propagation by adding a polynomial amount of redundant parity constraints to the problem. It is proven that without using additional variables, an exponential number of new parity constraints would be needed in the worst case. The presented classification and propagation methods are evaluated experimentally.

1 Introduction

Encoding a problem instance in conjunctive normal form (CNF) allows very efficient Boolean constraint propagation and conflict-driven clause learning (CDCL) techniques. This has contributed to the success of propositional satisfiability (SAT) solvers (see e.g. [1]) in a number of industrial application domains. On the other hand, an instance consisting only of parity (xor) constraints can be solved in polynomial time using Gaussian elimination but CNF-based solvers relying only on basic Boolean constraint propagation tend to scale poorly on the straightforward CNF-encoding of the instance. To handle CNF instances including parity constraints, common in application domains such as circuit verification, bounded model checking, and logical cryptanalysis, several approaches have been developed [2,3,4,5,6,7,8,9,10,11,12,13]. These approaches extend CNF-level SAT solvers by implementing different forms of constraint propagation for parity constraints, ranging from plain unit propagation via equivalence reasoning to Gaussian elimination. Compared to unit propagation, which has efficient implementation techniques, equivalence reasoning and Gaussian elimination allow stronger propagation but are computationally much more costly.

In this paper our main goal is not to design new inference rules and data structures for propagation engines, but to develop (i) methods for analyzing the structure of parity constraints in order to detect how powerful a parity reasoning engine is needed to achieve full forward propagation, and (ii) translations that allow unit propagation to simulate equivalence reasoning. We first present a method for detecting parity constraint

M. Milano (Ed.): CP 2012, LNCS 7514, pp. 357–372, 2012.

sets for which unit propagation achieves full forward propagation. For instances that do not fall into this category, we show how to extract easy-to-propagate parity constraint parts so that they can be handled by unit propagation and the more powerful reasoning engines can take care of the rest. We then describe a method for detecting parity constraint sets for which equivalence reasoning achieves full forward propagation. By analyzing the set of parity constraints as a constraint graph, we can characterize equivalence reasoning using the cycles in the graph. By enumerating these cycles and adding a new linear combination of the original constraints for each such cycle to the instance, we can achieve an instance in which unit propagation simulates equivalence reasoning. As there may be an exponential number of such cycles, we develop another translation to simulate equivalence reasoning with unit propagation. The translation is polynomial as new variables are introduced; we prove that if introduction of new variables is not allowed, then there are instance families for which polynomially sized simulation translations do not exist. This translation can be optimized significantly by adding only a selected subset of the new parity constraints. Even though the translation is meant to simulate equivalence reasoning with unit propagation, it can augment the strength of equivalence reasoning if equivalence reasoning does not achieve full forward propagation on the original instance. The presented detection and translation methods are evaluated experimentally on large sets of benchmark instances.

2   Preliminaries

An *atom* is either a propositional variable or the special symbol $\top$ which denotes the constant "true". A *literal* is an atom A or its negation $\neg A$; we identify $\neg\top$ with $\bot$ and $\neg\neg A$ with A. A traditional, non-exclusive *or-clause* is a disjunction $l_1 \vee \cdots \vee l_n$ of literals. Parity constraints are formally presented with xor-clauses: an *xor-clause* is an expression of form $l_1 \oplus \cdots \oplus l_n$, where $l_1, \ldots, l_n$ are literals and the symbol $\oplus$ stands for the exclusive logical or. In the rest of the paper, we implicitly assume that each xor-clause is in a *normal form* such that (i) each atom occurs at most once in it, and (ii) all the literals in it are positive. The unique (up to reordering of the atoms) normal form for an xor-clause can be obtained by applying the following rewrite rules in any order until saturation: (i) $\neg A \oplus C \rightsquigarrow A \oplus \top \oplus C$, and (ii) $A \oplus A \oplus C \rightsquigarrow C$, where C is a possibly empty xor-clause and A is an atom. For instance, the normal form of $\neg x_1 \oplus x_2 \oplus x_3 \oplus x_3$ is $x_1 \oplus x_2 \oplus \top$, while the normal form of $x_1 \oplus x_1$ is the empty xor-clause (). We say that an xor-clause is unary/binary/ternary if its normal form has one/two/three variables, respectively. We will identify $x \oplus \top$ with the literal $\neg x$. For convenience, we can represent xor-clauses in equation form $x_1 \oplus \ldots \oplus x_k \equiv p$ with $p \in \{\bot, \top\}$; e.g., $x_1 \oplus x_2$ is represented with $x_1 \oplus x_2 \equiv \top$ and $x_1 \oplus x_2 \oplus \top$ with $x_1 \oplus x_2 \equiv \bot$. The straightforward CNF translation of an xor-clause D is denoted by $\mathrm{cnf}(D)$; for instance, $\mathrm{cnf}(x_1 \oplus x_2 \oplus x_3 \oplus \top) = (\neg x_1 \vee \neg x_2 \vee \neg x_3) \wedge (\neg x_1 \vee x_2 \vee x_3) \wedge (x_1 \vee \neg x_2 \vee x_3) \wedge (x_1 \vee x_2 \vee \neg x_3)$. A *clause* is either an or-clause or an xor-clause.

A *truth assignment* τ is a set of literals such that $\top \in \tau$ and $\forall l \in \tau : \neg l \notin \tau$. We define the "satisfies" relation $\models$ between a truth assignment τ and logical constructs as follows: (i) if l is a literal, then $\tau \models l$ iff $l \in \tau$, (ii) if $C = (l_1 \vee \cdots \vee l_n)$ is an or-clause, then $\tau \models C$ iff $\tau \models l_i$ for some $l_i \in \{l_1, \ldots, l_n\}$, and (iii) if $C = (l_1 \oplus \cdots \oplus l_n)$ is an xor-clause, then $\tau \models C$ iff τ is total for C (i.e. $\forall 1 \leq i \leq n : l_i \in \tau \vee \neg l_i \in \tau$) and $\tau \models l_i$ for an odd number of literals of C. Observe that no truth assignment satisfies the empty or-clause () or the empty xor-clause (), i.e. these clauses are synonyms for $\bot$.

A *cnf-xor formula* ϕ is a conjunction of clauses, expressible as a conjunction

$$\phi = \phi_{\mathrm{or}} \wedge \phi_{\mathrm{xor}}, \tag{1}$$

where ϕ_{or} is a conjunction of or-clauses and ϕ_{xor} is a conjunction of xor-clauses. A truth assignment τ *satisfies* ϕ, denoted by $\tau \models \phi$, if it satisfies each clause in it; ϕ is called *satisfiable* if there exists such a truth assignment satisfying it, and *unsatisfiable* otherwise. The *cnf-xor satisfiability* problem studied in this paper is to decide whether a given cnf-xor formula has a satisfying truth assignment. A formula ϕ' is a *logical consequence* of a formula ϕ, denoted by $\phi \models \phi'$, if $\tau \models \phi$ implies $\tau \models \phi'$ for all truth assignments τ that are total for ϕ and ϕ'. The set of variables occurring in a formula ϕ is denoted by $\mathrm{vars}(\phi)$, and $\mathrm{lits}(\phi) = \{x, \neg x \mid x \in \mathrm{vars}(\phi)\}$ is the set of literals over $\mathrm{vars}(\phi)$. We use $C\,[A/D]$ to denote the (normal form) xor-clause that is identical to C except that all occurrences of the atom A in C are substituted with D once. For instance, $(x_1 \oplus x_2 \oplus x_3)\,[x_1/(x_1 \oplus x_3)] = x_1 \oplus x_3 \oplus x_2 \oplus x_3 = x_1 \oplus x_2$.

2.1 The DPLL(XOR) Framework

To separate parity constraint reasoning from the CNF-level reasoning, we apply the recently introduced DPLL(XOR) framework [10,12]. The idea in the DPLL(XOR) framework for satisfiability solving of cnf-xor formulas $\phi = \phi_{\mathrm{or}} \wedge \phi_{\mathrm{xor}}$ is similar to that in the DPLL(T) framework for solving satisfiability of quantifier-free first-order formulas modulo a background theory T (SMT, see e.g. [14,15]). In DPLL(XOR), see Fig. 1 for a high-level pseudo-code, one employs a conflict-driven clause learning (CDCL) SAT solver (see e.g. [1]) to search for a satisfying truth assignment τ over all the variables in $\phi = \phi_{\mathrm{or}} \wedge \phi_{\mathrm{xor}}$. The CDCL-part takes care of the usual unit clause propagation on the cnf-part ϕ_{or} of the formula (line 4 in Fig. 1), conflict analysis and non-chronological backtracking (line 15–17), and heuristic selection of decision literals (lines 19–20) which extend the current partial truth assignment τ towards a total one.

To handle the parity constraints in the xor-part ϕ_{xor}, an *xor-reasoning module* M is coupled with the CDCL solver. The values assigned in τ to the variables in $\mathrm{vars}(\phi_{\mathrm{xor}})$ by the CDCL solver are communicated as *xor-assumption literals* to the module (with the ASSIGN method on line 6 of the pseudo-code). If $\tilde{l}_1, ..., \tilde{l}_m$ are the xor-assumptions communicated to the module so far, then the DEDUCE method (invoked on line 7) of the module is used to deduce a (possibly empty) list of *xor-implied literals* $\hat{l}$ that are logical consequences of the xor-part ϕ_{xor} and xor-assumptions, i.e. literals for which $\phi_{\mathrm{xor}} \wedge \tilde{l}_1 \wedge ... \wedge \tilde{l}_m \models \hat{l}$ holds. These xor-implied literals can then be added to the current truth assignment τ (line 11) and the CDCL part invoked again to perform unit clause propagation on these. The conflict analysis engine of CDCL solvers requires that each implied (i.e. non-decision) literal has an *implying clause*, i.e. an or-clause that forces the value of the literal by unit propagation on the values of literals appearing earlier in the truth assignment (which at the implementation level is a sequence of literals instead of a set). For this purpose the xor-reasoning module has a method EXPLAIN that, for each xor-implied literal $\hat{l}$, gives an or-clause C of form $l'_1 \wedge ... \wedge l'_k \Rightarrow \hat{l}$,

solve($\phi = \phi_{\text{or}} \wedge \phi_{\text{xor}}$):
1. initialize xor-reasoning module M with ϕ_{xor}
2. $\tau = \langle \rangle$ /*the truth assignment*/
3. **while** true:
4. $(\tau', \mathit{confl}) = \text{UNITPROP}(\phi_{\text{or}}, \tau)$ /*unit propagation*/
5.   **if not** confl: /*apply xor-reasoning*/
6.   **for each** literal l in τ' but not in τ: $M.\text{ASSIGN}(l)$
7. $(\hat{l}_1, ..., \hat{l}_k) = M.\text{DEDUCE}()$
8.   **for** $i = 1$ **to** k:
9. $C = M.\text{EXPLAIN}(\hat{l}_i)$
10.   **if** $\hat{l}_i = \bot$ **or** $\neg\hat{l}_i \in \tau'$: $\mathit{confl} = C$, **break**
11.   **else if** $\hat{l}_i \notin \tau'$: add $\hat{l}_i^C$ to τ'
12.   **if** $k > 0$ **and not** confl:
13. $\tau = \tau'$; **continue**   /*unit propagate further*/
14.   **let** $\tau = \tau'$
15.   **if** confl: /*standard Boolean conflict analysis*/
16.   analyze conflict, learn a conflict clause
17.   backjump or return "unsatisfiable" if not possible
18.   **else**:
19.   add a heuristically selected unassigned literal in ϕ to τ
20.   or return "satisfiable" if no such variable exists

Fig. 1. The essential skeleton of the DPLL(XOR) framework

i.e. $\neg l'_1 \vee ... \vee \neg l'_k \vee \hat{l}$, such that (i) C is a logical consequence of ϕ_{xor}, and (ii) $l'_1, ..., l'_k$ are xor-assumptions made or xor-implied literals returned before $\hat{l}$. An important special case occurs when the "false" literal $\bot$ is returned as an xor-implied literal (line 10), i.e. when an *xor-conflict* occurs; this implies that $\phi_{\text{xor}} \wedge \tilde{l}_1 \wedge ... \wedge \tilde{l}_m$ is unsatisfiable. In such a case, the clause returned by the EXPLAIN method is used as the unsatisfied clause confl initiating the conflict analysis engine of the CDCL part (lines 10 and 15–17). *In this paper, we study the process of deriving xor-implied literals* and will not describe in detail how implying or-clauses are computed; the reader is referred to [10,12,13].

Naturally, there are many *xor-module integration strategies* that can be considered in addition to the one described in the above pseudo-code. For instance, if one wants to prioritize xor-reasoning, the xor-assumptions can be given one-by-one instead. Similarly, if CNF reasoning is to be prioritized, the xor-reasoning module can lazily compute and return the xor-implied literals one-by-one only when the next one is requested.

In addition to our previous work [10,12,13], also cryptominisat [9,11] can be seen to follow this framework.

3   Unit Propagation

We first consider the problem of deciding, given an xor-clause conjunction, whether the elementary unit propagation technique is enough for always deducing all xor-implied literals. As we will see, this is actually the case for many "real-world" instances.

$$\oplus\text{-Unit}^{+}:\frac{x \quad C}{C\,[x/\top]} \qquad \oplus\text{-Unit}^{-}:\frac{x \oplus \top \quad C}{C\,[x/\bot]}$$

Fig. 2. Inference rules of UP; The symbol x is variable and C is an xor-clause

The cnf-xor instances having such xor-clause conjunctions are probably best handled either by translating the xor-part into CNF or with unit propagation algorithms on parity constraints [8,9,13] instead of more complex xor-reasoning techniques.

To study unit propagation on xor-clauses, we introduce a very simple xor-reasoning system "UP" that can only deduce the same xor-implied literals as CNF-level unit propagation would on the straightforward CNF translation of the xor-clauses. To do this, UP implements the deduction system with the inference rules shown in Fig. 2. A *UP-derivation* from a conjunction of xor-clauses ψ is a sequence of xor-clauses $D_1, \ldots, D_n$ where each D_i is either (i) in ψ, or (ii) derived from two xor-clauses D_j, D_k with $1 \le j < k < i$ using the inference rule $\oplus\text{-Unit}^{+}$ or $\oplus\text{-Unit}^{-}$. An xor-clause D is *UP-derivable* from ψ, denoted $\psi \vdash_{\mathrm{up}} D$, if there exists a UP-derivation from ψ where D occurs. As an example, let $\phi_{\mathrm{xor}} = (a \oplus d \oplus e) \wedge (d \oplus c \oplus f) \wedge (a \oplus b \oplus c)$. Fig. 3(a) illustrates a UP-derivation from $\phi_{\mathrm{xor}} \wedge (a) \wedge (\neg d)$; as $\neg e$ occurs in it, $\phi_{\mathrm{xor}} \wedge (a) \wedge (\neg d) \vdash_{\mathrm{up}} \neg e$ and thus unit propagation can deduce the xor-implied literal $\neg e$ under the xor-assumptions (a) and $(\neg d)$.

Definition 1. *A conjunction ϕ_{xor} of xor-clauses is* UP-deducible *if for all $\tilde{l}_1, \ldots, \tilde{l}_k, \hat{l} \in$ $\mathrm{lits}(\phi_{\mathrm{xor}})$ it holds that (i) if $\phi_{\mathrm{xor}} \wedge \tilde{l}_1 \wedge \ldots \wedge \tilde{l}_k$ is unsatisfiable, then $\phi_{\mathrm{xor}} \wedge \tilde{l}_1 \wedge \ldots \wedge \tilde{l}_k \vdash_{\mathrm{up}} \bot$, and (ii) $\phi_{\mathrm{xor}} \wedge \tilde{l}_1 \wedge \ldots \wedge \tilde{l}_k \models \hat{l}$ implies $\phi_{\mathrm{xor}} \wedge \tilde{l}_1 \wedge \ldots \wedge \tilde{l}_k \vdash_{\mathrm{up}} \hat{l}$ otherwise.*

Unfortunately we do not know any easy way of detecting whether a given xor-clause conjunction is UP-deducible. However, as proven next, xor-clause conjunctions that are "tree-like", an easy to test structural property, are UP-deducible. For this, and also later, we use the quite standard concept of constraint graphs: the *constraint graph* of an xor-clause conjunction ϕ_{xor} is a labeled bipartite graph $G = \langle V, E, L \rangle$, where

- the set of vertices V is the disjoint union of (i) *variable vertices* $V_{\mathrm{vars}} = \mathrm{vars}(\phi_{\mathrm{xor}})$ which are graphically represented with circles, and (ii) *xor-clause vertices* $V_{\mathrm{clauses}} = \{D \mid D \text{ is an xor-clause in } \phi_{\mathrm{xor}}\}$ drawn as rectangles,
- $E = \{\{x, D\} \mid x \in V_{\mathrm{vars}} \wedge D \in V_{\mathrm{clauses}} \wedge x \in \mathrm{vars}(D)\}$ are the edges connecting the variables and the xor-clauses in which they occur, and
- L labels each xor-clause vertex $x_1 \oplus \ldots \oplus x_k \equiv p$ with the parity p.

A conjunction ϕ_{xor} is *tree-like* if its constraint graph is a tree or a union of disjoint trees.

Example 1. The conjunction $(a \oplus b \oplus c) \wedge (b \oplus d \oplus e) \wedge (c \oplus f \oplus g \oplus \top)$ is tree-like; its constraint graph is given in Fig. 3(b). On the other hand, the conjunction $(a \oplus b \oplus c) \wedge (a \oplus d \oplus e) \wedge (c \oplus d \oplus f) \wedge (b \oplus e \oplus f)$, illustrated in Fig. 3(c), is not tree-like.

Theorem 1. *If a conjunction of xor-clauses ϕ_{xor} is tree-like, then it is* UP-deducible.

Note that not all UP-deducible xor-clause constraints are tree-like. For instance, $(a \oplus b) \wedge (b \oplus c) \wedge (c \oplus a \oplus \top)$ is satisfiable and UP-deducible but not tree-like. No binary xor-clauses are needed to establish the same, e.g., $(a \oplus b \oplus c) \wedge (a \oplus d \oplus e) \wedge (c \oplus d \oplus f) \wedge (b \oplus e \oplus f)$ considered in Ex. 1 is satisfiable and UP-deducible but not tree-like.

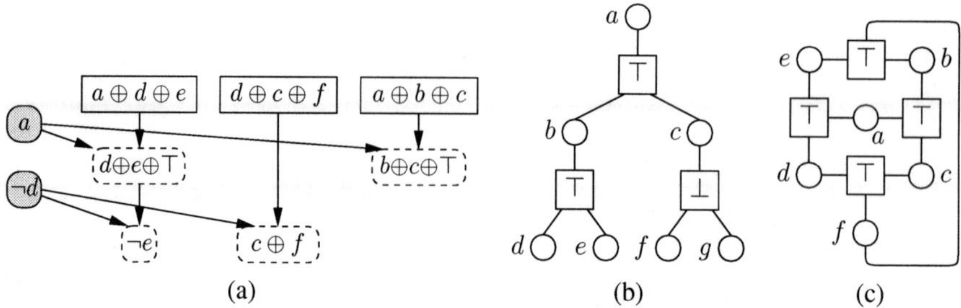

Fig. 3. A UP-derivation and two constraint graphs

3.1 Experimental Evaluation

To evaluate the relevance of this tree-like classification, we studied the benchmark instances in "crafted" and "industrial/application" categories of the SAT Competitions 2005, 2007, and 2009 as well as all the instances in the SAT Competition 2011 (available at http://www.satcompetition.org/). We applied the xor-clause extraction algorithm described in [11] to these CNF instances and found a large number of instances with xor-clauses. To get rid of some "trivial" xor-clauses, we eliminated unary clauses and binary xor-clauses from each instance by unit propagation and substitution, respectively. After this easy preprocessing, 474 instances (with some duplicates due to overlap in the competitions) having xor-clauses remained. Of these instances, 61 are tree-like.

As shown earlier, there are UP-deducible cnf-xor instances that are not tree-like. To find out whether any of the 413 non-tree-like cnf-xor instances we found falls into this category, we applied the following testing procedure to each instance: randomly generate xor-assumption sets and for each check, with Gaussian elimination, whether all xor-implied literals were propagated by unit propagation. For only one of the 413 non-tree-like cnf-xor instances the random testing could not prove that it is not UP-deducible; thus the tree-like classification seems to work quite well in practice as an approximation of detecting UP-deducibility. More detailed results are shown in Fig. 4(a). The columns "probably Subst" and "cycle-partitionable" are explained later.

As unit propagation is already complete for tree-like cnf-xor instances, it is to be expected that the more complex parity reasoning methods do not help on such instances. To evaluate whether this is the case, we ran cryptominisat 2.9.2 [9,11] on the 61 tree-like cnf-xor instances mentioned above in two modes: (i) parity reasoning disabled with CNF input, and (i) parity reasoning enabled with cnf-xor form input and full Gaussian elimination. The results in Fig. 4(b) show that in this setting it is beneficial to use CNF-level unit propagation instead of the computationally more expensive Gaussian elimination method.

3.2 Clausification of Tree-Like Parts

As observed above, a substantial number of real-world cnf-xor instances are not tree-like. However, in many cases a large portion of the xor-clauses may appear in

	SAT Competition			
	2005	2007	2009	2011
instances	857	376	573	1200
with xors	123	100	140	111
probably UP	19	10	18	15
tree-like	19	9	18	15
probably Subst	20	21	52	40
cycle-partitionable	20	13	24	40

(a)

(b)

Fig. 4. Instance classification (a), and cryptominisat run-times on tree-like instances (b)

tree-like parts of ϕ_{xor}. As an example, consider the xor-clause conjunction ϕ_{xor} having the constraint graph shown in Fig. 5(a). It is not UP-deducible as $\phi_{\mathrm{xor}} \wedge a \wedge \neg j \models e$ but $\phi_{\mathrm{xor}} \wedge a \wedge \neg j \not\models_{\mathrm{up}} e$. The xor-clauses (i), $(g \oplus h \oplus i \oplus \top)$, $(e \oplus f \oplus g)$, and $(d \oplus k \oplus m \oplus \top)$ form the tree-like part of ϕ_{xor}. Formally the *tree-like part of* ϕ_{xor}, denoted by $\mathrm{treepart}(\phi_{\mathrm{xor}})$, can be defined recursively as follows: (i) if there is a $D = (x_1 \oplus \ldots \oplus x_n \oplus p)$ with $n \geq 1$ in ϕ_{xor} and an $n-1$-subset W of $\{x_1, \ldots, x_n\}$ such that each $x_i \in W$ appears only in D, then $\mathrm{treepart}(\phi_{\mathrm{xor}}) = \{D\} \cup \mathrm{treepart}(\phi_{\mathrm{xor}} \setminus D)$, and (ii) $\mathrm{treepart}(\phi_{\mathrm{xor}}) = \emptyset$ otherwise.

One can exploit such tree-like parts by applying only unit propagation on them and letting the more powerful but expensive xor-reasoning engines take care only of the non-tree-like parts. Sometimes such a strategy can lead to improvements in run time. For example, consider a set of 320 cnf-xor instances modeling known-plaintext attack on Hitag2 cipher with 30–39 stream bits given. These instances typically have 2600–3300 xor-clauses, of which roughly one fourth are in the tree-like part. Figure 5(b) shows the result when we run cryptominisat 2.9.2 [9,11] on these instances with three configurations: (i) Gaussian elimination disabled, (ii) Gaussian elimination enabled, and (iii) Gaussian elimination enabled and the tree-like parts translated into CNF. On these instances, applying the relatively costly Gaussian elimination to non-tree-like parts only

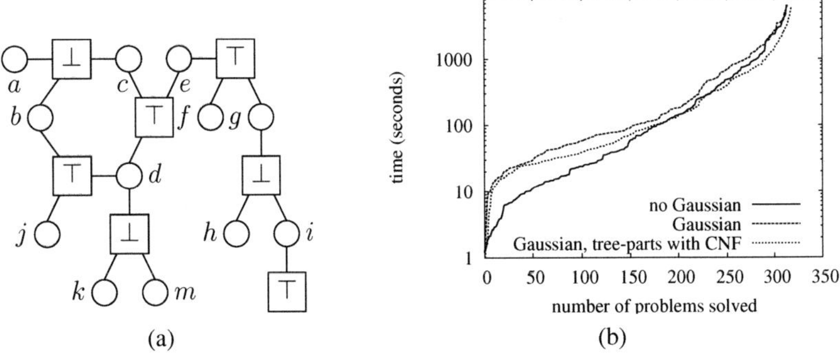

(a)

(b)

Fig. 5. A constraint graph (a), and run-times of cryptominisat on Hitag2 instances (b)

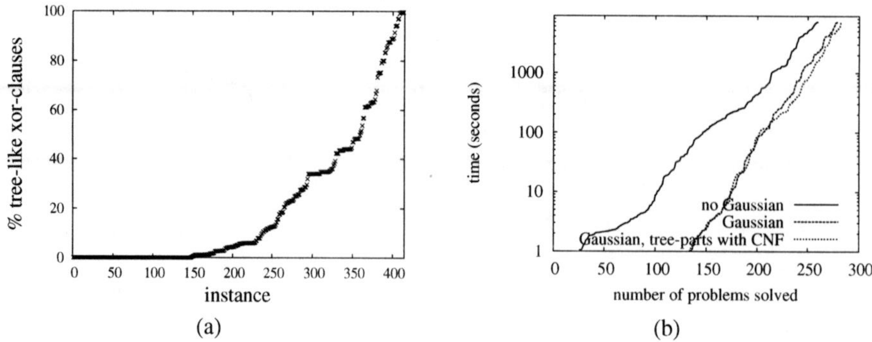

Fig. 6. Relative tree-like part sizes and run-times of non-tree-like instances

is clearly beneficial on the harder instances, probably due to the fact that the Gaussian elimination matrices become smaller. Smaller matrices consume less memory, are faster to manipulate, and can also give smaller xor-explanations for xor-implied literals. On some other benchmark sets, no improvements are obtained as instances can contain no tree-like parts (e.g. our instances modeling known-plaintext attack on TRIVIUM cipher) or the tree-like parts can be very small (e.g. similar instances on the Grain cipher). In addition, the effect is solver and xor-reasoning module dependent: we obtained no run time improvement with the solver of [10] applying equivalence reasoning.

We also ran the same cryptominisat configurations on all the 413 above mentioned non-tree-like SAT Competition benchmark instances. The instances have a large number of xor-clauses (the largest number is 312707) and Fig. 6(a) illustrates the relative tree-like part sizes. As we can see, a substantial amount of instances have a very significant tree-like part. Figure 6(b) shows the run-time results, illustrating that applying Gaussian elimination on non-tree-like instances can bring huge run-time improvements. However, one cannot unconditionally recommend using Gaussian elimination on non-tree-like instances as on some instances, especially in the "industrial" category, the run-time penalty of Gaussian elimination was also huge. Clausification of tree-like parts brings quite consistent improvement in this setting.

4   Equivalence Reasoning

As observed in the previous section, unit propagation is not enough for deducing all xor-implied literals on many practical cnf-xor instances. We next perform a similar study for a stronger deduction system, a form of equivalence reasoning [10,12]. Although it cannot deduce all xor-implied literals either, on many problems it can deduce more and has been shown to be effective on some instance families. The look-ahead based solvers EqSatz [2] and march_eq [7] apply same kind of, but not exactly the same, equivalence reasoning we consider here.

To study equivalence reasoning on xor-clauses, we introduce two equally powerful xor-reasoning systems: "Subst" [10] and "EC" [12]. The first is simpler to implement and to present while the second works here as a tool for analyzing the structure of xor-clauses with respect to equivalence reasoning. The "Subst" system simply adds two substitution rules to UP:

$$\oplus\text{-Eqv}^+: \frac{x \oplus y \oplus \top \quad C}{C\,[x/y]} \quad \text{and} \quad \oplus\text{-Eqv}^-: \frac{x \oplus y \quad C}{C\,[x/y \oplus \top]}$$

The "EC" system, standing for Equivalence Class based parity reasoning, has the inference rules shown in Fig. 7. As there are no inference rules for xor-clauses with more than three variables, longer xor-clauses have to be eliminated, e.g., by repeatedly applying the rewrite rule $(x_1 \oplus x_2 \oplus \ldots \oplus x_n) \rightsquigarrow (x_1 \oplus x_2 \oplus y) \wedge (\neg y \oplus x_3 \oplus \ldots \oplus x_n)$, where y is a fresh variable not occurring in other clauses. We define Subst- and EC-derivations, the relations $\vdash_{\mathsf{Subst}}$ and $\vdash_{\mathsf{ec}}$, as well as Subst- and EC-deducibility similarly to UP-derivations, $\vdash_{\mathsf{up}}$, and UP-deducibility, respectively.

Example 2. Figure 8 shows (parts of) Subst- and EC-derivations from $\phi_{\mathsf{xor}} \wedge (a) \wedge (\neg j)$, where ϕ_{xor} is the xor-clause conjunction shown in Fig. 5(a).

As shown in [12], on cnf-xor instances with xor-clauses having at most three variables, Subst and EC can deduce exactly the same xor-implied literals. Thus, such an instance ϕ_{xor} is Subst-deducible if and only if it is EC-deducible.

The EC-system uses more complicated inference rules than Subst, but it allows us to characterize equivalence reasoning as a structural property of constraint graphs. The EC rules Conflict, $\oplus$-Unit^2, and $\oplus$-Unit^3 are for unit propagation on xor-clauses with 1–3 variables, and the rules $\oplus$-Imply and $\oplus$-Conflict for equivalence reasoning. To simplify the following proofs and translations, we consider a restricted class of xor-clauses. A conjunction of xor-clauses ϕ_{xor} is in *3-xor normal form* if (i) every xor-clause in it has exactly three variables, and (ii) each pair of xor-clauses shares at most one variable. Given a ϕ_{xor}, an equi-satisfiable 3-xor normal form formula can be obtained by (i) eliminating unary and binary xor-clauses by unit propagation and substitution, (ii) cutting longer xor-clauses as described above, and (iii) applying the following rewrite rule: $(x_1 \oplus x_2 \oplus x_3) \wedge (x_2 \oplus x_3 \oplus x_4) \rightsquigarrow (x_1 \oplus x_2 \oplus x_3) \wedge (x_1 \oplus x_4 \oplus \top)$. In this normal form, $\oplus$-Conflict is actually a shorthand for two applications of $\oplus$-Imply and one application of Conflict, so the rule $\oplus$-Imply succinctly characterizes equivalence reasoning. We now prove that the rule $\oplus$-Imply is closely related to the cycles in the constraint graphs. An *xor-cycle* is an xor-clause conjunction of form $(x_1 \oplus x_2 \oplus y_1 \equiv p_1) \wedge \cdots \wedge (x_{n-1} \oplus x_n \oplus y_{n-1} \equiv p_{n-1}) \wedge (x_1 \oplus x_n \oplus y_n \equiv p_n)$, abbreviated with $XC(\langle x_1, \ldots, x_n \rangle, \langle y_1, \ldots, y_n \rangle, p)$ where $p = p_1 \oplus \ldots \oplus p_n$. We call $x_1, \ldots, x_n$ the *inner variables* and $y_1, \ldots, y_n$ the *outer variables* of the xor-cycle.

$$\frac{x \quad x \oplus \top}{\bot} \qquad \frac{x \oplus p_1 \quad x \oplus y \oplus p_2}{y \oplus (p_1 \oplus p_2 \oplus \top)} \qquad \frac{x_1 \oplus x_2 \oplus p_1 \oplus \top \quad \ldots \quad x_{n-1} \oplus x_n \oplus p_{n-1} \oplus \top \quad x_1 \oplus x_n \oplus y \oplus p}{y \oplus (p_1 \oplus p_2 \oplus \ldots \oplus p_{n-1} \oplus p)}$$

(a) Conflict   (b) $\oplus$-Unit^2 (d) $\oplus$-Imply

$$\frac{x \oplus p_1 \quad x \oplus y \oplus z \oplus p_2}{y \oplus z \oplus (p_1 \oplus p_2 \oplus \top)} \qquad \frac{x_1 \oplus x_2 \oplus p_1 \oplus \top \quad \ldots \quad x_{n-1} \oplus x_n \oplus p_{n-1} \oplus \top \quad x_n \oplus x_1 \oplus p_n \oplus \top}{\text{provided that } p_1 \oplus \ldots \oplus p_n = \top}{\bot}$$

(c) $\oplus$-Unit^3 (e) $\oplus$-Conflict

Fig. 7. Inference rules of EC; the symbols x, x_i, y, z are all variables while the p_i symbols are constants $\bot$ or $\top$

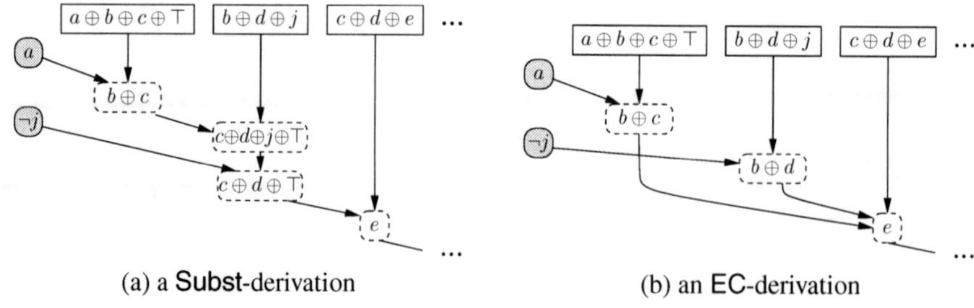

(a) a Subst-derivation   (b) an EC-derivation

Fig. 8. Subst- and EC-derivations from $\phi_{\text{xor}} \wedge (a) \wedge (\neg j)$, where ϕ_{xor} is given in Fig. 5(a)

Example 3. The cnf-xor instance shown in Fig. 5(a) has one xor-cycle $(a \oplus b \oplus c \oplus \top) \wedge$ $(c \oplus d \oplus e) \wedge (b \oplus d \oplus j)$, where b, c, d are the inner and a, e, j the outer variables.

A key observation is that the $\oplus$-Imply rule can infer a literal *exactly when there is an xor-cycle* with the values of the outer variables except for one already derived:

Lemma 1. *Assume an EC-derivation $\pi = D_1, \ldots, D_n$ from $\psi = \phi_{\text{xor}} \wedge \tilde{l}_1 \wedge \ldots \wedge \tilde{l}_k$, where ϕ_{xor} is a 3-xor normal form xor-clause conjunction. An xor-clause $(y \equiv p \oplus p_1' \oplus \ldots \oplus p_{n-1}')$ can be added to π after a number of other derivation steps using $\oplus$-Imply on the xor-clauses $\{(x_1 \oplus x_2 \equiv p_1 \oplus p_1'), \ldots, (x_{n-1} \oplus x_n \equiv p_{n-1} \oplus p_{n-1}'), (x_1 \oplus x_n \oplus y \equiv p_n)\}$ if and only if there is an xor-cycle $XC(\langle x_1, \ldots, x_n \rangle, \langle y_1, \ldots y_{n-1}, y \rangle, p) \subseteq \phi_{\text{xor}}$ where $p = p_1 \oplus \ldots \oplus p_n$ such that for each $y_i \in \{y_1, \ldots, y_{n-1}\}$ it holds that $\psi \vdash_{\text{ec}} (y_i \equiv p_i')$.*

4.1   Detecting When Equivalence Reasoning Is Enough

The presence of xor-cycles in the problem implies that equivalence reasoning might be useful, but does not give any indication of whether it is enough to always deduce all xor-implied literals. Again, we do not know any easy way to detect whether a given xor-clause conjunction is Subst-deducible (or equivalently, EC-deducible). However, we can obtain a very fast structural test for approximating EC-deducibility as shown and analyzed in the following.

We say that a 3-xor normal form xor-clause conjunction ϕ_{xor} is *cycle-partitionable* if there is a partitioning $(V_{\text{in}}, V_{\text{out}})$ of $\text{vars}(\phi_{\text{xor}})$ such that for each xor-cycle $XC(X, Y, p)$ in ϕ_{xor} it holds that $X \subseteq V_{\text{in}}$ and $Y \subseteq V_{\text{out}}$. That is, there should be no variable that appears as an inner variable in one xor-cycle and as an outer variable in another. For example, the instance in Fig. 5(a) is cycle-partitionable as $(\{b, c, d\}, \{a, e, f, \ldots, m\})$ is a valid cycle-partition. On the other hand, the one in Fig. 3(c) is not cycle-partitionable (although it is UP-deducible and thus EC-deducible). If such cycle-partition can be found, then equivalence reasoning is enough to always deduce all xor-implied literals.

Theorem 2. *If a 3-xor normal form xor-clause conjunction ϕ_{xor} is cycle-partitionable, then it is Subst-deducible (and thus also EC-deducible).*

Detecting whether a cycle-partitioning exists can be efficiently implemented with a variant of Tarjan's algorithm for strongly connected components.

To evaluate the accuracy of the technique, we applied it to the SAT Competition instances discussed in Sect. 3.1. The results are shown in the "cycle-partitionable" and "probably Subst" columns in Fig. 4(a), where the latter gives the number of instances for which our random testing procedure described in Sect. 3.1 was not able to show that the instance is not Subst-deducible. We see that the accuracy of the cycle-partitioning test is (probably) not perfect in practice although for some instance families it works very well.

4.2 Simulating Equivalence Reasoning with Unit Propagation

The connection between equivalence reasoning and xor-cycles enables us to consider a potentially more efficient way to implement equivalence reasoning. We now present three translations that add redundant xor-clauses in the problem with the aim that unit propagation is enough to always deduce all xor-implied literals in the resulting xor-clause conjunction. The first translation is based on the xor-cycles of the formula and does not add auxiliary variables, the second translation is based on explicitly communicating equivalences between the variables of the original formula using auxiliary variables, and the third translation combines the first two.

The redundant xor-clause conjunction, called an EC-*simulation formula* ψ, added to ϕ_{xor} by a translation should satisfy the following: (i) the satisfying truth assignments of ϕ_{xor} are exactly the ones of $\phi_{\mathrm{xor}} \wedge \psi$ when projected to $\mathrm{vars}(\phi_{\mathrm{xor}})$, and (ii) if $\hat{l}$ is EC-derivable from $\phi_{\mathrm{xor}} \wedge (\tilde{l}_1) \wedge ... \wedge (\tilde{l}_k)$, then $\hat{l}$ is UP-derivable from $(\phi_{\mathrm{xor}} \wedge \psi) \wedge (\tilde{l}_1) \wedge ... \wedge (\tilde{l}_k)$.

Simulation without Extra Variables. We first present an EC-simulation formula for a given 3-xor normal form xor-clause conjunction ϕ_{xor} without introducing additional variables. The translation adds one xor-clause with the all outer variables per xor-cycle:

$$cycles(\phi_{\mathrm{xor}}) = \bigwedge_{XC(\langle x_1,...,x_n\rangle, \langle y_1,...,y_n\rangle, p) \subseteq \phi_{\mathrm{xor}}} (y_1 \oplus ... \oplus y_n \equiv p)$$

For example, for the xor-clause conjunction ϕ_{xor} in Fig. 5(a) $cycles(\phi_{\mathrm{xor}}) = (a \oplus e \oplus j \oplus \top)$. Now $\phi_{\mathrm{xor}} \wedge cycles(\phi_{\mathrm{xor}}) \wedge (a) \wedge (\neg j) \vdash_{\mathrm{up}} e$ although $\phi_{\mathrm{xor}} \wedge (a) \wedge (\neg j) \nvdash_{\mathrm{up}} e$.

Theorem 3. *If ϕ_{xor} is a 3-xor normal form xor-clause conjunction, then $cycles(\phi_{\mathrm{xor}})$ is an EC-simulation formula for ϕ_{xor}.*

The translation is intuitively suitable for problems that have a small number of xor-cycles, such as the DES cipher. Each instance of our DES benchmark (4 rounds, 2 blocks) has 28–32 xor-cycles. We evaluated experimentally the translation on this benchmark using cryptominisat 2.9.2, minisat 2.0, minisat 2.2, and minisat 2.0 extended with the UP xor-reasoning module. The benchmark set has 51 instances and the clauses of each instance are permuted 21 times randomly to negate the effect of propagation order. The results are shown in Fig. 9(a). The translation manages to slightly reduce solving time for cryptominisat, but this does not happen for other solver configurations based on minisat, so the slightly improved performance is not completely due to simulation of equivalence reasoning using unit propagation. The xor-part (320 xor-clauses of which 192 tree-like) in DES is negligible compared to cnf-part (over 28000 clauses), so a great reduction in solving time is not expected.

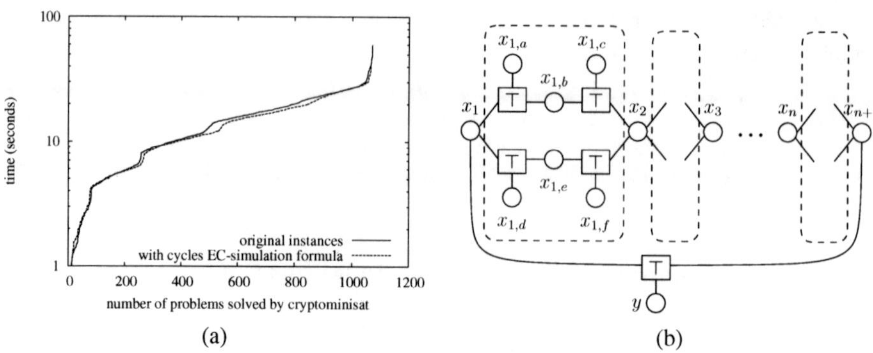

Fig. 9. The $cycles(\phi_{\mathrm{xor}})$ translation on DES instances (a), and the constraint graph of $D(n)$ (b)

Although equivalence reasoning can be simulated with unit propagation by adding an xor-clause for each xor-cycle, this is not feasible for all instances in practice due to the large number of xor-cycles. We now prove that, *without* using auxiliary variables, there are in fact families of xor-clause conjunctions for which all EC-simulation formulas, whether based on xor-cycles or not, are exponential. Consider the xor-clause conjunction $D(n) = (x_1 \oplus x_{n+1} \oplus y) \wedge \bigwedge_{i=1}^{n} (x_i \oplus x_{i,a} \oplus x_{i,b}) \wedge (x_{i,b} \oplus x_{i,c} \oplus x_{i+1}) \wedge (x_i \oplus x_{i,d} \oplus x_{i,e}) \wedge (x_{i,e} \oplus x_{i,f} \oplus x_{i+1})$ whose constraint graph is shown in Fig. 9(b). Observe that $D(n)$ is cycle-partitionable and thus Subst/EC-deducible. But all its EC-simulation formulas are at least of exponential size if no auxiliary variables are allowed:

Lemma 2. *Any* EC*-simulation formula ψ for $D(n)$ with* $\mathrm{vars}(\psi) = \mathrm{vars}(D(n))$ *contains at least 2^n xor-clauses.*

Simulation with Extra Variables: Basic Version. Our second translation $\mathrm{Eq}(\phi_{\mathrm{xor}})$ avoids the exponential increase in size by introducing a quadratic number of auxiliary variables. A new variable e_{ij} is added for each pair of distinct variables $x_i, x_j \in \mathrm{vars}(\phi_{\mathrm{xor}})$, with the intended meaning that e_{ij} is true when x_i and x_j have the same value and false otherwise. We identify e_{ji} with e_{ij}. Now the translation is

$$\mathrm{Eq}(\phi_{\mathrm{xor}}) = \bigwedge_{(x_i \oplus x_j \oplus x_k \equiv p) \in \phi_{\mathrm{xor}}} (e_{ij} \oplus x_k \oplus \top \equiv p) \wedge (e_{ik} \oplus x_j \oplus \top \equiv p) \wedge (x_i \oplus e_{jk} \oplus \top \equiv p) \wedge$$

$$\bigwedge_{x_i, x_j, x_k \in \mathrm{vars}(\phi_{\mathrm{xor}}), i < j < k} (e_{ij} \oplus e_{jk} \oplus e_{ik} \equiv \top)$$

where (i) the first line ensures that if we can deduce that two variables in a ternary xor-clause are (in)equivalent, then we can deduce the value of the third variable, and vice versa, and (ii) the second line encodes transitivity of (in)equivalences. The translation enables unit propagation to deduce all EC-derivable literals over the variables in the original xor-clause conjunction:

Theorem 4. *If ϕ_{xor} is an xor-clause conjunction in 3-xor normal form, then $\mathrm{Eq}(\phi_{\mathrm{xor}})$ is an* EC*-simulation formula for ϕ_{xor}.*

$\text{Eq}^\star(\phi_{\text{xor}})$: start with $\phi'_{\text{xor}} = \phi_{\text{xor}}$ and $V = \text{vars}(\phi_{\text{xor}})$

1.   while ($V \neq \emptyset$):
2. $x_j \leftarrow$ extract a variable v from V minimizing $|\,\text{vars}(\{C \in \phi'_{\text{xor}} \mid v \in \text{vars}(C)\}) \cap V|$
3.   for each $(x_i \oplus x_j \oplus e_{ij} \equiv p_{ij}),(x_j \oplus x_k \oplus e_{jk} \equiv p_{jk}) \in \phi'_{\text{xor}}$ such that $x_i, x_k \in V \wedge x_i \neq x_j \neq x_k$
4.   if $(x_i \oplus x_k \oplus y \equiv p'_{ik}) \in \phi'_{\text{xor}}$
5. $e_{ik} \leftarrow y; p_{ik} \leftarrow p'_{ik}$
6.   else
7. $e_{ik} \leftarrow$ new variable; $p_{ik} \leftarrow \top$
8. $\phi'_{\text{xor}} \leftarrow \phi'_{\text{xor}} \wedge (x_i \oplus x_k \oplus e_{ik} \equiv p_{ik})$
9. $\phi'_{\text{xor}} \leftarrow \phi'_{\text{xor}} \wedge (e_{ij} \oplus e_{jk} \oplus e_{ik} \equiv p_{ij} \oplus p_{jk} \oplus p_{ik})$
10. return $\phi'_{\text{xor}} \backslash \phi_{\text{xor}}$

Fig. 10. The $\text{Eq}^\star$ translation

Simulation with Extra Variables: Optimized Version. The translation $\text{Eq}(\phi_{\text{xor}})$ adds a cubic number of clauses with respect to the variables in ϕ_{xor}. This is infeasible for many real-world instances. The third translation combines the first two translations by implicitly taking into account the xor-cycles in ϕ_{xor} while adding auxiliary variables where needed. The translation $\text{Eq}^\star(\phi_{\text{xor}})$ is presented in Fig. 10. The xor-clauses added by $\text{Eq}^\star(\phi_{\text{xor}})$ are a subset of $\text{Eq}(\phi_{\text{xor}})$ and the meaning of the variable e_{ij} remains the same. The intuition behind the translation, on the level of constraint graphs, is to iteratively shorten xor-cycles by "eliminating" one variable at a time by adding auxiliary variables that "bridge" possible equivalences over the eliminated variable. The line 2 in the pseudo-code picks a variable x_j to eliminate. While the correctness of the translation does not depend on the choice, it is sensible to pick a variable that shares xor-clauses with fewest variables because the number of xor-clauses produced in lines 3–9 is then smallest. The loop in line 3 iterates over all possible xor-cycles where the selected variable x_j and two "neighboring" non-eliminated variables x_i, x_k may occur as inner variables. The line 4 checks if there already is an xor-clause that has both x_i and x_k. If so, then in line 5 an existing variable is used as e_{ik} capturing the equivalence between the variables x_i and x_k. If the variable p_{ik} is $\top$, then e_{ik} is true when the variables x_i and x_k have the same value. The line 9 adds an xor-clause ensuring that transitivity of equivalences between the variables x_i, x_j, and x_k can be handled by unit propagation.

Example 4. Consider the xor-clause conjunction $\phi_{\text{xor}} = (x_1 \oplus x_2 \oplus x_4) \wedge (x_2 \oplus x_3 \oplus x_5) \wedge (x_5 \oplus x_7 \oplus x_8) \wedge (x_4 \oplus x_6 \oplus x_7)$ shown in Fig. 11(a). The translation $\text{Eq}^\star(\phi_{\text{xor}})$ first selects one-by-one the variables in $\{x_1, x_3, x_6, x_8\}$ as each appears in only one xor-clause. The loop in lines 3–9 is not executed for any of them. The remaining variables are $V = \{x_2, x_4, x_5, x_7\}$. Assume that x_2 is selected. The loop in lines 3–9 is entered with values $x_i = x_4$, $x_j = x_2$, $e_{ij} = x_1$, $x_k = x_5$, $e_{jk} = x_3$, $p_{ij} = \top$, and $p_{jk} = \top$. The condition in line 4 fails, so the xor-clauses $(x_4 \oplus x_5 \oplus e_{45} \equiv \top)$ and $(x_1 \oplus x_3 \oplus e_{45} \equiv \top)$, where e_{45} is a new variable, are added. The resulting instance is shown in Fig. 11(b). Assume that x_5 is selected. The loop in lines 3–9 is entered with values $x_i = x_4$, $x_j = x_5$, $e_{ij} = e_{45}$, $x_k = x_7$, $e_{jk} = x_8$, $p_{ij} = \top$, and $p_{jk} = \top$. The condition in line 4 is true, so $e_{ik} = x_6$, and the xor-clause $(x_6 \oplus x_8 \oplus e_{45} \equiv \top)$ is added in line 9. The final result is shown in Fig. 11(c).

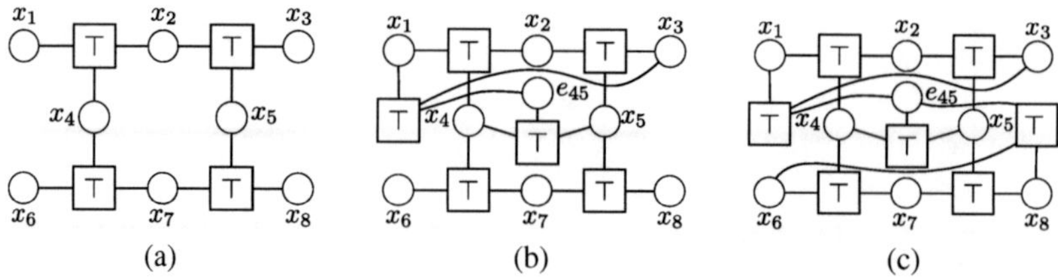

Fig. 11. Constraint graphs illustrating how the translation Eq* adds new xor-clauses

Benchmark	ϕ = Original		$\phi \wedge \mathrm{Eq}(\phi)$		$\phi \wedge \mathrm{Eq}^\star(\phi)$		$\frac{|\phi \wedge \mathrm{Eq}(\phi)|}{|\phi \wedge \mathrm{Eq}^\star(\phi)|}$
	vars	xor-clauses	vars	xor-clauses	vars	xor-clauses	
DES (4 rounds 2 blocks)	3781	320	7×10^6	9×10^9	3813	416	2.2×10^7
Grain (16 bit)	9240	6611	43×10^6	131×10^9	71670	3957571	33212.0
Hitag2 (33 bit)	6010	3747	18×10^6	36×10^9	21092	338267	106904.4
TRIVIUM (16 bit)	11485	8591	66×10^6	252×10^9	351392	30588957	8252.1

Fig. 12. Comparison of the translation sizes for Eq and Eq* on cipher benchmarks

Theorem 5. *If ϕ_{xor} is an xor-clause conjunction in 3-xor normal form, then* $\mathrm{Eq}^\star(\phi_{\mathrm{xor}})$ *is an* EC-*simulation formula for* ϕ_{xor}.

The translation Eq* usually adds fewer variables and xor-clauses than Eq. Fig. 12 shows a comparison of the translation sizes on four cipher benchmarks. The translation Eq yields an impractically large increase in formula size, while the translation Eq* adds still a manageable number of new variables and xor-clauses.

Experimental Evaluation. To evaluate the translation Eq*, we ran cryptominisat 2.9.2, and glucose 2.0 (SAT Competition 2011 application track winner) on the 123 SAT 2005 Competition cnf-xor instances preprocessed into 3-xor normal form with and without Eq*. The results are shown in Fig. 13. The number of decisions is greatly reduced, and this is reflected in solving time on many instances. Time spent computing Eq* is measured in seconds and is negligible compared to solving time. On some instances, the translation adds a very large number of xor-clauses (as shown in Fig. 14a) and the computational overhead of simulating equivalence reasoning using unit propagation becomes prohibitively large. For highly "xor-intensive" instances it is probably better to use more powerful parity reasoning; cryptominisat 2.9.2 with Gaussian elimination enabled solves majority of these instances in a few seconds. A hybrid approach first trying if Eq* adds a moderate number of xor-clauses, and if not, resorting to stronger parity reasoning could, thus, be an effective technique for solving cnf-xor instances.

Strengthening Equivalence Reasoning by Adding Xor-Clauses. Besides enabling unit propagation to simulate equivalence reasoning, the translation $\mathrm{Eq}^\star(\phi_{\mathrm{xor}})$ has an another interesting property: if ϕ_{xor} is not Subst-deducible, then even when $\phi_{\mathrm{xor}} \wedge \tilde{l}_1 \wedge \ldots \wedge \tilde{l}_n \nvdash_{\mathrm{Subst}} \hat{l}$ for some xor-assumptions $\tilde{l}_1, \ldots, \tilde{l}_n$, it might hold that $\phi_{\mathrm{xor}} \wedge \mathrm{Eq}^\star(\phi_{\mathrm{xor}}) \wedge \tilde{l}_1 \wedge \ldots \wedge \tilde{l}_n \vdash_{\mathrm{Subst}} \hat{l}$. For instance, let ϕ_{xor} be an xor-clause conjunction given in Fig. 14(b). It holds that $\phi_{\mathrm{xor}} \wedge (x) \models (z)$ but $\phi_{\mathrm{xor}} \wedge (x) \nvdash_{\mathrm{Subst}} (z)$. However, $\phi_{\mathrm{xor}} \wedge \mathrm{Eq}^\star(\phi_{\mathrm{xor}}) \wedge (x) \vdash_{\mathrm{Subst}} (z)$; the constraint graph of $\phi_{\mathrm{xor}} \wedge \mathrm{Eq}^\star(\phi_{\mathrm{xor}})$ is shown in Fig. 14(c).

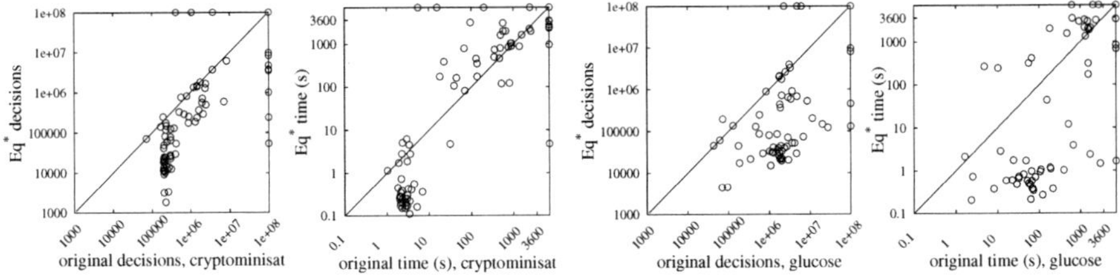

Fig. 13. Experimental results with/without $Eq^\star$ (cryptominisat on the left, glucose on the right)

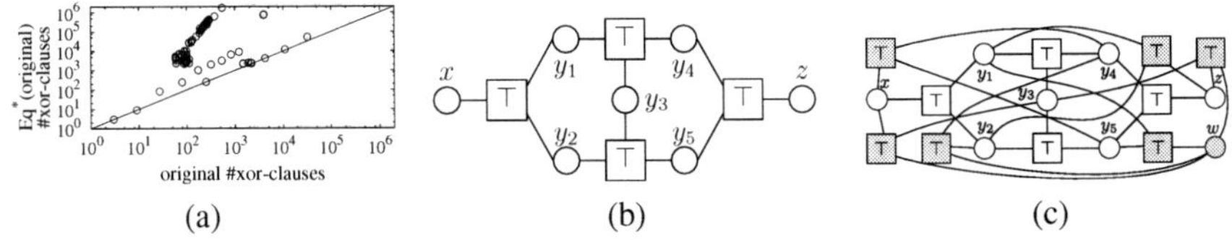

(a)   (b)   (c)

Fig. 14. (a) number of xor-clauses in our SAT'05 instances, (b) a non-**Subst**-deducible instance that becomes **Subst**-deducible with $Eq^\star$ in (c)

5   Conclusions

We have given efficient approximating tests for detecting whether unit propagation or equivalence reasoning is enough to achieve full propagation in a given parity constraint set. To our knowledge the computational complexity of exact versions of these tests is an open problem; they are certainly in co-NP but are they in P?

We have shown that equivalence reasoning can be simulated with unit propagation by adding a polynomial amount of redundant parity constraints to the problem. We have also proven that without introducing new variables, an exponential number of new parity constraints would be needed in the worst case. We have found many real-world problems for which unit propagation or equivalence reasoning achieves full propagation. The experimental evaluation of the presented translations suggests that equivalence reasoning can be efficiently simulated by unit propagation.

Acknowledgments. This work has been financially supported by the Academy of Finland under the Finnish Centre of Excellence in Computational Inference (COIN) and Hecse Helsinki Doctoral Programme in Computer Science.

References

1. Marques-Silva, J., Lynce, I., Malik, S.: Conflict-driven clause learning SAT solvers. In: Handbook of Satisfiability. IOS Press (2009)
2. Li, C.M.: Integrating equivalency reasoning into Davis-Putnam procedure. In: Proc. AAAI/IAAI 2000, pp. 291–296. AAAI Press (2000)
3. Li, C.M.: Equivalency reasoning to solve a class of hard SAT problems. Information Processing Letters 76(1-2), 75–81 (2000)

4. Baumgartner, P., Massacci, F.: The Taming of the (X)OR. In: Lloyd, J., Dahl, V., Furbach, U., Kerber, M., Lau, K.-K., Palamidessi, C., Moniz Pereira, L., Sagiv, Y., Stuckey, P.J. (eds.) CL 2000. LNCS (LNAI), vol. 1861, pp. 508–522. Springer, Heidelberg (2000)
5. Li, C.M.: Equivalent literal propagation in the DLL procedure. Discrete Applied Mathematics 130(2), 251–276 (2003)
6. Heule, M., van Maaren, H.: Aligning CNF- and Equivalence-Reasoning. In: Hoos, H.H., Mitchell, D.G. (eds.) SAT 2004. LNCS, vol. 3542, pp. 145–156. Springer, Heidelberg (2005)
7. Heule, M., Dufour, M., van Zwieten, J., van Maaren, H.: March_eq: Implementing Additional Reasoning into an Efficient Look-Ahead SAT Solver. In: Hoos, H.H., Mitchell, D.G. (eds.) SAT 2004. LNCS, vol. 3542, pp. 345–359. Springer, Heidelberg (2005)
8. Chen, J.: Building a Hybrid SAT Solver via Conflict-Driven, Look-Ahead and XOR Reasoning Techniques. In: Kullmann, O. (ed.) SAT 2009. LNCS, vol. 5584, pp. 298–311. Springer, Heidelberg (2009)
9. Soos, M., Nohl, K., Castelluccia, C.: Extending SAT Solvers to Cryptographic Problems. In: Kullmann, O. (ed.) SAT 2009. LNCS, vol. 5584, pp. 244–257. Springer, Heidelberg (2009)
10. Laitinen, T., Junttila, T., Niemelä, I.: Extending clause learning DPLL with parity reasoning. In: Proc. ECAI 2010, pp. 21–26. IOS Press (2010)
11. Soos, M.: Enhanced Gaussian elimination in DPLL-based SAT solvers. In: Pragmatics of SAT, Edinburgh, Scotland, GB, p. 1 (July 2010)
12. Laitinen, T., Junttila, T., Niemelä, I.: Equivalence class based parity reasoning with DPLL(XOR). In: Proc. ICTAI 2011, pp. 649–658. IEEE (2011)
13. Laitinen, T., Junttila, T., Niemelä, I.: Conflict-Driven XOR-Clause Learning. In: Cimatti, A., Sebastiani, R. (eds.) SAT 2012. LNCS, vol. 7317, pp. 383–396. Springer, Heidelberg (2012)
14. Nieuwenhuis, R., Oliveras, A., Tinelli, C.: Solving SAT and SAT modulo theories: From an abstract Davis-Putnam-Logemann-Loveland procedure to DPLL(T). Journal of the ACM 53(6), 937–977 (2006)
15. Barrett, C., Sebastiani, R., Seshia, S.A., Tinelli, C.: Satisfiability modulo theories. In: Handbook of Satisfiability. IOS Press (2009)

Consistencies for Ultra-Weak Solutions in Minimax Weighted CSPs Using the Duality Principle[*]

Arnaud Lallouet[1], Jimmy H.M. Lee[2], and Terrence W.K. Mak[2]

[1] Université de Caen, GREYC, Campus Côte de Nacre,
Boulevard du Maréchal Juin, BP 5186, 14032 Caen Cedex, France
Arnaud.Lallouet@unicaen.fr
[2] The Chinese University of Hong Kong, Shatin, N.T., Hong Kong
{jlee,wkmak}@cse.cuhk.edu.hk

Abstract. Minimax Weighted Constraint Satisfaction Problems (formerly called Quantified Weighted CSPs) are a framework for modeling soft constrained problems with adversarial conditions. In this paper, we describe novel definitions and implementations of node, arc and full directional arc consistency notions to help reduce search space on top of the basic tree search with alpha-beta pruning for solving ultra-weak solutions. In particular, these consistencies approximate the lower and upper bounds of the cost of a problem by exploiting the semantics of the quantifiers and reusing techniques from both Weighted and Quantified CSPs. Lower bound computation employs standard estimation of costs in the sub-problems used in alpha-beta search. In estimating upper bounds, we propose two approaches based on the Duality Principle: duality of quantifiers and duality of constraints. The first duality amounts to changing quantifiers from min to max, while the second duality re-uses the lower bound approximation functions on dual constraints to generate upper bounds. Experiments on three benchmarks comparing basic alpha-beta pruning and the six consistencies from the two dualities are performed to confirm the feasibility and efficiency of our proposal.

Keywords: constraint optimization, soft constraint satisfaction, minimax game search, consistency algorithms.

1   Introduction

The task at hand is that of a constraint optimization problem with *adversaries* controlling parts of the variables. As an example, we begin with a generalized version of the Radio Link Frequency Assignment Problem (RLFAP) [7] consisting of assigning frequencies to a set of radio links located between pairs of sites, with the goal of preventing interferences. The problem has two types of constraints. One type prevents radio links that are close together from interfering with one another, by restricting the links not to take frequencies with absolute differences smaller than a threshold. In practice, the

---

[*] We are grateful to the anonymous referees for their constructive comments. The work described in this paper was generously supported by grants CUHK413808 and CUHK413710 from the Research Grants Council of Hong Kong SAR.

M. Milano (Ed.): CP 2012, LNCS 7514, pp. 373–389, 2012.

threshold is measured depending on the physical environment, and is often overestimated. The second type of constraints are technological constraints, where each constraint ensures the distance between frequencies of a radio link from site A to B and its reverse radio link from site B to A must be equal to a constant. If the problem is unsatisfiable, one approach is to find assignments violating the first type of constraints as little as possible. Suppose now a certain set of links are placed in unsecured areas, and *adversaries* (e.g. terrorists/spies) may hijack/control these links. We are not able to re-adjust the frequencies for the other links immediately to minimize the interferences on the functioning ones. One interesting question for this type of scenarios is to find frequency assignments such that we can minimize the degree of radio links affected for the worst possible case (i.e. finding the best-worst case). The prime goal is to understand how well we can defend against the worst adversaries for planning purposes.

The example is optimization in nature, and the adversaries originate from the uncontrollable frequencies being assigned on the links in unsecured areas. The question can be translated to minimizing the interferences for all possible combinations of frequency adjustments the adversaries can control. One way to solve this problem is by tackling many COPs [2]/WCSPs [15], where each of them minimizes the interferences conditioned on a specific combination of frequency adjustments controlled by the adversaries. Another way is to model the problem as a QCSP [5] by finding whether there exists combinations of frequency adjustments for us for all frequency placements by the adversaries such that the total interferences is less than a cost k. To avoid solving multiple sub-problems, Minimax Weighted Constraint Satisfaction Problems (MWCSPs) (previously called Quantified Weighted Constraint Satisfaction Problems) [16] are proposed to tackle such problems, combining quantifier structures from QCSPs to model the adversaries and soft constraints from WCSPs to model costs information. Previous work defines a solution as a complete assignment representing the best-worst case, gives an introduction on how to adopt alpha-beta prunings to tackle the problem in branch and bound, and suggests two sufficient pruning conditions to achieve prunings and backtrackings.

When tackling game problems, more specifically two-person zero-sum games with perfect information [22,23], games can be solved at different levels. Allis [1,13] proposes three solving levels for games: *ultra-weakly solved*, *weakly solved*, and *strongly solved*. Ultra-weakly solved means the game-theoretic value of the initial position has been determined, which means we can determine the outcome of the scenario when both players are playing perfectly (i.e. best-worst case). Weakly solved means a strategy, noted as winning strategy [4] in QCSPs, has been determined for the initial position to achieve the game-theoretic value against any opposition. Strongly solved is being used for a game for which such a strategy has been determined for all legal positions. Once a game is solved at a stronger level, the game is automatically solved at weaker ones. Finding solutions at stronger levels, however, implies substantially higher computation requirements. In particular in terms of space, ultra-weak solutions are linear in size, while the other two stronger ones are exponential. In bi-level programs, there are cases in which we can assume there is a unique optimum for the follower or we are concerned with only the moves for the leader [11]. Finding ultra-weak solutions for these cases are sufficient, and the generalized RLFAP is an example. In adversarial game playing, many game search algorithms, e.g. minimax and alpha-beta [24], computes strategies assuming optimal plays to reduce computation costs. In fact, even determining just the ultra-weak solution in an offline manner is also an important and interesting line of research, e.g. a recent breakthrough on checkers [25].

The main focus of this paper is to further introduce novel consistency notions for solving ultra-weak solutions, by approximating the lower and upper bounds of the cost of the problem. Lower bound computation employs standard estimation of costs in the sub-problems used in alpha-beta search. In estimating upper bounds, we adopt the Principle of Duality in (integer) linear programming, which suggest to convert an original (primal) problem to its dual form and tackle the problem using both forms. We consider two dualities: duality of quantifiers and duality of constraints. The first approach allows us to formulate upper bound approximation functions by changing quantifiers in the lower bound functions from min to max, while the second approach re-uses the lower bound approximation functions on dual constraints to generate upper bounds. Discussions on whether our proposed techniques are applicable to the computation of the two stronger solutions will be given. Experimental evaluations on three benchmarks are performed to compare six consistencies defined using the two dualities to confirm the feasibility and efficiency of our proposal.

2 Background

In the first part, we give definitions and semantics of MWCSPs, followed by an example. In the second part, sufficient conditions allowing us to perform backtracking/prunings used in alpha-beta search are highlighted.

2.1 Definitions and Semantics

A *Minimax Weighted Constraint Satisfaction Problem* (MWCSP) [16] $\mathcal{P}$ is a tuple $(\mathcal{X}, \mathcal{D}, \mathcal{C}, \mathcal{Q}, k)$, where $\mathcal{X} = (x_1, \ldots, x_n)$ is defined as an ordered sequence of *variables*, $\mathcal{D} = (D_1, \ldots, D_n)$ is an ordered sequence of finite *domains*, $\mathcal{C}$ is a set of *soft constraints*, $\mathcal{Q} = (Q_1, \ldots, Q_n)$ is a *quantifiers sequence* where Q_i is either max or min associated with x_i, and k is the global upper bound. We denote $x_i = v_i$ an *assignment* assigning value $v_i \in D_i$ to variable x_i, and the set of assignments $l = \{x_1 = v_1, x_2 = v_2, \ldots, x_n = v_n\}$ a *complete assignment* on variables in $\mathcal{X}$, where v_i is the value assigned to x_i. A *partial assignment* $l[S]$ is a projection of l onto variables in $S \subseteq \mathcal{X}$. $\mathcal{C}$ is a set of *(soft) constraints*, each C_S of which represents a function mapping tuples corresponding to assignments on a subset of variables S, to a cost valuation structure $V(k) = ([0...k], \oplus, \leq)$. The structure $V(k)$ contains a set of integers $[0...k]$ with standard integer ordering $\leq$. Addition $\oplus$ is defined by $a \oplus b = \min(k, a+b)$. For any integer a and b where $a \geq b$, subtraction $\ominus$ is defined by $a \ominus b = a - b$ if $a \neq k$, and $a \ominus b = k$ if $a = k$. Without loss of generality, we assume the existence of $C_\varnothing$ denoting the lower bound of the minimum cost of the problem. If it is not defined, we assume $C_\varnothing = 0$. The *cost* of a complete assignment l in $\mathcal{X}$ is defined as: $cost(l) = C_\varnothing \oplus \bigoplus_{C_s \in \mathcal{C}} C_s(l[S])$.

In an MWCSP, ordering of variables is important. Without loss of generality, we assume variables are ordered by their indices. We define a variable with min (max resp.)

quantifier to be a minimization variable (maximization variable resp.). Let $\mathcal{P}[x_{i_1} = a_{i_1}][x_{i_2} = a_{i_2}]\ldots[x_{i_m} = a_{i_m}]$ be the *sub-problem* obtained from $\mathcal{P}$ by assigning value a_{i_1} to variable x_{i_i}, assigning value a_{i_2} to variable $x_{i_2},\ldots$, assigning value a_{i_m} to variable x_{i_m}. Let firstx($\mathcal{P}$) be a function returning the first unassigned variable in the variable sequence. If there are no such variables, it returns $\perp$. Suppose l is a complete assignment of $\mathcal{P}$. The *A-cost($\mathcal{P}$)* of an MWCSP $\mathcal{P}$ is defined recursively as follows:

$$
\text{A-cost}(\mathcal{P}) = \begin{cases} cost(l), \text{ if firstx}(\mathcal{P}) = \perp \\ \max(\mathbb{M}_i), \text{ if firstx}(\mathcal{P}) = x_i \text{ and } Q_i = \max \\ \min(\mathbb{M}_i), \text{ if firstx}(\mathcal{P}) = x_i \text{ and } Q_i = \min \end{cases}
$$

where l is the complete assignment of the completely assigned problem $\mathcal{P}$ (i.e. firstx($\mathcal{P}$) $= \perp$), and $\mathbb{M}_i = \{\text{A-cost}(\mathcal{P}[x_i = v])|v \in D_i\}$. An MWCSP $\mathcal{P}$ is *satisfiable* iff A-cost($\mathcal{P}$) $< k$.

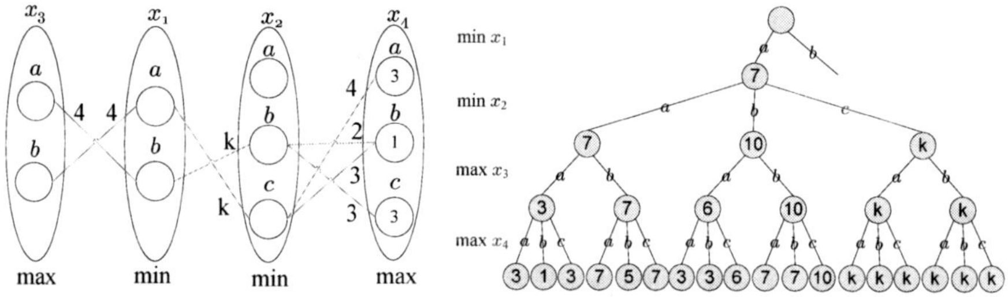

Fig. 1. Constraints for Example 1   Fig. 2. Labeling Tree for Example 1

We now define three solution concepts for MWCSPs based on the definition of A-costs. An *ultra-weak solution* of an MWCSP $\mathcal{P}$ is a complete assignment $\{x_1 = v_1,\ldots,x_n = v_n\}$ s.t. A-cost($\mathcal{P}$) $=$ A-cost($\mathcal{P}[x_1 = v_1]\ldots[x_i = v_i]$), $\forall 1 \leq i \leq n$. Solving an ultra-weak solution corresponds to finding the scenario when both players are playing perfectly. To capture weak (strong resp.) solutions, we re-use the concept of winning strategies [4]. Without loss of generality, we assume the max player is the adversary. A *weak solution* (*strong solution* resp.) is a

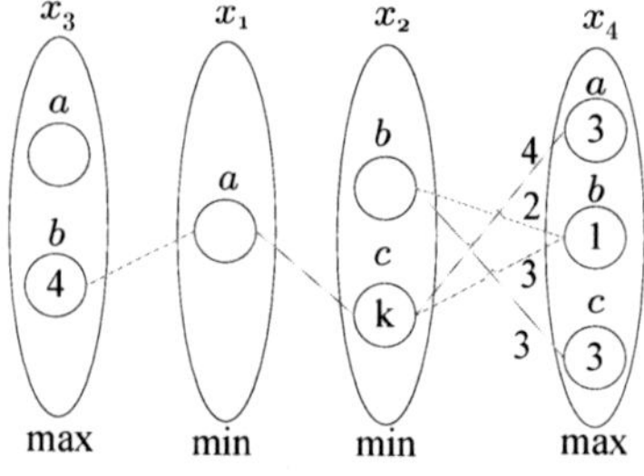

Fig. 3. Constraints for Example 2

set of functions $\mathcal{F}$, where each function $f_i \in \mathcal{F}$ corresponds to a min variable x_i. Let G_i be the set of domains of *max* variables (all variables resp.) preceding x_i, i.e. $G_i = \{D_j \in \mathcal{D}|Q_j = \max \wedge j < i\}$ ($G_i = \{D_j \in \mathcal{D}|j < i\}$ resp.). We define $f_i : \times_{D_j \in G_i} D_j \mapsto D_i$. If G_i is an empty set, then f_i is a constant function returning values from D_i. Let $\mathcal{P}'$ be a sub-problem of an MWCSP $\mathcal{P}$, where the next unassigned variable x_i is a min variable, and l be the set of assigned values for max variables (all variables resp.) x_j where $j < i$. For weak solutions, we further require the assigned values of min variables x_j where $j < i$ in $\mathcal{P}'$ follow f_j. We require all f_i to satisfy: A-cost($\mathcal{P}'[x_i = f_i(l)]$) $=$ A-cost($\mathcal{P}'$). In other words, we require $f_i(l)$ to return the best value for the min player, and the set of functions $\mathcal{F}$ will then be a best strategy for the min player. *This work focuses on ultra-weak solutions.* Note that computing ultra-weak solutions essentially computes the A-costs of an MWCSP, which are defined based on constraints and quantifiers, and in general, computing the A-costs of an MWCSP is PSPACE-hard [16]. A special case is that if all the quantifiers of an MWCSP are min quantifiers, finding an ultra-weak solution is equivalent to finding a complete assignment l with the minimum costs (i.e. $\mathrm{argmin}_l \, cost(l)$). The problem reduces [16] to a WCSP.

Example 1. We use the generalized Radio Link Frequency Assignment Problem introduced in the previous section as an example. The problem consists of four links $l_1, l_2, l_3,$ and l_4. Two of the links l_1 and l_2 connect sites A and B, and the other two links l_3 and l_4 connect sites B and C. Link l_2 (l_4 resp.) is the reverse link for l_1 (l_3 resp.). There is a variable x_i in the MWCSP $\mathcal{P}$ for each link l_i, which is used to represent the chosen frequency for link l_i. Site C is not secure and links l_3 and l_4 are subject to control. We need to pay costs if two links interfere with each other. Therefore, we want to find frequency assignments for l_1 and l_2 such that we can minimize the total costs for interference in the worst case. We set the quantifier sequence in $\mathcal{P}$ as $(Q_1 = \min, Q_2 = \min, Q_3 = \max, Q_4 = \max)$. For simplicity, we assume links l_1 and l_3 have two frequency choices, and the other two links have three. We measure the costs for interference only for links l_1 and l_3, and links l_2 and l_4. These costs will be modeled by constraints on variables x_1 and x_3, and also on variables x_2 and x_4. In addition, we maintain the technological constraint between links l_1 and l_2, which will be modeled by a binary constraint on variables x_1 and x_2. Figure 1 indicates there is one unary constraint C_4 and three binary constraints $C_{1,2}, C_{1,3},$ and $C_{2,4}$. For the unary constraint, non-zero unary costs are depicted inside a circle and domain values are placed above the circle. For binary constraints, non-zero binary costs are depicted as labels on edges connecting the corresponding pair of values. Only non-zero costs are shown. We set the global upper bound k to be 11. By following the partial labeling tree in Figure 2, we can easily infer the A-cost of the subproblem $\mathcal{P}' = \mathcal{P}[x_1 = a]$ is 7, and $\{x_1 = a, x_2 = a, x_3 = b, x_4 = a\}$ is one of the ultra-weak solutions for the sub-problem $\mathcal{P}'$.

2.2 Pruning Conditions in B and B

MWCSPs can be solved by applying alpha-beta pruning in branch and bound search [16] (Figure 4), by treating max and min variables as max and min players respectively. Alpha-beta pruning utilizes two bounds, α and β, for storing the current best costs for max and min players. We rename α and β as lower lb and upper ub bounds to fit with the common notations for bounds in constraint and integer programming. We initialize lb (ub resp.) to the lowest (largest resp.) possible costs, i.e. 0 (k resp.), and maintain the two bounds during assignments by the branch and bound. When a smaller costs (larger costs resp.) for min (max resp.) variable is found after exploring sub-trees, ub (lb resp.) will be updated (line 6 and 8). If $lb \geq ub$, then one of the previous branch must dominate over the current sub-tree, and we can perform backtrack (line 9).

Lee, Mak, and Yip [16] give pruning conditions that allow further derivation of consistency notions, and we introduce them as follows. Let $\mathcal{P}[x_{1..i-1} = v_{1..i-1}, x_i = v]$

```
1  function alpha_beta(P,lb,ub) :
2    if firstx(P) == ⊥: return cost(P)
3    i = firstx(P)
4    for v in Di:
5      if Qi == min:
6        ub = min(ub, alpha_beta(P[Xi=v],lb,ub))
7      else:
8        lb = max(lb, alpha_beta(P[Xi=v],lb,ub))
9      if ub <= lb: break
10   return (Qi == min) ?ub:lb
```

Fig. 4. Alpha-beta for MWCSPs

denote the subproblem $\mathcal{P}[x_1 = v_1][x_2 = v_2]...[x_{i-1} = v_{i-1}][x_i = v]$. Formally, we consider two conditions: $\exists v \in D_i$ s.t. $\forall v_1 \in D_1, ..., v_{i-1} \in D_{i-1}$:

$$\text{A-cost}(\mathcal{P}[x_{1..i-1} = v_{1..i-1}, x_i = v]) \geq ub \tag{1}$$

$$\text{A-cost}(\mathcal{P}[x_{1..i-1} = v_{1..i-1}, x_i = v]) \leq lb \tag{2}$$

where ub and lb are the upper and lower bounds in alpha-beta prunings respectively. When either of the above conditions is satisfied, we can apply prunings according to Table 1.

Checking Condition (1)/(2) by finding the *exact* value of the A-cost for each sub-problem is computationally expensive. Alternatively, we allow approximating functions to perform bounds approximations. Function $\text{ubaf}(\mathcal{P}, x_i = v)$ ($\text{lbaf}(\mathcal{P}, x_i = v)$ resp.) is

Table 1. When can we prune/backtrack

A-cost	$\geq ub$	$\leq lb$
$Q_i = \min$	prune v	backtrack
$Q_i = \max$	backtrack	prune v

an upper bound (a lower bound resp.) approximation function if it approximates the A-cost for the set S of sub-problems, where:

$$S = \{\mathcal{P}[x_{1..i-1} = v_{1..i-1}, x_i = v] | \forall v_1 \in D_1, ..., v_{i-1} \in D_{i-1}\}$$
$$\text{s.t.} \quad \forall \mathcal{P}' \in S, \text{A-cost}(\mathcal{P}') \leq \text{ubaf}(\mathcal{P}, x_i = v)$$
$$(\geq \text{lbaf}(\mathcal{P}, x_i = v) \text{ resp. })$$

From the definition, we can easily obtain:

$$\text{lbaf}(\mathcal{P}, x_i = v) \geq ub \implies \text{Condition (1)}$$
$$\text{ubaf}(\mathcal{P}, x_i = v) \leq lb \implies \text{Condition (2)}$$

By implementing lbaf()/ubaf() with good approximations, we can identify non-ultra-weak solution values from variable domains or perform backtracking earlier in search according to Table 1.

3   Consistency Techniques

In WCSPs, consistency notions [15,9] not only utilize constraint semantics, but also take the costs of constraints into account. This section discusses how we utilize costs information from unary constraints and binary constraints to formulate node and (full directional) arc consistencies. We start by giving an $lbaf()$ for node consistency called $nc_{lb}()$, which formulates lower bounds by gathering unary costs. We then further describe a stronger $lbaf()$ for (full directional) arc consistency called $ac_{lb}()$. To approximate upper bounds, we propose two approaches by utilizing the Duality Principle: duality of quantifiers and duality of constraints. In the last part, we discuss how to strengthen our consistency notions, by incorporating techniques in WCSPs. We write C_i for the unary constraint on variable x_i, $C_{i,j}$ for the binary constraint on variables x_i and x_j where $i < j$, $C_i(u)$ for the cost returned by the unary constraint when u is assigned to x_i, and $C_{i,j}(u,v)$ for the cost returned by the binary constraint when u and v are assigned to x_i and x_j respectively. To simplify our notations, we write the minimum costs $\min_{u \in D_j} C_j(u)$ and maximum costs $\max_{u \in D_j} C_j(u)$ of a unary constraint C_j as $\min C_j$ and $\max C_j$ respectively. We further write $Q_j C_j$ to mean $\min C_j$ if $Q_j = \min$, and $\max C_j$ if $Q_j = \max$.

3.1 Node Consistency: Lower Bound

We first give the definition for $nc_{lb}()$. We will then sketch the proof showing $nc_{lb}()$ is an lbaf() using a lemma. Without loss of generality, we now consider unary MWCSPs, which are MWCSPs with *unary constraints only*. We will show that computing A-costs for any sub-problems of unary MWCSPs are efficient (linear time), and therefore, computing the lower bound for these sub-problems are efficient. We then show using the same procedure on general MWCSPs, by viewing unary constraints only, the bound is still correct.

Definition 1. *The $nc_{lb}(\mathcal{P}, x_i = v)$ function approximates the* A-cost *for a set S of sub-problems $\{\mathcal{P}[x_{1..i-1} = v_{1..i-1}, x_i = v] | \forall v_1 \in D_1, \ldots, v_{i-1} \in D_{i-1}\}$. Define*

$$nc_{lb}(\mathcal{P}, x_i = v) \equiv C_\varnothing \oplus \left(\bigoplus_{j:j<i} \min C_j \right) \oplus (C_i(v)) \oplus \left(\bigoplus_{j:i<j} Q_j C_j \right)$$

where $Q_j \in \mathcal{Q}$ is the quantifier for variable x_j where $j > i$.

Lemma 1. *The* A-cost *of an MWCSP $\mathcal{P}$ with only unary constraints is equal to $\bigoplus_{i=1}^{n} Q_i C_i$.*

The proof of Lemma 1 follows directly from the definition of A-costs for MWCSPs.

Theorem 1. *The function $nc_{lb}(\mathcal{P}, x_i = v)$ is a lower bound approximating function* lbaf$(\mathcal{P}, x_i = v)$.

Lemma 1 suggests the computation of A-costs for unary MWCSPs can be done in $O(nd)$, where n is the number of variables and d is the maximum domain size. Therefore, computing the A-costs for any sub-problems is also efficient. The function $nc_{lb}()$ can be seen as a function extracting A-costs for the sub-problem in S with minimal A-costs following Lemma 1, by partitioning unary constraints into three groups: (a) $C_j, j < i$, (b) C_i, and (c) $C_j, j > i$. We skip the detailed reasoning on how to choose costs for these unary constraints. If $\mathcal{P}$ has only unary constraints, we can observe function $nc_{lb}()$ computes not only a correct lower bound for S, but also the exact A-cost for the sub-problem with minimum costs. Note that MWCSPs may have binary constraints and even high-arity constraints, but, these constraints must give positive costs to the problem. Therefore, by considering only unary constraints of general MWCSPs, $nc_{lb}()$ still returns a correct lower bound.

Example 2. We re-use Example 1. Suppose we are at sub-problem $\mathcal{P}' = \mathcal{P}[x_1 = a]$ and we have just visited the further sub-problem $\mathcal{P}'[x_2 = a]$ which have a new upper bound of 7. Before visiting $\mathcal{P}'[x_2 = b]$, we try to prune some values according to Table 1 using the new upper bound. Figure 3 shows the constraint graph for $\mathcal{P}'$. Suppose now $nc_{lb}()$ is applied and no unary costs for bounded variables, i.e. $C_\varnothing = 0$. We want to check if the value b can be pruned from D_2. In the sub-problem $\mathcal{P}'[x_2 = b]$, the quantifier Q_3 and Q_4 are both max, and they will take at least the maximum unary cost $\max C_3$ and $\max C_4$. We have $C_\varnothing + C_2(b) + \max C_3 + \max C_4 = 0 + 0 + 4 + 3 = 7 \geq ub$. The cost of any assignment in the sub-problem $\mathcal{P}'[x_2 = b]$ is at least 7. The value b can therefore be removed from domain D_2. Notice that such a node cannot be pruned by basic alpha-beta pruning.

3.2 Arc Consistency: Lower Bound

To obtain stronger lower bound, we further define function $ac_{lb}()$ based on $nc_{lb}()$. Without loss of generality, we restrict our attention to MWCSPs which have *only unary constraints and one binary constraint*. We will show that computing any sub-problems for these MWCSPs are efficient (polynomial time), and therefore, computing the lower bound for these sub-problems are again efficient. By similar argument, viewing unary constraints plus one binary constraint on general MWCSPs, the bound is still correct.

Definition 2. *The* $ac_{lb}[C_{i,j}](\mathcal{P}, x_i = v)$ *function approximates the* A-cost *for the set S of sub-problems* $\{\mathcal{P}[x_{1..i-1} = v_{1..i-1}, x_i = v] | \forall v_1 \in D_1, \ldots, v_{i-1} \in D_{i-1}\}$. *Define*

$$ac_{lb}[C_{i,j}](\mathcal{P}, x_i = v) \equiv C_\varnothing \oplus \left(\bigoplus_{k:k<i} \min C_k \right) \oplus (C_i(v))$$

$$\oplus \left(\bigoplus_{k:i<k \wedge j \neq k} Q_k C_k \right) \oplus \left(Q_j \{ C_j(u) \oplus C_{i,j}(v, u) \} \right)$$
$$\underset{u \in D_j}{}$$

where $Q_j \in \mathcal{Q}$ is the quantifier for variable x_j, and $Q_k \in Q$ is the quantifier for variable x_k where $k > i$ and $k \neq j$.

The first three terms are the same as in $nc_{lb}()$. The fourth term is equivalent to the last term in $nc_{lb}()$, except we do not consider costs for constraint C_j, which will be considered in the fifth term.

Lemma 2. *The* A-cost *of an MWCSP* $\mathcal{P} = (\mathcal{X}, \mathcal{D}, \mathcal{C}, \mathcal{Q}, k)$ *with only unary constraints and **one** binary constraint $C_{i,j}$ is equal to*

$$\bigoplus_{k \in [1 \ldots n] \backslash \{i,j\}} \underset{u \in D_k}{Q_k} C_k(u) \oplus \underset{u \in D_i}{Q_i} [\underset{v \in D_j}{Q_j} [C_i(u) \oplus C_j(v) \oplus C_{i,j}(u, v)]]$$

where $Q_i, Q_j, Q_k \in \mathcal{Q}$.

The proof of Lemma 2 follows from the definition of A-costs. Theorem 2 follows.

Theorem 2. *The function* $ac_{lb}[C_{i,j}](\mathcal{P}, x_i = v)$ *for binary constraint* $C_{i,j}$ *is a lower bound approximating function* $\mathrm{lbaf}(\mathcal{P}, x_i = v)$.

Note that Definition 2 is only one possible approach to define a lower bound approximation function for AC, following Lemma 2. It is designed in such a way that only **one** binary constraint is used in bounds calculation for costs estimation, and our approach is similar to AC in QCSPs [20,12]. It is natural for us to further ask for stronger/tighter functions which consider more than one binary constraint. Note that in classical local consistency enforcement such as: AC in CSPs [2]; AC* in WCSPs [15]; and (Q)AC [20] in QCSPs, we usually handle one (binary) constraint at a time. Consistency enforcement will be performed many times at each node of the search tree, and considering multiple constraints at a time may cause a huge increase in time complexity. We have to maintain a balance between amount of reasoning at each search node and amount of pruning achieved. There are stronger consistency notions with efficient algorithms which consider more than one binary constraint, e.g. Max Restriced Path Consistency [10] in CSPs and OSAC [8] in WCSPs/VCSPs. Investigations on stronger notions for MWCSPs is an interesting future work. One possibility to enhance ac_{lb} is to consider a subset of constraints that forms a tree, and employ a dynamic programming approach to enforce such stronger consistencies.

3.3   NC and AC Upper Bounds by the Duality Principle

In linear programming, duality [21,27] provides a standard way to obtain lower bounds (for minimization problems). In fact, the Principle/Theory of Duality [21] suggests that we can convert the original (primal) problem to its dual form, and tackle the problem by using both forms. In QCSPs, dual consistency [5] was defined by creating the dual QCSP problem, involving negation of the original constraints. We will now show how to implement upper bound approximation functions $nc_{ub}()$ and $ac_{ub}()$ by using the duality principle in MWCSPs.

Duality of Constraints.   One approach to create $nc_{ub}()/ac_{ub}()$ is to utilize the constraint duality property, which is similar to dual consistency [5] in QCSPs. We first define the dual problem of an MWCSP.

Definition 3. *Given an MWCSP* $\mathcal{P} = (\mathcal{X}, \mathcal{D}, \mathcal{C}, \mathcal{Q}, k)$. *The dual problem of* $\mathcal{P}$ *is an MWCSP* $\mathcal{P}^\dagger = (\mathcal{X}, \mathcal{D}, \mathcal{C}^\dagger, \mathcal{Q}^\dagger, k)$ *s.t. for a complete assignment* l,

$$C_\varnothing \oplus \bigoplus_{C_S \in \mathcal{C}} C_S(l[S]) = -1 \times \left(C_\varnothing^\dagger \oplus \bigoplus_{C_s^\dagger \in \mathcal{C}^\dagger} C_S^\dagger(l[S]) \right)$$

where the valuation structure of $\mathcal{P}^\dagger$ *is* $([-k...k], \oplus, \leq)$, $Q_i^\dagger = \min$ *if* $Q_i = \max$, *and* $Q_i^\dagger = \max$ *if* $Q_i = \min$.

We can observe that $\text{A-cost}(\mathcal{P}) = -1 \times \text{A-cost}(\mathcal{P}^\dagger)$, and a straightforward method to construct the **dual constraints** in the dual problem is to multiply costs for all constraints in the original problem by -1. We then show how we utilize the dual problem to check $\mathrm{ubaf}(\mathcal{P}, x_i = v) \leq lb$ (Condition 2) for an MWCSP $\mathcal{P}$.

Theorem 3. *Given an MWCSP $\mathcal{P}$ and its dual problem $\mathcal{P}^\dagger$. Suppose there is a lower bound approximation function* $\mathrm{lbaf}()$.

$$\mathrm{lbaf}(\mathcal{P}^\dagger, x_i = v) \geq -1 \times lb \implies \textit{Condition (2)}.$$

The proof of Theorem 3 can be shown by observing the labeling tree of the dual problem, and inferring $\mathrm{lbaf}(\mathcal{P}^\dagger, x_i = v) \times -1$ is an upper bound approximation function for the original problem. In fact, the upper bound $ub^\dagger$ (lower bound $lb^\dagger$ resp.) of $\mathcal{P}^\dagger$ is equal to -1 times the lower bound lb (upper bound ub resp.) of $\mathcal{P}$. Therefore, we further define $ub^\dagger = -1 \times lb$, and $lb^\dagger = -1 \times ub$. We then implement $nc_{ub}()$ and $ac_{ub}()$, via checking the $nc_{lb}()$ and $ac_{lb}()$ for the dual problem.

Definition 4. *An MWCSP $\mathcal{P}$ is dual constraint node consistent (DC-NC) iff $\forall x_i \in \mathcal{X}, \forall v \in D_i : nc_{lb}(\mathcal{P}, x_i = v) < ub \wedge nc_{lb}(\mathcal{P}^\dagger, x_i = v) < ub^\dagger$.*

Definition 5. *An MWCSP $\mathcal{P}$ is dual constraint arc consistent (DC-AC) iff $\mathcal{P}$ is DC-NC, $\forall C_{i,j} \in \mathcal{C}, \forall v \in D_i : ac_{lb}[C_{i,j}](\mathcal{P}, x_i = v) < ub$, and $\forall C_{ij}^\dagger \in \mathcal{C}^\dagger, \forall v \in D_i : ac_{lb}[C_{ij}^\dagger](\mathcal{P}^\dagger, x_i = v) < ub^\dagger$.*

Theorem 4. *DC-AC is strictly stronger than DC-NC.*

The proof follows from the definitions.

Duality of Quantifiers. Another way to check condition (2) for an MWCSP $\mathcal{P}$ is to scrutinize functions implementing $\mathrm{ubaf}(\mathcal{P}, x_i = v)$, by repeating similar reasonings for $nc_{lb}()$ on unary MWCSPs (plus a binary constraint). The idea is to use the duality of quantifiers, by replacing min quantifiers to max in the reasoning process. Recall we have three groups of unary constraints to consider. One direct way is to consider the maximum costs, instead of minimum costs from constraints in the first group (group (a)), hence changing quantifiers from min to max. However, using the resulting upper bound approximation functions, by reasoning on unary MWCSPs is incorrect for general MWCSPs. We cannot neglect costs given by high arity constraints. One way to make the bound correct is to add the maximum costs for constraints which will not be covered in the function, and we pre-compute these costs before search. Function $nc_{ub}(\mathcal{P}, x_i = v)$ and $ac_{ub}(\mathcal{P}, x_i = v)$ are given as follows, and we write $\max C^\star$ to mean the maximum costs for constraints which are not considered in the function.

Definition 6. *The $nc_{ub}(\mathcal{P}, x_i = v)$ function approximates the* A-cost *for a set S of sub-problems $\{P[x_{1..i-1} = v_{1..i-1}, x_i = v] | \forall v_1 \in D_1, v_2 \in D_2, \ldots, v_{i-1} \in D_{i-1}\}$. Define:*

$$nc_{ub}(\mathcal{P}, x_i = v) \equiv C_\varnothing \oplus \left(\bigoplus_{j:j<i} \max C_j \right) \oplus (C_i(v)) \oplus \left(\bigoplus_{j:i<j} Q_j C_j \right) \oplus (\max C^\star)$$

where $Q_j \in \mathcal{Q}$ is the quantifier for x_j, $j > i$.

We can easily observe $\max C^\star$ is equal to $\bigoplus_{j,k:j \neq k} \max C_{jk}$ if there are only unary and binary constraints.

Definition 7. *The function* $ac_{ub}[C_{i,j}](\mathcal{P}, x_i = v)$ *approximates the* A-cost *for the set* S *of sub-problems:* $\{P[x_{1..i-1} = v_{1..i-1}, x_i = v]|\forall v_1 \in D_1, v_2 \in D_2, \ldots, v_{i-1} \in D_{i-1}\}$. *Define:*

$$ac_{ub}[C_{i,j}](\mathcal{P}, x_i = v) \equiv C_{\varnothing} \oplus (\bigoplus_{j:j<i} \max C_j) \oplus (C_i(v)) \oplus (\bigoplus_{k:i<k \wedge j \neq k} Q_k C_k)$$

$$\oplus Q_{j_{u \in D_j}} \{C_j(u) \oplus C_{i,j}(v, u)\} \oplus (\max C^{\star})$$

where Q_k is the quantifier for variable x_k where $k > i$ and $k \neq j$, and Q_j is the quantifier for variable x_j.

If there are only unary and binary constraints, $\max C^{\star}$ is equal to $\bigoplus_{C_{k,l} \in B} \max C_{k,l}$, where $B = \{C_{k,l} \in \mathcal{C}|k \neq l\} - \{C_{i,j}\}$. We now define the node and arc consistencies by utilizing the constructed functions.

Definition 8. *An MWCSP $\mathcal{P}$ is dual quantifier node consistent (DQ-NC) iff $\forall x_i \in \mathcal{X}, \forall v \in D_i : nc_{lb}(\mathcal{P}, x_i = v) < ub \wedge nc_{ub}(\mathcal{P}, x_i = v) > lb$.*

Definition 9. *An MWCSP $\mathcal{P}$ is dual quantifier arc consistent (DQ-AC) iff $\mathcal{P}$ is DQ-NC, and $\forall C_{i,j} \in \mathcal{C}, \forall v \in D_i : ac_{lb}[C_{i,j}](\mathcal{P}, x_i = v) < ub \wedge ac_{ub}[C_{i,j}](\mathcal{P}, x_i = v) > lb$.*

Theorem 5. *DQ-AC is strictly stronger than DQ-NC.*

The proof follows from the definitions.

3.4   Consistency Enforcement

To enforce DC-NC and DC-AC, one major step is to compute $nc_{lb}()$ and $ac_{lb}()$, by computing costs from unary and binary constraints in both the original and dual MWC-SPs. For DQ-NC and DQ-AC, we compute $nc_{ub}()$ and $ac_{ub}()$ instead of the dual. To achieve these consistencies, we perform prunings/backtrackings according to Table 1. Similar to cascade propagation [2] in CSPs, a value of a variable being pruned may trigger prunings of other values in other variables and re-computation of the lbaf() and ubaf() functions. In addition, prunings caused by lower bound approximations may tighten upper bound approximations (and vice versa), and triggers extra prunings. Our propagation routine repeats until no values can be further pruned, or backtracks occur.

3.5   Strengthening Consistencies by Projection/Extension

Consistency algorithms for WCSPs use an equivalence preserving transformation called *projection* [9] to move costs from higher arity constraints to lower arity ones to extract and store bound information. Some further utilizes *extension* [9], which is the inverse of projection, to increase the consistency strength. We propose to re-use WCSP consistencies, especially the parts related to projections and extensions, to strengthen the approximating functions for MWCSPs.

WCSPs consistencies consist of two kinds of conditions: one for pruning and one for projection/extension. Since their pruning conditions are unsound w.r.t. MWCSPs, we adopt only their projection/extension conditions so as to strengthen DC-NC, DC-AC, DQ-NC, and DQ-AC. The projection/extension conditions for NC*, AC*, and FDAC* [15,14] are as follows:

$$\text{proj-NC* :} \forall C_i, \exists v \in D_i : C_i(v) = 0$$

$$\text{proj-AC* : proj-NC*} \wedge \forall C_{i,j}, \forall v_i \in D_i, \exists v_j \in D_j : C_{i,j}(v_i, v_j) = 0 \wedge$$

$$\forall C_{i,j}, \forall v_j \in D_j, \exists v_i \in D_i : C_{i,j}(v_i, v_j) = 0$$

$$\text{proj-FDAC* : proj-AC*} \wedge \forall C_{i,j} : i < j, \forall v_i \in D_i, \exists v_j \in D_j : C_{i,j}(v_i, v_j) \oplus C_j(v_j) = 0$$

Note that the enforcing algorithm for proj-FDAC* may decreases unary costs for max variables and increases unary costs for min variables; hence weakening the approximating functions. We tackle this issue by re-ordering the variables when enforcing proj-FDAC*, with max variables first. To further enforce these projecting conditions on the dual problem in DC-NC/DC-AC, we need to perform normalization, by transferring costs from $C_\varnothing$ to constraints with negative costs until all constraints except $C_\varnothing$ return non-negative costs. We now re-define DC-NC, DC-AC, DQ-NC, and DQ-AC, to allow users plugging in general projection/extension conditions τ.

Definition 10. *An MWCSP $\mathcal{P}$ is DC-NC[τ] (DC-AC[τ] resp.) iff $\mathcal{P}$ is DC-NC (DC-AC resp.), and all projection/extension conditions τ for both $\mathcal{P}$ and the dual problem $\mathcal{P}^\dagger$ are satisfied. An MWCSP $\mathcal{P}$ is DQ-NC[τ] (DQ-AC[τ] resp.) iff $\mathcal{P}$ is DQ-NC (DQ-AC resp.), and all the projection/extension conditions τ for $\mathcal{P}$ are satisfied.*

Previous work [16] shows experimental results on an implementation of DQ-NC[proj-NC*] and DQ-AC[proj-AC*], where DQ-NC[proj-NC*] and DQ-AC[proj-AC*] are named as node and arc consistency respectively.

3.6 Tackling Stronger Solution Definitions

This section discusses the scopes and limitations of our techniques on solving MWCSPs for the other two stronger solved levels: weakly solved and strongly solved.

In terms of space, the solution sizes for solving MWCSPs ultra-weakly, weakly, and strongly vary from $O(n)$, $O((n - m)d^m)$, to $O(d^n)$ respectively, where n is the total number of variables, $m \leq n$ is the number of variables owned by adversaries, and d is the maximum domain size of the MWCSP. A direct consequence is that we need exponential space to store weak/strong solutions during search, and most often, compact representations to represent weak/strong solutions are more desirable.

In terms of prunings in branch and bound tree search, a sound pruning condition when solving a weaker solution concept may not hold in stronger ones. This is caused by the removal of the assumption of optimal/perfect plays when dealing with stronger solution concepts. For example in alpha-beta prunings, when the min player obtains an A-costs which is lower than the lb (i.e. max player's last found best), we cannot immediately backtrack if we want to tackle weakly solved solutions, where we assume the max player is the adversary. The reason behind is that we cannot assume the max player must play a perfect move. We have to consider all moves for the max player. The situation is similar if we assume the min player is the adversary. By similar reasonings and inductions, we cannot perform prunings/backtrackings for the $\leq lb$ column ($\geq ub$

column resp.) in Table 1 if we want to tackle weakly solved solutions, assuming the max player (min player resp.) is the adversary. For solving strong solutions, the situation is even worse. We cannot assume optimal plays for both players. Therefore, we have to find A-costs for all sub-problems, and all prunings/backtrackings conditions in Table 1 cannot be used. In general, the fewer sound pruning/backtracking conditions available, the larger search space we have to search. By using tree search, we can observe finding stronger solutions is much harder than weaker ones.

When tackling real-life problems, one can ask for solutions which solve the problem in an intermediate level. For example, if the adversaries have multiple optimal strategies, we can require solutions containing responses to every different optimal choice the adversaries may choose. In this case, the solved level lies between ultra-weak and weak. One way to handle is to relax the bound updating procedure for the lower bound (upper bound resp.) in alpha-beta pruning (Line 6 and 8 in Figure 4), where we assume the max (min resp.) player is the adversary. When a larger lower bound lb (smaller upper bound ub resp.) is found, we update the lower bound to $lb - 1$ (upper bound to $ub + 1$ resp.). The major focus of this work is to give consistency notions to improve the search in finding the best-worst case, i.e. ultra-weak solutions, of a game.

4 Performance Evaluation

In this section, we compare our solver in seven modes: Alpha-beta pruning, DC-NC[proj-NC*], DQ-NC[proj-NC*], DC-AC[proj-AC*], DQ-AC[proj-AC*], DC-AC[proj-FDAC*], and DQ-AC[proj-FDAC*]. Values are labeled in static lexicographic order. We generate 20 instances for each benchmark's particular parameter setting. Results for each benchmark are tabulated with average time used (in sec.) and average number of tree nodes encountered. We take average for solved instances *only*. If there are any unsolved instances, we give the number of solved instances beside the average time (superscript in brackets). Winning entries are highlighted in bold. A symbol '-' represents all instances fail to run within the time limit. The experiment is conducted on a Core2 Duo 2.8GHz with 3.2GB memory. We have also performed experiments on

Table 2. Randomly Generated Problem

	Alpha-beta		DC-NC[proj-NC*]		DC-AC[proj-AC*]		DC-AC[proj-FDAC*]	
(n, d, p)	Time	#nodes	Time	#nodes	Time	#nodes	Time	#nodes
(12, 5, 0.4)	68.20	5,967,461	5.89	131,468	2.54	30,165	**2.13**	**20,397**
(12, 5, 0.6)	52.05	4,782,541	4.63	101,690	2.61	26,093	**2.24**	**16,178**
(14, 5, 0.4)	263.04[18]	19,770,953	52.72	948,783	19.33	198,476	**14.82**	**117,155**
(14, 5, 0.6)	271.72[17]	17,249,858	70.12	1,185,087	29.97	246,459	**23.11**	**143,197**
(16, 5, 0.4)	517.24[2]	26,269,025	332.65[19]	4,617,612	121.78	1,047,900	**102.82**	**706,913**
(16, 5, 0.6)	693.31[2]	36,315,673	461.68[16]	6,157,070	259.51	1,816,642	**208.52**	**1,054,326**
	QeCode		DQ-NC[proj-NC*]		DQ-AC[proj-AC*]		DQ-AC[proj-FDAC*]	
(n, d, p)	Time	#nodes	Time	#nodes	Time	#nodes	Time	#nodes
(12, 5, 0.4)	-	–	3.68	158,179	3.23	53,845	4.27	58,619
(12, 5, 0.6)	-	-	2.85	118,401	3.24	41,596	4.17	45,698
(14, 5, 0.4)	-	-	33.39	1,135,378	26.20	369,185	41.74	482,053
(14, 5, 0.6)	-	-	46.81	1,510,946	45.85	450,407	68.63	522,715
(16, 5, 0.4)	-	-	217.13	5,780,075	141.07	1,654,538	173.96	1,745,527
(16, 5, 0.6)	-	-	364.51[19]	9,401,844	341.71	3,071,036	362.12[17]	2,659,294

Table 3. Graph Coloring Game

(v, c, d)	Alpha-beta		DC-NC[proj-NC*]		DC-AC[proj-AC*]		DC-AC[proj-FDAC*]	
	Time	#nodes	Time	#nodes	Time	#nodes	Time	#nodes
(14, 4, 0.4)	19.88	1,572,978	6.71	122,266	3.20	37,252	**1.90**	**16,732**
(14, 4, 0.6)	24.12	1,730,473	10.38	185,111	5.88	59,359	**3.48**	**23,515**
(16, 4, 0.4)	167.75	10,050,800	48.37	688,200	22.67	221,484	**12.09**	**92,875**
(16, 4, 0.6)	166.83	9,213,029	45.71	625,944	27.03	212,934	**15.64**	**85,920**
(18, 4, 0.4)	784.47[3]	33,914,968	288.90	2,839,962	114.63	792,220	**65.58**	**357,457**
(18, 4, 0.6)	-	-	350.29	3,400,265	163.70	993,099	**80.06**	**343,146**
(v, c, d)	QeCode		DQ-NC[proj-NC*]		DQ-AC[proj-AC*]		DQ-AC[proj-FDAC*]	
	Time	#nodes	Time	#nodes	Time	#nodes	Time	#nodes
(14, 4, 0.4)	-	-	4.52	170,843	3.36	63,298	3.74	53,722
(14, 4, 0.6)	-	-	7.29	269,179	6.36	99,972	6.88	74,187
(16, 4, 0.4)	-	-	34.43	1,002,145	23.21	363,539	24.36	281,229
(16, 4, 0.6)	-	-	33.82	949,861	29.19	352,694	31.99	280,426
(18, 4, 0.4)	-	-	204.86	4,095,993	118.65	1,315,346	140.95	1,207,566
(18, 4, 0.6)	-	-	267.23	5,295,433	180.38	1,711,948	182.66	1,270,797

Table 4. Generalized Radio Link Frequency Assignment Problem

(i, n, d, r)	Alpha-beta		DC-NC[proj-NC*]		DC-AC[proj-AC*]		DC-AC[proj-FDAC*]	
	Time	#nodes	Time	#nodes	Time	#nodes	Time	#nodes
(1, 24, 4, 0.2)	-	-	86.38	442,362	50.54	74,182	53.85	**55,988**
(0, 24, 4, 0.4)	-	-	148.87	828,286	105.95	295,743	128.01	**286,122**
(1, 22, 6, 0.2)	-	-	618.93	3,580,885	**307.58**	352,439	309.63	**299,361**
(0, 24, 6, 0.2)	-	-	1230.33[19]	6,822,412	500.18	738,245	479.50	**651,762**
(i, n, d, r)	QeCode		DQ-NC[proj-NC*]		DQ-AC[proj-AC*]		DQ-AC[proj-FDAC*]	
	Time	#nodes	Time	#nodes	Time	#nodes	Time	#nodes
(1, 24, 4, 0.2)	-	-	**45.62**	449,164	50.75	77,286	47.08	62,734
(0, 24, 4, 0.4)	-	-	**96.55**	1,046,150	101.49	451,090	208.79	692,470
(1, 22, 6, 0.2)	-	-	338.42	3,719,348	374.34	374,385	309.96	368,643
(0, 24, 6, 0.2)	-	-	682.60[19]	7,224,677	539.69	803,087	**434.99**	812,048

QeCode, a solver for QCOPs [3], by transforming the instances to QCOPs according to the transformation in previous work [16].

4.1   Randomly Generated Problems and Graph Coloring Games

We re-use benchmark MWCSP instances and graph coloring game instances by Lee, Mak, and Yip [16]. The random MWCSP instances are generated with parameters (n, d, p), where n is the number of variables, d is the domain size for each variable, and p is the probability for a binary constraint to occur between two variables. There are no unary constraints which makes the instances harder, and the costs for each binary constraint are generated uniformly in [0..30]. Quantifiers are generated randomly with half probability for min (max resp.), and the number of quantifier levels vary from instances to instances. For the graph coloring game instances, numbers are used instead of colors, and the graph is numbered by two players. We partition the nodes into two sets A and B. Player 1 (Player 2 resp.) will number set A (B resp.). The goal of player 1 is to maximize the total difference between numbers of adjacent nodes, while player 2 wishes to minimize. The aim is to help player 1 extracting the best-worst case. We generate instances with parameters (v, c, d), where v is an even number of nodes in the graph, c is the range of numbers allowed to place, and d is the probability of an edge between two vertices. Player 1 (Player 2 resp.) is assigned to play the odd (even resp.)

numbered turns, and the node corresponding to each turn is generated randomly. Time limit for both benchmarks are 900 seconds. Table 2 and 3 show the results.

4.2 Generalized Radio Link Frequency Assignment Problem (GRLFAP)

We generate the GRLFAP according to two small but hard CELAR sub-instances [7], which are extracted from CELAR6. All GRLFAP instances are generated with parameters (i, n, d, r), where i is the index of the CELAR sub-instances ($\mathtt{CELAR6-SUB}_i$), n is an even number of links, d is an even number of allowed frequencies, and r is the ratio of links placed in unsecured areas, $0 \leq r \leq 1$. For each instance, we randomly extract a sequence of n links from $\mathtt{CELAR6-SUB}_i$ and fix a domain of d frequencies. We randomly choose $\lfloor (r \times n + 1)/2 \rfloor$ pairs of links to be unsecured. If two links are restricted not to take frequencies f_i and f_j with distance less than t, we measure the costs of interference by using a binary constraint with violation measure $\max(0, t - |f_i - f_j|)$. We set the time limit to 7200 seconds. Table 4 shows the results.

4.3 Results and Discussions

For all benchmarks, all six consistencies are significantly faster and stronger than alpha-beta pruning.

Comparing the two duality approaches, we observe duality of constraints (DC) is stronger than duality of quantifiers (DQ), and we conjecture for any projection/extension conditions τ, DC-NC[τ] (DC-AC[τ] resp.) is stronger than DQ-NC[τ] (DQ-AC[τ] resp.). Note that enforcing projection/extension conditions on DQ-NC/DQ-AC may strengthen one approximation function, and weaken the other at the same time. DC-NC/DC-AC extracts costs from different copies of constraints and resolve this issue.

For all benchmarks, DQ-NC[proj-NC*] runs faster than DC-NC[proj-NC*]. In randomly generated problems and the graph coloring game, DC-AC[proj-(FD)AC*] runs faster than DQ-AC[proj-(FD)AC*], with DC-AC[proj-FDAC] the fastest. In GRLFAP, DQ-NC[proj-NC*] runs faster than the others for smaller instances and stronger consistencies are faster for larger ones. Enforcing proj-FDAC* is more computational expensive than proj-AC* and proj-NC*, and implementing duality of constraints requires implementing two copies of constraints. Therefore, stronger consistencies are worthwhile for larger instances, but not for smaller ones due to the large computational over-head.

It is worth noting DQ[proj-FDAC*] prunes less than DQ[proj-AC*], suggested by the fact that adding stronger projection/extension conditions from WCSPs naively may not always strengthen our approximation functions. We have to further consider quantifier information.

All QCOP instances for even the smallest parameter settings for all benchmarks fail to run within the time limit. QCOPs are, in fact, more general [16] than MWCSPs. By viewing a more specific problem, it is natural for us to devise consistency techniques outperforming QeCode.

5 Concluding Remarks

We define and implement node and (full directional) arc consistency notions to reduce the search space of an alpha-beta search for MWCSPs, by approximating lower and upper bounds of the cost of the problem. Lower bound computation employs standard estimation of costs in the sub-problems and we propose two approaches: duality of quantifiers and duality of constraints, based on the Duality Principle in estimating upper bounds. Details on strengthening the approximation functions by re-using WCSPs consistencies are given. We also discuss capabilities and limitations of our approach on other stronger solution concepts. Experiments on comparing basic alpha-beta pruning and the six consistencies from the two dualities are performed.

There are two closely related frameworks, where both tackle constraint problems with adversaries. Brown et al. propose adversarial CSPs [6], which focuses on the case where two opponents take turns to assign variables, each trying to direct the solution towards their own objectives. Another related work is Stochastic CSPs [26], which can represent adversaries by known probability distributions. We seek actions to minimize/-maximize the expected cost for all the possible scenarios. Our work is similar in the sense that we are minimizing the cost for the worst case scenario.

Possible future work includes: consistency algorithms for high arity (soft) constraints similar to those for WCSPs [18,19,17], value/variable ordering heuristics, theoretical comparisons on different consistency notions, tackling stronger solutions, and online algorithms.

References

1. Allis, L.V.: Searching for solutions in games and artificial intelligence. Ph.D. thesis, University of Limburg (1994)
2. Apt, K.: Principles of Constraint Programming. Cambridge University Press, New York (2003)
3. Benedetti, M., Lallouet, A., Vautard, J.: Quantified Constraint Optimization. In: Stuckey, P.J. (ed.) CP 2008. LNCS, vol. 5202, pp. 463–477. Springer, Heidelberg (2008)
4. Bordeaux, L., Cadoli, M., Mancini, T.: CSP properties for quantified constraints: Definitions and complexity. In: AAAI 2005, pp. 360–365 (2005)
5. Bordeaux, L., Monfroy, E.: Beyond NP: Arc-Consistency for Quantified Constraints. In: Van Hentenryck, P. (ed.) CP 2002. LNCS, vol. 2470, pp. 371–386. Springer, Heidelberg (2002)
6. Brown, K.N., Little, J., Creed, P.J., Freuder, E.C.: Adversarial constraint satisfaction by game-tree search. In: ECAI 2004, pp. 151–155 (2004)
7. Cabon, B., de Givry, S., Lobjois, L., Schiex, T., Warners, J.: Radio link frequency assignment. Constraints 4, 79–89 (1999)
8. Cooper, M.C., de Givry, S., Schiex, T.: Optimal soft arc consistency. In: IJCAI 2007, pp. 68–73 (2007)
9. Cooper, M.C., de Givry, S., Sanchez, M., Schiex, T., Zytnicki, M., Werner, T.: Soft arc consistency revisited. Artificial Intelligence 174(7-8), 449–478 (2010)
10. Debruyne, R., Bessière, C.: From Restricted Path Consistency to Max-Restricted Path Consistency. In: Smolka, G. (ed.) CP 1997. LNCS, vol. 1330, pp. 312–326. Springer, Heidelberg (1997)
11. Dempe, S.: Foundations of Bilevel Programming. Kluwer Academic Publishers (2002)
12. Gent, I.P., Nightingale, P., Stergiou, K.: QCSP-Solve: A solver for quantified constraint satisfaction problems. In: IJCAI 2005, pp. 138–143 (2005)
13. van den Herik, H.J., Uiterwijk, J.W.H.M., van Rijswijck, J.: Games solved: Now and in the future. Artif. Intell. 134(1-2), 277–311 (2002)

14. Larrosa, J., Schiex, T.: In the quest of the best form of local consistency for weighted CSP. In: IJCAI 2003, pp. 239–244 (2003)
15. Larrosa, J., Schiex, T.: Solving weighted CSP by maintaining arc consistency. Artificial Intelligence 159(1-2), 1–26 (2004)
16. Lee, J.H.M., Mak, T.W.K., Yip, J.: Weighted constraint satisfaction problems with min-max quantifiers. In: ICTAI 2011, pp. 769–776 (2011)
17. Lee, J.H.M., Shum, Y.W.: Modeling soft global constraints as linear programs in weighted constraint satisfaction. In: ICTAI 2011, pp. 305–312 (2011)
18. Lee, J.H.M., Leung, K.L.: Consistency techniques for flow-based projection-safe global cost functions in weighted constraint satisfaction. JAIR 43, 257–292 (2012)
19. Lee, J.H.M., Leung, K.L., Wu, Y.: Polynomially decomposable global cost functions in weighted constraint satisfaction. In: AAAI 2012 (to appear, 2012)
20. Mamoulis, N., Stergiou, K.: Algorithms for Quantified Constraint Satisfaction Problems. In: Wallace, M. (ed.) CP 2004. LNCS, vol. 3258, pp. 752–756. Springer, Heidelberg (2004)
21. Murty, K.G.: Linear and Combinatorial Programming. R. E. Krieger (1985)
22. Neumann, J.V., Morgenstern, O.: Theory of Games and Economic Behavior. Princeton University Press (1944)
23. Nisan, N., Roughgarden, T., Tardos, E., Vazirani, V.V.: Algorithmic Game Theory. Cambridge University Press (2007)
24. Russell, S.J., Norvig, P.: Artificial Intelligence: A Modern Approach. Pearson Education (2003)
25. Schaeffer, J., Burch, N., Björnsson, Y., Kishimoto, A., Müller, M., Lake, R., Lu, P., Sutphen, S.: Checkers is solved. Science 317(5844), 1518–1522 (2007)
26. Walsh, T.: Stochastic constraint programming. In: ECAI 2002, pp. 111–115 (2002)
27. Wolsey, L.A.: Integer Programming. Wiley (1998)

Propagating Soft Table Constraints

Christophe Lecoutre, Nicolas Paris, Olivier Roussel, and Sébastien Tabary

CRIL - CNRS, UMR 8188,
Univ Lille Nord de France, Artois
F-62307 Lens, France
{lecoutre,paris,roussel,tabary}@cril.fr

Abstract. WCSP is a framework that has attracted a lot of attention during the last decade. In particular, many filtering approaches have been developed on the concept of equivalence-preserving transformations (cost transfer operations), using the definition of soft local consistencies such as, for example, node consistency, arc consistency, full directional arc consistency, and existential directional arc consistency. Almost all algorithms related to these properties have been introduced for binary weighted constraint networks, and most of the conducted experiments typically include networks with binary and ternary constraints only. In this paper, we focus on extensional soft constraints (of large arity), so-called soft table constraints. We propose an algorithm to enforce a soft version of generalized arc consistency (GAC) on such constraints, by combining both the techniques of cost transfer and simple tabular reduction, the latter dynamically maintaining the list of allowed tuples in constraint tables. On various series of problem instances containing soft table constraints of large arity, we show the practical interest of our approach.

1 Introduction

The Weighted Constraint Satisfaction Problem (WCSP) is an optimization framework used to handle soft constraints, which has been successfully applied in many applications of Artificial Intelligence and Operations Research. Each soft constraint is defined by a cost function that associates a violation degree, called cost, with every possible instantiation of a subset of variables. Using the (bounded) addition $\oplus$, these costs can be combined in order to obtain the overall cost of any complete instantiation. Finding a complete instantiation with a minimal cost is known to be NP-hard.

Several properties, such as Node Consistency (NC) and Arc Consistency (AC), introduced in the early 70's for the Constraint Satisfaction Problem (CSP), have been studied later in the context of WCSP. Since then, more and more sophisticated developments about the best form of soft arc consistency have been proposed over the years: full directional arc consistency (FDAC) [3], existential directional arc consistency (EDAC) [6], virtual arc consistency (VAC) and optimal soft arc consistency (OSAC) [4], among others. Cost transfer, which is the general principle behind the algorithms enforcing such properties, preserves the semantics of the soft constraints network while concentrating costs on domain values (unary constraints) and a global constant cost (nullary constraint). Quite interestingly, cost transfer algorithms have been shown to be particularly efficient to solve real-world problem instances, especially when soft constraints are binary or ternary (see http://costfunction.org for many such problems).

M. Milano (Ed.): CP 2012, LNCS 7514, pp. 390–405, 2012.
© Springer-Verlag Berlin Heidelberg 2012

For soft constraints of large arity, cost transfer becomes a serious issue because the risk of combinatorial explosion has to be controlled. A first solution is to postpone cost transfer operations until the number of unassigned variables (in constraint scopes) is sufficiently low. However, it may dramatically damage the filtering capability of the algorithms, in particular at the beginning of search. A second solution is to adapt soft local consistency algorithms to certain families of global soft constraints. This is the approach followed in [14,15] where the concept of projection-safe soft constraint is introduced. A third solution is to decompose soft constraints (cost functions) into soft constraints of smaller arity [7]. Decomposition of global soft constraints can also be envisioned [1]. Unluckily, not all soft constraints can be decomposed.

To enforce the property, known as Generalized Arc Consistency (GAC) on soft table constraints, i.e., soft constraints defined extensionally by listing tuples and their costs, we propose to combine two techniques, namely, Simple Tabular Reduction (STR) [16] and cost transfer. Basically, whenever some domain values are deleted during propagation or search, all tuples that become invalid are removed from constraint tables. This allows us to identify values that are no longer consistent with respect to GAC. Interestingly, because all valid tuples of tables are iterated over, it is easy and cheap to compute minimum costs of values. This is particularly useful for performing efficiently projection operations that are required to establish GAC.

The paper is organized as follows. In the first section, we present the technical background about WCSP. Next, we present the data structures and the algorithms used in our approach GAC^w-WSTR. We prove the correctness and discuss the complexity of our method. Finally, we introduce some new series of benchmarks and show the practical interest of our approach.

2   Technical Background

A *weighted constraint network* (WCN) P is a triplet $(\mathscr{X}, \mathscr{C}, k)$ where $\mathscr{X}$ is a finite set of n variables, $\mathscr{C}$ is a finite set of e soft (or weighted) constraints, and $k > 0$ is either a natural integer or ∞. Each variable x has a (current) domain, denoted by $dom(x)$, which is the finite set of values that can be (currently) assigned to x; the initial domain of x is denoted by $dom^{init}(x)$. d will denote the greatest domain size. An *instantiation* I of a set $X = \{x_1, \ldots, x_p\}$ of p variables is a set $\{(x_1, a_1), \ldots, (x_p, a_p)\}$ such that $\forall i \in 1..p, a_i \in dom^{init}(x_i)$; each a_i is denoted by $I[x_i]$. I is *valid* on P iff $\forall(x, a) \in I, a \in dom(x)$. Each soft constraint $c_S \in \mathscr{C}$ involves an ordered set S of variables, called its *scope*, and is defined as a cost function from $l(S)$ to $\{0, \ldots, k\}$ where $l(S)$ is the set of possible instantiations of S. When an instantiation $I \in l(S)$ is given the cost k, i.e., $c_S(I) = k$, it is said *forbidden*. Otherwise, it is permitted with the corresponding cost (0 being completely satisfactory). Costs are combined with the specific operator $\oplus$ defined as: $\forall \alpha, \beta \in \{0, \ldots, k\}, \alpha \oplus \beta = min(k, \alpha + \beta)$. The partial inverse of $\oplus$ is $\ominus$ defined by: if $0 \leq \beta \leq \alpha < k, \alpha \ominus \beta = \alpha - \beta$ and if $0 \leq \beta < k$, $k \ominus \beta = k$. A *unary* (resp., *binary*) constraint involves 1 (resp., 2) variable(s), and a *non-binary* one strictly more than 2 variables. For any constraint c_S, every pair (x, a) such that $x \in S \wedge a \in dom(x)$ is called a *value of* c_S.

For any instantiation I and any set of variables X, let $I_{\downarrow X} = \{(x, a) \mid (x, a) \in I \wedge x \in X\}$ be the projection of I on X. If c_S is a soft constraint and I is an instantiation of a set $X \supseteq S$, then $c_S(I)$ will be considered to be equal to $c_S(I_{\downarrow S})$ (in other words, projections will be implicit). For a WCN P and a complete instantiation I of P, the cost of I is $\bigoplus_{c_S \in \mathcal{C}} c_S(I)$. The usual (NP-hard) task of Weighted Constraint Satisfaction Problem (WCSP) [12] is, for a given WCN, to find a complete instantiation with a minimal cost.

Many forms of soft arc consistency have been proposed during the last decade (e.g., see [4]). We now briefly introduce some of them. Without any loss of generality, the existence of a nullary constraint $c_\emptyset$ (a constant) as well as the presence of a unary constraint c_x for every variable x is assumed. A variable x is node-consistent (NC) iff $\forall a \in dom(x), c_\emptyset \oplus c_x(a) < k$ and $\exists b \in dom(x) \mid c_x(b) = 0$. Some other consistencies introduced for WCSP are AC* [9,12], FDAC [3], EDAC [6], VAC and OSAC [4]. Algorithms enforcing such properties are based on equivalence-preserving transformations (EPT) that allow safe moves of costs among constraints: the cost of any complete instantiation is preserved. Two basic cost transfer operations are called **project** and **unaryProject** (see e.g., [4]). The former projects a given cost from a non-unary soft constraint to a unary constraint; for example, it is possible to project on $c_x(a)$ the *minimum* cost of a value (x, a) on a soft constraint c_S, which is $min_{I \in l(S) \land I[x]=a} c_S(I)$. The latter projects a given cost from a unary constraint to the nullary constraint $c_\emptyset$. We shall note $\phi(P)$ the enforcement of property ϕ (e.g., AC, EDAC, ...) on the (W)CN P. For non-binary soft constraints, generalized arc consistency (GAC), a well-known CSP property, has also been adapted to WCSP [5,4]. We first need to introduce the notion of extended cost. The *extended* cost of an instantiation $I \in l(S)$ on a soft constraint c_S, includes the cost of I on c_S as well as the nullary cost $c_\emptyset$ and the unary costs for I of the variables in S. It is defined by $ecost(c_S, I) = c_\emptyset + \sum_{x \in S} c_x(I[x]) + c_S(I)$; we shall say that I is *allowed* on c_S iff $ecost(c_S, I) < k$.

Definition 1. *A soft constraint c_S is GAC-consistent iff:*

- $\forall I \in l(S),\ c_S(I) = k\ if\ ecost(c_S, I) = k.$
- *for every value (x, a) of c_S, $\exists I \in l(S) \mid I[x] = a \land c_S(I) = 0.$*

Below, we propose an alternative to this definition of GAC for WCSP, and call it weak GAC (GAC^w for short).

Definition 2. *A value (x, a) of a soft constraint c_S is GAC^w-consistent on c_S iff $\exists I \in l(S) \mid I[x] = a \land c_S(I) = 0 \land ecost(c_S, I) < k$. A soft constraint c_S is GAC^w-consistent iff every value of c_S is GAC^w-consistent.*

GAC is stronger than GAC^w because it identifies instantiations of constraint scopes that are inconsistent. However, when domains are the only point of interest, we can observe that the set of values deleted when enforcing GAC on a soft constraint c_S is exactly the set of values deleted when enforcing GAC^w on c_S.

Finally, one alternative approach to cost transfer methods is the algorithm PFC-MRDAC [8,11,10]. This is a classical branch and bound algorithm that computes lower bounds at each node of the search tree and that is used in our experimentation.

3   GAC^w-WSTR

The algorithm we propose, called GAC^w-WSTR, can be applied to any soft table constraint c_S whose default cost is either 0 or k. Such constraints occur quite frequently in practice. For example, among the 31 packages of WCSP instances listed on `http://costfunction.org`, 19 packages contain instances where the default cost of all soft table constraints is 0, and 5 packages contain instances where a various proportion of soft table constraints have a default cost equal to 0. This means that our approach can be applied on more than 61% of the packages currently available on this website. In this section, we first describe the data structures, then we introduce the algorithm GAC^w-WSTR, and finally we study its properties (correctness and complexity).

3.1   Data Structures

A soft table constraint c_S is a constraint defined by a list $table[c_S]$ of t tuples[1] (built over S), a list $costs[c_S]$ of t integers, and an integer $default[c_S]$. The ith tuple in $table[c_S]$ is given as cost the ith value in $costs[c_S]$. Any *implicit* tuple, i.e., any tuple that is not present in $table[c_S]$, is given as (default) cost the value $default[c_S]$.

An important feature (inherited from STR) of the algorithm we propose is the cheap restoration of its structures when backtracking occurs. The principle is to split each constraint table into different sets such that each tuple is a member of exactly one set. One of these sets contains all tuples that are currently valid: tuples in this set constitute the content of the *current table*. For simplicity, data structures related to backtracking are not detailed in this paper (see [13]).

For a (soft) constraint table c_S, the following arrays provide access to the disjoint sets of valid and invalid tuples within $table[c_S]$:

- $position[c_S]$ is an array of size t that provides indirect access to the tuples of $table[c_S]$. At any given time, the values in $position[c_S]$ are a permutation of $\{1, 2, \ldots, t\}$. The ith tuple of c_S is $table[c_S][position[c_S][i]]$, and its cost is given by $costs[c_S][position[c_S][i]]$.
- $currentLimit[c_S]$ is the position of the last current tuple in $table[c_S]$. The current table of c_S is composed of exactly $currentLimit[c_S]$ tuples. The values in $position[c_S]$ at indices ranging from 1 to $currentLimit[c_S]$ are positions of the current tuples of c_S.

The top half of Figure 1 illustrates the use of our data structures for a given constraint. The array $table$ is composed of 7 tuples (ranging in lexicographic order from τ_0 to τ_6). For each tuple, an associated cost is given by the array $costs$. The array $position$ provides an indirect access to the tuples. The last valid tuple of the table is marked by a pointer: $currentLimit$. Initially all tuples of the table are valid and the current table is composed of exactly $currentLimit$ tuples. In our example the value of $c_\emptyset$ is set to 0 and the upper-bound k is set to 5. The data structure c_1 represents the unary cost of

---

[1] A tuple can be seen as an instantiation over the variables of the scope of a constraint.

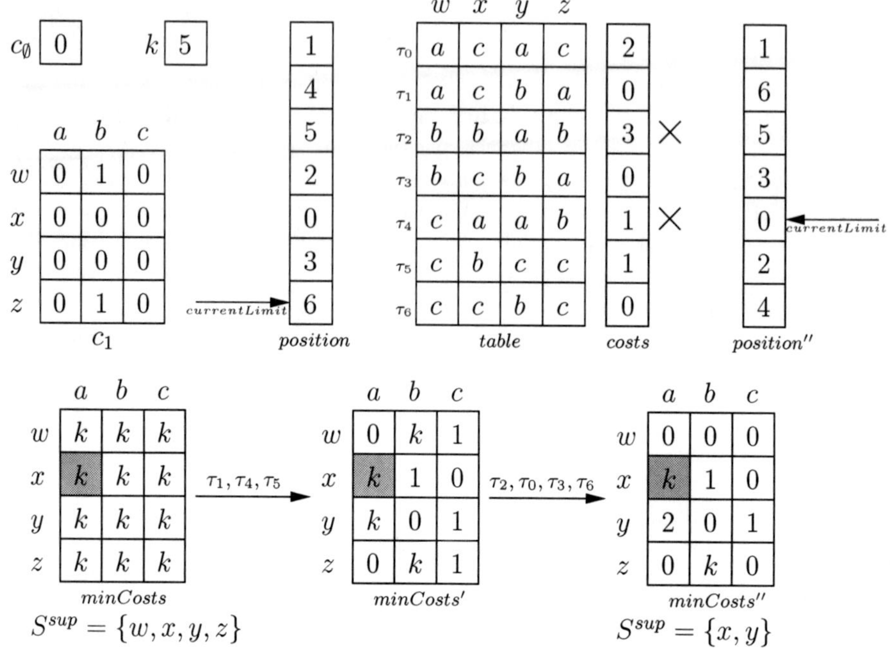

Fig. 1. Example of our data structures and their evolution after removing a from $dom(x)$

each value (x, a). The data structure *position″* represents the state of *position* after applying our algorithm. Changes in this data structure will be explained in Section 3.2.

As in [13], we also introduce two sets of variables, called S^{sup} and S^{val}. On the one hand, as soon as all values in the domain of a variable have been detected GAC^w-consistent, it is futile to continue to seek supports for values of this variable. We therefore introduce the set S^{sup} of uninstantiated variables (from the scope of the constraint) whose domains contain each at least one value for which a support has not yet been found. All main operations in our algorithm will only handle variables in S^{sup}. To update S^{sup}, we use an array to count the number $nbGacValues[c_S][x]$ of GAC^w-consistent values identified for each variable x.

On the other hand, at the end of an invocation of GAC^w-WSTR for a constraint c_S, we know that for every variable $x \in S$, every tuple τ such that $\tau[x] \notin dom(x)$ has been removed from the current table of c_S. If there is no backtrack and $dom(x)$ does not change between this invocation and the next invocation, then at the time of the next invocation it is certainly true that $\tau[x] \in dom(x)$ for every tuple τ in the current table of c_S. In this case, there is no need to check whether $\tau[x] \in dom(x)$; efficiency is gained by omitting this check. We implement this optimization by means of set S^{val}, which is the set of uninstantiated variables whose domain has been reduced since the previous invocation of GAC^w-WSTR. To set up S^{val}, we need to record the domain size of each variable $x \in S$ right after the execution of GAC^w-WSTR on c_S: this value is recorded in $lastSize[c_S][x]$.

For enforcing GAC^w on a given constraint c_S, we need to compute minimum costs of values on c_S. This can be achieved at low cost while traversing the current table of c_S. We just need an array $minCosts[c_S]$ to record those minimum costs; $minCosts[c_S][x][a]$ will denote the minimum cost of (x, a) on c_S Finally, when the default cost of c_S is 0, it is useful to count the number $nbTuples[c_S][x][a]$ of valid explicit tuples for each value (x, a), so as to determine whether a valid implicit tuple may exist.

To conclude this section, we briefly discuss how transfers of tuple costs can be implemented. Actually, to keep unchanged our table representation (i.e., keeping the same list of explicit tuples), we need to adopt the solution proposed in [5,12], which has a reasonable $O(|S|d)$ complexity. The principle is to keep original values in $costs[c_S]$ while recording in an auxiliary structure called $deltas[c_S]$ the cumulated cost of all projections performed with respect to each value. The current cost of a given tuple τ (at position $index$ in our representation) is then computed as follows: $c_S(\tau) = costs[c_S][index] \ominus_{x \in S} deltas[c_S][x][\tau[x]]$.

3.2 Algorithm

Whereas STR for crisp table constraints just requires simple iterations, for soft table constraints, we have to handle several potential iterations over table tuples (due to cost transfer operations). This is what we show now with Algorithm 10 that enforces GAC^w on any soft table constraint c_S whose default cost is either 0 or k. The first instruction is a call to Function initialize, Algorithm 5, which initializes both sets S^{sup} and S^{val}. More precisely, S^{sup} is initialized to contain future variables only, which is exactly $S \setminus past(P)$; the set $past(P)$ denotes the set of variables of the WCN P explicitly instantiated by the search algorithm. The set S^{val} contains the future variables whose domains have been changed since the last call to the algorithm (for the same constraint c_S). At line 6 of Algorithm 5, we have $|dom(x)|$ which is the size of the current domain of x, and $lastSize[c_S][x]$ which is the size of the domain of x, the last time the specific constraint c_S was processed; initially we have $lastSize[c_S][x] = -1$ for every pair composed of a constraint c_S and a variable x in S. Additionally, S^{val} also contains the last assigned variable, denoted by $lastPast(P)$ here, if it belongs to the scope of the constraint c_S. Indeed, after any variable assignment $x = a$, some tuples may become invalid due to the removal of values from $dom(x)$. The last assigned variable is the only instantiated variable for which validity operations must be performed.

First, let us assume that $default[c_S] = k$. In this case, a call to Function traverse-k, Algorithm 7, is performed at line 4 of Algorithm 10. Lines 1-4 of Algorithm 7 allow us to initialize the arrays $nbGacValues[c_S]$ and $minCosts[c_S]$: initially, no value has been proved to be GAC^w-consistent and no tuple with a cost lower than k has been found. Then the loop at lines $6 - 24$ successively processes all current tuples of the table of c_S. At line 10 of Algorithm 7, a validity check is performed on tuple τ when $cnt > 2$ (the operator 'or else' uses a short-circuit evaluation). This means that validity checks are only performed during the two first traversals of the table (i.e., two first calls to traverse-k) because after the second traversal, no other value can be deleted. The validity check is performed by Function isValidTuple, Algorithm 3, that deals only with variables in S^{val}. At line 11 of Algorithm 7, the extended cost of τ is computed (see Algorithm 4) and compared with k when $cnt > 1$. This means that such a computation is only performed during the first traversal of the table because the extended cost of any tuple on c_S remains constant after that traversal (projections do not modify extended costs as shown by Lemma 1). If tuple τ (whose cost is γ) is both valid and allowed, the array $minCosts[c_S]$ is subject to potential update (line 16). Besides, when the cost γ of τ is 0, we know that we have just found a support for (x, a) on c_S, so we can increment $nbGacValues[c_S][x]$ (line 18), and discard x from S^{sup} (line 20) in case (x, a) was the last value in $dom(x)$ without any proved support. In constant time at line 23 a tuple τ that is either invalid or forbidden is removed from the current table: actually it is moved to the end of the current table before the value of $currentLimit[c_S]$ is decremented.

The next instruction of Algorithm 10 is a call to Function pruneVars (Algorithm 6). This function allows us to remove all values proved to be inconsistent wrt GAC^w: they are the values (x, a) such that $minCosts[c_S][x][a] = k$. When at least one value is removed from the domain of a variable x, x is added to the set Y^{val} (line 6). Besides, after processing x, the value of $lastSize[c_S][x]$ is updated to record the new domain size (line 9). In Y^{sup}, we only keep variables for which at least one support must be sought (line 8). Finally, the sets Y^{sup} and Y^{val} become the new S^{sup} and S^{val}. Note that S^{sup} and S^{val} are not handled exactly as in STR2 [13], the difference residing mainly in Algorithm 6.

The main loop of Algorithm 10 aims at making successive projections in order to exhibit a support for each remaining value of c_S. All variables in S^{sup} require such operations. Each such variable is picked in turn at line 9, and projections (potentially followed by a unary projection) are performed at lines 10-16. If S^{sup} still contains at least one variable, the counter cnt is incremented, and traverse-k is called again. This new call permits to update the array $minCosts[c_S]$ as well as the set S^{sup}.

Now, let us assume that $default[c_S] = 0$. In that case, Function traverse-0 (Algorithm 9) is called instead of Function traverse-k, at lines 6 and 22 of Algorithm 10. The main difference between traverse-0 and traverse-k is that forbidden tuples must be kept in order to be able to count the number of valid tuples in the current table. Counting is managed at lines 5 and 16. Once the current table has been iterated over, we need to look for the existence of a valid implicit tuple for each value (lines 27-31). Function allowedImplicitTuple determines whether there exists such a valid implicit tuple containing the value (x, a).

The bottom half of Figure 1 illustrates the evolution of the data structure $minCosts$ during a call to Algorithm 7. We suppose that the event $x \neq a$ has triggered a reconsideration of the constraint. Note that before calling Algorithm 7 the structure $minCosts$ has been initialized (with value k) using Algorithm 5 and the set S^{sup} contains all the unassigned variables involved in the scope of the constraint. First, tuple τ_1 is considered. This tuple is valid (all values of the tuple belong to the current domains) and the tuple is also allowed since the extended cost of the tuple is equal to 0 (which is less than $k = 5$). Then the data structure $minCosts$ is updated for each value of τ_1. Next τ_4 is considered. Due to $x \neq a$, τ_4 is no more valid. So τ_4 is swapped with the last valid tuple and the pointer $currentLimit$ is decremented. In the figure, a cross identifies the removed tuple. The minimal cost of (x, a) remains k. The structure $minCosts'$ depicts the state of $minCosts$ after considering tuples τ_1, τ_4 and τ_5. Next when considering τ_2, this tuple is identified as not allowed because its extended cost is equal to $k = 5$. Hence,

---

Algorithm 1. project(c_S: soft constraint, x: variable, a: value, α: integer)

---

1   $c_x(a) \leftarrow c_x(a) \oplus \alpha$
2   $deltas[c_S][x][a] \leftarrow deltas[c_S][x][a] \oplus \alpha;$

Algorithm 2. unaryProject(x: variable, α: integer)

---

1   **foreach** *value* $a \in dom(x)$ **do**
2     $c_x(a) \leftarrow c_x(a) \ominus \alpha$

3   $c_\emptyset \leftarrow c_\emptyset \oplus \alpha$

Algorithm 3. isValidTuple(c_S: soft constraint, τ: tuple): Boolean

---

1   **foreach** *variable* $x \in S^{val}$ **do**
2   **if** $\tau[x] \notin dom(x)$ **then**
3   **return** *false*

4   **return** *true*

Algorithm 4. ecost(c_S: soft constraint, γ: integer, τ: tuple): integer

---

1   **return** $c_\emptyset \bigoplus_{x \in S} c_x(\tau[x]) \oplus \gamma$

Algorithm 5. initialize(c_S: soft constraint)

---

1   $S^{sup} \leftarrow \emptyset \, ; S^{val} \leftarrow \emptyset$
2   **if** $lastPast(P) \in S$ **then**
3     $S^{val} \leftarrow S^{val} \cup \{lastPast(P)\}$
4   **foreach** *variable* $x \in S \mid x \notin past(P)$ **do**
5     $S^{sup} \leftarrow S^{sup} \cup \{x\}$
6   **if** $|dom(x)| \neq lastSize[c_S][x]$ **then**
7       $S^{val} \leftarrow S^{val} \cup \{x\}$

Algorithm 6. pruneVars(c_S: soft constraint)

---

1   $Y^{sup} \leftarrow \emptyset, Y^{val} \leftarrow \emptyset$
2   **foreach** *variable* $x \in S^{sup}$ **do**
3   **foreach** $a \in dom(x)$ **do**
4   **if** $minCosts[c_S][x][a] = k$ **then**
5   remove a from $dom(x)$
6   add x to Y^{val}
7   **else if** $minCosts[c_S][x][a] > 0$ **then**
8   add x to Y^{sup}

9     $lastSize[c_S][x] \leftarrow |dom(x)|$
10   $S^{val} \leftarrow Y^{val} \, ; S^{sup} \leftarrow Y^{sup}$

Algorithm 7. traverse-k(c_S: soft constraint, cnt: integer)

---

1 **foreach** *variable* $x \in S^{sup}$ **do**
2     $nbGacValues[c_S][x] \leftarrow 0$
3   **foreach** $a \in dom(x)$ **do**
4        $minCosts[c_S][x][a] \leftarrow k$

5 $i \leftarrow 1$
6 **while** $i \leq currentLimit[c_S]$ **do**
7     $index \leftarrow position[c_S][i]$
8     $\tau \leftarrow table[c_S][index]$ `// current tuple`
9     $\gamma \leftarrow costs[c_S][index] \ominus_{x \in S} deltas[c_S][x][\tau[x]]$ `// tuple cost`
10    $valid \leftarrow cnt > 2$ **or else** isValidTuple(c_S, τ)
11    $allowed \leftarrow cnt > 1$ **or else** ecost$(c_S, \gamma, \tau) < k$
12   **if** $valid \wedge allowed$ **then**
13   **foreach** *variable* $x \in S^{sup}$ **do**
14          $a \leftarrow \tau[x]$
15   **if** $\gamma < minCosts[c_S][x][a]$ **then**
16            $minCosts[c_S][x][a] \leftarrow \gamma$
17   **if** $\gamma = 0$ **then**
18               $nbGacValues[c_S][x] + +$
19   **if** $nbGacValues[c_S][x] = |dom(x)|$ **then**
20                 $S^{sup} \leftarrow S^{sup} \setminus \{x\}$

21       $i \leftarrow i + 1$
22   **else**
23       $swap(position[c_S], i, currentLimit[c_S])$
24       $currentLimit[c_S] - -$

τ_2 is removed. The structure $minCosts''$ represents the state of $minCosts$ after considering all tuples. Finally, as all values of w have a minimal cost equal to 0, it means that all values have at least a support in this constraint. The variable w can then be safely removed from S^{sup}. One can apply a similar reasoning for the variable z. Actually, the value (z, b) will be removed when Algorithm 6 is called (since its minimal cost is equal to k), and since all the other values have a support then z can be safely removed from S^{sup}. Note that if τ_2 had not been removed (by omitting to compute its extended cost), the minimal cost of (z, b) would have been 3 (instead of k). Consequently this value would not have been removed. After the execution of Algorithms 7 and 6, the set S^{sup} only contains variables x and y.

3.3 Properties

Lemma 1. *Let c_S be a soft constraint, and (x, a) be a value of c_S. The extended cost of every tuple $\tau \in l(S)$ remains constant, whatever the operation* project(c_S, x, a, α) *or* unaryProject(x, α) *is performed (Proof omitted)*

Algorithm 8. allowedImplicitTuple(c_S: soft constraint, x: variable, a: value)

1 **foreach** $\tau \in l(S) \mid \tau[x] = a$ **do**
2   **if** $\neg binarySearch(\tau, table[c_S])$ **then**
3   **return** true

4 **return** false

Algorithm 9. traverse-0(c_S: soft constraint, cnt: integer)

1 **foreach** *variable* $x \in S^{sup}$ **do**
2 $nbGacValues[c_S][x] \leftarrow 0$
3   **foreach** $a \in dom(x)$ **do**
4 $minCosts[c_S][x][a] \leftarrow k$
5 $nbTuples[c_S][x][a] \leftarrow 0$

6 $i \leftarrow 1$
7 **while** $i \leq currentLimit[c_S]$ **do**
8 $index \leftarrow position[c_S][i]$
9 $\tau \leftarrow table[c_S][index]$ // current tuple
10 $\gamma \leftarrow costs[c_S][index] \ominus_{x \in S} deltas[c_S][x][\tau[x]]$ // tuple cost
11 $valid \leftarrow cnt > 2$ **or else** isValidTuple(c_S, τ)
12 $allowed \leftarrow$ ecost(c_S, γ, τ) $< k$
13   **if** $valid$ **then**
14   **foreach** *variable* $x \in S^{sup}$ **do**
15 $a \leftarrow \tau[x]$
16 $nbTuples[c_S][x][a] + +$
17   **if** $allowed \wedge \gamma < minCosts[c_S][x][a]$ **then**
18 $minCosts[c_S][x][a] \leftarrow \gamma$
19   **if** $\gamma = 0$ **then**
20 $nbGacValues[c_S][x] + +$
21   **if** $nbGacValues[c_S][x] = |dom(x)|$ **then**
22 $S^{sup} \leftarrow S^{sup} \setminus \{x\}$

23 $i \leftarrow i + 1$
24   **else**
25 $swap(position[c_S], i, currentLimit[c_S])$
26 $currentLimit[c_S] - -$

27 **foreach** *variable* $x \in S^{sup}$ **do**
28 $nb \leftarrow |\Pi_{y \in S | y \neq x} dom(y)|$
29   **foreach** $a \in dom(x)$ **do**
30   **if** $nbTuples[c_S][x][a] \neq nb \wedge minCosts[c_S][x][a] > 0$ **then**
31   **if** $allowedImplicitTuple(c_S, x, a)$ **then**
32 $minCosts[c_S][x][a] \leftarrow 0$

Algorithm 10. GAC^w-WSTR(c_S: soft constraint)

```
 1  initialize(c_S)
 2  cnt ← 1
 3  if default[c_S] = k then
 4  │   traverse-k(c_S, cnt)
 5  else
 6  └   traverse-0(c_S, cnt)                        // default[c_S] = 0
 7  pruneVars(c_S)
 8  while S^sup ≠ ∅ do
 9  │   pick and delete x from S^sup
10  │   α ← +∞
11  │   foreach a ∈ dom(x) do
12  │   │   if minCosts[c_S][x][a] > 0 then
13  │   │   └   project(c_S, x, a, minCosts[c_S][x][a])
14  │   └   α ← min(α, c_x(a))
15  │   if α > 0 then
16  │   └   unaryProject(x, α)
17  │   if S^sup ≠ ∅ then
18  │   │   cnt + +
19  │   │   if default[c_S] = k then
20  │   │   │   traverse-k(c_S, cnt)
21  │   │   else
22  │   │   └   traverse-0(c_S, cnt)                // default[c_S] = 0
```

Under our assumptions, a preliminary observation is that we do not have to keep track of the effect of projections $\mathsf{project}(c_S, x, a, \alpha)$ on the default cost. Indeed, if $default[c_S] = k$, we have $k \ominus \alpha = k$ and if $default[c_S] = 0$ a projection is only possible when no implicit tuple exists with $x = a$.

Proposition 1. *Algorithm 10 enforces GAC^w on any soft table constraint c_S such that $default[c_S] = k$.*

Proof. Let (x, a) be a value of c_S (before calling Algorithm 10), and let the value $\alpha = minCosts[c_S][x][a]$ be obtained (for the minimum cost of (x, a) on c_S) just before executing line 7 of Algorithm 10. On the one hand, if $\alpha = k$ then it means that there is no explicit valid tuple τ in the current table such that $\tau[x] = a \wedge ecost(c_S, \tau) < k$ (because all explicit tuples have just been iterated over by Function $\mathsf{traverse\text{-}k}$ called at line 4). Besides, as the default cost is k, there is no implicit tuple τ such that $ecost(c_S, \tau) < k$. We can conclude that (x, a) is inconsistent w.r.t. GAC^w. This is the reason why when $\mathsf{pruneVars}$ is called at Line 7, this value (x, a) is removed (see Line 5 of Algorithm 6). On the other hand, if $\alpha < k$, it means that there exists a non-empty set X of valid tuples τ such that $\tau[x] = a \wedge ecost(c_S, \tau) < k$. Let us first consider the call to Function $\mathsf{pruneVars}$ at line 7. For every value (y, b) removed at line 5 of Algorithm 6, we have $minCosts[c_S][y][b] = k$, which implies that for every $\tau \in X$, we have $\tau[y] \neq b$

(otherwise $minCosts[c_S][y][b]$ would have been α). Consequently, all tuples in X remain valid and allowed after the execution of pruneVars. Those tuples, present in X, will remain valid and allowed throughout the execution of the algorithm because after executing pruneVars, no more values can be deleted, and cost transfer operations do not modify extended costs (see Lemma 1). This guarantees that all values detected inconsistent by GAC^w are deleted during the call to pruneVars. Now, for (x, a), either $minCosts[c_S][x][a]$ incidentally becomes 0 by means of cost transfers concerning variables other than x, or $0 < minCosts[c_S][x][a] < k$ at the moment where x is picked at line 9 of Algorithm 10. When executing lines 11-14, all values of x are made GAC^w-consistent. So, this is the case for (x, a). We have just proved that every deleted value is inconsistent w.r.t. GAC^w, and that every remaining value is GAC^w-consistent. $\square$

Proposition 2. *Algorithm 10 enforces GAC^w on any soft table constraint c_S such that $default[c_S] = 0$.*

The proof (omitted here) is similar to that of Proposition 1, with the additional consideration of implicit valid tuples. Notice that Algorithm 10 enforces both GAC^w and NC on any soft table constraint whose default cost is either k or 0.

We now discuss the complexity of GAC^w-WSTR for a given constraint c_S. With $r = |S|$ being the arity of c_S, the space complexity is $O(tr)$ for structures $table[c_S]$ and $costs[c_S]$, $O(t)$ for $position[c_S]$, $O(r)$ for S^{sup}, S^{val}, $lastSize[c_S]$ and $nbGac\,Values[c_S]$, $O(rd)$ for $minCosts[c_S]$, $nbTuples[c_S]$ and $deltas[c_S]$. Overall, the worst-case space complexity is $O(tr + rd)$. The time complexity is $O(r)$ for initialize, $O(rd + tr)$ for traverse-k and $O(rd)$ for pruneVars. Importantly, the number of turns of the main loop starting at line 8 of Algorithm 10 is at most r because a variable can never be put two times in S^{sup}; the complexity for one iteration is $O(d)$ for lines 10-16 augmented with that of traverse-k or traverse-0. Overall, the worst-case time complexity of GAC^w-WSTR when $default[c_S] = k$ is $O(r^2(d + t))$. On the other hand, the time complexity of allowedImplicitTuple is $O(rt \log(t))$ because the loop at line 1 of Algorithm 8 is executed at most t times. Indeed, each call to binarySearch is $O(r \log(t))$ and the loop is stopped as soon as a valid tuple cannot be found in the table. traverse-0 is $O(rd + tr)$ for lines 1-26 and $O(r^2 dt \log(t))$. Finally, the worst-case time complexity of GAC^w-WSTR when $default[c_S] = 0$ is $O(r^3 dt \log(t))$.

4   Benchmarks

We have performed a first experimentation using a new series of Crossword instances called *crossoft*, which can be naturally represented by soft table constraints. Given a grid and a dictionary, the goal is to fill the grid with words present in the dictionary. To generate those instances, we used three series of grids (Herald, Puzzle, Vg) and one dictionary, called OGD, that contains common nouns (with a cost of 0) and proper nouns (with a cost r, where r is the length of the word). Penalties are inspired from the profits associated with words as described on the french web site

`http://ledefi.pagesperso-orange.fr.`

We have performed a second experimentation using random WCSP instances. We have generated different classes of instances by considering the CSP model RB [17]. With some well-chosen parameters, Theorem 2 in [17] holds: an asymptotic phase transition is guaranteed at a precise threshold point. CSP instances from model RB were translated into WCSP instances by associating a random cost (between 1 and k) with each forbidden CSP tuple, and considering a default cost for implicit tuples equal to 0. This guarantees the hardness of the generated random WCSP instances. Using $k = 10$, we generated 5 series of 10 WCSP instances of arity 3; rb-r-n-d-e-t-s is an instance of arity r with n variables, domain size d and e r-ary constraints of tightness t (generated with seed s). A second set was obtained by translating random CSP instances (with arity equal to 10) into WCSP. Such a translation was also used for the series *renault-mod*.

Next, we have performed experimental trials with a new series of instances called *poker* based on the version *Texas hold 'em* of poker. The goal is to fill an empty 5×5 board with cards taken from the initial set of cards so as to obtain globally the best hands in each row and column. The model used to generate the instances is the following: there is a variable per cell representing a card picked in the initial set of cards. In *Poker-n* the initial set of cards contains n cards of each suit and only combinations of at least 2 cards are considered. Of course, putting the same card several times on the grid is forbidden. The cost of each hand is given below:

Royal Flush	Straight Flush	Four of a Kind	Full House	Flush	Straight	Three of a Kind	Two Pairs	Pair	High Card
0	1	2	3	4	5	6	7	8	9

Finally, we have experimented our approach on real-world series from `http://costfunction.org/en/benchmark`. We have used the *ergo* and the *linkage* series which are structured WCSP instances with constraints of arity larger than 3.

5   Experimental Results

In order to show the practical interest of our approach to filter soft table constraints of large arity, we have conducted an experimentation (with our solver AbsCon) using a cluster of Xeon 3.0GHz with 1GiB of RAM under Linux. We have implemented a version of PFC-MRDAC, where minimum costs (required by the algorithm to compute lower bounds) are obtained by calling Functions traverse-0 and traverse-k. At each step of the search, only one call to either Function traverse-0 or Function traverse-k is necessary for each soft table constraint because PFC-MRDAC does not exploit cost transfer operations (due to lack of space, we cannot give full details). This version will be called PFC-MRDAC-WSTR, whereas the classical version will be called here PFC-MRDAC-GEN. We have also implemented the algorithm GAC^w-WSTR and embedded it in a backtrack search algorithm that maintains GAC^w. This search algorithm is also able to maintain AC* and FDAC, instead of GAC^w. Note that PFC-MRDAC-GEN, "maintaining AC*" and "maintaining FDAC" iterate over all valid tuples in order to compute lower bounds or minimum costs. To control such iterations (that are exponential with respect to the arity of constraints), we have pragmatically tuned a parameter

Table 1. Number of solved instances per series (a time-out of 1,200 seconds was set per instance)

| *Series* | #*Inst* | PFC-MRDAC- | | Maintaining- | | |
		WSTR	GEN	GAC^w-WSTR	AC*	FDAC
crossoft-herald	50	33	10	**47**	11	11
crossoft-puzzle	22	**22**	9	**22**	18	18
crossoft-vg	64	**14**	6	**14**	7	7
poker	18	**10**	2	**10**	5	5
rand-3 (rb)	48	20	29	20	**32**	30
rand-10	20	**20**	0	**20**	0	0
ergo	19	13	10	15	15	**17**
linkage	30	0	0	0	1	**9**
renault-mod	50	**50**	32	**50**	50	47

that delays the application of the algorithm until enough variables are assigned. We have conducted an experimentation on the benchmarks described in the previous section. A time-out of $1,200$ seconds was set per instance. The variable ordering heuristic was $wdeg/dom$ [2] and the value ordering heuristic selected the value with minimal cost.

The overall results are given in Table 1. Each line of this table corresponds to a series of instances: *crossoft-ogd, rand-3, ergo,...* The total number of instances for each series is given in the second column of the table. For each series, we provide the number of solved instances (optimum proved) by each method within 20 minutes. For series *crossoft*, the algorithms PFC-MRDAC-WSTR and "maintaining GAC^w-WSTR" solve more instances than the generic algorithms. We obtain the same kind of results with *poker*. Unsurprisingly, the STR approaches are not so efficient on RB series (*rand-3*), which can be explained by the low arity of the constraints (which are ternary) involved in these instances. On random problems with high arity (involving 10 variables) results are clearly better: generic algorithms can not solve any of these instances. Finally, for *ergo* and *linkage* series, results are not so significant. Indeed these instances have either constraints with low arity or variable domains with very few values (for example, the maximum domain size is 2 for the instance *cpcs422b*). When the size of variable domains is small, Cartesian products of domains grow slowly with the constraint arity, and so generic algorithms iterating over valid tuples can still be competitive.

Table 2 focuses on some selected instances with the same comparison of algorithms. We provide an overview of the results in terms of CPU time (in seconds). On instances of series *crossoft* and *poker*, our approach (PFC-MRDAC-WSTR and GAC^w-WSTR) outperforms the generic ones whatever the envisioned solving approach (i.e., with or without cost transfer) is. Note that results for maintaining AC* and FDAC are quite close. Instances of these two problems have constraints with high arity and variables with rather large domains. Therefore, the STR technique is well-adapted. Note that for various instances, generic approaches can not find and prove optimum solutions before the time limit whereas STR-based algorithms solve them in a few seconds.

Table 2. CPU time (in seconds) to prove optimality on various selected instances (a time-out of 1,200 seconds was set per instance)

| *Instances* | PFC-MRDAC- | | Maintaining- | | |
	WSTR	GEN	GAC^w-WSTR	AC*	FDAC
crossoft-ogd-15-09	26.5	$> 1,200$	**25.2**	273	269
crossoft-ogd-23-01	$> 1,200$	$> 1,200$	**565**	$> 1,200$	$> 1,200$
crossoft-ogd-puzzle-18	**6.29**	$> 1,200$	6.66	$> 1,200$	$> 1,200$
crossoft-ogd-vg-5-6	**0.4**	155	0.77	31.5	32.3
poker-5	0.26	92.4	**0.24**	1.39	1.5
poker-6	**0.38**	463	0.39	6.58	6.99
poker-9	0.79	$> 1,200$	**0.63**	782	1022
poker-12	1.51	$> 1,200$	**0.89**	$> 1,200$	$> 1,200$
rb-3-12-12-30-0.630-0	4.13	1.2	3.61	**0.79**	0.86
rb-3-16-16-44-0.635-2	94.3	7.41	51.6	**2.31**	3.11
rb-3-20-20-60-0.632-0	614	34.4	830	24.3	**23.3**
pedigree1	$> 1,200$	$> 1,200$	890	819	**35.0**
barley	$> 1,200$	$> 1,200$	40.7	23	**20.3**
cpcs422b	**7.4**	113	8.61	58.3	111
link	68.6	$> 1,200$	5.41	**4.55**	6.5
rand-10-20-10-5-9	3.94	$> 1,200$	**2.39**	$> 1,200$	$> 1,200$
rand-10-20-10-5-10	5.27	$> 1,200$	**2.67**	$> 1,200$	$> 1,200$
renault-mod-12	1.74	680	**1.39**	6.01	14.4
renault-mod-14	2.49	$> 1,200$	**1.49**	6.83	14.9

6 Conclusion

In this paper, we have introduced a filtering algorithm that enforces a form of generalized arc consistency, called GAC^w, on soft table constraints. This algorithm combines simple tabular reduction and cost transfer operations. The experiments that we have conducted show the viability of our approach when soft table constraints have large arity, whereas usual generic soft consistency algorithms are not applicable to their full extent. The algorithm we propose can be applied to any soft table constraint with a default cost of either 0 or k, which represents a large proportion of practical instances. A direct perspective of this work is to generalize our approach to soft table constraints with any default cost.

Acknowledgments. This work has been supported by both CNRS and OSEO within the ISI project 'Pajero'.

References

1. Allouche, D., Bessiere, C., Boizumault, P., de Givry, S., Gutierrez, P., Loudni, S., Métivier, J.-P., Schiex, T.: Decomposing global cost functions. In: Proceedings of AAAI 2012 (2012)
2. Boussemart, F., Hemery, F., Lecoutre, C., Sais, L.: Boosting systematic search by weighting constraints. In: Proceedings of ECAI 2004, pp. 146–150 (2004)
3. Cooper, M.C.: Reduction operations in fuzzy or valued constraint satisfaction. Fuzzy Sets and Systems 134(3), 311–342 (2003)
4. Cooper, M.C., de Givry, S., Sanchez, M., Schiex, T., Zytnicki, M., Werner, T.: Soft arc consistency revisited. Artificial Intelligence 174(7-8), 449–478 (2010)
5. Cooper, M.C., Schiex, T.: Arc consistency for soft constraints. Artificial Intelligence 154(1-2), 199–227 (2004)
6. de Givry, S., Heras, F., Zytnicki, M., Larrosa, J.: Existential arc consistency: Getting closer to full arc consistency in weighted CSPs. In: Proceedings of IJCAI 2005, pp. 84–89 (2005)
7. Favier, A., de Givry, S., Legarra, A., Schiex, T.: Pairwise decomposition for combinatorial optimization in graphical models. In: Proceedings of IJCAI 2011, pp. 2126–2132 (2011)
8. Freuder, E.C., Wallace, R.J.: Partial constraint satisfaction. Artificial Intelligence 58(1-3), 21–70 (1992)
9. Larrosa, J.: Node and arc consistency in weighted CSP. In: Proceedings of AAAI 2002, pp. 48–53 (2002)
10. Larrosa, J., Meseguer, P.: Partition-Based lower bound for Max-CSP. Constraints 7, 407–419 (2002)
11. Larrosa, J., Meseguer, P., Schiex, T.: Maintaining reversible DAC for Max-CSP. Artificial Intelligence 107(1), 149–163 (1999)
12. Larrosa, J., Schiex, T.: Solving weighted CSP by maintaining arc consistency. Artificial Intelligence 159(1-2), 1–26 (2004)
13. Lecoutre, C.: STR2: Optimized simple tabular reduction for table constraint. Constraints 16(4), 341–371 (2011)
14. Lee, J., Leung, K.: Towards efficient consistency enforcement for global constraints in weighted constraint satisfaction. In: Proceedings of IJCAI 2009, pp. 559–565 (2009)
15. Lee, J., Leung, K.: A stronger consistency for soft global constraints in weighted constraint satisfaction. In: Proceedings of AAAI 2010, pp. 121–127 (2010)
16. Ullmann, J.R.: Partition search for non-binary constraint satisfaction. Information Science 177, 3639–3678 (2007)
17. Xu, K., Boussemart, F., Hemery, F., Lecoutre, C.: Random constraint satisfaction: easy generation of hard (satisfiable) instances. Artificial Intelligence 171(8-9), 514–534 (2007)

WCSP Integration of Soft Neighborhood Substitutability

Christophe Lecoutre, Olivier Roussel, and Djamel E. Dehani

Université Lille-Nord de France, Artois
CRIL-CNRS UMR 8188
F-62307 Lens
{lecoutre,roussel,dehani}@cril.fr

Abstract. WCSP is an optimization problem for which many forms of soft local (arc) consistencies have been proposed such as, for example, existential directional arc consistency (EDAC) and virtual arc consistency (VAC). In this paper, we adopt a different perspective by revisiting the well-known property of (soft) substitutability. First, we provide a clear picture of the relationships existing between soft neighborhood substitutability (SNS) and a tractable property called *pcost* which allows us to compare the cost of two values (through the use of so-called cost pairs). We prove that under certain assumptions, *pcost* is equivalent to SNS but weaker than SNS in the general case since we show that SNS is coNP-hard. We also show that SNS preserves the property VAC but not the property EDAC. Finally, we introduce an algorithm to enforce *pcost* that benefits from several optimizations (early breaks, residues, timestamping). The practical interest of maintaining *pcost* together with AC*, FDAC or EDAC, during search. is shown on various series of WCSP instances.

1 Introduction

The Valued Constraint Satisfaction Problem (VCSP) [18] is a general optimization framework used to handle soft constraints, which has been successfully applied to many applications in Artificial Intelligence and Operations Research. A problem instance is modeled in this framework by means of a set of variables and a set of cost functions defined over a valuation structure. Each cost function determines a violation degree for each possible instantiation of a subset of variables. These degrees (or costs) can then be combined using the operator $\oplus$ of the valuation structure in order to obtain the overall cost of any complete instantiation. One can broadly classify the different VCSP instantiations according to the properties of the operator $\oplus$: those where $\oplus$ is idempotent (e.g., min) and those where $\oplus$ is monotonic (e.g., +).

Interchangeability is a general property of constraint networks introduced in [10]. Two values a and b for a variable x are interchangeable if for every solution I where x is assigned b, $I_{x=a}$ is also a solution, where $I_{x=a}$ means I with x set to a. Full interchangeability has been refined into several weaker forms such as neighborhood interchangeability, k-interchangeability, partial interchangeability, relational interchangeability and substitutability. Interchangeability and substitutability have been used in many contexts; see e.g. [11,3,7,14]. A partial taxonomy of these two properties can be found in [12].

M. Milano (Ed.): CP 2012, LNCS 7514, pp. 406–421, 2012.

A generalization of interchangeability and substitutability for soft constraints has been given in [1]. For example, a value a for a variable x is soft substitutable for another value b in the domain of x if for every complete instantiation I involving (x, a), the cost of I is less than or equal to the cost of $I_{x=b}$. Observation of this property can be used to delete value b for x whereas preserving optimality. Identifying full substitutability is not tractable, but is is known [1] that neighborhood substitutability, a limited form of substitutability where only constraints involving a given variable are considered, can be computed in polynomial time when $\oplus$ is idempotent (provided that the arity of constraints be bounded). However, when $\oplus$ is monotonic, as it is the case for the VCSP specialization called WCSP (Weighted CSP), there is no clear picture although dominance rules related to soft neighborhood substitutability have been used [13,8].

In this paper, we focus on soft neighborhood substitutability (SNS) for WCSP. We introduce a property based on cost pairs, called *pcost*, that allows us to identify efficiently soft substitutable values. We prove that under certain assumptions, *pcost* is equivalent to SNS. However, in the general case, and especially when the WCSP forbidden cost k is not ∞, *pcost* is weaker than SNS. Actually, identifying a soft neighborhood value is coNP-hard. We also study the relationships between SNS and known soft arc consistency properties such as Existential Directional Arc Consistency (EDAC) [9] and Virtual Arc Consistency (VAC) [5]. We prove that SNS preserves VAC but not necessarily EDAC. Finally, we develop a *pcost* algorithm that benefits from a very moderate best-case time complexity, and we show experimentally that it can be successfully combined with EDAC during search.

2   Technical Background

A *constraint network* (CN) P is a pair $(\mathscr{X}, \mathscr{C})$ where $\mathscr{X}$ is a finite set of n variables, also denoted by $vars(P)$, and $\mathscr{C}$ is a finite set of e constraints. Each variable x has a (current) domain, denoted by $dom(x)$, which is the finite set of values that can be (currently) assigned to x; the initial domain of x is denoted by $dom^{init}(x)$. The largest domain size will be denoted by d. Each constraint c_S involves an ordered set S of variables, called the *scope* of c_S, and represents a relation capturing the set of tuples allowed for the variables in S. A *unary* (resp., *binary*) constraint involves 1 (resp., 2) variable(s), and a *non-binary* one strictly more than 2 variables. An *instantiation* I of a set $X = \{x_1, \ldots, x_p\}$ of p variables is a set $\{(x_1, a_1), \ldots, (x_p, a_p)\}$ such that $\forall i \in 1..p, a_i \in dom^{init}(x_i)$; X is denoted by $vars(I)$ and each a_i is denoted by $I[x_i]$. An instantiation I on a CN P is an instantiation of a set $X \subseteq vars(P)$; it is *complete* if $vars(I) = vars(P)$. I is *valid* on P iff $\forall(x, a) \in I, a \in dom(x)$. I *covers* a constraint c_S iff $S \subseteq vars(I)$. I *satisfies* a constraint c_S with $S = \{x_1, \ldots, x_r\}$ iff (i) I covers c_S and (ii) the tuple $(I[x_1], \ldots, I[x_r]) \in c_S$. An instantiation I on a CN P is *locally consistent* iff (i) I is valid on P and (ii) every constraint of P covered by I is satisfied by I. A *solution* of P is a complete locally consistent instantiation on P; $sols(P)$ denotes the set of solutions of P.

A *weighted constraint network* (WCN) P is a triplet $(\mathscr{X}, \mathscr{W}, k)$ where $\mathscr{X}$ is a finite set of n variables, as for CSP, $\mathscr{W}$ is a finite set of e weighted constraints, also denoted by $cons(P)$, and $k > 0$ is a natural integer or ∞. Each weighted constraint $w_S \in \mathcal{W}$ involves an ordered set S of variables (its scope) and is defined as a cost function from $l(S)$ to $\{0, \ldots, k\}$ where $l(S)$ is the set of possible instantiations of S. When a constraint w_S assigns the cost k to an instantiation of S, it means that w_S forbids this instantiation. Otherwise, it is permitted with the corresponding cost (0 is completely satisfactory). Costs are combined with the specific operators $\oplus$ defined as: $\forall a, b \in \{0, \ldots, k\}, a \oplus b = min(k, a + b)$.

For any instantiation I and any set of variables X, let $I_{\downarrow X} = \{(x, a) \mid (x, a) \in I \wedge x \in X\}$ be the projection of I on X. We denote by $I_{x=a}$ the instantiation $I_{\downarrow X \setminus \{x\}} \cup \{(x, a)\}$, which is the instantiation obtained from I either by replacing the value assigned to x in I by a, or by extending I with (x, a). The set of neighbor constraints of x is denoted by $\Gamma(x) = \{c_S \in cons(P) \mid x \in S\}$. When $\Gamma(x)$ does not contain two constraints sharing at least two variables, we say that $\Gamma(x)$ is *separable*. If c_S is a (weighted) constraint and I is an instantiation of a set $X \supseteq S$, then $c_S(I)$ will be considered to be equal to $c_S(I_{\downarrow S})$ (in other words, projections will be implicit). If C is a set of constraints, then $vars(C) = \cup_{c_S} S$ is the set of variables involved in C; if I is an instantiation such that $vars(C) \subseteq vars(I)$, then $cost_C(I) = \oplus_{c_S \in C} c_S(I)$ is the cost of I obtained by considering all constraints in C. For a WCN P and a complete instantiation I of P, the cost of I is then $cost_{cons(P)}(I)$ which will be simplified into $cost(I)$. The usual (NP-hard) task of Weighted Constraint Satisfaction Problem (WCSP) is, for a given WCN, to find a complete instantiation with a minimal cost. CSP can be seen as specialization of WCSP (when only costs 0 and k are used) and WCSP [16] can be seen as a specialization of the generic framework of valued/semiring-based constraints [2].

Many forms of soft arc consistency have been proposed during the last decade. We briefly introduce them in the context of binary WCNs. Without any loss of generality, the existence of a zero-arity constraint $c_\emptyset$ (a constant) as well as the presence of a unary constraint c_x for every variable x is assumed. A variable x is node-consistent (NC*) iff $\forall a \in dom(x), c_\emptyset \oplus c_x(a) < k$ and $\exists b \in dom(x) \mid c_x(b) = 0$. A variable x is arc-consistent (AC*) iff x is NC* and $\forall a \in dom(x), \forall c_{xy} \in \Gamma(x), \exists b \in dom(y) \mid c_{xy}(a, b) = 0$ (b is called a simple support of a). A WCN is AC* iff each of its variable is AC* [15,16]. A WCN is full directional arc-consistent (FDAC) [4] with respect to an order $<$ on the variables if it is AC* and $\forall c_{xy} \mid x < y, \forall a \in dom(x), \exists b \in dom(y) \mid c_{xy}(a, b) = c_y(b) = 0$ (b is called a full support of a). A WCN is existential arc-consistent (EAC) [9] if it is NC* and $\forall x \in vars(P), \exists a \in dom(x) \mid c_x(a) = 0 \wedge \forall c_{xy} \in \Gamma(x), \exists b \in dom(y) \mid c_{xy}(a, b) = c_y(b) = 0$ (a is called the existential support of x). A WCN is existential directional arc-consistent (EDAC) with respect to an order $<$ on the variables if it is EAC and FDAC with respect to $<$. For any WCN P, we can derive an associated CN $Bool(P)$ by considering constraint relations that only allow tuples with cost zero in P. P is virtual arc-consistent (VAC) [5] if the arc consistency closure of $Bool(P)$ does not involve a variable with an empty domain, denoted by $AC(Bool(P)) \neq \perp$. A WCN is optimal soft arc-consistent (OSAC) if no SAC transformation (see [6]) applied to it increases $c_\emptyset$. OSAC is stronger than VAC, which itself is stronger than EDAC, when comparing the values of $c_\emptyset$. We shall note $\phi(P)$ the enforcement of property ϕ (e.g., AC, EDAC, ...) on the (W)CN P.

3   Soft Substitutability

In this section, we introduce soft neighborhood substitutability. Initially, *interchangeability* and *substitutability* are properties that have been introduced for CSP [10]. From now on, we consider given a (W)CN P.

Definition 1. *Let $x \in vars(P)$ and $\{a, b\} \subseteq dom(x)$,*

- *(x, a) is (fully) substitutable for (x, b) on P iff for every solution $I_{x=b}$ of P, $I_{x=a}$ is also a solution of P;*
- *(x, a) is (fully) interchangeable with (x, b) on P iff (x, a) is substitutable for (x, b) and (x, b) is substitutable for (x, a).*

For example, consider a CN P such that $vars(P) = \{x, y, z\}$ and $sols(P) = \{(a, a, a), (a, b, b), (b, a, a), (c, a, a), (c, b, b)\}$. The values (x, a) and (x, c) are interchangeable and both are substitutable for (x, b). When only a single solution is sought, we can remove a value that is interchangeable with another value (or for which a value is substitutable). Such removal preserves the satisfiability of the problem instance but not the full set of solutions.

From now on, we shall focus on substitutability that has been generalized [1] for WCSP as follows:

Definition 2. *Let $x \in vars(P)$ and $\{a, b\} \subseteq dom(x)$, (x, a) is soft substitutable for (x, b) on P iff for every complete instantiation I of P, $cost(I_{x=a}) \leq cost(I_{x=b})$.*

When (x, a) is soft substitutable for (x, b), b can be removed from $dom(x)$ without changing the cost of the optimal solution(s) of P. Indeed, possible solutions of P with (x, b) are lost, but it is guaranteed that solutions with (x, a) are at least as good.

Because identifying substitutable values involves handling complete instantiations, this is subject to combinatorial explosion. However, there is a form of local substitutability, called neighborhood substitutability [10,1], that may be helpful.

Definition 3. *Let $x \in vars(P)$ and $\{a, b\} \subseteq dom(x)$, (x, a) is soft neighborhood substitutable for (x, b) on P iff for every complete instantiation I of P, $cost_{\Gamma(x)}(I_{x=a}) \leq cost_{\Gamma(x)}(I_{x=b})$.*

We shall say that (x, b) is SNS-eliminable (on P) when there exists a value (x, a) such that (x, a) is soft neighborhood substitutable for (x, b). It is rather immediate that soft neighborhood substitutability implies soft (full) substitutability (but the reverse is not true). Interestingly enough, soft (neighborhood) substitutability allows compensation between constraint costs. Such compensation is made possible by the presence of all intermediate costs in the valuation structure, that is to say, the costs different from 0 and k. A simple illustration is given by Figure 1. There are two binary constraints c_{xy} and c_{xz} and three trivial unary constraints (all unary costs are equal to 0). Binary costs are depicted as labeled edges, and zero costs are not shown. Note that (x, a) is soft substitutable for (x, b).

Definition 3 requires to consider each instantiation of $vars(\Gamma(x))$, which has a high computational cost. Considering each constraint individually allows to reduce this cost,

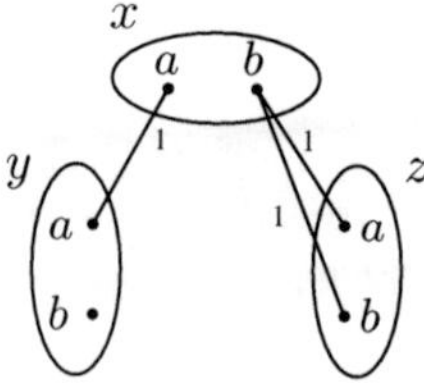

Fig. 1. (x, a) is soft substitutable for (x, b)

but in this case, it is highly desirable to be able to identify cost compensations between constraints. One straightforward way to do this is to compute a sum of minimal cost differences over all constraints, as mentioned in [13,8]. Unfortunately, these differences of costs introduce subtle problems when $k \neq \infty$. This is illustrated as follows:

Example 1. Consider the two families of constraints $C_i = \{c_i \mid i \in 1..n\}$ and $C'_i = \{c'_i \mid i \in 1..n'\}$ defined as:

x	y_i	c_i
a	c	0
b	c	1

x	z_i	c'_i
a	d	1
b	d	0

When $n = k$ and $n' = k+1$, (x, a) and (x, b) are interchangeable because both are forbidden (i.e., have maximum cost k). However, $\forall c_i \in C_i, cost_{c_i}(I_{x=b}) - cost_{c_i}(I_{x=a}) = 1$ and $\forall c'_i \in C'_i, cost_{c'_i}(I_{x=b}) - cost_{c'_i}(I_{x=a}) = -1$. In the summation over the constraints in $\Gamma(x) = C_i \cup C'_i$, the resulting value is -1 which would indicate that b has globally a lower cost than a, which is false since both a and b have a cost of k. To identify correctly this case of substitution when $k \neq \infty$, it is necessary to use a non commutative operator, which prevents us from using the usual -.

4   Computing Soft Substitutability

We now focus on soft neighborhood substitutability, and more precisely on the complexity of identifying SNS-eliminable values. We start with some related work. For the general framework VCSP, efficient algorithms for computing neighborhood substitutability exist [1] when the aggregation operator of the VCSP valuation structure is idempotent. For the framework FCSP (Fuzzy CSP), the notion of fuzzy neighborhood substitutability is proposed in [4]: it is shown that fuzzy neighborhood substitutable values can be identified efficiently when the aggregation operator of the FCSP valuation structure is strictly monotonic or when it is the operator max. More recently, the possibility of computing dominance forms weaker than soft neighborhood substitutability has been proposed in [13]. However, no qualitative study was led. This is what we propose now for WCSP.

First, we introduce cost pairs as our basic computation mechanism (this is related to what has been proposed in [4] for FCSP). Indeed, one way to circumvent the aforementioned problems with subtraction is to use only addition defined on pairs of costs.

This method is analogous to the construction of integers as equivalence classes of ordered pairs of natural numbers where a pair (β, α) represents the integer $\beta - \alpha$. We define $+$ (addition) on pairs of costs by $(\beta, \alpha) + (\beta', \alpha') = (\beta + \beta', \alpha + \alpha')$ (this is the regular $+$ and not $\oplus$) and the comparison of a pair of costs with zero by $(\beta, \alpha) \geq 0 \Leftrightarrow \beta \geq \alpha$. Pairs are ordered by the relation $\leq$ defined as $(\beta, \alpha) \leq (\beta', \alpha') \Leftrightarrow \beta - \alpha < \beta' - \alpha' \vee (\beta - \alpha = \beta' - \alpha' \wedge \alpha < \alpha')$. In a sense, the pair (β, α) conveys the difference $\beta - \alpha$ but also the information $min(\beta, \alpha)$ which is lost when a simple subtraction is used. Computing cost pairs on each constraint separately is sufficient for identifying certain soft neighborhood substitutable values: it suffices to reason from (sum) minimum differences of costs.

Definition 4. *Let $x \in vars(P)$ and $\{a, b\} \subseteq dom(x)$,*

- *the* cost pair *of (x, b) w.r.t. (x, a) on $c_S \in \Gamma(x)$ is defined as $pcost(c_S, x : a \to b) = min_{I \in l(S)}\{(c_S(I_{x=b}), c_S(I_{x=a}))\}$;*
- *the* cost pair *of (x, b) with respect to (x, a) on P is defined as $pcost(x : a \to b) = \sum_{c_S \in \Gamma(x)} pcost(c_S, x : a \to b)$.*

Proposition 1. *Let $x \in vars(P)$ and $\{a, b\} \subseteq dom(x)$. If $pcost(x : a \to b) \geq 0$ then (x, a) is soft neighborhood substitutable for (x, b) on P.*

Proof. For a constraint c_S, let I^{c_s} be the instantiation of $S - \{x\}$ which yields the minimal cost pair in $min_{I \in l(S)}\{(c_S(I_{x=b}), c_S(I_{x=a}))\}$. By definition, $pcost(c_S, x : a \to b) = (c_S(I^{c_s}_{x=b}), c_S(I^{c_s}_{x=a}))$. By definition of min on cost pairs, we have $\forall I, \forall c_S \in \Gamma(x), (c_S(I_{x=b}), c_S(I_{x=a})) \geq (c_S(I^{c_s}_{x=b}), c_S(I^{c_s}_{x=a}))$. By summing up, we obtain $\forall I$, $\sum_{c_S \in \Gamma(x)}(c_S(I_{x=b}), c_S(I_{x=a})) \geq \sum_{c_S \in \Gamma(x)}(c_S(I^{c_s}_{x=b}), c_S(I^{c_s}_{x=a}))$. By hypothesis, we have $pcost(x : a \to b) \geq 0$, so $\sum_{c_S \in \Gamma(x)}(c_S(I^{c_s}_{x=b}), c_S(I^{c_s}_{x=a})) \geq 0$, and consequently $\forall I, \sum_{c_S \in \Gamma(x)}(c_S(I_{x=b}), c_S(I_{x=a})) \geq 0$. From definition of $+$ and $\leq$ on cost pairs, we can derive $\forall I, \sum_{c_S \in \Gamma(x)} c_S(I_{x=b}) \geq \sum_{c_S \in \Gamma(x)} c_S(I_{x=a})$ which implies $\forall I, min(k, \sum_{c_S \in \Gamma(x)} c_S(I_{x=b})) \geq min(k, \sum_{c_S \in \Gamma(x)} c_S(I_{x=a}))$. Since $\forall a_i \in \{0, \ldots, k\}, a_1 \oplus \ldots \oplus a_n = min(k, a_1 + \ldots a_n)$, we can conclude that $\forall I, \bigoplus_{c_S \in \Gamma(x)} c_S(I_{x=b})) \geq \bigoplus_{c_S \in \Gamma(x)} c_S(I_{x=a}))$. Hence, (x, a) is soft neighborhood substitutable for (x, b) on P. $\square$

The converse of Proposition 1 is not true in the general case. One first case where it is false is when the scope of non-binary constraints intersect on more than one variable (non separable neighborhood). In this situation, constraints cannot be considered individually.

Example 2. Let's consider four variables x, y, z t such that $dom(x) = \{a, b\}, dom(y) = \{c, d\}, dom(z) = dom(t) = \{e\}$, and two ternary constraints c_{xyz}, c_{xyt} defined by the following cost table:

x y z t	c_{xyz}	c_{xyt}
a c e e	1	0
b c e e	0	1
a d e e	0	1
b d e e	1	0

It is easily seen that (x, a) is soft neighborhood substitutable for (x, b) but $pcost(c_{xyz}, x, a \to b) = pcost(c_{xyt}, x, a \to b) = (0, 1)$ and therefore $pcost(x, a \to b) = (0, 2) \not\geq 0$.

So a first condition for the converse of Proposition 1 to hold it that $\Gamma(x)$ be separable (which is the case of binary normalized networks).

Another case where the converse of Proposition 1 is false is when $k \neq \infty$. Considering the WCN of example 1 with $n = k$ and $n' = k + 1$, we can observe that $pcost(x : a \to b) = (n, n') = (k, k + 1) \not\geq 0$. However, both (x, a) and (x, b) imply cost k and therefore (x, a) is substitutable for (x, b) (and conversely). On this example, it might seem a good idea to use $\oplus$ instead of $+$ in the definition of the addition of pairs. However, Example 3 shows that this would lead to the incorrect identification of substitutable values.

Example 3. Consider the unary constraint c_x and the family of binary constraints $C_i = \{c_i \mid i \in 1..n\}$ defined by:

x	y_i	c_i
a	a	2
a	b	0
b	a	1
b	b	0

x	c_x
a	1
b	0

Clearly, (x, a) is not substitutable for (x, b). With the sum of pairs defined with $+$ and $n = k$, $pcost(x, a \to b) = (n, 2n + 1) \not\geq 0$. If the sum of pairs was defined with $\oplus$, we would obtain $(k, k) \geq 0$. Note that even if (k, k) was interpreted as $k - k = 0$ or as $k - k = k$ (absorbing element), the problem would remain: in both cases, we would have $(k, k) \geq 0$.

Interestingly, there are some situations where, even when $k \neq \infty$, using cost pairs allows us to identify exactly neighborhood substitutable values.

Proposition 2. *Let $x \in vars(P)$ and $\{a, b\} \subseteq dom(x)$ such that $\Gamma(x)$ is separable and $pcost(x, a \to b) = (\beta, \alpha)$ with $\alpha < k$. If (x, a) is soft neighborhood substitutable for (x, b) on P then $pcost(x : a \to b) \geq 0$.*

Proof. Since by hypothesis $\Gamma(x)$ is separable, one can define the instantiation I^{min} on $vars(\Gamma(x)) \setminus \{x\}$ as the union for each constraint $C_s \in \Gamma(x)$ of the instantiations I^{cs} defined in the proof of Proposition 1. I^{min} is such that $pcost(c_S, x : a \to b) = (c_S(I^{min}_{x=b}), c_S(I^{min}_{x=a}))$.

By hypothesis, $\forall I, cost_{\Gamma(x)}(I_{x=b}) \geq cost_{\Gamma(x)}(I_{x=a})$ which can be rewritten as $\forall I, \bigoplus_{c_S \in \Gamma(x)} c_S(I_{x=b}) \geq \bigoplus_{c_S \in \Gamma(x)} c_S(I_{x=a})$. This is especially true for $I = I^{min}$ therefore $\bigoplus_{c_S \in \Gamma(x)} c_S(I^{min}_{x=b}) \geq \bigoplus_{c_S \in \Gamma(x)} c_S(I^{min}_{x=a})$, giving $min(k, \beta) \geq min(k, \alpha)$ where $\beta = \sum_{c_S \in \Gamma(x)} c_S(I^{min}_{x=b}))$ and $\alpha = \sum_{c_S \in \Gamma(x)} c_S(I^{min}_{x=a})$. By definition, $pcost(x : a \to b) = (\beta, \alpha)$ and therefore $pcost(x : a \to b) \geq 0$ iff $\beta \geq \alpha$. Now, if $\alpha < k$, $min(k, \alpha) = \alpha$ and $min(k, \beta) \geq min(k, \alpha) \Rightarrow \beta \geq \alpha \Rightarrow pcost(x : a \to b) \geq 0$ (this is true for both $\beta < k$ and $\beta \geq k$). Note that when $\alpha \geq k$, $min(k, \beta) \geq min(k, \alpha) \not\Rightarrow \beta \geq \alpha$, a counter example being $\beta = k$ and $\alpha = k + 1$. $\qquad\square$

Corollary 1. *Let $x \in vars(P)$ and $\{a, b\} \subseteq dom(x)$ such that $\Gamma(x)$ is separable and $pcost(x : a \to b) = (\beta, \alpha)$ with $\alpha < k$. If $(\beta, \alpha) < 0$ then (x, a) is not soft neighborhood substitutable for (x, b) on P.*

When $pcost(x : a \to b) = (\beta, \alpha)$ with $\alpha \geq k$, deciding if (x, a) is soft neighborhood substitutable for (x, b) is much harder. Indeed, this problem is co-NP hard.

To prove this, we introduce the Multiple-choice Double Cost Problem (MCDP). We show that MCDP is NP-complete and exhibit a polynomial reduction of the MCDP problem to soft neighborhood substitutability.

Multiple-choice Double Cost Problem (MDCP) Given m sets E_1, E_2, ..., E_m of objects such that each object $o_j \in E_i$ has a cost value $r_{ij} \in \mathbb{Z}^+$ as well as a secondary cost value $s_{ij} \in \mathbb{Z}^+$. Given a maximal cost $C \in \mathbb{Z}^+$, the MDCP problem consists in deciding if it is possible to choose one object from each set such that the sum of the costs of these selected objects does not exceed C and does not exceed the sum of the secondary costs as well. This problem may be formulated as:

$$\sum_{i=1}^{m} \sum_{j \in E_i} r_{ij} x_{ij} \leq C,$$
$$\sum_{i=1}^{m} \sum_{j \in E_i} r_{ij} x_{ij} \leq \sum_{i=1}^{m} \sum_{j \in E_i} s_{ij} x_{ij},$$
$$\sum_{j \in E_i} x_{ij} = 1, i = 1, \ldots, m,$$
$$x_{ij} \in \{0, 1\}, i = 1, \ldots, m, j \in E_i.$$

Proposition 3. *The multiple-choice double cost problem is NP-complete.*

Proof. Membership to NP is immediate. For NP-hardness, we reduce Multiple-choice Knapsack problem (MCKP known to be NP-hard [17]) to Multiple-choice Double Cost Problem. For MCKP, we have also m sets, and each object is given a profit $p_{ij} \in \mathbb{Z}^+$ as well as a weight $w_{ij} \in \mathbb{Z}^+$. Given a minimal profit $P \in \mathbb{Z}^+$ and a maximal weight $W \in \mathbb{Z}^+$, the MCKP decision problem is formulated as:

$$\sum_{i=1}^{m} \sum_{j \in E_i} w_{ij} x_{ij} \leq W,$$
$$\sum_{i=1}^{m} \sum_{j \in E_i} p_{ij} x_{ij} \geq P,$$
$$\sum_{j \in E_i} x_{ij} = 1, i = 1, \ldots, m,$$
$$x_{ij} \in \{0, 1\}, i = 1, \ldots, m, j \in E_i.$$

To encode a MCKP instance into a MDCP instance, we keep the same structure (sets) and define:

$$C = qW - mP,$$
$$r_{ij} = qw_{ij} - P, i = 1, \ldots, m, j \in E_i,$$
$$s_{ij} = mp_{ij} + r_{ij} - P, i = 1, \ldots, m, j \in E_i.$$

with $q = 2mP$. With such a value for q, one can show that all values C, r_{ij} and s_{ij} belong to $\mathbb{Z}^+$. The first MDCP equation can be transformed as follows:

$\sum_{i=1}^{m} \sum_{j \in E_i} r_{ij} x_{ij} \leq C$

$\Rightarrow \sum_{i=1}^{m} \sum_{j \in E_i} (qw_{ij} - P) x_{ij} \leq qW - mP$

$\Rightarrow \sum_{i=1}^{m} \sum_{j \in E_i} qw_{ij} x_{ij} - mP \leq qW - mP$ because exactly m variables x_{ij} are set to 1

$\Rightarrow \sum_{i=1}^{m} \sum_{j \in E_i} w_{ij} x_{ij} \leq W$.

The second MDCP equation can be transformed as follows:

$\sum_{i=1}^{m} \sum_{j \in E_i} r_{ij} x_{ij} \leq \sum_{i=1}^{m} \sum_{j \in E_i} s_{ij} x_{ij}$,

$\Rightarrow 0 \leq \sum_{i=1}^{m} \sum_{j \in E_i} (m p_{ij} - P) x_{ij}$ by simplifying r_{ij} from both sides,

$\Rightarrow 0 \leq \sum_{i=1}^{m} \sum_{j \in E_i} m p_{ij} x_{ij} - mP$ because exactly m variables x_{ij} are set to 1,

$\Rightarrow P \leq \sum_{i=1}^{m} \sum_{j \in E_i} p_{ij} x_{ij}$. $\qquad\qquad\qquad\qquad\qquad\qquad\qquad$ $\square$

Proposition 2 states that *pcost* allows us to exactly identify soft neighborhood substitutable values when two conditions are verified: $\gamma(x)$ is separable and $pcost(x, a \rightarrow b) = (\beta, \alpha)$ with $\alpha < k$. In the following proposition, by construction, we deal with instances such that $\gamma(x)$ is separable. However, nothing is imposed on α, with the consequence that some SNS-eliminable values cannot be detected in polynomial time.

Proposition 4. *Deciding if a value is soft neighborhood substitutable for another on a WCN $(\mathscr{X}, \mathscr{W}, k)$ where $k \neq \infty$ is coNP-hard.*

Proof. Any MDCP instance can be reduced to the problem of deciding whether a value (x, a) is **not** soft substitutable for a value (x, b) on a WCN P. From the MDCP instance, we build the WCN P as follows. $vars(P)$ contains a variable x such that $dom(x) = \{a, b\}$ and a variable y_i for each set E_i ; the domain of y_i contains the objects $o_{i1}, o_{i2}, \ldots$ of E_i. $cons(P)$ contains exactly m binary soft constraints w_{xy_i}: we have $w_{xy_i}(\{(x, a), (y_i, o_{ij})\}) = s_{ij}$ and $w_{xy_i}(\{(x, b), (y_i, o_{ij})\}) = r_{ij}$. k is set to C. Determining if (x, a) is not soft substitutable for (x, b) on P is equivalent to finding an instantiation I of $vars(\Gamma(x))$ such that $cost_{\Gamma(x)}(I_{x=a}) > cost_{\Gamma(x)}(I_{x=b})$ which is equivalent to $\sum_{c_S \in \Gamma(x)} c_s(I_{x=b}) < k \wedge \sum_{c_S \in \Gamma(x)} c_s(I_{x=a}) > \sum_{c_S \in \Gamma(x)} c_s(I_{x=b})$. The first (resp. second) condition encodes the first (resp. second) inequality of the MDCP. Since variables x_{ij} of the MDCP instance correspond to the assignment of variables y_i in the WCSP ($x_{ij} = 1 \Leftrightarrow y_i = o_{ij}$), the third equation of the MDCP instance is directly encoded in the WCSP instance. $\qquad$ $\square$

5   Relationships with Soft Arc Consistency

After introducing soft neighborhood substitutability closure, this section presents some results relating soft neighborhood substitutability to various forms of soft arc consistency.

Definition 5. *The soft neighborhood substitutability closure (or SNS-closure) of a WCN P, denoted by $SNS(P)$, is any WCN obtained after iteratively removing SNS-eliminable values until convergence.*

Since this operation is not confluent, $SNS(P)$ is not unique. When we use the *pcost* approach to identify SNS-eliminable values, we note $PSNS(P)$.

Proposition 5. *Let P be an EDAC-consistent WCN. $SNS(P)$ is not necessarily EDAC-consistent.*

Proof. Consider the WCN P depicted in Figure 2(a). Note that P is EDAC-consistent w.r.t. the order $w > x > z > y$, and that (w, a) and (z, a) are respectively soft neighborhood substitutable for (w, b) and (z, b), since $pcost(w, a \rightarrow b) = (0, 0)$ and

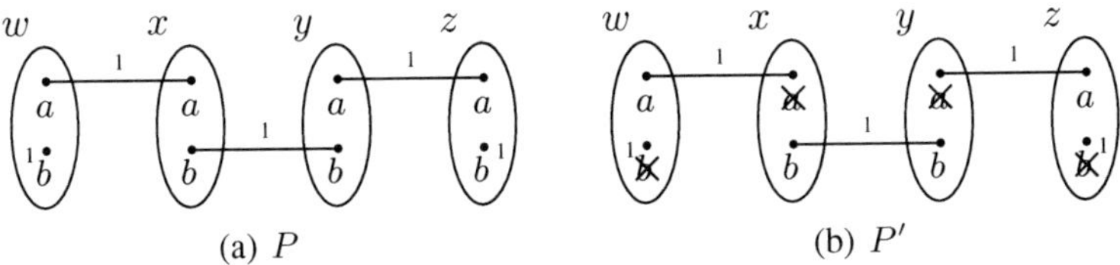

(a) P (b) P'

Fig. 2. EDAC versus SNS

$pcost(z, a \rightarrow b) = (0,0)$. There exists a unique SNS-closure of P, $P' = SNS(P)$, which is depicted in Figure 2(b). Clearly, P' is not EDAC-consistent since (x,b) and (y,b) have no support on c_{xy}. $\qquad\square$

Lemma 1. *Let P be a VAC-consistent WCN, $x \in S$ and $\{a,b\} \subseteq dom(x)$ such that (x,b) is AC-consistent in $Bool(P)$. If $pcost(x : a \rightarrow b) \geq 0$ on P then (x,a) is neighborhood substitutable (in the CSP sense [10]) for (x,b) on $Bool(P)$.*

Proof. We suppose that P is VAC-consistent and (x,b) is AC-consistent in $Bool(P)$. Because (x,b) is AC-consistent in $Bool(P)$, we know that for every constraint $c_S \in \Gamma(x)$, there exists an instantiation I of S such that $I[x] = b$ and $c_S(I) = 0$ (by construction of $Bool(P)$). This means that on every such constraint c_S, $pcost(c_S, x : a \rightarrow b) \leq 0$. As by hypothesis $pcost(x : a \rightarrow b) \geq 0$, for every constraint $c_S \in \Gamma(x)$, we have necessarily $pcost(c_S, x : a \rightarrow b) = 0$. We can deduce that for every constraint $c_S \in \Gamma(x)$, for every instantiation I of S such that $I[x] = b$ and $c_S(I) = 0$, the instantiation $I' = I_{x=a}$ is such that $c_S(I') = 0$. Finally, we can conclude that (x,a) is neighborhood substitutable for (x,b) on $Bool(P)$. $\qquad\square$

Proposition 6. *Let P be a VAC-consistent WCN, $x \in vars(P)$ and $\{a,b\} \subseteq dom(x)$. If $pcost(x : a \rightarrow b) \geq 0$ on P then $P \setminus \{(x,b)\}$ is VAC-consistent.*

Proof. On the one hand, suppose that (x,b) is AC-consistent in $Bool(P)$. From the previous lemma, we know that (x,a) is neighborhood substitutable for (x,b) on $Bool(P)$, and so we have $AC(Bool(P)) \neq \perp \Leftrightarrow AC(Bool(P \setminus \{(x,b)\})) \neq \perp$. Because P is VAC-consistent, necessarily $P \setminus \{(x,b)\}$ is VAC-consistent. On the other hand, suppose that (x,b) is not AC-consistent in $Bool(P)$. Clearly we have $AC(Bool(P)) = AC(Bool(P \setminus \{(x,b)\}))$, and so $P \setminus \{(x,b)\}$ is VAC-consistent (since P is VAC-consistent). $\qquad\square$

Corollary 2. *Let P be a WCN. If P is VAC-consistent then $SNS(P)$ is VAC-consistent.*

The previous corollary also holds for OSAC (because OSAC is stronger than VAC).

6 Algorithms

In this section, we introduce an algorithm to enforce AC*+PSNS (that can be easily adapted to EDAC+PSNS, for example). The main idea is always to start identifying SNS-eliminable values from a WCN that is AC*-consistent. This allows us to reduce the computation effort by using early breaks and residues.

The main procedure is Algorithm 1. As usual, we use a set, denoted by Q, to store the variables whose domain has been recently reduced. Initially, Q is initialized with all variables (line 4). Then, at line 6, a classical AC* algorithm, denoted here by W-AC*, is run (for example, this may be W-AC*2001 [16]), before soliciting a function called $PSNS^r$. The calls to W-AC* and $PSNS^r$ are interleaved until a fixed point is reached (i.e., $Q = \emptyset$).

Function $PSNS^r$, Algorithm 2, iterates over all variables in order to collect SNS-eliminable values into a set called Δ. This set is initialized at line 1 and updated at lines 6 and 8. Let us imagine that all SNS-eliminable values (that can be identified by means of pcost) for a variable x have been deleted, and that the domains of all variables in the neighborhood of x remain the same (while possible reductions happen elsewhere). Clearly, there is no need to consider x again for seeking SNS-eliminable values. This is the point of line 3. Here, timestamps are used. By introducing a global counter $time$ and by associating a time-stamp $stamp[x]$ with every variable x as well as a time-stamp $substamp$ with function $PSNS^r$, it is possible to determine which variables should be considered. The value of $stamp[x]$ indicates at which moment a value was most recently removed from $dom(x)$, while the value of $substamp$ indicates at which moment $PSNS^r$ was most recently called. Variables $time$, $stamp[x]$ for each variable x and $substamp$ are initialized at lines 1 to 3 of Algorithm 1. The value of $time$ is incremented whenever a variable is added to Q (line 13 of Algorithm 2 and this *must* also be performed *inside* W-AC*) and whenever $PSNS^r$ is called (line 9). All SNS-eliminable values collected in Δ are removed while updating Q for the next call to W-AC* (lines 10 to 12).

Algorithm 1. AC*-PSNS(P : WCN)

Output: P, made AC*-consistent and PSNS-closed

1 $time \leftarrow 0$;
2 $substamp \leftarrow -1$;
3 $stamp[x] \leftarrow 0, \forall x \in vars(P)$;
4 $Q \leftarrow vars(P)$;
5 **repeat**
6 $\quad\mid \quad PSNS^r(W\text{-}AC^*(P, Q))$;
7 **until** $Q \neq \emptyset$;

Algorithm 3 allows us to compute the pcost of (x, b) with respect to (x, a). Because we know that the WCN is currently AC*-consistent, we have the guarantee that the pcost of (x, b) with respect to (x, a) on any non-unary constraint c_S where $x \in S$ is less than or equal to 0. This means that we can never compensate a pair (β, α) s.t. $\alpha > \beta$ with a pair (β', α') s.t. $\alpha' < \beta$ (once the pair of unary costs has been taken into account). A strong benefit of that observation is the possibility of using early breaks during such computations. This is performed at lines 3, 6 and 13. Residues are another mechanism used to increase the performance of the algorithm. For every variable x, and every pair (a, b) of values in $dom(x)$, we store in $residues[x, a, b]$ the constraint c_S which guarantees that (x, b) is not SNS-eliminable by (x, a), if it exists. The residual constraint takes priority (lines 4 to 6); this way, if it compensates the initial unary cost,

---

Algorithm 2. $\text{PSNS}^r(P : \text{WCN AC*-consistent})$

---

1 $\Delta \leftarrow \emptyset$;
2 **foreach** $x \in vars(P)$ **do**
3 $\quad$ **if** $\exists y \in \Gamma(x) \mid stamp[y] > substamp$ **then**
4 $\quad\quad$ **foreach** $(a, b) \in dom(x)^2 \mid b > a$ **do**
5 $\quad\quad\quad$ **if** $pcost(x, a \to b) \geq 0$ **then**
6 $\quad\quad\quad\quad$ $\Delta \leftarrow \Delta \cup \{(x, b)\}$;
7 $\quad\quad\quad$ **else if** $pcost(x, b \to a) \geq 0$ **then**
8 $\quad\quad\quad\quad$ $\Delta \leftarrow \Delta \cup \{(x, a)\}$;

9 $substamp \leftarrow time\text{++}$;
10 **foreach** $(x, a) \in \Delta$ **do**
11 $\quad$ remove (x, a) from $dom(x)$;
12 $\quad$ $Q \leftarrow Q \cup \{x\}$;
13 $\quad$ $stamp[x] \leftarrow time\text{++}$;

Algorithm 3. $pcost(x, a \to b)$: cost pair

---

1 $pcst \leftarrow (c_x(b), c_x(a))$;
2 **if** $pcst < 0$ **then**
3 $\quad$ **return** $pcst$
4 $pcst \leftarrow pcst + pcost(residues[x, a, b], x, a \to b)$;
5 **if** $pcst < 0$ **then**
6 $\quad$ **return** $pcst$
7 **foreach** $c_S \in \Gamma(x) \mid c_S \neq residues[x, a, b]$ **do**
8 $\quad$ $d \leftarrow pcost(c_S, x, a \to b)$;
9 $\quad$ **if** $d < (c_x(a), c_x(b))$ **then**
10 $\quad\quad$ $residues[x, a, b] \leftarrow c_S$;
11 $\quad$ $pcst \leftarrow pcst + d$;
12 $\quad$ **if** $pcst < 0$ **then**
13 $\quad\quad$ **return** $pcst$
14 **return** $pcst$

Algorithm 4. $pcost(c_S, x : a \to b)$: cost pair

---

1 $pcst \leftarrow (1, 0)$;
2 **foreach** $I \in l(S \setminus \{x\})$ **do**
3 $\quad$ **if** $(c_S(I_{x=b}), c_S(I_{x=a})) < pcst$ **then**
4 $\quad\quad$ $pcst \leftarrow (c_S(I_{x=b}), c_S(I_{x=a}))$;
5 **return** $pcst$

---

we avoid unnecessary work. It is updated at lines 9-10. Note that we can initialize the array *residues* with any arbitrary constraints (not shown in the algorithm), and that Algorithm 4 necessarily returns a value less than or equal to 0 (this explains the initialization of $pcst$ to $(1, 0)$ at line 1).

We now discuss the complexity of $PSNS^r$ while assuming (for simplicity) that the WCN is binary. The space complexity is $O(nd^2)$ due to the use of the structure *residues*. The time complexity of Algorithm 4 is $O(d)$, and the time complexity of Algorithm 3 is $O(1)$ in the best case (if it is stopped at line 3) and $O(qd)$ in the worst case, where $q = |\Gamma(x)|$. Discarding line 3, the time complexity of Algorithm 2 is $O(nd^2)$ in the best case and $O(nd^3e)$ in the worst-case. Of course, Algorithm 2 can be called several times at line 6 of Algorithm 1, so we obtain a worst-case time complexity in $O(n^2d^4e)$. However, we have observed in our experiments that the number of successive calls to $PSNS^r$ is quite limited in practice (as we were predicting). Besides, imagine now that we call AC*-PSNS after the assignment of a value to a variable x (i.e., during search). In the best case (from a complexity point of view), no removal is made by W-$AC*$, and so, we just consider the neighbors of x, due to line 3 of Algorithm 2, which gives a best-case time complexity in $O(qd^2)$. This last result leans us toward experimenting maintaining AC*-PSNS during search.

7   Experimental Results

To show the practical value of removing SNS-eliminable values, we have conducted an experimentation using the series of WCSP instances available at `http://carlit.toulouse.inra.fr/cgi-bin/awki.cgi/BenchmarkS` and a cluster of Xeon

Table 1. Mean results obtained on various series (a time-out of 1,200 seconds was set per instance)

Series		AC*	AC* +PSNS	FDAC	FDAC +PSNS	EDAC	EDAC +PSNS
celar	#sol (cpu)	5 (337)	4 (231)	6 (316)	6 (341)	6 (344)	**7** (461)
#inst=7	avg-sub	0	3	0	6	0	4
driver	#sol (cpu)	18 (103)	18 (52)	19 (39.3)	19 (19.7)	19 (68.5)	**19 (56.8)**
#inst=19	avg-sub	0	1	0	1	0	1
geom	#sol (cpu)	5 (37.4)	5 (12.4)	5 (18.8)	**5 (11.0)**	5 (21.0)	5 (12.0)
#inst=5	avg-sub	0	0	0	1	0	0
mprime	#solv (cpu)	4 (9.82)	8 (17.4)	4 (11.6)	**8 (12.0)**	5 (206)	8 (21.0)
#inst=8	avg-sub	0	1	0	1	0	1
myciel	#sol (cpu)	3 (122)	3 (77.0)	3 (72.9)	**3 (34.3)**	3 (80.4)	3 (37.8)
#inst=3	avg-sub	0	2	0	3	0	3
scens+graphs	#sol (cpu)	1 (20.4)	3 (113)	**8 (242)**	7 (186)	6 (266)	5 (19.8)
#inst=9	avg-sub	0	8	0	8	0	8
spot5	#sol (cpu)	0	0	3 (20.5)	3 (12.6)	3 (20.1)	**3 (11.4)**
#inst=3	avg-sub	0	0	0	0	0	0
warehouse	#sol (cpu)	12 (286)	12 (46.2)	18 (166)	24 (139)	28 (53.2)	**29 (79.8)**
#inst=34	avg-sub	0	4	0	7	0	27
	#solved	48	53	66	75	75	**79**

Table 2. Illustrative results obtained on some problem instances

Instances		AC*	AC* +PSNS	FDAC	FDAC +PSNS	EDAC	EDAC +PSNS
cap101	cpu	232	42.1	1.6	1.62	1.48	**1.12**
	#nodes	242K	242K	835	835	75	75
	avg-sub	0	1	0	2	0	15
cap111	cpu	>1,200	>1,200	633	162	3.03	**2.74**
	#nodes	–	–	72,924	72,924	439	199
	avg-sub	0	2	0	3	0	12
capmo1	cpu	>1,200	>1,200	>1,200	>1,200	>1,200	**984**
	#nodes	–	–	–	–	–	–
	avg-sub	0	15	0	23	0	64
driverlog02ac	cpu	253	49.5	10.6	**7.99**	19.3	13.5
	#nodes	4,729K	701K	19,454	8,412	19,444	8,402
	avg-sub	0	0	0	0	0	0
driverlogs06	cpu	>1,200	>1,200	187	**61.0**	218	80.1
	#nodes	–	–	2,049K	609K	2,049K	609K
	avg-sub	0	0	0	0	0	0
mprime04ac	cpu	>1,200	30.9	>1,200	**15.7**	>1,200	43.7
	#nodes	–	189K	–	22,373	–	20,381
	avg-sub	0	0	0	1	0	1
myciel5g-3	cpu	38.1	19.6	**3.87**	4.18	4.36	4.57
	#nodes	518K	168K	10,046	9,922	6,159	6,128
	avg-sub	0	2	0	3	0	2
celar7-sub1	cpu	925	820	147	145	135	**86.4**
	#nodes	9,078K	1,443K	732K	91,552	796K	70,896
	avg-sub	0	6	0	9	0	6
graph07	cpu	>1,200	261	**3.11**	3.58	3.86	4.3
	#nodes	6,156K	145K	1,112	647	1,796	1,514
	avg-sub	0	23	0	9	0	4
scen06-24	cpu	>1,200	>1,200	**1,061**	>1,200	>1,200	>1,200
	#nodes	–	–	–	375K	–	–
	avg-sub	0	2	0	6	0	7
spot5-29	cpu	>1,200	>1,200	30.1	**18.0**	47.2	22.5
	#nodes	–	–	343K	174K	352K	185K
	avg-sub	0	0	0	0	0	0

3.0GHz with 1 GB of RAM under Linux. Our goal is to observe the relative efficiency of solving WCSP instances when *maintaining* AC*, AC*+PSNS, FDAC, FDAC+PSNS, EDAC, and EDAC+PSNS. For variable ordering during search, we use the simple static heuristic *max degree* which is independent of the pruning efficiency of the different algorithms, as in [8] where some experiments have been performed with a partial form of SNS enforced during a preprocessing stage.

Table 1 shows the average results obtained on various series. For each series, the number of considered instances (#inst) is given below the name of the series. We have discarded the instances that have not been solved by at least one of the algorithms, within 1,200 seconds. Here, a solved instance means that an optimal solution has been found and proved. In Table 1, the number (#sol) of instances solved within 1,200 seconds is given as well as the average CPU time over these solved instances. The average number (avg-sub) of SNS-eliminable values deleted during search (at each step) is also given (rounded to the nearest integer). On RLFAP instances (celar, scens, graphs), the benefit of using PSNS is rather erratic, but on planning (driver, mprime), coloring (myciel, geom), spot and warehouse instances, we can see a clear advantage of embedding PSNS. Overall, maintaining PSNS is cost-effective as it usually offers a benefit both in terms of solved instances (see last line of the table) and CPU time.

Table 2 presents the results obtained on some representative instances. It is interesting to note that on the warehouse instances (here, *cap101*, *cap111* and *capmo1*), enforcing PSNS does not entail a reduction of the size of the search tree (see the values of #nodes). However, PSNS permits to significantly reduce the size of the domains, making propagation of soft constraints quicker. On *celar7-sub1*, note that (maintaining) EDAC+PSNS is only about 50% faster than EDAC while the number of nodes has been divided by 10. This means that on such instances, PSNS is rather expensive, which maybe lets room for further improvements.

8   Conclusion

In this paper, we have investigated the property of soft neighborhood substitutability for weighted constraint networks, and have found a sufficient condition on substitutions that can be identified by an algorithm of reasonable complexity (restricted to the neighborhood and polynomial). We have proved that, even in simple cases, when $k \neq \infty$, the problem of deciding whether a value is soft neighborhood substitutable for another is coNP-hard. We have also given some properties linking substitutability and soft arc consistencies. Finally, we have proposed an algorithm that exploits early breaks and residues and shown experimentally that it can be helpful during search.

Acknowledgments. This work has been supported by both CNRS and OSEO within the ISI project 'Pajero'.

References

1. Bistarelli, S., Faltings, B., Neagu, N.: Interchangeability in Soft CSPs. In: O'Sullivan, B. (ed.) Constraint Solving and CLP. LNCS (LNAI), vol. 2627, pp. 31–46. Springer, Heidelberg (2003)
2. Bistarelli, S., Montanari, U., Rossi, F., Schiex, T., Verfaillie, G., Fargier, H.: Semiring-based CSPs and valued CSPs: Frameworks, properties, and comparison. Constraints 4(3), 199–240 (1999)
3. Choueiry, B.Y., Faltings, B., Weigel, R.: Abstraction by interchangeability in resource allocation. In: Proceedings of IJCAI 1995, pp. 1694–1710 (1995)

4. Cooper, M.C.: Reduction operations in fuzzy or valued constraint satisfaction. Fuzzy Sets and Systems 134(3), 311–342 (2003)
5. Cooper, M.C., de Givry, S., Sanchez, M., Schiex, T., Zytnicki, M.: Virtual arc consistency for weighted CSP. In: Proceedings of AAAI 2008, pp. 253–258 (2008)
6. Cooper, M.C., de Givry, S., Sanchez, M., Schiex, T., Zytnicki, M., Werner, T.: Soft arc consistency revisited. Artificial Intelligence 174(7-8), 449–478 (2010)
7. Cooper, M.C.: Fundamental properties of neighbourhood substitution in constraint satisfaction problems. Artificial Intelligence 90, 1–24 (1997)
8. de Givry, S.: Singleton consistency and dominance for weighted CSP. In: Proceedings of CP 2004 Workshop on Preferences and Soft Constraints (2004)
9. de Givry, S., Heras, F., Zytnicki, M., Larrosa, J.: Existential arc consistency: Getting closer to full arc consistency in weighted CSPs. In: Proceedings of IJCAI 2005, pp. 84–89 (2005)
10. Freuder, E.C.: Eliminating interchangeable values in constraint satisfaction problems. In: Proceedings of AAAI 1991, pp. 227–233 (1991)
11. Haselbock, A.: Exploiting interchangeabilities in constraint satisfaction problems. In: Proceedings of IJCAI 1993, pp. 282–287 (1993)
12. Karakashian, S., Woodward, R., Choueiry, B., Prestwhich, S., Freuder, E.: A partial taxonomy of substitutability and interchangeability. Technical Report arXiv:1010.4609, CoRR (2010)
13. Koster, A.M.: Frequency assignment: Models and Algorithms. PhD thesis, University of Maastricht, The Netherlands (1999)
14. Lal, A., Choueiry, B.Y., Freuder, E.C.: Neighborhood interchangeability and dynamic bundling for non-binary finite CSPs. In: Proceedings of AAAI 2005, pp. 397–404 (2005)
15. Larrosa, J.: Node and arc consistency in weighted CSP. In: Proceedings of AAAI 2002, pp. 48–53 (2002)
16. Larrosa, J., Schiex, T.: Solving weighted CSP by maintaining arc consistency. Artificial Intelligence 159(1-2), 1–26 (2004)
17. Martello, S., Toth, P.: Knapsack Problems: Algorithms and Computer Implementations. Wiley (1990)
18. Schiex, T., Fargier, H., Verfaillie, G.: Valued constraint satisfaction problems: Hard and easy problems. In: Proceedings of IJCAI 1995, pp. 631–639 (1995)

Increasing Symmetry Breaking by Preserving Target Symmetries

Jimmy H.M. Lee and Jingying Li

Department of Computer Science and Engineering,
The Chinese University of Hong Kong, Shatin, N.T.,Hong Kong
{jlee,jyli}@cse.cuhk.edu.hk

Abstract. Breaking the exponential number of all symmetries of a constraint satisfaction problem is often too costly. In practice, we often aim at breaking a subset of the symmetries efficiently, which we call target symmetries. In static symmetry breaking, the goal is to post a set of constraints to break these target symmetries in order to reduce the solution set and thus also the search space. Symmetries of a problem are all intertwined. A symmetry breaking constraint intended for a particular symmetry almost always breaks more than just the intended symmetry as a side-effect. Different constraints for breaking the same target symmetry can have different side-effects. Conventional wisdom suggests that we should select a symmetry breaking constraint that has more side-effects by breaking more symmetries. While this wisdom is valid in many ways, we should be careful where the side-effects take place. A symmetry σ of a CSP $\mathcal{P} = (\mathcal{V}, \mathcal{D}, \mathcal{C})$ is *preserved* by a set of symmetry breaking constraints C^{sb} iff σ is a symmetry of $\mathcal{P}' = (\mathcal{V}, \mathcal{D}, \mathcal{C} \cup C^{sb})$. We give theorems and examples to demonstrate that it is beneficial to post symmetry breaking constraints that preserve the target symmetries and restrict the side-effects to only non-target symmetries as much as possible. The benefits are in terms of the number of symmetries broken and the extent to which a symmetry is broken (or eliminated), resulting in a smaller solution set and search space. Extensive experiments are also conducted to confirm the feasibility and efficiency of our proposal empirically.

1   Introduction

Symmetries are common in Constraint Satisfaction Problems. Several methods [9,4] are proposed to avoid the exploration of search space segments with assignments that can be generated by representatives of symmetry classes. One common way is to add dedicated constraints statically at the modeling stage to eliminate symmetries [22], such as the LEXLEADER method [3]. A symmetry breaking constraint leaves only canonical solutions with their symmetrically-equivalent solutions eliminated. It prunes the search space in two ways: remove the symmetrically equivalent search branches, and trigger constraint propagation with other constraints and *vice versa*. They affect both the size of the solution set and the search tree of the problem.

There are some tractable classes of symmetries [7] and also methods to simplify the constraints [21]. However, in general more symmetry breaking constraints need to be posted in order to eliminate more symmetries. When the propagation overhead of

M. Milano (Ed.): CP 2012, LNCS 7514, pp. 422–438, 2012.

symmetry breaking constraints outweighs the time saved in exploring the search space, there is no longer better efficiency [20]. Eliminating all symmetries are too costly. Usually we only eliminate a subset of them, which we call *target symmetries*. Jefferson *et al.* [15] has given algorithms to generate a good set of target symmetries. While posting constraints for only the target symmetries, we show we can actually eliminate more symmetries and achieve a smaller solution set by carefully choosing the constraints.

When and how the choice of symmetry breaking constraints will affect the number of solutions are discussed systematically. We point out how the side-effects of symmetry breaking constraints in breaking other symmetries are common. To eliminate the same symmetry set, we have alternative choices which cause different side-effects, breaking or eliminating extra symmetries. We formally define the situation in which a symmetry is not removed by a symmetry breaking constraint: *preservation*. A symmetry σ of a CSP $\mathcal{P} = (\mathcal{V}, \mathcal{D}, \mathcal{C})$ is *preserved* by a set of symmetry breaking constraints C^{sb} iff σ is a symmetry of $\mathcal{P}' = (\mathcal{V}, \mathcal{D}, \mathcal{C} \cup C^{sb})$. Although, at first sight, constraints that break more symmetries as side-effects seem to be good choices, we should distinguish between side-effects on target symmetries and those on non-target symmetries.

We propose that symmetry breaking constraints aiming at some target symmetries should be selected to preserve other target symmetries and restrict the side-effects to non-target symmetries as much as possible. We analyze through the solution reduction ratio to show why preserving target symmetries can actually help us to eliminate more symmetries and thus are better choices. By carefully choosing symmetry breaking constraints to achieve specific side-effects, we achieve smaller solution set size and better efficiency. We also give observations on other factors that we should pay attention to when choosing symmetry breaking constraints.

A running example is given throughout the paper to demonstrate our ideas and results. Experimental results on four problems in the literature confirm empirically that models constructed using symmetry preservation achieve better efficiency up to one order of magnitude both in terms of runtime and search space.

2 Background

A *constraint satisfaction problem* (CSP) is a triple $\mathcal{P} = (\mathcal{V}, \mathcal{D}, \mathcal{C})$, consisting of a set of variables $\mathcal{V}$, each $v \in \mathcal{V}$ with a finite domain of possible values $\mathcal{D}(v)$ and a set of constraints $\mathcal{C}$, each defined over a subset of variables specifying the allowed combination of values. An *assignment* gives each variable a value from its domain. A *solution* α is an assignment that satisfies all constraints. We use $sol(\mathcal{P})$ to denote the set of all solutions of $\mathcal{P}$.

A *symmetry* for a CSP is a bijection on the set of all assignments that maps solutions to solutions, and thus also non-solutions to non-solutions. Two common types of symmetry are variable symmetry and value symmetry. A *variable symmetry* is a bijective mapping σ on the indices of variables. If $[X_1, ..., X_n] = [d_1, ..., d_n]$ is a solution then $[X_{\sigma(1)}, ..., X_{\sigma(n)}] = [d_1, ..., d_n]$ is also a solution. A *value symmetry* is a bijective mapping θ on the values. If $[X_1, ..., X_n] = [d_1, ..., d_n]$ is a solution then $[X_1, ..., X_n] = [\theta(d_1), ..., \theta(d_n)]$ is also a solution. There are also constraint symmetries [2] that act on both variables and values. Our results are general and work with every kind of symmetries.

The *symmetry group* G_Σ of a set of symmetries Σ is formed by closing Σ under composition. A *symmetry class* of G_Σ is a subset S of symmetrically-equivalent assignments of $\mathcal{P}$. If an assignment $\alpha \in S$, then $\forall \sigma \in G_\Sigma, \sigma(\alpha) \in S$. A symmetry class contains either all solutions or no solutions. Symmetry classes of G_Σ partition $sol(\mathcal{P})$, with solutions in the same symmetry class mapped to one another under symmetries in G_Σ. Without loss of generality, by symmetry classes, we refer to the ones in $sol(\mathcal{P})$. A simple running example with variable and value symmetries will be used to illustrate various concepts throughout the paper.

Example 1. The Diagonal Latin Square problem (DLS(n)) aims to assign numbers 1 to n to the cells of an $n \times n$ board with no numbers occurring more than once in each row, column and the 2 diagonals. For convenience, we call the one extending from left top diagonal 1, and that from left bottom diagonal 2. We use a matrix model $[X_{ij}]$ of n^2 variables, each representing a cell and with domain $\{1, ..., n\}$. Problem constraints consist of ALLDIFF [24] on each row, column and the 2 diagonals.

The variable symmetries of DLS include the geometric symmetry group G_{geo} of size 8: horizontal reflection $\sigma_{rx}(X_{ij}) = X_{i(n+1-j)}$, vertical reflection σ_{ry}, diagonal reflections $\sigma_{d1}(X_{ij}) = X_{ji}$ and σ_{d2}, rotational symmetries σ_{r90}, σ_{r180} and σ_{r270}, and the identity symmetry σ_{id}. The values in DLS are interchangeable, which means the permutation of values G_{val} preserves solution. The following 4 solutions of DLS(5) are in the same geometric symmetry class. Solution $\sigma_{rx}(\alpha)$ can be obtained by flipping α over the vertical axis.

α	$\sigma_{rx}(\alpha)$	$\sigma_{d1}(\alpha)$	$\sigma_{rx} \circ \sigma_{d1}(\alpha)$
1 2 3 4 5	5 4 3 2 1	1 2 5 3 4	4 3 5 2 1
2 4 5 3 1	1 3 5 4 2	2 4 3 1 5	5 1 3 4 2
5 3 2 1 4	4 1 2 3 5	3 5 2 4 1	1 4 2 5 3
3 1 4 5 2	2 5 4 1 3	4 3 1 5 2	2 5 1 3 4
4 5 1 2 3	3 2 1 5 4	5 1 4 2 3	3 2 4 1 5

Define $row([X_{ij}]) \equiv [X_{11}, ..., X_{1n}, X_{21}, ..., X_{2n}, ..., X_{n1}, ..., X_{nn}]$. To eliminate σ_{rx}, we can post the LEXLEADER constraint: $row([X_{ij}]) \leq_{lex} [X_{1n}, ..., X_{11}, X_{2n}, ..., X_{21}, ..., X_{nn}, ..., X_{n1}]$. From problem constraints we infer $X_{11} \neq X_{1n}$ and simplify the LEXLEADER constraint to:

$$X_{11} < X_{1n} \tag{1}$$

To eliminate σ_{d1}, we can post constraint $row([X_{ij}]) \leq_{lex} [X_{11}, ..., X_{n1}, X_{12}, ..., X_{n2}, ..., X_{1n}, ..., X_{nn}]$, which can be simplified since $X_{11} = X_{11}$ and $X_{1n} \neq X_{n1}$:

$$[X_{12}, ..., X_{1n}] \leq_{lex} [X_{21}, ..., X_{n1}] \tag{2}$$

3   Effects of Symmetry Breaking Constraints

This section reports our observations and views based on existing results from the literature. We first introduce some definitions related to symmetry breaking constraints. We are concerned with the actual set of symmetries out of the whole symmetry group on which we post symmetry breaking constraints. We systematically discuss the effects of symmetry breaking constraints on the final solution set size as split into two cases.

3.1   Properties of Symmetry Breaking Constraints

We first introduce some useful concepts for the rest of the paper. Suppose a CSP $\mathcal{P} = (\mathcal{V}, \mathcal{D}, \mathcal{C})$ contains symmetries Σ and C^{sb} is a set of symmetry breaking constraints such that $\mathcal{P}' = (\mathcal{V}, \mathcal{D}, \mathcal{C} \cup C^{sb})$. Here and throughout, $sol(\mathcal{P}, C^{sb})$ is a short hand for $sol(\mathcal{P}')$, which is $sol((\mathcal{V}, \mathcal{D}, \mathcal{C} \cup C^{sb}))$.

We adapt the following definitions from Katsirelos and Walsh [18]. A set of symmetry breaking constraints C^{sb} *breaks* a symmetry $\sigma \in \Sigma$ iff there exist a solution $\alpha \in sol(\mathcal{P}')$ such that $\sigma(\alpha) \notin sol(\mathcal{P}')$. A set of symmetry breaking constraints C^{sb} *eliminates* a symmetry $\sigma \in \Sigma$ iff for each solution $\alpha \in sol(\mathcal{P}')$, $\sigma(\alpha) \notin sol(\mathcal{P}')$. We define a stronger version of elimination. C^{sb} *fully eliminates* a symmetry $\sigma \in \Sigma$ iff for each solution $\alpha \in sol(\mathcal{P}')$, $\forall k \in \mathbb{Z}$, if $\sigma^k(\alpha) \neq \alpha$ then $\sigma^k(\alpha) \notin sol(\mathcal{P}')$. This definition is to ensure there is at most one solution left by C^{sb} in each symmetry class of $G_{\{\sigma\}}$. C^{sb} *breaks/(fully) eliminates* a symmetry set Σ iff C^{sb} breaks/(fully) eliminates each $\sigma \in \Sigma$. A set of symmetry breaking constraints is *sound/complete* to a symmetry set Σ iff it leaves at least/exactly one solution in each symmetry class of G_Σ.

In the following, we use $\mathcal{C}_\Sigma$ to denote a set of symmetry breaking constraints that eliminates Σ, and let $\mathcal{P}' = (\mathcal{V}, \mathcal{D}, \mathcal{C} \cup \mathcal{C}_\Sigma)$. We say that $\mathcal{C}_\Sigma$ is sound/complete to mean that $\mathcal{C}_\Sigma$ is sound/complete to Σ. Symmetry breaking constraints can be derived by pre-defining a *canonical variable ordering* [26] and forcing the canonical solution to be always smaller (bigger) than its symmetrically-equivalent counterpart. The canonical variable ordering is the row-wise ordering $row([X_{ij}])$ in the running example. Any permutation on $\text{row}([X_{ij}])$ can serve as a possible ordering. We consider other specific orderings by moving X_{1n} and X_{n1} forward in $row([X_{ij}])$ and their simplified symmetry breaking constraints to eliminate σ_{d1} are shown respectively as follows:

$$canonical\ ordering : [X_{11}, X_{1n}, X_{n1}, X_{12}, ..., X_{nn}] \to \mathcal{C}_{\sigma_{d1}} : X_{1n} < X_{n1} \quad (3)$$

$$canonical\ ordering : [X_{11}, X_{n1}, X_{1n}, X_{12}, ..., X_{nn}] \to \mathcal{C}_{\sigma_{d1}} : X_{n1} < X_{1n} \quad (4)$$

These two canonical orderings result in the same constraint to eliminate σ_{rx} as $row([X_{ij}])$. We are interested in whether the solution set size is affected by picking other canonical ordering or using other methods to eliminate symmetries Σ. There are two possibilities: a symmetry group is eliminated *entirely* or *partially*.

3.2   Eliminating a Symmetry Group Entirely

We can easily see that $\mathcal{C}_\Sigma$ is complete to Σ if $\mathcal{C}_\Sigma$ is sound to Σ and Σ is a symmetry group, i.e. $\Sigma = G_\Sigma$.

Theorem 1. *Given a CSP, any sound set of symmetry breaking constraints eliminating the same symmetry group results in exactly the same solution set size.*

Proof. A sound set of symmetry breaking constraints eliminating a symmetry group is complete. Symmetry classes formed in the solution space under a symmetry group are fixed. By picking exactly one solution from each symmetry class of G, we gain a solution set with the same size as the number of symmetry classes, no matter what constraints we use to eliminate the symmetries. $\qquad\square$

Many CSPs have exponentially-sized symmetry group. There exist efficient methods [6,7] that eliminate all symmetries in some tractable classes of symmetries such as piecewise variable and value symmetry. Also in special cases such as all different problems [21], a linear number of constraints can eliminate all symmetries. However, eliminating the whole symmetry group is intractable and too costly in general. We aim to eliminate a subset [20]. We call this subset of symmetries we intend to eliminate as *target symmetries*. In practice, target symmetries are usually a generator set [1] for which there exist efficient methods to eliminate and can be generated automatically [15].

3.3 Eliminating a Symmetry Group Partially

We demonstrate how the choice of symmetry breaking constraints affects the solution set size when only the target symmetries Σ are eliminated. Representatives selected from each symmetry class of $G_{\{\sigma_i\}}$ ($\sigma_i \in \Sigma$) intersects to form the remaining solutions in the symmetry classes of G_Σ. The final solution set is determined by the intersection of canonical solutions of each symmetry breaking constraints: $sol(\mathcal{P}, \cup_{\sigma_i \in \Sigma} \mathcal{C}_{\sigma_i}) = \cap_{\sigma_i \in \Sigma} sol(\mathcal{P}, \mathcal{C}_{\sigma_i})$. Even if each $\mathcal{C}_{\sigma_i}$ is complete, which means the size of each $sol(\mathcal{P}, \mathcal{C}_{\sigma_i})$ is fixed, picking different canonical solutions of each constraint makes the intersection significantly different.

Example 2. We give a simple example considering again σ_{rx} and σ_{d1} in the DLS problem. We compare the result of combining constraint (1) with either constraint (3) or constraint (4). Picking constraint (3) we obtain $\{X_{11} < X_{1n}, X_{1n} < X_{n1}\}$, a total ordering on the variable sequence $[X_{11}, X_{1n}, X_{n1}]$; picking constraint (4), we obtain $\{X_{11} < X_{1n}, X_{n1} < X_{1n}\}$, making X_{1n} the biggest value out of the three. Because X_{11} and X_{n1} are not ordered in $\{X_{11} < X_{1n}, X_{n1} < X_{1n}\}$, the size of $sol(\mathcal{P}, \{X_{11} < X_{1n}, X_{n1} < X_{1n}\})$ is twice as that of $sol(\mathcal{P}, \{X_{11} < X_{1n}, X_{1n} < X_{n1}\})$.

We want to formulate a set of symmetry breaking constraints to get a minimum intersection, but we should avoid selecting an unsound set that misses solutions. For example in the DLS problem, to eliminate σ_{rx}, we can post $X_{11} < X_{1n}$; to eliminate σ_{r90} (90 degree rotational symmetry), we can post $X_{11} > \max\{X_{1n}, X_{n1}, X_{nn}\}$. Combining the two constraints results in the empty solution set.

The number of symmetry classes of G_Σ is the minimum size for the intersection, since completeness is the best we can achieve for $\mathcal{C}_\Sigma = \cup_{\sigma_i \in \Sigma} \mathcal{C}_{\sigma_i}$. In other words, we can eliminate the symmetry group G_Σ by eliminating each σ_i. This is always possible if we can use table constraints to specify exactly which representatives to retain as solutions. In practice, however, table constraints are difficult to craft and problem-specific, and we are limited by the available symmetry breaking methods in existing constraint programming system. Thus, eliminating each σ_i is not always sufficient to eliminate all symmetries when $\Sigma \neq G_\Sigma$. We need to find practical ways to prove soundness and achieve as small a solution set as possible. Our proposal is based on the side-effects of symmetry breaking constraints.

4   Side-Effects of Symmetry Breaking Constraints

We assume that every symmetry breaking constraint aims at breaking a target symmetry. The *side-effect* of a symmetry breaking constraint is its effect in breaking symmetries other than the target symmetry. We show that side-effects are common in symmetry breaking constraints and study how strong and widespread side-effects could be. Different choices of symmetry breaking constraints aiming at the same target symmetry can have different side-effects. For symmetries that are not affected, we say they are *preserved*. Then we show that selecting constraints that restrict the side-effects to non-target symmetries and preserve other target symmetries can gain us smaller solution set. We support our claims with theoretical analysis, examples and experimental results.

4.1 Side-Effects Are Common

Our general study on side-effects are inspired by several examples in the literature. Katsirelos *et al.* [18,17] shows in examples a constraint eliminating symmetry σ_{rx} also breaks the symmetry σ_{r90}, and the DOUBLELEX constraint eliminates also the value interchangeability in the EFPA problem in addition to the row and column symmetries. We are going to show that these side-effects of symmetry breaking constraints are common and in many cases inevitable. Given a CSP $\mathcal{P}$ with symmetry G_Σ, each pair of solutions in a symmetry class are mapped to each other under some symmetry $\sigma \in G_\Sigma$.

Theorem 2. *Given a CSP $\mathcal{P}$ with symmetries G_Σ. Suppose C_{σ_i} eliminates symmetry $\sigma_i \in G_\Sigma$ and it reduces the number of solutions in one symmetry class of G_Σ to at least two. Then $\exists \sigma_j \in G_\Sigma$ such that $\sigma_i \neq \sigma_j$ and σ_j is broken by C_{σ_i}.*

Proof. Suppose $\alpha_1, \alpha_2 \in sol(\mathcal{P}, C_{\sigma_i})$ and $\sigma_i(\alpha_1) \neq \alpha_1$. Then $\sigma_i(\alpha_1)$ is not in $sol(\mathcal{P}, C_{\sigma_i})$ and there must exist a symmetry σ_j linking α_2 and $\sigma_i(\alpha_1)$. Since $\alpha_2 \in sol(\mathcal{P}, C_{\sigma_i})$ but $\sigma_j(\alpha_2)(= \sigma_i(\alpha_1))$ is not in $sol(\mathcal{P}, C_{\sigma_i})$, σ_j is broken by C_{σ_i} by definition. $\qquad\square$

As a symmetry breaking constraint can break other symmetries in addition to its target one, we analyze the possible number of symmetries broken or eliminated as side-effects with the help of Figure 1. Suppose the symmetry group G_Σ is of size m and a symmetry class S is of size n, $n \leq m$. We assume $n = m$ for ease of discussion. A symmetry class is represented as a directed graph and the following notions are used.

- circle node: solution.
- solid arrow directed edge with specific symmetry label: symmetry mapping. If an edge from node 1 to node 2 is labeled with σ_j and suppose node 1 is solution α, then node 2 is $\sigma_j(\alpha)$.
- dash line cut: symmetry breaking constraint. The effect of a symmetry breaking constraint C_{σ_i} on the symmetry class is a *cut* on the graph that partitions the nodes (solutions) into two parts, S_1 with n_1 nodes satisfying the constraint and S_2 with n_2 nodes violating it , as shown in Figure 1(a).
- dot-dash edge: edge labeled with target symmetry σ_i and removed by the cut C_{σ_i}.

Each pair of nodes have edges in between them in both directions. Each node has totally $n - 1$ out-going edges and $n - 1$ in-coming edges, each labeled with a distinct symmetry $\sigma \in G_\Sigma$, as shown in Figure 1(b). The cut by C_{σ_i} removes $2 \times n_1 \times n_2$ edges, among which $2 \times n_1$ edges are labeled with σ_i. $\frac{n_2-1}{n_2}$ of the removed edges are labeled with other symmetries. We quantify the number of symmetries broken or eliminated by analyzing the result of the cut in the graph.

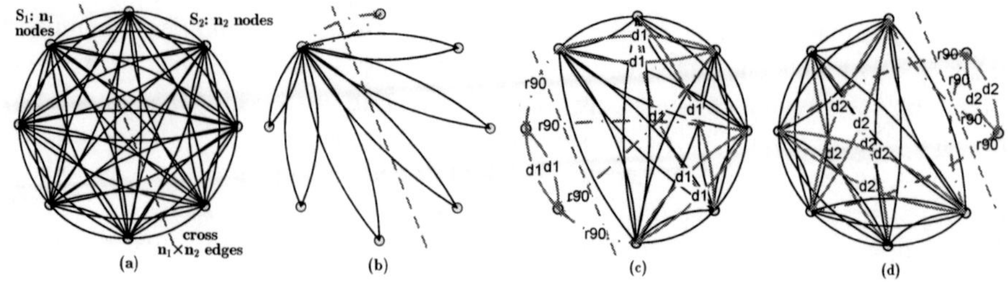

Fig. 1. A Symmetry Class of G_{geo}

- Symmetry σ_j is broken iff there exists an edge labeled with σ_j removed by the cut.
- Symmetry σ_j is eliminated iff no edges in S_1 are labeled with σ_j.

Broken Symmetries.

 # Broken Symmetries. Whenever an edge labeled with σ_j is cut, symmetry σ_j is broken by C_{σ_i}. If an edge labeled with σ_j is cut, there must exist an edge labeled with σ_j from S_1 to S_2 is cut. This is due to the closure of symmetries, which means there must exist a finite path from a node to itself consisting of only edges labeled with σ_j. All solutions in S_1 are in $sol(\mathcal{P}, C_{\sigma_i})$ while no solution in S_2 belong to $sol(\mathcal{P}, C_{\sigma_i})$. Therefore the symmetry σ_j linking a solution $\alpha \in S_1$ to $\sigma_j(\alpha) \in S_2$ is broken. For each node in S_1/S_2, there must exist n_2/n_1 edges labeled with distinct symmetries linking it to nodes in S_2/S_1. At least $\max\{n_1, n_2\}$ symmetries including the target symmetry are broken by the symmetry breaking constraint. *At least half the number of symmetries out of all are broken by a complete symmetry breaking constraint that aims at a single symmetry.*

 # Eliminated Symmetries. According to the definition of elimination, symmetries that do not label any edges in S_1 are eliminated. For each solution in S_1, there must exist $n_1 - 1$ symmetries linking it to the others S_1. *At least $n_1 - 1$ symmetries are not yet eliminated.* Take Figures 1(c) and (d) as examples, one to two symmetries are not yet eliminated. A smaller n_1 means smaller number of edges $n_1 \times (n_1 - 1)$ are left in S_1. A small size S_1 usually indicates that less symmetries are left un-eliminated.

If we choose different symmetry breaking constraints to eliminate the same target symmetry, the cut is different and respectively the side-effects are different. The side-effects can be different in the number of symmetries broken/eliminated or the specific symmetries broken/eliminated. We demonstrate it using a simple example. To eliminate symmetry σ_{r90} in the DLS example, we can use constraint $X_{11} < \min\{X_{1n}, X_{n1}, X_{nn}\}$ or $X_{1n} < \min\{X_{11}, X_{n1}, X_{nn}\}$. The effect of each constraint on the symmetry class is shown in Figures 1(c) and (d) respectively. Constraint $X_{11} < \min\{X_{1n}, X_{n1}, X_{nn}\}$ eliminates all but σ_{d1} while constraint $X_{1n} < \min\{X_{11}, X_{n1}, X_{nn}\}$ eliminates all but σ_{d2}. We say the constraint *preserves* symmetry σ_{d1}/σ_{d2}.

4.2   Symmetry Preservation

We formally define symmetry preservation in the following:

Definition 1. *Given a CSP $\mathcal{P} = (\mathcal{V}, \mathcal{D}, \mathcal{C})$ with a symmetry σ. The symmetry σ is preserved by a set of symmetry breaking constraints C^{sb} iff σ is a symmetry of $\mathcal{P}' = (\mathcal{V}, \mathcal{D}, \mathcal{C} \cup C^{sb})$.*

In other words, elements of each symmetry class of σ are either entirely removed from the solution set of $\mathcal{P}$ or retained as solutions by C^{sb}. Even if C^{sb} removes all solutions (i.e. making them non-solutions) of σ's symmetry class in $\mathcal{P}$, σ is not removed in the new problem $\mathcal{P}'$ since now σ still maps all such non-solutions to non-solutions, and is still a symmetry. A symmetry σ is preserved by C^{sb} iff the cut by C^{sb} does not remove any edges labeled with σ. For example in Figure 1(c), the edges labeled with symmetry σ_{d1} are not removed. A set of symmetries Σ is preserved iff each $\sigma \in \Sigma$ is preserved.

There are two nice properties of symmetry preservation. First, eliminating some symmetries can eliminate their composition under preservation.

Theorem 3. *Given a CSP $\mathcal{P}$ with symmetry σ_i and σ_j. If symmetry breaking constraint C_{σ_i} preserves symmetry σ_j, then $\sigma_i \circ \sigma_j$ and $\sigma_j \circ \sigma_i$ are eliminated by $C_{\sigma_i} \cup C_{\sigma_j}$.*

Proof. As shown in Figure 2, we have three disjoint sets $A = sol(\mathcal{P}, C_{\sigma_i} \cup C_{\sigma_j}), B = sol(\mathcal{P}, C_{\sigma_i}) - sol(\mathcal{P}, C_{\sigma_i} \cup C_{\sigma_j}), E = sol(\mathcal{P}) - sol(\mathcal{P}, C_{\sigma_i})$. Suppose $\alpha \in A$. By definition, symmetry σ_i links solutions (a) from A to E, (b) from B to E and (c) within E. Symmetry σ_j links solutions (a) within B, (b) within E and (c) from A to B. Therefore $\sigma_j(\alpha) \in B$ and $\sigma_i(\alpha) \in E$, $\sigma_i \circ \sigma_j(\alpha) \in E$ and $\sigma_j \circ \sigma_i(\alpha) \in E$. Since A and E are disjoint, both $\sigma_i \circ \sigma_j(\alpha)$ and $\sigma_j \circ \sigma_i(\alpha)$ are not in $sol(\mathcal{P}, C_{\sigma_i} \cup C_{\sigma_j})$. $\qquad\square$

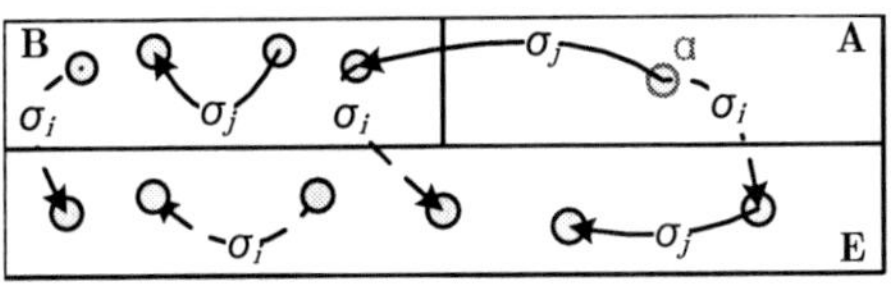

Fig. 2. σ_i preserve σ_j

Similarly, if a set of symmetry breaking constraints C_{G_1} preserves symmetry group G_2, then $\forall \sigma_i \in G_1, \sigma_j \in G_2$, $\sigma_i \circ \sigma_j$ and $\sigma_j \circ \sigma_i$ are eliminated by $C_{G_1} \cup C_{G_2}$. Second, when a symmetry set is preserved, if two symmetry breaking constraints are sound with respect to their target symmetries, their combination is also sound.

Theorem 4. *Given a set of symmetries $\Sigma = \Sigma_1 \cup \Sigma_2$. If C_{Σ_1} is sound and preserves Σ_2, then C_{Σ_2} being sound implies $C_{\Sigma_1} \cup C_{\Sigma_2}$ being sound.*

Proof. Based on the definition of preservation, if C_{Σ_1} preserves Σ_2 then Σ_2 is still a symmetry set of the new problem $\mathcal{P}' = (\mathcal{V}, \mathcal{D}, \mathcal{C} \cup C_{\Sigma_1})$. No matter what C_{Σ_2} is, as long as it is sound to $\mathcal{P}'$, it is able to regenerate all solutions of $sol(\mathcal{P}, C_{\Sigma_1})$ via symmetries in Σ_2. Then $sol(\mathcal{P})$ can be completely regenerated from $sol(\mathcal{P}, C_{\Sigma_1})$ via symmetries in Σ_1 since C_{Σ_1} is sound. $\qquad\square$

4.3 Solution Reduction by Symmetry Breaking Constraints

From the previous discussion, we can separate the effect of symmetry breaking constraints on other symmetries into three cases: *eliminate, break* and *preserve*. Our goal is to eliminate as many symmetries as possible and thus achieve smaller solution set by eliminating the target symmetries efficiently. What kind of effect is better?

The side-effects are common but we can choose to let the side-effects act on different symmetries. In the following, we show that symmetry breaking constraints that preserve other target symmetries and have more side-effects on non-target symmetries are better in the sense of symmetries eliminated and solution set size. We show how different side-effects may affect the final solution set size through analyzing the *solution reduction* by the symmetry breaking constraints.

The maximum solution reduction ratio $\frac{|sol(\mathcal{P})|}{|sol(\mathcal{P},\mathcal{C}_{\sigma_1})|}$ by a sound symmetry breaking constraint is achieved when it is complete to its target symmetry. Assume constraints are posted one by one and soundness must be ensured at each iteration. If the symmetry classes of σ_2 are left partially by $\mathcal{C}_{\sigma_1}$, the solution reduction ratio by $\mathcal{C}_{\sigma_2}$ in the new problem is smaller than that in the origin problem: $\frac{|sol(\mathcal{P},\mathcal{C}_{\sigma_1})|}{|sol(\mathcal{P},\mathcal{C}_{\sigma_1}\cup\mathcal{C}_{\sigma_2})|} < \frac{|sol(\mathcal{P})|}{|sol(\mathcal{P},\mathcal{C}_{\sigma_2})|}$. The origin solution set size of DLS(5) is 960. Posting constraint (1) $X_{11} < X_{1n}$ to eliminate σ_{ry} results in 480 solutions with reduction ratio 2. Posting it after symmetry σ_{rx} is eliminated by $X_{11} < X_{n1}$ results in 320 solutions out of 480 with reduction ratio 1.5.

This analysis also gives us insight into one reason why the efficiency we gained from eliminating more symmetries becomes weaker. When more constraints are added and their side-effects become stronger, not only the propagation overhead of constraints increases but also the available space that can be pruned becomes smaller. We consider this as an important factor behind partial symmetry breaking [20]. However, we can still achieve maximum solution reduction ratio when $\mathcal{C}_{\sigma_1}$ preserves σ_2. Without explicitly handling the composition symmetries, we are able to eliminate them as side-effects of eliminating the target symmetries when preserving other target symmetries.

A significant advantage of this approach is that, instead of introducing new constraints, we are able to eliminate more symmetries by selecting carefully the symmetry breaking constraints, which potentially entails better runtime.

4.4 Preservation Examples

Based on the previous theoretical analysis, we give three examples of achieving smaller solution set size by preserving symmetries. Two are from a big family of problems that can be modeled into matrix and contain a lot of symmetries.

Geometric Symmetries and Value Interchangeability. We consider again the DLS problem and show different side-effects on the geometric symmetry group when eliminating value symmetries. We compare 6 choices of distinct value symmetry breaking constraints: fixing the value of (a) the first row, (b) the first column, (c) middle row, (d) middle column, (e) diagonal 1 and (f) diagonal 2, as shown from left to right in Figure 3. All of the constraints result in the same solution set size according to Theorem 1. The side-effects of each choice is: (a) eliminate G_{geo}, (b) eliminate G_{geo}, (c) eliminate

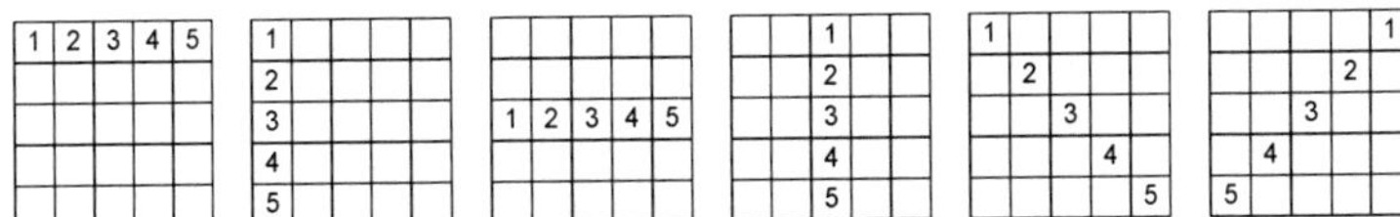

Fig. 3. Different Side-effect of $\mathcal{C}_{G_{val}}$

$G_{geo} \setminus \{\sigma_{ry}\}$, (d) eliminate $G_{geo} \setminus \{\sigma_{rx}\}$, (e) eliminate $G_{geo} \setminus \{\sigma_{d1}\}$ and (f) eliminate $G_{geo} \setminus \{\sigma_{d2}\}$. After posting consistent constraints to eliminate G_{geo} for each, the solution set sizes of (a) and (b) are twice of those of the rest. The reason is obvious: the preserved symmetry $\sigma_{ry}/\sigma_{rx}/\sigma_{d1}/\sigma_{d2}$ is further eliminated in (c)/(d)/(e)/(f) respectively and another half solution reduction is gained.

Matrix Symmetries and Value Interchangeability. A common symmetry group in CSP is the matrix symmetries [5]. Since the size of the matrix symmetry group is super-exponential, usually only row and column symmetries are considered as target symmetries. Methods like DOUBLELEX [5] and SNAKELEX [11], built upon LEXLEADER [3], are efficient in eliminating row and column symmetries, but they do not eliminate the matrix symmetries in general. In order to preserve target symmetries, we consider the multiset ordering constraint $\leq_m$ [8].

Enforcing lexicographical ordering in one dimension breaks the symmetries in the other dimension, but enforcing multiset ordering in one dimension preserves the permutation symmetries in the other dimension. However, multiset ordering may not force a unique ordering, and thus eliminates fewer symmetries and is weaker than lexicographical ordering. However, combining it with lex ordering may eliminate more symmetries than combining both lex orderings. Based on Theorem 3, *if multiset ordering determine a unique ordering in one dimension, combining it with lexicographical ordering in the other dimension eliminates the matrix symmetries*. We cannot guarantee multiset ordering achieve unique ordering in many problems, but we show in the following that the solution set left by the combination can be much smaller than the one left by both $\leq_{lex}$.

We take the Cover Array Problem $CA(t, k, g, b)$ from CSPLib prob045 as example and use the integrated model [14], which channels an original model and a compound model. The original model contains a $b \times k$ matrix X of integer variables with domain $\{1..g\}$. The compound model contains a $b \times \binom{k}{t}$ matrix of integer variable with domain $\{1..g^k\}$. The original model contains matrix symmetries and value interchangeability in each column. In the following, we consider tow sets of target symmetries and compare our approach with those in the literature respectively.

- Considering row and column symmetries as target symmetries, we compare with the popular approach DOUBLELEX, denoted as **dLex**. Now we explain our choice of constraints. The number of columns k is smaller than the number of rows b in satisfiable Cover Array instances $CA(t, k, g, b)$. Since multiset ordering is relatively weaker, we choose to post it between columns. We simulate the multiset ordering constraint [8] using the Global Cardinality Constraint [25]: $\forall i, 0 \leq i \leq k, gcc([X_{1i}, \ldots, X_{bi}], [1, \ldots, g], [O_{1i}, \ldots, O_{gi}])$. Multiset ordering between columns of the original model $\forall 1 \leq i < j \leq k, [X_{1i}, \ldots, X_{bi}] \leq_m [X_{1j}, \ldots, X_{bj}]$

is achieved by enforcing lex ordering on the cardinality variable sequences $\forall 1 \leq i < j \leq k, [O_{1i}, \ldots, O_{gi}] \leq_{lex} [O_{1j}, \ldots, O_{gj}]$. We denote it as **mLex**.

- Considering row, column and value symmetries as target symmetries, we compare with the combination of DOUBLELEX and PRECEDENCE [19] constraints on each column, denoted as **dLex-V**. PRECEDENCE [19] is a global constraint to eliminate value interchangeability. The value interchangeability of each column are corresponding to variable interchangeability of the cardinality variable sequence $[O_{1i}, \ldots, O_{gi}]$ of each column i. We can simply enforce an ordering on the cardinality variable sequence $O_{1i} \leq \cdots \leq O_{gi}$ for each column i to break the value symmetries. The combination of these value symmetry breaking constraints and mLex is denoted as **mLex-V**.

The value symmetry breaking constraints $\forall i, 1 \leq i \leq k, O_{1i} \leq \cdots \leq O_{gi}$ in mLex-V preserve both row and column permutations. For any solution obeying the constraints, permutating the rows does not change the number of occurrences of values $[O_{1i}, \ldots, O_{gi}]$ in each column i. Moreover, each column has the same ordering constraints on the cardinality variables and permutating the columns still obeys the constraints. When multiset ordering and value symmetry breaking constraint are complete respectively, the linear-size set of constraints mLex-V eliminates all symmetries of size $b! \times k! \times (g!)^k$.

Experiments are conducted on a Sun Blade 1000 (900MHz) running ILOG Solver 6.0. Variables are labeled in row-wise ordering. We conduct experiments on four instances of $CA(t, k, g, b)$ with different problem sizes. They vary in (1) the number of columns k in the original model, (2) the number of columns $\binom{k}{t}$ in the compound model and (3)

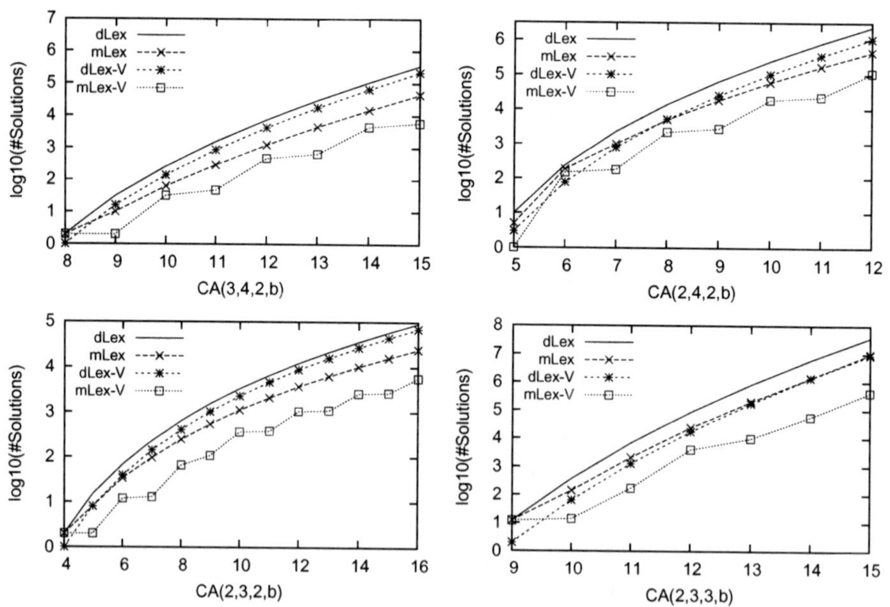

Fig. 4. Comparison of #Solutions in Cover Array Problem

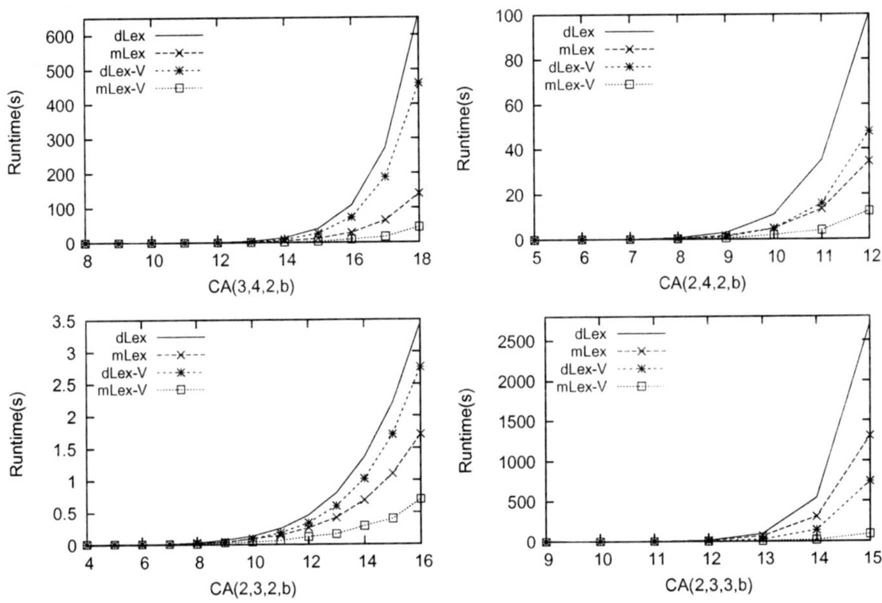

Fig. 5. Comparison of Runtime in Cover Array Problem

domain size g in the original model and g^t in the compound model. We show the growth of solution set size in *log scale* in base 10 in Figure 4 and runtime in Figure 5 as the number of rows b increases.

We can see that mLex/mLex-V always have fewer solutions than dLex/dLex-V, which supports that preservation can achieve smaller solution set size. The distance between the curve of dLex and dLex-V is relatively smaller than that between mLex and mLex-V. The value symmetry breaking constraints in mLex-V preserve the row and column permutations such that combining them achieves good solution reduction. In all instances, the growth of dLex and dLex-V are similar, but the growth of mLex is smaller than both. Although mLex has more solutions than dLex-V in small instances, the solution set size of mLex becomes smaller at certain point. Constraints selected under preservation may even break or eliminate more symmetries as side-effects than those intended for a bigger set of target symmetries.

Interestingly, mLex-V has zig-zag curves. When b is even in instances with $g = 2$, the solution set size is relatively bigger. We conjecture that the occurrence of each value in each column are equal with high probability when b has factor g and less value symmetries are broken.

Comparing the runtime of each approach in Figure 5, although the performance of mLex/mLex-V is not as good in small instances, it is better as b grows. The runtime of mLex is smaller than dLex-V when $g = 2$ but bigger when $g = 3$.

Piecewise Variable and Value Symmetries. Piecewise variable and value symmetries are identified as a tractable class of symmetries in CSPs [7]. There exists a linear-size set of constraints [6] that remove super-exponential number of symmetries. This set of constraints [6] is inclusively a case of symmetry preservation: the value symmetry breaking constraints preserve the variable symmetries and therefore all composition symmetries are eliminated as side-effects.

5   Interactions with Problem Constraints

We have discussed how to achieve smaller solution set and thus better efficiency when intending to eliminate the same target symmetries. In this section, we give observations on ways to further reduce the search space. This results in extra advantages even when we post symmetry breaking constraints that obtain the same solution set size.

Further Simplification. Sometimes, a symmetry breaking constraint can be simplified to an equivalent but cheaper one with the help of problem constraints, e.g. when there is an ALLDIFF constraint on all the variables [21]. Reduction rules [10] can be applied to reduce the number and arity of a conjunction of symmetry breaking constraints. We find that the degree of simplification can be different when choosing different alternatives. For example, to eliminate symmetry σ_{d1} in the DLS problem, we can post either LEXLEADER constraint (2) $[X_{12}, \ldots, X_{1n}] <_{lex} [X_{21}, \ldots, X_{n1}]$ or inequality constraints (3) $X_{1n} < X_{n1}$ or (4) $X_{n1} < X_{1n}$. Each choice cuts half of the solution set. Constraint (3) and (4) are preferred since it is cheaper and likely to entail better runtime.

To achieve simpler constraints, we can try different symmetry breaking constraints, especially those derived from canonical variable orderings that start with variables bounded by more problem constraints and other symmetry breaking constraints. The choice also affects how well symmetry breaking constraint interacts with problem constraints in terms of pruning.

Increasing Constraint Propagation. Symmetry breaking constraints are constraints and have interactions with other constraints. A number of new global constraints [13,16] that combines the symmetry breaking constraints with problem constraints are introduced to increase constraint propagation. We give an example of increasing constraint propagation by simply choosing another symmetry breaking constraint.

Example 3. Among various choices to eliminate the value symmetries of DLS(5), we pick three for discussion: fix the value of the first row, the middle column or the main diagonal. Variables in the center is constrained by four ALLDIFF constraints; variables in the diagonal line is constrained by three ALLDIFF constraints; and the rest by two ALLDIFF constraints. Figure 6 shows the remaining search space after the propagation.

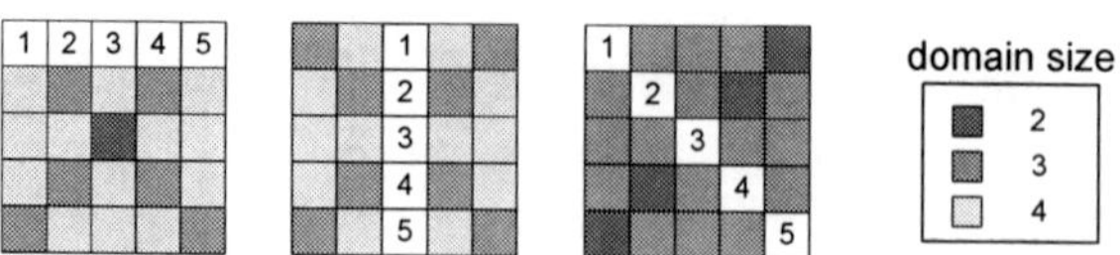

Fig. 6. remaining search space of DLS(5) - eliminate G_{val}

With the same solution set size, the remaining search space after propagation once when fixing the main diagonal is less than one hundredth of that when fixing the first row or middle column even in a small instance ($n = 5$). Even with the same number of solutions, different symmetry breaking constraints cooperate variously with problem constraints and other symmetry breaking constraints in terms of propagation. They may trigger different pruning power on variable domains when combined with problem constraints, some reducing dramatically more search space than others. Constraints that share variables with more or tighter problem constraints potentially trigger more propagation.

6 Experimental Results

This section evaluates how our choices of symmetry breaking constraints derived through preservation and the simple tips from Section 5 can influence solution set size and efficiency. In particular, we compare against the common practice reported in the literature. All problems can be formulated as matrix models containing both variable and value symmetries. We denote the symmetry breaking constraints derived from row-wise ordering as ROWWISE. The experiments are measured on a 3GHz Intel Core2 Duo PC with 3.2 GB RAM running Ubuntu 10.04.1 and Gecode-3.7.0. Depth first search is used and variables are searched in row-wise ordering with value tried in increasing order. We report the number of solutions, runtime and number of fails. Running time is expressed in seconds. Best results are highlighted in bold.

Our choices of symmetry breaking constraints for the first two benchmarks are as described in the first example of Section 4.4. Taking into account the interaction with problem constraints, constraints (5) and (6) in Figure 3 trigger more propagation and leave a smaller search tree. We choose (5) for experiments. Simplified constraint is posted for the un-eliminated symmetry σ_{d1}. The set of symmetry breaking constraint of Our Method is $\{[X_{11}, \ldots, X_{nn}] = [1, \ldots, n], X_{1n} < X_{n1}\}$. We compare with ROWWISE and VAR+OCC in the literature.

Diagonal Latin Square. Table 1 shows results of constraints derived from row-wise ordering and those we selected from preservation. *All solutions are searched above the double line and one solution for the rest.* Our method has half the number of solutions as that of ROWWISE. Benefiting from both preservation and the interaction with problem constraints, we have smaller number of fails that imply smaller search tree. With instance $n = 11$, the runtime of searching for one solution of our method is seven orders of magnitude better.

NNQueen. The problem is to place n colored queens on a $n \times n$ chessboard, such that no lines contain more than one queen with the same color. We model the problem as an n^2 variable matrix model and enforce an ALLDIFF constraint for each line. It also has geometric symmetries and value interchangeability. We compare with the symmetry breaking constraints by Puget [23]: VAR=$\{X_{11} < X_{1n}, X_{11} < X_{n1}, X_{11} < X_{nn}, X_{12} < X_{21}\}$ to eliminate geometric symmetries and a set of OCC constraints to eliminate value symmetries. Results presented in Table 2 show we gain better efficiency due to eliminating more symmetries.

Table 1. Diagonal Latin Square

n	RowWise			Our Method		
	#sol	time	#fails	#sol	time	#fails
5	8	0.001	7	**4**	**0.001**	**1**
6	128	0.029	3000	**64**	**0.004**	**652**
7	171200	12.981	1413K	**85600**	**1.954**	**163K**
8	1	0.002	140	1	**0.001**	**17**
9	1	40.04	4327K	1	**0.001**	**25**
10	1	0.031	2025	1	**0.002**	**175**
11	1	12052	1204124K	1	**0.005**	**339**

Table 2. NNQueen

n	VAR+OCC			Our Method		
	#sol	time	#fails	#sol	time	#fails
5	2	**0.001**	7	**1**	**0.001**	**0**
6	0	0.002	96	**0**	**0.001**	**55**
7	4	0.195	6201	**2**	**0.038**	**2496**
8	0	65.75	1824K	**0**	**16.47**	**1258K**
9	0	10660	232274K	**0**	**2349**	**153952K**

Error Correcting Code - Lee Distance (ECCLD). The problem is from CSPLib prob036. It requires to find the maximum number b of codes of length n drawn from 4 symbols $\{1, 2, 3, 4\}$ such that the Lee distance between any pair of codes is exactly c. The Lee distance between two symbols a and b is min $\{|a - b|, 4 - |a - b|\}$. We model it into a $b \times n$ matrix with domain $\{1..4\}$. In order to illustrate the effect on solution set size, we transform the optimization problem to a satisfaction one by setting b in advance. Due to limited runtime, we constrain the problem size by limiting b. In general, to which dimension multiset ordering should be applied is problem specific. Different from the Cover Array Problem, we find that the best performance of the combination of multiset ordering and lexicographical ordering is achieved by $\geq_m R, \leq_{lex} C$ (multiset ordering between rows and lexicographical ordering between columns). We compare with DoubleLex [5] and SnakeLex [11]. We can see from Table 3 that $\geq_m R, \leq_{lex} C$ is better in both solution set size and runtime when $b \leq n$ but worse roughly when $n < b$.

Table 3. Error Correcting Code - Lee Distance

(n, c, b)	DoubleLex			SnakeLex			$\geq_m R, \leq_{lex} C$		
	#sol	time	#fails	#sol	time	#fails	#sol	time	#fails
(4,4,8)	32469	**65.7**	**780K**	29384	86.9	996K	53972	81.6	928K
(5,2,10)	**87**	**7.98**	**42K**	107	10.9	78K	1040	11.7	80K
(5,6,4)	710731	45.2	426K	748248	29	660K	**213700**	**23.8**	**320K**
(5,6,5)	1441224	148	3378K	1468811	236	5661K	**379456**	79.6	1779K
(5,6,6)	297476	344	7090K	299821	602	11749K	**76528**	**204**	**3717K**
(6,4,4)	4698842	194	4036K	5061729	225	5341K	**1909044**	**115**	**2270K**
(6,4,5)	29345816	3340	72477K	29668229	3480	81624K	**11166072**	**1772**	**36960K**
(6,8,4)	59158	22.6	1175K	55618	29.8	1822K	**13163**	**12**	**588K**
(8,4,4)	35626714	2172	**48002K**	38629753	2554	63380K	**13403304**	**1108**	73211K

Conclusion. There are several methods [12,18,15] concerning the choice of the symmetry breaking constraints, such as model restarts [12] and dynamic posting [18] that reduce the conflict with search heuristic, and Jefferson *et al.*'s work [15] that chooses a better subset of symmetries to break, and use Crawford's [3] ordering constraints to generate the constraints. Our approach focuses on choosing better symmetry breaking constraints for a fixed set of target symmetries in the modeling stage so as to increase symmetry breaking or constraint interaction. The previous approach and ours are complementary to each other. Combining the approaches will be interesting future work.

Our goal is to find a set of symmetry breaking constraints that aims at only target symmetries but turns out to eliminate more symmetries. After analyzing the common side-effects of symmetry breaking constraints, we propose to select symmetry breaking constraints that preserve other target symmetries and restrict the side-effects to non-target symmetries. Unfortunately, our methods in the current form require the insight of a human modeler and automating it is non-trivial with our initial experience. The advantages of preserving target symmetries are demonstrated both analytically and empirically. We compare with approaches in the literature for problems in which eliminating all symmetries is not tractable. Without introducing new overhead, our approach can achieve smaller solution set and possibly better efficiency. To achieve symmetry preservation, different choices of symmetry breaking constraint need to be considered. It is interesting to investigate more flexible alternatives of symmetry breaking constraints that can preserve symmetries and propagate well with other constraints.

Acknowledgement. We are grateful to the anonymous referees for their comments. The work described in this paper was generously supported by grants CUHK413808 and CUHK413710 from the Research Grants Council of Hong Kong SAR.

References

1. Aloul, F.A., Ramani, A., Markov, I.L., Sakallah, K.A.: Solving difficult sat instances in the presence of symmetry. In: DAC 2002, pp. 731–736. ACM (2002)
2. Cohen, D., Jeavons, P., Jefferson, C., Petrie, K.E., Smith, B.M.: Symmetry Definitions for Constraint Satisfaction Problems. In: van Beek, P. (ed.) CP 2005. LNCS, vol. 3709, pp. 17–31. Springer, Heidelberg (2005)
3. Crawford, J., Ginsberg, M., Luks, E., Roy, A.: Symmetry breaking predicates for search problems. In: KR 1996, pp. 148–159 (1996)
4. Fahle, T., Schamberger, S., Sellmann, M.: Symmetry Breaking. In: Walsh, T. (ed.) CP 2001. LNCS, vol. 2239, pp. 93–107. Springer, Heidelberg (2001)
5. Flener, P., Frisch, A.M., Hnich, B., Kiziltan, Z., Miguel, I., Pearson, J., Walsh, T.: Breaking Row and Column Symmetries in Matrix Models. In: Van Hentenryck, P. (ed.) CP 2002. LNCS, vol. 2470, pp. 462–477. Springer, Heidelberg (2002)
6. Flener, P., Pearson, J., Sellmann, M., Van Hentenryck, P.: Static and Dynamic Structural Symmetry Breaking. In: Benhamou, F. (ed.) CP 2006. LNCS, vol. 4204, pp. 695–699. Springer, Heidelberg (2006)
7. Flener, P., Pearson, J., Sellmann, M., Van Hentenryck, P., Ågren, M.: Dynamic structural symmetry breaking for constraint satisfaction problems. Constraints 14(4), 506–538 (2009)
8. Frisch, A.M., Miguel, I., Kiziltan, Z., Hnich, B., Walsh, T.: Multiset ordering constraints. In: IJCAI 2003, pp. 221–226 (2003)
9. Gent, I.P., Smith, B.M.: Symmetry breaking in constraint programming. In: ECAI 2000, pp. 599–603 (2000)
10. Grayland, A., Jefferson, C., Miguel, I., Roney-Dougal, C.M.: Minimal ordering constraints for some families of variable symmetries. AMAI 57(1), 75–102 (2009)
11. Grayland, A., Miguel, I., Roney-Dougal, C.M.: Snake Lex: An Alternative to Double Lex. In: Gent, I.P. (ed.) CP 2009. LNCS, vol. 5732, pp. 391–399. Springer, Heidelberg (2009)

12. Heller, D., Panda, A., Sellmann, M., Yip, J.: Model Restarts for Structural Symmetry Breaking. In: Stuckey, P.J. (ed.) CP 2008. LNCS, vol. 5202, pp. 539–544. Springer, Heidelberg (2008)

13. Hnich, B., Kiziltan, Z., Walsh, T.: Combining symmetry breaking with other constraints: lexicographic ordering with sums. In: ISAIM 2004 (2004)

14. Hnich, B., Prestwich, S.D., Selensky, E., Smith, B.M.: Constraint models for the covering test problem. Constraints 11(2), 199–219 (2006)

15. Jefferson, C., Petrie, K.E.: Automatic Generation of Constraints for Partial Symmetry Breaking. In: Lee, J. (ed.) CP 2011. LNCS, vol. 6876, pp. 729–743. Springer, Heidelberg (2011)

16. Katsirelos, G., Narodytska, N., Walsh, T.: Combining Symmetry Breaking and Global Constraints. In: Oddi, A., Fages, F., Rossi, F. (eds.) CSCLP 2008. LNCS, vol. 5655, pp. 84–98. Springer, Heidelberg (2009)

17. Katsirelos, G., Narodytska, N., Walsh, T.: On the Complexity and Completeness of Static Constraints for Breaking Row and Column Symmetry. In: Cohen, D. (ed.) CP 2010. LNCS, vol. 6308, pp. 305–320. Springer, Heidelberg (2010)

18. Katsirelos, G., Walsh, T.: Symmetries of symmetry breaking constraints. In: ECAI 2010, pp. 861–866 (2010)

19. Law, Y.C., Lee, J.H.M.: Symmetry breaking constraints for value symmetries in constraint satisfaction. Constraints 11(2), 221–267 (2006)

20. McDonald, I., Smith, B.: Partial Symmetry Breaking. In: Van Hentenryck, P. (ed.) CP 2002. LNCS, vol. 2470, pp. 431–445. Springer, Heidelberg (2002)

21. Puget, J.-F.: Breaking symmetries in all different problems. In: IJCAI 2005, pp. 272–277 (2005)

22. Puget, J.-F.: On the Satisfiability of Symmetrical Constrained Satisfaction Problems. In: Komorowski, J., Raś, Z.W. (eds.) ISMIS 1993. LNCS, vol. 689, pp. 350–361. Springer, Heidelberg (1993)

23. Puget, J.-F.: Breaking All Value Symmetries in Surjection Problems. In: van Beek, P. (ed.) CP 2005. LNCS, vol. 3709, pp. 490–504. Springer, Heidelberg (2005)

24. Régin, J.C.: A filtering algorithm for constraints of difference in CSPs. In: AAAI 1994, pp. 362–367 (1994)

25. Régin, J.C.: Generalized arc consistency for global cardinality constraint. In: AAAI 1996, vol. 1, pp. 209–215 (1996)

26. Smith, B.M.: Sets of symmetry breaking constraints. In: SymCon 2005 (2005)

A Scalable Sweep Algorithm
for the *cumulative* Constraint

Arnaud Letort[1,*], Nicolas Beldiceanu[1], and Mats Carlsson[2]

[1] TASC team, (EMN-INRIA,LINA) Mines de Nantes, France
{arnaud.letort,nicolas.beldiceanu}@mines-nantes.fr
[2] SICS, P.O. Box 1263, SE-164 29 Kista, Sweden
matsc@sics.se

Abstract. This paper presents a sweep based algorithm for the *cumulative* constraint, which can operate in filtering mode as well as in greedy assignment mode. Given n tasks, this algorithm has a worst-case time complexity of $O(n^2)$. In practice, we use a variant with better average-case complexity but worst-case complexity of $O(n^2 \log n)$, which goes down to $O(n \log n)$ when all tasks have unit duration, i.e. in the bin-packing case. Despite its worst-case time complexity, this algorithm scales well in practice, even when a significant number of tasks can be scheduled in parallel. It handles up to 1 million tasks in one single *cumulative* constraint in both Choco and SICStus.

1 Introduction

In the 2011 Panel of the Future of CP [6], one of the identified challenges for CP was the need to handle large scale problems. Multi-dimensional bin-packing problems were quoted as a typical example [10], particularly relevant in the context of cloud computing. Indeed the importance of bin-packing problems was recently highlighted in [12] and is part of the topic of the 2012 Roadef Challenge [13].

Till now, the tendency is to use dedicated algorithms and metaheuristics [17] to cope with large instances. Following the line of research initiated with the *geost* constraint [2], our main objective is to provide global constraints that can handle a significant sub-problem while scaling well in a traditional CP solver. Typically, filtering algorithms focus on having the best possible deductions [9,20,21], rather than on scalability issues. This explains why all existing papers on *cumulative* [9,16,20,21] and *bin-packing* [18,7,14] usually focus on small size problems (i.e., typically less than 200 tasks up to 10000 tasks) but leave open the scalability issue. Like what was already done for the *geost* constraint, which handles up to 2 million boxes, our goal is to come up with a lean filtering algorithm for *cumulative*. In order to scale well in terms of memory, we design a lean filtering algorithm, which can also be turned into a greedy algorithm that benefits from the filtering of the lean filtering algorithm while fixing tasks. This approach

---

* Partially founded by the SelfXL project (contract ANR-08-SEGI-017).

M. Milano (Ed.): CP 2012, LNCS 7514, pp. 439–454, 2012.
© Springer-Verlag Berlin Heidelberg 2012

allows to avoid the traditional memory bottleneck problem of CP solvers due to trailing or copying data structures [15], while still benefitting from filtering. Moreover, like for *geost* our lean filtering algorithm and its derived greedy assignment mode are compatible in the sense that they can both be used at each node of the search tree, i.e., first call the greedy mode for trying to find a solution and, if that doesn't work, use the filtering mode to restrict the variables and continue the search.

This paper focuses on the *cumulative* constraint, originally introduced in [1] for modeling resource scheduling problems:

$$cumulative([s_0, \ldots, s_{n-1}], [d_0, \ldots, d_{n-1}], [e_0, \ldots, e_{n-1}], [h_0, \ldots, h_{n-1}], limit)$$

where $[s_0, \ldots, s_{n-1}]$, $[e_0, \ldots, e_{n-1}]$ are non-empty lists of domain variables,[1] and $[d_0, \ldots, d_{n-1}]$, $[h_0, \ldots, h_{n-1}]$ are lists of non-negative integers and *limit* is a non-negative integer. The *cumulative* constraint holds if (1-2) are true:

$$\forall t \in [0, n-1] \ : s_t + d_t = e_t \tag{1}$$

$$\forall i \in \mathbb{N} \ : \sum_{\substack{t \in [0,n-1]: \\ i \in [s_t, e_t)}} h_t \leq limit \tag{2}$$

Section 2 recalls the 2001 sweep algorithm for *cumulative* [3] and provides a critical analysis of its major bottlenecks. Then, Section 3 presents the new sweep based filtering algorithm and its greedy mode. Section 4 evaluates its implementations in both Choco [19] and SICStus [5] and compares them with the 2001 implementations [3] in both systems, as well as to a dedicated *bin-packing* constraint used in Entropy [8].

2   A Critical Analysis of the 2001 Sweep Algorithm

The algorithm is based on the *sweep* idea, which is widely used in computational geometry [4]. In constraint programming, sweep was used for implementing the *non-overlapping* constraint [2] as well as the *cumulative* constraint [3].

In 2 dimensions, a plane *sweep* algorithm solves a problem by moving a vertical line from left to right. The algorithm uses two data structures:

- The *sweep-line status*, which contains some information related to the current position δ of the vertical line.
- The *event point series*, which holds the events to process, ordered in increasing order according to the abscissa.

The algorithm initializes the sweep-line status for the starting position of the vertical line. Then the line "jumps" from event to event; each event is handled and inserted or removed from the sweep-line status. In our context, the sweep-line

---

[1] A *domain variable* v is a variable that ranges over a finite set of integers; $\underline{v}$ and $\overline{v}$ respectively denote the minimum and maximum value of variable v.

scans the time axis in order to build a pessimistic cumulated resource consumption profile (PCRCP) and to perform checks and pruning according to this profile and to *limit*. So the algorithm is a sweep variant of the *timetable* method [11]. Before defining the notion of PCRCP let us first introduce a running example that will be used throughout for illustrating the different algorithms.

Example 1. Consider four tasks t_0, t_1, t_2, t_3 which have the following *start, duration, end* and *height*:

- $t_0 : s_0 = 0,\ d_0 = 1,\ e_0 = 1,\ h_0 = 3,$
- $t_1 : s_1 \in [0, 2],\ d_1 = 2,\ e_1 \in [2, 4],\ h_1 = 3,$
- $t_2 : s_2 \in [2, 4],\ d_2 = 3,\ e_2 \in [5, 7],\ h_2 = 3,$
- $t_3 : s_3 \in [5, 7],\ d_3 = 1,\ e_3 \in [6, 8],\ h_3 = 3,$

subject to the constraint $cumulative([s_0, s_1, s_2, s_3], [d_0, d_1, d_2, d_3], [e_0, e_1, e_2, e_3],$ $[h_0, h_1, h_2, h_3], 5)$. Since task t_0 starts at instant 0 and since t_1 cannot overlap t_0 without exceeding the resource limit 5, the earliest start of t_1 is adjusted to 1. Since task t_1 occupies interval $[2, 3)$ and since t_1 and t_2 also cannot overlap for the same reason, the earliest start of t_2 is adjusted to 3. Since task t_2 occupies interval $[4, 6)$ and since t_2 and t_3 also cannot overlap, the earliest start of t_3 is adjusted to 6. The purpose of the sweep algorithm is to perform such filtering in an efficient way.   □

Given a set of tasks $\mathcal{T}$, the PCRCP of the set $\mathcal{T}$ consists of the aggregation of the compulsory parts of the tasks in $\mathcal{T}$, where the *compulsory part* of a task is the intersection of all its feasible instances. On the one hand, the height of the compulsory part of a task t at a given time point i is defined by h_t if $i \in [\overline{s_t}, \underline{e_t})$ and 0 otherwise. On the other hand, the height of the PCRCP at a given time point i is given by $\sum_{\substack{t \in \mathcal{T}, \\ i \in [\overline{s_t}, \underline{e_t})}} h_t.$

Continuation of Example 1 (Compulsory Part of a Task). Task t_0 and t_2 initially have a non-empty compulsory part: task t_0 uses 3 resource units on interval $[0, 1)$, while task t_2 uses 3 resource units on interval $[4, 5)$. After reaching the fixpoint, task t_1 also has a non-empty compulsory part: task t_1 uses 3 resource units on interval $[2, 3)$ while the compulsory part of t_2 now occupies interval $[4, 6)$.   □

Event Point Series. In order to build the PCRCP and to prune the start of the tasks, the sweep algorithm considers the following types of events:

- *Profile events* for building the PCRCP correspond to the latest starts and the earliest ends of the tasks for which the latest start is strictly less than the earliest end (i.e. the start and the end of a non-empty compulsory part).
- *Pruning events* for recording the tasks to prune, i.e. the not yet fixed tasks that intersect δ.

Table 1 (top) describes the different types of events, where each event corresponds to a quadruple ⟨*event type, task generating the event, event date, available space increment*⟩. These events are sorted by increasing date.

Continuation of Example 1 (Generated Events). The following events are generated and sorted by increasing date: ⟨$SCP, 0, 0, -3$⟩, ⟨$PR, 1, 0, 0$⟩, ⟨$ECP, 0, 1, 3$⟩, ⟨$PR, 2, 2, 0$⟩, ⟨$SCP, 2, 4, -3$⟩, ⟨$ECP, 2, 5, 3$⟩, ⟨$PR, 3, 5, 0$⟩.   □

Table 1. Event types for the 2001 sweep (top) and the dynamic sweep (bottom) with corresponding condition for generating them. The last event attribute is only relevant for event types *SCP*, *ECP* and *ECPD*.

Generated Events (2001 algo.)		Conditions
$\langle SCP, t, \overline{s_t}, -h_t \rangle$ and $\langle ECP, t, \underline{e_t}, +h_t \rangle$		$\overline{s_t} < \underline{e_t}$
$\langle PR, t, \underline{s_t}, 0 \rangle$		$\underline{s_t} \neq \overline{s_t}$

New Events	Events (2001 algo.)	Conditions
$\langle SCP, t, \overline{s_t}, -h_t \rangle$	$\langle SCP, t, \overline{s_t}, -h_t \rangle$	$\overline{s_t} < \underline{e_t}$
$\langle ECPD, t, \underline{e_t}, +h_t \rangle$	$\langle ECP, t, \underline{e_t}, +h_t \rangle$	$\overline{s_t} < \underline{e_t}$
$\langle CCP, t, \overline{s_t}, 0 \rangle$		$\overline{s_t} \geq \underline{e_t}$
$\langle PR, t, \underline{s_t}, 0 \rangle$	$\langle PR, t, \underline{s_t}, 0 \rangle$	$\underline{s_t} \neq \overline{s_t}$

Sweep-Line Status. The sweep-line maintains three pieces of information:

- The current sweep-line position δ, initially set to the date of the first event.
- The amount of available resource at instant δ, denoted by *gap*, i.e., the difference between the resource limit and the height of the PCRCP.
- A list of tasks $\mathcal{T}_{prune}$, recording all tasks that potentially can overlap δ.

The sweep algorithm first creates and sorts the events wrt. their date. Then, the sweep-line moves from one event to the next event, updating *gap* and $\mathcal{T}_{prune}$. Once all events at δ have been handled, the sweep algorithm tries to prune all tasks in $\mathcal{T}_{prune}$ wrt. *gap* and interval $[\delta, \delta')$ where δ' is the next sweep-line position, i.e. the date of the next event. More precisely, given a task $t \in \mathcal{T}_{prune}$ such that $h_t > gap$, the interval $[\delta - d_t + 1, \delta')$ is removed from the start of task t.

Continuation of Example 1 (Illustrating the 2001 Sweep Algorithm). The sweep algorithm reads the two events $\langle SCP, 0, 0, -3 \rangle$, $\langle PR, 1, 0, 0 \rangle$ and sets *gap* to $5 - 3$ and $\mathcal{T}_{prune}$ to $\{t_1\}$. During a first sweep, the compulsory part of task t_0 (see Part (A) of Figure 1) permits to prune the start of t_1 since the *gap* on $[0, 1)$ is strictly less than h_1. The pruning of the earliest start of t_1 during the first sweep causes the creation of a compulsory part for task t_1 which is not immediately used to perform more pruning (see Part (B)). It is necessary to wait for a second sweep to take advantage of this new compulsory part to adjust the earliest start of task t_2. This last adjustment causes the extension of the compulsory part of t_2 on $[4, 6)$ (see Part (C)). A third sweep adjusts the earliest start of task t_3 which cannot overlap t_2. A fourth and last sweep is performed to find out that the fixpoint was reached (see Part (D)). $\qquad\square$

Weakness of the 2001 Sweep Algorithm

① [Too static] The potential increase of the PCRCP during a single sweep is not dynamically taken into account. In other words, creations and extensions of compulsory parts during a sweep are not immediately used to perform more pruning while sweeping. Example 1 illustrates this point since the sweep needs to be run four times before reaching its fixpoint.

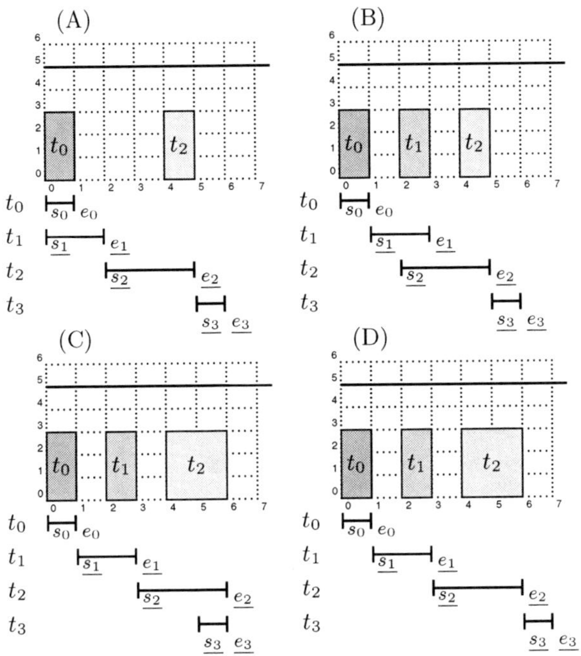

Fig. 1. Parts (A), (B), (C) and (D) respectively represent the earliest positions of the tasks and the PCRCP, of the initial problem described in Example 1, after a first sweep, after a second sweep and after a third sweep

② **[Often reaches its worst-case time complexity]** The worst-case time complexity of the 2001 sweep algorithm is $O(n^2)$ where n is the number of tasks. This complexity is often reached in practice when most of the tasks can be placed everywhere on the time line. The reason is that it needs at each value of δ to systematically re-scan all tasks that overlap δ. Profiling the 2001 implementation indicates that the sweep algorithm spends up to 45% of its overall running time scanning the list of potential tasks to prune.

③ **[Creates holes in the domains]** The 2001 sweep algorithm removes intervals of consecutive values from domain variables. This is a weak point, which prevents handling large instances since a variable cannot just be compactly represented by its minimum and maximum values.

④ **[Does not take advantage of bin-packing]** For instances where all tasks have duration one, the worst time complexity $O(n^2)$ is left unchanged.

3   The Dynamic Sweep Algorithm

This section presents our contribution, a new sweep algorithm that handles the four performance issues of the 2001 sweep algorithm raised at the end of Sect.2, i.e., points ① to ④. We first introduce some general design decisions of the new sweep algorithm as well as the property the algorithm maintains, and then describe it in a similar way the 2001 original sweep algorithm was presented in Sect. 2. We first present the new *event point series*, then the new *sweep-line status*, and the overall algorithm. Finally we prove that the property initially introduced is maintained by the new algorithm and give its complexity in the general case as well as in the case where all task durations are fixed to one.

The first difference from the 2001 sweep is that our algorithm only deals with domain bounds, which is a good way to reduce the memory consumption for the representation of domain variables (see Point ③ of Sect. 2).[2] Consequently, we need to change the 2001 algorithm, which creates holes in the domain of task origins. The new sweep algorithm filters the task origins in two distinct sweep stages. A first stage, called *sweep_min*, tries to adjust the earliest starts of tasks by performing a sweep from left to right, and a second stage tries to adjust the latest ends by performing a sweep from right to left. The greedy mode of the new sweep algorithm will be derived from *sweep_min*, in the sense that it takes advantage of the propagation performed by *sweep_min* and fixes the start of tasks rather than adjusting them. W.l.o.g, we focus from now on the first stage *sweep_min* since the second stage is completely symmetric.

As illustrated by Example 1, the 2001 sweep algorithm needs to be re-run several times in order to reach its fixpoint (i.e., 4 times in our example). This is due to the fact that, during one sweep, restrictions on task origins are not immediately taken into account. Our new algorithm, *sweep_min*, dynamically uses these deductions to reach its fixpoint in one single sweep. To deal with this aspect, our new sweep algorithm introduces the concept of *conditional events*, i.e., events that are created while sweeping over the time axis.

We first give the property that holds when *sweep_min* reaches its fixpoint. This property will be proved at the end of this section.

Property 1. Given a *cumulative* constraint with its set of tasks $\mathcal{T}$ and its resource limit *limit*, *sweep_min* ensures that:

$$\forall t \in \mathcal{T}, \forall i \in [\underline{s_t}, \underline{e_t}) \; : \; h_t + \sum_{\substack{t' \in \mathcal{T} \backslash \{t\}: \\ i \in [\overline{s_{t'}}, \underline{e_{t'}})}} h_{t'} \leq limit \tag{3}$$

Property 1 ensures that, for any task t of the *cumulative* constraint, one can schedule t at its earliest start without exceeding the resource limit wrt. the PCRCP for the tasks of $\mathcal{T} \backslash \{t\}$. We now present the different parts of the new sweep algorithm.

3.1 Event Point Series

In order to address point ① [Too static] of Sect. 2, *sweep_min* should handle the extension and the creation of compulsory parts caused by the adjustment of earliest starts of tasks in one single sweep. We therefore need to modify the

---

[2] Note that most Operation Research scheduling algorithms only adjust the earliest start and latest ends of tasks.

events introduced in Table 1. The bottom part presents the events of *sweep_min* and their relations with the events of the 2001 algorithm.

- The event type $\langle SCP, t, \overline{s_t}, -h_t \rangle$ for the start of compulsory part of task t is left unchanged. Note that, since *sweep_min* only adjusts earliest starts, the start of a compulsory part (which corresponds to a latest start) can never be extended to the left.
- The event type $\langle ECP, t, \underline{e_t}, h_t \rangle$ for the end of the compulsory part of task t is converted to $\langle ECPD, t, \underline{e_t}, h_t \rangle$ where D stands for *dynamic*. The date of such event corresponds to the earliest end of t (also the end of its compulsory part) and may increase due to the adjustment of the earliest start of t.
- A new event type $\langle CCP, t, \overline{s_t}, 0 \rangle$, where CCP stands for *conditional compulsory part*, is created for each task t that does not have any compulsory part. At the latest, once the sweep-line reaches position $\overline{s_t}$, it adjusts the earliest start of t. Consequently the conditional event can be transformed into an SCP and an $ECPD$ events, reflecting the creation of compulsory part.
- The event type $\langle PR, t, \underline{s_t}, 0 \rangle$ for the earliest start of t is left unchanged.

On the one hand, some of these events have their dates modified (see $ECPD$). On the other hand, some events create new events (see CCP). Consequently, rather than just sorting all events initially, we insert them by increasing date into a heap called h_{events}.

Continuation of Example 1 (New Generated Events for sweep_min*).* The following events are generated and sorted according to their date: $\langle SCP, 0, 0, -3 \rangle$, $\langle PR, 1, 0, 0 \rangle$, $\langle \mathbf{ECPD}, \mathbf{0}, \mathbf{1}, \mathbf{3} \rangle$, $\langle \mathbf{CCP}, \mathbf{1}, \mathbf{2}, \mathbf{0} \rangle$, $\langle PR, 2, 2, 0 \rangle$, $\langle SCP, 2, 4, -3 \rangle$, $\langle \mathbf{ECPD}, \mathbf{2}, \mathbf{5}, \mathbf{3} \rangle$, $\langle PR, 3, 5, 0 \rangle$, $\langle \mathbf{CCP}, \mathbf{3}, \mathbf{7}, \mathbf{0} \rangle$.   □

3.2   Sweep-Line Status

The sweep-line maintains the following pieces of information:

- The current sweep-line position δ, initially set to the date of the first event.
- The amount of available resource at instant δ, denoted by *gap*, i.e., the difference between the resource limit and the height of the PCRCP.
- Two heaps $h_{conflict}$ and h_{check} for partially avoiding point ② [Often reaches its worst-case time complexity] of Sect. 2. W.l.o.g. assume that the sweep-line is at its initial position and that we handle an event of type PR (i.e., we try to find out the earliest possible start of a task t).

  - If the height of task t is strictly greater than the available gap at δ, we know that we have to adjust the earliest start of t. In order to avoid re-checking each time we move the sweep-line whether or not the gap is big enough wrt. h_t, we say that t is in conflict with δ. We insert task t in the heap $h_{conflict}$, which records all tasks that are in conflict with δ, sorted by increasing height, i.e. the top of the heap $h_{conflict}$ corresponds to the smallest value. This order is induced by the fact that, if we need to adjust the earliest start of a task t, all earliest task starts with a height greater than or equal to h_t also need to be adjusted.

- If the height of task t is less than or equal to the available gap at δ, we know that the earliest start of task t could be equal to δ. But to be sure, we need to check Property 1 for t (i.e., $\mathcal{T} = \{t\}$). For this purpose we insert t in the heap h_{check}, which records all tasks for which we currently check Property 1. Task t stays in h_{check} until a conflict is detected (i.e., h_t is greater than the available gap, and t goes back in $h_{conflict}$) or until the sweep-line passes instant $\delta + d_t$ (and we have found a feasible earliest start of task t wrt. Property 1). In the heap h_{check}, tasks are sorted by decreasing height, i.e. the top of the heap h_{check} corresponds to the largest value, since if a task t is not in conflict with δ, all other tasks of h_{check} of height less than or equal to h_t are also not in conflict with δ. In the following, $empty(h)$ returns $true$ if the heap h is empty, $false$ otherwise. Function $get\_top\_key(h)$ returns the key of the top element in the heap h. We introduce an array of integers $mins$, which stores for each task t in h_{check} the value of δ when t was added into h_{check}.

3.3   Algorithm

The $sweep\_min$ algorithm performs one single sweep over the event point series in order to adjust the earliest start of the tasks wrt. Property 1. It consists of a main loop, a filtering part and a synchronization part. This last part is required in order to directly handle the deductions attached to the creation or increase of compulsory parts in one single sweep. In addition to the heaps h_{check} and $h_{conflict}$ we introduce an array of booleans $evup$ for which the tth entry indicates whether events related to the compulsory part of t were updated or not. It is set to true once we have found the final values of the start and end of the compulsory part of t. We introduce a list $newActiveTasks$, which records all tasks that have their PR event at δ. The primitive $adjust\_min\_start(t, v)$ adjusts the minimum value of the start variable of task t to value v.

Main Loop. The main loop (Algorithm 1) consists of:

- [INITIALIZATION] (lines 3 to 5). The events are generated and inserted into h_{events} according to the conditions given in Table 1. The h_{check} and $h_{conflict}$ heaps are initialized as empty heaps. The list $newActiveTasks$ is initialized as an empty list. δ is set to the date of the first event.
- [MAIN LOOP] (lines 7 to 24). For each date the main loop processes all the corresponding events. It consists of the following parts:
  - [HANDLING A SWEEP-LINE MOVE] (lines 9 to 16). Each time the sweep-line moves, we update the sweep-line status (h_{check} and $h_{conflict}$) wrt. the new $active\ tasks$, i.e. the tasks for which the earliest start is equal to δ. All the new active tasks that are in conflict with δ in the PCRCP are added into $h_{conflict}$ (lines 9 and 10). For tasks that are not in conflict we check whether the sweep interval $[\delta, \delta')$ is big enough wrt. their durations. Tasks for which the sweep interval is too small are added into h_{check} (line 10).

```
 1: function sweep_min(n, s_{[0..n-1]}, s̄_{[0..n-1]}, e_{[0..n-1]}, d_{[0..n-1]}, h_{[0..n-1]}) : boolean
 2:   [INITIALIZATION]
 3:   h_events ← generation of events wrt. n, s_t, s̄_t, d, e_t and h and Table 1.
 4:   h_check, h_conflict ← ∅;  newActiveTasks ← ∅
 5:   δ ← get_top_heap(h_events);  δ' ← δ;  gap ← limit
 6:   [MAIN LOOP]
 7:   while ¬empty(h_events) do
 8:     [HANDLING A SWEEP-LINE MOVE]
 9:     if δ ≠ δ' then
10:       while ¬empty(newActiveTasks) do
11:         extract first task t from newActiveTasks
12:         if h_t > gap then add ⟨h_t, t⟩ in h_conflict
13:         else if d_t > δ' − δ then add ⟨h_t, t⟩ in h_check;  mins_t ← δ
14:         else evup_t ← true
15:       if ¬filter_min(δ, δ', gap) then return false
16:       δ ← δ'
17:     [HANDLING CURRENT EVENT]
18:     δ ← synchronize(h_events, δ)
19:     extract ⟨type, t, δ, dec⟩ from h_events
20:     if type = SCP ∨ type = ECPD then gap ← gap + dec
21:     else if type = PR then newActiveTasks ← newActiveTasks ∪ {t}
22:     [GETTING NEXT EVENT]
23:     if empty(h_events) ∧ ¬filter_min(δ, +∞, gap) then return false
24:     δ' ← synchronize(h_events, δ)
25: return true
```

Algorithm 1. False if a resource overflow is detected, true otherwise.

Then *filter_min* (see Alg. 2) is called to update h_{check} and $h_{conflict}$ and to adjust the earliest start of tasks for which a feasible position was found.

- [HANDLING CURRENT EVENT] (lines 18 to 21). Conditional events (CCP) and dynamic events ($ECPD$) at the top of h_{events} are processed (see Alg. 3). The top event is extracted from the heap h_{events}. Depending of its type (i.e., SCP or $ECPD$), the gap of the available resource is updated, or (i.e., PR), the task is added into the list of new active tasks.
- [GETTING NEXT EVENT] (lines 23 to 24). If there is no more event in h_{events}, *filter_min* is called in order to empty the heap h_{check}, which may generate new compulsory part events.

The Filtering Part. Algorithm 2 processes tasks in h_{check} and $h_{conflict}$ in order to adjust the earliest start of the tasks. The main parts of the algorithm are:

- [CHECK RESOURCE OVERFLOW] (line 3). If the available resource *gap* is negative on the sweep interval $[\delta, \delta')$, Alg. 2 returns false meaning a failure (i.e. the resource capacity limit is exceeded).
- [UPDATING TOP TASKS OF h_{check}] (lines 5 to 11). All tasks in h_{check} of height greater than the available resource *gap* are extracted.

```
 1: function filter_min(δ, δ′, gap) : boolean
 2: [CHECK RESOURCE OVERFLOW]
 3: if gap < 0 then return false
 4: [UPDATING TOP TASKS OF h_check ]
 5: while ¬empty(h_check) ∧ (empty(h_events) ∨ get_top_key(h_check) > gap) do
 6:     extract ⟨h_t, t⟩ from h_check
 7:     if δ ≥ s̄_t ∨ δ − mins_t ≥ d_t ∨ empty(h_events) then
 8:         adjust_min_start(t, mins_t)
 9:         if ¬evup_t then update events of the compulsory part of t; evup_t ← true
10:     else
11:         add ⟨h_t, t⟩ in h_conflict
12: [UPDATING TOP TASKS OF h_conflict ]
13: while ¬empty(h_conflict) ∧ get_top_key(h_conflict) ≤ gap do
14:     extract ⟨h_t, t⟩ from h_conflict
15:     if δ ≥ s̄_t then
16:         adjust_min_start(t, s̄_t)
17:         if ¬evup_t then update events of the compulsory part of t; evup_t ← true
18:     else
19:         if δ′ − δ ≥ d_t then
20:             adjust_min_start(t, δ)
21:             if ¬evup_t then update events of the compulsory part of t; evup_t ← true
22:         else
23:             add ⟨h_t, t⟩ in h_check; mins_t ← δ
24: return true
```

Algorithm 2. Tries to adjust earliest starts of tasks in h_{check} and $h_{conflict}$ wrt. the sweep interval $[\delta, \delta')$ and the available resource gap and returns false if a resource overflow is detected, true otherwise.

- A first case to consider is when task t has been in h_{check} long enough (i.e. $\delta - mins_t \geq d_t$, line 7), meaning that the task is not in conflict on interval $[mins_t, \delta)$, whose size is greater than or equal to d_t. Consequently, we adjust the earliest start of task t to value $mins_t$.
- A second case to consider is when δ has passed the latest start of task t (i.e. $\delta \geq \overline{s_t}$, line 7). That means task t was not in conflict on interval $[mins_t, \delta)$ either, and we can adjust its earliest start to $mins_t$.
- A third case is when there is no more event in the heap h_{events} (i.e. empty(h_{events}), line 7). It means that the height of the PCRCP is equal to zero and we need to empty h_{check}.
- Otherwise, the task is added into $h_{conflict}$ (line 11).

- [UPDATING TOP TASKS OF $h_{conflict}$] (lines 13 to 23). All tasks in $h_{conflict}$ that are no longer in conflict with δ are extracted. If δ has passed the latest start of task t, we know that t cannot be scheduled before its latest position. Otherwise, we compare the duration of t with the sweep interval and decide whether to adjust the earliest start of t or to add it into h_{check}.

```
 1: function synchronize(h_events, δ) : integer
 2:    [UPDATING TOP EVENTS]
 3:    repeat
 4:       if empty(h_events) then return −1
 5:       sync ← true; ⟨date, t, type, dec⟩ ← consult top event of h_events
 6:       [PROCESSING DYNAMIC EVENT]
 7:       if type = ECPD ∧ ¬evup_t then
 8:          if t ∈ h_check then update event date to mins_t + d_t
 9:          else update event date to s̄_t + d_t
10:          evup_t ← true; sync ← false;
11:          [PROCESSING CONDITIONAL EVENT]
12:       else if type = CCP ∧ ¬evup_t ∧ date = δ then
13:          if t ∈ h_check ∧ mins_t + d_t > δ then
14:             add ⟨SCP, t, δ, −h_t⟩ and ⟨ECPD, t, mins_t + d_t, h_t⟩ into h_events
15:          else
16:             add ⟨SCP, t, δ, −h_t⟩ and ⟨ECPD, t, ē_t, h_t⟩ into h_events
17:          evup_t ← true; sync ← false;
18:    until sync
19:    return date
```

Algorithm 3. Checks that the event at the top of h_{events} is updated and returns the date of the next event or *null* if h_{events} is empty.

The Synchronization Part. Before each extraction or access to h_{events}, Alg. 3 checks and updates the top event and returns the next event date. The main parts of the algorithm are:

- [UPDATING TOP EVENTS] (lines 3 to 18). Dynamic and conditional events require to check whether the next event to be extracted by Alg. 1 needs to be updated or not. The repeat loop updates the next event if necessary until the top event is up to date.
- [PROCESSING DYNAMIC EVENT] (lines 7 to 10). An event of type $ECPD$ must be updated if the related task t is in h_{check} or in $h_{conflict}$. If t is in $h_{conflict}$, it means that t cannot start before its latest starting time $\overline{s_t}$. Consequently, its $ECPD$ event is pushed back to the date $\overline{s_t} + d_t$ (line 9). If t is in h_{check}, it means that its earliest start can be adjusted to $mins_t$. Consequently, its $ECPD$ event is updated to the date $mins_t + d_t$ (line 8).
- [PROCESSING CONDITIONAL EVENT] (lines 12 to 17). When the sweep-line reaches the position of a CCP event for a task t, we need to know whether or not a compulsory part for t is created. As $evup_t$ is set to *false*, we know that t is either in h_{check} or in $h_{conflict}$. If t is in $h_{conflict}$ the task is fixed to its latest position and related events are added into h_{events} (line 16). If t is in h_{check}, a compulsory part is created iff $mins_t + d_t > δ$ (lines 13-14).

Continuation of Example 1 (Illustrating the Dynamic Sweep Algorithm). The sweep algorithm first reads the two events $\langle SCP, 0, 0, -3\rangle$, $\langle PR, 1, 0, 0\rangle$ and sets *gap* to 2. Since the height of task t_1 is greater than the available resource *gap*, t_1 is added into

$h_{conflict}$ (see Alg. 1 line 12 and Fig. 2 Part (A)). The call of *filter_min* only checks that the gap is non-negative. Then, the sweep-line moves to the position 1, reads the event $\langle ECPD, 0, 1, +3 \rangle$ and sets gap to 5. The call of *filter_min* with $\delta = 1$, $\delta' = 2$ and $gap = 5$ retrieves t_1 from $h_{conflict}$ and inserts it into h_{check} (see Alg. 2, line 23). In *synchronize* (called in Alg. 1 line 24), the next event $\langle CCP, 1, 2, 0 \rangle$ is converted into two events $\langle SCP, 1, 2, -3 \rangle$ and $\langle ECPD, 1, 3, +3 \rangle$ standing for the creation of a compulsory part on interval $[2, 3)$ for the task t_1 (see Fig. 2 Part (B)). Note that the creation of the compulsory part occurs after the sweep-line position, which is key to ensuring Property 1. □

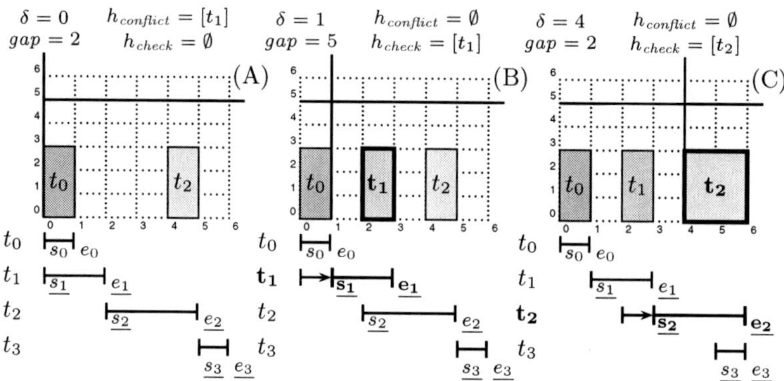

Fig. 2. Parts (A), (B) and (C) represent the earliest positions of the tasks and the PCRCP at different values of δ. Part (A) is when $\delta = 0$ just before the call of *filter_min* (Alg. 1 line 15). Part (B) is when $\delta = 1$ just after the call of *synchronize* (Alg. 1 line 24). Part (C) is when $\delta = 4$ just after the call of *synchronize* (line 24).

3.4   Correctness and Property Achieved by *sweep_min*

We now prove that after the termination of *sweep_min*(Alg. 1), Property 1 holds. For this purpose, we first introduce the following lemma.

Lemma 1. *At any point of its execution,* sweep_min*(Alg. 1) cannot generate a new compulsory part that is located before δ.*

Proof. Since the start of the compulsory part of a task t corresponds to $\overline{s_t}$, which is indicated by its CCP or SCP event, and since *sweep_min* only prunes earliest starts, the compulsory part of t cannot start before this event. Consequently, the latest value of δ to know whether the compulsory part of t is created is $\overline{s_t}$. This case is processed by Alg. 3, lines 12 to 17.

The end of the compulsory part of a task t corresponds to $\underline{e_t}$ and is indicated by its $ECPD$ event. To handle its potential extension to the right, the earliest start of t must be found before the sweep extracts its $ECPD$ event. This case is processed by Alg. 3, lines 7 to 10. □

Proof (of Property 1). Given a task t, let δ_t and min_t respectively denote the position of the sweep-line when the earliest start of t is adjusted by *sweep_min*, and the new earliest start of t. We successively show the following points:

① When the sweep-line is located at instant δ_t we can start task t at min_t without exceeding *limit*, i.e.

$$\forall t' \in \mathcal{T}\backslash\{t\}, \forall i \in [min_t, \delta_t) : h_t + \sum_{\substack{t'\in\mathcal{T}\backslash\{t\}: \\ i\in[\overline{s_{t'}},e_{t'})}} h_{t'} \leq limit$$

The adjustment of the earliest start of task t to min_t implies that t is not in conflict on the interval $[min_t, \delta_t)$ wrt. the PCRCP. Condition $get\_top\_key(h_{check}) > gap$ (Alg. 2 line 5) ensures that the adjustment in line 8 does not induce a resource overflow on $[min_t, \delta_t)$, otherwise t should have been added into $h_{conflict}$. Condition $get\_top\_key(h_{conflict}) \leq gap$ (Alg. 2 line 13) implies that task t is in conflict until the current sweep-line position δ. If $\delta \geq \overline{s_t}$ (line 15) the conflict on $[\overline{s_t}, \delta_t)$ is not "real" since the compulsory part of t is already taken into account in the PCRCP. Alg. 2 (line 20), the earliest start of task t is adjusted to the current sweep-line position, consequently the interval $[min_t, \delta_t)$ is empty.

② For each value of δ greater than δ_t, *sweep_min* cannot create a compulsory part before instant δ_t. This is implied by Lemma 1, which ensures that *sweep_min* cannot generate any compulsory part before δ.

Consequently once *sweep_min* is completed, any task t can be fixed to its earliest start without exceeding the resource limit *limit*.   □

Property 2. [Correctness.] For any task t, there is no feasible position before its earliest start min_t wrt. the PCRCP.

Proof. By contradiction. Given a task t, let $omin_t$ be its earliest start before the execution of *sweep_min*. If the earliest start of t is pruned during the sweep, i.e. $min_t > omin_t$, then t is in conflict at a time point in the interval $[omin_t, min_t)$ (see Alg. 2, line 11). Consequently condition $\delta - mins_t > d_t$ (Alg. 2, line 7) is false, which ensures that there is no earliest feasible position before min_t.   □

3.5   Complexity

Given a *cumulative* constraint involving n tasks, the worst-case time complexity of the dynamic sweep algorithm is $O(n^2 \log n)$. First note that the overall worst-case complexity of synchronize over a full sweep is $O(n)$ since conditional and dynamic events are updated at most once. The worst-case $O(n^2 \log n)$ can be reached in the special case when the PCRCP consists of a succession of high peaks and deep, narrow valleys. Assume that one has $O(n)$ peaks, $O(n)$ valleys, and $O(n)$ tasks to switch between h_{check} and $h_{conflict}$ each time. A heap operation costs $O(\log n)$. The resulting worst-case time complexity is $O(n^2 \log n)$.[3] For bin-packing, the two heaps $h_{conflict}$ and h_{check} permit to reduce the worst-case time complexity down to $O(n \log n)$. Indeed, the earliest start of the tasks of duration one that exit $h_{conflict}$ can directly be adjusted (i.e. h_{check} is unused).

---

[3] Note that when the two heaps are replaced by a list where for each active task we record its status (in checking mode or in conflict mode), we get an $O(n^2)$ worst-case time complexity which, like the 2001 algorithm, is reached in practice.

3.6   Greedy Mode

The motivation for a greedy assignment mode is to handle larger instances in a CP solver. This propagation mode reuses the *sweep_min* part of the filtering algorithm in the sense that once the minimum value of a start variable is found, the greedy mode directly fixes the start to its earliest feasible value wrt. Property 1 rather than adjusting it. Then, the sweep-line is reset to this start and the process continues until all tasks get fixed or a resource overflow occurs. Thus the greedy mode directly benefits from the propagation performed while sweeping.

4   Evaluation

We implemented the dynamic sweep algorithm on Choco [19] and SICStus [5]. Benchmarks were run with an Intel i7 at 2.93 GHz processor on one single core, memory limited to 13GB under Mac OS X 64 bits.

In a first experiment, we ran random instances of cumulative and bin-packing problems. Instances were randomly generated with a density close to 0.7. For a given number of tasks, we generated two different instances with the average number of tasks overlapping a time point equal to 10 (denoted by *ttu1*) resp. 100 (denoted by *ttu2*). For cumulative problems, we compared the time needed to find a first solution using fail-first search with the 2001 sweep algorithm (denoted by *sweep*), the dynamic sweep (denoted by *dynamic*) and the greedy mode (denoted by *greedy*). For bin-packing problems, we also tested a dedicated filtering algorithm (denoted by *fastbp*) coming from Entropy [7], an open-source autonomous virtual machine manager. The Choco results are shown in Fig. 3 (top and middle). The SICStus results (omitted) paint a similar picture. We notice a significant difference from the 2001 algorithm due to an inappropriate design of the code for large instances in Choco (iterating over objects). SICStus is up to 8 times faster than Choco on *sweep* and twice as fast on *dynamic* and *greedy*. The dynamic sweep is always faster than the 2001 sweep with a speedup increasing with the number of tasks (e.g., for 8000 tasks up to 7 times in Choco and 5 times in SICStus). The dynamic sweep algorithm is also more robust than the 2001 algorithm wrt. different heuristics. For the bin-packing case (ttu2), *greedy* could handle up to 10 million tasks in one *cumulative* constraint in SICStus in 8h20m.

In a second experiment, we ran the J30 single-mode resource-constrained project scheduling benchmark suite from PSPLib [4], comparing *sweep* with *dynamic*. Each instance involves four *cumulative* constraints on 30 tasks and several precedence constraints. The variable with the smallest minimal value was chosen during search. The SICStus results are shown in Fig. 3 (bottom). The left hand scatter plot compares run times for instances that were solved to within 5% of the optimal makespan within a 1 minute time-out by both algorithms. The right hand scatter plot compares backtrack counts for instances that timed out in at least one of the two algorithms. Search trees are the same for *sweep* and *dynamic*, and a higher backtrack count means that propagation is faster, and so

---

[4] http://129.187.106.231/psplib/, (480 instances)

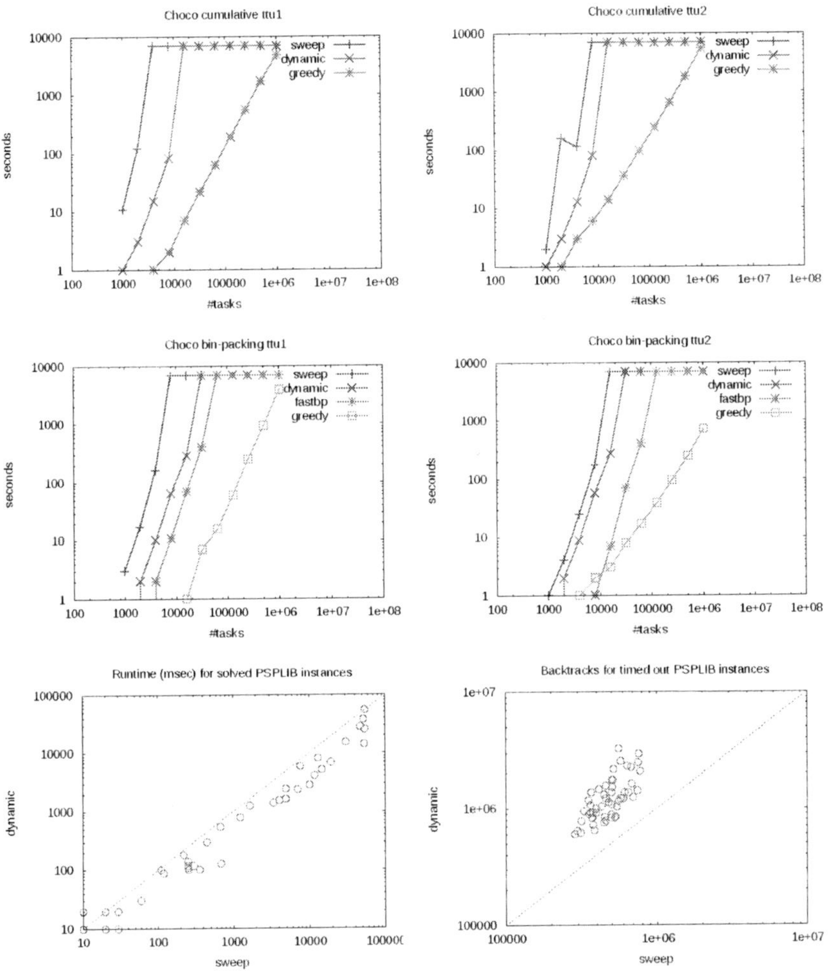

Fig. 3. Runtimes on random instances (top and middle). Comparing runtimes and backtrack counts on PSPLib instances (bottom).

both plots confirm the finding that the dynamic sweep outperforms the 2001 one by a factor up to 3, and not just for problems stated with a single constraint.

5   Conclusion

We have presented a new sweep based filtering algorithm, which dynamically handles deductions while sweeping. In filtering mode, the new algorithm is up to 8 times faster than the 2001 implementation. In assignment mode, it allows to handle up to 1 million tasks in both Choco and SICStus. Future work will focus on the adaptation of this algorithm to multiple resources.

Acknowledgments. Thanks to S. Demassey for providing the *fastbp* constraint.

References

1. Aggoun, A., Beldiceanu, N.: Extending CHIP in order to solve complex scheduling and placement problems. Mathl. Comput. Modelling 17(7), 57–73 (1993)
2. Beldiceanu, N., Carlsson, M., Poder, E., Sadek, R., Truchet, C.: A Generic Geometrical Constraint Kernel in Space and Time for Handling Polymorphic k-Dimensional Objects. In: Bessière, C. (ed.) CP 2007. LNCS, vol. 4741, pp. 180–194. Springer, Heidelberg (2007)
3. Beldiceanu, N., Carlsson, M.: A New Multi-resource *cumulatives* Constraint with Negative Heights. In: Van Hentenryck, P. (ed.) CP 2002. LNCS, vol. 2470, pp. 63–79. Springer, Heidelberg (2002)
4. de Berg, M., van Kreveld, M., Overmars, M., Schwarzkopf, O.: Computational geometry - algorithms and Applications. Springer (1997)
5. Carlsson, M., et al.: SICStus Prolog User's Manual. SICS, 4.2.1 edn. (2012), http://www.sics.se/sicstus
6. Freuder, E., Lee, J., O'Sullivan, B., Pesant, G., Rossi, F., Sellman, M., Walsh, T.: The future of CP. Personal communication (2011)
7. Hermenier, F., Demassey, S., Lorca, X.: Bin Repacking Scheduling in Virtualized Datacenters. In: Lee, J. (ed.) CP 2011. LNCS, vol. 6876, pp. 27–41. Springer, Heidelberg (2011)
8. Hermenier, F., Lorca, X., Menaud, J.M., Muller, G., Lawall, J.: Entropy: a consolidation manager for clusters. In: VEE 2009, pp. 41–50. ACM (2009)
9. Kameugne, R., Fotso, L.P., Scott, J., Ngo-Kateu, Y.: A Quadratic Edge-Finding Filtering Algorithm for Cumulative Resource Constraints. In: Lee, J. (ed.) CP 2011. LNCS, vol. 6876, pp. 478–492. Springer, Heidelberg (2011)
10. O'Sullivan, B.: CP panel position - the future of CP. Personal communication (2011)
11. Pape, C.L.: Des systèmes d'ordonnacement flexibles et opportunistes. Ph.D. thesis, Université Paris IX (1988) (in French)
12. Régin, J.C., Rezgui, M.: Discussion about constraint programming bin packing models. In: AI for Data Center Management and Cloud Computing. AAAI (2011)
13. ROADEF: Challenge 2012 machine reassignment (2012), http://challenge.roadef.org/2012/en/index.php
14. Schaus, P., Deville, Y.: A global constraint for bin-packing with precedences: application to the assembly line balancing problem. In: AAAI 2008, pp. 369–374. AAAI Press (2008)
15. Schulte, C.: Comparing trailing and copying for constraint programming. In: Schreye, D.D. (ed.) ICLP 1999, pp. 275–289. The MIT Press (1999)
16. Schutt, A., Feydy, T., Stuckey, P.J., Wallace, M.G.: Why *Cumulative* Decomposition Is Not as Bad as It Sounds. In: Gent, I.P. (ed.) CP 2009. LNCS, vol. 5732, pp. 746–761. Springer, Heidelberg (2009)
17. Shaw, P.: Using Constraint Programming and Local Search Methods to Solve Vehicle Routing Problems. In: Maher, M.J., Puget, J.-F. (eds.) CP 1998. LNCS, vol. 1520, pp. 417–431. Springer, Heidelberg (1998)
18. Shaw, P.: A Constraint for Bin Packing. In: Wallace, M. (ed.) CP 2004. LNCS, vol. 3258, pp. 648–662. Springer, Heidelberg (2004)
19. Team, C.: Choco: an open source Java CP library. Research report 10-02-INFO, Ecole des Mines de Nantes (2010), http://choco.emn.fr/
20. Vilím, P.: Edge Finding Filtering Algorithm for Discrete Cumulative Resources in $O(kn \log n)$. In: Gent, I.P. (ed.) CP 2009. LNCS, vol. 5732, pp. 802–816. Springer, Heidelberg (2009)
21. Vilím, P.: Timetable Edge Finding Filtering Algorithm for Discrete Cumulative Resources. In: Achterberg, T., Beck, J.C. (eds.) CPAIOR 2011. LNCS, vol. 6697, pp. 230–245. Springer, Heidelberg (2011)

A New Encoding from MinSAT into MaxSAT[*]

Zhu Zhu[1], Chu-Min Li[1], Felip Manyà[2], and Josep Argelich[3]

[1] MIS, Université de Picardie Jules Verne, Amiens, France
`{zhu.zhu,chu-min.li}@u-picardie.fr`
[2] Artificial Intelligence Research Institute (IIIA, CSIC), Bellaterra, Spain
`felip@iiia.csic.es`
[3] Dept. of Computer Science, Universitat de Lleida, Lleida, Spain
`jargelich@diei.udl.cat`

Abstract. MinSAT is the problem of finding a truth assignment that minimizes the number of satisfied clauses in a CNF formula. When we distinguish between hard and soft clauses, and soft clauses have an associated weight, then the problem, called Weighted Partial MinSAT, consists in finding a truth assignment that satisfies all the hard clauses and minimizes the sum of weights of satisfied soft clauses. In this paper we define a novel encoding from Weighted Partial MinSAT into Weighted Partial MaxSAT, which is also valid for encoding Weighted Partial MaxSAT into Weighted Partial MinSAT. Moreover, we report on an empirical investigation that shows that our encoding significantly outperforms existing encodings on weighted and unweighted Min2SAT and Min3SAT instances.

1 Introduction

Solving NP-complete decision problems by reducing them to the propositional satisfiability problem (SAT) is a powerful solving strategy that is widely used to tackle both academic and industrial problems. Recently, the success of SAT has led to explore MaxSAT formalisms such as Weighted MaxSAT and Weighted Partial MaxSAT [10] for solving practical optimization problems. Nowadays, MaxSAT formalisms are quite competitive on certain domains, and we believe that the development of new solving techniques and the annual celebration of a MaxSAT Evaluation [1–3] will act as a driving force to incorporate MaxSAT technology in industrial environments.

In this paper we focus on MinSAT, which is close to MaxSAT but the goal now is to minimize the cost of satisfied clauses instead of maximizing that cost. Specifically, we focus on the Weighted Partial MinSAT problem, where instances are formed by a set of clauses, each clause is declared to be either hard or soft, and each soft clause has an associated weight. Solving a Weighted Partial MinSAT instance amounts to finding an assignment that satisfies all the hard clauses, and minimizes the sum of the weights of satisfied soft clauses. On the one hand, we distinguish between constraints that are

---

[*] Research partially funded by French ANR UNLOC project: ANR-08-BLAN-0289-03, National Natural Science Foundation of China (NSFC) grant No. 61070235, the Generalitat de Catalunya under grant AGAUR 2009-SGR-1434, the Ministerio de Economía y Competividad research projects AT CONSOLIDER CSD2007-0022, INGENIO 2010, ARINF TIN2009-14704-C03-01, TASSAT TIN2010-20967-C04-01/03, and Newmatica INNPACTO IPT-2011-1496-310000 (funded by the Ministerio de Ciencia y Tecnología until 2011).

M. Milano (Ed.): CP 2012, LNCS 7514, pp. 455–463, 2012.

compulsory (hard) and constraints that can be relaxed (soft). On the other hand, we establish a priority among soft constraints by assigning them a weight that represents the significance of the constraint.

Although MinSAT and MaxSAT are both extensions of SAT, their solving techniques are quite different as well as complementary in the sense that problems that are beyond the reach of current MaxSAT solvers can be solved with MinSAT solvers, and vice versa [14, 15]. Because of that, we believe that it makes sense to define suitable encodings from MinSAT (MaxSAT) into MaxSAT (MinSAT). To the best of our knowledge, there exist two papers addressing this kind of encodings [8, 13]. In [13], a number of encodings were defined to reduce (unweighted) MinSAT to Partial MaxSAT. One drawback of that work is that the defined encodings do not generalize to Weighted Partial MinSAT. More recently, Kügel [8] has defined the so-called natural encoding (c.f. Section 3), which inspired the present work.

In this paper we define a novel encoding, called natural flow network encoding, from Weighted Partial MinSAT into Weighted Partial MaxSAT which, as Kügel's encoding, is also valid for encoding Weighted Partial MaxSAT into Weighted Partial MinSAT. Our encoding improves Kügel's encoding by allowing to detect earlier more contradictions, and by performing a more efficient treatment of weights. Moreover, we report on an empirical investigation that shows that our encoding significantly outperforms existing encodings on weighted and unweighted Min2SAT and Min3SAT instances.

The paper is structured as follows. Section 2 introduces basic concepts that are used in the rest of the paper. Section 3 presents first the natural encoding, and then the natural flow network encoding. Section 4 reports on the empirical investigation, and Section 5 presents the conclusions.

2 Preliminaries

A literal is a propositional variable or a negated propositional variable. A clause is a disjunction of literals. A weighted clause is a pair (c, w), where c is a clause and w, its weight, is a natural number or infinity. A clause is hard if its weight is infinity; otherwise it is soft. A Weighted Partial MinSAT (MaxSAT) instance is a multiset of weighted clauses $\phi = \{(h_1, \infty), \ldots, (h_k, \infty), (c_1, w_1), \ldots, (c_m, w_m)\}$, where the first k clauses are hard and the last m clauses are soft. For simplicity, in what follows, we omit infinity weights, and write $\phi = \{h_1, \ldots, h_k, (c_1, w_1), \ldots, (c_m, w_m)\}$. Notice that a soft clause (c, w) is equivalent to having w copies of the clause $(c, 1)$,[1] and that $\{(c, w_1), (c, w_2)\}$ is equivalent to $(c, w_1 + w_2)$.

A truth assignment is a mapping that assigns to each propositional variable either 0 or 1. The cost of a truth assignment I for ϕ is the sum of the weights of the soft clauses satisfied by I. The Weighted Partial MinSAT problem for an instance ϕ consists in finding an assignment with minimum cost that satisfies all the hard clauses (i.e, an optimal assignment), while the Weighted Partial MaxSAT problem consists in finding an assignment with maximum cost that satisfies all the hard clauses. The Weighted MinSAT (MaxSAT) problem is the Weighted Partial MinSAT (MaxSAT) problem when there are no hard clauses. The Partial MinSAT (MaxSAT) problem is the Weighted Partial

---

[1] From a complexity point of view, this transformation can be exponential in the worst case.

MinSAT (MaxSAT) problem when all soft clauses have weight 1. The (Unweighted) MinSAT (MaxSAT) problem is the Partial MinSAT (MaxSAT) problem when there are no hard clauses. The SAT problem is the Partial MaxSAT or the Partial MinSAT problem when there are no soft clauses.

A flow network is a directed graph $G = (V, E)$ with two distinguished vertices: the source s, with no incoming edges, and the sink t, with no outgoing edges. We will assume that every vertex $v \in V$ is in some path from s to t. Every edge (u, v) has a non-negative integer weight, called the capacity $c(u, v)$ of (u, v); if $(u, v) \notin E$, then $c(u, v) = 0$. The capacity represents the maximum amount of flow that can pass through an edge. A flow is a real-valued function $f : V \times V \rightarrow R^+$ subject to the following constraints: (i) for all $u, v \in V$, $0 \leq f(u, v) \leq c(u, v)$ (capacity constraint: the flow of an edge cannot exceed its capacity), and (ii) for all $u \in V \setminus \{s, t\}$, $\sum_{v \in V} f(u, v) - \sum_{v \in V} f(v, u) = 0$ (conservation of flows: the sum of the flows entering a vertex must equal the sum of the flows exiting a vertex, except for the source and the sink).

The value of a flow $|f|$ is defined as follows: $|f| = \sum_{v:(s,v) \in E} f(s, v) = \sum_{u:(u,t) \in E} f(u, t)$. It represents the amount of flow passing from the source to the sink. The maximum flow problem for a flow network G consists in finding a flow of maximum value on G (i.e., to route as much flow as possible from the source to the sink).

3   Encodings from MinSAT (MaxSAT) into MaxSAT(MinSAT)

3.1   The Natural Encoding (NE)

Kügel has recently defined an encoding, called Natural Encoding (NE), from Weighted Partial MinSAT into Weighted Partial MaxSAT, which is also valid for encoding Weighted Partial MaxSAT into Weighted Partial MinSAT. Encoding NE does not introduce auxiliary variables and just increases the number of clauses by a factor of the clause size. Furthermore, any optimal assignment of the MaxSAT instance is an optimal assignment of the MinSAT instance, and vice versa.

Given a Weighted Partial MinSAT instance, encoding NE consists in creating a Weighted Partial MaxSAT instance with the same hard clauses, and replacing every soft clause with its negation in clausal form. Since negation has to preserve the number of unsatisfied clauses, the negation in conjunctive normal form of the soft clause $(c, w) = (l_1 \vee l_2 \vee \cdots \vee l_k, w)$, denoted by $\mathrm{CNF}\overline{(c, w)}$, is defined as follows:[2]

$$\mathrm{CNF}\overline{(c, w)} = (\neg l_1, w) \wedge (l_1 \vee \neg l_2, w) \wedge \cdots \wedge (l_1 \vee l_2 \vee \cdots \vee \neg l_k, w)$$

Example 1. Let $\phi = \{x_1 \vee x_2, (x_1 \vee x_3, 2), (\neg x_2 \vee x_3, 3)\}$ be a Weighted Partial MinSAT instance. Encoding NE generates the following Weighted Partial MaxSAT instance:

$$\{x_1 \vee x_2, (\neg x_1, 2), (x_1 \vee \neg x_3, 2), (x_2, 3), (\neg x_2 \vee \neg x_3, 3)\}.$$

---

[2] Recall that this negation was first used for defining resolution-like inference in MaxSAT [4, 9]. Also recall that if an assignment falsifies (c, w), then it satisfies all the clauses in $\mathrm{CNF}\overline{(c, w)}$; and if an assignment satisfies (c, w), then it falsifies exactly one clause in $\mathrm{CNF}\overline{(c, w)}$.

Encoding NE relies on the following equations:

$$\mathrm{wpmin}(\{h_1, \ldots, h_k, (c_1, w_1), \ldots, (c_m, w_m)\}) = \\ \mathrm{wpmax}(\{h_1, \ldots, h_k, \mathrm{CNF}(c_1, w_1), \ldots, \mathrm{CNF}(c_m, w_m)\})$$

$$\mathrm{wpmax}(\{h_1, \ldots, h_k, (c_1, w_1), \ldots, (c_m, w_m)\}) = \\ \mathrm{wpmin}(\{h_1, \ldots, h_k, \mathrm{CNF}(c_1, w_1), \ldots, \mathrm{CNF}(c_m, w_m)\})$$

where $\mathrm{wpmin}(\phi)$ is the maximum sum of weights of falsified soft clauses in the MinSAT instance ϕ, and $\mathrm{wpmax}(\phi)$ is the minimum sum of weights of falsified soft clauses in the MaxSAT instance ϕ. Recall that branch-and-bound MinSAT (MaxSAT) solvers find an optimal solution by maximizing (minimizing) the sum of weights of falsified clauses.

3.2   The Natural Flow Network Encoding (NFNE)

In this section we define a novel encoding, called Natural Flow Network Encoding (NFNE), that improves Kügel's encoding. Encoding NFNE relies on the following observation: the way of definining $\mathrm{CNF}(c, w)$ is not unique, it depends on the ordering in which we consider the literals of the clause, and this ordering is important because contradictions are detected if complementary unit clauses are derived. For example, $\mathrm{CNF}(x_1 \vee x_2 \vee x_3, w)$ can be defined as either $(\neg x_1, w), (x_1 \vee \neg x_2, w), (x_1 \vee x_2 \vee \neg x_3, w)$, or $(\neg x_2, w), (x_2 \vee \neg x_1, w), (x_2 \vee x_1 \vee \neg x_3, w)$, or ... Furthermore, there are cases where it is advantageous to split a weighted clause (c, w) into several clauses in such a way that each one of these clauses generates a different unit clause when we derive $\mathrm{CNF}(c, w)$.

For example, in order to encode the clauses $\{(x_1 \vee x_2, 5), (\neg x_1 \vee x_3, 2), (\neg x_2 \vee \neg x_4, 3)\}$ into MaxSAT, we split $(x_1 \vee x_2, 5)$ into the clauses $(x_1 \vee x_2, 2)$ and $(x_1 \vee x_2, 3)$, derive the unit clause $(\neg x_1, 2)$ in $\mathrm{CNF}(x_1 \vee x_2, 2)$, derive the unit clause $(\neg x_2, 3)$ in $\mathrm{CNF}(x_1 \vee x_2, 3)$, derive the unit clause $(x_1, 2)$ in $\mathrm{CNF}(\neg x_1 \vee x_3, 2)$, and derive the unit clause $(x_2, 3)$ in $\mathrm{CNF}(\neg x_2 \vee \neg x_4, 3)$, then we derive an empty clause with weight 5 when the complementary unit clauses of the obtained Partial MaxSAT instance are resolved. If $(x_1 \vee x_2, 5)$ is not split, we derive an empty clause with weight at most 3.

Our aim is to define an encoding that splits weighted clauses and selects the literals in such a way that the number of complementary unit clauses that are derived is maximized. To this end, we represent the soft clauses in a flow network G and then compute a maximum value flow on G using the Ford-Fulkerson algorithm [7].

Given a Weighted Partial MinSAT instance $\phi = \{h_1, \ldots, h_k, (c_1, w_1), \ldots, (c_m, w_m)\}$, without tautological clauses, over the set of propositional variables $\{x_1, \ldots, x_n\}$, we create a flow network G with set of vertices $\{s, t, c_1^+, \ldots, c_m^+, c_1^-, \ldots, c_m^-, x_1, \ldots, x_n, \neg x_1, \ldots, \neg x_n\}$, and having the following edges:

1. There is an edge (s, c_i^+) with capacity w_i for all $0 \leq i \leq m$.
2. There is an edge (c_i^-, t) with capacity w_i for all $0 \leq i \leq m$.
3. There is an edge $(x_i, \neg x_i)$ with capacity $\sum_{i=0}^{m} w_i$ for all $0 \leq i \leq m$.
4. There is an edge (c_i^+, x_j) with capacity w_i for each clause (c_i, w_i), $0 \leq i \leq m$, containing the literal x_j for all $0 \leq j \leq n$.

5. There is an edge $(\neg x_j, c_i^-)$ with capacity w_i for each clause (c_i, w_i), $0 \le i \le m$, containing the literal $\neg x_j$ for all $0 \le j \le n$.

Given a flow f of maximum value on G, we derive a Weighted Partial MaxSAT instance ϕ' as follows:

1. The hard clauses of ϕ' and ϕ are the same.
2. For each path P from s to t of the form $s \to c_i^+ \to x_j \to \neg x_j \to c_k^- \to t$ that does not contain edges with flow 0, add the soft clauses of $\text{CNF}\overline{(c_i, f(c_i^+, x_j))}$ having as unit clause $(\neg x_j, f(c_i^+, x_j))$, and the soft clauses of $\text{CNF}\overline{(c_k, f(\neg x_j, c_k^-))}$ having as unit clause $(x_j, f(\neg x_j, c_k^-))$. Clauses corresponding to edges belonging to several paths are added only once.
3. For each soft clause $(c_i, w_i) \in \phi$, add the soft clauses of $\text{CNF}\overline{(c_i, w_i - w_i^s)}$ if $w_i > w_i^s$, where $w_i^s = \sum_{k=1}^{n} f(c_i^+, x_k) + \sum_{k=1}^{n} f(\neg x_k, c_i^-)$. In this case, there is no constraint on the literal used to generate the unit clause. We use the first literal in lexicographical ordering.

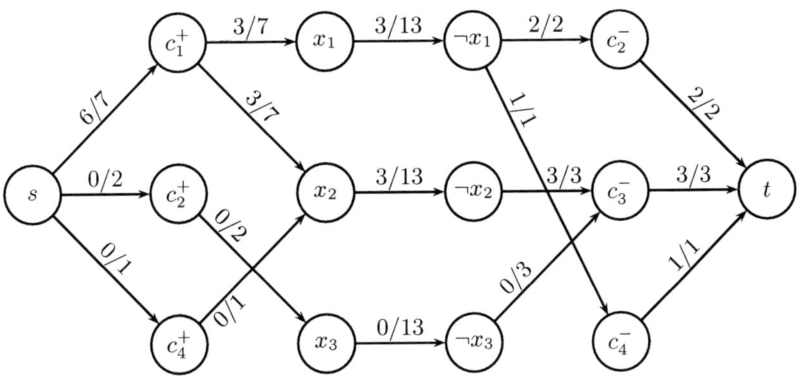

Fig. 1. Flow network for the MinSAT instance $\phi = \{(c_1, w_1), (c_2, w_2), (c_3, w_3), (c_4, w_4)\} = \{(x_1 \lor x_2, 7), (\neg x_1 \lor x_3, 2), (\neg x_2 \lor \neg x_3, 3), (\neg x_1 \lor x_2, 1)\}$

Example 2. Let $\phi = \{(c_1, w_1), (c_2, w_2), (c_3, w_3), (c_4, w_4)\} = \{(x_1 \lor x_2, 7), (\neg x_1 \lor x_3, 2), (\neg x_2 \lor \neg x_3, 3), (\neg x_1 \lor x_2, 1)\}$ be a Weighted Partial MinSAT instance. For deriving the NFNE encoding, we first build the graph $G = (V, E)$ of Figure 1, and compute a flow of maximum value. Each edge of G has a label of the form f/c, where f is the edge flow value and c is the edge capacity.

We have 3 paths where there is no edge with flow value 0:

1. $P_1 = s \to c_1^+ \to x_1 \to \neg x_1 \to c_2^- \to t$. From this path, we derive the soft clauses $\phi_1' = \{\text{CNF}\overline{(x_1 \lor x_2, 3)}, \text{CNF}\overline{(\neg x_1 \lor x_3, 2)}\}$, deriving unit clauses $(\neg x_1, 3)$ and $(x_1, 2)$.
2. $P_2 = s \to c_1^+ \to x_1 \to \neg x_1 \to c_4^- \to t$. From this path, we derive the soft clauses $\phi_2' = \text{CNF}\overline{(\neg x_1 \lor x_2, 1)}$, deriving unit clause $(x_1, 1)$. We do not add

$\mathrm{CNF}(\overline{x_1 \vee x_2, 3})$ because the edge associated with these clauses was considered in path P_1.

3. $P_3 = s \rightarrow c_1^+ \rightarrow x_2 \rightarrow \neg x_2 \rightarrow c_3^- \rightarrow t$. From this path, we derive the soft clauses $\phi_3' = \{\mathrm{CNF}(\overline{x_1 \vee x_2, 3}), \mathrm{CNF}(\overline{\neg x_2 \vee \neg x_3, 3})\}$, deriving unit clauses $(\neg x_2, 3)$ and $(x_2, 3)$.

Moreover, we should add the clauses $\mathrm{CNF}(\overline{x_1 \vee x_2, 1})$ because $w_1 = 7$ and $w_1^s = 6$. In summary, the resulting encoding ϕ' is formed by $\phi_1' \cup \phi_2' \cup \phi_3' \cup \mathrm{CNF}(\overline{x_1 \vee x_2, 1})$.

4 Experimental Results

We conducted experiments for comparing the performance of encoding NFNE with the performance of the encoding NE defined in [8], where it was shown that NE is significantly better than the encodings defined in [13]. As benchmarks we used weighted and unweighted Min2SAT and Min3SAT instances as in [8]. We do not use the more applied instances in [14] (maxclique and combinatorial auctions) because all their soft clauses are unit and, in this case, there is no difference between the unit clauses generated with NE and NFNE. Benchmarks were solved on a 2.70 Ghz intel XEON E5-2680 CPU with Linux and 8 Gb memory, using a cutoff time of 1800 seconds.

The solvers used in our empirical investigation are:

- Maxsatz [11, 12]: MaxSatz is a representative branch-and-bound Weighted Partial MaxSAT solver that is particularly efficient on random MaxSAT instances. We used the version submitted to the 2011 MaxSAT Evaluation.
- MinSatz: The implementation of the Weighted Partial MinSAT solver used in [14]. It is the only publicly available exact Weighted Partial MinSAT solver.

We generated weighted and unweighted Min2SAT instances with 160, 180 and 200 variables, the clause-to-variable ratio being 4, 5 and 6 for each number of variables. The Min3SAT instances have 70, 80 and 90 variables, and the clause-to-variable ratio also ranges from 4 to 6. For each number of variables, and for each ratio, 50 instances were generated and solved. The weight of each weighted clause is randomly generated between 1 and 10. These instances are generated in the same way as in the MaxSAT Evaluation.

Table 1 and Table 2 compare the performance of MinSatz and MaxSatz when solving the unweighted random Min2SAT and Min3SAT instances. The MinSAT instances solved with MaxSatz were encoded using encoding NE of Kügel and our new encoding NFNE. Each entry gives the number of instances solved within 1800 seconds (in brackets), and the average run time of the instances solved within the cutoff time (including the time needed to generate the encodings), as well as the average search tree size of these solved instances.

As Table 1 shows, MaxSatz using encoding NE is slower than the dedicated MinSAT solver MinSatz on Min2SAT instances, essentially because the search tree is too large for these instances. On the contrary, encoding NFNE is substantially better than NE, and makes MaxSatz even faster than MinSatz on Min2SAT instances. We observe that NFNE especially allows to reduce the search tree size of MaxSatz for these instances, by deriving more empty clauses more quickly than NE.

Table 1. Number of instances solved within 30 minutes (in brackets), average run times in seconds, and search tree size of MaxSatz and MinSatz for random unweighted Min2SAT instances. C/V is the clause-to-variable ratio.

instance		$MinSatz$		$MaxSatz(NE)$		$MaxSatz(NFNE)$	
#var	C/V	time	size	time	size	time	size
160	4	**0.01(50)**	3511	1(50)	17953	**0.01(50)**	455
180	4	**0.01(50)**	5191	4(50)	44210	**0.01(50)**	790
200	4	4(50)	10227	32(50)	228807	**0.01(50)**	2143
160	5	10(50)	17005	35(50)	212903	**0.01(50)**	3372
180	5	36(50)	47446	138(49)	718034	**1(50)**	7348
200	5	95(50)	99360	509(46)	2143832	**4(50)**	13502
160	6	96(49)	104737	184(50)	769490	**2(50)**	11047
180	6	354(44)	313081	759(38)	2623855	**19(50)**	55947
200	6	703(30)	519333	820(5)	2249294	**75(50)**	193008

Table 2. Number of instances solved within 30 minutes (in brackets), average run times in seconds, and search tree size of MaxSatz and MinSatz for random unweighted Min3SAT instances. C/V is the clause-to-variable ratio.

instance		$MinSatz$		$MaxSatz(NE)$		$MaxSatz(NFNE)$	
#var	C/V	time	size	time	size	time	size
70	4	**0.01(50)**	4906	3(50)	55817	**0.01(50)**	15081
80	4	**2(50)**	22031	23(50)	252116	7(50)	77994
90	4	**6(50)**	46174	82(50)	735247	21(50)	184906
70	5	**2(50)**	19330	15(50)	145294	6(50)	64111
80	5	**15(50)**	84226	77(50)	584794	36(50)	279558
90	5	**103(50)**	481784	443(50)	2961189	215(50)	1401195
70	6	**12(50)**	66989	42(50)	314221	23(50)	178947
80	6	**88(50)**	374512	286(50)	1814812	154(50)	947687
90	6	**528(47)**	1944530	1078(28)	5458486	707(38)	3580599

Table 2 shows that encoding NFNE is always substantially better than encoding NE on Min3SAT instances, although the gain is less spectacular than for Min2SAT. Observe that NFNE could be further improved for Min3SAT (or Min$k$SAT for $k \geq 3$), since after determining the first literal to derive the unit clause, we have the choice of the second literal for the binary clause. For example, in order to encode the soft clauses $\{(x_1 \vee x_2 \vee x_3, 1), (\neg x_1 \vee x_3 \vee x_4, 1)\}$, we can derive the unit clause $(\neg x_1, 1)$ and the binary clause $(x_1 \vee \neg x_3, 1)$ (by choosing x_3 for the binary clause) from $\text{CNF}(x_1 \vee x_2 \vee x_3, 1)$, and the unit clause $(x_1, 1)$ and the binary clause $(\neg x_1 \vee \neg x_3, 1)$ from $\text{CNF}(\neg x_1 \vee x_3 \vee x_4, 1)$, so that MaxSAT resolution can derive an empty clause from the two unit clauses, and an unit clause $(\neg x_3, 1)$ from the two binary clauses.

Table 3 and Table 4 show the results for weighted Min2SAT and Min3SAT. Encoding NFNE is always significantly better than encoding NE. The performance difference between NFNE and NE for weighted instances appears to be larger than for unweighted instances. In fact, MaxSatz using NFNE solves 73 weighted Min2SAT instances more than MaxSatz using NE, while MaxSatz using NFNE solves 62 unweighted Min2SAT instances more than MaxSatz using NE. In addition, MaxSatz using NFNE solves 21 weighted Min3SAT instances more than MaxSatz using NE, while MaxSatz using NFNE solves 10 unweighted Min3SAT instances more than MaxSatz using NE. Recall that, in the weighted case, NFNE allows to split the weight of clauses to maximize the weight of complementary unit clauses, while NE does not.

Table 3. Number of instances solved within 30 minutes (in brackets), average run times in seconds, and search tree size of MaxSatz and MinSatz for random weighted Min2SAT instances. C/V is the clause-to-variable ratio.

$instance$		$MinSatz$		$MaxSatz(NE)$		$MaxSatz(NFNE)$	
#var	C/V	$time$	$size$	$time$	$size$	$time$	$size$
160	4	**0.01(50)**	9037	3(50)	18045	**0.01(50)**	385
180	4	1(50)	11363	14(50)	46142	**0.01(50)**	529
200	4	3(50)	13661	64(50)	193995	**0.01(50)**	772
160	5	2(50)	11299	46(50)	124140	**0.01(50)**	730
180	5	6(50)	15170	193(50)	398049	**0.01(50)**	1049
200	5	23(50)	24030	544(37)	979496	**3(50)**	2201
160	6	19(50)	18011	348(50)	647277	**1(50)**	1438
180	6	69(50)	31774	876(33)	1427495	**5(50)**	2653
200	6	168(48)	55038	1218(7)	1873148	**17(50)**	5402

Table 4. Number of instances solved within 30 minutes (in brackets), average run times in seconds, and search tree size of MaxSatz and MinSatz for random weighted Min3SAT instances. C/V is the clause-to-variable ratio.

$instance$		$MinSatz$		$MaxSatz(NE)$		$MaxSatz(NFNE)$	
#var	C/V	$time$	$size$	$time$	$size$	$time$	$size$
70	4	**0.01(50)**	2214	5(50)	21095	1(50)	3121
80	4	**0.01(50)**	4198	28(50)	83732	5(50)	7764
90	4	**2(50)**	8283	101(50)	245622	19(50)	20496
70	5	**1(50)**	4394	20(50)	51106	7(50)	9219
80	5	**6(50)**	14310	110(50)	209113	43(50)	41907
90	5	**29(50)**	42583	445(48)	723592	181(50)	129093
70	6	**6(50)**	12013	61(50)	109430	35(50)	33545
80	6	**34(50)**	45771	343(50)	492617	210(50)	144586
90	6	**145(50)**	155974	1002(26)	1166419	689(45)	389266

5   Conclusions

Observing that MaxSAT and MinSAT solving techniques are very different, and complementary in the sense that problems beyond the reach of the current MaxSAT solvers can be solved using MinSAT solvers and vice versa, we have proposed a new encoding from MinSAT into MaxSAT, which can also be used to encode MaxSAT into MinSAT, allowing to solve MinSAT problems using MaxSAT solvers, and vice versa. The new encoding is based on modeling the relationships between clauses and literals via flow networks, in which a maximum flow can be computed using the Ford-Fulkerson algorithm no matter how clauses and literals are ordered. This approach allows to maximize the sum of weights of complementary unit clauses in the resulting MaxSAT instance, so that empty clause with the maximum weight can be derived quickly. Moreover, we conducted experiments that show that the new encoding is significantly better on weighted and unweighted Min2SAT and Min3SAT instances.

In the future, we plan to improve the new encoding for Min$k$SAT ($k \geq 3$), because we could yet derive additional unit clauses after deriving the empty clauses. Another possible future research direction, pointed out by a reviewer, is to investigate how the reformulation techniques proposed by Boros & Hammer for Max2SAT [5], and Cooper et al. for the valued CSP framework [6] compare with the flow network techniques we have defined in the present paper.

References

1. Argelich, J., Li, C.M., Manyà, F., Planes, J.: The first and second Max-SAT evaluations. Journal on Satisfiability, Boolean Modeling and Computation 4, 251–278 (2008)
2. Argelich, J., Li, C.M., Manyà, F., Planes, J.: Analyzing the Instances of the MaxSAT Evaluation. In: Sakallah, K.A., Simon, L. (eds.) SAT 2011. LNCS, vol. 6695, pp. 360–361. Springer, Heidelberg (2011)
3. Argelich, J., Li, C.M., Manyà, F., Planes, J.: Experimenting with the instances of the MaxSAT Evaluation. In: Proceedings of the 14th International Conference of the Catalan Association for Artificial Intelligence, CCIA 2011, Lleida, Catalonia, Spain, pp. 31–40 (2011)
4. Bonet, M.L., Levy, J., Manyà, F.: Resolution for Max-SAT. Artificial Intelligence 171(8-9), 240–251 (2007)
5. Boros, E., Hammer, P.L.: Pseudo-boolean optimization. Discrete Applied Mathematics 123(1-3), 155–225 (2002)
6. Cooper, M.C., de Givry, S., Sanchez, M., Schiex, T., Zytnicki, M., Werner, T.: Soft arc consistency revisited. Artif. Intell. 174(7-8), 449–478 (2010)
7. Ford, L.R., Fulkerson, D.R.: Flows in Networks. Princeton University Press, Princeton (1962)
8. Kügel, A.: Natural Max-SAT encoding of Min-SAT. In: Proceedings of the Learning and Intelligent Optimization Conference, LION 6, Paris, France (2012)
9. Larrosa, J., Heras, F., de Givry, S.: A logical approach to efficient Max-SAT solving. Artificial Intelligence 172(2-3), 204–233 (2008)
10. Li, C.M., Manyà, F.: MaxSAT, hard and soft constraints. In: Biere, A., van Maaren, H., Walsh, T. (eds.) Handbook of Satisfiability, pp. 613–631. IOS Press (2009)
11. Li, C.M., Manyà, F., Mohamedou, N.O., Planes, J.: Resolution-based lower bounds in MaxSAT. Constraints 15(4), 456–484 (2010)
12. Li, C.M., Manyà, F., Planes, J.: New inference rules for Max-SAT. Journal of Artificial Intelligence Research 30, 321–359 (2007)
13. Li, C.M., Manyà, F., Quan, Z., Zhu, Z.: Exact MinSAT Solving. In: Strichman, O., Szeider, S. (eds.) SAT 2010. LNCS, vol. 6175, pp. 363–368. Springer, Heidelberg (2010)
14. Li, C.M., Zhu, Z., Manyà, F., Simon, L.: Minimum satisfiability and its applications. In: Proceedings of the 22nd International Joint Conference on Artificial Intelligence, IJCAI 2011, Barcelona, Spain, pp. 605–610 (2011)
15. Li, C.M., Zhu, Z., Manyà, F., Simon, L.: Optimizing with minimum satisfiability. Artificial Intelligence (2012), http://dx.doi.org/10.1016/j.artint.2012.05.004

Solving Minimal Constraint Networks
in Qualitative Spatial and Temporal Reasoning

Weiming Liu and Sanjiang Li

Centre for Quantum Computation and Intelligent Systems, Faculty of Engineering and
Information Technology, University of Technology Sydney, Australia

Abstract. The *minimal label problem* (MLP) (also known as the deductive closure problem) is a fundamental problem in qualitative spatial and temporal reasoning (QSTR). Given a qualitative constraint network Γ, the minimal network
of Γ relates each pair of variables (x, y) by the minimal label of (x, y), which is
the minimal relation between x, y that is entailed by network Γ. It is well-known
that MLP is equivalent to the corresponding consistency problem with respect to
polynomial Turing-reductions. This paper further shows, for several qualitative
calculi including Interval Algebra and RCC-8 algebra, that deciding the minimality of qualitative constraint networks and computing a solution of a minimal
constraint network are both NP-hard problems.

1 Introduction

Spatial and temporal information is pervasive and increasingly involved in both industry
and everyday life. Many tasks in real or virtual world demand sophisticated spatial and
temporal reasoning systems. The qualitative approach to spatial and temporal reasoning
(QSTR) provides a promising framework and has boosted research and applications in
areas such as natural language processing, geographical information systems, robotics,
content-based image retrieval (see e.g. [5]).

Concentrating on different aspects of the space and/or time, dozens of qualitative relation models have been proposed, which are called qualitative calculi in QSTR. While
Point Algebra [2] and Interval Algebra [1] are two most popular qualitative temporal
calculi, the Cardinal Relation Algebra [13] and the RCC-8 algebra [18] are two popular
qualitative spatial calculi which model directional and, respectively, topological spatial
information. Roughly speaking, a qualitative calculus has a fixed (usually infinite) universe of entities (e.g., temporal points in the real line) and uses a finite vocabulary (viz.,
basic relations, e.g., $<, >$ and $=$) to model the relationship between the entities. In a
qualitative calculus, basic relations are used to represent the knowledge of which we
are certain, while non-basic relations (e.g., $\leq, \geq$) are used for uncertain knowledge.

In QSTR, spatial or temporal information is usually represented in terms of basic
or non-basic relations in a qualitative calculus, and reasoning tasks are formulated as
solving a set of qualitative constraints (called a *qualitative constraint network*). A qualitative constraint is a formula of the form xRy, which asserts that variables x and y
should be interpreted by two entities in the universe such that relation R holds between
them, where R could be a basic or non-basic relation. The *consistency problem* is to

M. Milano (Ed.): CP 2012, LNCS 7514, pp. 464–479, 2012.

decide whether a set of constraints can be satisfied simultaneously, i.e., whether there is an interpretation of all the variables such that the constraints are all satisfied by this interpretation. Note that the universe of a qualitative calculus is usually infinite, which implies that the general techniques developed in the community of classical CSP can hardly be applied directly. Even though, researches have successfully solved the consistency problem in a number of qualitative calculi including Point Algebra [2], Interval Algebra [1,17] and the RCC-8 algebra [19].

The *minimal label problem* (MLP) (also known as the deductive closure problem) is another fundamental problem in QSTR, which is equivalent to the corresponding consistency problem with respect to polynomial Turing-reductions. A qualitative constraint network Γ is called *minimal* if for each constraint, say xRy, R is the minimal label of (x, y), i.e., R is the minimal relation between x, y that is entailed by Γ.

Since its introduction in 1974, minimal constraint network [16] has drawn attention from both classical CSP (see [10] and references therein) and QSTR [4,9] researchers. In a recent paper [10], Gottlob proved that, in classical CSP, it is NP-hard to compute a solution of a minimal constraint network, which confirms a conjecture proposed by Dechter [6]. Inspired by this work, we investigate the same problem for minimal constraint networks in the context of QSTR. For partially ordered Point Algebra (in which two points can be incomparable), Interval Algebra, Cardinal Relation Algebra, and RCC-8 algebra, we find that deciding the minimality of constraint networks is in general NP-hard. Furthermore, we prove that it is also NP-hard to compute solutions of minimal RCC-8 constraint networks, though it was assumed in [4] that this can be easily accomplished. In fact, for all qualitative calculi mentioned above, it is NP-hard to compute even a single solution of a minimal constraint network.

The remainder of this paper proceeds as follows. Section 2 introduces basic notions as well as the qualitative calculi discussed in this paper. Section 3 shows that computing a solution of a minimal constraint network in partially ordered Point Algebra and RCC-8 algebra is NP-hard, while Section 4 proves the same result for Cardinal Relation Algebra and Interval Algebra. The last section concludes the paper.

2   Preliminaries

In this section, we introduce basic notions in qualitative constraint solving and the four important qualitative calculi to be discussed later.

2.1   Qualitative Calculi

QSTR is mainly based on qualitative calculi. Suppose U is the universe of spatial or temporal entities. Write $\mathbf{Rel}(U)$ for the Boolean algebra of binary relations on U. Binary *qualitative calculi* (cf. e.g.[14]) on U are finite Boolean subalgebras of $\mathbf{Rel}(U)$. Note that $\mathcal{M}$ contains the universal relation (denoted by ?) which is defined as $U \times U$.

Let $\mathcal{M}$ be a qualitative calculus on U. We call a relation α in $\mathcal{M}$ a *basic* relation if it is an atom in $\mathcal{M}$. The *converse* of a binary relation α, denoted by $\alpha^{\sim}$, is defined as $\alpha^{\sim} = \{(x, y) : (y, x) \in \alpha\}$. We next recall the well-known Point Algebra (PA) [20,2,3], Cardinal Relation Algebra (CRA) [8,13], Interval Algebra (IA) [1], and RCC-8 algebra [18], which are all closed under converse.

Definition 1 (partially ordered Point Algebra [3]). *Let $(U, \geq)$ be a partial order. The following three relations on U (where $a \not\geq b$ denotes that (a,b) is not in relation $\geq$),*

$$> = \{(a,b) \in U \times U : a \geq b, b \not\geq a\},$$
$$< = \{(a,b) \in U \times U : a \not\geq b, b \geq a\},$$
$$|| = \{(a,b) \in U \times U : a \not\geq b, b \not\geq a\},$$

together with the identity relation $=$, are a jointly exhaustive and pairwise disjoint (JEPD) set of binary relations on U. The Point Algebra is the Boolean subalgebra generated by $\{<, >, =, ||\}$.

The following definition is a special case where $\geq$ is a total order.

Definition 2 (totally ordered Point Algebra [20]). *Let $(U, \geq)$ be a totally ordered set. The (totally ordered) Point Algebra is the Boolean subalgebra generated by the JEPD set of relations $\{<, >, =\}$, where $<, >, =$ are defined as in Definition 1.*

Note that there are eight relations for totally ordered Point Algebra, viz. the three basic relations $<, >, =$, the empty relation, and the four non-basic nonempty relations $\leq, \geq$, $\neq, ?$, where $?$ is the universal relation.

Definition 3 (Cardinal Relation Algebra [8,13]). *Let U be the real plane. The Cardinal Relation Algebra (CRA) is generated by the nine JEPD relations $NW, N, NE, W,$ EQ, E, SW, S, SE defined in Table 1 (a).*

Table 1. (a) Definitions of basic CRA relations between points (x, y) and (x', y') (b) Illustrations of CRA relations, where P_1 NW Q and P_2 E Q

Relation	Converse	Definition
NW	SE	$x < x', y > y'$
N	S	$x = x', y > y'$
NE	SW	$x > x', y > y'$
W	E	$x < x', y = y'$
EQ	EQ	$x = x', y = y'$

(a)   (b)

The Cardinal Relation Algebra can be viewed as the Cartesian product of two totally ordered Point Algebras (with $\mathbb{R}$ as their domains).

Definition 4 (Interval Algebra [1]). *Let U be the set of closed intervals on the real line. Thirteen binary relations between two intervals $x = [x^-, x^+]$ and $y = [y^-, y^+]$*

Table 2. Basic IA relations and their converses between intervals $[x^-, x^+]$ and $[y^-, y^+]$

Relation	Symbol	Converse	Meaning
before	p	pi	$x^- < x^+ < y^- < y^+$
meets	m	mi	$x^- < x^+ = y^- < y^+$
overlaps	o	oi	$x^- < y^- < x^+ < y^+$
starts	s	si	$x^- = y^- < x^+ < y^+$
during	d	di	$y^- < x^- < x^+ < y^+$
finishes	f	fi	$y^- < x^- < x^+ = y^+$
equals	eq	eq	$x^- = y^- < x^+ = y^+$

are defined by the order of the four endpoints of x and y (see below). The Interval Algebra is generated by these JEPD relations.

Definition 5 (RCC-8 Algebra[18][1]). *Let U be the set of nonempty regular closed sets on the real plane. The RCC-8 algebra is generated by the eight topological relations*

$$\mathbf{DC, EC, PO, EQ, TPP, NTPP, TPP^{\sim}, NTPP^{\sim}},$$

where $\mathbf{EQ, DC, EC, PO, TPP}$ and $\mathbf{NTPP}$ are defined as in Table 5, and $\mathbf{TPP^{\sim}}$ and $\mathbf{NTPP^{\sim}}$ are the converses of $\mathbf{TPP}$ and $\mathbf{NTPP}$ respectively.

Table 3. Definitions for basic RCC-8 relations between plane regions a and b, where a°, b° are the interiors of a, b respectively

Relation	Meaning	Relation	Meaning	Relation	Meaning
EQ	$a = b$	**EC**	$a \cap b \neq \varnothing, a^\circ \cap b^\circ = \varnothing$	**TPP**	$a \subset b, a \not\subset b^\circ$
DC	$a \cap b = \varnothing$	**PO**	$a \not\subseteq b, b \not\subseteq a, a^\circ \cap b^\circ \neq \varnothing$	**NTPP**	$a \subset b^\circ$

Illustrations for basic RCC-8 relations are provided in Figure 2.1. Note that regions in general may have multiple pieces or holes.

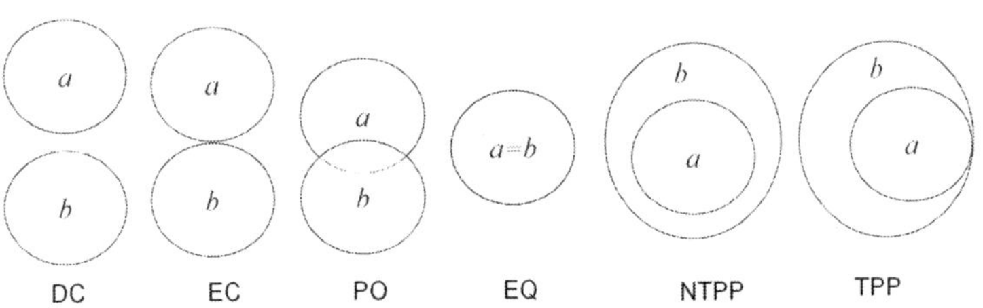

Fig. 1. Illustration for basic RCC-8 relations

---

[1] We note that the RCC algebras have interpretations in arbitrary topological spaces. In this paper, we only consider the interpretation in the real plane.

2.2   Constraint Networks and Minimal Networks

A qualitative calculus $\mathcal{M}$ provides a constraint language by using formulas of the form $v_i \alpha v_j$, where v_i, v_j are variables and α is a relation in $\mathcal{M}$. Formulas of the form $v_i \alpha v_j$ are called *constraints* (in $\mathcal{M}$). If α is a basic relation in $\mathcal{M}$, $v_i \alpha v_j$ is called a *basic constraint*. The consistency problem over $\mathcal{M}$ can then be formulated as below.

Definition 6. *[5] Let $\mathcal{M}$ be a qualitative calculus on universe U, and $\mathcal{S}$ be a subset of $\mathcal{M}$. The consistency problem* CSPSAT($\mathcal{S}$) *(in qualitative calculus $\mathcal{M}$) is defined as:*

> ***Instance**: A 2-tuple (V, Γ). Here V is a finite set of variables $\{v_1, \cdots, v_n\}$, and Γ is a finite set of binary constraints of the form $x\alpha y$, where $\alpha \in \mathcal{S}$ and $x, y \in V$.* [2]
> ***Question**: Is there an assignment $\nu : V \to U$ s.t. all constraints in Γ are satisfied?*

If an assignment ν satisfies all constraints in Γ, we say ν is a solution of Γ and Γ is satisfiable *or* consistent.

Note that Point Algebras rely on the underlying (partial or total) orders. Consequently, the consistency of an instance in PA depends on both the underlying (partial or total) order and the constraint network. However, we are usually more interested in the constraint network than in the particular underlying structure. Meanwhile, for the totally ordered Point Algebra, it is clear that if an instance is consistent in *some* Point Algebra, then it is also consistent in the PA generated by the total order $(\mathbb{R}, \leq)$, i.e., any finite total order can be embedded into $(\mathbb{R}, \leq)$. Therefore, we fix the underlying total order as $(\mathbb{R}, \leq)$ for all totally ordered PA constraint networks. Similarly, we fix the underlying partial order for partially ordered PA as $(\mathbb{N}^+, \succeq)$, where $a \succeq b$ if there is an integer k such that $a = bk$, as any finite partial order can be embedded into $(\mathbb{N}^+, \succeq)$.

The consistency problem as defined in Dfn. 6 has been investigated for many different qualitative calculi (see e.g. [1,2,17,19,15,3]). In particular it is shown in [2] that the consistency problem for the totally ordered Point Algebra can be solved in $O(n^2)$, where n is the number of variables. For most other qualitative calculi including the partially ordered PA, IA, CRA, and RCC-8, the consistency problems are NP-hard. Nonetheless, researchers have proved that the consistency problems CSPSAT($\mathcal{S}$) are tractable for some subsets $\mathcal{S}$ in the qualitative calculi. Such a set $\mathcal{S}$ is called a *tractable subclass*. A tractable subclass is called *maximal* if it has no proper superset which is tractable. Maximal subclasses of IA and RCC-8 have been identified, see [17,7,19].

A set of constraints Γ is called a *basic constraint network* if Γ contains exactly one basic constraint for each pair of variables. When only basic constraint networks are considered, the consistency problems of all qualitative calculi mentioned in Section 2 can be decided in $O(n^3)$ time by enforcing path-consistency (cf. [12]).

Definition 7 (refinement, scenario). *Let $\mathcal{M}$ be a qualitative calculus. Suppose (V, Γ) and (V, Γ') are two constraint networks over the same variable set V in $\mathcal{M}$, where $\Gamma = \{v_i \alpha_{ij} v_j\}_{i,j=1}^n$ and $\Gamma' = \{v_i \beta_{ij} v_j\}_{i,j=1}^n$. We say (V, Γ') is a* refinement *of (V, Γ), if for any $1 \leq i, j \leq n$ it holds that $\beta_{ij} \subseteq \alpha_{ij}$. We say (V, Γ') is a* scenario *of (V, Γ) if it is a basic constraint network.*

---

[2] We may simply denote the instance by Γ when V is clear or less important.

A refinement of a constraint network is a network with stronger constraints. Scenarios are the finest refinements. It is clear that a constraint network is consistent iff it has a consistent scenario. As we have mentioned, for all qualitative calculi considered in this paper, the consistency of a scenario can be determined by checking path-consistency.

Next we introduce the concept of minimal network (cf. [16]).

Definition 8 (minimal network). *Let $\Gamma = \{v_i \alpha_{ij} v_j\}_{i,j=1}^{n}$ be a constraint network in qualitative calculus $\mathcal{M}$. We say Γ is* minimal, *if for any $1 \leq i, j \leq n$ and any basic relation $b \subseteq \alpha_{ij}$ in $\mathcal{M}$, the refinement of Γ obtained by refining α_{ij} to b is consistent.*

There are two important problems regarding minimal networks. First, how to decide whether a network is minimal and how to compute the equivalent minimal network (i.e., the minimal constraint network with the same solution set). Second, how to get one (or all) solution(s) of a minimal constraint network. In what follows, we call the problem of deciding whether a constraint network in qualitative calculus $\mathcal{M}$ is minimal the *minimality problem* in $\mathcal{M}$.

For the minimality problem, we note that the problem can be decided in polynomial time for *tractable* subclasses, as we only need to check the consistency of at most $n^2 B$ networks, where n is the number of variables and B is the number of basic relations. Meanwhile, if the equivalent minimal network of an input network can be computed in polynomial time, then the minimality problem is also tractable (we may simply compare a network to its equivalent minimal network). In this sense, the minimality problem is simpler than the consistency problem and the problem of computing equivalent network, which are both known to be NP-hard in general. However, it is not clear before this paper whether a polynomial algorithm exists for the minimality problem. This paper proves that the problem is NP-hard for all qualitative calculi introduced above except the totally ordered Point Algebra.

The main interest of this paper is the second problem. We prove that, for all qualitative calculi mentioned above except the totally ordered Point Algebra, it is also NP-hard to compute a single solution of a minimal constraint network.[3] We do not distinguish the semantic difference between 'computing a solution' and 'computing a consistent scenario'. This is because, on one hand, polynomial algorithms to construct solutions for consistent scenarios have been proposed for all qualitative calculi discussed in this paper (see e.g.[11]); and on the other hand, a consistent scenario can always be computed in polynomial time from a solution.

A strategy has been introduced in [10] to prove the NP-hardness of computing a solution of a minimal constraint network in classical CSP. We use the same strategy in QSTR. The general framework is as follows, where $\mathcal{M}$ is a qualitative calculus in which the consistency of any basic network (scenario) can be decided in polynomial time.

- Construct a polynomial reduction R from an NP-hard problem $\mathcal{N}$ (variants of the SAT problem in this paper) to the consistency problem CSPSAT($\mathcal{M}$) for $\mathcal{M}$.
- Show any CSPSAT($\mathcal{M}$) instance generated by R is either inconsistent or minimal.

---

[3] Note that computing a solution of a constraint network is not a decision problem (i.e., problem with answer either 'yes' or 'no'). So the 'NP-hardness' here means that, if we have a polynomial algorithm that computes a solution of a minimal network, then we can provide a polynomial algorithm that solves an NP-complete problem [10].

We claim that the existence of such a reduction R implies the NP-hardness of computing a solution of a minimal constraint network in $\mathcal{M}$. This is because, if some polynomial algorithm $\mathcal{A}$ is able to compute a solution (or a consistent scenario) of a minimal constraint network with upper bounding time $p(x)$ (x is the size of input instance), then the following polynomial algorithm $\mathcal{A}^*$ would solve the NP-hard problem $\mathcal{N}$, where i is an instance of $\mathcal{N}$.

- Compute the CSPSAT($\mathcal{M}$) instance $R(i)$ by the reduction R.
- Call Algorithm $\mathcal{A}$ with input $R(i)$.
- If $\mathcal{A}$ does not halt in $p(size(R(i)))$ time, return 'No'.
- Verify whether the output of $\mathcal{A}$ is a solution of $R(i)$. If not, return 'No'.
- Return 'Yes'.

Because the consistency of any scenario in $\mathcal{M}$ can be determined in polynomial time, we know $\mathcal{A}^*$ is a polynomial algorithm. We next show that Algorithm $\mathcal{A}^*$ is sound. First, suppose i is a positive instance of $\mathcal{N}$. Then $R(i)$ is a minimal constraint network by the assumption of R. So Algorithm $\mathcal{A}$ should return a solution of $R(i)$ in $p(size(R(i)))$ time, and thus $\mathcal{A}^*$ returns the correct output 'Yes'. Second, suppose i is a negative instance of $\mathcal{N}$, in which case $R(i)$ is inconsistent, and Algorithm $\mathcal{A}$ gets an invalid input. So Algorithm $\mathcal{A}$ may either not halt in $p(size(R(i)))$ time, or halt with some output which is not a solution of $R(i)$ (as $R(i)$ is inconsistent). In both cases, Algorithm $\mathcal{A}^*$ returns 'No'. Therefore, $\mathcal{A}^*$ is sound and we conclude that computing a solution of a minimal network in qualitative calculus $\mathcal{M}$ is NP-hard.

Note that the reduction R above (if it exists) is also a polynomial reduction from the NP-hard problem $\mathcal{N}$ to the minimality problem in $\mathcal{M}$. Therefore the existence of R also implies the NP-hardness of the minimality problem in $\mathcal{M}$.

Theorem 1. *Let $\mathcal{M}$ be a qualitative calculus in which the consistency of basic networks can be decided in polynomial time. Suppose there exists a polynomial reduction R from an NP-hard problem to the consistency problem in $\mathcal{M}$ such that any CSPSAT($\mathcal{M}$) instance generated by R is either inconsistent or minimal. Then the minimality problem in $\mathcal{M}$ is NP-complete. Furthermore, it is also NP-complete to compute a solution of a minimal constraint network in $\mathcal{M}$.*

2.3   Variants of the SAT Problem

The NP-hardness results provided in the following sections are achieved by polynomial reductions from special variants of the SAT problem introduced below. Note that Definition 9 is original while Definition 10 comes from [10].

Definition 9 (symmetric SAT). *We say a SAT instance ϕ is symmetric, if for any truth value assignment $\nu : Var(\phi) \to \{\textbf{true}, \textbf{false}\}$, ν satisfies ϕ iff $\overline{\nu}$ satisfies ϕ, where $\overline{\nu}$ is defined by $\overline{\nu}(p) = \textbf{true}$ if $\nu(p) = \textbf{false}$, and $\overline{\nu}(p) = \textbf{false}$ if $\nu(p) = \textbf{true}$, for $p \in Var(\phi)$.*

It is clear that any unsatisfiable SAT instance is a symmetric instance.

Lemma 1. *There exists a polynomial-time transformation that transforms each SAT instance ϕ into a symmetric instance ϕ', such that ϕ is satisfiable iff ϕ' is satisfiable.*

Proof. Suppose $\phi = \bigwedge_{j=1}^{m} c_j$ is a SAT instance with propositional variables $\mathrm{Var}(\phi) = \{p_1, p_2, \cdots, p_n\}$ and clauses $c_j = \bigvee_{k=1}^{t_j} l_{j,k}$, where $l_{j,k}$ are literals. For a literal l, its *negation*, denoted by $\bar{l}$, is defined to be $\neg p_i$ if $l = p_i$, or p_i if $l = \neg p_i$. Define $\overline{c_j} = \bigvee_{k=1}^{t_j} \overline{l_{j,k}}$. Clearly, a truth value assignment ν satisfies c_j iff $\bar{\nu}$ satisfies $\overline{c_j}$.

Now we define a SAT instance ϕ' with $\mathrm{Var}(\phi') = \mathrm{Var}(\phi) \cup \{q\}$ and $2m$ clauses,

$$\phi' = \bigwedge_{i=1}^{m} (c_i \vee q) \wedge \bigwedge_{i=1}^{m} (\overline{c_i} \vee \neg q).$$

It is straightforward to verify that ϕ is satisfiable iff ϕ' is satisfiable. $\square$

Gottlob introduced the following concept of k-supersymmetry of SAT instances ($k \geq 1$) and proved that any SAT instance can be transformed into a k-supersymmetric instance while preserving its satisfiability.

Definition 10 (k-supersymmetric SAT, [10]). *Let ϕ be a SAT instance. We say ϕ is k-supersymmetric, if either it is unsatisfiable, or for any set of k propositional variables $V_k \subseteq Var(\phi)$, and any partial truth value assignment ν to V_k, there is an extension of ν which satisfies ϕ.*

Lemma 2 ([10]). *For each fixed integer $k \geq 1$, there exists a polynomial-time transformation that transforms a SAT instance ϕ into a k-supersymmetric instance ϕ^k, such that ϕ is satisfiable iff ϕ^k is satisfiable.*

It is clear that a symmetric SAT instance is also 1-supersymmetric and a $(k + 1)$-supersymmetric instance is also k-supersymmetric. The following lemma synthesizes Lemmas 1 and 2.

Lemma 3. *For each fixed integer $k \geq 1$, there exists a polynomial-time transformation that transforms a SAT instance ϕ into a symmetric and k-supersymmetric instance ϕ^*, such that ϕ is satisfiable iff ϕ^* is satisfiable.*

Proof. We first transform ϕ into a symmetric SAT instance ϕ' by Lemma 1, then transform ϕ' into a k-supersymmetric instance ϕ^* using the second transformation described in [10, Lemma 2]. It can be straightforwardly verified that this transformation preserves symmetry. Therefore ϕ^* is symmetric and k-supersymmetric. This procedure is polynomial as both of the two transformations are polynomial. $\square$

3   Partially Ordered Point Algebra and RCC-8

We first prove that computing a solution of a minimal constraint network in partially ordered Point Algebra is NP-hard. We achieve this by devising a reduction from the symmetric and 3-supersymmetric SAT problem to the consistency problem in partially ordered PA. Let $\phi = \bigwedge_{j=1}^{m} c_j$ be a symmetric and 3-supersymmetric SAT instance with $\mathrm{Var}(\phi) = \{p_1, p_2, \cdots, p_n\}$ and clauses $c_j = \bigvee_{k=1}^{t_j} l_{j,k}$. We construct a constraint network (V_ϕ, Γ_ϕ) in partially ordered PA.

The variable set is $V_\phi = V_0 \cup V_1 \cup \cdots \cup V_m$, where $V_0 = \{x_i, y_i : 1 \le i \le n\}$ and $V_j = \{w_{j,k} : 1 \le k \le t_j\}$. The constraints in Γ_ϕ are as follows, where $w_{j,t_j+1} = w_{j,1}$.

- For $x_i, y_i \in V_0$, $x_i \{<,>\} y_i$,
- For $k \ne k'$, $w_{j,k} \{<,>\} w_{j,k'}$,
- If $l_{j,k} = p_i$, then $x_i \{<,>\} w_{j,k+1}, y_i \{<,>\} w_{j,k}, x_i \parallel w_{j,k}, y_i \parallel w_{j,k+1}$,
- If $l_{j,k} = \neg p_i$, then $x_i \{<,>\} w_{j,k}, y_i \{<,>\} w_{j,k+1}, x_i \parallel w_{j,k+1}, y_i \parallel w_{j,k}$,
- For $j \ne j'$ and any $w \in V_j, w' \in V_{j'}, w \parallel w'$,
- For any other pair of variables $(u, v) \in V_\phi, u \{<,>,\parallel\} v$.

We next provide a brief explanation. We use variables $x_i, y_i \in V_0$ to simulate propositional variable p_i. The case that $x_i < y_i$ ($x_i > y_i$, resp.) corresponds to p_i being assigned **true** (**false**, resp.). Variables in V_j simulate clause c_j in ϕ, where the relationship ($<$ or $>$) between $w_{j,k}$ and $w_{j,k+1}$ corresponds to literal $l_{j,k}$ in c_j. Suppose $l_{j,k} \in \{p_i, \neg p_i\}$. Then the constraints between $x_i, y_i, w_{j,k}, w_{j,k+1}$ as specified above establish the connection between the relation of x_i and y_i and that of $w_{j,k}$ and $w_{j,k+1}$, which simulates the dependency of $l_{j,k}$ on p_i. In detail, if $l_{j,k} = p_i$ ($\neg p_i$, resp.), then the relation between $w_{j,k}$ and $w_{j,k+1}$ should be the same as (opposite to, resp.) that between x_i and y_i. For example, if $l_{j,k} = p_i$, then $x_i < y_i$ implies $x_i < w_{j,k+1}$ (otherwise, we shall have $w_{j,k+1} < x_i < y_i$ and $y_i \parallel w_{j,k+1}$, which is inconsistent). Similarly we have $y_i > w_{j,k}$ and $w_{j,k} < w_{j,k+1}$ in such case. See Figure 2 for illustration. By these constraints, we relate the case $w_{j,k} < w_{j,k+1}$ ($w_{j,k} > w_{j,k+1}$, resp.) to that $l_{j,k}$ being assigned **true** (**false**, resp.).

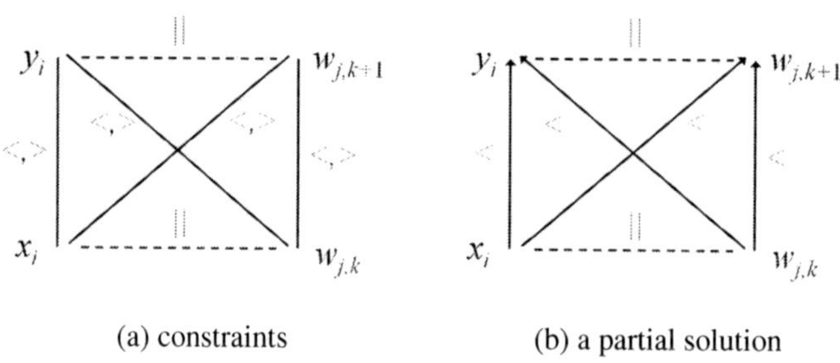

(a) constraints   (b) a partial solution

Fig. 2. Passing the relation between x_i and y_i to that between $w_{j,k}$ and $w_{j,k+1}$

Clause c_j rules out the assignments that assign all the literals $l_{j,k}$ **false**. Because ϕ is a symmetric SAT instance, any assignment ν that assigns all literals $l_{j,k}$ **true** would also falsify ϕ (otherwise, ϕ should also be satisfied by $\bar{\nu}$, which fails to satisfy c_j). Correspondingly, (V_ϕ, Γ_ϕ) is inconsistent if $w_{j,1} > w_{j,2} > \cdots > w_{j,t_j} > w_{j,1}$ or $w_{j,1} < w_{j,2} < \cdots < w_{j,t_j} < w_{j,1}$. Note that we only constrain variables in V_j by enforcing a total order and propagating the configuration of V_0 to them. In summary, by introducing of variables in V_j (and related constraints), we forbid two certain kinds of configurations of V_0, which respectively correspond to assignments of $\mathrm{Var}(\phi)$ that assign all literals in c_j **true** or **false**.

Proposition 1. *Let ϕ be a symmetric SAT instance, and (V_ϕ, Γ_ϕ) be the constraint network as constructed above. Then ϕ is satisfiable iff (V_ϕ, Γ_ϕ) has a consistent scenario.*

Proof (sketch). Suppose ϕ is satisfiable and $\nu : \mathrm{Var}(\phi) \to \{\textbf{true}, \textbf{false}\}$ is a truth value assignment that satisfies ϕ. We construct a consistent scenario for (V_ϕ, Γ_ϕ).

The constraints between variables in V_0 are as follows, see Figure 3 for illustration.

- If $\nu(p_i) = \textbf{true}$, then $x_i < y_i$; otherwise $x_i > y_i$,
- Let big_i ($small_i$ resp.) be the big one (small one resp.) in $\{x_i, y_i\}$. Then for $i \neq i'$,

$$big_i \,\|\, big_{i'}, \quad small_i \,\|\, small_{i'}, \quad small_i < big_{i'}.$$

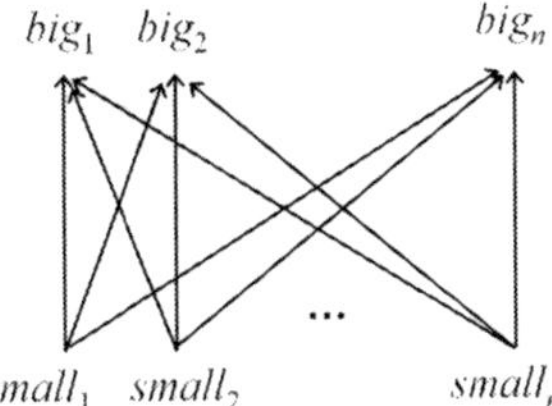

Fig. 3. Constraints between variables in V_0 in the scenario, where $\{big_i, small_i\} = \{x_i, y_i\}$

Assume $l_{j,k} \in \{p_i, \neg p_i\}$. The constraints between $x_i, y_i, w_{j,k}$ and $w_{j,k+1}$ are decided by the constraint between x_i and y_i, as we have explained before the proof. The scenario should provide a total order on V_j consistent to the determined orders of $w_{j,k}$ and $w_{j,k+1}$. Such a total order always exists, because the determined orders contain no circles (due to the fact that ν satisfies clause c_j). Pick arbitrary one if such total orders are not unique.

It remains to refine constraints of the form $u \,\{<, >, \|\}\, v$, where $u \in V_0$ and $v \in V_j$ for some j. Part of these constraints need to be refined to $<$ (or $>$) due to the transitivity of $<$ (or $>$). The rest are refined to $\|$.

We now get a scenario, the consistency of which can be verified by checking its path-consistency. We omit the details here.

Now suppose (V_ϕ, Γ_ϕ) has a consistent scenario. We define a truth value assignment ν by $\nu(p_i) = \textbf{true}$ if $x_i < y_i$ in the scenario, or $\nu(p_i) = \textbf{false}$ otherwise. For each clause c_j in ϕ, it can be checked that ν satisfies c_j because the constraints about variables in V_j in the scenario are consistent. Therefore, ϕ is satisfiable. $\square$

The PA network (V_ϕ, Γ_ϕ) in the reduction above has $2n + T$ variables, where $n = |\mathrm{Var}(\phi)|$, and $T = t_1 + t_2 + \cdots + t_m$ is the number of all literals in the SAT instance. Therefore the reduction is polynomial. We next prove, by exploiting the 3-supersymmetry of the SAT instance ϕ, that (V_ϕ, Γ_ϕ) is minimal if ϕ is satisfiable.

Proposition 2. *Suppose ϕ is a symmetric and 3-supersymmetric SAT instance. Let (V_ϕ, Γ_ϕ) be the constraint network as constructed above. If ϕ is satisfiable, then (V_ϕ, Γ_ϕ) is a minimal constraint network.*

Proof (sketch). First note that (V_ϕ, Γ_ϕ) contains three forms of constraints, viz. $u \parallel v$, $u\{<,>\}v$ and $u\{<,>,\parallel\}v$. We aim to show the latter two forms of non-basic constraints can be refined to any basic constraint without violating the consistency. We have proved that (V_ϕ, Γ_ϕ) has a consistent scenario, say, (V_ϕ, Γ_0). Let (V_ϕ, Γ_1) be the scenario of (V_ϕ, Γ_ϕ) which refines non-basic constraints in the opposite way, i.e.,

- Constraint $u\{<,>\}v$ is refined to $u < v$ ($u > v$ resp.) if it is refined to $u > v$ ($u < v$ resp.) in Γ_0.
- Constraint $u\{<,>,\parallel\}v$ is refined to $u < v$ ($u > v$, $u \parallel v$ resp.) if it is refined to $u > v$ ($u < v$, $u \parallel v$ resp.) in Γ_0.

It can be proved that scenario (V_ϕ, Γ_1) is path-consistent (otherwise (V_ϕ, Γ_0) would not be path-consistent) and hence consistent.

Now we need only to show that constraint $u\{<,>,\parallel\}v$ in (V_ϕ, Γ_ϕ) can be refined to any of $u < v$, $u > v$ and $u \parallel v$. There are only two possible cases of such u and v:

- $u \in \{x_i, y_i\}$ and $v \in \{x_{i'}, y_{i'}\}$ for some $i \neq i'$.
- $u \in \{x_i, y_i\}$ and $v = w_{j,k}$ for some i, j, k such that $l_{j,k}, l_{j,k+1} \notin \{p_i, \neg p_i\}$.

For the first case, w.l.o.g., we may suppose $u = x_i$ and $v = x_{i'}$. Because ϕ is 3-supersymmetric and hence 2-supersymmetric, there exists a truth value assignment ν which satisfies ϕ and $\nu(p_i) = \mathbf{true}$, $\nu(p_{i'}) = \mathbf{true}$. So we may get a consistent scenario of (V_ϕ, Γ_ϕ) as in the proof of Proposition 1 according to ν. In this scenario, we have $x_i < y_i, x_{i'} < y_{i'}$ and $x_i \parallel x_{i'}$. Therefore, there exists a consistent scenario of (V_ϕ, Γ_ϕ) in which $x_i \parallel x_{i'}$. Similarly, we may obtain consistent scenarios in which $x_i < x_{i'}$ or $x_i > x_{i'}$, by other truth value assignments on p_i and $p_{i'}$.

The second case is slightly complicated and is briefly described here. W.o.l.g., suppose we want to refine constraint $x_i\{<,>,\parallel\}w_{j,k}$ to $x_i < w_{j,k}$, $x_i > w_{j,k}$ and $x_i \parallel w_{j,k}$ respectively. For $x_i < w_{j,k}$, we need an assignment ν which satisfies ϕ and $\nu(p_i) = \mathbf{true}$, $\nu(l_{j,k-1}) = \mathbf{true}$ and $\nu(l_{j,k}) = \mathbf{false}$. Such assignment ν exists as ϕ is 3-supersymmetric. We are able to get a scenario as in the proof of Proposition 1 in which we further require $w_{j,k}$ to be the maximal element in V_j (note $w_{j,k} > w_{j,k-1}, w_{j,k+1}$ as $\nu(l_{j,k-1}) = \mathbf{true}$ and $\nu(l_{j,k}) = \mathbf{false}$). In this scenario either $x_i < w_{j,k}$ or $x_i \parallel w_{j,k}$. In the latter case, we replace it with $x_i < w_{j,k}$, which does not jeopardize the consistency as x_i is in the smaller part of V_0 and $w_{j,k}$ is the maximal element in V_j. So we get a consistent scenario in which $x_i < w_{j,k}$. The other two cases are similar, where for $x_i > w_{j,k}$ we need $\nu(p_i) = \mathbf{false}$ and $w_{j,k}$ to be the minimal element in V_j, and for $x_i \parallel w_{j,k}$ we need $\nu(p_i) = \mathbf{true}$ and $w_{j,k}$ to be the minimal element. $\square$

Now we have the following conclusion for partially ordered Point Algebra.

Theorem 2. *Computing a solution of a minimal network in partially ordered Point Algebra is NP-complete.*

Proof. By Propositions 1, 2 and Theorem 1, we know that computing a solution (or a consistent scenario) of a minimal network in partially ordered PA is NP-hard. The problem is in NP as we may guess a consistent scenario by indeterminism. $\square$

The above technique can be directly applied to the RCC-8 algebra.

Theorem 3. *Computing a solution of a minimal network in RCC-8 is NP-complete.*

Proof. Given a symmetric and 3-supersymmetric 3-SAT instance ϕ, we may construct an instance of the consistency problem in RCC-8 by substituting PA relations $<, >, \|$ in the reduction provided above with RCC-8 relations $\mathbf{NTPP}, \mathbf{NTPP}^{\sim}, \mathbf{PO}$ respectively. Propositions 1 and 2 still hold and can be proved in the same way. Therefore, computing a solution of a minimal RCC-8 network is NP-complete. $\qquad\qquad\square$

4   Cardinal Relation Algebra and Interval Algebra

This section shows that computing a solution for a minimal CRA or IA constraint network is also NP-hard. The proof is similar to but simpler than that in previous section, as 3-supersymmetry is no longer necessary. We first introduce some abbreviations that may clarify the specification of constraints. For a CRA relation α, We use

$$a|b \;\; \alpha \;\; c|d \quad \text{to denote constraints} \quad a\,\alpha\,c, a\,\alpha\,d \text{ and } b\,\alpha\,c, b\,\alpha\,d,$$

$$\frac{a}{b} \;\; \alpha \;\; \frac{c}{d} \quad \text{to denote constraints} \quad a\,\alpha\,c \text{ and } b\,\alpha\,d.$$

Now we discuss the Cardinal Relation Algebra. Suppose $\phi = \bigwedge_{j=1}^{m} c_j$ is a symmetric SAT instance with $\mathrm{Var}(\phi) = \{p_1, p_2, \cdots p_n\}$ and clauses $c_j = \bigvee_{k=1}^{t_j} l_{j,k}$. We now construct a CRA constraint network (V_ϕ, Γ_ϕ). The spatial variables for propositional variable p_i are still x_i and y_i, while $2t_j$ spatial variables $c_{j,k}, d_{j,k}\,(k = 1, 2, \cdots, t_j)$ are introduced for clause c_j (which has t_j literals). So the variable set is $V_\phi = V_0 \cup V_1 \cup \cdots \cup V_m$, where $V_0 = \{x_i, y_i : 1 \le i \le n\}$ and $V_j = \{c_{j,k}, d_{j,k} : 1 \le k \le t_j\}$.

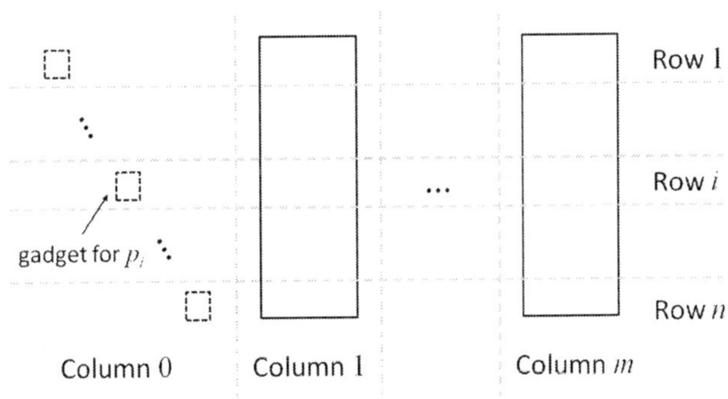

Fig. 4. Overview of the configuration of (V_ϕ, Γ_ϕ)

We describe the relative locations of variables which are implied by the CRA constraints in Γ_ϕ. The variables in V_0 will be located in the leftmost column (Column 0) in Figure 4. Among them, x_i and y_i will be located in the dashed small box which is in the

i-th row. Meanwhile, x_i is either to the northwest of or to the southeast of y_i. Formally, we impose the following CRA constraints, where $1 \leq i < i' \leq n$,

$$x_i \; \{\text{NW, SE}\} \; y_i, \quad x_i | y_i \; \text{NW} \; x_{i'} | y_{i'}.$$

The case that x_i is to the northwest (southeast resp.) of y_i corresponds to that p_i is assigned **true** (**false** resp.).

The $2t_j$ variables $c_{j,k}, d_{j,k} (1 \leq k \leq t_j)$ in V_j are all located in Column j in Figure 4. For the vertical positions, suppose $l_{j,k} \in \{p_i, \neg p_i\}$, then $c_{j,k}$ and $d_{j,k}$ are located in the i-th row. Precisely, we impose

$$\frac{x_i}{y_i} \; \text{W} \; \frac{c_{j,k}}{d_{j,k}}, \quad \frac{x_i}{y_i} \; \{\text{NW, SW}\} \; \frac{d_{j,k}}{c_{j,k}},$$

and

$$c_{j,k} \; \{\text{NW, SE}\} \; d_{j,k} \quad \text{if } l_{j,k} = p_i, \quad \text{or} \quad c_{j,k} \; \{\text{NE, SW}\} \; d_{j,k} \quad \text{if } l_{j,k} = \neg p_i.$$

That is to say, $c_{j,k}$ ($d_{j,k}$ resp.) is to the east of x_i (y_i resp.). By these constraints, the CRA relation between $c_{j,k}$ and $d_{j,k}$ is decided by the relation between x_i and y_i and literal $l_{j,k}$ (being positive or negative). In fact, it is straightforward to check that the horizontal relation (i.e., west or east) between $c_{j,k}$ and $d_{j,k}$ is in accordance with the truth value assigned to literal $l_{j,k}$ (**true** or **false**), see Figure 5. The relative positions (or constraints) between $c_{j,k}$, $d_{j,k}$ and $x_{i'}$, $y_{i'}$ where $i \neq i'$ can be completely decided by the constraints above.

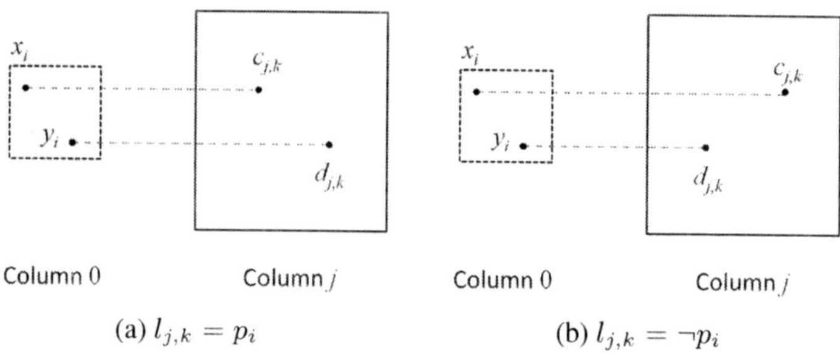

Column 0   Column j Column 0   Column j

(a) $l_{j,k} = p_i$ (b) $l_{j,k} = \neg p_i$

Fig. 5. Passing the relation between x_i and y_i to that between $c_{j,k}$ and $d_{j,k}$, assuming x_i NW y_i

We now discuss the constraints between variables in V_j. Suppose $u \in \{c_{j,k}, d_{j,k}\}$ and $v \in \{c_{j,k'}, d_{j,k'}\}$ are two variables in V_j, where $k \neq k'$. Suppose $l_{j,k} \in \{p_i, \neg p_i\}$ and $l_{j,k'} \in \{p_{i'}, \neg p_{i'}\}$ for some $i \neq i'$ (if $i = i'$ then either one literal can be removed, or the clause is unsatisfiable and we may simply construct an inconsistent CRA instance). We consider the constraint between u and v in the vertical direction and in the horizontal direction separately. The vertical relation between u and v is determined by i and i', as u is in the i-th row and v is in the i'-th row.

The horizontal relation between u and v is the key point which connects the SAT instance ϕ and the CRA constraint network (V_ϕ, Γ_ϕ). Note that the j-th clause c_j in ϕ rules out the assignments that assign all literals $l_{j,k}$ **false**. Therefore, we expect that (V_ϕ, Γ_ϕ) forbids the case that all $d_{j,k}$ are to the left (i.e., northwest or southwest) of $c_{j,k}$ for $k \in \{1, 2, \cdots, t_j\}$. To this end, $d_{j,k}$ and $c_{j,k+1}$ are required to lie on the same vertical line, while any other pair of variables in V_j are required not to, where $t_j + 1$ is considered as 1. It is clear that $d_{j,k}$ (or equivalently $c_{j,k+1}$) being to the west of $c_{j,k}$ for all $k \in \{1, 2, \cdots, t_j\}$ is not realisable.

Note that the above constraints also forbid the case that all $d_{j,k}$ are to the right (i.e., northeast of southeast) of $c_{j,k}$ for $k \in \{1, 2, \cdots, t_j\}$. This does not cause problems since ϕ rules out the assignments that assign all literals $l_{j,k}$ in c_j **true** by its symmetry.

Now we consider the constraint between $u \in V_j$ and $v \in V_{j'}$ for $j \neq j'$. Suppose $u \in \{c_{j,k}, d_{j,k}\}$, $v \in \{c_{j',k'}, d_{j',k'}\}$, where $l_{j,k} \in \{p_i, \neg p_i\}$, $l_{j',k'} \in \{p_{i'}, \neg p_{i'}\}$. The horizontal relation between u and v is determined by j and j' (u is in Column j and v is in Column j'). For the vertical constraint, note that u and v are located in the i-th and i'-th rows respectively. Therefore the case that $i \neq i'$ is clear. If $i = i'$, $c_{j,k}$ and $c_{j',k'}$ are both to the east of x_i, while $d_{j,k}$ and $d_{j',k'}$ are both to the east of y_i. So we specify the following constraints if $j < j'$. The case when $j > j'$ is similar.

$$\frac{c_{j,k}}{d_{j,k}} \; \text{E} \; \frac{c_{j',k'}}{d_{j',k'}}, \quad \frac{c_{j,k}}{d_{j,k}} \; \{\text{NW,SW}\} \; \frac{d_{j',k'}}{c_{j',k'}}.$$

Proposition 3. *Given a symmetric SAT instance ϕ, suppose (V_ϕ, Γ_ϕ) is the CRA instance as constructed above. Then ϕ is satisfiable iff (V_ϕ, Γ_ϕ) has a consistent scenario.*

Proof. This part can be straightforwardly proved by the connection between ϕ and (V_ϕ, Γ_ϕ) described above. We omit the details here. $\qquad\qquad\square$

Therefore, we have a reduction from symmetric SAT to CSPSAT(CRA). Note that for a symmetric SAT instance ϕ with n propositional variables and T literals, the CRA instance (V_ϕ, Γ_ϕ) consists $2n + 2T$ spatial variables. So the reduction is polynomial.

Proposition 4. *Suppose ϕ is a satisfiable symmetric SAT instance. Then the CRA constraint network (V_ϕ, Γ_ϕ) constructed above is minimal.*

Proof. We need to prove that, after refining a non-basic constraint to any basic relation it contains, the network is still consistent. Note that the CRA relation in any non-basic constraints of (V_ϕ, Γ_ϕ) is exactly the union of two basic relations. Suppose Γ_0 is a consistent scenario of (V_ϕ, Γ_ϕ). Let Γ_1 be the scenario obtained by refining non-basic constraints in (V_ϕ, Γ_ϕ) to the basic constraint different from the one in Γ_0. Scenario Γ_1 is also consistent by path-consistency (otherwise, Γ_0 can not be path-consistent). Therefore, each non-basic constraint in (V_ϕ, Γ_ϕ) can be refined to any basic constraint (either in scenario Γ_0 or in scenario Γ_1). So (V_ϕ, Γ_ϕ) is a minimal CRA network. $\quad\square$

Theorem 4. *Computing a solution of a minimal CRA network is NP-complete.*

Proof. The theorem can be proved in the same way as Theorem 2.

For Interval Algebra, observing that interval $[a, b]$ (where $a < b$) naturally corresponds to the point (a, b) on the half plane, we can prove the following theorem.

Theorem 5. *Computing a solution of a minimal IA network is NP-complete.*

Proof (sketch). We may translate the reduction for CRA into a reduction for IA by replacing CRA relations with IA relations as in the following table.

CRA relation	NW	N	NE	W	EQ	E	SW	S	SE
IA relation	di	si	oi	fi	eq	f	o	s	d

We then may prove Propositions 3 and 4 for IA in the same way.   □

5   Conclusion and Future Work

In this paper we have discussed the minimal constraint networks in qualitative spatial and temporal reasoning. We have proved for four major qualitative calculi (viz., partially ordered Point Algebra, Cardinal Relation Algebra, Interval Algebra and RCC-8 algebra) that deciding the minimality of networks and computing solutions of minimal constraint networks are both NP-complete problems. We have provided a polynomial reduction from a specialized SAT problem to the consistency problem of each qualitative calculus, which maps positive SAT instances to minimal constraint networks. The reduction exploits the symmetry of qualitative calculi, as it uses 'symmetric' intractable subclasses of relations, for example, $\{||, \{<, >\}, \{<, >, ||\}\}$ in partially ordered Point Algebra, and $\{\mathbf{PO}, \{\mathbf{NTPP}, \mathbf{NTPP}^{\sim}\}, \{\mathbf{NTPP}, \mathbf{NTPP}^{\sim}, \mathbf{PO}\}\}$ in RCC-8.

The work of Gottlob [10] reveals the intractability of solving minimal constraint networks in classical CSP (with finite domains). This paper discussed the same problem, but in the context of QSTR. In this situation, the domains are infinite, and the constraints are all taken from a fixed and finite set of relations in a qualitative calculus. The minimality problem in such a qualitative calculus can also be regarded as a 'special' problem in classical CSP. The NP-hardness of the general problem proved by Gottlob (where 2-supersymmetry is enough for the binary case) does by no means imply the NP-hardness of the special problems considered in this paper, where symmetry (and 3-supersymmetry for CRA and IA) are required.

Acknowledgements. This work was partially supported by an ARC Future Fellowship (FT0990811) and an ARC Discovery Project (DP120104159). We thank the reviewers for their helpful suggestions.

References

1. Allen, J.F.: Maintaining knowledge about temporal intervals. Commun. ACM 26(11), 832–843 (1983)
2. van Beek, P.: Reasoning about qualitative temporal information. Artif. Intell. 58(1-3), 297–326 (1992)
3. Broxvall, M., Jonsson, P.: Point algebras for temporal reasoning: Algorithms and complexity. Artif. Intell. 149(2), 179–220 (2003)
4. Chandra, P., Pujari, A.K.: Minimality and convexity properties in spatial CSPs. In: ICTAI, pp. 589–593. IEEE Computer Society (2005)

5. Cohn, A.G., Renz, J.: Qualitative spatial reasoning. In: van Harmelen, F., Lifschitz, V., Porter, B. (eds.) Handbook of Knowledge Representation. Elsevier (2007)
6. Dechter, R.: Constraint processing. Elsevier Morgan Kaufmann (2003)
7. Drakengren, T., Jonsson, P.: A complete classification of tractability in Allen's algebra relative to subsets of basic relations. Artif. Intell. 106(2), 205–219 (1998)
8. Frank, A.U.: Qualitative spatial reasoning with cardinal directions. In: Kaindl, H. (ed.) ÖGAI. Informatik-Fachberichte, vol. 287, pp. 157–167. Springer (1991)
9. Gerevini, A., Saetti, A.: Computing the minimal relations in point-based qualitative temporal reasoning through metagraph closure. Artif. Intell. 175(2), 556–585 (2011)
10. Gottlob, G.: On Minimal Constraint Networks. In: Lee, J. (ed.) CP 2011. LNCS, vol. 6876, pp. 325–339. Springer, Heidelberg (2011)
11. Li, S.: On topological consistency and realization. Constraints 11(1), 31–51 (2006)
12. Li, S., Ying, M.: Region Connection Calculus: Its models and composition table. Artif. Intell. 145(1-2), 121–146 (2003)
13. Ligozat, G.: Reasoning about cardinal directions. J. Vis. Lang. Comput. 9(1), 23–44 (1998)
14. Ligozat, G., Renz, J.: What Is a Qualitative Calculus? A General Framework. In: Zhang, C., Guesgen, H.W., Yeap, W.-K. (eds.) PRICAI 2004. LNCS (LNAI), vol. 3157, pp. 53–64. Springer, Heidelberg (2004)
15. Liu, W., Zhang, X., Li, S., Ying, M.: Reasoning about cardinal directions between extended objects. Artif. Intell. 174(12-13), 951–983 (2010)
16. Montanari, U.: Networks of constraints: Fundamental properties and applications to picture processing. Inf. Sci. 7, 95–132 (1974)
17. Nebel, B., Bürckert, H.J.: Reasoning about temporal relations: A maximal tractable subclass of Allen's Interval Algebra. J. ACM 42(1), 43–66 (1995)
18. Randell, D.A., Cui, Z., Cohn, A.G.: A spatial logic based on regions and connection. In: KR, pp. 165–176 (1992)
19. Renz, J., Nebel, B.: On the complexity of qualitative spatial reasoning: A maximal tractable fragment of the Region Connection Calculus. Artif. Intell. 108(1-2), 69–123 (1999)
20. Vilain, M.B., Kautz, H.A.: Constraint propagation algorithms for temporal reasoning. In: AAAI, pp. 377–382 (1986)

Containment, Equivalence and Coreness
from CSP to QCSP and Beyond

Florent Madelaine[1,2] and Barnaby Martin[3,*]

[1] Clermont Université, Université d'Auvergne, Clermont-Ferrand, France
[2] Laboratoire d'Informatique de l'École Polytechnique (CNRS, UMR 7161), France
[3] School of Engineering and Computing Sciences, Durham University
Science Laboratories, South Road, Durham DH1 3LE, UK

Abstract. The constraint satisfaction problem (CSP) and its quantified extensions, whether without (QCSP) or with disjunction (QCSP$_\vee$), correspond naturally to the model checking problem for three increasingly stronger fragments of positive first-order logic. Their complexity is often studied when parameterised by a fixed model, the so-called template. It is a natural question to ask when two templates are equivalent, or more generally when one "contain" another, in the sense that a satisfied instance of the first will be necessarily satisfied in the second. One can also ask for a smallest possible equivalent template: this is known as the core for CSP. We recall and extend previous results on containment, equivalence and "coreness" for QCSP$_\vee$ before initiating a preliminary study of cores for QCSP which we characterise for certain structures and which turns out to be more elusive.

1   Introduction

We consider the following increasingly stronger fragments of first-order logic:

1. primitive positive first-order ($\{\exists, \wedge\}$-FO)
2. positive Horn ($\{\exists, \forall, \wedge\}$-FO)
3. positive equality-free first-order ($\{\exists, \forall, \wedge, \vee\}$-FO); and,
4. positive first-order logic ($\{\exists, \forall, \wedge, \vee, =\}$-FO)

The *model checking problem* for a logic $\mathscr{L}$ takes as input a sentence of $\mathscr{L}$ and a structure $\mathcal{B}$ and asks whether $\mathcal{B}$ models $\mathcal{L}$. The structure $\mathcal{B}$ is often assumed to be a fixed parameter and called the template; and, unless otherwise stated, we will assume implicitly that we work in this so-called *non-uniform* setting.

For the above first three fragments, the model checking problem is better known as the *constraint satisfaction problem* CSP($\mathcal{B}$), the *quantified constraint satisfaction problem* QCSP($\mathcal{B}$) and its extension with disjunction which we shall denote by QCSP$_\vee$($\mathcal{B}$). Much of the theoretical research into CSPs is in respect of a large complexity classification project – it is conjectured that CSP($\mathcal{B}$) is

* Author supported by EPSRC grant EP/G020604/1.

M. Milano (Ed.): CP 2012, LNCS 7514, pp. 480–495, 2012.
© Springer-Verlag Berlin Heidelberg 2012

always either in P or NP-complete [10]. This *dichotomy* conjecture remains unsettled, although dichotomy is now known on substantial classes (e.g. structures of size ≤ 3 [18,3] and smooth digraphs [12,1]). Various methods, combinatorial (graph-theoretic), logical and universal-algebraic have been brought to bear on this classification project, with many remarkable consequences. A conjectured delineation for the dichotomy was given in the algebraic language in [4].

Complexity classifications for QCSPs appear to be harder than for CSPs. Just as $\mathrm{CSP}(\mathcal{B})$ is always in NP, so $\mathrm{QCSP}(\mathcal{B})$ is always in Pspace. No overarching polychotomy has been conjectured for the complexities of $\mathrm{QCSP}(\mathcal{B})$, as $\mathcal{B}$ ranges over finite structures, but the only known complexities are P, NP-complete and Pspace-complete (see [2,17] for some trichotomies). It seems plausible that these complexities are the only ones that can be so obtained.

Distinct templates may give rise to the same model-checking-problem or preserve acceptance,

($\mathscr{L}$-**equivalence**) for any sentence φ of $\mathscr{L}$, $\mathcal{A}$ models $\varphi \Leftrightarrow \mathcal{B}$ models φ
($\mathscr{L}$-**containment**) for any sentence φ of $\mathscr{L}$, $\mathcal{A}$ models $\varphi \Rightarrow \mathcal{B}$ models φ.

We will see that containment and therefore equivalence is decidable, and often quite effectively so, for the four logics we have introduced.

For example, when $\mathscr{L}$ is $\{\exists, \wedge\}$-FO, any two bipartite undirected graphs that have at least one edge are equivalent. Moreover, there is a canonical *minimal* representative for each equivalence class, the so-called *core*. For example, the core of the class of bipartite undirected graphs that have at least one edge is the graph $\mathcal{K}_2$ that consists of a single edge. The core enjoys many benign properties and has greatly facilitated the classification project for CSPs (which corresponds to the model-checking for $\{\exists, \wedge\}$-FO): it is unique up to isomorphism and sits as an induced substructure in all templates in its equivalence class. A core may be defined as a structure all of whose endomorphisms are automorphisms. To review, therefore, it is well-known that two templates $\mathcal{A}$ and $\mathcal{B}$ are equivalent iff there are homomorphisms from $\mathcal{A}$ to $\mathcal{B}$ and from $\mathcal{B}$ to $\mathcal{A}$, and in this case there is an (up to isomorphism) unique core $\mathcal{C}$ equivalent to both $\mathcal{A}$ and $\mathcal{B}$ such that $\mathcal{C} \subseteq \mathcal{A}$ and $\mathcal{C} \subseteq \mathcal{B}$.

The situation for $\{\exists, \forall, \wedge\}$-FO and QCSP is somewhat murkier. It is known that non-trivial $\mathcal{A}$ and $\mathcal{B}$ are equivalent iff there exist integers r and r' and surjective homomorphisms from $\mathcal{A}^r$ to $\mathcal{B}$ and from $\mathcal{B}^{r'}$ to $\mathcal{A}$ (and one may give an upper bound on these exponents) [7]. However, the status and properties of "core-ness" for QCSP were hitherto unstudied.

We **might** call a structure $\mathcal{B}$ a *Q-core* if there is no $\{\exists, \forall, \wedge\}$-FO-equivalent $\mathcal{A}$ of strictly smaller cardinality. We will discover that **this** Q-core is a more cumbersome beast than its cousin the core; it need not be unique nor sit as an induced substructure of the templates in its class. However, in many cases we shall see that its behaviour is reasonable and that – like the core – it can be very useful in delineating complexity classifications.

The erratic behaviour of Q-cores sits in contrast not just to that of cores, but also that of the *U-X-cores* of [15], which are the canonical representatives of the equivalence classes associated with $\{\exists, \forall, \wedge, \vee\}$-FO, and were instrumental in deriving a full complexity classification – a tetrachotomy – for $QCSP_{\vee}$ in [15]. Like cores, they are unique and sit as induced substructures in all templates in their class. Thus, primitive positive logic and positive equality-free logic behave genially in comparison to their wilder cousin positive Horn.

Continuing to add to our logics, in restoring equality, we might arrive at positive logic. Two finite structures agree on all sentences of positive logic iff they are isomorphic – so here every finite structure satisfies the ideal of "core". When computing a/the smallest substructure with the same behaviour with respect to the four decreasingly weaker logics – positive logic, positive equality-free, positive Horn, and primitive positive – we will obtain possibly decreasingly smaller structures. In the case of positive equality-free and primitive positive logic, as pointed out, these are unique up to isomorphism; and for the U-X-core and the core, these will be induced substructures. **A** Q-core will necessarily contain the core and be included in the U-X-core. This phenomenon is illustrated on Table 1 and will serve as our running example.

Table 1. different notions of "core" (the circles represent self-loops)

$\{\exists, \forall, \wedge, \vee, =\}$-FO	$\{\exists, \forall, \wedge, \vee\}$-FO	$\{\exists, \forall, \wedge\}$-FO	$\{\exists, \wedge\}$-FO
$\mathcal{A}_4$	$\mathcal{A}_3$	$\mathcal{A}_2$	$\mathcal{A}_1$
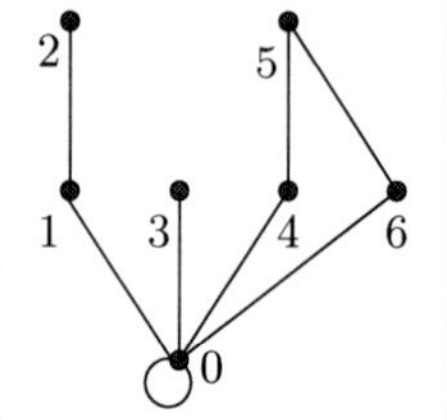	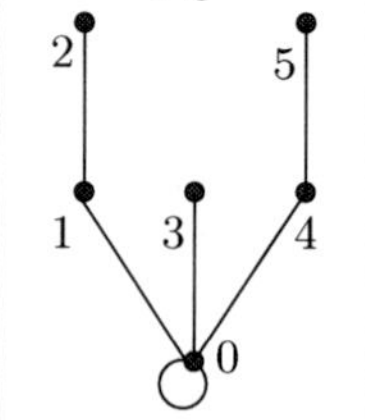	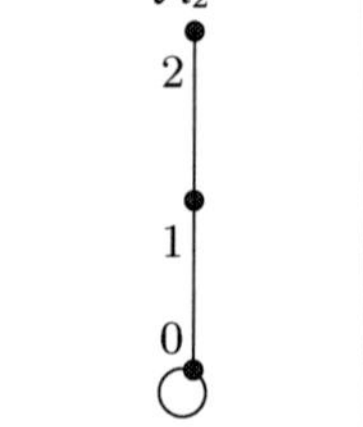	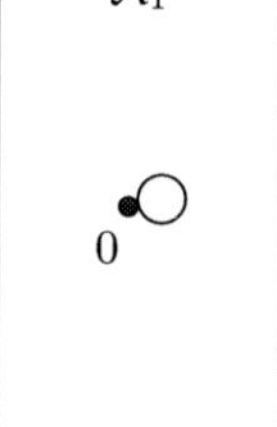
isomorphism	U-X-Core	Q-core	Core

The paper is organised as follows. In § 2, we recall folklore results on CSP. We move on to $QCSP_{\vee}$ in § 3, where we recall results on coreness and spell out containment for $\{\exists, \forall, \wedge, \vee\}$-FO that were only implicit in [15]. In § 4, we move on to QCSP and recall results on the decidability of containment from [7] together with new lower bounds before initiating a study of Q-cores. In § 5, we show that Q-cores behaves well for a number of classes and in § 6 that it can be used to delineate complexity classification. In § 7, we propose a method to compute Q-cores. Due to space restriction, several proofs are missing or are only sketched, they can be found *in extenso* in the full version of this paper [14].

2 The Case of CSP

Unless otherwise stated, we consider structures over a fixed relational signature σ. We denote by A the domain of a structure $\mathcal{A}$ and for every relation symbol R in σ of arity r, we write $R^{\mathcal{A}}$ for the interpretation of R in $\mathcal{A}$, which is a

r-ary relation that is $R^{\mathcal{A}} \subseteq A^r$. We write $|A|$ to denote the cardinality of the set A. A homomorphism (resp., strong homomorphism) from a structure $\mathcal{A}$ to a structure $\mathcal{B}$ is a function $h : A \to B$ such that $(h(a_1), \ldots, h(a_r)) \in R^{\mathcal{B}}$, if (resp., iff) $(a_1, \ldots, a_r) \in R^{\mathcal{A}}$. We will occasionally consider signatures with constant symbols. We write $c^{\mathcal{A}}$ for the interpretation of a constant symbol c and homomorphisms are required to preserve constants as well, that is $h(c^{\mathcal{A}}) = c^{\mathcal{B}}$.

Containment for $\{\exists, \wedge\}$-FO is a special case of conjunctive query containment from databases [5]. Given a finite structure $\mathcal{A}$, we write $\varphi_{\mathcal{A}}$ for the so-called *canonical conjunctive query* of $\mathcal{A}$, the quantifier-free formula that is the conjunction of the positive facts of $\mathcal{A}$, where the variables $v_1, \ldots, v_{|A|}$ correspond to the elements $a_1, \ldots, a_{|A|}$ of $\mathcal{A}$.

Theorem 1 (Containment). *Let $\mathcal{A}$ and $\mathcal{B}$ be two structures. The following are equivalent.*

 (i) for every sentence φ in $\{\exists, \wedge\}$-FO, if $\mathcal{A} \models \varphi$ then $\mathcal{B} \models \varphi$.
 (ii) The exists a homomorphism from $\mathcal{A}$ to $\mathcal{B}$.
 (iii) $\mathcal{B} \models \exists v_1 \exists v_2 \ldots v_{|A|} \varphi_{\mathcal{A}}$.

where $\varphi_{\mathcal{A}}$ denotes the canonical conjunctive query[1] *of $\mathcal{A}$.*

It is well known that the core is unique up to isomorphism and that it is an induced substructure [13]. It is usually defined via homomorphic equivalence, but because of the equivalence between (i) and (ii) in the above theorem, we may define the core as follows.

Definition 2. *The* core $\mathcal{B}$ *of a structure $\mathcal{A}$ is a minimal substructure of $\mathcal{A}$ such that for every sentence φ in $\{\exists, \wedge\}$-FO, $\mathcal{A} \models \varphi$ if and only if $\mathcal{B} \models \varphi$.*

Corollary 3 (equivalence). *Let $\mathcal{A}$ and $\mathcal{B}$ be two structures. The following are equivalent.*

 (i) for every sentence φ in $\{\exists, \wedge\}$-FO, $\mathcal{A} \models \varphi$ if and only if $\mathcal{B} \models \varphi$.
 (ii) There are homomorphisms from $\mathcal{A}$ to $\mathcal{B}$ and from $\mathcal{B}$ to $\mathcal{A}$.
 (iii) The core of $\mathcal{A}$ and the core of $\mathcal{B}$ are isomorphic.

As a preprocessing step, one could replace the template $\mathcal{A}$ of a CSP by its core $\mathcal{B}$ (see Algorithm 6.1 in [8]). However, the complexity of this preprocessing step would be of the same order of magnitude as solving a constraint satisfaction problem.[2] This drawback, together with the uniform nature of the instance in constraints solvers, means that this preprocessing is not exploited in practice to the best of our knowledge.

The notion of a core can be extended and adapted suitably to solve important questions related to data exchange and query rewriting in databases [9]. It is also very useful as a simplifying assumption when classifying the complexity: with the algebraic approach, it allows to study only idempotent algebras [4].

---

[1] Most authors consider the canonical query to be the sentence which is the existential quantification of $\varphi_{\mathcal{A}}$.

[2] Checking that a graph is a core is coNP-complete [11]. Checking that a graph is the core of another given graph is DP-complete [9].

3   The Case of QCSP with Disjunction

For $\{\exists, \forall, \wedge, \vee\}$-FO, it is no longer the homomorphism that is the correct concept to transfer winning strategies.

Definition 4. *A surjective hypermorphism f from a structure $\mathcal{A}$ to a structure $\mathcal{B}$ is a function from the domain A of $\mathcal{A}$ to the power set of the domain B of $\mathcal{B}$ that satisfies the following properties.*

- *(**total**) for any a in A, $f(a) \neq \emptyset$.*
- *(**surjective**) for any b in B, there exists a in A such that $f(a) \ni b$.*
- *(**preserving**) if $R(a_1, \ldots, a_i)$ holds in $\mathcal{A}$ then $R(b_1, \ldots, b_i)$ holds in $\mathcal{B}$, for all $b_1 \in f(a_1), \ldots, b_i \in f(a_i)$.*

Lemma 5 (strategy transfer). *Let $\mathcal{A}$ and $\mathcal{B}$ be two structures such that there is a surjective hypermorphism from $\mathcal{A}$ to $\mathcal{B}$. Then, for every sentence φ in $\{\exists, \forall, \wedge, \vee\}$-FO, if $\mathcal{A} \models \varphi$ then $\mathcal{B} \models \varphi$.*

Example 6. Consider the structures $\mathcal{A}_4$ and $\mathcal{A}_3$ from Table 1. The map f given by $f(x) := \{x\}$ for $0 \leq x \leq 5$ and $f(6) := \{4\}$ is a surjective hypermorphism from $\mathcal{A}_4$ to $\mathcal{A}_3$. The map g given by $g(x) := \{x\}$ for $x \in \{0, 1, 2, 3, 5\}$ and $g(4) := \{4, 6\}$ is a surjective hypermorphism from $\mathcal{A}_3$ to $\mathcal{A}_4$. The two templates are equivalent w.r.t. $\{\exists, \forall, \wedge, \vee\}$-FO.

We can define $\Theta_{\mathcal{A}, |\mathcal{B}|}$ a canonical sentence of $\{\exists, \forall, \wedge, \vee\}$-FO that is defined in terms of $\mathcal{A}$ and $|\mathcal{B}|$ and that is modelled by $\mathcal{A}$ by construction.

Lemma 7. *Let $\mathcal{A}$ and $\mathcal{B}$ be two structures. If $\mathcal{B} \models \theta_{\mathcal{A}, |\mathcal{B}|}$ then there is a surjective hypermorphism from $\mathcal{A}$ to $\mathcal{B}$.*

Theorem 8 (Containment for $\{\exists, \forall, \wedge, \vee\}$-FO). *Let $\mathcal{A}$ and $\mathcal{B}$ be two structures. The following are equivalent.*

- *(i) for every sentence φ in $\{\exists, \forall, \wedge, \vee\}$-FO, if $\mathcal{A} \models \varphi$ then $\mathcal{B} \models \varphi$.*
- *(ii) The exists a surjective hypermorphism from $\mathcal{A}$ to $\mathcal{B}$.*
- *(iii) $\mathcal{B} \models \theta_{\mathcal{A}, |\mathcal{B}|}$*

where $\Theta_{\mathcal{A}, |\mathcal{B}|}$ is a canonical sentence of $\{\exists, \forall, \wedge, \vee\}$-FO that is defined in terms of $\mathcal{A}$ and $|\mathcal{B}|$ and that is modelled by $\mathcal{A}$ by construction.

Let U and X be two subsets of A and a surjective hypermorphism h from $\mathcal{A}$ to $\mathcal{A}$ that satisfies $h(U) = \bigcup_{u \in U} h(u) = A$ and $h^{-1}(X) = A$ where $h^{-1}(X)$ stands for $\{a \in A : \exists x \in X \text{ s. t. } x \in h(a)\}$. Let $\mathcal{B}$ be the substructure of $\mathcal{A}$ induced by $B := U \cup X$. Then f and g, the range and domain restriction of h to B, respectively, are surjective hypermorphisms between $\mathcal{A}$ and $\mathcal{B}$ witnessing that $\mathcal{A}$ and $\mathcal{B}$ satisfy the same sentence of $\{\exists, \forall, \wedge, \vee\}$-FO.[3] Note that in particular h induces a retraction of $\mathcal{A}$ to a subset of X; and, dually a retraction of the

---

[3] $f := h_{|B}$ (the usual function restriction) and for any x in A, $g(x) := h(x) \cap B$.

complement structure [4] of $\mathcal{B}$ to a subset of U. Additional minimality conditions on U, X and $U \cup X$ ensure that $\mathcal{B}$ is minimal.[5] It is also *unique* up to isomorphism and within $\mathcal{B}$ the set U and X are *uniquely determined*. Consequently, $\mathcal{B}$ is called the U-X-core of $\mathcal{A}$ (for further details see [15]) and may be defined as follows.

Definition 9. *The U-Xcore $\mathcal{B}$ of a structure $\mathcal{A}$ is a minimal substructure of $\mathcal{A}$ such that for every sentence φ in $\{\exists, \forall, \wedge, \vee\}$-FO, $\mathcal{A} \models \varphi$ if and only if $\mathcal{B} \models \varphi$.*

Example 10. The map $h(0) := \{0\}$, $h(1) := \{1\}$, $h(2) := \{0, 2\}$, $h(3) := \{0, 3\}$, $h(4) := \{4\}$, $h(5) := \{0, 5\}$, $h(6) := \{4, 6\}$, is a surjective hypermorphism from $\mathcal{A}_4$ to $\mathcal{A}_4$ with $U = \{2, 3, 5\}$ and $X := \{0, 1, 4\}$ (in general U and X need not be disjoint). The substructure induced by $U \cup X$ is $\mathcal{A}_3$. It can be checked that it is minimal.

The U-X-core is just like the core an induced substructure. There is one important difference in that U-X-cores should be genuinely viewed as a minimal equivalent substructure induced by *two* sets. Indeed, when evaluating a sentence of $\{\exists, \forall, \wedge, \vee\}$-FO, we may assume w.l.o.g. that all $\forall$ variables range over U and all $\exists$ variables range over X. This is because we can "pull" any move of $\forall$ into U and "push" any successful move of $\exists$ into X in the spirit of Lemma 5. Ultimately, we may extract a winning strategy for $\exists$ that can restrict herself to play only on X, even if $\forall$ plays arbitrarily [15, Lemma 5]. Hence, as a *preprocessing step*, one could compute U and X and restrict the domain of each universal variable to U and the domain of each universal variable to X. *The complexity of this processing step is no longer of the same magnitude* and is in general much lower than solving a QCSP$_\vee$.[6]   Thus, even when taking into account the uniform nature of the instance in a quantified constraints solver, this preprocessing step might be exploited in practice. This could turn out to be ineffective when there are few quantifier alternation (as in bilevel programming), but should be of particular interest when the quantifier alternation increases. Another interesting feature is that storing a winning strategy over U and X together with the surjective hypermorphism h from $\mathcal{A}$ to $\mathcal{A}$, allows to recover a winning strategy even when $\forall$ plays in an unrestricted manner. This provides a *compression mechanism* to store certificates.

4   The Case of QCSP

In primitive positive and positive Horn logic, one normally considers equalities to be permitted. From the perspective of computational complexity of CSP and QCSP, this distinction is unimportant as equalities may be propagated out by

---

[4] It has the same domain as $\mathcal{A}$ and a tuple belongs to a relation R iff it did not in $\mathcal{A}$.

[5] This is possible since given h_1 s.t. $h_1(U) = A$ and h_2 such that $h_2^{-1}(X) = A$, their composition $h = h_2 \circ h_1$ (where $h(x) := \bigcup_{y \in h_1(x)} h_2(y)$) satisfies both $h(U) = A$ and $h^{-1}(X) = A$.

[6] The question of U-X-core identification is DP-complete, whereas QCSP$_\vee$ is Pspace-complete in general.

substitution. In the case of positive Horn and QCSP, though, equality does allow the distinction of a trivial case that can not be recognised without it. The sentence $\exists x \forall y\ x = y$ is true exactly on structures of size one. The structures $\mathcal{K}_1$ and $2\mathcal{K}_1$, containing empty relations over one element and two elements, respectively, are therefore distinguishable in $\{\exists, \forall, \wedge, =\}$-FO, but not in $\{\exists, \forall, \wedge\}$-FO. Since we disallow equalities, many results from this section apply only to *non-trivial* structures of size ≥ 2. Note that equalities can not be substituted out from $\{\exists, \forall, \wedge, \vee, =\}$-FO, thus it is substantially stronger than $\{\exists, \forall, \wedge, \vee\}$-FO.

For $\{\exists, \forall, \wedge\}$-FO, the correct concept to transfer winning strategies is that of *surjective homomorphism from a power*. Recall first that the *product* $\mathcal{A} \times \mathcal{B}$ of two structures $\mathcal{A}$ and $\mathcal{B}$ has domain $\{(x, y) : x \in A, y \in B\}$ and for a relation symbol R, $R^{\mathcal{A} \times \mathcal{B}} := \{((a_1, b_1), \ldots, (a_r, b_r)) : (a_1, \ldots, a_r) \in R^{\mathcal{A}}, (b_1, \ldots, b_r) \in R^{\mathcal{B}}\}$; and, similarly for a constant symbol c, $c^{\mathcal{A} \times \mathcal{B}} := (c^{\mathcal{A}}, c^{\mathcal{B}})$. The *mth power* $\mathcal{A}^m$ of $\mathcal{A}$ is $\mathcal{A} \times \ldots \times \mathcal{A}$ (m times).

Lemma 11 (strategy transfer). *Let $\mathcal{A}$ and $\mathcal{B}$ be two structures and $m \geq 1$ such that there is a surjective homomorphism from $\mathcal{A}^m$ to $\mathcal{B}$. Then, for every sentence φ in $\{\exists, \forall, \wedge\}$-FO, if $\mathcal{A} \models \varphi$ then $\mathcal{B} \models \varphi$.*

Example 12. Consider an undirected bipartite graph with at least one edge $\mathcal{G}$ and $\mathcal{K}_2$ the graph that consists of a single edge. There is a surjective homomorphism from $\mathcal{G}$ to $\mathcal{K}_2$. Note also that $\mathcal{K}_2 \times \mathcal{K}_2 = \mathcal{K}_2 + \mathcal{K}_2$ (where $+$ stands for disjoint union) which we write as $2\mathcal{K}_2$. Thus, $\mathcal{K}_2{}^j = 2^{j-1}\mathcal{K}_2$ (as $\times$ distributes over $+$). Hence, if $\mathcal{G}$ has no isolated element and m edges there is a surjective homomorphism from $\mathcal{K}_2^{1+\log_2 m}$ to $\mathcal{G}$.

This examples provides a lower bound for m which we can improve.

Proposition 13 (lower bound). *For any $m \geq 2$, there are structures $\mathcal{A}$ and $\mathcal{B}$ with $|A| = m$ and $|B| = m+1$ such that there is only a surjective homomorphism from $\mathcal{A}^j$ to $\mathcal{B}$ provided that $j \geq |A|$.*

Proof (sketch). We consider a signature that consists of a binary symbol E together with a monadic predicate R. Consider for $\mathcal{A}$ an oriented cycle with m vertices, for which R holds for all but one. Consider for $\mathcal{B}$ an oriented cycle with

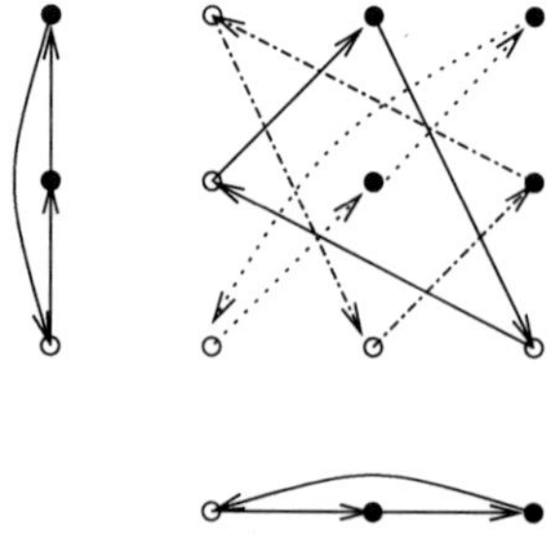

Fig. 1. the power of oriented cycles is a sum of oriented cycles

m vertices, for which R does not hold, together with a self-loop on which R holds. The square of $\mathcal{A}$ will consists of $|A| = m$ oriented cycles with m vertices: one cycle will be a copy of $\mathcal{A}$, all the other will be similar but with two vertices on which R does not hold (this is depicted on Figure 1 in the case $m = 3$: white vertices do not satisfy R while black ones do). It is only for $j = m$ that we will get as an induced substructure of $\mathcal{A}^j$ one copy of an oriented cycle on which R does not hold as in $\mathcal{B}$.

There is also a *canonical* $\{\exists, \forall, \wedge\}$-*FO-sentence* which turns out to be in Π_2-form, that is with a quantifier prefix of the form $\forall^\star \exists^\star$.

Theorem 14 (Containment for $\{\exists, \forall, \wedge\}$-FO [7]). *Let $\mathcal{A}$ and $\mathcal{B}$ be two nontrivial structures. The following are equivalent.*

  (i) for every sentence φ in $\{\exists, \forall, \wedge\}$-FO, if $\mathcal{A} \models \varphi$ then $\mathcal{B} \models \varphi$.
  (ii) There exists a surjective homomorphism from $\mathcal{A}^r$ to $\mathcal{B}$, with $r \leq |A|^{|B|}$.
 (iii) $\mathcal{B} \models \psi_{\mathcal{A}, |B|}$

where $\psi_{\mathcal{A}, |B|}$ is a canonical sentence of $\{\exists, \forall, \wedge\}$-FO with quantifier prefix $\forall^{|B|} \exists^\star$ that is defined in terms of $\mathcal{A}$ and modelled by $\mathcal{A}$ by construction.

Following our approach for the other logics, we now define a minimal representative as follows.

Definition 15. *A* Q-core $\mathcal{B}$ *of a structure* $\mathcal{A}$ *is a minimal substructure of* $\mathcal{A}$ *such that for every sentence* φ *in* $\{\exists, \forall, \wedge\}$-FO, $\mathcal{A} \models \varphi$ *if and only if* $\mathcal{B} \models \varphi$.

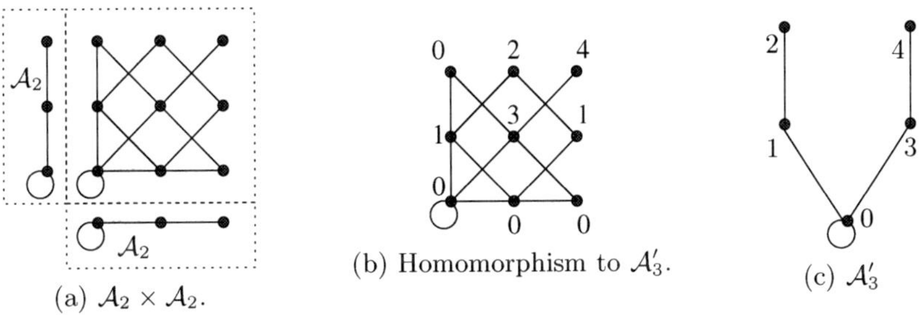

(a) $\mathcal{A}_2 \times \mathcal{A}_2$.

(b) Homomorphism to $\mathcal{A}_3'$.

(c) $\mathcal{A}_3'$

Fig. 2. surjective homomorphism from a power

Example 16. Consider $\mathcal{A}_3$ and $\mathcal{A}_2$ from Table 1. We consider the subgraph $\mathcal{A}_3'$ of $\mathcal{A}_3$ as depicted on Figure 2c. The map $f(0) := 0$, $f(1) := 1$, $f(2) := 2$, $f(3) := 0$, $f(4) := 0$ is a surjective homomorphism from $\mathcal{A}_3'$ to $\mathcal{A}_2$. The square of $\mathcal{A}_2$ is depicted on Figure 2a; and, a surjective homomorphism from it to $\mathcal{A}_3'$ is depicted on Figure 2b. Thus $\mathcal{A}_3'$ and $\mathcal{A}_2$ are equivalent w.r.t. $\{\exists, \forall, \wedge\}$-FO. In a similar fashion but using a cube rather than a square, one can check that $\mathcal{A}_3$ and $\mathcal{A}_2$ are equivalent w.r.t. $\{\exists, \forall, \wedge\}$-FO. One can also check that $\mathcal{A}_2$ is minimal and is therefore a Q-core of $\mathcal{A}_3$, and *a posteriori* of $\mathcal{A}_4$.

The behaviour of the Q-core differs from its cousins the core and the U-X-core.

Proposition 17. *The Q-core of a 3-element structure $\mathcal{A}$ is not always an induced substructure of $\mathcal{A}$.*

Proof. Consider the signature $\sigma := \langle E, R, G \rangle$ involving a binary relation E and two unary relations R and G. Let $\mathcal{A}$ and $\mathcal{B}$ be structures with domain $\{1, 2, 3\}$ with the following relations.

$$E^{\mathcal{A}} := \{(1,1), (2,3), (3,2)\} \; R^{\mathcal{A}} := \{1,2\} \; G^{\mathcal{A}} := \{1,3\}$$
$$E^{\mathcal{B}} := \{(1,1), (2,3), (3,2)\} \; \; R^{\mathcal{B}} := \{1\} \; \; \; G^{\mathcal{B}} := \{1\}$$

Since $\mathcal{B}$ is a substructure of $\mathcal{A}$, we have $\mathcal{B} \twoheadrightarrow \mathcal{A}$. Conversely, the square of $\mathcal{A}^2$ contains an edge that has no vertex in the relation R and G, which ensures that $\mathcal{A}^2 \twoheadrightarrow \mathcal{B}$ (see Figure 3). Observe also that no two-element structure $\mathcal{C}$, and *a fortiori* no two-element substructure of $\mathcal{A}$ agrees with them on $\{\exists, \forall, \wedge\}$-FO.

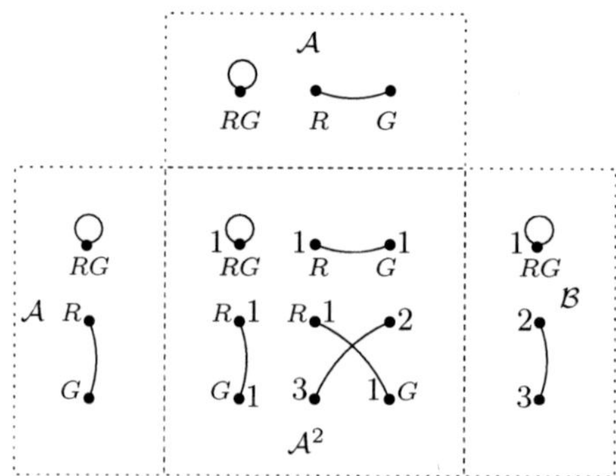

Fig. 3. example of two distinct 3-element structures (signature, E binary and two unary predicates R and G) that are equivalent w.r.t. $\{\exists, \forall, \wedge\}$-FO

We do not know whether the Q-core of a structure is unique. We will explore in the following section Q-cores over some special classes and show that it behaves well in these cases.

5   Q-cores over Classes

5.1   The Boolean Case

A Boolean structure $\mathcal{B}$ has domain $B := \{0, 1\}$. The results of this section apply to arbitrary (not necessarily finite) signatures.

Theorem 18. *Let $\mathcal{A}$ and $\mathcal{B}$ be Boolean structures that are equivalent w.r.t. $\{\exists, \forall, \wedge\}$-FO. Then $\mathcal{A}$ and $\mathcal{B}$ are isomorphic.*

In the extended logic $\{\exists, \forall, \wedge, =\}$-FO, it follows that every structure of size at most 2 satisfies the ideal of core. For $\{\exists, \forall, \wedge\}$-FO we can only say the following.

Proposition 19. *Every Boolean structure $\mathcal{B}$ is either a Q-core, or its Q-core is the substructure induced by either of its elements. In particular, **the** Q-core of $\mathcal{B}$ is unique up to isomorphism and is an induced substructure of $\mathcal{B}$.*

5.2   Unary Structures

Let σ be a fixed relational signature that consists of n unary relation symbols $M_1, M_2, \ldots, M_n$. A structure over such a signature is deemed *unary*. Let w be a string of length n over the alphabet $\{0, 1\}$. We write $w(x)$ as an abbreviation for the quantifier-free formula $\bigwedge_{1 \leq i \leq n, w[i]=1} M_i(x)$. An element a of a unary structure $\mathcal{A}$ corresponds naturally to a word w, namely the word with maximal Hamming weight s.t. $w(a)$ holds. The unary structure $\mathcal{A}$ satisfies the *canonical universal sentence* $\forall y \, w_\forall(y)$, where $w_\forall$ is the conjunction of the words associated with each element.

Proposition 20. *The Q-core of a unary structure $\mathcal{A}$ is the unique substructure of $\mathcal{A}$ defined as follows. The Q-core of $\mathcal{A}$ is the core $\mathcal{A}'$ of $\mathcal{A}$ if they share the same canonical universal sentence and the disjoint union of $\mathcal{A}'$ with a single element corresponding to $w_\forall$ where $\forall y \, w_\forall(y)$ is the canonical universal sentence of $\mathcal{A}$, otherwise.*

5.3   Structures with an Isolated Element

We say that φ is a *proper $\{\exists, \forall, \wedge\}$-FO-sentence*, if it has at least one universally quantified variable x that occurs in some atom in the quantifier-free part.

Let σ be a signature that consists of finitely many relation symbols R_i of respective arity r_i. We will consider the set of minimal proper $\{\exists, \forall, \wedge\}$-FO-sentences w.r.t. σ, that is all formulae of the form $\forall x_1 \exists x_2 \ldots \exists x_{r_i} R_i(\bar{x})$, where the tuple $\bar{x}$ is a permutation of the variables $x_1, x_2, \ldots x_{r_i}$ where x_1 has been transposed with some other variable. There are $r(\sigma) = \Sigma_{R_i \in \sigma} r_i$ such formulae.

Theorem 21. *Let $\mathcal{A}$ be a σ-structure. The following are equivalent.*

1. *$\mathcal{A}$ does not satisfy any proper $\{\exists, \forall, \wedge\}$-FO-sentence.*
2. *$\mathcal{A}$ does not satisfy any of the $r(\sigma)$ minimal proper $\{\exists, \forall, \wedge\}$-FO-sentences w.r.t. σ.*
3. *$\mathcal{A}^{r(\sigma)}$ contains an isolated element (that does not occur in any relation).*

Proof (sketch for digraphs). Assume that the signature σ consists of a single binary symbol E. The minimal proper sentences are $\forall x_1 \exists x_2 E(x_1, x_2)$ and $\forall x_1 \exists x_2 E(x_2, x_1)$ and we have $r(\sigma) = 2$.

A directed graph which does not satisfy them will satisfy their negation $\exists x_1 \forall x_2 \neg E(x_1, x_2)$ and $\exists x_1 \forall x_2 \neg E(x_2, x_1)$. A witness for the existential x_1 in the first sentence will be a *source*, and in the second sentence a *sink*, respectively.

The first point implies trivially the second. The converse is essentially trivial for digraphs (less so in general). The second point is equivalent to the last: the isolated element in the square $\mathcal{A}^2$ is of the form "source $\times$ sink".

Corollary 22. *The Q-core of a structure $\mathcal{A}$ that does not satisfy any proper $\{\exists, \forall, \wedge\}$-FO-sentence is the unique substructure of $\mathcal{A}$ that may be found as follows. The Q-core of $\mathcal{A}$ is the core $\mathcal{A}'$ of $\mathcal{A}$, if $\mathcal{A}'^{r(\sigma)}$ contains an isolated element, and the disjoint union of $\mathcal{A}'$ and an isolated element, otherwise.*

Remark 23. Checking whether a structure with an isolated element is a Q-core is of the same complexity as checking whether it is a core, it is co-NP-complete [12].

6   The Usefulness of Q-Cores

Graphs are relational structures with a single symmetric relation E. We term them *reflexive* when any vertex has a self-loop; *partially reflexive* (p.r.) to emphasise that any vertex may or may not have a self-loop; and, *irreflexive* when they have none. A *p.r. tree* may contain self-loops but no larger cycle C_n for $n \geq 3$. A *p.r. forest* is the disjoint union of p.r. trees.

Since p.r. forests are closed under substructures, we can be assured that a Q-core of a p.r. forest is a p.r. forest. It is clear from inspection that the Q-core of p.r. forest is unique up to isomorphism, but we do not prove this as it does not shed any light on the general situation. The doubting reader may substitute "a/ all" for "the" in future references to Q-cores in this section.

The complexity classifications of [16] were largely derived using the properties of equivalence w.r.t. $\{\exists, \forall, \wedge\}$-FO. This will be the central justification for the following propositions.

Let $\mathcal{K}_i^\star$ and $\mathcal{K}_i$ be the reflexive and irreflexive i-cliques, respectively. Let $[n] := \{1, \ldots, n\}$. For $i \in [n]$ and $\alpha \in \{0,1\}^n$, let $\alpha[i]$ be the ith entry of α. For $\alpha \in \{0,1\}^*$, let $\mathcal{P}_\alpha$ be the path with domain $[n]$ and edge set $\{(i,j) : |j - i| = 1\} \cup \{(i,i) : \alpha[i] = 1\}$. For a tree $\mathcal{T}$ and vertex $v \in T$, let $\lambda_T(v)$ be the shortest distance in $\mathcal{T}$ from v to a looped vertex (if $\mathcal{T}$ is irreflexive, then $\lambda_T(v)$ is always infinite). Let λ_T be the maximum of $\{\lambda_T(v) : v \in T\}$. A tree is *loop-connected* if the self-loops induce a connected subtree. A tree $\mathcal{T}$ is *quasi-loop-connected* if either 1.) it is irreflexive, or 2.) there exists a connected reflexive subtree $\mathcal{T}_0$ (chosen to be **maximal**) such that there is a walk of length λ_T from every vertex of $\mathcal{T}$ to $\mathcal{T}_0$.

6.1   Partially Reflexive Forests

It is not true that, if $\mathcal{H}$ is a p.r. forest, then either $\mathcal{H}$ admits a majority polymorphism, and QCSP($\mathcal{H}$) is in NL, or QCSP($\mathcal{H}$) is NP-hard. However, the notion of Q-core restores a clean delineation.

Proposition 24. *Let $\mathcal{H}$ be a p.r. forest. Then either the Q-core of $\mathcal{H}$ admits a majority polymorphism, and QCSP($\mathcal{H}$) is in NL, or QCSP($\mathcal{H}$) is NP-hard.*

Proof. We assume that graphs have at least one edge (otherwise the Q-core is $\mathcal{K}_1$). Irreflexive forests are a special case of bipartite graphs, which are all equivalent w.r.t. $\{\exists, \forall, \wedge\}$-FO, their Q-core being $\mathcal{K}_2$ when they have no isolated vertex (see example 12) and $\mathcal{K}_2 + \mathcal{K}_1$ otherwise.

We assume from now on that graphs have at least one edge and one self-loop. The one vertex case is $\mathcal{K}_1^\star$. We assume larger graphs from now on. If the graph contains an isolated element then its Q-core is $\mathcal{K}_1 + \mathcal{K}_1^\star$. Assume from now on that the graph does not have an isolated element.

We deal with the disconnected case first. If the graph is reflexive, then its Q-core is $\mathcal{K}_1^\star + \mathcal{K}_1^\star$. Otherwise, the graph is properly partially reflexive in the sense that it embeds both $\mathcal{K}_1^\star$ and $\mathcal{K}_1$. If the graph has an irreflexive component then its Q-core is $\mathcal{K}_2 + \mathcal{K}_1^\star$. If the graph has no irreflexive component, then its Q-core is $\mathcal{K}_1^\star + \mathcal{P}_{10^\lambda}$ where λ is the longest walk from any vertex to a self-loop. The equivalence follows from analysing surjective homomorphism from suitable powers and requires some work. The minimality follows from the fact that the Q-core must not satisfy $\forall x \exists y_1, \ldots, y_{\lambda-1} \; E(x, y_1) \wedge E(y_1, y_2) \wedge \ldots \wedge E(y_{\lambda-2}, y_{\lambda-1})$ and must be disconnected .

We now follow the classification of [16]. If a p.r. forest contains more than one p.r. tree, then the Q-core is among those formed from the disjoint union of exactly two (including the possibility of duplication) of $\mathcal{K}_1$, $\mathcal{K}_1^\star$, $\mathcal{P}_{10^\lambda}$, $\mathcal{K}_2$. Each of these singularly admits a majority polymorphism, therefore so does any of their disjoint unions.

We now move on to the connected case, i.e. it remains to consider p.r. trees $\mathcal{T}$. If $\mathcal{T}$ is irreflexive, then its Q-core is $\mathcal{K}_2$ or $\mathcal{K}_1$, which admit majority polymorphisms. If $\mathcal{T}$ is loop-connected, then it admits a majority polymorphism [16]. If $\mathcal{T}$ is quasi-loop-connected, then it is QCSP-equivalent to one of its subtrees that is loop-connected [16] which will be its Q-core. In all other cases QCSP($\mathcal{T}$) is NP-hard, and $\mathcal{T}$ does not admit majority [16].

6.2   Irreflexive Pseudoforests

A *pseudotree* is a graph that involves at most one cycle. A *pseudoforest* is the disjoint union of a collection of pseudotrees.

Proposition 25. *Let $\mathcal{H}$ be an irreflexive pseudoforest. Then either the Q-core of $\mathcal{H}$ admits a majority polymorphism, and QCSP($\mathcal{H}$) is in NL, or QCSP($\mathcal{H}$) is NP-hard.*

Proof. We follow the classification of [17]. If $\mathcal{H}$ is bipartite, then its Q-core is either $\mathcal{K}_2$, $\mathcal{K}_1$, $\mathcal{K}_2 + \mathcal{K}_1$ (see [7]) and this admits a majority polymorphism. Otherwise its Q-core contains an odd cycle, which does not admit a majority polymorphism, and QCSP($\mathcal{H}$) is NP-hard.

7   Computing a Q-Core

We may use Theorem 14 to provides a first algorithm (Algorithm 1). This does not appear very promising if we wish to use Q-cores as a preprocessing step.

Algorithm 1. a naive approach to compute the Q-cores

> **input**   : A structure $\mathcal{A}$
> **output**   : The list L of Q-cores of $\mathcal{A}$
> **initialisation**: set $L := \{\mathcal{A}\}$
> **forall the** *substructure $\mathcal{B}$ of $\mathcal{A}$* **do**
> > **if** *there exists a surjective homomorphism from $\mathcal{A}^{|A|^{|B|}}$ to $\mathcal{B}$* **then**
> > > **if** *there exists a surjective homomorphism from $\mathcal{B}^{|B|^{|A|}}$ to $\mathcal{A}$* **then**
> > > > Remove any structure containing $\mathcal{B}$ in L;
> > > > Add $\mathcal{B}$ to L;
> > >
> > > **end**
> >
> > **end**
>
> **end**
> **output**   : List of Q-cores L

We will propose and illustrate a general and less naive method to compute Q-cores by computing U-X-core and cores first.

Another nice feature of cores and U-X-cores which implies their uniqueness is the following: any substructure $\mathcal{C}$ of $\mathcal{A}$ that agrees with it on $\{\exists, \wedge\}$-FO (respectively on $\{\exists, \forall, \wedge, \vee\}$-FO) will contain the core (respectively the U-X-core). Consequently, the core and the U-X-core may be computed in a *greedy fashion*. Assuming that the Q-core would not satisfy this nice property, why should this concern the Q-core? Well, we know that **any** Q-core will lie somewhere between the U-X-core and the core that are induced substructures: this is a direct consequence of the inclusion of the corresponding fragments of first-order logic and their uniqueness. Moreover, according to our current knowledge, checking for equivalence appears, at least on paper, much easier for $\{\exists, \forall, \wedge, \vee\}$-FO than $\{\exists, \forall, \wedge\}$-FO: compare the number of functions from A to the power set of B $(2^{|B|^{|A|}} = 2^{|B| \times |A|})$ with the number of functions from A^r to B $(|B|^{|A|^r})$ where r could be as large as $|A|^{|B|}$ and can certainly be greater than $r \approx |A|$ (see Proposition 13). So it make sense to bound the naive search for Q-cores.

Furthermore, we know that the U-X-core can be identified by specific surjective hypermorphisms that act as the identity on X and contain the identity on U [15] which makes the search for the U-X-core somewhat easier than its definition suggest (see Algorithm 2).

Observe also that X must contain the core $\mathcal{C}$ of the U-X-core $\mathcal{B}$, which is also the core of the original structure $\mathcal{A}$ (this is because h induces a so-called retraction of $\mathcal{A}$ to the substructure $\mathcal{A}_{|X}$ induced by X). Thus we may compute the core greedily from X. Next, we do a little bit better than using our naive algorithm, by interleaving steps where we find a substructure that is $\{\exists, \forall, \wedge\}$-FO-equivalent, with steps where we compute its U-X-core (one can find a sequence of distinct substructures $\mathcal{B}, \mathcal{D}$ and $\mathcal{B}'$ such that $\mathcal{B}$ is a U-X-core, which is $\{\exists, \forall, \wedge\}$-FO-equivalent to $\mathcal{D}$, whose U-X-core $\mathcal{B}'$ is strictly smaller than $\mathcal{B}$, see Example 26). Algorithm 3 describes this proposed method informally when we want to compute **one** Q-core (of course, we would have no guarantee that we

---

Algorithm 2. a greedy approach to compute the U-X-core

---

input	: a structure $\mathcal{A}$.
output	: the U-X-core of $\mathcal{A}$.
variables	: U and X two subsets of A.
variable	: h a surj. hypermorphism from $\mathcal{A}$ to $\mathcal{A}$ s.t. $h(U) = A$ and $h^{-1}(X) = A$.
variable	: $\mathcal{B}$ an induced substructure of $\mathcal{A}$ such that $B = U \cup X$.

initialisation: set $U := A$, $X := A$, $\mathcal{B} := \mathcal{A}$, h the identity
repeat
    guess a subset U' of U and a subset X' of X;
    let h' be a map from B to B;
    forall the x' *in* X' **do set** $h'(x') := \{x'\}$;
    forall the u' *in* $U' \setminus X'$ **do guess** x' in X' **set** $h'(u') := \{u', x'\}$;
    forall the z' *in* $B \setminus (U' \cup X')$ **do**
        guess x' in X' **set** $h(z') := \{x'\}$;
        guess u' in U' **set** $h(u') := h(u') \cup \{z'\}$;
    end
    if h' *is a surj. hypermorphism from* B *to* B **then**
        set $\mathcal{B}$ to be the substructure of $\mathcal{B}$ induced by $U' \cup X'$;
        set $U := U'$, $X := X'$ and $h := h' \circ h$;
    end
until U *and* X *are minimal*;
output   : $\mathcal{B}$.

---

get the smallest Q-core, unless the Q-core can be also greedily computed, which holds for all cases we have studied so far).

In Algorithm 3, we have purposely not detailed line 3. We could use the characterisation of $\{\exists, \forall, \wedge\}$-FO-containment via surjective homomorphism from a power of Theorem 14 as in Algorithm 1. Alternatively, we can use a refined form of (iii) in this Theorem and use the canonical sentences in Π_2-form $\psi_{\mathcal{B},m_1}$ and $\psi_{\mathcal{D},m_2}$, with $m_1 := \min(|D|, |U|)$ and $m_2 := |U|$.

The test would consists in checking that $\mathcal{B}$ satisfies $\psi_{\mathcal{D},m_2}$ (where we may relativise to universal variables to $\mathcal{U}$ and existential variables to X) and $\mathcal{D}$ satisfies $\psi_{\mathcal{B},m_1}$. This is correct because we know that we may relativise every universal variable to U within $\mathcal{B}$. Thus, it suffices to consider Π_2-sentences with at most $|U|$ universal variables.

Example 26. We describe a run of Algorithm 3 on input $\mathcal{A} := \mathcal{A}_4$. During the initialisation, we compute its U-X-core $\mathcal{B} := \mathcal{A}_3$ and discover that $U = \{2, 3, 5\}$ and $X = \{0, 1, 4\}$. We compute $\mathcal{C} := \mathcal{A}_1$, the core of the substructure induced by X. Next the algorithm guesses a substructure $\mathcal{D}$ of $\mathcal{B}$ that contains $\mathcal{C}$: e.g. it takes for $\mathcal{D}$ the substructure induced by $\{0, 1, 2, 3, 4\}$. It checks succesfully equivalence w.r.t. $\{\exists, \forall, \wedge\}$-FO (trivially, there is a surjective homomorphism from $\mathcal{A}_3$ to $\mathcal{D}$; and, one may compose the natural surjective homomorphism from $\mathcal{D} \times \mathcal{D}$ to $\mathcal{P}_{0100} \times \mathcal{P}_{100}$ with a suitable surjective homomorphism from the latter to $\mathcal{A}_3$).

Algorithm 3. bounded Search for a Q-core

    input   : a structure $\mathcal{A}$
    output   : a Q-core $\mathcal{B}$ of $\mathcal{A}$
    initialisation: compute the U-X-core of $\mathcal{A}$ as in Algorithm 2;
    set $\mathcal{B}$ to be the U-X-core;
    set $\mathcal{C}$ to be the core $\mathcal{C}$ of the substructure of $\mathcal{B}$ induced by X;

1 **repeat**
2 | **guess** $\mathcal{D}$ a substructure of $\mathcal{B}$ that contains $\mathcal{C}$;
3 | **check** that $\mathcal{B}$ and $\mathcal{D}$ are equivalent w.r.t. $\{\exists, \forall, \wedge\}$-FO;
4 | **set** $\mathcal{B}$ to be the U-X-core of $\mathcal{D}$;
5 **until** $\mathcal{B}$ *is minimal*;
    output   : $\mathcal{B}$.

Next the algorithm computes the U'-X'-core $\mathcal{B}'$ of $\mathcal{D}$ which is the substructure induced by the union of $U' = \{2, 3\}$ with $X' = \{0, 1\}$ (witnessed by $h'(0) = \{0\}$, $h'(1) = \{1\}$, $h'(2) = \{0, 2\}$, $h'(3) = \{1, 3, 4\}$, $h'(4) = \{0\}$). This substructure $\mathcal{B}'$ is isomorphic to $\mathcal{P}_{0100}$. The algorithm sets $\mathcal{B} := \mathcal{B}'$ and starts over. It stops eventually and outputs $\mathcal{A}_2$ as it is minimal.

8   Conclusion

We have introduced a notion of Q-core and demonstrated that it does not enjoy all of the properties of cores and U-X-cores. In particular, there need not be a unique minimal element w.r.t. size in the equivalence class of structures agreeing on $\{\exists, \forall, \wedge\}$-FO-sentences. However, we suspect that the notion of Q-core we give is robust, in that the Q-core of any structure $\mathcal{B}$ is unique up to isomorphism; and, that it sits inside any substructure of $\mathcal{B}$ that satisfies the same sentence of $\{\exists, \forall, \wedge\}$-FO, making it computable in a greedy fashion. Thus, the nice behaviour of Q-cores is almost restored, but "induced substructure" in the properties of core or U-X-core must be replaced by the weaker "substructure".

Generalising the results about Q-cores of structures with an isolated element to disconnected structures is already difficult. Just as the $\{\exists, \forall, \wedge\}$-FO-theory of structures with an isolated element is essentially determined by their $\{\exists, \wedge\}$-FO-theory, so the $\{\exists, \forall, \wedge\}$-FO-theory of disconnected structures is essentially determined by its $\forall^1 \exists^\star$ fragment (see [17]).

We hope that a better understanding of Q-cores will permit progress in the classification of the complexity of the QCSP, in particular to dispense from the frequent assumption in the papers by Chen *et. al* that all constants are present [6]. Indeed this assumption implies that only cores are considered, which can be assumed without loss of generality in the classification of the CSP but not provably so in the classification of the QCSP.

Acknowledgements. The authors thank David Savourey and Shwetha Raghuraman for helping correct mistakes in earlier versions of this paper, and the anonymous referees for their helpful comments.

References

1. Barto, L., Kozik, M., Niven, T.: The CSP dichotomy holds for digraphs with no sources and no sinks (a positive answer to a conjecture of Bang-Jensen and Hell). SIAM Journal on Computing 38(5), 1782–1802 (2009)
2. Börner, F., Bulatov, A.A., Chen, H., Jeavons, P., Krokhin, A.A.: The complexity of constraint satisfaction games and QCSP. Inf. Comput. 207(9), 923–944 (2009)
3. Bulatov, A.: A dichotomy theorem for constraint satisfaction problems on a 3-element set. J. ACM 53(1), 66–120 (2006)
4. Bulatov, A., Krokhin, A., Jeavons, P.G.: Classifying the complexity of constraints using finite algebras. SIAM Journal on Computing 34, 720–742 (2005)
5. Chandra, A.K., Merlin, P.M.: Optimal implementation of conjunctive queries in relational data bases. In: Proceddings of STOC 1977, pp. 77–90 (1977)
6. Chen, H.: Meditations on quantified constraint satisfaction. CoRR abs/1201.6306 (2012)
7. Chen, H., Madelaine, F., Martin, B.: Quantified constraints and containment problems. In: 23rd Annual IEEE Symposium on Logic in Computer Science, pp. 317–328 (2008)
8. Cohen, D., Jeavons, P.: The complexity of constraint languages. Appears in: Handbook of Constraint Programming (2006)
9. Fagin, R., Kolaitis, P.G., Popa, L.: Data exchange: getting to the core. ACM Trans. Database Syst. 30(1), 174–210 (2005)
10. Feder, T., Vardi, M.: The computational structure of monotone monadic SNP and constraint satisfaction: A study through Datalog and group theory. SIAM Journal on Computing 28, 57–104 (1999)
11. Hell, P., Nesetril, J.: The core of a graph. Discrete Math. 109, 117–126 (1992)
12. Hell, P., Nešetřil, J.: On the complexity of H-coloring. Journal of Combinatorial Theory, Series B 48, 92–110 (1990)
13. Hell, P., Nešetřil, J.: Graphs and Homomorphisms. Oxford University Press (2004)
14. Madelaine, F., Martin, B.: Containment, equivalence and coreness from CSP to QCSP and beyond. CoRR abs/1204.5981 (2012)
15. Madelaine, F.R., Martin, B.: A tetrachotomy for positive first-order logic without equality. In: LICS, pp. 311–320 (2011)
16. Martin, B.: QCSP on Partially Reflexive Forests. In: Lee, J. (ed.) CP 2011. LNCS, vol. 6876, pp. 546–560. Springer, Heidelberg (2011)
17. Martin, B., Madelaine, F.: Towards a Trichotomy for Quantified H-Coloring. In: Beckmann, A., Berger, U., Löwe, B., Tucker, J.V. (eds.) CiE 2006. LNCS, vol. 3988, pp. 342–352. Springer, Heidelberg (2006)
18. Schaefer, T.J.: The complexity of satisfiability problems. In: Proceedings of STOC 1978, pp. 216–226 (1978)

An Optimal Filtering Algorithm for Table Constraints

Jean-Baptiste Mairy[1], Pascal Van Hentenryck[2], and Yves Deville[1]

[1] ICTEAM, Université catholique de Louvain, Belgium
{Jean-Baptiste.Mairy,Yves.Deville}@uclouvain.be
[2] Optimization Research Group, NICTA, University of Melbourne, Australia
pvh@nicta.com.au

Abstract. Filtering algorithms for table constraints are constraint-based, which means that the propagation queue only contains information on the constraints that must be reconsidered. This paper proposes four efficient value-based algorithms for table constraints, meaning that the propagation queue also contains information on the removed values. One of these algorithms (AC5TC-Tr) is proved to have an optimal time complexity of $O(r \cdot t + r \cdot d)$ per table constraint. Experimental results show that, on structured instances, all our algorithms are two or three times faster than the state of the art STR2+ and MDD^c algorithms.

1 Introduction

Domain-consistency algorithms are usually classified as constraint-based (i.e., the propagation queue only contains information on the constraints that must be reconsidered) or value-based (i.e., removed values are also stored in the propagation queue). For table constraints, which have been the focus of much research in recent years, all existing algorithms (except in [14]) are constraint-based. This paper proposes four original value-based algorithms for table constraints, which are all instances of the AC5 generic algorithm. The proposed propagators maintain, for every value of the variables, the index of its first current support in the table. They also use, for each variable of a tuple, the index of the next tuple sharing the same value for this variable. The algorithms differ in their use of information on the validity of the tuples. Three of the proposed algorithms have a time complexity of $O(r^2 \cdot t + r \cdot d)$ per table constraint and one of them (AC5TC-Tr) has the optimal time complexity of $O(r \cdot t + r \cdot d)$, where r is the arity of the table, d the size of the largest domain and t the number of tuples in the table. One of the proposed algorithms, AC5TC-Recomp, is the (unpublished) propagator of the Comet system.

Experimental results show that, on structured instances, our algorithms improve upon the state-of-the-art STR2+ [11] and MDD^c [3]: The speedup is between 1.95 and 3.66 over STR2+ and between 1.83 and 4.57 over MDD^c. Our (theoretically) optimal algorithm is not always the fastest in practice. Interestingly, on purely random tables, our algorithms do not compete with STR2+ and MDD^c. Since most real problems are structured, we expect our algorithms to be an interesting contribution to the field. The rest of this paper is organized as follows. Section 2 presents background information and related work. Section 3 describes the first two table-constraint propagators. Section 4 presents our optimal propagator, while Section 5 proposes an efficient variant of our first algorithms. Section 6 describes the experimental results.

M. Milano (Ed.): CP 2012, LNCS 7514, pp. 496–511, 2012.

2   Background

A CSP $(X, D(X), C)$ is composed of a set of n *variables* $X = \{x_1, \ldots, x_n\}$, a set of *domains* $D(X) = \{D(x_1), \ldots, D(x_n)\}$ where $D(x)$ is the set of possible *values* for variable x, and a set of *constraints* $C = \{c_1, \ldots, c_e\}$, with $Vars(c_i) \subseteq X \ (1 \leq i \leq e)$. We let $d = max_{1 \leq i \leq n}(\#D(x_i))$, and $D(X)_{x_i=a}$ be the set of tuples $\mathbf{v}$ in $D(X)$ with $\mathbf{v}_i = a$. Given $Y = \{x_1, \ldots, x_k\} \subseteq X$, the set of tuples in $D(x_1) \times \ldots \times D(x_k)$ is denoted by $D(X)[Y]$ or simply $D(Y)$. A support in a constraint c for a variable value pair (x, a) is a tuple $\mathbf{v} \in D(Vars(c))$ such that $c(\mathbf{v})$ and $\mathbf{v}[x] = a$. The following (Inconsistent and Consistent) sets are useful for specifying domain consistency and propagation methods. Let c be a constraint of a CSP $(X, D(X), C)$ with $y \in Vars(c)$, and $B(X)$ be some domain.

$$Inc(c, B(X)) = \{(x, a)| \ x \in Vars(c) \wedge a \in D(x) \wedge \forall \mathbf{v} \in B(Vars(c))_{x=a} : \neg c(\mathbf{v})\}$$
$$Cons(c, y, b) = \{(x, a)|x \in Vars(c) \wedge \ a \in D(x) \wedge \exists \mathbf{v} : \mathbf{v}[x] = a \wedge \mathbf{v}[y] = b \wedge c(\mathbf{v})\}$$
$$Inc(c) = Inc(c, D(X))$$

A constraint c in a CSP $(X, D(X), C)$ is *domain-consistent* iff $Inc(c) = \emptyset$. A CSP $(X, D(X), C)$ is *domain-consistent* iff all its constraints are domain-consistent.

Table Constraints. Given a set of tuples T of arity r, a table constraint c over T holds if $(x_1, \ldots, x_r) \in T$. The size t of a table constraint c is its number of tuples, which is denoted by $c.length$. We assume an implicit ordering of the tuples: $\sigma_{c,i}$ denotes the ith element of the table in c and $\sigma_{c,i}[x]$ is the value of $\sigma_{c,i}$ for variable x. We introduce a top value $\top$ (resp. bottom value $\bot$) greater (resp. smaller) than any other value. We also introduce a universal tuple $\sigma_{c,\top}$, with $\sigma_{c,\top}[x] = *$ forall $x \in X$ and abuse notations in postulating that $\forall a \in D(x), * = a$. This implies that, for any table T, $\sigma_{c,\top} \in T$. Given a table constraint, we say that a tuple σ is *allowed* if it belongs to the table. A tuple σ is *valid* if all its values belong to the domain of the corresponding variables. To achieve domain consistency, one must at least check the validity of each tuple and, in the worst case, remove all the values from the domains. Hence a domain-consistency algorithm has a complexity $\Omega(r \cdot t + r \cdot d)$ per table constraint in the worst case. An AC5-like algorithm with a complexity $O(r \cdot t + r \cdot d)$ per table constraint is thus optimal. As usual for such algorithms, if a domain-consistency algorithm has a time complexity of $O(f)$, then the time complexity of the aggregate executions of this algorithm along any path in the search tree is also $O(f)$.

Related Work. A lot of research effort has been spent on table constraints. The existing propagators can be categorized in 3 classes: index-based, compression-based, and based on a dynamic table. The index-based approaches use an indexing of the table to speed up its traversal. Examples of such propagators are GAC3-allowed and other constraint-based variants (GAC3$_{rm}$-allowed, GAC2001-allowed) [10,1,12,6]. For each variable value pair (x, a), the index data structure has an array of the indexes of the tuples with value a for x. The space complexity of the data structure is $O(r \cdot t)$. The time complexity of GAC3-allowed is $O(r^3 \cdot d \cdot t + r \cdot d^2)$ per table constraint. GAC2001-allowed has a time complexity of $O(r^3 \cdot d \cdot t + r^2 \cdot t)$ per table constraint. Indexing can also be used in value-based propagators. In [14], the authors propose a value-based propagator for table constraints implementing GAC6. It uses a structure which indexes, for each variable value pair (x, a) and each tuple, the next tuple in the table with value a for x. The space complexity of the data structure is $O(r \cdot d \cdot t)$. This space usage can be reduced by using a data structure called hologram [13]. Another index type, proposed in [7], indexes, for each tuple and variable, the next tuple having a different value for the variable. Compression-based propagators compress the table in a form that allows a fast traversal. One of such compressed forms uses a trie for each variable [7]. Another example of compression-based techniques [3,2] uses a *Multi Valued Decision Diagram* (MDD) to represent the table more efficiently. During propagation, the tries or MDD are traversed using the current domains to perform the pruning. These algorithms are constraint-based and have a time complexity of $O(r^2 \cdot d \cdot t)$ per table constraint. Compression and faster traversal can also be achieved by using compressed tuples, which represent a set of tuples [8,16]. Propagators based on dynamic tables maintain the table by suppressing invalid tuples from it. The STR algorithm [17] and its refined version, STR2 [11], are constraint-based and scan only the previously valid tuples to extract the valid values. The time complexity of STR2 is $O(r^2 \cdot d^2 + r^2 \cdot d \cdot t)$ per table constraint. The *or-tool* propagator [15] also maintains a dynamic table. It uses a bitset on the tuples of the table to maintain their validity. One bitset per variable value (x, a) is also used for easy access of the tuples with value a for variable x. This propagator has a $O(r \cdot d \cdot t)$ time complexity per table constraint.

The AC5 Algorithm. AC5 [18,4] is a generic value-based domain-consistency algorithm. In a value-based approach, information on the removed values is also stored in the queue for the propagation. Specification 1 describes the main methods of AC5 which uses a queue Q of triplets (c, x, a) stating that the domain consistency of constraint c should be reconsidered because value a has been removed from $D(x)$. When a value is removed from a domain, the method `enqueue` puts the necessary information on the queue. In the postcondition, Q_o represents the value of Q at call time. The method `post(c,`$\triangle$`)` is called once when posting the constraint. It computes the inconsistent values of the constraint c and initializes specific data structures required for the propagation of the constraint. As long as (c, x, a) is in the queue, it is algorithmically desirable to consider that value a is still in $D(x)$ from the perspective of constraint c. This is captured by the following definition.

Definition 1. *The local view of a domain $D(x)$ wrt a queue Q for a constraint c is defined as $D(x, Q, c) = D(x) \cup \{a | (c, x, a) \in Q\}$.*

For table constraints, a tuple σ is *Q-valid* if all its values belong to $D(X, Q, c)$. The central method of AC5 is the `valRemove` method, where the set $\triangle$ is the set of values becoming inconsistent because b is removed from $D(y)$. In this specification, b is a value that is no longer in $D(y)$ and `valRemove` computes the values (x, a) no longer supported in the constraint c because of the removal of b from $D(y)$. Note that values in the queue are still considered in the potential supports as their removal has not yet been reflected in this constraint. The minimal pruning $\triangle_1$ only deals with variables and values previously supported by (y, b). However, we give `valRemove` the possibility of achieving more pruning ($\triangle_2$), which is useful for table constraints.

```
1   enqueue(in x: Variable; in a: Value;  inout Q: Queue)
2   // Pre: x ∈ X, a ∉ D(x)
3   // Post: Q = Q₀ ∪ {(c, x, a)|c ∈ C, x ∈ Vars(c)}
4   post(in c: Constraint; out △: Set of Values)
5   // Pre: c ∈ C
6   // Post: △ = Inc(c) + initialization of specific data structures
7   valRemove(in c: Constraint; in y: Variable; in b: Value;
8                 out △: Set of Values)
9   // Pre: c ∈ C,  b ∉ D(y, Q, c)
10  // Post:  △₁ ⊆ △ ⊆ △₂  with  △₁ = Inc(c, D(X, Q, c)) ∩ Cons(c, y, b)
11  //                         and  △₂ = Inc(c)
```

Specification 1. The enqueue, post, and valRemove Methods for AC5

3   Efficient Value-Based Algorithms for Table Constraints

Our value-based approaches use a data structure FS memorizing first supports. Intuitively $FS[x, a]$ is the index of the first Q-valid support of the variable value pair (x, a). To speed up the table traversal, our algorithms use a second data structure called $next$ that links all the elements of the table sharing the same value for a given variable. The $next$ data structure is semantically equivalent to the index of [12]. More formally, for a given table constraint c, FS and $next$ satisfy the following invariant (called FS-invariant) before dequeuing an element from Q.

$$\forall x \in Vars(c) \, \forall a \in D(x, Q, c) : FS[x, a] = i \Leftrightarrow$$
$$\sigma_{c,i}[x] = a \wedge i \neq \top \wedge \sigma_{c,i} \in D(Vars(c), Q, c) \wedge$$
$$\forall j < i : \sigma_{c,j}[x] = a \Rightarrow \sigma_j \notin D(Vars(c), Q, c)$$
$$\forall x \in Vars(c) \, \forall 1 \leq i \leq c.length : next[x, i] = Min\{j|i < j \wedge \sigma_{c,j}[x] = \sigma_{c,i}[x]\}$$

The $next$ data structure, illustrated in Figure 1, is static as it does not depend on the domain of the variables. However, FS must be trailed during the search. Methods `postTC` and `valRemoveTC` are given in Algorithms 1 and 2. They use the `seekNextSupportTC` method (Algorithm 3) which searches the next Q-valid tuple. Abstract method `isQValidTC(c,i)` tests whether $\sigma_{c,i}$ is Q-valid (i.e., $\sigma_{c,i} \in D(X, Q, c)$) and can be implemented in many ways. One simple way is to record the Q-validity of tuples in some data structure, initialized in method `initSpecStructTC` and updated in method `setQInvalidTC`. Method `postTC` initializes the FS and $next$ data structures and returns the set of inconsistent values. Method `valRemoveTC` has only to consider the tuples in the $next$ chain starting at $FS[y, b]$. When one of these tuples $\sigma_{c,i}$ is the first support of an element $a = \sigma_{c,i}[x]$, a new support $FS[x, a]$ must be found. If such a support does not exist, then (x, a) belongs to the set $\triangle_1$. Method `valRemoveTC` thus computes the set $\triangle_1$ and maintains the FS-invariant.

Not considering `initSpecStructTC`, method `postTC` has a time complexity of $O(rt + rd)$. After the `postTC` method, the domain size of x is $O(t)$. We now establish the complexity of all executions of `valRemoveTC` for a given table constraint,

```
1    postTC(in c: Constraint;out △: Set of Values){
2    // Pre: c ∈ C, c is a table constraint
3    // Post: △ = Inc(c) + initialization of the next, FS and specific data structures
4        △ = ∅;
5        initSpecStructTC(c);
6        forall(x in Vars(c), a in D(x)) c.FS[x,a]=⊤;
7        forall(x in Vars(c), i in 1..c.length)   c.next[x,i] = ⊤;
8        forall(i in c.length..1)
9            if (σc,i in D(Vars(c))) {
10               forall(x in Vars(c)) {
11                   c.next[x,i] = FS[x,σc,i[x]];
12                   c.FS[x,σc,i[x]] = i;
13               }
14           }
15           else setQInvalidTC(c,i);
16       forall(x in Vars(c),a in D(x))
17           if(c.FS[x,a]==⊤) △ += (x,a);
18   }
```

Algorithm 1. Method `postTC` for Table Constraints

assuming this table constraint is one of the constraints of the CSP on which domain consistency is achieved. Consider first all executions of `valRemoveTC` without line 13. For a given variable y, these executions follow the different *next* chains of the variable y. The chains for all values of y have a total number of t elements. The complexity of lines 9–16 (without line 13) is $O(r)$. Since the table has r variables, the complexity of all `valRemoveTC` executions during the fixed point (without line 13) is thus $O(r^2 \cdot t)$, assuming a $O(1)$ complexity of `setQInvalidTC`. Consider now all executions of line 13 in `valRemoveTC` for a variable x. Since line 13 always increases the value of $FS[x,a]$ in the *next* chain of (x,a), we have a global complexity of $O(V \cdot t)$ for the variable x, where V is the time complexity of `isQValidTC`. All executions of line 13 in `valRemoveTC` thus take time $O(V \cdot r \cdot t)$. The time complexity of all executions of `valRemoveTC` is then $O(r^2 \cdot t + V \cdot r \cdot t)$. Even with a $O(1)$ the time complexity of `isQValidTC`, the algorithm is not optimal but it turns out to be more efficient than state-of-the-art algorithms on different classes of problems. The AC5 algorithm with the `postTC` and `valRemoveTC` implementation for table constraint is called AC5TC (AC5 for Table Constraints).

Proposition 1. *Assuming that* `initSpecStructTC` *and* `setQInvalidTC` *have a time complexity of* $O(r \cdot t + r \cdot d)$ *and* $O(1)$ *respectively and allow a correct implementation of* `isQValidTC` *to have a complexity of* $O(r)$, *then AC5TC is correct and has a time complexity of* $O(r^2 \cdot t + r \cdot d)$ *per table constraint.*

We now present two implementations of AC5TC. They differ in the implementations of methods `isQValidTC`, `setQInvalidTC` and `initSpecStructTC`. AC5TC-Bool, the first implementation of AC5TC, is shown in Algorithm 4. It uses a data structure $isQValid[i]$ to record the Q-validity of the element $σ_{c,i}$. It satisfies invariant $isQValid[i] \Leftrightarrow σ_{c,i} \in D(X,Q,c)$ before dequeuing an element from Q $(1 \leq i \leq$

```
 1  valRemoveTC(in c: Constraint; in y: Variable;  in b: Value;
 2              out △: Set of Values) {
 3  // Pre: c ∈ C, c is a table constraint and b ∉ D(y, Q, c)
 4  // Post: △₁ ⊆ △ ⊆ △₂  with  △₁ = Inc(c, D(X, Q, c)) ∩ Cons(c, y, b)
 5  //                     and  △₂ = Inc(c, x)
 6      △ = ∅;
 7      i = c.FS[y, b];
 8    while(i!=⊤){
 9        setQInvalidTC(c, i);
10        forall(x in Vars(c): x!=y){
11            a = σ_{c,i}[x];
12            if (c.FS[x, a]==i){
13                c.FS[x, a] = seekNextSupportTC(c, x, i);
14                if(c.FS[x, a]==⊤ && a in D(x))  △ += (x, a);
15            }
16        }
17        i = c.next[y, i];
18    }
19  }
```

Algorithm 2. Method `valRemoveTC` for Table Constraints.

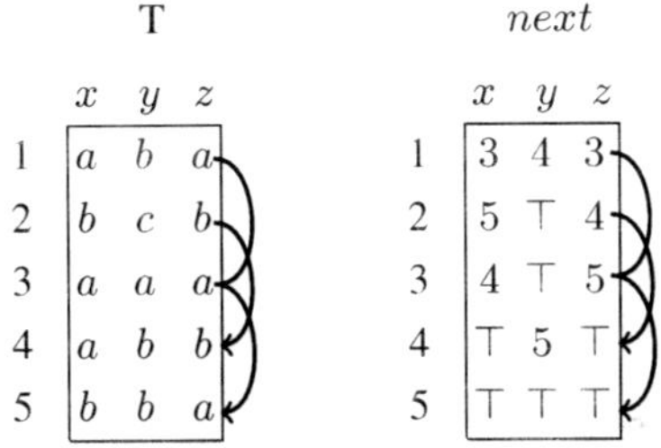

Fig. 1. Example of a *next* data structure of a table T (arrow pointers for variable z only)

c.length). The data structure must be trailed as it depends on the domains. The methods for Q-validity are given in Algorithm 4. As the methods `isQValidTC-Bool` is correct, AC5TC-Bool is correct. The time complexity of `isQValidTC-Bool` and `setQInvalidTC-Bool` is $O(1)$ and `initSpecStructTC-Bool` is $O(t)$. The time complexity of AC5TC-Bool is then $O(r^2 \cdot t + r \cdot d)$ per table constraint.

AC5TC-Bool must trail the *isQValid* boolean array. We now propose an implementation that only trails one integer, building upon an idea in STR and STR2 [17,11]. The implementation simply keeps invalid elements at the end of the table, with a single variable *size* representing the boundary between valid (before position *size*) and invalid elements (after position *size*). When an element becomes invalid, it is swapped with the element at position *size* and *size* is decremented by one. The *size* variable must be trailed but the table does not need to: The valid elements are automatically restored, albeit at a different position in the table. This is sometimes called semantic backtracking [19]. Instead of swapping tuples, our implementation uses two arrays

```
1   function seekNextSupportTC(in c: Constraint; in x: Variable;
2                              in i: Index)  : Index {
3   // Pre: c ∈ C, c is a table constraint, x ∈ Vars(c), 1 ≤ i ≤ c.length
4   // Post: return the first index greater than i which is Q−valid
5       i = c.next[x,i];
6       while(i!=⊤){
7           if(isQValidTC(c,i))   return i;
8           i = c.next[x,i];
9       }
10      return ⊤;
11  }
```

Algorithm 3. Function `seekNextSupportTC` for Table Constraints.

```
1   initSpecStructTC-Bool(in c: Constraint) {
2       forall(i in 1..c.length)   c.isQValid[i] = true;
3   }
4   function isQValidTC-Bool(in c: Constraint;in i: Index) {
5   // Pre: c ∈ C, c is a table constraint and 1 ≤ i ≤ c.length
6   // Post: returns σc,i ∈ D(X, Q, c)
7       return c.isQValid[i];
8   }
9   setQInvalidTC-Bool(in c: Constraint;in i: Index) {
10  // Pre: c ∈ C, c is a table constraint and 1 ≤ i ≤ c.length
11      c.isQValid[i] = false;
12  }
```

Algorithm 4. Implementation of the specific methods of AC5TC-Bool

Map and *Dyn* that give the virtual position of the tuples and the positions of the virtual tuples in the table. These arrays do not have to be trailed. For a given table constraint c, the data structures satisfy the following invariants before dequeuing an element from Q ($1 \leq i \leq c.length$): $Map[i] \leq size \Leftrightarrow \sigma_i \in D(X, Q, c) \land Dyn[Map[i]] = i$

The implementations are given in Algorithm 5 and the algorithm is called AC5TC-CutOff. The time complexity of `isQValidTC-CutOff` and `setQInvalidTC-CutOff` is $O(1)$ and `initSpecStructTC-CutOff` takes time $O(t)$. The time complexity of AC5TC-CutOff is thus $O(r^2 \cdot t + r \cdot d)$.

4   An Optimal Algorithm

In method `valRemoveTC`, executions of the `seekNextSupportTC` (line 13 of Algorithm 2) take $O(r \cdot t)$ assuming `isQValidTC` takes constant time. However, the method revisits Q-invalid tuples because the *next* data structure is static. To remedy this situation, the idea is to make the *next* data structure dynamic and to always ensure that the element following a Q-valid element in a *next* chain is also Q-valid. This avoids unnecessary Q-validity checks and can be easily implemented using a doubly-linked list.

```
 1   initSpecStructTC-CutOff(in c: Constraint){
 2       forall(i in 1..c.length){
 3           c.Map[i] = i;
 4           c.Dyn[i] = i;
 5       }
 6       c.size = c.length;
 7   }
 8   function isQValidTC-CutOff(in c: Constraint; in i: Index; out b: Bool){
 9   // Pre: c ∈ C, c is a table constraint and 1 ≤ i ≤ c.length
10   // Post: return (σ_{c,i} ∈ D(X, Q, c))
11       return (c.Map[i] <= c.size);
12   }
13   setQInvalidTC-CutOff(in c: Constraint; in i: Index){
14   // Pre: c ∈ C, c is a table constraint and 1 ≤ i ≤ c.length
15       c.Dyn[c.Map[i]] = c.Dyn[c.size];
16       c.Dyn[c.size[c]] = i;
17       c.Map[c.Dyn[c.Map[i]]] = c.Map[i];
18       c.Map[i] = c.size;
19       c.size--;
20   }
```

Algorithm 5. Implementation of the specific methods of AC5TC-CutOff

More formally, for a given table constraint c, the data structure satisfies the following invariant before dequeuing an element from Q

$$\forall x \in Vars(c) \; \forall 1 \leq i \leq c.length : \sigma_{c,i} \in D(X, Q, c) \Rightarrow$$
$$nextTr[x, i] = Min\{j | i < j \wedge \sigma_{c,j}[x] = \sigma_{c,i}[x] \wedge \sigma_{c,j} \in D(X, Q, c)\} \wedge$$
$$predTr[x, nextTr[x, i]] = i$$

The *nextTr* and *predTr* data structures should be trailed as they depend on the current domains. The algorithm also uses the FS data structure with its original invariant. No other data structures are necessary.

Methods `postTC-Tr` and `valRemoveTC-Tr` are given in Algorithms 6 and 7. Method `postTC-Tr` now initializes *predTr* as well. Method `valRemoveTC-Tr` (c, y, b) does not need to search for a support as the next element in the *nextTr* chain is necessarily Q-valid. However, if the first support for (x, a) is before $FS[y, b]$ (it cannot be after because of the FS invariant), *nextTr* and *predTr* must be updated to ensure that the new invalid tuples are no longer in the *next* chains. It will thus never be visited twice. Method `valRemoveTC-Tr` computes the set $\triangle_1$ and maintains the invariants on FS and on $nextTr / predTr$.

Method `postTC-Tr` has a time complexity of $O(r \cdot t + r \cdot d)$. We establish the complexity of all executions of `valRemoveTC-Tr` for a given table constraint during the fixed point algorithm, assuming the presence of other constraints on which domain consistency is also enforced. We first show that all executions of lines 7 and 21 lead to different values of i in $\{1, \ldots, t\}$ (except when $i == \top$). By the FS invariant, we never

```
1   postTC-Tr(in c: Constraint; out △: Set of Values){
2   // Pre: c ∈ C, c is a table constraint
3   // Post: △ = Inc(c) + initialization of the next, pred and FS data structures
4       △ = ∅;
5       forall(x in Vars(c), a in D(x))  c.FS[x,a]=⊤;
6       forall(x in Vars(c), i in 1..c.length)
7           c.nextTr[x,i] = ⊤; c.predTr[x,i] = ⊥;
8       forall(i in c.length..1: σc,i in D(Vars(c)))
9           forall(x in Vars(c)){
10              c.nextTr[x,i] = c.FS[x,σc,i[x]];
11              if (c.FS[x,σc,i[x]]!=⊤) c.predTr[x,FS[x,σc,i[x]] = i;
12              c.FS[x,σc,i[x]] = i;
13          }
14      forall(x in Vars(c),a in D(x))
15          if(c.FS[x,a]==⊤) △ += (x,a);
16  }
```

Algorithm 6. An optimal `postTC-Tr` method for Table Constraints

have $i == \top$ at line 7. For a given value $i \neq \top$, by the FS invariant, $FS[x,a] == i$ or $FS[x,a] < i$ holds at line 11. If $FS[x,a] == i$, $FS[x,a]$ is incremented and the tuple i will never be reconsidered by removing a from $D(x)$. If $FS[x,a] < i$, the tuple i is removed from the $nextTr$ chain and will never be reconsidered. This holds for all variables $x \neq y$. Hence the tuple i will never be reconsidered in future executions of `valRemoveTC-Tr`. Hence, lines 9-20 are executed $O(t)$ times. Since the complexity of lines 9-20 is $O(r)$, the aggregate complexity of all executions of `valRemoveTC-Tr` is $O(r \cdot t)$. The AC5 algorithm with the `postTC-Tr` and `valRemoveTC-Tr` implementation for table constraint is called AC5TC-Tr (AC5 for Table Constraints with Trailing).

Proposition 2. *AC5TC-Tr is correct and has an optimal time complexity of $O(r \cdot t + r \cdot d)$ per table constraint.*

5   A Variation Based on Recomputation

We now propose a variation of the AC5TC algorithm, called AC5TC-Recomp, that does not require any data structure to maintain the Q-validity of tuples. AC5TC-Recomp is the (unpublished) table constraint algorithm of the Comet system. It replaces the Q-validity test by a function `isValidTC` that tests the validity of the tuples. The straight-forward implementation of the `isValid` function is given in Algorithm 8. Method `initSpecStructTC` and `setQInvalidTC` are just empty. Since AC5TC-Recomp tests validity instead of Q-validity, method `valRemoveTC` must be slightly modified; the test $a \in D(x)$ should be moved from line 14 to line 10 which becomes[1]

```
forall(x in Vars(c): x!=y && σc,i[x] in D(x)){
```

---

[1] This modification also maintains the correctness of our generic AC5TC algorithm but requires a more sophisticated FS-invariant. With this change, our earlier algorithms would have the same theoretical complexity but are less efficient in practice.

```
 1  valRemoveTC-Tr(in c: Constraint; in y: Variable; in b: Value;
 2              out △: Set of Values) {
 3  // Pre: c ∈ C, c is a table constraint and b ∉ D(y, Q, c)
 4  // Post: △₁ ⊆ △ ⊆ △₂ with △₁ = Inc(c, D(X, Q, c)) ∩ Cons(c, y, b)
 5  //                    and △₂ = Inc(c)
 6     △ = ∅;
 7     i = c.FS[y,b];
 8     while(i!=⊤){
 9       forall(x in Vars(c): x!=y){
10         a = σ_{c,i}[x];
11         if(c.FS[x,a] == i){
12           c.FS[x,a] = c.nextTr[x,i];
13           if(c.FS[x,a]==⊤ && a in D(x))   △ += (x,a);
14         } else { //c.FS[x,a] < i
15           if (c.predTr[x,i]!=⊥)
16               c.nextTr[x,c.predTr[x,i],c] = c.nextTr[x,i,c];
17           if (c.nextTr[x,i]!=⊤)
18               c.predTr[x,c.nextTr[x,i],c] = c.predTr[x,i,c];
19         }
20       }
21       i = c.next[y,i];
22     }
23  }
```

Algorithm 7. An Optimal `valRemoveTC-Tr` method for Table Constraints

This modified version of `valRemoveTC` exploits the flexibility of its specification by computing a set $\triangle$ between $\triangle_1$ and $\triangle_2$. AC5TC-Recomp has a runtime complexity of is $O(r^2 \cdot t + r \cdot d)$ and per table constraint and improves state-of-the-art algorithms on some classes of problems.

6   Experimental Results

All proposed algorithms have been implemented on top of Comet, including AC5TC-Recomp. For comparison, classical constraint-based algorithms have also been implemented on top of Comet. The GAC3-Allowed algorithm has been chosen because it is the standard GAC3 algorithm for the table constraints [10]. The two state-of-the-art methods were also reimplemented: The STR2+ algorithm from [11] and the MDD^c algorithm from [3]. They are respectively called STR and MDD in the experimental results. All experiments were conducted on an Intel Xeon 2.53GHz using Comet 2.1.1. The algorithms are compared within a MAC search. This section presents results on fully random instances, on the geometric problem, on Langford problem, and on the Traveling Salesman Problem.

For each instance set, the experimental results report the mean execution times in seconds (*totTime*), the mean "posting" times in seconds (*postT*), the number of propagator calls (*nProp*), the percentage to the best with respect to execution time (*%best*), the mean of percentage to the best algorithm in terms of execution time (μ*%best*), the

```
1   function isValid(in c: Constraints;in i: Index) : Bool {
2   // Pre: c ∈ C, c is a table constraint and 1 ≤ i ≤ c.length
3   // Post: return (σ_{c,i} ∈ D(X))
4       forall(x in Vars(c))
5           if(!σ_{c,i}[x] in D(x)) return false;
6       return true;
7   }
```

Algorithm 8. The `isValid` Function of AC5TC-Recomp

number of validity checks (*valChk*), Q-validity checks (*QvalChk*), and the number of pointers followed (*pFollow*). The difference between the *%best* and μ*%best* is the following: for *%best*, the execution times are averaged before computing the quantity. There is thus one best algorithm. For μ*%best*, the percentages are computed instance by instance and aggregated with a geometrical mean at the end. This measure takes into account that different instances may have different best algorithms. The μ*%best* measure uses a geometrical mean as suggested in [5]. The last reported quantity, *pFollow*, has different meanings for different algorithms. For GAC3-Allowed, it corresponds to the number of times the tuples are accessed. For the AC5TC algorithms, it is defined as the number of times the *next* or *nextTr* structures are used to traverse the table. For MDD, it corresponds to the number of edges followed in the MDD structure. Although referring to different quantities, *pFollow* is useful for comparing the behavior of the propagators as it reflects the usage of their specific structures.

Random Instances. These instances contain random table constraints of random scope generated by the RD-model [21]. Parameters are chosen to generate instances close to the phase transition, using Theorems 1 and 2 from [21]. The instances have 10 variables, a uniform domain size of 10, and 15 table constraints of arity 5. The expected number of tuples in each table is thus 20000. 10 instances were generated with those settings. The search strategy is the *dom* heuristic with lexicographic value ordering.

Table 1. Results of the propagators on fully random instance set (times in seconds)

propagator	totTime	postT	nProp	%best	μ%best	valChk	QvalChk	pFollow
GAC3-Allowed	3 000	1.5	614 k	2 725	2 660	523 M	0	523 M
AC5TC-Bool	4 636	1.0	2.8 M	4 211	4 070	19 k	257 M	481 M
AC5TC-CutOff	3 991	0.8	2.8 M	3 626	3 538	19 k	257 M	481 M
AC5TC-Tr	994	5.2	2.8 M	903	930	19 k	0	16 M
AC5TC-Recomp	3 874	0.8	2.4 M	3 519	3 357	98 M	0	305 M
STR2	483	0.7	614 k	439	455	22 M	0	0
MDD	**110**	12.4	614 k	100	100	0	0	12 M

Table 1 summarizes the results, which remain similar for other parameter settings. The standard STR2 and MDD algorithms outperform our value-based propagators.

Observe the large number of validity checks of AC5TC-Recomp and Q-validity checks of AC5TC-Bool and AC5TC-CutOff, as well as the number of times they follow a pointer. AC5TC-Tr, the best value-based propagator, follows far less pointers than our other propagators because it does not follow pointers to a previously inspected tuple. Due to the lack of structure of the constraint set, the first three AC5TC propagators check multiple times the same tuples. Also, those random instances have large tables, which makes the cost of the trailable *nextTr* structure in AC5TC-Tr too high.

The Geometric Problem. Instances of the geometric problem are random instances generated following a specific structure proposed by Rick Wallace [20]. Each variable is randomly placed in the unit square. A fixed distance (less than $\sqrt{2}$) is randomly chosen. For each pair of variables (x, y), if the distance between their associated points is less than or equal to this fixed distance, the arc (x, y) is added to the constraint graph. Constraint relations are then created like in fully random CSP instances. We use the instance set from [9] which counts 100 instances. The search strategy uses the heuristic *dom/deg* with lexicographic value ordering. A timeout of 5 minutes has been used. The quantity *%solv* gives the percentage of solved instances.

Table 2. Results of the propagators on the geom instances (times in seconds)

propagator	totTime	postT	nProp	%best	μ%best	%solv	valChk	QvalChk	pFollow
GAC3-Allowed	10.1	0.3	288 k	128	138	86	28 k	0	28 k
AC5TC-Bool	12.5	0.3	867 k	158	159	84	300	25 k	50 k
AC5TC-CutOff	10.8	0.2	867 k	137	131	86	300	25 k	50 k
AC5TC-Tr	9.6	0.8	867 k	122	200	87	300	0	13 k
AC5TC-Recomp	**7.9**	0.2	831 k	100	100	87	6 k	0	29 k
STR2	24.9	0.3	288 k	315	316	82	26 k	0	0
MDD	14.7	1.6	288 k	186	337	86	0	0	65 k

Table 2 presents the experimental results. The quantities are computed on instances for which none of the techniques timeout. All our propagators outperform the state-of-the-art STR and MDD. AC5TC-Tr and AC5TC-Recomp are also better than the classical AC3-Allowed. AC5TC-Recomp is the fastest on these instances. which are relatively easy and contain only binary tables. Checking the validity (not costly for binary tables) allows AC5TC-Recomp to follow less pointers than AC5TC-Bool and AC5TC-CutOff by performing longer jumps in the table. The cost of the data structures in AC5TC-Tr is too expensive and outweights its benefits. The large difference between *%best* and μ*%best* for AC5TC-Tr is due to the easiest instances, where the propagator is even more disadvantaged due to the cost of its data structures. AC5TC-Recomp also performs less validity checks than STR2 and the number of pointers followed by our propagators are less than those of MDD.

Langford Number Problem. Langford number problem $L(k, n)$ amounts to arranging k sets of numbers 1 to n into a sequence of numbers, so that each occurrence of a number m is m numbers apart from its previous occurrence. Those problems are modeled with binary (positive) table constraints only. The search strategy used is *dom/deg* with lexicographic value ordering. Problems where all the propagators take more than 5 minutes are removed from the sets. For $k = 2$, 12 instances are used: $n \in \{5..12, 15, 16, 19, 20\}$, for $k = 3$, 8 instances: $n \in \{3..10\}$ and for $k = 4$, 9 instances: $n \in \{3..11\}$. The results for k of 2, 3 and 4 can respectively be found in Table 3.

Table 3. Experimental Results on Langford instances (times in seconds)

propagator	totTime	postT	nProp	%best	μ%best	valChk	QvalChk	pFollow
				$k = 2$				
GAC3-Allowed	16.3	0.6	1 M	173	172	166 k	0	166 k
AC5TC-Bool	18.6	0.8	2 M	197	182	576	178 k	316 k
AC5TC-CutOff	16.8	0.5	2 M	178	147	576	178 k	316 k
AC5TC-Tr	**9.4**	2.5	2 M	100	260	576	0	42 k
AC5TC-Recomp	10.1	0.4	2 M	107	106	27 k	0	154 k
STR2	26.7	1.3	1 M	283	342	46 k	0	0
MDD	26.6	3.7	1 M	282	517	0	0	307 k
				$k = 3$				
GAC3-Allowed	2.5	0.3	75 k	162	148	12 k	0	12 k
AC5TC-Bool	3.5	0.3	242 k	227	184	380	10 k	21 k
AC5TC-CutOff	2.5	0.2	242 k	163	147	380	10 k	21 k
AC5TC-Tr	2.2	0.9	242 k	140	198	380	0	4 k
AC5TC-Recomp	**1.5**	0.2	239 k	100	107	2 k	0	12 k
STR2	3.7	0.6	75 k	240	243	5 k	0	0
MDD	3.9	1.5	75 k	249	360	0	0	22 k
				$k = 4$				
GAC3-Allowed	23.4	1.3	419 k	137	155	19 k	0	19 k
AC5TC-Bool	42.5	1.6	1.6 M	250	215	677	20 k	36 k
AC5TC-CutOff	29.8	1.0	1.6 M	175	157	677	20 k	36 k
AC5TC-Tr	21.8	5.0	1.6 M	128	254	677	0	5 k
AC5TC-Recomp	**17.0**	0.8	1.58 M	100	100	3 k	0	18 k
STR2	33.2	3.3	419 k	195	277	10 k	0	0
MDD	31.2	7.3	419 k	183	392	0	0	35 k

Except for AC5TC-Bool on the $k = 4$ set of instances, all our propagators improve the state-of-the-art STR and MDD. AC5TC-Tr and AC5TC-Recomp are also better than the classical AC3-Allowed. AC5TC-Tr is the fastest propagator for $k = 2$ and AC5TC-Recomp is the fastest on the other instance sets. Observe that the number of followed pointers is globally higher for the first instance set ($k = 2$), due to inclusion of instances with larger n. The number of calls to the propagators during the search is also higher for the $k = 2$ set. This suggests that AC5TC-Tr requires harder instances (found in the $k = 2$ set) for amortizing the cost of its data structures. For the last two sets, AC5TC-Tr is the second fastest propagator in terms of mean solving time. AC5TC-Tr is closer to AC5TC-Recomp on the $k = 4$ instance set. The $k = 4$ instance set includes an instance for which $n = 11$ (for $k = 3$: $n \leq 10$). Here again, the large difference between *%best* and μ*%best* for AC5TC-Tr can be attributed to the easiest instances.

Traveling Salesman Problems. We conclude with results of the propagators on the Traveling Salesman Problem (TSP) constraint satisfaction instances. We used the set of instances *tsp-20* and *tsp-25* [9]. Those structured instances are composed of very different table constraints. Their arity varies between 2 and 3 and they may count up to 20 000 tuples but also as few as 20. The variables also have quite different domains: Some have small domains, while others feature domains containing up to 1000 values. There are 61 variables and 230 table constraints in *tsp-20* instances. The *tsp-25* instances count 76 variables and 350 constraints. The negative table constraints found in those instances have been transformed into positive ones. The search strategy used here is *dom/deg* with lexicographic value ordering.

Tables 4 and 5 present the results. We first observe that STR2 and MDD perform worse than our propagators. AC5TC-Recomp is the winning strategy on *tsp-20* instances while AC5TC-Tr is faster on the *tsp-25* ones. The latter instances are more

Table 4. Results of the propagators for instance set TSP-20 (times in seconds)

propagator	totTime	postT	nProp	%best	μ%best	valChk	QvalChk	pFollow
GAC3-Allowed	797	1.7	6.7 M	733	587	11 M	0	11 M
AC5TC-Bool	186	0.8	21.2 M	171	187	2 k	1 M	2 M
AC5TC-CutOff	153	0.5	21.2 M	141	144	2 k	1 M	2 M
AC5TC-Tr	120	3.3	21.2 M	111	164	2 k	0	466 k
AC5TC-Recomp	**109**	0.3	20.9 M	100	104	391 k	0	1 M
STR2	398	1.4	6.7 M	366	353	803 k	0	0
MDD	456	19.0	6.7 M	419	769	0	0	7 M

Table 5. Results of the propagators for instance set TSP-25 (times in seconds)

propagator	totTime	postT	nProp	%best	μ%best	valChk	QvalChk	pFollow
GAC3-Allowed	6 607	2.4	73 M	606	509	23 M	0	23 M
AC5TC-Bool	2 625	1.3	198 M	241	233	2 k	11 M	19 M
AC5TC-CutOff	1 937	0.7	198 M	178	175	2 k	11 M	19 M
AC5TC-Tr	**1 089**	5.2	198 M	100	100	2 k	0	3 M
AC5TC-Recomp	1 315	0.5	196 M	121	120	3 M	0	10 M
STR2	3 740	2.9	73 M	343	333	5 M	0	0
MDD	4 974	25.2	73 M	457	425	0	0	28 M

difficult. We can also see that checking the validity instead of the Q-validity allows AC5TC-Recomp to follow less pointers and perform fewer validity checks than the Q-validity checks of AC5TC-Bool and AC5TC-CutOff. Moreover, on these instances, the small arity makes the validity check ($O(r)$) cheap compared to Q-validity. Again, on those instances, the light-weight trailable structures of AC5TC-CutOff make it faster than AC5TC-Bool.

When we merge binary tables in tsp-20 instances into higher arity tables, we observed that AC5TC-Tr, our optimal algorithm, solves more instances than STR2, and with a smaller total execution time on the instances solved by both solvers. MDD does not compete on these instances. On simple instances, STR2 is more efficient than AC5TC-Recomp which is also more efficient than AC5TC-Tr. However, on hard instances, the optimality of AC5TC-Tr pays off and it becomes the best algorithm.

Summary. We conclude that, for the fully random instances, the lack of structure in the tables prevents our propagators from competing with state-of-the-art algorithms. However, for structured instances, our propagators are faster. Globally, AC5TC-Bool and AC5TC-CutOff are slower than AC5TC-Recomp since they are testing Q-validity, not validity, and hence they perform smaller jumps in the table. Moreover, maintaining their data structures is costly. Only the optimal AC5TC-Tr outperforms AC5TC-Recomp on difficult instances while using Q-validity. However, on easier instances, the cost of its trailable *nextTr* data-structure makes it slower than AC5TC-Recomp.

7   Conclusion

This paper proposed four different value-based, domain-consistency algorithms for table constraints, all using the AC5 generic framework. The new propagators record, for every value of the variables, the index of its first current support in the table. They also use, for each variable of a tuple, the index of the next tuple sharing the same value for this variable. They differ in their use of information on the validity of the tuples. AC5TC-Tr and AC5TC-Recomp are the two best value-based algorithms: AC5TC-Recomp does not maintain any validity information and recomputes it on demand and AC5TC-Tr embeds the Q-validity information into the indexing structure, avoiding unnecessary visits of invalid tuples and leading to an optimal algorithm with a time complexity of $O(r{\cdot}t+r{\cdot}d)$ per table constraint. Our other algorithms have a time complexity of $O(r^2{\cdot}t+r{\cdot}d)$ per table constraint. Experimental results show that on, purely random tables, our algorithms do not compete with the state-of-the-art STR2+ and MDD^c algorithms. On structured instances, our propagators outperform STR2+ and MDD^c, with a speed up varying between 1.83 and 4.57. As future work, it would be interesting to extend AC5TC to handle negative tables through its disallowed tuples and to integrate the compressed representation of tuples introduced in [16].

Acknowledgments. The authors want to thank the anonymous reviewers for their helpful comments. The first author is supported as a Research Assistant by the Belgian FNRS (National Fund for Scientific Research). This research is also partially supported by the Interuniversity Attraction Poles Program (Belgian State, Belgian Science Policy) and the FRFC project 2.4504.10 of the Belgian FNRS.

References

1. Bessière, C., Régin, J.-C.: Arc consistency for general constraint networks: Preliminary results. In: IJCAI (1), pp. 398–404 (1997)
2. Carlsson, M.: Filtering for the case constraint. Talk given at the advanced school on global constraints (2006)
3. Cheng, K., Yap, R.: An mdd-based generalized arc consistency algorithm for positive and negative table constraints and some global constraints. Constraints 15, 265–304 (2010)
4. Deville, Y., Van Hentenryck, P.: Domain Consistency with Forbidden Values. In: Cohen, D. (ed.) CP 2010. LNCS, vol. 6308, pp. 191–205. Springer, Heidelberg (2010)
5. Fleming, P.J., Wallace, J.J.: How not to lie with statistics: the correct way to summarize benchmark results. Commun. ACM 29(3), 218–221 (1986)
6. Gent, I.P., Jefferson, C., Miguel, I.: Watched Literals for Constraint Propagation in Minion. In: Benhamou, F. (ed.) CP 2006. LNCS, vol. 4204, pp. 182–197. Springer, Heidelberg (2006)
7. Gent, I.P., Jefferson, C., Miguel, I., Nightingale, P.: Data structures for generalised arc consistency for extensional constraints. In: Proceedings of the Twenty Second Conference on Artificial Intelligence, pp. 191–197. AAAI Press (2007)
8. Katsirelos, G., Walsh, T.: A Compression Algorithm for Large Arity Extensional Constraints. In: Bessière, C. (ed.) CP 2007. LNCS, vol. 4741, pp. 379–393. Springer, Heidelberg (2007)
9. Lecoutre, C.: Instances of the Constraint Solver Competition, http://www.cril.fr/~lecoutre/
10. Lecoutre, C.: Constraint Networks: Techniques and Algorithms. ISTE/Wiley (2009)
11. Lecoutre, C.: Str2: optimized simple tabular reduction for table constraints. Constraints 16, 341–371 (2011)
12. Lecoutre, C., Szymanek, R.: Generalized Arc Consistency for Positive Table Constraints. In: Benhamou, F. (ed.) CP 2006. LNCS, vol. 4204, pp. 284–298. Springer, Heidelberg (2006)
13. Lhomme, O.: Arc-Consistency Filtering Algorithms for Logical Combinations of Constraints. In: Régin, J.-C., Rueher, M. (eds.) CPAIOR 2004. LNCS, vol. 3011, pp. 209–224. Springer, Heidelberg (2004)
14. Lhomme, O., Régin, J.-C.: A fast arc consistency algorithm for n-ary constraints. In: Proceedings of the Nationnal Conference on Artificial Intelligence, pp. 405–410. AAAI Press (2005)
15. Perron, L., Furnon, V.: or-tools, http://code.google.com/p/or-tools
16. Régin, J.-C.: Improving the expressiveness of table constraints. In: Proceedings of Workshop ModRef 2011 at CP 2011 (2011)
17. Ullmann, J.R.: Partition search for non-binary constraint satisfaction. Inf. Sci. 177(18), 3639–3678 (2007)
18. Van Hentenryck, P., Deville, Y., Teng, C.-M.: A generic arc-consistency algorithm and its specializations. Artif. Intell. 57(2-3), 291–321 (1992)
19. Van Hentenryck, P., Ramachandran, V.: Backtracking without Trailing in CLP($\Re_{lin}$). ACM Transactions on Programming Languages and Systems 17(4), 635–671 (1995)
20. Wallace, R.: Factor Analytic Studies of CSP Heuristics. In: van Beek, P. (ed.) CP 2005. LNCS, vol. 3709, pp. 712–726. Springer, Heidelberg (2005)
21. Xu, K., Boussemart, F., Hemery, F., Lecoutre, C.: Random constraint satisfaction: Easy generation of hard (satisfiable) instances. Artif. Intell. 171(8-9), 514–534 (2007)

Parallel SAT Solver Selection and Scheduling

Yuri Malitsky[1], Ashish Sabharwal[2], Horst Samulowitz[2], and Meinolf Sellmann[2]

[1] Cork Constraint Computation Centre, University College Cork, Ireland
y.malitsky@4c.ucc.ie
[2] IBM Watson Research Center, Yorktown Heights, NY 10598, USA
{ashish.sabharwal,samulowitz,meinolf}@us.ibm.com

Abstract. Combining differing solution approaches by means of solver portfolios has proven as a highly effective technique for boosting solver performance. We consider the problem of generating parallel SAT solver portfolios. Our approach is based on a recently introduced sequential SAT solver portfolio that excelled at the last SAT competition. We show how the approach can be generalized for the parallel case, and how obstacles like parallel SAT solvers and symmetries induced by identical processors can be overcome. We compare different ways of computing parallel solver portfolios with the best performing parallel SAT approaches to date. Extensive experimental results show that the developed methodology very significantly improves our current parallel SAT solving capabilities.

1 Introduction and Related Work

In the past decade, solver portfolios have boosted our capability to solve hard combinatorial problems. Portfolios of existing solution algorithms have excelled in competitions in satisfiability (SAT), constraint programming (CP), and quantified Boolean formulae (QBF) [8, 12, 14].

In the past years, a new trend has emerged, namely the development of parallel solver portfolios. The gold-winning ManySAT solver [6] is, when we ignore features like clause-sharing, a static parallel portfolio of the MiniSAT solver [5] with different parameterizations. At the SAT Competition 2011, an extremely simple static parallel portfolio, ppfolio [10], dominated the wall-clock categories on random and crafted SAT instances and came very close to winning the applications category as well.

The obvious next step is to consider dynamic parallel portfolios, i.e., portfolios that are composed based on the features of the given problem instance. Traditionally, sequential portfolios simply *select* one of the constituent solvers which appears best suited for the given problem instance. At least since the invention of CP-Hydra [8] and SatPlan [13], sequential portfolios also *schedule* solvers. That is, they may select more than just one constituent solver and assign each one a portion of the time available for solving the given instance. Yun and Epstein [15] introduced a heuristic method to build dynamic parallel portfolios. The method relies heavily on the observation that running a deterministic solver on more than one processor is a waste of time and is thus limited to the use of

M. Milano (Ed.): CP 2012, LNCS 7514, pp. 512–526, 2012.

sequential SAT solvers only. A similar restriction applies to the work by Petrik and Zilberstein [9]. They introduced a method to compute static parallel schedules that are optimal with respect to the training instances, based on formulating the problem as a non-linear optimization problem and considering only sequential constituent solvers.

The best-performing *sequential* dynamic portfolio at the SAT Competition 2011 was 3S [7] where it won gold medals in the CPU-time category on random and crafted instances. 3S combines a fixed-time static solver schedule with the dynamic selection of *one* long-running SAT solver. To compute the static schedule offline and to select the long-running solver online, 3S combines low-bias nearest neighbor regression with integer programming optimization. In this paper, we augment this methodology to devise dynamic parallel SAT portfolios which include parallel SAT solvers.

2   SAT Solver Selector (3S)

Before considering the challenges of parallel portfolios, let us first review sequential 3S in more detail. 3S works in two phases, an offline learning phase, and an online execution phase.

- At Runtime: In the execution phase, as all dynamic solver portfolios, 3S first computes *features* of the given problem instance. In particular, 3S uses the same 48 core features as SATzilla [14]. Then, 3S selects $k \in \mathbb{N}$ instances that are most "similar" to the given instance in a training set of SAT instances for which 3S knows all runtimes of all solvers. Similarity in 3S is determined by the Euclidean distance of the (normalized) feature vectors of the given instance and the training instances. 3S selects the solver that can solve most of these k instances within the given time limit (ties are broken by shorter runtime). Finally, 3S first runs a fixed schedule of solvers for 10% of the time limit and then runs the selected solver for the remaining 90% of the available time.
- Offline: In the learning phase, which takes place during the development of the portfolio solver, 3S considers three tasks:

  1. Computation of features and simulation of solvers on all instances to determine their runtime on all training instances.
  2. To compute a desirable size k of the neighborhood, 3S employs a cross validation by random subsampling. That is, 3S repeatedly splits the training set into a base and a validation set and determines which size of k results in the best average validation set performance when using only the base set training instances to determine the long running solver.
  3. Lastly, 3S computes the fixed schedule of solvers that are run for 10% of the competition runtime. The objective when producing this schedule is to maximize the number of instances that can be solved within this reduced time limit. Among schedules that can solve the same number

of instances, 3S selects one that minimizes the runtime of the schedule and then scales this shorter schedule back to the 10% time limit by increasing the runtime of each solver in the schedule proportionally.

2.1 Schedule Computation

This last step deserves our special attention as it is at the core of what we will need to do to generalize the 3S methodology for the case of parallel solver execution. The problem again is to select a schedule of solvers – that is, a sequence of solvers with associated runtimes – that maximizes the number of instances solved within the reduced time limit. This problem is obviously an optimization problem, and it actually resembles a bit the set covering problem.

3S considers the following integer program (IP) to compute a solver schedule. Let $V_{S,t}$ denote the set of instances i that can be solved by solver S within time limit t.

Solver Scheduling IP

$$\min \ (C+1) \sum_i y_i + \sum_{S,t} t x_{S,t}$$

$$s.t. \ y_i + \sum_{(S,t) \mid i \in V_{S,t}} x_{S,t} \geq 1 \qquad \forall i$$

$$\sum_{S,t} t x_{S,t} \leq C$$

$$y_i, x_{S,t} \in \{0,1\} \qquad \forall i, S, t$$

For all pairs of solvers S and time limits t, there is one variable $x_{S,t}$. Note that there are a number of solvers times the number of training instances of such variables as for each solver S 3S only considers time limits t where the solver just solves an instance in the training set. $x_{S,t}$ will be equal one if and only if, in our schedule, we will run solver S for t seconds.

The second set of variables are the y_i, one for each training instance i. Variable y_i will be one if and only if the solver schedule cannot solve instance i.

The first set of constraints ensure that each instance is covered – either because one of the selected solver/time pairs means that the respective solver can solve the instance in the allocated time, or because the instance is counted as not covered by y_i. The final knapsack constraint simply ensures that the total schedule time does not exceed the reduced time limit C.

The objective is first to minimize the number of uncovered instances. The second criterion is to minimize the time of the schedule. Both is achieved simultaneously by minimizing the term $(C+1) \sum_i y_i + \sum_{S,t} t x_{S,t}$. The latter summand obviously minimizes the total time scheduled. The first summand minimizes the number of uncovered instances. Note that the factor $C+1$ ensures that the objective will always be less for schedules that solve at least one more instance, even when the scheduled time would increase from 0 seconds to the maximum of C seconds.

For further details on 3S the reader is referred to [7] which also contains a in-depth comparison to CP-Hydra [8].

2.2 Reducing the Number of Integer Variables

3S has 26 constituent SAT solvers, 11 of which are considered with two parameterizations each. Consequently, the scheduler considers 37 solvers. Moreover, unlike prior SAT portfolios, the 3S portfolio is identical for all categories in the SAT competition (application, crafted, and random). Consequently, it is based on a vast set of training instances, almost 5,500. Therefore, the IP above has more than 200,000 variables and more than 5,000 constraints. Although solved offline, to solve this IP more quickly, 3S uses a heuristic in [7]. 3S first solves the continuous relaxation of the solver scheduling IP. That is, it considers the linear program (LP) where constraints $y_i, x_{S,t} \in \{0,1\}$ are replaced by $0 \leq y_i, x_{S,t} \leq 1$.

When solving this relaxed LP the simplex algorithm [4] will only consider variables where, at some point during the optimization, it is beneficial to introduce these variables. In practice, during the optimization the vast majority of the 200,000 variables will never be set to a value greater than zero. Without going into the theory of linear programming, the important aspect is that the simplex algorithm has a *precise necessary condition* to determine whether a variable can improve the objective or not. Namely, in each step of the optimization, the simplex algorithm prices each constraint with a so-called "dual value."

For each variable, it then computes a "reduced cost." The latter is defined as the actual cost factor in the objective of the variable, minus the sum of the variable's coefficients in each constraint times the dual price of that constraint. Formally, when c_j is the cost coefficient, A_{ij} is the matrix coefficient for variable z_j on constraint i, and when π_i is the dual price for constraint i, then the reduced costs $\bar{c}_j$ for variable z_j are defined as:

$$\bar{c}_j = c_j - \sum_i A_{ij}\pi_i.$$

Now, the simplex algorithm will only consider setting variable z_j to a value different from 0 when $\bar{c}_j < 0$. We then say, that the variable (or the respective column in the matrix) has *negative reduced costs*.

So 3S solves the relaxation LP by introducing one (potentially new) variable in each iteration. Then, to solve the actual integer problem, 3S removes all variables from the solver scheduling IP which have never been introduced during the optimization. This reduction in the number of integer variables speeds up the solution to the integer program. However, Kadioglu et al. [7] showed that the solutions found in this manner are near-optimal in practice and, on average, work just as well on the test set as the optimal schedule would.

3 Parallel Solver Portfolios

The objective of this work is to generalize the 3S technology for the development of parallel SAT solver portfolios. At the core of 3S lie two optimization problems. The first is the selection of the long running solver primarily based on the maximum number of instances solved. The second is the solver scheduling problem.

Consider the first problem when there are $p > 1$ processors available. The objective is to select p solvers that, as a set, will solve the most number of instances. Note that this problem can no longer be solved by simply choosing the one solver that solves most instances in time. Moreover, we will now need to decide how to integrate the newly chosen solvers with the ones from the static schedule. The second problem is the solver scheduling problem discussed before, with the additional problem that solvers need to be assigned to processors so that the total makespan is within the allowed time limit.

A major obstacle in solving these problems efficiently is the symmetry induced by the identical processors to which we can assign each solver. Symmetries can hinder optimization very dramatically as equivalent (partial) schedules (which can be transformed into one another by permuting processor indices) will be considered again and again by a systematic solver. For example, when there are 8 processors, for each schedule there exist over 40,000 (8 factorial) equivalent versions. An optimization that used to take about half a second may now easily take 6 hours.

Another consideration is the fact that a parallel solver portfolio may obviously include parallel solvers as well. Assuming there are 8 processors and a parallel solver employs 4 of them, there are 70 different ways to allocate processors for this solver. The portfolio that we will develop later will have 37 sequential and 2 4-core parallel solvers. The solver scheduling IP that needs to be solved for this case has over 1.5 million variables. Eventually, we will apply our technology to a set of 72 of the latest SAT solvers from 2011, among them four parallel solvers which we will consider to run with 1, 2, 3, and 4 processors. Note that, in the parallel case, we will need to solve these IPs *at runtime*. Consequently, where 3S could afford to price out all variables at each step, we will need a more sophisticated method to speed up the optimization time – which directly competes with the remaining time to solve the actual SAT problem that was given.

3.1 Parallel Solver Scheduling

Recall again that we need to solve two different optimization problems. The first is to compute a static schedule for the first 10% of the allowed runtime. This problem is solved once, offline. The second optimization problem schedules solvers for the remaining 90% of the allowed time. This is done instance-specifically, taking into account the specific features of the SAT instance that is given.

We will address both optimization problems by considering the following IP. Let $t_S \geq 0$ denote the minimum time that solver S must run in the schedule, let $M = \{S; |; t_S > 0\}$ denote the set of solvers that have a minimal runtime, let p be the number of processors, and let $n_S \leq p$ denote the number of processors that solver S requires.

Parallel Solver Scheduling IP - CPU time

$$\min \ (pC+1) \sum_i y_i + \sum_{S,t,P} tn_S x_{S,t,P}$$

$$s.t. \ y_i + \sum_{(S,t) \ | \ i \in V_{S,t}, P \subseteq \{1,\dots,p\}, |P|=n_S} x_{S,t,P} \geq 1 \qquad \forall i$$

$$\sum_{S,t,P \subseteq \{1,\dots,p\} \cup \{q\}, |P|=n_S} tx_{S,t,P} \leq C \qquad \forall q \in \{1,\dots,p\}$$

$$\sum_{S,t,P \subseteq \{1,\dots,p\}, |P|=n_S, t \geq t_S} x_{S,t,P} \geq 1 \qquad \forall S \in M$$

$$\sum_{S,t,P \subseteq \{1,\dots,p\}, |P|=n_S} x_{S,t,P} \leq N$$

$$y_i, x_{S,t,P} \in \{0,1\} \qquad \forall i, S, t, P \subseteq \{1,\dots,p\}, |P| = n_S$$

Variables y_i are exactly what they were before. There are now variables $x_{S,t,P}$ for all solvers S, time limits t, *and subsets of processors* $P \subseteq \{1,\dots,p\}$ with $|P| = n_S$. $x_{S,t,P}$ is 1 if an only if solver S is run for time t on the processors in P in the schedule.

The first constraint is again to solve all instances with the schedule or count them as not covered. There is now a time limit constraint *for each processor*. The third set of constraints ensures that all solvers that have a minimal solver time are included in the schedule, with an appropriate time limit. The last constraint finally places a limit on the number of solvers that can be included in the schedule.

The objective is again to minimize the number of uncovered instances. The secondary criterion is to minimize the total CPU time of the schedule.

Remark 1. Note that the IP above needs to be solved both offline to determine the static solver schedule (for this problem $M = \emptyset$ and the solver limit is infinite) and *during the execution phase* (when M and the solver limit are determined by the static schedule computed offline). Therefore, we absolutely need to be able to solve this problem quickly, despite its huge size and its inherent symmetry caused by the multiple processors.

Note also that the parallel solver scheduling IP does not directly result in an executable solver schedule. Namely, the IP does not specify the actual start times of solvers. In the sequential case this does not matter as solvers can be sequenced in any way without affecting the total schedule time or the number of instances solved. In the parallel case, however, we need to ensure that the parallel processes are in fact run in parallel. We omit this aspect from the IP above to avoid further complicating the optimization. Instead, after solving the parallel solver IP, we heuristically schedule the solvers in a best effort approach, whereby we may preempt solvers and eventually even lower the runtime of the solvers to obtain a legal schedule. In our experiments presented later it turned out that in practice the latter was never necessary. Hence, the quality of the schedule was never diminished by the necessity to schedule processes that belong to the same parallel solver at the same time.

3.2   Solving the Parallel Solver Scheduling IP

We cannot afford to solve the parallel solver scheduling IP exactly during the execution phase. Each second spent on solving this problem is one second less for solving the actual SAT instance. Hence, we revert to solving the problem heuristically by employing *column generation*, whereby the generation of IP variables is limited to the root node.

While 3S prices all columns in the IP during each iteration, fortunately we actually do not need to do this here. Consider the reduced costs of a variable. Denote with $\mu_i \leq 0$ the dual prices for the instance-cover constraints, $\pi_q \leq 0$ the dual prices for the processor time limits, $\nu_S \geq 0$ the dual prices for the minimum time solver constraints, and $\sigma \leq 0$ the dual price for the limit on the number of solvers. Finally, let $\bar{\nu}_S = \nu_S$ when $S \in M$ and 0 otherwise. Then:

$$\bar{c}_{S,t,P} = n_S t - \sum_{i \in V_{S,t}} \mu_i - \sum_{q \in P} t \pi_q - \bar{\nu}_S - \sigma.$$

The are two important things to note here: First, the fact that we only consider variables introduced during the column generation process means that we *reduce the processor symmetry* in the final IP. While it is not impossible, it is unlikely that the variables that would form a symmetric solution to a schedule that can already be formed from the variables already introduced would have negative reduced costs.

Second, to find a new variable that has the most negative reduced costs, we do not need to iterate through all $P \subseteq \{1, \ldots, p\}$ for all solver/time pairs (S, t). Instead, we order the processors by their decreasing dual prices. The next variable introduced will use the first n_S processors in this order as all other selections of processors would result in higher reduced costs.

3.3   Minimizing Makespan and Post Processing the Schedule

We now have everything in place to develop our parallel SAT solver portfolio. In the offline training phase we compute a static solver schedule based on all training instances for 10% of the available time. We use this schedule to determine a set M of solvers that must be run for at least the static scheduler time at runtime. During the execution phase, given a new SAT instance we compute its features, determine the k closest training instances, and compute a parallel schedule that will solve as many of these k instances in the shortest amount of CPU time possible.

In our experiments we consider a second variant of the parallel solver scheduling IP where the secondary criterion is not to minimize CPU time but the *makespan* of the schedule. The corresponding IP is given below, where variable

m measure the minimum idle time for all processors. The reduced cost computation changes accordingly.

Parallel Solver Scheduling IP - Makespan

$$\min\ (C+1) \sum_i y_i - m$$

$$s.t.\ y_i + \sum_{(S,t)\ \mid\ i \in V_{S,t},\, P \subseteq \{1,\ldots,p\},\, |P|=n_S} x_{S,t,P} \geq 1 \qquad \forall i$$

$$m + \sum_{S,t,P \subseteq \{1,\ldots,p\} \cup \{q\},\, |P|=n_S} t x_{S,t,P} \leq C \qquad \forall q \in \{1,\ldots,p\}$$

$$\sum_{S,t,P \subseteq \{1,\ldots,p\},\, |P|=n_S,\, t \geq t_S} x_{S,t,P} \geq 1 \qquad \forall S \in M$$

$$\sum_{S,t,P \subseteq \{1,\ldots,p\},\, |P|=n_S} x_{S,t,P} \leq N$$

$$y_i, x_{S,t,P} \in \{0,1\} \qquad \forall i, S, t, P \subseteq \{1,\ldots,p\}, |P| = n_S$$

Whether we minimize CPU time or makespan, as remarked earlier, we post-process the result by assigning actual start times to solvers heuristically. We also scale the resulting solver times to use as much of the available time as possible. For low values of k, we often compute schedules that solve all k instances in a short amount of time without utilizing all available processors. In this case, we assign new solvers to the unused processors in the order of their ability to solve the highest number of the k neighboring instances.

4   Experimental Results

Using the methodology above, we built two parallel portfolios. The first based on the 37 constituent solvers of 3S [7]. We refer to this portfolio as p3S-37. The second portfolio that we built includes two additional solvers, 'Cryptominisat (2.9.0)' [11] and 'Plingeling (276)' [2], both executed on four cores. We refer to this portfolio as p3S-39. It is important to emphasize that *all solvers that are part of our portfolio were available before the SAT Competition 2011*. We would have liked to compare our portfolio builder with other parallel portfolios. However, existing works on parallel portfolios do not accommodate parallel solvers. Consequenlty, in our experiments we will compare p3S-37 and p3S-39 with the state of the art in parallel SAT solving. The winners in the parallel tracks at the 2011 SAT Competition were the parallel solver portfolio 'ppfolio' [10] and 'Plingeling (587f)' [3], both executed on eight cores. Note that these competing solvers are new solvers that were introduced for the SAT Competition 2011.

As our benchmark set of SAT instances, to the $5,464$ instances from all SAT Competitions and Races between 2002 and 2010 [1], we added the $1,200$ (300

Table 1. Average performance comparison parallel portfolio when optimizing CPU time and varying neighborhood size k based on 10-fold cross validation

CPU Time	10	25	50	100	200
Average (σ)	320 (45)	322 (43.7)	329 (42.2)	338 (43.9)	344 (49.9)
Par 10 (σ)	776 (241)	680 (212)	694 (150)	697 (156)	711 (221)
# Solved (σ)	634 (2.62)	636 (2.22)	636 (1.35)	637 (1.84)	636 (2.37)
% Solved (σ)	99.0 (0.47)	99.2 (0.39)	99.2 (0.27)	99.2 (0.28)	99.2 (0.41)

Table 2. Average performance comparison parallel portfolio when optimizing Makespan and varying neighborhood size k based on 10-fold cross validation

Makespan	10	25	50	100	200
Average (σ)	376.1 (40.8)	369.2 (42.9)	374 (40.7)	371 (40.8)	366 (36.9)
Par 10 (σ)	917 (200)	777 (192)	782 (221)	750 (153)	661 (164)
# Solved (σ)	633 (2.16)	635 (2.28)	634.9 (2.92)	635 (1.89)	637 (2.01)
% Solved (σ)	98.8 (0.39)	99.1 (0.39)	99.1 (0.46)	99.2 (0.32)	99.3 (0.34)

application, 300 crafted, 600 random) instances from last years SAT Competition 2011. Based on this large set of SAT instances, we created a number of benchmarks. Based on all SAT instances that can be solved by at least one of the solvers considered in p3S-39 within 5,000 seconds, we created an equal 10 partition. We use this partition to conduct a ten-fold cross validation, whereby in each fold we use nine partitions as our training set (for building the respective p3S-37 and p3S-39 portfolios), and evaluate the performance on the partition that was left out before. For this benchmark we report average performance over all ten splits. On top of this cross-validation benchmark, we also consider the split induced by the SAT Competition 2011. Here we use all instances prior to the competition as training set, and the SAT Competition instances as test set. Lastly, we also created a competition split based on application instances only.

As performance measures we consider the number of instances solved, average runtime, and PAR10 score. The PAR10 is a penalized average runtime where instances that time out are penalized with 10 times the timeout. Experiments were run on dual Intel Xeon 5540 (2.53 GHz) quad-core Nehalem processors with 24 GB of DDR-3 memory.

Impact of the IP Formulation and Neighborhood Size. In Tables 1 and 2 we show the average cross-validation performance of p3S-39 when using different neighborhood sizes k and the two different IP formulations (tie breaking by minimum CPU time or minimizing schedule makespan). As we can see, the size of the neighborhood k affects the most important performance measure, the number of instances solved, only very little. There is a slight trend towards larger k's working a little bit better. Moreover, there is also not a great difference between the two IP formulations, but on average we find that the version that breaks ties by minimizing the makespan solves about 1 instance more per split. Based on

Table 3. Performance of 10-fold cross validation on all data. Results are averages over the 10 folds.

Cross	p3S-37		p3S-39	
Validation	4 core	8 core	4 core	8 core
Average (*sigma*)	420 (22.1)	**355 (31.3)**	435 (48.5)	366 (36.9)
Par 10 (*sigma*)	991 (306)	679 (176)	1116 (256)	**661 (164)**
Solved (*sigma*)	630 (4.12)	633 (2.38)	631 (2.75)	**637 (2.01)**
% Solved (*sigma*)	98.3 (0.63)	98.8 (0.35)	98.5 (0.49)	**99.3 (0.34)**

these results, p3S in the future refers to the portfolio learned on the respective training benchmark using $k = 200$ and the IP formulation that minimizes the makespan.

4.1 Impact of Parallel Solvers and the Number of Processors

Next we investigate the impact of employing parallel solvers in the portfolio. In Tables 3 and 4 we compare the performance of p3S-37 (without parallel solvers) and p3S-39 (which employs two 4-core parallel solvers) on the cross-validation and on the competition split. We observe a small difference in the number of solved instances in the cross-validation, and a significant gap in the competition split.

Two issues are noteworthy about that competition split. First, since this was the latest competition and these instances were also used for the parallel track, the instances in the test set of this split are significantly harder than the instances from earlier years. The relatively low percentage of instances solved even by the best solvers at the SAT Competition 2011 is an indication for this. Second, some instance families in this test set are completely missing in the training partition. That is, for a good number of instances in the test set there may be no training instance that is very similar. These features of any competition-induced split (which is the realistic split scenario!) explain why the average cross-validation performance is often significantly better than the competition performance. Moreover, they explain why p3S-39 has a significant advantage over p3S-37: When a lot of the instances are out of reach of the sequential solvers within the competition timeout *then the portfolio must necessarily include parallel solvers to perform well.*

As a side remark: the presence of parallel solvers is what makes the computation of parallel portfolios challenging in the first place. Not only do parallel solvers complicate the optimization problems that we considered earlier. In the extreme case, if all solvers were sequential, we could otherwise have as many processors as solvers, and then a trivial portfolio would achieve the performance of the virtual best solver. That is to say: The more processors we have, the easier solver selection becomes. We were curious to see what would happen when we made the selection harder than it actually is under the competition setting and reduced the number of available processors to 4. For both p3S-37 and p3S-39, the cross-validation performance decreases only moderately while, under the

Table 4. Performance of the solvers on all 2011 SAT Competition data.

Competition	3S		//3S		VBS
	4 cores	8 cores	4 cores	8 cores	
Average	1907	1791	1787	1640	1317
Par 10	12,782	12,666	11,124	10,977	10,580
Solved	843	865	853	892	953
% Solved	70.3	72.1	71.1	74.3	79.4

competition split, performance decays significantly. At the same time, the advantages of p3S-39 over p3S-37 shrink a lot. As one would expect, the advantage of employing parallel solvers decays with a shrinking number of processors.

4.2 Parallel Solver Selection and Scheduling vs. State-of-the-Art

The dominating parallel portfolio to date is 'ppfolio' [10]. In the parallel track at the SAT Competition 2011, it won gold in the crafted and random categories and came in just shy to winning the application category as well where it was beat by just one instance. In the application category, the winning solver was 'Plingeling (587f)' run on 8 cores. We compare against both competing approaches in Figures 1 and 2.

Instances from All SAT Categories. The top plot in Figure 1 shows the scaling behavior in the form of a "cactus plot" for 8-core runs of ppfolio, p3S-37, and p3S-39, for the competition split containing all 1,200 instances used in the 2011 SAT Competition. This plot shows that p3S-39 (whose curve stays the lowest as we move to the right) can solve significantly more instances than the other two approaches for any given time limit larger than around 800 sec. We also see that p3S-37, based solely on sequential constituent solvers, performs similar to ppfolio for time limits up to 3,000 sec, and begins to outperform it for larger time limits.

This comparative performance profile is by no means accidental. It is well known in SAT that an instance that is exceedingly difficult to solve for one solver poses almost no problem at all for another. This is the deeper reason why the 10% static schedules are well motivated, because there exists a realistic chance that one of the solvers scheduled for a short period of time will solve the instance.

At a higher level, we are observing the same here. Sequential SAT solvers have, for a good number of instances, the chance to solve an instance within some time. Consequently, a portfolio of sequential solvers only can, up to a point, compete with a solver portfolio that incorporates parallel solvers as well. However, a realistic set of hard instances, as the one considered at the SAT Competition, also contains instances that are very hard to solve, even by the best solver for that instance. Some of the instances will not be solvable by any sequential algorithm within the available time. This is why it is so important to be able to include parallel solvers in a parallel portfolio.

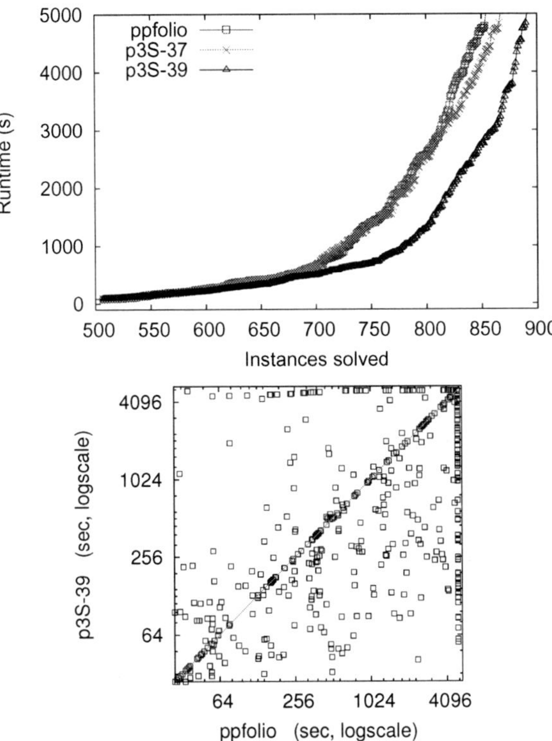

Fig. 1. Comparison on all 1200 instances used in the 2011 SAT Competition, across all categories. Left: cactus plot depicting the scaling behavior of solvers. Right: per-instance comparison between ppfolio and p3S-39.

The bottom plot in Figure 1 shows the per-instance performance of p3S-39 vs. ppfolio, with runtimes in log-scale on both axes. More points being below the diagonal red line signifies that p3S-39 is faster than ppfolio on a large majority of the instances. ppfolio also times out on many instances that p3S-39 can solve, as evidenced the large number of points on the right margin of the plot.

Overall, p3S-39 was able to solve 892 instances, 47 more than ppfolio. p3S-37 was somewhere in-between, solving 20 more than ppfolio. In fact, even with only 4 cores, p3S-37 and p3S-39 solved 846 and 850 instances, respectively, more than the 845 ppfolio solved on 8 cores.

Industrial Instances. Traditionally, portfolios did not excel in the category for industrial SAT instances. In part, this is because there are much fewer representative training instances available than in the random or crafted categories. That is to say, at the competition there is a much better chance to encounter an application instance that is very much different from anything that was considered during training. Moreover, progress on solvers that work well on industrial instances is commonly much more pronounced. Since competition portfolios are

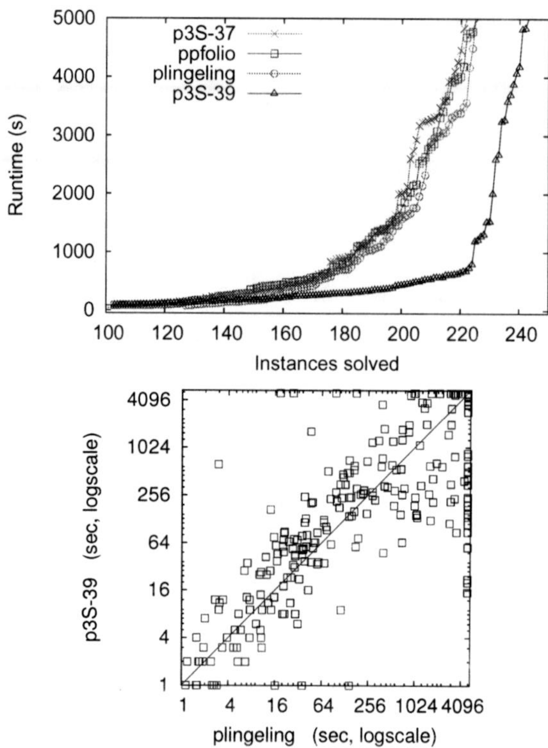

Fig. 2. Comparison on the 300 application category instances used in the 2011 SAT Competition. Left: cactus plot depicting the scaling behavior of solvers. Right: per-instance comparison between Plingeling and p3S-39.

based on older solvers, even their intelligent selection may not be enough to make up for the progress on the solvers themselves.

Figure 2 shows similar comparisons, but on the competition split restricted to the application category, and with Plingeling as one of the competing solvers. The cactus plot on top still shows a significantly better scaling behavior of p3S-39 than both Plingeling and ppfolio. The scatter plot shows that Plingeling, not surprisingly, is able to solve several easy instances within just a few seconds (as evidenced by the points on the bottom part of the left edge of the plot), but begins to take more time than p3S-39 on challenging instances and also times out on many more instances (shown as points on the right edge of the plot).

Overall, with 8 cores, p3S-39 solved 248 application category instances, 23 more than ppfolio and 22 more than Plingeling. Moreover, p3S-37, based only on sequential constituent solvers, was only two instances shy of matching Plingeling's performance. This performance improvement is quiete significant. In the application category, the best performing algorithms usually lie just a couple of instances solved apart. In 2011, the top ten solvers solved between 200 and 215

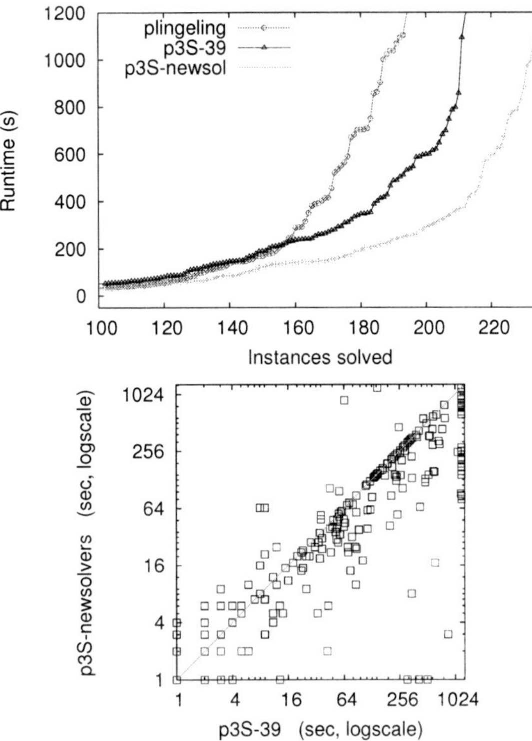

Fig. 3. Performance of p3S built using latest solvers, on all the 300 application category instances used in the 2011 SAT Competition. Left: cactus plot depicting the scaling behavior of solvers. Right: per-instance comparison between p3S-39 and p3S-newsolvers.

instances. An improvement of over 20 instances over the best-performing solver from nine months ago is much more than we had expected.

2012 Competition Portfolio. Finally, to demonstrate the efficacy of the method presented here, we trained a parallel portfolio based on 40 of the latest available parallel and sequential SAT solvers. Two of them were run on 1, 2, 3, and 4 processors. For all solvers, we consider a secondary setting where the given instance is first simplified by the Satelite program. In total, we have thus 92 solvers, 6 of them run in parallel on 2, 3, or 4 processors.

In Figure 3 we compare the performance of this portfolio against Plingeling, the winning parallel solver in the 2011 SAT Competition nine months ago, and p3S-39, our portfolio of solvers from 2010 and before. We observe that our method of devising parallel portfolios continues to result in strong performance and generalizes well to this extended set of solvers and corresponding training data. The parallel portfolio based on the latest SAT solvers currently competes in the 2012 SAT Challenge.

5 Conclusion

We presented the first method for devising *drynamic parallel solver portfolios that accommodate parallel solvers*. Our approach is based on the recently introduced SAT Solver Selector and Scheduler (3S). We combine core methods from machine learning, such as nearest neighbor regression, with methods from optimization, in particular integer programming and column generation, to produce parallel solver schedules *at runtime*. We compared different formulations of the underlying optimization problems and found that minimizing makespan as a tie breaking rule works slightly better than minimizing CPU time.

We compared the resulting portfolio, p3S-39, with the current state-of-the-art parallel solvers on instances from all SAT categories and from the application category only. We found that p3S-39 marks a very significant improvement in our ability to solve SAT instances.

Acknowledgements. This research has been partially supported by EU FET grant ICON (project number 284715).

References

[1] SAT Competition, http://www.satcomptition.org
[2] Biere, A.: Lingeling, plingeling, picosat and precosat at sat race 2010. Technical report, Johannes Kepler University, Linz, Austria (2010)
[3] Biere, A.: Lingeling and friends at the sat competition 2011. Technical report, Johannes Kepler University, Altenbergerstr. 69, 4040 Linz, Austria (2011)
[4] Dantzig, G.: Linear programming and extensions. Princeton University Press, Princeton (1963)
[5] Een, N., Sorensson, N.: An extensible sat-solver [ver 1.2] (2003)
[6] Hamadi, Y., Jabbour, S., Lakhdar, S.: Manysat: a parallel sat solver. Journal on Satisfiability, Boolean Modeling and Computation 6, 245–262 (2009)
[7] Kadioglu, S., Malitsky, Y., Sabharwal, A., Samulowitz, H., Sellmann, M.: Algorithm Selection and Scheduling. In: Lee, J. (ed.) CP 2011. LNCS, vol. 6876, pp. 454–469. Springer, Heidelberg (2011)
[8] O'Mahony, E., Hebrard, E., Holland, A., Nugent, C., O'Sullivan, B.: Using case-based reasoning in an algorithm portfolio for constraint solving. In: Irish Conference on Artificial Intelligence and Cognitive Science (2008)
[9] Petrik, M., Zilberstein, S.: Learning static parallel portfolios of algorithms. In: Ninth International Symposium on Artificial Intelligence and Mathematics (2006)
[10] Roussel, O.: Description of ppfolio (2011),
http://www.cril.univ-artois.fr/~roussel/ppfolio/solver1.pdf
[11] Soos, M.: Cryptominisat 2.9.0 (2011)
[12] Stern, D., Samulowitz, H., Herbrich, R., Graepel, T., Pulina, L., Tacchella, A.: Collaborative expert portfolio management. In: AAAI (2010)
[13] Streeter, M., Smith, S.: Using decision procedures efficiently for optimization. In: ICAPS, pp. 312–319 (2007)
[14] Xu, L., Hutter, F., Hoos, H., Leyton-Brown, K.: Satzilla: Portfolio-based algorithm selection for sat. JAIR 32(1), 565–606 (2008)
[15] Yun, X., Epstein, S.: Learning algorithm portfolios for parallel execution. In: Workshop on Learning and Intelligent Optimization (2012)

Constraint Satisfaction over Bit-Vectors

Laurent D. Michel[1] and Pascal Van Hentenryck[2]

[1] University of Connecticut, Storrs, CT 06269-2155
[2] Optimization Research Group, NICTA and The University of Melbourne

Abstract. Reasoning over bit-vectors arises in a variety of applications in verification and cryptography. This paper presents a bit-vector domain for constraint programming and its associated filtering algorithms. The domain supports all the traditional bit operations and correctly models modulo-arithmetic and overflows. The domain implementation uses bit operations of the underlying architecture, avoiding the drawback of a bit-blasting approach that associates a variable with each bit. The filtering algorithms implement either domain consistency on the bit-vector domain or bit consistency, a new consistency notion introduced in this paper. Filtering algorithms for logical and structural constraints typically run in constant time, while arithmetic constraints such as addition run in time linear in the size of the bit-vectors. The paper also discusses how to channel bit-vector variables with an integer variable.

1 Introduction

A number of applications in cryptography and verification require reasoning over bit-vectors. For instance, cryptographic hash functions form an active research area where researchers aim at creating secure hash algorithms that produce a short fixed-length digest from an arbitrary message. A good hash function f makes it difficult to find collisions, i.e., two messages m and m' such that $f(m) = f(m')$. It should be equally difficult to find a preimage given a digest d, i.e., a message m' satisfying $f(m') = d$. Cryptanalysts faced with the task of comparing potential hash functions need tools to assess their resistance against collision or preimage attacks. A constraint solver capable of using a hash function f and a partial message m (where some bits are unknown) to recover the lost information would be invaluable to assess the value of potential hash functions. Bit-vectors also enjoy a significant popularity in the verification community (e.g., [5,2,11,4,3,12]).

Existing solvers *encode* bit-vectors into their native languages (e.g., SAT, LP, CP, or MIP), all of which bring some strengths and some weaknesses (see Section 7). It is often the case that a representation which is appropriate for reasoning about bit-vectors is less adequate when the vectors are interpreted as numbers, and vice versa. The goal of this paper is to study whether constraint programming can unify these strengths and avoid (some of) the weaknesses.

To address this challenge, the paper investigates *a bit-vector domain for constraint programming* that is particularly suitable for verification and cryptography applications. The domain supports all the traditional bit operations and

M. Milano (Ed.): CP 2012, LNCS 7514, pp. 527–543, 2012.

correctly models modulo-arithmetic and overflows. Its implementation uses bit operations of the underlying architecture, avoiding the drawback of a bit-blasting approach that associates a variable with each bit. The filtering algorithms implement either domain consistency on the bit-vector domain or bit consistency, a new consistency notion introduced in this paper. Filtering algorithms for logical and structural constraints typically run in constant time (as long as the size of the bit-vector is not greater than the size of the machine registers), while arithmetic constraints such as addition run in time linear in the size of the bit-vectors. The paper also discusses how to channel bit-vector and variables, combining the inference strengths of both representations.

The paper gives an overview of the domain and its filtering algorithms. A short companion paper [18] presents an application in cryptography. For space reasons, the presentation of the bit-vector domain itself cannot be comprehensive and the paper only aims at presenting the main conceptual ideas and filtering algorithms. In particular, the paper only focuses on unsigned number, although similar algorithms also exist for signed numbers.

The paper is organized as follows: Section 2 introduces the bit-vector domain. Section 3 presents three classes of constraints and two consistency notions. Section 4 discusses the implementation of representative constraints, Section 5 discusses the channeling mechanism, and Section 6 briefly discusses how to extend the results to signed numbers. Section 7 covers the related work and Section 8 finally concludes the paper.

2 Bit-Vector Variables

Definition 1 (Bit-Vectors). *A bit-vector $b_{[k]}$ denotes a sequence of $k \geq 1$ binary digits (bits). b_i ($i \in 0..k - 1$) denotes the ith bit in the sequence. Bit b_0 is called the least significant bit and b_{k-1} the most significant bit. The most significant bit comes first in the sequence and the least significant bit comes last.*

For instance, 001 is a bit-vector of size 3 with its least significant bit equal to 1. In this paper, we assume that bit-vectors are of length k unless specified otherwise and we use b to denote $b_{[k]}$ for simplicity. Bit-vectors are typically denoted by the letters b, l, and u, possibly superscripted.[1] Bit-vectors are often used to represent (a finite subset of) natural numbers.

Definition 2 (Natural Number Interpretation). *The natural number represented by a bit-vector b is given by the formula*

$$\mathcal{I}(b) = \sum_{i=0}^{k-1} b_i \cdot 2^i$$

Bit-vectors of size k can represent natural numbers in the range $[0, 2^k - 1]$.

---

[1] Superscripts are used, since subscripts are used to access the individual bits.

Function $\mathcal{I}$ defines a total order on bit-vectors, which also coincides with the lexicographic ordering on the sequence.

Definition 3 (Ordering on Bit-Vectors). *Let b_1 and b_2 be bit-vectors. Then*

$$b_1 \leq b_2 \ \textit{iff} \ \mathcal{I}(b_1) \leq \mathcal{I}(b_2).$$

Definition 4 (Bit-Vector Domain). *A bit-vector domain is a pair $\langle l, u \rangle$ of bit-vectors such that $l_i \leq u_i$ ($0 \leq i < k$). The bit-vector domain represents the set of bit-vectors*

$$\{b \mid l \leq b \leq u \wedge \forall i \in F(\langle l, u \rangle) : b_i = l_i\}$$

where $F(\langle l, u \rangle)$ represents the fixed bits of the domain, i.e.,

$$F(\langle l, u \rangle) = \{i \in 0..k - 1 \mid l_i = u_i\}.$$

The free bits $V(\langle l, u \rangle)$ of domain $\langle l, u \rangle$ complements the fixed bits, i.e.,

$$V(\langle l, u \rangle) = \{i \in 0..k - 1 \mid l_i < u_i\}.$$

Example 1. The bit-vector domain $D = \langle 010, 111 \rangle$ denotes $\{010, 011, 110, 111\}$, $F(D) = \{1\}$, and $V(D) = \{0, 2\}$. When the bit-vector domain is viewed as representing a set of natural numbers, the domain can be interpreted as $\{2, 3, 6, 7\}$.

Definition 5 (Bit Domain in a Bit-Vector Domain). *Let D be a bit-vector domain. The domain of bit i in D, denoted by D_i, is the set $\{b_i \mid b \in D\}$.*

Definition 6 (Bit-Vector Variable). *A bit-vector variable x is associated with a domain $D = \langle l, u \rangle$, in which it takes its value. We use l^x and u^x to denote the bit-vectors l and u defining the domain of x and F^x and V^x to denote $F(\langle l^x, u^x \rangle)$ and $V(\langle l^x, u^x \rangle)$.*

Bit-vector variables can be viewed as a sequence of binary 0-1 variables. Making them first-class objects however has significant benefits, not only from a modeling, but also from a computational standpoint as will become clear later in the paper. The primary purpose of bit-vector variables is to reason about which bits in the vector are fixed. Although they can be used for representing sets of natural numbers, bit-vector variables are not always able to represent arbitrary sets and intervals accurately.

Example 2. Consider the interval 3..6. It cannot be represented exactly by a bit-vector domain. Indeed, the domain $D = \langle 011, 110 \rangle$ is not valid since $l_0 > u_0$. The bit-vectors associated with integers in 3..6 are $\{011, 100, 101, 110\}$ and there is no bit which has the same value in all these bit-vectors. The most precise approximation of 3..6 by a bit-vector domain is $D = \langle 000, 111 \rangle$ which captures no information.

Bit-vector variables are denoted by letters x, y, z, w, possibly superscripted, in the following.

3   The Constraint System

Constraints over bit-vector variables contain logical, arithmetic, and structural constraints. This section gives an overview of some of the constraints useful in practical applications. It is not intended to be comprehensive for space reasons.

3.1   The Constraints

Logical Constraints Logical constraints include equality $x = y$, bitwise negation $x = \neg y$, bitwise conjunction $x \wedge y = z$, bitwise disjunction $x \vee y = z$, bitwise exclusive or $x \oplus y = z$, conditional $\mathtt{IF}(x, y, z) = w$, and reified equality also known as equivalence $(x = y) = z$. The semantics of these constraints is the natural bit-wise semantics. For instance, $x = y$ holds iff $\forall i \in 0..k - 1 : x_i = y_i$ and $x \wedge y = z$ holds iff $\forall i \in 0..k - 1 : x_i \wedge y_i = z_i$. The constraint $\mathtt{IF}(x_{[k]}, y_{[k]}, z_{[k]}) = w_{[k]}$ is a quaternary constraint with the following semantics

$$\forall i \in 0..k - 1 \ : \ w_i = y_i \wedge x_i = 1 \ \vee \ w_i = z_i \wedge x_i = 0.$$

Finally, the reified equality $(x = y) = z$ is equivalent to

$$\forall i \in 0..k - 1 \ : \ z_i = 1 \wedge x_i = y_i \ \vee \ z_i = 0 \wedge x_i \neq y_i$$

and could be recast as $\neg(x \oplus y) = z$.

Arithmetic Constraints. Since bit-vector variables can be used to represent natural numbers, it is natural to include a collection of arithmetic constraints, which includes membership, inequalities, addition $x + y = (z, c)$, and unsigned multiplication $x \cdot y = z$. Constraint $x \in [L, U]$, where L and U are integers and $L \leq U$, holds if $L \leq \mathcal{I}(x) \leq U$. Constraint $x \leq y$ holds if $\mathcal{I}(x) \leq \mathcal{I}(y)$. Constraint $x + y = (z, c)$ holds if $(\mathcal{I}(x) + \mathcal{I}(y)) \bmod 2^k = \mathcal{I}(z)$ and c denotes the carry bit-vector variable, while $x \cdot y = z$ if $(\mathcal{I}(x) \cdot \mathcal{I}(y)) \bmod 2^k = \mathcal{I}(z)$. It is important to point out that the addition constraint models actual computer architectures where the addition can overflow. The carry vector c has $k + 1$ bits and the last carry-out bit c_k is 1 when the addition overflows.

Structural Constraints. Structural constraints are useful to extract sub-sequences from bit-vectors, concatenate, shift, rotate, or even perform some extension operations. Some of the constraints in our implementation are listed below.

$\mathtt{SHL}(x, n) = y$: The constraint holds if y shifts x by n positions to the left. The n bits introduced to the right are 0.

$\mathtt{LSHR}(x, n) = y$: The constraint holds if y is the logical right shift of x by n positions to the right. The n bits introduced to the left are all 0.

$\mathtt{ROTL}(x, n) = y$: The constraint holds if y is the rotation of x by n positions to the left, i.e., $\forall i \in 0..k - 1 : x_i = y_{(i+n) \bmod k}$.

$\mathtt{ROTR}(x, n) = y$: The constraint holds if y is the rotation of x by n positions to the right, i.e., $\forall i \in 0..k - 1 : y_i = x_{(i+n) \bmod k}$.

$\texttt{EXTU}(x_{[k]}, n) = y_{[n]}$: Assuming that $n > k$, the constraint holds if $\forall i \in 0..k - 1$: $y_i = x_i \wedge \forall i \in k..n : y_i = 0$.

$\texttt{EXTRACT}(x_{[k]}, f, n) = y_{[n]}$: Assuming that $n \leq k$ and $0 \leq f \leq k - 1$, the constraint holds when $\forall i \in 0..n - 1$: $y_i = x_{f+i}$, i.e., when y denotes the sub-sequence of n bits extracted from position f in x.

$\texttt{CONCAT}(x_{[k]}, y_{[l]}) = z_{[k+l]}$: The constraint holds if z is the concatenation of the two bit sequences x and y, i.e., $\forall i \in 0..l - 1 : z_i = y_i \wedge \forall i \in 0..k - 1 : z_{l+i} = x_i$.

3.2 Consistency Notions

The constraint implementation considers two consistency notions: Domain consistency and bit consistency. Domain consistency is the standard definition lifted to bit-vector variables.

Definition 7 (Domain Consistency). *A bit-vector constraint $C(x^1, \ldots, x^k)$ is domain-consistent with respect to $D^1 \times \ldots \times D^k$ and variable x_j iff*

$$\forall b^j \in D^j \; \exists b^1 \in D^1, \ldots, b^{j-1} \in D^{j-1}, b^{j+1} \in D^{j+1}, \ldots, b^k \in D^k : C(b^1, \ldots, b^k).$$

It is domain-consistent with respect to $D^1 \times \ldots \times D^k$ iff it is domain-consistent with respect to these domains for all variables.

Bit consistency is a weaker consistency notion that only guarantees that each free bit for a variable is not fixed to the same value in all solutions.

Definition 8 (Bit Consistency). *A bit-vector constraint $C(x^1, \ldots, x^k)$ is bit-consistent with respect to $D^1 \times \ldots \times D^k$, variable x^j, and bit i iff*

$$\forall v \in D_i^j \; \exists b^1 \in D^1, \ldots, b^k \in D^k : \; b_i^j = v \wedge C(b^1, \ldots, b^k).$$

It is bit-consistent with respect to $D^1 \times \ldots \times D^k$ iff it is bit-consistent with respect to these domains for all variables and all bits.

Example 3. Constraint $x \geq y$ is bit-consistent wrt $D(x) = \langle 000, 111 \rangle$ and $D(y) = \langle 011, 111 \rangle$. Indeed, the potential solutions for x are $\{011, 100, 101, 110, 111\}$ and all three bits appear in a solution with values 0 and 1. Constraint $x \geq y$ is not bit-consistent wrt $D(x) = \langle 000, 111 \rangle$ and $D(y) = \langle 100, 111 \rangle$, since bit 2 is fixed to 1 in all solutions. It is bit-consistent wrt $D(x) = \langle 100, 111 \rangle$ and $D(y) = \langle 100, 111 \rangle$.

4 Constraint Implementation

4.1 Overview

The key insight of the implementation is that many constraints[2] can be implemented through bit operations of the underlying computer architecture, making the implementation constant time when the length of the bit-vector variables

---

[2] Arithmetic constraints are different and need to be handled differently.

```
1  function  propagate (C(x, y))   :   Bool
2       low^x    = f^x_low(x, y);
3       up^x     = f^x_up(x.y);
4       low^y    = f^y_low(x, y);
5       up^y     = f^y_up(x, y);
6       ⟨l^x, u^x⟩ = ⟨low^x, up^x⟩;
7       ⟨l^y, u^y⟩ = ⟨low^y, up^y⟩;
8       return  valid (⟨l^x, u^x⟩) ∧ valid (⟨l^y, u^y⟩);
```

Fig. 1. The Overall Implementation Structure of Logical Constraints

does not exceed the word size. Indeed, under these circumstances, an expression over bit-vectors l, u, and v such as $\neg(l \oplus u) \wedge (l \vee v)$ can be evaluated in four instructions of the underlying architecture. This is particularly appealing on recent Intel architectures which feature extended register sets, i.e., the so-called SSE registers with 128 bits. These SSE registers are larger than our bit-vectors for our targeted applications.

Consider now a constraint $C(x, y)$.[3] The domain of x is $\langle l^x, u^x \rangle$ and the domain of y is $\langle l^y, u^y \rangle$. The goal of the constraint propagation is to determine new domains $\langle low^x, up^x \rangle$ and $\langle low^y, up^y \rangle$ for x and y in terms of expressions over l^x, u^x, l^y, and u^y. As mentioned above, these expressions are then evaluated in constant time. In addition to specifying low^x, up^x, low^y, up^y, the constraint should also indicate if it fails or succeeds. In the implementation, this check is performed by verifying that the domains $\langle low^x, up^x \rangle$ and $\langle low^y, up^y \rangle$ are valid. Recall that a bit-vector domain $\langle l, u \rangle$ must satisfy $\forall i \in 0..k - 1 : l_i \leq u_i$. Such a validity test can be performed in constant time using the following expression

$$valid(\langle l, u \rangle) = \neg(l \oplus u) \vee u.$$

Indeed, this expression checks that, whenever $l_i \neq u_i$, we have $u_i = 1$. As a result, the implementation of a constraint $C(x, y)$ typically looks like the schema depicted in Figure 1. Hence the implementation in the paper only specifies low^x, up^x, low^y, and up^y. These constraints are propagated each time a bit is fixed in one of the arguments.

4.2   Logical Constraints

This section presents the implementation of logical constraints. Because they operate bitwise, logical constraints are domain-consistent iff the propagation algorithm for a single bit is domain-consistent. Indeed, a logical constraint $C(x, y)$ can be seen as an abbreviation of

$$\forall i \in 0..k - 1 : C_b(x_i, y_i)$$

---

[3] This overview focuses on binary constraints but generalizes to n-ary constraints.

where k is the word width, x_i and y_i represent bit variables associated with bit i in bit-vectors x and y, and C_b is the propagation algorithm for a single bit. Since all these constraints are independent, $C(x, y)$ is domain-consistent iff the propagation algorithm for a single bit is bound-consistent.

Theorem 1. *The propagation of a logical constraint over bit-vectors is domain-consistent iff the propagation algorithm for a single bit is domain-consistent.*

In the following, we derive the propagation algorithm for logical constraint from Boolean propagation rules as those defined in [13].

Equality: The propagator for the $x = y$ constraint is defined as follows:

$$low^x = low^y = l^x \vee l^y$$
$$up^x = up^y \quad = u^x \wedge u^y$$

The propagators of logical constraints are derived systematically from inference rules and we illustrate the process on this simple case. The two inference rules for the new domain of x are

$$x_i = 1 \text{ if } y_i = 1$$
$$x_i = 0 \text{ if } y_i = 0$$

The inference rule for $x_i = 1$ defines the expression for low_i^x which is obtained by the disjunction of l_i^x and a rewriting of the right-hand side. In the right-hand side, an expression $y_i = 1$ is rewritten as l_i^y and an expression $y_i = 0$ is rewritten into $\neg u_i^y$. In this case, we obtain

$$low_i^x = l_i^x \vee l_i^y.$$

The inference rule for $x_i = 0$ defines the expression for up_i^x which is obtained by a conjunction of u_i^x and the negation of the rewriting of the right-hand side. The conjunction and the negation are necessary, since bit i is fixed to zero when $u_i^x = 0$. We then obtain

$$up_i^x = u_i^x \wedge \neg(\neg u_i^y) = u_i^x \wedge u_i^y.$$

The final result is obtained by lifting the expression from bits to bit-vectors.

Bitwise Conjunction: The propagation for $x \wedge y = z$ is defined as follows:

$$low^z = l^z \vee (l^x \wedge l^y) \qquad up^x = u^x \wedge \neg(\neg u^z \wedge l^y)$$
$$up^z = u^z \wedge u^x \wedge u^y \qquad low^y = l^y \vee l^z$$
$$low^x = l^x \vee l^z \qquad up^y = u^y \wedge \neg(\neg u^z \wedge l^x)$$

The new domain of z is computed by the two inference rules

$$z_i = 1 \text{ if } x_i = 1 \wedge y_i = 1$$
$$z_i = 0 \text{ if } x_i = 0 \vee y_i = 0$$

The new domain of x is computed by the two inference rules

$$x_i = 1 \text{ if } z_i = 1$$
$$x_i = 0 \text{ if } z_i = 0 \wedge y_i = 1$$

Bitwise Negation: The propagation for $x = \neg y$ is defined as follows:

$$low^y = l^y \vee \neg u^x \qquad low^x = l^x \vee \neg u^y$$
$$up^y = u^y \wedge \neg l^x \qquad up^x = u^x \wedge \neg l^y$$

Bitwise Disjunction: The propagation for $x \vee y = z$ is defined as follows:

$$low^z = l^z \vee l^x \vee l^y \qquad up^x = u^x \wedge u^z$$
$$up^z = u^z \wedge (u^x \vee u^y) \qquad low^y = l^y \vee (\neg u^x \wedge l^z)$$
$$low^x = l^x \vee (\neg u^y \wedge l^z) \qquad up^y = u^y \wedge u^z$$

The new domain of z is computed by the two inference rules

$$z_i = 1 \text{ if } x_i = 1 \vee y_i = 1$$
$$z_i = 0 \text{ if } x_i = 0 \wedge y_i = 0$$

The new domain of x is computed by the two inference rules

$$x_i = 1 \text{ if } z_i = 1 \wedge y_i = 0$$
$$x_i = 0 \text{ if } z_i = 0$$

Bitwise Exclusive Or: The propagation for $x \oplus y = z$ is defined as follows:

$$low^z = l^z \vee (\neg u^x \wedge l^y) \vee (l^x \wedge \neg u^y) \qquad up^x = u^x \wedge (u^z \vee u^y) \wedge \neg(l^y \wedge l^z)$$
$$up^z = u^z \wedge (u^x \vee u^y) \wedge \neg(l^x \wedge l^y) \qquad low^y = l^y \vee (\neg u^z \wedge l^x) \vee (l^z \wedge \neg u^x)$$
$$low^x = l^x \vee (\neg u^z \wedge l^y) \vee (l^z \wedge \neg u^y) \qquad up^y = u^y \wedge (u^z \vee u^x) \wedge \neg(l^x \wedge l^z))$$

The new domain of z is computed by the two inference rules

$$z_i = 1 \text{ if } (x_i = 0 \wedge y_i = 1) \vee (x_i = 1 \wedge y_i = 0)$$
$$z_i = 0 \text{ if } (x_i = 0 \wedge y_i = 0) \vee (x_i = 1 \wedge y_i = 1)$$

The new domain of x is computed by the two inference rules

$$x_i = 1 \text{ if } (z_i = 1 \wedge y_i = 0) \vee (z_i = 0 \wedge y_i = 1)$$
$$x_i = 0 \text{ if } (z_i = 1 \wedge y_i = 1) \vee (z_i = 0 \wedge y_i = 0)$$

Conditional: The propagation of $\text{IF}(x, y, z) = w$ relies, in essence, on a conditional pair of mutually exclusive assignments. In the following, $a \Rightarrow b$ is a bitwise implication which can be rewritten with DeMorgan's law into $\neg a \vee b$. The implementation is as follows:

$$low^w = l^w \vee (l^x \wedge l^y) \vee (\neg u^x \wedge l^z) \vee (l^y \wedge l^z)$$
$$up^w = u^w \wedge \neg((l^x \wedge \neg u^y) \vee (\neg u^x \wedge \neg u^z) \vee (\neg u^y \wedge \neg u^z))$$
$$low^y = l^y \vee (l^w \wedge (l^x \vee \neg u^z))$$
$$up^y = u^y \wedge \neg(\neg u^w \wedge (l^x \vee l^z))$$
$$low^z = l^z \vee (l^w \wedge (l^x \vee \neg u^y))$$
$$up^y = u^z \wedge \neg(\neg u^w \wedge (l^x \vee l^y))$$
$$low^x = l^x \vee (l^w \wedge \neg u^z) \vee (\neg u^w \wedge l^z)$$
$$up^x = u^x \wedge \neg((l^w \wedge \neg u^y) \vee (\neg u^w \wedge l^y))$$

The new domain of w is computed by the two inference rules

$$w_i = 1 \text{ if } (x_i = 1 \wedge y_i = 1) \vee (x_i = 0 \wedge z_i = 1) \vee (y_i = 1 \wedge z_i = 1)$$
$$w_i = 0 \text{ if } (x_i = 1 \wedge y_i = 0) \vee (x_i = 0 \wedge z_i = 0) \vee (y_i = 0 \wedge z_i = 0)$$

The new domain of y is computed by the two inference rules

$$y_i = 1 \text{ if } (w_i = 1) \wedge (x_i = 1 \vee z_i = 0)$$
$$y_i = 0 \text{ if } (w_i = 0) \wedge (x_i = 1 \vee z_i = 1)$$

The new domain of x is computed by the two inference rules

$$x_i = 1 \text{ if } w_i \neq z_i$$
$$x_i = 0 \text{ if } w_i \neq y_i$$

4.3   Structural Constraints

We briefly review structural constraints which enforce domain consistency.

Shifts and Rotations: The propagation of $\mathtt{SHL}(x,n) = y$ is defined as follows:

$$low^y = ((l^x \ll n) \vee l^y) \wedge 1_{[k-n]}.0_{[n]}$$
$$up^y = (u^x \ll n) \wedge u^y$$
$$low^x = l^x \vee (l^y \gg n)$$
$$up^x = ((u^y \gg n) \wedge u^x) \vee 1_{[n]}.0_{[k-n]})$$

For low^y, the implementation first shifts the bits of l^x by n position (with a logical left shift implemented in hardware and denoted by $\ll$), then applies a disjunction with l^y as in the equality constraint, and finally force the n low-order bits down to 0 since a left shift introduces 0 bits. The last part is achieved through a conjunction with a mask whose $k-n$ high-order bits are 1 (in order to preserve the part just constructed) and the n low-order bits are 0. Similarly, up^x starts with a logical right shift by n position of u^y and takes the conjunction with the old value of u^x. Since a logical left shift introduces 0 to the left, the result is disjuncted with a mask whose n high-order bits are 1. The implementation runs in $\mathcal{O}(1)$ if the required masks are available (there are only $\mathcal{O}(k)$ of them).

Extension: $\mathtt{EXTU}(x_{[k]}, n) = y_{[n]}$ expects $x_{[k]}$ to be a $k-$bit long sequence and extends it to a $n-$bit long sequence $y_{[n]}$ with padding to the left. The implementation runs in $\mathcal{O}(1)$ time with a suitable mask.

Extraction: $\mathtt{EXTRACT}(x_{[k]}, f, n) = y_{[n]}$ is yet another variant of equality which is even more direct and also runs in $\mathcal{O}(1)$. Indeed, both sequences (l^x, u^x) must be shifted right by f to discard the low-order bits and a mask must be applied to relax the $k-n$ high-order bits. Once this is done, it is a normal equality.

4.4   Arithmetic Constraints

Finally, we focus on arithmetic constraints that are particularly interesting.

```
1  function propagate (x ∈ [L, U])  :  Bool
2      i = k − 1
3      while i ≥ 0 do
4          if I(uˣ) < L ∨ I(lˣ) > U
5              return false ;
6          if i ∈ Vˣ then
7              if I(uˣ) − 2ⁱ < L then lᵢˣ = 1
8              else if I(lˣ) + 2ⁱ > U then uᵢˣ = 0
9              else break ;
10         i = i − 1
11     return true ;
```

Fig. 2. Propagating the Membership Constraint

Membership Constraint. Figure 2 depicts the implementation of the membership constraint. Given a bit-vector variable x and a free bit i in x, the algorithm uses two propagation rules (line 7 and line 8). If $\mathcal{I}(u^x) - 2^i < L$, then bit i must be fixed to 1 since otherwise x cannot reach the lower bound. Similarly, if $\mathcal{I}(l^x) + 2^i > U$, then bit i must be fixed to 0 since otherwise x would exceed the upper bound. The algorithm applies these two rules, starting with the most significant free bits. If, at some point, none of these two rules apply, subsequent, less significant, free bits do not have to be considered and the algorithm terminates early (line 9). Note also that the algorithm tests for feasibility at every iteration of the loop (lines 4–5), since the constraint may become infeasible even if $\mathcal{I}(l^x) \leq L \leq U \leq \mathcal{I}(u^x)$ holds initially. The algorithm runs in $\mathcal{O}(k)$ in the worst case since $\mathcal{I}(l^x)$ and $\mathcal{I}(u^x)$ are constant time lookup and 2^i can be maintained in $\mathcal{O}(1)$ via shifts.

Example 4 (Membership Constraint (1)). Let x be a bit-vector variable with domain $D^x = \langle 000010, 111110 \rangle$. In the following, we associate $I(l^x)$ and $I(u^x)$ with the domain for clarity. So D^x is denoted by $\langle 000010 : 2, 111110 : 62 \rangle$. Observe that the last two bits are fixed and all others are free. Consider the constraint $x \in [42, 45]$. The algorithm proceeds left to right. Since $62 - 32 < 42$ in line 7, bit 5 must be fixed to 1 and the domain becomes $\langle 100010 : 34, 111110 : 62 \rangle$. The algorithm now considers bit 4. Since $34 + 16 > 45$, bit 4 is fixed to zero by lines (9–10) and the domain becomes $\langle 100010 : 34, 101110 : 46 \rangle$. Bit 3 is now examined. Since $46 - 8 < 42$, bit 3 must be fixed to one and the domain is updated to $\langle 101010 : 42, 101110 : 46 \rangle$. The algorithm now looks at bit 2. Since $42 + 4 > 45$, bit 2 must be fixed to zero, producing the final domain $\langle 101010 : 42, 101010 : 42 \rangle$.

Example 5 (Membership Constraint (2)). With the same variable x as in Example 4, consider the constraint $x \in [43, 45]$. The first couple of steps are similar, producing a domain $\langle 101010 : 42, 101110 : 46 \rangle$. Bit 2 is now considered. Since $46 - 4 < 43$, the new domain becomes $\langle 101110 : 46, 101110 : 46 \rangle$ and a failure is detected in line 4.

Theorem 2. *The propagation of $x \in [L, U]$ achieves bit consistency.*

```
1  function propagate (x ≤ y)  :  Bool
2      do
3          ⟨ol^x, ou^x⟩ = ⟨I(l^x), I(u^x)⟩
4          ⟨ol^y, ou^y⟩ = ⟨I(l^y), I(u^y)⟩
5          if  ! propagate (x ∈ [I(l^x), I(u^y)])
6                  return false ;
7          if  ! propagate (y ∈ [I(l^x), I(u^y)])
8                  return false ;
9      while  ⟨ol^x, ou^x⟩ ≠ ⟨l^x, u^x⟩ ∨ ⟨ol^y, ou^y⟩ ≠ ⟨l^y, u^y⟩ ;
10     return true ;
```

Fig. 3. The Propagation Algorithm for $x \leq y$

Proof. We first show that the algorithm can terminate early. If the algorithm terminates in line 11, we know that $\mathcal{I}(u^x) - 2^i \geq L$ and $\mathcal{I}(l^x) + 2^i \leq U$. It follows that $\mathcal{I}(u^x) - 2^j \geq L$ and $\mathcal{I}(l^x) + 2^j \leq U$ for all $0 \leq j < i$. We next show that, if more than one rule applies, the constraint is infeasible and will fail. Assume that $\mathcal{I}(u^x) - 2^i < L$ holds for bit i. The algorithm fixes bit i to 1 and the lower bound of the domain becomes $\mathcal{I}(l^x) + 2^i$. If the second rule applies, this would mean that $\mathcal{I}(l^x) + 2^i > U$ and this is detected at the next iteration of the loop in line 4. As a result, the two propagation rules do not apply to any of the free bits at the end of the propagation.

Consider now the most significant free bit i in the resulting domain. Observe first that the bit-vector representations of L and U must have bit i equal to 0 and 1 respectively since otherwise bit i would be fixed. Since the propagation rules do not apply, we have that $\mathcal{I}(u^x) - 2^i \geq L$ and $\mathcal{I}(l^x) + 2^i \leq U$. Moreover, the value $\mathcal{I}(u^x) - 2^i$ belongs to D^x by definition of the bit-vector domain, satisfies $L \leq \mathcal{I}(u^x) - 2^i \leq U$, and has bit i set to 0. Similarly, the value $\mathcal{I}(l^x) + 2^i$ belongs to D^x by definition of the bit-vector domain, satisfies $L \leq \mathcal{I}(l^x) + 2^i \leq U$, and has bit i set to 1. Hence, bit i is bit-consistent.

Finally, consider another free bit $j < i$ and the bit-vectors associated with $\mathcal{I}(u^x) - 2^i$ and $\mathcal{I}(l^x) + 2^i$. We know that these bit-vectors are in the domain and satisfy the constraint. Moreover, $\mathcal{I}(u^x) - 2^i$ has bit j assigned to 1 and $\mathcal{I}(l^x) + 2^i$ has bit j assigned to 0. Hence, bit j is bit-consistent. $\qquad\square$

Inequalities Figure 3 depicts the propagation algorithm for the inequality constraint $x \leq y$. The algorithm uses the membership constraint and implements a fixpoint computation.

Theorem 3. *The propagation of $x \leq y$ achieves bit consistency.*

Proof. At the end of the fixpoint, $x \in [\mathcal{I}(l^x), \mathcal{I}(u^y)]$ and $y \in [\mathcal{I}(l^x), \mathcal{I}(u^y)]$ are bit-consistent. Consider variable x. For every free bit of x, we can select the bit-vector for $\mathcal{I}(u^y)$ in D^y and the bit-vectors in D^x selected in the proof of Theorem 2. Hence, x is bit-consistent with respect to the updated domains. The reasoning for y is similar. $\qquad\square$

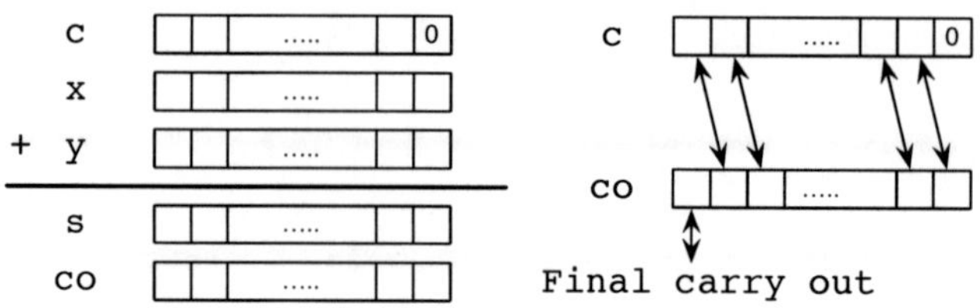

Fig. 4. Illustrating the Sum Constraint

Theorem 4. *The propagation of $x \leq y$ runs in time $O(k)$.*

Proof. By definition of the constraint, l^x and u^y cannot be updated. Hence, only one iteration is needed to reach the fixpoint.

Addition. The implementation of addition also aims at exploiting bit operations on the underlying architecture. This is however more challenging than for logical constraints, since the bits are not independent: The carry-out of bit i is the carry-in of bit $i + 1$. We could implement the constraint by associating a propagator with each bit i but such an algorithm would not exploit bitwise operations of the underlying machine. Instead, our implementation decouples the carries and uses constraints to connect them. Our implementation contains two constraints:

1. A logical full-adder constraint;
2. The constraint connecting the carry-in and the carry-out.

The constraints are decoupled by using a brand new bit-vector variable co for the carry-out's. Let us present the intuition bottom-up, starting from the bits (See also Figure 4 for an illustration). Constraint $fulladder(x_i, y_i, c_i, s_i, co_i)$ holds if

$$s_i = c_i \oplus x_i \oplus y_i$$
$$co_i = (x_i \wedge y_i) \vee (c_i \wedge (x_i \oplus y_i))$$

which can be found in any introductory class on computer architecture. The link between the carry-out of bit i and the carry-in of bit $i + 1$ is achieved with

$$c_{i+1} = co_i \tag{1}$$

Now observe that the full-adder constraint only involves i of the underlying bit-vector variables. Hence, the implementation can process all the bits simultaneously using the bitwise operations of the underlying architecture, like all the logical constraints proposed earlier. Moreover, equation 1 can be recast as a shift constraint, leading to the constraint system over bit-vector variables

$$fulladder(x, y, c, s, co)$$
$$c = co \ll 1$$
$$c_0 = 0$$

```
1  function  propagate (x + y = (z, c))  :  Bool
2     co ∈ ⟨0...0, 1...1⟩:  new  bit-vector  variable  with  k  bits
3     return  propagate ({  fulladder(x, y, c, s, co),  c = co ≪ 1,  c_0 = 0  });
```

Fig. 5. The Propagation Algorithm for the Addition Constraint

The resulting propagation algorithm is depicted in Figure 5. To propagate the *sum* constraint, the algorithm propagates the system of constraints until a fixpoint is reached. The propagation may fix the carry-out co_i which means that the half-adder for bit $i+1$ must be reconsidered. Similarly, the propagation may fix the carry-in c_{i+1} which means that the half-adder for bit i must reconsider. Of course, the propagation for all bits takes place simultaneously.

Theorem 5. *The propagation of $x + y = (z, c)$ enforces bit-consistency.*

Proof. Consider a finite-domain constraint system containing the full-adder constraint for each bit and the equations 1 for linking the carries. The constraint graph is Berge-acyclic. Hence, if the half-adder constraint enforces domain-consistency, the constraint system is domain-consistent when the fixpoint is reached. The constraint system in Figure 5 enforces the same propagation and hence is bit-consistent.

Since there are 5 bit-vectors with k bits in the implementation, the number of iterations in the fixpoint algorithm is $O(k)$, since each bit can be fixed at most once. Moreover, the propagation of every constraint takes constant time, since they consist of a constant number of bitwise operations. Hence, the overall propagation runs in time $O(k)$.

Theorem 6. *The propagation of $x + y = (z, c)$ runs in time $O(k)$.*

5   Channelling

Example 2 indicated that bit-vector domains cannot always precisely capture a set of integers. A common remedy is to use a dual representation consisting of a bit-vector variable and an interval variable connected by a channelling constraint.

The channelling constraint is defined in Figure 6 where (x, X) is a pair of channelled variables: x is a bit-vector variable and X is its interval counterpart. The implementation first updates X given D^x, then x given the updated D^X and then process a final update of X given the resulting domain of x. No further step is needed, since D^X is updated exactly to $[l^x, u^x]$ at that stage.

Theorem 7. *The propagation of the channeling constraint runs in time $O(k)$.*

It is also interesting to observe that an update to the lower bound of X may produce a narrowing in the lower and upper bounds of x. Thus, this constraint is propagated every time a bit is fixed in x or a bound is updated in variable X.

```
1  function  propagate ( channel ( x, X ) )  :  Bool
2       if  ! propagate ( X ∈ [𝓘(l^x), 𝓘(u^x)] )
3            return  false ;
4       if  ! propagate ( x ∈ D^X )
5            return  false ;
6       return  propagate ( X ∈ [𝓘(l^x), 𝓘(u^x)] ) ;
```

Fig. 6. The Propagation Algorithm for the Channelling Constraint

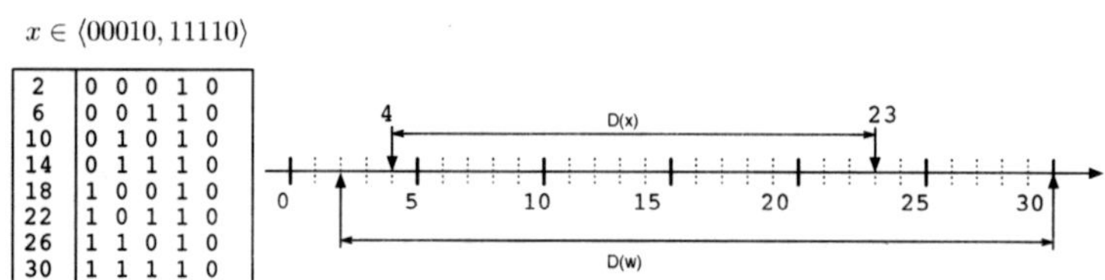

Fig. 7. Illustrating the Channelling Constraint

Example 6. Consider a variable x with domain $\langle 00010 : 2, 11110 : 30 \rangle$ and its channelled counterpart X with domain $[2, 30]$. The situation is depicted in Figure 7[left] which specifies the domain of x. First, observe that, if the domain of X is updated to $[4, 23]$, the domain of x does not shrink, highlighting the need for the dual representation (indeed, each bit has supports both for 0 and 1). Figure 7[right] depicts the relationship between the domains of x and X. If the lower bound of X is updated to 15, the propagation of $x \in [15, 23]$ produces a domain $\langle 10010 : 18, 10110 : 22 \rangle$ and the domain of X is updated to $[18,22]$. It is interesting to observe that a tightening of the inner domain narrows the outer domain which, in turn, shrinks the inner domain.

6   Extension to Signed Numbers

This paper focused on an unsigned interpretation of bit-vectors. Similar techniques can be used for signed numbers, in which case the interpretation becomes

$$\mathcal{S}(b) = -b_{k-1} \cdot 2^k + \sum_{i=0}^{k-2} b_i \cdot 2^i.$$

The bit-vector variables do not change but various constraints would interpret them as signed or unsigned variables, like various computer instructions would interpret memory words as signed or unsigned numbers. For instance, our constraint system in fact includes an signed operation **EXTS** in addition to its unsigned counterpart **EXTU**. A dual representation and a channeling representation can also be defined for the signed case. Moreover, the two interpretations can be used simultaneously for different bit-vector variables in the same way as computer programs can use both signed and unsigned integers.

7   Related Work

The satisfiability community features a number of SMT solvers [9,8,10,14,16,21] for bit-vectors (including BOOLECTOR or BEAVER). They rely on SAT literals to encode bit-vector constraints into a CNF formula. This technique is often referred to as *bit-blasting* as each bit of each bit-vector is encoded as a boolean variable and the bit-vector constraints are encoded as propositional clauses.

Mixed integer programming [7,23], constraint programming [20,22,4], or their hybridizations [1] have also been considered to side-step the scaling issues arising from bit-blasting. The approaches are also tempting given their potential to model and efficiently reason with arithmetic constraints. Difficulties with MIP and CP encodings typically stem from modulo arithmetic. Modular arithmetic was considered in [15].

In a 2006 note [6], Bordeaux et al. described a channeling constraint between a vector of 0-1 variables and an arithmetic constraint. The inference rules are similar to those used in the membership constraint and they prove a result similar to Theorem 3. This paper however proposes an algorithm that runs in time $\mathcal{O}(k)$ in the worst case and stops at the first free bit (from the most significant bit downwards) for which the inference rules do not apply. Bardin [4] outlines a bit-list domain $\mathcal{BL}$ and provides $\mathcal{O}(k)$ algorithms for propagators. Their paper also combines $\mathcal{BL}$ with interval lists domains. However, their implementation does not exploit bit operations of the underlying architecture which allows for many operations to run in time $\mathcal{O}(1)$ rather than $\mathcal{O}(k)$. Finally, while their propagator for the sum runs in $\mathcal{O}(k)$, it does not achieve bit-consistency (See [4], page 9). Papers that make use of bit-vectors to implement traditional finite domains, e.g., [19,17], are not related to the ideas developed here.

8   Conclusion

The paper introduced a new bit-vector domain for constraint programming motivated by applications in cryptography and verification. The domain supports all the traditional bit operations and correctly models modulo-arithmetic and overflows. Its implementation uses bit operations of the underlying architecture, avoiding the drawback of a bit-blasting approach that associates a variable with each bit. The filtering algorithms implement either domain consistency or bit consistency, a new consistency notion introduced in this paper. Filtering algorithms for logical and structural constraints run in constant time (when the size of the bit-vector is not greater than the size of the largest machine registers), while arithmetic constraints such as addition run in time linear in the size of the bit-vectors. The paper also discusses how to channel bit-vector variables with an integer variable. Future work on the domain will also study how to integrate lazy clause generation in the domain, by explaining each constraint. This will ensure that the domain combines the advantages of existing approaches.

References

1. Achterberg, T., Berthold, T., Koch, T., Wolter, K.: Constraint Integer Programming: A New Approach to Integrate CP and MIP. In: Trick, M.A. (ed.) CPAIOR 2008. LNCS, vol. 5015, pp. 6–20. Springer, Heidelberg (2008)
2. Baray, F., Codognet, P., Díaz, D., Michel, H.: Code-Based Test Generation for Validation of Functional Processor Descriptions. In: Garavel, H., Hatcliff, J. (eds.) TACAS 2003. LNCS, vol. 2619, pp. 569–584. Springer, Heidelberg (2003)
3. Bardin, S., Herrmann, P.: Structural testing of executables. In: ICST 2008, pp. 22–31. IEEE Computer Society, Washington, DC (2008)
4. Bardin, S., Herrmann, P., Perroud, F.: An Alternative to SAT-Based Approaches for Bit-Vectors. In: Esparza, J., Majumdar, R. (eds.) TACAS 2010. LNCS, vol. 6015, pp. 84–98. Springer, Heidelberg (2010)
5. Biere, A., Cimatti, A., Clarke, E., Zhu, Y.: Symbolic Model Checking without BDDs. In: Cleaveland, W.R. (ed.) TACAS 1999. LNCS, vol. 1579, pp. 193–207. Springer, Heidelberg (1999)
6. Bordeaux, L., Hamadi, Y., Quimper, C.-G.: The bit-vector constraint. Technical Report 86, Microsoft Research (2006)
7. Brinkmann, R., Drechsler, R.: RTL-datapath verification using integer linear programming. In: ASP-DAC 2002, pp. 741–746. IEEE Computer Society, Washington, DC (2002)
8. Brummayer, R., Biere, A.: Boolector: An Efficient SMT Solver for Bit-Vectors and Arrays. In: Kowalewski, S., Philippou, A. (eds.) TACAS 2009. LNCS, vol. 5505, pp. 174–177. Springer, Heidelberg (2009)
9. Brummayer, R., Biere, A., Lonsing, F.: Btor: bit-precise modelling of word-level problems for model checking. In: SMT 2008/BPR 2008, pp. 33–38. ACM, New York (2008)
10. Bruttomesso, R., Cimatti, A., Franzén, A., Griggio, A., Sebastiani, R.: The MATH-SAT 4 SMT solver. In: Gupta, A., Malik, S. (eds.) CAV 2008. LNCS, vol. 5123, pp. 299–303. Springer, Heidelberg (2008)
11. Bryant, R.E., Kroening, D., Ouaknine, J., Seshia, S.A., Strichman, O., Brady, B.: Deciding Bit-Vector Arithmetic with Abstraction. In: Grumberg, O., Huth, M. (eds.) TACAS 2007. LNCS, vol. 4424, pp. 358–372. Springer, Heidelberg (2007)
12. Clarke, E., Kroning, D., Lerda, F.: A Tool for Checking ANSI-C Programs. In: Jensen, K., Podelski, A. (eds.) TACAS 2004. LNCS, vol. 2988, pp. 168–176. Springer, Heidelberg (2004)
13. Dincbas, M., Van Hentenryck, P., Simonis, H., Aggoun, A., Graf, T., Berthier, F.: The Constraint Logic Programming Language CHIP. In: International Conference on 5th Generation Computer Systems, Tokyo, Japan (December 1988)
14. Ganesh, V., Dill, D.L.: A Decision Procedure for Bit-Vectors and Arrays. In: Damm, W., Hermanns, H. (eds.) CAV 2007. LNCS, vol. 4590, pp. 519–531. Springer, Heidelberg (2007)
15. Gotlieb, A., Leconte, M., Marre, B.: Constraint solving on modular integers. In: ModRef Worksop, associated to CP 2010, Saint-Andrews, Royaume-Uni (September 2010)
16. Jha, S., Limaye, R., Seshia, S.A.: Beaver: Engineering an Efficient SMT Solver for Bit-Vector Arithmetic. In: Bouajjani, A., Maler, O. (eds.) CAV 2009. LNCS, vol. 5643, pp. 668–674. Springer, Heidelberg (2009)
17. Lecoutre, C., Vion, J.: Enforcing arc consistency using bitwise operations. Constraint Programming Letters (2008)

18. Michel, L., Van Hentenryck, P., Johnson, G.: Cryptanalysis of SHA-1 With Constraint Programming Over Bit-Vectors. In: CP 2012 (submitted, 2012)
19. Ullmann, J.R.: Bit-vector algorithms for binary constraint satisfaction and subgraph isomorphism. J. Exp. Algorithmics 15, 1.6:1.1–1.6:1.64 (2011)
20. Vemuri, R., Kalyanaraman, R.: Generation of design verification tests from behavioral VHDL programs using path enumeration and constraint programming. IEEE Trans. Very Large Scale Integr. Syst. 3(2), 201–214 (1995)
21. Wille, R., Fey, G., Grobe, D., Eggersgluss, S., Drechsler, R.: Sword: A sat like prover using word level information. In: VLSI - SoC 2007, pp. 88–93 (October 2007)
22. Zeng, Z., Ciesielski, M.J., Rouzeyre, B.: Functional test generation using constraint logic programming. In: VLSI-SOC 2001, pp. 375–387. Kluwer, B.V., Deventer (2002)
23. Zeng, Z., Kalla, P., Ciesielski, M.: Lpsat: a unified approach to RTL satisfiability. In: DATE 2001, pp. 398–402. IEEE Press, Piscataway (2001)

Towards Solver-Independent Propagators[*]

Jean-Noël Monette, Pierre Flener, and Justin Pearson

Uppsala University, Department of Information Technology, Uppsala, Sweden
{jean-noel.monette,pierre.flener,justin.pearson}@it.uu.se

Abstract. We present an extension to indexicals to describe propagators for global constraints. The resulting language is compiled into actual propagators for different solvers, and is solver-independent. In addition, we show how this high-level description eases the proof of propagator properties, such as correctness and monotonicity. Experimental results show that propagators compiled from their indexical descriptions are sometimes not significantly slower than built-in propagators of Gecode. Therefore, our language can be used for the rapid prototyping of new global constraints.

1 Introduction

One of the main assets of constraint programming (CP) is the existence of numerous filtering algorithms, called propagators, that are tailored specially for global constraints and allow one to solve efficiently hard combinatorial problems. Successful CP solvers enable the definition of new constraints and their associated propagators. However, it may be a tedious task to implement a propagator that is correct, efficient, compliant with the specific interface of the solver, and hand-coded in the solver's implementation language, such as C++.

In this paper, we propose a solver-independent language to describe a large class of propagators. Our contribution towards such a language is twofold.

First, we ease the implementation and sharing of propagators for constraints. The propagators are described concisely and without reference to the implementation details of any solver. Implementations of propagators are generated from their description. This allows one to prototype rapidly a propagator for a new constraint, for which one has no built-in propagator or no (good enough) decomposition; the generated propagator may even serve as a baseline for further refinement. This also allows one to integrate an existing constraint into another solver, as each solver can be equipped with its own back-end for the propagator description language. We believe that such a language can play the same role for sharing propagators between solvers as solver-independent modelling languages play for sharing models between solvers.

Second, we ease the proof of propagator properties, such as correctness and monotonicity. The higher level of abstraction of our propagator description language allows us to design tools to analyse and transform a propagator. This provides a method to apply systematically theoretical results on propagators.

---

[*] This work is supported by grant 2011-6133 of the Swedish Research Council (VR). We thank the anonymous reviewers and Christian Schulte for their useful comments.

M. Milano (Ed.): CP 2012, LNCS 7514, pp. 544–560, 2012.

Our approach is based on the seminal work on indexicals [27]. In a nutshell, an indexical defines a restriction on the domain of a decision variable, given the current domains of other decision variables. Indexicals have been used to implement user-defined constraints in various finite-domain systems, such as SICStus Prolog [6]. While indexicals can originally only deal with constraints of fixed arity, we extend them to deal with constraints of non-fixed arity (often referred to as global constraints) by handling arrays of decision variables and operations on such arrays (iteration and n-ary operators). Also, in contrast to classical implementations of indexicals, indexicals are not interpreted here, but compiled into instructions in the source language of the targeted solver.

The paper is structured as follows. After defining the relevant background (Section 2) and presenting motivating examples of propagator descriptions in our language (Section 3), we list our design decisions behind the language (Section 4). Next we describe our language (Section 5), show how to analyse propagators written in it (Section 6), discuss our current implementation (Section 7), and experimentally evaluate it (Section 8). We end the paper with a review of related work and a look at future research directions (Section 9).

2   Background

Let X be a set of integer decision variables that take values in some universe $\mathcal{U}$, where $\mathcal{U}$ can be $\mathbb{Z}$ but is in practice a subset thereof. In a finite-domain (FD) solver, a *store* is a mapping $S\colon X \to \mathcal{P}(\mathcal{U})$, where $\mathcal{P}(\mathcal{U})$ is the power set of $\mathcal{U}$. For a variable $x \in X$, the set $S(x)$ is called the *domain* of x and is the set of possible values of x. A store S is an *assignment* if every variable has a singleton domain; we say that such variables are *ground*. A store S is *failed* if some variable has an empty domain. A store S is *stronger* than a store T (denoted by $S \sqsubseteq T$) if the domain under S of each variable is a subset of its domain under T.

A *constraint* $C(Y)$ on a sequence Y over X is a restriction on the possible values that these variables can take at the same time. The constraint $C(Y)$ is a subset of $\mathcal{U}^n$, where n is the length of Y. An assignment A *satisfies* a constraint $C(Y)$ if the sequence $[v \mid \{v\} = A(y) \wedge y \in Y]$ is a member of $C(Y)$. Given a store S, a value $v \in S(y)$ for some $y \in Y$ is *consistent* with a constraint C if there exists an assignment $A \sqsubseteq S$ with $A(y) = \{v\}$ that satisfies C. A constraint C is *satisfiable* in a store S if there exists an assignment $A \sqsubseteq S$ that satisfies C. A constraint C is *entailed* in a store S if all assignments $A \sqsubseteq S$ satisfy C.

A *checker* for a constraint C is a function that tells if an assignment satisfies C or not. A *propagator* for a constraint C is a function from stores to stores whose role is to remove domain values that are not consistent (or *inconsistent*) with C. We are here interested in writing propagators. In actual solvers, the propagators are not returning a store but modifying the current store. There are a number of desirable properties of a propagator:

- *Correct*: A correct propagator never removes values that are consistent with respect to its constraint. This property is mandatory for all propagators.

- *Checking*: A checking propagator decides if an assignment satisfies its constraint. This divides into *singleton correctness* (accept all satisfying assignments) and *singleton completeness* (reject all non-satisfying assignments).
- *Contracting*: A propagator P is contracting if $P(S) \sqsubseteq S$ for all stores S.
- *Monotonic*: A propagator P is monotonic if $S \sqsubseteq T$ implies $P(S) \sqsubseteq P(T)$ for all stores S and T.
- *Domain consistent* (DC): A DC propagator removes all inconsistent values. Weaker consistencies exist, such as bounds consistency and value consistency.
- *Idempotent*: A propagator P is idempotent if $P(S) = P(P(S))$ for all stores S. This allows an improved scheduling by the solver [22].

In addition, and for efficiency reasons, one aims at propagators with a low time complexity. This is the reason why domain consistency is often replaced by weaker consistencies. Another concern is to avoid executing a propagator that cannot remove any value from the current store. Several mechanisms may be implemented by propagators to avoid such executions: report idempotency, report entailment, subscribe only to relevant modification events of the domains.

Indexicals were introduced in [27] to describe propagators for the cc(FD) solver. An *indexical (expression)* is of the form x **in** σ, meaning that the domain of the decision variable x must be restricted to the intersection of its current domain with the set σ; the set σ may depend on other variables, and if so is computed based on their domains in the *current* store. An indexical is executed whenever the domain of one of the variables appearing in σ is modified. A propagator is typically described by several indexicals, namely one for each decision variable of the constraint. Indexicals have been included in several other systems, featuring extensions such as checking indexicals [6], conditional expressions [23], [6], and guards [28].

3   Examples of Propagator Descriptions

In this paper, we extend the syntax of indexicals to deal with arrays of decision variables and operations on such arrays (iteration and n-ary operators). We first present a few examples of propagator descriptions that showcase the main features of our language. This will lead us to explain the decisions we made in the design of the language, and a more precise definition of the language.

Figure 1 presents the $\textsc{Sum}(X, N)$ constraint, which holds if the sum of the values in the array X is equal to N. It is possible to describe several propagators for a constraint, and to write an optional checker. In particular, we have here two propagators: `v1` uses the entire domains of the variables (e.g., `dom(N)`), while `v2` only uses their bounds (e.g., `min(N)`). A propagator description is comprised of a set of indexicals. The domains of variables can be accessed using the four functions `dom`, `min`, `max`, and `val`. Arithmetic operators can be applied on integers as well as on sets. The `sum` operator is n-ary, as it operates on a sequence of values of arbitrary length. The `rng` operator denotes the range of indices of an array, and $\ell\,.\,.\,u$ the range of integers from ℓ to u included. Line 9 reads as: "The domain of N must be intersected with the range whose lower bound is the sum of the

```
1   def SUM(vint[] X,vint N){
2     propagator(v1){
3       N in sum(i in rng(X))(dom(X[i]));
4       forall(i in rng(X)){
5         X[i] in dom(N) - sum(j in {k in rng(X):k!=i})(dom(X[j]));
6       }
7     }
8     propagator(v2){
9       N in sum(i in rng(X))(min(X[i])) .. sum(i in rng(X))(max(X[i]));
10      forall(i in rng(X)){
11        X[i] in min(N) - sum(j in {k in rng(X):k!=i})(max(X[j])) ..
12                  max(N) - sum(j in {k in rng(X):k!=i})(min(X[j]));
13      }
14    }
15    checker{ val(N) = sum(i in rng(X))(val(X[i])) }
16  }
```

Fig. 1. Code for the SUM constraint, with two propagators

smallest values in the domains of all the variables in X, and whose upper bound is the sum of the largest values in the domains of all the variables in X." This example also shows how to write loops (`forall` in lines 4–6 and 10–13).

Figure 2 presents the EXACTLY(X, N, v) constraint, which holds if exactly N variables of the array X are equal to the given value v. This example illustrates the use of conditions (`when`), boolean-to-integer conversion (b2i($false$) $= 0$ and b2i($true$) $= 1$), and reference to other constraints (`EQ` and `NEQ`, constraining a variable to be respectively equal to, and different from, a given value). The functions `entailed` and `satisfiable` check the status of a constraint given the current domains of the variables, while `post` invokes a propagator of the given constraint. Lines 3–4 restrict the domain of N to be between two bounds. The lower bound is computed as the number of variables in X that *must* be assigned to v, and the upper bound as the number of variables that *may* be assigned to v. The body of the loop removes v from the domain of a variable (lines 6–9), or fixes a variable to v (lines 10–13), when some conditions involving the other variables are respected. The domain modifications are performed by invoking the propagation of other constraints.

4   Language Design Decisions

The language, as showcased in the previous section and defined more precisely in the next section, has been designed according to the following decisions.

The language is based on indexicals. Indexicals have already been used successfully in several solvers, and a fair amount of work has been done to deal with their use and properties, e.g. in [5] and [9]. Also, indexicals are very simple to understand and are often very close to the first reasoning one might come up

```
 1   def EXACTLY(vint[] X, vint N, int v){
 2     propagator{
 3       N in sum(i in rng(X))(b2i(entailed(EQ(X[i], v)))) ..
 4                   sum(i in rng(X))(b2i(satisfiable(EQ(X[i], v))));
 5       forall(i in rng(X)){
 6         once(val(N) <=
 7           sum(j in {j in rng(X):i!=j})(b2i(entailed(EQ(X[j], v))))){
 8             post(NEQ(X[i], v));
 9         }
10         once(val(N) >
11           sum(j in {j in rng(X):i!=j})(b2i(satisfiable(EQ(X[j], v))))){
12             post(EQ(X[i], v));
13         }
14       }
15     }
16     checker{ val(N) = sum(i in rng(X))(b2i(val(X[i]) = v)) }
17   }
```

Fig. 2. Code for the EXACTLY constraint

with when thinking about a propagator. One of the restrictions that we currently preserve is that indexicals are *stateless*, hence we cannot describe advanced propagators, such as a DC propagator for the ALLDIFFERENT constraint [19]. This is quite a strong limitation, but a choice must be made between the simplicity of the language and the intricacy of the propagators. However, a large number of constraints have efficient enough stateless propagators.

The language is strongly typed, in order to simplify the understanding and compilation of propagators. This requires the addition of the b2i operator.

We introduce arrays and n-ary operators to deal effectively with global constraints. For example, the expression sum(i in rng(X))(val(X[i])) is parametrised by the looping index (i here), its domain (rng(X) here), and the index-dependent expression (val(X[i]) here) that must be aggregated (summed here).

We also introduce meta-constraints, constraint invocation, and local variables in order to help write concise propagators. See Section 5 for details.

For simplicity of use, we want only a few different operators and language constructs. This also allows us to have a relatively simple compilation procedure. However it is not impossible that some new constructs will be added in the future, but with care.

For generality (in the FD approach), we refrain from adding solver-specific hooks. In particular, our language only has four accessors (see Section 5) to the domain of a variable. The other communication channels with a solver are domain narrowing functions, which are provided by any FD solver, and a fail mechanism (which can be mimicked by emptying a domain).

Our language is also missing some constructs that are found in the implementation of constraints, such as entailment detection, watched literals [13], and fine-grained events [15]. Part of our future work will be dedicated to studying how these can be included without overcomplicating the language.

5   Definition of the Language

In Section 5.1, we define the syntax and semantics of our language. It is strongly typed and has five basic types: integers (`int`), booleans (`bool`), sets of integers (`set`), integer decision variables (`vint`), and constraints (`cstr`). This last type is discussed in Section 5.2. We support arrays of any basic type (but currently not arrays of arrays). Identifiers of (arrays of) decision variables start with an uppercase letter. Identifiers of constants denoting integers, booleans, sets, and arrays thereof start with a lowercase letter.

5.1   Syntax and Semantics

Figure 3 presents the grammar of our language. We now review the different production rules. The main rule (**CSTR**) defines a constraint. A *constraint* is defined by its name and list of arguments. A constraint definition also contains the description of one or more propagators and an optional checker.

A *propagator* has an optional identifier and contains a list of instructions. An instruction (**INSTR**) can be an *indexical, x in σ*, whose meaning is that the domain of the decision variable x must be restricted to the intersection of its current domain with the set σ. Other instructions are `fail` and `post`. The effect of `fail` is to transform the current store into a failed store. The instruction $\mathtt{post}(C, P)$ invokes the propagator P of constraint C; if P is not specified, then the first (or only) propagator of C is invoked. There are two control structures. The `forall` control structure creates an iteration over a set, and `once` creates a conditional block; the reason for not naming the latter `if` is to stress that propagators should be monotonic, and that once the condition becomes true, it should remain true. See Section 6 for a discussion on monotonicity.

Most of the rules on sets, integers, and booleans do not need any explanations or were already explained in Section 3. Some constants are defined: `univ` denotes the universe $\mathcal{U}$, `inf` its infimum, `sup` its supremum, and `emptyset` the empty set. Arithmetic operations on integers are lifted as point-wise operations to sets.

There are four accessors to the domain of a decision variable: $\mathtt{dom}(x)$, $\mathtt{min}(x)$, $\mathtt{max}(x)$, and $\mathtt{val}(x)$ denote respectively the domain of decision variable x, its minimum value, its maximum value, and its unique value. As $\mathtt{val}(x)$ is only determined when the decision variable x is ground, the compiler must add guards to ensure a correct treatment when x is not ground.

While the instruction `post` invokes the propagator of another constraint, the functions `entailed`, `satisfiable`, and `check` query the status of another constraint. Let S be the current store: $\mathtt{entailed}(c)$ and $\mathtt{satisfiable}(c)$ decide whether the constraint c is entailed (respectively, satisfiable) in S; if S is an assignment, then the function $\mathtt{check}(c)$ can be called and decides whether S satisfies the constraint c (an example will be given in the next sub-section).

```
CSTR     ::= def CNAME(ARGS){ PROPAG+ CHECKER?)}
PROPAG   ::= propagator(PNAME?){ INSTR* }
CHECKER  ::= checker{ BOOL }
INSTR    ::= VAR in SET ; | post(CINVOKE,PNAME?); | fail; |
             once(BOOL){ INSTR* } | forall(ID in SET){ INSTR* }
SET      ::= univ | emptyset | ID | INT..INT | rng(ID) | dom(VAR) |
             NSETOP(ID in SET)(SET) | -SET | SET BSETOP SET |
             {INT+} | {ID in SET:BOOL}
INT      ::= inf | sup | NUM | ID | card(SET)| min(SET) | max(SET) |
             min(VAR) | max(VAR) | val(VAR) | - INT | INT BINTOP INT |
             b2i(BOOL) | NINTOP(ID in SET)(INT)
BOOL     ::= true | false | ID | INT INTCOMP INT | INT memberOf SET |
             SET SETCOMP SET | not BOOL | BOOL BBOOLOP BOOL |
             NBOOLOP(ID in SET)(BOOL) |
             entailed(CINVOKE) | satisfiable(CINVOKE) | check(CINVOKE)
BINTOP   ::= + | - | * | / | mod
NINTOP   ::= sum | min | max
BSETOP   ::= union | inter | minus | + | - | * | / | mod
NSETOP   ::= union | inter | sum
INTCOMP  ::= == | != | <= | < | >= | >
SETCOMP  ::= == | subseteq
BBOOLOP  ::= and | or | =
NBOOLOP  ::= and | or
CINVOKE  ::= CNAME | CNAME(ARGS)
```

Fig. 3. BNF-like grammar of our language. Constructions in grey were already in previous definitions of indexicals [27], [6]. The rules corresponding to `ARGS` (list of arguments), `CNAME`, `PNAME`, `ID`, `VAR` (respectively identifier of a constraint, propagator, constant, and variable), and `NUM` (integer literal) are not shown.

5.2 Meta-constraints

A new feature of our language is what we call a meta-constraint, which is a constraint that takes other constraint(s) as argument(s). Meta-constraints allow one to write more concise propagators by encapsulating common functionalities.

For example, the $\textsc{Among}(X, N, s)$ constraint, which holds if there are N elements in array X that take a value in set s, would be described almost identically to the $\textsc{Exactly}(X, N, v)$ constraint in Figure 2. The common code can be factored out in the meta-constraint `COUNT(vint[] X, vint N, cstr C, cstr NC)`, whose full description is not shown here, but whose meaning is defined by its checker: `val(N) = sum(i in rng(X))(b2i(check(C(X[i]))))`, that is exactly N variables of the array `X` satisfy constraint `C`. The argument `NC` is the negation of constraint `C` (see [3] for how to negate even global constraints), and is used in the propagator of $\textsc{Count}$ (in the way `NEQ` is used on line 8 of Figure 2). We can then describe $\textsc{Exactly}$ and $\textsc{Among}$ as shown in Figure 4. The $\textsc{Count}$ meta-constraint is closely related to the cardinality operator [25] but we allow the user to describe more meta-constraints.

```
def EXACTLY(vint[] X, vint N, int v){   def AMONG(vint N, vint[] X, set s){
  propagator{                             propagator{
    cstr EQv(vint V):= EQ(V,v);             cstr INs(vint V):= INSET(V,s);
    cstr N_EQv(vint V):= NEQ(V,v);          cstr NINs(vint V):= NOTINSET(V,s);
    post(COUNT(X,N,EQv,N_EQv));             post(COUNT(X,N,INs,NINs));
  }                                       }
}                                       }
```

Fig. 4. EXACTLY and AMONG, described using the COUNT meta-constraint

6   Syntactic Analysis and Tools

One of our objectives is to ease the proof of propagator properties. Before turning
to the actual compilation, we show how our language helps with this, and with
other propagator-writing related functionalities.

6.1   Analysis

Among the properties of a propagator presented in Section 2, most are difficult
to prove for a given propagator (except contraction, which indexicals satisfy
by definition). However, as has been shown in [5], it is possible to prove the
monotonicity of indexicals. This result can be lifted to our more general language.
We show further how to prove the correctness of some propagators with respect
to their constraints.

Monotonicity. The procedure to check the monotonicity of indexicals is combined with the addition of guards for **val** accesses. The syntax tree representing
the indexicals is traversed by a set of mutually recursive functions. To ensure
monotonicity of the whole propagator, each function verifies an *expected* behaviour of the subtree it is applied on. The recursive functions are labelled
monotonic, anti-monotonic, and *fixed* for boolean expressions; *increasing, decreasing,* and *fixed* for integer expressions; *growing, shrinking,* and *fixed* for set
expressions; and *monotonic* for instructions. For instance, the *increasing* function verifies that its integer expression argument is non-strictly increasing when
going from a store to a stronger store. These functions return two values: whether
the expression is actually respecting its expected behaviour, and the set of variables that need to be ground to ensure a safe use of the **val** accessor. This set of
variables is used to add guards in the generated propagator. Those guards are
added not only to instructions, but also inside the body of **b2i** expressions.

For lack of space, we cannot exhibit all the rules that make up the recursive
functions (there are about 200 rules). Instead we show a few examples. As an
example of a rule, consider the call of the function *increasing* on an expression
of the form $\min(i \text{ in } \sigma)(e)$. For this expression to be increasing, σ must be
shrinking and e must be *increasing*. In addition, the set of variables to guard
is the union of the variables that must be ground for those two subexpressions.

The table below shows some examples of the results of the procedure. In this table, $\texttt{ground}(x)$ represents an operator (not part of our indexical language) that decides if variable x is ground. The second column shows where guards are added to indexicals. The third column reports if the indexical is proven monotonic or not.

Original expression	Guarded expression	Mono
`X in {val(Y)}`	`once(ground(Y)) X in {val(Y)}`	true
`B in b2i(val(X)=v) ..` ` b2i(v memberOf dom(X))`	`B in b2i(ground(X) and val(X)=v) ..` `                b2i(v memberOf dom(X))`	true
`once(min(B)=1) X in {v}`	`once(min(B)=1) X in {v}`	false
`once(min(B)>=1)X in {v}`	`once(min(B)>=1) X in {v}`	true
`once(val(B)=1) X in {v}`	`once(ground(B) and val(B)=1)X in {v}`	true

The first line just adds a guard to the indexical. The second line shows that a guard can be added inside a `b2i` expression so that it returns 0 while `X` is not ground. The three last lines show how small variations change the monotonicity of an expression. The condition `min(B)=1` is (syntactically) not monotonic, because in general this condition might be true in some store and become false in a stronger store; however if we know that `B` represents a boolean (with domain `0..1`), then we can replace the equality by an inequality as done in the fourth example. The last line shows another way to get monotonicity, namely by replacing the `min` accessor by `val`; this requires the addition of a guard.

The soundness of the monotonicity checking procedure can be shown by induction on the recursive rules, as suggested in [5]. However, as this procedure is syntactical, it is incomplete. An example of propagator for $\text{EQ}(X, Y)$ that is monotonic but not recognised as such is given on the left of Figure 5. The procedure does not recognise the monotonicity because it requires the sets over which the `forall` loops iterate to be growing (because the indexical expressions that are applied on a store must also be applied on a stronger store), while $\texttt{dom}(x)$ can only shrink. However, this is not the simplest way to describe propagation for this constraint: a 2-line propagator can be found on the upper right of Figure 5.

Correctness. To prove algorithmically that a propagator is correct with respect to its constraint, we use the known fact that a propagator is correct if it is singleton-correct and monotonic. We devise an incomplete but sound procedure to prove that a propagator P is singleton-correct with respect to its checker C, and hence with respect to its constraint (assuming the checker is correct). We need to prove that if an assignment satisfies C, then it is not ruled out by P. To this end, from the indexical description of P, we derive a formula $C(P)$ that defines which assignments are accepted by the propagator. Singleton correctness then holds if the formula $C \wedge \neg C(P)$ is unsatisfiable (i.e., if $C \Rightarrow C(P)$). To derive $C(P)$, we transform the indexicals into an equivalent checking formula using the following rewrite rules:

```
1   def EQ(vint X1, vint X2){
2     propagator{
3       forall(i in dom(X1)){
4         once(not i memberOf dom(X2)){
5           X1 in univ minus {i};
6         }
7       }
8       forall(i in dom(X2)){
9         once(not i memberOf dom(X1)){
10          X2 in univ minus {i};
11        }
12      }
13    }
14  }
```

```
1   def EQ(vint X, vint Y){
2     propagator{
3       X in dom(Y);
4       Y in dom(X);
5     }
6     checker{ val(X) = val(Y) }
7   }
```

```
1   def EQ(vint X, int cY){
2     propagator{
3       X in {cY};
4       once(not cY
5           memberOf dom(X))
6         fail;
7     }
8     checker{ val(X) = cY }
9   }
```

Fig. 5. Variations of the EQ constraint

```
dom(x) → {val(x)}          x in σ → val(x) memberOf σ
min(x) → val(x)            once(b){y} → (not b) or y
max(x) → val(x)            forall(i in σ){y} → and(i in σ)(y)
fail → false
```

Our current procedure to prove the unsatisfiability of $C \wedge \neg C(P)$ tries to simplify the formula to `false` using rewrite rules. We implemented around 240 rewrite rules, ranging from boolean simplification (e.g., `false or` $b \to b$) to integer and set simplification (e.g., `min(`i `in` $\ell\,..\,u$`)(`i`)` $\to \ell$) and partial evaluation (e.g., $2 + x + 3 \to x + 5$). As this procedure is incomplete and able to prove singleton correctness only for a small portion of the propagators (see Section 7.2), we plan to improve it by calling an external prover.

As an example, applying the propagator-to-checker transformation on the propagator on the upper right of Figure 5 results in the formula `val(X) memberOf {val(Y)} and val(Y) memberOf {val(X)}`. This formula can be shown equivalent to the checker of the constraint (using the following rules: x `memberOf` $\{y\}$ $\to x = y$, b `and` $b \to b$, and b `and not` $b \to$ `false`). In summary, this propagator can be automatically proven monotonic, singleton correct, singleton complete (see below), and therefore correct and checking.

Checking. The approach to proving correctness can also be used to prove that a propagator is checking. Indeed, singleton completeness is shown by proving the implication $C(P) \Rightarrow C$ (the converse of singleton correctness), and a propagator that is singleton-correct and singleton-complete is checking (i.e., $C(P) \Leftrightarrow C$).

6.2   Transformation

In addition to the analysis, we can algorithmically transform a propagator. We have devised two first transformations that seem of interest: changing the level of reasoning, and grounding some decision variables.

Changing the level of reasoning. As shown in the SUM example (Figure 1), a propagator can be described to use different levels of reasoning according to the amount of data it uses:

- Under domain reasoning, the whole domains of decision variables may be used to perform propagation.
- Under bounds reasoning, only the bounds of the domains are used (i.e., the `dom` accessor does not appear).
- Under value reasoning, no propagation is performed until some variables are ground (i.e., only the `val` accessor is used).

A few remarks are necessary here. First, the level of reasoning can be distinct for the different variables of a constraint. Second, those levels of reasoning are not directly linked to the usual notions of consistency (domain, bounds, or value consistency). Indeed, using the `dom` accessor does not provide any guarantee of domain consistency. Conversely some propagators that do not use `dom` may achieve domain consistency. Third, the level of reasoning is also distinct from the level of narrowing of variables, which is how propagation affects the domain of variables, i.e., if it only updates the bounds, or if it can create holes.

It is possible to change the level of reasoning from a strong level to a weaker one, i.e., from domain reasoning to bounds reasoning, and from bounds reasoning to value reasoning. All one has to do is to replace the appearances of $\mathtt{dom}(x)$ by $\mathtt{min}(x)\,.\,.\,\mathtt{max}(x)$, and of $\mathtt{min}(x)$ and $\mathtt{max}(x)$ by $\mathtt{val}(x)$. This allows one to describe one propagator, and have for free up to three different implementations that one may try and compare.

Variable grounding. Another propagator transformation is the grounding of some variables, that is the replacement of a variable argument by a constant (or of an array of variables by an array of constants). Again, this allows one to describe only one propagator for the general case and then specialise it to specific cases. For instance, one might want to derive a propagator for the `EQ(vint,int)` constraint from the one of `EQ(vint,vint)`. The transformation is close in spirit to the computation of the equivalent checking formula of a propagator presented in Section 6.1. The transformation is however only applied to one variable (the one being grounded). The most interesting rewrite rule is that if variable x is replaced by a constant c, then x in σ is replaced by `once(not` c `memberOf` σ`)` `fail`. Most of the time this check is redundant with the other instructions of the propagator (as in the example on the lower right of Figure 5, where lines 4–6 check a condition enforced by line 3). However, we have not found yet a general and cheap way to tell when this instruction is indeed redundant. Currently, it is the responsibility of the user to remove it if he wants to.

7   Compilation

We now discuss our compiler design decisions and our current compiler.

7.1   Compiler Design Decisions

Instead of interpretation, we made the decision to compile our language into the language in which propagators are written for a particular solver. This has the double advantage of having an infrastructure that is relatively independent of the solvers (only the code generation part is solver-dependent), and of having compiled code that is more efficient than interpreted code. The generated code is also self-contained (it can be distributed without the compiler).

The compiled propagators are currently stateless (as are indexical propagators) and use coarse-grained wake-up events. This choice is meant to simplify the compilation. However, upon a proper analysis, it should be possible to produce propagators that incorporate some state or use more fine-grained events.

The compilation produces one propagator for each `propagator` description. The compilation does not alter the order of the indexicals inside a propagator. Furthermore, to get idempotency, the full propagator is repeated until it reaches its internal fixpoint. Another valid choice would have been to create a propagator for each indexical, and let the solver perform the scheduling. We have not evaluated all the potential trade-offs of this choice. An intermediate approach would be to analyse the internal structure of the propagator description to generate a good scheduling policy of the indexicals inside the propagator. This requires substantially more work and is left as future work.

The invocations of propagators (`post`) or checkers (`check`) are replaced by the corresponding code. This is similar to function inlining in classical programming languages. For the functions `entailed` and `satisfiable`, the description of the corresponding checker is first transformed in order to deal with stores that are not assignments. This is done by about 130 rewrite rules forming a recursive procedure similar to the one for monotonicity checking. As a difference, the `val` accessors are replaced by `min`, `max`, or `dom` when possible, or are properly guarded otherwise. For example, calling the *shrinking* function on the singleton $\{$`val`$(x)\}$ returns `dom`(x), but calling the *growing* function adds a guard instead. Entailment requires a *monotonic* boolean formula, and satisfiability an *antimonotonic* formula. For example, the generated satisfiability checker of `EQ(X,Y)` is `not dom(X) inter dom(Y) = emptyset`. In turn, the generated entailment checker of `EQ(X,Y)` is `ground(X) and ground(Y) and val(X)=val(Y)`.

The choice of inlining constraint invocations greatly simplifies the compilation process. However, this means that all referenced constraints must be described by indexicals. We plan to explore how we can remove this limitation in order to be able to invoke propagators built into the targeted solver.

7.2   Implementation and Target Solvers

We have written a prototype compiler in Java. It uses Antlr [18] for the parsing and StringTemplate [17] for the code generation. Currently, we compile into propagators for Comet [10], Gecode [12], and Scampi [21]. A big part of the compilation process amounts to rewriting the n-ary operators as loops. Some optimisations are performed, such as a dynamic programming pre-computation of arrays (replacing nested loops by successive loops) [24, Section 9] and the factorisation of repeated expressions. The compiler detects the events that should wake up the propagator; this is performed by walking the syntax tree and gathering the variable accessors. The compiler also adds entailment detection to the propagator of constraint c, by testing if `entailed(c)` is true.

Currently, we have written about 700 lines of indexicals for describing 76 propagators of 48 constraints, of which 14 are meta-constraints, 17 are global constraints, and 17 are binary or ternary constraints. Out of the 76 propagators, 69 are proven monotonic, of which 16 are proven singleton-correct and 29 are proven singleton complete, making 16 propagators provably correct. These numbers could be improved with a better unsatisfiability proof procedure.

To get an idea of the conciseness of the language, note that our compiler produces from the 17-line description of EXACTLY in Figure 2 a propagator for Comet that is about 150 lines of code, and one for Gecode of about 170 lines. We estimate the code for the built-in propagator of this constraint to be around 150 lines of code in Gecode.

The current prototype, as well as the propagator descriptions, are available on demand from the first author.

8   Experimental Evaluation

To assess that propagators described by indexicals behave reasonably well, we compare a few generated propagators with built-in propagators of Gecode and simple constraint decompositions. We do not expect the generated propagators to be as efficient as the hand-crafted ones, but the goal is to show that they are a viable alternative when one has little time to develop a propagator for a constraint.

Our experimental setting is as follows. We use Gecode 3.7.3. For each constraint, we search for all its solutions. We repeat the search using several branching heuristics to try and exercise as many parts of the propagators as possible.

The studied constraints are SUM, MAXIMUM, EXACTLY, and ELEMENT. Their indexical descriptions are representative of the other constraints we implemented. In addition, they share the property that one of the variables is functionally dependent on the other ones. This allows us to compare the different propagators of a constraint with a dummy problem where the constraint is absent but the functionally dependent variable is instead fixed to an arbitrary value. As the considered constraint is only defining the functional dependency, the number of solutions is the same and the size of the search tree is the same, but the time spent by propagation is null. We can then compute the runtime of a propagator by subtracting the total runtimes.

For SUM and MAXIMUM, we use bounds-reasoning versions of the indexicals and built-in propagators; for EXACTLY and ELEMENT, we use domain-reasoning

Table 1. Relative runtimes (in percent)

	Maximum	Sum	Exactly	Element
Built-in	100	100	100	100
Indexicals	125	269	252	118
Decomposition	195	296	313	204
Automaton	675	n/a	n/a	487

versions. The indexical descriptions of Sum and Exactly are shown in Figures 1 (v2) and 2 respectively. For space reasons, Maximum and Element are not shown. The decompositions of $\text{Sum}(X, N)$ and $\text{Maximum}(X, N)$ introduce an array A of $n = \|X\|$ auxiliary variables. The decomposition of Sum is expressed as $A[1] = X[1] \wedge \forall_{i \in 2..n} (A[i-1] + X[i] = A[i]) \wedge A[n] = N$, and the one of Maximum is $A[1] = X[1] \wedge \forall_{i \in 2..n} (\max(A[i-1], X[i]) = A[i]) \wedge A[n] = N$. The decomposition of $\text{Exactly}(X, N, v)$ introduces an array B of boolean variables and is defined as $\forall_{i \in 1..n} (B[i] \equiv X[i] = v) \wedge N = \sum_{i \in 1..n} B[i]$; the sum of boolean variables is implemented by a built-in propagator. The $\text{Element}(X, Y, Z)$ constraint (holding if $X[Y] = Z$) is decomposed into $Y \in 1..n \wedge \forall_{i \in 1..n} (Y = i \Rightarrow X[i] = Z)$. Additionally, we use the automaton formulations of Maximum and Element, given in the Global Constraint Catalogue [4]. The automata of Maximum and Element and the decomposition of Element do not perform all the possible pruning (while the other decompositions do so, at least under the used heuristics). This incurs an overhead of about 15% more nodes visited for the automata, and 7% for the decomposition of Element.

Table 1 presents the relative runtimes of the different implementations of the constraints for arrays of 9 variables over domains of 9 values. The used search heuristics are some combinations of the variable ordering (order of the arguments of the constraints, and within an array the smallest or the largest domain first) and of the value ordering (assign the minimum, split in two, assign the median). For each constraint and each propagator, the runtimes are summed over the different search heuristics. Then the sum of the times to explore the search tree is subtracted. Finally, for each constraint, the sum of the times of each propagator is divided by the sum of the times of the Gecode built-in propagator. Compared to the built-in propagators, the generated propagators induce only a small overhead for Maximum and Element, but do not behave so well for Sum and Exactly. However, in all cases, the indexicals have a better runtime than the decomposition, though sometimes only slightly. In particular, for the Exactly constraint, the decomposition has a better runtime for some smaller instances (not shown). We explain the behaviour of the indexicals on this constraint by the fact that it is awakened each time the domain of a variable changes, even though some variables might not affect the constraint status anymore. The built-in propagator and the decomposition are more clever and ignore the variables that cannot take the given value anymore. The indexicals have a much better runtime than the automata.

The propagators compiled from indexicals have an average runtime per call that (necessarily) increases linearly with the number of variables. The runtimes of the built-in propagators increase also but with a much gentler slope.

These experiments show that indexical descriptions of global constraints are useful, but that there is still room for improvement in the compilation.

9 Conclusion

We have presented a solver-independent language to describe propagators. The aim is to ease the writing and sharing of propagators and to make proving their formal properties much easier. The resulting language, based on indexicals, is high-level enough to abstract away implementation details and to allow some analyses and transformations. It is compiled into source code for target solvers.

The idea of letting the user write his own propagators is not new. The system cc(FD) [27] is one of the first to have proposed this, through the use of indexicals. Since then, most CP solvers claim to be open, in the sense that any user can add new propagators (especially for global constraints) to the kernel of built-in propagators. In such systems, the user writes code in the host programming language of the solver and integrates it with the solver through an interface defining mainly how to interact with the variables and the core of the solver. Some solvers take a quite different approach and propose a language to define propagation, examples include constraint handling rules [11] and action rules [28]. Our language is a level of abstraction above those approaches, as it can be translated into code for *any* solver. From this point of view, it is close in spirit to what a solver-independent modelling language is to solvers: a layer above solvers to describe easily problem models, which are then compiled into the solver input language. In our case, propagators are described, not models.

Propagators have also been described using atomic constraints and propagation rules [8]. However, that description language has been devised to reason about propagators, and it is not practical for actually implementing propagators in FD solvers, except for approaches based on SAT solvers that use such low-level constraints (e.g., lazy-clause generation [16]).

The pluggable constraints of [20] also decouple the implementation of constraints from the solver architecture, using a solver-independent interface for the communication between the two components. Our approach aims at a higher level of abstraction, possibly losing some fine control.

This paper opens several interesting research directions, in addition to those already listed in Sections 4 and 7.1 for overcoming initial design decisions: the language needs to allow better control of the propagation algorithm while staying simple and general. Simplicity is important, as we believe that having *automated* propagator analysis tools eases the writing of propagators.

Future compilation targets are other FD solvers (e.g., Choco [7] and Ja-CoP [14]), lazy-clause generators in SAT [16], cutting plane generators in MIP [1], and penalty functions or invariants in constraint-based local search [26], [2].

References

1. Achterberg, T.: SCIP: Solving constraint integer programs. Mathematical Programming Computation 1, 1–41 (2009)
2. Ågren, M., Flener, P., Pearson, J.: Inferring Variable Conflicts for Local Search. In: Benhamou, F. (ed.) CP 2006. LNCS, vol. 4204, pp. 665–669. Springer, Heidelberg (2006)
3. Beldiceanu, N., Carlsson, M., Flener, P., Pearson, J.: On the reification of global constraints. Tech. Rep. T2012:02, Swedish Institute of Computer Science (February 2012), `http://soda.swedish-ict.se/view/sicsreport/`
4. Beldiceanu, N., Carlsson, M., Rampon, J.X.: Global constraint catalog, 2nd edn. (revision a). Tech. Rep. T2012:03, Swedish Institute of Computer Science (February 2012), `http://soda.swedish-ict.se/view/sicsreport/`
5. Carlson, B., Carlsson, M., Diaz, D.: Entailment of finite domain constraints. In: Proceedings of ICLP 1994, pp. 339–353. MIT Press (1994)
6. Carlsson, M., Ottosson, G., Carlson, B.: An Open-Ended Finite Domain Constraint Solver. In: Glaser, H., Hartel, P., Kuchen, H. (eds.) PLILP 1997. LNCS, vol. 1292, pp. 191–206. Springer, Heidelberg (1997)
7. CHOCO: An open source Java CP library, `http://www.emn.fr/z-info/choco-solver/`
8. Choi, C.W., Lee, J.H.M., Stuckey, P.J.: Removing propagation redundant constraints in redundant modeling. ACM Transactions on Computational Logic 8(4) (2007)
9. Dao, T.B.H., Lallouet, A., Legtchenko, A., Martin, L.: Indexical-Based Solver Learning. In: Van Hentenryck, P. (ed.) CP 2002. LNCS, vol. 2470, pp. 541–555. Springer, Heidelberg (2002)
10. Dynadec, Dynamic Decision Technologies Inc.: Comet tutorial, v2.0 (2009), `http://dynadec.com/`
11. Frühwirth, T.W.: Theory and practice of constraint handling rules. Journal of Logic Programming 37(1-3), 95–138 (1998)
12. Gecode Team: Gecode: A generic constraint development environment (2006), `http://www.gecode.org/`
13. Gent, I.P., Jefferson, C., Miguel, I.: Watched Literals for Constraint Propagation in Minion. In: Benhamou, F. (ed.) CP 2006. LNCS, vol. 4204, pp. 182–197. Springer, Heidelberg (2006)
14. JaCoP: Java constraint programming solver, `http://jacop.osolpro.com/`
15. Mohr, R., Henderson, T.: Arc and path consistency revisited. Artificial Intelligence 28, 225–233 (1986)
16. Ohrimenko, O., Stuckey, P., Codish, M.: Propagation via lazy clause generation. Constraints 14, 357–391 (2009)
17. Parr, T.J.: Enforcing strict model-view separation in template engines. In: Proceedings of the 13th International Conference on the World Wide Web, pp. 224–233. ACM (2004)
18. Parr, T.J.: The Definitive ANTLR Reference: Building Domain-Specific Languages. The Pragmatic Bookshelf (2007)
19. Régin, J.C.: A filtering algorithm for constraints of difference in CSPs. In: Hayes-Roth, B., Korf, R.E. (eds.) Proceedings of AAAI 1994, pp. 362–367. AAAI Press (1994)
20. Richaud, G., Lorca, X., Jussien, N.: A portable and efficient implementation of global constraints: The tree constraint case. In: Proceedings of CICLOPS 2007, pp. 44–56 (2007)

21. Scampi (2011), `https://bitbucket.org/pschaus/scampi/`
22. Schulte, C., Stuckey, P.J.: Speeding Up Constraint Propagation. In: Wallace, M. (ed.) CP 2004. LNCS, vol. 3258, pp. 619–633. Springer, Heidelberg (2004)
23. Sidebottom, G., Havens, W.S.: Nicolog: A simple yet powerful cc(FD) language. Journal of Automated Reasoning 17, 371–403 (1996)
24. Tack, G., Schulte, C., Smolka, G.: Generating Propagators for Finite Set Constraints. In: Benhamou, F. (ed.) CP 2006. LNCS, vol. 4204, pp. 575–589. Springer, Heidelberg (2006)
25. Van Hentenryck, P., Deville, Y.: The cardinality operator: A new logical connective for constraint logic programming. In: Proceedings of ICLP 1991, pp. 745–759 (1991)
26. Van Hentenryck, P., Michel, L.: Differentiable Invariants. In: Benhamou, F. (ed.) CP 2006. LNCS, vol. 4204, pp. 604–619. Springer, Heidelberg (2006)
27. Van Hentenryck, P., Saraswat, V., Deville, Y.: Design, implementation, and evaluation of the constraint language cc(FD). Tech. Rep. CS-93-02, Brown University, Providence, USA (January 1993), revised version in Journal of Logic Programming 37(1-3), 293–316 (1998). Based on the unpublished manuscript Constraint Processing in cc(FD) (1991)
28. Zhou, N.F.: Programming finite-domain constraint propagators in action rules. Theory and Practice of Logic Programming 6, 483–507 (2006)

Interactive Algorithm for Multi-Objective Constraint Optimization

Tenda Okimoto[1], Yongjoon Joe[1], Atsushi Iwasaki[1], Toshihiro Matsui[2],
Katsutoshi Hirayama[3], and Makoto Yokoo[1]

[1] Kyushu University, Fukuoka 8190395, Japan
[2] Nagoya Insutitute of Technology, Nagoya 4668555, Japan
[3] Kobe University, Kobe 6580022, Japan
{tenda@agent.,yongjoon@agent.,iwasaki@,yokoo@}inf.kyushu-u.ac.jp,
matsui.t@nitech.ac.jp, hirayama@maritime.kobe-u.ac.jp

Abstract. Many real world problems involve multiple criteria that should be considered separately and optimized simultaneously. A Multi-Objective Constraint Optimization Problem (MO-COP) is the extension of a mono-objective Constraint Optimization Problem (COP). In a MO-COP, it is required to provide the most preferred solution for a user among many optimal solutions. In this paper, we develop a novel Interactive Algorithm for MO-COP (MO-IA). The characteristics of this algorithm are as follows: (i) it can guarantee to find a Pareto solution, (ii) it narrows a region, in which Pareto front may exist, gradually, (iii) it is based on a pseudo-tree, which is a widely used graph structure in COP algorithms, and (iv) the complexity of this algorithm is determined by the induced width of problem instances. In the evaluations, we use an existing model for representing a utility function, and show empirically the effectiveness of our algorithm. Furthermore, we propose an extension of MO-IA, which can provide the more detailed information for Pareto front.

1   Introduction

Many real world optimization problems involve multiple criteria that should be considered separately and optimized simultaneously. A Multi-Objective Constraint Optimization Problem (MO-COP) [7, 8, 9, 16] is the extension of a mono-objective Constraint Optimization Problem (COP) [4, 18]. A COP is a problem to find an assignment of values to variables so that the sum of the resulting rewards is maximized. A MO-COP is a COP involves multiple criteria. In a MO-COP, generally, since trade-offs exist among objectives, there does not exist an ideal assignment, which maximizes all objectives simultaneously. Therefore, we characterize the optimal solution of a MO-COP using the concept of Pareto optimality. Solving a MO-COP is to find the Pareto front. The Pareto Front is a set of reward vectors obtained by Pareto solutions. An assignment is a Pareto solution, if there does not exist another assignment that improves all of the criteria. A COP and a MO-COP can be represented using a constraint graph, in which a node represents a variable and an edge represents a constraint.

M. Milano (Ed.): CP 2012, LNCS 7514, pp. 561–576, 2012.

Various complete algorithms have been developed for solving a MO-COP, e.g., Russian Doll Search algorithm (MO-RDS) [17], Multi-objective AND/OR Branch-and-Bound search algorithm (MO-AOBB) [8], and MultiObjective Bucket Elimination (MO-BE) [16]. In a MO-COP, even if a constraint graph has the simplest tree structure, the size of the Pareto front, i.e., the number of Pareto solutions, is often exponential in the number of reward vectors. In such MO-COP problems, finding all Pareto solutions is not real. On the other hand, several incomplete algorithms have been developed for solving a MO-COP, e.g., Multi-Objective Mini-Bucket Elimination (MO-MBE) [16], Multi-objective Best-First AND/OR search algorithm (MO-AOBF) [9], and Multiobjective A* search algorithm (MOA*) [15]. MO-MBE computes a set of lower bounds of MO-COPs. MO-AOBF and MOA* compute a relaxed Pareto front using ϵ-dominance [14].

Various algorithms have been developed for solving a Multi-Objective Optimization Problem (MOOP) [1, 2, 3, 5, 11]. In a MOOP, a variable takes its value from a continuous domain, while a variable takes its value from a discrete domain in a MO-COP. In this paper, we focus on a MO-COP.

An Aggregate Objective Function (AOF) [12, 13] is the simplest and the most widely used classical method to find the Pareto solutions of a MOOP. This method scalarizes the set of objective functions into a weighted mono-objective function, and find an optimal solution. It is well known that an optimal solution obtained by AOF is a Pareto solution of the original MOOP problem [13]. If Pareto front is convex, AOF guarantees to find all Pareto solutions. Otherwise, it cannot find Pareto solutions in non-convex region. In our research, we use AOF to find the Pareto solutions of a MO-COP.

In this paper, we develop a novel Interactive Algorithm for MO-COPs (MO-IA). Our algorithm finds a set of Pareto solutions and narrows a region, in which Pareto front may exist, gradually. Our algorithm utilizes a graph structure called a pseudo-tree, which is widely used in COP algorithms. The complexity of our algorithm is determined by the induced width of problem instances. Induced width is a parameter that determines the complexity of many COP algorithms. We evaluate our algorithm using a Constraint Elasticity of Substitution (CES) utility function [10], which is widely used in many economic textbooks representing utility functions, and show empirically the effectiveness of our algorithm. Furthermore, we propose an extension of MO-IA, which finds several Pareto solutions so that we can provide a narrower region, in which Pareto front may exist, i.e., we can provide the more detailed information for Pareto front. As far as the authors aware, there exists virtually no work on interactive algorithms for a MO-COP, although various MO-COP algorithms have been developed [8, 9, 15, 16, 17].

Our proposed algorithm is similar to Physical Programming (PP) [11] and Directed Search Domain algorithm (DSD) [5]. However, these are interactive algorithms for MOOPs, while our algorithm is for MO-COPs. If we apply PP and DSD to MO-COPs, there is no guarantee to find a Pareto solution. On the other hand, our algorithm can always find a Pareto solution. Furthermore, compared to evolutionary algorithms [1, 3] for solving a MOOP, the advantage of our algorithm is that our algorithm guarantees to find a Pareto solution.

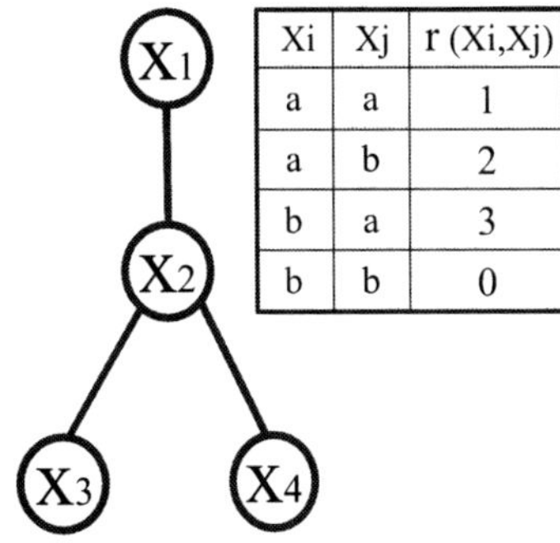
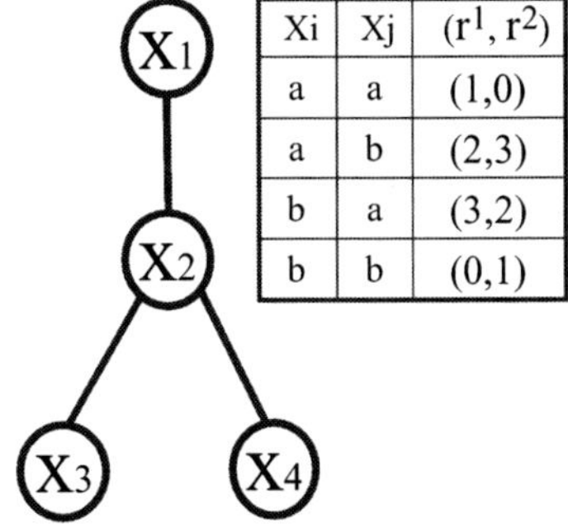

Fig. 1. A mono-objective COP with four variables. The optimal solution of this problem is $\{(x_1, a), (x_2, b), (x_3, a), (x_4, a)\}$ and the optimal value is eight.

Fig. 2. A bi-objective COP with four variables. The Pareto solutions of this problem are $\{\{(x_1, b), (x_2, a), (x_3, b), (x_4, b)\}, \{(x_1, a), (x_2, b), (x_3, a), (x_4, a)\}\}$, and the Pareto front is $\{(7, 8), (8, 7)\}$.

About application domains of MO-COP, we believe design/configuration tasks would be promising. For a simple toy example, we can consider Build to Order Custom Computers, where one can configure their own PC by choosing various options, e.g., CPU clock, memory size, hard drive size, operating system, and monitor size, considering multiple criteria, e.g., cost, performance, and required space.

The remainder of this paper is organized as follows. Section 2 provides some preliminaries on COPs, MO-COPs, AOF, and an existing model for representing a utility function. Section 3 introduces our interactive algorithm for MO-COPs, and Section 4 evaluates our algorithm using an existing model described in Section 2. Furthermore, we provides the extension of our algorithm. Section 5 concludes this paper and gives future works.

2   Preliminaries

In this section, we briefly describe the formalizations of Constraint Optimization Problems (COPs) and Multi-objective Constraint Optimization Problems (MO-COPs), which is the extension of a mono-objective COP. Also, we show an Aggregate Objective Function (AOF), which is the most widely used classical method to find a Pareto solution. Furthermore, we introduce a Constraint Elasticity of Substitution (CES) utility function and an indifference curve that are widely used in many economic textbooks representing utility functionsD

2.1   Mono-Objective Constraint Optimization Problem

A Constraint Optimization Problem (COP) [4, 18] is a problem to find an assignment of values to variables so that the sum of the resulting rewards is maximized.

A COP is defined by a set of variables X, a set of binary constraint relations C, and a set of binary reward functions F. A variable x_i takes its value from a finite, discrete domain D_i. A binary constraint relation (i, j) means there exists a constraint relation between x_i and x_j. For x_i and x_j, which have a constraint relation, the reward for an assignment $\{(x_i, d_i), (x_j, d_j)\}$ is defined by a binary reward function $r_{i,j} : D_i \times D_j \to \mathbb{R}$. For a value assignment to all variables A, let us denote

$$R(A) = \sum_{(i,j)\in C, \{(x_i,d_i),(x_j,d_j)\}\subseteq A} r_{i,j}(d_i, d_j). \tag{1}$$

Then, an optimal assignment A^* is given as $\arg\max_A R(A)$, i.e., A^* is an assignment that maximizes the sum of the value of all reward functions, and an optimal value is given by $R(A^*)$. A COP can be represented using a constraint graph, in which nodes correspond to variables and edges correspond to constraints.

A pseudo-tree is a special graph structure, which is widely used in COP algorithms. In a pseudo-tree, there exists a unique root node, and each non-root node has a parent node. The pseudo-tree contains all nodes and edges of the original constraint graph, and the edges are categorized into tree edges and back edges. There are no edges between different subtrees. For each node x_i, we denote the parent node, ancestors, and children of x_i as follows:

- p_i: the parent node, which is connected to x_i through a tree edge.
- PP_i: a set of the ancestors which are connected to x_i through back edges.
- C_i: a set of children which are connected to x_i through tree and back edges.

Example 1 (COP). Figure 1 shows a mono-objective COP with four variables x_1, x_2, x_3 and x_4. $r(x_i, x_j)$ is a binary reward function where $i < j$. Each variable takes its value assignment from a discrete domain $\{a, b\}$. The optimal solution of this problem is $\{(x_1, a), (x_2, b), (x_3, a), (x_4, a)\}$, and the optimal value is eight.

2.2 Multi-Objective Constraint Optimization Problem

A Multi-Objective Constraint Optimization Problem (MO-COP) [7, 8, 9, 16] is the extension of a mono-objective COP. A MO-COP is defined by variables $X = \{x_1, \ldots, x_n\}$, multi-objective constraints $C = \{C^1, \ldots, C^m\}$, i.e., a set of sets of binary constraint relations, and multi-objective functions $O = \{O^1, \ldots, O^m\}$, i.e., a set of sets of objective functions (binary reward functions). A variable x_i takes its value from a finite, discrete domain D_i. A binary constraint relation (i, j) means there exists a constraint relation between x_i and x_j. For an objective l $(1 \le l \le m)$, variables x_i and x_j, which have a constraint relation, the reward for an assignment $\{(x_i, d_i), (x_j, d_j)\}$ is defined by a binary reward function $r_{i,j}^l : D_i \times D_j \to \mathbb{R}$. For an objective l and a value assignment to all variables A, let us denote

$$R^l(A) = \sum_{(i,j)\in C^l, \{(x_i,d_i),(x_j,d_j)\}\subseteq A} r_{i,j}^l(d_i, d_j). \tag{2}$$

Then, the sum of the values of all reward functions for m objectives is defined by a reward vector, denoted $R(A) = (R^1(A), \ldots, R^m(A))$. To find an assignment that maximizes all objective functions simultaneously is ideal. However, in general, since trade-offs exist among objectives, there does not exist such an ideal assignment. Therefore, we characterize the optimal solution of a MO-COP using the concept of Pareto optimality.

Definition 1 (Dominance). *For a MO-COP and two reward vectors $R(A)$ and $R(A')$, we call that $R(A)$ dominates $R(A')$, denoted by $R(A') \prec R(A)$, iff $R(A')$ is partially less than $R(A)$, i.e., (i) it holds $R^l(A') \leq R^l(A)$ for all objectives l, and (ii) there exists at least one objective l, such that $R^l(A') < R^l(A)$.*

Definition 2 (Pareto solution). *For a MO-COP and an assignment A, we call that A is the Pareto solution, iff there does not exist another assignment A', such that $R(A) \prec R(A')$.*

Definition 3 (Pareto Front). *For a MO-COP, we call a set of reward vectors obtained by Pareto solutions as the Pareto front.*

Solving a MO-COP is to find the Pareto front. A MO-COP can be also represented using a constraint graph as a COP. In this paper, we assume that all reward values are non-negative.

Example 2 (MO-COP). Figure 2 shows a bi-objective COP, which is an extension of a mono-objective COP in Fig. 1. Each variable takes its value from a discrete domain $\{a, b\}$. The Pareto solutions of this problem are $\{\{(x_1, b), (x_2, a), (x_3, b), (x_4, b)\}, \{(x_1, a), (x_2, b), (x_3, a), (x_4, a)\}\}$, and the Pareto front is $\{(7, 8), (8, 7)\}$, which is a set of reward vectors obtained by these Pareto solutions.

2.3 Aggregate Objective Function

An Aggregate Objective Function (AOF) [12, 13] is the simplest and the most widely used classical method to find the Pareto solutions of a MOOP. This method scalarizes the set of objective functions into a weighted mono-objective function and find an optimal solution. For objective functions $o^1, \ldots, o^m$ of a MOOP, we define a weight denoted by $\alpha = (\alpha_1, \ldots, \alpha_m)$, $where \sum_{1 \leq i \leq m} \alpha_i = 1, \alpha_i > 0$. Next, we make a weighted mono-objective function $\alpha_1 o^1 + \ldots + \alpha_m o^m$, and find the optimal solution. Then, the following theorem holds:

Theorem 1 (AOF). *For a MOOP, an optimal solution A^* obtained by AOF is a Pareto solution of the original problem.*

It is well known that AOF can guarantee to find all Pareto solutions, if Pareto front is convex. Otherwise, it cannot find all Pareto solutions. In this paper, we use this method to find the Pareto solutions of a MO-COP. Theorem 1 holds also for MO-COPs. We omit the proof due to space limitations.

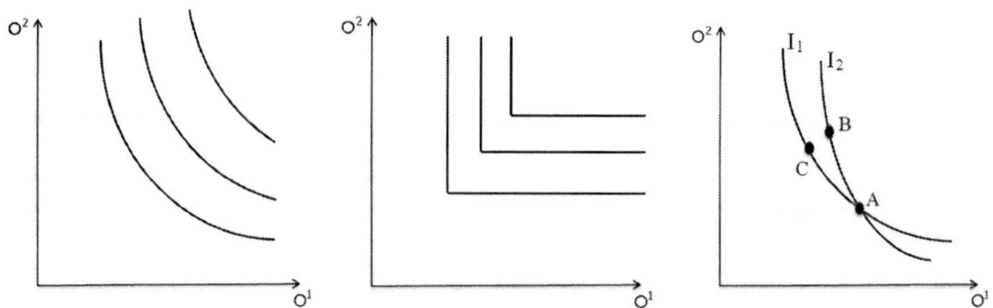

Fig. 3. Indifference curves of Cobb-Douglas function for a bi-objective COP

Fig. 4. Indifference curves of Leontief function for a bi-objective COP

Fig. 5. Indifference curves I_1 and I_2 intersect at point A

2.4 Constraint Elasticity of Substitution Utility Function

A Constraint Elasticity of Substitution (CES) utility function [10] is a function which is widely used in many economic textbooks representing utility functionsD A CES utility function has the form

$$u(x_1, \ldots, x_m) = (\alpha_1 x_1^p + \ldots + \alpha_m x_m^p)^{1/p}, \tag{3}$$

where $\sum_{1 \leq i \leq m} \alpha_i = 1$, $\alpha_i > 0$, $p < 1$. Linear, Cobb-Douglas and Leontief functions are special cases of the CES utility function. For example, as $p \to 1$, the CES utility function becomes a linear function

$$u(x_1, \ldots, x_m) = \alpha_1 x_1 + \ldots + \alpha_m x_m. \tag{4}$$

As $p \to 0$, the CES utility function becomes Cobb-Douglas function

$$u(x_1, \ldots, x_m) = x_1^{\alpha_1} \times \ldots \times x_m^{\alpha_m}. \tag{5}$$

As $p \to -\infty$, the CES utility function becomes Leontief function

$$u(x_1, \ldots, x_m) = min(x_1, \ldots, x_m). \tag{6}$$

2.5 Indifference Curves

The indifference curve [10, 19] shows the various combinations of goods that make a person equally satisfied. [1] For example, two different pairs, e.g., a pair of 10 compact discs and 150 candy bars, and a pair of 12 compact discs and 130 candy bars, are on the same indifference curve means that a person has a same utility, whichever pair he/she chooses.

Example 3 (Indifference curves). Figure 3 and 4 show indifference curves of Cobb-Douglas and Leontief functions for a bi-objective COP, respectively. On the graphs, o_1 represents quantity of goods, e.g., compact discs, while o_2 represents quantity of goods, e.g., candy bars. A person is equally satisfied at any point along a given curve, i.e., each point brings the same utility.

---

[1] For $m \geq 3$ goods (objectives), we can consider an indifference surface.

The following are the typical properties of indifference curves:

- Indifference curves are convex to the origin.
- Indifference curves cannot intersect each other.
- Higher indifference curves represents higher utility.

The first property is derived from the principle called diminishing marginal rate of substitution [19]. As a person substitutes good o_1 for good o_2, the marginal rate of substitution diminishes as o_1 for o_2 along an indifference curve. The slope of the curve is referred as the marginal rate of substitution. The marginal rate of substitution is the rate at which a person must sacrifice units of one good to obtain one more unit of another good.

We show the second property by contradiction. Assume that the indifference curves I_1 and I_2 intersect at point A (see Fig. 5). That would mean that a person is indifferent between A and all points on I_1. In particular, he/she would be indifferent between A and B, between A and C, and accordingly between B and C. However, since B involves higher values of both objective functions than C, B is clearly preferred to C. Thus, indifference curves cannot intersect each other. Furthermore, for the third property, since the combination of goods which lies on a higher indifference curve will be preferred by a person to the combination which lies on a lower indifference curve, the higher indifference curve represents a higher utility/satisfaction.

3 Interactive Algorithm for MO-COP

In this section, we develop a novel Interactive Algorithm for MO-COP (MO-IA). This algorithm finds a set of Pareto solutions and narrows a region, in which Pareto front may exist, gradually. First, we find optimal solutions of weighted mono-objective functions using AOF. Then, we provide a user with an optimal value and a region, in which Pareto front may exist. A user determines whether he/she is satisfied by the Pareto solution. If he/she is satisfied, our algorithm terminates. Otherwise, he/she chooses a preference point in the region, in which Pareto front may exist. Next, we find a point (a reward vector obtained by a Pareto solution) in the region that is closest to the user's preference point by using a distance defined in our algorithm, and update the region, in which Pareto front may exist. Then, as a new information, we provide the user with a set of Pareto solutions and a new narrower region, in which Pareto front may exist. We continue this process until the user will be satisfied by at least one of the provided Pareto solutions.

3.1 Interactive Algorithm

Our algorithm has three phases:

Phase 1: For each objective function, find an optimal solution.
Phase 2: For a weighted mono-objective function, find the optimal solution.

Phase 3: Find a point in a region that is closest to a user's preference point.

Let us describe Phase 1. We use AOF to find an optimal solution for each objective function, respectively. Specifically, for m objective functions of a MO-COP, we give the following m weights $(1, 0, \ldots, 0), (0, 1, 0, \ldots, 0), \ldots, (0, \ldots, 0, 1)$ and make the m weighted objective functions $o^1, \ldots, o^m$. Then, we find an optimal solution for each weighted mono-objective function o^i $(1 \leq i \leq m)$, respectively, i.e., it is equivalent to solve m COP problems independently. In this paper, we denote the obtained m optimal values as $R^1_{max}, \ldots, R^m_{max}$.

In Phase 2, we use AOF and make a weighted mono-objective function where each weight has a same value. Then, we find the optimal solution. Specifically, for m objective functions of a MO-COP, we make the following weighted mono-objective function, denoted π, giving the weights $\alpha_1 = \frac{1}{m}, \ldots, \alpha_m = \frac{1}{m}$, and find the optimal solution.

$$\pi : \frac{1}{m} o^1 + \ldots + \frac{1}{m} o^m \tag{7}$$

Let A^* be an optimal solution of a weighted mono-objective function π. By Theorem 1, A^* is a Pareto solution of the original problem. In this paper, we call this Pareto solution as a candidate solution. For optimal values $R^1_{max}, \ldots, R^m_{max}$ obtained by Phase 1 and a candidate solution A^* obtained by Phase 2, let A be an another Pareto solution, and let $R(A)$ be a reward vector obtained by A which is different from $R(A^*)$. Then, the following theorem holds.

Theorem 2. *For reward vectors $R(A^*)$ and $R(A)$, it holds:*
(1) $\sum_{l=1}^{m} R^l(A) \leq \sum_{l=1}^{m} R^l(A^*)$.
(2) $\exists l : R^l(A^*) < R^l(A) \leq R^l_{max}$.

Proof. Since A^* is an optimal solution of a weighted mono-objective function π, it holds

$$\frac{1}{m} R^1(A) + \ldots + \frac{1}{m} R^m(A) \leq \frac{1}{m} R^1(A^*) + \ldots + \frac{1}{m} R^m(A^*). \tag{8}$$

Also, there exists no reward vector that dominates a reward vector on π.

Next, we show that it holds $\exists l : R^l(A^*) < R^l(A) \leq R^l_{max}$. Since R^l_{max} is a reward vector obtained by an optimal solution of the objective function o^l, it holds $R^l(A) \leq R^l_{max}$. Furthermore, we show that it holds $\exists l : R^l(A^*) < R^l(A)$ by contradiction. Assume that $\forall l : R^l(A^*) \geq R^l(A)$ holds. Since A^* is a candidate solution, i.e., Pareto solution, and $R(A)$ is a reward vector which is different from $R(A^*)$, there exists at least one objective l, such that $R^l(A^*) > R^l(A)$. Then, it holds $R(A) \prec R(A^*)$ by Definition 1, i.e., $R(A^*)$ dominates $R(A)$. However, since A is a Pareto solution, i.e., there exist no reward vector that dominates $R(A)$, this is a contradiction. Thus, it holds $\exists l : R^l(A^*) < R^l(A)$. $\qquad\square$

Example 4. Figure 6 shows a region, in which the Pareto front of a bi-objective COP may exist. The x-axis represents the rewards for objective 1 and the y-axis represents those for objective 2. R^1_{max} and R^2_{max} are reward vectors obtained in

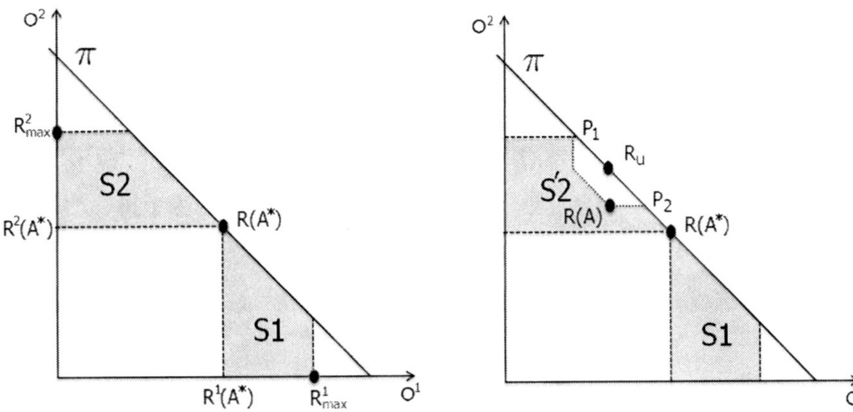

Fig. 6. A region, in which the Pareto front of a bi-objective COP may exist

Fig. 7. A new region which is narrower compared to the region in Fig. 6

Phase 1. The line π represents a weighted mono-objective function, and $R(A^*)$ is a reward vector obtained by a candidate solution A^* in Phase 2. Let $R(A)$ be a reward vector obtained by a Pareto solution, which is different from $R(A^*)$. By Theorem 2, $R(A)$ exists in the region under a function π. Furthermore, it holds $R^1(A^*) < R^1(A)$ or $R^2(A^*) < R^2(A)$. Also, since $R^1(A) \leq R^1_{max}$ and $R^2(A) \leq R^2_{max}$ must be hold, $R(A)$ exists in the region $S1$ or $S2$.

A user determines whether he/she is satisfied by a candidate solution obtained in Phase 2. If he/she is satisfied, our algorithm terminates. Otherwise, he/she chooses a point in a region, in which Pareto front may exist. We call the point as a user's preference point and denote it by $R_u(= (R^1_u, \ldots, R^m_u))$.

Let us describe Phase 3. We find a point in a region that is closest to a user's preference point. Specifically, for a reward vector $R(A)$ obtained by an assignment A and a user's preference point R_u, we define the distance between $R(A)$ and R_u as follows:

$$dis(R(A), R_u) = f(R^1(A), R^1_u) + \ldots + f(R^m(A), R^m_u),$$

$$\forall l : f(R^l(A), R^l_u) = \begin{cases} R^l_u - R^l(A) & (R^l_u \geq R^l(A)) \\ -\epsilon_l(R^l(A) - R^l_u) & (R^l_u < R^l(A)) \end{cases}, \tag{9}$$

where ϵ_l is small enough. In Phase 3, we find an assignment A so that the distance between a reward vector $R(A)$ and a user's preference point is minimal. Strictly speaking, the distance function is non-linear. However, the term $-\epsilon_l(R^l(A) - R^l_u)$ is used only for a tie-breaker. We can easily encode this metric in standard MO-COP algorithms.

We show the procedure of MO-IA (Phase 3) in Algorithm 1. In our algorithm, we assume that a pseudo-tree based on total ordering $x_1, \ldots, x_n$ is given, where

---

Algorithm 1. MO-IA (Phase 3)

MO-IA(X,D,O)

1 Given : R_u // user's preference point on π

2 $JOIN_1 = \text{null}, \ldots, JOIN_n = \text{null}$

3 **for each** $i = n, \ldots, 1$

4 **if** i is a leaf **then**

5    $JOIN_i = R_i^{p_i} \oplus (\bigoplus_{h \in PP_i} R_i^h)$ // join all reward tables

6   Compute $\underset{a}{\operatorname{argmin}}\, dis(R(a), R_u)$ for each a in combination of assignments of

     p_i and PP_i

7    $JOIN_i = JOIN_i \perp_{x_i}$ // use projection to eliminate x_i

8 **else**

9    $JOIN_i = R_i^{p_i} \oplus (\bigoplus_{h \in PP_i} R_i^h) \oplus (\bigoplus_{j \in C_i} JOIN_j)$

10   Compute $\underset{a}{\operatorname{argmin}}\, dis(R(a), R_u)$ for each a in combination of assignments of

     p_i and PP_i

11    $JOIN_i = JOIN_i \perp_{x_i}$ // use projection to eliminate x_i

12 **end if**

13 **end for**

---

x_1 is a root node. The $\oplus$ operator is the operator to join two reward tables and the $\bigoplus$ operator is the operator to join all reward tables. The $\perp_x$ operator is the projection to eliminate x. $JOIN_i$ represents a reward table maintained by a node x_i. Also, $R_i^{p_i}$ and R_i^h represent the reward tables between a node x_i and its parent node p_i, and its ancestor $h \in PP_i$, respectively. Our algorithm processes bottom-up, which starts from the leaves and propagates upwards only through tree edges (line 3). If x_i is a leaf node, x_i joins all reward tables it has with its ancestor using $\bigoplus$ operator, and the reward table it has with its parent using $\oplus$ operator (line 5). Then, for each combination of assignments of p_i and PP_i, we compute an assignment so that the distance from a user's preference point is minimal (line 6). We use the $\perp$ operator to eliminate x_i from the reward table $JOIN_i$ (line 7). If x_i is not a leaf node, we access the reward tables of its children, and join the following reward tables $R_i^{p_i}$, R_i^h, and all $JOIN_{j \in C_i}$ (line 9). Then, we conduct the same process we did for a leaf node (line 10 and 11). In this algorithm, each node chooses an assignment so that the distance from a user's preference point is minimal. Thus, for an assignment to all variables A, the distance between the reward vector $R(A)$ obtained by A and a user's preference point is minimal. This is because we deal with maximization MO-COPs. We omit the proof due to space limitations. For an assignment obtained by our algorithm, the following theorem holds.

Theorem 3. *An assignment obtained by MO-IA is a Pareto solution.*

Proof. Let A^* be an assignment obtained by our algorithm and $R(A^*)$ be a reward vector obtained by A^*. We show that there exists no assignment A, such that $R(A^*) \prec R(A)$. Assume that $\exists A : R(A^*) \prec R(A)$ holds. By Definition 1, it holds (i) $\forall l : R^l(A^*) \leq R^l(A)$ and (ii) $\exists l : R^l(A^*) < R^l(A)$. Let R_u be a user's preference point. By (i), when $R_u^l \geq R^l(A)$, the following holds for all objectives:

$$f(R^l(A), R_u^l) = R_u^l - R^l(A) \leq R_u^l - R^l(A^*) = f(R^l(A^*), R_u^l). \qquad (10)$$

Otherwise, i.e., when $R_u^l < R^l(A)$, it holds

$$f(R^l(A), R_u^l) = -\epsilon_l(R^l(A) - R_u^l) \leq -\epsilon_l(R^l(A^*) - R_u^l) = f(R^l(A^*), R_u^l). \quad (11)$$

By (ii), when $R_u^l \geq R^l(A)$, the following holds at least one objective:

$$f(R^l(A), R_u^l) < f(R^l(A^*), R_u^l). \qquad (12)$$

Otherwise, i.e., when $R_u^l < R^l(A)$, it holds

$$f(R^l(A), R_u^l) < f(R^l(A^*), R_u^l). \qquad (13)$$

Thus, it holds $dis(R(A), R_u) < dis(R(A^*), R_u)$. However, $dis(R(A^*), R_u)$ is minimal. This is a contradiction. Thus, A^* is a Pareto solution. $\qquad \square$

In Phase 3, we can obtain a Pareto solution that gives the closest point to a user's preference point. It means that there exists no Pareto front within the distance from a user's preference point to a point obtained by Phase 3. Thus, the new region, in which the Pareto front may exist, is the remaining region obtained from the original region removing the region within this distance.

Example 5. Figure 7 shows a new region, in which the Pareto front may exist. $R(A)$ represents a reward vector obtained by our algorithm (Phase 3). Since there exists no Pareto solution within the distance $dis(R(A), R_u)$, a new region is the region obtained by removing the region enclosed by π and $S'2$ from the original region. The new region is narrower compared to that in Fig. 6.

Let a Pareto solution obtained by Phase 3 be a new candidate solution. A user determines whether he/she is satisfied by at least one of the two candidate solutions, i.e., the first candidate solution obtained by Phase 2 or a new candidate solution. If he/she is satisfied, our algorithm terminates. Otherwise, he/she chooses a new preference point in the new narrower region, in which Pareto front may exist. We conduct the Phase 3 repeatedly, i.e., we compute a set of candidate solutions and the narrowed regions, in which Pareto front may exist, until the user is satisfied by at least one of the candidate solutions. Since our algorithm repeatedly narrows the region where the Pareto front can exist, we can expect that it converges after a finite number of iterations. However, there exists a pathological case where the algorithm repeats infinitely. This happens when the user's preference is Leontief, which is very different from our distance function. To guarantee the terminating of this algorithm, we need to set a threshold value, where the user terminates the iteration when a possible maximal improvement becomes less than the threshold.

Complexity

Our algorithm MO-IA is time $O(e \times m \times |D|^{w^*+1})$ and space $O(n \times m \times |D|^{w^*})$, where n is the number of variables, m is the number of objectives, $|D|(= |D_1| = ,\ldots,= |D_n|)$ is the domain size, w^* is the induced width, e is the number of constraints. The complexity of MO-IA is determined by the induced width of a problem instance. Induced width is a parameter that determines the complexity of many COP algorithms. Specially, if a problem instance has the tree structure, i.e., the induced width is one, the complexity of MO-IA is constant.

4   Evaluations

In this section, we evaluate our algorithm using CES utility functions. Specifically, we define the following four users that have different utility functions, and examine the number of the required iterations for each user until our algorithm terminates.

User 1: Linear utility function.
User 2: CES utility function where the parameter p is 0.5.
User 3: Cobb-Douglas utility function.
User 4: Leontief utility function.

Let us explain how we examine the number of the required iterations. First, we compute a candidate solution and a region, in which Pareto front may exist. Then, we determine a user's preference point, which is the intersection of a utility function and a weighted mono-objective function. If he/she is satisfied by the candidate solution, our algorithm terminates. Then, the number of the required iterations is one. Otherwise, we find a Pareto solution that gives the closest point to a user's preference point, and let this solution be a new candidate solution. Next, using indifference curves of the user, we determine a new user's preference point in the region, in which Pareto front may exist, i.e., Pareto front without the computed candidate solutions. If he/she is satisfied by at least one of the candidate solutions, our algorithm terminates, and the number of the required iterations is increased by one. We continue this process until he/she will be satisfied, and examine how many iterations are required for each user.

Let us describe termination conditions of our algorithm. For a MO-COP, a user's preference point $R_u(= (R_u^1, \ldots, R_u^m))$, and a reward vector $R(A)$ obtained by one of the candidate solutions, a user is satisfied, if the following holds:

$$\exists A : u(R_u^1, \ldots, R_u^m) \leq u(R^1(A), \ldots, R^m(A)), \tag{14}$$

i.e., termination conditions for user 1, 2, 3 and 4 are as follows:

Termination conditions for user 1

$$\exists A : \sum_{l=1}^{m} \alpha_l R_u^l \leq \sum_{l=1}^{m} \alpha_l R^l(A) \tag{15}$$

Table 1. Number of required iterations in bi-objective COPs

Nodes	User 1	User 2	User 3	User 4
10	2.5	1.6	2.1	25.3
20	2.5	2.0	1.9	17.4
30	2.6	2.2	1.8	12.8
40	2.8	2.3	1.6	13.4
50	2.8	2.5	1.7	12.9
60	2.9	2.3	1.5	11.1
70	2.9	2.3	1.5	11.9
80	2.9	2.5	1.4	11.5
90	2.8	2.5	1.2	12.5
100	2.8	2.6	1.3	10.4

Table 2. Number of required iterations in tri-objective COPs

Nodes	User 1	User 2	User 3	User 4
10	2.4	2.3	2.3	44.4
20	2.2	2.1	2.4	33.0
30	2.2	2.2	2.3	46.8
40	2.2	2.7	2.2	36.7
50	2.4	2.3	2.0	43.6
60	2.2	2.6	2.1	38.8
70	2.3	2.6	2.0	40.7
80	2.4	2.5	2.0	30.5
90	2.3	2.3	2.0	41.8
100	2.3	2.4	2.0	42.9

Termination conditions for user 2

$$\exists A : \sum_{l=1}^{m} \alpha_l \sqrt{R_u^l} \leq \sum_{l=1}^{m} \alpha_l \sqrt{R^l(A)} \tag{16}$$

Termination conditions for user 3

$$\exists A : \prod_{l=1}^{m} (R_u^l)^{\alpha_l} \leq \prod_{l=1}^{m} (R^l(A))^{\alpha_l} \tag{17}$$

Termination conditions for user 4

$$\exists A \ \forall l : R_u^l \leq R^l(A) \tag{18}$$

In our evaluations, the domain size of each variable is two, and we chose the reward value uniformly at random from the range $[0,\ldots,10]$ for all objectives. We generate bi/tri-objective COP problem instances randomly, and determine the parameter α of CES utility functions random for each problem instance. For each objective, we generate the same constraint graph. The number of constraints is given by $|X| * |O|$, where $|X|$ and $|O|$ are the number of variables and objectives. The results represent an average of 50 problem instances. For the parameter of the distance in Phase 3, we set that ϵ_l is 0.001 for all l.

The experimental results for bi-objective COPs are summarized in Table 1. For 10 nodes, the number of required iterations for user 1, 2 and 3 are 2.5, 1.6 and 2.1, respectively. These results are almost unchanged, when the number of nodes increases. For 100 nodes, the number of required iterations for user 1, 2 and 3 are 2.8, 2.6 and 1.3, respectively. We can see that our algorithm satisfies the preferences of user 1, 2 and 3 with few iterations. We consider that this is because the "closest" solution defined by our algorithm are almost same as the "closest" solution that user 1, 2 and 3 think. For user 4, the number of required iterations are significantly increased compared to those for other users.

In Table 1, the number of required iterations are 25.3 for 10 nodes and 10.4 for 100 nodes. We consider that this is because there exists a divergence between the "closest" solution defined by our algorithm and that user 4 thinks. Furthermore, for user 4, the number of the required iterations decreases, when the number of nodes increases. The number of the required iterations 10.4 for 100 nodes are less than half of that for 10 nodes. We consider that this is because the solution space of bi-objective COPs becomes dense, when the number of nodes increases.

We confirmed the similar results for tri-objective COPs. The experimental results are summarized in Table 2. For 10 nodes, the number of required iterations for user 1, 2 and 3 are 2.4, 2.3 and 2.3, respectively. These results are almost unchanged, when the number of nodes increases. For user 4, the number of required iterations are significantly increased compared to those for other users. The number of required iterations for user 4 increases compared to those for bi-objective COPs. We consider that this is because the Pareto solutions are sparse in tri-objective COPs compared to that in bi-objective COPs. Furthermore, we do not see any direct relationship between the number of nodes and the required iterations in Table 2. We consider that this is because Pareto solutions in three dimensional tri-objective COPs are still sparse for 100 nodes, while Pareto solutions in two dimensional bi-objective COPs becomes dense.

In summary, these experimental results reveal that our algorithm is effective for user 1, 2 and 3, i.e., CES utility functions where the parameter p is between 0 and 1. However, for user 4, the number of the required iterations are significantly increased compared to those for other users. Our future works include performing more detailed analysis, e.g., examining the relationships between the size of Pareto front and the number of nodes/objectives. Furthermore, we hope to examine the performance of our algorithm based on the utilities of real people by experiments with human subjects.

Let us propose a method to reduce the number of the required iterations for a user who has Leontief utility function. In the evaluations, our algorithm required a large number of iterations for user 4. We propose the following method to improve the results for user 4. First, we estimate a coefficient α of an utility function from a user's preference point. Next, if our algorithm does not terminate in a constant number of the required iterations, we assume that a user has Leontief utility function using the estimated α. Then, we compute Pareto front repeatedly without asking a user until the required iterations converge. We examined the number of the required iterations for user 4 using this method. We used the same problem instances in section 4, i.e., 50 bi-objective COP problem instances and 50 tri-objective COP problem instances. We set a constant number of the required iterations to three. Our algorithm terminated, when the number of the required iterations was four.

Extended MO-IA

We propose an extension of our algorithm, which finds several Pareto solutions so that we can provide a narrower region, in which Pareto front may exist, i.e., more detailed information for Pareto front. When we consider an interaction in the real world, it is natural to provide several candidate solutions. Also, a narrower region is desirable. We extend the Phase 3 of our algorithm as follows.

Phase 3' : Determine additional virtual (preference) points which are different from a user's preference point, and find a Pareto solution that is closest to each point, respectively.

In our original algorithm, we find a candidate solution and provide a region, in which Pareto front may exist, gradually. On the other hand, the extended algorithm finds several candidate solutions and provides a narrower region. The narrower region is obtained from the original region removing a region within each distance between a preference point and the corresponding candidate solution. For the virtual preference points, for example, we choose the intersections (P_1 and P_2) of function π and the border of the removed region in Fig. 7.

5   Conclusions

We developed a novel interactive algorithm for a MO-COP. This algorithm finds a set of Pareto solutions and narrows a region, in which Pareto front exist, gradually. Furthermore, we showed that the complexity of our algorithm is determined by the induced width of problem instances. In the evaluations, we defined four users using a CES utility function, and examined the number of required iterations for each user. We showed empirically that our algorithm is effective for the users, who have linear and Cobb-Douglas utility functions. Finally, we proposed a method that can reduce the number of the required iterations for a user, who has Leontief utility function. Also, we proposed an extension of MO-IA, which finds several Pareto solutions so that we can provide a narrower region, in which Pareto front may exist. As future works, we intend to apply our algorithm on challenging real world problems. Furthermore, we will develop an interactive algorithm for a multi-objective DCOP, which is formalized in [6].

References

[1] Bringmann, K., Friedrich, T., Neumann, F., Wagner, M.: Approximation-guided evolutionary multi-objective optimization. In: Proceedings of the 22nd International Joint Conference on Artificial Intelligence, pp. 1198–1203 (2011)

[2] Das, I., Dennis, J.E.: Normal-boundary intersection: a new method for generating the Pareto surface in nonlinear multicriteria optimization problems. SIAM Journal on Optimization 8(3), 631–657 (1998)

[3] Deb, K., Agrawal, S., Pratap, A., Meyarivan, T.: A fast and elitist multiobjective genetic algorithm: NSGA-II. IEEE Trans. Evolutionary Computation 6(2), 182–197 (2002)

[4] Dechter, R.: Constraint Processing. Morgan Kaufmann Publishers (2003)

[5] Erfani, T., Utyuzhnikov, S.V.: Directed search domain: a method for even generation of the Pareto frontier in multiobjective optimization. Engineering Optimization 43(5), 467–484 (2010)

[6] Fave, F.M.D., Stranders, R., Rogers, A., Jennings, N.R.: Bounded decentralised coordination over multiple objectives. In: Proceedings of the 10th International Conference on Autonomous Agents and Multiagent Systems, pp. 371–378 (2011)

[7] Junker, U.: Preference-based inconsistency proving: When the failure of the best is sufficient. In: Proceedings of the 17th European Conference on Artificial Intelligence, pp. 118–122 (2006)

[8] Marinescu, R.: Exploiting Problem Decomposition in Multi-objective Constraint Optimization. In: Gent, I.P. (ed.) CP 2009. LNCS, vol. 5732, pp. 592–607. Springer, Heidelberg (2009)

[9] Marinescu, R.: Best-first vs. depth-first and/or search for multi-objective constraint optimization. In: Proceedings of the 22nd IEEE International Conference on Tools with Artificial Intelligence, pp. 439–446 (2010)

[10] Mas-Colell, A., Whinston, M.D., Green, J.R.: Microeconomic Theory. Oxford University Press (1995)

[11] Messac, A., Mattson, C.: Generating well-distributed sets of Pareto points for engineering design using physical programming. Optimization and Engineering 3(4), 431–450 (2002)

[12] Messac, A., Puemi-sukam, C., Melachrinoudis, E.: Aggregate objective functions and Pareto frontiers: Required relationships and practical implications. Optimization and Engineering 1(2), 171–188 (2000)

[13] Miettinen, K.: Nonlinear Multiobjective Optimization. Kluwer Academic Publishers, Boston (1999)

[14] Papadimitriou, C.H., Yannakakis, M.: On the approximability of trade-offs and optimal access of web sources. In: Proceedings of the 41st Annual Symposium on Foundations of Computer Science, pp. 86–92 (2000)

[15] Perny, P., Spanjaard, O.: Near admissible algorithms for multiobjective search. In: Proceedings of the 18th European Conference on Artificial Intelligence, pp. 490–494 (2008)

[16] Rollon, E., Larrosa, J.: Bucket elimination for multiobjective optimization problems. Journal of Heuristics 12(4-5), 307–328 (2006)

[17] Rollon, E., Larrosa, J.: Multi-objective Russian doll search. In: Proceedings of the 22nd AAAI Conference on Artificial Intelligence, pp. 249–254 (2007)

[18] Schiex, T., Fargier, H., Verfaillie, G.: Valued constraint satisfaction problems: Hard and easy problems. In: Proceedings of the 14th International Joint Conference on Artificial Intelligence, pp. 631–639 (1995)

[19] Stiglitz, J.E.: Economics. W.W.Norton & Company (1993)

FOCUS: A Constraint for Concentrating High Costs

Thierry Petit

TASC (Mines Nantes, LINA, CNRS, INRIA),
4, Rue Alfred Kastler, FR-44307 Nantes Cedex 3, France
Thierry.Petit@mines-nantes.fr

Abstract. Many Constraint Programming models use integer cost variables aggregated in an objective criterion. In this context, some constraints involving exclusively cost variables are often imposed. Such constraints are complementary to the objective function. They characterize the solutions which are acceptable in practice. This paper deals with the case where the set of costs is a sequence, in which high values should be concentrated in a few number of areas. Representing such a property through a search heuristic may be complex and overall not precise enough. To solve this issue, we introduce a new constraint, $\text{FOCUS}(X, y_c, len, k)$, where X is a sequence of n integer variables, y_c an integer variable, and len and k are two integers. To satisfy FOCUS, the minimum number of distinct sub-sequences of consecutive variables in X, of length at most len and that involve exclusively values strictly greater than k, should be less than or equal to y_c. We present two examples of problems involving FOCUS. We propose a complete filtering algorithm in $O(n)$ time complexity.

1 Introduction

Encoding optimization problems using Constraint Programming (CP) often requires to define cost variables, which are aggregated in an objective criterion. To be comparable, those variables have generally a totally ordered domain. They can be represented by integer variables. In this context, some constraints on cost variables are complementary to the objective function. They characterize the solutions which are acceptable in practice. For instance, to obtain balanced solutions several approaches have been proposed: Balancing constraints based on statistics [6,11], as well as classical or dedicated cardinality constraints when the set of costs is a sequence [9,8]. Some applications of these techniques are presented in [12,7]. Representing such constraints, as well as solving efficiently the related problems, form an important issue because real-life problems are rarely "pure". In this context, CP is a well-suited technique. CP is generally robust to the addition of constraints, providing that they come up with filtering algorithms which impact significantly the search process.

Conversely to balancing constraints, in some problems involving a sequence of cost variables, the user wishes to minimize the number of sub-sequences of consecutive variables where high cost values occur.

Example 1. We consider a problem where some activities have to be scheduled. Each activity consumes an amount of resource. The total amount of consumption at a given time is limited by the capacity of the machine that produces the resource. If the time

M. Milano (Ed.): CP 2012, LNCS 7514, pp. 577–592, 2012.

window where activities have to be scheduled is fixed, in some cases not all the activities can be scheduled, because there is not enough quantity of resource to perform all the activities on time. Assume that, in this case, we rent a second machine to solve the problem. In practice, it is often less costly to rent such a machine within a package, that is, during consecutive periods of time. If you rent the machine during three consecutive days, the price will be lower than the price of three rentals of one day in three different weeks. Moreover, such packages are generally limited, e.g., the maximum duration of one rental is one week. If you exceed one week then you need to sign two separate contracts. Thus, to satisfy the end-user, a solution should both limit and concentrate the exceeds of resource consumption, given a maximum rental duration.   ⊛

In Example 1, a solution minimizing the exceeds with many short and disjoint rental periods will be more expensive for the end-user than a non-minimum solution where rentals are focused on a small number of periods. Such a constraint cannot be easily simulated with a search strategy, a fortiori when the duration of packages is limited. Furthermore, to solve instances, search heuristics are generally guided by the underlying problem (in Example 1, the cumulative problem). Our contribution is a generic, simple and effective way to solve this issue. It comes in two parts.

1. A new constraint, FOCUS(X, y_c, len, k), where X is a sequence of integer variables, y_c an integer variable, and len and k two integer values. y_c limits the number of distinct sub-sequences in X, each of length at most len, involving exclusively values strictly greater than k. More precisely, the minimum possible number of such sub-sequences should be less than or equal to y_c, while any variable in X taking a value $v > k$ belongs to exactly one sub-sequence.
2. A $O(n)$ Generalized Arc-Consistency (GAC) filtering algorithm for FOCUS.

Section 2 defines the FOCUS constraint. Section 3 presents two examples of use. In section 4, we present the $O(n)$ complete filtering algorithm for FOCUS. Section 5 introduces some variations of FOCUS, namely the case where len is a variable and the case where the constraint on the variable y_c is more restrictive. We discuss the related work and propose an automaton-based reformulation of FOCUS. Our experiments, in Section 7, show the importance of providing FOCUS with a complete filtering algorithm.

2   The FOCUS Constraint

Given a sequence of integer variables $X = \langle x_0, x_1, \ldots, x_{n-1} \rangle$ of length $|X| = n$, an instantiation of X is a valid assignment, *i.e.*, a sequence of values $I[X] = \langle v_0, v_1, \ldots v_{n-1} \rangle$ such that $\forall j \in \{0, 1, \ldots, n-1\}$, v_j belongs to $D(x_j)$, the domain of x_j.

Definition 1 (FOCUS). *Given $X = \langle x_0, x_1, \ldots, x_{n-1} \rangle$, let y_c be an integer variable such that $0 \leq y_c \leq |X|$, len be an integer such that $1 \leq len \leq |X|$, and $k \geq 0$ be an integer. Given an instantiation $I[X] = \langle v_0, v_1, \ldots v_{n-1} \rangle$, and a value v_c assigned to y_c, FOCUS$(I[X], v_c, len, k)$ is satisfied if and only if there exists a set S of **disjoint** sequences of consecutive variables in X such that three conditions are all satisfied:*

1. Number of sequences: $|S| \leq v_c$
2. One to one mapping of all values strictly greater than k:
$\quad \forall j \in \{0, 1, \ldots, n-1\},\ v_j > k \Leftrightarrow \exists s_i \in S$ *such that* $x_j \in s_i$
3. Length of a sequence in S: $\forall s_i \in S,\ 1 \leq |s_i| \leq len.$

If $len = |X|$, y_c limits the number of disjoint maximum length sequences where all the variables are assigned with a value strictly greater than k. Otherwise, len limits the length of the sequences counted by y_c. Example 2 illustrates the two cases.

Example 2. Let $I[X] = \langle 1, 3, 1, 0, 1, 0 \rangle$. FOCUS($I[X], \langle 2 \rangle, 6, 0$) is satisfied since we can have 2 disjoint sequences of length ≤ 6 of consecutive variables with a value > 0, *i.e.*, $\langle x0, x1, x2 \rangle$, and $\langle x4 \rangle$. FOCUS($I[X], \langle 2 \rangle, 2, 0$) is violated since it is not possible to include all the strictly positive variables in X with only 2 sequences of length ≤ 2.   ⊛

3   Examples of Use

Constraints and Music. An important field in musical problems is automatic composition and harmonization. In many cases, the end-user wishes to obtain the maximum length sequences of measures where her rules are minimally violated. We consider the example of the *sorting chords* problem [5,13]. The goal is to sort n distinct chords. A chord is a set of at most p notes played simultaneously. p can vary from one chord to another. The sort should reduce as much as possible the number of notes changing between two consecutive chords. The musician may be particularly interested by large sub-sequences of consecutive chords where there is at most *nchange* different notes between two consecutive chords, and thus she aims at concentrating high changes in a few number of areas. We represent the sequence by n variables $Chords = \langle ch_0, ch_1, \ldots, ch_{n-1} \rangle$, such as each variable can be instantiated with any of the chords. The constraint ALLDIFF($Chords$) [10] imposes that all ch_i's are pairwise distinct. *nchange* is at least 1. Therefore, we define the cost between two consecutive chords in the sequence as the number of changed notes less one. It is possible to compute that cost for each pair of chords (the number of costs is $n \times (n-1)/2$), and link this value with the chords through a ternary table constraint. We call such a constraint COSTC$_i$($ch_i, ch_{i+1}, cost_i$), where $cost_i \in X$ is the integer variable representing the cost of the pair (ch_i, ch_{i+1}). Its domain is the set of distinct cost values considered when COSTC$_i$($ch_i, ch_{i+1}, cost_i$) is generated. FOCUS is imposed on $X = \langle cost_0, cost_1, \ldots, cost_{n-2} \rangle$, in order to concentrate high costs (for instance costs > 2, that is $nchange = 3$) in a few number of areas. If their length is not constrained $len = |X|$, otherwise the end-user can fix a smaller value. The constraint model is:
ALLDIFF($Chords$) $\land\ \forall i \in \{0, 1, \ldots, n-2\}$ COSTC$_i$($ch_i, ch_{i+1}, cost_i$)
$\land$ FOCUS($X, y_c, len, 2$) $\land\ sum = \sum_{i \in \{0,1,\ldots,n-2\}} cost_i$
Two objectives can be defined: $minimize(sum)$ and $minimize(y_c)$.

Over-Loaded Cumulative Scheduling. In Example 1 of the Introduction, the core of the problem can be represented using the SOFTCUMULATIVE constraint [2]. The time window starts at time 0 and ends at a given strictly positive integer, the *horizon* (*e.g.*, 160 points in times which are, for instance, the total amount of hours of 4 weeks of work). Activities $a_k \in A$ are represented by three variables: starting time, duration, resource consumption. Using SOFTCUMULATIVE, some intervals of time $I_i \in \mathcal{I}$ (e.g., one day of 8 hours), one to one mapped with cost variables $cost_i \in X$, are given by the user. A cost measures how much the capacity $capa$ is exceeded within the interval I_i. The maximum value in the domain of each variable $cost_i$ expresses the maximum allowed excess. In [2], several definitions of costs are proposed. We can for instance define $cost_i$ as the exceed of the maximum over-loaded hour in the interval I_i.

The constraint related to the additional machine is FOCUS($X, y_c, len, 0$), where $X = \langle cost_0, cost_1, \ldots, cost_{|I|-1} \rangle$. len is the maximum duration of one rental, e.g., 5 days, that is, $len = 40$ (if the time unit is one hour). The constraint model is:

$$\text{SOFTCUMULATIVE}(A, X, \mathcal{I}, horizon) \wedge \text{FOCUS}(X, y_c, len, 0)$$
$$\wedge \ sum = \sum_{i \in \{0,1,\ldots,n-2\}} cost_i$$

Two objectives can be defined: $minimize(sum)$ and $minimize(y_c)$.

4   Linear Filtering Algorithm

4.1   Characterization of Sequences

Notation 1 (Status of a variable). *Let $X = \langle x_0, x_1, \ldots, x_{n-1} \rangle$ be a sequence of integer variables and k an integer. According to k, a variable $x_i \in X$ is:* Penalizing (P_k) *if and only if the minimum value in its domain $min(x_i)$ is such that $min(x_i) > k$.* Neutral (N_k) *if and only if the maximum value in its domain $max(x_i)$ is such that $max(x_i) \leq k$.* Undetermined (U_k) *if $min(x_i) \leq k$ and $max(x_i) > k$.*

Definition 2 (Maximum σ-sequence). *Let $X = \langle x_0, x_1, \ldots, x_{n-1} \rangle$ be a sequence of integer variables, k an integer, and $\sigma \subseteq \{P_k, N_k, U_k\}$. A σ-sequence $\langle x_i, x_{i+1}, \ldots, x_j \rangle$ of X is a sequence of consecutive variables in X such that all variables have a status in σ and for all status $s \in \sigma$ there exists at least one variable in the sequence having the status s. It is* maximum *if and only if the two following conditions are satisfied:*

1. If $i > 0$ then the status of x_{i-1} is not in σ.
2. If $j < n - 1$ then the status of x_{j+1} is not in σ.

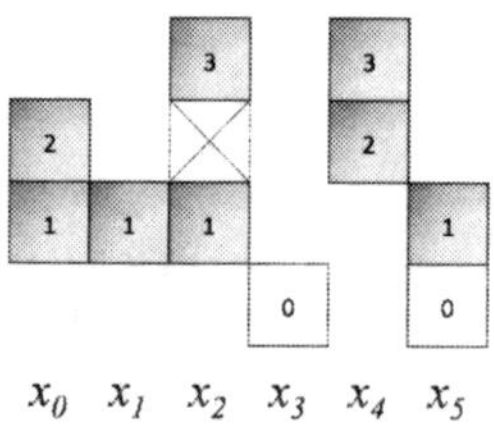

Fig. 1. $X = \langle x_0, x_1 \ldots, x_5 \rangle$ is a maximum-length $\{N_0, P_0, U_0\}$-sequence, which contains one maximum-length $\{N_0, P_0\}$-sequence $\langle x_0, x_1, \ldots, x_4 \rangle$, two maximum-length $\{P_0\}$-sequences $\langle x_0, x_1, x_2 \rangle$ and $\langle x_4 \rangle$, one maximum length $\{P_0, U_0\}$-sequence $\langle x_4, x_5 \rangle$, one maximum-length $\{N_0\}$-sequence $\langle x_3 \rangle$ and one maximum-length $\{U_0\}$-sequence $\langle x_5 \rangle$

Figure 1 illustrates Definition 2. The picture shows the domains in a sequence of $n = 6$ variables, with $k = 0$. Grey squares are values strictly greater than k, while the white ones correspond to values less than or equal to k.

Definition 3 (Focus cardinality of a σ-sequence). *Given a sequence of variables X and len and k two integer values, the focus cardinality $card(X, len, k)$ is the minimum value v_c such that* FOCUS(X, v_c, len, k) *has a solution.*

We can evaluate the focus cardinality according to the different classes of sequences.

Property 1. Given a $\{P_k\}$-sequence Y, $card(Y, len, k) = \lceil \frac{|Y|}{len} \rceil$.

Proof. $\lfloor \frac{|Y|}{len} \rfloor$ is the minimum number of distinct sequences of consecutive variables of length *len* within Y, and the remainder r of $\frac{|Y|}{len}$ is such that $0 \leq r < len$. $\square$

Notation 2. *Given a $\{N_k, P_k\}$-sequence X, $P_k(X)$ denotes the set of disjoint maximum $\{P_k\}$-sequences extracted from X.*

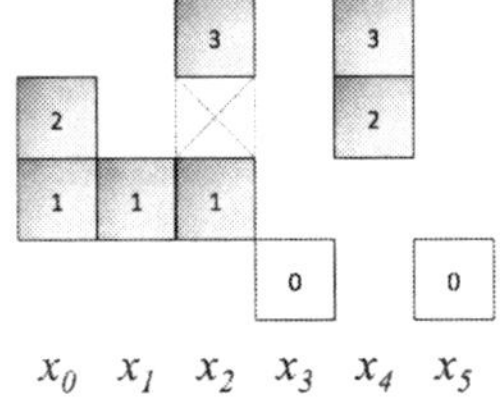

Fig. 2. A $\{N_0, P_0\}$-sequence $X = \langle x_0, x_1, \ldots, x_5 \rangle$. $card(X, 1, 0) = \sum_{Y \in P_0(X)} \lceil \frac{|Y|}{1} \rceil = 3 + 1 = 4$. $card(X, 2, 0) = \sum_{Y \in P_0(X)} \lceil \frac{|Y|}{2} \rceil = 2 + 1 = 3$. $card(X, 4, 0) = \sum_{Y \in P_0(X)} \lceil \frac{|Y|}{4} \rceil = 2$

Property 2. Given   a $\{N_k, P_k\}$-sequence X, $card(X, len, k)$ $=$ $\sum_{Y \in P_k(X)} card(Y, len, k)$.

Proof. By definition of a $\{N_k, P_k\}$-sequence, variables outside these sequences take a value less than or equal to k. From Property 1, the property holds. $\square$

Figure 2 illustrates Property 2. When X is a $\{N_k, P_k\}$-sequence, for instance an instantiation $I[X]$, and y_c is fixed to a value v_c, we can encode a checker for FOCUS, based on the computation of the focus cardinality of X.[1]

The correctness of Algorithm 1 is proved by Properties 1 and 2. Its time complexity is obviously $O(n)$. The computation for of a $\{N_k, P_k, U_k\}$-sequence requires to prove some properties. In Figure 1, we have $len = 1$. Depending whether x_5 is assigned to 0 or to 1, the value of y_c satisfying FOCUS$(X, y_c, len, 0)$ is either 4 or 5.

4.2   Feasibilty and Filtering Algorithm

Definition 4. *Given $x_i \in X$, $i \in \{0, 1, \ldots, n - 1\}$, and $v \in D(x_i)$,*

- $\underline{p}(x_i, v)$ *is the focus cardinality $card(\langle x_0, x_1, \ldots x_i \rangle, len, k)$ of the prefix sequence $\langle x_0, x_1, \ldots x_i \rangle$ when $x_i = v$.*

---

[1] For a end-user, we can provide a set of sub-sequences corresponding to the focus cardinality: the algorithm is similar to Algorithm 1 (we store the sequences instead of counting them).

Algorithm 1: ISSATISFIED($\{N_k, P_k\}$-sequence $X = \langle x_0, x_1, \ldots, x_{n-1} \rangle$, v_c, *len*, k): boolean

---

1 Integer $nb := 0$;
2 Integer $size := 0$;
3 Boolean *prevpk* := false;
4 **for** Integer $i := 0$; $i < n$; $i := i + 1$ **do**
5   **if** $\min(x_i) > k$ **then**
6          $size := size + 1$;
7   *prevpk* := true;
8   **else**
9   **if** *prevpk* **then** $nb := nb + \lceil \frac{size}{len} \rceil$;
10         $size := 0$;
11   *prevpk* := false;
12 **if** *prevpk* **then** $nb := nb + \lceil \frac{size}{len} \rceil$;
13 return $nb \leq v_c$; // focus cardinality of X

- $\underline{s}(x_i, v)$ *is the focus cardinality* $card(\langle x_i, x_{i+1}, \ldots x_{n-1} \rangle, len, k)$ *of the suffix sequence* $\langle x_i, x_{i+1}, \ldots x_{n-1} \rangle$ *when* $x_i = v$.

The remaining of this section is organized as follows. First, we show how we can check the feasibility of FOCUS and enforce a complete filtering of domains of variables in X and $D(y_c)$, provided we have the data of Definition 4. Then, we explain how such a data and the filtering algorithm can be obtained in $O(n)$.

Given $x_i \in X$, the two quantities of Definition 4 can have, each, at most two distinct values: one for the values in $D(x_i)$ strictly greater than k, one for the values in $D(x_i)$ less than or equal to k. This property holds by the definition of the constraint FOCUS itself (Definition 1): From the point of view of FOCUS, value $k + 1$ or value $k + 1000$ for x_i are equivalent. We use a new notation, which groups values of Definition 4.

Notation 3. *Given* $x_i \in X$,

- $\underline{p}(x_i, v_>)$ *is the value of* $\underline{p}(x_i, v)$ *for all* $v \in D(x_i)$ *such that* $v > k$, *equal to* $n + 1$ *if there is no value* $v > k$ *in* $D(x_i)$.
- $\underline{p}(x_i, v_\leq)$ *is the value of* $\underline{p}(x_i, v)$ *for all* $v \in D(x_i)$ *such that* $v \leq k$, *equal to* $n + 1$ *if there is no value* $v \leq k$ *in* $D(x_i)$.

Similarly, we use the notations $\underline{s}(x_i, v_>)$ *and* $\underline{s}(x_i, v_\leq)$ *for suffix sequences.*

Given such quantities for the last variable (or the first if we consider suffixes), we obtain a feasibility check for FOCUS. Their computation is explained in next section.

Algorithm 2: ISSATISFIED($X = \langle x_0, x_1, \ldots, x_{n-1} \rangle$, y_c, *len*, k): boolean

---

1 return $\min(\underline{p}(x_{n-1}, v_>), \underline{p}(x_{n-1}, v_\leq)) \leq \max(y_c)$;

We use the following notation: $minCard(X) = \min(\underline{p}(x_{n-1}, v_>), \underline{p}(x_{n-1}, v_\leq))$. With that data, we can update $\min(y_c)$ to $\min(\overline{minCard}(X), \min(y_c))$. Then from Definition 1, all the values in $D(y_c)$ have a valid support on FOCUS (by definition any value of y_c greater than $minCard(X)$ satisfies the constraint). By applying $O(n)$ times Algorithm 2, in order to study each variable x_i in X successively restricted to the range of values $\leq k$ as well as the range of values $> k$, we perform a complete filtering.

Lemma 1. . *Given a* U_k *variable* x_i, *let* $X_i^> = \{x_0^>, x_1^>, \ldots, x_{n-1}^>\}$ *be the set of variables derived from* X *such that* $\forall j \in \{0, 1, \ldots, i-1, i+1, \ldots, n-1\}$, $D(x_j^>) = D(x_j)$ *and* $D(x_i^>) = D(x_i) \cap [k+1, \max(x_i)]$. *If* $minCard(X_i^>) > \max(y_c)$ *then the range* $[k+1, \max(x_i)]$ *can be removed from* $D(x_i)$.

Lemma 2. *Given a* U_k *variable* x_i, *let* $X_i^> = \{x_0^>, x_1^>, \ldots, x_{n-1}^>\}$ *be the set of variables derived from* X *such that* $\forall j \in \{0, 1, \ldots, i-1, i+1, \ldots, n-1\}$, $D(x_j^\leq) = D(x_j)$ *and* $D(x_i^\leq) = D(x_i) \cap [\min(x_i), k]$. *If* $minCard(X_i^\leq) > \max(y_c)$ *then the range* $[\min(x_i), k]$ *can be removed from* $D(x_i)$.

Proof (Lemmas 1 and 2). Direct consequence of Definitions 1 and 3. $\qquad\qquad\square$

Given $O(\Phi)$ the time complexity of an algorithm computing $minCard(X)$, we can perform the complete filtering of variables in $X \cup \{y_c\}$ in $O(n \times \Phi)$, where $n = |X|$. We now show how to decrease the whole time complexity to $O(n)$, both for computing $minCard(X)$ and shrink the domains of all the variables in X. Given $x_i \in X$, the first idea is to compute $\underline{p}(x_i, v_>)$ from $\underline{p}(x_{i-1}, v_>)$ and $\underline{p}(x_{i-1}, v_\leq)$. To do so, we have to estimate the minimum length of a $\{P_k\}$-sequence containing x_i, within an instantiation of $\langle x_0, x_1, \ldots, x_i \rangle$ of focus cardinality $\underline{p}(x_i, v_>)$. We call this quantity $\underline{plen}(x_i)$. Next lemmas provide the values of $\underline{p}(x_i, v_>)$, $\underline{p}(x_{i-1}, v_\leq)$ and $\underline{plen}(x_i)$, from x_{i-1}.

Lemma 3 (case of x_0).

- If x_i is a $\{P_k\}$-variable, $\underline{p}(x_0, v_\leq) = n+1$, $\underline{p}(x_0, v_>) = 1$ and $\underline{plen}(x_0) = 1$.
- If x_i is a $\{N_k\}$-variable, $\underline{p}(x_0, v_\leq) = 0$, $\underline{p}(x_0, v_>) = n+1$ and $\underline{plen}(x_0) = 0$.
- If x_i is a $\{U_k\}$-variable, $\underline{p}(x_0, v_\leq) = 0$, $\underline{p}(x_0, v_>) = 1$ and $\underline{plen}(x_0) = 1$.

Proof. If x_0 takes a value $v > k$ then by Definition 4 $\underline{p}(x_0, v_>) = 1$ and $\underline{plen}(x_0) = 1$. Otherwise, there is no $\{P_k\}$-sequence containing x_0 and $\underline{plen}(x_0) = 0$: We use the convention $\underline{p}(x_0, v_>) = n+1$ (an absurd value: the max. number of sequences in X is n). If x_0 belongs to a $\{P_k\}$-sequence then $\underline{p}(x_0, v_\leq) = n+1$. $\qquad\square$

Lemma 4 (computation of $\underline{p}(x_i, v_\leq)$, $0 < i < n$). *If* x_i *is a* $\{P_k\}$-*variable then* $\underline{p}(x_i, v_\leq) = n+1$. *Otherwise,* $\underline{p}(x_i, v_\leq) = \min(\underline{p}(x_{i-1}, v_>), \underline{p}(x_{i-1}, v_\leq))$.

Proof. If x_i belongs to a $\{P_k\}$-sequence then x_i does not take a value $v \leq k$, thus $\underline{p}(x_0, v_\leq) = n+1$. If there exists some values less than or equal to k in $D(x_i)$, assigning one such value to x_i leads to a number of $\{P_k\}$-sequences within the prefix sequence $\langle x_0, x_1, \ldots, x_i \rangle$ which does not increase compared with the sequence $\langle x_0, x_1, \ldots, x_{i-1} \rangle$. Thus, $\underline{p}(x_i, v_\leq) = \min(\underline{p}(x_{i-1}, v_>), \underline{p}(x_{i-1}, v_\leq))$. $\qquad\square$

Lemma 5 (computation of $\underline{p}(x_i, v_>)$ and $\underline{plen}(x_i)$, $0 < i < n$). *We have:*

- If x_i is a $\{N_k\}$-variable then $\underline{p}(x_i, v_>) = n+1$ and $\underline{plen}(x_i) = 0$.
- Otherwise,
    - If $\underline{plen}(x_{i-1}) = len \vee \underline{plen}(x_{i-1}) = 0$ then $\underline{p}(x_i, v_>) = \min(\underline{p}(x_{i-1}, v_>) + 1, \underline{p}(x_{i-1}, v_\leq) + 1)$ and $\underline{plen}(x_i) = 1$.
    - Otherwise $\underline{p}(x_i, v_>) = \min(\underline{p}(x_{i-1}, v_>), \underline{p}(x_{i-1}, v_\leq) + 1)$ and:

Algorithm 3: MINCARDS($X = \langle x_0, x_1, \ldots, x_{n-1} \rangle$, *len*, k): Integer matrix

---

```
 1  cards := new Integer[||X||][3] ;
 2  if min(x₀) ≤ k ∧ max(x₀) > k then
 3  │   cards[0][0] := 0;
 4  │   cards[0][1] := 1;
 5  │   cards[0][2] := 1;
 6  else
 7  │   if min(x₀) > k then
 8  │   │   cards[0][0] := n + 1;
 9  │   │   cards[0][1] := 1;
10  │   │   cards[0][2] := 1;
11  │   else
12  │   │   cards[0][0] := 0;
13  │   │   cards[0][1] := n + 1;
14  │   │   cards[0][2] := 0;
15  for Integer i := 1; i < n; i := i + 1 do
16  │   if max(xᵢ) > k then
17  │   │   if min(xᵢ) > k then  cards[i][0] := n + 1;
18  │   │   else  cards[i][0] := min(cards[i − 1][0], cards[i − 1][1]);
19  │   │   if cards[i − 1][2] = 0 ∨ cards[i − 1][2] = len then
20  │   │   │   cards[i][1] := min(cards[i − 1][0] + 1, cards[i − 1][1] + 1);
21  │   │   │   cards[i][2] := 1;
22  │   │   else
23  │   │   │   cards[i][1] := min(cards[i − 1][0] + 1, cards[i − 1][1]);
24  │   │   │   if cards[i − 1][1] < cards[i − 1][0]+1 then  cards[i][2] := cards[i−1][2]+1;
25  │   │   │   else  cards[i][2] := 1;
26  │   else
27  │   │   cards[i][0] := min(cards[i − 1][0], cards[i − 1][1]);
28  │   │   cards[i][1] := n + 1;
29  │   │   cards[i][2] := 0;
30  return cards;
```

$\quad*$ If $\underline{p}(x_{i-1}, v_>) \leq \underline{p}(x_{i-1}, v_\leq)$ then $\underline{plen}(x_i) = \underline{plen}(x_{i-1}) + 1$.
$\quad*$ Else $\underline{plen}(x_i) = 1$.

Proof. If x_i is a $\{N_k\}$-variable then it cannot take a value $> k$. By convention $\underline{p}(x_i, v_>) = n + 1$ and $\underline{plen}(x_i) = 0$. Otherwise, recall that from Definition 3, the focus cardinality is the minimum possible number of $\{P_k\}$-sequences. If $0 < \underline{plen}(x_{i-1}) < len$ the last $\{P_k\}$-sequence can be extended by variable x_i within an assignment having the same focus cardinality than the one of $\langle x_0, x_1, \ldots, x_{i-1} \rangle$, thus $\min(\underline{p}(x_{i-1}, v_>), \underline{p}(x_{i-1}, v_\leq) + 1)$ and $\underline{plen}(x_i)$ is updated so as to remain the minimum length of a $\{P_k\}$-sequence containing x_i in an instantiation of $\langle x_0, x_1, \ldots, x_i \rangle$ of focus cardinality $\underline{p}(x_i, v_>)$. Otherwise, the focus cardinality will be increased by one if x_i takes a value $v > k$. We have $\underline{p}(x_i, v_>) = \min(\underline{p}(x_{i-1}, v_>) + 1, \underline{p}(x_{i-1}, v_\leq) + 1)$. Since we have to count a new $\{P_k\}$-sequence starting at x_i, $\underline{plen}(x_i) = 1$. □

Given $X = \langle x_0, x_1, \ldots, x_{n-1} \rangle$, Algorithm 3 uses the Lemmas to computes in $O(n)$ the quantities. It returns a matrix $cards$ of size $n \times 3$, such that at each index i:

$$cards[i][0] = \underline{p}(x_i, v_\leq); \ cards[i][1] = \underline{p}(x_i, v_>); \ cards[i][2] = \underline{plen}(x_i)$$

We can then compute for each x_i $\underline{s}(x_i, v_>)$, $\underline{s}(x_{i-1}, v_\leq)$ and $\underline{slen}(x_i)$ (the equivalent of $\underline{plen}(x_i)$ for suffixes), by using Lemmas 3, 4 and 5 with X sorted in the reverse order: $\langle x_{n-1}, x_{n-2}, \ldots, x_0 \rangle$. To estimate for each variable $minCard(X_i^\leq)$ and $minCard(X_i^>)$, we have to aggregate the quantities on prefixes and suffixes.

Property 3. $minCard(X_i^\leq) = \underline{p}(x_i, v_\leq) + \underline{s}(x_i, v_\leq)$ and $minCard(X_i^>)$ is equal to:

- $\underline{p}(x_i, v_>) + \underline{s}(x_i, v_>) - 1$ if and only if $\underline{plen}(x_i) + \underline{slen}(x_i) - 1 \leq len$.
- $\underline{p}(x_i, v_>) + \underline{s}(x_i, v_>)$ otherwise.

Proof. Any pair of instantiations of focus cardinality respectively equal to $\underline{p}(x_i, v_\leq)$ and $\underline{s}(x_i, v_\leq)$ correspond to disjoint $\{P_k\}$-sequences (which do not contain x_i). Thus the quantities are independent and can be summed. With respect to $minCard(X_i^>)$, the last current $\{P_k\}$-sequence taken into account in $\underline{p}(x_i, v_>)$ and $\underline{s}(x_i, v_>)$ contains x_i. Thus, their union (of length $\underline{plen}(x_i) + \underline{slen}(x_i) - 1$) forms a unique $\{P_k\}$-sequence, from which the maximum-length sub-sequence containing x_i should not be counted twice when it is not strictly larger than len. $\square$

Changing one value in an instantiation modifies its focus cardinality of at most one.

Property 4. Let $I[X] = \langle v_0, v_1, \ldots, v_{n-1} \rangle$ be an instantiation of focus cardinality v_c and $x_i \in X$, and $I'[X] = \langle v'_0, v'_1, \ldots, v'_{n-1} \rangle$ be the instantiation such that $\forall j \in \{0, 1, \ldots, n-1\}, j \neq i, v_j = v'_j$ and: *(1)* If $v_i > k$ then $v'i \leq k$. *(2)* If $v_i \leq k$ then $v'_i > k$. The focus cardinality v'_c of $I'[X]$ is such that $|v_c - v'_c| \leq 1$.

Proof. Assume first that $v_i > k$. x_i belongs to a $\{P_k\}$-sequence p. Let s be the length of this $\{P_k\}$-sequence within $I[X]$. We can split p into $p_1 = \langle x_k, x_{k+1}, \ldots, x_{i-1} \rangle$, $p_2 = \langle x_i \rangle$, $p_3 = \langle x_{i+1}, x_{i+1}, \ldots, x_l \rangle$ ($p1$ and/or $p3$ can be empty). Let q_1, q_3 and r_1, r_3 be positive or null integers such that $r_1 < len, r_3 < len$ and $s = q_1 \times len + r_1 + 1 + q_3 \times len + r_3$. By construction, the maximum contribution of the variables in p to the focus cardinality of $I'[X]$ (that is, with $x_i \leq k$), is equal to $q_1 + 1 + q_3 + 1 = q_1 + q_3 + 2$. With respect to $I[X]$, the contribution is then equal to $q_1 + q_3 + \lceil \frac{r_1 + r_3 + 1}{len} \rceil$. The minimum value of $\lceil \frac{r_1 + r_3 + 1}{len} \rceil$ is 1. In this case the property holds. The minimum contribution of the variables in p to the focus cardinality of $I'[X]$ is equal to $q_1 + q_3$. In this case, with respect to $I[X]$ the maximum value of $\lceil \frac{r_1 + r_3 + 1}{len} \rceil$ is $\lceil \frac{1}{len} \rceil = 1$, the property holds. The last intermediary case is when the contribution of the variables in p to the focus cardinality of $I'[X]$ is equal to $q_1 + q_3 + 1$. The minimum value of $\lceil \frac{r_1 + r_3 + 1}{len} \rceil$ is 1 and its maximum is 2, the property holds. The reasoning for $v_i \leq k$ is symmetrical. $\square$

From Property 4, we know that domains of variables in X can be pruned only once y_c is fixed since the variation coming from a single variable in X is at most one.

Algorithm 4 shrinks $D(y_c)$ and all the variables in X in $O(n)$. It first calls Algorithm 3 to obtain $minCard(X)$ from $\min(\underline{p}(x_{n-1}, v_>)$ and $\underline{p}(x_{n-1}, v_\leq))$ and eventually shrinks $D(y_c)$. Then, it computes the data for suffixes, and uses Property 3 to reduce domains of variables in X according to $\max(y_c)$. Since removed values of variables in X cannot lead to a focus cardinality strictly less than $\max(y_c)$, it enforces GAC. Algorithm 4 does not directly modify domains: X and y_c are locally copied, and the filtered copies are returned. The reason is that we will use this algorithm in an extension of FOCUS in Section 5. To improve the readability, we assume that the solver raises an exception FAILEXCEPTION if a domain of one copy becomes empty.

Algorithm 4: FILTER($X = \langle x_0, x_1, \ldots, x_{n-1} \rangle, y_c, len, k$): Set of variables

---

1 $cards := \text{MINCARDS}(X, len, k)$;
2 Integer $lb := \min(cards[n-1][0], cards[n-1][1])$;
3 **if** $\min(y_c) < lb$ **then** $D(y_c) := D(y_c) \setminus [\min(y_c), lb[$;
4 **if** $\min(y_c) = \max(y_c)$ **then**
5 $\quad$ $sdrac := \text{MINCARDS}(\langle x_{n-1}, x_{n-2}, \ldots, x_0 \rangle, len, k)$;
6 $\quad$ **for** Integer $i := 0; i < n; i := i+1$ **do**
7 $\quad\quad$ **if** $cards[i][0] + sdrac[n-1-i][0] > \max(y_c)$ **then**
8 $\quad\quad\quad$ $D(x_i) := D(x_i) \setminus [\min(x_i), k[$;
9 $\quad\quad$ Integer $regret := 0$;
10 $\quad\quad$ **if** $cards[i][2] + sdrac[n-1-i][2] - 1 \leq len$ **then** $regret := 1$;
11 $\quad\quad$ **if** $cards[i][1] + sdrac[n-1-i][1] - regret > \max(y_c)$ **then**
12 $\quad\quad\quad$ $D(x_i) := D(x_i) \setminus]k, \max(x_i)]$;

13 return $X \cup \{y_c\}$;

Example 3. Consider FOCUS($X = \langle x_0, x_1, \ldots, x_4 \rangle, y_c, len, 0$) and $D(y_c) = \{1, 2\}$. *(1)* Assume $len = 2$ and $D(x_0)=D(x_2)=D(x_3)=\{1,2\}, D(x_1) = \{0\}$ and $D(x_4) =\{0,1,2\}$. Line 3 of Algorithm 4 removes $[1,2[$ from $D(y_c)$. Since the length of the $\{P_k\}$-sequence $\langle x_2, x_3 \rangle$ is equal to len, $cards[4][1]=3$ and $cards[4][2]=1$. $sdrac[5-1-4][1]=1$ and $sdrac[5-1-4][2]=1$. $regret=1$ and thus $3+1-regret=3>\max(y_c)$, $]0,2]$ is removed from $D(x_4)$ (line 12). *(2)* Assume now $len=3$ and $D(x_0)=D(x_2)=D(x_4)=\{1,2\}, D(x_1)=\{0\}$ and $D(x_3)=\{0,1,2\}$. Line 3 of Algorithm 4 removes $[1,2[$ from $D(y_c)$. Since value 0 for x_3 leads to a focus cardinality of 3 ($cards[3][0] =2$ and $sdrac[5-1-3][0]=1$), strictlty greater than $\max(y_c)$, $\langle x_2, x_3, x_4 \rangle$ must be a $\{P_k\}$-sequence (of length $3 \leq len$). Algorithm 4 removes value 0 from $D(x_3)$ (line 8).   ✳

5 Constraints Derived from FOCUS

5.1 Using a Variable for *len*

Assume that, in Example 1 of the Introduction, several companies offer leases with different maximum duration. We aim at computing the best possible configuration for each different offer, based on each maximum duration. To deal with this case, we can extend FOCUS so as to define *len* as a variable, with a discrete domain since the maximum durations of rentals are proper to each company. Another use of this extension is the case where the end-user wishes to compare for the same company several maximum package duration, enumerate several solutions, etc.

Algorithm 5: FILTERVARLEN($X = \langle x_0, x_1, \ldots, x_{n-1} \rangle, y_c, len, k$): Set of variables

---

1   IntegerVariable[][] $vars$:= new IntegerVariable[$|D(len)|$][];
2   Integer $j := 0$;
3   **foreach** $v_l \in D(len)$ **do**
4   **try** $vars[j]$:= FILTER(X, y_c, v_l, k); // the last variable is y_c
5   **catch** FAILEXCEPTION: $D(len) := D(len) \setminus \{v_l\}$; // in this case $vars[j] = null$
6      $j := j + 1$;
7   Integer $min_c := \max(y_c) + 1$;
8   $j := 0$;
9   **foreach** $v_l \in D(len)$ **do**
10   **if** $vars[j] \neq null$ **then** $min_c := \min(min_c, \min(vars[j][n]))$;
11     $j := j + 1$;
12   $D(y_c) := D(y_c) \setminus [\min(y_c), min_c[$;
13   **for** Integer $i := 0; i < n; i := i + 1$ **do**
14   Integer $min_i := \max(x_i) + 1$;
15   Integer $max_i := \min(x_i) - 1$;
16     $j := 0$;
17   **foreach** $v_l \in D(len)$ **do**
18   **if** $vars[j] \neq null$ **then**
19         $min_i := \min(min_i, \min(vars[j][i]))$;
20         $max_i := \max(max_i, \max(vars[j][i]))$;
21       $j := j + 1$;
22     $D(x_i) := D(x_i) \setminus ([\min(x_i), min_i[\, \cup \,]max_i, \max(x_i)])$;
23   return $X \cup \{y_c\} \cup \{len\}$;

The filtering algorithm of this extension of FOCUS uses following principle: For each value v_l in $D(len)$, we call FILTER(X, y_c, v_l, k) (Algorithm 4). If an exception FAILEXCEPTION is raised, v_l is removed from $D(len)$. Otherwise, we store the result of the filtering. At the end of the process, value $v \in D(x_i)$, $x_i \in X$, is removed from its domain if and only if it was removed by all the calls to FILTER(X, y_c, v_l, k) that did not raised an exception. Algorithm 5 implements this principle. Since it calls Algorithm 4 for each value in $D(len)$, it enforces GAC. Its time complexity is $O(n \times |D(len)|)$.

5.2   Harder Constraint on y_c

In Definition 1, the number of sequences in S could be constrained by an equality: $|S| = v_c$. When the maximum value for the variable y_c is taken, the number of counted disjoint sequences of length at most len is maximized. This maximum is equal to the number of $\{P_k, U_k\}$-sequences (we consider sequences of length one). The filtering is obvious. If, in addition, we modify the condition 3 of Definition 1 to make it stronger, for instance $3 \leq |s_i| \leq len$, it remains possible to compute recursively the maximum possible number of disjoint sequences by traversing X, similarly to the Lemmas used for the focus cardinality. Conversely, the aggregation of prefix and suffix data to obtain an algorithm in $O(n)$ is different. We did not investigate this point because we are not convinced of the practical significance of this variation of FOCUS.

6 Discussion: Related Work and Decomposition

Although it seems to be similar to a specialization of GROUP,[2] the FOCUS constraint cannot be represented using GROUP because of *len*: Using FOCUS, the variable in the sequence which directly precedes (or succeeds) a counted group of values strictly greater than the parameter k can also take itself a value strictly greater than k, which violates the notion of group.

To remain comparable with a filtering algorithm having a time complexity linear in the number of variables, the automaton-based reformulation of FOCUS should directly manipulate an automaton with counters, which are used to estimate the focus cardinality. We selected the paradigm presented in [1]. Under some conditions, this framework leads to a reformulation where a complete filtering is achieved, despite the counters.[3] This paradigm is based on automata derived from constraint checkers. We propose an automaton $\mathcal{A}$ representing FOCUS deduced from Algorithm 1, depicted by Figure 3.

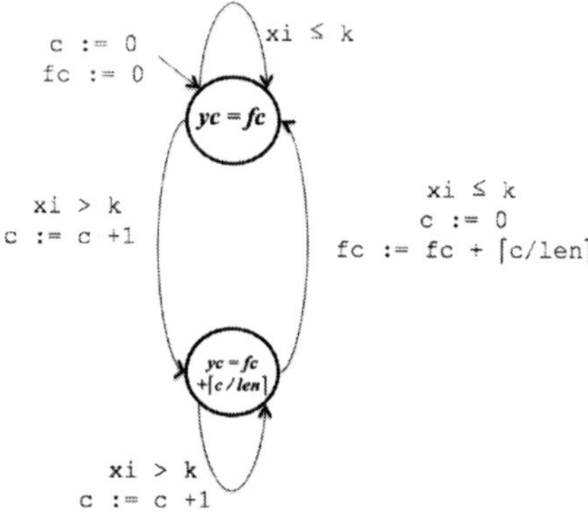

Fig. 3. Automaton with two counters c and fc, representing FOCUS. The two states are terminal. x_i denotes a variable in X and we consider the variable y_c and the value *len* of FOCUS.

As it is shown by figure 3, the automaton has two terminal states and maintains two counters c and fc, which represent respectively the size of the current traversed $\{P_k\}$-sequence and the focus cardinality. Therefore, when $i = n - 1$, fc is compared to the value of y_c. The principle is the same than in Algorithm 1: Each time a $\{P_k\}$-sequence ends, its contribution is added to the counter representing the focus cardinality. The automaton has two states and two counters. The constraint network $\mathcal{N}$ (hypergraph) encoding the automaton [1] is not Berge-Acyclic [4]. Propagating directly the constraints in $\mathcal{N}$ does not necessarily entail a complete filtering. However, within the automaton $\mathcal{A}$, the choice of the next transition only depends on the value of $x_i \in X$. Therefore, in $\mathcal{N}$, no signature constraints share a variable. From [1, p. 348-349], by enforcing pairwise consistency on the $O(n)$ pairs of transition constraints sharing more than one

---

[2] http://www.emn.fr/z-info/sdemasse/gccat/Cgroup.html
[3] Conversely to the COSTREGULAR constraint [3], for instance.

variable, we obtain a complete filtering. Three variables are shared: one representing the possible next states, and the variables for the two counters. In the worst case, pairwise consistency considers all the tuples for the shared variables. Counter variables (fc and c) have initially a domain of order $O(n)$ while the third domain is in $O(1)$, which leads to a time complexity in $O(n^3)$ for filtering FOCUS. When len and $\max(len)$ are small, this decomposition could be used. Thus, it is a contribution. The behavior with generic search strategies could be different for the decomposition and for the linear algorithm. However, using the decomposition requires to enforce pairwise consistency, which is not simpler and less generic than a dedicated filtering algorithm, because of events propagation. The decomposition is dependent on the priority rules of the solver.

7 Experiments

Since our filtering algorithm is complete, in $O(n)$, and without heavy data structures, performing benchmarks of the constraint isolated from a problem is not relevant. This section analyzes the impact of FOCUS on the solutions of the *sorting chords* problem described in Section 3. We selected this example rather than, for instance, the cumulative problem of Section 3, because all the other constraints (ALLDIFF as well as the ternary table constraints) can be provided with a complete filtering algorithm. With respect to the variable sum, we define it as the objective to minimize. Thus, instead of an equality we enforce the constraint $sum \leq \sum cost_i$, which is also tractable. We use the solver Choco (`http://www.emn.fr/z-info/choco-solver/`) with the default variable and value search strategies (DomOverWDeg and MinValue), on a Mac OSX 2.2 GHz Intel Core i7 with 8GB of RAM memory. We are interested in two aspects:

1. Our constraint is used for refining an objective criterion and, from Property 4, we know that domains of variables in X can be pruned only once y_c is fixed. An important question is the following: Is our filtering algorithm significantly useful during the search, compared with a simple checker?
2. Are the instances harder and/or the value of the objective variable sum widely increased when we restrict the set of solutions by imposing FOCUS?

7.1 Pruning Efficacy

In a first experiment, we run sets of 100 random instances of the *sorting chords* problem, in order to compare the complete filtering of FOCUS with a simple checker. For each set, the maximum value of y_c is either 1 or 2, in order to consider instances where FOCUS as a important influence. We compare the pruning efficacy of FOCUS with a light version of FOCUS, where the propagation is reduced to the checker.

Table 1 summarizes the results on 8 and 9 chords (respectively 17 and 19 variables in the problem). The instances are all satisfiable and optima were always proved. With larger instances ($\geq$ 10 chords), optimum cannot be proved in a reasonable number of backtracks using the checker. In the table, $\overline{y_c} = \max(y_c)$, and len and k are the parameters of FOCUS. $nmax$ is the maximum possible number of changing notes between any two chords. Average values are given as integers. Table 1 clearly shows that,

Table 1. Comparison of Algorithm 4 with a checker, on the *sorting chords* problem. Each row represents 100 randomly generated instances. $\overline{y_c}$ is $\max(y_c)$. *nmax* indicates the maximum possible common notes between two chords. Optimum solutions were found for all the considered instances. "#backtracks" means the number of backtracks, "#fails" the number of fails, "#optimum with $sum > 0$" is the number of solutions with a non null objective value.

Instances			FOCUS(X, y_c, len, k)				CHECKER(X, y_c, len, k)			
nb. of chords	$\overline{y_c}$-*len* -k -$nmax$	#optimum with $sum > 0$	average #backtracks (of 100)	average #fails (of 100)	max. #backtracks (of 100)	average time (ms) (of 100)	average #backtracks (of 100)	average #fails (of 100)	max. #backtracks (of 100)	average time (ms) (of 100)
8	1-4-0-3	66	61	46	462	11	1518	951	15300	17
8	1-4-1-3	66	45	33	342	1	91	59	1724	2
8	2-4-0-3	66	47	34	247	1	58	42	455	1
8	2-4-1-3	66	45	33	301	1	44	32	349	1
8	1-6-0-4	84	198	141	2205	12	15952	10040	86778	213
8	1-6-1-4	84	114	79	767	3	1819	1117	45171	24
8	2-6-0-4	84	127	88	620	3	2069	1361	64509	28
8	2-6-1-4	84	118	81	724	3	250	168	7027	4
8	1-8-0-5	98	261	184	1533	6	39307	25787	167575	566
8	1-8-1-5	98	148	103	662	3	11821	7642	94738	168
8	2-8-0-5	98	164	113	803	4	21739	14400	173063	317
8	1-8-0-5	98	183	127	882	4	10779	6939	92560	153
8	1-8-0-6	99	290	203	1187	18	46564	30488	130058	690
8	1-8-1-6	99	238	166	1167	11	29256	19150	134882	438
8	2-8-0-6	99	221	152	1458	6	29455	19607	123857	445
8	2-8-1-6	99	209	144	1118	9	21332	14095	117768	329
9	1-9-0-4	88	415	299	4003	18	214341	133051	1095734	3244
9	1-9-1-4	88	268	185	2184	6	12731	7988	751414	203
9	2-9-0-4	88	270	188	2714	6	22107	14065	374121	337
9	2-9-1-4	88	266	182	3499	6	1364	941	92773	23
9	1-9-0-5	97	574	407	2437	26	360324	230167	1355934	6584
9	1-9-1-5	97	404	273	1677	11	62956	40277	881441	1150
9	2-9-0-5	97	451	309	3327	12	228072	147007	1124630	4263
9	2-9-1-5	97	386	260	1698	10	58421	37589	989900	1079

even with a GAC filtering algorithm for ALLDIFF and for the ternary table constraints COSTC($ch_i, ch_{i+1}, cost_i$), we can only solve small instances without propagating FOCUS, some of them requiring more than one million backtracks. Conversely, the number of backtracks for proving optimality is small and stable when Algorithm 4 is used for propagating FOCUS. Using the filtering algorithm of FOCUS is mandatory.

7.2   Impact on the Objective Value

In a second experiment, we run sets of 100 random instances of the *sorting chords* problem, in order to compare problems involving FOCUS with the same problems where FOCUS is removed from the model. The goal of this experiment is to determine, with respect to the sorting chords problem, whether the optimum objective value increases widely or not when FOCUS is imposed, as well as the time and backtracks required to obtain such an optimum value. Table 2 summarizes the results with a number of chords varying from 6 to 20 (13 to 51 variables in the problem). The instances are all satisfiable and optima were always found and proved. In each set, we count the number of instances with distinct objective values. For sake of space, we present the results for $k = 0$ and $len = 4$. Similar results were found for closed values of k and len. Despite the instances with FOCUS are more constrained with respect to the variables involved in the objective, Table 2 does not reveals any significant difference, both with respect

Table 2. Comparison of instances of the *sorting chords* problem with and without FOCUS. The column "#optimum equal with and without FOCUS" indicates the number of instances for which, with and without FOCUS, the optimum objective value is equal. "max. value of *sum*" indicates the maximum objective value among the 100 instances.

Instances				with FOCUS				without FOCUS			
nb. of chords	$\overline{y_c}$ -*len* -*k* -*nmax*	#optimum with $sum > 0$	#optimum equal with / without FOCUS	max. value of *sum*	average #backtracks (of 100)	max. #backtracks (of 100)	average time (ms) (of 100)	max. value of *sum*	average #backtracks (of 100)	max. #backtracks (of 100)	average time (ms) (of 100)
6	2-4-0-4	88	99	7	23	76	3	7	22	76	1
8	2-4-0-4	84	95	8	131	618	4	7	115	449	3
10	2-4-0-4	78	98	6	457	4579	11	6	360	3376	9
12	2-4-0-4	69	98	5	952	12277	28	5	1010	10812	27
16	2-4-0-4	43	100	4	4778	132019	153	4	6069	95531	189
20	2-4-0-4	7	100	3	15650	1316296	747	3	15970	1095399	679
6	2-4-0-6	97	96	13	37	113	1	13	37	121	1
8	2-4-0-6	99	93	11	218	1305	5	11	198	860	4
10	2-4-0-6	97	77	10	1247	5775	32	9	1159	10921	26
12	2-4-0-6	96	75	12	5092	34098	155	11	4844	54155	145
16	2-4-0-6	88	84	9	45935	724815	2002	9	73251	2517570	3407
20	2-4-0-6	79	91	8	264881	4157997	14236	6	174956	2918335	8284

Table 3. Comparison of the results for providing a first solution on large instances of the *sorting chords* problem, with and without FOCUS. The "average gap in *sum*" is the average of the difference, for each of the 100 instances, between the objective value without FOCUS and the objective value with FOCUS. This latter one (with FOCUS) is smaller or equal for all the instances.

Instances			with FOCUS				without FOCUS			
nb. of chords	$\overline{y_c}$ -*len* -*k* -*nmax*	average gap in *sum*	min(*sum*) / max(*sum*) (of 100)	average #backtracks (of 100)	max. #backtracks (of 100)	average time (ms) (of 100)	min(*sum*) / max(*sum*) (of 100)	average #backtracks (of 100)	max. #backtracks (of 100)	average time (ms) (of 100)
50	4-6-2-4	14	45/73	0	16	84	55/94	0	0	80
100	8-6-2-4	27	100/145	1	57	1168	119/175	0	0	1413
50	5-6-1-4	30	19/69	212	12698	101	55/94	0	0	85
100	10-6-1-4	63	40/113	93	3829	1002	119/175	0	0	1407

to the number of backtracks (and solving time) and the optimum objective value. We finally have differences in objective values at most equal to 2.

To complete our evaluation, we search for the first solution of larger problems, in order to compare the scale of size that can be reached with and without using FOCUS. The value heuristic assigns first the smaller value, which is semantically suited to the goal of the problem although we do not search for optimum solutions.

Results are compared in Table 3. They show that, with FOCUS, the number of backtracks grows for some instances when the parameters of FOCUS are shrunk (one of the instance with 50 chords required 12698 backtracks). However, the objective value of the first solution is systematically significantly better when FOCUS is set in the model. One explanation of these results is the following: Focusing the costs on a small number of areas within the sequence semantically tends to limit the value of their total sum.

8   Conclusion

We presented FOCUS, a new constraint for concentrating high costs within a sequence of variables. In the context of scheduling, many use cases for the constraint make sense. We proposed a $O(n)$ complete filtering algorithm, where n is the number of variables.

We proposed an automaton-based decomposition and we discussed extensions of our constraint. Our experiments investigated the importance of propagating FOCUS and the impact of FOCUS on the solutions of problems. Our results demonstrated that the complete filtering algorithm of FOCUS is mandatory for solving instances.

Acknowledgments. We thank Jean-Guillaume Fages for the helpful comments he provided on the paper.

References

1. Beldiceanu, N., Carlsson, M., Debruyne, R., Petit, T.: Reformulation of global constraints based on constraints checkers. Constraints 10(4), 339–362 (2005)
2. De Clercq, A., Petit, T., Beldiceanu, N., Jussien, N.: Filtering Algorithms for Discrete Cumulative Problems with Overloads of Resource. In: Lee, J. (ed.) CP 2011. LNCS, vol. 6876, pp. 240–255. Springer, Heidelberg (2011)
3. Demassey, S., Pesant, G., Rousseau, L.-M.: A cost-regular based hybrid column generation approach. Constraints 11(4), 315–333 (2006)
4. Janssen, P., Vilarem, M.-C.: Problèmes de satisfaction de contraintes: techniques de résolution et application à la synthèse de peptides. Research Report C.R.I.M. 54 (1988)
5. Pachet, F., Roy, P.: Musical harmonization with constraints: A survey. Constraints 6(1), 7–19 (2001)
6. Pesant, G., Régin, J.-C.: SPREAD: A Balancing Constraint Based on Statistics. In: van Beek, P. (ed.) CP 2005. LNCS, vol. 3709, pp. 460–474. Springer, Heidelberg (2005)
7. Petit, T., Poder, E.: Global propagation of side constraints for solving over-constrained problems. Annals of Operations Research, 295–314 (2011)
8. Petit, T., Régin, J.-C.: The ordered distribute constraint. International Journal on Artificial Intelligence Tools 20(4), 617–637 (2011)
9. Petit, T., Régin, J.-C., Bessière, C.: Meta constraints on violations for over constrained problems. In: Proc. IEEE-ICTAI, pp. 358–365 (2000)
10. Régin, J.-C.: A filtering algorithm for constraints of difference in CSPs. In: Proc. AAAI, pp. 362–367 (1994)
11. Schaus, P., Deville, Y., Dupont, P., Régin, J.-C.: The Deviation Constraint. In: Van Hentenryck, P., Wolsey, L.A. (eds.) CPAIOR 2007. LNCS, vol. 4510, pp. 260–274. Springer, Heidelberg (2007)
12. Schaus, P., Van Hentenryck, P., Régin, J.-C.: Scalable Load Balancing in Nurse to Patient Assignment Problems. In: van Hoeve, W.-J., Hooker, J.N. (eds.) CPAIOR 2009. LNCS, vol. 5547, pp. 248–262. Springer, Heidelberg (2009)
13. Truchet, C., Codognet, P.: Musical constraint satisfaction problems solved with adaptive search. Soft Comput. 8(9), 633–640 (2004)

Refining Abstract Interpretation Based Value Analysis with Constraint Programming Techniques*

Olivier Ponsini, Claude Michel, and Michel Rueher

University of Nice–Sophia Antipolis, I3S/CNRS
BP 121, 06903 Sophia Antipolis Cedex, France
`firstname.lastname@unice.fr`

Abstract. Abstract interpretation based value analysis is a classical approach for verifying programs with floating-point computations. However, state-of-the-art tools compute an over-approximation of the variable values that can be very coarse. In this paper, we show that constraint solvers can significantly refine the approximations computed with abstract interpretation tools. We introduce a hybrid approach that combines abstract interpretation and constraint programming techniques in a single static and automatic analysis. RAiCP, the system we developed is substantially more precise than FLUCTUAT, a state-of-the-art static analyser. Moreover, it could eliminate 13 false alarms generated by FLUCTUAT on a standard set of benchmarks.

Keywords: Program verification, Floating-point computation, Constraint solvers over floating-point numbers, Constraint solvers over real number intervals, Abstract interpretation-based approximation.

1 Introduction

Programs with floating-point computations control complex and critical physical systems in various domains such as transportation, nuclear energy, or medicine. Floating-point computations are an additional source of errors and famous computer bugs are due to errors in floating-point computations, e.g., the Patriot missile failure. Floating-point computations are usually derived from mathematical models on real numbers [14]. However, real and floating-point computation models are different: for the same sequence of operations, floating-point numbers do not behave identically to real numbers. For instance, with binary floating-point numbers, some decimal real numbers are not representable (e.g., 0.1 has no exact representation), arithmetic operators are not associative and may be subject to phenomena such as absorption (e.g., $a + b$ is rounded to a when a is far greater than b) or cancellation (subtraction of nearly equal operands after rounding that only keeps the rounding error).

---

* This work was partially supported by ANR VACSIM (ANR-11-INSE-0004), ANR AEOLUS (ANR-10-SEGI-0013), and OSEO ISI PAJERO projects.

M. Milano (Ed.): CP 2012, LNCS 7514, pp. 593–607, 2012.

Value analysis is often used to check the absence of run-time errors, such as invalid integer or floating-point operations, as well as simple user assertions [8]. Value analysis can also help with estimating the accuracy of floating-point computations with respect to the same sequence of operations in an idealized semantics of real numbers. Existing automatic tools are mainly based on abstract interpretation techniques. For instance, FLUCTUAT [9], a state-of-the-art static analyzer, computes an over-approximation of the domains of the variables for a C program considered with a semantics on real numbers. It also computes an over-approximation of the error due to floating-point operations at each program point. However, these over-approximations may be very coarse even for usual programming constructs and expressions. As a consequence, numerous false alarms[1]—also called false positives—may be generated.

In this paper, we introduce a hybrid approach for the value analysis of floating-point programs that combines abstract interpretation (AI) and constraint programming techniques (CP). We show that constraint solvers over floating-point and real numbers can significantly refine the over-approximations computed by abstract interpretation. RAICP, the system we developed, uses both FLUCTUAT and the following constraint solvers:

- REALPAVER [17], a safe and correct solver for constraints over real numbers,
- FPCS [21,20], a safe and correct solver for constraints over floating-point numbers.

Experiments show that RAICP is substantially more precise than FLUCTUAT, especially on C programs that are difficult to handle with abstract interpretation techniques. This is mainly due to the refutation capabilities of filtering algorithms over the real numbers and the floating-point numbers used in RAICP. RAICP could also eliminate 13 false alarms generated by FLUCTUAT on a set of 57 standard benchmarks proposed by D'Silva et al [12] to evaluate CDFL, a program analysis tool that embeds an abstract domain in the conflict driven clause learning algorithm of a SAT solver. Moreover, RAICP is on average at least 5 times faster than CDFL on this set of benchmarks.

Section 2 illustrates our approach on a small example. Basics on the techniques and tools we use are introduced in Section 3. Next section is devoted to related work. Section 5 details our approach whereas experiments are analysed in Section 6.

2   Motivation

In this section, we illustrate our approach on a small example. The program in Fig. 1 is mentioned in [13] as a difficult program for abstract interpretation based analyses. On floating-point numbers, as well as on real numbers, this function

---

[1] A false alarm corresponds to the case when the abstract semantics intersects the forbidden zone, i.e., erroneous program states, while the concrete semantics does not intersect this forbidden zone. So, a potential error is signaled which can never occur in reality (see http://www.di.ens.fr/~cousot/AI/IntroAbsInt.html).

```
1 /* Pre-condition :  x ∈ [0, 10] */
2 double conditional(double x) {
3   double y = x*x - x;
4   if (y >= 0)
5     y = x/10;
6   else
7     y = x*x + 2;
8   return y; }
```

Fig. 1. Example 1

returns a value in the interval $[0, 3]$. Indeed, from the conditional statement of line 4, we can derive the following information:

- `if` branch: $x = 0$ or $x \geq 1$, and thus $y \in [0, 1]$ at the end of this branch;
- `else` branch: $x \in]0, 1[$, and thus $y \in]2, 3[$ at the end of this branch.

However, classical abstract domains (e.g., intervals, polyhedra), as well as the abstract domain of *zonotopes* used in FLUCTUAT, fail to obtain a good approximation of this value. The best interval obtained with these abstractions is $[0, 102]$, both over the real numbers and the floating-point numbers. The difficulty for these analyses is to intersect the abstract domains computed for y at lines 3 and 4. Actually, they are unable to derive from these statements any constraint on x. As a consequence, in the `else` branch, they still estimate that x ranges over $[0, 10]$.

We propose here to compute an approximation of the domains in both execution paths. On this example, CSP filtering techniques are strong enough to reduce the domains of the variables. Consider for instance the constraint system over the real numbers $\{y_0 = x_0 * x_0 - x_0, y_0 < 0, y_1 = x_0 * x_0 + 2, x_0 \in [0, 10]\}$ which corresponds to the execution path[2] through the `else` branch of the function `conditional`. From the constraints $y_0 = x_0 * x_0 - x_0$ and $y_0 < 0$, the interval solver over the real numbers we use can reduce the initial domain of x_0 to $[0, 1]$. This reduced domain is then used to compute the one of y_1 *via* the constraint $y_1 = x_0 * x_0 + 2$, which yields $y_1 \in [2, 3.001]$. Likewise, our constraint solver over the floating-point numbers will reduce x_0 to $[4.94 \times 10^{-324}, 1.026]$ and y_1 to $[2, 3.027]$.

To sum up, we explore the control flow graph (CFG) of a program and stop each time two branches join. There, we build one constraint system per branch that reaches the join point. Then, we use filtering techniques on these systems to reduce the domains of the variables computed by FLUCTUAT at this join point. Exploration goes on with the reduced domains. CFG exploration is performed on-the-fly. Branches are cut as soon as an inconsistency of the constraint system is detected by a local filtering algorithm. Table 1 collects the results obtained by the different techniques on the example of the function `conditional`. On this

---

[2] Statements are converted into DSA (Dynamic Single Assignment) form where each variable is assigned exactly once on each program path [2].

Table 1. Return domain of the `conditional` function

	Domain	Time
Exact real and floating-point domains	$[0, 3]$	–
FLUCTUAT (real and floating-point domains)	$[0, 102]$	0.1 s
FPCS (floating-point domain)	$[0, 3.027]$	0.2 s
REALPAVER (real domain)	$[0, 3.001]$	0.3 s

example, contrary to FLUCTUAT, our approach computes very good approximations. Analysis times are very similar. In [13], the authors proposed an extension to the zonotopes—named *constrained zonotopes*—which attempts to overcome the issue of program conditional statements. This extension is defined for the real numbers and is not yet implemented in FLUCTUAT. The approximation computed with *constrained zonotopes* is better than the one of FLUCTUAT (the upper bound is reduced to 9.72) but remains less precise than the one computed with REALPAVER.

3   Background

Before going into the details, we recall basics on abstract interpretation and FLUCTUAT, as well as on the constraint solvers REALPAVER and FPCS used in our implementation.

Abstract interpretation[3] consists in considering an abstract semantics, that is a super-set of the concrete program semantics. The abstract semantics covers all possible cases, thus, if the abstract semantics is safe (i.e. does not intersect the forbidden zone) then so is the concrete semantics.

FLUCTUAT is a static analyzer for C programs specialized in estimating the precision of floating-point computations[4] [9]. FLUCTUAT compares the behavior of the analyzed program over real numbers and over floating-point numbers. In other words, it allows to specify ranges of values for the program input variables and computes for each program variable v:

- bounds for the domain of variable v considered as a real number;
- bounds for the domain of variable v considered as a floating-point number;
- bounds for the maximum error between real and floating-point values;
- the contribution of each statement to the error associated with variable v ;
- the contribution of the input variables to the error associated with variable v.

---

[3] See `http://www.di.ens.fr/~cousot/AI/IntroAbsInt.html` for a nice informal introduction.

[4] FLUCTUAT is developed by CEA-LIST (`http://www-list.cea.fr/validation _en.html`) and was successfully used for industrial applications of several tens of thousands of lines of code in transportation, nuclear energy, or avionics areas.

FLUCTUAT proceeds by abstract interpretation. It uses the weakly relational abstract domain of zonotopes [15]. Zonotopes are sets of affine forms that preserve linear correlations between variables. They offer a good trade-off between performance and precision for floating-point and real number computations. Indeed, the analysis is fast and scales well, processes accurately linear expressions, and keeps track of the statements involved in the loss of accuracy of floating-point computations. To increase the analysis precision, FLUCTUAT allows to use arbitrary precision numbers or to subdivide up to two input variable intervals. However, over-approximations computed by FLUCTUAT may be very large because the abstract domains do not handle well conditional statements and non-linear expressions.

REALPAVER is an interval solver for numerical constraint systems over the real numbers[5] [17]. Constraints can be non-linear and can contain the usual arithmetic operations and transcendental elementary functions.

REALPAVER computes reliable approximations of continuous solution sets using correctly rounded interval methods and constraint satisfaction techniques. More precisely, the computed domains are closed intervals bounded by floating-point numbers. REALPAVER implements several partial consistencies: box, hull, and $3B$ consistencies. An approximation of a solution is described by a box, i.e., the Cartesian product of the domains of the variables. REALPAVER either proves the unsatisfiability of the constraint system or computes small boxes that contains all the solutions of the system.

The REALPAVER modeling language does not provide strict inequality and not-equal operators, which can be found in conditional expressions in programs. As a consequence, in the constraint systems generated for REALPAVER, strict inequalities are replaced by non strict ones and constraints with a not-equal operator are ignored. This may lead to over-approximations, but this is safe since no solution is lost.

FPCS is a constraint solver designed to solve a set of constraints over floating-point numbers without losing any solution [21,20]. It uses $2B$-consistency [19] along with projection functions adapted to floating-point arithmetic [22,4].

The main difficulty lies in computing inverse projection functions that keep all the solutions. Indeed, direct projections only requires a slight adaptation of classical results on interval arithmetic, but inverse projections do not follow the same rules because of the properties of floating-point arithmetic. More precisely, each constraint is decomposed into an equivalent binary or ternary constraint by introducing new variables if necessary. A ternary constraint $x = y \odot_f z$, where $\odot_f$ is an arithmetic operator over the floating-point numbers, is decomposed into three projection functions:

- the direct projection, $\Pi_x(x = y \odot_f z)$;
- the first inverse projection, $\Pi_y(x = y \odot_f z)$;
- the second inverse projection, $\Pi_z(x = y \odot_f z)$.

A binary constraint of the form $x \odot_f y$, where $\odot_f$ is a relational operator (among ==, !=, <, <=, >, and >=), is decomposed into two projection functions: $\Pi_x(x \odot_f y)$ and $\Pi_y(x \odot_f y)$. The computation of the approximation of these projection functions is mainly inspired from interval arithmetic and benefits from floating-point numbers being a totally ordered finite set.

FPCS also implements stronger consistencies—e.g., kB-consistencies [19]—to deal with the classical issues of multiple occurrences and to reduce more substantially the bounds of the domains of the variables.

The floating-point domains handled by FPCS also include infinities. Moreover, FPCS handles all the basic arithmetic operations, as well as most of the usual mathematical functions. Type conversions are also correctly processed.

4   Related Work

Different methods address static validation of programs with floating-point computations: abstract interpretation based analyses, proofs of programs with proof assistants or with decision procedures in automatic solvers.

Analyses based on abstract interpretation capture rounding errors due to floating-point computation in their abstract domains. They are usually fast, automatic, and scalable. However, they may lack of precision and they are not tailored for automatically generating a counter-example, that is to say, input variable values that violate some assertion in a program. ASTRÉE [8] is probably one of the most famous tool in this family of methods. The tool estimates the value of the program variables at every program point and can show the absence of run-time errors, that is the absence of behavior not defined by the programming language, e.g., division by zero, arithmetic overflow. As said before, FLUCTUAT estimates in addition the accuracy of the floating-point computations, that is, a bound on the difference between the values taken by variables when the program is given a real semantics and when it is given a floating-point semantics [9].

Proof assistants like Coq [3] or HOL [18] allow their users to formalize floating-point arithmetic. Proofs of program properties are done manually in the proof assistants which guarantee proof correctness. Even though some parts of the proofs may be automatized, these tools usually require a lot of user interaction. Moreover, when a proof strategy fails to prove a property, the user often does not know whether the property is false or another strategy could prove it. Like abstract interpretation, proof assistants usually do not provide automatic generation of counter-examples for false properties. The Gappa tool [11] combines interval arithmetic and term rewriting from a base of theorems. The theorems rewrite arithmetic expressions so as to compensate for the shortcomings of interval arithmetic, e.g., loss of dependency between variables. Whenever the computed intervals are not precise enough, theorems can be manually introduced or the input domains can be subdivided. The cost of this semi-automatic method is then considerable. In [1], the authors propose axiomatizing floating-point arithmetic within first-order logic to automate the proofs conducted in proof assistants such as Coq by calling external SMT (Satisfiability Modulo

Theories) solvers and Gappa. Their experiments show that human interaction with the proof assistant is still required.

The classical bit-vector approach of SAT solvers is ineffective on programs with floating-point computations because of the size of the domains of floating-point variables and the cost of bit-vector operations. An abstraction technique was devised for CBMC in [5]. It is based on under and over-approximation of floating-point numbers with respect to a given precision expressed as a number of bits of the mantissa. However, this technique remains slow. D'Silva et al [12] developed recently CDFL, a program analysis tool that embeds an abstract domain in a conflict driven clause learning algorithm of a SAT solver. CDFL is based on a sound and complete analysis for determining the range of floating-point variables in control software. In [12] the authors state that CDFL is more than 200 times faster than CBMC. In Section 6 we compare the performances of CDFL and rAiCp on a set of benchmarks proposed by D'Silva et al.

Links between abstract interpretation and constraint logic programming have been studied at a theoretical level (e.g., [6]) and recent work investigate the use of abstract interpretation and abstract domains in the context of constraint programming. In [10], the authors introduce a new global constraint to model iterative arithmetic relations between integer variables. The associated filtering algorithm is based on abstract interpretation over polyhedra. In [23], the authors propose to use the octagonal abstract domain, which proved efficient in abstract interpretation, to represent the variable domains in a continuous constraint satisfaction problem. Then, they generalize local consistency and domain splitting to this octagonal representation. In this paper, we show how abstract interpretation and constraint programming techniques can complement each other for the static analysis of floating-point programs.

5   rAiCp, a Hybrid Approach

The approach we propose here is based on successive explorations and merging steps. More precisely, we call FLUCTUAT to compute a first approximation of the variable values at the first program node of the CFG where two branches join. Then, we build one constraint system per branch and use filtering techniques to reduce the domains of the variables computed by FLUCTUAT. Reduced domains obtained for each branch are merged and exploration goes on with the result of the merge.

5.1   Control Flow Graph Exploration

The CFG of a program is explored using a forward analysis going from the beginning to the end of the program. Statements are converted into DSA (Dynamic Single Assignment) form where each variable is assigned exactly once on each program path [2]. Lengths of the paths are bounded since loops are unfolded a bounded number of times, after which they are abstracted by the domains computed by abstract interpretation. At any point of an execution path, the possible states of a program are represented by a constraint system over the program variables. Domains of the variables are intervals over the real numbers in the constraint store of REALPAVER; domains are intervals over the floating-point numbers that correspond to the `int`, `float` and `double` machine types of the C language[6] in the constraint store of FPCS. Each program statement adds new constraints and variables to these constraint stores. This technique for representing programs by constraint systems was introduced for bounded verification of programs in CPBPV [7]. The implementation of the approach proposed in this paper relies on libraries developed for CPBPV.

CFG exploration is performed on-the-fly and unreachable branches are interrupted as soon as an inconsistency is detected in the constraint store. We collect constraints between two join points in the CFG. If, for all executable paths between these points, the constraint systems are inconsistent for some interval I of an output variable x, then we can remove the interval I from the domain of x. Note that we differentiate between program *input* variables, whose domains cannot be reduced, and program *output* variables, whose domains depend on the program computations and input variable domains, and thus can be reduced.

Merging program states at each join point not only allows a tight cooperation between FLUCTUAT and the constraint solvers but also limits the number of executable paths to explore.

5.2   Filtering Techniques

We use constraint filtering techniques for two different purposes in RAICP:

- elimination of unreachable branches during CFG exploration;
- reduction of the domain of the variables at CFG join points.

On floating-point numbers constraint systems, we perform $3B(w)$-consistency filtering with FPCS; on real numbers constraint systems, we perform a $BC5$-consistency filtering in paving mode with REALPAVER[7].

6   Experiments

In this section, we compare in detail FLUCTUAT and RAICP on programs that are representative of FLUCTUAT limitations. We also compare RAICP to a state-of-the-art tool, CDFL on the benchmarks provided by the authors of the latter system.

All results were obtained on an Intel Core 2 Duo at 2.8 GHz with 4 GB of memory running Linux using FLUCTUAT version 3.8.73, REALPAVER version 0.4 and the downloadable version of CDFL. All the programs are available at `http://users.polytech.unice.fr/~rueher/Benchs/RAICP`.

---

[6] Note that the behavior of programs containing floating-point computations may vary with the programming language or the compiler, but also, with the operating system or the hardware architecture. We consider here C programs, compiled with GCC without any optimization option and intended to be run on an x86 architecture managed by a 32-bit Linux operating system.

[7] $BC5$-consistency is a combination of interval Newton method, hull-consistency and box-consistency.

Table 2. Domains of the roots of the `quadratic` function

		conf. #1: $a \in [-1, 1]$ $b \in [0.5, 1]$ $c \in [0, 2]$			conf. #2: $a, b, c \in [1, 1 \times 10^6]$		
		x0	x1	Time	x0	x1	Time
$\mathbb{R}$	Fluctuat	$[-\infty, \infty]$	$[-\infty, \infty]$	0.14 s	$[-2 \times 10^6, 0]$	$[-1 \times 10^6, 0]$	0.14 s
	rAiCp	$[-\infty, 0]$	$[-8.006, \infty]$	1.55 s	$[-1 \times 10^6, 0]$	$[-5.186 \times 10^5, 0]$	0.58 s
$\mathbb{F}$	Fluctuat	$[-\infty, \infty]$	$[-\infty, \infty]$	0.13 s	$[-2 \times 10^6, 0]$	$[-1 \times 10^6, 0]$	0.13 s
	rAiCp	$[-\infty, 0]$	$[-8.125, \infty]$	0.39 s	$[-1 \times 10^6, 0]$	$[-3\,906.26, 0]$	0.39 s

6.1 Improvements over Fluctuat

We show here how our approach improves the approximations computed by
Fluctuat on programs with conditionals, non-linearities, and loops.

Conditionals: The first benchmark concerns conditional statements, for which
abstract domains need to be intersected with the condition of the conditional
statement. The function `gsl_poly_solve_quadratic` comes from the GNU scientific library and contains many of these conditional statements. It computes
the real roots of a quadratic equation $ax^2 + bx + c$ and puts the results in variables
x0 and x1.

Table 2 shows analysis times and approximations of the domains of variables
x0 and x1 for two configurations of the input variables. The first two rows present
the results of Fluctuat and rAiCp (with RealPaver) over the real numbers.
The next two rows present the results of Fluctuat and rAiCp (with FPCS)
over the floating-point numbers.

In the first configuration, Fluctuat's over-approximation is so large that
it does not give any information on the domain of the roots, whereas rAiCp
drastically reduce these domains both over $\mathbb{R}$ and $\mathbb{F}$. However, intersection of
abstract domains has not always such a significant impact on the bounds of all
domains as illustrated by the domain over $\mathbb{F}$ of x0 in the second configuration.

To increase analysis precision, Fluctuat allows to divide the domains of at
most two input variables into a given number of sub-domains. Analyses are then
run over each combination of sub-domains and the results are merged. Finding
appropriate subdivisions of the domains is a critical issue: subdividing may not
improve the analysis precision, but it always increases the analysis time. Table 3
reports the results with 50 subdivisions when only one domain is divided, and 30
when two domains are divided. Over $\mathbb{R}$, in the first configuration, the subdivisions
yield no improvement and, in the second configuration, the results are identical
to those over $\mathbb{F}$.

Subdividing domains can be quite time consuming with little gains in precision:

- In the first configuration, subdivisions of the domain of a lead to a significant
  reduction of the domain of x0. No subdivision combination could reduce the
  domain of x1.

Table 3. Domains over $\mathbb{F}$ for the `quadratic` function with input domains subdivided

	conf. #1		conf. #2	
	x0	Time	x1	Time
FLUCTUAT a subdivided	$[-\infty, \text{-}0]$	> 1 s	$[-1 \times 10^6, 0]$	> 1 s
FLUCTUAT b subdivided	$[-\infty, \infty]$	> 1 s	$[-5 \times 10^5, 0]$	> 1 s
FLUCTUAT c subdivided	$[-\infty, \infty]$	> 1 s	$[-1 \times 10^6, 0]$	> 1 s
FLUCTUAT a & b subdivided	$[-\infty, \text{-}0]$	> 10 s	$[-1.834 \times 10^5, 0]$	> 10 s
FLUCTUAT a & c subdivided	$[-\infty, \text{-}0]$	> 10 s	$[-1 \times 10^6, 0]$	> 10 s
FLUCTUAT b & c subdivided	$[-\infty, \infty]$	> 10 s	$[-5 \times 10^5, 0]$	> 10 s

Table 4. Domains of the return value of `sinus` and `rump` functions

		sinus $x \in [-1, 1]$		rump $x \in [7 \times 10^4, 8 \times 10^4]$ $y \in [3 \times 10^4, 4 \times 10^4]$	
		Domain	Time	Domain	Time
$\mathbb{R}$	FLUCTUAT	$[-1.009, 1.009]$	0.12 s	$[-1.168 \times 10^{37}, 1.992 \times 10^{37}]$	0.13 s
	RAICP	$[-0.842, 0.843]$	0.34 s	$[-1.144 \times 10^{36}, 1.606 \times 10^{37}]$	1.26 s
$\mathbb{F}$	FLUCTUAT	$[-1.009, 1.009]$	0.12 s	$[-1.168 \times 10^{37}, 1.992 \times 10^{37}]$	0.13 s
	RAICP	$[-0.853, 0.852]$	0.22 s	$[-1.168 \times 10^{37}, 1.992 \times 10^{37}]$	0.22 s

- In the second configuration, the best reduction of the domain of `x1` is obtained by subdividing the domains of both a and b. The gain remains however quite small and no subdivision combination could reduce the domain of `x0`.

RAICP turns out to be more efficient: it often improves the precision of the approximation and requires less time than the subdividing process of FLUCTUAT. Moreover, RAICP could also take advantage of the subdivision technique.

Non-linearity: The abstract domain used by FLUCTUAT is based on affine forms that do not allow an exact representation of non-linear operations: the image of a zonotope by a non-linear function is not a zonotope in general. Non-linear operations are thus over-approximated. FPCS handles the non-linear expressions better. This is illustrated on the 7th-order Taylor series of function sinus (see Table 4, column `sinus`).

FPCS and REALPAVER also use approximations to handle non-linear terms, and thus, are not always more precise than FLUCTUAT. The second row of Table 4 shows that RAICP could not reduce the domain computed by FLUCTUAT for the `rump` polynomial program [24], a very particular polynomial designed to outline a catastrophic cancellation phenomenon.

Table 5. Domain of the return value of the `sqrt` and `bigLoop` functions

		sqrt #1: $x \in [4.5, 5.5]$		sqrt #2: $x \in [5, 10]$		bigLoop	
		Domain	Time	Domain	Time	Domain	Time
$\mathbb{R}$	Fluctuat	$[2.116, 2.354]$	0.13 s	$[2.098, 3.435]$	0.2 s	$[-\infty, \infty]$	0.15 s
	RAiCp	$[2.121, 2.346]$	0.35 s	$[2.232, 3.165]$	0.57 s	$[0, 10]$	0.8 s
$\mathbb{F}$	Fluctuat	$[2.116, 2.354]$	0.13 s	$[-\infty, \infty]$	0.2 s	$[-\infty, \infty]$	0.15 s
	RAiCp	$[2.121, 2.347]$	0.81 s	$[2.232, 3.168]$	1.59 s	$[0, 10]$	0.7 s

Loops: FLUCTUAT unfolds loops a bounded number of times[8] before applying the widening operator of abstract interpretation. The widening operator allows to find a fixed point for a loop without unfolding it completely. In RAiCp, we also unfold loops a user-defined number of times, after which the loop is abstracted by the invariant computed by abstract interpretation. Note that we can also use FLUCTUAT to estimate an upper bound on the number of necessary unfoldings [16].

`sqrt` is a program based on the so-called Babylonian method that computes an approximate value, with an error of 1×10^{-2}, of the square root of a number greater than 4. For the analysis of this program with two different input domains (see Table 5), ten unfoldings are sufficient to exit the loop. Both FLUCTUAT and RAiCp obtain accurate results over $\mathbb{R}$. Over $\mathbb{F}$, in the second configuration RAiCp shrinks the domain to $[2.232, 3.168]$ whereas FLUCTUAT couldn't achieve any reduction.

Program `bigLoop` contains non-linear expressions followed by a loop that iterates one million times. On such programs, it is not possible to completely unfold loops. FLUCTUAT fails to analyze accurately the loop in this program because of over-approximations of the non-linear expressions. RAiCp refines significantly the over-approximations computed by FLUCTUAT, even without any initial unfoldings. This example shows that a tight cooperation between CP and AI techniques can be very efficient.

Contributions of AI and CP: FLUCTUAT often yields a first approximation that is tight enough to allow efficient filtering with partial consistencies. Even though the same domain reductions can sometimes be achieved without starting from the approximation computed by FLUCTUAT (i.e., starting from $[-\infty, \infty]$), our experiments show that our approach usually benefits from the approximation computed by FLUCTUAT.

$3B$-consistency filtering works well with FPCS. $2B$-consistency is not strong enough to reduce the domains computed by FLUCTUAT whereas a stronger kB-consistency is too time-consuming. We experimented also with various consistencies implemented in REALPAVER: $BC5$, a combination of hull and box consistencies with interval Newton method, $HC4$, $3B$-consistency. $3B$-consistency was in general too time-consuming. $BC5$-consistency provided the best trade-off between time cost and domain reduction.

---

[8] Default value is ten times.

Table 6. Execution times (s) of CDFL, Fluctuat and rAiCp

	CDFL	Fluctuat	rAiCp		CDFL	Fluctuat	rAiCp
newton.1.1	0.5	0.12	0.62	eps_line1	0.12	0.11	0.28
newton.1.2	1.64	0.13	0.68	muller	0.13	0.11	0.2
newton.1.3	4.6	0.21	1.89	sac.10	2.49	1.25	1.6
newton.2.1	0.95	0.11	1.47	sac.20	2.46	1.38	1.75
newton.2.2	3.44	0.14	0.82	sac.30	2.49	1.39	1.68
newton.2.3	9.32	0.21	1.79	sac.40	2.47	1.38	1.68
newton.3.1	1.95	0.12*	1.3	sac.50	2.46	1.38	1.71
newton.3.2	5.61	0.13	1.13	sac.60	2.48	1.4	1.76
newton.3.3	15.9	0.22	2.35	sac.70	2.46	1.37	1.7
newton.4.1	1.07	0.12	1.74	sac.80	2.48	1.37	1.7
newton.4.2	8.4	0.13	1.82	sac.90	2.47	1.37	1.67
newton.4.3	23.63	0.22	2.49	sine.1	0.68	0.12	0.31
newton.5.1	1.76	0.12	1.83	sine.2	0.96	0.11	0.28
newton.5.2	14.61	0.13*	2.68	sine.3	0.5	0.11	0.28
newton.5.3	38.19	0.23*	4.01	sine.4	7.89	0.12*	0.3
newton.6.1	1.28	0.12	2.15	sine.5	0.68	0.12*	0.23
newton.6.2	2.33	0.13	8.85	sine.6	0.3	0.12*	0.26
newton.6.3	3.59	0.15	4.76	sine.7	0.13	0.12*	0.22
newton.7.1	1.8	0.12	2.23	sine.8	0.08	0.12	0.23
newton.7.2	1.57	0.14	1.59	square.1	0.16	0.12	0.26
newton.7.3	19.45	0.15	1.68	square.2	0.32	0.12	0.25
newton.8.1	0.41	0.11	0.86	square.3	0.7	0.11	0.25
newton.8.2	1.67	0.12	0.88	square.4	1.05	0.12*	0.22
newton.8.3	7.49	0.12	1.05	square.5	0.68	0.12*	0.22
GC4	0.04	0.14	0.23	square.6	0.55	0.11*	0.23
Poly	0.16	0.11	0.23	square.7	0.36	0.12*	0.23
Rump	0.02	0.11	0.21	square.8	0.06	0.12	0.21
Sterbenz	0	0.12	0.2	**Total**	**208.99**	**18.37**	**40.55**

6.2   Comparison with CDFL

CDFL [12] is a program analysis tool designed for proving the absence of run-time errors in critical programs. In [12], the authors show that CDFL is much more efficient than CBMC and much more precise than Astrée [8] for determining the range of floating-point variables on various programs.

We compare here rAiCp and CDFL on the set of benchmarks[9] proposed in [12]. The set consists of 57 benchmarks made from 12 programs by varying the input variable domains, the loop bounds, and the constants in the properties to check. We discarded two benchmarks as they are related to integer computations which are not the focus of this work. All the programs are based on academic numerical algorithms, except Sac which is generated from a Simulink controller model. The program properties are simple assertions on program variable domains.

---

[9] These benchmarks are available at http://www.cprover.org/cdfpl

Table 6 provides the runing time of RAICP, FLUCTUAT and CDFL. RAICP was only run with FPCS since the properties and the programs are both defined over the floating-point numbers.

All three analyses may report false alarms: i.e., they may answer a property is false while it is not. Actually, RAICP and CDFL correctly reported all the 33 true properties. FLUCTUAT gave 11 false alarms that are noted with * in FLUCTUAT columns of Table 6. The domain refinements performed by RAICP successfully eliminated the false alarms produced by FLUCTUAT.

On average, RAICP is 5 times faster than CDFL for the same precision. On some benchmarks, we observe a speed-up factor of 25. On average, RAICP is 2.2 times slower than FLUCTUAT used alone but this is largely compensated by the gain in precision.

7   Conclusion

In this paper, we introduced a new approach for computing tight intervals of floating-point variables of C programs. RAICP, the prototype we developed, relies on the static analyser FLUCTUAT and on FPCS and REALPAVER, two constraint solvers which are respectively correct over floating-point and real numbers. So, RAICP can exploit the refutation capabilities of partial consistencies to refine the domains computed by FLUCTUAT.

We showed that RAICP is fast and efficient on programs that are representative of the difficulties of FLUCTUAT (conditional constructs and non-linearities). Experiments on a significant set of benchmarks showed also that RAICP is as precise and faster than CDFL, a state-of-the-art tool for bound analysis and assertion checking on programs with floating-point computations.

This integration of AI and CP works well because often the first approximation of variable bounds computed by AI is small enough to allow efficient filtering with partial consistencies. In the case of FLUCTUAT, sets of affine forms abstract non-linear expressions and constraints. These sets constitute better approximations of linear constraint systems than the boxes used in interval-based constraint solvers. Nonetheless, they are less adapted for non-linear constraint systems where filtering techniques used in numeric CSP solving offer a more flexible and extensible framework.

Further work concerns a tighter integration of abstract interpretation and constraint solvers, for instance, at the abstract domain level instead of the interval domain level.

Acknowledgments. The authors gratefully acknowledge Sylvie Putot, Éric Goubault and Franck Védrine for their advice and help on using FLUCTUAT.

References

1. Ayad, A., Marché, C.: Multi-Prover Verification of Floating-Point Programs. In: Giesl, J., Hähnle, R. (eds.) IJCAR 2010. LNCS, vol. 6173, pp. 127–141. Springer, Heidelberg (2010)

2. Barnett, M., Leino, K.R.M.: Weakest-precondition of unstructured programs. Information Processing Letters 93(6), 281–288 (2005)
3. Boldo, S., Filliâtre, J.C.: Formal verification of floating-point programs. In: 18th IEEE Symposium on Computer Arithmetic, pp. 187–194. IEEE (2007)
4. Botella, B., Gotlieb, A., Michel, C.: Symbolic execution of floating-point computations. Software Testing, Verification and Reliability 16(2), 97–121 (2006)
5. Brillout, A., Kroening, D., Wahl, T.: Mixed abstractions for floating-point arithmetic. In: 9th International Conference on Formal Methods in Computer-Aided Design, pp. 69–76. IEEE (2009)
6. Codognet, P., Filé, G.: Computations, abstractions and constraints in logic programs. In: International Conference on Computer Languages (ICCL 1992), pp. 155–164. IEEE (1992)
7. Collavizza, H., Rueher, M., Hentenryck, P.V.: A constraint-programming framework for bounded program verification. Constraints Journal 15(2), 238–264 (2010)
8. Cousot, P., Cousot, R., Feret, J., Miné, A., Mauborgne, L., Monniaux, D., Rival, X.: Varieties of static analyzers: A comparison with ASTRÉE. In: 1st Joint IEEE/IFIP Symposium on Theoretical Aspects of Software Engineering, pp. 3–20. IEEE (2007)
9. Delmas, D., Goubault, E., Putot, S., Souyris, J., Tekkal, K., Védrine, F.: Towards an Industrial Use of FLUCTUAT on Safety-Critical Avionics Software. In: Alpuente, M., Cook, B., Joubert, C. (eds.) FMICS 2009. LNCS, vol. 5825, pp. 53–69. Springer, Heidelberg (2009)
10. Denmat, T., Gotlieb, A., Ducassé, M.: An Abstract Interpretation Based Combinator for Modelling While Loops in Constraint Programming. In: Bessière, C. (ed.) CP 2007. LNCS, vol. 4741, pp. 241–255. Springer, Heidelberg (2007)
11. de Dinechin, F., Lauter, C.Q., Melquiond, G.: Certifying the floating-point implementation of an elementary function using Gappa. IEEE Transactions on Computers 60(2), 242–253 (2011)
12. D'Silva, V., Haller, L., Kroening, D., Tautschnig, M.: Numeric Bounds Analysis with Conflict-Driven Learning. In: Flanagan, C., König, B. (eds.) TACAS 2012. LNCS, vol. 7214, pp. 48–63. Springer, Heidelberg (2012)
13. Ghorbal, K., Goubault, E., Putot, S.: A Logical Product Approach to Zonotope Intersection. In: Touili, T., Cook, B., Jackson, P. (eds.) CAV 2010. LNCS, vol. 6174, pp. 212–226. Springer, Heidelberg (2010)
14. Goldberg, D.: What every computer scientist should know about floating point arithmetic. ACM Computing Surveys 23(1), 5–48 (1991)
15. Goubault, E., Putot, S.: Static Analysis of Numerical Algorithms. In: Yi, K. (ed.) SAS 2006. LNCS, vol. 4134, pp. 18–34. Springer, Heidelberg (2006)
16. Goubault, E., Putot, S.: Static Analysis of Finite Precision Computations. In: Jhala, R., Schmidt, D. (eds.) VMCAI 2011. LNCS, vol. 6538, pp. 232–247. Springer, Heidelberg (2011)
17. Granvilliers, L., Benhamou, F.: Algorithm 852: RealPaver: an interval solver using constraint satisfaction techniques. ACM Transactions on Mathematical Software 32(1), 138–156 (2006)
18. Harrison, J.: A Machine-Checked Theory of Floating Point Arithmetic. In: Bertot, Y., Dowek, G., Hirschowitz, A., Paulin, C., Théry, L. (eds.) TPHOLs 1999. LNCS, vol. 1690, pp. 113–130. Springer, Heidelberg (1999)
19. Lhomme, O.: Consistency techniques for numeric CSPs. In: 13th International Joint Conference on Artificial Intelligence, pp. 232–238 (1993)
20. Marre, B., Michel, C.: Improving the Floating Point Addition and Subtraction Constraints. In: Cohen, D. (ed.) CP 2010. LNCS, vol. 6308, pp. 360–367. Springer, Heidelberg (2010)

21. Michel, C.: Exact projection functions for floating-point number constraints. In: 7th International Symposium on Artificial Intelligence and Mathematics (2002)
22. Michel, C., Rueher, M., Lebbah, Y.: Solving Constraints over Floating-Point Numbers. In: Walsh, T. (ed.) CP 2001. LNCS, vol. 2239, pp. 524–538. Springer, Heidelberg (2001)
23. Pelleau, M., Truchet, C., Benhamou, F.: Octagonal Domains for Continuous Constraints. In: Lee, J. (ed.) CP 2011. LNCS, vol. 6876, pp. 706–720. Springer, Heidelberg (2011)
24. Rump, S.M.: Verification methods: Rigorous results using floating-point arithmetic. Acta Numerica 19, 287–449 (2010)

Time-Dependent Simple Temporal Networks

Cédric Pralet and Gérard Verfaillie

ONERA – The French Aerospace Lab, F-31055, Toulouse, France
{cedric.pralet,gerard.verfaillie}@onera.fr

Abstract. Simple Temporal Networks (STN) allow conjunctions of minimum and maximum distance constraints between pairs of temporal positions to be represented. This paper introduces an extension of STN called Time-dependent STN (TSTN), which covers temporal constraints for which the minimum and maximum distances required between two temporal positions x and y are not necessarily constant but may depend on the assignments of x and y. Such constraints are useful to model problems in which the transition time required between two activities may depend on the time at which the transition is triggered. Properties of the new framework are analyzed, and standard STN solving techniques are extended to TSTN. The contributions are applied to the management of temporal constraints for so-called "agile" satellites.

1   Motivations

Managing temporal aspects is crucial when solving planning and scheduling problems. Indeed, the latter generally involve constraints on the earliest start times and latest end times of activities, precedence constraints between activities, no-overlapping constraints over sets of activities, or constraints over the minimum and maximum temporal distance between activities. In many cases, these constraints can be expressed as *simple* temporal constraints, written as $x - y \in [\alpha, \beta]$ with x, y two variables corresponding to temporal positions and α, β two constants. Such simple temporal constraints can be represented using the STN framework (Simple Temporal Networks [1]). This framework is appealing in practice due to the polynomial complexity of important operations such as determining the consistency of an STN or computing the earliest/latest times associated with each temporal variable of an STN, which is useful to maintain a schedule offering temporal flexibility. Another feature of STN is that they are often used as a basic element when solving more complex temporal problems such as DTN (Disjunctive Temporal Networks [2]).

In this paper, we propose an extension of the STN framework and of STN algorithms. This extension is illustrated on an application from the space domain. The latter corresponds to the management of Earth observation satellites such as those of the *Pleiades* system. Such satellites are moving around the Earth on a circular, low-altitude orbit (several hundreds of kilometers). They are said to be *agile*, which means that they have the capacity to move around the three axes (roll, pitch, and yaw). This agility allows them to point to the right, left, in front of, or behind of the Earth point at the vertical of the satellite at each time

M. Milano (Ed.): CP 2012, LNCS 7514, pp. 608–623, 2012.
© Springer-Verlag Berlin Heidelberg 2012

(Nadir). The mission of these satellites is to perform acquisitions of polygons at the Earth surface. These polygons are split into strips which must be scanned using an observation instrument fixed on the satellite. Scanning a given strip requires at any time a particular configuration of the satellite called an *attitude*, defined by a pointing direction and by a speed on each of the three axes.

In the agile satellite context, contrary to the simplified version of the 2003 ROADEF Challenge [3], the minimum transition time taken by a maneuver between the end of an acquisition i and the start of an acquisition j is not constant and depends on the precise time at which the first acquisition ends [4]. Transition times may vary of about ten seconds on the examples provided in Fig. 1, duration during which the satellite covers between 50 and 100 kilometers on the ground. Fig. 1 also shows how diverse minimum transition times evolution schemes can be. They are obtained by solving a continuous command optimization problem which takes into account the movement of the satellite on its orbit, the movement of points on the ground due to the rotation of Earth, and kinematic constraints restricting the possible attitude moves of the satellite.

This context motivates the need for a new modeling framework for problems in which the minimum transition time between two activities can depend on the precise time at which the transition is triggered. This aspect is close to work on *time-dependent scheduling* [5,6], where transition times take particular forms, piecewise constant or piecewise linear (these forms cannot be directly reused here). It also appears in applications such as congestion-aware logistics, in which traveling times depend on the hour of the day, due to traffic. The framework proposed, called Time-dependent STN, is first introduced (Sect. 2). Techniques are then defined for computing the earliest and latest times associated with each temporal variable (Sect. 3 and 4). These techniques are used for scheduling activities of an agile satellite, in the context of a local search algorithm (Sect. 5).

2   Towards Time-Dependent STN

2.1   Simple Temporal Networks (STN)

We first recall some definitions associated with STN. In the following, the domain of values of a variable x is denoted $\mathbf{d}(x)$.

Definition 1. *An STN is a pair (V, C) with V a finite set of continuous variables whose domain is a closed interval $[l, u] \subset \mathbb{R}$, and C a finite set of binary constraints of the form $x - y \in [\alpha, \beta]$ with $x, y \in V$, $\alpha \in \mathbb{R} \cup \{-\infty\}$, and $\beta \in \mathbb{R} \cup \{+\infty\}$. Such constraints are called* simple temporal constraints. *A solution to an STN (V, C) is an assignment of all variables in V satisfying all constraints in C. An STN is consistent iff it has at least one solution.*

Unary constraints $x \in [\alpha, \beta]$, including those defining the domains of possible values of variables, can be formulated as simple temporal constraints $x - x_0 \in [\alpha, \beta]$, with x_0 a variable of domain $[0, 0]$ playing the role of a temporal reference. Moreover, as $x - y \in [\alpha, \beta]$ is equivalent to $(x - y \leq \beta) \wedge (y - x \leq -\alpha)$, it is possible to use only constraints of the form $y - x \leq c$ with c some constant.

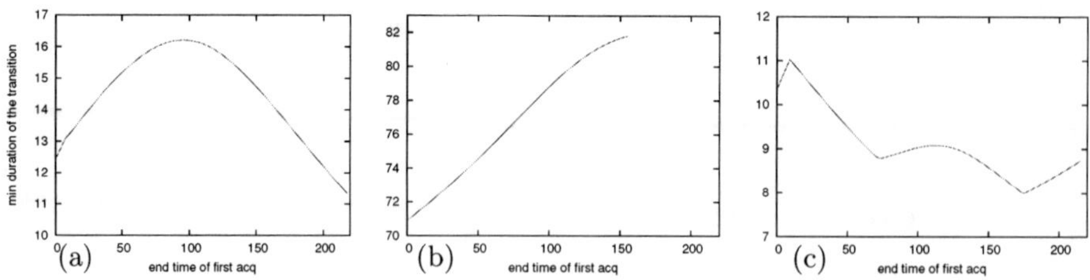

Fig. 1. Minimum durations, in seconds, for a satellite maneuver from a strip i ending at point of latitude-longitude $41°17'48''$N-$2°5'12''$E to a strip j starting at point of latitude-longitude $42°31'12''$N-$2°6'15''$E, for different scanning angles with regard to the trace of the satellite on the ground: (a) scan of i at $40°$ and scan of j at $20°$; (b) scan of i at $40°$ and scan of j at $-80°$; (c) scan of i at $90°$ and scan of j at $82°$

An important element associated with an STN is its *distance graph*. This graph contains one node per variable of the STN and, for each constraint $y - x \le c$ of the STN, one arc from x to y weighted by c. Based on this distance graph, the following results can be established [1] (some of these results are similar to earlier work on PERT and critical path analysis):

1. an STN is consistent iff its distance graph has no cycle of negative length;
2. if d_{0i} (resp. d_{i0}) denotes the length of the shortest path in the distance graph from the reference node labeled by x_0 to a node labeled by temporal variable x_i (resp. from x_i to x_0), then interval $[-d_{i0}, d_{0i}]$ gives the set of consistent assignments of x_i; the shortest paths can be computed for every i using Bellman-Ford's algorithm or arc-consistency filtering [7,8,9,10];
3. if d_{ij} (resp. d_{ji}) denotes the length of the shortest path from x_i to x_j (resp. x_j to x_i) in the distance graph, then interval $[-d_{ji}, d_{ij}]$ corresponds to the set of all possible temporal distances between x_i and x_j; shortest paths can be computed for every i, j using Floyd-Warshall's algorithm or path-consistency filtering [1,11,12,13], which produces the *minimal network* of the STN [14].

Example. Let us consider a simplified satellite scheduling problem. This problem involves 3 acquisitions acq_1, acq_2, acq_3 to be realized in order $acq_3 \rightarrow acq_1 \rightarrow acq_2$. For every $i \in [1..3]$, $Tmin_i$ and $Tmax_i$ denote the earliest start time and latest end time of acq_i, and Da_i denotes the duration of acq_i. The minimum durations of the transitions between the end of acq_3 and the start of acq_1, and between the end of acq_1 and the start of acq_2, are denoted $Dt_{3,1}$ and $Dt_{1,2}$ respectively. These durations are considered as constant in this first simplified version. We also consider two temporal windows $w_1 = [Ts_1, Te_1]$, $w_2 = [Ts_2, Te_2]$ during which data download to ground stations is possible. The satellite must download acq_2 followed by acq_3 in window w_1, before downloading acq_1 in window w_2. The duration taken by the download of acq_i is denoted Dd_i.

This problem can be modeled as an STN containing, for every acquisition acq_i ($i \in [1..3]$), (a) two variables sa_i and ea_i denoting respectively the start time and end time of the acquisition, with domains of values $\mathbf{d}(sa_i) = \mathbf{d}(ea_i) =$

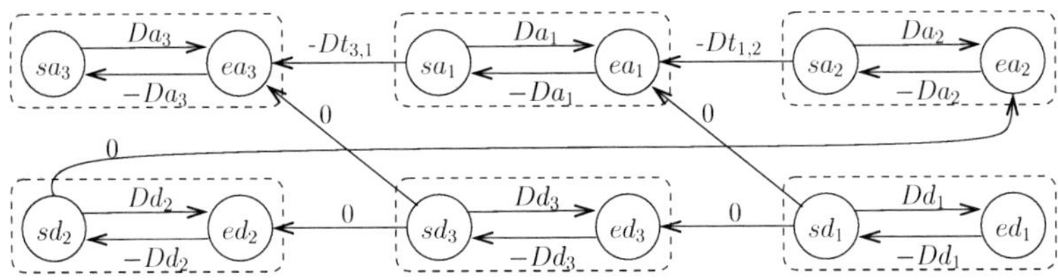

Fig. 2. Distance graph (reference temporal position x_0 is not represented)

$[Tmin_i, Tmax_i]$; (b) two variables sd_i and ed_i, denoting respectively the start time and end time of the download of the acquisition, with domains of values $[Ts_1, Te_1]$ for $i = 2, 3$ and $[Ts_2, Te_2]$ for $i = 1$.

Simple temporal constraints in Eq. 1 to 4 are imposed over these variables. Eq. 1 defines the duration of acquisitions and data downloads. Eq. 2 imposes minimum transition times between acquisitions. Eq. 3 enforces no-overlap between downloads. Eq. 4 expresses that an acquisition can start being downloaded only after its realization. Fig. 2 gives the distance graph of the obtained STN.

$$\forall i \in [1..3], \; (ea_i - sa_i = Da_i) \land (ed_i - sd_i = Dd_i) \tag{1}$$

$$(sa_1 - ea_3 \geq Dt_{3,1}) \land (sa_2 - ea_1 \geq Dt_{1,2}) \tag{2}$$

$$(sd_3 - ed_2 \geq 0) \land (sd_1 - ed_3 \geq 0) \tag{3}$$

$$\forall i \in [1..3], \; sd_i - ea_i \geq 0 \tag{4}$$

2.2 T-Simple Temporal Constraints and TSTN

We now introduce a new class of temporal constraints which can be used to model transitions whose minimum duration depends on the precise time at which the transition is triggered. These constraints are called *t-simple temporal constraints* for "time-dependent"-simple temporal constraints.

Definition 2. *A t-simple temporal constraint is a triple $(x, y, dmin)$ composed of two temporal variables x and y, and of one function $dmin : \mathbf{d}(x) \times \mathbf{d}(y) \to \mathbb{R}$ called* minimum distance function *(function not necessarily continuous). A t-simple temporal constraint $(x, y, dmin)$ is also written as $y - x \geq dmin(x, y)$. The constraint is satisfied by $(a, b) \in \mathbf{d}(x) \times \mathbf{d}(y)$ iff $b - a \geq dmin(a, b)$.*

Informally, $dmin(x, y)$ specifies a minimum temporal distance between the events associated with temporal variables x and y respectively.

To illustrate why having a minimum distance function $dmin$ depending on both x and y is useful, consider the example of agile satellites. Let x be a variable representing the end time of an acquisition acq. Let $Att(x)$ denote the attitude obtained when finishing acq at time x. Let y be a variable representing the start time of an acquisition acq', to be performed just after acq. Let $Att'(y)$ denote the attitude required for starting acq' at time y. Let $minAttTransTime$ be the function (available in our agile satellite library) such that $minAttTransTime(att, att')$ gives the minimum transition time required by a satellite maneuver to move from attitude att to attitude att'. Then, t-simple temporal constraint $y - x \geq dmin(x,y)$ with $dmin(x,y) = minAttTransTime(Att(x), Att'(y))$ expresses that the duration between the end of acq and the start of acq' must be greater than the minimum duration required to move from attitude $Att(x)$ to attitude $Att'(y)$.

In some cases, function $dmin(x,y)$ does not depend on y. This concerns *time-dependent scheduling* [5,6], for which the processing time of a task only depends on the start time of this task (t-simple temporal constraint $y - x \geq dmin(x)$ with $dmin(x)$ the processing time of the task when this task starts at time x). T-simple temporal constraints also cover simple temporal constraints $y - x \geq c$, by using a constant minimum distance function $dmin = c$. They also cover constraints of maximum temporal distance between two temporal variables $y - x \leq dmax(x,y)$, since the latter can be rewritten as $x - y \geq dmin(y,x)$ with $dmin(y,x) = -dmax(x,y)$.

Note that a t-simple temporal constraint only refers to the *minimum* duration of a transition. Such an approach can be used for handling agile satellites under the (realistic) assumption that any maneuver which can be made in duration δ is also feasible in duration $\delta' \geq \delta$. This assumption of feasibility of a "lazy maneuver" is not necessarily satisfied by every physical system.

On this basis of t-simple temporal constraints, a new framework called TSTN for Time-dependent STN can be introduced.

Definition 3. *A TSTN is a pair (V, C) with V a finite set of continuous variables of domain $[l, u] \subset \mathbb{R}$, and C a finite set of t-simple temporal constraints $(x, y, dmin)$ with $x, y \in V$. A solution to a TSTN is an assignment of variables in V that satisfies all constraints in C. A TSTN is said to be consistent iff it admits at least one solution.*

Example Let us reconsider the example involving 3 acquisitions acq_1, acq_2, acq_3 and remove the unrealistic assumption of constant minimum transition durations between acquisitions. In the TSTN model obtained, the only difference with the initial STN model is that simple temporal constraints of Eq. 2 are replaced by the t-simple temporal constraints given in Eq. 5 and 6, in which given an acquisition acq_i, $Satt_i(t)$ and $Eatt_i(t)$ respectively denote the attitudes required at the start and at the end of acq_i if this start/end occurs at time t. The definition of the distance graph associated with a TSTN is similar to the definition of the distance graph associated with an STN (see Fig. 3).

$$sa_1 - ea_3 \geq minAttTransTime(Eatt_3(ea_3), Satt_1(sa_1)) \tag{5}$$

$$sa_2 - ea_1 \geq minAttTransTime(Eatt_1(ea_1), Satt_2(sa_2)) \tag{6}$$

3   Arc-Consistency of t-Simple Temporal Constraints

A first important element for establishing arc-consistency is the *delay function*.

Definition 4. *The* delay function *associated with a t-simple temporal constraint $ct : (x, y, dmin)$ is function $delay_{ct} : \mathbf{d}(x) \times \mathbf{d}(y) \rightarrow \mathbb{R}$ defined by $delay_{ct}(a, b) = a + dmin(a, b) - b$.*

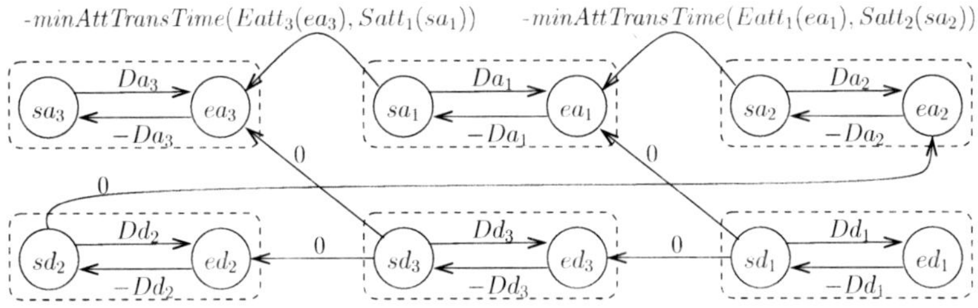

Fig. 3. TSTN distance graph (temporal reference x_0 is not represented)

Informally, $delay_{ct}(a, b)$ is the delay obtained in b if a transition in minimum time from x to y is triggered at time a. This delay corresponds to the difference between the minimum arrival time associated with the transition $(a + dmin(a, b))$ and the required arrival time (b). A strictly negative delay corresponds to a transition ending before deadline b. A strictly positive delay corresponds to a violation of constraint ct. A null delay corresponds to an arrival right on time.

Definition 5. *A t-simple temporal constraint ct : $(x, y, dmin)$ is said to be delay-monotonic iff its delay function $delay_{ct}(.,.)$ satisfies the conditions below:*

$$\forall a, a' \in \mathbf{d}(x), \forall b \in \mathbf{d}(y), \ (a \leq a') \rightarrow (delay_{ct}(a, b) \leq delay_{ct}(a', b))$$

$$\forall a \in \mathbf{d}(x), \forall b, b' \in \mathbf{d}(y), \ (b \leq b') \rightarrow (delay_{ct}(a, b) \geq delay_{ct}(a, b'))$$

Definition 5 means that for being delay-monotonic, a t-simple temporal constraint $(x, y, dmin)$ must verify that on one hand, the later the transition is triggered in x, the greater the delay in y, and on the other hand the earlier the transition must end in y, the greater the delay. When monotonicities over the two arguments are strict, we speak of a strictly delay-monotonic t-simple temporal constraint. The notion of delay-monotonicity can be related to the notion of monotonic constraints, defined for instance in [15]. One difference is that in TSTN, domains considered are continuous.

We now introduce the functions of earliest arrival time and latest departure time associated with a t-simple temporal constraint. In the following, given a function $F : \mathbb{R} \rightarrow \mathbb{R}$ and a closed interval $I \subset \mathbb{R}$, we denote by (1) $firstNeg(F, I)$ the smallest $a \in I$ such that $F(a) \leq 0$ (value $+\infty$ if such a value does not exist); (2) $lastNeg(F, I)$ the greatest $a \in I$ such that $F(a) \leq 0$ (value $-\infty$ if such a value does not exist).[1]

Definition 6. *The functions of earliest arrival time and latest departure time associated with a t-simple temporal constraint ct : $(x, y, dmin)$ are functions denoted $earr_{ct}$ and $ldep_{ct}$ and defined over $\mathbf{d}(x)$ and $\mathbf{d}(y)$ respectively, by:*

$$\forall a \in \mathbf{d}(x), \ earr_{ct}(a) = firstNeg(delay_{ct}(a, .), \mathbf{d}(y))$$

$$\forall b \in \mathbf{d}(y), \ ldep_{ct}(b) = lastNeg(delay_{ct}(., b), \mathbf{d}(x))$$

---

[1] Quantities $firstNeg(F, I)$ and $lastNeg(F, I)$ are mathematically not necessarily well-defined if function F has discontinuities; we implicitly use the fact that all operations are done on computers with finite precision.

Informally, $earr_{ct}(a)$ gives the smallest arrival time in y without delay if the transition from x is triggered at time a. $ldep_{ct}(b)$ gives the latest triggering time of the transition in x for an arrival in b without delay.

Prop. 1 shows that these two functions help establishing bound arc-consistency.

Proposition 1. *Bound arc-consistency for a t-simple temporal constraint* ct : $(x, y, dmin)$ *can be enforced using the following domain modification rules:*

$$\mathbf{d}(y) \leftarrow \mathbf{d}(y) \cap [earr_{ct}(\min(\mathbf{d}(x))), +\infty[\tag{7}$$

$$\mathbf{d}(x) \leftarrow \mathbf{d}(x) \cap]-\infty, ldep_{ct}(\max(\mathbf{d}(y)))] \tag{8}$$

Proof. Assume that $earr_{ct}(\min(\mathbf{d}(x))) \neq +\infty$ and $ldep_{ct}(\max(\mathbf{d}(y))) \neq -\infty$. By definition of $earr_{ct}$ and $ldep_{ct}$, we then have $delay(\min(\mathbf{d}(x)), earr_{ct}(\min(\mathbf{d}(x)))) \leq 0$ and $delay(ldep_{ct}(\max(\mathbf{d}(y))), \max(\mathbf{d}(y))) \leq 0$. Hence min and max bounds of x and y all have a support after application of Rules 7-8 if domains obtained are not empty.

Rule 7 updates the earliest time associated with y. Rule 8 updates the latest time associated with x. These domain modification rules are such that current domains $\mathbf{d}(x)$ and $\mathbf{d}(y)$ remain closed intervals. Prop. 2 below establishes the equivalence between bound arc-consistency and arc-consistency for delay-monotonic constraints.

Proposition 2. *Let* ct : $(x, y, dmin)$ *be a t-simple temporal constraint with monotonic delay. Establishing bound arc-consistency for* ct *using Rules 7 and 8 is equivalent to establishing arc-consistency over the whole domains of* x *and* y.

Proof. Let x^-, x^+, y^-, y^+ denote the min/max bounds of x and y before application of the rules. Let $b \in [y^-, y^+]$. If $b < earr_{ct}(x^-)$, then b has no support over x for ct because $\forall a \in [x^-, x^+]$, $delay_{ct}(a, b) \geq delay_{ct}(x^-, b) > 0$ (by delay-monotonicity and by definition of $earr_{ct}(x^-)$). Conversely, if $b \geq earr_{ct}(x^-)$, then $delay_{ct}(x^-, b) \leq delay_{ct}(x^-, earr_{ct}(x^-)) \leq 0$, hence b is supported by x^-. Therefore, y-values pruned by Rule 7 are those that have no support over x. Similarly, it can be shown that x-values pruned by Rule 8 are those that have no support over y.

When delay-monotonicity is violated, Rules 7-8 can be applied but they do not necessarily establish arc-consistency. Prop. 3 generalizes a STN result to TSTN and shows why maintaining bound arc-consistency is useful.

Proposition 3. *If all constraints of a TSTN are made bound arc-consistent using Rules 7-8, then the schedule which assigns to each variable its earliest (resp. latest) possible time is a solution of the TSTN.*

Proof. Let ct : $(x, y, dmin)$ be a constraint of the TSTN. As shown in the proof of Prop. 1, the min bounds of x and y after application of Rules 7-8 form a consistent pair of values for ct, as well as their max bounds.

Concerning the way $earr$ and $ldep$ can be computed in practice, for simple temporal constraints $y - x \geq c$, an analytic formulation of $earr$ and $ldep$ can be given. However, in the general case, $firstNeg(F, I)$ and $lastNeg(F, I)$ must be computed, which corresponds to an optimization problem in itself. An iterative method for approximating $firstNeg(F, I = [a_1, a_2])$ is given in Algorithm 1. This method generalizes the *false position method*, used to find a zero of an arbitrary function. Applied to the case of t-simple temporal constraints, the method works as follows. If leftmost point $P_1 = (a_1, F(a_1))$ has a negative delay ($F(a_1) \leq 0$), then a_1 is directly returned. Otherwise, if rightmost point $P_2 = (a_2, F(a_2))$ has a strictly positive delay ($F(a_2) > 0$), then $+\infty$ is returned. Otherwise, points P_1 and P_2 have opposite delay-signs ($F(a_1) > 0$ and $F(a_2) \leq 0$), and the method computes delay $F(a_3)$ in a_3, the x-value of the intersection between segment (P_1, P_2) and the x-axis. If the delay in $P_3 = (a_3, F(a_3))$ is positive (resp. negative), then the mechanism is applied again by taking $P_1 = P_3$ (resp. $P_2 = P_3$). If the t-simple temporal constraint considered has a strictly monotonic delay, the convergence to $firstNeg(F, I)$ is ensured; otherwise, the method may return a value $a > firstNeg(F, I)$, but in this case a still satisfies $F(a) \leq 0$ (with a given precision). It can be observed in practice that the convergence speed is particularly good for the delay function associated with agile satellites.

Algorithm 1. Possible way of computing $firstNeg(F, I)$, with I=$[a_1, a_2]$, maxIter a maximum number of iterations, and prec a desired precision

---

```
1  firstNeg(F, [a₁, a₂], maxIter, prec)
2  begin
3      f₁ ← F(a₁); if f₁ ≤ 0 then return a₁
4      f₂ ← F(a₂); if f₂ > 0 then return +∞
5      for i = 1 to maxIter do
6          a₃ = (f₁ * a₂ − f₂ * a₁)/(f₁ − f₂)
7          f₃ = F(a₃)
8          if |f₃| < prec then return a₃
9          else if f₃ > 0 then (a₁, f₁) ← (a₃, f₃)
10         else (a₂, f₂) ← (a₃, f₃)
11     return a₂
```

---

4   Solving TSTN

The problem considered hereafter is to determine the consistency of a TSTN and to compute the earliest and latest possible times associated with each temporal variable. We also consider a context in which temporal constraints can be successively added and removed from the problem. This dynamic aspect is useful for instance when using local search for solving scheduling problems. In this kind of search, local moves are used for modifying a current schedule. They may correspond to additions and removals of activities, which are translated into additions and removals of temporal constraints. The different techniques used, which generalize existing STN resolution techniques, are successively presented.

4.1   Constraint Propagation

We first use constraint propagation for computing min and max bounds of temporal variables. This standard method is inspired by approaches defined in [8,9,10]. The latter correspond to maintaining a list of variables for which constraints holding over these variables must be revised with, for each variable z of the list, the nature of the revision(s) to be performed: (a) if z had its min bound updated, then the min bound of every variable t linked to z by a constraint $t - z \geq c$ must be revised; (b) if z had its max bound updated, then the max bound of every variable t linked to z by a constraint $z - t \geq c$ must be revised.

Compared to standard STN approaches, we choose for TSTN a constraint propagation scheme in which a list containing constraints to be revised is maintained, instead of a list containing variables. This list is partitioned into two sub-lists, the first one containing constraints to be revised which may modify a min bound (constraints $y - x \geq dmin(x, y)$ awoken following a modification of $\min x$, which may modify $\min y$), and the second one containing constraints to be revised which may modify a max bound (constraints $y - x \geq dmin(x, y)$ awoken following a modification of $\max y$, which may modify $\max x$). Compared to the version maintaining lists of variables, maintaining lists of constraints allows some aspects to be more finely handled (more details below).

Last, a t-simple temporal constraint is revised using Rules 7 and 8 of Prop. 1.

4.2   Negative Cycle Detection

With bounded domains of values, the establishment of arc-consistency for STN is able to detect inconsistency. However, the number of constraint revisions required for deriving inconsistency may be prohibitive compared to STN approaches defined in [7,8], which use the fact that STN inconsistency is equivalent to the existence of a cycle of negative length in the distance graph.

The basic idea of these existing STN approaches consists in detecting such negative cycles on the fly by maintaining so-called *propagation chains*. The latter can be seen as explanations for the current min and max bounds of the different variables. A constraint $y - x \geq c$ is said to be active with regard to min bounds (resp. max bounds) if and only if the last revision of this constraint is responsible for the last modification of the min of y (resp. the max of x). It is shown in [7] that if there exists a cycle in the directed graph where an arc is associated with each active constraint with regard to min bounds, then the STN is inconsistent. The intuition is that if a propagation cycle $x_1 \rightarrow x_2 \rightarrow \ldots \rightarrow x_n \rightarrow x_1$ is detected for min bounds, then this means that the min value of x_1 modified the min value of x_2... which modified the min value of x_n which modified the min value of x_1. By traversing this propagation cycle a sufficient number of times, the domain of x_1 can be entirely pruned. The same result holds for the directed graph containing one arc per active constraint with regard to max bounds.

These results cannot however be directly reused for t-simple temporal constraints, since for TSTN in general, the existence of a propagation cycle does not necessarily imply inconsistency, as shown in the example below.

Example. Let $dmin$ be the minimum distance function defined by $dmin(a, b) = 1 - a/2$. Let $(V, C) = (\{x, y\}, \{ct_1 : x - y \geq -0.5, ct_2 : y - x \geq dmin(x, y)\})$ be a TSTN containing two temporal variables of domains $\mathbf{d}(x) = \mathbf{d}(y) = [0.5, 2]$ and two constraints. The delay functions associated with ct_1 and ct_2 are strictly monotonic (for ct_2, it equals $delay_{ct_2}(a, b) = a + dmin(a, b) - b = 1 + a/2 - b$).

Propagating ct_2 using Rule 7 updates the min of y and gives $\mathbf{d}(y) = [1+1/4, 2]$. Propagating ct_1 using the same rule then updates the min of x and gives $\mathbf{d}(x) = [1 - 1/4, 2]$. The result obtained is a cycle of propagation since the min value of x modified the min of y which itself modified the min of x. In the context of STN, the existence of such a cycle means inconsistency. In the context of TSTN, such a conclusion does not always hold because for instance assignment $x = 1$, $y = 1.5$ is consistent.

The reason is that in TSTN, domain reductions obtained by traversing cycles again and again may become smaller and smaller. This is what happens here, where we get $\mathbf{d}(x) = [1 - 1/2^n, 2]$ after n traversals of the propagation cycle between x and y. The finite computer precision implies that cycle traversals stop at some step, but potentially only after many iterations.

The example also shows that the strict monotonicity of the delay function does not suffice for deriving inconsistency in case of cycle detection. A sufficient condition satisfied for standard STN is given in Prop. 4. This condition ensures that a cycle does not become "less negative" when traversed again and again.

Definition 7. *A t-simple temporal constraint $ct : (x, y, dmin)$ is said to be shift-monotonic iff it satisfies:*

$$\forall a, a' \in \mathbf{d}(x), \forall b \in \mathbf{d}(y), (a \leq a') \to (delay_{ct}(a', b) \geq delay_{ct}(a, b) + (a' - a))$$
$$\forall a \in \mathbf{d}(x), \forall b, b' \in \mathbf{d}(y), (b \leq b') \to (delay_{ct}(a, b) \geq delay_{ct}(a, b') + (b' - b))$$

Informally, shift-monotonicity means that on one hand, when the start time of a transition is shifted forward, the arrival time is shifted forward by at least the same amount, and on the other hand when the arrival time of the transition is shifted backward, the delay is increase by at least the same amount.

Proposition 4. *If a propagation cycle involving only shift-monotonic constraints is detected in a TSTN, then the TSTN is inconsistent.*

Proposition 5. *In particular, (1) for TSTN containing only shift-monotonic constraints, the existence of a propagation cycle implies inconsistency; (2) for TSTN whose distance graph does not contain cycles involving non shift-monotonic constraints, the existence of a propagation cycle implies inconsistency.*

Proof. For Prop. 4, assume that propagation cycle $x_1 \to x_2 \to \ldots \to x_n \to x_1$ is detected for min bounds, following the revision of a constraint linking x_n and x_1. Let $\delta > 0$ be the increase in the min bound of x_1 following this last constraint revision. It can be shown that shift-monotonicity implies that $(a \leq a') \to (earr_{ct}(a') \geq earr_{ct}(a) + (a' - a))$. Therefore, if the cycle is traversed again, the min bounds of $x_2, \ldots, x_n$ will be increased again by at least δ. After a sufficient number of cycle traversals, the domain of one variable of the cycle becomes empty. Prop. 5 is a direct consequence of Prop. 4.

In the agile satellite application which motivates this work, the minimum distance functions used are not necessarily shift-monotonic, as can be seen in Fig. 1, but point 2 of Prop. 5 applies for case studies considered. Inferring inconsistency due to propagation cycle detection is correct in this case. Checking the satisfaction of the condition given in point 2 of Prop. 5 is easy (linear in the number of variables and constraints).

If none of the sufficient conditions given in Prop. 5 is satisfied, several options can be considered. The first one consists in not considering non shift-monotonic constraints in propagation chains; this approach is correct but may lose time in propagation cycles. The second option consists in considering a TSTN as inconsistent as soon as a propagation cycle is detected, even if it contains non shift-monotonic constraints; this may be incorrect in the sense that it may wrongly conclude to inconsistency. A possible trade-off is to keep the first option but to stop propagating constraints when some time-limit or some precision is reached.

In terms of complexity, Prop. 6 below generalizes polynomial complexity results available on STN to TSTN, and therefore to time-dependent scheduling.

Proposition 6. *Given a TSTN (V, C), if the existence of a propagation cycle implies inconsistency, then the algorithm using Rules 7-8 for propagation plus a FIFO ordering on the propagation queue plus propagation cycle detection establishes bound arc-consistency in $O(|V||C|)$ constraint revisions (bound independent of the size of the variable domains).*

Proof. Similar to the result stating that the number of arc revisions in the Bellman-Ford's FIFO label-correcting algorithm is $O(|V||C|)$.

In terms of implementation, we perform on the fly detection of propagation cycles based on an efficient data structure introduced in [16]. The latter is used for maintaining a topological order of nodes in the graphs of propagation of min and max bounds. When no topological order exists, the graph contains a cycle.

Prop. 8 and 9 show that the two monotonicity properties considered in this paper (delay- and shift-monotonicity) are satisfied by simple temporal constraints and by several constraints used in time-dependent scheduling (see [5]).

Proposition 7. *Shift-monotonicity implies strict delay-monotonicity.*

Proposition 8. *Simple temporal constraints $y - x \geq c$ are shift-monotonic (and therefore also strictly delay-monotonic).*

Proposition 9. *Let x, y be two temporal variables corresponding to the start time and end time of a task respectively. Monotonicity results of Table 1 hold.*

Proof. Prop. 7 is straightforward. Prop. 8 holds because if $dmin$ is constant, then $delay_{ct}(a, b) - delay_{ct}(a', b) = a - a'$ and $delay_{ct}(a, b) - delay_{ct}(a, b') = b' - b$. For Prop. 9, some intermediate results can be used: (a) if $dmin(x, y) = dmin(x)$, shift-monotonicity holds iff $dmin(x)$ is a non-decreasing function; (b) if $dmin(x)$ decreases at some step, then delay-monotonicity holds provided that the decrease slope is ≥ -1.

Table 1. Monotonicity of some distance functions used in time-dependent scheduling, with x a variable whose domain is not reduced to a singleton, and A, B, D constants such that $A \geq 0$, $B > 0$, and $D > \min(\mathbf{d}(x))$

Distance $dmin(x,y) = dmin(x)$	form	shift-monotonic	delay-monotonic
$A + Bx$		yes	yes (strict)
$A - Bx$		no	yes iff $B \leq 1$ (strict iff $B < 1$)
$\max(A, A + B(x - D))$		yes	yes (strict)
A if $x < D$, $A + B$ otherwise		yes	yes (strict)
$A - B \min(x, D)$		no	yes iff $B \leq 1$ (strict iff $B < 1$)

4.3 Constraint Depropagation for Dynamic TSTN

Constraint propagation techniques are directly able to handle constraint addition or constraint strengthening. As for constraint removal or constraint weakening, constraint depropagation strategies defined in [10] for STN can be directly reused. These strategies allow min and max bounds of temporal variables to be recomputed at minimum cost. They avoid reinitializing all variable domains and repropagating all constraints from scratch when a constraint is removed or weakened. The basic idea is to use propagation chains in order to determine which variable domains must be reinitialized and which constraints need to be revised. More precisely, when a constraint $y - x \geq dmin(x, y)$ is removed or weakened, if this constraint is active with regard to the min bound of y (resp. the max bound of x), then the min bound of y (resp. the max bound of x) is reinitialized to the value it had before any propagation. This reinitialization may trigger other reinitializations. TSTN constraints of the form $y - z \geq dmin(z, y)$ (resp. $z - x \geq dmin(x, z)$) are then added to the list of constraints to be revised from the point of view of min bounds (resp. max bounds).

The only difference when compared to standard STN techniques is the use of lists of constraints to be revised instead of lists of variables. This allows constraint depropagation to be slightly less costly: on the example of reinitialization of the min bound of y, the standard STN version would add to a list of variables to be propagated every variable z linked to y by some constraint $y - z \geq dmin(z, y)$, and doing so would repropagate in the end all constraints of the form $u - z \geq dmin(z, u)$, even those with $u \neq y$.

4.4 Constraint Revision Ordering

A last technique is used for minimizing the number of constraint revisions. This can be particularly useful for TSTN, for which revising one constraint can be significantly more costly than for STN. The proposed approach extends a technique developed for STN$^-$[9], a sub-class of STN in which every constraint must be rewritable as $y - x \geq c$ with $c \geq 0$. The idea consists in building the strongly connected components of the distance graph, in ordering them in topological order, and in using this order to determine which constraint to propagate first. We first recall definitions concerning strongly connected components.

Definition 8. *Let $G = (V, A)$ be a directed graph with V the set of nodes and A the set of arcs. A* Strongly Connected Component *(SCC) of G is a maximum sub-graph G' of G such that there exists in G' a path from every node to every other node.*

The DAG *(Directed Acyclic Graph) of SCCs of G is the directed graph whose nodes are the SCCs of G and which contains an arc from SCC c_1 to SCC c_2 iff there exists in G an arc from one of the nodes of c_1 to one of the nodes of c_2.*

A topological order of SCCs is an order $\preceq$ where each SCC c is put strictly after each of its parents c' in the DAG of SCCs ($c' \prec c$). Given a node x in graph G, $scc(x)$ denotes the unique SCC of G that contains x.

Propagating temporal constraints following a topological order of SCCs of the distance graph boils down to using the fact that solving shortest path problems is easier for acyclic graphs than for arbitrary graphs. To apply this result, constraints to be propagated are ordered according to a topological order of SCCs. More precisely, concerning the propagation of min bounds, we propagate first constraints $y - x \geq dmin(x, y)$ such that $scc(y)$ is maximum in the order of SCCs and, in case of equality, we propagate first constraints such that $scc(x) \neq scc(y)$, to postpone as much as possible the propagation of "internal" constraints in an SCC. To break remaining ties, a FIFO ordering strategy is used. Concerning the propagation of max bounds, constraints are ordered by increasing $scc(x)$ and, in case of equality, we propagate first constraints such that $scc(y) \neq scc(x)$, and break remaining ties using a FIFO ordering strategy. In the example of Fig. 3, SCCs are represented as dotted boxes. A bad propagation order for min bounds would consist in propagating first the constraint between sa_2 and ea_1, and then the constraint between sa_1 and ea_3. A good order, consistent with the order of SCCs, would consist in using the opposite strategy.

Compared to the way SCCs are used in [9] for STN$^-$, the method we propose is adapted not only to general STN, but also to TSTN. In terms of implementation, in order to avoid recomputing the DAG of SCCs from scratch after each constraint addition or removal, we use recent algorithms proposed for maintaining SCCs in a dynamic graph [17,18].

5 Experiments

All techniques presented in Section 4 (constraint propagation, propagation cycle detection, constraint depropagation, SCC ordering) are integrated and simultaneously used in a scheduling tool based on local search. The local search aspect entails that doing/undoing a local move is fast, similarly to constraint-based local search tools Comet [19] and LocalSolver [20]. Our STN/TSTN solver is implemented in Java. Results are obtained on an Intel i5-520 1.2GHz, 4GBRAM.

Experiments not detailed here were first performed on STN obtained from scheduling problems of the SMT-LIB. The objective was to evaluate the propagation heuristics based on a topological ordering of the SCCs. This heuristics appears to be a robust strategy, which significantly decreases the number of constraint revisions on some problems. More precisely, for consistent STN, the

SCC heuristics is always at least as good as a pure FIFO heuristics, but for inconsistent STN, it is not always the fastest strategy for proving inconsistency.

We detail below experiments realized on TSTN in the context of agile satellites. The problem considered here is a simple no overlapping constraint over an ordered sequence of n acquisitions $acq_1 \rightarrow \ldots \rightarrow acq_n$, with n varying between 5 and 13. These acquisitions correspond to ground strips located between the north of Spain and the north of France. The no-overlapping constraint between acquisitions can be written as a set of t-simple temporal constraints of the form $s_{i+1} - e_i \geq minAttTransTime(Eatt_i(e_i), Satt_{i+1}(s_{i+1}))$ with, for an acquisition j, s_j/e_j the start/end time of this acquisition, and $Satt_j(t)/Eatt_j(t)$ the attitudes required to start/end j at time t. In addition, simple temporal constraints are used to define the constant duration of each acquisition.

Two methods are compared: (1) a TSTN approach in which exact transition times between acquisitions are used, and (2) an STN approach in which upper bounds on transition times are pre-computed, by sampling on the different possible start times of the transitions. The schedule obtained in both cases is flexible in the sense that the domains of values after propagation over STN/TSTN are generally not reduced to singletons. The criterion considered for comparing the two approaches is the mean temporal flexibility $mtf = \frac{1}{|V|} \sum_{x \in V} (max(x) - min(x))$, measured as the mean, over all temporal variables $x \in V$, of the difference between the earliest and latest possible times associated with x. Such a flexibility is important in practice to offer as much freedom as possible concerning the choice of an angle of acquisition of ground strips, which influences image quality.

Three scenarios are considered. In the first one, acquisitions correspond to strips of length about 80km, to be observed with a scanning direction of 0 degrees (angle between the trace of the satellite on the ground and the direction in which the strip must be scanned). Fig. 4(a) shows that in this case, the temporal flexibility obtained with TSTN only slightly improves the flexibility obtained with STN. The reason is that if all acquisitions are realized with a scanning direction of 0 degrees, the minimum transition times between acquisitions considered are almost independent of the precise triggering time of transitions: they are only time-dependent when the rotation on the pitch axis is the most constraining from a temporal point of view, compared to the rotation on the roll axis.

In the second scenario, the scanning direction becomes 30 degrees. Fig. 4(b) shows that the temporal flexibility obtained with TSTN is better than with STN (improvement of about 20 seconds in flexibility), and that the flexibility gap between STN and TSTN increases with the number of acquisitions planned.

In the third and last scenario, the length of the strips considered becomes approximately 40km, and the scanning direction is chosen at random for each strip. In this case, Fig. 4(c) shows that the STN approach only allows sequences of length 5 and 6 to be scheduled. It concludes to an inconsistency of the problem for $n \geq 7$. On the other hand, the TSTN approach schedules all 13 acquisitions considered. One reason explaining these results is that the more distinct the scanning directions are, the more the minimum transition times between acqui-

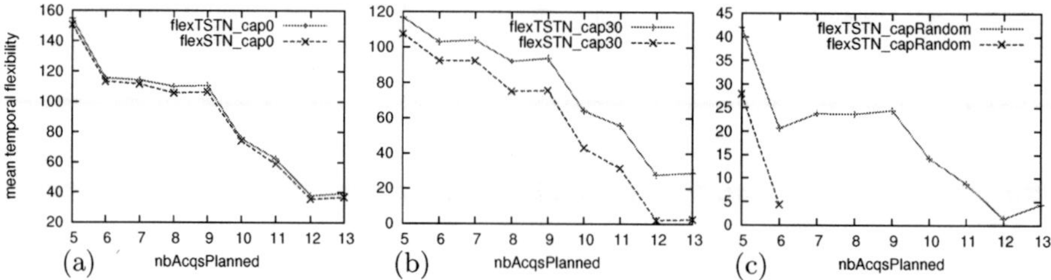

Fig. 4. Comparison of temporal flexibilities, in seconds, obtained with precomputed upper bound on transition times (flexSTN) and with exact transition times (flexTSTN)

sitions depend on the triggering time of the transitions. The possibility to have distinct scanning directions is important in practice. It indeed allows acquisitions defined as polygons to be split into strips whose orientation can be freely chosen, which can reduce the number of strips to be scanned.

To give an idea of computation times, for 13 acquisitions added one by one to the current schedule, a precision of one second on dates, and a maximum number of iterations equal to 10^4 for computing *firstNeg* and *lastNeg*, the TSTN approach takes about 2ms per acquisition addition. With STN, the computation time is less than 0.1ms per addition. For precisions of 10^{-1}, 10^{-2}, and 10^{-3} second on dates, computation times with TSTN respectively become 3ms, 12ms, and 66ms per addition. A typical technique can consist in first searching for schedules with a fast coarse-grained approach, before using a finer precision.

As a conclusion, this paper introduced TSTN, their properties, resolution techniques, and their application to agile satellites. It would be interesting to extend other features of STN to TSTN, e.g. concerning *decomposability* issues [1], and to test TSTN on other applications, e.g. from the logistics domain.

References

1. Dechter, R., Meiri, I., Pearl, J.: Temporal constraint networks. Artificial Intelligence 49, 61–95 (1991)
2. Stergiou, K., Koubarakis, M.: Backtracking algorithms for disjunctions of temporal constraints. Artificial Intelligence 120, 81–117 (2000)
3. Challenge ROADEF-03: Handling the mission of Earth observation satellites (2003), http://challenge.roadef.org/2003/fr/
4. Lemaître, M., Verfaillie, G., Jouhaud, F., Lachiver, J.M., Bataille, N.: Selecting and scheduling observations of agile satellites. Aerospace Science and Technology 6, 367–381 (2002)
5. Cheng, T., Ding, Q., Lin, B.: A concise survey of scheduling with time-dependent processing times. European Journal of Operational Research 152, 1–13 (2004)
6. Gawiejnowicz, S.: Time-dependent scheduling. Springer (2008)
7. Cervoni, R., Cesta, A., Oddi, A.: Managing dynamic temporal constraint networks. In: Proc. of AIPS 1994, pp. 13–18 (1994)
8. Cesta, A., Oddi, A.: Gaining efficiency and flexibility in the simple temporal problem. In: Proc. of TIME 1996, pp. 45–50 (1996)

9. Gerevini, A., Perini, A., Ricci, F.: Incremental algorithms for managing temporal constraints. In: Proc. of ICTAI 1996, pp. 360–365 (1996)
10. Shu, I., Effinger, R., Williams, B.: Enabling fast flexible planning through incremental temporal reasoning with conflict extraction. In: Proc. of ICAPS 2005, pp. 252–261 (2005)
11. Xu, L., Choueiry, B.: A new efficient algorithm for solving the simple temporal problem. In: Proc. of TIME-ICTL 2003, pp. 210–220 (2003)
12. Planken, L., de Weerdt, M., van der Krogt, R.: P3C: a new algorithm for the simple temporal problem. In: Proc. of ICAPS 2008, pp. 256–263 (2008)
13. Planken, L., de Weerdt, M., Yorke-Smith, N.: Incrementally solving STNs by enforcing partial path consistency. In: Proc. of ICAPS 2010, pp. 129–136 (2010)
14. Montanari, U.: Networks of constraints: fundamental properties and applications to picture processing. Information Sciences 7(2), 95–132 (1974)
15. Hentenryck, P.V., Deville, Y., Teng, C.: A generic arc-consistency algorithm and its specializations. Artificial Intelligence 57(2-3), 291–321 (1992)
16. Bender, M.A., Cole, R., Demaine, E.D., Farach-Colton, M., Zito, J.: Two Simplified Algorithms for Maintaining Order in a List. In: Möhring, R.H., Raman, R. (eds.) ESA 2002. LNCS, vol. 2461, pp. 152–164. Springer, Heidelberg (2002)
17. Haeupler, B., Kavitha, T., Mathew, R., Sen, S., Tarjan, R.: Incremental cycle detection, topological ordering, and strong component maintenance. ACM Transactions on Algorithms 8(1) (2012)
18. Roditty, L., Zwick, U.: Improved dynamic reachability algorithms for directed graphs. SIAM Journal on Computing 37(5), 1455–1471 (2008)
19. Hentenryck, P.V., Michel, L.: Constraint-based local search. The MIT Press (2005)
20. Benoist, T., Estellon, B., Gardi, F., Megel, R., Nouioua, K.: Localsolver 1.x: a black-box local-search solver for 0-1 programming. 4OR: A Quarterly Journal of Operations Research 9(3), 299–316 (2011)

Improved Bounded Max-Sum
for Distributed Constraint Optimization

Emma Rollon and Javier Larrosa

Departament de Llenguatges i Sistemes Informàtics
Universitat Politècnica de Catalunya, Spain

Abstract. Bounded Max-Sum is a message-passing algorithm for solving Distributed Constraint Optimization Problems able to compute solutions with a guaranteed approximation ratio. Although its approximate solutions were empirically proved to be within a small percentage of the optimal solution on low and moderately dense problems, in this paper we show that its theoretical approximation ratio is overestimated, thus overshadowing its good performance. We propose a new algorithm, called Improved Bounded Max-Sum, whose approximate solutions are at least as good as the ones found by Bounded Max-Sum and with a tighter approximation ratio. Our empirical evaluation shows that the new approximation ratio is significantly tighter.

1   Introduction

Decentralised coordination techniques are a very important topic of research. A common approach is to cast the problem as a *multi-agent distributed constraint optimization problem* (DCOP), where the possible actions that agents can take are associated with *variables* and the utility for taking joint actions are encoded with (soft) *constraints* [8]. The set of constraints define a global utility function $F(x)$ to be optimized via decentralised coordination of the agents. In general, complete algorithms [6,5,7] (i.e. algorithms that find the true optimum) exhibit an exponentially increasing coordination overhead, which makes them useless in many practical situations.

Approximate algorithms constitute a very interesting alternative. They require little computation and communication at the cost of sacrificing optimality. There are several examples showing that they can provide solutions which are very close to optimality [3,4]. However, this observation can only be verified on small toy instances, because it requires the computation of the true optimal to compare with, and it is not available in real-size real-world situations.

A significant breakthrough along this line of work was the Bounded Max-Sum algorithm (BMS) [8]. This algorithm comes with a guarantee approximation ratio $\tilde{\rho}$, meaning that its approximate solution $\tilde{\mathbf{x}}$ has a utility $F(\tilde{\mathbf{x}})$ which is no more than a factor $\tilde{\rho} \geq 1$ away from the optimum (i.e, $F(\tilde{\mathbf{x}}) \leq F(\mathbf{x}^*) \leq \tilde{\rho}F(\tilde{\mathbf{x}})$). Clearly, large values of $\tilde{\rho}$ reflect lack of confidence in the solution $\tilde{\mathbf{x}}$. There are two possible reasons for a large $\tilde{\rho}$: i) the algorithm failed in finding a solution close to the optimal, ii) the approximation ratio is not tight. Clearly, if we want $\tilde{\rho}$ to be our measure of confidence about the quality of $\tilde{\mathbf{x}}$, we want a tight $\tilde{\rho}$ (i.e, $F(\mathbf{x}^*) \approx \tilde{\rho}F(\tilde{\mathbf{x}})$). Thus, the quality of the approximation ratio is a matter of the utmost importance.

M. Milano (Ed.): CP 2012, LNCS 7514, pp. 624–632, 2012.

In this paper we propose an improvement of BMS with approximation ratio ρ. We theoretically show that it is always better than the previous one (i.e., $\rho \leq \widetilde{\rho}$). Moreover, our experiments show that, in practice, ρ is much tighter than $\widetilde{\rho}$.

2 Preliminaries

In this Section we review the main elements to contextualize our work. Definitions and notation are borrowed almost directly from [8]. We urge the reader to visit that reference for more details and examples.

2.1 DCOP

A *Distributed Constraint Optimization Problem* (DCOP) is a quadruple $P = (\mathbf{A}, \mathbf{X}, \mathbf{D}, \mathbf{F})$, where $\mathbf{A} = \{\mathcal{A}_1, \ldots, \mathcal{A}_r\}$ is a set of agents, and $\mathbf{X} = \{x_1, \ldots, x_n\}$ and $\mathbf{D} = \{\mathbf{d}_1, \ldots, \mathbf{d}_n\}$ are variables and domains. $\mathbf{F} = \{f_1, \ldots, f_e\}$ is a set of cost functions. The objective function is,

$$F(x) = \sum_{j=1}^{e} f_j(x^j)$$

where $x^j \subseteq \mathbf{X}$ is the scope of f_j. A *solution* is a complete assignment $\mathbf{x}$. An *optimal solution* is a complete assignment $\mathbf{x}^*$ such that $\forall \mathbf{x}, \ F(\mathbf{x}^*) \geq F(\mathbf{x})$. The usual task of interest is to find $\mathbf{x}^*$ through the coordination of the agents.

In the applications under consideration, the agents search for the optimum via decentralised coordination. We assume that each agent can control only its local variable(s) and has knowledge of, and can directly communicate with, a few neighboring agents. Two agents are neighbors if there is a relationship connecting variables and functions that the agents control.

The structure of a DCOP problem $P = (\mathbf{A}, \mathbf{X}, \mathbf{D}, \mathbf{F})$ can be transformed into a factor graph. A *factor graph* is a bipartite graph having a variable node for each variable $x_i \in \mathbf{X}$, a factor node for each local function $f_j \in \mathbf{F}$, and an edge connecting variable node x_i to factor node f_j if and only if x_i is an argument of f_j.

2.2 Max-Sum Algorithm

The *Max-Sum* algorithm [2,1] is a message-passing algorithm for solving DCOP problems. It operates over a factor graph by sending functions (a.k.a., messages) along its edges. Edge (i, j) has associated two messages $q_{i \rightarrow j}$, from variable node x_i to function node f_j, and $r_{j \rightarrow i}$, from function node f_j to variable node x_i. These messages are defined as follows:

– **From variable to function:**

$$q_{i \rightarrow j}(x_i) = \alpha_{ij} + \sum_{k \in \mathcal{M}_i \setminus j} r_{k \rightarrow i}(x_i)$$

where $\mathcal{M}_i$ is a vector of function indexes, indicating which function nodes are connected to variable node x_i, and α_{ij} is a normalizing constant to prevent the messages from increasing endlessly in cyclic graphs.

- **From function to variable:**

$$r_{j \to i}(x_i) = \max_{x^j \setminus x_i} \{ f_j(x^j) + \sum_{k \in \mathcal{N}_j \setminus i} q_{k \to i}(x_i) \}$$

where $\mathcal{N}_j$ is a vector of variable indexes, indicating which variable nodes are connected to function node f_j and $x^j \setminus x_i = \{ x_k \mid k \in \mathcal{N}_j \setminus i \}$

Max-Sum is a distributed synchronous algorithm, since the agent controlling node i has to wait to receive messages from all its neighbors but j, to be able to compute (and send) its message to j. When the factor graph is cycle free, the algorithm is guaranteed to converge to the global optimal solution. Once the convergence is reached, each variable node can compute function,

$$z_i(x_i) = \max_{x_i} \sum_{k \in \mathcal{M}_i} r_{k \to i}(x_i)$$

The optimal solution is $\max_{x_i} \{ z_i(x_i) \}$ and the optimal assignment $\mathbf{x}_i^* = \arg\max_{x_i} \{ z_i(x_i) \}$. When the factor graph is cyclic, the algorithm may not converge to the optimum and only provides an approximation.

3   Bounded Max-Sum Algorithm

The *Bounded Max-Sum* algorithm (BMS) [8], is an approximation algorithm built on the Max-Sum algorithm. From a possibly cyclic problem P, the idea is to remove cycles in its factor graph by ignoring dependencies between functions and variables which have the least impact on the solution quality, producing a new acyclic problem $\widetilde{P}$. Then, Max-Sum is used to optimally solve $\widetilde{P}$ while simultaneously computing the approximation ratio $\widetilde{\rho}$. A more detailed description follows. For the sake of simplicity, we will restrict ourselves to the case of binary functions $f_j(x_i, x_k)$. The extension to general functions is direct. The algorithm works in three phases, each one implementable in a decentralised manner (see [8] for further details):

- **Relaxation Phase:** First, the algorithm weights each edge (i, j) of the original factor graph as,

$$w_{ij} = \max_{x_k} \{ \max_{x_i} f_j(x_i, x_k) - \min_{x_i} f_j(x_i, x_k) \}$$

Then, it finds a maximum spanning tree T. Let W be the sum of weights of the removed edges (i.e, $W = \sum_{(i,j) \notin T} w_{ij}$). Next, the original problem P is transformed into an acyclic one $\widetilde{P}$ having the spanning tree T as factor graph. This is done as follows: for each edge (i, j) in the original graph that does not belong to the tree, the cost function $f_j(x_i, x_k)$ is transformed into another function $\widetilde{f}_j(x_k)$ defined as,

$$\widetilde{f}_j(x_k) = \min_{x_i} f_j(x_i, x_k)$$

Note that the objective function of $\widetilde{P}$ is

$$\widetilde{F}(x) = \sum_{(i,j),(k,j)\in T} f_j(x_i, x_k) + \sum_{(i,j)\notin T} \widetilde{f}_j(x_k)$$

- **Solving Phase:** BMS solves $\widetilde{P}$ with Max-Sum. Let $\widetilde{\mathbf{x}}$ be the solution of this problem. Since the factor graph of $\widetilde{P}$ is acyclic, $\widetilde{\mathbf{x}}$ is its optimal assignment.
- **Bounding Phase:** In [8], it is proved that,

$$F(\widetilde{\mathbf{x}}) \le F(\mathbf{x}^*) \le \widetilde{F}(\widetilde{\mathbf{x}}) + W$$

We can rewrite the previous upper bound expression as,

$$F(\mathbf{x}^*) \le \frac{\widetilde{F}(\widetilde{\mathbf{x}}) + W}{F(\widetilde{\mathbf{x}})} F(\widetilde{\mathbf{x}})$$

Therefore, the algorithm computes $\widetilde{\rho} = \frac{\widetilde{F}(\widetilde{\mathbf{x}})+W}{F(\widetilde{\mathbf{x}})}$, which is a guarantee approximation ratio.

4 Improved BMS

4.1 Theoretical Elements

Consider an edge (i, j) in the original factor graph that does not belong to the spanning tree. We define $\widehat{f}_j(x_k)$ as,

$$\widehat{f}_j(x_k) = \max_{x_i} f_j(x_i, x_k)$$

Let $\widehat{P}$ denote the problem containing the unmodified functions $f_j(x_i, x_k)$ (for $(i, j), (k, j) \in T$) and the $\widehat{f}_j(x_k)$ functions (for $(i, j) \notin T$). Note that $\widehat{P}$ and $\widetilde{P}$ have the same acyclic factor graph. Note as well that the objective function of $\widehat{P}$ is

$$\widehat{F}(x) = \sum_{(i,j),(k,j)\in T} f_j(x_i, x_k) + \sum_{(i,j)\notin T} \widehat{f}_j(x_k)$$

We can solve $\widehat{P}$ with Max-Sum. Let $\widehat{\mathbf{x}}$ be the optimal solution of this problem. It is obvious that $F(\widehat{\mathbf{x}})$ is a lower bound of $F(\mathbf{x}^*)$. Furthermore, as we prove next, $\widehat{F}(\widehat{\mathbf{x}})$ is an upper bound of $F(\mathbf{x}^*)$. Therefore, $\widehat{\rho} = \frac{\widehat{F}(\widehat{\mathbf{x}})}{F(\widehat{\mathbf{x}})}$ is a guarantee approximation ratio.

Theorem 1. $F(\mathbf{x}^*) \le \widehat{F}(\widehat{\mathbf{x}})$.

Proof. By definition, $F(\mathbf{x}^*) = \sum_{(i,j),(k,j)\in T} f_j(\mathbf{x}_i^*, \mathbf{x}_k^*) + \sum_{(i,j)\notin T} f_j(\mathbf{x}_i^*, \mathbf{x}_k^*)$. Since for all f_j we have that $f_j(x_i, x_k) \le \max_{x_i} f_j(x_i, x_k)$, then

$$F(\mathbf{x}^*) \le \sum_{(i,j),(k,j)\in T} f_j(\mathbf{x}_i^*, \mathbf{x}_k^*) + \sum_{(i,j)\notin T} \max_{x_i} f_j(x_i, \mathbf{x}_k^*) = \widehat{F}(\mathbf{x}^*)$$

From the optimality of $\widehat{\mathbf{x}}$, we know that $\widehat{F}(\mathbf{x}^*) \le \widehat{F}(\widehat{\mathbf{x}})$, which proves the theorem.

Next, we show that $\widehat{F}(\widehat{\mathbf{x}})$ is a tighter upper bound than $\widetilde{F}(\widetilde{\mathbf{x}}) + W$.

Theorem 2. $\widehat{F}(\widehat{\mathbf{x}}) \leq \widetilde{F}(\widetilde{\mathbf{x}}) + W$.

Proof. The proof is direct once it has been noted that for all $f_j(x_i, x_k)$,

$$\widehat{f}_j(x_k) \leq \widetilde{f}_j(x_k) + w_{ij}$$

which we prove next. By definition, the previous equation corresponds to,

$$\max_{x_i} f_j(x_i, x_k) \leq \min_{x_i} f_j(x_i, x_k) + \max_{x_k}\{\max_{x_i} f_j(x_i, x_k) - \min_{x_i} f_j(x_i, x_k)\}$$

which can be rewritten as,

$$\max_{x_i} f_j(x_i, x_k) - \min_{x_i} f_j(x_i, x_k) \leq \max_{x_k}\{\max_{x_i} f_j(x_i, x_k) - \min_{x_i} f_j(x_i, x_k)\}$$

which clearly holds.

We cannot establish any dominance relation between $\widetilde{\rho}$ and $\widehat{\rho}$ because there is no dominance between $F(\widetilde{\mathbf{x}})$ and $F(\widehat{\mathbf{x}})$. However, one way to circumvent this situation is to take $\rho = \frac{\widehat{F}(\widehat{\mathbf{x}})}{\max\{F(\widetilde{\mathbf{x}}),F(\widehat{\mathbf{x}})\}}$. The new ratio ρ dominates $\widetilde{\rho}$.

Theorem 3. $\rho \leq \widetilde{\rho}$.

Proof. Direct from Theorem 2 and the fact that $\max\{F(\widetilde{\mathbf{x}}), F(\widehat{\mathbf{x}})\} \geq F(\widetilde{\mathbf{x}})$.

4.2 IBMS

Improved BMS (IMBS) works, as BMS, in three phases:

- **Relaxation Phase:** IBMS computes the spanning tree T and the relaxed problem $\widetilde{P}$ exactly as BMS does. Additionally, IBMS computes the relaxed problem $\widehat{P}$.
- **Solving Phase:** IBMS solves $\widetilde{P}$ and $\widehat{P}$ with Max-Sum. Let $\widetilde{\mathbf{x}}$ and $\widehat{\mathbf{x}}$ be the solutions of these problems. The agents will act according to the best solution $(\max\{F(\widetilde{\mathbf{x}}), F(\widehat{\mathbf{x}})\})$.
- **Bounding Phase:** IBMS computes the approximation ratio $\rho = \frac{\widehat{F}(\widehat{\mathbf{x}})}{\max\{F(\widetilde{\mathbf{x}}),F(\widehat{\mathbf{x}})\}}$.

The computation, storage and communication effort of IBMS is essentially twice that of BMS, because it requires solving two relaxed problems with Max-Sum. Given the low cost of BMS, doubling it seems acceptable. However, when it is not the case, one can always run a weaker version of IBMS ignoring $\widetilde{P}$. This weaker version will be exactly as costly as BMS. Its disadvantage is that $\widehat{\mathbf{x}}$ is not guaranteed to be better than $\widetilde{\mathbf{x}}$. In fact, our experiments show that there is no clear winner among them. Interestingly, the approximation ratio of the weaker version $\widehat{\rho}$ is systematically better than the approximation ratio of BMS $\widetilde{\rho}$.

Example 1. Consider the problem P given in Figure 1 with two variables $\{x_1, x_2\}$ and two functions $\{f_1, f_2\}$. The spanning tree of its factor graph is given with solid lines (i.e., edge (x_2, f_1) has been removed, shown as a dashed line). Thus, $W = 10$. Functions $\widetilde{f}_1$ and $\widehat{f}_1$ in $\widetilde{P}$ and $\widehat{P}$, respectively, are given in the figure. Max-Sum finds assignments $\widetilde{\mathbf{x}} = \widehat{\mathbf{x}} = (x_1 = a, x_2 = a)$, with utility $\widetilde{F}(\widetilde{\mathbf{x}}) = \widehat{F}(\widehat{\mathbf{x}}) = 20$. Their evaluation on the original problem P is $F(\widetilde{\mathbf{x}}) = F(\widehat{\mathbf{x}}) = 20$. The approximation ratios are $\widetilde{\rho} = 1.5$, $\widehat{\rho} = 1$, and $\rho = 1$.

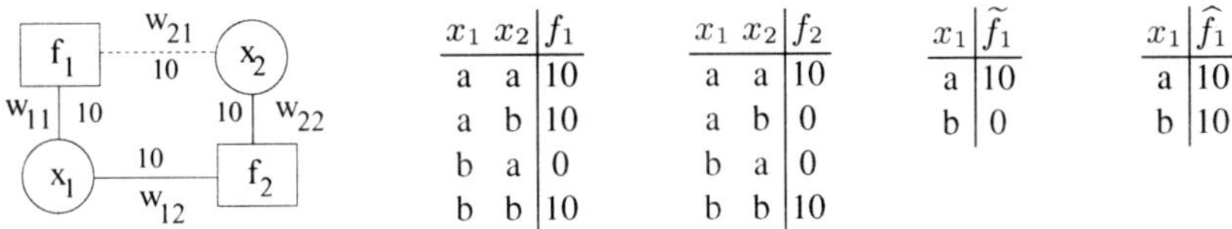

Fig. 1. Example of a factor graph containing cycles and a spanning tree formed by removing the edge between variable node x_2 and function node f_1

5   Empirical Evaluation

The purpose of the experiments is to evaluate the improvement of our upper bound $\widehat{F}(\widehat{\mathbf{x}})$ and approximation ratios ρ and $\widehat{\rho}$ over the BMS upper bound $\widetilde{F}(\widetilde{\mathbf{x}}) + W$ and approximation ratio $\widetilde{\rho}$, respectively. We consider the same set of problems from the ADOPT repository[1] used in [8]. These problems represent graph coloring problems with two different link densities (i.e., the average connection per agent) and different number of nodes. Each agent controls one node (i.e., variable), with domain $|\mathbf{d}_i| = 3$, and each edge of the graph represents a pairwise constraint between two agents. Each edge is associated with a random payoff matrix, specifying the payoff that both agents will obtain for every possible combination of their variables' assignments. Each entry of the payoff matrix is a real number sampled from two different distributions: a gamma distribution with $\alpha = 9$ and $\beta = 2$, and a uniform distribution with range $(0, 1)$. For each configuration, we report average values over 25 repetitions. For the sake of comparison, we compute the optimal utility by a complete centralized algorithm, although this value can only be computed up to 12 agents by a complete decentralized algorithm, as shown in [8].

Figure 2 (first and second rows) shows the upper and lower bound obtained by IBMS (i.e., $\widehat{F}(\widehat{\mathbf{x}})$ and $\max\{F(\widehat{\mathbf{x}}), F(\widetilde{\mathbf{x}})\}$, respectively) and BMS (i.e., $\widetilde{F}(\widetilde{\mathbf{x}})$ and $F(\widetilde{\mathbf{x}})$, respectively), along with the optimal utility (i.e., $F(\mathbf{x}^*)$), for the different link densities and payoff distributions. The behavior of both algorithms is very similar across all link densities and payoff distributions. IBMS always computes an upper bound tighter than the one computed by BMS. The improvement is slightly better for the uniform distribution. The lower bounds computed by both algorithms are very close, although IBMS lower bound is slightly better.

Figure 2 (bottom row) shows a detail on the lower bounds $F(\widehat{\mathbf{x}})$ and $F(\widetilde{\mathbf{x}})$ obtained on each instance of a given parameter configuration. Since the behavior across all number of agents, link densities and payoff distributions is very similar, we only report results on instances with 25 agents and gamma distribution. Both lower bounds are very close, and none of them is consistently better than the other.

Figure 3 shows the percentage of improvement of the approximation ratio of IBMS ρ and the weaker version of IBMS $\widehat{\rho}$ over the approximation ratio of MBS $\widetilde{\rho}$ (left y-axe). The figure also reports the percentage of deterioration of the approximation ratio of the 7-size-bounded-distance criteria introduced in [9] according to the minimum maximum reward bound ($S7_r$) and the minimum fraction bound ($S7_f$) presented in [10] over the

---

[1] http://teamcore.usc.edu/dcop

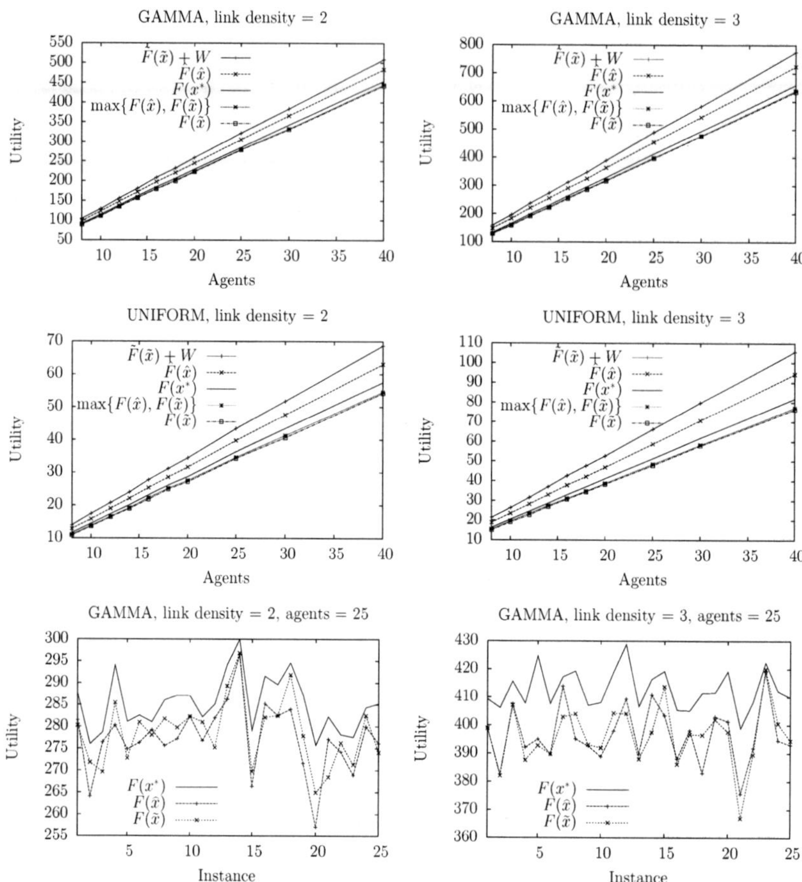

Fig. 2. First and second row, bounds obtained by algorithms IBMS and BMS varying the number of agents; third row, lower bound detail for instances with 25 agents and gamma distribution

approximation ratio of MBS $\widetilde{\rho}$ (right y-axe). Since the relation between the optimal solution of the problem $F(\mathbf{x}^*)$ and an approximation ratio ρ of a given solution $\mathbf{x}$ is $1 \leq \frac{F(\mathbf{x}^*)}{F(\mathbf{x})} \leq \rho$, we compute the improvement of an approximation ratio ρ over $\widetilde{\rho}$ as,

$$\frac{(\widetilde{\rho} - 1) - (\rho - 1)}{\widetilde{\rho} - 1} * 100$$

The improvement of ρ is always higher than 37%, and up to almost 50%. Its mean improvement for the gamma and uniform distributions is higher than 40% and 45%, respectively. The improvement of $\widehat{\rho}$ is always higher than 32%, and up to almost 46%. Its mean improvement for the gamma and uniform distributions is higher than 35% and 37%, respectively. Therefore, both IBMS and its weaker version always significantly outperforms BMS. Recall that the weaker version of IBMS has the same communication demands as BMS. Both approximation ratios $S7_r$ and $S7_f$ are worse than the

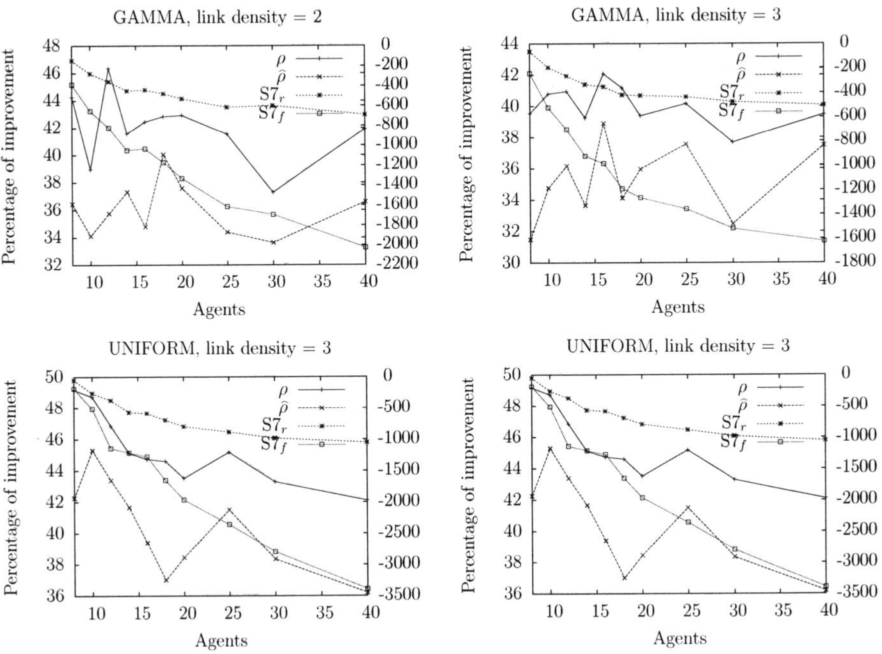

Fig. 3. Percentage of improvement of the approximation ratio of IBMS ρ and weaker version of IBMS $\widehat{\rho}$ (left y-axe), and percentage of decrease of the 7-size-bounded-distance criteria using the the minimum maximum reward bound ($S7_r$) and the minimum fraction bound ($S7_f$) (right y-axe) over the approximation ratio of BMS $\widetilde{\rho}$

approximation ratio of BMS $\widetilde{\rho}$ (the percentage is always negative). Their quality decreases as the number of agents increases for both distributions.

6   Conclusions

In this paper we introduced a new algorithm, called Improved Bounded Max-Sum (IBMS), based on the Bounded Max-Sum algorithm. We theoretically proved that its approximation ratio is always better than the previous one, at the only cost of doubling the communication requirements. We also introduced a weaker version of IBMS having the same communication demands as Bounded Max-Sum. Our experiments show that the approximation ratio of both algorithms is significantly tighter.

Acknowledgements. The authors are thankful to the authors of [8,9] for providing us with their implementations, and to the authors of [8] for sharing their benchmarks with us. This work was supported by project TIN2009-13591-C02-01.

References

1. Aji, S.M., McEliece, R.J.: The generalized distributive law. IEEE Transactions on Information Theory 46(2), 325–343 (2000)

2. Farinelli, A., Rogers, A., Petcu, A., Jennings, N.R.: Decentralised coordination of low-power embedded devices using the max-sum algorithm. In: AAMAS, pp. 639–646 (2008)
3. Fitzpatrick, S., Meetrens, L.: Distributed coordination through anarchic optimization. In: Distributed Sensor Networks A Multiagent Perspective, pp. 257–293. Kluwer Academic (2003)
4. Maheswaran, R.J., Pearce, J., Tambe, M.: A family of graphical-game-based algorithms for distributed constraint optimization problems. In: Coordination of Large-Scale Multiagent Systems, pp. 127–146. Springer (2005)
5. Mailler, R., Lesser, V.R.: Solving distributed constraint optimization problems using cooperative mediation. In: AAMAS, pp. 438–445 (2004)
6. Modi, P.J., Shen, W.M., Tambe, M., Yokoo, M.: Adopt: asynchronous distributed constraint optimization with quality guarantees. Artif. Intell. 161(1-2), 149–180 (2005)
7. Petcu, A., Faltings, B.: A scalable method for multiagent constraint optimization. In: IJCAI, pp. 266–271 (2005)
8. Rogers, A., Farinelli, A., Stranders, R., Jennings, N.R.: Bounded approximate decentralised coordination via the max-sum algorithm. Artif. Intell. 175(2), 730–759 (2011)
9. Vinyals, M., Shieh, E., Cerquides, J., Rodriguez-Aguilar, J.A., Yin, Z., Tambe, M., Bowring, E.: Quality guarantees for region optimal dcop algorithms. In: AAMAS, pp. 133–140 (2011)
10. Vinyals, M., Shieh, E., Cerquides, J., Rodriguez-Aguilar, J.A., Yin, Z., Tambe, M., Bowring, E.: Reward-based region optimal quality guarantees. In: OPTMAS Workshop (2011)

A Hybrid MIP/CP Approach
for Multi-activity Shift Scheduling

Domenico Salvagnin[1] and Toby Walsh[2]

[1] DEI, University of Padova
[2] NICTA and UNSW, Sydney

Abstract. We propose a hybrid MIP/CP approach for solving multi-activity shift scheduling problems, based on regular languages that partially describe the set of feasible shifts. We use an aggregated MIP relaxation to capture the optimization part of the problem and to get rid of symmetry. Whenever the MIP solver generates a integer solution, we use a CP solver to check whether it can be turned into a feasible solution of the original problem. A MIP-based heuristic is also developed. Computational results are reported, showing that the proposed method is a promising alternative compared to the state-of-the-art.

1 Introduction

A shift scheduling problem assigns a feasible working shift to a set of employees, in order to satisfy the demands for a given set of activities at each period in a given time horizon. The set of feasible shifts that can be assigned to employees is often defined by a complex set of work regulation agreements and other rules. Assigning a shift to an employee means specify an activity for each period, which may be a working activity or a rest activity (e.g., lunch). The objective is usually to minimize the cost of the schedule, which is usually a linear combination of working costs plus some penalties for undercovering/overcovering the demands of the activities in each time period. If the set of working activities W is made by a single activity, we talk of *single-activity* shift scheduling, while if there are several working activities we talk of *multi-activity* shift scheduling. In this paper we consider the latter case, with the additional constraint that all employees are identical.

In particular, suppose we are given a planning horizon divided into a set of periods T, a set of activities A, a subset $W \subset A$ of working activities, and a set of employees E. For each period $t \in T$ and for each working activity $a \in W$, we are given a demand d_{at}, an assignment cost c_{at}, an undercovering cost c_{at}^- and an overcovering cost c_{at}^+. Introducing the set of integer variables y_{at}, which count the number of employees assigned to activity a at period t, and integer variables s_{at}^-, s_{at}^+ that count the appropriate under/over covering, we can formulate the

M. Milano (Ed.): CP 2012, LNCS 7514, pp. 633–646, 2012.

problem as:

$$\min \sum_a \sum_t c_{at} y_{at} + \sum_a \sum_t c_{at}^+ s_{at}^+ + \sum_a \sum_t c_{at}^- s_{at}^- \tag{1}$$

$$y_{at} - s_{at}^+ + s_{at}^- = d_{at} \quad \forall a \in W, \forall t \in T \tag{2}$$

$$\sum_e x_{eat} = y_{at} \quad \forall a \in W, \forall t \in T \tag{3}$$

$$\langle x \text{ define a feasible shift } \forall e \in E \rangle \tag{4}$$

$$y_{at}, s_{at}^+, s_{at}^- \in \mathbb{Z}^+ \tag{5}$$

$$x_{eat} \in \{0, 1\} \tag{6}$$

Depending on how we formulate the constraints (4), we may end up with very different models. A convenient way to define the set of feasible shifts that can be assigned to a given employee is to use a regular or a context-free language, i.e., the set of feasible shifts can be viewed as the words accepted by a finite automaton or more generally by a push-down automaton. It has been shown in [14,3] that it is possible to derive a polyhedron that describes a given regular/context-free language. Such representations are compact (in an appropriate extended space, i.e., introducing additional variables) and thus lead directly to a MIP formulation of the problem. In particular, the extended formulation for a regular language is essentially a network flow formulation based on the expanded graph associated with the accepting automaton (see [13,3] for details). The extended formulation for the context-free language, on the other hand, is based on an and-or graph built by the standard CYK parser [10] for the corresponding grammar [14,16].

Note that it is not necessary to describe completely the set of feasible shifts by a regular/context-free language. The formal language may capture only some of the constraints defining a feasible shift, with the remaining ones described as linear inequalities. This may simplify the corresponding automaton considerably. For example, regular languages are notoriously bad at handling counting arguments, and an automaton describing the set of feasible shifts completely in the presence of even a few cardinality constraints may require thousands of states. Such large automaton are not trivial to generate. This also has a direct influence on the size of the model and the efficiency of reasoning about it. The same holds for context free languages. It is true that they can be enriched considerably by adding constraints that limit the applicability of productions rules, without even increasing the size of the model. However, certain cardinality constraints may overly complicate the language. Finally, depending on the application, the model derived using a context free language may be much bigger than an equivalent one derived using a regular language.

However, describing the set of feasible shifts with formal languages alone has some important implications. First of all, it has been proven for both the regular and context-free languages that the derived polyhedron is integral [14], and thus, if the are no other constraints, it is possible to optimize a linear function over the set of feasible shifts by solving just a linear program. Even more importantly, these results have been extended also to polyhedra describing sets of feasible shifts [4]. It is then possible to consider an aggregated (*implicit*) model and reconstruct an optimal solution of the original one with a polynomial post-processing phase. This gives the current state-of-the-art for solving multi-activity shift scheduling problems. Finally, if the formal language completely describes the set of feasible shifts, then it is possible to apply some very effective large neighborhood search heuristics to find quickly high quality solutions [17].

2   A Hybrid MIP/CP Approach

The explicit MIP model based on formal languages mentioned in the previous section has two drawbacks. First of all, its size is directly proportional to the number of employees in the instance; given that thousands of variables may be needed to completely describe the set of feasible shifts for a single employee, the linear programming relaxation can quickly become the bottleneck for branch-and-cut algorithm solving the instance. Second, with interchangeable employees, the enumeration itself explodes because of symmetry issues. In other words, the explicit model scales very badly as the number of employees increases.

A recently developed technique in the MIP community to deal with symmetric instances is called *orbital shrinking* [6], which is closely related to the implicit model mentioned earlier. The basic idea behind orbital shrinking is, given an orbit partitioning of the variables of a problem, to aggregate the variables within any orbit and consider the derived shrinked model on these aggregated variables, which is at the same time smaller and symmetry free. In the case of scheduling, where the symmetries are due to the interchangeable employees, this procedure automatically produces the implicit model used in [4]. Note that orbital shrinking is an exact reformulation only for convex optimization problems. On the one hand, this means that the LP relaxation of the shrinked model yields the same dual bound as the LP relaxation of the explicit model. On the other hand, given an arbitrary MIP, such reformulation is in general only a relaxation, although it can be tighter and/or faster to compute than the LP relaxation (more on this in Section 3).

Interestingly, for some special cases, the orbital shrinking reformulation is exact also for the MIP problem. This happens, for example, whenever it can be proven that an optimal solution of the aggregated model can always be turned into a solution of the original model of the same cost (and thus optimal). Examples of this behaviour are the assignment polytope and the regular/context-free language polytopes.

Consider for example the regular polytope in its extended form: the optimal solution is always a flow of integral value, say k, and basic network flow theory guarantees that it can be decomposed into k paths of unitary flow (and since each path in the expanded graph corresponds to a word in the language, this is a feasible solution for the original explicit problem). Similar reasoning applies to the grammar polytope (although it is not a flow model), as successfully shown in [4].

Unfortunately, it is not always reasonable to describe the set of feasible shifts completely with a formal language. While it is true that formal languages can be extended without changing the complexity of the corresponding MIP encoding (this is particularly true for context-free languages [16]), still some cardinality constraints may be very awkward to express, as shown in the following example:

Example 1. Let's consider a time horizon of 18 hours, divided into 18 periods. A feasible shift is a word of length 18 build from the alphabet $\Sigma = \{a, b, r\}$ (where a denotes the only working activity, r is a rest period, while b is a break period) that follows the pattern *rest-work-break-work-break-work-break-work-rest.* Suppose that the breaks are constrained to be one period long, and the number of working periods must be between 6 and 8. Then, a very simple grammar encoding the set of feasible shifts, ignoring the cardinality constraint, is:

$$S \rightarrow RFR \quad F \rightarrow PBP \quad P \rightarrow WBW$$

$$R \rightarrow Rr|r \quad W \rightarrow Wa|a \quad B \rightarrow b$$

In this particular case, since the number of breaks in the shift is fixed (3), it is very easy to extend the grammar to deal with the cardinality constraint by restricting the production rule $F \rightarrow PBP$ to be applied only with substring of length between 9 and 11. This can be handled very well by the CYK parser, and thus the cardinality constraint can be added essentially at no cost.

However, let's consider a slightly more complicated case. The pattern of a feasible shift is the same, but now the length of breaks is *not* fixed to one. In particular, the number of break periods is constrained to be between 4 and 6. The best we can do keeping approximately the same grammar as before is the following (we use the notation of [4] to indicate restrictions on production rules):

$$S \rightarrow RFR \quad F_{[10,14]} \rightarrow PBP \quad P_{[3,10]} \rightarrow WBW$$

$$R \rightarrow Rr|r \quad W \rightarrow Wa|a \quad B \rightarrow Bb|b$$

It is easy to see that the restrictions cannot be tightened any further, otherwise we may lose feasible shifts. However, the grammar also accepts the substring $rrababababbbbbbbaarr$, which violates both cardinality constraints.   □

Of course the issues of the previous example are not theoretical. As the set of feasible shifts is finite, there always exists a regular/context-free language that describes that set. However, the corresponding automaton may be unreasonably large in practice.

In order to turn the orbital shrinking approach into a complete method for the multi-activity shift scheduling problem when the formal language does *not* completely describe the set of feasible shifts, we propose a hybrid MIP/CP approach based on decomposition. In particular, whenever the MIP solver generates an integer feasible solution of the aggregated model, we must check whether it can be turned into a feasible solution of the explicit model. Because orbital shrinking always aggregates variables with the same costs (otherwise they would not be on the same orbit), this is indeed a pure feasibility problem. As such, we propose to formulate the check as a CSP problem, to be solved with a CP solver. In this way, we not only avoid solving the LP relaxations (that would provide no meaningful bounds), but we can explicitly state symmetry breaking constraints. Note that this is essentially a master/slave decomposition similar to a generalized Benders method, where the master problem is a MIP model, while the slave is a CP model.

2.1 MIP Model

The MIP model that we use is a simple modification of the general model hinted at in Section 1. The main difference is that we partition the set of feasible shifts Ω into k subsets Ω_k, each of which is described by a potentially different deterministic finite automaton (DFA) and cardinality constraints. This partition can simplify a lot the structure of the DFAs, and in general makes the implicit model more accurate, since the cardinality constraints are aggregated only within employees of the same "kind". This of course increases the size of the relaxation, but since the aggregated model is quite compact, this is usually well worth it. For each shift type Ω_k, the MIP model decides how many employees are assigned a shift in Ω_k, and then computes an aggregated integer flow of the same value. In details:

$$\min \sum_{k} \sum_{a} \sum_{t} c_{at} y_{at}^{k} + \sum_{a} \sum_{t} c_{at}^{+} s_{at}^{+} + \sum_{a} \sum_{t} c_{at}^{-} s_{at}^{-} \tag{7}$$

$$\sum_{k} y_{at}^{k} - s_{at}^{+} + s_{at}^{-} = d_{at} \quad \forall a \in W, \forall t \in T \tag{8}$$

$$\mathrm{regular}(y^{k}, w^{k}, \mathrm{DFA}^{k}) \quad \forall k \in K \tag{9}$$

$$\langle \text{cardinality constraints for } y^{k} \rangle \quad \forall k \in K \tag{10}$$

$$\sum_{k} w^{k} \leq E \tag{11}$$

$$w^{k}, y_{at}^{k}, s_{at}^{+}, s_{at}^{-} \in \mathbb{Z}^{+} \tag{12}$$

Note that we use the notation of constraint (9) to refer to the extended MIP formulation of the regular constraint involving flow variables. The constraint ensures that variables y^{k} can be decomposed into w^{k} words accepted by the automaton DFA^{k}. Constraints (10) refers to the cardinality constraints expressed as linear constraints that complete the description of sets Ω_k. Finally, if an upper bound E is given on the number of employees that can scheduled, it can be imposed in constraint (11).

2.2 CP Checker

The decision to partition the set of feasible shifts into k subsets Ω_k has an important consequence on the structure of the CP checker: the model actually decomposes into k separate CP models, one for each type of shift. Given an index k, suppose the master (MIP) model assigns $\overline{w}^{k}$ employees, with their aggregated shifts described by $\overline{y}_{at}$. Then the corresponding CP model, which is similar to the one proposed in [5], reads:

$$\mathrm{gcc}(x^e, \sigma^e, A) \quad \forall e \in 1, \ldots, \overline{w}^k \tag{13}$$

$$\tau^e = \sum_{a \in W} \sigma_a^e \quad \forall e \in 1, \ldots, \overline{w}^k \tag{14}$$

$$\langle \text{cardinality constraints for } \sigma^e, \tau^e \rangle \quad \forall e \in 1, \ldots, \overline{w}^k \tag{15}$$

$$\mathrm{regular}(x^e, \mathrm{DFA}^k) \quad \forall e \in 1, \ldots, \overline{w}^k \tag{16}$$

$$\mathrm{gcc}(x_t, \overline{y}_t, A) \quad \forall t \in T \tag{17}$$

$$x^e \preceq x^{e+1} \quad \forall e \in 1, \ldots, \overline{w}^k - 1 \tag{18}$$

Variables x_t^e denote the activity assigned to employee e at time t. Variables σ_a^e count the number of periods assigned to each activity for employee e, while τ^e gives the sum over all working activities. Both are needed to specify the cardinality constraints (15). Constraints (17) link the variables in the CP model to the master solution $\overline{y}_{at}$. Finally, we impose a lexicographic order among the shifts of the employees with constraints (18).

The CP model above is usually extremely fast in proving whether the aggregated solution can be turned into a solution of the original. However, as the number of activities and employees increases, it can occasionally become very time consuming. The main reason for this behavior is the weak interaction between the cardinality constraints (17) and the symmetry breaking constraints (18). The issue can be easily explained with an example:

Example 2. Let's consider a vector of 5 binary variables $x_1, \ldots, x_5$, each with initial domain $\{0, 1\}$, and aggregate variables $y_0 = 2$ and $y_1 = 3$, linked with the x by a cardinality constraint. In addition, there are symmetry breaking constraints of the form $x_1 \leq x_2 \leq \cdots \leq x_5$. From the cardinality constraint point of view, any permutation of the solution $(0, 0, 1, 1, 1)$ is feasible, so no reductions are possible. The same happens from the symmetry breaking point of view, because symmetry breaking alone cannot exclude the assignments where all x variables take the same value (0 or 1). However, it is clear that the only solution feasible for both constraints together is indeed $(0, 0, 1, 1, 1)$. □

To overcome this issue, we implemented an ad-hoc propagator that implements a custom symmetry breaking strategy based on the cardinality constraints. This propagator handles the case in which there is a matrix of variables, with cardinality constraints in each column and symmetry on the rows (i.e., any permutation of the rows is feasible). As such, although it is very related to our CP model, it is not specific to our particular instances. The propagator works by partitioning the rows of the matrix into sets of identical rows (given the current domains). Then, for each set, it considers the first column with unassigned variables. For each possible value v in the domains of this subset of variables, it computes a lower bound l_v and an upper bound u_v on the number of variables that must be assigned to v, by taking into account the cardinality constraints on the column and the domains of the variables. Finally, it assigns, for each value v, l_v

variables to v. The complexity of such propagation is linear in the size of the matrix/domains. Note that this symmetry breaking strategy does *not* enforce constraints (18) on the rows of the matrix, and is not guaranteed to remove all possible row symmetries from the model, as shown in the following example.

Example 3. Suppose we have a cell with 5 rows and let $x_1, \ldots, x_5$ be the first 5 unassigned variables (one for each row). Suppose that we have 2 possible values, a and b, and that we can compute the following lower/upper bounds on the number of occurrences of these 2 values in the 5 vars:

$$l_a = 1 \quad u_a = 3 \quad l_b = 2 \quad u_b = 5$$

Before propagation, the domains of the 5 variables are all equal to $\{a, b\}$. After symmetry breaking we can reduce the domains to $\{a\}, \{b\}, \{b\}, \{a, b\}, \{a, b\}$. However, since it is possible to have both *abbab* and *abbba*, the propagator does not enforce a lexicographic order on the rows, and does not eliminates all symmetry.   □

Another issue with the CP model above is that the minimum/maximum length of a working shift (i.e., the number of periods, breaks included, between the first and last working period) is constrained only implicitly by the regular constraints. Again, we implemented a custom propagator that deals with that. The combined effect of these propagators is impressive: we often observed reductions of $2-3$ orders of magnitude in both the number of nodes and the running times on hard instances. On occasion, we observed even higher savings. For example, on one instance with 9 full-time employees and 3 activities, we reduced the running times from 6 minutes to 10^{-5} seconds, with a number of nodes dropping from 765,026 to 1.

Finally, the same model could be solved repeatedly if the aggregated solutions are too similar, i.e., if the values of the variables y^k coincide for some k. So, we implemented a "caching" mechanism that stores the last CP models and their status, in order to avoid solving the same model twice. According to our computational experience, the custom propagators and the cache were sufficient to keep the time spent within the CP solver negligible.

2.3   MIP Repair/Improve Heuristic

Finding a good quality feasible solution early in the process is often crucial to effectively solve an optimization problem. However, dual decomposition strategies (such as generalized Benders decomposition and the hybrid approach presented here) are usually quite weak at finding (good) feasible solutions. In order to solve this issue, we devised an ad-hoc heuristic procedure for the shift scheduling problem.

The procedure is a simple generalization of the large neighborhood search developed in [17]. Suppose the set Ω can be completely described by a regular language. Then every feasible shift is a path in the expanded graph associated to the DFA and a solution to the problem is just a set of paths in this graph. Given a solution $S = \{s_1, \ldots, s_n\}$, let N_S be the set of solutions that can obtained by replacing a shift s_k with a different shift r. Given a choice s_k for the shift to remove, searching an improving replacement shift r can be formulated as a shortest path problem on the expanded graph, and N_S can be used as a neighborhood in the large neighborhood search heuristic (see [17] for details).

If Ω is not described completely by a regular language (the main assumption in the present work), then the search for an improving shift r cannot be formulated anymore as a shortest path. However, it can still be formulated as a MIP, to be solved by a black box MIP solver. This is the basic step of our heuristic, to be used in a scheme akin to the one in [17]. Note that basically the same MIP can be used to:

- find a feasible shift r to replace another feasible shift s_k.
- find a feasible shift r to replace an *infeasible* shift s_k.
- find a feasible shift r to add to current solution S (if doing so reduces the cost).

As such, the same heuristic can be used to (i) construct an initial feasible solution, (ii) improve a feasible solution, and (iii) repair an infeasible solution.

3   Computational Results

We tested our method on the multi-activity instances used in [3,4,17]. This testbed is derived from a real-world store, and contains instances with 1 to 10 working activities (each class has 10 instances). A basic description of the problem is as follows:

- The planning horizon of 1 day is divided into 96 slots of 15 minutes.
- A part-time employee must work a minimum of 3 hours and less than 6 hours, and is entitled to one break of 15 minutes.
- A full-time employee can work between 6 and 8 hours, and is entitled to two breaks of 15 minutes plus a lunch break of 1 hour (in any order).
- When an employee starts working on one activity, it must do it for at least 1 hour. In addition, a break/lunch is needed before changing activity.
- A break cannot be scheduling at the beginning/end of the shift.
- At specific times of the day (e.g., when the store is closed), no employee is allowed to work.
- Overcovering/undercovering is allowed, with an associated cost.
- The cost of a shift is the sum of the costs of all working activities performed in the shift.

We implemented our method in C++, using IBM ILOG Cplex 12.2 [11] as black box MIP solver, and Gecode 3.7.1 [7] as CP solver. All tests have been performed on a PC with an Intel Core i5 CPU running at 2.66GHz, with 8GB of RAM (only one core was used by each process). Every method was given a time limit of 1 hour per instance. Concerning the set of feasible shifts Ω, we simply partitioned it into full-time and part-time shifts. We could have partitioned the full-time shifts further (depending on the relative order of breaks and lunch), but it seemed overkill because all full-time shifts share the same cardinality constraints (this was confirmed by some preliminary tests). In general, disaggregating shifts depending on the cardinality constraints seems to work well in practice.

From the implementation point of view, our hybrid method is made of the following phases:

- First, the aggregated model is solved with Cplex, using the default settings. The outcome of this (usually fast) first phase is a dual bound (potentially stronger than the LP bound) and the set of aggregated solutions collected by the MIP solver during the solution process (not necessarily feasible for the original model).
- We apply our MIP repair/improve heuristic to each aggregated solution which is within 20% of the aggregated model optimal solution. The outcome of this phase is always a feasible solution for the original model, thus a primal bound. Note that if the gap between the two is already below the 1% threshold, we are done.
- We solve the aggregated model again, this time implementing the hybrid MIP/CP approach. This means that we disable dual reductions (otherwise the decomposition would not be correct) and use Cplex callbacks framework to implement the decomposition.

Here is a more detailed description of the last phase. Whenever the MIP solver finds an integer solution, either with its own heuristics or because the LP relaxation happens to be integer, we build the corresponding CP models and solve them with Gecode DFS algorithm. As far as the branching strategy of the CP solver is concerned, after some trial-and-error we found that ranking the variables by increasing time period was the most successful policy. If the check is successful, then we update the incumbent, otherwise the solution is rejected. In both cases, we apply the MIP repair/improve heuristic on it to try to find a new incumbent. If the solution was the optimal solution of an LP relaxation, then we force a branching on a integer variable and keep going. As for branching inside the MIP solver, we let Cplex apply its own powerful strategy whenever the relaxation has some fractional variables. If this is not the case, we branch first on the w variables and then, if all w variables are already fixed, on the y, again ranking them by increasing time period. The rationale behind this strategy is that if the w variables are not fixed to some value, then we cannot even formulate the CP checking model, so the sooner we fix them the better. Note that as soon as the w variables are fixed, we can build a CP model akin to (13)-(18) where the y variables are not necessarily fixed but just take the domains of the current node. In this case, we let the CP solver run with a strict fail limit (1000 in our code) and, if it detects infeasibility, then we prune the node.

Table 1 reports a comparison between the proposed method and others in the literature, for a number of activities from 1 to 10. As far as the number of employees is concerned, we put an upper bound of 12 for instances with up to 2 activities, of 24 for instances with 3 to 8 activities and of 30 for instances

Table 1. Average computing times between the different methods to solve to near-optimality (gap $\leq 1\%$) the instances with up to 10 activities

# act.	# solved (10)			time(s)		
	cpx-reg	hybrid	grammar	cpx-reg	hybrid	grammar
1	10	10	10	41.3	9.1	283.7
2	9	10	9	707.9	194.5	379.9
3	4	5	9	2957.3	1996.4	205.4
4	3	6	10	2970.2	1827.9	300.5
5	0	8	10	3600.0	1438.4	146.2
6	1	4	10	3530.6	2340.6	213.8
7	1	6	10	3438.7	2399.0	230.9
8	0	5	10	3600.0	2201.5	257.1
9	0	4	10	3600.0	2444.0	289.1
10	0	2	10	3600.0	3275.6	516.7

with 9 or 10 activities. cpx-reg refers to the explicit model based on the regular constraint in [3], while grammar refers to the implicit model based on the grammar constraint in [4]. Note that for grammar we are reporting the results from [4], which were obtained on a different machine and, more importantly, with an older version of Cplex, so the numbers are meant to give just a reference. All methods were run to solve the instances to near-optimality, stopping when the final integrality gap dropped below 1%.

According to Table 1, hybrid outperforms significantly the explicit model cpx-reg, which, as already noted, scales very poorly because of symmetry issues and slow LPs. When compared to grammar, hybrid is very competitive only for up to 2 activities, while after that threshold grammar clearly takes the lead. This is somewhat expected: the set of feasible shifts in these instances can indeed be described without too much effort with an extended grammar, and it is no surprise that the pure implicit MIP model outperforms our decomposition approach. However, hybrid is likely to be the best approach if the extended grammar is not a viable option.

Table 2 reports a closer comparison between cpx-reg and hybrid, reporting the average final gap, average number of variables in the model and average node throughput for each category. According to the table, hybrid consistently yields very small gaps (always below 3% on average), while for cpx-reg is always above 60% with more than 4 activities. As far as the number of variables of the models is concerned, hybrid needs approximately 1/10 of the number of variables of cpx-reg, which promptly turns into a much faster node throughput: hybrid is more than two order of magnitude faster in exploring nodes than cpx-reg. Note that, according to [4], grammar models range from 70,000 variables for instances with 1 activity to 96,000 for instances with 10 activities, so the hybrid model based on regular languages is significantly smaller.

Table 2. Comparison of average final gap between `cpx-reg` and `hybrid`

# act.	gap(%) cpx-reg	gap(%) hybrid	#vars cpx-reg	#vars hybrid	node/sec cpx-reg	node/sec hybrid
1	0.72	0.24	9,956	1,908	21.99	10.36
2	0.78	0.61	13,608	2,925	3.52	20.60
3	3.74	3.00	34,903	4,152	0.59	8.42
4	25.18	1.39	43,005	5,291	0.20	3.92
5	62.55	1.01	52,979	6,828	0.05	3.32
6	75.89	1.59	62,442	8,364	0.03	1.86
7	90.00	0.90	73,693	9,936	0.01	1.64
8	100.00	1.92	78,809	10,603	0.01	1.22
9	100.00	1.52	104,561	11,509	0.01	1.05
10	100.00	2.76	120,049	13,302	0.01	0.86

Finally, Table 3 shows the gap just before the beginning of the last phase (but after the aggregated model has been solved and its solutions have been used to feed the MIP repair/improve heuristic). On almost all categories the average final gap is below 10%, with an average running time of 1 minute. This heuristic alone significantly outperforms `cpx-reg` for a number of activities greater than 3. It is also clear from the table that solving the orbital shrinking relaxations with a black box MIP solver is usually very fast. Interestingly, solving these MIPs turn out to be often faster than solving the LP relaxations of the original models, while providing better or equal dual bounds. For example, on one instance with 1 activity, the LP relaxation of the original model that 0.26 seconds to solve, yielding a dual bound of 142.48, while the shrinked MIP takes 0.12 seconds and yields a dual bound of 182.54 (in this case, equal to the value of the optimal solution). On another instance with 10 activities, the LP relaxation takes 269.55, while the shrinked MIP takes only 52.77 seconds, both yielding the same dual bound in this case.

Table 3. MIP repair/improve heuristics standalone results

# act.	time(s)	gap(%)
1	6.2	1.5
2	46.5	6.5
3	24.7	20.3
4	30.3	7.1
5	34.5	5.9
6	33.5	10.5
7	63.2	7.1
8	69.3	7.7
9	89.8	6.7
10	65.9	8.0

4   Related Work

We divide the related work into two parts: previous work on using regular and context-free languages to specify constraints, especially those constraints occurring in shift scheduling problems, and previous work on hybridizations of CP and MIP solving.

Pesant introduced the global regular constraint in which constraints are specified by regular constraints [13]. He gave a complete propagation algorithm based on dynamic programming. Coincidently Beldiceanu, Carlsson and Petit proposed specifying global constraints by means of finite automata augmented with counters [1]. Regular languages are precisely those accepted by (deterministic) finite automata. Propagators for such an automaton are constructed automatically from the specification of the automaton by means of a decomposition into simpler constraints. Quimper and Walsh proposed a closely related decomposition of the regular constraint based on transition constraints and variables introduced to represent the states of the unfolded automaton which recognizes the language [15]. They showed that such decomposition was effective and efficient in practice. Demassay et al. [5] used a column generation technique to solve a shift scheduling problem in which the columns are generated with a CP solver using the cost regular constraint, a variation of the regular constraint, whilst the optimization process is driven by the simplex method. Côté et al. [2] encoded the regular constraint into a MIP and efficiently solved some instances of the shift scheduling problem using the same automaton as Demassay et al.

Quimper and Walsh proposed the context-free grammar constraint in which constraints are specified by a context free grammar [15]. They gave two different propagators, one based on the CYK and the other on the Earley parser. At the same time and independently, Sellmann proposed the same global constraint and gave a similar propagator based on the CYK parser [18]. In [16,3], context-free grammar constraints have been used to model complex shift-scheduling problems. More recently, Côté, Gendron, Quimper and Rousseau have proposed mixed-integer programming (MIP) encodings of the regular and context-free grammar constraints [3]. The MIP encoding of the regular constraint introduces linear inequalities to model the flow constructed by unfolding the automaton into a layered transition graph. When this is the only constraint in a problem, this can be solved with a specialized path finding algorithm. However, when there are other constraints in the problem, it needs to be solved with a more general 0/1 MIP solver. The MIP encoding has one significant difference with the CYK propagator. If there is more than one parsing for a sequence, it picks one arbitrarily whilst the CYK propagator keeps all. This simplifies the MIP encoding without changing the set of solutions since only one parsing is needed to show membership in a context-free grammar. Experiments on a shift scheduling problem show that such MIP encodings are highly effective.

The last couple of decades have seen many hybrid approaches to solving optimization and decision problems that exploit both MIP and CP technqiues [12,20]. There are several different approaches to such hybridization including:

Double modeling: We use both CP and MIP models and exchange information while solving. This may include bounds, infeasibility, nogoods, etc.

Search-inference duality [8]: We view CP and MIP methods as special cases of a search/inference duality. Search methods can be complete methods like branching or incomplete methods like local search. Inference methods can be CP based methods like domain reduction or MIP based methods like cutting plane inference.

Decomposition [9]: We decompose problems into a CP part and a MIP part using, for example, a Benders style scheme. The master problem searches over some of these variables. Given an instantiation for these variable, we get a subproblem which we can, for instance, solve using CP. The MIP/CP hybridization proposed here has this form.

Relaxation [19]: We combine CP based search methods like branching with relaxation techniques from MIP which solve a simpler approximated form of the problem like Langrangian relaxation.

5   Conclusions

We have proposed a hybrid MIP/CP approach for solving multi-activity shift scheduling problems, based on regular languages that *partially* describe the set of feasible shifts. This choice is justified by the fact that it may be much easier from the modeling point of view to define the appropriate formal language, and the corresponding MIP models may be much smaller. Computational results show that the method is a promising alternative compared to the state-of-the-art, being the fastest method when there are not too many activities. When the number of activities increases, the method cannot compete, in its present state, with an implicit formulation where the formal language completely describes the set of feasible shifts. However, it outperforms significantly explicit MIP based formulations, thus becoming the method of choice when it is not practical to describe the set of feasible shifts with a formal language alone. Future work may address the interesting question of how to derive Benders style cutting planes from the CP model (when infeasible), in order to speed up the enumeration of the MIP.

References

1. Beldiceanu, N., Carlsson, M., Petit, T.: Deriving Filtering Algorithms from Constraint Checkers. In: Wallace, M. (ed.) CP 2004. LNCS, vol. 3258, pp. 107–122. Springer, Heidelberg (2004)
2. Côté, M.-C., Gendron, B., Rousseau, L.-M.: Modeling the Regular Constraint with Integer Programming. In: Van Hentenryck, P., Wolsey, L.A. (eds.) CPAIOR 2007. LNCS, vol. 4510, pp. 29–43. Springer, Heidelberg (2007)
3. Côté, M.C., Gendron, B., Quimper, C.G., Rousseau, L.M.: Formal languages for integer programming modeling of shift scheduling problems. Constraints 16(1), 54–76 (2011)

4. Côté, M.C., Gendron, B., Rousseau, L.M.: Grammar-based integer programming models for multiactivity shift scheduling. Management Science 57(1), 151–163 (2011)
5. Demassey, S., Pesant, G., Rousseau, L.M.: A cost-regular based hybrid column generation approach. Constraints 11(4), 315–333 (2006)
6. Fischetti, M., Liberti, L.: Orbital Shrinking. In: Mahjoub, A.R., Markakis, V., Milis, I., Paschos, V.T. (eds.) ISCO 2012. LNCS, vol. 7422, pp. 48–58. Springer, Heidelberg (2012)
7. Gecode Team: Gecode: Generic constraint development environment (2012), http://www.gecode.org
8. Hooker, J.N.: Integrated Methods for Optimization. Springer (2006)
9. Hooker, J.N., Ottosson, G.: Logic-based Benders decomposition. Mathematical Programming 96(1), 33–60 (2003)
10. Hopcroft, J.E., Ullman, J.D.: Introduction to Automata Theory, Languages and Computation. Addison Wesley (1979)
11. IBM ILOG: CPLEX 12.2 User's Manual (2011)
12. Milano, M.: Constraint and Integer Programming: Toward a Unified Methodology. Kluwer Academic Publishers (2003)
13. Pesant, G.: A Regular Language Membership Constraint for Finite Sequences of Variables. In: Wallace, M. (ed.) CP 2004. LNCS, vol. 3258, pp. 482–495. Springer, Heidelberg (2004)
14. Pesant, G., Quimper, C.-G., Rousseau, L.-M., Sellmann, M.: The Polytope of Context-Free Grammar Constraints. In: van Hoeve, W.-J., Hooker, J.N. (eds.) CPAIOR 2009. LNCS, vol. 5547, pp. 223–232. Springer, Heidelberg (2009)
15. Quimper, C.-G., Walsh, T.: Global Grammar Constraints. In: Benhamou, F. (ed.) CP 2006. LNCS, vol. 4204, pp. 751–755. Springer, Heidelberg (2006)
16. Quimper, C.-G., Walsh, T.: Decomposing Global Grammar Constraints. In: Bessière, C. (ed.) CP 2007. LNCS, vol. 4741, pp. 590–604. Springer, Heidelberg (2007)
17. Quimper, C.G., Rousseau, L.M.: A large neighbourhood search approach to the multi-activity shift scheduling problem. Journal of Heuristics 16, 373–392 (2010)
18. Sellmann, M.: The Theory of Grammar Constraints. In: Benhamou, F. (ed.) CP 2006. LNCS, vol. 4204, pp. 530–544. Springer, Heidelberg (2006)
19. Sellmann, M.: Theoretical Foundations of CP-Based Lagrangian Relaxation. In: Wallace, M. (ed.) CP 2004. LNCS, vol. 3258, pp. 634–647. Springer, Heidelberg (2004)
20. Van Hentenryck, P., Milano, M.: Hybrid Optimization: The Ten Years of CPAIOR. Springer Optimization and Its Applications. Springer (2010)

Contributions to the Theory of Practical Quantified Boolean Formula Solving

Allen Van Gelder

University of California, Santa Cruz
http://www.cse.ucsc.edu/~avg

Abstract. Recent solvers for quantified boolean formulas (QBFs) use a clause learning method based on a procedure proposed by Giunchiglia et al. (JAIR 2006), which avoids creating tautological clauses. The underlying proof system is Q-resolution. This paper shows an exponential worst case for the clause-learning procedure. This finding confirms empirical observations that some formulas take mysteriously long times to solve, compared to other apparently similar formulas.

Q-resolution is known to be refutation complete for QBF, but not all logically implied clauses can be derived with it. A stronger proof system called QU-resolution is introduced, and shown to be complete in this stronger sense. A new procedure called QPUP for clause learning without tautologies is also described.

A generalization of pure literals is introduced, called effectively depth-monotonic literals. In general, the variable-elimination resolution operation, as used by Quantor, sQueezeBF, and Bloqqer is unsound if the existential variable being eliminated is not at innermost scope. It is shown that variable-elimination resolution is sound for effectively depth-monotonic literals even when they are not at innermost scope.

1 Introduction

Solvers for quantified boolean formulas (QBFs) are rapidly increasing in strength, partly due to increased understanding of how to incorporate conflict-driven clause learning (CDCL), which found great practical success in propositional satisfiability. Several current solvers are patterned after the Q-resolution method described by Giunchiglia *et al.* [13]. A thorough survey of the field through 2005 may be found in this paper.

It is starting to be recognized that simply delivering a 0 or 1 answer is not good enough, for a solver. There has to be some formal proof system to back up claimed answers. More than just verifying a claimed answer, users want additional information relevant to their applications. To accommodate these needs, solvers need to be implemented in an organized way. This paper addresses a few issues found in current solvers and suggests improvements.[1]

---

[1] See http://www.cse.ucsc.edu/~avg/QPUP/qpup-cp12-long.pdf for a longer version of this paper.

M. Milano (Ed.): CP 2012, LNCS 7514, pp. 647–663, 2012.

First, in Section 3 we motivate the study by showing that current clause-learning methods based on [13] might take exponential time to learn one clause. We describe a family of QBF formulas such that the first clause to be learned takes exponential time, although it has a linear-length Q-resolution derivation. Exponential time to learn one clause does not occur with a properly implemented propositional CDCL procedure.

In Section 4 we describe a clause-learning system that runs in polynomial time per clause learned. This avoids the exponential-time worst cases mentioned above. Although it may be impractical in its current form, it might provide the basis for a practical adaptation into the QDPLL framework.

In Section 5 we introduce QU-resolution, a resolution system for QBF that is capable of deriving any logically implied clause (Definition 2.5), a characteristic that is absent from all presently known QBF solvers that utilize the QDPLL framework of [13]. We show that QU-resolution can produce exponentially shorter refutations than Q-resolution on some QBF families. In related work, Egly recently showed that certain sequent systems can produce exponentially shorter refutations than Q-resolution on some QBF families [7]. These sequent systems, in which introduction of new variables is central, can produce exponentially shorter refutations than QU-resolution on the same formulas.

Like Q-resolution, QU-resolution preserves tree models. Tree models represent winning strategies for true formulas. These strategies, also called Skolem functions, contain the extra information needed by many practical applications to achieve the goal that is encoded in the QBF.

Effectively depth-monotonic literals (Section 6) allow QBF variable-elimination resolution in more cases. The paper concludes with Section 7.

2   Preliminaries

In general, **quantified boolean formulas** (QBFs) generalize propositional formulas by adding operations consisting of universal and existential quantification of boolean variables. A **closed QBF** is one in which every variable is quantified. See [15,5,17] for thorough introductions. This paper uses standard notation as much as possible, but some terms have no standard version. We consider resolution and universal reduction separately, although some papers combine them. Also, we use tree models, which are not found in all QBF papers, to define *super-sound* and *safe* operations, and to distinguish between them. Also, we define ordered assignments.

We say that a QBF is in *prenex conjunctive normal form* if all the quantifiers are outermost operators (the prenex, or quantifier prefix), and the quantifier-free portion (also called the matrix) is in CNF; i.e., $\Psi = \vec{Q} . \mathcal{F}$ consists of prenex $\vec{Q}$ and matrix $\mathcal{F}$. For this paper QBFs are in prenex conjunctive normal form. If p precedes q in the quantifier prefix, we say p **is outer to** q and q **is inner to** p. Clauses in $\mathcal{F}$ are called *input clauses*.

A *closed* QBF evaluates to either *invalid* (false) or *valid* (true), as defined by induction on its principal operator. We use 0 and 1 for truth values of literals and use true and false for semantic values of formulas.

1. $(\exists\, x\, \Phi(x))$ is true if and only if ($\Phi(0)$ is true or $\Phi(1)$ is true).
2. $(\forall\, x\, \Phi(x))$ is false if and only if ($\Phi(0)$ is false or $\Phi(1)$ is false).
3. Other operators have the same semantics as in propositional logic.

This definition emphasizes the connection of QBF to two-person games, in which player E (Existential) tries to set existential variables to make the QBF evaluate to true, and player A (Universal) tries to set universal variables to make the QBF evaluate to false. Players set their variable when it is outermost, or for non-prenex, when it is the root of a subformula (see [16] for more details). Only one player has a winning strategy.

For this paper a **clause** is a *disjunctively* connected set of literals. If the term **cube** is used, it refers to a *conjunctively* connected set of literals. Literals are variables or negated variables, with overbar denoting negation. We also use $\perp$ as a literal representing false when it makes more uniform notation. Clauses may be written as literals enclosed in square brackets (e.g., $[p, q, \overline{r}]$), and $[]$ denotes the empty clause. Where the context permits, letters e and others near the beginning of the alphabet denote existential literals, while letters u and others near the end of the alphabet denote universal literals. Letters like p, q, r denote literals of unspecified quantifier type. The variable underlying a literal p is denoted by $|p|$ where necessary. Free variables or free literals e and u may be indicated by $\Psi(e, u)$.

The quantifier prefix is partitioned into maximal contiguous subsequences of variables of the same quantifier type, called **quantifier blocks**. Each quantifier block has a unique **qdepth**, with the outermost block having qdepth $= 1$. The **scope** of a quantified variable is the qdepth of its quantifier block. We say scopes are *outer* or *inner* to another scope to avoid any confusion about the direction, since there are varying conventions in the literature for numbering scopes.

Definition 2.1. An **assignment** is a partial function from variables to truth values, usually represented as the set of literals mapped to 1. A **total assignment** is an assignment to all variables. Assignments are denoted by ρ, σ, τ. Application of an assignment σ to a logical expression is called a **restriction** and is denoted by $q\lceil_\sigma$, $C\lceil_\sigma$, $\mathcal{F}\lceil_\sigma$, etc. Quantifiers for assigned variables are deleted in $\Psi\lceil_\sigma$.

An **ordered assignment** is a special term that denotes a total assignment that is represented by a sequence of literals that are assigned 1 and are in the same order as their variables appear in the quantifier prefix. $\qquad\square$

A **winning strategy** can be presented as an unordered directed tree. If it is a winning strategy for the E player, it is also called a **tree model**, which we now describe. We shorten *unordered directed tree* to *tree* throughout this paper. The qualifier "unordered" means that the children of a node do not have a specified order; they are a set. Recall that a **branch** in a tree is a path from the root node to some leaf node. A tree can be represented as the set of its branches. We also define a **branch prefix** to be a path from the root node that might terminate before reaching a leaf.

Definition 2.2. Let a QBF $\Phi = \vec{Q} \cdot \mathcal{F}$ be given. In this definition, σ denotes a (possibly empty) branch prefix of some ordered assignment for Φ. A ***tree model*** M for Φ is a nonempty set of ordered assignments for Φ that defines a tree, such that

1. Each ordered assignment makes $\mathcal{F}$ true, i.e., *satisfies* $\mathcal{F}$ in the usual propositional sense.
2. If e is an existential literal in Φ and some branch of M has the prefix (σ, e), then *no* branch has the prefix $(\sigma, \overline{e})$; that is, treating σ as a tree node in M, it has only one child and the edge to that child is labeled e.
3. If u is an universal literal in Φ and some branch of M has the prefix (σ, u), then *some* branch of M has the prefix $(\sigma, \overline{u})$; that is, treating σ as a tree node in M, it has two children and the edges to those children are labeled u and $\overline{u}$.

If τ is a partial assignment to all variables outer to existential variable e, then requirement (2) ensures that the "Skolem function" $e(\tau)$ is well defined as the unique literal following τ in a branch of M. If the formula evaluates to **false**, the set of tree models is empty. □

Definition 2.3. A ***tree countermodel*** R for Φ is essentially the dual of a tree model. That is, a tree node has two children with the edges labeled e and $\overline{e}$ when e is existential and has one child when the edge is labeled with universal literal u, and each branch *falsifies* some clause of $\mathcal{F}$.

If τ is a partial assignment to all variables outer to universal variable u, then (the dual of) requirement (3) ensures that the "Herbrand function" $u(\tau)$ is well defined as the unique literal following τ in a branch of R. If the formula evaluates to **true**, the set of tree countermodels is empty. □

Definition 2.4. For our purposes, an operation on a closed QBF is said to be ***safe*** if it does not change the truth value of the formula. An operation on a closed QBF is said to be ***super sound*** if it *preserves* the set of tree models (i.e., does not *add* or *delete* tree models). Clearly, preserving the set of tree models is a sufficient condition for safety. □

Definition 2.5. Let $\Psi = \vec{Q} \cdot \mathcal{F}$ be a closed prenex QBF. Another quantifier-free formula $\mathcal{G}$ (usually a clause or a set of clauses) is said to be ***logically implied*** by Ψ if $\vec{Q} \cdot (\mathcal{F} \wedge \mathcal{G})$ has the same set of tree models as Ψ; that is, the operation of adding $\mathcal{G}$ to $\mathcal{F}$ is super sound. In other words, $\mathcal{G}$ evaluates to **true** in every tree-model of Ψ. □

The proof system known as **Q-resolution** consists of two operations, *resolution* and *universal reduction,* defined next. Q-resolution is of central importance for QBFs because it is a *refutationally* complete proof system [14]. Unlike resolution for propositional logic, Q-resolution is not *inferentially* complete. That is, a new (non-tautological) clause C might be logically implied by a closed QBF Ψ, yet no subset of C is derivable by Q-resolution (see Example 5.5).

Definition 2.6. ***Resolution*** is defined as usual. Let clauses $C_1 = [q, \alpha]$ and $C_2 = [\overline{q}, \beta]$, where q is called the *clashing literal*. Let α be a literal sequence without conflicting literals and without q and $\overline{q}$. Let the same be true of β. Either or both of α and β may be empty. Then $\mathsf{res}_q(C_1, C_2) = [\alpha \cup \beta]$ is the *resolvent*. For Q-resolution, the clashing literal q is required to be existential, and the resolvent is not permitted to be tautologous.

Universal reduction is special to QBF. Let clauses $C_1 = [u, \alpha]$, where u is called the *reduction literal* and is universal. Let α be a literal sequence without conflicting literals and without u and $\overline{u}$. Further, let u be ***tailing for*** α, which means that the quantifier depth of u is greater than (u is inner to) that of any existential literal in α. Then $\mathsf{unrd}_u(C_1) = [\alpha]$. $\qquad\square$

Lemma 2.7. Resolution and universal reduction are super-sound operations. Also, resolution preserves the set of tree countermodels.

Proof: Straightforward application of the definitions. $\blacksquare$

We are not aware of a prior definition of tree countermodel, which explicitly encodes a winning strategy for the universal player when the QBF is false. However, there has been work on extracting a winning strategy for the universal player from a Q-resolution refutation [2,10].

It is worth observing that universal reduction might *add* tree countermodels, as shown by the following example. This is another reason to treat it as a separate operation from resolution.

Example 2.8. Consider $\Phi = \exists\, d\, \forall\, u\, \exists\, e\, \big\{[d,\, u]\,, [\overline{u},\, e]\,, [\overline{d},\, \overline{e}]\big\}$. The only tree countermodel is

$$\big\{(\overline{d},\, \overline{u},\, \overline{e}),\, (\overline{d},\, \overline{u},\, e),\, (d,\, u,\, \overline{e}),\, (d,\, u,\, e)\big\},$$

where parentheses enclose branches (ordered assignments).

If universal reduction is applied on the clause $[d,\, u]$ giving $[d]$, there is a second tree countermodel in which u is positive on all four branches. $\qquad\square$

3   QDPLL Exponential Case

QDPLL is the name commonly used for a family of QBF solving procedures with clause learning similar to those described by Giunchiglia *et al.* [11,12], [13]. Letz describes similar ideas that are used in `SemProp`, but with a different learning procedure [21]. For QDPLL, it is assumed that the input clauses are non-tautological and that any tailing universal literals have been reduced away (Definition 2.6).

When a conflict occurs, QDPLL derives a *learned clause* using Q-resolution on clauses in the conflict graph. We begin by showing a case in which QDPLL spends time that is exponential in the size of the conflict graph to derive its first learned clause. This is remarkable because the learned clause has a linear Q-resolution derivation.

At a high level, the clause learning procedure is similar to that found in CDCL SAT solvers. We assume the reader is generally familiar with the SAT version.

1. Begin with the working clause W_0 being the clause that became falsified.
2. At each derivation step i, choose a literal $\overline{q}$ in the current working clause W_{i-1} and Q-resolve with the antecedent of q in the conflict graph to produce the next working clause W_i. The **antecedent** clause is the clause that became unit to imply q.
3. Stop when W_i is an asserting clause that satisfies some additional conditions specific to QBF. Learn W_i. A clause is **asserting** when it has a unique literal at the highest decision level among the decision levels represented in the clause.

The procedure is called `Rec-C-Resolve` in [13]. The algorithm in which it is used is called *Q-DLL-LN*.

The central point of the above procedure is the choice of literal $\overline{q}$ in W_{i-1}, which will be used as the clashing literal. The policy in the cited paper is this: Choose the existential literal that was implied *most recently* unless that that choice would produce a tautologous resolvent (in which case Q-resolution is not defined). Otherwise choose an existential literal in W_{i-1} whose quantifier block has innermost scope. This policy is implemented in early versions of `QuBE`. Essentially the same policy is used in `depQBF` [18,19]. A slight variant is used in `CirQit` [9,8].

Example 3.1. We now describe a family named *qdpllexp* whose run time increases exponentially with instance length for the three solvers just mentioned.[2] The essence of `qdpllexp_06` is shown in Figure 1. The generation pattern simply varies the number of middle sections and should be self-evident.

The computation begins by assuming outermost existential b_1 is true, implying e_{99} and f_{99} at innermost scope. Now u_{10} is tailing, allowing e_9 and f_9 to be implied. This pattern continues until $\left[\overline{d_1}, \overline{c_1}\right]$ is falsified. In each four-literal clause the two negative existential literals "block" the universal literal. After they are falsified by unit-clause propagation, the universal literal can be reduced, yielding a new implied existential literal.

After $\left[\overline{d_1}, \overline{c_1}\right]$ is falsified, it becomes W_0 for the learning procedure outlined above. This is resolved with $\left[d_1, \overline{f_3}, \overline{e_3}, \overline{u_2}\right]$ (either choice gives similar results) to give $W_1 = \left[\overline{c_1}, \overline{f_3}, \overline{e_3}, \overline{u_2}\right]$. Now the most recently implied literal is c_1, but resolving W_1 with the antecedent of c_1 would be tautologous, so an innermost existential, say f_3, is used instead. Thus W_2 contains $\overline{u_4}$, preventing resolution on $\overline{e_3}$. Also, $\overline{e_3}$ still prevents reduction of $\overline{u_2}$. This pattern continues down the conflict graph. Eventually, e_{99} and f_{99} are introduced in W_5 by resolution on the antecedent of f_9. After e_{99} and f_{99} are resolved out for the first time (allowing $\overline{u_{10}}$ to be reduced), they are *re-introduced* by resolution on the antecedent of e_9. Then they get resolved out for the second time. By the conclusion of the procedure e_{99} and f_{99} are resolved out 32 times.

---

[2] See `www.cse.ucsc.edu/~avg/ProofChecker/QdpllexpSimple.tar` for some instances.

Prefix: $\exists\, b_1, c_1, d_1 \,\forall\, u_2 \,\exists\, e_3, f_3 \,\forall\, u_4 \,\exists\, e_5, f_5 \,\forall\, u_6 \,\exists\, e_7, f_7 \,\forall\, u_8 \,\exists\, e_9, f_9 \,\forall\, u_{10} \,\exists\, e_{99}, f_{99}\, \ldots$

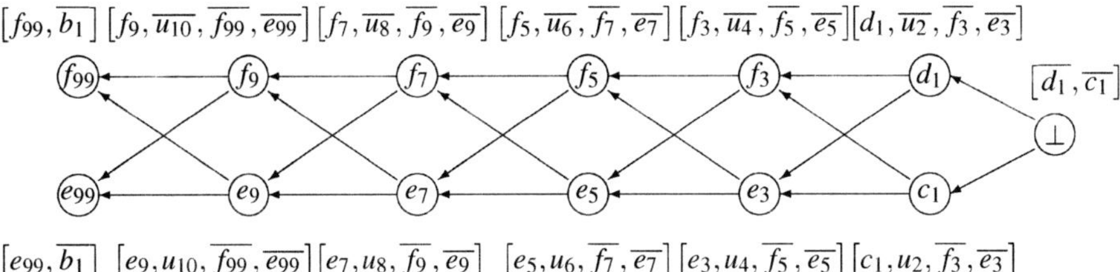

Fig. 1. Excerpt of exponential family of QBF formulas; see Example 3.1. The implied literal is shown inside each circle, which represents a vertex in the conflict graph. Antecedent clauses are shown above or below each circle. Arrows point to reason literals. E. g., f_9 is implied because $\overline{f_{99}}$ and $\overline{e_{99}}$ are falsified, then $\overline{u_{10}}$ is reduced away.

Table 1. Running times in seconds on *qdpllexp* family. See Example 3.1

family index	18	19	20	21	22	23
`QuBE 1.3`	10	22	47	105	segv	segv
`depQBF 0.1`	8	16	32	69	140	298
`CirQit3.15`	1	1	3	5	11	21

"segv" denotes "segmentation violation".

Further study shows that all paths in the graph are traversed, and it is well known that this family of graphs has exponentially many paths.

Table 1 provides empirical confirmation that the running time doubles for each increase of one level in the family. A one-level increase produces 10 additional variables and four more four-literal clauses. The clauses seen in Figure 1 are essentially cloned with fresh variables, except that b_1 appears instead of $\overline{b_1}$.

Cases of unexpectedly bad performance on application instances might be due to getting trapped in a similar structure during some derivations. Next, Section 4 proposes a modified clause learning procedure that avoids both exponential worst cases and tautologous resolvents. Example 4.2 shows that $\boxed{b_1}$ can be learned with a linear-length derivation. Hence, there is an opportunity for significant improvement in the efficiency of QBF clause learning. □

Performance problems of this kind can be avoided by accepting clauses with contradictory universal literals, a scheme dubbed *long-distance resolution* by Zhang and Malik [28]. Although they argued that the practice was sound *in the context of their particular solver*, there is no proof theory for it, in the sense of Cook and Reckhow [6]. According to Narizzano *et al.* [22], the version of `QuBE` used as the basis for `QuBE-cert` uses (some version of) long-distance resolution.[3]

Algorithm 1. Compute qpup[f]. Upon conflict, call computeQpup($\perp$).

/* *Precondition:* f *is* **implied** *(hence is existential or* $\perp$*).*   */
/* *Postcondition:* **qpup** [f] *is computed.*   */

1 **computeQpup** (f)
2 workCl = copy(antecedent[f])
3 **for** (each **implied** q such that $\overline{q} \in$ antecedent[f], in order of implication) **do**
4   **if** ($\overline{q} == f$) **then**
5   **continue**
6   **if** (**qpup** [q] == **NULL**) **then**
7   **computeQpup** (q)
8   workCl = **res** ($\overline{q}$, workCl, **qpup** [q])
9   **for** each **tailing** universal $u_i \in$ workCl **do**
10   workCl = **unrd** (u_i, workCl)
11 **qpup** [f] = workCl
12 **return**

---

4   QBF Pseudo-Unit Propagation

Pseudo-unit propagation (PUP) was introduced for propositional clause learning [26] and found empirically to produce longer proofs than the 2005 version of zchaffSE, which uses the more common first UIP technique.

This section introduces ***QBF pseudo-unit propagation*** (**QPUP**). QPUP can be used to derive a learned clause from a conflict graph. To focus on the main ideas, this section assumes that the pure literal rule is not in use. The combination of the pure literal rule with clause learning for QBF solving raises issues that are discussed in [13].

For QDPLL, it is assumed that the input clauses are non-tautological and that any tailing universal literals have been reduced away (Definition 2.6).

Definition 4.1. Let a QDPLL search be given, comprising a sequence of assumptions and unit-clause propagations. Each assumption is made upon a variable in the outermost quantifier block that contains a currently unassigned variable.

An ***implied literal*** is an existential literal that is assigned as a result of unit-clause propagation, including universal reductions. We call $\perp$ the implied literal when the search ends with a conflict. Each implied literal has an associated ***antecedent*** clause, which implied it. At the time e is implied, let qpup[e] be initialized to NULL.

In conjunction with the QDPLL search, suppose f is an implied literal with antecedent clause C. That is, all literals in C other than f are either assigned **false**, or they are universal literals that are tailing with respect to f (all universal literals are tailing with respect to $\perp$). Then the single-parameter partial function $qpup(f)$ is defined inductively for this search, as shown in Algorithm 1.

1. If C has no negations of implied literals, then $qpup(f) = C$. Note that C contains no unassigned universal literals tailing to f in this case.
2. If C contains $\overline{e_i}$ for $i = 1, \ldots, k$ and $k \geq 1$, where e_i are earlier implied literals in the same order as they were implied, then $qpup(f)$ is the result of successively resolving C with $qpup(e_i)$ for $i = 1, \ldots, k$. Applicable universal reductions are performed after each resolution step.

Note that f may be $\bot$. Then $qpup(\bot)$ contains only negated assumptions and non-tailing universal literals that have been assigned false by the search procedure. This is a conflict clause in the usual sense: this set of literals is inconsistent with the given formula. $\qquad\square$

The $qpup$ clauses may be computed lazily when a conflict occurs, or eagerly, as soon as a literal is implied. If computed eagerly, no recursive calls occur at lines 6 and 7 in `computeQpup`.

Example 4.2. The idea of QPUP is illustrated by an example, referring to the clauses in Figure 1. For simplicity, we assume $qpup$ is computed eagerly. As in Example 3.1, b_1 is assumed. Next, b_1 implies f_{99}, but we define $qpup(f_{99}) = \left[f_{99}, \overline{b_1}\right]$ because b_1 is only an assumption. Similarly, b_1 implies e_{99} and we define $qpup(e_{99}) = \left[e_{99}, \overline{b_1}\right]$. Now f_9 is implied by unit-clause propagations on f_{99} and e_{99}, to start.

Unit-clause propagation can also be thought of as unit-clause resolution. If we had really derived f_{99} and e_{99} as unit clauses, we could resolve them with $\left[f_9, \overline{f_{99}}, \overline{e_{99}}, \overline{u_{10}}\right]$ to shorten that clause. Instead, we resolve $qpup(f_{99})$ and $qpup(e_{99})$ with this clause, then apply universal reduction, to get $qpup(f_9) = \left[f_9, \overline{b_1}\right]$. Notice that the introduced literal $\overline{b_1}$ cannot block the universal reduction because b_1 was an assumption, so it must have outer scope to any unassigned universal literals.

Continuing in this way, $qpup(p)$ and $qpup(\overline{p})$ are derived successively for $p = f_9, e_9, \ldots, d_1, c_1$. Finally, $\left[\overline{b_1}\right]$ is learned with a linear number of steps. $\qquad\square$

Lemma 4.3. With $qpup(f)$ as in Definition 4.1, let q be the most recent assumption before f was implied. Then every existential literal in $qpup(f)$ *other than f* has scope outer to or equal to the scope of q.

Proof: The existential literals of $qpup(f)$ other than f are negations of assumptions, and assumptions are made in outer to inner order. $\qquad\blacksquare$

If cube processing is intermixed with unit-clause propagation, $qpup(f)$ can contain universal literals whose scopes are inner to q (the most recent assumption) and outer to f. These literals were assigned false by the search procedure. However, if $f = \bot$, there are no such literals in $qpup(f)$. When $f \neq \bot$, these universal literals cannot cause tautologous resolvents, because unit-clause propagation would satisfy any clause that might produce a tautologous resolvent. The results of QPUP are soundly derived clauses, each containing an implied literal plus negations of some earlier-assigned literals.

For a formula with n variables, there are at most n *qpup* clauses at any point in the search, and each *qpup* clause is derived with at most n steps. Only *qpup* clauses are used to derive $qpup(\perp)$, so the procedure to learn one clause is polynomial in n. The problem with Example 3.1 is that most derived clauses are not *qpup* clauses.

In summary, QPUP shows that conflict clauses can be learned by QDPLL with polynomially long derivations, although current implementations of QDPLL sometimes carry out exponentially long derivations. We hope this motivates the search for a better system of clause learning in QBF.

5   QU-Resolution

This section introduces **QU-Resolution**. This is a natural extension of Q-resolution. Although Q-Resolution is refutationally complete, it cannot derive all logically implied clauses (see Definition 2.5). Since clause learning involves deriving logically implied clauses, and we saw in Section 3 that such derivations might be exponentially long, in terms of the overall number of clauses needed to derive the learned clause, it makes sense to look at variants of Q-resolution.

This section shows that QU-resolution is **inferentially complete**, in the sense that if some (non-tautological) clause C is a super-sound addition to a closed QBF Φ, then some subset of C is derivable by QU-resolution. We also show that it provides a theoretical underpinning for previously reported failed-literal preprocessing in QBF [20,27].

For propositional conflict-driven clause learning, it is known that the learned clause has a quadratically long derivation, in terms of the overall number of clauses needed to derive it, even when clause minimization and "volunteers" are included [23,24]. (See cited papers; minimization and volunteers are not important to this paper.)

Definition 5.1. A **QU-derivation** is the same as a Q-resolution derivation (see Section 2), except that it includes resolutions in which the clashing literal is universal. Tautologous resolvents are still prohibited. A **QU-refutation** is a QU-derivation of the empty clause.

A **regular** QU-derivation is one such that no variable appears twice as a clashing literal or reduction literal on any directed path through the derivation DAG. An **ordered** QU-derivation is one such that variables appear in the same order on every directed path through the derivation DAG. (Not all variables need appear on a path; ordered derivations are necessarily regular). A **prefix-ordered** QU-derivation is an ordered QU-derivation such that variables appear in outer to inner order of the quantifier prefix on paths directed away from the root of the derivation DAG. □

Definition 5.2. If D is a clause, and clause $D' \subseteq D$ as a set of literals, then we say D' **subsumes** D. The notation $D^{(-)}$ means "*some* clause that subsumes D". That is, a statement "$D^{(-)} \ldots$" should be read as "For *some* clause D' that subsumes D, $D' \ldots$". In general, the statement will not hold for all subsets of D. Recall that the empty clause and D itself are subsets of D. □

Lemma 5.3. If clause D is derived from $\Psi = \vec{Q}.\mathcal{F}$ by QU-Resolution, then D is logically implied by Ψ.

The proof of the next theorem employs the framework first published by Anderson and Bledsoe [1]. The original proof that Q-resolution is refutationally complete uses the same idea [14], but that paper does not mention that the refutation may be required to have the additional properties of being prefix-ordered and regular.

Theorem 5.4 If (non-tautological) clause D is logically implied by $\Psi = \vec{Q}.\mathcal{F}$, then $D^{(-)}$ can be derived from Ψ by prefix-ordered QU-Resolution. If D is the empty clause then the QU-resolution can also be a Q-resolution.

Proof: Although many prefix orders may be equivalent, fix one for the proof. The proof is by induction on n, the number of variables in the quantifier prefix. The base case is $n = 0$. The conclusion is immediate, as D must be the empty clause and $\mathcal{F}$ must contain the empty clause to by hypotheses of the theorem. For $n > 0$, assume the claim holds for $m < n$ variables. There are several cases.

If literal $p \in D$ and $|p|$ is outermost, consider $\Psi_1 = \Psi\lceil_{\overline{p}}$ and $D_1 = D - \{p\}$. By the inductive hypothesis, $D_1^{(-)}$ has a QU-derivation from Ψ_1; call it π_1. Let π do the same proof operations as π_1, except starting with clauses in $\mathcal{F}$ instead of $\mathcal{F}\lceil_{\overline{p}}$. All operations remain correct because p is outermost. Each clause in π has at most an extra literal p, compared to the corresponding clause in π_1. Therefore π derives $D^{(-)}$.

If literals p and $\overline{p}$ are *not* in D and $|p|$ is outermost, consider $\Psi_0 = \Psi\lceil_p$ and $\Psi_1 = \Psi\lceil_{\overline{p}}$. If M_0 is any tree model of Ψ_0, then prepending $p = 1$ to every ordered assignment in M_0 gives a subset of some model of Ψ; call it M. By hypothesis of the theorem, D is true on every branch of M, so D is true on every branch of M_0. By the inductive hypothesis, $D^{(-)}$ has a QU-derivation from Ψ_0; call it π_0. (If Ψ_0 has no tree model, let π_0 be a Q-refutation of Ψ_0.) Let π do the same proof operations as π_0, except starting with clauses in $\mathcal{F}$ instead of $\mathcal{F}\lceil_p$. All operations remain correct because p is outermost. Each clause in π has at most an extra literal $\overline{p}$, compared to the corresponding clause in π_0. Therefore π derives $D_0 = (D \cup \{\overline{p}\})^{(-)}$ from Ψ. (If Ψ_0 has no tree model, $D_0 = (\{\overline{p}\})^{(-)}$.) Similarly, $D^{(-)}$ has a QU-derivation from Ψ_1; call it π_1; and $D_1 = (D \cup \{p\})^{(-)}$ can be derived from Ψ. If D_0 lacks $\overline{p}$, it serves as $D^{(-)}$ derived from Ψ. If D_1 lacks p, it serves as $D^{(-)}$ derived from Ψ. Otherwise, resolving D_0 and D_1 on p derives $D^{(-)}$.

If D is the empty clause and universal variable u is outermost, the refutation can be achieved with prefix-ordered Q-resolution. Let Ψ_0 and Ψ_1 be as defined in the previous paragraph with u in the place of p. By hypothesis of the theorem and the semantics of QBF (see Definition 2), either Ψ_0 is false or Ψ_1 is false. By the inductive hypothesis, either π_0 or π_1 derives the empty clause with Q-resolution. Then either the empty clause or $[u]$ or $[\overline{u}]$ can be derived from Ψ with Q-resolution. Universal reduction, if needed, completes the refutation.   ∎

The proof shows that the part of the derivation inner to the variables that occur in the target clause D can always use universal reduction in preference to resolution with a universal clashing literal (if preserving tree countermodels is not required).

Since QU-resolution is able to eliminate universal literals in some cases without using universal reduction and Example 2.8 showed that universal reduction can change the set of tree countermodels, QU-resolution has a greater capability than Q-resolution to preserve tree countermodels, which might be important for certain applications that want to extract strategies for the A player.

Example 5.5. The following QBF family is given by Kleine Büning and Lettman [15] in the proof of their Theorem 7.4.8. The k-th QBF is:

$$\exists\, d_0\, d_1\, e_1 \,\forall\, x_1 \,\exists\, d_2\, e_2 \,\forall\, x_2 \cdots \exists\, d_k\, e_k \,\forall\, x_k \,\exists\, f_1 \cdots \exists\, f_k.$$

$$
\begin{array}{ll}
\left[\,d_0\,\right] & \left[d_j,\overline{x_j},\overline{d_{j+1}},\overline{e_{j+1}}\right] \text{ for } 1 \le j < k \\
\left[d_0,\overline{d_1},\overline{e_1}\right] & \left[e_j,x_j,\overline{d_{j+1}},\overline{e_{j+1}}\right] \text{ for } 1 \le j < k \\
\left[d_k,\overline{x_k},\overline{f_1},\ldots,\overline{f_k}\right] & \left[\overline{x_j},f_j\right] \text{ for } 1 \le j \le k \\
\left[e_k,x_k,\overline{f_1},\ldots,\overline{f_k}\right] & \left[x_j,f_j\right] \text{ for } 1 \le j \le k
\end{array}
$$

That proof states that every Q-refutation of the k-th formula has at least 2^k steps.

With QU-resolution, begin by resolving the binary clauses on universal literals to produce unit clauses $[f_j]$ for $1 \le j \le k$. Then derive unit clauses $[d_k]$ and $[e_k]$ (using universal reduction on x_k and $\overline{x_k}$). The remainder is straightforward with Q-resolution. The overall number of steps is linear. $\qquad\square$

Lonsing and Biere introduced *abstractions* of QBF for preprocessing purposes [20]. The idea is developed further in SAT 2012 [27]. Essentially the "abstraction" with respect to a variable p treats all universal variables outer to p as though they were existential. For inferential purposes, this amounts to using QU-resolution on these variables.

6   Depth-Monotonic Literals

This section addresses the question of when **QBF variable-elimination resolution (QVER)** is safe. QVER is the "resolve" operation in `Quantor`, described by Biere [3]. Variable-elimination resolution is the basic DP operation for CNF. For Q-resolution, add operation (4):

1. Find all resolvents with existential clashing literal q and add them to $\mathcal{F}$.
2. Discard tautologous resolvents.
3. Delete all clauses with q or $\overline{q}$.
4. Perform all universal reductions that are possible.

When is QVER safe? `Quantor` uses QVER for variables at innermost scope, but Biere did not prove that it is safe [3]. However, a proof may be found in [5, prop.

2.5.2]. In the cases that q is *not* in the innermost scope, the problem is step 3 above, deletion of clauses. Removal of constraints might change the value of a formula from false to true. We take a detour and return to this question.

A complementary question was considered by Bubeck and Kleine Büning. They considered universal expansion (the "expand" operation in `Quantor`, which eliminates a *universal* variable). They showed in their Theorem 1 [4] and Theorem 5.6.1 [5] that the operation can be simplified by not expanding existential variables that meet a certain criterion called a one-sided dependency.

Nondecisive clauses have been defined for propositional formulas. We extend this idea to QBF.

Definition 6.1. Let $\Phi = \vec{Q}.\mathcal{F}$ be a closed QBF with matrix $\mathcal{F}$, and let C be a clause in $\mathcal{F}$ that contains the literal q. C is **depth-nondecisive** on q with respect to $\mathcal{F}$ if for all clauses $D \in \mathcal{F}$ that contain $\overline{q}$ it is the case that at least one of the following conditions holds:

1. $\mathsf{res}_q(C, D)$ contains *no* literal with scope inner to q, even if the resolvent is tautologous; **or**
2. $\mathsf{res}_q(C, D)$ is tautologous due to the presence of literals $r \in C$ and $\overline{r} \in D$, where r is *not* inner to the scope of q; **or**
3. $\mathsf{res}_q(C, D)$ is subsumed by some clause C_1, where C_1 contains *no* literal with scope inner to q and either $C_1 \in \mathcal{F}$ or C_1 is a resolvent created under case (1) above.

The idea is that for any resolvent $R = \mathsf{res}_q(C, D)$, either R has no literals inner to q or R has some literals inner to q but even if those literals were deleted from R giving R', then R' could still be safely discarded because it is tautologous or subsumable. $\qquad\qquad\square$

Example 6.2. In case (3) the subsumption by another resolvent using C does not imply that $\mathcal{F}$ already contains a subsumable clause. For example, let $C = [c, e]$, $D_1 = [c, d, \overline{e}]$, and $D_2 = [d, \overline{e}, f]$. Assume each variable has a different scope and literals are listed from outer to inner scope, which is also alphabetical order. No clauses are subsumed, and $\mathsf{res}_e(C, D_2)$ contains f, yet C is depth-nondecisive on e. $\qquad\qquad\square$

Definition 6.3. Let $\Phi = \vec{Q}.\mathcal{F}$ be a closed QBF with matrix $\mathcal{F}$, which partitions into $\mathcal{G}_1 + \mathcal{G}_2 + \mathcal{G}_3$ ("$+$" denotes **disjoint union** in this section), and let q be an existential literal in Φ. Assume that:

1. $\mathcal{G}_1$ consists of clauses with q or $\overline{q}$ and *no* deeper variables. Deeper means strictly greater qdepth.
2. $\mathcal{G}_2$ consists of clauses with q or $\overline{q}$ and *some* deeper variables.
3. $\mathcal{G}_3$ consists of clauses without q or $\overline{q}$.

If every clause C in $\mathcal{G}_2$ that contains $\overline{q}$ is depth-nondecisive on $\overline{q}$ with respect to $\mathcal{F}$, then we say that q is **effectively depth-monotonic** in Φ. We say that the variable $|q|$ is effectively depth-monotonic in Φ if it holds for q or $\overline{q}$.

Clearly, if $|q|$ is pure in $\mathcal{G}_2$, then $|q|$ is effectively depth-monotonic in Φ. In this case we say that $|q|$ is ***depth-monotonic*** in Φ. $\square$

Theorem 6.4 With q, Φ, $\mathcal{F}$, $\mathcal{G}_1$, $\mathcal{G}_2$ and $\mathcal{G}_3$ defined as in Definition 6.3, let $|q|$ be the clashing existential variable being contemplated for QVER on Φ, and let $|q|$ be effectively depth-monotonic in Φ. Then QVER on $|q|$ is safe.

Furthermore, let $\Phi_1 = \vec{Q}_1 . (\mathcal{G}_3 \cup \mathcal{G}_4)$ be the result of QVER on $|q|$ and Φ. That is, $\vec{Q}_1$ is $\vec{Q}$ with $|q|$ deleted, and $\mathcal{G}_4$ consists of all non-tautologous resolvents with $|q|$ as the clashing variable using clauses in $\mathcal{G}_1 + \mathcal{G}_2$. Then every tree model M_1 for Φ_1 can be extended to a tree model M for Φ with a one-to-one correspondence between the branches of M_1 and the branches of M.

Proof: (Sketch[4]) By super-soundness of Q-resolution, it suffices to consider only the case that Φ_1 is true. For notation, let $A = \{\alpha\}$, $B = \{\beta\}$, $\Gamma = \{\gamma\}$, $\Delta = \{\delta\}$ denote the clause sets in $\vec{Q}_1$ corresponding to $\mathcal{G}_1$ and $\mathcal{G}_2$ as follows:

$$\mathcal{G}_1 = \{[q, \alpha] \mid \alpha \in A\} + \{[\overline{q}, \beta] \mid \beta \in B\}$$
$$\mathcal{G}_2 = \{[q, \gamma] \mid \gamma \in \Gamma\} + \{[\overline{q}, \delta] \mid \delta \in \Delta\}$$

Some α or β might be empty, but all γ and δ are nonempty. W.l.o.g., assume that q is effectively depth-monotonic in Φ.

Let M_1 be any tree model for Φ_1. Let (σ, τ) be any branch in M_1 where σ assigns values to all variables outer to $|q|$ and τ assigns values to all variables inner to $|q|$. Construct the corresponding branch in M as follows:

Condition on σ	Corresponding Branch in M
$\exists\, \beta \in B$ such that $\beta\lceil_\sigma \neq 1$, or $\exists\, \delta \in \Delta$ such that $\delta\lceil_\sigma \neq 1$	$(\sigma, \overline{q}, \tau) \in M$
$\forall\, \beta \in B\ \ \beta\lceil_\sigma = 1$, and $\forall\, \delta \in \Delta\ \ \delta\lceil_\sigma = 1$	$(\sigma, q, \tau) \in M$

(Note that $\delta\lceil_\sigma$ has some unassigned literals if it does not simplify to 1.)

By construction, M satisfies the requirement for tree models that, if (ρ, q) is a branch prefix of M, then there is no branch prefix of the form $(\rho, \overline{q})$, and *vice versa*. M satisfies $\mathcal{G}_3$, since M_1 satisfies $\mathcal{G}_3$. M satisfies all clauses in $\mathcal{G}_1 + \mathcal{G}_2$ that contain $\overline{q}$ by construction.

It remains to show that M satisfies all clauses in $\mathcal{G}_1 + \mathcal{G}_2$ that contain q. Assume $(\sigma, \overline{q}, \tau)$ is the branch in M corresponding to (σ, τ) in M_1, otherwise satisfaction is immediate.

First, consider an arbitrary clause $C = [q, \alpha] \in \mathcal{G}_1$. There is some $\beta \in B$ or $\delta \in \Delta$ that caused the assignment $q = 0$ in the construction of M.

(Subcase β) If $\beta\lceil_\sigma \neq 1$ and $[\alpha, \beta] \in \mathcal{G}_4$, then $\alpha\lceil_\sigma = 1$. If $\beta\lceil_\sigma \neq 1$ and $[\alpha, \beta] \notin \mathcal{G}_4$, it must be due to being tautologous. Then there is some literal r that is not inner to q such that $r \in \alpha$ and $\overline{r} \in \beta$. It follows that $\alpha\lceil_\sigma = 1$.

(Subcase δ) Suppose some δ causes the assignment $q = 0$. If $\delta\lceil_\sigma \neq 1$ and $[\alpha, \delta] \in \mathcal{G}_4$, then it must be subsumed by some $D_1 \in (\mathcal{G}_3 \cup \mathcal{G}_4)$ because $[\alpha, \delta]$ must have some literal inner to q. To qualify, D_1 must have no literals inner

---

[4] See http://www.cse.ucsc.edu/~avg/QPUP/qpup-cp12-long.pdf for full proofs.

to q (case (3) in Definition 6.1). Some literal $p \in D_1$ is assigned 1 by σ. By the subcase hypothesis $p \notin \delta$, so again $\alpha\lceil_\sigma = 1$. If $\delta\lceil_\sigma \neq 1$ and $[\alpha, \delta] \notin \mathcal{G}_4$, the argument in subcase β applies with δ in the place of β.

Second, consider an arbitrary clause $C_2 = [q, \gamma] \in \mathcal{G}_2$. There is some $\beta \in B$ or $\delta \in \Delta$ that caused the assignment $q = 0$ in the construction of M. By arguments similar to the first case, $\gamma\lceil_\sigma = 1$. ∎

Corollary 6.5. With q, Φ, $\mathcal{F}$, $\mathcal{G}_1$, $\mathcal{G}_2$ and $\mathcal{G}_3$ defined as of Theorem 6.3, if $|q|$ is depth-monotonic in Φ and the universal player has a winning strategy, then for each universal variable u_i that has inner scope to q, the winning strategy for u_i can be expressed as a function that is independent of q.

Proof: (Sketch) Place $|q|$ innermost in its quantifier block. W.l.o.g., let the pure literal be q for $\mathcal{G}_2$. All occurrences of u_i are in $\mathcal{G}_2$ or $\mathcal{G}_3$. Now consider the modified formula Φ_1 that differs from Φ only in that $|q|$ is outermost within the innermost quantifier block. It remains to show that Φ_1 is false.

Apply the quadrangle dependency theory of [25], Theorem 4.7. In Φ_1, $|q|$ does not have a quadrangle dependency with any u_i that is inner to $|q|$ in Φ, because paths from u_i or $\overline{u_i}$ to $\overline{q}$ require connections outer to u_i. Therefore, starting from Φ_1, $|q|$ can repeatedly be transposed with the next outer variable in the quantifier prefix without changing the value of the formula, until $|q|$ reaches the position it has in Φ, which is false by hypothesis. Since $|q|$ is inner to all u_i in Φ_1, if Φ_1 is false, then the universal player has a winning strategy such that the winning value for each u_i is independent of $|q|$. ∎

Example 6.6. This example illustrates several topics from this section. Φ is shown in chart form.

Φ	$\exists q$	$\exists r$	$\forall u$	$\exists x$	$\exists y$
C_1	$\overline{q}$			x	
C_2			$\overline{u}$	x	
C_3	q	$\overline{r}$			
C_4	q	r			y
C_5			u		$\overline{y}$

The winning strategy for u depends on q but not r; i.e., $u(q, r) = q$.

QVER(r) is sound. QVER(q) is not sound. Notice that q has no clauses in common with u, and neither does r. The difference lies in the paths from u or $\overline{u}$ to the various existential literals. Thus the two paths $C_2 - C_1$ and $C_5 - C_4$ establish a quadrangle dependency of q upon u. However, the only path from u or $\overline{u}$ to $\overline{r}$ involves a connection through q, but q is not inner to r, so this path does not qualify for establishing a quadrangle dependency.

By quadrangle dependency theory, r and u can be transposed without changing the truth value of Φ. After the transposition, r is innermost, so QVER(r) is sound and might be selected by `Quantor`. Since the formula that results after QVER(r) is the same, whether the transposition is carried out or not, an implementation need not actually perform the transposition. □

7   Conclusion

This paper presents theoretical analysis of several issues closely related to modern QBF solvers. Avenues for improvement in the state of the art are suggested. No QBF solvers with publicly available code were found to be very amenable to research experimentation, in contrast to MiniSat in the SAT domain. Therefore we leave it to various implementers to decide how to incorporate this paper's results into their own solvers.

Acknowledgment. We thank Florian Lonsing, Mikolas Janota, Uwe Bubeck, and the reviewers for their careful reading of an earlier draft.

References

1. Anderson, R., Bledsoe, W.W.: A linear format for resolution with merging and a new technique for establishing completeness. JACM 17 (1970)
2. Balabanov, V., Jiang, J.-H.R.: Resolution Proofs and Skolem Functions in QBF Evaluation and Applications. In: Gopalakrishnan, G., Qadeer, S. (eds.) CAV 2011. LNCS, vol. 6806, pp. 149–164. Springer, Heidelberg (2011)
3. Biere, A.: Resolve and Expand. In: Hoos, H.H., Mitchell, D.G. (eds.) SAT 2004. LNCS, vol. 3542, pp. 59–70. Springer, Heidelberg (2005)
4. Bubeck, U., Kleine Büning, H.: Bounded Universal Expansion for Preprocessing QBF. In: Marques-Silva, J., Sakallah, K.A. (eds.) SAT 2007. LNCS, vol. 4501, pp. 244–257. Springer, Heidelberg (2007)
5. Bubeck, U.: Model-based Transformations for Quantified Boolean Formulas. PhD thesis, University of Paderborn (2010) (DISKI, 329, IOS Press; also at http://www.ub-net.de/bubeck-qbf-transformations-2010.pdf)
6. Cook, S., Reckhow, R.: The relative efficiency of propositional proof systems. J. Symbolic Logic 44, 36–50 (1979)
7. Egly, U.: On Sequent Systems and Resolution for QBFs. In: Cimatti, A., Sebastiani, R. (eds.) SAT 2012. LNCS, vol. 7317, pp. 100–113. Springer, Heidelberg (2012)
8. Goultiaeva, A., Bacchus, F.: Exploiting QBF duality on a circuit representation. In: AAAI (2010)
9. Goultiaeva, A., Iverson, V., Bacchus, F.: Beyond CNF: A Circuit-Based QBF Solver. In: Kullmann, O. (ed.) SAT 2009. LNCS, vol. 5584, pp. 412–426. Springer, Heidelberg (2009)
10. Goultiaeva, A., Van Gelder, A., Bacchus, F.: A uniform approach for generating proofs and strategies for both true and false QBF formulas. In: Proc. IJCAI (2011)
11. Giunchiglia, E., Narizzano, M., Tacchella, A.: Learning for Quantified Boolean Logic Satisfiability. In: AAAI/IAAI, pp. 649–654 (2002)
12. Giunchiglia, E., Narizzano, M., Tacchella, A.: Monotone Literals and Learning in QBF Reasoning. In: Wallace, M. (ed.) CP 2004. LNCS, vol. 3258, pp. 260–273. Springer, Heidelberg (2004)
13. Giunchiglia, E., Narizzano, M., Tacchella, A.: Clause/term resolution and learning in the evaluation of quantified boolean formulas. JAIR 26 (2006)
14. Kleine Büning, H., Karpinski, M., Flögel, A.: Resolution for quantified boolean formulas. Information and Computation 117, 12–18 (1995)
15. Kleine Büning, H., Lettmann, T.: Propositional Logic: Deduction and Algorithms. Cambridge University Press (1999)

16. Klieber, W., Sapra, S., Gao, S., Clarke, E.: A Non-prenex, Non-clausal QBF Solver with Game-State Learning. In: Strichman, O., Szeider, S. (eds.) SAT 2010. LNCS, vol. 6175, pp. 128–142. Springer, Heidelberg (2010)
17. Lonsing, F.: Dependency Schemes and Search-Based QBF Solving: Theory and Practice. PhD thesis, Johannes Kepler University (2012)
18. Lonsing, F., Biere, A.: A Compact Representation for Syntactic Dependencies in QBFs. In: Kullmann, O. (ed.) SAT 2009. LNCS, vol. 5584, pp. 398–411. Springer, Heidelberg (2009)
19. Lonsing, F., Biere, A.: Integrating Dependency Schemes in Search-Based QBF Solvers. In: Strichman, O., Szeider, S. (eds.) SAT 2010. LNCS, vol. 6175, pp. 158–171. Springer, Heidelberg (2010)
20. Lonsing, F., Biere, A.: Failed Literal Detection for QBF. In: Sakallah, K.A., Simon, L. (eds.) SAT 2011. LNCS, vol. 6695, pp. 259–272. Springer, Heidelberg (2011)
21. Letz, R.: Lemma and Model Caching in Decision Procedures for Quantified Boolean Formulas. In: Egly, U., Fermüller, C. (eds.) TABLEAUX 2002. LNCS (LNAI), vol. 2381, pp. 160–175. Springer, Heidelberg (2002)
22. Narizzano, M., Peschiera, C., Pulina, L., Tacchella, A.: Evaluating and certifying QBFs: A comparison of state-of-the-art tools. AI Commun. 22(4), 191–210 (2009)
23. Van Gelder, A.: Improved Conflict-Clause Minimization Leads to Improved Propositional Proof Traces. In: Kullmann, O. (ed.) SAT 2009. LNCS, vol. 5584, pp. 141–146. Springer, Heidelberg (2009)
24. Van Gelder, A.: Generalized Conflict-Clause Strengthening for Satisfiability Solvers. In: Sakallah, K.A., Simon, L. (eds.) SAT 2011. LNCS, vol. 6695, pp. 329–342. Springer, Heidelberg (2011)
25. Van Gelder, A.: Variable Independence and Resolution Paths for Quantified Boolean Formulas. In: Lee, J. (ed.) CP 2011. LNCS, vol. 6876, pp. 789–803. Springer, Heidelberg (2011)
26. Van Gelder, A.: Producing and verifying extremely large propositional refutations: Have your cake and eat it too. AMAI (2012) (accepted subject to revisions)
27. Van Gelder, A., Wood, S.B., Lonsing, F.: Extended Failed-Literal Preprocessing for Quantified Boolean Formulas. In: Cimatti, A., Sebastiani, R. (eds.) SAT 2012. LNCS, vol. 7317, pp. 86–99. Springer, Heidelberg (2012)
28. Zhang, L., Malik, S.: Conflict driven learning in a quantified boolean satisfiability solver. In: Proc. ICCAD, pp. 442–449 (2002)

Breaking Variable Symmetry
in Almost Injective Problems

Philippe Vismara[1,2] and Remi Coletta[1]

[1] LIRMM, UMR5506 Université Montpellier II - CNRS, Montpellier, France
`{coletta,vismara}@lirmm.fr`
[2] MISTEA, UMR729 Montpellier SupAgro - INRA, Montpellier, France

Abstract. Lexicographic constraints are commonly used to break variable symmetries. In the general case, the number of constraint to be posted is potentially exponential in the number of variables. For injective problems (AllDiff), Puget's method[12] breaks all variable symmetries with a linear number of constraints.

In this paper we assess the number of constraints for "almost" injective problems. We propose to characterize them by a parameter μ based on Global Cardinality Constraint as a generalization of the AllDiff constraint. We show that for almost injective problems, variable symmetries can be broken with no more than $\binom{n}{\mu}$ constraints which is XP in the framework of parameterized complexity. When only ν variables can take duplicated values, the number of constraints is FPT in μ and ν.

Keywords: Variable Symmetry, Global Cardinality Constraint, Parameterized complexity.

1   Introduction

The importance of symmetry is now widely recognized in Constraint Satisfaction Problems. Many methods have been devised to break symmetry especially for variables. Most of them require to post a number of constraints which is equal to the number of symmetries. This is the case of lexicographic constraints [6] or dynamic constraints in SBDS [9].

Unfortunately, the number of variable symmetries can be exponential in the general case. Even so, Puget [12] has shown that, for injective problems, symmetry can be broken with a linear number of constraints.

Between these two extreme, the aim of this paper is to use a parameterized complexity approach in order to evaluate the number of constraints required to break variable symmetries.

Parameterized complexity [7] offers a measure of complexity for NP-complete problems which is based on an isolate parameter k independent from the size n. The complexity of a problem is XP if it is in the form $\mathcal{O}(n^k)$ or FPT if it is in $\mathcal{O}(f(k)n^c)$ where c is a constant and f any function, even exponential.

Many problems in Artificial Intelligence, especially CSP, have been analyzed with parameterized complexity [10]. For instance, breaking value symmetries has

M. Milano (Ed.): CP 2012, LNCS 7514, pp. 664–671, 2012.

a fixed-parameter complexity in $\mathcal{O}(2^k nd)$, where k is the potentially exponential number of value symmetries [2].

By analogy, we can claim that variable symmetry can be broken with a number of lexicographic constraints which is FTP in the number of symmetries. In this paper we try to propose new parameters to measure the number of required constraints.

2 Breaking Variable Symmetry

A variable symmetry is a permutation σ on the set of variables $\{x_i\}_{i \in 1..n}$ that maps solutions onto solutions. More formally, any assignment $(x_i = v_i)_{i \in 1..n}$ is a solution if and only if $(x_{\sigma(i)} = v_i)_{i \in 1..n}$ is also a solution.

The general method for breaking variable symmetry is to add lexicographical constraints [6]. Given an order $x_{i_1}, \ldots x_{i_n}$ on variables, we post, for any symmetry σ, the lexicographical ordering constraint:

$$x_{i_1}, \ldots, x_{i_n} \leq_{lex} x_{\sigma(i_1)}, \ldots, x_{\sigma(i_n)} \tag{1}$$

Unfortunately, the number of variable symmetries can be exponential. It is equal to the size of the corresponding permutation group $\mathcal{G}$. This group is generally given as a set S of generators. For instance, Nauty [11] or Saucy are well known programs to compute the permutation group of a graph (automorphisms). Even if their theoretical worst-case time complexity is exponential, they are very efficient in practice.

In the special case of "piecewise symmetry" – where the set of variables (resp. values) is partitioned into subsets of interchangeable variables (resp. values) – both variable and value symmetry can be broken with a linear number of constrains [8].

For injective problems, Puget[12] has shown that equation (1) can be simplified into $x_{i_k} \leq x_{\sigma(i_k)}$, where i_k is the smallest index such that $\sigma(i_k) \neq i_k$. Hence there are less than n^2 such constraints for all variable symmetries.

Given a variable index i_k, any symmetry σ for which we post the constraint $x_{i_k} \leq x_{\sigma(i_k)}$ is a stabilizer of $i_1, .., i_{k-1}$. So σ belongs to the stabilizer subgroup $\mathcal{G}^{[i_k]} = \{\sigma \in \mathcal{G} \mid \forall z < k, \ \sigma(i_z) = i_z\}$. More precisely, we need to compute the image (orbit) of i_k by any symmetry that leaves $i_1, .., i_{k-1}$ invariant. This set is defined by $\Delta^{[i_k]} = \{\sigma(i_k) | \sigma \in \mathcal{G}^{[i_k]}\}$

This approach is interesting because it is possible to compute all sets $\Delta^{[i_k]}$ without enumerating $\mathcal{G}$, thanks to Schreier-Sims' algorithm.

2.1 Schreier-Sims' Algorithm

In 1970, Sims introduced the notion of *base* for a permutation group. A sequence $B = (\beta_1, \beta_2, ..., \beta_m)$, where $m \leq n$, is a base for $\mathcal{G}$ if the only permutation which fixes each of the points in B is the identity.

The word "base" is used because an element σ of the group $\mathcal{G}$ is uniquely determined by the image $\sigma(B) = (\sigma(\beta_1), \sigma(\beta_2), ..., \sigma(\beta_m))$.

A base B induces a stabilizer chain $\mathcal{G} = \mathcal{G}^{[\beta_1]} \geq \mathcal{G}^{[\beta_2]} \geq \ldots \geq \mathcal{G}^{[\beta_m]} \geq \{id.\}$ since $\mathcal{G}^{[\beta_{k+1}]}$ (the permutations that fix $\beta_1, .., \beta_k$) is a subgroup of $\mathcal{G}^{[\beta_k]}$.

If every $\mathcal{G}^{[\beta_{k+1}]}$ is a proper (not included) subgroup of $\mathcal{G}^{[\beta_k]}$, the base B is called *nonredundant* and $2^{|B|} \leq |\mathcal{G}|$ thus $|B| \leq \log_2(|\mathcal{G}|)$.

Given any set S of generators of the group $\mathcal{G}$, the Schreier-Sims's algorithm computes incrementally a nonredundant base B. This approach is analogous to Gaussian elimination in Algebra. The algorithm adds new generators to S such that $S \cap \mathcal{G}^{[\beta_k]}$ is a generator of $\mathcal{G}^{[\beta_k]}$. The resulting base is called a *strong generating set*.

For each β_k in B, the algorithm computes the orbit $\Delta^{[\beta_k]}$ and chooses, for each value γ in $\Delta^{[\beta_k]}$, one representative permutation $u_{\beta_k}^{\gamma} \in \mathcal{G}^{[\beta_k]}$ such that $u_{\beta_k}^{\gamma}(\beta_k) = \gamma$. Thanks to this set of representatives, one can easily enumerate $\mathcal{G}$.

Let $U^{[k]} = \{u_{\beta_k}^{\gamma} | \gamma \in \Delta^{[\beta_k]}\}$ be the set of representatives for β_k.

Then $\mathcal{G} = \{v_{\beta_1} \circ v_{\beta_2} \circ \ldots \circ v_{\beta_m} \mid \forall k \in 1..m, v_{\beta_k} \in U^{[k]}\}$.

There exists several variants of Schreier-Sims' algorithm and a vast literature on this topic[5,14]. The simplest deterministic version has a time complexity in $\mathcal{O}(n^2 \log^3 |\mathcal{G}| + |S|n^2 \log |\mathcal{G}|)$ and $\mathcal{O}(n^2 \log |\mathcal{G}| + |S|n)$ in space.

If we choose $B = (1, 2, ..., n)$, the Jerrum's variant has a time complexity in $\mathcal{O}(n^5)$ and $\mathcal{O}(n^2)$ in space.

2.2   A Polynomial Number of Constraints

With Schreier-Sims' algorithm one can enumerate in polynomial time all orbits $\Delta^{[\beta_k]}$ from a given[1] generating set of $\mathcal{G}$.

Since any symmetry σ must belong to a subgroup $\mathcal{G}^{[\beta_k]}$, each constraint (1) is equivalent to an inequality $x_{\beta_k} < x_{\sigma(\beta_k)}$ with $\sigma(\beta_k) \in \Delta^{[\beta_k]}$. Hence all variable symmetries can be broken with the following constraints:

$$\forall \beta_i \in B, \ \forall \gamma \in \Delta^{[\beta_i]}, \gamma \neq \beta_i, \text{ we post } x_{\beta_i} < x_{\gamma} \tag{2}$$

The total number of inequalities is $\sum_{\beta_i \in B}(|\Delta^{[\beta_i]}| - 1)$. So it is in $\mathcal{O}(n \log_2 |\mathcal{G}|)$ or in $\mathcal{O}(n^2)$.

In [12], Puget has shown that equation (2) can be reduced to one inequality for each γ in $\Delta^{[\beta_i]}$. The principle is to associate each γ to the larger β_i (different from γ) such that $\gamma \in \Delta^{[\beta_i]}$.

Formally, let $\mathcal{R}_{\gamma} = \{\beta_i \in B \setminus \{\gamma\} \mid \gamma \in \Delta^{[\beta_i]}\}$. If $\mathcal{R}_{\gamma} \neq \emptyset$ let us define $r(\gamma) = \max(\mathcal{R}_{\gamma})$ otherwise $r(\gamma) = \gamma$.

One can prove[2] that equation (2) is equivalent to the following linear number of constraints:

$$\forall \gamma \in 1..n \text{ such that } r(\gamma) \neq \gamma, \text{ we post } x_{r(\gamma)} < x_{\gamma} \tag{3}$$

---

[1] Or computed by *Nauty* with potentially exponential time.

[2] The proof given in [12] can be sketched as follows: by transitivity, $x_{\beta_i} < x_{\gamma}$ is derived from $x_{\beta_i = r(...r(r(\gamma)))} < \cdots < x_{r(r(\gamma))} < x_{r(\gamma)} < x_{\gamma}$.

3   Gcc: A Relaxation of Alldiff for Almost Injective Problems

A constraint network that involves an Alldiff constraint is injective and we showed in Section 2 that the number of constraints required to break all variable symmetries is polynomial. The Alldiff imposes that each value be taken at most 1 time, which is quite restrictive in terms of modelisation. In many practical applications, we may want to impose that a value should appear at least l and/or at most u times. With this purpose, the Global Cardinality Constraint Gcc has been introduced in [13] to deal globally with a conjunction of AtLeast (stating that a value has to appear at least a given number of times) and AtMost (stating that a value has to appear at most a given number of times). The Gcc is widely used to model industrial problems and is available in almost all existing constraints solvers.

Definition 1 (Global Cardinality Constraint). *A Gcc constraint, denoted $Gcc(X, lb, up)$ involves a set of variables X and two functions lb, ub : $\bigcup_{x \in X} D(x) \to \mathcal{N}$. The Gcc constraint is satisfied if for each value $v \in \bigcup_{x \in X} D(x)$ the number of variables in X assigned to v is between $lb(v)$ and $ub(v)$.*

This constraint does not guarantee the problem to be injective, but in the remainder of this section we show that a problem with Gcc constraint may be *almost injective*. To measure the distance of a given problem N to a perfectly injective problem, we introduce the parameter $\mu(N)$, maximum number of variables that can be equal simultaneously. If $\mu(N) = 0$, then the problem N is perfectly injective. If $\mu(N)$ is *small* we call the problem N is *almost injective*.

Example 1. Consider a problem $N = <X, D, C>$ with n variables $X = \{x_i\}_{i \in 1..n}$, uniform domains $D(x_1) = \ldots = D(x_n) = \{v_1, \ldots, v_d\}$ and a unique constraint $C = Gcc(X, lb, ub)$, with $lb(v_i) = 0$ and $ub(v_i) = 1$ for all value, except for the special value v_d that can be taken at most 3 times $(ub(v_d) = 3)$. This problem is almost injective and we have $\mu(N) \leq 3$.

We provide below a first upper bound of $\mu(N)$ for a constraint network N which contains a Gcc constraint on all variables involved in the symmetries, by considering in a static way the different upper bounds of the values.

Property 1. Let $N = < X, D, C >$ be a constraint network, with $Gcc(X, lb, ub) \in C$. $\mu \leq \sum_{v \in D \ s.t. \ ub(v) > 1} ub(v)$.

This bound takes into account neither the overlap between variables domains nor the lower bound of values. To deal with both the lower bounds and the variables domains and then achieve a better upper bound of μ, we have to look at the implementation of the Gcc.

The Gcc constraint propagation algorithm proposed in [13] consists in finding a flow in a bipartite graph, such that one of Figure 1. The set of the set of

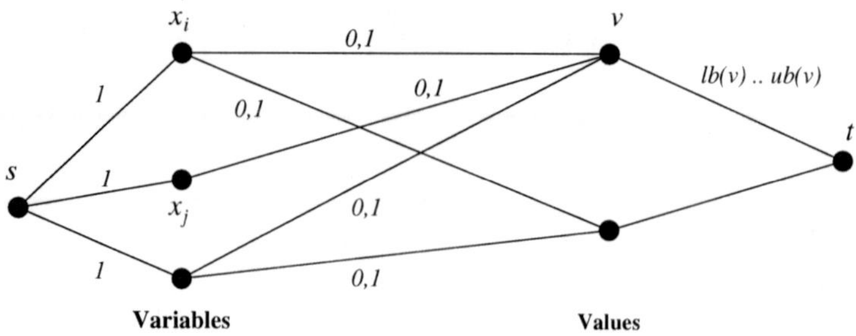

Fig. 1. The network used for GCC propagation

nodes is the union of the variables involved in the Gcc and their values and the two special nodes s and t. The flow between the source s and a variable x is exactly 1, stating that each variable has to be assigned to a single value. The flow between a variable x and a value v in its domain is either 0 or 1, depending if x is assigned to v or not. The flow between a value v and s is constrained by the lower and upper bounds of the value in the Gcc.

Property 2. Given a constraint network $N =< X, D, C >$ with $Gcc(X, lb, ub) \in C$, $\mu(N)$ is bounded by the number of variables minus the minimal number of values which can be assigned to exactly one variable while respecting the Gcc.

Computing this minimal number of values is more complex than establishing the upper bound of Prop. 1. We propose to compute it using the CSP encoding described below. As described above the underlying algorithm of the Gcc relies on a flow problem. But, the flow problem itself has been encapsulated in a constraint in [4], allowing to solve problems involving flow and additional constraints, for instance large workforce scheduling problems [1]. In this flow constraint, the flow allowed to an edge (i, j) is expressed as a CSP variable $x_{flow(i,j)}$.

To compute the minimal number of values which can be assigned to exactly one variable while respecting the Gcc, we propose to reuse the encoding of the Gcc enriched with extra variables $nb\_occ(v_i)$ to count the number of occurrences of a value. $nb_occ(v_i)$ is the variable expressing the flow between v_i and t. Minimize $\sum_{v_i \in D}(nb\_occ(v_i) = 1)$ is equivalent to compute the minimal number of values which can be assigned to exactly one variable while respecting the Gcc.

Lemma 1. *The problem of determining if $\sum_{v_i \in D}(nb\_occ(v_i) = 1) \leq N$ is NP-hard.*

Proof sketch: Consider a $Gcc(X, lb, up)$ with $lb(v) = 0, ub(v) = |X|$, finding $\sum_{v_i \in D}(nb\_occ(v_i) \geq 1) \leq N$ exactly fits the definition of the $atmostN values(X, N)$ constraint, which was shown NP-hard in [3] by a reduction to $3 - SAT$. For $\sum_{v_i \in D}(nb\_occ(v_i) = 1) \leq N$ the proof is identical except the point 3-SAT is replaced by exactly-1 3SAT (also called 1-in-3 SAT).

4 Generalization to "almost injective" Problems

Lets consider an almost injective problem where no more than μ variables can be equal simultaneously.

For each variable symmetry $\sigma \in \mathcal{G}^{[\beta_k]}$, constraint (1) is no longer equivalent to $x_{\beta_k} < x_{\sigma(\beta_k)}$ because x_{β_k} can be equal to $x_{\sigma(\beta_k)}$. In such a case, checking the lexicographical constraint involves to find the next $\beta_z > \beta_k$ such that $x_{\beta_z} \neq x_{\sigma(\beta_z)}$ and so $\sigma(\beta_z) \neq \beta_z$.

Given any symmetry $\sigma \in \mathcal{G}^{[\beta_k]}$, consider the increasing sequence $i_\sigma^1, i_\sigma^2, \ldots i_\sigma^t$ of the elements[3] of B for which $\sigma(i) \neq i$. Since $\sigma \in \mathcal{G}^{[\beta_k]}$ we have $i_\sigma^1 = \beta_k$.

For injective problems we have seen that lexicographic constraint (1) can be replaced by the constraint $x_{i_\sigma^1} < x_{\sigma(i_\sigma^1)}$. For almost injective problem with no more than μ simultaneous pairs of equal variables, constraint (1) simplifies to:

$$x_{i_\sigma^1}, x_{i_\sigma^2}, \ldots, x_{i_\sigma^\rho} \leq_{lex} x_{\sigma(i_\sigma^1)}, x_{\sigma(i_\sigma^2)}, \ldots, x_{\sigma(i_\sigma^\rho)} \tag{4}$$

where $\rho = \min(\mu + 1, t)$.

Because there are no more than $\binom{n}{2\mu}$ constraints (4) for all symmetries in $\mathcal{G}$ and $\binom{n}{k} < n^k$, we have the following lemma:

Lemma 2. *Given a CSP where no more than μ variables can be equal simultaneously, all variable symmetries can be broken with a number of constraints which is XP in μ.*

To compute constraints (4), we have to enumerate the whole group $\mathcal{G}$. This can be done in $\mathcal{O}(n|\mathcal{G}|)$ thanks to sets $U^{[k]}$ computed by Schreier-Sims' algorithm.

By definition, constraint (4) is equivalent to the following constraints:

$$x_{i_\sigma^1} \leq x_{\sigma(i_\sigma^1)} \tag{5a}$$

$$x_{i_\sigma^1} = x_{\sigma(i_\sigma^1)} \rightarrow x_{i_\sigma^2} \leq x_{\sigma(i_\sigma^2)} \tag{5b}$$

$$\ldots$$

$$x_{i_\sigma^1} = x_{\sigma(i_\sigma^1)} \wedge \ldots \wedge x_{i_\sigma^{\rho-1}} = x_{\sigma(i_\sigma^{\rho-1})} \rightarrow x_{i_\sigma^\rho} \leq x_{\sigma(i_\sigma^\rho)} \tag{5c}$$

There are $\sum_{\beta_i \in B}(|\Delta^{[\beta_i]}| - 1)$ constraints of type (5a), no more than $\binom{n}{4}$ constraints (5b) and finally no more than $\binom{n}{\mu+2}$ constraints (5c). It is a crude upper bound in the worst case. In practice some constraints can be discarded. For instance, if we post a constraint (5a) like $x_a \leq x_b$, it is unnecessary to post any constraint that ends with "$\rightarrow x_a \leq x_b$". Moreover, constraints (5a) are equivalent to a linear number of constraints of the form $x_{r(j)} \leq x_j$.

The total number of constraints required to break symmetry also depends on the ordering of the variables in base B. For instance, suppose that δ is the first index in B such that x_δ takes a duplicate value. For each $\sigma \in \mathcal{G}^{[\beta_k]}$ such that $i_\sigma^1 = \beta_k < \delta$, the lexicographic constraint is equivalent to $x_{i_\sigma^1} < x_{\sigma(i_\sigma^1)}$. All these inequalities can be reduced to a linear number of constraints of the form $x_{r(j)} \leq x_j$. The special case when $\delta > |B|$ (with $|B| < n$) is completely equivalent to injective problems.

---

[3] If $|B| < n$ we can add missing values at the end of the base. It does not affect the strong generating set computed with Schreier-Sims' algorithm.

Therefore, we can try to rearrange base B in order to set, at the end of the list, the indexes of the variables that take duplicate values.

A dynamic rearrangement could be expensive because permuting a single pair of contiguous indexes has a complexity in $\mathcal{O}(n^4)$[5].

Let us assume that there are only ν variables that can take a duplicate value. We can choose B in order that the indexes of these variables are placed at the end of the list. Then all the symmetries in $\mathcal{G} \setminus \mathcal{G}^{[\beta_n - \nu]}$ can be broken with a linear number of constraints as in injective problems. Lexicographic constraints (4) are only required for symmetries that belong to $\mathcal{G}^{[\beta_n - \nu]}$. All the scopes of these constraints are included in a set of ν variables. Hence there are at most $\binom{\nu}{\mu}$ and the total number of constraints is in $\mathcal{O}(\binom{\nu}{\mu} + n)$. This proves the following:

Lemma 3. *Given a CSP where no more than μ simultaneous variables can be equal, and only a subset of ν variables can take duplicates values, all variable symmetries can be broken with a number of constraint which is FPT in ν and μ.*

The Case of Heterogeneous Domains

In Section 4, we provide a theoretical bound of number of constraints to be posted. As shown in example 2, this bound does not take into account initial domains of variables, when they are not all equal.

Example 2. Let $x_{i_\sigma^1}$ and $x_{\sigma(i_\sigma^1)}$ be two variables involved in a symmetry. If $D(x_{i_\sigma^1}) \cap D(x_{\sigma(i_\sigma^1)}) = \emptyset$, the constraint 5b is useless and can be discarded (not posted) as well as all constraints containing $x_{i_\sigma^1} = x_{\sigma(i_\sigma^1)}$ in their left part.

The conjunction of the Gcc constraint and the initial domains of variables may also forbid combinations of equalities between pairs of variables:

Example 3. Let $D(x_1) \cap D(x_2) = \{v\} = D(x_3) \cap D(x_4)$ and $ub(v) = 3$. The Gcc forbids to have simultaneously $x_1 = x_2$ and $x_3 = x_4$. Then, all constraints 5c, within $x_1 = x_2 \wedge x_3 = x_4$ in their left part may be discarded.

More generally, during the generation of symmetry breaking constraints of the constraint network $N =< X, D, C >$, we propose the following technique: Before posting the constraint $x_{i_\sigma^1} = x_{\sigma(i_\sigma^1)} \wedge \ldots \wedge x_{i_\sigma^{\rho-1}} = x_{\sigma(i_\sigma^{\rho-1})} \rightarrow x_{i_\sigma^\rho} \leq x_{\sigma(i_\sigma^\rho)}$ we propose to solve the subproblem $N =< X, D, C' >$ with $C' = \{x_{i_\sigma^1} = x_{\sigma(i_\sigma^1)}, \ldots, x_{i_\sigma^{\rho-1}} = x_{\sigma(i_\sigma^{\rho-1})}\} \cup \{Gcc\}$, the sub-problem of N restricted to both the Gcc constraint and the equality constraints of the left part of this constraint. If N' is not soluble, the constraint is discarded.

Lemma 4. *The satisfiability problem of the class of constraint networks involving only a Gcc and some equality constraints is polynomial.*

The idea of the proof relies on a slight modification of the Gcc encoding as flow problem. For each pair of variable (x_i, x_j) involved in an equality constraint, we replace them by a merged node $x_{i,j}$ The flow between the source s and this new node variable $x_{i,j}$ is exactly 2 to enforce both x_i and x_j to be assigned. The flow between $x_{i,j}$ and any value v in $D(x_i) \cap D(x_j)$ is either 0 or 2 to count 2 uses of v when x_{ij} (in fact both x_i and x_j) is assigned to v.

In addition of theoretical results in terms parametrized complexity, this technique provides a practical way to restrict the number of constraints to be posted for breaking symmetries in almost injective problems.

5   Conclusion

We have introduced a characterization of "almost injective" problems which is based on the number μ of variables that can be equal simultaneously. We showed that variable symmetry can be broken with no more than $\binom{n}{\mu}$ constraints which is XP in the framework of parameterized complexity.

When only ν variables can take duplicated values, the number of constraints is FPT in μ and ν.

In case of heterogeneous domains, we presented a polynomial method (based on Gcc) for eliminating unnecessary constraints.

References

1. Benoist, T., Gaudin, E., Rottembourg, B.: Constraint Programming Contribution to Benders Decomposition: A Case Study. In: Van Hentenryck, P. (ed.) CP 2002. LNCS, vol. 2470, pp. 603–617. Springer, Heidelberg (2002)
2. Bessiere, C., Hebrard, E., Hnich, B., Kiziltan, Z., Quimper, C.G., Walsh, T.: The parameterized complexity of global constraints. In: Proc. AAAI, pp. 235–240 (2008)
3. Bessière, C., Hebrard, E., Hnich, B., Kiziltan, Z., Walsh, T.: Filtering Algorithms for the NVALUE Constraint. In: Barták, R., Milano, M. (eds.) CPAIOR 2005. LNCS, vol. 3524, pp. 79–93. Springer, Heidelberg (2005)
4. Bockmayr, A., Pisaruk, N., Aggoun, A.: Network Flow Problems in Constraint Programming. In: Walsh, T. (ed.) CP 2001. LNCS, vol. 2239, pp. 196–210. Springer, Heidelberg (2001)
5. Butler, G.: Fundamental Algorithms for Permutation Groups. LNCS, vol. 559. Springer, Heidelberg (1991)
6. Crawford, J., Ginsberg, M., Luks, E., Roy, A.: Symmetry-breaking predicates for search problems. In: Proc. KR 1996, pp. 148–159 (1996)
7. Downey, R.G., Fellows, M.R., Stege, U.: Parameterized complexity: A framework for systematically confronting computational intractability. In: DIMACS Series in Discrete Mathematics and Theoretical Computer Science, vol. 49, pp. 49–99 (1997)
8. Flener, P., Pearson, J., Sellmann, M.: Static and dynamic structural symmetry breaking. Ann. Math. Artif. Intell. 57(1), 37–57 (2009)
9. Gent, I.P., Harvey, W., Kelsey, T.: Groups and Constraints: Symmetry Breaking during Search. In: Van Hentenryck, P. (ed.) CP 2002. LNCS, vol. 2470, pp. 415–430. Springer, Heidelberg (2002)
10. Gottlob, G., Szeider, S.: Fixed-parameter algorithms for artificial intelligence, constraint satisfaction and database problems. Comput. J. 51(3), 303–325 (2008)
11. McKay, B.D.: nauty user's guide. Tech. rep., Australian National University (2009), `http://cs.anu.edu.au/~bdm/nauty/`
12. Puget, J.F.: Breaking symmetries in all different problems. In: Proc. IJCAI 2005, pp. 272–277 (2005)
13. Régin, J.C.: Generalized arc consistency for global cardinality constraint. In: Proc. AAAI 1996, pp. 209–215 (1996)
14. Seress, Á.: Permutation group algorithms. Cambridge tracts in mathematics. Cambridge University Press (2003)

Understanding, Improving and Parallelizing MUS Finding Using Model Rotation

Siert Wieringa⋆

Aalto University, School of Science
Department of Information and Computer Science
P.O. Box 15400, FI-00076 Aalto, Finland

Abstract. Recently a new technique for improving algorithms for extracting Minimal Unsatisfiable Subsets (MUSes) from unsatisfiable CNF formulas called "model rotation" was introduced [Marques-Silva et. al. SAT2011]. The technique aims to reduce the number of times a MUS finding algorithm needs to call a SAT solver. Although no guarantees for this reduction are provided the technique has been shown to be very effective in many cases. In fact, such model rotation algorithms are now arguably the state-of-the-art in MUS finding.

This work analyses the model rotation technique in detail and provides theoretical insights that help to understand its performance. These new insights on the operation of model rotation lead to several modifications and extensions that are empirically evaluated. Moreover, it is demonstrated how such MUS extracting algorithms can be effectively parallelized using existing techniques for parallel incremental SAT solving.

1 Introduction

Despite the theoretical hardness of the satisfiability problem (SAT) current state-of-the-art decision procedures for SAT, so called SAT solvers, have proven to be efficient problem solvers for many real life applications. For instances of SAT that are unsatisfiable the notion of a Minimal Unsatisfiable Subset (MUS) can be defined as follows. A MUS of an unsatisfiable formula is a subset of its constraints that is minimal in the sense that removing any constraint will make it satisfiable.

In recent years there has been a lot of research into algorithms for MUS finding [12]. Such algorithms have several useful applications in for example product configuration [18], electronic design automation [15] and algorithms for maximum satisfiability [11]. Deciding whether a formula is minimally unsatisfiable is a DP-complete problem [16]. SAT was the first problem ever proven NP-complete [5]. As the complexity class DP is not believed to be included in the class NP an algorithm for MUS finding using an NP oracle, such as a SAT solver, will include repeated calls to that oracle.

MUS finding algorithms have been categorized as either *constructive*, *destructive* or *dichotomic* [8]. Destructive algorithms start by approximating the MUS

⋆ Financially supported by the Academy of Finland project 139402.

contains all m clauses of the input formula, and then iteratively remove clauses from this approximation, requiring $O(m)$ calls to a SAT solver. Constructive algorithms such as [10] start by under-approximating the MUS as an empty set and adding clauses from the input formula until this approximation becomes unsatisfiable. In a straightforward constructive algorithm a clause is proven critical if its addition to the MUS approximation causes this approximation to become unsatisfiable, after which the algorithm restarts using only clauses already proven critical as its MUS approximation. Such an algorithm requires $O(m \times k)$ calls to a SAT solver where k is the number of clauses in the largest MUS [8]. Dichotomic algorithms [9] have not received much attention in recent publications. Those algorithms use a binary search to construct the MUS and require $O(k \ log \ m)$ solver calls.

Recent work [13] blurs the distinction between constructive and destructive algorithm by presenting a constructive style algorithm that requires $O(m)$ SAT solver calls. Arguably the most significant recent contribution to the field of MUS finding is *model rotation*, which was also introduced in [13]. Model rotation is presented as a heuristic technique for reducing the required number of SAT solver calls, and it seems to be very effective in practice. Several other MUS algorithms make use of the *resolution proof* that can be generated by an extended SAT solver, but [13] provides evidence that such algorithms can be outperformed by algorithms that do not require a proof logging SAT solver.

This paper contributes new insights on what it is that makes model rotation such a powerful technique. Using this insight several improvements are suggested and empirically evaluated. Moreover, it is demonstrated how model rotation algorithms can be parallelized using existing techniques for parallelizing incremental SAT.

2   Basic Definitions

A literal l is a Boolean variable $l = x$ or its negation $l = \neg x$. For any literal l it holds that $\neg\neg l = l$. A clause $c = \{l_1, l_2, \cdots, l_{|c|}\}$ is a non-empty set of literals, representing the disjunction $l_1 \vee l_2 \vee \cdots \vee l_{|c|}$. A formula $\mathcal{F}$ is a set of clauses. An assignment a is a set of literals such that if $l \in a$ then $\neg l \notin a$. If $l \in a$ then it is said that literal l is assigned the value **true**, if $\neg l \in a$ then l it is said that l assigned value **false**. Assignment a satisfies clause c if there exists a literal $l \in a$ such that $l \in c$. An assignment satisfies a formula if it satisfies all clauses in the formula. An assignment a is a *complete assignment* for a formula $\mathcal{F}$ if for all $c \in \mathcal{F}$ and all $l \in c$ either $l \in a$ or $\neg l \in a$. A formula that has no satisfying assignments is called unsatisfiable. A formula $\mathcal{F}$ is *minimal unsatisfiable* if it is unsatisfiable and any subformula $\mathcal{F}' \subset \mathcal{F}$ is satisfiable.

Definition 1 (assoc). *An associated assignment (assoc) [10] for a clause $c \in \mathcal{F}$ is a complete assignment a for the formula $\mathcal{F}$ that satisfies the formula $\mathcal{F} \setminus \{c\}$ and does not satisfy c. Let $A(c, \mathcal{F})$ be the set of all assocs for $c \in \mathcal{F}$.*

Note that an unsatisfiable formula $\mathcal{F}$ is minimal unsatisfiable iff for all clauses $c \in \mathcal{F}$ it holds that c has an assoc, i.e. $A(c, \mathcal{F}) \neq \emptyset$. Such clauses are referred to in

---

Algorithm 1. MUS finder with recursive model rotation [2,13]

Given an unsatisfiable formula $\mathcal{F}$:

1. $M = \emptyset$
2. `while` $\mathcal{F} \neq M$
3.   `pick` a clause $c \in \mathcal{F} \setminus M$
4.   `if` $\mathcal{F} \setminus \{c\}$ is satisfiable `then`
5. $M = M \cup \{c\}$
6.   `modelRotate`(c, a) where a is a compl. satisfying assign. for $\mathcal{F} \setminus \{c\}$
7.   `else`
8. $\mathcal{F} = \mathcal{F} \setminus \{c\}$
9. `return` $\mathcal{F}$

function `modelRotate(clause` c`, assignment` a`)`   `//` such that $a \in A(c, \mathcal{F})$

  I `for all` $l \in c$ `do`
 II $a' = $ `rotate`$(a, \neg l)$
III   `if` exactly one clause $c' \in \mathcal{F}$ is not satisfied by a' and $c' \notin M$ `then`
IV $M = M \cup \{c'\}$
 V   `modelRotate`(c', a')

---

other work as *critical clauses* (e.g. [10]) or *transition clauses* (e.g [13]). Because of the special interest in the assocs, this work will however always explicitly refer to "a clause for which an assoc exists". Note that the problem of finding a MUS in the unsatisfiable formula $\mathcal{F}$ is equivalent to *proving an assoc exists for every clause in an unsatisfiable formula* $\mathcal{F}' \subseteq \mathcal{F}$.

Clearly, for any unsatisfiable formula $\mathcal{F}$ and clause $c \in \mathcal{F}$ a single assoc $a \in A(c, \mathcal{F})$ or proof that $A(c, \mathcal{F}) = \emptyset$ can be obtained by testing the satisfiability of the formula $\mathcal{F} \setminus \{c\}$ using a SAT solver. Given an assoc for a clause c the technique proposed in [13] and improved to its recursive version in [2] attempts to obtain an assoc for another clause c' by *model rotation*, which is replacing a single literal in the assoc by its negation.

Definition 2 (`rotate`(a, l)). *Let* `rotate`(a, l) *be a function that negates literal* l *in assignment* a, *i.e.:* `rotate`$(a, l) = (a \setminus \{l\}) \cup \{\neg l\}$

Algorithm 1 gives the pseudocode for a MUS finder using model rotation. The only difference from the classical 'destructive' MUS finding algorithm is the addition of the call to the function `modelRotate` on Line 6. A SAT solver is used to determine the satisfiability of the formula $\mathcal{F} \setminus \{c\}$ on Line 4. If the formula is satisfiable then the solver has found a satisfying assignment and this assignment is an assoc $a \in A(c, \mathcal{F})$. Any clause for which an assoc is found is added to M, the MUS under construction, and the algorithm continues until an assoc is found for every clause remaining in $\mathcal{F}$. On Line 8 the algorithm removes clause c from $\mathcal{F}$. In typical implementations of MUS finding algorithms instead of removing c from $\mathcal{F}$ the formula is replaced by an unsatisfiable subset $\mathcal{F}' \subseteq \mathcal{F} \setminus \{c\}$ that can be cheaply calculated by the SAT solver.

3   Understanding Model Rotation

In this section the model rotation technique will be studied by thinking of it as an algorithm that traverses a graph.

Definition 3 (flip graph). *For a CNF formula $\mathcal{F}$ let the flip graph $G = (V, E)$ be a graph in which there is a vertex for every clause, i.e. $V = \mathcal{F}$. Each edge $(c_i, c_j) \in E$ is labelled with the set of literals $L(c_i, c_j)$ such that:*

$$L(c_i, c_j) \quad = \quad \{l \mid l \in c_i \text{ and } \neg l \in c_j\}$$

The set of edges E of the flip graph is defined by $(c_i, c_j) \in E$ iff $L(c_i, c_j) \neq \emptyset$

Even though $(c_i, c_j) \in E$ iff $(c_j, c_i) \in E$ in this work the flip graph is considered to be a directed graph. This is because of the interest in the sets labelling the edges, which contain the same literals in opposite polarity for the same edge in different directions, and because a subset of edges will be defined that imposes a truly directed graph.

In some existing literature (e.g. [17]) the flip graph is defined as an undirected graph and referred to as the *resolution graph*. That name was apparently chosen because resolution can only be performed on clauses that are neighbors in the graph, but it is slightly confusing because the name resolution graph is more commonly used (e.g. in [7]) to denote the directed acyclic graph (DAG) describing a resolution proof.

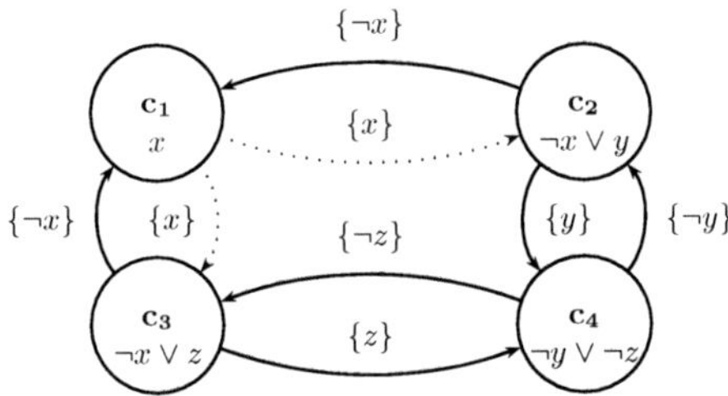

Fig. 1. The *flip graph* for the formula $\mathcal{F}_{fig1} = \{\{x\}, \{\neg x, y\}, \{\neg x, z\}, \{\neg y, \neg z\}\}$

Model rotation can be thought of as an algorithm that traverses a path in the flip graph. The intuition behind the following negatively stated lemma is that model rotation can only successfully traverse edges labelled with exactly one literal by rotation of that same literal.

Lemma 1. *Let $\mathcal{F}$ be an unsatisfiable formula, c_i and c_j two clauses $c_i, c_j \in \mathcal{F}$, and l a literal $l \in c_i$. If $L(c_i, c_j) \neq \{l\}$ then for any assoc $a_i \in A(c_i, \mathcal{F})$ the assignment $a_j = \mathtt{rotate}(a_i, \neg l)$ satisfies c_j (and thus $a_j \notin A(c_j, \mathcal{F})$).*

Proof. Note that because a_i is an assoc $a_i \in A(c_i, \mathcal{F})$ it satisfies c_j. Note also that the only literal that is in a_i and not in a_j is the literal $\neg l$. If $L(c_i, c_j) = \emptyset$

then $\neg l \notin c_j$ and thus a_j satisfies c_j. If $L(c_i, c_j) \neq \emptyset$ and $L(c_i, c_j) \neq \{l\}$ then there exists a literal $l' \neq l$ such that $l' \in L(c_i, c_j)$. Because a_i does not satisfy c_i and $l' \in c_i$ it must hold that $\neg l' \in a_i$. But then also $\neg l' \in a_j$ and because $\neg l' \in c_j$ it holds that c_j is satisfied by a_j.

Clearly by Lemma 1 for any two $c_i, c_j \in \mathcal{F}$ negating a single literal in an assoc $a_i \in A(c_i, \mathcal{F})$ can never result in an assoc $a_j \in A(c_j, \mathcal{F})$ if $|L(c_i, c_j)| \neq 1$. This leads to the definition of a subset of the edges E of the flip graph of $\mathcal{F}$ called the *possible rotation* edges[1] $E_P \subseteq E$. Moreover, a set of *guaranteed rotation* edges $E_G \subseteq E_P$ is defined.

Definition 4 (rotation edges). *Given a formula $\mathcal{F}$, let:*

$$E_P = \{(c_i, c_j) \mid c_i, c_j \in \mathcal{F} \text{ and } |L(c_i, c_j)| = 1\}$$
$$E_G = \{(c_i, c_j) \mid c_i, c_j \in \mathcal{F} \text{ and } |L(c_i, c_j)| = 1 \text{ and for all } c_k \in \mathcal{F}$$
$$\text{it holds that } L(c_i, c_j) \neq L(c_i, c_k) \text{ if } c_k \neq c_j\}$$

In Fig. 1 the flip graph for an example formula $\mathcal{F}_{fig1}$ is given. Because there are no two clauses $c_i, c_j \in \mathcal{F}_{fig1}$ such that $|L(c_i, c_j)| > 1$ it holds that the set of possible rotation edges E_P is equal to the set of all edges E in the flip graph. However, only the solid edges in the figure belong to the set of guaranteed rotation edges E_G. The dotted edges are not in the set E_G because the two outgoing edges from vertex c_1 have the same label $L(c_1, c_2) = L(c_1, c_3) = \{x\}$.

Theorem 1. *Let $\mathcal{F}$ be an unsatisfiable formula and E_G the set of guaranteed rotation edges it induces. If $(c_i, c_j) \in E_G$ then for any assoc $a_i \in A(c_i, \mathcal{F})$ an assignment $a_j = \mathtt{rotate}(a_i, \neg l)$ such that $L(c_i, c_j) = \{l\}$ is an assoc $a_j \in A(c_j, \mathcal{F})$.*

Proof. By the definition of E_G for all clauses $c_k \in \mathcal{F}$ such that $c_k \neq c_j$ it holds that $L(c_i, c_k) \neq \{l\}$. It follows from Lemma 1 that all such clauses c_k are satisfied by a_j. As $\mathcal{F}$ is unsatisfiable and a_j satisfies $\mathcal{F} \setminus \{c_j\}$ it must hold that a_j does not satisfy c_j. Thus a_j is an assoc $a_j \in A(c_j, \mathcal{F})$

From Th. 1 it follows that an assoc exists for every clause for which there is a path over edges in E_G from a clause for which an assoc exists. Thus for formula $\mathcal{F}_{fig1}$ presented in Fig. 1 obtaining any assoc $a \in A(c_i, \mathcal{F}_{fig1})$ such that $i \in \{2, 3, 4\}$ is sufficient to determine that the formula is minimal unsatisfiable. Obtaining an assoc for clause c_1 may however be less effective. Note that:

$$A(c_1, \mathcal{F}_{fig1}) = \{ \{\neg x, \neg y, \neg z\}, \{\neg x, \neg y, z\}, \{\neg x, y, \neg z\} \}$$

Although by replacing $\neg x$ by x the second and third assoc in this set can be rotated into a valid assoc for c_2 and c_3 respectively, no negation of a single literal will make the first assoc into a valid assoc for any other clause in the formula $\mathcal{F}_{fig1}$.

---

[1] Note that the set E_P also corresponds to all pairs of clauses (c_i, c_j) on which resolution $c_i \otimes c_j$ can be performed without creating a tautology.

Corollary 1. *Let c_i and c_j be two clauses $c_i, c_j \in \mathcal{F}$ for an unsatisfiable formula $\mathcal{F}$. Let E_G be the set of guaranteed rotation edges induced by $\mathcal{F}$. If in graph $G = (\mathcal{F}, E_G)$ there exists a path from c_i to c_j, and also a path from c_j to c_i, then $A(c_i, \mathcal{F}) \neq \emptyset$ iff $A(c_j, \mathcal{F}) \neq \emptyset$.*

Recall that a single assoc $a \in A(c, \mathcal{F})$, or proof that no such assoc exists, can be obtained by testing the satisfiability of the formula $\mathcal{F} \setminus \{c\}$ using a SAT solver.

Corollary 2. *Given the unsatisfiable formula $\mathcal{F}$ an assoc for every clause in a minimal unsatisfiable subformula $\mathcal{F}' \subseteq \mathcal{F}$ can be obtained using at most as many solver calls as there are strongly connected components (SCCs)[2] in the graph $G = (\mathcal{F}, E_G)$.*

Let a *root* SCC of a graph be an SCC that has no incoming edges from different SCCs. In other words, a root SCC of the graph $G = (V, E)$ is a strongly connected component containing vertices $V' \subseteq V$ such that for all $v' \in V'$ and $(v, v') \in E$ it holds that $v \in V'$. Observe that every directed graph has at least one root SCC. Because the existence of an assoc $a \in A(c, \mathcal{F})$ for a single clause c implies an assoc for every clause reachable from c the following holds:

Corollary 3. *Given the minimal unsatisfiable formula $\mathcal{F}$ an assoc for every clause in $\mathcal{F}$ can be obtained using at most as many solver calls as there are root SCCs in the graph $G = (\mathcal{F}, E_G)$.*

Given in Alg. 2 is the pseudo-code for a MUS finding algorithm that respects the upper bounds on the number of solver calls established by Corollaries 2 and 3. Note that the set of clauses added to M on Line 7 of Alg. 2 is a superset of the clauses in the SCC $\mathcal{F}_i$. In an implementation of this algorithm once again an improvement for the unsatisfiable case can be made if the solver is capable of calculating an unsatisfiable subset $\mathcal{F}' \subseteq \mathcal{F}_i \setminus \{c\}$. Given such an unsatisfiable subset $\mathcal{F}'$ any SCC $\mathcal{F}_j$ such that for some $c \in \mathcal{F}_j$ it holds that $c \notin \mathcal{F}'$ can be removed from S, and its clauses removed from $\mathcal{F}$. Moreover, Alg. 2 can be extended with model rotation to also discover assocs for clauses reachable over edges only in E_P.

3.1   Benchmark Statistics

The benchmark set used in [13] is a set of 500 *trimmed* benchmarks. Trimming means reducing an unsatisfiable formula to a unsatisfiable subset that is not necessarily minimal. Reducing the size of the formula without proving it minimal reduces the amount of work left for the expensive MUS finding algorithm, and is thus typically beneficial for the overall performance. A second benchmark set for MUS finders was found in the MUS finding track of the SAT competition

---

[2] A directed graph is strongly connected if there exists a path between every two of its vertices. The SCCs of a directed graph are its maximal strongly connected subgraphs.

Algorithm 2. MUS finder respecting upper bounds of Cor. 2 and Cor. 3

Given an unsatisfiable formula $\mathcal{F}$ and the set E_G it induces:

1. Let $S = \{\mathcal{F}_1, \mathcal{F}_2, \cdots, \mathcal{F}_{|S|}\}$ the division of $\mathcal{F}$ into the SCCs of $G = (\mathcal{F}, E_G)$
2. $M = \emptyset$
3. **while** $\mathcal{F} \neq M$
4.   **pick** $\mathcal{F}_i \in S$ corresponding to a root SCC such that $M \cap \mathcal{F}_i = \emptyset$
5.   **pick** a clause $c \in \mathcal{F}_i$
6.   **if** $\mathcal{F} \setminus \{c\}$ is satisfiable **then**
7. $M = M \cup \{c' \mid c' = c \text{ or } c' \text{ is reachable from } c \text{ in } G = (\mathcal{F}, E_G)\}$
8.   **else**
9. $\mathcal{F} = \mathcal{F} \setminus \mathcal{F}_i$
10. $S = S \setminus \{\mathcal{F}_i\}$
11. **return** $\mathcal{F}$

2011^3. It consists of 300 benchmarks, 150 of which are the original formulas for which the trimmed version appear in the benchmark set of [13].

Table 1 gives statistics on these benchmarks. Note that the statistics concern properties of the benchmarks themselves, rather than the result of some empirical evaluation of an algorithm running on those benchmarks. Looking at the statistics in the columns labelled 'original' in the table gives some insight in why model rotation performs so well for these benchmarks. The table provides statistics on the average number of outgoing guaranteed rotation edges from any clause ('avg. out-degree E_G'). As for the near minimal formulas described here this number is greater than 1 any single execution of the model rotation function is probable to result in finding more than one assoc. For the larger formulas from the SAT competition the average out-degree of guaranteed edges is still substantial at 0.89, and it should be noted that during the execution of the MUS finding algorithm this out-degree will increase (towards 1.59 on average) as the formula shrinks to a MUS.

In the set of trimmed benchmarks from [13] 148 benchmarks have only exactly one root SCC. This means that algorithm Alg. 2 requires only one call to a SAT solver to establish an assoc for all clauses in that benchmark. The last two rows in the table state the average upper bounds as established in Corollaries 2 and 3 as a percentage of the number of clauses in the input formula. The statistics thus demonstrate that for the benchmark set from [13] Alg. 2 requires at most 43.6% of the number of solver calls used in the worst case by a classical destructive algorithm without model rotation. The column in the table labelled 'MUSes' provide statistics for the 'original' benchmarks after reduction using a MUS finding algorithm. The 'MUSes' benchmark sets are slightly smaller because they do not include a MUS for formulas for which one could not be established in reasonable time. To prove that the MUSes found in the formulas from [13] are indeed MUSes Alg. 2 requires at most 27% of the number of solver calls required in the worst case without model rotation.

---

^{3} http://www.satcompetition.org/2011

Table 1. Statistics on $G = (\mathcal{F}, E_G)$ for formulas from various benchmark sets

	From [13]		SAT11 competition	
	original	MUSes	original	MUSes
# benchmarks	500	491	298	262
# with single root SCC	148	148	0	51
avg. # clauses	6874.4	6204.2	404574	8162.7
avg. # SCCs	3000.4	2484.1	327815	3355.6
of size 1	2051.1	1635.2	276341	2254.2
avg. # root SCCs	2124.7	1680.5	258350	1891.1
of size 1	1613.4	1240.5	227129	1420.8
avg. clause length	2.32	2.35	2.53	2.42
avg. out-degree E_P	12.30	11.24	86.84	14.16
avg. out-degree E_G	1.60	1.63	0.89	1.59
avg. Cor. 2 bound	$\frac{3000.4}{6874.4} = 43.6\%$	$\frac{2484.1}{6204.2} = 40\%$	$\frac{327815}{404574} = 81\%$	$\frac{3355.6}{8162.7} = 41\%$
avg. Cor. 3 bound	n/a	$\frac{1680.5}{6204.2} = 27\%$	n/a	$\frac{1891.1}{8162.7} = 23\%$

What Alg. 2 can be proven to do serves to illustrate what the model rotation algorithm Alg. 1 can do. This is because Lines 4 and 5 of Alg. 2 can be thought of as a good strategy for picking a clause on Line 3 in Alg. 1. If the resulting formula $\mathcal{F} \setminus \{c\}$ is satisfiable then model rotation will find an assoc for at least all clauses reachable over guaranteed rotation edges E_G in c, so the set of clauses added to M by Alg. 1 is a superset of the set of clauses added to M by Alg. 2. Hence, Cor. 3 can be thought of as establishing an upper bound on the number of solver calls in the most efficient execution sequence possible for Alg. 1. For non-minimal input formulas following a solver call with unsatisfiable result Alg. 2 removes a complete SCC rather than a single clause on Line 9. This means that in theory executing Alg. 1 may result in more solver calls with result unsatisfiable than Alg. 2. However, as stated before in practice Alg. 1 is implemented such that more than one clause at the time is removed by obtaining an unsatisfiable core from the solver. As a result even if the input formula is far from minimal the number of solver calls with result unsatisfiable is typically small compared to the number of calls with result satisfiable.

4   Improving Model Rotation

Studying model rotation as an algorithm traversing the flip graph revealed a possible algorithmic improvement which is best explained by example. In Fig. 2 the graph $G = (\mathcal{F}_{ph3}, E_G)$ for the formula $\mathcal{F}_{ph3}$ representing an encoding of the *pigeon hole principle* for three pigeons and two holes is given. For each of three pigeons x, y and z there are two variables. The assignment of x_1 to the value **true** means that pigeon x is in hole 1, the assignment y_2 to **true** means pigeon y is in hole 2 etcetera. The clauses c_2, c_4 and c_6 for which the vertices are drawn with double circles in Fig. 2 are representing for pigeons x, y and z respectively the constraint that the pigeon is either in hole 1 or in hole 2. The other vertices

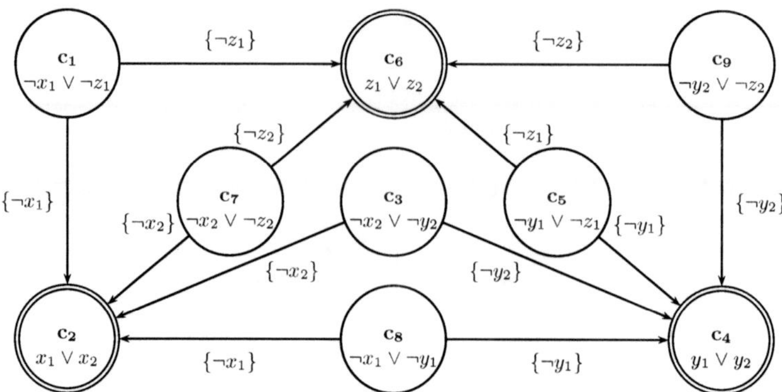

Fig. 2. The graph $G = (\mathcal{F}_{ph3}, E_G)$ for formula $\mathcal{F}_{ph3}$

represent clauses that state for each combination of two pigeons and a hole that the two pigeons are not both in that hole. Obviously this formula is unsatisfiable as there is no placement of three pigeons into two holes such that each pigeon is in a hole and no two pigeons are in the same hole. Observe that it is also minimal unsatisfiable.

As none of the vertices in $G = (\mathcal{F}_{ph3}, E_G)$ is reachable from itself there are no SCCs of size larger than one. Nevertheless the pigeon hole does not represent a worst-case for model rotation. Let us assume that $\mathcal{F}_{ph3}$ is given as an input to Alg. 1 and the first time line 3 is reached the algorithm chooses clause $\neg x_1 \vee \neg z_1$. The formula $\mathcal{F}_{ph3} \setminus \{\neg x_1, \neg z_1\}$ has a single satisfying assignment representing the case where pigeon x and z both sit in hole 1, and pigeon y sits in hole 2. The resulting execution of the `modelRotate` function is depicted in Table 2 by the recursive function call sequence above the double horizontal line.

The last call in the sequence of recursive function calls is the one performed for clause c_7. As $(c_7, c_2) \in E_G$ rotation of the assoc for c_7 is guaranteed to give an assoc for clause c_2, but as $c_2 \in M$ the condition on Line III of the pseudo-code of function `modelRotate` is not satisfied. Note that the assoc for c_2 that can be obtained by rotation from the assoc for c_7 is different from the assoc that was obtained in the first execution step by the rotation of the assoc for c_1. The steps under the double horizontal line in Table 2 depict how rotation of the new assoc for c_2 would allow finding an assoc for the two remaining clauses c_8 and c_9.

Obviously, if the pseudo-code of the `modelRotate` function is modified by simply removing the subexpression "**and** $c' \notin M$" from Line III then the function no longer terminates. One possible solution would be to store with each clause a list containing each assoc found for that clause, and perform model rotation for assocs that have not been rotated before. This would guarantee termination, but as one clause may have exponentially many assocs the length of the execution sequence may be non-linear in the size of the formula, and the amount of storage space required for the assocs would be substantial. A better approach is illustrated in the pseudo-code given in Alg. 3. Intuitively, the improved model

Table 2. Possible execution of function `modelRotate` for $\mathcal{F}_{ph3}$

$|M|$	l	Function call
1		`modelRotate`$(c_1, \{x_1, \neg x_2, \neg y_1, y_2, z_1, \neg z_2\})$
2	x_1	`modelRotate`$(c_2, \{\neg x_1, \neg x_2, \neg y_1, y_2, z_1, \neg z_2\})$
3	x_2	`modelRotate`$(c_3, \{\neg x_1, x_2, \neg y_1, y_2, z_1, \neg z_2\})$
4	y_2	`modelRotate`$(c_4, \{\neg x_1, x_2, \neg y_1, \neg y_2, z_1, \neg z_2\})$
5	y_1	`modelRotate`$(c_5, \{\neg x_1, x_2, y_1, \neg y_2, z_1, \neg z_2\})$
6	z_1	`modelRotate`$(c_6, \{\neg x_1, x_2, y_1, \neg y_2, \neg z_1, \neg z_2\})$
7	z_2	`modelRotate`$(c_7, \{\neg x_1, x_2, y_1, \neg y_2, \neg z_1, z_2\})$
7	x_2	`modelRotate`$(c_2, \{\neg x_1, \neg x_2, y_1, \neg y_2, \neg z_1, z_2\})$
8	x_1	`modelRotate`$(c_8, \{x_1, \neg x_2, y_1, \neg y_2, \neg z_1, z_2\})$
8	y_1	`modelRotate`$(c_4, \{x_1, \neg x_2, \neg y_1, \neg y_2, \neg z_1, z_2\})$
9	y_2	`modelRotate`$(c_9, \{x_1, \neg x_2, \neg y_1, y_2, \neg z_1, z_2\})$

Algorithm 3. Improved recursive model rotation

function `improvedModelRotate`(**clause** c, **assignment** a, **literal** l_r)

1. **if** $l_r \neq$ **undefined then** $seen[c, l_r] = $ **true**;
2. **for all** $l \in c$ such that $l \neq \neg l_r$ **do**
3. $a' = $ **rotate**$(a, \neg l)$
4.   **if** exactly one clause $c' \in \mathcal{F}$ is not satisfied by a' and $seen[c', l] = $ **false then**
5. $M = M \cup \{c'\}$ // **Has no effect if** c' **is already in** M
6.   `improvedModelRotate`(c', a', l)

---

rotation function allows a path that passes through the same clause multiple times as long as that clause is reached over edges with different labels. To use this improved model rotation function the MUS finding algorithm should initialize $seen[c, l] = $ **false** for all clauses $c \in \mathcal{F}$ and literals $l \in c$. Moreover, the call to `modelRotate`(c, a) on Line 6 of the MUS finding algorithm Alg. 1 should be replaced by `improvedModelRotate`$(c, a, $ **undefined**$)$. Note that for formula $\mathcal{F}_{ph3}$ the execution sequence of the improved model rotation function is exactly the sequence of all steps in Table 2.

During the review process of this work an article by the developers of model rotation appeared [1] in which they also observed that the termination condition of the model rotation function can be weakened. They proposed a parameterized function called *extended model rotation* which stores a limited number of assocs with every visited clause, but they were unable to achieve good performance using this function.

5   Parallelizing Model Rotation

Alg. 4 is an implementation of a MUS finding algorithm that can be used with an external incremental solver such as Tarmo [20]. The idea behind the Tarmo solver

---

Algorithm 4. Parallelized MUS finding algorithm

---

1. **generating thread**
   1.1 $i = 0$
   1.2 $M = \emptyset$
   1.3 **forever do**
   1.4   **if** $\mathcal{F} \neq M$ **then** $\mathcal{F}_i = \mathcal{F} \setminus \{c_i\}$ for some $c_i \in \mathcal{F} \setminus M$ **else return** $\mathcal{F}$
   1.5   **submit** formula $\mathcal{F}_i$ to solver
   1.6 $i = i + 1$

2. **result handling thread**
   2.1 **forever do**
   2.2   **if** solver reports result for some formula $\mathcal{F}_j$ **then**
   2.3   **if** formula $\mathcal{F}_j$ is **satisfiable then**
   2.4 $M = M \cup \{c_j\}$; `modelRotate`(c_j, a) where $a \in A(c_j, \mathcal{F}_j)$
   2.5   **else if** $\mathcal{F}_j \subset \mathcal{F}$ **then**
   2.6 $\mathcal{F} = \mathcal{F}_j$

---

is that given a sequence of related formulas encoded incrementally parallelization can be performed by solving multiple formulas from that sequence at the same time in parallel. This can be efficient for applications of incremental SAT solvers like Bounded Model Checking (BMC) [4] where the hardness of formulas in the sequence is varying. Inside Tarmo information gathered by each individual solver thread in the form of *conflict clauses* [14] is shared with the other solver threads. The clause sharing database design of Tarmo makes sure that this is performed correctly even when those solver threads are working on different formulas at the same time.

Algorithm 4 uses two threads that operate simultaneously in parallel. Thread 1 generates formulas and submits them to the external incremental SAT solver, Thread 2 handles the results reported by the SAT solver as they come in. The results may be reported out of order by the external solver, allowing parallelization by using the solver Tarmo.

A single line of pseudocode for a thread constitutes an atomic operation, which means that a thread can not observe state changes caused by the other thread during the execution of a single line. In fact, it is natural to think of the execution of the algorithm as a sequential *interleaving* of the execution steps of both threads. This means that both threads take turns and in each turn they execute one or more lines of their pseudocode. The execution order is assumed to be *fair* which means that no single thread executes forever without allowing the other thread to execute. The execution of the algorithm ends if and only if 'return $\mathcal{F}$' is reached on Line 1.4. Typical modern SAT solvers are complete, meaning that for any input formula they will eventually report an answer. Given those preconditions Alg. 4 is guaranteed to terminate.

If at most one formula at the time is submitted at the time to the external solver, i.e. after each formula submission Thread 1 does not execute until Thread 2 has handled the result for the submitted formula, then Alg. 4 behaves exactly like Alg. 1. The only expression in the pseudo-code of Alg. 4 that has no analogue in the pseudo-code for Alg. 1 is the constraint $\mathcal{F}_j \subset \mathcal{F}$ on Line 2.5. The extra constraint is required in this parallel algorithm because the input formula may contain more than one distinct MUS. As a consequence it is possible that two formulas $\mathcal{F}_1 \subset \mathcal{F}$ and $\mathcal{F}_2 \subset \mathcal{F}$ are both unsatisfiable, while $\mathcal{F}_1 \cap \mathcal{F}_2$ is not. The extra constraint ensures that the working formula $\mathcal{F}$ shrinks monotonically while remaining unsatisfiable. Of course also in an implementation of Alg. 4 the size of a non-minimal formula can be reduced faster if an unsatisfiable core is obtained from the solver.

Note that if two formulas are solved in parallel then the result of the formula that is solved first may imply that the effort made solving the second formula is wasted. This can happen either if both formulas are satisfiable in case the clauses that can be added to M due to the second result is a subset of those added before due to the first result. Or, if both formulas are unsatisfiable and as a result of solving the first formula the condition $F_j \subset \mathcal{F}$ is not met once the second formula is solved. The parallelization will thus have to perform well enough to make up for these "wasted" solver calls.

6 Experimental Results

In this section the performance of implementations of the algorithms Alg. 1 and Alg. 2 are evaluated. Both make use of the unsatisfiable core returned by the SAT solver to reduce the size of a non-minimal formula by more than one clause or SCC at the time. The implementation of Alg. 2 uses Tarjan's algorithm [19] to compute the SCCs.

Table 3. Instances solved and average solver calls over benchmarks solved by all

Model rotation	none			standard			improved		
	#	SAT	UNSAT	#	SAT	UNSAT	#	SAT	UNSAT
Alg. 1	-	-	-	136	2696.8	193.4	136	2346.1	188.3
Alg. 2	106	4389.2	162.5	123	3610.8	163.6	131	2341.9	172.4
Alg. 1 Tarmo 4	-	-	-	-	-	-	134	2660.9(88%)	436.0(42%)
Alg. 2 Tarmo 4	-	-	-	-	-	-	135	2448.4(95%)	377.2(43%)
Alg. 1 Tarmo 8	-	-	-	-	-	-	136	2963.0(79%)	796.5(23%)
Alg. 2 Tarmo 8	-	-	-	-	-	-	139	2605.7(90%)	700.6(23%)

The non-parallelized versions of the MUS finding algorithms use the SAT solver MiniSAT[4] version 2.2.0 [6]. The implementation of the MUS finding algorithm and the SAT solver are compiled at once into a single executable process. Parallelization of both algorithms was implemented using the scheme of Alg. 4 in combination with Tarmo as an external solver. Each of the parallel solver threads

---

[4] http://www.minisat.se

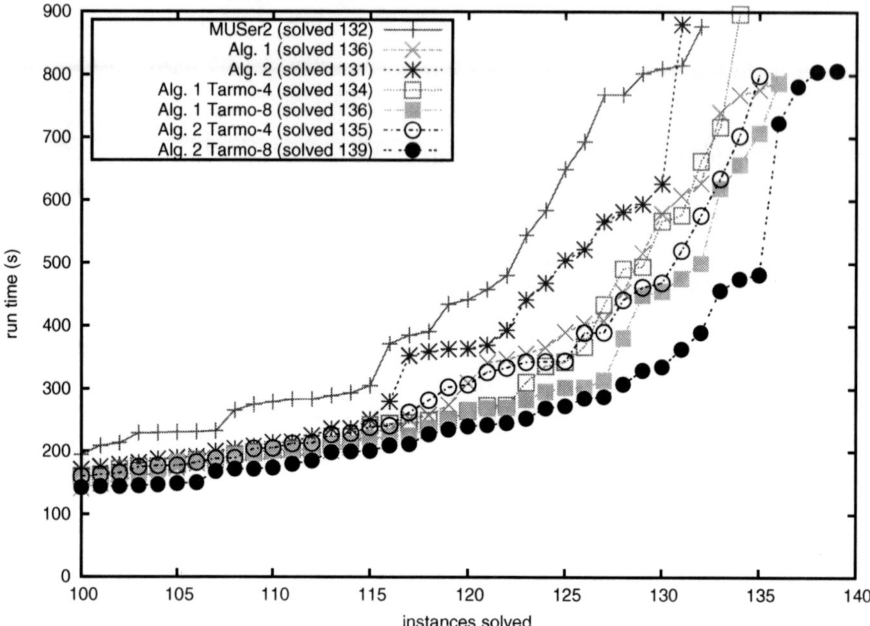

Fig. 3. Comparison of multiple versions

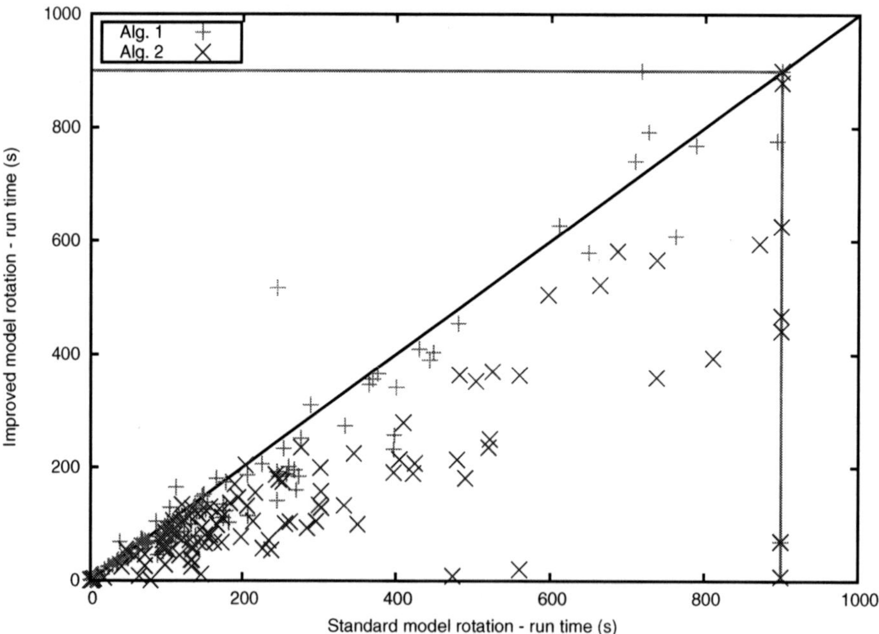

Fig. 4. Effect of improved model rotation

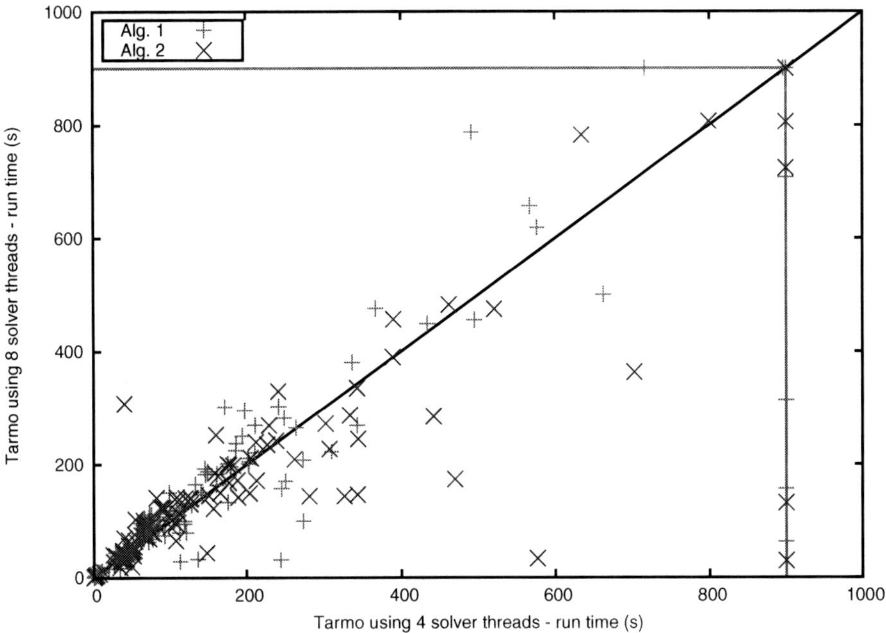

Fig. 5. Performance of 4 versus 8-threads

in Tarmo uses the same version of MiniSAT as the non-parallelized MUS finding algorithms. The MUS finder and the Tarmo solver are not linked together,
but rather Tarmo is executed as a separate process and communication is performed through UNIX pipes. This interface to Tarmo had been implemented
before to illustrate that also incremental SAT solvers can be used as stand-alone
"black-box" solvers. The downside is a rather substantial loss of performance[5].

All experiments were run in a computing cluster in which each computing
node has two six core Intel Xeon X5650 processors. Each separate result has
been obtained on a single such machine that was not performing any other
computation tasks at the same time. The chosen timeout for all experiments was
900 seconds, and the memory was limited to 2500MB per core. The benchmarks
used were a subset of the 800 benchmarks described in Sec. 3.1. This subset
contained the 178 benchmarks for which the classical destructive algorithm takes
more than two minutes (73 from [13], 104 from the competition).

In Table 3 the columns marked '#' denote the number of benchmarks solved.
Note that the performance of algorithm Alg. 2 is worse than that of the simpler
Alg. 1, at least for non-parallel versions. Fig. 4 shows the effect of the improve-
ment for model rotation function described in Sec. 4. Although this improvement
reduces the number of calls to the solver the impact on the execution time for
Alg. 1 is rather minimal. The improvement seems to however be important for the

---

[5] Using Tarmo with a single core Alg. 2 with improved model rotation solves 123
benchmarks instead of 131.

performance of Alg. 2, possibly as it explores path over edges in E_G before paths in E_P. It seems that picking the first arbitrary clause as Alg. 1 does on average leads to more effective executions than picking the first clause in an arbitrary root SCC. It is possible that this is caused by the existence of root SCCs of size 1, leading Alg. 2 to select a clause with no outgoing edges in E_G regularly.

Fig. 3 is a cactus plot, as used for example in the SAT competitions for comparing solver performance. This cactus plot shows a comparison between the implementations of both algorithms using different numbers of solver threads against the model rotation MUS finder MUSer2 [2,3]. The average number of solver calls used by each version is provided in Table 3, were calls with satisfiable and unsatisfiable result are listed separately in their respective columns. To allow fair comparison these averages are computed only over the results for the 103 benchmarks solved by all versions listed in the table. The percentages given for the parallel versions correspond to the percentage of solver calls that were actually effective to the progress of the algorithm.

Although Alg. 2 is slower than Alg. 1 in the single threaded case the SCC calculation is important for the effectiveness of their parallelizations. This is because the parallel version of Alg. 2 chooses clauses from different root SCCs to construct the formulas to be submitted in parallel. This has a much higher probability of leading to independently effective results than the strategy of Alg. 1 which constructs formulas testing the existing of an assoc for multiple arbitrary clauses in parallel. Fig. 5 illustrates this by showing that using 8 solver threads instead of 4 has more impact on the performance of the parallelization of Alg. 2 than on the performance of the parallel version of Alg. 1.

7 Conclusion

This paper studies a recently introduced technique [13,2] for improving MUS finding algorithms called *model rotation*. Model rotation was presented as a heuristic technique, but in this work it is proven that, when the input formula posses certain common properties, model rotation is in fact guaranteed to reduce the number of solver calls required by a destructive MUS finding algorithm. The presented statistics for a set of benchmarks designed for testing MUS finding algorithms illustrates why model rotation performs so well in practice.

The theoretical insights were followed by the presentation of an improvement for the model rotation technique. The third contribution of the paper is the presentation of a parallelization for MUS finding algorithms using model rotation that builds upon existing work for parallel incremental SAT solving.

Finally, the presented improvements and parallelization were empirically evaluated and shown to be effective.

References

1. Belov, A., Lynce, I., Marques-Silva, J.P.: Towards efficient MUS extraction (2012) (to appear in AI Communications)
2. Belov, A., Marques-Silva, J.P.: Accelerating MUS extraction with recursive model rotation. In: FMCAD, pp. 37–40 (2011)

3. Belov, A., Marques-Silva, J.P.: MUSer2: An efficient MUS extractor (2012), to appear in proceedings of Pragmatics of SAT (POS)
4. Biere, A., Cimatti, A., Clarke, E., Zhu, Y.: Symbolic Model Checking without BDDs. In: Cleaveland, W.R. (ed.) TACAS 1999. LNCS, vol. 1579, pp. 193–207. Springer, Heidelberg (1999)
5. Cook, S.A.: The complexity of theorem-proving procedures. In: Harrison, M.A., Banerji, R.B., Ullman, J.D. (eds.) STOC, pp. 151–158. ACM (1971)
6. Eén, N., Sörensson, N.: An Extensible SAT-solver. In: Giunchiglia, E., Tacchella, A. (eds.) SAT 2003. LNCS, vol. 2919, pp. 502–518. Springer, Heidelberg (2004)
7. Goldberg, E.I., Novikov, Y.: Verification of proofs of unsatisfiability for CNF formulas. In: DATE, pp. 10886–10891. IEEE Computer Society (2003)
8. Grégoire, É., Mazure, B., Piette, C.: On approaches to explaining infeasibility of sets of boolean clauses. In: ICTAI (1), pp. 74–83. IEEE Computer Society (2008)
9. Junker, U.: QUICKXPLAIN: Preferred explanations and relaxations for over-constrained problems. In: McGuinness, D.L., Ferguson, G. (eds.) AAAI, pp. 167–172. AAAI Press/The MIT Press (2004)
10. van Maaren, H., Wieringa, S.: Finding Guaranteed MUSes Fast. In: Kleine Büning, H., Zhao, X. (eds.) SAT 2008. LNCS, vol. 4996, pp. 291–304. Springer, Heidelberg (2008)
11. Marques-Silva, J.P., Planes, J.: Algorithms for maximum satisfiability using unsatisfiable cores. In: DATE, pp. 408–413. IEEE (2008)
12. Marques-Silva, J.P.: Minimal unsatisfiability: Models, algorithms and applications (invited paper). In: ISMVL, pp. 9–14. IEEE Computer Society (2010)
13. Marques-Silva, J., Lynce, I.: On Improving MUS Extraction Algorithms. In: Sakallah, K.A., Simon, L. (eds.) SAT 2011. LNCS, vol. 6695, pp. 159–173. Springer, Heidelberg (2011)
14. Marques-Silva, J.P., Sakallah, K.A.: GRASP: A search algorithm for propositional satisfiability. IEEE Trans. Computers 48(5), 506–521 (1999)
15. Oh, Y., Mneimneh, M.N., Andraus, Z.S., Sakallah, K.A., Markov, I.L.: AMUSE: a minimally-unsatisfiable subformula extractor. In: Malik, S., Fix, L., Kahng, A.B. (eds.) DAC, pp. 518–523. ACM (2004)
16. Papadimitriou, C.H., Wolfe, D.: The complexity of facets resolved. J. Comput. Syst. Sci. 37(1), 2–13 (1988)
17. Sinz, C.: Visualizing SAT instances and runs of the DPLL algorithm. J. Autom. Reasoning 39(2), 219–243 (2007)
18. Sinz, C., Kaiser, A., Küchlin, W.: Formal methods for the validation of automotive product configuration data. AI EDAM 17(1), 75–97 (2003)
19. Tarjan, R.E.: Depth-first search and linear graph algorithms. SIAM J. Comput. 1(2), 146–160 (1972)
20. Wieringa, S., Niemenmaa, M., Heljanko, K.: Tarmo: A framework for parallelized bounded model checking. In: Brim, L., van de Pol, J. (eds.) PDMC. EPTCS, vol. 14, pp. 62–76 (2009)

Revisiting Neighborhood Inverse Consistency on Binary CSPs

Robert J. Woodward[1], Shant Karakashian[1], Berthe Y. Choueiry[1], and
Christian Bessiere[2]

[1] Constraint Systems Laboratory, University of Nebraska-Lincoln, USA
{rwoodwar,shantk,choueiry}@cse.unl.edu
[2] LIRMM-CNRS, University of Montpellier, France
bessiere@lirmm.fr

Abstract. Our goal is to investigate the definition and application of
strong consistency properties on the dual graphs of binary Constraint
Satisfaction Problems (CSPs). As a first step in that direction, we study
the structure of the dual graph of binary CSPs, and show how it can
be arranged in a triangle-shaped grid. We then study, in this context,
Relational Neighborhood Inverse Consistency (RNIC), which is a consistency property that we had introduced for non-binary CSPs [17]. We
discuss how the structure of the dual graph of binary CSPs affects the
consistency level enforced by RNIC. Then, we compare, both theoretically and empirically, RNIC to Neighborhood Inverse Consistency (NIC)
and strong Conservative Dual Consistency (sCDC), which are higher-
level consistency properties useful for solving difficult problem instances.
We show that all three properties are pairwise incomparable.

1   Introduction

Enforcing consistency properties on Constraint Satisfaction Problems (CSPs)
allows us to effectively prune the exponential search space of these problems,
and constitutes one of the most significant contributions of Constraint Programming (CP). While lower level consistencies, such as Arc Consistency (AC) [15],
are often sufficient for solving easy problems, solving difficult problems often
requires enforcing higher consistency levels. To facilitate solving difficult CSPs,
we propose, as a research goal, to investigate the effectiveness of enforcing higher
levels of consistency on the *dual* graphs of binary CSPs.

   To this end, we first study the structure of the *dual graph* of binary CSPs
and show that it exhibits an interesting triangle-shaped grid that, in general,
may affect the 'level' of the consistency property enforced and the operation
of the algorithms for enforcing it. Then, we focus our attention on Relational
Neighborhood Inverse Consistency (RNIC) [17], a consistency property that we
had proposed and evaluated as an extension of Neighborhood Inverse Consistency (NIC) introduced by Freuder and Elfe [8]. We show how the structure of
the dual graph of a binary CSP affects the consistency level enforced by RNIC,

M. Milano (Ed.): CP 2012, LNCS 7514, pp. 688–703, 2012.

characterizing the conditions where RNIC cannot be stronger than another relational consistency property that we had defined in [11] when both properties are enforced on binary CSPs. In order to characterize the effectiveness of RNIC on binary CSPs despite the identified limitation imposed by the structure, we turn our attention back to 'strong' consistency properties defined for binary CSPs, and compare RNIC, both theoretically and empirically, to NIC and strong Conservative Dual Consistency (sCDC) [14], showing that all three properties are incomparable.

This paper is structured as follows. Section 2 reviews background information about CSPs. Section 3 discusses the structure of the dual graph of a binary CSP, mainly, that it is a triangle-shaped grid. Section 4 discusses RNIC on binary CSPs. Section 5 reviews the state of the art in relational consistency. Section 6 discusses experimentally the filtering power of NIC, sCDC, and RNIC on binary CSPs. Finally, Section 7 concludes this paper.

2   Background

A Constraint Satisfaction Problem (CSP) is defined by $\mathcal{P} = (\mathcal{V}, \mathcal{D}, \mathcal{C})$ where $\mathcal{V}$ is a set of variables, $\mathcal{D}$ is a set of domains, and $\mathcal{C}$ is a set of constraints. Each variable $V_i \in \mathcal{V}$ has a finite domain $D_i \in \mathcal{D}$, and is constrained by a subset of the constraints in $\mathcal{C}$. Each constraint $C_i \in \mathcal{C}$ is specified by a relation R_i defined on a subset of the variables, called the scope of the relation and denoted $scope(R_i)$. Given a relation R_i, a tuple $\tau_i \in R_i$ is a vector of allowed values for the variables in the scope of R_i. Solving a CSP corresponds to finding an assignment of a value to each variable such that all the constraints are satisfied. The dual encoding of a CSP, $\mathcal{P}$, is a binary CSP whose variables are the relations of $\mathcal{P}$, their domains are the tuples of those relations, and the constraints enforce *equalities* over the shared variables.

2.1   Graphical Representations

A binary CSP is graphically represented by its *constraint graph* where the vertices are the variables of the CSP and the edges represent the relations [6]. $\text{Neigh}(V_i)$ denotes the set of variables adjacent to a variable V_i in the constraint graph. The *dual graph* of a CSP is a graph whose vertices represent the relations of the CSP, and whose edges connect two vertices corresponding to relations whose scopes overlap. $\text{Neigh}(R_i)$ denotes the set of relations adjacent to a relation R_i in the dual graph. Janssen et al. [10] and Dechter [6] observed that, in the dual graph, an edge between two vertices is *redundant* if there exists an alternate path between the two vertices such that the shared variables appear in every vertex in the path. Redundant edges can be removed without modifying the set of solutions. Janssen et al. introduced an efficient algorithm for computing the *minimal dual graph* [10]. Many minimal graphs may exist, but all are guaranteed to have the same number of edges. Figure 1 shows the constraint, dual graph, and a minimal dual graph of a small CSP.

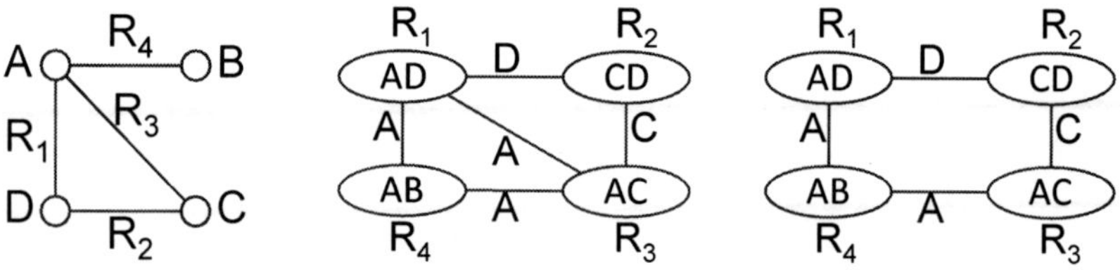

Fig. 1. A constraint graph, dual graph, and minimal dual graph

2.2 Consistency Properties and Algorithms

CSPs are in general $\mathcal{NP}$-complete and solved by search. To reduce the severity of the combinatorial explosion, they are usually 'filtered' by enforcing a given local consistency property [2]. One common property is Arc Consistency (AC). A CSP is arc consistent iff for every binary constraint, any value in the domain of one variable can be extended to the domain of the other variable while satisfying the constraint, and vice versa. The more difficult the CSP, the larger is its search space, and the more advantageous it is to enforce consistency properties of higher levels. In fact, Freuder provided a sufficient condition for guaranteeing a backtrack-free search that links the level of consistency to a structural parameter of the CSP [7]. However, enforcing higher-level consistencies may add constraints and modify the structure of the problem. For this reason, we focus, in this paper, on higher-level consistency properties for binary CSPs that do not modify the graphical representations of a problem.

Freuder and Elfe introduced Neighborhood Inverse Consistency (NIC), which ensures that each value in the domain of a variable can be extended to a solution of the subproblem induced by the variable and all the variables in its neighborhood in the constraint graph [8]. NIC is defined on binary CSPs and the algorithm for enforcing it operates on the constraint graph. RNIC ensures that any tuple in any relation can be extended in a consistent assignment to all the relations in its neighborhood in the dual graph. Enforcing NIC (RNIC) does not modify the constraint graph (dual graph). Further, it exploits the structure of the problem to focus the pruning on where a variable (relation) most tightly interacts with the problem. Thus, the topology of the constraint graph (dual graph) of a problem can determine the level of consistency enforced.

As extensions to RNIC, we also proposed wRNIC, triRNIC, and wtriRNIC, which modifies the structure of the dual graph but not the CSP solution set [17]. wRNIC is defined on a minimal dual graph; triRNIC is defined on a triangulated dual graph;[1] and wtriRNIC is defined on a triangulation of a minimal dual graph. We gave a selection strategy, selRNIC, for automatically determining which RNIC variation to use based on the density of the dual graph.[2] We

---

[1] Graph triangulation adds an edge (a chord) between two non-adjacent vertices in every cycle of length four or more [9]. While minimizing the number of edges added by the triangulation process is NP-hard, MINFILL is an efficient heuristic commonly used for this purpose [12,6].

[2] The density of a graph $G = (V, E)$ is considered to be $\frac{2|E|}{|V|(|V|-1)}$.

showed that selRNIC statistically dominates all other RNIC properties. We have also studied m-wise consistency, which we denoted R$(*,m)$C, and proposed the first algorithm for enforcing it [11]. R$(*,m)$C ensures that every tuple in every relation can be extended in a consistent assignment to every combination of $m-1$ relations in the problem. In this paper, we use the knowledge about the structure of the dual graph of binary CSPs to formally characterize the relationship between RNIC and R$(*,m)$C.

Strong consistency properties for binary CSPs that do not affect the topology of the constraint graph have been carefully reviewed, studied, and compared to each others [5,14]. Such properties include maxRPC [3], SAC [4], CDC [13], and, the strongest of them all, sCDC [14]. Further, Lecoutre et al. show that, on binary CSPs, strong Conservative Dual Consistency (sCDC) is equal SAC+CDC [14].[3] While NIC was shown to be incomparable to SAC [5], its relationship to sCDC has not yet been addressed [13]. In this paper, we complete the comparison of NIC, RNIC, and sCDC, and show that they are, both theoretically and empirically, pairwise incomparable. Thus, our results contribute to the characterization of strong consistency properties for binary CSPs.

When enforcing a relational consistency property, we always terminate the process by filtering the variables' domains by projecting on them the filtered relations. For RNIC, we call the resulting consistency property RNIC+DF (domain filtering). To compare a consistency property p_i defined on the relations of a CSP to another one defined on the variables, we always consider p_i+DF. To compare two consistency properties p and p', we use the following terminology [4]:

- p is *stronger* than p' if, in any CSP where p holds, p' also holds.
- p is *strictly stronger* than p' if p is stronger than p' and there exists at least one CSP in which p' holds but p does not.
- p and p' are *equivalent* when p is stronger than p' and vice versa.
- Finally, p and p' are *incomparable* when there exists at least one CSP in which p holds but p' does not, and vice versa.

In practice, when a consistency property is stronger (respectively, weaker) than another, enforcing the former never yields less (respectively, more) pruning than enforcing the latter on the same problem.

3   Structure of the Dual Graph of Binary CSPs

The structure of the dual graph determines the neighborhoods of its vertices (i.e., CSP relations) and may affect the level of relational consistency that can be enforced on the CSP. We first discuss the case of a binary CSP with a complete constraint graph, showing that the structure of its dual graph can be arranged

---

[3] Singleton Arc Consistency (SAC) ensures that a binary CSP remains AC after instantiating any single variable to any value in the variable's domain [4]. Conservative Dual Consistencies (CDC) ensures that for every instantiation of two variables in the scope of some constraint, $\{(V_i, a), (V_j, b)\}$, that b remain in the domain of V_j in the arc-consistent CSP where a is assigned to V_i and vice versa [13].

in a triangle-shaped grid. We show that redundant edges can be removed in a way to maintain the grid structure. We then discuss the case of a binary CSP with a non-complete constraint graph, and show that its dual graph can also be arranged in a triangle-shaped grid but with fewer vertices and a less regular shape than that of a CSP with complete constraint graph. We discuss the effects of the dual-graph structure on RNIC in Section 4.

3.1   Binary CSP with a Complete Constraint Graph

Theorem 1. *The $\frac{n(n-1)}{2}$ vertices of the dual graph of a binary CSP of n variables whose constraint graph is complete such as the one shown in Figure 2 (i.e., forms a clique of n vertices, K_n), can be arranged in an $(n-1) \times (n-1)$ triangle-shaped grid where:*

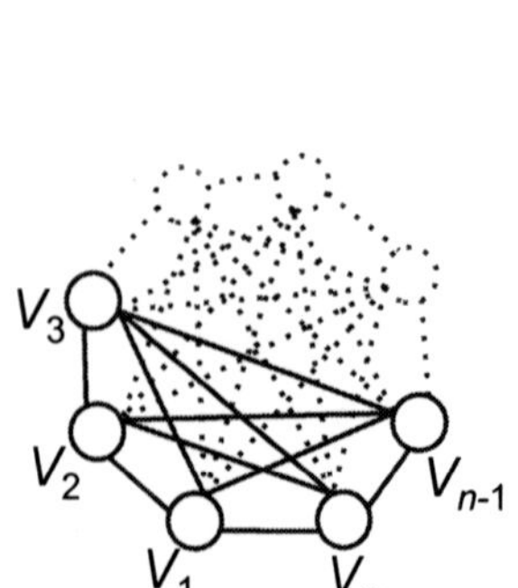

Fig. 2. A complete constraint graph of n vertices

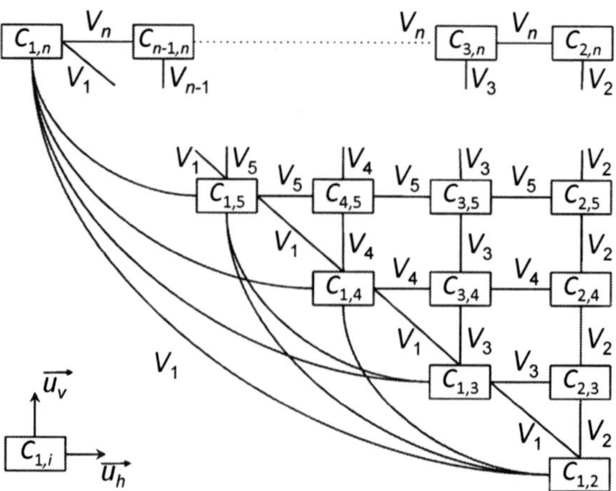

Fig. 3. Dual graph corresponding to the CSP in Figure 2

1. *The $n-1$ vertices on the diagonal of the triangle correspond to the constraints over the variable V_1. They are denoted $C_{1,i}$ where $i \in [2,n]$ and completely connected. The connecting edges are labeled with V_1.*
2. *The $n-1$ vertices corresponding to the constraints over variable $V_{i\geq2}$ are located along the path in the grid shown in Figure 4 and specified as follows:*
   - *Considering the coordinate system defined by the horizontal and vertical unit vectors $\boldsymbol{u_h}$, $\boldsymbol{u_v}$ and centered on $C_{1,i}$,*
   - *$i-2$ vertices are lined up along the horizontal axis $\boldsymbol{u_h}$, and*
   - *$n-i$ vertices are lined up along the vertical axis $\boldsymbol{v_h}$.*
   - *Those $n-1$ vertices are completely connected, and the connecting edges are labeled with V_i. (For the sake of clarity, Figure 3 does not show all the edges of the dual graph: only all the edges labeled V_1 are shown on the diagonal of the grid.)*

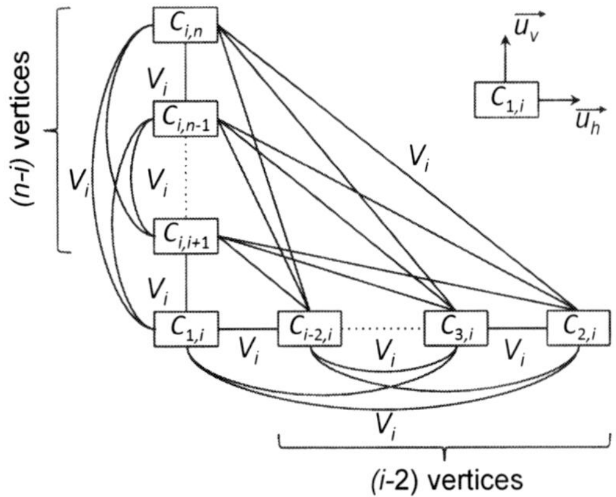

Fig. 4. The path for the constraints over variables $V_{i \geq 2}$ of the grid of Figure 3

Proof: See Appendix A.   □

Corollary 1. *After the removal of redundant edges, the dual graph of a binary CSP of n variables whose constraint graph is complete can be arranged in a $(n-1) \times (n-1)$ triangle-shaped grid, where every CSP variable annotates the edges of a chain of length $n-2$.*

Proof: See Appendix B.   □

Because redundancy removal is not unique, not all minimal dual graphs necessarily yield a triangle-shaped grid as we show using a counter-example. One possible minimal dual graph for the complete constraint graph of five vertices of Figure 5 is shown in Figure 6. In this example, there is a cycle of size six in

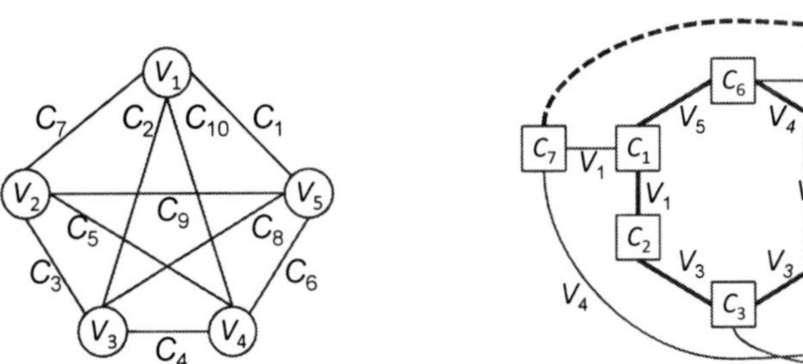

Fig. 5. A complete graph with five variables

Fig. 6. A minimal dual graph of Figure 5, which does not form a grid

the dual graph, indicated by the bold lines in Figure 6. Thus, the dual graph is not a grid. Further, the variable V_2 does not annotate a chain, but a star, as indicated by the dotted lines in the dual graph.

3.2   Binary CSP with a Non-complete Constraint Graph

In a binary CSP with a non-complete constraint graph, the dual graph can be thought of as the complete binary constraint graph with some missing vertices. Because, in the dual graph of any complete constraint graph, all the vertices corresponding to the constraints that apply to a given CSP variable are completely connected, it is always possible, even in the case of a CSP with a non-complete constraint graph, to form, in its corresponding dual graph, a chain connecting vertices related to the same variable. However, the length of such a chain may be less than $n-2$. Thus, the triangle-shaped grid can be preserved. For example, consider the binary CSP with $n = 5$ variables given in Figure 7. A minimal dual

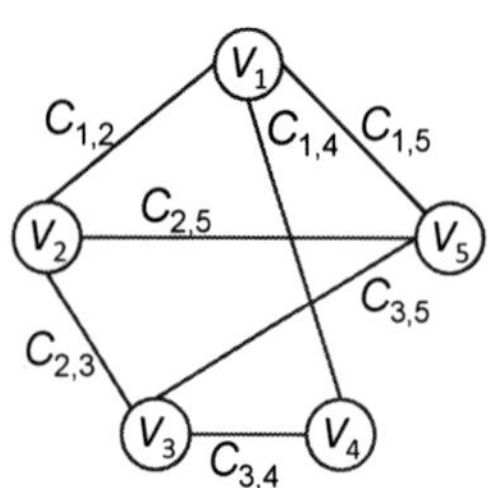

Fig. 7. A constraint graph with five variables

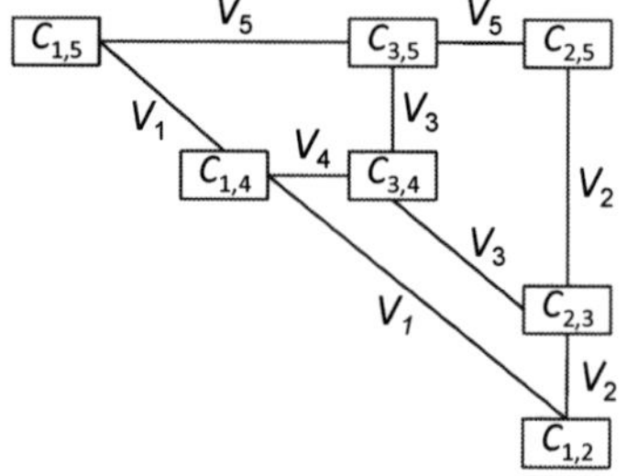

Fig. 8. The minimal dual graph of Figure 7

graph for that binary CSP is given in Figure 8, which was constructed from the dual graph for the complete CSP by removing the vertices corresponding to the constraints that are not in the CSP. Again, because the minimal dual graph is not unique, there exists minimal dual graphs that do not favor the chains, and thus, are not triangle-shaped grids.

4   RNIC on Binary CSPs

Knowing the structure of the dual graph of a binary CSP, we first prove limitations of the filtering power of RNIC and wRNIC (RNIC enforced on a minimal dual graph). Then we theoretically compare sCDC, RNIC, and NIC, showing that they are pairwise incomparable.

4.1   Effects of the Dual-Graph's Structure on RNIC

Theorem 2. *RNIC, R(*,2)C, and R(*,m)C are equivalent on any dual graph that is tree structured or is a cycle of length $\geq maximum(4, m + 1)$.*

Proof: By straightforward generalization of Theorem 5 in [17].   □

Any redundancy-free dual graph of an arbitrary binary CSP can contain only one or more of the following configurations:

1. A cycle of length four, on a grid-shaped dual graph
2. A cycle of length larger than four as shown in Figure 6.
3. A triangle along the diagonal.

In the first two cases above, enforcing RNIC on the minimal dual graph (wRNIC) is equivalent to R(*,2)C by Theorem 2. On the third case, wRNIC is equivalent to R(*,3)C.

Theorem 3. *On a binary CSP, wRNIC is never strictly stronger than R(*,3)C.*

Proof: See Appendix C. $\square$

Using an algorithm for enforcing RNIC to enforce R(*,3)C is wasteful of resources. Indeed, the former executes more consistency-checking operations than needed to enforce R(*,3)C given that the neighborhoods considered by the former are supersets of those considered by the latter.

4.2   Comparing sCDC, RNIC and NIC

Theorem 4. *On binary CSPs, sCDC and RNIC+DF are incomparable.*

Proof: In Figure 9, the CSP is RNIC+DF but not sCDC. sCDC empties all

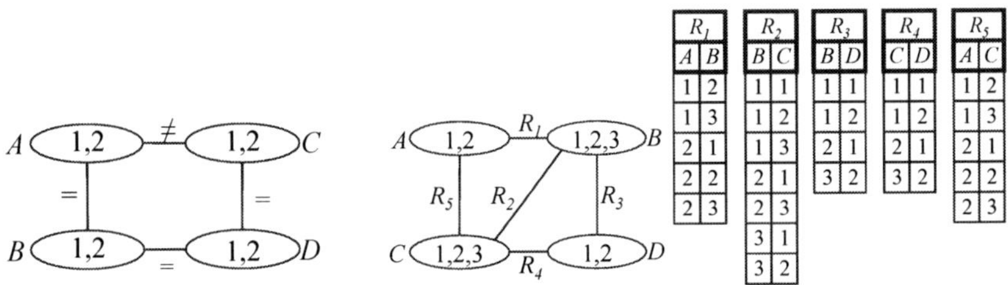

R_1		R_2		R_3		R_4		R_5	
A	B	B	C	B	D	C	D	A	C
1	2	1	1	1	1	1	1	1	2
1	3	1	2	1	2	1	2	1	3
2	1	1	3	2	1	2	1	2	1
2	2	2	1	3	2	3	2	2	2
2	3	2	3					2	3
		3	1						
		3	2						

Fig. 9. RNIC+DF but not sCDC	**Fig. 10.** sCDC but not RNIC+DF

variables domains. In Figure 10, borrowed from Debruyne and Bessière [5], the CSP is sCDC but not RNIC+DF. RNIC removes $\{(2,3),(3,2)\}$ from R_2, $\{(1,2),(1,3)\}$ from R_1, and $\{(1,2),(1,3)\}$ from R_5. Therefore, RNIC+DF removes the value 1 from A. $\square$

Theorem 5. *On binary CSPs, sCDC and NIC are incomparable.*

Proof: In Figure 9, the CSP is NIC but not sCDC. In Figure 11, borrowed from Debruyne and Bessière [5], the CSP is sCDC but not NIC. NIC removes the value 1 from A. $\square$

Theorem 6. *On binary CSPs, NIC and RNIC+DF are incomparable.*

Proof: In Figure 12, the CSP is NIC but not RNIC+DF. RNIC removes the tuples in $\{(0,2),(2,2)\}$ from R_0, $\{(0,0),(1,2)\}$ from R_1, $\{(0,2)\}$ from R_2, $\{(0,2)\}$ from R_3, and $\{(0,1),(2,1)\}$ from R_4. Therefore, RNIC+DF removes the value 0 from A. In Figure 13, the CSP is RNIC+DF but not NIC. NIC removes the value 0 from D. $\square$

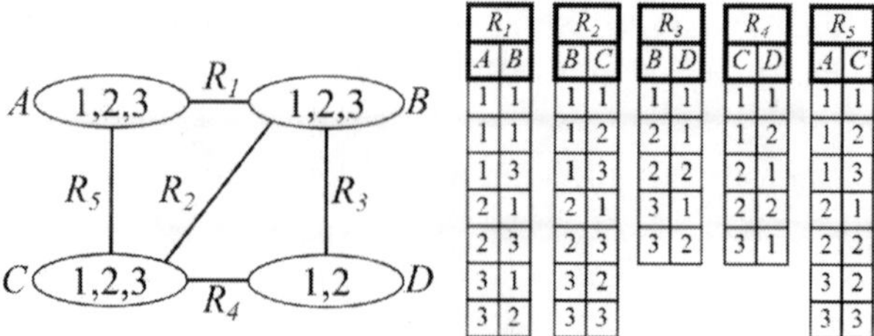

Fig. 11. sCDC but not NIC

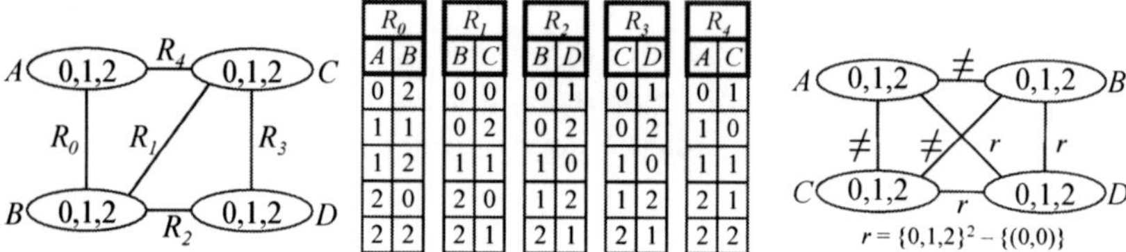

Fig. 12. NIC but not RNIC+DF   **Fig. 13.** RNIC+DF but not NIC

5 Related Work

Most of the related work is discussed in Section 2.2. To the best of our knowledge, no prior work has investigated the structure of the dual graph of binary CSPs.

Bacchus et al. studied the application of NIC to the dual encoding of a CSP, which they denoted *nic(dual)* [1]. While this property is identical to RNIC, the paper did not go beyond stating that *nic(dual)* is strictly stronger than *ac(dual)*.

Despite its pruning power and light space overhead, NIC received relatively little attention in the literature, likely because of the prohibitive cost of the algorithm for enforcing it. Freuder and Elfe tested their algorithm in a preprocessing step to backtrack search for solving binary instances whose constraint density did not exceed 4.25% [8]. Debruyne and Bessière showed that NIC is ineffective on sparse graphs and too costly on dense graphs [5].

6 Experimental Results

We study the impact of enforcing consistency on binary CSPs in a pre-processing step, prior to search. We then find the first solutions of the CSPs with backtrack search using a dynamic domain/weighted-degree variable (dom/wdeg) ordering and MAC [16] for full lookahead. We consider the four consistency properties: AC, sCDC, NIC and selRNIC. We measure the number of visited during search and the total CPU time (i.e., pre-processing and search). In practice, out of the five RNIC consistency algorithms, *the performance of only selRNIC matters* because it enforces the algorithm chosen by a selection policy [17].

We ran our experiments on benchmarks from the CSP Solver Competition.[4] We limited the processing time per instance to 90 minutes. We tested 571

---

[4] All constraints are normalized http://www.cril.univ-artois.fr/CPAI09/.

Table 1. CPU time: Search (MAC+dom/wdeg) with pre-processing by AC3.1, sCDC1, NIC, and selRNIC

Benchmark	#Instances	AC3.1	sCDC1	NIC	selRNIC
		CPU Time (msec)			
		NIC Quickest			
bqwh-15-106	100/100	3,505	3,860	**1,470**	3,608
bqwh-18-141	100/100	68,629	82,772	**38,877**	77,981
graphColoring-sgb-queen	12/50	680,140	(+3) -	(+9) **57,545**	634,029
graphColoring-sgb-games	3/4	41,317	33,307	(+1) **860**	41,747
rand-2-23	10/10	1,467,246	1,460,089	**987,312**	1,171,444
rand-2-24	3/10	**567,620**	677,253	(+7) 3,456,437	677,883
		sCDC1 Quickest			
driver	2/7	(+5) 70,990	(+5) **17,070**	358,790	(+4) 185,220
ehi-85	87/100	(+13) 27,304	(+13) **573**	513,459	(+13) 75,847
ehi-90	89/100	(+11) 34,687	(+11) **605**	713,045	(+11) 90,891
frb35-17	10/10	41,249	**38,927**	179,763	73,119
		selRNIC Quickest			
composed-25-1-25	10/10	226	335	1,457	**114**
composed-25-1-2	10/10	233	283	1,450	**88**
composed-25-1-40	9/10	(+1) 288	(+1) 357	120,544	(+1) **137**
composed-25-1-80	10/10	223	417	(+1) -	**190**
composed-75-1-25	10/10	2,701	1,444	363,785	**305**
composed-75-1-2	10/10	2,349	1,733	48,249	**292**
composed-75-1-40	7/10	(+3) 1,924	(+3) 1,647	631,040	(+3) **286**
composed-75-1-80	10/10	1,484	1,473	(+1) -	**397**

instances of binary CSPs, with density in [1%,100%] with an average of 18%. We report the following measures:

- *#Instances*: We report two numbers for each benchmark. The first number is the number of instances completed by all four algorithms. The second number is the total number of instances in the benchmark.
- *CPU Time (msec)*: The average CPU time in milliseconds for pre-processing and search, computed over only the instances completed by all algorithms. The number of additional instances solved above the number completed by all is given in parenthesis.
- *BT-Free*: The number of instances that were solved backtrack-free.
- *#NV*: the average number of nodes visited for search, computed over only the instances that were completed by all algorithms.

In the results tables, we highlight in boldface the best values. When one of the algorithms does not complete any instances within the time threshold, no averages can be computed. To obtain averages over these instances, we compute the averages over only the algorithms that 'completed,' and mark in gray the box corresponding to the ignored algorithm. Table 1 shows the CPU times for the

Table 2. Number of nodes visited (#NV): Search (MAC+dom/wdeg) with pre-processing by AC3.1, sCDC1, NIC, and selRNIC

Benchmark	#Instances	AC3.1	sCDC1	NIC	selRNIC	AC3.1	sCDC1	NIC	selRNIC
		BT-Free				#NV			
		NIC Quickest							
bqwh-15-106	100/100	0	3	**8**	5	1,807	1,881	**739**	1,310
bqwh-18-141	100/100	0	0	**1**	0	25,283	25,998	**12,490**	22,518
graphColoring-sgb-queen	12/50	1	-	**16**	1	91,853	-	**15,798**	91,853
graphColoring-sgb-games	3/4	1	1	**4**	1	14,368	14,368	**40**	14,368
rand-2-23	10/10	0	0	**10**	0	471,111	471,111	**12**	471,111
rand-2-24	3/10	0	0	**10**	0	222,085	222,085	**24**	222,085
		sCDC1 Quickest							
driver	2/7	**1**	**2**	**1**	**1**	3,893	**409**	3,763	3,763
ehi-85	87/100	0	**100**	87	**100**	1,425	**0**	**0**	**0**
ehi-90	89/100	0	**100**	89	**100**	1,298	**0**	**0**	**0**
frb35-17	10/10	**0**	**0**	**0**	**0**	24,491	24,491	24,491	**24,346**
		selRNIC Quickest							
composed-25-1-25	10/10	0	**10**	**10**	**10**	153	**0**	**0**	**0**
composed-25-1-2	10/10	0	**10**	**10**	**10**	162	**0**	**0**	**0**
composed-25-1-40	9/10	0	**10**	9	**10**	172	**0**	**0**	**0**
composed-25-1-80	10/10	0	**10**	-	**10**	112	**0**	-	**0**
composed-75-1-25	10/10	0	**10**	**10**	**10**	345	**0**	**0**	**0**
composed-75-1-2	10/10	0	**10**	**10**	**10**	346	**0**	**0**	**0**
composed-75-1-40	7/10	0	**10**	7	**10**	335	**0**	**0**	**0**
composed-75-1-80	10/10	0	**10**	-	**10**	199	**0**	-	**0**

tested benchmarks, and Table 2 shows the number of instances solved backtrack-free and the number of nodes visited. We split the results into three sections based on average CPU time: those where NIC performs well, those where sCDC1 performs well, and those where selRNIC performs well.

While NIC may be too costly to use in general [5], there are difficult problems that benefit from higher level consistency. On instances where NIC performs the quickest in terms of CPU time, its search has orders of magnitude lower nodes visited than the other algorithms. Interestingly, NIC performs well on the rand-2-23/24 benchmarks, where the density is 100% (there is a constraint between every pair of variables). This result is interesting, because NIC is solving the instance during pre-processing. Therefore, it is solving every instance backtrack-free. Note, that on rand-2-24, despite taking a large amount of CPU time, NIC solves 7 additional instances that no other tested algorithm solved. The average density for the other benchmarks where NIC performs well is 14%.

On instances where sCDC1 performs the quickest in terms of CPU time, it is able to filter the instances very quickly. The frb35-17 benchmark has an average density of 44%. This large density explains why NIC is performing poorly on these instances. An interesting benchmark is ehi-85/90 whose instances are all unsatisfiable. Interestingly, sCDC1 detects unsatisfiability at pre-processing by a domain wipe-out of the very first variable that it checks. Thus, its speed. selRNIC and (to a lesser extent) NIC, detect unsatisfiability at pre-processing, but cost more effort than sCDC1. Note that AC3.1 is too weak to uncover inconsistency at pre-precessing.

Interestingly, selRNIC automatically selected RNIC, and not wRNIC, on all tested benchmarks, except rand-2-23/24 where it selected wRNIC, thus not hindering itself as predicted by Theorem 3. selRNIC performs well on the composed benchmarks, where all of the algorithms, except AC3.1, were able to detect unsatisfiability at pre-processing. For the composed-25 benchmarks, the average density of the CSP is 50% (the average dual graph density is 12%), and for the composed-75 benchmarks, the average density of the CSP is 20% (the average dual graph density is 5%). The large densities of the composed-75 benchmark explain the poor performance of NIC.

7   Conclusions and Future Work

An important contribution of this paper is the understanding of the structure of the dual graph on binary CSPs, which should impact the development of future consistency algorithms that operate on the dual graph of binary CSPs. We also theoretically showed that NIC, sCDC, and RNIC are incomparable. Despite previous work showing that NIC may be too costly to use in general [5], our experimental results show that there are instances that benefit from higher level consistency.

The algorithm we use to remove redundant edges from a dual graph generates triangle-shaped grids for binary CSPs [10]. However, there may also be non-grid shaped minimal dual graphs. We propose to investigate why this algorithm favors the triangle-shaped grids. We also propose to develop a portfolio-based algorithm that measure the structure of a constraint graph and of its dual graph to select the appropriate consistency property to enforce.

Acknowledgments. Experiments were conducted on the equipment of the Holland Computing Center at the University of Nebraska-Lincoln. Robert Woodward was partially supported by a National Science Foundation (NSF) Graduate Research Fellowship grant number 1041000. This research is supported by NSF Grant No. RI-111795.

A   Proof of Theorem 1

(By induction of number of variables.)

Base Step: Stated for $n = 3$.

For $n = 3$, the constraint graph is shown in Figure 14 and the corresponding dual graph in Figure 15. The dual graph is obviously a triangle.

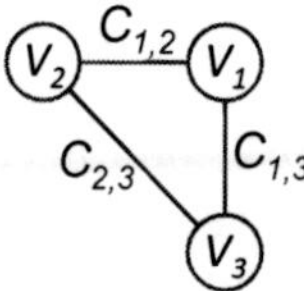

Fig. 14. A complete constraint graph with 3 variables

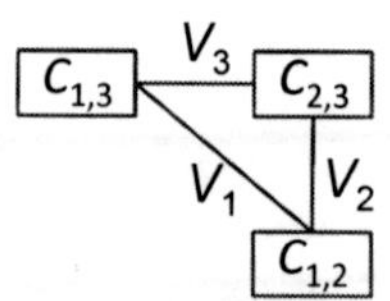

Fig. 15. The dual graph of a complete constraint graph with 3 variables

- The two vertices corresponding to the constraints over the variable V_1 form the diagonal.
- The two vertices corresponding to the constraints over V_2 start at $C_{1,2}$ and have 0 vertices along the horizontal axis, and one vertex along the vertical axis. Also, the two vertices corresponding to the constraints over V_3 start at $C_{1,3}$ have 0 vertices along the horizontal axis, and one vertex along the vertical axis.

Inductive Step: Assume that the theorem holds for a CSP with k variables (inductive hypothesis). Show the theorem holds for a CSP with $k + 1$ variables (inductive step).

Consider the complete constraint graph of a CSP with k variables, which is the clique K_k. By the inductive hypothesis, the dual graph can be arranged in the triangle-shaped grid. Now, add the variable V_{k+1} to the CSP. In order to connect V_{k+1} to all k variables, k constraints are added to the constraint graph of the CSP, as shown in Figure 16. Namely, these k constraints are $C_{i,k+1}, \forall i \leq k$. Place the dual variables as follows, going from right to left in Figure 17:

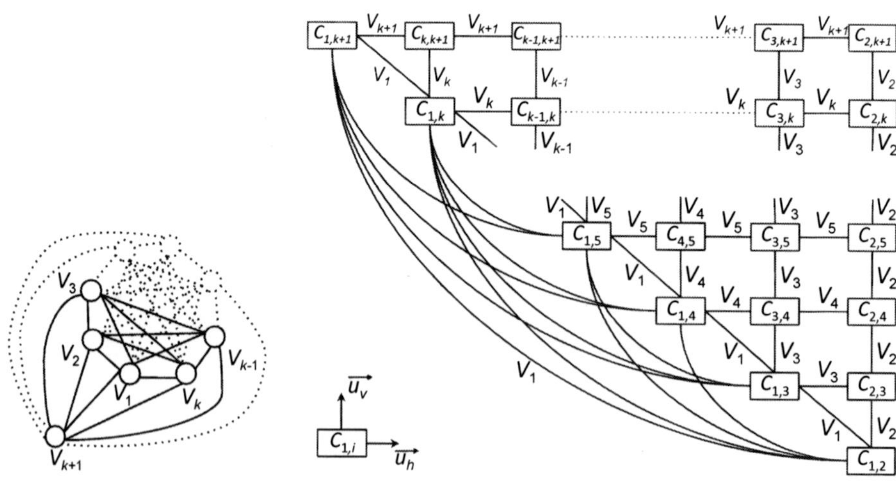

Fig. 16. A complete graph with $k + 1$ variables

Fig. 17. The dual graph of Figure 16

- $C_{i,k+1}, i \in [2, k-1]$ is placed above $C_{i,k}$,
- $C_{k,k+1}$ is placed above $C_{1,k}$, and
- $C_{1,k+1}$ is placed to the left of $C_{k,k+1}$.

This arrangement yields a dual graph that is a triangle-shaped grid because:

- The vertices corresponding to the constraints over the variable V_1 are located on the diagonal of the triangle because $C_{k+1,1}$ is to the left of $C_{k+1,k}$,
- The coordinate system centered on $C_{1,i\in[2,k]}$ increases by one vertical unit for vertex $C_{k+1,i}$ and labeled with variable V_i.
- The coordinate system centered on $C_{1,k+1}$ has $(k+1) - 2 = k - 1$ vertices on the horizontal axis and 0 vertices in the vertical axis. The k vertices on the top row of the triangle form a clique whose edges are labeled with V_{k+1} (shown partially, for readability).

Consequently, this new dual graph of a complete constraint graph of $k + 1$ variables has the topology of a triangle-shaped grid. $\qquad\square$

B   Proof of Corollary 1

Let us consider the $n - 1$ vertices corresponding to the constraints that apply on variable V_i and the coordinate system defined by the horizontal and vertical unit vectors $\boldsymbol{u_h}$, $\boldsymbol{u_v}$ and centered on $C_{1,i}$. All edges between the $i - 2$ horizontal vertices and the $n - i$ vertical vertices that link two non-consecutive vertices are redundant and can be removed, leaving a path linking the $n - 1$ vertices along the horizontal and vertical axis. As for V_1, a similar operation can be applied to the vertices along the diagonal of the triangle. $\qquad\square$

C   Proof of Theorem 3

(By contradiction) Assume that wRNIC is strictly stronger than R($*$,3)C, that is, enforcing the former can result in more filtering more than enforcing the latter. To filter more, wRNIC has to consider simultaneously four constraints. Therefore, there must be a configuration of the minimal dual graph where a given constraint, C_1, has three adjacent constraints C_2, C_3, and C_4, and where C_1 is not an articulation point (otherwise, wRNIC would have the same filtering power as R($*$,3)C). The only redundancy-free configuration is the one shown in Figure 18. We show that this configuration is not possible.

1. Given the topology of the graph shown in Figure 18, the three edges incident to C_1 cannot have the same labeling, for example variable V_1, because C_1 becomes a unary constraint. There cannot be three different labeling, for example variables V_1, V_2, and V_3, otherwise C_1 becomes a ternary constraint. Thus, they must be labeled with two variables, V_1 and V_2, as shown in Figures 19 and 20.

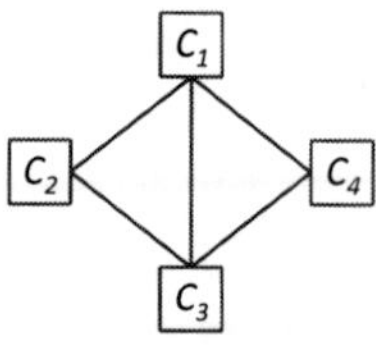
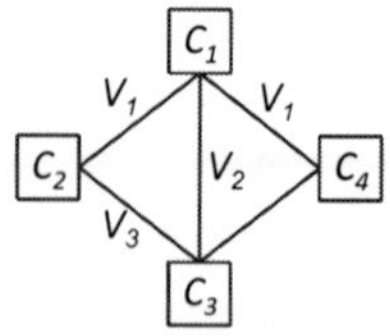
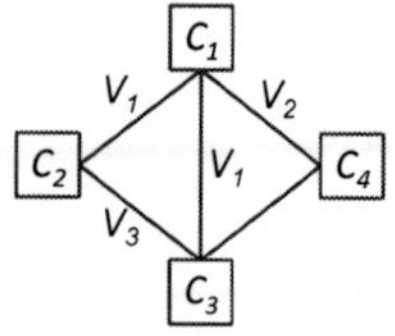

Fig. 18. A redundancy-free configuration of four binary constraints

Fig. 19. One possible labeling of the edges incident to C_1

Fig. 20. The other possible labeling of the edges incident to C_1

2. In Figure 19, the edge between C_2 and C_3 cannot be labeled V_1 (otherwise, C_2 becomes a unary constraint); cannot be labeled V_2 (otherwise, the scopes of C_2 and C_1 become equal, and we assume that the CSP is normalized); therefore, it must be labeled V_3. The edge between C_3 and C_4 cannot be labeled V_1 or V_4 (otherwise, C_3 becomes a ternary constraint); cannot be labeled V_2 (otherwise, the scopes of C_1 and C_3 become equal); cannot be labeled V_3 (otherwise, the scopes of C_2 and C_4 become equal). Therefore, no possible labeling for the edge between C_3 and C_4 exists, and this configuration is impossible.

3. In Figure 20, the edge between C_2 and C_3 cannot be labeled V_1 (otherwise, C_2 would be a unary constraint); cannot be labeled V_2 (otherwise, the scopes of C_1 and C_2 become equal); cannot be labeled V_3 (otherwise, the scopes of C_2 and C_3 become equal). Therefore, no possible labeling for the edge between C_2 and C_3 exist, and this configuration is impossible.

Consequently, no redundancy-free dual graph of a binary CSP can have a configuration of its vertices for enforcing R(∗,4)C. □

References

1. Bacchus, F., Chen, X., Beek, P.V., Walsh, T.: Binary vs. Non-Binary Constraints. Artificial Intelligence 140, 1–37 (2002)
2. Bessière, C.: Constraint Propagation. In: Handbook of Constraint Programming. Elsevier (2006)
3. Debruyne, R., Bessière, C.: From Restricted Path Consistency to Max-Restricted Path Consistency. In: Smolka, G. (ed.) CP 1997. LNCS, vol. 1330, pp. 312–326. Springer, Heidelberg (1997)
4. Debruyne, R., Bessière, C.: Some Practicable Filtering Techniques for the Constraint Satisfaction Problem. In: Proceedings of the 15th International Joint Conference on Artificial Intelligence, pp. 412–417 (1997)
5. Debruyne, R., Bessière, C.: Domain Filtering Consistencies. Journal of Artificial Intelligence Research 14, 205–230 (2001)
6. Dechter, R.: Constraint Processing. Morgan Kaufmann (2003)
7. Freuder, E.C.: A Sufficient Condition for Backtrack-Free Search. JACM 29(1), 24–32 (1982)
8. Freuder, E.C., Elfe, C.D.: Neighborhood Inverse Consistency Preprocessing. In: Proceedings of AAAI 1996, Portland, Oregon, pp. 202–208 (1996)
9. Golumbic, M.C.: Algorithmic Graph Theory and Perfect Graphs. Annals of Discrete Mathematics, vol. 75. Elsevier (2004)

10. Janssen, P., Jégou, P., Nougier, B., Vilarem, M.C.: A Filtering Process for General Constraint-Satisfaction Problems: Achieving Pairwise-Consistency Using an Associated Binary Representation. In: IEEE Workshop on Tools for AI, pp. 420–427 (1989)
11. Karakashian, S., Woodward, R., Reeson, C., Choueiry, B.Y., Bessière, C.: A First Practical Algorithm for High Levels of Relational Consistency. In: 24th AAAI Conference on Artificial Intelligence (AAAI 2010), pp. 101–107 (2010)
12. Kjærulff, U.: Triagulation of Graphs - Algorithms Giving Small Total State Space. Research Report R-90-09, Aalborg University, Denmark (1990)
13. Lecoutre, C., Cardon, S., Vion, J.: Conservative Dual Consistency. In: Proceedings of AAAI 2007, pp. 237–242 (2007)
14. Lecoutre, C., Cardon, S., Vion, J.: Second-Order Consistencies. Journal of Artificial Intelligence Research 40, 175–219 (2011)
15. Mackworth, A.K.: Consistency in Networks of Relations. Artificial Intelligence 8, 99–118 (1977)
16. Sabin, D., Freuder, E.C.: Contradicting Conventional Wisdom in Constraint Satisfaction. In: Proceedings of the 11th European Conference on Artificial Intelligence, Amsterdam, The Netherlands, pp. 125–129 (1994)
17. Woodward, R., Karakashian, S., Choueiry, B.Y., Bessière, C.: Solving Difficult CSPs with Relational Neighborhood Inverse Consistency. In: 25th AAAI Conference on Artificial Intelligence (AAAI 2011), pp. 112–119 (2011)

Syntactically Characterizing Local-to-Global Consistency in ORD-Horn

Michał Wrona*

Department of Computer and Information Science
Linköpings universitet
SE-581 83 Linköping, Sweden
michal.wrona@liu.se

Abstract. Establishing *local consistency* is one of the most frequently used algorithmic techniques in constraint satisfaction in general and in spatial and temporal reasoning in particular. A collection of constraints is *globally consistent* if it is completely explicit, that is, every partial solution may be extended to a full solution by greedily assigning values to variables one at a time. We will say that a structure **B** has *local-to-global consistency* if establishing local-consistency yields a globally consistent instance of **CSP(B)**.

This paper studies local-to-global consistency for ORD-Horn languages, that is, structures definable over the ordered rationals $(\mathbb{Q}; <)$ within the formalism of ORD-Horn clauses. This formalism has attracted a lot of attention and is of crucial importance to spatial and temporal reasoning. We provide a syntactic characterization (in terms of first-order definability) of all ORD-Horn languages enjoying local-to-global consistency.

1   Introduction

A constraint satisfaction problem is a computational problem whose instance consists of a domain, a set of variables and a set of constraints imposed on the variables. The goal is to answer whether there is an assignment to variables satisfying all the constraints. As it is common, we consider here constraint satisfaction problems, **CSP(B)**, parametrized by relational structures **B** over finite signatures. (All constraints in an instance of **CSP(B)** come from **B**.) Such structures **B** are typically referred to as languages.

It is very well known that a number of natural problems can be formulated in this way. In this paper we focus on CSPs arising from a subdiscipline of Artificial Intelligence called spatial and temporal reasoning. In this case, a domain is a set of intervals, time-points, spatial regions, or the like; and constraints reflect temporal or spatial dependencies on the elements of the domain [1,2,3]. A prominent example of a temporal constraint calculi, which plays a central role in qualitative reasoning in general, is Allen's interval algebra [4].

---

* This work is partially supported by the *National Graduate School in Computer Science* (CUGS), Sweden.

A primary algorithmic technique for solving CSPs in temporal and spatial reasoning is the process of establishing k-consistency [5], which converts an instance Φ of $\mathbf{CSP}(\mathbf{B})$ into Φ' that has the same set of solutions as Φ and is k-consistent. According to the definitions in [6,7] the instance Φ' is k-consistent if every partial solution to $(k-1)$ variables may be extended to any other variable. An instance of $\mathbf{CSP}(\mathbf{B})$ involving n variables is strongly k-consistent if it is consistent for every $i \leq k$, and globally consistent if it is strongly n-consistent. In this paper, we say that $\mathbf{B}$ has *local-to-global consistency* if there exists k such that every strongly k-consistent instance Φ' of $\mathbf{CSP}(\mathbf{B})$ is also globally consistent. It is equivalent to saying that $\mathbf{CSP}(\mathbf{B})$ has bounded strict width [8].

As indicated in the abstract, a globally consistent set of constraints represents explicitly all its solutions. Since local-to-global consistency of $\mathbf{B}$ implies the existence of a polynomial algorithm that converts an instance Φ of $\mathbf{CSP}(\mathbf{B})$ into a globally consistent Φ' equivalent to Φ, local-to-global consistency is certainly a very desirable property. Therefore it makes sense to examine its applicability to important families of languages such as *ORD-Horn (OH) languages*, which were introduced in order to define the most well-known subclass of Allen's interval algebra called ORD-Horn [9]. An OH language is a relational structure such that each of its relations is definable in $(\mathbb{Q}; <)$ by a conjunction of *OH clauses* each of which is in one of the following forms:

- $((x_1 \neq y_1 \vee \cdots \vee x_k \neq y_k) \vee (x_0 \circ y_0))$,
- $(x_1 \neq y_1 \vee \cdots \vee x_k \neq y_k)$, or
- $x_k \circ y_k$,

where $\circ \in \{<, \leq, =\}$.

In this paper, we give a complete syntactic characterization of ORD-Horn languages with local-to-global consistency, summarized in Theorem 1. It is well-known that $\mathbf{CSP}(\mathbf{B})$ for every OH language $\mathbf{B}$ can be solved by establishing local consistency, however, as we will see, not every OH language has local-to-global consistency. We say that an OH language is *basic OH* if each of its relations is definable by a conjunction of *basic OH clauses*, that is, OH clauses in one of the following forms:

- $((x_1 \neq x_2 \vee \cdots \vee x_1 \neq x_k) \vee (y_1 \neq y_2 \vee \cdots \vee y_1 \neq y_l) \vee x_1 < y_1)$,
- $(x_1 \neq y_1 \vee \cdots \vee x_k \neq y_k)$, or
- $(x \circ y)$,

where $\circ \in \{<, \leq, =\}$.

Theorem 1. *Let $\mathbf{B}$ be an ORD-Horn language. Then, $\mathbf{B}$ has local-to-global consistency if and only if it is basic ORD-Horn.*

1.1   The Outline of the Proof of Theorem 1

Positive Results. To show that basic OH languages have local-to-global consistency we follow the proof strategy in [10] and use the algebraic characterization of this property for ω-categorical structures [11], which is a generalization of that for finite structures [8,12]. It says that a structure $\mathbf{B}$ of the domain B has local-to-global consistency if and only if there exists $k \geq 3$ such that for each finite subset A of B there is a homomorphism from $\mathbf{B}^k$ to $\mathbf{B}$ satisfying property $f(x, \ldots, x, y, x, \ldots, x) = f(x, \ldots, x)$ for all $x, y \in A$, or equivalently, $\mathbf{B}$ is preserved by an oligopotent quasi near-unanimity function (in short, an oligopotent QNUF) (see Sect. 2). A construction of an oligopotent QNUF preserving a fixed basic OH language is provided in Sect. 3.

Negative Results. In Sect. 4 we provide a characterization of all OH languages that are not basic. These structures are called *complex*. In Sect. 5 we show that complex OH structures are not preserved by oligopotent QNUFs. First we select certain 3/4-ary relations called *domino*, *split*, and *windmill* relations. To show that these simple relations are not preserved by oligopotent QNUFs we use the characterization of local-to-global consistency by a generalization of the classical result by Baker and Pixley to ω-categorical structures [13]. To obtain the same for all complex OH structures we prove that every such structure primitively positively defines a domino, a split, or a windmill relation; and use the well-known fact that a relation primitively positively defined in a structure $\mathbf{B}$ is preserved by all the functions preserving $\mathbf{B}$.

1.2   Related Work

Point algebra [14] and extensions thereof by disjunctions of disequalities whose local-to-global consistency was studied by Koubarakis [15] are special cases of basic OH languages. Furthermore, the oligopotent QNUF constructed in Sect. 3 reproves local-to-global consistency for languages studied in Theorems 23–27 in [10]. Similar in spirit research may be found also in [16] and [17], where authors identify subclasses of Allen's Interval Algebra for which strong path-consistency implies global consistency.

Although the complexity of CSPs for structures with a first-order definition in $(\mathbb{Q}, <)$, called also temporal languages, was classified in [19], the corresponding classification of QCSPs [18] proved to be much harder to obtain (for partial results see [20,21,22]). One of the difficulties is that OH structures, whose CSPs are solvable in polynomial time, give rise to QCSPs of varying complexity [20]. By the result in [10], the fact that an oligopotent QNUF constructed in Sect. 3 is surjective, implies that QCSPs for basic OH languages are tractable. Since they strictly contain the family of negative equality languages, which form the only polynomial class in [20], this paper significantly contributes to understanding the complexity of temporal QCSPs.

2   Preliminaries

The logical and algebraic terminology and notions that we use are fairly standard. Here, we provide a brief review of the concepts most central to this article. By $[n]$, we denote the set $\{1, \ldots, n\}$.

Structures. We study relational structures. A *signature* is a set of relation symbols, where each symbol has an associated arity $k \geq 1$. In this paper, we assume that each signature is finite. A *structure* $\mathbf{B}$ over signature σ consists of a *universe* B and an interpretation $R^{\mathbf{B}} \subseteq B^k$ for each symbol $R \in \sigma$; here, k denotes the arity of R. Two or more structures are said to be *similar* if they are over the same signature. When $\mathbf{A}, \mathbf{B}$ are similar structures, a mapping $h : A \to B$ is considered to be a *homomorphism* from $\mathbf{A}$ to $\mathbf{B}$ if for each symbol R and each tuple $(a_1, \ldots, a_k) \in R^{\mathbf{A}}$, it holds that $(h(a_1), \ldots, h(a_k)) \in R^{\mathbf{B}}$. A *bijective homomorphism* is a homomorphism that is both surjective and injective. An isomorphism is a bijective homomorphism, whose inverse is also a homomorphism. An automorphism of $\mathbf{B}$ is an isomorphism between $\mathbf{B}$ and itself. The set of automorphisms of a structure $\mathbf{B}$ is denoted by $Aut(\mathbf{B})$.

In this paper we say that a relational structure $\mathbf{B}$ is *first-order definable* in $\mathbf{A}$ if $\mathbf{B}$ has the same domain as $\mathbf{A}$, and for every relation R of $\mathbf{B}$ there is a first-order formula ϕ in the signature of $\mathbf{A}$ such that ϕ holds exactly on those tuples that are contained in R. Likewise, we say that $\mathbf{B}$ is *primitively positively definable* (pp-definable) in $\mathbf{A}$ if there is ϕ that is a primitive positive formula (first-order formula built exclusively from conjunction, existential quantifiers, equality and relation symbols).

A first-order theory is ω-categorical if all of its countable models are isomorphic, and a countable structure is ω-categorical if its theory is ω-categorical. A relation R over a domain B is ω-categorical if the structure $(B; R)$ is ω-categorical. The theorem of Engeler, Ryll- Nardzewski, and Svenonius provides characterizations of ω-categoricity for structures [23]. In particular it implies that every n-ary relation that is first-order definable in an ω-categorical structure $\mathbf{B}$ is ω-categorical and is the union of a finite set of orbits of $Aut(\mathbf{B})$, that is, sets of the form $\{(\alpha(x_1), \ldots, \alpha(x_n)) \in B^n \mid \alpha \in Aut(\mathbf{B})\}$ for some $x_1, \ldots, x_n \in B$.

Algebra. Let $\mathbf{B}$ be a structure. A finitary operation $h : B^k \to B$ is a *polymorphism* of $\mathbf{B}$ if h is a homomorphism from $\mathbf{B}^k$ to $\mathbf{B}$; it is a *polymorphism* of a relation $T \subseteq B^k$ if it is a polymorphism of the structure $(B; T)$. When h is a polymorphism of a structure (or relation) we also say that the structure (or relation) is *preserved* by h. We will make use of the following well-known and straightforwardly verified fact.

Proposition 1. *Let $\mathbf{B}$ be a structure. Let ϕ be a primitive positive formula having k free variables. Each polymorphism of $\mathbf{B}$ is a polymorphism of the relation $T \subseteq B^k$ defined by ϕ over $\mathbf{B}$.*

This proposition says that preservation of a relation by all polymorphisms of a structure $\mathbf{B}$ is necessary for the relation to be pp-definable; it is certainly worth mentioning that for ω-categorical structures, this preservation condition is also sufficient [24].

A function f on a domain B is called a *quasi near-unanimity function* (short, a *QNUF*), if it satisfies $f(x, \ldots, x, y) = f(x, \ldots, y, x) = \ldots f(y, \ldots, x, x)$ for all $x, y \in B$. We say that a polymorphism g of a structure $\mathbf{B}$ is *oligopotent* if $f(x) := g(x, \ldots, x)$ preserves all relations that are first-order definable in $\mathbf{B}$.

ORD-Horn Languages. Recall from the introduction the definitions of OH and basic OH languages. We say that a relation R over a domain B is OH or basic OH if $(B; R)$ is OH or basic OH, respectively. We will write that two OH clauses (or in general formulas) are equivalent if they are equivalent in $(\mathbb{Q}; <)$.

It is well known that $(\mathbb{Q}; <)$ is ω-categorical and hence, by the remark above, every OH structure is ω-categorical and is the union of a finite number of orbits of $Aut((\mathbb{Q}; <))$. In the following an orbit of $Aut((\mathbb{Q}; <))$ is simply called an orbit. For example, a relation $\{(x, y, z) \mid (x \leq y \leq z) \wedge x \neq z\}$ is the union of orbits: $\{(x, y, z) \mid x < y < z\}$, $\{(x, y, z) \mid x = y < z\}$, and $\{(x, y, z) \mid x < y = z\}$.

In the following we say that an orbit O satisfies a formula ϕ if ϕ holds on all tuples from O. Likewise, we say that an orbit O violates a formula ϕ if ϕ does not hold on tuples from O.

The following lemmas may be proved using the algebraic characterization of ORD-Horn structures from [25]. The proof of the first one as well as other omitted proofs will be included in the full version of the paper. The second one is proved here using more straightforward methods.

Lemma 1. *The set of all OH languages is closed under pp-definitions, that is, every structure pp-defined in some OH language is also an OH language.*

Lemma 2. *Let R be an n-ary OH relation and π some permutation of $[n]$. Then R contains an orbit*

$$O = \{(v_1, \ldots, v_n) \mid v_{\pi(1)} < \ldots < v_{\pi(n)}\}$$

if and only if for all $a < b$ in $[n]$ the relation R contains an orbit $O_{a,b}$ satisfying $(v_{\pi(a)} < v_{\pi(b)})$.

Proof. The left-to-right implication is obvious. To show the reverse implication assume on the contrary that O is not contained in R and all of $O_{a,b}$ are. Then a definition Φ_R of R contains a clause C that violates O. Since all the tuples in O are injective it is easy to see that C involves two variables. Every such clause is of the form $(v_{\pi(a)} \circ v_{\pi(b)})$, where $1 \leq a < b \leq n$ and $\circ \in \{\geq, >, =\}$. But the fact that the clause $(v_{\pi(a)} \circ v_{\pi(b)})$ is in Φ_R contradicts the assumption that $O_{a,b}$ is contained in R. Hence we have proved the lemma. $\qquad\square$

Local-to-Global Consistency. For a structure $\mathbf{B}$, the *constraint satisfaction problem* on $\mathbf{B}$, denoted by $\mathbf{CSP}(\mathbf{B})$, is the computational problem of deciding, given a primitive positive formula over a signature of $\mathbf{B}$ if it is satisfiable in $\mathbf{B}$.

The process of establishing (strong) k-consistency, see the definitions in the introduction, permits the detection of some unsatisfiable instances of $\mathbf{CSP}(\mathbf{B})$. If $\mathbf{B}$ is ω-categorical, then it can be performed in polynomial time [11]. We say that a structure $\mathbf{B}$ has *local-to-global consistency with respect to k* if every strongly k-consistent instance of $\mathbf{CSP}(\mathbf{B})$ is globally consistent. A structure $\mathbf{B}$ has local-to-global consistency if it has local-to-global consistency wrt. some k. In the terminology of [11], a structure $\mathbf{B}$ has local-to-global consistency wrt. k if and only if $\mathbf{CSP}(\mathbf{B})$ has strict with k. The following theorem that characterizes local-to-global consistency is a corollary of Theorem 13 in [11] and Theorem 19 in [13].

Theorem 2. *Let $\mathbf{B}$ be an ω-categorical structure. Then for every $k \geq 2$, the following are equivalent.*

1. *The structure $\mathbf{B}$ has an oligopotent $(k + 1)$-ary quasi near-unanimity polymorphism.*
2. *For every n, every n-ary relation pp-definable in Γ is k-decomposable, that is, it contains all tuples t such that for every subset I of $\{1, \ldots, n\}$ with $|I| \leq k$ there is a tuple $s \in R$ such that $t[i] = s[i]$ for all $i \in I$.*
3. *The structure $\mathbf{B}$ has local-to-global consistency with respect to k.*

3 QNUF Construction

In this section, we demonstrate that an oligopotent QNUF with certain desirable properties can be constructed for basic ORD-Horn structures. We make use of the following notions from [10, Sect. 4] (adapted to the current context). Let $t \in \mathbb{Q}^k$ be a tuple, where $k \geq 3$, and let $\pi : [k] \to [k]$ be a permutation such that $t_{\pi(1)} \leq \cdots \leq t_{\pi(k)}$. If it holds that $t_{\pi(2)} = \cdots = t_{\pi(k-1)}$, we say that this value is the *main value* of t. Not every tuple $t \in \mathbb{Q}^k$ has a main value, but when a tuple has a main value, it is unique. We define the equivalence relation $\equiv_m$ on $\mathbb{Q}^k$ as follows: $t \equiv_m t'$ if and only if $t = t'$ or t, t' have the same main value. We say that a function $h : \mathbb{Q}^k \to \mathbb{Q}$ is *main-injective* if for all $t, u \in \mathbb{Q}^k$, it holds that $h(t) = h(u)$ if and only if $t \equiv_m u$.

Theorem 3. *Let $\mathbf{B}$ be a basic ORD-Horn structure. Then there exists a main-injective, surjective, oligopotent QNUF that is a polymorphism of $\mathbf{B}$, and consequently, $\mathbf{B}$ has local-to-global consistency.*

We devote the rest of this section to the proof of this theorem. Observe first that by Proposition 1, we can assume that $\mathbf{B}$ contains only relations definable by single basic OH clauses. Let F be the set that for every relation R in $\mathbf{B}$ contains exactly one basic OH clause defining R.

We will make use of the following notation: for some relation $R \subseteq \mathbb{Q}^2$, tuples $t, u \in \mathbb{Q}^k$, and nonnegative integer n, we write $[tRu, n]$ if tRu holds in at least n coordinates, that is, if there exists a subset $S \subseteq [k]$ of size greater than or equal to n such that for all $i \in S$, it holds that $t_i R u_i$. In place of tuples $t, u \in \mathbb{Q}^k$, we may use this notation with a value $b \in \mathbb{Q}$, which will be understood to represent the k-tuple $(b, \ldots, b) \in \mathbb{Q}^k$.

We choose r to be a sufficiently large positive integer so that (1) for each formula $(x_1 \neq x_2 \vee \cdots \vee x_1 \neq x_q) \vee (x_1 < y_1) \vee (y_1 \neq y_2 \vee \cdots \vee y_1 \neq y_{q'})$ contained in F, it holds that $r \geq q + q'$, and (2) $4r + 3$ exceeds $2s + 1$ where s is the maximum number of variables occurring in a disjunction of disequalities $(x_1 \neq y_1 \vee \cdots \vee x_p \neq y_p)$ in F. We set k to be equal to $4r + 3$. We will construct a k-ary QNUF operation.

We now define a binary relation $\trianglelefteq$ on the set $\mathbb{Q}^k$. For $t, u \in \mathbb{Q}^k$, we define $t \trianglelefteq u$ if and only if one of the following holds:

- $[t \leq u, k]$
- one (or both) of t, u has a main value and $[t < u, 2r + 3]$.

We will first observe that there are no cycles in $(\mathbb{Q}^k; \trianglelefteq)/ \equiv_m$ where the notation $(\mathbb{Q}^k; \trianglelefteq)/ \equiv_m$ indicates the structure whose universe consists of $\equiv_m$-equivalence classes, and where $T \trianglelefteq T'$ (for two such equivalence classes T, T') if and only if there exist $t \in T$, $t' \in T'$ such that $t \trianglelefteq t'$. An equivalence class of $(x_1, \ldots, x_k)$ wrt. $\equiv_m$ will be denoted by $(x_1, \ldots, x_k)/ \equiv_m$.

The following lemma is similar in spirit to Lemma 20 in [10].

Lemma 3. *There are no $\equiv_m$-equivalence classes $T_1, \ldots, T_c$ such that $T_1 \trianglelefteq T_2 \trianglelefteq \cdots \trianglelefteq T_c \trianglelefteq T_1$ and that the T_i are pairwise distinct.*

Now, we show that there is a bijective homomorphism h from $(\mathbb{Q}^k; \trianglelefteq)/ \equiv_m$ to $(\mathbb{Q}; \leq)$, and then argue that the resulting natural function $f : \mathbb{Q}^k \to \mathbb{Q}$ satisfying $f(x_1, \ldots, x_k) = h((x_1, \ldots, x_k)/ \equiv_m)$ for all $x_1, \ldots, x_k \in \mathbb{Q}^k$ is a polymorphism of $\mathbf{B}$ with all of the desired properties. Let it be implicitly understood that the relation $\trianglelefteq$ and the relation $\leq$ have the same relation symbol in both structures.

To show that there is a bijective homomorphism h from $(\mathbb{Q}^k; \trianglelefteq)/ \equiv_m$ to $(\mathbb{Q}; \leq)$ we show that there is a *homomorphic back-and-forth system* defined as follows.

Let $\mathbf{A}$ and $\mathbf{B}$ be τ-structures. A *homomorphic back-and-forth system* from $\mathbf{A}$ to $\mathbf{B}$ is a non-empty set I of pairs $(\bar{a}, \bar{b})$ of tuples, with $\bar{a}$ from A and $\bar{b}$ from B, such that all entries in $\bar{a}$ and $\bar{b}$ are pairwise different and all of the following hold.

1. If $(\bar{a}, \bar{b}) \in I$ then $\bar{a}$ and $\bar{b}$ have the same length and every atomic sentence (that is, in our case, a first-order formula built from the relational symbol and constants only) that holds in $(\mathbf{A}, \bar{a})$ also holds in $(\mathbf{B}, \bar{b})$.
2. (Going Forth.) For every pair $(\bar{a}, \bar{b}) \in I$ and every element c of $\mathbf{A}$ which is not in $\bar{a}$ there is an element d of $\mathbf{B}$ such that the pair $(\bar{a}c, \bar{b}d) \in I$.
3. (Going Back.) For every pair $(\bar{a}, \bar{b}) \in I$ and every element d of $\mathbf{B}$ which is not in $\bar{b}$ there is an element c of $\mathbf{A}$ such that the pair $(\bar{a}c, \bar{b}d) \in I$.

Homomorphic back-and-forth system is a variant of a back-and-forth system, as defined in [23], Sect. 3.2. The first difference is that in Item 1 of the standard definition we require an atomic sentence to hold in $(A, \bar{a})$ if and only if it holds in $(B, \bar{b})$. Here we need less. Indeed, since we want a bijective homomorphism rather than an isomorphism, its inversion does not have to be a homomorphism. The second difference is that we require entries of $\bar{a}$ and $\bar{b}$ to be pairwise different. This is necessary by the first modification and the fact that our goal is to construct an injective homomorphism.

There is a back-and-forth system from $\mathbf{A}$ to $\mathbf{B}$ if and only if there is a bijective homomorphism from $\mathbf{A}$ to $\mathbf{B}$ (this follows from a straightforward modification of Lemma 3.2.2. and Theorem 3.2.3 (b) in [23]).

Let I be the set of *all* pairs $(\bar{a}, \bar{b})$, where $\bar{a}$ and $\bar{b}$ are tuples of equal length with pairwise different entries over A and B, respectively, such that every atomic formula that holds in $(A, \bar{a})$ also holds in $(B, \bar{b})$. We show that I is a homomorphic back-and-forth system from $(\mathbb{Q}^k; \trianglelefteq)/ \equiv_m$ to $(\mathbb{Q}; \leq)$. For going forth, let $(\bar{a}, \bar{b}) \in I$

and let c be an element of $(\mathbb{Q}^k; \trianglelefteq)/\equiv_m$. We have to find an element d of $(\mathbb{Q}; \leq)$ such that the pair $(\bar{a}c, \bar{b}d) \in I$. Let A_1 be the set of indices i of $\bar{a}$ such that $a_i \trianglelefteq c$, and let A_2 be the set of indices i of $\bar{a}$ such that $c \trianglelefteq a_i$ in $(\mathbb{Q}^k; \trianglelefteq)/\equiv_m$. By Lemma 3, the sets A_1 and A_2 are disjoint. It is then clear that we can find a $d \in \mathbb{Q}$ such that for all $i \in A_1$ we have $b_i \leq d$, and for all $i \in A_2$ we have $d \leq b_i$.

For going back, let $(\bar{a}, \bar{b}) \in I$ and let d be an element of $\mathbb{Q}$. We have to find an element c of $(\mathbb{Q}^k; \trianglelefteq)/\equiv_m$ such that the pair $(\bar{a}c, \bar{b}d) \in I$. We claim that there is an element c of $(\mathbb{Q}^k; \trianglelefteq)/\equiv_m$ not comparable to any entry in $\bar{a}$; clearly, we then have $(\bar{a}c, \bar{b}d) \in I$. To prove the claim, recall that $k = 4r + 3$. Partition the k coordinates into two parts P, Q with $|P| = 2r + 1$, $|Q| = 2r + 2$. Define $t \in \mathbb{Q}^k$ where, at all entries $i \in P$, $t_i < v_i$ in case that $a_i = \{(v_1, \ldots, v_k)\}$ is an equivalence class of size 1, and $t_i < v$ when v is the main value of a_i. Similarly, we require that at all entries $i \in Q$, $t_i > v_i$ in case that $a_i = \{(v_1, \ldots, v_k)\}$ is an equivalence class of size 1, and $t_i > v$ when v is the main value of a_i. It is straightforward to verify that the equivalence class of t is incomparable to any entry in $\bar{a}$.

So we have shown that the existence of a bijective homomorphism h from $(\mathbb{Q}^k; \trianglelefteq)/\equiv_m$ to $(\mathbb{Q}; \leq)$. Let $f : \mathbb{Q}^k \to \mathbb{Q}$ be the natural function induced by h.

Clearly, by the definition of $\trianglelefteq$, it holds that f preserves the relation $\leq$. Also, by the choice of k, it holds that f preserves all disjunctions of disequalities in F, by the argumentation in [10, proof of Theorem 17]. We argue that f preserves all formulas $(x_1 \neq x_2 \vee \cdots \vee x_1 \neq x_q) \vee (x_1 < y_1) \vee (y_1 \neq y_2 \vee \cdots \vee y_1 \neq y_{q'})$ in F, as follows. Let $t_1, \ldots, t_q, u_1, \ldots, u_{q'} \in \mathbb{Q}^k$ be tuples such that $(t_1 \neq t_2 \vee \cdots \vee t_1 \neq t_q) \vee (t_1 < u_1) \vee (u_1 \neq u_2 \vee \cdots \vee u_1 \neq u_{q'})$ holds in k coordinates. Suppose that $f(t_1) = \cdots = f(t_q)$ and that $f(u_1) = \cdots = f(u_{q'})$. We need to prove that $f(t_1) < f(u_1)$. We consider two cases:

- If neither t_1 nor u_1 has a main value, then all of the t_i are equal and, likewise, all of the u_i are equal. It then follows that $[t_1 < u_1, k]$. Since f preserves both $\neq$ and $\leq$, it also preserves $<$ and it follows that $f(t_1) < f(u_1)$.
- If one of the tuples t_1, u_1 has a main value, then the set of coordinates $S \subseteq [k]$ where $(t_1 \neq t_2 \vee \cdots \vee t_1 \neq t_q) \vee (u_1 \neq u_2 \vee \cdots \vee u_1 \neq u_{q'})$ is less than or equal to $2(q + q')$, which in turn is less than or equal to $2r$. It thus follows that $[t_1 < u_1, 2r + 3]$, implying that $t_1 \trianglelefteq u_1$ and that $f(t_1) \leq f(u_1)$. As $t_1 \not\equiv_m u_1$, we conclude that $f(t_1) < f(u_1)$.

The function f is certainly surjective and main-projective. The latter implies that it is a QNUF. That f is oligopotent follows from the fact that it preserves the relation $<$. This concludes the proof of Theorem 3.

4   Complex OH Languages

This section is devoted to provide a smooth definition of OH languages that are not basic. They will be called *complex OH languages*. Such definition should in particular cover all OH structures with a single relation definable by an OH clause which is not basic such as $(x \neq y \vee x = z)$ or $(x \neq y \vee y \leq z)$. Indeed, it is easy to verify that by Definition 4 all such languages are complex.

First, we provide some additional tools.

A-subpartitions. An *A-subpartition of a set S* is: either (i) a subpartition (a partition of a subset) of S into sets of size at least 2, or (ii) $\{\emptyset\}$. The set of A-subpartitions of $[n]$ will be denoted by $\mathbf{P}_n$.

For instance, if $n = 4$, then the set $\mathbf{P}_4$ contains $\{S\}$ for every $S \subseteq [4]$ such that $|S| \neq 1$ as well as $\{\{1,2\},\{3,4\}\}, \{\{1,3\},\{2,4\}\}, \{\{1,4\},\{2,3\}\}$.

For a given n, we define an order $(\mathbf{P}_n; \preceq)$ in the following way. We have $P_1 \preceq P_2$ for $P_1, P_2 \in \mathbf{P}_n$ if for each $S_1 \in P_1$ there is $S_2 \in P_2$ such that $S_1 \subseteq S_2$. If $P_1 \preceq P_2$ and $P_1 \neq P_2$, then we write $P_1 \prec P_2$. It is easy to see that for every $n \in N$, the order $(\mathbf{P}_n; \preceq)$ is a partial order and $\{\emptyset\}$ is its least element.

Normalized OH Clauses. For the purposes of this paper, we provide another way of defining OH relations.

Definition 1. *We say that an OH clause is* normalized *if it is of the form*

$$\bigvee_{S \in P} \bigvee_{a,b \in S; a \neq b} v_a \neq v_b \vee (v_i \circ v_j) \tag{1}$$

for some $P \in \mathbf{P}_n$, $i, j \in [n]$, and $\circ \in \{\leq, <, =\}$.

In particular, a normalized OH clause may be of the form $(\bigvee_{S \in P} \bigvee_{a,b \in S; a \neq b} v_a \neq v_b)$ or $(v_i \circ v_j)$. For the sake of simplicity we shorten (1) and write $(\Phi[P] \vee v_i \circ v_j)$ or $(\Phi[P])$. A *normalized OH formula* is a conjunction of normalized OH clauses.

As an example of a clause that is not normalized, consider $(v_1 \neq v_2 \vee v_2 \neq v_3 \vee v_3 \neq v_4)$. Observe that it is equivalent to $(\bigvee_{S \in \{[4]\}} \bigvee_{a,b \in S; a \neq b} v_a \neq v_b)$, or $\Phi[\{[4]\}]$ for short.

Lemma 4. *Every OH relation is definable by a normalized OH formula.*

If not stated otherwise we will assume that Φ_R is a normalized OH formula over variables $v_1, \ldots, v_n$ such that R is equal to $\{(v_1, \ldots, v_n) \mid \Phi_R(v_1, \ldots, v_n)\}$. Thus, Φ_R defines R so that a variable v_i in Φ_R corresponds exactly to the i-th coordinate of R.

Entailment. We adopt the standard definition of *entailment* to our needs. Let Φ and Ψ be first-order formulas over the signature $\{<, \leq, \neq, =\}$ and variables $v_1, \ldots, v_n$. We say that Φ *entails* Ψ if

$$\forall v_1 \ldots \forall v_n (\Phi \rightarrow \Psi),$$

is true in $(\mathbb{Q}; <, \leq, \neq, =)$. We also say that an OH relation R entails Ψ if Φ_R entails Ψ. The set of all normalized OH clauses entailed by R is denoted by $\mathcal{C}_R$.

Lemma 5. *If $\bigwedge_{i=1}^n D_i$ entails C and $\bigwedge_{i=1}^m E_i$ entails D_1, then $\bigwedge_{i=1}^m E_i \wedge \bigwedge_{i=2}^n D_i$ entails C.*

Proper Entailment. We will first define a partial order $(\mathcal{C}_R; \precsim^R)$, and then use it to refine the notion of entailment.

Definition 2. *Let R be an n-ary OH relation R, and $<_1^R$ a binary relation on the set $\mathcal{C}_R$ such that for all $C_1, C_2 \in \mathcal{C}_R$ we have $C_1 <_1^R C_2$ if and only if:*

1. *C_1 is basic OH, and C_2 is not basic OH;*
2. *or both are not basic OH and one of the following holds:*
   - *(a) C_1 is $(\Phi[P_1] \vee v_i \circ_1 v_j)$, C_2 is $(\Phi[P_2] \vee v_k \circ_2 v_l)$ and $P_1 \prec P_2$;*
   - *(b) C_1 is $(\Phi[P_1] \vee v_i \circ_1 v_j)$, C_2 is $(\Phi[P_1] \vee v_k \circ_2 v_l)$ and $|Var(C_1)| < |Var(C_2)|$; or*
   - *(c) C_1 is $(\Phi[P_1] \vee v_i \circ_1 v_j)$, C_2 is $(\Phi[P_1] \vee v_i \circ_2 v_j)$ and $(v_i \circ_1 v_j)$ entails $(v_i \circ_2 v_j)$;*

   where $P_1, P_2 \in \mathbf{P}_n$, $i, j, k, l \in [n]$ and $\circ_1, \circ_2 \in \{<, =, \leq\}$.

We define $\precsim^R$ to be the reflexive and transitive closure of $<_1^R$ and write $C_1 \precnsim^R C_2$ if $C_1 \precsim^R C_2$ and $C_1 \neq C_2$. We omit R and write $\precsim$ or $\precnsim$ if R is obvious from the context.

Lemma 6. *Let R be an n-ary OH relation. Then $(\mathcal{C}_R; \precsim^R)$ is a partial order.*

We are now ready to define proper entailment.

Definition 3. *Let R be an OH relation that entails a normalized OH clause C. We say that C is* not properly entailed *by R if there exist $n \in \mathbb{N}$ and normalized OH clauses $C_1, \ldots, C_n$ such that all of the following holds:*

- *R entails $C_1 \wedge \cdots \wedge C_n$;*
- *$C_1, \ldots, C_n \precnsim C$; and*
- *$C_1 \wedge \cdots \wedge C_n$ entails C.*

Otherwise, we say that C is properly entailed *by R.*

As an example, consider an OH relation R defined by $C \wedge C_1 \wedge C_2$, where $C := (v_1 \neq v_2 \vee v_2 = v_3)$, $C_1 := (v_1 \leq v_3)$ and $C_2 := (v_3 \leq v_2)$. Certainly, each of these clauses is entailed by R. Furthermore, observe that $C_1, C_2 \precnsim C$ by applying either Item 1 or Item 2a in Definition 2 and that $C_1 \wedge C_2$ entails C. It implies that C is not properly entailed by R; on the other hand, it is straightforward to check that C_1 and C_2 are properly entailed by R and that $C_1 \wedge C_2$ defines R.

In general every OH relation R can be defined as the conjunction Φ_R^p of all normalized OH clauses properly entailed by R. The following lemma can be proved using Lemmas 5 and 6.

Lemma 7. *The conjunction Φ_R^p, defines R.*

Basic and Complex OH Languages. An OH clause is *complex* if it is not basic.

Definition 4. *We say that an OH relation R is* complex *if it properly entails a complex, normalized OH clause. An OH language is* complex *if it contains a complex relation.*

We will now prove that an OH relation cannot be both basic and complex.

Lemma 8. *An OH relation R is complex if and only if it is not basic.*

Proof. If R is not complex, then Φ_R^p contains basic OH clauses only, and hence R is basic. Assume now that R is both complex and basic. If it is complex then it properly entails a complex, normalized OH clause C. If R is basic, then it has a definition Φ_R consisting of basic, normalized OH clauses only. By Item 1 of Definition 2, for every clause D in Φ_R, we have $D \underset{\sim}{\lesssim} C$. Hence by Definition 3, the clause C cannot be properly entailed by R. This contradicts the assumption, and thus we have proved the lemma. $\qquad\square$

5   OH Languages without Local-to-Global Consistency

In this section we show that complex OH relations, and therefore languages, are not preserved by oligopotent QNUFs, and hence by Theorem 2, do not enjoy local-to-global consistency. First we show this statement for certain classes of ternary and four-ary OH relations called domino, split and windmill relations. The next step is to show that an arbitrary complex OH relation pp-defines some relation from one of these three classes.

5.1   Domino, Split, and Windmill Relations

To prove that a domino, a split or a windmill relation R is not preserved by an oligopotent QNUF of arity $k+1$, we use the equivalence of Item 1 and Item 2 in Theorem 2, and show that R pp-defines a relation, which is not k-decomposable.

Definition 5. *Let R be a ternary OH relation. We say that R is a* domino *relation if R entails $(v_1 \neq v_2 \vee v_2 = v_3)$ and contains*

1. *$\{(v_1, v_2, v_3) \mid v_1 = v_2 = v_3 \vee v_1 < v_2 < v_3\}$, or*
2. *$\{(v_1, v_2, v_3) \mid v_1 = v_2 = v_3 \vee v_1 > v_2 > v_3\}$.*

An obvious example of a domino relation is $\{(v_1, v_2, v_3) \mid v_1 \neq v_2 \vee v_2 = v_3\}$.

Lemma 9. *Let R be a domino relation. Then it is not preserved by any oligopotent QNUF.*

Proof. We restrict ourselves to domino relations satisfying Item 1 in Definition 5. The other case is analogous. Let $k \geq 2$. To show that R is not preserved by any $(k+1)$-ary oligopotent QNUF, we prove that a relation defined by a formula ϕ of the form:

$$\exists y_1 \ldots \exists y_k\; R(z_1, z_2, y_1) \wedge R(x_1, y_1, y_2) \wedge \cdots \wedge R(x_{k-1}, y_{k-1}, y_k) \wedge R(x_k, y_k, x_{k+1}),$$

whose quantifier-free part will be denoted by ψ, is not k-decomposable. Observe that it is enough to indicate an assignment a from $V := \{z_1, z_2, x_1, \ldots, x_{k+1}\}$ to $\mathbb{Q}$ that violates ϕ, but each of its restriction to the subset $V \setminus \{v\}$, where $v \in V$, satisfies $\exists v\, \phi$.

Consider $a(z_1) = a(z_2) = a(x_1) = \cdots = a(x_k) < a(x_{k+1})$, which clearly violates ϕ. To complete the proof we will show that for every $v \in V$, the restriction a' of a to variables in $V \setminus \{v\}$ can be extended to b on variables $\{z_1, z_2, x_1, \ldots, x_{k+1}, y_1, \ldots, y_k\}$ satisfying ψ.

1. If $v \in \{z_1, z_2\}$, then we choose b such that $b(z_1) < b(z_2) < b(y_1) < \cdots < b(y_k) < b(x_{k+1})$. Since b is an extension of a' we have that $b(z_1) < b(z_2) < b(y_1)$, $b(x_i) < b(y_i) < b(y_{i+1})$ for each $i \in [k-1]$, and $b(x_k) < b(y_k) < b(x_{k+1})$. Hence b satisfies ψ.
2. In the case where $v = x_i$ for some $i \in [k]$ we claim that b satisfying $b(x_i) < b(z_1)$, $b(z_1) = b(y_1) = \cdots = b(y_i)$ and $b(y_i) < b(y_{i+1}) < \cdots < b(y_k) < b(x_{k+1})$ does the job. Indeed, observe that $b(z_1) = b(z_2) = b(y_1)$, $b(x_j) = b(y_j) = b(y_{j+1})$ for $j \in [i-1]$, $b(x_j) < b(y_j) < b(y_{j+1})$ for $j \in \{i, \ldots, k-1\}$, and $b(x_k) < b(y_k) < b(x_{k+1})$.
3. Finally, if $v = x_{k+1}$, then we choose b so that the values of all the variables in ψ are equal. It certainly satisfies ψ. $\qquad\square$

We now present split and windmill relations.

Definition 6. *Let R be a four-ary OH relation. We say that R is a split relation if R entails $(v_1 \neq v_2 \vee v_3 = v_4)$ and contains*

1. $\{(v_1, v_2, v_3, v_4) \mid v_1 = v_2 < v_3 = v_4 \vee v_1 < v_2 < v_3 < v_4\}$, or
2. $\{(v_1, v_2, v_3, v_4) \mid v_1 = v_2 > v_3 = v_4 \vee v_1 > v_2 > v_3 > v_4\}$

As an example of a split relation consider $\{(v_1, v_2, v_3, v_4) \mid (v_1 \neq v_2 \vee v_3 = v_4) \wedge v_1 < v_3 \wedge v_1 < v_4 \wedge v_2 < v_3 \wedge v_2 < v_4\}$.

Definition 7. *Let R be a four-ary OH relation. We say that R is a windmill relation if R entails $(v_1 \neq v_2 \vee v_3 < v_4)$ and contains*

1. $\{(v_1, v_2, v_3, v_4) \mid v_1 = v_2 < v_3 < v_4 \vee v_1 < v_2 < v_4 < v_3\}$, or
2. $\{(v_1, v_2, v_3, v_4) \mid v_1 = v_2 > v_4 > v_3 \vee v_1 > v_2 > v_3 > v_4\}$.

For example, $\{(v_1, v_2, v_3, v_4) \mid v_1 \neq v_2 \vee v_3 < v_4\}$ is a windmill relation.

Lemma 10. *Let R be a split relation. Then it is not preserved by any oligopotent QNUF.*

Lemma 11. *Let R be a windmill relation. Then it is not preserved by any oligopotent QNUF.*

5.2   Complex OH Languages

In this section we show that every complex OH relation pp-defines a domino, a split, or a windmill relation. An n-ary OH relation R is complex if it properly entails a complex OH clause C of the form: $(\Phi[P] \vee v_i = v_j)$, $(\Phi[P] \vee v_i \leq v_j)$, or $(\Phi[P] \vee v_i < v_j)$. We will say that R is =-complex, $\leq$-complex, or <-complex, respectively. We treat each case separately, however, we use the same strategy every time. We select some $g, h \in S$ for some $S \in P$ and then consider a formula:

$$\Theta := \exists w_1 \ldots \exists w_k \; (R(v_1, \ldots, v_n) \wedge \bigwedge_{S \in P'} \bigwedge_{a,b \in S} v_a = v_b), \tag{2}$$

where $\{w_1, \ldots, w_k\} = \{v_1, \ldots, v_n\} \setminus \{v_g, v_h, v_i, v_j\}$, and $P' \in \mathbf{P}_n$ is either $\{\emptyset\}$ if $P = \{\{g, h\}\}$ or $(P \setminus \{S\}) \cup \{S \setminus \{h\}\}$ otherwise. In every case, using Lemmas 12, 13, and 2, we show that the formula Θ defines a domino, a split, or a windmill relation R_Θ. Observe that by Lemma 1, this relation is an OH relation; and therefore we can apply Lemma 2 to it.

Lemma 12. *Let R be an n-ary OH relation and $P \neq \{\emptyset\}$ be in $\mathbf{P}_n$. Let Θ and P' be as defined above. Then all of the following holds:*

- *R entails $(\Phi[P] \vee v_k \circ_1 v_l)$ for some $k, l \in [n]$ and $\circ_1 \in \{=, \leq, <\}$ if and only if Θ entails $(v_g \neq v_h \vee v_k \circ_1 v_l)$;*
- *R entails $(\Phi[P'] \vee v_k \circ_1 v_l)$ for some $k, l \in [n]$ and $\circ_1 \in \{=, \leq, <\}$ if and only if Θ entails $(v_k \circ_1 v_l)$;*
- *R entails $(\Phi[P])$ if and only if Θ entails $(v_g \neq v_h)$.*

Lemma 13. *Let R be an n-ary OH relation that properly entails a complex OH clause $(\Phi[P] \vee v_i \circ v_j)$. Then the relation R_Θ defined by Θ as in (2) has orbits:*

- *O_1 satisfying $(v_g = v_h \wedge v_i \circ v_j)$, and*
- *O_2 satisfying $(v_g \neq v_h \wedge v_i \circ_1 v_j)$, where $v_i \circ_1 v_j$ is equivalent to $\neg(v_i \circ v_j)$.*

Assume first that R properly entails $(\Phi[P] \vee v_i = v_j)$.

Lemma 14. *Let R be an n-ary OH relation that is =-complex. Then it pp-defines a domino relation or a split relation.*

Proof. Since R properly entails C, there is no S in P containing both i and j.

First, we consider the case where there is $S \in P$ that contains either i or j. Without loss of generality, we assume the former. We choose g to be equal to i, and h to be another element in S. In this case Θ is over variables v_h, v_i, v_j. We will show that $R_\Theta = \{(v_h, v_i, v_j) \mid \Theta(v_h, v_i, v_j)\}$ is, up to a permutation of coordinates, a domino relation. By Lemma 12, the formula Θ entails $(v_h \neq v_i \vee v_i = v_j)$, and by Lemma 13, the relation R_Θ contains $O_1 = \{(v_h, v_i, v_j) \mid v_h = v_i = v_j\}$ and an orbit O_2 satisfying $(v_h \neq v_i \wedge v_i \neq v_j)$. We assume without loss of generality that O_2 satisfies $(v_h < v_i \wedge v_i \neq v_j)$ and show that in this case R_Θ or $R_\Theta(v_i, v_h, v_j)$, which is R_Θ with a permutation of the first two arguments, is a domino relation. By Definition 5, it suffices to prove that R_Θ

contains an orbit $\{(v_h, v_i, v_j) \mid v_h < v_i < v_j\}$ or $\{(v_h, v_i, v_j) \mid v_j < v_h < v_i\}$. Assume the contrary. Then, by Lemma 2, an orbit satisfying (i) $v_h < v_i$, (ii) $v_i < v_j$, or (iii) $v_h < v_j$ and an orbit satisfying (iv) $v_j < v_h$, (v) $v_h < v_i$, or (vi) $v_j < v_i$ is not in R_Θ. Since O_2 is in R_Θ cases (i) and (v) cannot hold. Hence, we have that Θ entails $v_h = v_j$ (by (iii) and (iv)); $v_i = v_j$ (by (ii) and (vi)); $v_h \geq v_j \geq v_i$ (by (iii) and (vi)); or $v_i \geq v_j \geq v_h$ (by (ii) and (iv)). We claim that each of these cases contradicts the assumption that C is properly entailed by R. Indeed, if Θ entails $v_h = v_j$ or $v_i = v_j$, then, by Lemma 12, R entails $(\Phi[P'] \vee v_h = v_j)$ or $(\Phi[P'] \vee v_i = v_j)$, respectively. Futhermore, if Θ entails $(v_h \geq v_j \geq v_i)$ or $(v_i \geq v_j \geq v_h)$, then R entails $D_1 := (\Phi[P'] \vee v_h \geq v_j)$ and $D_2 := (\Phi[P'] \vee v_j \geq v_i)$; or $E_1 := (\Phi[P'] \vee v_i \geq v_j)$ and $E_2 := (\Phi[P'] \vee v_j \geq v_h)$. Since $D_1 \wedge D_2$ as well as $E_1 \wedge E_2$ entail C and $D_1, D_2, E_1, E_2 \nleq C$, we are done.

We will now consider the situation where every $S \in P$ contains neither i nor j. In this case Θ defines a four-ary relation. We will show that $R_\Theta = \{(v_g, v_h, v_i, v_j) \mid \Theta(v_g, v_h, v_i, v_j)\}$ is, up to a permutation of coordinates, a split relation. By Lemma 12, the formula Θ entails $(v_g \neq v_h \vee v_i = v_j)$. By Lemma 13, R_Θ contains an orbit O_1 satisfying $(v_g = v_h \wedge v_i = v_j)$ and an orbit O_2 satisfying $(v_g \neq v_h \wedge v_i \neq v_j)$. In the next step, we show that we can choose O_1 so that it satisfies $(v_g = v_h \neq v_i = v_j)$. Assume the contrary. Then Θ entails both $(v_g \neq v_h \vee v_h = v_i)$ and $(v_g \neq v_h \vee v_h = v_j)$. By Lemma 12, we have that R entails $D_1 := (\Phi[P] \vee v_h = v_i)$ as well as $D_2 := (\Phi[P] \vee v_h = v_j)$. Note that v_j and v_i do not occur in D_1 and D_2, resp. Thus, $D_1, D_2 \nleq C$. Since $D_1 \wedge D_2$ entails C, we have the contradiction with the assumption that C is properly entailed by R. Thus, O_1 satisfies either $\{(v_g, v_h, v_i, v_j) \mid v_g = v_h < v_i = v_j\}$ or $\{(v_g, v_h, v_i, v_j) \mid v_g = v_h > v_i = v_j\}$. In the following, we assume the first possibility. The other case is symmetric. Since R_Θ contains an orbit satisfying $(v_g \neq v_h \wedge v_i \neq v_j)$, it contains $(v_g < v_h \wedge v_i < v_j)$, $(v_g < v_h \wedge v_i > v_j)$, $(v_g > v_h \wedge v_i < v_j)$, or $(v_g > v_h \wedge v_i > v_j)$. By Lemma 2, since $O_1 \subseteq R_\Theta$, it contains $\{(v_g, v_h, v_i, v_j) \mid v_g < v_h < v_i < v_j\}$, $\{(v_g, v_h, v_i, v_j) \mid v_g < v_h < v_j < v_i\}$, $\{(v_g, v_h, v_i, v_j) \mid v_h < v_g < v_i < v_j\}$, or $\{(v_g, v_h, v_i, v_j) \mid v_h < v_g < v_j < v_i\}$. Thus $R_\Theta(v_g, v_h, v_i, v_j)$, $R_\Theta(v_g, v_h, v_j, v_i)$ $R_\Theta(v_h, v_g, v_i, v_j)$, or $R_\Theta(v_h, v_g, v_j, v_i)$ is a split relation. $\qquad\square$

The second case, that is, if R properly entails $(\Phi[P] \vee v_i \leq v_j)$ can be proved in the similar manner. We first show that R pp-defines $\{(x, y) \mid x \leq y\}$; and then that $(\Theta \wedge v_i \geq v_j)$ pp-defines a domino, or a split relation.

Lemma 15. *Let R be a $\leq$-complex OH relation, then R pp-defines a domino or a split relation.*

The third case requires an auxiliary lemma.

Lemma 16. *Let R be an OH relation that is $<$-complex but neither $=$-complex nor $\leq$-complex. Then there is an OH clause of the form $(\Phi[P_1] \vee v_k < v_l)$ properly entailed by R and $m \in \bigcup_{S \in P_1} S$ such that R does not entail $(\Phi[P_1] \vee v_k \leq v_m)$ or $(\Phi[P_1] \vee v_m \leq v_l)$.*

Lemma 17. *Let R be an OH relation that is $<$-complex, then it pp-defines a split, a domino, or a windmill relation.*

Proof. By Lemmas 14 and 15, we can assume that R is neither $=$-complex nor $\leq$-complex. Hence, by Lemma 16, we can assume that there is a clause C of the form $(\Phi[P] \vee v_i < v_j)$ and $h \in \bigcup_{S \in P} S$ such that R does not entail $(\Phi[P] \vee v_i \leq v_h)$ or R does not entail $(\Phi[P] \vee v_h \leq v_j)$. In the following, we assume the former; observe that by Lemma 12, it implies that Θ does not entail $(v_g \neq v_h \vee v_i \leq v_h)$. Let S be an element of P that contains h and $g \neq h$ be another element of S. In this case Θ, see (2), defines a four-ary relation $R_\Theta := \{(v_g, v_h, v_i, v_j) \mid \Theta(v_g, v_h, v_i, v_j)\}$. We will show that R_Θ is, up to a permutation of coordinates, a windmill relation. By Lemma 12, the formula Θ entails $(v_g \neq v_h \vee v_i < v_j)$. By Lemma 13, we have that R_Θ contains O_1 satisfying $(v_g = v_h \wedge v_i < v_j)$. Since Θ does not entail $(v_g = v_h \wedge v_i \leq v_h)$, we can choose O_1 equal to $\{(v_g, v_h, v_i, v_j) \mid v_g = v_h < v_i < v_j\}$. To complete the proof, we will show that R_Θ contains $\{(v_g, v_h, v_i, v_j) \mid v_g < v_h < v_j < v_i\}$ or $\{(v_g, v_h, v_i, v_j) \mid v_h < v_g < v_j < v_i\}$. It implies that R_Θ, or $R_\Theta(v_h, v_g, v_i, v_j)$ is a windmill relation.

Since R entails C, it entails a normalized OH clause D_1 equivalent to $(\Phi[P] \vee v_i \neq v_j)$. Let D_2 be $(\Phi[P'] \vee v_i \leq v_j)$. Recall that C is properly entailed by R. Since $D_1 \wedge D_2$ entails C and $D_1, D_2 \not\lesssim C$, we have that R does not entail D_2. By Lemma 12, the formula Θ does not entail $v_i \leq v_j$, and hence R_Θ contains an orbit satisfying $(v_g < v_h \wedge v_j < v_i)$ or $(v_h < v_g \wedge v_j < v_i)$. Since $O_1 \subseteq R_\Theta$, by Lemma 2, R_Θ contains $\{(v_g, v_h, v_i, v_j) \mid v_g < v_h < v_j < v_i\}$ or $\{(v_g, v_h, v_i, v_j) \mid v_h < v_g < v_j < v_i\}$. $\qquad\square$

We summarize this section with the following theorem.

Theorem 4. *Let $\mathbf{B}$ be a complex OH language. Then it does not have local-to-global consistency.*

Proof. If $\mathbf{B}$ is complex, then it contains a relation R that is complex. If R is complex, then by Lemmas 14, 15 and 17, we have that R pp-defines $R_{3/4}$ which is a domino, a split, or a windmill relation. Furthermore, by Lemmas 9, 10, and 11, the relation $R_{3/4}$ is not preserved by any oligopotent QNUF. Thus, the fact that $\mathbf{B}$ does not have an oligopotent QNUF follows from Proposition 1. The proof is completed by applying Theorem 2. $\square$

Acknowledgements. The author would like to thank Hubie Chen and Manuel Bodirsky for many inspiring discussion on the subject of this paper, and the referees for their helpful comments.

References

1. Fisher, M., Gabbay, D., Vila, L. (eds.): Handbook of Temporal Reasoning in Artificial Intelligence. Elsevier (2005)
2. Renz, J., Nebel, B.: Qualitative spatial reasoning using constraint calculi. In: Aiello, M., Pratt-Hartmann, I., van Benthem, J. (eds.) Handbook of Spatial Logics, pp. 161–215. Springer, Berlin (2007)

3. Duentsch, I.: Relation algebras and their application in temporal and spatial reasoning. Artificial Intelligence Review 23, 315–357 (2005)
4. Allen, J.F.: Maintaining knowledge about temporal intervals. Communications of the ACM 26(11), 832–843 (1983)
5. Mackworth, A.K.: Consistency in networks of relations. AI 8, 99–118 (1977)
6. Freuder, E.: A sufficient condition for backtrack-free search. Journal of the ACM 29(1), 24–32 (1982)
7. Dechter, R.: From local to global consistency. Artificial Intelligence 55(1), 87–108 (1992)
8. Feder, T., Vardi, M.: The computational structure of monotone monadic SNP and constraint satisfaction: A study through Datalog and group theory. SIAM Journal on Computing 28, 57–104 (1999)
9. Nebel, B., Bürckert, H.J.: Reasoning about temporal relations: A maximal tractable subclass of Allen's interval algebra. Journal of the ACM 42(1), 43–66 (1995)
10. Bodirsky, M., Chen, H.: Qualitative temporal and spatial reasoning revisited. In: CSL 2007, pp. 194–207 (2007)
11. Bodirsky, M., Dalmau, V.: Datalog and Constraint Satisfaction with Infinite Templates. In: Durand, B., Thomas, W. (eds.) STACS 2006. LNCS, vol. 3884, pp. 646–659. Springer, Heidelberg (2006)
12. Jeavons, P., Cohen, D., Gyssens, M.: Closure properties of constraints. Journal of the ACM 44(4), 527–548 (1997)
13. Bodirsky, M., Chen, H.: Oligomorphic clones. Algebra Universalis 57(1), 109–125 (2007)
14. Vilain, M., Kautz, H., van Beek, P.: Constraint propagation algorithms for temporal reasoning: A revised report. Reading in Qualitative Reasoning about Physical Systems, 373–381 (1989)
15. Koubarakis, M.: From local to global consistency in temporal constraint networks. Theor. Comput. Sci. 173(1), 89–112 (1997)
16. van Beek, P., Cohen, R.: Exact and approximate reasoning about temporal relations. Computational Intelligence 6, 132–144 (1990)
17. Bessi, C., Isli, A.: Global consistency in interval algebra networks: Tractable subclasses. In: Proc. ECAI 1996, pp. 3–7. Wiley (1996)
18. Börner, F., Bulatov, A., Jeavons, P., Krokhin, A.: Quantified Constraints: Algorithms and Complexity. In: Baaz, M., Makowsky, J.A. (eds.) CSL 2003. LNCS, vol. 2803, pp. 58–70. Springer, Heidelberg (2003)
19. Bodirsky, M., Kára, J.: The complexity of temporal constraint satisfaction problems. Journal of the ACM 57(2) (2009); An extended abstract appeared in the proceedings of STOC 2008
20. Bodirsky, M., Chen, H.: Quantified equality constraints. SIAM J. Comput. 39(8), 3682–3699 (2010)
21. Charatonik, W., Wrona, M.: Quantified Positive Temporal Constraints. In: Kaminski, M., Martini, S. (eds.) CSL 2008. LNCS, vol. 5213, pp. 94–108. Springer, Heidelberg (2008)
22. Charatonik, W., Wrona, M.: Tractable Quantified Constraint Satisfaction Problems over Positive Temporal Templates. In: Cervesato, I., Veith, H., Voronkov, A. (eds.) LPAR 2008. LNCS (LNAI), vol. 5330, pp. 543–557. Springer, Heidelberg (2008)
23. Hodges, W.: A shorter model theory. Cambridge University Press, Cambridge (1997)
24. Bodirsky, M., Nešetřil, J.: Constraint satisfaction with countable homogeneous templates. Journal of Logic and Computation 16(3), 359–373 (2006)
25. Bodirsky, M., Kára, J.: A fast algorithm and Datalog inexpressibility for temporal reasoning. ACM Transactions on Computational Logic 11(3) (2010)

A Hybrid Paradigm for Adaptive Parallel Search

Xi Yun[1] and Susan L. Epstein[1,2]

[1] Department of Computer Science,
The Graduate Center of the City University of New York, New York, NY 10016, USA
[2] Department of Computer Science,
Hunter College of the City University of New York, New York, NY 10065, USA
xyun@gc.cuny.edu, susan.epstein@hunter.cuny.edu

Abstract. Parallelization offers the opportunity to accelerate search on constraint satisfaction problems. To parallelize a sequential solver under a popular message passing protocol, the new paradigm described here combines portfolio-based methods and search space splitting. To split effectively and to balance processor workload, this paradigm adaptively exploits knowledge acquired during search and allocates additional resources to the most difficult parts of a problem. Extensive experiments in a parallel environment show that this paradigm significantly improves the performance of an underlying sequential solver, outperforms more naive approaches to parallelization, and solves many difficult problems left open after recent solver competitions.

1 Introduction

SPREAD (Search by Probing and REcursive Adaptive Domain-splitting) is an adaptive paradigm that harnesses parallel computation to enhance an underlying sequential constraint solver (henceforward, a *solver*). Because SPREAD does not alter its solver, only minimal programming for message passing is required for use with modern solvers. Our thesis is that, on difficult problems, parallelization that combines efficient task assignment with effective exploitation of information can significantly improve performance. The principal results reported here are that SPREAD significantly improves the performance of its solver, outperforms a variety of reasonable alternatives, and solves many difficult constraint satisfaction problems left open after recent solver competitions.

SPREAD makes only two assumptions about its solver. First, the solver directs search with a variable-ordering heuristic toward *contention*, variables whose constraints are more likely to cause wipeout [1]. (Here, we used learned variable weights [2], but variable impact would be an alternative [3].) Second, the solver uses a restart strategy to extricate search from early unproductive assignments [4]. Most modern solvers satisfy both conditions.

SPREAD uses a *manager-worker* framework, where a *manager* assigns tasks and coordinates messages among all the other processors (the *workers*). SPREAD has two phases: a time-limited portfolio phase followed by a splitting phase. In the *portfolio phase*, SPREAD's multiple workers search in parallel from random

M. Milano (Ed.): CP 2012, LNCS 7514, pp. 720–734, 2012.

seeds; if any worker reports a solution or proves that there is none, the problem is solved. Otherwise, once the portfolio phase exhausts its time allocation, SPREAD begins its *splitting phase*, where the manager partitions the original problem into subproblems based on weights learned thus far. The manager distributes the subproblems to the workers with search limits based on the search effort during the portfolio phase. If any worker reports a solution, or if all the subproblems are proved to have no solution, the problem is solved. Any subproblem returned unsolved to the manager undergoes further partitioning. Those new subproblems are enqueued and eventually re-distributed with larger search limits as workers become available. This recursive partitioning mechanism naturally directs computational power to difficult subproblems.

SPREAD facilitates parallelization. It accepts any constraint supported by its solver. To partition problems, SPREAD manipulates domains rather than constraints, so that users need not learn propagators provided by the solver or implement new ones. Because its domain splitting method is general, SPREAD could be extended to continuous variable domains. SPREAD's portfolio phase solves easy problems quickly and stably. For more difficult problems, the portfolio phase also learns weights that determine how the manager in the subsequent splitting phase generates subproblems. Moreover, workers can exploit those same weights during their search on subproblems. In practice, the variables used to generate subproblems can be statically chosen before the splitting phase (*SPREAD-S*), or determined dynamically from weights learned during search on the corresponding subproblem (*SPREAD-D*). (For clarity, we refer to the paradigm here as SPREAD, and the individual implementations as SPREAD-S and SPREAD-D.) After relevant background and related work in the next section, we describe SPREAD, offer some reasonable alternatives, evaluate SPREAD-S and SPREAD-D against them, and discuss their advantages and limitations.

2   Background and Related Work

A constraint satisfaction problem (*CSP*) $P = \langle \mathcal{X}, \mathcal{D}, \mathcal{C} \rangle$ is defined by a set of variables $\mathcal{X} = \{X_1, ..., X_n\}$, each with an associated domain $\mathcal{D} = \{d_1, ..., d_n\}$, and a set of constraints $\mathcal{C} = \{c_1, ..., c_m\}$. A *solution* to P assigns a value to each variable in $\mathcal{X}$ from its respective domain so that it satisfies $\mathcal{C}$. If P has a solution, it is *satisfiable*; otherwise it is *unsatisfiable*. The solver here is assumed complete; it executes systematic backtracking, traditionally envisioned as a search tree. There, after each value assignment, inference removes from the domains of the as-yet-unbound variables all values that it shows inconsistent with $\mathcal{C}$. If a domain becomes empty (a *wipeout*), *weight learning* increases the weight of the constraint that removed the last value. Once search stops, the weight of a variable is taken as the sum of the weights on the constraints that restrict it [2].

Parallelization seeks to exploit the massive computing resources increasingly available on multicore computers, and in clusters, grids, and clouds. Research on parallelization for CSP solvers includes a broad spectrum of parallel programming models (e.g., OpenMP [5,6,7], Message Passing Interface (*MPI*) [8,9]) and a variety of platforms (e.g., single node [5,7], cluster [9,10], and grid [11]), on a scale from a few processors to thousands. In particular, MPI is intended for high-performance parallel computing on platforms without shared memory. Its convenience and portability have made it the *de facto* standard for a variety of technological platforms. This work uses MPI on a cluster and executes extensive experiments on up to 256 processors, a number widely available in modern computing environments.

Search space splitting can explore different search subspaces on different processors. For SAT instances, with their boolean domains, search space splitting usually relies on a guiding path (e.g., Fig. 1(a)). A boolean flag δ_i indicates whether a node is *closed* (both values attempted, $\delta_i = 0$, black circle) or *open* (one value attempted, $\delta_i = 1$, white circle) [8]. Although identification of a helpful guiding path is non-trivial (as in [5]), variables with particular properties have proved effective for splitting [7]. Iterative partitioning with clause learning, where search spaces of SAT subproblems may overlap, can also be an effective strategy [11].

Given non-boolean domains and various kinds of constraints, search space splitting for CSPs becomes more complex. A SAT solver can conveniently partition its search space by adding new clauses (e.g., parity constraints [12]), without any modification to its search strategy or propagation methods. A CSP solver that tries to add new constraints to split a search space, however, might confront constraints it could not directly process. Even splitting only with already existing constraints (as in [13]) might have to contend with different formulations and different models for the same problem under different solvers. Partitioning by domain manipulation avoids such difficulties. One approach, *network extraction* (NE), performs a sequence of domain splits on a subset of $\mathcal{X}$ under a given variable ordering for a single processor [14], as in Fig. 1(b). A split on the ith variable produces subproblems P_i^1 and P_i^2 that differ only in the ith variable's

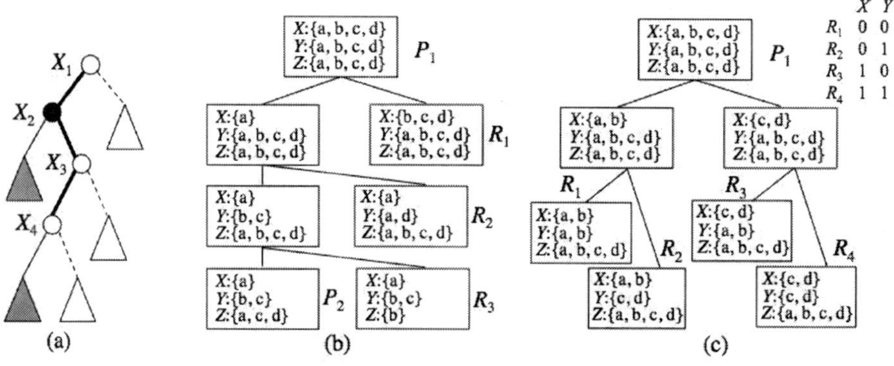

Fig. 1. (a) A guiding path with open nodes at X_1, X_3, and X_4. (b) Extraction of subproblem P_2 from P_1 (under variable order X, Y, Z) produces subproblems R_1, R_2, R_3. (c) Iterative bisection partitioning on X and Y creates a virtual binary search tree of subproblems, shown with their bit-string representations.

domain: P_i^1 has some values for the ith variable, and P_i^2 has the rest. NE was developed to avoid duplicate search on visited search spaces after restart [14].

Another prevalent parallelization approach for CSP solvers uses an algorithm portfolio [15]. A *portfolio-based* method schedules a set of algorithms (its *portfolio*) on one or more processors, hoping to outperform any of its constituent algorithms [12,16]. Although this approach can benefit from information shared among processors, as when parallel SAT solvers share clauses [17], most portfolio-based methods for CSPs do not share information [12].

Additional parallelization methods include various workload-balancing mechanisms, such as work stealing [5,6,10] and work sharing [9]. The SAT-solver parallelization methods most relevant to SPREAD are described in [7] and [11]. The first passes information from a portfolio phase to a subsequent splitting phase for effective partitioning; the second iteratively partitions a SAT problem with learned clauses. SPREAD combines and extends them to parallelize adaptive search for CSPs. (An earlier version of SPREAD appeared as a modification to an explore-and-follow parallelization paradigm [18].) The CSP work most relevant to SPREAD is [13]. SPREAD, however, better utilizes its computing resources initially, splits the search space by domain manipulation, avoids nogood learning, and proves convenient for systematic experimental evaluation.

3   The SPREAD Paradigm

SPREAD uses its manager to partition and distribute tasks, and leaves search entirely to its workers. Each worker executes the solver exactly as it would on a single processor, but may receive different parameter values from the manager. SPREAD starts with a weak portfolio, where different workers execute the same solver from different random seeds on the full problem. The subsequent splitting phase formulates subproblems, and recursively partitions those that prove difficult to direct additional computing resources to them.

Given problem P with search limit ℓ and restart schedule *policy*, the SPREAD-S manager executes the portfolio phase with Algorithm 1 on workers under the control of Algorithm 2. The manager then uses Recursive Splitting with Iterative Bisection Partitioning (*RS-IBP*) during the splitting phase with Algorithm 3. As is traditional in MPI, the manager executes on processor 0, and the other k processors are workers. We extend SPREAD-S to SPREAD-D at the end of this section.

Portfolio Phase. To begin, the manager sends the initialization signal 0 to each of the k workers (Algorithm 1, lines 2-4). On receipt of that message, workers (Algorithm 2) execute a weak algorithm portfolio, attempting to solve P within ℓ with different random seeds. Any proof of either P's satisfiability or unsatisfiability within ℓ leads to an immediate report as well as the termination of the MPI environment, including execution on all workers (Algorithm 2, lines 5-6). Otherwise, worker i has exhausted ℓ, and reports to the manager its learned weights w_l and backtrack count b_l (Algorithm 2, line 8). The manager receives w_p and b_p from worker p (Algorithm 1, line 6), possibly in non-numerical order.

Finally, the manager forwards to the splitting phase *weights*, the average of the variable weights w_i received from all the workers, and *base*, the average of the backtrack counts b_i received from them, where they will be used to guide search on the subproblems (Algorithm 1, lines 8-9).

Algorithm 1 Portfolio (Manager)
Input: P, *policy*
Output: weights, backtrack counts
1: $signal \leftarrow 0$
2: **for** $i = 1$ to k
3:   $w_i \leftarrow 0$, $b_i \leftarrow 0$
4:   MPI.Send($signal$, i)
5: **while** $i > 0$
6:   MPI.Recv($\langle w_p, b_p \rangle$, p)
7:   $i \leftarrow i - 1$
8: Compute *weights* from all w_i
   and *base* from all b_i
9: **return** $\langle weights, base \rangle$

Algorithm 2 Worker
Input: P, ℓ, *policy*
Output: result of search on P
1: **while** TRUE
2:   MPI.Recv($signal$, 0)
3:   **if** $signal = $ -1 **break**
4:   **if** $signal = 0$ // portfolio phase
5:     **if** solve(P, ℓ, *policy*, *rand_seed*)
6:       Output result and abort MPI
7:     **else**
8:       MPI.Send($\langle w_l, b_l \rangle$, 0)
9:   **else**   // splitting phase
10:    MPI.Recv($\langle P^S, \ell_r, weights \rangle$, 0)
11:    Initialize variable weights of P^S
12:    **if** solve(P^S, ℓ_r, *policy*, *rand_seed*)
13:      **if** P^S is satisfiable
14:        Output *sol* and abort MPI
15:      **else do** MPI.Send($\langle 0, P^S \rangle$, 0)
16:    **else do** MPI.Send($\langle 1, P^S \rangle$, 0)

Algorithm 3 RS-IBP (Manager)
Input: P, weights w, *base*, *policy*
Output: solution of P
1: $v \leftarrow get\_splitting\_number(k)$
2: $threshold \leftarrow compute\_threshold(v)$
3: $splits \leftarrow choose(v, P, w)$
4: **for** P^S in IBP(P, $splits$)
5:   Q.push(P^S)
6: **for** $i = 1$ to 2^v **do** L.push($base$)
7: $\#enqueued \leftarrow Q.size()$
8: $\#feedback \leftarrow 0$, $signal \leftarrow 1$
9: **for** $i = 1$ to k **do** $distribute(i)$
10: **while** $\#feedback < \#enqueued$
11:   MPI.Recv($\langle feedback, P^S \rangle$, p)
12:   $\#feedback \leftarrow \#enqueued + 1$
13:   **if** $feedback = 0$
14:     **if** $!Q.empty()$
15:       $distribute(p)$
16:   **else**
17:     **if** $Q.size() < threshold$
18:       $splits \leftarrow choose(\xi, P^S, w)$
19:       **for** P_i^S in IBP(P^S, $splits$)
20:         Q.push(P_i^S)
21:         L.push($get\_limit(P_i^S, base)$)
22:         $\#enqueued \leftarrow \#enqueued+1$
23:     **else**
24:       Q.push(P^S)
25:       L.push($get\_limit(P^S, base)$)
26:       $\#enqueued \leftarrow \#enqueued+1$
27:     **for** $i = 1$ to k
28:       **if** i is idle && $!Q.empty()$
29:         $distribute(i)$
30: Send $signal$ -1 to all processors

A portfolio-based method that shares information must address the trade-off between diversification and intensification [19]. Diversification uses dramatically different search strategies, and expects its searchers to proceed independently. In contrast, intensification explores with relatively small variations around a single strategy, and expects to share the information it gathers among all its searchers. Because SPREAD is intended to solve difficult CSPs, it does intensification in its portfolio phase, as recommended in [19]. Indeed, the primary purpose of SPREAD's portfolio phase is to glean information to support search space splitting, not to solve P.

Iterative Bisection Partitioning. A *bisection partition* (BP) on variable X_i with domain d_i replaces X_i with two variables, X_i' and X_i'', whose respective domains d_i' and d_i'' partition d_i. To generate subproblems with search spaces that may have similar sizes, without bias toward particular domain values, we adopt an (almost) even bisection partition where $d_i' = \{v_1, ..., v_\chi\}$, $d_i'' = \{v_{\chi+1}, ..., v_{|d|}\}$, and $\chi = \lceil |d|/2 \rceil$. *Iterative bisection partitioning* (IBP) repeats BP on v ordered *splitting variables* to generate 2^v subproblems. Fig. 1 (c) illustrates IBP on variables X and Y of P_1 to generate subproblems R_1, R_2, R_3, and R_4. Intuitively, overall search performance on P_1 could be improved by processing such subproblems on different processors in parallel.

In SPREAD, the manager chooses as splitting variables those with the highest weights. This conserves the promising variable ordering already found effective in the portfolio phase by a solver that exploits those weights (e.g., variable-ordering heuristic *dom/wdeg* [2]). Moreover, since IBP splits domains of CSP instances much the way a guiding path splits $\{0,1\}$ for SAT problems, an IBP-generated subproblem can analogously be represented by a guiding path, where L_i indicates whether the ith splitting variable X_i is associated with d_i' ($L_i = 0$) or with d_i'' ($L_i = 1$). (See Fig. 1(c).) This simple bit-string representation reduces the communication effort required to pass subproblems to workers.

Splitting Phase. In its splitting phase, SPREAD recursively splits the search space with IBP (Algorithm 3). Initially, the manager partitions P into several subproblems, each represented as a bit string for the partition that gave rise to it, and allocates *base* backtracks to each one. Subproblems and their backtrack limits are maintained in queues $\mathcal{Q}$ and $\mathcal{L}$, respectively. For k workers, the manager determines how many initial splitting variables to use (here, $v = \lceil \log_2 2k \rceil$), computes the queue length *threshold* (here, 2^v), and then chooses *splits*, the v variables with the highest weights in *weights* learned for P during the portfolio phase (lines 1-3). Next the manager partitions P on *splits* in descending order of weight, and tracks the resultant subproblems and their respective backtrack limits (lines 4-6). Before it distributes subproblems to workers with backtrack limits and variable weights (line 9), the manager notifies the ith worker with signal 1 that it is about to do so. The manager then dequeues and sends the first k subproblems on $\mathcal{Q}$ with their corresponding weights and backtrack limits from $\mathcal{L}$, and awaits feedback.

As in the portfolio phase, a worker immediately reports any detected solution to the manager, and terminates the MPI environment (Algorithm 2, line 14). If a worker proves its subproblem P^S unsatisfiable, however, it notifies the manager with message 0 (Algorithm 2, line 15). The manager replies with a new subproblem from $\mathcal{Q}$ (if any is waiting, Algorithm 3, lines 14-15). Otherwise, the worker has exhausted its resources ℓ_r and returns its subproblem to the manager with message 1 (Algorithm 2, line 16). If the subproblem queue has fewer than *threshold* subproblems, the manager recursively partitions the returned subproblem on new splitting variables, and enqueues the resultant subproblems with their resource limits (Algorithm 3, lines 18-22). If there is insufficient space on the queue to repartition the subproblem, the manager re-enqueues it as it was, but with a larger resource limit (Algorithm 3, lines 24-26). Whether or not it repartitions returned subproblems, the manager continues to distribute subproblems from Q to any idle worker (Algorithm 3, lines 27-29). RS-IBP terminates when some worker finds a solution, or when all subproblems are proved unsatisfiable.

When eventually distributed, an unresolved subproblem (even without repartitioning) will break ties with a random seed, and may therefore have a different search experience. To bound the size of Q, for each split, SPREAD-S here chooses ξ as $\max\{\lceil \log_2(threshold - Q.size())\rceil, 1\}$ (Algorithm 3, line 18). This bounds the length of Q at $2^v + 2^{v-1} - 1$, which happens only when an unresolved subproblem confronts a queue of length $2^{v-1} - 1$. Nonetheless, IBP's concise bit-string representation makes it space-efficient, and in practice allows large queues.

Spread-D. SPREAD-S always uses the same splitting variables in the same order, determined by the weights first learned during its portfolio phase. Intuitively, for a returned subproblem, it could be more accurate to determine splitting variables dynamically, with weights learned during search on that subproblem. SPREAD-D is an extension of SPREAD-S that dynamically chooses its splitting variables. When a SPREAD-D worker fails to solve a subproblem within the allocated resource, it returns to the manager both the subproblem and the weights learned on it (i.e., received initially from the manager and modified during this search). This requires modification of only Algorithm 2, line 16 and Algorithm 3, line 11. The manager then chooses, in line 18, additional splitting variables with the highest weights acquired during search on the returned subproblem. Because SPREAD-D never changes *splits*, which originally designated the subproblem returned in line 2, it guarantees mutually exclusive subproblems.

In the portfolio phase, SPREAD-S and SPREAD-D terminate only when some worker finds a solution or proves the problem unsatisfiable. Otherwise they enter the splitting phase where, without a search limit, they terminate only when a solution is found or all subproblems are proved unsatisfiable. SPREAD is complete, because IBP generates subproblems with mutually exclusive, collectively exhaustive search spaces, and a subproblem is always partitioned or receives larger search limits. Section 5 demonstrates that SPREAD is also effective.

4   Experimental Design

The experiments reported here evaluate parallelization methods on their ability to solve both problems difficult for the underlying solver and problems difficult for all the solvers in the two most recent international CSP solver competitions [20,21]. From the repository of more than 7000 problems in those competitions, we selected 51 representative classes that cover a broad variety of CSPs with relatively uniform population distribution, shown in Fig. 2. To avoid any bias toward large classes, we stratified selection from each class to reflect any pre-specified subclasses and naming conventions, and chose a subset from each class in proportion to its original subclass sizes. This produced 1765 problems in classes of sizes from 7 to 65.

The experimental platform was a Cray XE6m system with 160 dual-socket compute nodes. Each node contains two 8-core AMD Magny-Cours processors running at 2.3 GHz. (Here a SPREAD processor corresponds to a Cray core.) Without a readily-available parallel CSP solver as a benchmark, this paper compares the performance of SPREAD-S and SPREAD-D to a variety of parallelization methods inspired by relevant work. We chose to parallelize the solver Mistral-1.331 (with C++ source code from [20]) because it is compatible with MPI on the Cray, and allows us to curate sets of difficult problems from recent CSP solver competitions and to evaluate the performance improvement under SPREAD-S and SPREAD-D. Mistral can be compiled to run sequentially on the Cray XE6m either under the GCC compiler (*Mistral-GCC*) or the CC compiler (*Mistral-CC*), but Mistral-GCC runs about two to three times faster than Mistral-CC. This gives the Mistral-GCC benchmark a considerable advantage over all our parallel solvers, which require the CC compiler for MPI.

We solved each of the 1765 problems with Mistral-GCC, and eliminated the 1398 problems solved by Mistral-GCC in less than one minute. The 119 that could be solved by sequential Mistral-GCC within 1 to 30 minutes on the Cray became the *hard set*; the remaining 248 became the *harder set*. Finally, the *challenge set* consists of the 133 problems never solved by any solver within 30 minutes in either competition, and not already included in the harder set.

We tested Mistral-GCC and Mistral-CC alone, as well as SPREAD-S and SPREAD-D with Mistral-CC, and the following parallelization approaches:

- *Naive Random* (*NR*) races 63 copies of Mistral-CC with random seeds.
- *Parallel Portfolio* (*PP*) races 63 combinations of heuristics and restart policies. The heuristics were *impact, dom/wldeg, dom/wdeg*, and *impact/wdeg*. The restart policies were Luby-k (k (backtracks per unit) $\in \{128, 256, 512, 1024, 2048, 4096\}$), geometric (restart limit $x(n) = 100p^n$ at step n, where

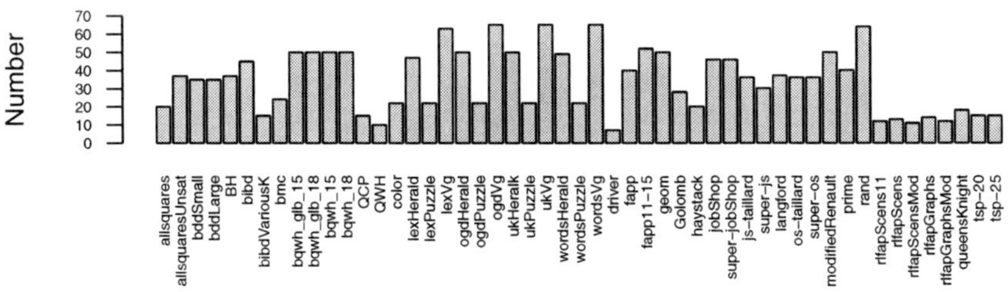

Fig. 2. Population distribution of the 1765 problems in the hard and harder problem sets, identified under stratified selection from 51 CSP competition classes

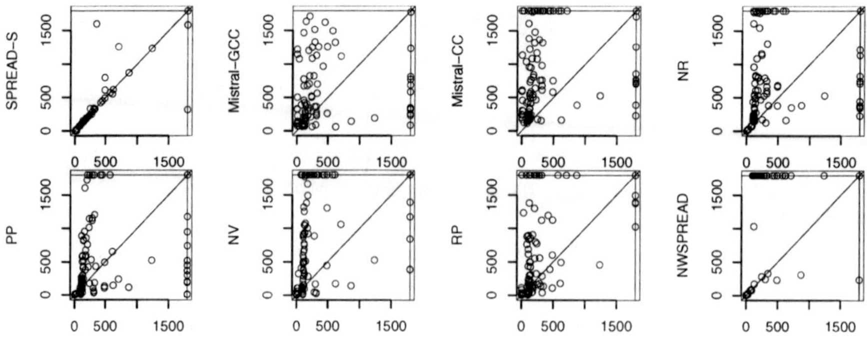

Fig. 3. On the 119-problem hard set, solution time in seconds for SPREAD-D (x-axis) plotted against that for other methods (y-axis). Each circle represents a problem; black areas indicate a heavy concentration of problems. Circles at the top and far right represent unsolved problems.

$p = 1.3$, 1.5 or 2.0), arithmetic ($x(n) = 16000n$, $8000n$, $1000n^2$, and $500n^2$), and dynamic. Dynamic adaptively determines whether to execute geometric restart with exponent 1.3, 1.5, or 2.0 based on the problem formulation, and restarts on the minimum of 1000 and the number of variables. Given these $4{\times}16$ possibilities but only 63 workers, PP did not execute *impact/wdeg* with dynamic restart and exponent 2.0.

- *Naive Variable* (*NV*) partially fixes the variable orders, as suggested in [16]. NV races 63 copies of Mistral, each of which randomly selects and orders the first 3 variables it assigns (but not their values) and reuses those variables on every restart.
- *Random Partitioning* (*RP*) splits on 7 randomly-chosen splitting variables, and enqueues those 128 subproblems for distribution to 63 workers, which run them to completion. Some workers process more than one subproblem.
- *No-Weight SPREAD* (*NWSPREAD*) is an ablated version of SPREAD intended to gauge the impact of learned weights. NWSPREAD does not use the weights from the portfolio phase for the workers, either to split or to search.

5   Experimental Results

Unless otherwise stated, all results reported here use the median of the values from three runs (as in recent parallel SAT solver competitions [22]), under a 30-minute per problem time limit, with the portfolio phase in both versions of SPREAD limited to 100 seconds. The initial backtrack limit was *base*, the average generated in the portfolio phase (Algorithm 1, line 8). When a subproblem was partitioned on ξ additional splitting variables, this limit was multiplied by $(1.5)^\xi$.

On the Hard Problem Set. Fig. 3 compares SPREAD-D's runtime to that of the other approaches in Section 4. Although a few instances (along the right margin) went unsolved under SPREAD-D, Fig. 3 shows that both versions of SPREAD

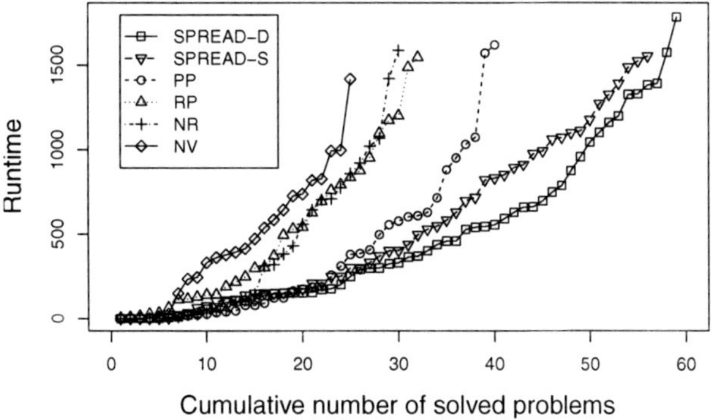

Fig. 4. On the 248 problems in the harder problem set, cumulative number of solved problems across 1800 seconds for SPREAD-D, SPREAD-S, and four competitors

clearly outperform most of the other benchmark methods. Indeed, on the problems solved both by SPREAD-D and each competitor, SPREAD-D achieved average speedups of 19.08 ($\sigma = 79.41$) over Mistral-GCC, 27.91 ($\sigma = 148.32$) over Mistral-CC, 2.65 ($\sigma = 2.77$) over NR, 1.98 ($\sigma = 1.92$) over PP, 4.03 ($\sigma = 3.89$) over NV, 3.34 ($\sigma = 6.48$) over RP, and 1.59 ($\sigma = 1.61$) over NW. Both SPREAD-S and SPREAD-D solved 43.70% of the hard set within 100 – 200 seconds. This is the time when both SPREAD versions have just begun to use critical splitting variables, while PP tries a complementary algorithm portfolio instead. The plot for PP on the lower left is a clear demonstration that search space splitting is essential. Moreover, search space splitting without the knowledge from the portfolio phase (NWSPREAD, on the lower right) was dramatically inferior; it could not solve 75.63% of these problems in 30 minutes, even though NR solved 17.65% of them within 100 seconds. Given their performance, NWSPREAD, sequential Mistral-CC, and Mistral-GCC were excluded from further comparisons.

On the Harder Problem Set. Fig. 4 compares both versions of SPREAD to the remaining parallelization methods. Given 30 minutes per problem, SPREAD-S solved 56 problems (44 satisfiable), 16 more (40.00% improvement) than the best benchmark method PP (which solved 40), and 31 more (124.00%) than the worst, NV (which solved 25). In addition, SPREAD-D solved 59 (46 satisfiable), 3 more than SPREAD-S. As one would expect, both versions of SPREAD behaved early on much like the portfolio-based methods NR and PP. SPREAD-S solved 10 (all satisfiable) within the first 100 seconds, its portfolio phase, while SPREAD-D solved 12 (all satisfiable). Both versions of SPREAD also solved more problems that required more time. In the last 800 seconds, SPREAD-S solved 18 (11 satisfiable) and SPREAD-D solved 12 (6 satisfiable).

On the Challenge Set. SPREAD-S and SPREAD-D significantly outperformed the other parallelization methods. Table 1 compares runtimes for SPREAD-S, SPREAD-D, and RP on the 35 problems solved by at least one of them more than once. (The other approaches from Fig. 4, NR, NV, and PP, solved 1, 1,

Table 1. Challenge problem solution times for SPREAD-S (S-S) and SPREAD-D (S-D), with best in boldface. 10 denotes a 10-second (rather than 100-second) portfolio phase. – denotes failure to solve in 30 minutes.

Problem	SAT	RP	S-S-10	S-S-100	S-D-10	S-D-100
costasArray-20	yes	**721.84**	–	–	876.13	1120.20
crossword-m1-words-21-10	yes	–	–	846.01	**746.36**	795.21
crossword-m1c-ogd-vg10-13_ext	no	–	744.89	**583.22**	748.62	750.28
crossword-m1c-ogd-vg10-14_ext	no	–	1302.41	**402.33**	1264.02	1280.17
crossword-m1c-ogd-vg12-12_ext	no	–	461.62	586.02	450.51	**450.22**
crossword-m1c-uk-vg11-12_ext	no	–	–	**1081.71**	–	–
frb53-24-2-mgd_ext	yes	–	749.33	**329.85**	749.25	330.59
frb53-24-5_ext	yes	748.17	63.04	255.86	**62.74**	256.10
frb56-25-2-mgd_ext	yes	–	661.94	822.12	**661.32**	822.18
graphcoloring-myciel6-6	no	–	–	–	–	**1185.96**
graphcoloring-myciel7-6	no	–	–	–	–	**1178.54**
langford-2-14	no	485.81	187.39	401.30	**150.96**	–
langford-3-16	no	567.92	659.73	446.50	537.89	**129.86**
rand-3-24-24-76-632-17_ext	yes	358.80	240.02	326.82	240.18	**239.56**
rand-3-24-24-76-632-fcd-47_ext	yes	823.91	693.89	**207.30**	691.45	697.14
rand-3-24-24-76-632-fcd-50_ext	yes	692.31	59.71	168.40	59.63	**58.01**
rand-3-28-28-93-632-16_ext	yes	–	1551.52	–	1551.47	**1541.81**
rand-3-28-28-93-632-23_ext	yes	–	551.02	758.57	**550.41**	592.62
rand-3-28-28-93-632-25_ext	yes	–	448.20	464.58	**448.14**	449.93
rand-3-28-28-93-632-3_ext	yes	–	1306.23	**648.04**	1305.86	1304.37
rand-3-28-28-93-632-30_ext	yes	–	**893.93**	1061.22	894.57	897.32
rand-3-28-28-93-632-35_ext	no	–	**1186.84**	1321.97	1192.83	1189.95
rand-3-28-28-93-632-37_ext	yes	–	–	**238.10**	–	–
rand-3-28-28-93-632-8_ext	no	–	1126.08	–	1126.21	**1118.44**
rand-3-28-28-93-632-fcd-16_ext	yes	–	1531.64	**530.76**	1529.82	1519.74
rand-3-28-28-93-632-fcd-20_ext	yes	**24.79**	299.64	314.52	299.73	295.269
rand-3-28-28-93-632-fcd-21_ext	yes	–	1322.25	–	1322.01	**1221.21**
rand-3-28-28-93-632-fcd-24_ext	yes	–	1122.49	–	1116.40	**1116.19**
rand-3-28-28-93-632-fcd-27_ext	yes	–	–	1349.44	–	–
rand-3-28-28-93-632-fcd-31_ext	yes	–	700.22	**211.54**	690.09	684.804
rand-3-28-28-93-632-fcd-35_ext	yes	–	494.40	616.61	492.14	**489.88**
rand-3-28-28-93-632-fcd-40_ext	yes	–	**137.29**	219.16	137.32	197.24
rand-3-28-28-93-632-fcd-42_ext	yes	–	144.94	**124.55**	152.84	138.62
rand-3-28-28-93-632-fcd-46_ext	yes	1410.23	168.30	**159.21**	171.66	166.41
super-js-taillard-20-20	no	–	–	–	**1142.61**	–
Problems solved least twice	–	9	27	27	30	30
Problems solved at least once	–	20	31	30	33	32

and 2 problems, respectively, all among these 35.) SPREAD-S and SPREAD-D are shown with both 10-second and 100-second portfolio phases; neither ever solved a problem during the portfolio phase. Although SPREAD did best with *rand* problems, it also solved problems in such categories as *Langford*, *crossword*, *super-jobshop*, and *graph-coloring*.

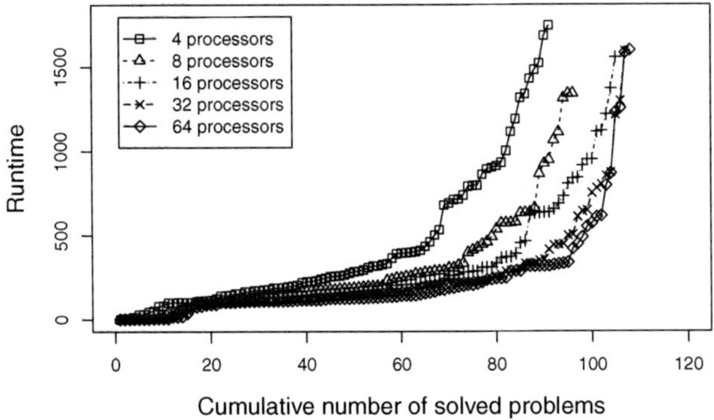

Fig. 5. Cumulative number of problems from the hard set solved by SPREAD-S

SPREAD's search is influenced by the variables it splits on and by their order, but the portfolio-phase search limit also has a strong effect. (Recall that the splitting-phase search limits are proportional to the backtracks consumed in the portfolio phase.) Because we report a median of three runs, to record a problem on any but the last line in Table 1, a program must have solved it at least twice. Both versions of SPREAD actually solved more problems; the last row indicates how many different problems they solved at least once in the three runs. Solved problems not listed in Table 1 include the satisfiable queenAttacking-8 and tdsp-C5-3-91, and the unsatisfiable pseudo-par-32-3-c, super-js-taillard-20-12, and super-js-taillard-20-22. Were the splitting-phase search limit infinite, SPREAD would partition only once and would probably benefit from a longer portfolio phase, but could readily be modified to search for all solutions.

Scalability. Fig. 5 shows that, given more processors, SPREAD consistently solved more problems from the hard set. More than 64 processors, however, introduced only marginal improvement on these problems. (Data omitted.) Because we did not tune SPREAD specifically for Mistral, we would expect similar improvement with other CSP solvers. Recall that, among our curated problems, the hard set contains the easiest ones, where further improvement by SPREAD is relatively difficult. In contrast, Fig. 6 shows how SPREAD scales on two typical problems from the harder problem set, given one hour. With more processors, SPREAD was significantly more likely to succeed within the time limit, and its runtime variance decreased, which produced more stable performance.

Other Statistics. When a worker completes its subproblem but no subproblems remain in the queue, that worker becomes *idle*. To investigate how well SPREAD uses its computing resources, let the *idle ratio* of the ith worker be the fraction of overall runtime that it was idle. SPREAD-S's average idle ratio on problems solved during the splitting phase rose as high as 0.8251 on hard, 0.8793 on harder, and 0.5015 on challenge problems. Large idle ratios were likely caused by a high backtrack limit on an extremely unbalanced search tree, which forced most other workers to wait for a new assignment. The idle ratio could be improved by a backtrack limit tailored to a particular problem class. Overall,

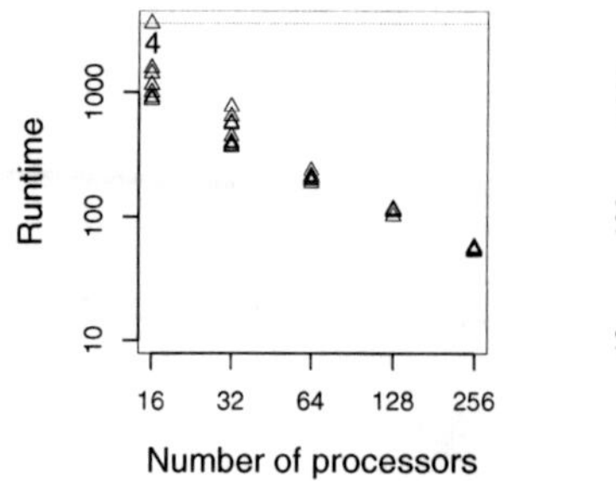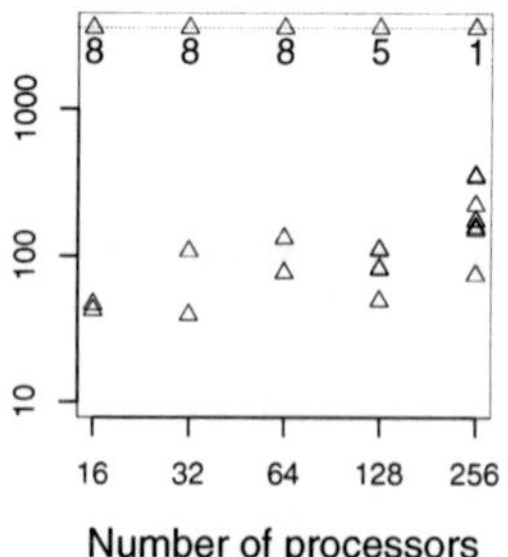

Fig. 6. Runtimes to solution within 1 hour across 10 runs of Spread-S with different numbers of processors for (a) rlfapScens11-f1 (unsatisfiable) (b) js-taillard-20-15-105-4 (satisfiable). Numbers with the uppermost triangles count failed runs.

however, Spread's idle ratio was under 0.1 on 56.92% of the hard, 69.92% of the harder, and 75.00% of the challenge problems. Spread-D's idle ratio was similar: 58.50%, 76.81%, and 75.00% under 0.1, respectively. Finally, Table 2 provides data on subproblems generated during the splitting phase.

6   Discussion

There are many plausible ways to parallelize a solver. One might perturb initial assignments, to vary the top of the search tree, using the same variables with different values. That was tested here as NV, and shown adequate only for the easiest of our test problems. Given the success of restarts and the ability of solvers to learn about contention, one might race the solver against copies of itself with different seeds. That was tested here as NR, and shown only slightly more effective. Given the success of some splitting and portfolio approaches, one might execute random partitioning, or race different solvers against one another. That was tested here as RP and PP, respectively, and shown adequate for some problems, but significantly less so for more difficult ones.

Spread could define its phases' search limits in number of backtracks, consistency checks, or search tree size. In the portfolio phase, time is the limiting factor because it forces all the workers to finish at once. In the splitting phase, however, there is a backtrack limit, to reduce the likelihood that all the workers will communicate with the manager at once.

To split a search space, Spread uses IBP, which, for generality, assumes no knowledge about problem domains. It could, however, be profitable to exploit domain characteristics. For example, one might partition the large domain of a critical variable into more subproblems, or partition extremely small domains (e.g., binary, as in SAT) with parity constraints [12].

Our work now proceeds along three lines. First, IBP may be misled by information collected during the portfolio phase. A typical example comes from the *queens-knights* (QK) problems. Although the contention in QK lies with the knights, weight-based variable-ordering heuristics prefer the queens variables at

Table 2. During the splitting phase, mean split subproblems (#), average maximum subproblem queue length (Max), and average maximum split variable number (μ)

Implementation	Hard			Harder			Challenge		
	#	Max	μ	#	Max	μ	#	Max	μ
SPREAD-S	156	129	7.6	518	136	13.6	244	132	9.6
SPREAD-D	146	129	7.5	374	137	11.3	211	130	8.6

the beginning of search, when weight-based heuristics (e.g., *dom/wdeg*) are close to those not based on weights (e.g., *dom/deg*). We are exploring adaptive methods that dynamically choose duration for the portfolio phase. Second, SPREAD-D did not always outperform SPREAD-S. In SPREAD-D, weights emphasize the local perspective of the subproblem, and preserve the portfolio phase's global perspective on the full problem only at the top of the search tree. We suspect that the initial partitioning is effective because it is based on parallel probing, and that repartitioning is less effective because it lacks the benefit of restart within the subproblem. We are therefore exploring restart strategies for SPREAD. Finally, nogood learning (as clause learning) has proved crucial in SAT, but has thus far received relatively little attention in parallel CSP solvers, including SPREAD. Future work includes combinations of adaptive splitting variable selection with nogood learning to avoid the loss of useful information.

Meanwhile, SPREAD offers a complete and effective method to parallelize a CSP solver. As a parallelization paradigm, SPREAD makes no assumption about domains or constraint types, and so accepts any class of CSPs that its solver can handle. Its bit-string representation permits programmers to ignore the implementation details of the solver, significantly simplifying parallelization, and dramatically reduces communication effort. Its portfolio phase solves easy problems quickly, and informs the splitting phase for effective partitioning. By recursively partitioning difficult subproblems with RS-IBP, it gradually allocates more computing cycles to the difficult parts of a problem, and thereby adaptively balances processor workload. Finally, SPREAD provides a natural way to embed restart policies into an MPI environment without recoding its underlying solver.

Acknowledgements. This work was supported in part by the National Science Foundation under IIS-0811437, CNS-0958379 and CNS-0855217, and the City University of New York's High Performance Computing Center. The authors thank the referees for their thoughtful suggestions.

References

1. Grimes, D., Wallace, R.J.: Sampling Strategies and Variable Selection in Weighted Degree Heuristics. In: Bessière, C. (ed.) CP 2007. LNCS, vol. 4741, pp. 831–838. Springer, Heidelberg (2007)
2. Boussemart, F., Hemery, F., Lecoutre, C., Sais, L.: Boosting systematic search by weighting constraints. In: Proceedings of Sixteenth European Conference on Artificial Intelligence, pp. 146–149. IOS Press, Amsterdam (2004)

3. Refalo, P.: Impact-Based Search Strategies for Constraint Programming. In: Wallace, M. (ed.) CP 2004. LNCS, vol. 3258, pp. 557–571. Springer, Heidelberg (2004)
4. Gomes, C.P., Selman, B., Crato, N.: Heavy-Tail Distributions in Combinatorial Search. In: Smolka, G. (ed.) CP 1997. LNCS, vol. 1330, pp. 121–135. Springer, Heidelberg (1997)
5. Chu, G., Schulte, C., Stuckey, P.J.: Confidence-Based Work Stealing in Parallel Constraint Programming. In: Gent, I.P. (ed.) CP 2009. LNCS, vol. 5732, pp. 226–241. Springer, Heidelberg (2009)
6. Michel, L., See, A., Van Hentenryck, P.: Parallelizing Constraint Programs Transparently. In: Bessière, C. (ed.) CP 2007. LNCS, vol. 4741, pp. 514–528. Springer, Heidelberg (2007)
7. Martins, R., Manquinho, V., Lynce, I.: Improving search space splitting for parallel SAT solving. In: Proceedings of Twenty-Second International Conference on Tools with Artificial Intelligence, pp. 336–343 (2010)
8. Zhang, H., Bonacina, M.P., Hsiang, J.: PSATO: a distributed propositional prover and its application to quasigroup problems. Journal of Symbolic Computation 21, 543–560 (1996)
9. Xie, F., Davenport, A.: Massively Parallel Constraint Programming for Supercomputers: Challenges and Initial Results. In: Lodi, A., Milano, M., Toth, P. (eds.) CPAIOR 2010. LNCS, vol. 6140, pp. 334–338. Springer, Heidelberg (2010)
10. Schubert, T., Lewis, M.D.T., Becker, B.: PaMiraXT: Parallel SAT solving with threads and message passing. Journal of Satisfiability, Boolean Modeling and Computation 6, 203–222 (2009)
11. Hyvärinen, A.E.J., Junttila, T., Niemelä, I.: Grid-Based SAT Solving with Iterative Partitioning and Clause Learning. In: Lee, J. (ed.) CP 2011. LNCS, vol. 6876, pp. 385–399. Springer, Heidelberg (2011)
12. Bordeaux, L., Hamadi, Y., Samulowitz, H.: Experiments with massively parallel constraint solving. In: Proceedings of Twenty-First International Joint Conference on Artificial Intelligence, pp. 443–448 (2009)
13. Kotthoff, L., Moore, N.: Distributed solving through model splitting. In: Proceedings of Third Workshop on Techniques for Implementing Constraint Programming Systems, pp. 26–34 (2010)
14. Mehta, D., O'Sullivan, B., Quesada, L., Wilson, N.: Search Space Extraction. In: Gent, I.P. (ed.) CP 2009. LNCS, vol. 5732, pp. 608–622. Springer, Heidelberg (2009)
15. Gomes, C., Selman, B.: Algorithm portfolio design: theory vs. practice. In: Proceedings of Thirteenth Conference on Uncertainty in Artificial Intelligence, pp. 190–197 (1997)
16. Kadioglu, S., Malitsky, Y., Sabharwal, A., Samulowitz, H., Sellmann, M.: Algorithm Selection and Scheduling. In: Lee, J. (ed.) CP 2011. LNCS, vol. 6876, pp. 454–469. Springer, Heidelberg (2011)
17. Hamadi, Y., Sais, L.: ManySAT: a parallel SAT solver. Journal on Satisfiability, Boolean Modeling and Computation 6, 245–262 (2009)
18. Yun, X., Epstein, S.: Adaptive parallelization for constraint satisfaction search. Accepted by SoCS (2012)
19. Guo, L., Hamadi, Y., Jabbour, S., Sais, L.: Diversification and Intensification in Parallel SAT Solving. In: Cohen, D. (ed.) CP 2010. LNCS, vol. 6308, pp. 252–265. Springer, Heidelberg (2010)
20. CSP Competition (CPAI 2008), http://www.cril.univ-artois.fr/CPAI08/
21. CSP Competition (CSC 2009), http://www.cril.univ-artois.fr/CPAI09/
22. SAT Competitions, http://www.satcompetition.org

A Constraint Programming Approach
for the Traveling Purchaser Problem

Hadrien Cambazard and Bernard Penz

G-SCOP
Université de Grenoble / Grenoble-INP / UJF-Grenoble 1 / CNRS
{hadrien.cambazard,bernard.penz}@grenoble-inp.fr

Abstract. We present a novel approach to the *Traveling Purchaser Problem* (TPP), based on constraint programming and Lagrangean relaxation. The TPP is a generalization of the Traveling Salesman Problem involved in many real-world applications. Given a set of markets providing products at different prices and a list of products to be purchased, the problem is to determine the route minimizing the sum of the traveling and purchasing costs. We propose in this paper an efficient approach when the number of markets visited in an optimal solution is low. We believe that the real-world applications of this problem often assume a bounded number of visits when they involve a physical routing. It is an actual requirement from our industrial partner which is developing a web application to help their customers' shopping planning. The approach is tested on academic benchmarks. It proves to be competitive with a state of the art branch-and-cut algorithm and can provide in some cases new optimal solutions for instances with up to 250 markets and 200 products.

1   Introduction and Industrial Context

The *Traveling Purchaser Problem* (TPP) introduced by Ramesh [18], is a generalization of the *Traveling Salesman Problem* (TSP) and occurs in many real-world applications related to routing, wharehousing and scheduling [23]. Given a hometown for the traveler, a set of markets providing products at different prices and a list of products to be purchased, the problem is to determine the route minimizing the sum of the traveling and purchasing costs. The TPP was brought to our attention by a startup ("Le Bon Côté des Choses")[1] developing a web application to help their customers' shopping planning. A customer enters his location, a list of products, a maximum number of markets to visit in the application and is told the most profitable shopping plan. The original question faced by the startup was therefore a TPP with a side constraint bounding the number of markets in the route.

Such an application requires very short response times and a heuristic was previously designed to cope with this requirement. It constructs a feasible solution by greedily adding markets, then attempts to improve it using a two-opt technique. The startup is now in the process of gathering data and extending their approach with additional features. Two main extensions (not revealed here for confidentiality reasons) currently

---

[1] http://www.leboncotedeschoses.fr/

M. Milano (Ed.): CP 2012, LNCS 7514, pp. 735–749, 2012.

figure on top of their priorities. We note here that a classical extension of the TPP found in the literature [15] is to consider a limited supply of products in each market. However, this feature is not considered by the startup for the moment since the stock levels are not available online (unlike the catalogues of products). To handle the two extensions mentioned, the initial heuristic must now be deeply restructured. The need for flexibility in extending and maintaining the solver lead us to consider in parallel the development of a constraint programming approach.

The purpose of this paper is to propose a new exact algorithm, based on Constraint Programming (CP), for the TPP with a bounded number of visits in the tour. Our approach takes advantage of three key sub-problems of the TPP, and the propagation algorithms are based on dynamic programming and Lagrangean relaxation. Due to the lack of mature industrial benchmark at this stage, we tackle an academic benchmark and compare to a state of the art exact algorithm [15]. Although the approach was initially designed for a small number of visits, it proves to be surprisingly competitive when applied in the unbounded case.

The rest of the paper is organized as follows. Section 2 precisely defines the Traveling Purchaser Problem and briefly presents the literature, focusing on exact approaches. Section 3 describes the constraint programming model. The details of the main constraints are given in the following sections (4, 5 and 6). The branching strategy is detailed in section 7. Finally computational experiments are reported in section 8.

2   Problem Definition and State of the Art

Notations are similar to the ones of [15]. Let $K = \{p_1, \ldots, p_m\}$ be the set of products, $M = \{v_1, \ldots, v_n\}$ the set of markets and $M_k \subseteq M$ the set of markets where the product p_k is available. The problem is to determine the route starting from a depot v_0 (purchaser's hometown), and minimizing the sum of the traveling and purchasing costs to acquire the products of K. The price of product p_k in market v_i ($v_i \in M_k$) is z_{ki} and the traveling cost between two nodes v_i and v_j of $V = \{v_0\} \cup M$ is c_{ij} (we assume that the costs satisfy the triangular inequality). Moreover, each p_k has to be bought in a specific amount and this demand is denoted d_k. In this paper, we deal with the unrestricted TPP *i.e* the problem where the supply in each shop is unlimited. The unrestricted TPP was extended in [15] by bounding the amount of p_k available in each market. In an optimal solution of the unrestricted TPP, all the demand of a product p_k is bought in a single shop. Therefore, we simplify our notations by introducing b_{ki}, the cost for buying all the demand of p_k in market v_i: $b_{ki} = d_k z_{ki}$. Finally, we add a parameter, B, to bound the number of markets visited. This last constraint was also considered in [11] and is often a reasonable assumption made in routing problems as discussed in [5].

The TPP [18] is NP-hard in the strong sense since it generalizes two classical strongly NP-hard problems: the Uncapacited Facility Location Problem (UFLP) and the Traveling Salesman Problem (TSP) [8]. Any TSP can indeed be seen as a TPP where each product (one per market) is available in only one market (so that all markets have to be visited). The UFLP can also be seen as a TPP by mapping products to clients and markets to facilities.

The Traveling Purchaser problem has been largely studied during the last two decades. Numerous heuristics were developed, starting with [10,17] and more recently by [21]. The first exact algorithm based on a lexicographic search was proposed by [18] and was able to solve optimally problems up to 12 markets and 10 products. A branch and bound algorithm was designed later on by [23]. They used a bound based on the simple plant location problem and managed to solve efficiently problems with 20 markets and 50 products. Laporte et al. [15] developed an efficient branch-and-cut algorithm for the undirected TPP. Riera-Ledesma and Salazar-Gonzalez proposed to extend the previous branch-and-cut algorithm to solve the asymmetric case [20]. To our knowledge, this is the best known exact algorithm for the TPP, designed to handle both cases of unlimited and limited supply. It seems reasonable since branch-and-cut is the state of art method for tackling TSP and the best known exact algorithms for facility locations problems are based on linear programming. In Laporte et al. [15], valid inequalities are identified based on the *cycle* (generalization of the TSP where only a subset of vertices must be visited) and the *set-covering* polytopes. The branch-and-cut algorithm generates four types of constraints (four separation procedures) but also variables (pricing procedure) to keep the size of the model reasonable. Finally it uses a primal heuristic at each node (including a 2-opt mechanism for the TSP) to improve the upper bound. It is able to solve optimally instances up to 250 markets and 200 products.

3   CP Model

Our constraint programming approach is built on three core sub-problems at the heart of the TPP: the *Traveling Salesman Problem* (TSP), the *P-Median* problem and the *Hitting Set* problem. It strongly relies on the fact that the number of visited markets is bounded (by B) so that the following observations make sense:

- whether all products can be bought in less than B markets is a minimum *Hitting Set* problem (one set per product p_k containing the markets where p_k is available);
- finding the cheapest way to buy all products in less than B markets is a *P-Median* problem where each facility is a market, each client is a product and the cost of connecting a client to a facility is the cost of buying the corresponding product in the given market;
- once the markets visited are known, we are left with a TSP problem on the corresponding set of markets.

We will derive propagation mechanisms taking advantage of these three core sub-problems and achieve a strong level of consistency. One key idea of our model is to exclude the routing problem (TSP) from the search space by performing exponential time propagation in B *i.e* by encapsulating the TSP inside a constraint.

Variables. We use the variables $Ct \geq 0$ and $Cs \geq 0$ to respectively denote the total traveling and shopping cost. Variables $Cs_k \geq 0$ represent the cost of buying each product p_k. Boolean variables $y_i \in \{0, 1\}$ indicates whether market v_i is visited in the tour of the purchaser. The finite domain variables $s_k \in \{i | v_i \in M_k\}$ give the market where product k is bought. Finally $Nvisit \in \{1, \ldots, B\}$ represents the number of markets visited in the tour.

Model. The model is written as follows:

$$\text{Minimize } Ct + Cs$$

(1) $\quad Cs = \sum_{k=1}^{m} Cs_k$

(2) $\quad Cs_k = \text{ELEMENT}([b_{k1}, \ldots, b_{ki}, \ldots, b_{kn}], s_k) \qquad (\forall \, p_k \in K)$

(3) $\quad \text{OCCURRENCE}(i, [s_1, \ldots, s_m]) \geq 1 \Leftrightarrow y_i = 1 \qquad (\forall \, v_i \in M)$

(4) $\quad Nvisit = \sum_{v_i \in M} y_i$

(5) $\quad \text{NVALUE}([s_1, \ldots, s_m], Nvisit)$

(6) $\quad \text{TSP}([y_1, \ldots, y_n], Ct, Nvisit, \{c_{ij} | v_i, v_j \in M\})$

(7) $\quad \text{PMEDIAN}([y_1, \ldots, y_n], [s_1, \ldots, s_m], Cs, Nvisit,$
$\qquad\qquad \{b_{ki} | p_k \in K, v_i \in M\})$

$$s_k \in \{i | v_i \in M_k\} \qquad\qquad\qquad (\forall \, p_k \in K)$$
$$Cs_k \geq 0 \qquad\qquad\qquad\qquad\quad (\forall \, p_k \in K)$$
$$y_i \in \{0, 1\} \qquad\qquad\qquad\qquad (\forall \, v_i \in M)$$
$$Ct \geq 0, \;\; Cs \geq 0$$

The domains of the variables s_k, y_i and $Nvisit$ are finite *enumerated* domains (each value is maintained in the domain representation) whereas Cs, Cs_k, Ct are represented only by their lower and upper bounds. In the following we denote by $D(x)$ the domain of variable x and by $\underline{x}$ (resp. $\overline{x}$) the lower (resp. upper) bound of x so that x takes a value from $D(x) = [\underline{x}, \ldots, \overline{x}]$.

Constraints. Constraints (2) relate the shopping cost of a product to the market where it is bought. It states that $Cs_k = b_{ks_k}$. The ELEMENT constraint allows to index a table of values by an integer variable. The lower bound of Cs_k is thus maintained as the minimum price of product p_k among all currently possible markets: $\underline{Cs_k} = \min_{v_i | i \in D(s_k)} b_{ki}$.

Constraints (3) are channeling constraints linking y_i and s_k variables. Note that at least one product must be bought in a visited market. Our solver does not support this constraint directly but it can be easily decomposed or implemented directly.

Constraint (4) links the number of visits ($Nvisit$) to the y_i variables.

Constraint (5) is a redundant constraint enforcing the number of visits to be equal to the number of different values of the s_k variables ($Nvisit = |\{s_k | 1 \leq k \leq m\}|$). Our solver only provides ATMOSTNVALUE whose focus is on the lower bound of $Nvisit$: the minimum number of markets that must be visited to get all products *i.e* is the aforementioned *Hitting Set* problem. Typically, this constraint efficiently detects an unfeasible problem where B is too restrictive compared to the domains of the s_k (for example in case of rare products). This constraint is NP-hard and is propagated with a greedy algorithm [3].

Constraint (6) is a dedicated global constraint enforcing Ct to be equal to the optimal tour visiting all the markets i such as $y_i = 1$. *SM* denotes the set of sure markets *i.e* $SM = \{v_i \mid y_i = 1\}$. The scope of (6) encompasses $Nvisit$ because it can be used to derive a stronger lower bound on Ct. Indeed, $\underline{Nvisit}$ might be greater than $|SM|$ in which case the problem of propagating the constraint is known as the *k-TSP*: find the optimal tour visiting k (in our case $\underline{Nvisit}$) cities out of the set of original cities (in our case the set of markets associated to y_i variables not yet fixed to 0).

Constraint (7) is redundant with constraints (1) and (2) together. It propagates a lower bound of Cs by solving the corresponding *P-Median* problem by Lagrangian relaxation.

The three core constraints (5), (6) and (7) are presented in details in the next three sections but we outline now a few elements of this model.

Firstly, note that the exact route followed by the purchaser is not explicitly represented in the model but is present as a support of $\underline{Ct}$ in the TSP global constraint. This has two main drawbacks for a user: the solution is not readily available in the variables and side constraints on the tour are difficult to add. Typically, side constraints to enforce a partial order for visiting the markets or a maximum distance (a resource constraint), must be defined as new parameters of the TSP constraint. The model can also be extended to explicitly represent the route similarly to [5].

Secondly, observe that once the markets are known, it is only a matter of buying each product at the cheapest price among all the markets in the tour so that the s_k variables can be excluded from the search space. Dominance and search strategy are described in section 7.

Finally, the NVALUE and PMEDIAN constraints are redundant. They help strengthening the connection with the TSP sub-problem by increasing $\underline{Nvisit}$ which in turn helps the TSP constraint to derive a sharper $\underline{Ct}$. Similarly, NVALUE and TSP contribute to the decrease of $\overline{Nvisit}$ which helps the PMEDIAN to derive a sharper $\underline{Cs}$. The use of *P-Median* is similar to [23] whose bound is based the *Simple Plant Location Problem*. Our approach is different since the two bounds obtained for Cs and Ct can be added to give the global lower bound. The PMEDIAN provides a very strong filtering but at a high cost. Since it is redundant it can be easily removed from the model when very fast response times are needed for low B.

4   The TSP Global Constraint

We start the description of the TSP global constraint by recalling its scope:

$$\text{TSP}([y_1, \ldots, y_n], Ct, Nvisit, \{c_{ij} | v_i, v_j \in M\})$$

It is a dedicated global constraint enforcing Ct to be equal to the optimal tour visiting all the selected markets (a market i such that $y_i = 1$). We recall that $SM = \{v_i \,|\, y_i = 1\}$ is the set of currently sure markets and denote $PM = \{v_i \,|\, y_i \text{ unknown}\}$ the set of potential markets that can still be included in the tour. We recall that $|SM|$ is assumed to be bounded by a small constant B. Efficient approaches for solving small TSPs are based on dynamic programming or constraint programming [5]. Solving the optimal TSP by dynamic programming can be done in $O(B^2 \times 2^B)$ [13]. The recursive formulation of the TSP is easily written with $f^*(S, x)$, the value of the optimal path starting from the hometown of the traveler, visiting all cities/markets in S and finishing in city x ($x \in S$):

$$f^*(S, x) = \min_{y \in S}(f^*(S - \{x\}, y) + d(y, x))$$

We denote by *optSM* the value of the optimal tour of the sure markets *i.e optSM* $= f^*(SM, v_0)$. Dynamic programming can be used to compute efficiently the optimal tour for instances with around 15 cities. It still fits in memory for 21-22 cities but requires several seconds.

Lower bound of Ct. The lower bound of Ct can be updated to $optSM$ since the costs satisfy the triangular inequality (adding markets in the set SM can not introduce short-cuts and thus only increase the traveling cost). This lower bound can be refined by taking into account $\underline{Nvisit}$. A minimum of $k = \underline{Nvisit} - |SM|$ number of additional markets must be part of the final tour. We describe a simple lower bound of this quantity. Let $ci(a, b, c)$ be the increase of cost for inserting market a between b and c so that $ci(a, b, c) = d_{ba} + d_{ac} - d_{bc}$ and $ci(a, S)$ the best insertion cost of a in the set of markets $S : ci(a, S) = \min_{b,c \in S \times S | b \neq c, a \neq b, a \neq c} ci(a, b, c)$. Let $< \sigma_1, \ldots, \sigma_{|PM|} >$ be the sequence of markets in PM sorted by increasing $ci(a, SM \cup PM)$ *i.e* the best insertion cost in the set $SM \cup PM$. We use the following rule for updating $\underline{Ct}$:

$$\underline{Ct} = optSM + \sum_{i=1}^{k} ci(\sigma_i, SM \cup PM)$$

The bound is summing the value of the optimal tour on the sure markets and the k best insertion costs. Alternatively the propagation of this constraint is exactly a *k-TSP* problem with a number of mandatory cities. Good approximation algorithms exist for this problem based on a primal-dual scheme initially described in [9]. A more recent paper [1] presents such a scheme with the presence of mandatory cities. A classical linear formulation with an exponential number of constraints (the sub-tour constraints) of the *k-TSP* is given. Its dual has an exponential number of variables but a feasible solution can be obtained greedily without considering all the variables explicitly and provides a lower bound. We experimented with this approach, but the bounds obtained are very weak and only improve the previous bound when k is very large (small $|SM|$ and large $\underline{Nvisit}$). Note finally that the *k-MST* (minimum spanning tree where only k nodes have to be spanned) is also NP-hard [7].

Upper bound of $Nvisit$. A simple upper bound of $Nvisit$ can be derived from the previous reasonings. Let $kmax$ be the smallest integer such that $optSM + \sum_{i=1}^{kmax} ci(\sigma_i, SM \cup PM) > \overline{Ct}$, we have:

$$\overline{Nvisit} = |SM| + kmax - 1$$

Filtering of unreachable markets. All markets $v_i \in PM$ such that $optSM + ci(v_i, SM) > \overline{Ct}$ can be eliminated from the tour by setting y_i to 0.

Scaling up with B. To ensure that the constraint can still be applied when $|SM|$ is not bounded, the classical Held and Karp bound [14] based on Lagrangian relaxation can be used instead of dynamic programming. This bound is often extremely close to the optimal value especially for small TSPs and can turn out to be faster than dynamic programming.

Implementation. The propagator of the constraint is finally implemented as follows. At any update of a domain (mainly $\underline{Ct}$) a contradiction might be raised if it leads to an inconsistency (typically by overloading $\overline{Ct}$) immediately interrupting the propagation:

- Compute a minimum spanning tree of SM and update $\underline{Ct}$ accordingly.
- If $|SM| < 15$, use dynamic programming to solve optimally the corresponding TSP and get $optSM$. Otherwise, set $optSM$ to the value of the Held and Karp bound. Update $\underline{Ct}$ accordingly.
- Compute $ci(x, SM \cup PM)$ for all $x \in PM$. Update $\underline{Ct}$ to $optSM + \sum_{i=1}^{k} ci(\sigma_i, SM \cup PM)$ as explained above along with $\overline{Nvisit}$.
- If $PM = \emptyset$ we need to solve the TSP optimally to instantiate Ct. This has already been done if $|SM| < 15$. Otherwise, an exact TSP solver is called. We use for this purpose a CP model designed on top of the Held and Karp bound such as the one of [5] and also inspired by [2].

5   The PMEDIAN Global Constraint

The idea of using Lagrangean relaxation for *P-Median* problem in order to perform variables fixings (as a pre-processing) and reduce the size of a linear programming model was proposed in [4]. We intend to go a step further and design a PMEDIAN global constraint to achieve propagation during search. We recall its scope :

$$\text{PMEDIAN}([y_1, \ldots, y_n], [s_1, \ldots, s_m], Cs, Nvisit, \{b_{ki} | p_k \in K, v_i \in M\})$$

The lower bound of Cs obtained from the propagation of the ELEMENT constraints does not take advantage of the limitation enforced by $Nvisit$. Computing a sharp lower bound on Cs by using the information from $Nvisit$ is an NP-hard problem, it is exactly a *P-Median* problem where p can be set to $\overline{Nvisit}$ (visiting more markets is always cheaper so using the upper bound of $Nvisit$ ensures a lower bound). The classical formulation is the following:

$$\text{Minimize} \qquad Cs = \sum_{i=1}^{n} \sum_{k=1}^{m} b_{ki} x_{ki}$$

$$
\begin{array}{llll}
(1) & \sum_{i=1}^{n} x_{ki} & = 1 & (\forall \, p_k \in K) \\
(2) & \sum_{i=1}^{n} y_i & = \overline{Nvisit} & \\
(3) & x_{ki} - y_i & \leq 0 & (\forall \, v_i \in M, p_k \in K) \\
(4) & x_{ki} \in \{0,1\} & & (\forall \, v_i \in M, p_k \in K) \\
(5) & y_i \in \{0,1\} & & (\forall \, v_i \in M)
\end{array}
$$

A traditional technique to compute a lower bound $\underline{Cs}$ is to use Lagrangian relaxation [16]. Lagrangian relaxation [25] is a technique that moves the "complicating constraints" into the objective function with a multiplier, $\lambda \in \mathbb{R}$, to penalize their violation. For a given value of λ, the resulting problem is the Lagrangian sub-problem and, in the context of minimization, provides a lower bound on the objective of the original problem. The Lagrangian dual is to find the set of multipliers that provide the best possible lower bound.

A lower bound $\underline{Cs}$ can be computed by relaxation of the assignment constraints (1) [16]. Relaxations based on constraints (3) (see [6]) or both constraints (1) and (2) (see [12]) are also possible. The latter seems to be a "standard" relaxation for *P-Median*. We chose to relax (1) since keeping constraints (2) does not make the sub-problem much harder (it only adds a log factor) and seems in practice to greatly improve the convergence. Thus our Lagrangian sub-problem is (λ is unrestricted in sign since we are relaxing an equality constraint) :

$$\begin{aligned}
\text{Minimize } w_\lambda(x,y) &= \sum_{i=1}^{n} \sum_{k=1}^{m} b_{ki} x_{ki} + \sum_{k=1}^{m} (\lambda_k (1 - \sum_{i=1}^{n} x_{ki})) \\
&= \sum_{i=1}^{n} \sum_{k=1}^{m} (b_{ki} - \lambda_k) x_{ki} + \sum_{k=1}^{m} \lambda_k
\end{aligned}$$

$$\text{subject to } (2) - (5)$$
$$\lambda_k \in \mathbb{R} \qquad (\forall\, p_k \in K)$$

The objective function basically amounts to minimizing $\sum_{i=1}^{n} \sum_{k=1}^{m} (b_{ki} - \lambda_k) x_{ki}$ and the Lagrangian dual is: $\max_\lambda (\min_{x,y} w_\lambda(x,y))$. The Lagrangian sub-problem can be solved directly by inspection:

- For each market v_i, we evaluate the change of the objective function when setting y_i to 1 by computing

$$\alpha(i) = \sum_{k=1}^{m} min(b_{ki} - \lambda_k, 0)$$

- The optimal solution is made of the $\overline{Nvisit}$ markets with smallest $\alpha(i)$. More formally, let $<\delta_1, \ldots, \delta_m>$ be the sequence of markets sorted by increasing value of alpha so that $\alpha(\delta_1) \leq \alpha(\delta_2) \leq \ldots \leq \alpha(\delta_m)$. We have

$$w_\lambda^*(x,y) = \sum_{k=1}^{m} \lambda_k + \sum_{i=1}^{\overline{Nvisit}} \alpha(\delta_i)$$

Solving the sub-problem therefore takes $O(nm + nlog(n))$ if we solve the second steps by sorting. Note that relaxing constraint (2) would not change the quality of the bound of the relaxation since both have the integrality property (so the global bound is the one of the linear relaxation of formulation (2)). It would make the sub-problem easier but we find that it significantly increases the number of iterations in practice and does not pay off.

Solving the Lagrangian Dual. We followed the classical approach and used the subgradient method. The algorithm iteratively solves w_λ for different values of λ, initialised to 0 at the first iteration. The values of λ are updated by following the direction of a supergradient of w at the current value λ for a given step length μ.
A supergradient is given by the violation of the assignment constraints so that:

$$\lambda_k^{t+1} = \lambda_k^t + \mu_t (1 - \sum_{i=1}^{n} x_{ki})$$

The step lengths have to be chosen to guarantee convergence. In particular μ_t must converge toward 0 and not too quickly. We used $\mu_t = \mu_0 \times \epsilon^t$ with $\epsilon < 1$ ($\epsilon = 0.99$) and $\mu_0 = 10^5$. We refer the reader to [25] for more details.

Filtering. If a value is proven inconsistent in at least one Lagrangian sub-problem then it is inconsistent in the original problem [22]. We therefore try to identify infeasible values at each iteration of the algorithm. Let us consider a value i in the domain of s_k such that $i \notin [\delta_1, \ldots, \delta_{\overline{Nvisit}}]$. To establish the feasibility of i we replace $\delta_{\overline{Nvisit}}$ (the least profitable market) by i and recompute the bound by enforcing the assignment x_{ki} to 1. This is done immediately using the benefits computed previously and value i is pruned if:

$$w_\lambda^*(x, y) - \alpha(\delta_{\overline{Nvisit}}) + \alpha(\delta_i) + \max(b_{ki} - \lambda_k, 0) > \overline{Cs}$$

We can notice that $b_{ki} - \lambda_k$ is already counted in $\alpha(\delta_i)$ if it is negative thus the term $\max(b_{ki} - \lambda_k, 0)$. This Lagrangian filtering is performed in $O(nm)$. Note that infeasible markets (y_i that must be set to 0) are detected as a result of the previous filtering if all products are removed from their domain. Such markets i would indeed satisfy:

$$w_\lambda^*(x, y) - \alpha(\delta_{\overline{Nvisit}}) + \alpha(\delta_i) > \overline{Cs}$$

A market i can be proved mandatory by considering the next most beneficial market $\delta_{\overline{Nvisit+1}}$. For all $i \in [\delta_1, \ldots, \delta_{\overline{Nvisit}}]$, we can set y_i to 1 if:

$$w_\lambda^*(x, y) + \alpha(\delta_{\overline{Nvisit+1}}) - \alpha(\delta_i) > \overline{Cs}$$

Finally, a simple lower bound of $Nvisit$ can be derived from the previous reasonings. Let $kmax$ be the largest integer such that $\sum_{k=1}^{m} \lambda_k + \sum_{i=1}^{i=kmax} \alpha(\delta_i) > \overline{Cs}$, we have:

$$\underline{Nvisit} = kmax + 1$$

Implementation. We report here a number of observations that we believe are important when implementing the global constraint.

Optimal values of the dual variables λ are stored after resolution at a given node and restored upon backtracking. When going down in the tree the optimal λ found at the father node are used to initialize the new λ. When going up, the optimal λ previously found at this node are re-used to start the algorithm.

It is also important to stop the algorithm as soon as the lower bound becomes greater than $\overline{Cs}$. This can save many calls to the sub-problem.

All previous reasonings have to be adjusted to take into account the current domains of the y and s variables. This quickly reduces the size of the sub-problem as one moves down in the search tree.

Two pre-conditions are used to avoid starting the costly computations of the relaxation. They provide necessary conditions for any improvement of $\underline{Cs}$ by the use of Lagrangean relaxation compared to the bound given by the ELEMENT constraints. The first one is very cheap to compute and very simple:

$$\overline{Nvisit} < |\{v_i | y_i \neq 0\}|$$

The second one is based on the greedy resolution of the following *Hitting Set* problem. A set is associated to each market v_i and contains the products p_k that can be bought in v_i at their current overall best possible price given by $\underline{Cs_k}$. If a hitting set of cardinality less than $\overline{Nvisit}$ exists then it is a support of the current lower bound of Cs. In this case, we know the bound will not increase by solving the relaxation. The set provides a solution where each product can be bought at its minimum cost. In any of these two cases we do not call the relaxation since it would not lead to any improvement of $\underline{Cs}$ (note that filtering on y and s_k can be lost nonetheless, but this is marginal compared to increasing $\underline{Cs}$).

6   The ATMOSTNVALUE Global Constraint: *Hitting Set*

This constraint is NP-hard and is propagated with a greedy algorithm described in [3]. Bessiere and al. [3] also shows that the best bound for this constraint is the linear relaxation of the Linear Programming formulation of the problem. The formulation is based on variables y_i to know whether value i (market v_i) is included in the set:

$$\text{Minimize } \sum_{i=1}^n y_i$$

$$\begin{array}{ll} (1) & \sum_{i \in M_k} y_i \geq 1 \ (\forall\, p_k \in K) \\ (2) & y_i \in \{0, 1\} \qquad (\forall\, v_i \in M) \end{array}$$

Similarly to the *P-Median* problem, we can use Lagrangian relaxation to obtain the bound of this linear program by relaxing the covering constraints (1):

$$\begin{aligned} \text{Minimize } w_\lambda(x, y) &= \sum_{i=1}^n y_i + \sum_{k=1}^m \left(\lambda_k \left(1 - \sum_{i \in M_k} y_i\right)\right) \\ &= \sum_{i=1}^n y_i\left(1 - \sum_{k | i \in M_k} \lambda_k\right) + \sum_{k=1}^m \lambda_k \end{aligned}$$

$$\begin{array}{ll} y_i \in \{0, 1\} & (\forall\, v_i \in M) \\ \lambda_k \geq 0 \quad (\forall\, p_k \in K) & \end{array}$$

The objective function basically amounts at minimizing $\sum_{i=1}^n y_i(1 - \sum_{k | i \in M_k} \lambda_k)$. The Lagrangian sub-problem can again be solved directly by inspection similarly to section 5. This approach could be used to strengthen the propagation of our ATMOSTNVALUE constraint without the need of a simplex algorithm. We intend to do so in the future and only use it in this current paper to get an initial lower bound of $Nvisit$.

7   Dominance and Branching

We end the description of the CP model with an observation about dominance and the search strategy.

Dominance. A simple form of dominance can be added to enforce buying the products in the cheapest market of the tour. Let's consider a product p_k, when a market t is added to the tour (when y_t is set to one), all values $j \in s_k$ such that $j \neq t$ and $b_{kj} \geq b_{kt}$ can be removed from the domain of s_k by dominance. The reasoning is implemented in a dedicated constraint and stated for each product p_k.

Search. Our branching strategy proceeds in two steps, it branches first on *Nvisit* and then operates on the y_i variables. Once all y_i are instantiated, the dominance ensures that all s_k variables are also grounded and a solution is reached. Let h^* refers to the lower bound of *Nvisit* obtained by solving the minimum *Hitting Set* problem by Lagrangean relaxation described in section 6. Four branches start from the root node, constraining *Nvisit* to be within the following intervals (from the left branch to the right branch): $[h^*, h^*+1], [h^*+2, 15], [16, 25], [26, m]$. The rational behind this branching is to ensure that good upper bounds involving small number of markets are obtained early in the search (this can be seen as a form of iterative deepening) and before facing large TSP problems. The hope is that we will be able to rule them out without having to solve them explicitly once good upper bounds are known. The values 15 and 25 are chosen simply because they correspond to the limit of efficient solving of the TSP by dynamic programming and our CP complete solver respectively. The branching continues on the y_i variables using *Impact Based Search* [19].

8   Experimental Results

Benchmark. Our approach is tested on the benchmark generated by [15]. We are using their "class 3" instances where the n markets are generated randomly in the plan (x and y coordinates are taken with a uniform distribution in $[0, 1000] \times [0, 1000]$), each product is available in a number of markets randomly chosen in $[1, n]$ and product prices are generated in $[1, 500]$ with a uniform discrete distribution. 5 instances are generated for each $n \in \{50, 100, 150, 200, 250\}$ and $m \in \{50, 100, 150, 200\}$ leading to 100 instances in total (20 instances for each value of n) and each instance is identified by $n.m.z$ where $z \in [1, 5]$. We chose this benchmark because the number of markets visited in the optimal solutions are relatively small (up to 28 markets among the known optima) and because it was the hardest benchmark (with unlimited supply) for the branch-and-cut[2].

Set-up. The experiments ran as a single thread on a Dual Quad Core Xeon CPU, 2.66GHz with 12MB of L2 cache per processor and 16GB of RAM overall, running Linux 2.6.25 x64. A time limit of 2 hours was used for each run.

Algorithms. We evaluate two algorithms: CP refers to the algorithm without the redundant PMEDIAN constraint whereas CP+PM is the version including it. In both case, we start by computing the lower bound of *Nvisit* as explained in section 6. We also apply a heuristic at the root node before starting the search to get an upper bound (we recall that [15] uses a similar heuristic at each node). It builds a feasible solution and improves it by inserting, removing or swapping markets in the pool of visited markets until a local optimum is reached. Its performances are indicated in Table 1 with the average gap to the best known value (in percentage), the number of optimal values identified (column $\#Opt$) and its CPU times.

---

[2] The benchmark and the detailed results of [15] can be found online:
    `http://webpages.ull.es/users/jriera/TPP.htm`

Table 1. Performances of the heuristic applied at the root node for the 100 instances

Average gap to opt (%)	#Opt	Time (s)			
		Avg	Median	Min	Max
10,8	14	2,15	1,3	0	13,03

Table 2. Results of the three approaches on each class of problem (20 instances per class)

n	Branch-and-Cut			CP+PM				CP			
	Time (s)			Time (s)				Time (s)			
	Avg	Median	#F	Avg (s)	Median (s)	#F	Avg Back	Avg (s)	Median (s)	#F	Avg Back
50	23,9	17,5	0	7,1	3,9	0	574,4	11,9	2,7	0	1162,4
100	299,8	295,0	0	187,3	16,3	2	34327,6	214,2	7,0	3	57579,9
150	1734,5	1515,5	0	997,7	231,9	0	41563,6	778,7	110,3	4	108435,5
200	4983,2	3275,0	2	518,5	89,1	6	83670,8	1329,3	43,3	7	201362,0
250	9720,2	10211,0	9	728,2	266,9	2	53741,6	1057,5	357,0	4	145697,2

Table 3. Cpu times of the two algorithms (CP and CP+PM) on the 100 instances of the benchmark and for $B \in \{5, 10\}$

	CP+PM (B = 5)	CP (B = 5)	CP+PM (B = 10)	CP (B = 10)
NbInfeasible	80	80	9	9
Avg Time (s)	0,68	1,22	17,31	122,37
Median Time (s)	0,50	0,48	3,96	1,99
StDev on Time (s)	0,83	4,48	29,23	674,00
Min Time (s)	0,06	0,06	0,14	0,07
Max Time (s)	5,55	43,88	156,83	5496,67

Results. The aim of these experiments is threefold:

Firstly, we show the interest of the PMEDIAN global constraint in Table 2. It reports CPU times (average and median computed only on instances that did not reach the time limit) for each value of n (20 instances in each sub-class), the number of problems where an algorithm fails to prove optimality (reported in column #F) and the average number of backtracks (column Avg Back). CP+PM is clearly more efficient. It fails to prove optimality on 10 instances whereas CP fails on 18 instances.

Secondly, we compare the results with the algorithm of [15]. The branch and cut was run with a timelimit of 5 hours (18000 seconds) on a very old machine (Pentium 500 Mhz with CPLEX 6.0) and the code is not available anymore. Therefore, we do not perform a direct comparison with the CPU times reported for [15] in Tables 2 and 4. They are indicative and only serve to draw general trends. Table 4 gives detailed results for each instance (N and Obj are respectively the number of visits and the value of the objective function in the best solution found). In any case, the branch and cut is more robust since it can handle efficiently large TSP problems. CP+PM fails on 10 instances whose optimal number of visits is between 19 and 28 (instance 200.100.5). Surprisingly, it can find solutions including up to 25 markets (150.200.3) and prove their optimality. It also shows a very good complementarity to the branch-and-cut (see for example 250.50.5 where it needs less than 1s when the branch and cut nearly reaches

Table 4. Details of the results on the class 3 instances of [15]

Instance name	N	Obj	Time(s)	N	Obj	Time(s)	back	Instance name	N	Obj	Time(s)	N	Obj	Time(s)	Back
		Branch-and-cut			CP+PM					Branch-and-cut			CP+PM		
50.50.1	8	1856	3	8	1856	0.54	16	50.50.2	8	1070	0	8	1070	0.48	33
50.50.3	9	1553	9	9	1553	0.41	192	50.50.4	6	1394	10	5	1394	0.08	8
50.50.5	2	1536	10	2	1536	0.09	5	50.100.1	12	2397	15	12	2397	4.31	289
50.100.2	11	2138	23	11	2138	4.39	562	50.100.3	10	1852	10	10	1852	0.43	8
50.100.4	15	3093	39	15	3093	4.73	917	50.100.5	12	2603	21	12	2603	1.77	159
50.150.1	15	2784	32	15	2784	16.7	1465	50.150.2	11	2137	12	11	2137	0.64	74
50.150.3	14	2308	49	14	2308	7.3	570	50.150.4	14	2524	15	14	2524	3.73	327
50.150.5	16	3150	53	16	3150	26.56	2411	50.200.1	19	3354	18	20	3354	28.28	1260
50.200.2	15	2397	17	15	2397	8.17	604	50.200.3	11	2319	62	11	2319	1.87	103
50.200.4	16	2858	27	16	2858	4.01	252	50.200.5	17	3575	53	17	3575	28.27	2232
100.50.1	9	1468	139	9	1468	1.17	316	100.50.2	7	971	66	7	971	1.67	306
100.50.3	10	1623	159	10	1623	0.92	206	100.50.4	10	1718	80	10	1718	1.68	349
100.50.5	11	2494	168	10	2494	1.31	695	100.100.1	16	2121	130	15	2121	8.74	1258
100.100.2	13	1906	405	13	1906	17.0	2926	100.100.3	13	1822	199	13	1822	15.28	1037
100.100.4	9	1649	55	9	1649	0.73	170	100.100.5	15	2925	758	14	2925	75.39	16289
100.150.1	14	2195	534	14	2195	132.73	11660	100.150.2	16	2806	423	16	2806	141.41	7148
100.150.3	18	2257	440	18	2257	98.84	7974	100.150.4	19	2625	234	18	2625	118.09	9820
100.150.5	18	3150	484	18	3150	390.29	21056	100.200.1	10	1883	166	10	1883	15.64	680
100.200.2	24	3077	369	24	3087	> 7200s	182021	100.200.3	23	2791	356	23	2791	2284.13	76395
100.200.4	19	3409	455	19	3409	> 7200s	342591	100.200.5	18	2732	376	17	2732	66.64	3655
150.50.1	12	1658	498	12	1658	7.58	335	150.50.2	7	1383	512	7	1383	0.27	25
150.50.3	7	821	516	7	821	1.22	308	150.50.4	10	1676	1612	10	1676	13.75	5045
150.50.5	12	1823	1647	12	1823	8.34	1408	150.100.1	17	1717	1243	17	1717	512.73	38148
150.100.2	11	1798	1419	11	1798	27.45	3259	150.100.3	16	1959	2304	15	1959	258.08	32622
150.100.4	14	1609	3408	14	1609	34.42	4179	150.100.5	14	1585	1216	14	1585	275.86	33068
150.150.1	19	1669	367	19	1669	117.76	4264	150.150.2	24	2526	1801	22	2526	2218.69	99598
150.150.3	19	2456	3092	19	2456	5037.65	245597	150.150.4	16	1761	1268	15	1761	205.76	11488
150.150.5	18	2355	3155	16	2355	1152.24	87066	150.200.1	19	1760	365	19	1760	331.94	8090
150.200.2	24	2312	1732	22	2312	1037.48	28630	150.200.3	25	2594	1317	25	2594	5677.0	122773
150.200.4	15	1889	3431	15	1889	154.14	17961	150.200.5	22	2472	3787	22	2472	2881.27	87408
200.50.1	13	1102	1644	12	1102	22.3	5631	200.50.2	6	607	337	6	607	0.45	23
200.50.3	6	530	550	6	530	0.5	71	200.50.4	9	908	849	9	908	0.63	158
200.50.5	11	1067	2248	11	1067	2.78	495	200.100.1	12	949	490	12	949	4.03	237
200.100.2	18	2271	8188	16	2271	381.32	39562	200.100.3	13	1611	2515	13	1611	37.36	4420
200.100.4	17	1799	4697	17	1799	176.97	18758	200.100.5	28	3161	8599	23	3178	> 7200s	310494
200.150.1	16	1730	1574	15	1730	140.87	16012	200.150.2	27	2745	15951	15	2790	> 7200s	212333
200.150.3	20	1861	2613	20	1861	682.82	17643	200.150.4	23	2460	18024	15	**2441**	> 7200s	238821
200.150.5	18	2079	11505	18	2079	446.03	48271	200.200.1	18	1736	3937	18	1736	578.88	21584
200.200.2	22	2352	10647	22	2359	> 7200s	156492	200.200.3	23	2505	7873	22	2505	> 7200s	179516
200.200.4	20	3314	18053	21	**2344**	4783.84	173311	200.200.5	23	2427	5481	23	2462	> 7200s	229584
250.50.1	6	533	556	6	533	0.98	145	250.50.2	10	1103	5451	10	1103	15.67	1316
250.50.3	12	1295	14030	12	1295	12.75	1349	250.50.4	13	1553	15487	12	1553	16.43	3377
250.50.5	5	1142	17399	5	1142	0.66	45	250.100.1	11	2447	18057	15	**1301**	257.56	30983
250.100.2	12	932	2771	12	932	17.22	382	250.100.3	17	1361	16376	17	1361	111.45	10874
250.100.4	16	1759	18029	16	**1673**	284.68	43921	250.100.5	14	1708	18059	15	**1641**	256.96	17568
250.150.1	15	1168	1453	15	1168	367.23	15596	250.150.2	22	2205	7999	22	2205	2036.28	169070
250.150.3	15	1582	10211	15	1582	276.17	11033	250.150.4	18	2636	18078	17	**1836**	1203.97	129473
250.150.5	15	2121	18074	18	**1531**	417.35	38506	250.200.1	20	1677	15189	20	1677	1455.16	61277
250.200.2	25	2787	18019	25	2856	> 7200s	149248	250.200.3	25	3555	18065	20	**1924**	5665.07	140904
250.200.4	17	2432	18045	15	**2139**	> 7200s	230347	250.200.5	18	2771	18071	17	**1797**	712.4	19417

the five hours). Overall it improves 10 solutions (in bold in the table) and manage to close 8 instances among the 11 that were still open.

Thirdly, Table 3 shows the results obtained when bounding B to 5 and 10. 80 instances are proved infeasible for $B = 5$ and 9 for $B = 10$. All instances are solved to optimality and CP+PM clearly outperforms CP. It shows that the approach can be very effective for low B. The CPU times reported include the proof of optimality which is irrelevant in the industrial case. We plan to analyze the quality of the solution found within 1s of time limit and the time to obtain the best solution without time limit to get more insights on the possibility of using this algorithm in the industrial case.

9   Conclusions and Future Works

We proposed a new exact algorithm based on Constraint Programming for the Traveling Purchaser Problem. The TSP and PMEDIAN global constraints are introduced to efficiently handle two core structures of the TPP and rely on Lagrangean relaxation. The proposed algorithm is designed for problems involving a bounded number of visited markets (around 5 in the industrial application) but is robust enough to be applied on academic benchmark in the unbounded case. It proves to be very complementary to the state of the art exact algorithm and manages to close 8 instances out the 11 open on this particular benchmark.

We intend to investigate further how propagation could be strengthened. Firstly, by using Lagrangean relaxation to propagate and filter the ATMOSTNVALUE constraint. Secondly by adding the ATLEASTNVALUE constraint which would propagate an upper bound of $Nvisit$ based on a maximum matching. We plan to investigate further how to efficiently implement global constraints with Lagrangean relaxation. In particular the work of [2] for the TSP and [4] for the *P-Median* are very good starting points. Finally the use of a state of the art TSP solver might allow the approach to scale further and overcome its current limitation.

References

1. Arora, S., Karakostas, G.: A 2+epsilon approximation algorithm for the k-mst problem. In: Shmoys, D.B. (ed.) SODA, pp. 754–759. ACM/SIAM (2000)
2. Benchimol, P., Régin, J.-C., Rousseau, L.-M., Rueher, M., van Hoeve, W.-J.: Improving the Held and Karp Approach with Constraint Programming. In: Lodi, A., Milano, M., Toth, P. (eds.) CPAIOR 2010. LNCS, vol. 6140, pp. 40–44. Springer, Heidelberg (2010)
3. Bessiere, C., Hebrard, E., Hnich, B., Kiziltan, Z., Walsh, T.: Filtering algorithms for the nvalue constraint. Constraints 11, 271–293 (2006)
4. Briant, O., Naddef, D.: The optimal diversity management problem. Operations Research 52(4), 515–526 (2004)
5. Caseau, Y., Laburthe, F.: Solving small tsps with constraints. In: ICLP, pp. 316–330 (1997)
6. Christofides, N., Beasley, J.E.: A tree search algorithm for the p-median problem. European Journal of Operational Research 10(2), 196–204 (1982)
7. Fischetti, M., Hamacher, H.W., Jørnsten, K., Maffioli, F.: Weighted k-cardinality trees: Complexity and polyhedral structure. Networks 24(1), 11–21 (1994)
8. Garey, M.R., Johnson, D.S.: Computers and Intractability: A Guide to the Theory of NP-Completeness. W. H. Freeman & Co., New York (1979)
9. Goemans, M., Williamson, D.P.: A general approximation technique for constrained forest problems. SIAM Journal on Computing 24, 296–317 (1992)
10. Golden, B., Levy, L., Dahl, R.: Two generalizations of the traveling salesman problem. Omega 9(4), 439–441 (1981)
11. Gouveia, L., Paias, A., Voí, S.: Models for a traveling purchaser problem with additional side-constraints. Computers and Operations Research 38, 550–558 (2011)

12. Hanjoul, P., Peeters, D.: A comparison of two dual-based procedures for solving the p-median problem. European Journal of Operational Research 20(3), 387–396 (1985)
13. Held, M., Karp, R.M.: A dynamic programming approach to sequencing problems. In: Proceedings of the 1961 16th ACM National Meeting, ACM 1961, pp. 71.201–71.204. ACM, New York (1961)
14. Held, M., Karp, R.M.: The traveling-salesman problem and minimum spanning trees: Part ii. Mathematical Programming 1, 6–25 (1971)
15. Laporte, G., Riera-Ledesma, J., Salazar-González, J.-J.: A branch-and-cut algorithm for the undirected traveling purchaser problem. Operations Research 51, 940–951 (2003)
16. Narula, S.C., Ogbu, U.I., Samuelsson, H.M.: An algorithm for the p-median problem. Operations Research 25(4), 709–713 (1977)
17. Pearn, W.L., Chien, R.C.: Improved solutions for the traveling purchaser problem. Computers and Operations Research 25(11), 879–885 (1998)
18. Ramesh, T.: Travelling purchaser problem. Opsearch 2(18), 78–91 (1981)
19. Refalo, P.: Impact-based search strategies for constraint programming. In: Wallace (ed.) [24], pp. 557–571
20. Riera-Ledesma, J., Salazar-Gonzalez, J.-J.: Solving the asymmetric traveling purchaser problem. Annals of Operations Research 144(1), 83–97 (2006)
21. Riera-Ledesma, J., Salazar-González, J.J.: A heuristic approach for the travelling purchaser problem. European Journal of Operational Research 162(1), 142–152 (2005)
22. Sellmann, M.: Theoretical foundations of cp-based lagrangian relaxation. In: Wallace (ed.) [24], pp. 634–647
23. Singh, K.N., van Oudheusden, D.L.: A branch and bound algorithm for the traveling purchaser problem. European Journal of Operational Research 97(3), 571–579 (1997)
24. Wallace, M. (ed.): CP 2004. LNCS, vol. 3258. Springer, Heidelberg (2004)
25. Wolsey, L.A.: Integer programming. Wiley-Interscience series in discrete mathematics and optimization. Wiley (1998)

Constraint-Based Register Allocation and Instruction Scheduling

Roberto Castañeda Lozano[1], Mats Carlsson[1], Frej Drejhammar[1],
and Christian Schulte[2,1]

[1] SICS (Swedish Institute of Computer Science), Sweden
{rcas,matsc,frej,cschulte}@sics.se
[2] School of ICT, KTH Royal Institute of Technology, Sweden

Abstract. This paper introduces a constraint model and solving techniques for code generation in a compiler back-end. It contributes a new model for global register allocation that combines several advanced aspects: multiple register banks (subsuming spilling to memory), coalescing, and packing. The model is extended to include instruction scheduling and bundling. The paper introduces a decomposition scheme exploiting the underlying program structure and exhibiting robust behavior for functions with thousands of instructions. Evaluation shows that code quality is on par with LLVM, a state-of-the-art compiler infrastructure.

The paper makes important contributions to the applicability of constraint programming as well as compiler construction: essential concepts are unified in a high-level model that can be solved by readily available modern solvers. This is a significant step towards basing code generation entirely on a high-level model and by this facilitates the construction of correct, simple, flexible, robust, and high-quality code generators.

1 Introduction

Compilers consist of a front-end and a back-end. The *front-end* analyzes the input program, performs architecture-independent optimizations, and generates an intermediate representation (IR) of the input program. The *back-end* takes the IR and generates assembly code for a particular processor. This paper introduces a constraint model and solving techniques for substantial parts of a compiler back-end and contributes an important step towards compiler back-ends that exclusively use a constraint model for code generation.

Today's back-ends typically generate code in stages: instruction selection (choose appropriate instructions for the program being compiled) is followed by register allocation (assign variables to registers or memory) and instruction scheduling (order instructions to improve their throughput). Each stage commonly executes a heuristic algorithm as taking optimal decisions is considered either too complex or computationally infeasible. Both staging and heuristics compromise the quality of the generated code and by design preclude optimal code generation. Capturing common architectural features and adapting to new architectures and frequent processor revisions is difficult and error-prone with

M. Milano (Ed.): CP 2012, LNCS 7514, pp. 750–766, 2012.
© Springer-Verlag Berlin Heidelberg 2012

heuristic algorithms. The use of a constraint model as opposed to staged and heuristic algorithms facilitates the construction of correct, simple, flexible, and robust code generators with the potential to generate optimal code.

Approach. The paper uses LLVM [1] for the compiler front-end and assumes that instruction selection has already been done yielding a representation of the input program in SSA (static single assignment). The paper extends SSA by introducing LSSA (linear SSA), which represents programs as *basic blocks* (blocks of instructions without control flow, *blocks* for short) and relations of *temporaries* (program variables) among blocks.

A function in LSSA (the compilation unit in this paper) is used to generate a model for global register allocation (assign temporaries to registers for an entire function). The model combines several advanced aspects:

Multiple register banks also capture the spilling of temporaries to memory as just another register bank due to lack of available processor registers.

Temporary coalescing attempts to assign related temporaries to the same register in order to save move instructions.

Register packing can assign several small unrelated temporaries to the same register. For example, two 16-bit temporaries can be assigned to the upper and lower half of a 32-bit register.

Both multiple register banks and coalescing are modeled by optional copy instructions between temporaries identified as related in the LSSA representation. The model is extended to include instruction scheduling and instruction bundling (for executing several instructions in parallel). The single model accurately captures the interdependencies between register allocation and instruction scheduling. Hence, it faithfully reflects the trade-off between conflicting decisions during code generation.

The model is solved by decomposition, exploiting the underlying program structure as explicated in LSSA. First, the relation of temporaries among blocks is established followed by solving constraints for each block. The code generator exhibits robust behavior for functions with thousands of instructions, where we choose the `bzip2` program as part of the standard SPECint 2006 benchmark suite. Evaluation shows that code quality is on par with LLVM.

Key contributions. The paper makes the following contributions: − LSSA as a new program form explicating program structure used for modeling; − a constraint model unifying register allocation and instruction scheduling; − in particular, the register allocation model unifies multiple register banks, spilling, coalescing, and packing; and − a code generator based on a problem decomposition showing promising code quality and robustness.

Related approaches. Optimal register allocation and instruction scheduling have been approached with different optimization techniques, both in isolation and as an integrated problem.

Register allocation has been approached as an integer linear programming (ILP) problem [2,3] and as a partitioned Boolean quadratic problem [4]. To the best of our knowledge, there has been no attempt to solve global register allocation with constraint programming (CP).

Instruction scheduling has typically been modeled as a resource-constrained scheduling problem. Local instruction scheduling has been approached with both CP [5,6] and ILP [7,8]. Proposed solutions for the global case include the use of CP [9], ILP [10], and special-purpose optimization algorithms [11,12].

Different ILP and CP approaches have addressed the integration of both problems in a single model [13,14,15,16]. To the best of our knowledge, none of these approaches deals with all essential aspects of global register allocation such as coalescing and spilling.

The integration of instruction selection with these problems has also been considered, using both CP [17] and ILP [18,19]. These approaches are limited to single blocks, and it is unclear how to extend them to handle entire functions robustly.

Plan of the Paper. Section 2 reviews SSA-based program representation whereas Sect. 3 introduces LSSA used for the constraint model in the paper. A constraint model for register allocation is introduced in Sect. 4 which is refined in Sect. 5 to integrate instruction scheduling. Section 6 discusses model limitations. Section 7 introduces a decomposition scheme which is evaluated in Sect. 8. Section 9 concludes the paper and presents future work.

2   SSA-Based Program Representation

This section describes SSA (static single assignment) as a state-of-the-art representation for programs where processor instructions have already been selected. The factorial function, whose C source code is shown to the right, is used as a running example throughout the paper.

```
int factorial(int n) {
  int f = 1;
  while (n > 0) {
    f = f * n; n--;
  }
  return f;
}
```

A function is represented by its control-flow graph (CFG). The CFG's vertices correspond to blocks and its arcs define the control flow (jumps and branches) between blocks. A block consists of instructions and temporaries. Temporaries are storage locations corresponding to program variables after instruction selection. Other types of operands such as immediate values are not of interest in this context. An instruction is a triple represented as $D \leftarrow \text{op } U$, where D and U are the sets of *defined* and *used* temporaries and **op** is the *processor operation* that implements the instruction. For example, an instruction that uses a temporary t to define a temporary t' by executing the operation **neg** is represented as $t' \leftarrow \text{neg } t$. The remainder of the paper uses operations from MIPS32 [20], a simple instruction set chosen for ease of illustration.

A *program point* is located between two consecutive statements. A temporary is *live* at a program point if it holds a value that might be used by an instruction in the future. The *live range* of a temporary is the set of program points where it is live. Two temporaries *interfere* if their live ranges overlap. A temporary is *live-in* (respectively *live-out*) in a block if it is live at its entry (exit) point.

Architectural constraints and ABIs (application binary interfaces) predetermine the registers to which certain temporaries are assigned. A temporary t that is *pre-assigned* to a register $\mathbf{r}$ is represented by $t \triangleright \mathbf{r}$.

SSA is a program form where temporaries are only defined once [21] (as is common, SSA in this paper means conventional SSA [22]). For programs where temporaries are redefined, SSA inserts ϕ-*functions* among the *natural instructions* derived from program statements and expressions. ϕ-functions disambiguate definitions of temporaries that depend on program control flow. For instance, in the factorial function, the return value might be defined by `int f = 1` or within the `while`-loop. In SSA, a ϕ-function is inserted defining a new return value that is equal to either `1` or to the the value of `f` computed in the loop. ϕ-functions define a congruence among temporaries, where two temporaries are ϕ-*congruent* if they are accessed by the same ϕ-function. Since ϕ-functions are not provided by processor instruction sets, their resolution is a prerequisite for generating executable code.

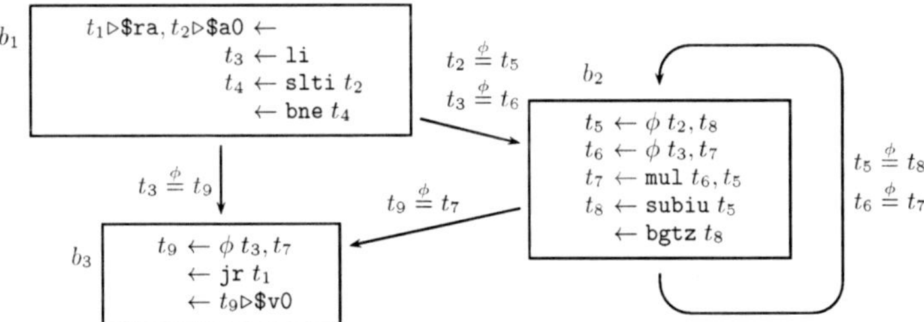

Fig. 1. MIPS32 instruction-selected function in SSA

SSA simplifies the computation of liveness and interference. Since this simplification eases register allocation [23], SSA form is used as input to the code generator. Fig. 1 shows the control-flow graph of the running example transformed to SSA form with MIPS32 operations. Arc labels show the ϕ-congruence connecting temporaries related by ϕ-functions. Temporaries t_1, t_2 and t_9 are pre-assigned to the return address (`$ra`), first argument (`$a0`) and first return value (`$v0`) registers.

3   Program Representation for Register Allocation

This section introduces the program representation on which the model is based.

3.1   Register Allocation

Register allocation is the problem of assigning temporaries to either processor registers (hereafter called *registers*) or memory. Since access to registers is orders of magnitude faster than access to memory, it is desirable to assign all temporaries to registers. As the number of registers is limited, not all temporaries can be assigned a register. A first way to improve register utilization is to store temporaries only while they are live. This allows register allocation to assign several temporaries to the very same register provided the temporaries do not interfere with each other.

In general, even optimal register utilization does not guarantee the availability of processor registers for all temporaries. In this case, some temporaries must be stored in memory (that is, *spilled*). Since access to memory is costly, the decision of which temporaries are assigned to memory and at which program point they are assigned has high impact on the efficiency of the generated code.

The input program to register allocation may contain temporaries related by copy instructions (that is, instructions that copy the content of one temporary into another). Assigning these temporaries to the same register (that is, *coalescing* them) allows the removal of the corresponding copy instructions, improving the efficiency and size of the code.

Each temporary has a certain bit width (hereafter just called *width*) which is determined by the source data type that it represents. Many processors allow temporaries of different widths to be assigned to different parts of the same physical register. For example, Intel's x86 has 16-bit registers (**AX**) that combine pairs of 8-bit registers (**AH**, **AL**). For these processors, the ability to pack non-interfering temporaries into different parts of the same physical register is a key technique to improve register utilization [24].

Register allocation can be *local* or *global*. Local register allocation deals with one block at a time, spilling all temporaries that are live at block boundaries. Global register allocation yields better code by considering entire functions and can keep temporaries in the same register across blocks.

3.2   Linear Static Single Assignment Form

The live range of a temporary depends on where it is defined and used by instructions. In a model that captures simultaneous register allocation and instruction scheduling, the live ranges and their interferences are mutually dependent. Although SSA makes live range and interference computation easier than in a general program form, it is unclear how to model interference of temporaries with variable live ranges that can span block boundaries and follow branches.

To overcome this limitation and enable simple and direct modeling of live ranges, this paper introduces linear SSA (LSSA). LSSA is stricter than SSA in that each temporary is only defined and used within a single block. This property leads to simple, linear live ranges which do not span block boundaries and can be directly modeled as in Sect. 4. Furthermore, this simplification enables a problem decomposition that can be exploited for robust code generation, as Sect. 7 explains.

This paper is the first to take advantage of the explicit congruence structure that LSSA defines even though this structure appears in some SSA construction approaches [25,26].

To restrict live ranges to single blocks, SSA ϕ-functions are generalized to *delimiter instructions* (hereafter just called *delimiters*). These instructions are placed at the block boundaries and are not part of the generated code. Each block contains two delimiters: an *in-delimiter* which defines the live-in temporaries and an *out-delimiter* which uses the live-out temporaries.

In LSSA, the live range of a temporary cannot span the boundaries of the block where it is defined. Liveness across blocks in the original program is captured by a new definition of temporary congruence, relating temporaries that represent the same original temporary in different blocks. This congruence generalizes the ϕ-congruence defined for SSA.

Fig. 2 shows the control-flow graph of the running example transformed to LSSA. Arc labels show the generalized congruence. The temporary t_1 which is live across all branches in Fig. 1 corresponds to the congruent temporaries t_1, t_5 and t_{10} in Fig. 2, each of them having a linear live range. The figure shows that the only link between blocks in LSSA is given by temporary congruences.

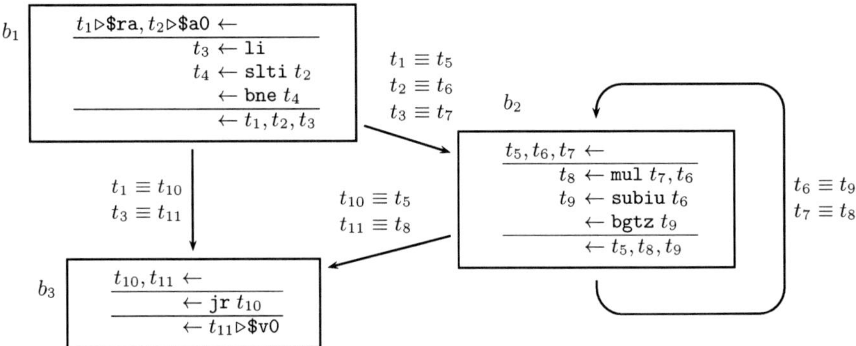

Fig. 2. MIPS32 instruction-selected function in LSSA

LSSA can be easily constructed from SSA by applying the following steps to each block:

1. introduce delimiters;
2. for each live-in temporary, introduce a new definition by the in-delimiter; for each live-out temporary, introduce a new use by the out-delimiter (applying the liveness definition by Sreedhar et al. [22]);
3. remove all ϕ-functions;
4. connect the new definitions and uses with their corresponding live-in and live-out temporaries by defining congruences.

3.3   Copies

Spilling requires copying the contents of temporaries to new temporaries that can then be assigned to different processor registers or to memory. This is captured in the model by extending the program representation with the *copy* instruction type, similarly to Appel and George's approach [3]. A copy replicates the content of a temporary t_s to a new temporary t_d. To allow t_s and t_d to be assigned to different types of locations such as registers or memory, the copy is implemented by the execution of one of a set of alternative operations $\{o_1, o_2, \ldots, o_n\}$ and represented as $t_d \leftarrow \{o_1, o_2, \ldots, o_n\}\ t_s$.

The way in which a program is extended with copies depends on the processor. In load/store processors such as MIPS32, arithmetic/logic operations define and use temporaries in registers. In this setting, register allocation needs to be able to copy a temporary defined in a register to another register or to memory. If a temporary has been copied to memory (that is, spilled), it must be copied back to a register before its use by an arithmetic/logic operation. In MIPS32, the program is extended with copies of the form $t_d \leftarrow \{\texttt{move}, \texttt{sw}\}\ t_s$ after the definition of t_s in a register and $t_d \leftarrow \{\texttt{move}, \texttt{lw}\}\ t_s$ before the use of t_d in a register, where the operations $\texttt{move}$, $\texttt{sw}$ and $\texttt{lw}$ implement register-to-register, register-to-memory and memory-to-register copies. Fig. 3 shows how the function in Fig. 2 is extended with such copies.

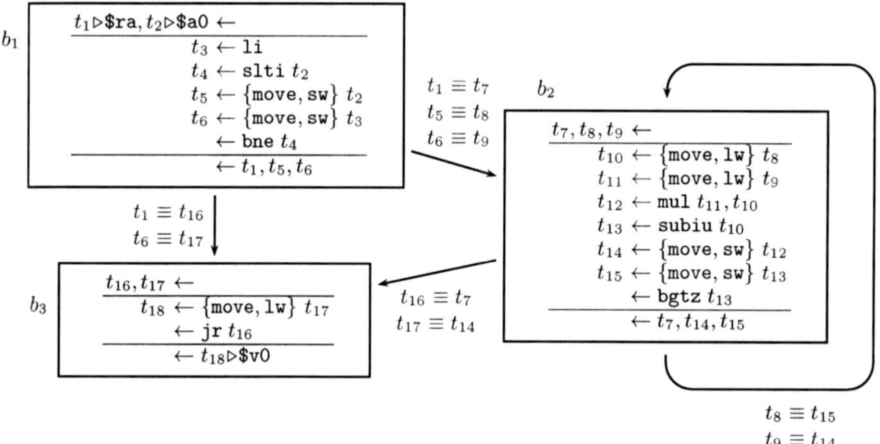

Fig. 3. Function in LSSA extended with copies

4   Register Allocation and Packing

This section describes the constraint model for register allocation and packing. The constraint model is parameterized with respect to a program in LSSA and a processor. The entire model, extended with instruction scheduling, is shown in Fig. 4. The main text contains references to the formulas (1-11) from the figure. Sections 4.1 and 4.2 discuss local and global register allocation. Section 4.3 refines the model to handle register packing.

4.1   Local Register Allocation

The variables in the constraint model are described in Fig. 4. A valid assignment of the register (r_t) and operation (o_i) variables constitutes a solution to the register allocation problem. When solving this problem in isolation, the issue cycle (c_i) and live range (ls_t, le_t) variables (1) are pre-assigned and can be seen as program parameters. They act as variables when the model is extended with instruction scheduling as described in Sect. 5.

Natural instructions and delimiters define the meaning of a program and must be active (2). Unlike natural instructions and delimiters, a copy i might be implemented by different operations or be inactive. To support the latter case, the domain of its variable o_i is extended with a virtual operation null (3). Delimiters are implemented by virtual in and out operations. A solution to the register allocation problem implies deciding whether a copy i is active (a_i) and which operation implements it (o_i).

A processor typically contains one or more register banks which can be directly accessed by instructions. Traditional register allocation treats memory and different register banks as separate entities, which leads to specialized methods for different aspects of register allocation such as spilling and dealing with multiple register banks. We consider a unified register array that fully integrates these aspects. A unified register array is a sequence of register *spaces*. A space is a sequence of related registers. Spaces capture different processor register banks as well as *memory registers* (representing memory locations on the runtime stack). A memory space contains a practically infinite sequence of memory registers $(\texttt{m1}, \texttt{m2}, \ldots)$. Fig. 5a shows the unified register array for MIPS32.

To the best of our knowledge, this paper presents the first application of a unified register array to integrate different aspects of register allocation in the context of native code generation.

Local register allocations can be projected onto a rectangular area. The horizontal dimension represents the registers in the unified register array, whereas the vertical dimension represents time in clock-cycles (see Fig. 5a). In this projection, each temporary t is represented as a rectangle with $\text{width}(t) = 1$. The top and bottom coordinates of the rectangle reflect the issue cycle of its *definer* and the last issue cycle of its *users*. The horizontal coordinate represents the register to which the temporary is assigned (see Fig. 5b).

In this representation, two temporaries interfere when their rectangles overlap vertically. The non-overlapping rectangles constraint disjoint2 [27] enforces interfering temporaries to be assigned to different registers (4). Fig. 6a shows a register allocation for a given instruction schedule of block b_1 from Fig. 2. In this schedule, all temporaries interfere with each other and must be assigned to different processor registers.

Due to architectural constraints, operations can only access their operands in certain spaces. The operation that implements an instruction determines the space to which its temporaries are allocated (5).

A copy i from a temporary $\text{src}(i)$ to a temporary $\text{dst}(i)$ is active when these temporaries are not coalesced (6). Fig. 6b shows a register allocation for the

Program parameters

B, I, T	sets of blocks, instructions and temporaries
$\mathrm{ins}(b)$	set of instructions of block b
$\mathrm{tmp}(b)$	set of temporaries defined and used within block b
$\mathrm{tmp}(i)$	set of temporaries defined and used by instruction i
$\mathrm{definer}(t)$	instruction that defines temporary t
$\mathrm{users}(t)$	set of instructions that use temporary t
$t \equiv t'$	whether temporaries t and t' are congruent
$\mathrm{dep}(b)$	dependency graph of the instructions of block b
$\mathrm{ops}(i)$	set of alternative operations implementing i (singleton for non-copies)
$\mathrm{width}(t)$	number of register atoms that temporary t occupies
$t \triangleright \mathbf{r}$	whether temporary t is pre-assigned to register $\mathbf{r}$
$\mathrm{src}(i)$	source temporary of copy instruction i
$\mathrm{dst}(i)$	destination temporary of copy instruction i
$\mathrm{freq}(b)$	estimation of the execution frequency of block b

Processor parameters

$\mathrm{space}(i, \mathbf{op}, t)$	register space in which instruction i implemented by $\mathbf{op}$ accesses t
$\mathrm{forbidden}(t)$	forbidden first assigned register atoms for temporary t
$\mathrm{lat}(\mathbf{op})$	latency of operation $\mathbf{op}$
R	set of processor resources
$\mathrm{cap}(r)$	capacity of processor resource r
$\mathrm{use}(\mathbf{op}, r)$	units of processor resource r used by operation $\mathbf{op}$
$\mathrm{dur}(\mathbf{op}, r)$	duration of usage of processor resource r by operation $\mathbf{op}$

Variables

$r_t \in \mathbb{N}_0$	register to which temporary t is assigned
$o_i \in \mathbb{N}_0$	operation that implements instruction i
$c_i \in \mathbb{N}_0$	issue cycle of instruction i relative to the beginning of its block
$a_i \in \{0, 1\}$	whether instruction i is active
$ls_t \in \mathbb{N}_0$	start of live range of temporary t
$le_t \in \mathbb{N}_0$	end of live range of temporary t

Register allocation constraints

$$ls_t = c_{\mathrm{definer}(t)} \quad \forall t \in T; \qquad le_t = \max_{u \in \mathrm{users}(t)} c_u \quad \forall t \in T \tag{1}$$

$$a_i \quad \forall i : (i \text{ is a natural instruction}) \vee (i \text{ is a delimiter}) \tag{2}$$

$$o_i \in \mathrm{ops}(i) \cup \{\texttt{null}\} \quad \forall i : i \text{ is a copy} \tag{3}$$

$$\mathrm{disjoint2}(\{\langle r_t, r_t + \mathrm{width}(t), ls_t, le_t \rangle : t \in \mathrm{tmp}(b)\}) \quad \forall b \in B \tag{4}$$

$$o_i = \mathbf{op} \implies r_t \in \mathrm{space}(i, \mathbf{op}, t) \quad \forall t \in \mathrm{tmp}(i), \forall \mathbf{op} \in \mathrm{ops}(i), \forall i \in I \tag{5}$$

$$r_{\mathrm{src}(i)} \neq r_{\mathrm{dst}(i)} \iff a_i \quad \forall i : i \text{ is a copy} \tag{6}$$

$$r_t = \mathbf{r} \quad \forall t \in T : t \triangleright \mathbf{r} \tag{7}$$

$$r_t = r_{t'} \quad \forall t, t' \in T : t \equiv t' \tag{8}$$

$$r_t \notin \mathrm{forbidden}(t) \quad \forall t \in T \tag{9}$$

Instruction scheduling constraints

$$a_i \wedge a_j \implies c_j \geq c_i + \mathrm{lat}(o_i) \quad \forall (i, j) \in \mathrm{edges}(\mathrm{dep}(b)), \forall b \in B \tag{10}$$

$$\mathrm{cumulative}(\{\langle c_i, \mathrm{dur}(o_i, r), a_i \times \mathrm{use}(o_i, r) \rangle : i \in \mathrm{ins}(b)\}, \mathrm{cap}(r)) \quad \forall b \in B, \forall r \in R \tag{11}$$

Fig. 4. Model parameters, variables, and constraints

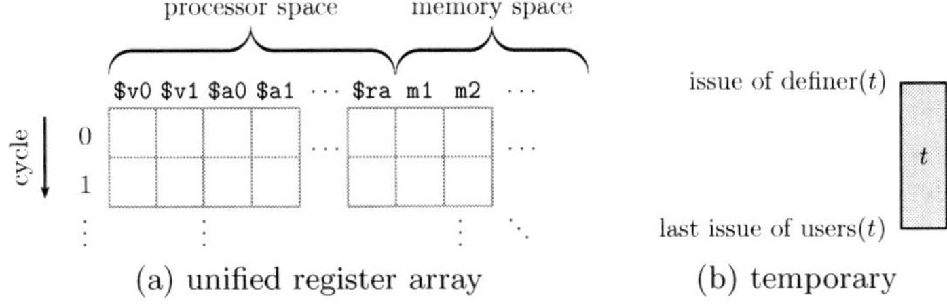

(a) unified register array   (b) temporary

Fig. 5. Geometric interpretation of local register allocation

block b_1 in Fig. 3, where dotted arrows connect copy-related temporaries. t_2 and t_5 are coalesced, rendering their copy inactive. The copy from t_3 to t_6 is implemented by `sw` and represents a spill to `m1`. This frees `$v0` from cycle 2 onwards and allows t_4 to reuse it, reducing the set of used processor registers from $\{\$v0, \$v1, \$a0, \$ra\}$ in Fig. 6a to $\{\$v0, \$a0, \$ra\}$ in Fig. 6b.

Some temporaries are pre-assigned to registers (7).

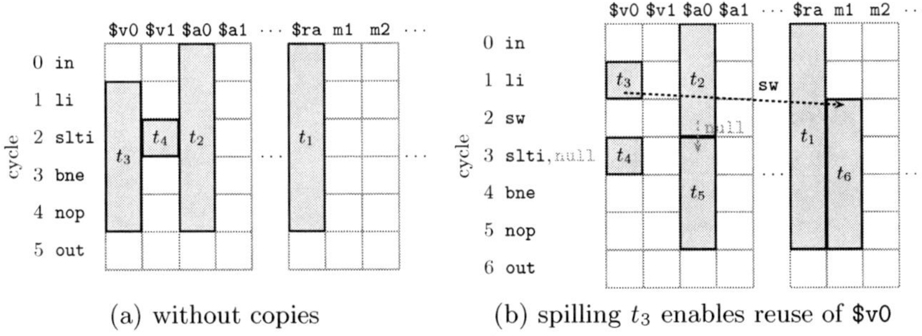

(a) without copies   (b) spilling t_3 enables reuse of `$v0`

Fig. 6. Two register allocations for block b_1

4.2 Global Register Allocation

In LSSA, the global relation between temporaries is solely captured by temporary congruences, which leads to a direct extension of the local problem. Temporaries whose live ranges span block boundaries are decomposed in LSSA into congruent temporaries with linear live ranges. Congruent temporaries resulting from this decomposition represent the same original temporary and are assigned to the same register (8).

4.3 Register Packing

The register allocation model described in this section is extended with register packing following the approach of Pereira and Palsberg [28]. Registers are decomposed into register *atoms*. An atom is the minimum part of a physical register that can be referenced by an operation (for example, `AH` in x86). A space is a sequence of atoms, each of which corresponds to a column in the unified register array. width(t) becomes a program parameter giving the number of atoms that the temporary t occupies. The variable r_t represents the first of the atoms to which t is assigned. Enforcing non-interference among temporaries assigned to the same register (4) thus becomes isomorphic to rectangle packing.

Most processors restrict the combinations of atoms out of which wider registers can be formed. For example, double-width temporaries such as t_2 and t_4 in the figure to the right cannot be assigned to $\{AL, BH\}$ or to $\{BL, CH\}$ in the 16-bit x86 register array. Such forbidden atoms are removed from the register variable domains (9).

5 Instruction Scheduling and Bundling

This section describes how to extend the register allocation model with instruction scheduling and bundling.

Local instruction scheduling. To model instruction scheduling, the issue cycle c_i of each instruction i and the (derived) live range (ls_t, le_t) of each temporary t become variables (1).

Flow of data and control cause dependencies among instructions. The dependencies in a block b form a *dependency graph* dep(b) with instructions as vertices. For example, the figure to the right shows dep(b_1) from Fig. 2. Solid, dashed and dotted arcs respectively represent data, control and artificial dependencies. The latter are added to make delimiters first and last in any topological ordering. These dependencies dictate precedence constraints among instructions. The minimum issue distance in a precedence (i, j) is equal to the latency of the parent $\mathrm{lat}(o_i)$. Precedence constraints are only effective when both instructions are active (10).

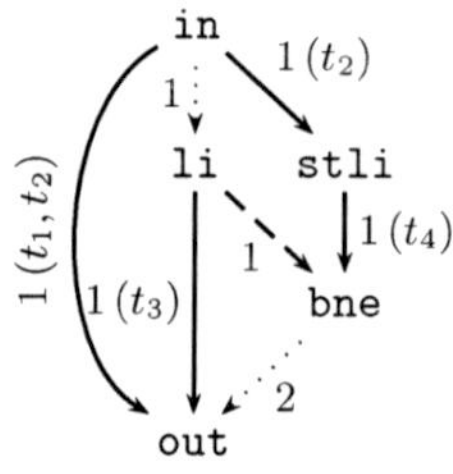

Operations share limited processor resources such as functional units and data buses. This is naturally captured as a task-resource model with a cumulative constraint [29] for each processor resource and block. These constraints include a task for each active instruction in the block (11).

Instruction bundling. Very Long Instruction Word and Explicitly Parallel Instruction Computing processors can issue several instructions every clock cycle. To exploit this capability, instructions must be combined into valid bundles satisfying precedence (10) and processor resource (11) constraints [30]. The presented scheduling model already subsumes bundling, by interpreting sets of instructions issued in the same cycle as bundles and giving the appropriate processor resource configuration.

6 Model Limitations

This section discusses some model limitations to be addressed in the future.

Limited coalescing. The constraint model employs a simple definition of interference with a direct geometrical interpretation: two temporaries *interfere* when the rectangles representing their live ranges overlap. Copy-related temporaries that do not interfere can often be coalesced, rendering their copy instructions inactive and saving execution cycles. That is the case, for example, for t_2 and t_5 in Fig. 6b. Relaxed definitions of interference have been proposed which expose more coalescing opportunities [31]. In the example to the right, t_i and t_j can be coalesced into a single temporary since both hold the same value. However, in the constraint model this makes their rectangles overlap, which is not allowed. This limitation, shared by the related register allocation approaches mentioned in the introduction, is significantly mitigated in the constraint model by the possibility of rearranging live ranges through instruction scheduling.

$$\cdots$$
$$t_j \leftarrow t_i$$
$$\cdots$$
$$\leftarrow \ldots, t_i, t_j, \ldots$$

Spilling reused temporaries. Once a temporary is spilled, it must be loaded into a register before every use. If the temporary is used multiply, it might be desirable to load it to a register once and keep it there for the remaining uses. The example to the right illustrates this limitation: if t_1 is spilled to memory by a `sw` operation, t_2 must be loaded into a register twice by `lw` operations, once for each operation `op`. Fortunately, in SSA most of the temporaries are only used once [32], and this percentage is even larger in LSSA since SSA temporaries are further decomposed.

$$\cdots$$
$$t_2 \leftarrow \{\texttt{move}, \texttt{sw}\}\ t_1$$
$$\cdots$$
$$t_3 \leftarrow \{\texttt{move}, \texttt{lw}\}\ t_2$$
$$\cdots \leftarrow \texttt{op}\ t_3$$
$$\cdots$$
$$t_4 \leftarrow \{\texttt{move}, \texttt{lw}\}\ t_2$$
$$\cdots \leftarrow \texttt{op}\ t_4$$
$$\cdots$$

7 Decomposition-Based Code Generation

This section introduces a decomposition scheme and a code generator that exploit the properties of LSSA.

The main property of LSSA is that temporaries are live in single blocks only. All temporaries accessed by delimiters (*global* temporaries) are pre-assigned or congruent to temporaries in other blocks. For example, the global temporaries in Fig. 3 are $\{t_1, t_2, t_5, t_6, t_7, t_8, t_9, t_{14}, t_{15}, t_{16}, t_{17}, t_{18}\}$. The only link between different blocks in the model is given by the congruence constraints (8), which relate the register variables r_t of global temporaries. Thus, once these variables are assigned, the rest of the register allocation and instruction scheduling problem can be solved independently for each block. For example, assuming that $\{t_2, t_5\}$ and $\{t_3, t_6\}$ have been coalesced, the problem variables that remain to be assigned for block b_1 in Fig. 3 are the register $\{r_{t_3}, r_{t_4}\}$ and issue cycle variables.

Based on this observation, we devise a decomposition scheme that significantly reduces the search space. It proceeds by first solving the global problem (assigning the r_t variables of global temporaries) and then solving a local problem for each block (assigning the remaining variables).

The objective is to minimize execution cycles according to an estimate of block execution frequency:

$$\text{minimize} \sum_{b \in B} \text{freq}(b) \times \max_{i \in \text{ins}(b)} c_i \tag{12}$$

Fig. 7 depicts the architecture of the code generator. SSA functions are transformed to LSSA and extended with copies as described in Sect. 3. Then, a satisfaction problem with all constraints is solved, assigning global temporaries to registers. This assignment constitutes a *global solution*. As variable selection heuristic, the global solver branches first on the largest temporary congruence class ($\{t_6, t_9, t_{14}, t_{17}\}$ in Fig. 3). As value selection heuristic, it performs a cost-benefit analysis to determine the most effective register for each temporary. The benefit component estimates the saved spilling overhead, while the cost component is based on an estimate of the increased space occupation. This analysis is parameterized with an aggressiveness factor to direct the heuristic towards either the benefit or the cost component. In the example, all temporaries are assigned to processor registers regardless of this parameter, since spilling is costly and the processor register space occupation is low.

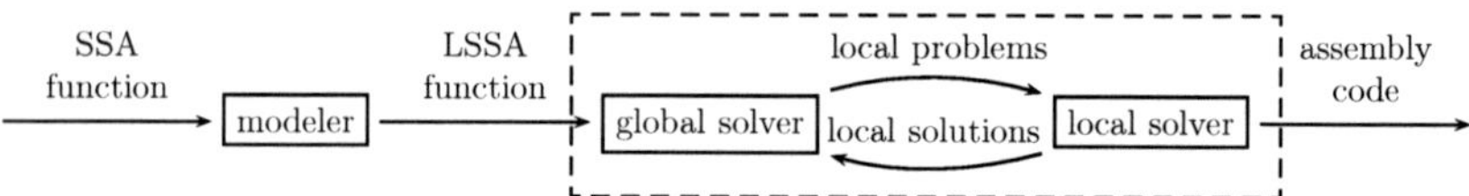

Fig. 7. Architecture of the code generator

Once a global solution is found, an optimization problem with all but the congruence constraints (8) is solved for each block, seeking a locally optimal assignment of the remaining variables. This assignment constitutes a *local solution*. The search starts by branching on the copy activation variables a_i, trying first to inactivate the copy. Then, the solver branches on the c_i variables in topological dependency order, selecting the earliest cycle first. Finally, local temporaries are assigned to registers by assigning their r_t variables. The global and local solutions are then combined and the total cost is computed. This process is repeated, increasing the aggressiveness of the global solver in each iteration until it proves optimality or reaches a time limit.

8   Evaluation

Two essential characteristics of the code generator are examined by the experiments: the quality of the generated code and its solving time. The global and local solvers are implemented with the constraint programming system Gecode 3.7.3 [33]. As input, we have used functions from the C program `bzip2` as a representative of the standard SPECint 2006 benchmark suite, optimized the intermediate code of each function and selected MIPS32 instructions using the LLVM 3.0 compiler infrastructure. The instruction-selected functions are passed as input to the code generator for this experiment. The global and local solvers are run with a time limit of 10 and 3 seconds respectively. All experiments are run using sequential search on a Linux machine with a quad-core Intel Core i5-750 processor and 4 GB of main memory.

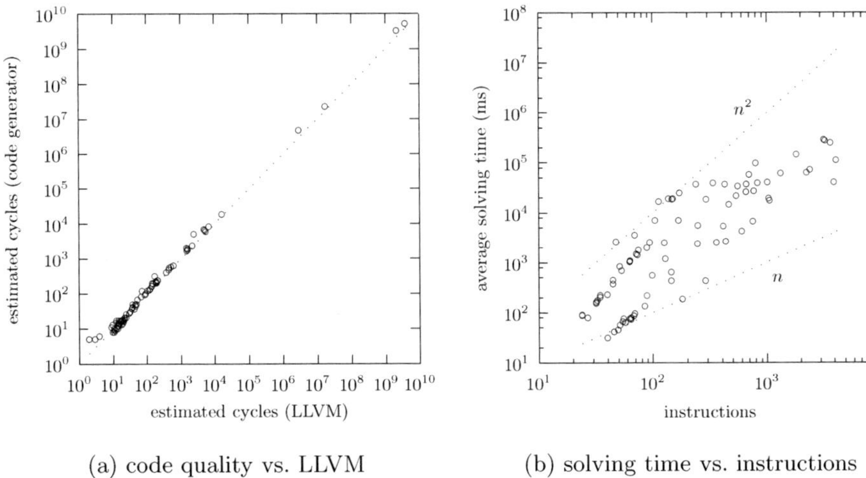

(a) code quality vs. LLVM   (b) solving time vs. instructions

Fig. 8. Evaluation of the decomposition-based code generator

Code quality. Lack of post-code-generation support (including generation of assembly directives) prevents running the generated code. Therefore, to measure its quality, we estimate the number of execution cycles by computing the value of the objective function (12). This measure is computed for the code generated by both our system and LLVM's register allocator (based on priority-based coloring [34]) and instruction scheduler (based on list-scheduling [35]). These LLVM components are run with the following flags: `-O3 -enable-mips-delay-filler -disable-post-ra -disable-tail-duplicate -disable-branch-fold`. This comparison is meaningful since a) the input to the LLVM components is the same as to our system, and b) the optimization flags given to LLVM are aligned with the objective function (12). For example, `-O3` emphasizes speed at the possible expense of code size. Furthermore, the same optimization level is enforced after code generation by disabling *tail duplication* and *branch folding* in LLVM.

From a total of 106 functions in `bzip2`, the code generator pre-processes and solves 86 functions. The remaining ones cannot yet be pre-processed by the `modeler` module (see Fig. 7) because of lack of support for the MIPS32 floating-point extension and incompleteness of the interface to LLVM's instruction selector. Fig. 8a shows the estimated execution cycles of each solved function. The figure shows that the code generator is competitive with LLVM, a state-of-the-art compiler infrastructure, in terms of code quality. The cases in which LLVM generates better code are due to a) few of the 86 functions being solved optimally because of global and local time-outs, and b) the limitations discussed in Sect. 6, which in particular prevent coalescing certain copies in blocks that belong to deeply nested loops.

Solving time. The global and local solvers dominate the execution time of the code generator. Fig. 8b shows, for each function in the first experiment, the average time to solve the constraint problems and the number of instructions. The average is calculated on 10 iterations, where the maximum coefficient of variation per function is 10%. The figure reveals a sub-quadratic relation between solving time and size of the compiled functions, confirming the robustness of the code generator for functions with thousands of instructions. This robust behavior is due partially to the effect of solver time-outs on all non-trivial functions.

9   Conclusion and Future Work

This paper introduces a constraint model capturing global register allocation and local instruction scheduling as two main tasks of code generation. In particular, the model of register allocation combines all essential aspects in this problem: generalized spilling, coalescing and register packing. The paper introduces LSSA to enable a direct model of global register allocation and a problem decomposition. A code generator is presented that exploits this decomposition to achieve robust behavior. Experiments show that it generates code that is competitive with a state-of-the-art compiler infrastructure for functions of thousands of instructions.

Future work. This paper presents an important step towards basing code generation entirely on a high-level model as opposed to limited heuristic algorithms. There is considerable future work in this direction. A first step is to address the limitations from Sect. 6. Also, although the code generator shows robust behavior, there is still a significant efficiency gap with respect to state-of-the-art compiler infrastructure such as LLVM. We have identified several opportunities to improve the efficiency of the code generator by applying standard modeling techniques: breaking symmetries in the dependency graph [6] and in the register array, strengthening the geometric reasoning on the register array by coalescing-aware custom propagators and inferring implied constraints to improve propagation are some examples. Another possible improvement is to refine the decomposition described in Sect. 7 into a Benders-like scheme [36], where the local solver feeds back problem knowledge to the global solver.

We plan to study how to extend the model from this paper with instruction selection and other code generation problems to further improve code quality. Finally, we intend to evaluate the generality and flexibility of the model by targeting more challenging architectures such as digital signal processors and Intel's x86. We conjecture that the additional difficulties imposed by these architectures will only but highlight the advantages of a constraint-based approach.

Acknowledgments. The research reported in this paper has been partially funded by LM Ericsson AB and the Swedish Research Council (VR 621-2011-6229). The authors are grateful for helpful comments from Gabriel Hjort Blindell and the anonymous reviewers.

References

1. Lattner, C., Adve, V.: LLVM: A compilation framework for lifelong program analysis & transformation. In: CGO (March 2004)
2. Goodwin, D.W., Wilken, K.D.: Optimal and near-optimal global register allocations using 0-1 integer programming. Software – Practice and Experience 26, 929–965 (1996)
3. Appel, A.W., George, L.: Optimal spilling for CISC machines with few registers. SIGPLAN Not. 36, 243–253 (2001)
4. Scholz, B., Eckstein, E.: Register allocation for irregular architectures. SIGPLAN Not. 37, 139–148 (2002)
5. Anton Ertl, M., Krall, A.: Optimal Instruction Scheduling Using Constraint Logic Programming. In: Małuszyński, J., Wirsing, M. (eds.) PLILP 1991. LNCS, vol. 528, pp. 75–86. Springer, Heidelberg (1991)
6. Malik, A.M., McInnes, J., van Beek, P.: Optimal basic block instruction scheduling for multiple-issue processors using constraint programming. International Journal on Artificial Intelligence Tools 17(1), 37–54 (2008)
7. Leupers, R., Marwedel, P.: Time-constrained code compaction for DSP's. IEEE Transactions on Very Large Scale Integration Systems 5, 112–122 (1997)
8. Wilken, K., Liu, J., Heffernan, M.: Optimal instruction scheduling using integer programming. SIGPLAN Not. 35, 121–133 (2000)
9. Malik, A.M., Chase, M., Russell, T., van Beek, P.: An Application of Constraint Programming to Superblock Instruction Scheduling. In: Stuckey, P.J. (ed.) CP 2008. LNCS, vol. 5202, pp. 97–111. Springer, Heidelberg (2008)
10. Winkel, S.: Exploring the performance potential of Itanium processors with ILP-based scheduling. In: CGO, pp. 189–200. IEEE (2004)
11. Chou, H.C., Chung, C.P.: An optimal instruction scheduler for superscalar processor. IEEE Transactions on Parallel and Distributed Systems 6, 303–313 (1995)
12. Shobaki, G., Wilken, K.: Optimal superblock scheduling using enumeration. In: MICRO, pp. 283–293. IEEE (2004)
13. Gebotys, C.H., Elmasry, M.I.: Simultaneous scheduling and allocation for cost constrained optimal architectural synthesis. In: DAC, pp. 2–7. ACM (1991)
14. Kästner, D.: PROPAN: A Retargetable System for Postpass Optimisations and Analyses. In: Davidson, J., Min, S.L. (eds.) LCTES 2000. LNCS, vol. 1985, pp. 63–80. Springer, Heidelberg (2001)
15. Kuchcinski, K.: Constraints-driven scheduling and resource assignment. ACM Trans. Des. Autom. Electron. Syst. 8, 355–383 (2003)
16. Nagarakatte, S.G., Govindarajan, R.: Register Allocation and Optimal Spill Code Scheduling in Software Pipelined Loops Using 0-1 Integer Linear Programming Formulation. In: Adsul, B., Vetta, A. (eds.) CC 2007. LNCS, vol. 4420, pp. 126–140. Springer, Heidelberg (2007)
17. Bashford, S., Leupers, R.: Phase-coupled mapping of data flow graphs to irregular data paths. Design Automation for Embedded Systems, 119–165 (1999)
18. Wilson, T., Grewal, G., Halley, B., Banerji, D.: An integrated approach to retargetable code generation. In: ISSS, pp. 70–75. IEEE (1994)
19. Eriksson, M.V., Skoog, O., Kessler, C.W.: Optimal vs. heuristic integrated code generation for clustered VLIW architectures. In: SCOPES, pp. 11–20 (2008)
20. Sweetman, D.: See MIPS Run, 2nd edn. Morgan Kaufmann (2006)
21. Cytron, R., Ferrante, J., Rosen, B.K., Wegman, M.N., Zadeck, F.K.: Efficiently computing static single assignment form and the control dependence graph. ACM TOPLAS 13(4), 451–490 (1991)

22. Sreedhar, V.C., Ju, R.D.-C., Gillies, D.M., Santhanam, V.: Translating Out of Static Single Assignment Form. In: Cortesi, A., Filé, G. (eds.) SAS 1999. LNCS, vol. 1694, pp. 194–210. Springer, Heidelberg (1999)

23. Hack, S., Grund, D., Goos, G.: Register Allocation for Programs in SSA-Form. In: Mycroft, A., Zeller, A. (eds.) CC 2006. LNCS, vol. 3923, pp. 247–262. Springer, Heidelberg (2006)

24. Smith, M.D., Ramsey, N., Holloway, G.: A generalized algorithm for graph-coloring register allocation. SIGPLAN Not. 39, 277–288 (2004)

25. Appel, A.W.: SSA is functional programming. SIGPLAN Not. 33(4), 17–20 (1998)

26. Aycock, J., Horspool, N.: Simple Generation of Static Single-Assignment Form. In: Watt, D.A. (ed.) CC 2000. LNCS, vol. 1781, pp. 110–124. Springer, Heidelberg (2000)

27. Beldiceanu, N., Carlsson, M.: Sweep as a Generic Pruning Technique Applied to the Non-overlapping Rectangles Constraint. In: Walsh, T. (ed.) CP 2001. LNCS, vol. 2239, pp. 377–391. Springer, Heidelberg (2001)

28. Pereira, F., Palsberg, J.: Register allocation by puzzle solving. SIGPLAN Not. 43, 216–226 (2008)

29. Aggoun, A., Beldiceanu, N.: Extending CHIP in order to solve complex scheduling and placement problems. Mathematical and Computer Modelling 17(7), 57–73 (1993)

30. Kessler, C.W.: Compiling for VLIW DSPs. In: Handbook of Signal Processing Systems, pp. 603–638. Springer (2010)

31. Chaitin, G.J., Auslander, M.A., Chandra, A.K., Cocke, J., Hopkins, M.E., Markstein, P.W.: Register allocation via coloring. Computer Languages 6(1), 47–57 (1981)

32. Boissinot, B., Hack, S., Grund, D., Dupont de Dinechin, B., Rastello, F.: Fast liveness checking for SSA-form programs. In: CGO, pp. 35–44. ACM (2008)

33. Gecode Team: Gecode: generic constraint development environment (2006), http://www.gecode.org

34. Chow, F., Hennessy, J.: Register allocation by priority-based coloring. SIGPLAN Not. 19(6), 222–232 (1984)

35. Rau, B.R., Fisher, J.A.: Instruction-level parallel processing: history, overview, and perspective. J. Supercomput. 7, 9–50 (1993)

36. Hooker, J.N., Ottosson, G.: Logic-based Benders decomposition. Mathematical Programming 96(1), 33–60 (2003)

Maximising the Net Present Value
of Large Resource-Constrained Projects

Hanyu Gu[1], Peter J. Stuckey[2], and Mark G. Wallace[3]

[1] National ICT Australia, Victoria Research Laboratory,
The University of Melbourne, Victoria 3010, Australia
`Hanyu.Gu@nicta.com.au`
[2] National ICT Australia, Department of Computing and Information Systems,
The University of Melbourne, Victoria 3010, Australia
`pstuckey@unimelb.edu.au`
[3] Monash University, Victoria 3100, Australia
`mark.wallace@monash.edu`

Abstract. The Resource-constrained Project Scheduling Problem (RCPSP), in which a schedule must obey resource constraints and precedence constraints between pairs of activities, is one of the most studied scheduling problems. An important variation of this problem (RCPSPDC) is to find a schedule which maximises the net present value (discounted cash flow). Large industrial applications can require thousands of activities to be scheduled over a long time span. The largest case we have (from a mining company) includes about 11,000 activities spanning over 20 years. We apply a Lagrangian relaxation method for large RCPSPDC to obtain both upper and lower bounds. To overcome the scalability problem of this approach we first decompose the problem by relaxing as fewer as possible precedence constraints, while obtaining activity clusters small enough to be solved efficiently. We propose a hierarchical scheme to accelerate the convergence of the subgradient algorithm of the Lagrangian relaxation. We also parallelise the implementation to make better use of modern computing infrastructure. Together these three improvements allow us to provide the best known solutions for very large examples from underground mining problems.

1 Introduction

The Resource-constrained Project Scheduling Problem (RCPSP) is one of the most studied scheduling problems. It consists of scarce resources, activities and precedence constraints between pairs of activities where a precedence constraint expresses that an activity can be run after the execution of its preceding activity is finished. Each activity requires some units of resources during their execution. The aim is to build a schedule that satisfies the resource and precedence constraints. Here, we assume renewable resources (*i.e.*, their supply is constant during the planning period) and non-preemptive activities (*i.e.*, once started their execution cannot be interrupted). Usually the objective in solving

M. Milano (Ed.): CP 2012, LNCS 7514, pp. 767–781, 2012.

RCPSP problems is to minimise the makespan, *i.e.*, to complete the entire project in the minimum total time. But another important objective is to maximise the net present value (*npv*), because it better captures the financial aspects of the project. In this formulation each activity has an associated cash flow which may be a payment (negative cash flow) or a receipt (positive cash flow). These cash flows are discounted with respect to a discount rate, which makes it, in general, beneficial for the *npv* to execute activities with a positive (negative) cash flow as early (late) as possible. The problem is to maximise the *npv* for a given RCPSP problem. We denote the problem RCPSPDC, *i.e.*, RCPSP with discounted cash flows. It is classified as $m, 1|cpm, \delta_n, c_j|npv$ [13] or $PS|prec| \sum_j C_j^F \beta^{C_j}$ [3].

Optimisation of the net present value for project scheduling problems was first introduced in [21]. Different complete and incomplete methods for RCPSPDC with or without generalised precedence constraints have been proposed, the reader is referred to [12] for a more extensive literature overview of solution approaches for RCPSPDC and other variants or extensions of RCPSP.

Most complete methods for RCPSPDC use a branch-and-bound algorithm to maximise the *npv*. The approaches in [14,24,19] are based on the branch-and-bound algorithm in [6,7] for RCPSP. These algorithms use a scheduling generation scheme which resolves resource conflicts by adding new precedence constraints between activities in conflict. The method in [24] improves upon the one in [14] whereas the work [19] considers RCPSPDC with generalised precedence constraints. Recently we developed lazy clause generation approaches to RCPSPDC [22] which provide the state of the art complete methods for RCPSPDC.

But complete methods can only solve problems up to a hundred activities. To cope with large industrial applications with thousands of activities, various rules based heuristics [2,20] are used in practice. However these approaches are not robust especially for problems with tight constraints. Decomposition methods are widely used for large-scale combinatorial optimisation problems. Lagrangian relaxation was successfully applied on RCPSP for up to 1000 jobs [18]. It has also been applied to RCPSPDC with good results [17] and is more robust than rule based heuristics. However our experience shows that this method scales badly for example underground mining problems with over a few thousand activities. The main obstacle is that the Lagrangian relaxation problem, which is a max-flow problem, cannot be solved efficiently for large problems. Furthermore the Lagrangian dual problem requires many more iterations to converge. We address these two problems in this paper to improve the Lagrangian relaxation method to generate tight upper and lower bounds for RCPSPDC problems with up to 10,000 activities.

The contributions of this paper are: *(i)* We propose a general approach to relax as few as possible of the precedence constraints but still obtain activity clusters small enough to be solved efficiently. *(ii)* We define a hierarchical scheme to accelerate the convergence of the Lagrangian multipliers for the precedence constraints when using subgradient algorithm. *(iii)* We parallelise the algorithm to make more effective use of modern multi-core desktop computers. *(iv)* We produce highly competitive results on very large underground mining problems within a reasonable computing time.

2   Lagrangian Relaxation of Resource Constraints

Let T be the deadline of the project, α the discount rate, R_k be the capacity of resource $k \in R$, r_{jk} be the resource requirement of activity $j \in J$ on resource k, p_j be the processing time of activity j, c_j is the discounted cash flow for activity j to start at time 0, and precedence constraint $(i,j) \in L$ if activity j cannot start before activity i completes. The RCPSPDC problem can be stated as follows:

$$NPV = \text{maximise } \sum_j e^{-\alpha s_j} c_j \tag{1}$$

$$\text{subject to} \qquad s_i + p_i \leq s_j \qquad\qquad (i,j) \in L \tag{2}$$

$$\texttt{cumulative}(s, p, [r_{jk}|j \in J], R_k) \quad k \in R, \tag{3}$$

$$0 \leq s_j \leq T - p_j \qquad\qquad j \in J. \tag{4}$$

where $s_j, j \in J$ is the start time of activity j. This model uses the global constraint `cumulative`, and hence requires a CP technology to solve directly.

The time-indexed formulation for the RCPSPDC problem, breaks the start time variables into binary form, and encodes the resource constraints explicitly for each time point, giving a binary program. Let w_{jt} be the discounted cash flow of activity j when starting at time t, i.e. $w_{jt} = e^{-\alpha t} c_j$. The time-indexed formulation is [8]:

$$NPV = \text{maximise } \sum_j \sum_t w_{jt} x_{jt} \tag{5}$$

$$\text{subject to} \qquad \sum_t x_{jt} = 1 \qquad\qquad j \in J \tag{6}$$

$$\sum_{s=t}^{T} x_{is} + \sum_{s=0}^{t+p_j-1} x_{js} \leq 1 \qquad \forall (i,j) \in L, t = 0, \cdots, T \tag{7}$$

$$\sum_j r_{jk} \left(\sum_{s=t-p_j+1}^{t} x_{js} \right) \leq R_k \qquad k \in R, t = 0, \cdots, T \tag{8}$$

$$\text{all variables binary} \tag{9}$$

where the binary variable $x_{jt} = 1$ if activity j starts at t ($s_j = t$), and $x_{jt} = 0$ otherwise. The assignment constraints (6) ensure that each activity has exactly one start time. The precedence constraints (7) imply that activity j cannot start before $t + p_j$ if activity i starts at or after time t for each $(i,j) \in L$. The resource constraints (8) enforce that the resource consumption of all activities processed simultaneously must be within the resource capacity.

The Lagrangian relaxation method identifies "hard" constraints in the optimisation problem, and removes these "hard" constraints by putting them in the objective function as penalties for the violation of these relaxed constraints [10]. The Lagrangian Relaxation Problem (LRP) obtained by relaxing the resource constraints (8) with Lagrangian multipliers λ_{kt}, $k \in R$, $t = 0, \cdots, T$ is

$$Z_{LR}(\lambda) = \text{maximise } LRP(x) \tag{10}$$

$$\text{subject to} \quad (6), (7), (9) \tag{11}$$

where

$$LRP(x) = \sum_j \sum_t w_{jt} x_{jt} + \sum_{k \in R} \sum_t \lambda_{kt} \left(R_k - \sum_j r_{jk} \left(\sum_{s=t-p_j+1}^{t} x_{js} \right) \right) \quad (12)$$

By rearranging the terms in (12) we have

$$LRP(x) = \sum_j \sum_t z_{jt} x_{jt} + \sum_{k \in R} \sum_t \lambda_{kt} R_k \quad (13)$$

where

$$z_{jt} = w_{jt} - \sum_{s=t}^{t+p_j-1} \sum_{k \in R} \lambda_{ks} r_{jk} \quad (14)$$

The Lagrangian multiplier λ_{ks} can be interpreted as the unit price for using resource k at time period s. The discounted cash flow of activity j starting at time t is then further reduced in (14) by the amount paid for all the resources used from the start to the completion of this activity. It is well-know [10] that $Z_{LR}(\lambda)$ is a valid upper bound of RCPSPDC for $\lambda \geq 0$.

The polytope described by (6), (7), and (9) is integral [4]. However, it is inefficient to solve LRP using a general LP solver. Instead, it can be transformed into a minimum cut problem [18] and solved efficiently by a general max-flow algorithm. We briefly explain the network flow model of [18], denoted by $G = (V, A)$ in order to be self-contained. The node set $V = \{a, b\} \cup \bigcup_{i \in J} V_i$ is defined as

- a is the source node and b is the sink node.
- $V_i = \{v_{it} | t = e(i), \cdots, l(i)+1\}$ where $e(i)$ and $l(i)$ are the earliest and latest start time of activity i respectively.

The directed edge set $A = A^a \cup A^b \cup A_i^A \cup A_{i,j}^P$ is defined as

- $A^a = \{(a, v_{j,e(j)}) | \forall j \in J\}$ are the auxiliary edges connecting the source and the first node of each activity. All edges in A^a have infinite capacity.
- $A^b = \{(v_{j,l(j)+1}, b) | \forall j \in J\}$ are the auxiliary edges connecting the last node of each activity and the sink. All edges in A^b have infinite capacity.
- $A_i^A = \{(v_{it}, v_{i,t+1}) | t = e(i), \cdots, l(i)\}$ are the assignment edges that forms a chain for each activity. $(v_{it}, v_{i,t+1})$ has capacity z_{it}.
- $A_{i,j}^P = \{(v_{it}, v_{j,t+p_i}) | e(i) + 1 \leq t \leq l(i), e(j) + 1 \leq t + p_i \leq l(j)\}$ are the precedence edges for $(i, j) \in L$. All the precedence edges have infinite capacity.

Since the network flow model has $O(|J|T)$ nodes and $O((|J| + |L|)T)$ edges, a state of the art max-flow solver [5] can solve it in $O(|J||L|T^2 \log(T))$. We use the push-relabel implementation in c++ BOOST BGL [23] which is based on the source code of the authors of [5]. This implements the highest-label version of the push-relabel method with global relabeling and gap relabeling heuristics.

With just 1400 activities and $T = 4000$ the network has about 5 million nodes and it takes on average 4 minutes to solve the maximum flow problem. For some larger cases we could not even set up the network model on a desktop computer with 16GB memory.

3   Lagrangian Relaxation of Precedence Constraints

To overcome the scalability problem of the max-flow algorithm we further relax some precedence constraints so that activities can form clusters that are independent from each other. We partition the set of activities J into $J = J_1 \cup J_2 \cup \cdots \cup J_U$ where U is the given number of clusters. The multi-cut of this partition is defined as the set of precedence constraints $E = \{(i,j) \in L | i \in J_p, j \in J_q, p \neq q\}$. Denote the set of precedence relations that hold on cluster J_i by $L_i = \{(k,j) \in L | k \in J_i, j \in J_i\}$. Obviously we have $L = E \cup L_1 \cup \cdots \cup L_U$. As an example given if Figure 1, the set of nine activities is partitioned into $J_1 = \{1,3,5\}$, $J_2 = \{2,4,6\}$ and $J_3 = \{7,8,9\}$. The multi-cut of this partition is $E = \{e1, e2, e3, e4\}$.

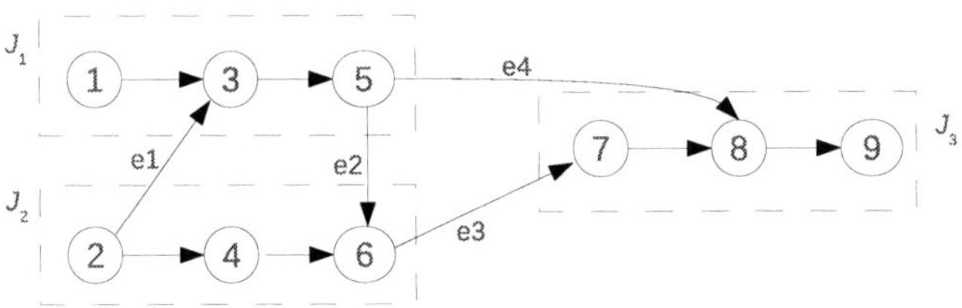

Fig. 1. Example of multi-cut for 9 activities and 3 clusters

To reduce the number of Lagrangian multipliers introduced for the relaxed precedence relations, we use the weak form of the precedence constraints [18]

$$\sum_t t(x_{jt} - x_{it}) \geq p_i, \quad \forall (i,j) \in E \tag{15}$$

We give the details here how (15) can be derived by (6) and (7) even when x_{jt} is allowed to be fractional. By rewriting (6) as

$$\sum_0^{s=t+p_j-1} x_{js} + \sum_{s=t+p_j} x_{js} = 1$$

and replacing the common terms in (7) we get

$$\sum_{s=t}^{T} x_{is} + 1 - \sum_{s=t+p_j} x_{js} \leq 1$$

$$\Rightarrow x_{it} + \sum_{s=t+1}^{T} x_{is} - \left(\sum_{s=t+1}^{T} x_{js} - \sum_{s=t+1}^{t+p_i-1} x_{js}\right) \leq 0$$

$$\Rightarrow \sum_{s=t+1}^{T}(x_{js} - x_{is}) \geq \sum_{s=t+1}^{t+p_i-1} x_{js} + x_{it} \tag{16}$$

By summing up (16) over all t we have

$$\sum_t t(x_{jt} - x_{it}) \geq \sum_t \sum_{s=t+1}^{t+p_i-1} x_{js} + \sum_t x_{it}$$

$$= \sum_{k=1}^{p_i-1} \sum_{s=k} x_{jk} + 1$$

$$= \sum_{k=1}^{p_i-1} \sum_{s=0} x_{jk} + 1 = p_i$$

which exploits that $x_{jt} = 0$, $t = 0, \cdots, p_i - 1$.

By relaxing the precedence constraints (15) with Lagrangian multipliers μ we can obtain a Decomposable Lagrangian Relaxing Problem (DLRP)

$$Z_{LR}(\lambda, \mu) = \text{maximise } LRP(x) + \sum_{v=(i,j)\in E} \mu_v(\sum_t t(x_{jt} - x_{it}) - p_i) \quad (17)$$

$$\text{subject to} \qquad\qquad (6), (9) \qquad\qquad\qquad\qquad (18)$$

$$\sum_{s=t}^{T} x_{is} + \sum_{s=0}^{t+p_j-1} x_{js} \leq 1 \quad \forall(i,j) \in L\backslash E, t = 0, \cdots, T \quad (19)$$

Let the set of precedence constraints in the multi-cut that have activity i as predecessor be $P_i = \{(i,j) \in E\}$, the set of precedence constraints in the multi-cut that have activity i as successor be $S_i = \{(j,i) \in E\}$. For the example in Figure 1, we have empty P_i and S_i except that $P_2 = \{e1\}$, $S_3 = \{e1\}$, $P_5 = \{e2, e4\}$, $S_6 = \{e2\}$, $P_6 = \{e3\}$, $S_7 = \{e3\}$, and $S_8 = \{e4\}$.

By rearranging the items in the objective function of DLRP in (17), we get

$$LRP(x) + \sum_j \sum_t (\sum_{e\in S_j} t\mu_e - \sum_{e\in P_j} t\mu_e)x_{jt} - \sum_{e=(i,j)\in E} \mu_e p_i \qquad (20)$$

By ignoring the constant terms the DLRP can be decomposed into U independent subproblems on each of the clusters J_k, $k = 1, \cdots, U$

$$\text{maximise} \qquad \sum_{j\in J_k} \sum_t (z_{jt} + \sum_{e\in S_j} t\mu_e - \sum_{e\in P_j} t\mu_e)x_{jt} \qquad (21)$$

$$\text{subject to} \qquad\qquad \sum_t x_{jt} = 1 \quad j \in J_k \qquad\qquad\qquad (22)$$

$$\sum_{s=t}^{T} x_{is} + \sum_{s=0}^{t+p_j-1} x_{js} \leq 1 \quad \forall(i,j) \in L_k, t = 0, \cdots, T \quad (23)$$

$$\text{all variables binary} \qquad\qquad\qquad (24)$$

Since each subproblem has smaller size, the max-flow solver can solve DLRP much faster than LRP. Also these subproblems can be solved in parallel utilising the multi-core computers that are now very popular. If main memory of the computer is a bottleneck we can construct the network flow model of each cluster on the fly. In this way we have solved the DLRP with over a hundred million variables within 500M memory.

The upper bound will become worse, i.e., $Z_{LR}(\lambda) \leq Z_{LR}(\lambda, \mu)$ since the weak form of the precedence constraints is used. Our goal therefore is to relax as few as possible the precedence constraints but still obtain activity clusters small enough to solve efficiently as a maximum flow problem. This can be formulated as the Min-Cut Clustering problem (MCC) as in [15]

$$\text{minimise} \quad \sum_{g=1}^{U} \sum_{e \in L} z_{eg} \tag{25}$$

$$\text{subject to} \quad \sum_{g=1}^{U} x_{ig} = 1 \qquad i \in J \tag{26}$$

$$x_{ig} - x_{jg} \leq z_{eg} \qquad \forall e = (i,j) \in L, g = 1, \cdots, U \tag{27}$$

$$l \leq \sum_{i \in J} x_{ig} \leq u \qquad g = 1, \cdots, U \tag{28}$$

$$\text{all variables binary} \tag{29}$$

where U is the upper bound of the number of clusters, x_{ig} is 1 if activity i is included in the cluster g, and otherwise 0. The set partitioning constraints (26) make sure that each activity is contained in only one cluster; constraints (27) and the minimisation of (25) imply that the binary variable z_{eg} is 1 if the predecessor activity of e is included in the cluster g but the successor activity is in a different cluster, and otherwise 0; constraints (28) ensure that the cluster size is within the specified range $[l, u]$.

MCC is also NP-hard, and only small problems can be solved to optimality. For our purpose the cluster size constraints (28) are just soft constraints. We can use heuristics to generate good partitions very quickly. Our experimentation with METIS [16] shows that the project with 11,000 activities can be partitioned into 100 balanced parts within 0.1 second and only 384 precedence constraints need to be relaxed.

4   Hierarchical Subgradient Algorithm (HSA)

The upper bound obtained by solving DLRP can be tightened by optimising the Lagrangian Dual Problem(LDP) as

$$\min_{\lambda \geq 0, \mu \geq 0} \quad Z_{LR}(\lambda, \mu) \tag{30}$$

We use the *standard subgradient algorithm* (SSA) [10] which updates the Lagrangian multipliers at the ith iteration (λ^i, μ^i) according to

$$(\lambda^{i+1}, \mu^{i+1}) = \left[(\lambda^i, \mu^i) - \delta^i \frac{Z_{LR}(\lambda^i, \mu^i) - LB^*}{||g_\lambda^i||^2 + ||g_\mu^i||^2} (g_\lambda^i, g_\mu^i) \right]^+ \tag{31}$$

where $[\cdot]^+$ denotes the non-negative part of a vector, δ^i is a scalar step size, LB^* is the best known lower bound, and (g_λ^i, g_μ^i) is a subgradient calculated as

$$g_\lambda^i(k, t) = R_k - \sum_j r_{jk} \left(\sum_{s=t-p_j+1}^{t} x_{js}^i \right) \tag{32}$$

and

$$g_\mu^i(j, k) = \sum_t t(x_{kt}^i - x_{jt}^i) - p_j \qquad \forall (j, k) \in E \tag{33}$$

where x^i is the optimal solution of DLRP at the ith iteration.

In practice δ^i is reduced by a factor ρ if Z_{LR} is not improved by at least Δ^i after p iterations. The algorithm can terminate when δ^i is small enough to avoid excessive iterations.

The subgradient algorithm tends to converge slowly for problems of high dimensions due to the zig-zag phenomenon [25]. For large RCPSPDC problems we observed that the convergence of the precedence multipliers μ was extremely slow using the updating rule (31). The reasons could be

- The contribution of the precedence multipliers in the objective function value Z_{LR} is trivial. It can be even smaller than Δ^i which is used to test if the upper bound is improved. Too small Δ^i can only lead to excessive iterations before δ^i can be reduced.
- In (31) the resource component of the subgradient $||g_\lambda^i||^2$ is much larger than the precedence component $||g_\mu^i||^2$, which may lead to steps too small for the convergence of μ.

Good μ can lead to near-feasible solution with respect to the precedence constraints, which is important for the Lagrangian relaxation based heuristics to produce good lower bound. We will demonstrate this in Section 6.

To accelerate the convergence of precedence multipliers we introduce a *hierarchical subgradient algorithm* (HSA) which has two levels. At the first level we update the multipliers according to (31) for a certain number of iterations i_1 and then move to the second level by just updating the precedence multipliers as

$$\mu^{i+1} = \left[\mu^i - \delta^i \delta_\mu^i \frac{Z_{LR}(\lambda^i, \mu^i)}{||g_\mu^i||^2} g_\mu^i\right]^+ \tag{34}$$

Only δ_μ^i is reduced at the second level if Z_{LR} is not improved after p iterations. After a certain number of iterations i_2 the algorithm will switch back to the first level. This process is repeated until some stopping criterion are met.

5   Lagrangian Relaxation Based Heuristic

The Lagrangian relaxation DLRP produces upper bounds for the original NPV problem. But in practice we are interested in finding feasible solutions of high value. We can use the Lagrangian relaxation solution to create a heuristic which created strong solutions. Combining Lagrangian relaxation with list scheduling has been previously successfully applied to different variants of RCPSP [18] [17] problems.

The basic idea is motivated by the intuition that violation of relaxed constraints tend to be reduced in the course of the subgradient optimization. Therefore, a near-feasible solution of the Lagrangian relaxation contains valuable information on how conflicts on constraints can be resolved. In [18] the activities are sorted in the increasing order of the so called $\alpha-$point. Assume the Lagrangian relaxation solution assigned start time $l_j, j \in J$ to each activity. The $\alpha-$point is calculated as $l_j + \alpha * p_j$ which is the earliest time that at least α of the activity has completed if the activity starts at l_j. α is normally evenly chosen between 0 and 1. A parallel list scheduling scheme [11] is then employed to produce feasible solutions. For RCPSPDC left and right shifting techniques are used to further improve the solution quality [17] of the parallel list scheduling using just the start time in the Lagrangian relaxation solution.

We use the parallel list scheduling scheme to generate feasible solution at each iteration of the subgradient algorithm. The list scheduling scheme starts at time $t = 0$, and determines the subset of candidate activities $j \in A$ for scheduling as those whose predecessors have all completed $s_i + p_i \leq t, (i,j) \in L$, and whose start time $l_j \leq t$. The candidate activities $j \in A$ are then scheduled, so s_j is set, in increasing order of l_j, where the resource requirements are satisfied. Candidate activities $j \in A$ with positive cash flow ($c_j > 0$) are moved as early as possible (so it may be the case that $s_j < l_j$). After an activity i is scheduled, its successor j where $(i,j) \in L$ activities may become eligble for scheduling and are added to the candidates A. The process continues until no activity is schedulable at or before time t. We then set $t := t + 1$ and repeat.

A schedule created in this way will almost always have a makespan ms larger than the deadline T. To overcome this we simply modify the Lagrangian relaxation solution $l_j, j \in J$ used to drive the heuristic. We set $l_j := l_i - T + ms$ for all $j \in J$. This means there is many more candidate activities at the start of the search, and also allows activities with negative cash flows to be scheduled earler.

6   Experiments

We implemented the algorithms in C++. BOOST version 1.49.0 [1] is used for the max-flow solver and multi-threading. METIS version 5.0 [16] is used to solve the MCC problem. All tests were run on a Dell PowerEdge R415 computer with two 2.8GHz AMD 6-Core Opteron 4184 cpu which has 3M L2 and 6M L3 Cache, and 64 GB memory. We use the same set of parameters for all the tests. For the subgradient algorithm we use $\delta^0 = 1$ in (31), $\delta^0_\mu = 0.01$ in (34), the threshold for significant objective value improvement is $\Delta^i = 0.0001 Z_{LR}(\lambda^i, \mu^i)$, the number of iterations for reducing δ^i is $p = 3$ and the factor $\rho = 0.5$, the maximal number of iterations for each level of HSA is $i_1 = i_2 = 10$, and the maximal number of iterations for the list scheduling heuristic is $i_3 = 20$.

6.1   Test Cases

Table 1 shows six of the eight test cases we obtained from a mining consulting company. The other two have no resource constraints and can be solved to optimality within 2 minutes. The first column in the table is the name of each case. The next two columns are the number of activities $|J|$ and the average number of successors of each activity in the precedence constraints L. The fourth column is the number of resources $|R|$. The fifth column is the number of Natural Clusters (NC) which is the number of clusters without relaxing any precedence

Table 1. Test cases for large RcpspDc

case name	$|J|$	$|L|/|J|$	$|R|$	T	NC	$(U,|E|)$		
caNZ_def	1410	1.18	7	3040	14	(10,38)	(50,126)	(100,233)
caW	2817	1.26	2	3472	1	(10,120)	(50,314)	(100,578)
caGL	3838	1.16	5	2280	17	(10,59)	(50,174)	(100,269)
caZ	5032	1.36	5	8171	1	(50,274)	(100,427)	
caCH	8284	1.24	4	7251	1	(100,623)	(200,866)	
caZF	11769	1.16	5	6484	1	(100,384)	(200,595)	

Table 2. Test results for relaxing resource constraints only

case name	makespan	ub	lb	gap	ite	time	$|V|$	$|A|$
canNZ_def	3040	**1.199E9**	**1.140E9**	0.0496	81	10370	2874917	5854335
caW	3471	**6.014E8**	4.681E8	0.2217	72	12336	6178671	13449822
caGL	2269	**1.055E9**	**1.021E9**	0.0318	86	60885	6371416	13224808

constraints. The remaining columns give the pairs of the number of obtained clusters (U) after relaxing the number of precedence constraints (E) by solving MCC. These test cases ranges from about 1400 activities to about 11,000 activities. The average number of successors per activity is small for all of these test cases. However only the smaller canNZ_def and caGL have natural clusters. Even for these two cases the natural clusters are not balanced in size. For example 38 precedence constraints have to be relaxed for canNZ_def to have 10 balanced clusters which is smaller than the number of natural clusters. It can be seen that more precedence constraints have to be relaxed when the number of clusters required increases. We omit here the running times of METIS because all MCC instances for our six test cases in Table 1 can be solved within 0.1 seconds.

6.2　Relaxing Resource Constraints Only

Without relaxing precedence constraints we can only solve the three smaller test cases and the results are shown in Table 2. The makespan of the feasible solution with the best *npv*, the upper bound(ub) and lowerbound(lb) are reported in columns 2 to 4. The fifth column is the optimality gap calculated as $(ub - lb)/ub$. The next two columns are the total number of iterations for the subgradient algorithm, and the total cpu time. The last two columns are the number of nodes $|V|$ and number of edges $|A|$ in the network model for solving the Lagrangian relaxation problem. All times are in seconds. Entries in bold are the best over entries in Tables 2–5. It can be seen that caNZ_def and caGL have very good optimality gaps which are under 5%, while caW has quite a large gap. Although caW and caGL have similar sizes of network flow model, caGL is much slower to solve. The reason could be that caGL has larger edge capacities which can also affect the performance of max-flow algorithm.

Table 3. Test results of SSA for comparison with HSA

case name	U	makespan	ub	lb	gap	ite	time
canNZ_def	10	3035	1.202E9	1.047E9	0.128	96	2261
caW	10	—	6.036E8	—	—	62	2103
caGL	10	2237	1.058E9	1.010E9	0.046	89	7887
caZ	100	7969	3.003E8	1.259E8	0.581	100	19357
caCH	200	**7251**	3.031E9	2.326E9	0.232	100	18885
caZF	200	**6337**	3.979E8	2.091E8	0.474	100	11371

Table 4. Test results of HSA for comparison with SSA

case name	U	makespan	ub	lb	gap	ite	time
canNZ_def	10	3033	**1.200E9**	1.136E9	0.054	100	6330
caW	10	**3456**	6.017E8	**4.803E8**	0.202	100	6657
caGL	10	2254	1.056E9	1.019E9	0.035	100	9523
caZ	100	7931	**2.919E8**	**1.735E8**	0.406	100	23076
caCH	200	**7251**	3.060E9	**2.449E9**	0.200	100	44353
caZF	200	6368	3.952E8	2.394E8	0.394	100	17318

6.3 Relaxing Both Resource and Precedence Constraints

We first study how convergence can be affected after relaxing precedence constraints by comparing the standard subgradient algorithm (SSA) and the heirarchical subgradient algorithm (HSA). The maximal number of iterations is set to be 100 for all tests. We use 10 cores to speed up the tests.

The results for SSA and HSA are reported in Table 3 and Table 4 respectively. For the three smaller test cases it can be seen clearly that the upper bounds are trivially worsened by relaxing precedence constraints. However the lower bounds become significantly worse if SSA is used. The test case caW could not even find a feasible solution. In sharp contrast HSA finds a better lower bound for caW than without precedence constraint relaxation. This shows that HSA makes the Lagrangian multipliers associated with precedence constraints converge much faster. For the three larger cases HSA produces much better lower bounds than SSA especially for caZ. However SSA got a better upper bound for caCH. The reason is that HSA only uses half the number of iterations on updating the Lagrangian multipliers associated with the resource constraints. All the following tests will use HSA because of the overwhelming advantages on lower bounds over SSA.

We study how the quality of bounds are affected by the number of clusters. The results are reported in Tables 4 and 5. For the three smaller test cases, both upper bounds and lower bounds consistently become worse with the number of clusters increased. The lower bounds are more adversely affected than the upper bounds. For the three larger test cases the worsening of upper bounds is more than 1% after the number of clusters is doubled. However the lower bounds

Table 5. Test results for effects of the number of clusters

case name	U	makespan	ub	lb	gap	ite	time
canNZ_def	10	3033	**1.200E9**	1.136E9	0.054	100	6330
canNZ_def	50	3030	1.202E9	1.125E9	0.064	100	1349
canNZ_def	100	**3025**	1.203E9	1.100E9	0.086	100	967
caW	10	**3456**	6.017E8	**4.803E8**	0.202	100	6657
caW	50	3457	6.025E8	4.615E8	0.234	100	4390
caW	100	3460	6.036E8	4.380E8	0.274	100	3699
caGL	10	2254	1.056E9	1.019E9	0.035	100	9523
caGL	50	2228	1.060E9	1.017E9	0.040	100	3574
caGL	100	**2218**	1.060E9	1.010E9	0.047	100	2337
caZ	50	**7909**	2.856E8	1.714E8	0.400	100	31180
caZ	100	7931	**2.919E8**	**1.735E8**	0.406	100	23076
caCH	100	**7251**	**3.023E9**	2.365E9	0.218	100	64758
caCH	200	**7251**	3.060E9	**2.449E9**	0.200	100	44353
caZF	100	6427	**3.860E8**	**2.398E8**	0.379	100	27663
caZF	200	6368	3.952E8	2.394E8	0.394	100	17318

become significantly better on caZ and caCH. The reason could be that list scheduling is not robust. But it can also be that the HSA does not converge well on the Lagrangian multipliers related to the precedence constraints. By setting $i_1 = i_2 = 50$ and keeping the maximal number of iterations the same we get lower bound 1.86E8 for caZ with $U = 50$. This suggests more work need to be done to adaptively tune parameters of HSA.

We next study the impact on solution time from using the techniques in the paper. We calculate the speedup factor f_d due to solving DLRP instead of LRP as

$$f_d = ATL/ATD$$

where ATD is the average solution time of DLRP while ATL is the average solution time of LRP. We also calculate the speedup factor f_p due to solving DLRP using multi-cores as

$$f_p = ATD/ATM$$

where ATM is the average solution time of DLRP with multi-cores. We implemented a thread pool using the BOOST thread library. If the number of clusters is larger the number of cores available the longest solution time first rule [9] is used to schedule the threads. Solution time is estimated based on the previous iterations. If the estimation of the solution time produces the same list of threads as using the real solution time, this rule has performance guarantee 4/3 .

The results for smaller cases are reported in Table 6 using 10 threads. It can be seen that the benefit from increasing the number of clusters is small. The reason is that the complexity of the max-flow algorithm depends not just on the size of the cluster but also on the number of precedence constraints on it. The capacities of the edges can also be an important factor. It is interesting to see that f_d is even less than 1. We show the solution time of caW for each iteration

Table 6. Test results for speedups due to decomposition and parallelization as an effect of the number of clusters

case name	$U = 10$			$U = 50$			$U = 100$		
	f_d	f_p	$f_d * f_p$	f_d	f_p	$f_d * f_p$	f_d	f_p	$f_d * f_p$
canNZ_def	0.94	2.16	2.02	1.68	5.63	9.49	2.46	5.38	13.24
caW	0.92	2.81	2.57	0.63	6.18	3.90	0.76	6.11	4.63
caGL	1.50	4.97	7.43	3.32	5.97	19.81	5.11	5.93	30.29

with $U = 50$ in Figure 6.3(a). The solution times at the first few solutions are quite small. After the capacities of edges are updated according to the HSA, the solution times increase quickly. We also show the solution time of caW for each cluster with $U = 50$ in Figure 6.3(b). It can be seen that some clusters have much larger solution time than others.

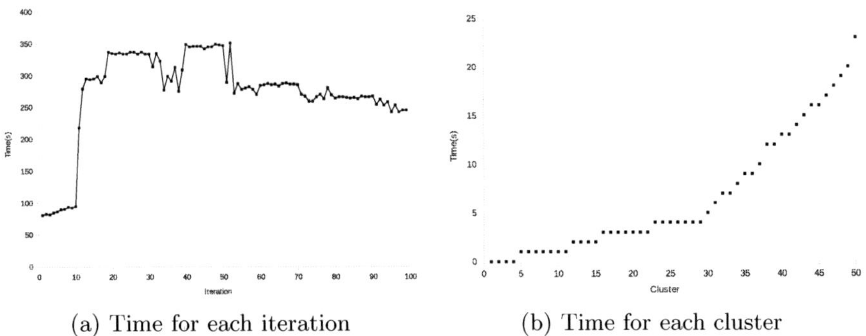

(a) Time for each iteration   (b) Time for each cluster

Fig. 2. Solutions times for caW with $U = 50$

We cannot calculate f_d for the larger cases. f_p is similar to those of the smaller cases.

The mining consulting company also provides us with the results from two other research teams J and D. J sets a time limit of 5 hours but does not disclose details of their method. D sets a time limit of 1 hour and is based on Ant Colony Optimisation. We report our results with a time limit of 5h in Table 7 along with J and D's results. Although D's method is the fastest, it has difficulty in finding good solutions when the deadline is tight. It has the largest makespans for all the cases except for caZ. We don't know how it performs if running time is extended to 5 hours. For the three smaller cases we cannot produce solutions with the same makespan as J's. However our *npv* are significantly larger with small increase of the makespan. For the three larger cases our method produces solutions with tightest makespan and best *npv*. For caCH our *npv* is almost 40% higher than J's.

Table 7. Comparison with J and D's methods in 5 hours limit

| | LR | | | | J | | D | |
case name	Makespan	ub	npv	$|U|$	makespan	npv	Makespan	npv
canNZ_def	3040	1.199E9	**1.140E9**	1	**3014**	1.040E9	3559	9.810E8
caW	3456	6.017E8	**4.803E8**	10	**3417**	4.670E8	3472	4.740E8
caGL	2254	1.056E9	**1.019E9**	10	**2196**	9.920E8	2283	1.010E9
caZ	**7931**	2.997E8	**1.735E8**	100	8171	1.670E8	8078	1.670E8
caCH	**7251**	3.215E9	**2.449E9**	200	**7251**	1.750E9	8097	1.700E9
caZF	**6368**	3.952E8	**2.394E8**	200	6384	2.290E8	6904	2.290E8

7   Conclusion

We have applied a Lagrangian relaxation method for large RCPSPDC problems
by relaxing both resource and precedence constraints. By minimising the number of relaxed precedence constraints we can still achieve relatively tight upper
and lower bounds. Together with a parallel implementation and a hierarchical subgradient algorithm this approach produced highly competitive results on
very large underground mining problems within reasonable computing time. Further work is needed to combine with the CP technologies to improve solution
qualities.

Acknowledgements. NICTA is funded by the Australian Government as represented by the Department of Broadband, Communications and the Digital
Economy and the Australian Research Council through the ICT Centre of Excellence program.

References

1. Boost.org, `http://www.boost.org/`
2. Baroum, S., Patterson, J.: The development of cash flow weight procedures for maximizing the net present value of a project. Journal of Operations Management 14,
209–227 (1996)
3. Brucker, P., Drexl, A., Möhring, R., Neumann, K., Pesch, E.: Resource-constrained
project scheduling: Notation, classification, models, and methods. European Journal of Operational Research 112(1), 3–41 (1999)
4. Chaudhuri, S., Walker, R., Mitchell, J.: Analyzing and exploiting the structure of
the constraints in the ilp approach to the scheduling problem. IEEE Trans. Very
Large Scale Integration (VLSI) Systems 2, 456–471 (1994)
5. Cherkassky, B., Goldberg, A.: On implementing push-relabel method for the maximum flow problem. In: IPCO 1995, pp. 157–171 (1995)
6. Demeulemeester, E.L., Herroelen, W.S.: A branch-and-bound procedure for
the multiple resource-constrained project scheduling problem. Management Science 38(12), 1803–1818 (1992)
7. Demeulemeester, E.L., Herroelen, W.S.: New benchmark results for the resource-
constrained project scheduling problem. Management Science 43(11), 1485–1492
(1997)

8. Doersch, R., Patterson, J.: Scheduling a project to maximize its present value: A zero-one programming approach. Management Science 23, 882–889 (1977)
9. Dósa, G.: Graham's example is the only tight one for $P||C_{max}$. Annales Universitatis Scientarium Budapestiensis de Rolando Eötös nominatae, Sectio Mathematica 47, 207–210 (2004)
10. Fisher, M.: The lagrangian relaxation method for solving integer programming problems. Management Science 27, 1–18 (1981)
11. Graham, R.L.: Bounds for certain multiprocessing anomalies. Bell System Tech. J. (45), 1563–1581 (1966)
12. Hartmann, S., Briskorn, D.: A survey of variants and extensions of the resource-constrained project scheduling problem. European Journal of Operational Research 207(1), 1–14 (2010)
13. Herroelen, W.S., Demeulemeester, E.L., De Reyck, B.: A classification scheme for project scheduling. In: Weglarz, J. (ed.) Project Scheduling. International Series in Operations Research and Management Science, vol. 14, pp. 1–26. Kluwer Academic Publishers (1999)
14. Icmeli, O., Erengüç, S.S.: A branch and bound procedure for the resource constrained project scheduling problem with discounted cash flows. Management Science 42(10), 1395–1408 (1996)
15. Johnson, E., Mehrotra, A., Nemhauser, G.: Min-cut clustering. Mathematical Programming 62, 133–151 (1993)
16. Karypis, G.: METIS, A Software Package for Partitioning Unstructured Graphs, Partitioning Meshes, and Computing Fill-Reducing Orderings of Sparse Matrices, Version 5.0 (2011), `http://glaros.dtc.umn.edu/gkhome/views/metis`
17. Kimms, A.: Maximizing the net present value of a project under resource constraints using a lagrangian relaxation based heuristic with tight upper bounds. Annals of Operations Research 102, 221–236 (2001)
18. Möhring, R.H., Schulz, A.S., Stork, F., Uetz, M.: Solving project scheduling problems by minimum cut computations. Management Science 49(3), 330–350 (2003)
19. Neumann, K., Zimmermann, J.: Exact and truncated branch-and-bound procedures for resource-constrained project scheduling with discounted cash flows and general temporal constraints. Central European Journal of Operations Research 10(4), 357–380 (2002)
20. Padman, R., Smith-Daniels, D.E., Smith-Daniels, V.L.: Heuristic scheduling of resource-constrained projects with cash flows. Naval Research Logistics 44, 365–381 (1997)
21. Russell, A.H.: Cash flows in networks. Management Science 16(5), 357–373 (1970)
22. Schutt, A., Chu, G., Stuckey, P.J., Wallace, M.G.: Maximising the Net Present Value for Resource-Constrained Project Scheduling. In: Beldiceanu, N., Jussien, N., Pinson, É. (eds.) CPAIOR 2012. LNCS, vol. 7298, pp. 362–378. Springer, Heidelberg (2012)
23. Siek, J.G., Lee, L.Q., Lumsdaine, A.: The Boost Graph Library: User Guide and Reference Manual. Addison-Wesley (2001)
24. Vanhoucke, M., Demeulemeester, E.L., Herroelen, W.S.: On maximizing the net present value of a project under renewable resource constraints. Management Science 47, 1113–1121 (2001)
25. Zhao, X., Luh, P.B.: New bundle methods for solving lagrangian relaxation dual problems. J. Optim. Theory Appl. 113, 373–397 (2002), `http://dx.doi.org/10.1023/A:1014839227049`

Comparing Solution Methods
for the Machine Reassignment Problem

Deepak Mehta, Barry O'Sullivan, and Helmut Simonis

Cork Constraint Computation Centre, University College Cork, Ireland
{d.mehta,b.osullivan,h.simonis}@4c.ucc.ie

Abstract. The machine reassignment problem is defined by a set of machines and a set of processes. Each machine is associated with a set of resources, e.g. CPU, RAM etc., and each process is associated with a set of required resource values and a currently assigned machine. The task is to reassign the processes to machines while respecting a set of hard constraints in order to improve the usage of the machines, as defined by a complex cost function. We present a natural Constraint Programming (CP) formulation of the problem that uses a set of well-known global constraints. However, this formulation also requires new global constraints. Additionally, the instances are so large that systematic search is not practical especially when the time is limited. We therefore developed a CP formulation of the problem that is especially suited for a large neighborhood search approach (LNS). We also experimented with a mixed-integer programming (MIP) model for LNS. We first compare our formulations on a set of ROADEF'12 problem instances where the number of processes and machines are restricted to 1000 and 100 respectively. Both MIP and CP-based LNS approaches find solutions that are significantly better than the initial ones and compare well to other submitted entries. We further investigate their performances on larger instances where the number of processes and machines can be up to 50000 and 5000 respectively. The results suggest that our CP-based LNS can scale to large instances and is superior in memory use and the quality of solutions that can be found in limited time.

1   Introduction

The management of data centres provides a rich domain for constraint programming, and combinatorial optimisation in general. The *EU Stand-by Initiative* recently published a Code of Conduct for Energy Efficiency in Data Centres.[1] This report stated that in 2007 Western European data centres consumed 56 Tera-Watt Hours (TWh) of power, which is expected to almost double to 104 TWh per year by 2020. A typical optimisation challenge is workload consolidation whereby one tries to keep servers well utilised such that the power costs are minimised. A naïve model for the problem of loading servers to a desired utilisation level for each resource can be regarded as a multi-dimensional bin packing problem [6]. In such a model the servers are represented by bins and the set of resources define the set of dimensions. Many related optimisation

---

[1] http://re.jrc.ec.europa.eu/energyefficiency/html/
standby_initiative_data_centers.htm

M. Milano (Ed.): CP 2012, LNCS 7514, pp. 782–797, 2012.

problems arise in this domain, such as the adaptive control of virtualised resources [3], and cluster resource management [5,9]. A MIP approach to dynamically configuring the consolidation of multiple services or applications in a virtualised server cluster has been proposed [4]. That work both focuses on power efficiency, and considers the costs of turning on/off the servers. Power and migration cost-aware application placement has also been considered [8,7].

Given the growing level of interest from the optimisation community in data centre optimisation and virtualisation, the 2012 ROADEF Challenge is focused on the topic of machine reassignment, a common task in virtualisation and service configuration on data centres.[2] We present a mixed integer programming and a constraint programming solution for the problem. We also present a large neighbourhood search scheme for both. Our results show that as the problem size grows, the CP approach is significantly more scalable than the MIP approach. This demonstrates the applicability of CP, and in particular LNS, as a promising optimisation approach for this application domain.

2   Problem Description and Notations

In this section, we describe the 2012 machine reassignment problem of ROADEF/EURO Challenge, which was in collaboration with Google. For more details the reader is referred to the specification of the problem, which is available online[3]. We also introduce some notations that will be used throughout the paper. Let M be the set of machines and P be the set of processes. A solution of the machine reassignment problem is an assignment of each process to a machine subject to a set of constraints. The objective is to find a solution that minimizes the cost of the assignment.

2.1   Constraints

Capacity Constraints. Let R be the set of all resources (e.g., CPU, RAM, etc.). Let c_{mr} be the capacity of machine m for resource r and let r_{pr} be the requirement of process p for resource r. The usage by a machine m of resource r, denoted by u_{mr}, is equal to the sum of the amount of resource required by processes which are assigned to machine m. The usage by a machine of a resource should not exceed the capacity of the resource.

Conflict Constraints. A service is a set of processes. Let S be the set of services that partition P. The processes of a service should be assigned to different machines.

Spread Constraints. A location is a set of machines. Let L be the set of all locations that partition the machines. We abuse the notation by using $L(m)$ to denote the location of machine m. The processes of a service s should be assigned to the machines such that their corresponding locations are spread over at least $spreadmin_s$ number of locations.

---

[2] http://challenge.roadef.org/2012/en/index.php
[3] http://challenge.roadef.org/2012/files/
   problem_definition_v1.pdf

Neighborhood Constraints. A neighborhood is a set of machines that partition the machines. Let N be the set of all neighborhoods. We abuse the notation by using $N(m)$ to denote the neighborhood of machine m. If service s depends on service s' then it is denoted by $\langle s, s' \rangle$. Let D be the set of such dependencies. If $\langle s, s' \rangle \in D$ then the set of the neighborhoods of the machines assigned to the processes of s must be a subset of the set of the neighborhoods of the machines assigned to the processes of service s'.

Transient Usage Constraints. Let o_p be the original machine assigned to a process p. When a process is moved from o_p to another machine, some resources, e.g., hard disk, are required in both source and target machines during the move. These resources are called transient resources. Let $T \subseteq R$ be the set of transient resources. The transient usage of a machine m for a resource $r \in T$ is equal to sum of the amount of resource required by processes whose original or current machine is m. The transient usage of a machine for a given resource should not exceed the capacity of the resource.

2.2   Costs

Load Cost. It is usually not a good idea to use the full capacity of a resource. The safety capacity limit provides a soft limit, any use above that limit incurs a cost. Let sc_{mr} be the safety capacity of machine m for resource r. The load cost for a resource r is equal to $\sum_{m \in M} \max(0, u_{mr} - sc_{mr})$.

Balance Cost. In order to balance the availability of resources, a balance b can be defined by a triple of resources r_b^1 and r_b^2, and a multiplier t_b. Let B be the set of all such triples. For a given triple b, the balance cost is $\sum_{m \in M} \max(0, t_b \times A(m, r_b^1) - A(m, r_b^2))$ with $A(m, r) = c_{mr} - u_{mr}$.

Process Move Cost. Let pmc_p be the cost of moving a process p from its original machine to any other machine. The process move cost is the sum of all pmc_p such that p is moved from its original machine.

Service Move Cost. To balance the movement of processes among services, a service move cost is defined as the maximum number of moved processes for only services.

Machine Move Cost. Let $mmc(m_1, m_2)$ be the cost of moving a process from a machine m_1 to a machine m_2. Obviously, if m_1 is equal to m_2 then this cost is 0. The machine move cost is the sum of these costs for all processes.

Objective Function. The objective is to minimize the weighted sum of all load-costs, balance-costs, service-move cost, process-move and machine-move cost of each process. Let w_r be the weight of the load-cost for $r \in R$, v_b be the weight of the balance cost for $b \in B$, w^{pmc} be the weight for process move cost, w^{smc} be the weight for service move cost, and w^{mmc} be the weight for machine move cost.

3   Finite Domain Model

In this section, we present a natural finite domain constraint programming formulation of the machine reassignment problem.

3.1 Variables

For each process p, we use three integer variables: (1) a variable x_p which indicates the machine to which process p is assigned, (2) a variable l_p indicating the location of the machine to which process p is assigned, and (3) a variable n_p indicating the neighborhood of the machine to which process p is assigned.

3.2 Constraints

We use the constraint format of the *Global Constraint Catalog* [1]. Of course, the names and arguments of constraints for specific constraint solvers may differ.

Projection. Let LT and NT be two integer arrays for mapping a machine to its location and its neighborhood. We need element constraints to project from the decision variable x_p to the location and the neighborhood respectively:

$$\forall_{p\in P}: \quad \texttt{element}(x_p, \text{LT}, l_p), \tag{1}$$

$$\forall_{p\in P}: \quad \texttt{element}(x_p, \text{NT}, n_p). \tag{2}$$

Capacity. The hard capacity constraints can be expressed with a $\texttt{bin\_packing\_capa}$ global constraint for each resource:

$$\forall_{r\in R}: \quad \texttt{bin\_packing\_capa}([(m, c_{mr})|m \in M], [(x_p, r_{pr})|p \in P]). \tag{3}$$

Conflict. All processes in the same service must be allocated to different machines, i.e. they must be alldifferent:

$$\forall_{s\in S}: \quad \texttt{alldifferent}([x_p|x \in s]). \tag{4}$$

Spread. A service s must run in at least $spreadmin_s$ locations:

$$\forall_{s\in S}: \quad \texttt{atleast\_nvalue}(spreadmin_s, [l_p|p \in s]). \tag{5}$$

Dependency. If a service s depends on another s', then that service must run in all neighborhoods where s is running:

$$\forall\langle s, s'\rangle \in D: \texttt{uses}([n_{p'}|p' \in s'], [n_p|p \in s]). \tag{6}$$

Transient. For the transient resource, we have to add some optional tasks, which only occur if the process is not assigned to its original value o_p. Different constraint systems will require different techniques to implement this. In the best case, optional tasks are available, or resource height can be a variable. In the worst case, dummy tasks with escape values should be considered:

$$\forall_{r\in T}: \quad \texttt{bin\_packing\_capa}([(m, c_{mr})|m \in M], \\ [(x_p, r_{pr})|p \in P] \cup [(o_p, r_{pr})|x_p \neq o_p]). \tag{7}$$

3.3   Objective

The five different costs described in Section 2 depend on the assignment of a machine to a process. A standard way of modelling the objective function would be an additive, weighted sum of these costs. However, it seems unlikely that such a sum of the different cost factors would allow any significant constraint propagation. Furthermore, systematic search does not seem to be a feasible approach especially when the size of the instance is very large and the solution-time is very limited. Therefore it did not lead to a working system.

4   A Mixed Integer Linear Programming Model

In this section we present a mixed integer linear programming formulation of the machine reassignment problem. We use the notation $a := b$ to abbreviate expression b with name a. We may, but do not have to, create a new variable for a.

4.1   Variables

x_{pm} A binary variable that indicates if process p is running on machine m. Thus, x_{po_p} indicates that the process is not moving, while $(1 - x_{po_p})$ indicates that the process is moving from its original assignment o_p.

y_{sl} A binary variable that indicates if service s is running in location l.

z_{sn} A binary variable that states that service s is running in neighborhood n.

lc_{mr} A continuous (non-negative) variable for resource r on machine m that shows how much the usage exceeds the safety capacity.

bc_{bm} A continuous (non-negative) variable for balance b on machine m that links resources r_b^1 and r_b^2 with multiplier t_b.

smc A continuous (non-negative) variable for cost of moving processes of the services.

The following may be used as abbreviations or variables depending on the system that is used: (1) u_{mr} denotes resource usage of machine m for resource r, (2) a_{mr} denotes free capacity of machine m for resource r, (3) lc_r denotes load cost of resource r, (4) bc_b denotes balance cost of balance b, (5) pmc denotes process move cost, (6) smc_s contribution of service s to service move cost, (7) mmc_p contribution of process p to machine move cost, and (8) mmc denotes machine move cost.

4.2   Constraints

Assignment. Every process must be allocated to exactly one machine.

$$\forall_{p \in P} : \quad \sum_{m \in M} x_{pm} = 1 \tag{8}$$

Capacity. For each resource, the resource use for all processes on one machine must be below the resource limit of that machine. We denote with u_{mr} the resource use on machine m for resource r by all process which run on this machine.

$$\forall_{r \in R} \forall_{m \in M} : \quad u_{mr} := \sum_{p \in P} r_{pr} x_{pm} \leq c_{mr} \tag{9}$$

Conflict. Processes belonging to the same service can not be run on the same machine.

$$\forall_{s \in S} \forall_{m \in M} : \quad \sum_{p \in s} x_{pm} \leq 1 \tag{10}$$

Spread. A service must run in at least spreadmin locations. We introduce y_{sl} to denote if one of the processes in service s is running on one of the machines in location l.

$$\forall_{l \in L} \forall_{s \in S} : \quad \sum_{m \in l} \sum_{p \in s} x_{pm} \leq |l||s| y_{sl} \tag{11}$$

$$\forall_{l \in L} \forall_{s \in S} : \quad \sum_{m \in l} \sum_{p \in s} x_{pm} \geq y_{sl} \tag{12}$$

$$\forall_{s \in S} : \quad \sum_{l \in L} y_{sl} \geq \text{spreadmin}_s \tag{13}$$

Dependency (I). If a service depends on another service, then that service must run in the same neighborhood.

$$\forall_{\langle s,s' \rangle \in D} \forall_{n \in N} \forall_{p \in s} \forall_{m \in n} : \quad x_{pm} \leq \sum_{p' \in s'} \sum_{m' \in n} x_{p'm'} \tag{14}$$

Dependency (II). Dependency (I) seems to require too many constraints, we can introduce binary variables z_{sn} which indicate that service s is run in neighborhood n. We then have the constraints to link the x_{pm} and the z_{sn} variables:

$$\forall_{n \in N} \forall_{s \in S} : \quad \sum_{m \in n} \sum_{p \in s} x_{pm} \leq |n||s| z_{sn} \tag{15}$$

$$\forall_{n \in N} \forall_{s \in S} : \quad \sum_{m \in n} \sum_{p \in s} x_{pm} \geq z_{sn} \tag{16}$$

For every service dependency and every neighborhood, we have an implication:

$$\forall_{\langle s,s' \rangle \in D} \forall_{n \in N} : \quad z_{sn} \leq z_{s'n} \tag{17}$$

Dependency (III). We can reduce the number of constraints also by reorganizing the first version. On the left hand side, we can add all x_{pm} variables for the same process, as their sum is always less than or equal to 1. On the right rand side, we can reuse the sum for multiple left hand sides, so don't have to create these sums over and over again. This seems to make approach (III) feasible for first set of instances of ROADEF.

$$\forall_{\langle s,s' \rangle \in D} \forall_{n \in N} \forall_{p \in s} : \quad \sum_{m \in n} x_{pm} \leq \sum_{p' \in s'} \sum_{m' \in n} x_{p'm'} \tag{18}$$

In our implementation, the third alternative for the dependency constraint is used.

788   D. Mehta, B. O'Sullivan, and H. Simonis

Transient. For transient resources, the resource use must include the original and the new machine, not just the new machine. It is important to avoid double counting if the process does not move.

$$\forall_{r \in T} \forall_{m \in M} : \quad \sum_{p \in P, o_p = m} r_{pr} + \sum_{p \in P, o_p \neq m} r_{pr} x_{pm} \leq c_{mr} \tag{19}$$

Note that the transient resource type only affects this capacity constraint, not how the load cost is calculated.

4.3   Objective

Load Cost. We need continuous, non-negative variables lc_{mr} to express this cost. Recall that u_{mr} denotes the resource use of resource r on machine m, and integer sc_{mr} is the safety capacity limit for resource r on machine m.

$$\forall_{r \in R} \forall_{m \in M} : \quad lc_{mr} \geq u_{mr} - sc_{mr} \tag{20}$$

$$\forall_{r \in R} : \quad lc_r := \sum_{m \in M} lc_{mr} \tag{21}$$

Balance Cost. We define the free capacity for resource r on machine m as

$$\forall_{m \in M} \forall_{r \in R} : \quad a_{mr} := c_{mr} - u_{mr} \tag{22}$$

We express the contribution of balance cost b on machine m as

$$\forall_{b \in B} \forall_{m \in M} : \quad bc_{bm} \geq t_b \times a_{mr_b^1} - a_{mr_b^2} \tag{23}$$

The total contribution of balance cost b is summed over all machines as follows:

$$\forall_{b \in B} : \quad bc_b := \sum_{m \in M} bc_{bm} \tag{24}$$

Process Move Cost. This considers if a process is moved from its original assignment. Each process p can be weighted individually with constants pmc_p.

$$pmc := \sum_{p \in P} (1 - x_{po_p}) pmc_p \tag{25}$$

Service Move Cost. We require a continuous, non-negative variable smc to express this cost:

$$\forall_{s \in S} : \quad smc_s := \sum_{p \in s} (1 - x_{po_p}) \tag{26}$$

$$\forall_{s \in S} : \quad smc \geq smc_s \tag{27}$$

smc_s denotes how many processes in service s are assigned to new machines, smc is bounded by those limits. This requires smc to be used inside the cost function for minimization with a non-negative coefficient.

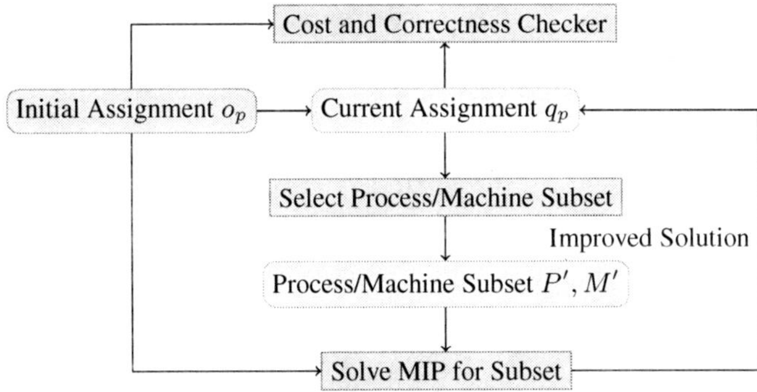

Fig. 1. Principles of the LNS-based approach

Machine Move Cost. This considers the cost matrix $mmc()$ for moving processes between machines. This gives different weights for from-to machine pairs, but does not distinguish processes.

$$\forall_{p \in P}: \quad mmc_p := \sum_{m \in M} x_{pm} mmc(o_p, m) \tag{28}$$

$$mmc := \sum_{p \in P} mmc_p \tag{29}$$

Overall Cost. The overall cost is a weighted sum of all the different cost elements defined above. The weight factors are defined as part of the input data of the problem instance. The objective is to minimize the following:

$$Cost = \sum_{r \in R} w_r lc_r + \sum_{b \in B} v_b bc_b + w^{pmc} pmc + w^{smc} smc + w^{mmc} mmc \tag{30}$$

4.4 Solution Method

We use a large neighborhood search starting from the MIP formulation of the problem. The overall solution method is shown in Figure 1. We maintain a current assignment, which is initialized by the initial solution given as input. At every iteration step, we select a subset of the processes to be reassigned, and setup the MIP model, freezing all other processes. We solve the resulting, small MIP with a timeout, and keep the best solution found as our new current assignment. At each step, we check the correctness and cost of our current solution by a separate checker to avoid isssues with accuracy and tolerances in the MIP solver.

A key observation was that selecting all processes from only some machines for reassignment works better than selecting only a few processes from many machines. Our subproblem selection therefore is based on selecting a subset of the machines, and allowing all processes on those machines to be reassigned. A set of machines are selected by solving a small combinatorial optimization problem. We compare the current usage, denoted by cu_{mr}, of the selected machines (when the Boolean variable $u_m = 1$) with an ideal solution where we can distribute the utilization over all machines such that the area above the safety capacity is minimal. The objective is to maximize the following:

$$\sum_{m \in M} (\sum_{r \in R} \max(0, cu_{mr} - sc_{mr}))u_m - \sum_{r \in R} \max(0, \sum_{m \in M} (cu_{mr} - sc_{mr})u_m)$$

This is optimistic, as we normally will not be able to achieve this totally balanced spread, but this will guide the subset selection to find promising machine subsets. For some sub-problems this works perfectly. At every step the promised improvement is nearly fully realized by the subsequent MIP run, until the promise (and cost) converge to 0. For problems with significant transient resource elements, the approach works less well, but is still much better than a random selection. We also use a tabu list and a progressive parameter weighting to improve results over all problem instances.

5   CP Model for LNS

We now present a CP model of the problem that facilitates incrementality in the large neighbourhood search method. This helps in creating subproblems efficiently as explained later.

5.1   Variables

Let x_p be an integer variable that indicates which machine is assigned to process p. The domain of x_p is $[0, |M| - 1]$. Let u_{mr} be an integer variable that denotes the usage of resource r on machine m. The domain of u_{mr} is $[0, c_{mr}]$. Let t_{mr} be an integer variable that denotes the transient usage of machine m for transient-resource r. The domain of t_{mr} is $[0, c_{mr}]$. The following may be used as abbreviations or variables depending on the system that is used: (1) um_s denotes a set of machines assigned to the processes of service s, (2) yu_{sl} denotes the number of machines belonging to location l that are assigned to the processes of the service s, (3) nul_s denotes the number of locations used by the processes of the service s, (4) zu_{sn} denotes the number of machines of neighborhood n that are used by service s, (5) zm_{sn} denotes the number of services that enforce that n should be a neighborhood of s, (6) nmp_s denotes how many processes in service s are assigned to new machines, and (7) mmp denotes the maximum number of processes of services that are moved to new machines.

5.2   Constraints

Load Constraints. The usage of resource $r \in R$ on machine m is the sum of the resources required by those processes p that are assigned to machine m:

$$\forall m \in M, r \in R : u_{mr} = \sum_{\forall p \in P \wedge x_p = m} r_{pr}. \tag{31}$$

Transient Constraints. The transient-usage of transient-resource $r \in T$ on machine m is the sum of u_{mr} and the resources required by those processes p whose original machine is m but whose current machine is not m.

$$\forall m \in M, r \in T : t_{mr} = u_{mr} + \sum_{o_p=m \wedge x_p \neq o_p} r_{pr}. \tag{32}$$

Conflict Constraints. The set of machines used by the processes of service s is:

$$um_s := \{x_p : p \in s\}.$$

Processes belonging to the same service s cannot be run on the same machine:

$$\forall s \in S, |s| = |um_s|. \tag{33}$$

Spread Constraints. The number of machines at a location l used by a service s is:

$$\forall s \in S, l \in L, yu_{sl} := |\{x_p | p \in s \wedge L(x_p) = l\}|.$$

The number of locations used by the processes of a service s is:

$$nul_s := |\{l | l \in L \wedge yu_{sl} > 0\}|.$$

The number of locations used by service s should be greater than or equal to the minimum spread of service s:

$$nul_s \geq spreadmin_s. \tag{34}$$

Neighborhood Constraints. The number of machines used by a service s in a neighborhood n is:

$$zu_{sn} := |\{x_p | p \in s \wedge N(x_p) = n\}|.$$

The number of services that want n to be a (mandatory) neighborhood of service s is:

$$zm_{sn} := |\{s' | \langle s', s \rangle \in D \wedge zu_{s'n} > 0\}|.$$

The number of mandatory neighborhoods of service s that are not used by service s is:

$$nmn_s := |\{n \in N \wedge zm_{sn} > 0 \wedge zu_{sn} = 0\}|$$

Each mandatory neighborhood of service s should be used by service s:

$$nmn_s = 0. \tag{35}$$

5.3   Objective

We define the cost of a machine as the sum of the weighted load costs of all resources and the sum of all the weighted balance costs of all the balances:

$$cost_m := \sum_{r \in R} \max(0, u_{mr} - sc_{mr}) \cdot w_r + \sum_{b \in B} \max(0, t_b \cdot a_{mr_b^1} - a_{mr_b^2}) \cdot v_b. \tag{36}$$

We define the cost of a process as the sum of the weighted process move cost and weighted machine move cost:

$$cost_p := \min(1, |x_p - o_p|) \cdot pmc_p \cdot w^{pmc} + mmc(o_p, x_p) \cdot w^{mmc}. \qquad (37)$$

We define the cost of a service as the weighted sum of the number of moved processes of the service:

$$cost_s := \sum_{p \in s \wedge x_p \neq o_p} w^{sms}. \qquad (38)$$

The overall cost, to be minimized, is:

$$Cost = \sum_{m \in M} cost_m + \sum_{p \in P} cost_p + \max_{s \in S}(cost_s). \qquad (39)$$

5.4   Solution Method

We use large neighborhood search for the CP model described above. Using the LNS of MIP for the CP model is a non-starter. Therefore we use a different LNS. In the following we describe how a subproblem is selected and created, and how it is solved.

Subproblem Selection. We first select a number of machines, denoted by k_m, whose processes we want to reassign and then for each selected machine we select a number of processes, bounded by k_p, which we want to reassign. Both k_m and k_p are non-zero positive integers. Initially k_m is set to 1, it is incremented as search progresses, and it is re-initialized to 1 when it exceeds 10. We also compute an ordering on the machines based on Equation (36) after regular intervals. We progressively select one machine from this pre-computed ordering and select the remaining $k_m - 1$ machines randomly.

Depending on the value of k_m a fixed value of k_p is used. If all the processes of a machine are selected then many of them may select the same machine again. Therefore, we restrict that the number of processes selected for a given machine is less than or equal to the half of the average number of processes on a machine. We further restrict that the maximum value of k_p is 10. The total number of processes selected for reassignment is bounded by $k_m \times k_p \leq 40$. Hence, the number of processes selected from a given machine is determined by $\min(40/k_m, \min(|P|/(2 \cdot |M|), 10))$.

Subproblem Creation. For solving large sized problems using LNS one may need to select, create and solve many subproblems. When the time for solving the problem is limited, one challenge is to create a subproblem efficiently. One approach is to create the full problem in memory and reinitialize/recompute the domains of the required variables at each iteration. Depending on the model and the solution technique it may not always be possible to allocate the required memory for this approach, especially considering that the number of processes could be up to 50,000 and the number of machines could be up to 5,000. Another approach could be to allocate the memory and create a subproblem and reinitialize/recompute the domains of the required variables at each iteration, which may not be efficient in terms of time. Creating a subproblem in the context of machine-reassignment problem can be seen as unassigning a set of machines

---

Algorithm 1. removeProcessFromMachine(p)

1: $m \leftarrow x_p; s \leftarrow S_p; n \leftarrow N_m; l \leftarrow L_m$
2: $um_s \leftarrow um_s - \{m\};$
3: $yu_{sl} \leftarrow yu_{sl} - 1$
4: **if** $yu_{sl} = 0$ **then**
5:      $nul_s \leftarrow nul_s - 1$
6: $zu_{sn} \leftarrow zu_{sn} - 1$
7: **if** $zu_{sn} = 0$ **then**
8:      **if** $zm_{sn} > 0$ **then**
9:           $nmn_s \leftarrow nmn_s + 1$
10:     **for all** $\langle s, s' \rangle \in D$ **do**
11:        $zm_{s'n} \leftarrow zm_{s'n} - 1$
12:        **if** $zu_{s'n} = 0 \wedge zm_{s'n} = 0$ **then**
13:           $nmn_{s'} \leftarrow nmn_{s'} - 1$
14: $Cost \leftarrow Cost - \sum_{r \in R} \max(0, u_{mr} - sc_{mr}) \times w_r$
15: $Cost \leftarrow Cost - \sum_{b \in B} \max(0, t_b \times (c_{mr_b^1} - u_{mr_b^1}) - (c_{mr_b^2} - u_{mr_b^2})) \times v_b$

16: **for all** $r \in R$ **do**
17:      $u_{mr} \leftarrow u_{mr} - r_{pr}$
18:      **if** $r \in T$ and $m \neq o_p$ **then**
19:          $t_{mr} \leftarrow t_{mr} - r_{pr}$
20: $Cost \leftarrow Cost + \sum_{r \in R} \max(0, u_{mr} - sc_{mr}) \times w_r$
21: $Cost \leftarrow Cost + \sum_{b \in B} \max(0, t_b \times (c_{mr_b^1} - u_{mr_b^1}) - (c_{mr_b^2} - u_{mr_b^2})) \times v_b$

22: **if** $x_p \neq o_p$ **then**
23:      $Cost \leftarrow Cost - pmc_p \times w^{pmc}$
24:      $nmp_s \leftarrow nmp_s - 1$
25:      **if** $mmp = nmp_s + 1$ **then**
26:         $mmp \leftarrow \max_{s \in S} nmp_s$
27:         **if** $nmp_s + 1 \neq mmp$ **then**
28:            $Cost \leftarrow Cost - w^{smc}$
29: $Cost \leftarrow Cost - mmc(o_p, x_p) \times w^{mmc}$

---

from a set of processes and updating the domains accordingly. We remark that removing values from the domains due to constraint-propagation after an assignment is generally done efficiently in constraint-solvers. However restoring values to the domains after undoing an assignment may not always be straightforward, when no assumption is made on the ordering of assignments. The presented CP-based LNS approach facilitates that domains can be updated efficiently for creating a subproblem.

The two main operations for CP-based LNS for solving machine-reassignment problem are: removing a process from a machine (unassigning a machine from a process), and adding a process to a machine (assigning a machine to a process). When a process p is removed from machine x_p, the domains are updated as shown in Algorithm 1. x_p is removed from the set of machines used by service s (Line 2). The number of machines of location l used by s is decremented (Line 3) and if that becomes 0 (Line 4) the number of locations used by s is also decremented (Line 5). If the number of machines of neighborhood n used by s becomes 0 (Line 6), and if the number of services that want n to be included is greater than 0, then the number of mandatory neighborhoods of s is incremented. If n is no longer used by s then the the number of services that want n to be included is decremented for each service s', on which s depends on. For each s' if n is not used anymore then the number of mandatory neighborhoods of s' is also decremented. Lines 14–29 update the cost. First the load and balance costs for all resources and balances are subtracted from the cost (Lines 14–15). Then, the usage and transient usage are updated (Lines 16–19). The load and balance costs are recomputed

---

Algorithm 2. addProcessToMachine(p, m)

1: $s \leftarrow S_p; n \leftarrow N_m; l \leftarrow L_m$
2: $x_p \leftarrow m$
3: $Cost \leftarrow Cost + mmc(o_p, x_p) \times w^{mmc}$
4: **if** $x_p \neq o_p$ **then**
5:         $Cost \leftarrow Cost + pmc_p \times w^{pmc}$
6:         $nmp_s \leftarrow nmp_s + 1$
7:         **if** $mmp < nmp_s$ **then**
8:               $mmp \leftarrow nmp_s$
9:               $Cost \leftarrow Cost + w^{smc}$
10: $Cost \leftarrow Cost - \sum_{r \in R} \max(0, u_{mr} - sc_{mr}) \times w_r$
11: $Cost \leftarrow Cost - \sum_{b \in B} \max(0, t_b \times (c_{mr_b^1} - u_{mr_b^1}) - (c_{mr_b^2} - u_{mr_b^2})) \times v_b$

12: **for all** $r \in R$ **do**
13:         $u_{mr} \leftarrow u_{mr} + r_{pr}$
14:         **if** $r \in T$ and $x_p \neq o_p$ **then**
15:               $t_{mr} \leftarrow t_{mr} + r_{pr}$
16: $Cost \leftarrow Cost + \sum_{r \in R} \max(0, u_{mr} - sc_{mr}) \times w_r$
17: $Cost \leftarrow Cost + \sum_{b \in B} \max(0, t_b \times (c_{mr_b^1} - u_{mr_b^1}) - (c_{mr_b^2} - u_{mr_b^2})) \times v_b$

18: **if** $zu_{sn} = 0$ **then**
19:         **if** $zm_{sn} > 0$ **then**
20:               $nmn_s \leftarrow nmn_s - 1$
21:         **for all** $\langle s, s' \rangle \in D$ **do**
22:               **if** $zu_{s'n} = 0 \wedge zm_{s'n} = 0$ **then**
23:                     $nmn_{s'} \leftarrow nmn_{s'} + 1$
24:               $zm_{s'n} \leftarrow zm_{s'n} + 1$
25: $zu_{sn} \leftarrow zu_{sn} + 1$
26: **if** $yu_{sl} = 0$ **then**
27:         $nul_s \leftarrow nul_s + 1$
28: $yu_{sl} \leftarrow yu_{sl} + 1$
29: $um_s \leftarrow um_s \cup \{m\};$

---

and added to the cost (Lines 20–21). The process and service move costs are subtracted (Lines 22-28) followed by the subtraction of machine-move cost (Line 29).

Re-optimizing a Subproblem. We use systematic search for solving a given subproblem. However, the search is stopped when the number of failures exceed a given threshold. Three important components of a CP-based systematic search are: variable ordering heuristics, value ordering heuristics, and filtering rules. The variable ordering heuristic we use is based on an aggregation of the following information that we maintain for each process p: (a) maximum increment in the objective cost when assigning a best machine to process p, (b) total weighted requirement of a process which is the sum of the weighted requirements of all resources, and (c) the number of machines available for p. The value ordering ordering heuristic that is used to select a machine for a given process is based on the minimum cost while ties are broken randomly.

At each node of the search tree constraint propagation is performed to reduce the search space. Whenever a machine m is assigned to a process p, the domains are filtered as shown in Algorithm 2, and the affected constraints are checked. Additionally, during subproblem solving, usage and cost based filtering is also applied for removing machines from the domains of the processes. We omit the details of these pruning rules due to lack of space. Let Q be a set of unassigned processes. The minimum load-cost that will be incurred as a result of assigning machines to processes in Q is denoted by lc^{bound}. The minimum balance-cost that will be incurred as a result of assigning

Table 1. Properties of Instances in Datasets A and B

Set	No.	$|P|$	$|R|$	$|T|$	$|M|$	$|S|$	$|L|$	$|N|$	$|D|$	$|B|$	w_r	v_b	w^{pmc}	w^{smc}	w^{mmc}	mmc^{max}
a1	1	100	2	0	4	79	4	1	0	1	10	10	1	10	100	2
a1	2	1000	4	1	100	980	4	2	40	0	10	-	1	10	100	2
a1	3	1000	3	1	100	216	25	5	342	0	10	-	1	10	100	2
a1	4	1000	3	1	50	142	50	50	297	1	10	10	1	10	100	2
a1	5	1000	4	1	12	981	4	2	32	1	10	10	1	10	100	2
a2	1	1000	3	0	100	1000	1	1	0	0	10	-	1	10	100	0
a2	2	1000	12	4	100	170	25	5	0	0	10	-	1	10	100	2
a2	3	1000	12	4	100	129	25	5	577	0	10	-	1	10	100	2
a2	4	1000	12	0	50	180	25	5	397	1	10	10	1	10	100	2
a2	5	1000	12	0	50	153	25	5	506	0	10	-	1	10	100	2
b	1	5000	12	4	100	2512	10	5	4412	0	10	-	1	10	100	2
b	2	5000	12	0	100	2462	10	5	3617	1	10	10	1	10	100	2
b	3	20000	6	2	100	15025	25	5	16560	0	10	-	1	10	100	2
b	4	20000	6	0	500	1732	50	5	40485	1	10	10	1	10	100	2
b	5	40000	6	2	100	35092	10	5	14515	0	10	-	1	10	100	2
b	6	40000	6	0	200	14680	50	5	42081	1	10	10	1	10	100	0
b	7	40000	6	0	4000	15050	50	5	43873	1	10	10	1	10	100	2
b	8	50000	3	1	100	45030	10	5	15145	0	10	-	1	10	100	2
b	9	50000	3	0	1000	4609	100	5	4337	1	10	10	1	10	100	2
b	10	50000	3	0	5000	4896	100	5	47260	1	10	10	1	10	100	2

machines to the processes in Q is denoted by bc^{bound}. We use the following to compute lc^{bound} and bc^{bound}, during search:

$$lc^{bound} = \sum_{r \in R} \left(\sum_{p \in Q} r_{pr} - \sum_{m \in M} \max(0, sc_{mr} - u_{mr}) \right) \times w_r.$$

$$bc^{bound} = \sum_{b \in B} \left(-t_b \times \sum_{p \in Q} r_{pr_b^1} + \sum_{p \in Q} r_{pr_b^2} \right) \times v_b.$$

The value of lc^{bound} is obtained by adding the total demand for each resource and comparing it to the total safety capacity for that resource on all machines, and multiplying with the appropriate cost-factor. Similarly, the value of bc^{bound} is obtained by considering the weighted difference of total resources and multiplying with the appropriate cost factor. The lower-bound of the objective function is obtained by adding lc^{bound} and bc^{bound} to the current cost of the partial solution.

6   Experimental Results

In this section we present some results to demonstrate the effectiveness of our approaches, and in particular the scalability of the CP approach. We experimented with the instances of Datasets A and B of EURO/ROADEF'12 Challenge. Table 1 shows the properties of these instances. In set A (B) the maximum number of processes is 1,000 (50,000) and the maximum number of machines is 100 (5,000). For MIP we used CPLEX and the algorithms were implemented in Java. For CP we did not use any existing local search solver, e.g., COMET [2], in order to get more freedom in controlling different aspects of LNS approach. All the algorithms were implemented in C.

Table 2. Cost Results for set A obtained within 300 seconds

Problem	Initial	LB	Best ROADEF	MIP-LNS	CP-LNS
a1-1	49,528,750	44,306,501.00	44,306,501	**44,306,501**	**44,306,501**
a1-2	1,061,649,570	777,531,000.00	777,532,896	792,813,766	**778,654,204**
a1-3	583,662,270	583,005,717.00	583,005,717	583,006,527	**583,005,829**
a1-4	632,499,600	242,397,000.00	252,728,589	258,135,474	**251,189,168**
a1-5	782,189,690	727,578,309.00	727,578,309	**727,578,310**	727,578,311
a2-1	391,189,190	103.87	198	273	**196**
a2-2	1,876,768,120	526,244,000.00	816,523,983	836,063,347	**803,092,387**
a2-3	2,272,487,840	1,025,730,000.00	1,306,868,761	1,393,648,719	**1,302,235,463**
a2-4	3,223,516,130	1,680,230,000.00	1,681,353,943	1,725,846,815	**1,683,530,845**
a2-5	787,355,300	307,041,000.00	336,170,182	359,546,818	**331,901,091**

Table 3. Cost Results for set B obtained within 300 seconds for CP approach

Name	Initial	LB	CP-LNS	Iterations
b-1	7,644,173,180	3,290,754,940	3,337,329,571	813,519
b-2	5,181,493,830	1,015,153,860	1,022,043,596	477,375
b-3	6,336,834,660	156,631,070	157,273,705	1,271,094
b-4	9,209,576,380	4,677,767,120	4,677,817,475	226,561
b-5	12,426,813,010	922,858,550	923,335,604	968,840
b-6	12,749,861,240	9,525,841,820	9,525,867,169	618,878
b-7	37,946,901,700	14,833,996,360	14,838,521,000	36,886
b-8	14,068,207,250	1,214,153,440,	1,214,524,845	1,044,842
b-9	23,234,641,520	15,885,369,400	15,885,734,072	115,054
b-10	42,220,868,760	18,048,006,980	18,049,556,324	31,688

Table 2 shows results for dataset A. For each instance, we give the initial cost, the best lower bound found (LB), the best solution found by any solver submitted to the ROADEF competition (Best ROADEF) and the best result obtained within 300s for one overall set of parameter settings for our MIP and CP approaches. The full sized MIP problem could not be solved for all the instances of set A. The approach that outperforms another approach in terms of cost is made bold for each instance of set A. Overall, CP-based LNS outperforms MIP-based LNS.

Table 3 presents results for dataset B. For each instance, we give the initial cost, the lower bound (LB) and the result obtained within 300s for the CP approach, and the number of iterations (Iterations). The results for MIP-based LNS are not presented. The reason is that in each iteration 10 machines are selected, which requires roughly 10 seconds to find an improving solution. This is feasible with 100 machines, but with 5000 not enough combinations are explored in 300 seconds. The results suggest that our CP-based LNS can scale to very large problem instances and is superior both in memory use and the quality of solutions that can be found in limited time. This is mainly because of three factors. First, *selecting a subproblem* by selecting a set of processes from only a few selected machines works better than selecting a few processes from many machines. Second, *creating a subproblem* efficiently by removing a set of processes from their corresponding machines and updating the domains incrementally increases the number of iterations resulted in a greater level of exploration of the search space. Finally, *solving a subproblem* by incrementally maintaining the domains, using cheap constraint checks and using a heuristic that selects a machine that incurs the lowest cost for a given process resulted in efficient exploration of the search-space of the subproblem.

7   Conclusions and Future Work

We presented MIP and CP-based LNS approaches for solving the machine reassignment problem. Results shows that our CP-based LNS approach is scalable, thus suited for solving large instances, and has better anytime-behaviour which is important when solutions must be reported subject to a time limit. The incrementality aspect of our approach allows to create and solve subproblems efficiently, which is a key-factor in finding good quality solutions in a limited time. In the future, we plan to exploit multi-cores that might be available while solving the problem.

Acknowledgments. This work is supported by Science Foundation Ireland Grant No. 10/IN.1/I3032.

References

1. Beldiceanu, N., Carlsson, M., Rampon, J.: Global constraint catalog, 2nd edn. (revision a). Technical Report T2012:03, SICS (2012)
2. Van Hentenryck, P., Michel, L.: Constraint-Based Local Search. The MIT Press (2009)
3. Padala, P., Shin, K.G., Zhu, X., Uysal, M., Wang, Z., Singhal, S., Merchant, A., Salem, K.: Adaptive control of virtualized resources in utility computing environments. In: EuroSys 2007: Proceedings of the 2nd ACM SIGOPS/EuroSys European Conference on Computer Systems 2007, pp. 289–302. ACM, New York (2007)
4. Petrucci, V., Loques, O., Mosse, D.: A dynamic configuration model for power-efficient virtualized server clusters. In: Proceedings of the 11th Brazilian Workshop on Real-Time and Embedded Systems (2009)
5. Shen, K., Tang, H., Yang, T., Chu, L.: Integrated resource management for cluster-based internet services. SIGOPS Oper. Syst. Rev. 36(SI), 225–238 (2002)
6. Srikantaiah, S., Kansal, A., Zhao, F.: Energy aware consolidation for cloud computing. In: Proceedings of HotPower (2008)
7. Steinder, M., Whalley, I., Hanson, J.E., Kephart, J.O.: Coordinated management of power usage and runtime performance. In: NOMS, pp. 387–394. IEEE (2008)
8. Verma, A., Ahuja, P., Neogi, A.: pMapper: Power and Migration Cost Aware Application Placement in Virtualized Systems. In: Issarny, V., Schantz, R. (eds.) Middleware 2008. LNCS, vol. 5346, pp. 243–264. Springer, Heidelberg (2008)
9. Wang, X., Chen, M.: Cluster-level feedback power control for performance optimization. In: HPCA, pp. 101–110. IEEE Computer Society (2008)

A Boolean Model for Enumerating Minimal Siphons and Traps in Petri Nets

Faten Nabli, François Fages, Thierry Martinez, and Sylvain Soliman

EPI Contraintes, Inria Paris-Rocquencourt, France
{Faten.Nabli,Francois.Fages,Thierry.Martinez,Sylvain.Soliman}@inria.fr

Abstract. Petri nets are a simple formalism for modeling concurrent computation. Recently, they have emerged as a promising tool for modeling and analyzing biochemical interaction networks, bridging the gap between purely qualitative and quantitative models. Biological networks can indeed be large and complex, which makes their study difficult and computationally challenging. In this paper, we focus on two structural properties of Petri nets, siphons and traps, that bring us information about the persistence of some molecular species. We present two methods for enumerating all minimal siphons and traps of a Petri net by iterating the resolution of Boolean satisfiability problems executed with either a SAT solver or a CLP(B) program. We compare the performances of these methods with respect to a state-of-the-art algorithm from the Petri net community. On a benchmark with 80 Petri nets from the Petriweb database and 403 Petri nets from curated biological models of the Biomodels database, we show that miniSAT and CLP(B) solvers are overall both faster by two orders of magnitude with respect to the dedicated algorithm. Furthermore, we analyse why these programs perform so well on even very large biological models and show the existence of hard instances in Petri nets with unbounded degrees.

1 Introduction

Petri nets were introduced in the 60's as a simple formalism for describing and studying information processing systems that are characterized as being concurrent, asynchronous, non-deterministic and possibly distributed [21].

The use of Petri nets for representing biochemical reaction models, by mapping molecular species to places and reactions to transitions, was introduced quite late in [22], together with some Petri net concepts and tools for the analysis of metabolic networks [28]. In [24], a Constraint Logic Program over finite domains (CLP(FD)) is proposed for computing place invariants, which in turn provides structural conservation laws that can be used to reduce the dimension of the Ordinary Differential Equations (ODE) associated to a biochemical reaction model.

In this paper, we consider the Petri net concepts of siphons and traps. A siphon is a set of places that, once it is unmarked, remains so. A trap is a set of places that, once it is marked, can never loose all its tokens. Thus, siphons

M. Milano (Ed.): CP 2012, LNCS 7514, pp. 798–814, 2012.

and traps have opposing effects on the token distribution in a Petri net. These structural properties provide sufficient conditions for reachability (whether the system can reach a given state) and liveness (freedom of deadlocks) properties. It is proved that in order to be live, it is necessary that each siphon remains marked. Otherwise (i.e. once it is empty), transitions having their input places in a siphon can not be live. One way to keep each siphon marked is to have a marked trap inside it. In fact, this condition is necessary and sufficient for a free-choice net to be live [21]. Mixed integer linear programs have been proposed in [19,4] and a state-of-the-art algorithm from the Petri net community has been described later in [6] to compute minimal sets of siphons and traps in Petri nets.

In this article, we present a simple Boolean model capturing these notions and two methods for enumerating the set of all minimal siphons and traps of a Petri net. The first method iterates the resolution of the Boolean model executed with a SAT solver while the second proceeds by backtracking with a CLP(B) program.

On a benchmark composed of the 80 Petri nets of Petriweb[1] [10] and the 403 curated biological models of the biomodels.net[2] repository [16], we show that miniSAT and CLP(B) solvers are both faster by two orders of magnitude than the dedicated algorithms and can in fact solve all instances. Furthermore, we analyse why these programs perform so well on even very large biological models and show the existence of hard instances in Petri nets with unbounded degrees.

2 Preliminaries

2.1 Petri Nets

A *Petri net graph PN* is a weighted bipartite directed graph $PN = (P, T, W)$, where P is a finite set of vertices called *places*, T is a finite set of vertices (disjoint from P) called *transitions* and $W : ((P \times T) \cup (T \times P)) \to \mathbf{N}$ represents a set of directed arcs weighted by non-negative integers (the weight zero represents the absence of arc). Places are graphically represented by circles and transitions by boxes. Unlabeled edges are implicitly labeled with weight 1. A *marking* for a Petri net graph is a mapping $m : P \to \mathbf{N}$ which assigns a number of tokens to each place. A place p is marked by a marking m iff $m(p) > 0$. A subset $S \subseteq P$ is marked by m iff at least one place in S is marked by m. A *Petri net* is a 4-tuple (P, T, W, m_0) where (P, T, W) is a Petri net graph and m_0 is an *initial marking*.

The set of *predecessors* (resp. *successors*) of a transition $t \in T$ is the set of places ${}^{\bullet}t = \{p \in P \mid W(p,t) > 0\}$ (resp. $t^{\bullet} = \{p \in P \mid W(t,p) > 0\}$). Similarly, the set of predecessors (resp. successors) of a place $p \in P$ is the set of transitions ${}^{\bullet}p = \{t \in T \mid W(t,p) > 0\}$ (resp. $p^{\bullet} = \{t \in T \mid W(p,t) > 0\}$).

For every two markings $m, m' : P \to \mathbf{N}$ and every transition $t \in T$, there is a transition step $m \xrightarrow{t} m'$, if for all $p \in P$, $m(p) \geq W(p,t)$ and $m'(p) =$

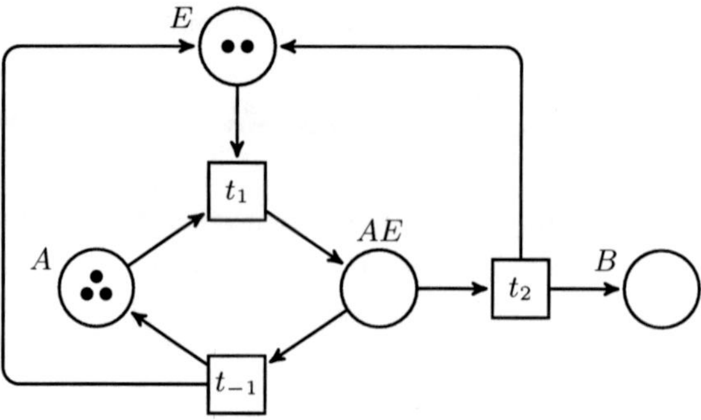

Fig. 1. Petri net associated to the biochemical reaction model of Example 1, displayed here with an arbitrary marking that enables the transition t_1

$m(p) - W(p,t) + W(t,p)$. This notation extends to sequence of transitions $\sigma = (t_0 \ldots t_n)$ by writing $m \xrightarrow{\sigma} m'$ if $m \xrightarrow{t_0} m_1 \xrightarrow{t_1} \ldots \xrightarrow{t_{n-1}} m_n \xrightarrow{t_n} m'$ for some markings $m_1, \ldots, m_n$.

The classical Petri net view of a reaction model is to associate biochemical *species* to *places* and biochemical *reactions* to *transitions*.

Example 1. The system known as Michaelis-Menten enzymatic reactions can be represented by the Petri net depicted in Figure 1. It consists of three enzymatic reactions that take place in two discrete steps: the first involves reversible formation of a complex (AE) between the enzyme (E) and substrate (A) and the second step involves breakdown of the (AE) to form product (B) and to regenerate the enzyme.

$$A + E \leftrightarrows AE \rightarrow B + E$$

2.2 Siphons and Traps

Let $PN = (P, T, W)$ be a Petri net graph.

Definition 1. *A* trap *is a non-empty set of places $P' \subseteq P$ whose successors are also predecessors, $P'^\bullet \subseteq {}^\bullet P'$.*

A siphon *is a non-empty set of places $P' \subseteq P$ whose predecessors are also successors: ${}^\bullet P' \subseteq P'^\bullet$.*

A siphon (resp. a trap) is proper *if its predecessor set is strictly included in its successor set, ${}^\bullet P' \subsetneq P'^\bullet$ (resp. $P'^\bullet \subsetneq {}^\bullet P'$).*

A siphon (resp. a trap) is minimal *if it does not contain any other siphon (resp. trap).*

It is worth remarking that a siphon in PN is a trap in the dual Petri net graph, obtained by reversing the direction of all arcs in PN. Note also that since predecessors and successors of an union are the union of predecessors (resp. successors), the union of two siphons (resp. traps) is a siphon (resp. a trap).

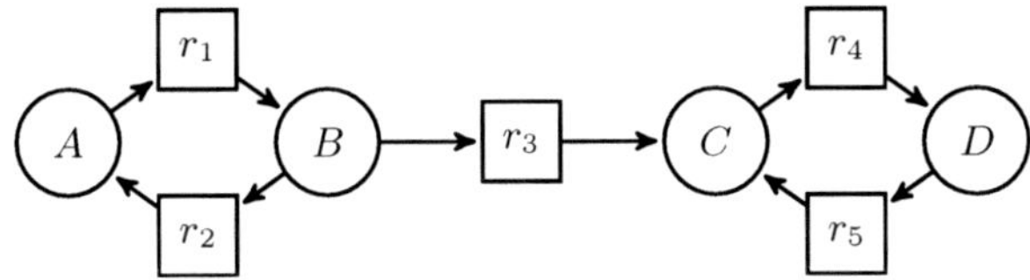

Fig. 2. Petri net graph of Example 2

Example 2. In the Petri net graph depicted in Figure 2, $\{A, B\}$ is a minimal proper siphon, since $^\bullet\{A, B\} = \{r_1, r_2\} \subset \{^\bullet A, B\} = \{r_1, r_2, r_3\}$. $\{C, D\}$ is a minimal proper trap, scine $\{C, D\}^\bullet = \{r_4, r_5\} \subset {}^\bullet\{C, D\} = \{r_3, r_4, r_5\}$.

The following propositions show that traps and siphons provide a structural characterization of some particular dynamical properties on markings.

Proposition 1. *[21] For every subset $P' \subseteq P$ of places, P' is a trap if and only if for any marking $m \in \mathbf{N}^P$ with $m_p \geq 1$ for some place $p \in P'$, and any marking $m' \in \mathbf{N}^P$ such that $m \xrightarrow{\sigma} m'$ for some sequence σ of transitions, there exists a place $p' \in P'$ such that $m'_{p'} \geq 1$.*

Proposition 2. *[21] For every subset $P' \subseteq P$ of places, P' is a siphon if and only if for any marking $m \in \mathbf{N}^P$ with $m_p = 0$ for all $p \in P'$, and any marking $m' \in \mathbf{N}^P$ such that $m \xrightarrow{\sigma} m'$ for some sequence σ of transitions, we have $m'_{p'} = 0$ for all $p' \in P'$.*

Although siphons and traps are stable under union, it is worth noting that minimal siphons do not form a generating set of all siphons. A siphon is called a *basis* siphon if it can not be represented as a union of other siphons [19]. Obviously, a minimal siphon is also a basis siphon, however, not all basis siphons are minimal. For instance, in Example 2, there are two basis siphons, $\{A, B\}$ and $\{A, B, C, D\}$, but only the former is minimal, the latter cannot be obtained by union of minimal siphons.

2.3   Application to Deadlock Detection

One reason to consider minimal siphons is that they provide a sufficient condition for the non-existence of deadlocks.

It has been shown indeed that in a deadlocked Petri net (i.e. where no transition can fire), all unmarked places form a siphon [3]. The siphon-based approach for deadlock detection checks if the net contains a proper siphon that can become unmarked by some firing sequence. A proper siphon does not become unmarked if it contains an initially marked trap. If such a siphon is identified, the initial marking is modified by the firing sequence and the check continues for the remaining siphons until a deadlock is identified, or until no further progress can be done. Considering only the set of minimal siphons is sufficient because if any

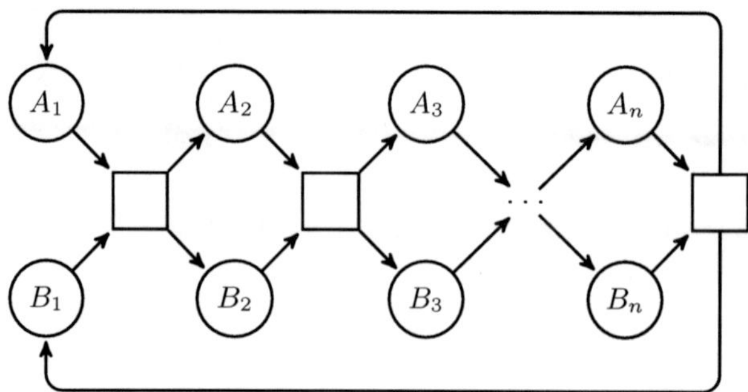

Fig. 3. Petri net representation of the model of Example 3

siphon becomes unmarked during the analysis, then at least one of the minimal siphons must be unmarked.

The relevance of siphons and traps for other liveness properties is summarized in [11].

2.4   Complexity

Deciding whether a Petri net contains a siphon or a trap and exhibiting one if it exists is polynomial [5]. However, the decision problem of the existence of a minimal siphon containing a given place is NP-hard [26].

Furthermore, there can be an exponential number of minimal siphons and traps in a Petri net, as shown by the following:

Example 3. In the Petri net depicted in Figure 3, defined by the transitions: $A_1 + B_1 \rightarrow A_2 + B_2, A_2 + B_2 \rightarrow A_3 + B_3, \ldots, A_n + B_n \rightarrow A_1 + B_1$, there are 2^n minimal siphons and 2^n minimal traps, each one including either A_i or B_i but not both of them, for all i's.

2.5   Application to Systems Biology

One example of the relevance of traps and siphons in biology was given in [28] for the analysis of the potato plant that produces starch and accumulates it in the potato tubers during growth, while starch is consumed after the tubers are deposited after the harvest. The starch and several of its precursors then form traps in the reaction net during growth, while starch and possible intermediates of degradation form siphons after the harvest.

The underlying Petri net is shown in Figure 4, where G_1 stands for glucose-1-phosphate, G_u is UDP-glucose, S is the starch, I stands for intermediary species and P_1 and P_2 represent external metabolites [25]. In this model, either the branch producing starch (t_3 and t_4) or the branch consuming it (t_5 and t_6) is operative. Two Petri nets are derived from this model: one Petri net where t_5 and t_6 are removed (in this Petri net, t_3 and t_4 are said to be operative) and one Petri net where t_3 and t_4 are removed.

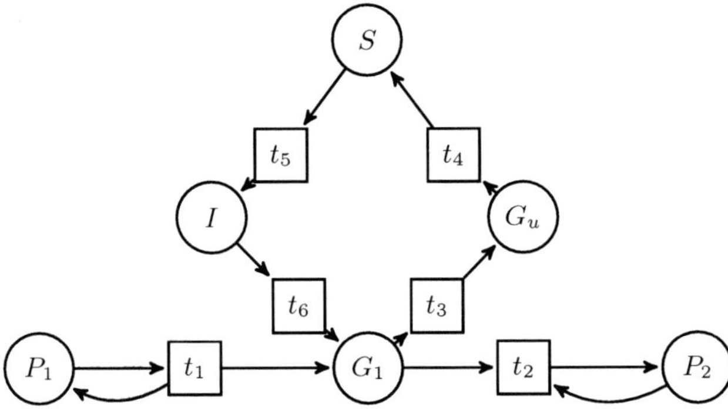

Fig. 4. Petri net graph modeling the growth metabolism of the potato plant [28]

It can be easily observed that the set $\{G_u, S\}$ is a trap when t_3 and t_4 are operative: once a token arrives in S, no transition can be fired and the token remains there independently of the evolution of the system. Dually, $\{S, I\}$ is a siphon when t_5 and t_6 are operative: once the last token is consumed from S and I, no transition can generate a new token in these places, so they remain empty.

In most cells containing starch, starch and specific predecessors form traps, whereas starch and specific successors form siphons. This provides a very simple explanation for the fact that either the branch producing starch or the branch degrading it is operative. This is realized by complete inhibition of the appropriate enzymes by the gene regulatory network.

Another interesting example, also from [28], deals with the analysis of the role of the triosephosphate isomerase (TPI) in Trypanosoma brucei metabolism by detecting solely siphons and traps. At the beginning, Helfert et al. [12] supposed that glycolysis could proceed without TPI. But unexpected results where all system fluxes (Pyruvate, Glycerol) decrease were found so that the authors built a kinetic model for explaining that phenomenon. Then a purely structural explanation for the necessary presence of TPI in glycolysis and glycerol production was provided in [28] by simply considering the presence of siphons and traps in the model.

3   Boolean Model

In the literature, many algorithms have been proposed to compute minimal siphons and traps of Petri nets. Since a siphon in a Petri net N is a trap of the dual net N', it is enough to focus on siphons, the traps are obtained by duality. Some algorithms are based on linear programming [19,4], Horn clause satisfaction [13,17] or algebraic approaches [15]. More recent state-of-the-art methods are presented in [5,6].

Here we present two Boolean methods for enumerating minimal siphons. First, siphons can be straightforwardly characterized with a boolean model representing the belonging or not of each place to the siphon. For a Petri net with n

places and m transitions, a siphon S is a set of places whose predecessors are also successors. S can be represented with a vector V of $\{0,1\}^n$ such that for all $i \in \{1, 2, .., n\}$, $V_i = 1$ if and only if $p_i \in S$. The siphon constraint can then be formulated as

$$\forall i, V_i = 1 \Rightarrow {}^\bullet p_i \subseteq \left(\bigcup_{V_j=1} \{p_j\} \right)^\bullet$$

which is equivalent to

$$\forall i, V_i = 1 \Rightarrow \left(\forall t \in T, t \in {}^\bullet p_i \Rightarrow t \in \left(\bigcup_{V_j=1} \{p_j\} \right)^\bullet \right)$$

which is equivalent to

$$\forall i, V_i = 1 \Rightarrow \left(\forall t \in T, t \in {}^\bullet p_i \Rightarrow \exists p_j \in {}^\bullet t, V_j = 1 \right)$$

which can be rewritten in clausal form as:

$$\forall i, V_i = 1 \Rightarrow \bigwedge_{t \in {}^\bullet p_i} \left(\bigvee_{p_j \in {}^\bullet t} V_j = 1 \right)$$

To exclude the case of the empty set, the following constraint is added:

$$\bigvee_i V_i = 1$$

These clauses are Horn-dual clauses (i.e. clauses with at most one negative literal). They are trivially satisfied by taking all variables true.

Second, the enumeration of all minimal siphons (*w.r.t.* set inclusion) can be ensured by a search strategy and the addition of new Boolean constraints during search. One strategy is to find siphons in set inclusion order, and to add a new constraint

$$\bigvee_{p_i \in S} V_i = 0$$

each time a siphon S is found to disallow any superset of this siphon to be found in the continuation of the search. It is worth remarking that this clause is not the dual of a Horn clause. The whole clauses are thus now non-Horn.

In a previous approach based on Constraint Logic Programming [20], the enumeration by set inclusion order was ensured by labeling a cardinality variable in increasing order. Labeling directly on the Boolean variables, with increasing value selection (first 0, then 1), reveals however much more efficient and in fact easier to enforce. The following proposition shows that this strategy correctly finds siphons in set inclusion order.

Proposition 3. *Given a binary tree such that, in each node instantiating a variable X, the left sub-edge posts the constraint $X = 0$ and the right sub-edge posts the constraint $X = 1$, then for all distinct leaves A and B, leaf A is on the left of leaf B only if the set represented by B is not included in the set represented by A (that is to say, there exists a variable X such that $X_B > X_A$, where X_A and X_B denote the values instantiated to X in the paths leading to A and B respectively).*

Proof. A and B have a least common ancestor node instantiating a variable X. If leaf A is on the left of leaf B, the sub-edge leading to A is the left one, with the constraint $X = 0$ and the sub-edge leading to B is the right one, with the constraint $X = 1$, therefore $X_B > X_A$. □

In a post-processing phase, the computed set of minimal siphons can be filtered for only keeping the minimal siphons that contain a given set of places, and hence solve the above mentioned NP-hard decision problem. It is worth remarking that posting the inclusion of the selected places first would not ensure that the siphons found are indeed minimal w.r.t. set inclusion.

4 Boolean Algorithms

This section describes two implementations of the above model and search strategy, one using an iterated SAT procedure and the other based on Constraint Logic Programming with Boolean constraints.

4.1 Iterated SAT Algorithm

The Boolean model can be directly interpreted using a SAT solver to check the existence of a siphon or trap. We use sat4j[3], an efficient library of SAT solvers in Java for Boolean satisfaction and optimization. It includes an implementation of the MiniSAT algorithm in Java.

The example of the enzymatic reaction of example 1 is encoded as follows: each line is a space-separated list of variables representing a clause; a positive value means that corresponding variable is under positive form (so 2 means V_2), and a negative value means the negation of that variable (so -3 means $\neg V_3$). In this example, variables 1, 2, 3 and 4 correspond respectively to E, A, AE and B. In the first iteration, the problem amounts to solve the following encoding of Horn-dual clauses:

```
-2 3
-3 1 2
-1 3
-1 3
-4 3
```

The problem is satisfied with the values: $-1, 2, 3, -4$ which means that $\{A, AE\}$ is a minimal siphon.

To ensure minimality, the (non Horn-dual) clause $-2\ -3$ is added and the program iterates an other time. The problem is satisfied with $1, -2, 3, -4$, meaning that $\{E, AE\}$ is also a minimal siphon. A new clause is added stating that either E or AE does not belong to the siphon and no more variable assignment can satisfy the problem.

---

[3] `http://www.sat4j.org/`

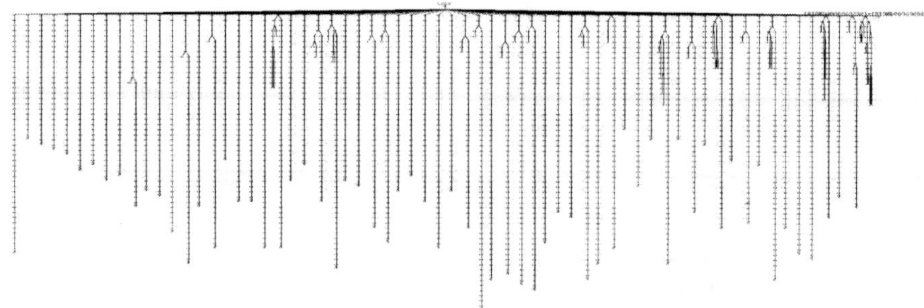

Fig. 5. Search tree developed with the backtrack replay strategy for enumerating the 64 minimal siphons of model 239 of biomodels.net (described in Section 5.1)

Therefore, this model contains 2 minimal siphons: $\{A, AE\}$ and $\{E, AE\}$. The enzyme E is a catalyst protein for the transformation of the substrate E in a product B. Such a catalyst increases the rate of the reaction but is conserved in the reaction.

4.2 Backtrack Replay CLP(B) Algorithm

The search for siphons can also be implemented with a Constraint Logic Program with Boolean constraints (CLP(B)). We use GNU-Prolog[4] [8] for its efficient low-level implementation of Boolean constraint propagators.

The enumeration strategy is a variation of *branch-and-bound*, where the search is restarted to find a non-superset siphon each time a new siphon is found. We tried two variants of the branch-and-bound: with restart from scratch and by backtracking.

In the branch-and-bound with restart method, it is essential to choose a variable selection strategy which ensures diversity. Indeed, an enumeration method with a fixed variable order accumulates failures by always trying to enumerate the same sets first and these failures are only lately pruned by the non-superset constraints. As a consequence, the developed search tree gets more and more dense after each iteration since the previous forbidden sets are repeatedly tried again. This phenomenon does not exist in SAT solvers thanks to no-good recording. In CLP, this problem can be compensated for however, by using a random selection strategy for variables. This provides a good diversity and performs much better than any uniform heuristics.

However, branch-and-bound by backtracking gives better performance when care is taken for posting the non-superset constraint only once, since reposting it at each backtrack step proved to be inefficient. Our backtrack replay strategy is implemented as follows:

1. each time a siphon is found, the path leading to this solution is memorized,
2. then the search is fully backtracked in order to add to the model the new non-superset constraint,

---

[4] `http://www.gprolog.org/`

3. and then the memorized path is rolled back to continue the search at the point it was stopped.

Figure 5, generated with CLPGUI[5] [9] depicts the search tree that is developed for enumerating the 64 minimal siphons of a biological model of 51 species and 72 reactions. Each sub-tree immediately connected to the root corresponds to the replay of the path with a minimality constraint added. It is remarkable that with the backtrack replay strategy, very few backtracking steps are necessary to search for all solutions.

5   Evaluation

5.1   Benchmark

Petriweb. Our first benchmark of Petri nets is Petriweb [10], a benchmark of 80 Petri nets from the Petri net community. The most difficult instances of this benchmark come from case studies in process refinement, namely problems 1454, 1479 and 1516.

Biomodels.net. We also consider the Petri nets associated to biochemical reaction models of the biomodels.net repository of 403 models [16] and some other complex biochemical models. The most difficult models are the following ones:

- Kohn's map of the mammalian cell cycle control [14,2], a model of 509 species and 775 reactions;
- Model BIOMD0000000175 of biomodels.net, a model of 118 species and 194 reactions involved in ErbB signaling;
- Model BIOMD0000000205 of biomodels.net, a model of 194 species and 313 reactions involved in the regulation of EGFR endocytosis and EGFR-ERK signaling by endophilin-mediated RhoA-EGFR crosstalk;
- Model BIOMD0000000239, a core model of 51 species and 72 reactions representing the glucose-stimulated insulin secretion of pancreatic beta cells.

5.2   Results and Comparison

In this section, we compare the two Boolean methods described in the previous section with the state-of-the-art dedicated algorithm of [6]. This algorithm uses a recursive problem partitioning procedure to reduce the original search problem to multiple simpler search subproblems. Each subproblem has specific additional place constraints with respect to the original problem. This algorithm can be applied to enumerate minimal siphons, place-minimal siphons, or even siphons that are minimal with respect to a given subset of places.

Table 1 presents the CPU times for enumerating all minimal siphons of the Petri nets in Petriweb and biomodels.net. All times are in milliseconds and have been obtained on a PC with an intel Core processor 2.20 GHz and 8 GB of

---

[5] http://contraintes.inria.fr/~fages/CLPGUI

Table 1. Performance on the whole benchmark

Database	# model	# siphons min–max (avg.)	siphons size min–max (avg.)	total time dedicated algorithm	SAT	GNU Prolog
Biomodels.net	403	0–64 (4.21)	1–413 (3.10)	19734	611	195
Petriweb	80	0–11 (2.85)	0–7 (2.03)	2325	156	6

Table 2. Performance on the hardest instances

model	# siphons	# places	# transitions	dedicated algorithm	sat	GNU Prolog
Kohn's map of cell cycle	81	509	775	28	1	221
BIOMD000000175	3042	118	194	∞	137000	∞
BIOMD000000205	32	194	313	21	1	34
BIOMD000000239	64	51	72	2980	1	22

memory. For each benchmark, we provide the total number of models, the minimal, maximal and average numbers of siphons and the total computation time for enumerating all of them.

Surprisingly, but happily, on all these practical instances, except one instance detailed below, the SAT and CLP(B) programs solve the minimal siphon enumeration problem, in less than one millisecond in average, with a better performance for the CLP(B) program over the SAT solver, and by two orders of magnitude over the dedicated algorithm.

However, one particular model, number 175 in biomodels.net, was excluded from this table because its computational time is very high. Table 2 presents the performance figures obtained on this model and on the three other hardest instances for which we also provide the number of places and transitions. On these hard instances, the SAT solver is faster than the CLP(B) program by one to two orders of magnitude, and is the only algorithm to solve the problem for model 175, in 137 seconds.

That model 175 represents a quantitative model that relates EGF and HRG stimulation of the ErbB receptors to ERK and AKt activation in MCF-7 breast cancer cells [1]. This is the first model to take into account all four ErbB receptors, simultaneous stimulation with two ligands, and both the ERK and AKt pathways. Previous models of ErbB (e.g. the model developed in [23]) were limited to a single ErbB because of combinatorial complexity. It is well known that the ErbB signaling network is highly connected and indeed the underlying Petri net contains the highest number of arcs of the biomodels.net repository.

5.3 Hard Instances

MiniSAT and CLP(B) outperform the specialized algorithm by at least one order of magnitude and the computation time is extremely short on our practical examples. Even if the model is quite large, e.g. for Kohn's map of the cell cycle control with 509 species and 775 reactions, the computation time for enumerating

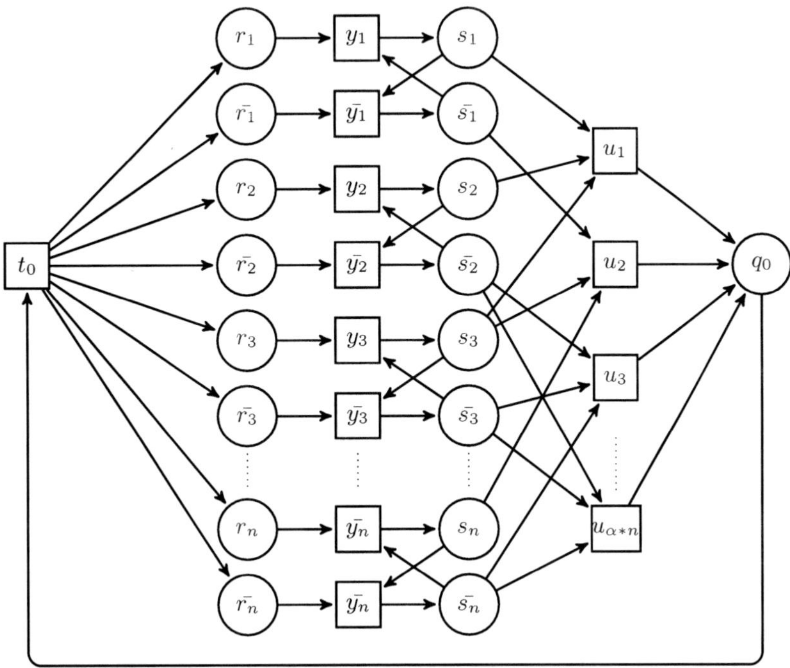

Fig. 6. Petri net graphs considered for the reduction of 3-SAT to the existence of a minimal siphon containing place q_0

its 81 minimal siphons is astonishingly short: one millesecond only. However, this enumeration of all minimal siphons solves the decision problem of the existence of a minimal siphon containing a given set of places which has been proved NP-hard by reduction of 3-SAT in [27], and the question is: why the existing benchmarks from systems biology and petriweb are so easy?

We can provide some hints of explanation by considering the well-known phase transition phenomenon in 3-SAT. The probability that a random 3-SAT problem is satisfiable has been shown to undergo a sharp phase transition as the ratio α of the number of clauses over the number of variables crosses the critical value of about 4.26 [18,7], going from satisfiability to unsatisfiability with probability one when the number of variables grows to the infinity.

The reduction of three-satisfiability (3-SAT) to the problem of existence of a minimal siphon containing a given place has been shown in [27] with the Petri net structure illustrated in Figure 6. It is worth noticing that in this encoding, the Petri net has a maximum place indegree (for q_0) which is linear in the number of clauses, and a maximum place outdegree (for t_0) which is linear in the number of variables.

Not surprisingly, this family of Petri nets provides a hard benchmark for enumerating minimal siphons[6]. Table 3 contains experimental results on these

---

[6] All benchmarks of this section are available at
`http://contraintes.inria.fr/~nabli/indexhardinstances.html`.

Table 3. Computational results for the enumeration of minimal siphons in Petri nets encoding 3-SAT

model	# siphons	Petri net view			3-SAT view			time
		# places	# transitions	density	# variables	# clauses	α	(ms)
pn0.2.xml	>129567	801	441	0.56	200	40	0.2	60000
pn0.6.xml	>32392	801	521	0.65	200	120	0.6	60000
pn1.xml	>1075	801	601	0.751	200	200	1	60000
pn2.xml	>74462	801	801	1	200	400	2	60000
pn3.xml	>63816	801	1001	1.24	200	600	3	60000
pn4.xml	>59827	801	1201	1.49	200	800	4	60000
pn4.2.xml	>41415	801	1241	1.54	200	840	4.2	60000
pn4.4.xml	200	801	1281	1.59	200	880	4.4	1596
pn4.6.xml	200	801	1321	1.64	200	920	4.6	1411
pn5.xml	200	801	1401	1.74	200	1000	5	370
pn6.xml	200	801	1601	1.99	200	1200	6	175
pn7.xml	200	801	1801	2.24	200	1400	7	157
pn8.xml	200	801	2001	2.49	200	1600	8	157
pn9.xml	200	801	2201	2.74	200	1800	9	133
pn10.xml	200	801	2401	2.99	200	2000	10	137

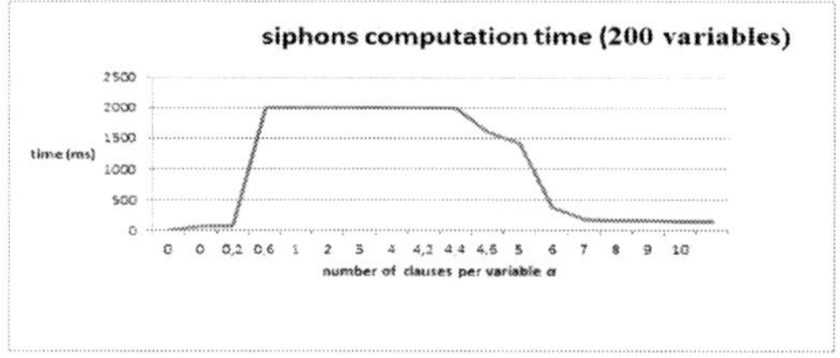

Fig. 7. CPU time for computing all minimal siphons in Petri nets encoding 3-SAT problems of density ranging from 0 to 10 with a time-out of 2 seconds

Petri nets associated to random 3-SAT problems. The table gives the number of minimal siphons and the time to compute them with a timeout of 60 seconds. The table also provides information concerning both the 3-SAT problem and its corresponding Petri net. For each 3-SAT problem, we provide the number of Boolean variables, the number of clauses and the ratio α. For the corresponding Petri net, we provide the number of places, the number of transitions and the density (ratio of the number of transitions over the number of places). The computation time of all minimal siphons as a function of α, the density of the initial 3-SAT problem, is represented in Figure 7.

The reason for the timeout obtained for all 3-SAT problems of density below the threshold value 4.26 is that for small values of α, the clause is satisfiable with an exponential number of valuations which gives rise to an exponential number of minimal siphons to compute. On the other hand, for values of $alpha$ above the threshold, the clause are unsatisfiable and there is indeed no minimal siphon containing q_0 (only the 200 minimal siphons without q_0 are computed).

Table 4. Computational results for the enumeration of minimal siphons in random Petri nets with varying density and linear versus bounded degrees

	nb places	nb transitions	density	nb siphons	time
random-degree-pn1	801	80	0.099875156	>77918	60000
random-degree-pn2	801	320	0.399500624	>69246	60000
random-degree-pn3	801	560	0.699126092	>45782	60000
random-degree-pn4	801	801	1	>28285	60000
random-degree-pn5	801	1041	1.299625468	0	7473
random-degree-pn6	801	1281	1.599250936	0	11233
random-degree-pn7	801	1521	1.898876404	0	15040
random-degree-pn8	801	1762	2.199750312	0	9548
random-degree-pn9	801	2242	2.799001248	0	13807
bounded-degree-pn1	801	80	0.099875	377	120
bounded-degree-pn2	801	320	0.399501	250	55
bounded-degree-pn3	801	560	0.699126	146	32
bounded-degree-pn4	801	801	1	66	14
bounded-degree-pn5	801	1041	1.299625	29	11
bounded-degree-pn6	801	1281	1.599251	13	3
bounded-degree-pn7	801	1521	1.898876	4	5
bounded-degree-pn8	801	1762	2.199750312	1	1
bounded-degree-pn9	801	2242	2.799001248	0	2

Table 5. In and out degrees for places and transitions in the Petri nets of the biomodels.net benchmark with model 175 apart

	minimum	maximum	average	model 175
number of arcs	1	913	92	1125
Avg-indegree-places	0	8	1.89	5
Avg-indegree-transitions	0	3	1.06	3
Max-indegree-places	1	54	5.94	31
Max-indegree-transitions	1	14	2.72	11
Avg-outdegree-places	0	8	1.93	5
Avg-outdegree-transitions	0	3	0.99	3
Max-outdegree-places	0	36	5.53	32
Max-outdegree-transitions	1	14	2.87	11

Now, Table 4 shows that similar bad performance figures are obtained with randomly generated Petri nets with a number of in and out degrees that is *linear* in the number of places and transitions, while on random Petri nets with a *bounded degree* (less than 5), the enumeration of minimal siphons is easy. This is the situation encountered in our practical application. As shown in Table 5, the Petri nets associated to the biochemical reaction models of biomodels.net have small in and out degrees for places and transitions even in very large models. Model 175 mentioned in Section 5.2 appears as an exception combining a large size with a high connectivity on places, with some species that are both the reactants of 32 reactions and the products of 31 reactions.

6   Conclusion

Siphons and traps in Petri nets define meaningful pools of places that display a specific behavior during the dynamical evolution of a Petri net, or of a system of biochemical reactions whatever kinetic parameters are.

We have described a Boolean model for the problem of enumerating all minimal siphons in a Petri net and have compared two Boolean methods to a state-of-the-art algorithm from the Petri net community [6]. The miniSAT solver and the CLP(B) program both solve our large benchmark of real-size problems and outperform the dedicated algorithm by two orders of magnitude. On the benchmark of 403 biological models in biomodels.net, the Boolean method for enumerating all minimal siphons using miniSAT is very efficient. It also scales very well in the size of the net. The CLP(B) program also solves all but one instances of the benchmark, with a better performance than miniSAT in average, but does not scale-up as well to large size models like Kohn's map with 509 species and 775 reactions.

The surprising efficiency of the miniSAT and CLP(B) methods for solving the practical instances of this NP-hard problem has been analyzed in connection to the well-known phase transition phenomenon in 3-SAT, and to the fact that the degree of Petri nets associated to even very large models of several hundreds of biochemical species and reactions remains limited to small values in practice. This explains why these Boolean methods perform so well in the practical context of systems biology applications.

These results militate for the analysis of biochemical networks with Petri net concepts and Constraint Programming tools.

Acknowledgment. This work is supported by the French OSEO project Biointelligence.

References

1. Birtwistle, M.R., Hatakeyama, M., Yumoto, N., Ogunnaike, B.A., Hoek, J.B., Kholodenko, B.N.: Ligand-dependent responses of the ErbB signaling network: experimental and modeling analysis. Molecular Systems Biology 3(144) (September 2007)
2. Chabrier-Rivier, N., Chiaverini, M., Danos, V., Fages, F., Schächter, V.: Modeling and querying biochemical interaction networks. Theoretical Computer Science 325(1), 25–44 (2004)
3. Chu, F., Xie, X.-L.: Deadlock analysis of petri nets using siphons and mathematical programming. IEEE Transactions on Robotics and Automation 13(6), 793–804 (1997)
4. Cordone, R., Ferrarini, L., Piroddi, L.: Characterization of minimal and basis siphons with predicate logic and binary programming. In: Proceedings of IEEE International Symposium on Computer-Aided Control System Design, pp. 193–198 (2002)
5. Cordone, R., Ferrarini, L., Piroddi, L.: Some results on the computation of minimal siphons in petri nets. In: Proceedings of the 42nd IEEE Conference on Decision and Control, Maui, Hawaii, USA (December 2003)

6. Cordone, R., Ferrarini, L., Piroddi, L.: Enumeration algorithms for minimal siphons in petri nets based on place constraints. IEEE Transactions on Systems, Man and Cybernetics. Part A, Systems and Humans 35(6), 844–854 (2005)

7. Crawford, J.M., Auton, L.D.: Experimental results on the crossover point in satisfiability problems. In: Proceedings of the 11th National Conference on Artificial Intelligence, pp. 21–27. AAAI Press (1993)

8. Diaz, D., Codognet, P.: Design and implementation of the GNU Prolog system. Journal of Functional and Logic Programming 6 (October 2001)

9. Fages, F., Soliman, S., Coolen, R.: CLPGUI: a generic graphical user interface for constraint logic programming. Journal of Constraints, Special Issue on User-Interaction in Constraint Satisfaction 9(4), 241–262 (2004)

10. Goud, R., van Hee, K.M., Post, R.D.J., van der Werf, J.M.E.M.: Petriweb: A Repository for Petri Nets. In: Donatelli, S., Thiagarajan, P.S. (eds.) ICATPN 2006. LNCS, vol. 4024, pp. 411–420. Springer, Heidelberg (2006)

11. Heiner, M., Gilbert, D., Donaldson, R.: Petri Nets for Systems and Synthetic Biology. In: Bernardo, M., Degano, P., Zavattaro, G. (eds.) SFM 2008. LNCS, vol. 5016, pp. 215–264. Springer, Heidelberg (2008)

12. Helfert, S., Estevez, A., Bakker, B., Michels, P., Clayton, C.: Roles of triosephosphate isomerase and aerobic metabolism in trypanosoma brucei. Biochem. J. 357, 117–125 (2001)

13. Kinuyama, M., Murata, T.: Generating siphons and traps by petri net representation of logic equations. In: Proceedings of 2nd Conference of the Net Theory SIG-IECE, pp. 93–100 (1986)

14. Kohn, K.W.: Molecular interaction map of the mammalian cell cycle control and DNA repair systems. Molecular Biology of the Cell 10(8), 2703–2734 (1999)

15. Lautenbach, K.: Linear algebraic calculation of deadlocks and traps. In: Voss, G., Rozenberg (eds.) Concurrency and Nets Advances in Petri Nets, pp. 315–336. Springer, New York (1987)

16. le Novère, N., Bornstein, B., Broicher, A., Courtot, M., Donizelli, M., Dharuri, H., Li, L., Sauro, H., Schilstra, M., Shapiro, B., Snoep, J.L., Hucka, M.: BioModels Database: a free, centralized database of curated, published, quantitative kinetic models of biochemical and cellular systems. Nucleic Acid Research 1(34), D689–D691 (2006)

17. Minoux, M., Barkaoui, K.: Deadlocks and traps in petri nets as horn-satisfiability solutions and some related polynomially solvable problems. Discrete Applied Mathematics 29, 195–210 (1990)

18. Mitchell, D., Selman, B., Levesque, H.: Hard and easy distributions of SAT problems. In: Proceedings of the 10th National Conference on Artificial Intelligence, pp. 459–465. AAAI Press (1992)

19. Murata, T.: Petri nets: properties, analysis and applications. Proceedings of the IEEE 77(4), 541–579 (1989)

20. Nabli, F.: Finding minimal siphons as a CSP. In: CP 2011: The Seventeenth International Conference on Principles and Practice of Constraint Programming, Doctoral Program, pp. 67–72 (September 2011)

21. Peterson, J.L.: Petri Net Theory and the Modeling of Systems. Prentice Hall, New Jersey (1981)

22. Reddy, V.N., Mavrovouniotis, M.L., Liebman, M.N.: Petri net representations in metabolic pathways. In: Hunter, L., Searls, D.B., Shavlik, J.W. (eds.) Proceedings of the 1st International Conference on Intelligent Systems for Molecular Biology (ISMB), pp. 328–336. AAAI Press (1993)

23. Schoeberl, B., Eichler-Jonsson, C., Gilles, E., Muller, G.: Computational modeling of the dynamics of the map kinase cascade activated by surface and internalized egf receptors. Nature Biotechnology 20(4), 370–375 (2002)
24. Soliman, S.: Finding minimal P/T-invariants as a CSP. In: Proceedings of the Fourth Workshop on Constraint Based Methods for Bioinformatics, WCB 2008, associated to CPAIOR 2008 (May 2008)
25. Stryer, L.: Biochemistry. Freeman, New York (1995)
26. Tanimoto, S., Yamauchi, M., Watanabe, T.: Finding minimal siphons in general petri nets. IEICE Trans. on Fundamentals in Electronics, Communications and Computer Science, 1817–1824 (1996)
27. Yamauchi, M., Watanabe, T.: Time complexity analysis of the minimal siphon extraction problem of petri nets. IEICE Trans. on Fundamentals of Electronics, Communications and Computer Sciences, 2558–2565 (1999)
28. Zevedei-Oancea, I., Schuster, S.: Topological analysis of metabolic networks based on petri net theory. Silico Biology 3(29) (2003)

Cardinality Reasoning for Bin-Packing Constraint: Application to a Tank Allocation Problem

Pierre Schaus[1], Jean-Charles Régin[2], Rowan Van Schaeren[3], Wout Dullaert[4], and Birger Raa[5]

[1] ICTEAM, Université catholique de Louvain, Belgium
pschaus@gmail.com
[2] University of Nice
gcregin@gmail.com
[3] Antwerp Maritime Academy
rowan.van.schaeren@hzs.be
[4] VU University Amsterdam and University of Antwerp
wout.dullaert@vu.nl
[5] University of Gent
birger.raa@ugent.be

Abstract. Flow reasoning has been successfully used in CP for more than a decade. It was originally introduced by Régin in the well-known Alldifferent and Global Cardinality Constraint (GCC) available in most of the CP solvers. The BinPacking constraint was introduced by Shaw and mainly uses an independent knapsack reasoning in each bin to filter the possible bins for each item. This paper considers the use of a cardinality/flow reasoning for improving the filtering of a bin-packing constraint. The idea is to use a GCC as a redundant constraint to the BinPacking that will count the number of items placed in each bin. The cardinality variables of the GCC are then dynamically updated during the propagation. The cardinality reasoning of the redundant GCC makes deductions that the bin-packing constraint cannot see since the placement of all items into every bin is considered at once rather than for each bin individually. This is particularly well suited when a minimum loading in each bin is specified in advance. We apply this idea on a Tank Allocation Problem (TAP). We detail our CP model and give experimental results on a real-life instance demonstrating the added value of the cardinality reasoning for the bin-packing constraint.

Keywords: Tank Allocation, Constraint Programming, Load Planning.

1 Bin-Packing Constraint

The BinPacking constraint was introduced in [1]:

$$BinPacking([X_1, \ldots, X_n], [w_1, \ldots, w_n], [L_1, \ldots, L_m]).$$

M. Milano (Ed.): CP 2012, LNCS 7514, pp. 815–822, 2012.

This constraint enforces the relation $L_j = \sum_i (X_i = j) \cdot w_i, \forall j$. It makes the link between n weighted items (item i has a weight w_i) and the m different capacitated bins in which they are to be put. Only the weights of the items are integers, the other arguments of the constraints are finite domain (f.d.) variables. Note that in this formulation, Lj is a variable which is bounded by the maximal capacity of the bin j. Without loss of generality we assume the item variables and their weights are sorted such that $w_i \leq w_{i+1}$. Example: $BinPacking([1, 4, 1, 2, 2], [2, 3, 3, 3, 4], [5, 7, 0, 3])$.

Classical formulation The traditional way to model a BinPacking constraint is to introduce a binary variable $B_{i,j}$ for each pair (item, bin) which is 1 (true) if item i is placed into bin j, 0 (false) otherwise. Then for each bin j, we add the constraint $L_j = \sum_i B_{i,j} \cdot w_i$. As noted in [1] one important redundant constraint to add is $\sum_j L_j = \sum_i w_i$ allowing a better communication between the other constraints.

Existing Filtering Algorithms. A specific filtering algorithm for the BinPacking constraint in addition to its classical formulation has been first proposed in [1]. This algorithm essentially filters the domains of the X_i's using a knapsack-like reasoning to detect if forcing an item into a particular bin j would make it impossible to reach a load L_j for that bin. This procedure is very efficient but can return false positive saying that an item is OK for a particular bin while it is not. Shaw [1] also introduced a failure detection algorithm computing a lower bound on the number of bins necessary to complete the partial solution. This last consistency check has been extended by [2]. Finally, Cambazard and O'Sullivan [3] propose to filter the domains using an LP arc-flow formulation.

The existing filtering algorithms use the upper bounds of the loading variables $\max(L_j)$ (i.e. capacity of the bins). They do not focus much on the lower bounds of these variables $\min(L_j)$. In classical bin-packing problems, the capacity of the bins $\max(L_j)$ are constrained while the lower bounds $\min(L_j)$ are usually set to 0 in the model. The additional cardinality/flow based filtering we introduce is well suited when those lower bounds are also constrained initially $\min(L_j) > 0$.

2   Cardinality Reasoning for Bin-Packing

Existing filtering algorithms for the *BinPacking* constraint do not make use of the cardinality information inside each bin (*i.e..* the number of items packed inside each bin). However this information can be very valuable in some situations. Consider the extreme case where every item has an equal weight (assume a weight of 1) such that the *BinPacking* constraint reduces to a *GCC*. It is clear that the filtering algorithms for the *BinPacking* are very weak compared to the global arc consistent filtering for the GCC in such a situation. Of course this situation rarely happens in practice but in many applications the weights of the items to place are not so different and it is preferable not to lose completely the reasoning offered by a cardinality constraint (the flow reasoning). Our idea is to introduce one redundant *GCC* in the modelling of the *BinPacking*:

$$GCC([X_1, \ldots, X_n], [C_1, \ldots, C_m])$$

with C_j a variable that represents the number of items placed into bin j. Initially $Dom(C_j) = \{0, \ldots, n\}$. The cardinality variables are pruned dynamically during the search as the bounds of the L_j's and the domains of the X_i's change. Let $bound(X_i, j)$ be equal to true if the variable X_i is instantiated to value j (*i.e.* item i is placed into bin j), false otherwise. Let l_j be the load of the packed items into bin j: $l_j = \sum_{i:bound(X_i,j)} w_i$ and c_j be the number of packed items into bin j: $c_j = \sum_{i:bound(X_i,j)} 1$. Note that it is possible that $\min(L_j) \geq l_j$ because of the filtering from [1]. Furthermore let $poss_j$ be the set of possible items into bin j: $poss_j = \{i \mid |Dom(X_i)| > 1 \ \wedge \ j \in Dom(X_i)\}$. Given a subset of items S , let $sum(S) = \sum_{i \in S} w_i$. The rules to update the lower and upper bounds of C_j are obtained by combining cardinalities and capacity information:

$$\min(C_j) \leftarrow \max(\min(C_j), c_j + |A_j|) \tag{1}$$
$$\max(C_j) \leftarrow \min(\max(C_j), c_j + |B_j|) \tag{2}$$

where $A_j \subseteq poss_j$ is the minimum cardinality set of items such that $l_j + sum(A_j) \geq \min(L_j)$ and $B_j \subseteq poss_j$ is the maximum cardinality set of items such that $l_j + sum(B_j) \leq \max(L_j)$. Since items $w_1, \ldots, w_n$ are sorted increasingly, both rules (1) and (2) can be implemented in $O(n)$ by scanning the items from right to left for rule (1) and from left to right for rule (2).

Example 1. Five items with weights 3,3,4,5,7 can be placed into bin 1 having a possible load $L_1 \in [20..22]$. Two other items are already packed into that bin with weights 3 and 7 ($c_1 = 2$ and $l_1 = 10$). Clearly we have that $|A_1| = 2$ obtained with weights 5,7 and $|B_1| = 3$ obtained with weights 3,3,4. The domain of the cardinality variable C_1 is thus set to $[4..5]$.

3   Tank Allocation for Liquid Bulk Vessels

The tank allocation problem involves the assignment of different cargoes (volumes of chemical products to be shipped by the vessel) to the available tanks of the vessel. The loading plans of bulk vessels are generally generated manually by the vessel planners although it is difficult to generate high quality solutions. The constraints to satisfy are mainly segregation constraints:

1. prevent chemicals from being loaded into certain types of tanks because
   - the chemical may need to have its temperature managed and the tank needs to be equipped with a heating system,
   - the tank must be resistant to the chemical,
   - a tank may still be contaminated by previous cargoes incompatible with the chemical.
2. prevent some pairs of cargoes to be placed next to each other: not only the chemical interactions between the different cargoes need to be considered but also the temperature at which they need to be transported. Too different temperature requirements for adjacent tanks cause the second one to solidify due to cooling off by the first cargo or the first may become chemically unstable due to heating up of the second cargo.

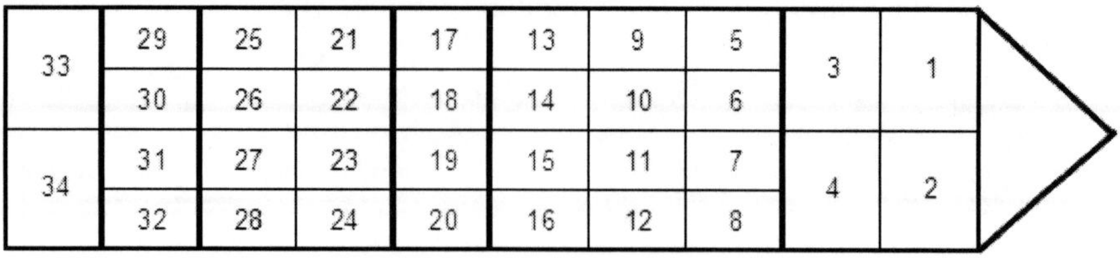

818   P. Schaus et al.

33	29	25	21	17	13	9	5	3	1
	30	26	22	18	14	10	6		
34	31	27	23	19	15	11	7	4	2
	32	28	24	20	16	12	8		

Fig. 1. Layout of the vessel

In order to minimize the costs and inconvenience of tank cleaning, an ideal loading plan should maximize the total volume of unused tanks (i.e. free space).

Instance. The characteristics of the real instance[1] that we received from a major chemical tanker company:

- 20 cargoes with volumes ranging from 381 to 1527 tons.
- The vessel has 34 tanks with capacities from 316 to 1017 tons.
- There are 5 pairs of cargoes that cannot be placed into adjacent tanks.
- Each tank has between 1 to 3 cargoes that cannot be assigned to it.

4   A CP Model

The whole tank allocation problem is a mixed integer programming problem since the decision of which cargo is assigned to each tank is discrete but the exact volume to assign to each tank is a continuous decision. This paper only deals with the discrete problem by assigning each cargo to a set of tanks having a total capacity large enough to accommodate the whole cargo volume. The subsequent decision of the distribution of the cargo volume among those tanks must take the stability constraints into account and is beyond the scope of this paper. We just assume here that all cargo must be completely loaded. The Scala model of the problem implemented in OscaR [4] is given in Listing 1.1. In this model, two sets of variables are introduced:

- $cargo_t$: represents the type of cargo assigned to cargo tank t (type 0 represents the empty cargo). The domain of $cargo_t$ only contains cargo identifiers that can be placed into that specific cargo tank (remember that not every tank can accommodate every cargo).
- $load_c$: represents the total tank capacity available for shipping cargo c. The minimum value of $load_c$ is set to the total volume of cargo c: $volume_c$. For $load_0$ the minimum is set to 0 since there is no need to have empty cargo tanks.

---

[1] Available upon request

Listing 1.1. Scala/OscaR CP Model

```scala
class Cargo(val volume: Int)
class Tank(val id: Int, val capa: Int, val neighbors: Set[Int], val possibleCargos: Set[Int])
val cargos: Array[Cargo] // all the cargo data
val tanks: Array[Tanks] // all the tanks data
val compatibles: Set[(Int, Int)] // compatibles neighbor cargos

val cp = CPSolver()
// the cargo type for each tank (dummy if empty tank)
val cargo = Array.tabulate(tanks.length)(t => CPVarInt(cp, tanks(t).possibleCargos))
// the total capacity allocated to cargo (at least the volume to place)
val load = Array.tabulate(cargos.size)(c => CPVarInt(cp, cargos(c).volume to totCapa))
// objective = the total empty space = volume allocated to dummy
val freeSpace = load(0)
// tanks allocated to cargo c in current partial solution
def tanksAllocated(c: Int) =
    (0 until tanks.size). filter (t => (cargo(t).isBound && cargo(t).getValue == c))
// volume allocated to cargo c in current partial solution
def volumeAllocated(c: Int) = tanksAllocated(c).map(tanks(_).capa).sum
// the objective, the constraints and the search
cp.maximize(freeSpace) subjectTo {
    // links cargo and load vars with binpacking constraint
    cp.add(binpacking(cargo, tanks.map(_.capa), load), Strong)
    // new cardinality redundant constraints
    cp.add(binpackingCardinality(cargo, tanks.map(_.capa), load))
    // dominance rules constraints
    for (i <- 1 until cargos.size) {
      cp.add(new DominanceRules(cargos(i),tanks,cargo))
    }
    // two neighbor tanks, must contain compatible cargo types
    for (t <- tanks; t2 <- t.neighbors; if (t2 > t.id)) {
      cp.add(table(cargo(t.id-1),cargo(t2-1),compatibles))
    }
} exploration {
    while(!allBounds(cargo)) {
      val volumeLeft = Array.tabulate(cargos.size) (c => cargos(c).volume -
          volumeAllocated(c))
      // unbounds cargo with their index
      val unboundTanks = cargo.zipWithIndex.filter{case (x,c) => !x.isBound}
      // unbound cargo (and its index) with the largest capa, tie break on domain size
      val (tankVar,tank) = unboundTanks.maxBy{case (x,c) => (tanks(c).capa,-x.getSize)}
      // cargo with largest volume still to place that can be used in the selected tank
      val cargoToPlace = (0 until
          cargos.size). filter (tankVar.hasValue(_)).maxBy(volumeLeft(_))
      cp.branch(cp.post(tankVar == cargoToPlace)) // left branch
            (cp.post(tankVar != cargoToPlace)) // right branch
    }
  }
}
```

Example 2. Consider the Figure 2. The vessel is divided into four different tanks with capacities 500, 400, 640, 330. There are two cargoes to load (A and B). The quantity of cargo A to load is 1000 and of cargo B is 790. One bin is introduced for each of them with lower bounds $\min(load_A) = 1000$ and $\min(load_B) = 790$. The objective is to assign the tanks (items) to them such that this minimum load is met meaning that all the cargo volumes can be loaded into the tanks. An assignment of the tanks to the cargoes satisfying this requirement is given on the picture: $1040 \geq 1000$ and $830 \geq 790$.

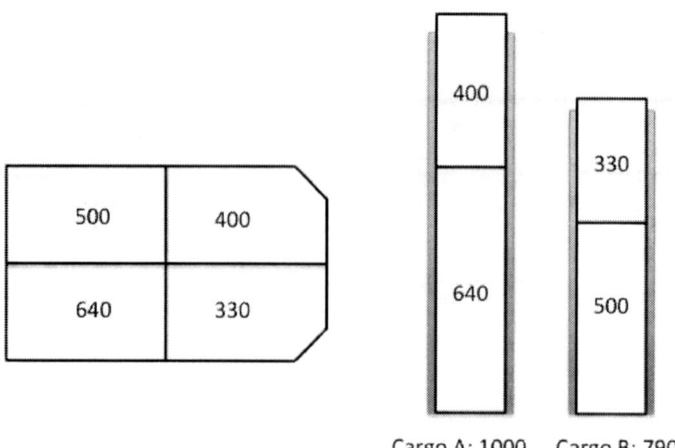

Fig. 2. Tank assignment

Assigning cargo to tanks (or tanks to cargoes in our model), is handled with the *BinPacking* constraint to express the volume requirements of each cargo. This global constraint enforces the following relation:

$$load_c = \sum_{t=1}^{nbTanks} capa_t \cdot (cargo_t = c) \ , \forall c$$

linking the two sets of variables $cargo_t$, $load_c$ and the tank capacities $capa_t$. The segregation constraints require the layout of the cargo vessels: which cargo tanks are considered adjacent and which are not. Let $\mathcal{A} \subset [1..nbTanks] \times [1..nbTanks]$ be the set of pairs of adjacent tanks and let $\mathcal{C} \subset [1..nbCargoes] \times [1..nbCargoes]$ be the set of pairs of cargoes which are compatible. We must have that

$$(cargo_{t_i}, cargo_{t_j}) \in \mathcal{C}, \quad \forall(t_i, t_j) \in \mathcal{A}.$$

These constraints are enforced with classical table constraints. The objective function of the problem is the maximization total capacity in the unused tanks: $maximize(load_0)$.

Dominance rules. Let T_c be the set of tanks allocated to a cargo c in a solution. If there exists a subset of those tanks having enough capacity to accommodate the cargo volume, then the solution is not optimal since it can be improved by allocating the subset of tanks to the cargo. More formally, a solution is not dominated if for every cargo $c \in \{1, \ldots, nbCargoes\}$, $\forall\ t' \in T_c : \sum_{t \in (T_c - t')} capa_t < volume_c$. Avoiding to generate dominated solutions can easily be achieved by implementing a dedicated propagator. As soon as the propagator realizes there is enough capacity to accommodate a cargo, this cargo value is removed from the domain of every unbound tank variable.

Heuristic. Let us define as $left_c = volume_c - \sum_{t:bound(cargo_t) \wedge cargo_t = c} capa_t$ the difference between the volume of a cargo c and the current total tank volume allocated to it. If it is positive it means that the cargo does not have enough tank volume to transport it on the vessel. If it is negative it means there is a surplus of volume for that cargo.

The variable heuristic selects the unbound variable $cargo_t$ corresponding to the tank with the largest capacity breaking ties by preferring the variable with smallest domain size.

The value heuristic tries on the left branch to assign to $cargo_t$, the cargo $c \in Dom(cargo_t)$ having the largest $left_c$ value. On the right branch, this value is removed. The idea is to use first the tanks with large capacities for the large cargo volumes, finishing down a branch with a finer granularity of tank capacities allowing more flexibility to find good feasible solutions.

Strengthening the model with lower bounds (preliminary ideas). For every cargo c, a set of tanks is allocated to it in the final solution. Let us define as $surplus_c = \sum_{t:cargo_t=c} capa_t - volume_c$ the difference between the final total tank volume allocated to a cargo c and the volume of this cargo. An interesting question is the possibility to compute a lower bound on $surplus_c$ in the final solution. A lower bound for $surplus_c$ can be found for every cargo c by solving the following sub-problem:

$$\underline{surplus}_c = \min \left(\sum_t X_t \cdot capa_t \right) - volume_c \tag{3}$$

$$s.t. : \sum_t X_t \cdot capa_t \geq volume_c \tag{4}$$

$$X_t \in \{0, 1\} \tag{5}$$

The summations in the above model are done only on the tanks that can accommodate the cargo c i.e. $\{t : c \in Dom(cargo_t)\}$. This is indeed a relaxation since a same tank can be selected for different cargo and the segregation constraints are not considered. These sub-problems can be solved with dynamic programming in pseudo-polynomial time. The resulting values can be used to strengthen the model by adding the constraints: $\forall c : load_c - volume_c \geq \underline{surplus}_c$ and also to compute an upper bound on the empty tanks volume: $\overline{load_0} \leq \sum_t capa_t - \sum_c (volume_c + \underline{surplus}_c)$.

5   Experimental Results

With the new redundant bin-packing cardinality constraint [2], the first feasible solution is easily found with just 28 backtracks (this solution uses all the tanks). Without these redundant constraints, we were not able to find any feasible solution in one hour of computation.

---

[2] The flow based propagator for the GCC should be used. A forward checking propagation for the GCC does not help on this problem.

The best solution using the Depth First Search (DFS) branch and bound (empty space = 1811) was found after 5 minutes and 1,594,159 backtracks but it was not possible to prove its optimality.

Using a Large Neighbourhood Search (LNS) on top of our model fixing 50% of the tanks randomly from the current best solution and restarting every 1000 backtracks, we were able to find a solution with an empty space of 2296 within 3 seconds and after a dozen of restarts.

We also experimented with two MIP solvers to solve this problem:

- lp_solve was not able to find any feasible solution.
- CPLEX could find and prove the optimum after 3 seconds confirming that the solution found with CP+LNS is optimal.

We plan to extend the TAP problem and also to consider (i) the maximization of the total volume to place in the case it exceeds the capacity of the vessel (ii) the integrated routing problem of a single vessel servicing multiple ports, and (iii) the stability constraints of the vessel which are non linear. We believe that those last two constraints will make it more difficult to build a MIP model. This is the reason why we developed a CP approach.

6   Conclusion

We introduced a new additional filtering algorithm for BinPacking constraint based on cardinality reasoning to count the number of items placed in each bin. This new filtering is particularly useful when a lower bound on the capacity is specified in the bins as in the TAP problem since it can immediately deduce a minimum number of items to place inside each bin. This new filtering was experimented and showed to be crucial to solve a real-life instance of a tank allocation problem with CP.

References

1. Shaw, P.: A Constraint for Bin Packing. In: Wallace, M. (ed.) CP 2004. LNCS, vol. 3258, pp. 648–662. Springer, Heidelberg (2004)
2. Dupuis, J., Schaus, P., Deville, Y.: Consistency Check for the Bin Packing Constraint Revisited. In: Lodi, A., Milano, M., Toth, P. (eds.) CPAIOR 2010. LNCS, vol. 6140, pp. 117–122. Springer, Heidelberg (2010)
3. Cambazard, H., O'Sullivan, B.: Propagating the Bin Packing Constraint Using Linear Programming. In: Cohen, D. (ed.) CP 2010. LNCS, vol. 6308, pp. 129–136. Springer, Heidelberg (2010)
4. OscaR (Scala in OR) Solver, `https://bitbucket.org/oscarlib/oscar`

The Offshore Resources Scheduling Problem: Detailing a Constraint Programming Approach

Thiago Serra, Gilberto Nishioka, and Fernando J.M. Marcellino

PETROBRAS – Petróleo Brasileiro S.A.
Avenida Paulista, 901, 01311-100, São Paulo – SP, Brazil
{thiago.serra,nishioka,fmarcellino}@petrobras.com.br

Abstract. The development of maritime oil wells depends on the availability of specialized fleet capable of performing the required activities. In addition, the exploitation of each well can only start when it is connected through pipes to a producing unit. The Offshore Resources Scheduling Problem (ORSP) combines such restrictions with the aim to prioritize development projects of higher return in oil production. In this work, an extended description of the ORSP is tackled with Constraint Programming (CP). Such extension refers to the customization of pipes and to the scheduling of fleet maintenance periods. The use of CP is due to the noted ability of the technique to model and solve large and complex scheduling problems. For the type of scenarios faced by Petrobras, the reported approach was able to quickly deliver good-quality solutions.

Keywords: Scheduling, Time-interval Variables, Well Developments.

1   Introduction

As Daniel Yergin observed, "the competition for oil and the struggle for energy security seem to never end" [24]. More than 80% of the world primary energy consumption is currently supplied by fossil fuels. Among them, oil is considered the scarcest source because it has the smallest reserves-to-production ratio [6]. Once an opportunity is detected, the exploitation of a new oil reservoir usually requires large investments, and sometimes incurs a number of technological challenges. Such is the case currently faced by Petrobras with the Tupi field, which is located in the pre-salt layer along the Brazilian coast. Announced in 2007 as the largest oil discovery of the last 30 years in the Western hemisphere, it was estimated to contain from 5 to 8 billion recoverable barrels of oil equivalent [7]. On account of enabling its exploitation, Petrobras had the largest public share offering ever seen [5]. The company now possesses a 5-year investment budget of 225 billion US dollars [20], which is mainly focused on exploiting the pre-salt layer. In this context, the optimized use of resources is critical to leverage the success of such ventures and to help attending the increasing demand of the society for energy.

One of the challenges to exploit new oil fields consists of mitigating the shortage of machinery. Resources such as a specialized fleet are highly demanded but also scarce and very expensive. That is the case of oil platforms – usually

M. Milano (Ed.): CP 2012, LNCS 7514, pp. 823–839, 2012.

referred as oil rigs – to drill and complete wells, and pipelay vessels to carry pipes to connect developed wells to producing units. The use of such resources has been approached in a variety of ways along the life cycle of oil reservoirs. For instance, the scheduling of exploratory activities for preliminary assessment was addressed by Glinz and Berumen [8]; the integrated planning and scheduling of well developments in oil fields was approached by Iyer et al. [12]; and the scheduling of well maintenance activities was focused by Aloise et al. [2]. The development of wells consists of a midpoint between the latter two cases. It is possible to address it by means of the Offshore Resources Scheduling Problem (ORSP), which is aimed at prioritizing developments with higher combined return of oil production.

Hasle et al. [9] introduced this problem, which was shown to be NP-hard by Nascimento [18]. Since then, many approaches have been refining its extent and solving methodology [17, 19, 21–23]. The decisions consist of selecting well development projects, and then assigning and sequencing resources to perform each of their activities. The optimization criterion is to maximize the incurred short-term production of oil. In addition, its constraints avoid the assignment of an unfit resource to any activity, provide time for routine maintenance of those resources, and limit their presence at any site to a maximum allowed for security reasons. Hence, solving the ORSP can aid the operational decision-making in a process that is directly related to the return on investment of oil companies.

There is a long record of approaches to the ORSP at Petrobras. Most of it is related to a system named ORCA, which stands for Optimization of Resources of Critical use in Activities for exploitation. Its development started with the work of Accioly et al. [1], which focused mostly on well maintenance activities. Their motivation was the fact that fewer resources are usually available for such type of activity, thus creating a demand-pull for the use of optimization. Pereira et al. [19] and Moura et al. [17] broadened the focus of the system to well developments. More recently, new efforts have been reported by Serra et al. [21–23] to tackle larger instances and to cope with new bottlenecks that have been recognized. Resources regarded as critical to well developments were split between rigs and vessels, and the inventory of pipes on vessels was included in [21]. The model introduced in [21] was refined and extended in [23], when the selection of development projects became a problem decision. The first attempt to provide an upper bound to the production of each instance was made in [21] and then tightened in [22]. Each of those efforts has contributed to improve the quality of the achieved solutions and to reduce their gap with respect to operational needs.

This work reports an extended description of the ORSP as well as a model to tackle it using Constraint Programming (CP). The current refinement refers to the remaining details that were included after the development of ORCA was resumed. They are related to the availability of specific pipes for connecting each well, and to the scheduling of resource maintenance activities. We aim to present a detailed description of the developed CP model and to evaluate its performance on a benchmark of instances based on a past scenario of the company.

The paper is organized as follows. Section 2 contains a definition of the ORSP. The rationale for using CP is presented in section 3, while section 4 describes the developed model and its evaluation. The benefits attributed to CP are discussed in section 5. Final remarks are presented afterwards.

2   The Offshore Resources Scheduling Problem

The ORSP consists of scheduling oil rigs and pipelay vessels to develop oil wells while targeting to maximize the incurred short-term production of oil. Rigs and vessels are required to attend to a large geographical area, in which displacements may take a considerable time. They are scheduled to perform development activities at wells, auxiliary loading activities at harbors and their own maintenance. Well development and loading activities must be assigned to a resource compatible with their needs, thus comprising requirements such as if the rig is able to operate at the depth of a given well or if the vessel is able to unload at once a given weight of pipes. Some of the development activities require pipes which, in turn, must have been previously loaded at harbors; and the amount of loading activities varies according to how many pipes are loaded at each time.

The problem decisions are *if*, *when* and *how* to perform each of those activities. Time is represented in days, starting from a date set as 0. The following convention is used to describe the problem. Resources will be denoted by index $\mathbf{i}$ assuming values in the set $\mathbf{I}$, where $\mathbf{I_R} \subseteq I$ denotes the set of rigs and $\mathbf{I_V} = I \setminus I_R$ the set of vessels. Activities will be denoted by index $\mathbf{j}$ assuming values in the set $\mathbf{J}$, where $\mathbf{J_W}$ denotes the set of well development activities, $\mathbf{J_H}$ the set of loading activities, $\mathbf{J_M}$ the set of resource maintenance activities, $J_W \cup J_H \cup J_M = J$ and these subsets are pairwise disjoint. Let $\mathbf{J_{M'}} \subseteq J_M$ be the set of resource maintenances allowing concurrency of other activities. The compatibility of assignments is represented by matrix $\mathbf{C}$, where $c_{ij} = 1$ if, and only if, resource i is compatible with activity j. Locations will be denoted by index $\mathbf{k}$ assuming values in the set $\mathbf{K}$, where $\mathbf{K_W} \subseteq K$ denotes the set of wells and $\mathbf{K_H} = K \setminus K_W$ the set of harbors. Pipes will be denoted by index $\mathbf{p}$ assuming values in the set $\mathbf{P}$.

2.1   Optimization Criterion

The short-term production is measured as how much each developed well would produce until a time horizon $\mathbf{H}$. The conclusion of each activity j induces a daily production rate $\mathbf{pr_j}$, which is nonzero only if it finishes a well development.

2.2   Resource Constraints

- Each resource performs at most one development or load at a time.
- Only one resource can be placed at a well at any time.
- Each harbor k supports up to $\mathbf{s_k}$ resources at the same time.
- If a resource i performs consecutive activities on distinct locations k_1 and k_2, there must be a minimum displacement time $\mathbf{dt_{ik_1k_2}}$ between them.
- Each resource i has a contractual period of use, ranging from the release date $\mathbf{rr_i} \geq 0$ to the deadline $\mathbf{rd_i}$.

2.3 Activity Constraints

- Activities are non-preemptive, i.e., they are performed without interruption.
- Each development or maintenance j can only be scheduled between release date $\mathbf{ar_j} \geq 0$ and deadline $\mathbf{ad_j} \geq ar_j$. It is mandatory if $ad_j < H$ or $j \in J_M$.
- Each activity j is associated with a location $\mathbf{loc_j}$.
- Either all activities of a well are performed or none is.
- Each development or maintenance j requires $\mathbf{p_j}$ days to be performed.
- The length of a loading activity on resource i ranges from $\mathbf{mil_i}$ to $\mathbf{mal_i}$ days.
- If a well development activity j_2 must be preceded by another activity j_1, then $\mathbf{pc_{j_1 j_2}} = 1$, and $\mathbf{pd_{j_1 j_2}}$ denotes the minimum delay between both.
- Each well development activity j belongs to a cluster of activities with index $\mathbf{cl_j}$, and all activities of a cluster must be assigned to a single resource.
- Each resource maintenance activity j is associated with a resource $\mathbf{rm_j}$.
- For each maintenance activity $j \in J_{M'}$, which only makes a resource partially unavailable, let $\mathbf{J'_j} \subseteq J_W$ be the set of activities that cannot concur with it.

2.4 Inventory Constraints

- Only vessels may perform loading activities at harbors.
- Each vessel has an empty inventory at time 0.
- Each vessel i has an inventory varying between 0 and $\mathbf{ic_i}$ along time.
- Each loading activity takes one or more pipes at a harbor.
- The number of loading activities is variable. However, an upper limit can be set with as much activities at each harbor as the number of pipes it has.
- A pipe p is only loaded at harbor $\mathbf{hp_p}$ from the release date $\mathbf{rp_p}$ onwards.
- The load of a pipe p implies an inventory increase of $\mathbf{wp_p}$.
- The increase of the onboard inventory due to a loading activity is limited by the time it lasts, with such a rate that it can achieve ic_i after mal_i days.
- The unload of a pipe p is made by a development activity of connection $\mathbf{ca_p}$, which can only be assigned to a resource that has previously loaded pipe p.

3 Why CP

The selection of CP in the current project stems from a balance between performance and maintainability. Many stakeholders were leaning towards the use of a commercial solver in order to benefit from a modeling environment favoring fast prototyping. It was also seen as beneficial that such solvers use state-of-art algorithms that have been broadly tested elsewhere. Since many solvers support CP or Mathematical Programming (MP), there was a strong preference for either of those techniques. There was also a great concern with the timely deliver of good solutions. A successful use of CP was reported by early approaches both inside [1] and outside [9] the company. Metaheuristics were considered as well by Nascimento [18] and for a while at Petrobras due to Pereira et al. [19] and Moura et al. [17]. However, the favorable results of a metaheuristic against CP in these latter works are somewhat related to the fact that the tested instances were not much constrained. In particular, such prevalence was not as strong when large clusters of activities had to be assigned to a single resource. Due to the introduction of loading activities, higher levels of constrainedness were expected. In addition, the observed development of propagation mechanisms for resource constraints [3, 4, 13] favored using CP to model inventory on vessels. There was also some skepticism regarding the possibility of using MP due to the increase in the size of the expected instances. In such cases, CP solvers have a competitive edge because their models are more concise and feasible solutions are found earlier in the search. Hence, we pondered that CP was a reasonable choice.

Nevertheless, both CP and MP were employed in the developments that followed. To highlight the observed progress, we will use the results achieved with the largest available instance. In accordance to Hooker [10], equal hardware conditions were observed in all cases. The first solution was obtained by a CP model described in [21], where an ad-hoc algorithm assessed a worst-case optimality gap of 38%. Such gap was then reduced to almost 18% with a continuous-time Mixed-Integer Linear Programming (MILP) model without loading activities in [22]. Such experiment was helpful to assess the scalability issues of a straightforward MP approach to a scheduling problem: compared to CP, time and memory requirements increased at a much faster step. In the following, an improved CP model with project selection in [23] was able to find a solution 11% better. Thus, the worst-case optimality gap of the largest instance tested in previous works is below 6%. Such results increased our confidence on CP, which was selected to tackle the ORSP in the full extent currently required at Petrobras.

4 How CP

The ORSP was modeled with abstractions based on conditional time-interval variables. Such concept was introduced by Laborie and Rogerie [14] and further extended by Laborie et al. [15]. It specializes the abstract modeling targeted with the Optimization Programming Language (OPL) [16] to handle resource-constrained scheduling problems. For that sake, abstractions from the specialized CP subfield of Constraint-Based Scheduling (CBS) [3, 4] such as activities and resources are defined by means of intervals and other auxiliary elements.

Each interval variable depicts an event through a collection of interdependent properties such as its presence, start time, length, and end time. Such intervals can be used to model complex relations according to a hierarchical structure imposed by one-to-many constraints as well as sequence variables, cumulative and state functions, and the constraints that can be imposed on them. Further details about those elements are presented along the model description.

4.1 Modeling the ORSP

The model described in this section is an extension of the model M_2, which was introduced in [23]. It is depicted below with separate sections for the definition of variables and domains, constraints, objective function and search phases.

Variables and Domains. Interval variables are used to represent each of the problem activities in vector **a**, and each combination of a resource and an activity requiring assignment in matrix **M**. To each activity $j \in J$, there is a corresponding interval variable a_j in vector a. In the case of well development and resource maintenance activities, the following domain restrictions apply:

$$start(a_j) \geq ar_j, \qquad \forall j \in J_W \cup J_M \quad (1)$$
$$end(a_j) \leq ad_j, \qquad \forall j \in J_W \cup J_M \quad (2)$$
$$length(a_j) = p_j, \qquad \forall j \in J_W \cup J_M \quad (3)$$
$$presence(a_j) = 1, \qquad \forall j \in J_W, ad_j < H \quad (4)$$
$$presence(a_j) = 1, \qquad \forall j \in J_M \quad (5)$$

Each cell m_{ij} of matrix M corresponds to the combination of resource $i \in I$ and activity $j \in J_W \cup J_H$. Thus, the presence of an interval m_{ij} implies the assignment of resource i to activity j. Since most of the resource-activity pairs are incompatible, the actual implementation of M is aimed to leverage such sparsity: separate tuple-indexed vectors are employed for activities on wells and activities on harbors; and each of them contains only intervals corresponding to compatible combinations, i.e., m_{ij} is there if, and only if, $c_{ij} = 1$. However, since such details are more of an implementation issue than a modeling design choice, the matrix notation was kept for simplicity. The following domain restrictions are imposed to intervals of M related to well developments:

$$start(m_{ij}) \geq MAX(rr_i, ar_j), \qquad \forall i \in I, j \in J_W, c_{ij} = 1 \quad (6)$$
$$end(m_{ij}) \leq MIN(rd_i, ad_j), \qquad \forall i \in I, j \in J_W, c_{ij} = 1 \quad (7)$$
$$presence(m_{ij}) = 0, \qquad \forall i \in I, j \in J_W, c_{ij} \neq 1 \quad (8)$$

For loading activities, the corresponding cells have the following restrictions:

$$start(m_{ij}) \geq rr_i, \qquad \forall i \in I, j \in J_H, c_{ij} = 1 \quad (9)$$
$$end(m_{ij}) \leq rd_i, \qquad \forall i \in I, j \in J_H, c_{ij} = 1 \quad (10)$$
$$length(m_{ij}) \geq mil_i, \qquad \forall i \in I, j \in J_H, c_{ij} = 1 \quad (11)$$
$$length(m_{ij}) \leq mal_i, \qquad \forall i \in I, j \in J_H, c_{ij} = 1 \quad (12)$$

Since we have considered the existence of a loading activity for each pipe, we can refer explicitly to such correspondence with a bijection $\mathbf{f} : P \rightarrow J_H$ mapping each pipe $p \in P$ to a loading activity $j \in J_H$. Without loss of generality, we can use it to postpone the early start of each loading activity according to the release date of its corresponding pipe, and thus reduce the search space:

$$start(a_j) \geq rp_p, \qquad \forall j \in J_H, p \in P, f(p) = j \quad (13)$$

Such premise also incurs that $loc_j = hp_p$, $\forall j \in J_H, p \in P, f(p) = j$, i.e., it associates each loading activity with the harbor of its corresponding pipe.

The use of each resource and the concurrency of resources on each location are represented by the vectors of cumulative functions **u** and **x**, respectively.

Cumulative functions. Cumulative functions are piecewise time-domain functions whose discrete changes in value are associated with the start and end of interval variables. They represent a generalization of cumulative resources [15]; for which many types and levels of consistency have been proposed and refined, and a number of filtering algorithms has been studied [3]. The composition of a cumulative function consists of a linear combination of steps, pulses and other cumulative functions. A step can be associated with the start and end of an interval with functions *stepAtStart* and *stepAtEnd*, respectively; and a pulse, which corresponds to equal but opposite steps at the start and at the end of an interval, is declared with function *pulse*. In either case, the first argument is an interval variable, which can be followed by one argument, a, if the variation is fixed or two arguments, a and b, if the variation must lay in the range $[a, b]$.

To each resource $i \in I$ there is an associated cumulative function u_i composed of unitary pulses associated with intervals of M, and to each location $k \in K$ there is a cumulative function x_k composed of unitary pulses on intervals of a:

$$u_i = \sum_{j \in J_W \cup J_H : c_{ij} = 1} pulse(m_{ij}, 1), \qquad \forall i \in I \quad (14)$$

$$x_k = \sum_{j \in J_W \cup J_H : loc_j = k} pulse(a_j, 1), \qquad \forall k \in K \quad (15)$$

The balance of inventory on the resources is modeled with the vector of cumulative functions **b**. There is one cumulative function b_i corresponding to each vessel $i \in I_V$. It is composed of the sum of cumulative functions related to each resource-harbor pair in the matrix **BH**. In turn, each cumulative function bh_{ik} from BH is composed of positive steps at the end of each loading activity performed by resource i at harbor k and negative steps for each activity on resource i that is associated with the release of a pipe from harbor k. The increase of inventory due to each load is variable, non-negative and limited to ic_i. The decrease of inventory is fixed and given by the weight of the unloaded pipes. Thus, we have the following definitions of the elements of BH and b:

$$bh_{ik} = \sum_{j \in J_H : c_{ij} = 1 \wedge loc_j = k} stepAtEnd(m_{ij}, 0, ic_i)$$

$$- \sum_{j \in J_W, p \in P : c_{ij} = 1 \wedge ca_p = j \wedge hp_p = k} stepAtEnd(m_{ij}, wp_p),$$

$$\forall i \in I_V, k \in K_H \quad (16)$$

$$b_i = \sum_{k \in K_H} bh_{ik}, \qquad \forall i \in I_V \quad (17)$$

Finally, the location of the resources along time is represented with the vector of state functions l. A state function represents a qualitative property that varies in time. It can be accompanied by an auxiliary function that defines the minimum transition time between its states. In the current case, let $d_i : K \times K \to \mathbb{N}$

be a function of resource $i \in I$ such that $d_i(k_1, k_2) = dt_{ik_1k_2}$, $\forall k_1 \in K, k_2 \in K, k_1 \neq k_2$. Thus, d_i corresponds to the displacement function of resource i. It can be used in combination with l_i to guarantee the minimum displacement time between activities on different locations that were assigned to resource i:

$$l_i : \mathbb{N} \to K \cup \{0\}, \qquad\qquad\qquad \forall i \in I \quad (18)$$

$$[l_i(t_1) = k_1 \wedge l_i(t_2) = k_2] \to t_1 + d_i(k_1, k_2) \le t_2,$$
$$\forall i \in I, k_1 \in K, k_2 \in K, k_1 \neq k_2,$$
$$t_1 \in \mathbb{N}, t_2 \in \mathbb{N}, t_1 < t_2 \quad (19)$$

The value 0 represents the state at which the resource is not being used anywhere.

Constraints. Vector a and matrix M are bound by constraint *alternative*, which is employed to state that at most one interval of the j-th column of M occurs and that it corresponds to interval a_j:

$$alternative(a_j, M_j), \qquad\qquad\qquad \forall j \in J_W \cup J_H \quad (20)$$

The clustering constraints are represented by logical implications, which force the presence of all development intervals of a cluster in a single line of M:

$$presence(m_{i_1j_1}) \to \neg\, presence(m_{i_2j_2}), \qquad \forall i_1 \in I, i_2 \in I, i_1 \neq i_2,$$
$$j_1 \in J_W, j_2 \in J_W, j_1 \neq j_2,$$
$$c_{i_1j_1} = 1, c_{i_2j_2} = 1, cl_{j_1} = cl_{j_2} \quad (21)$$

In order to prevent resources from being assigned to more than one activity at a time, an upper limit of 1 is set to the functions of u. Besides, the functions of l are set as equal to the location of the activities performed on each resource:

$$u_i \le 1, \qquad\qquad\qquad\qquad \forall i \in I \quad (22)$$

$$[presence(m_{ij}) = 1] \to [l_i(t) = loc_j], \qquad \forall i \in I, j \in J_W \cup J_H,$$
$$c_{ij} = 1, t \in [start(m_{ij}), end(m_{ij})) \quad (23)$$

With constraints on vector u, we can also prevent any resource from performing an activity during a period of full unavailability. That is made by constraining the value of the functions of u to 0 when such type of maintenance occurs:

$$alwaysIn(u_i, a_j, 0, 0), \qquad\qquad \forall i \in I, j \in J_M \setminus J_{M'}, rm_j = i \quad (24)$$

The constraint $alwaysIn(f, v, a, b)$ states that the value of a cumulative function f during the occurrence of an interval v must lay in the range $[a, b]$.

In the case of resource maintenance activities demanding only a partial unavailability, an auxiliary vector of cumulative functions $\mathbf{u'}$ is used to represent the use of the associated resources by a conflicting activity. Each of such functions is constrained to be 0 during the associated interval of partial unavailability:

$$u'_{j_m} = \sum_{j \in J'_{j_m} : c_{ij}=1} pulse(m_{ij}, 1), \qquad \forall i \in I, j_m \in J_{M'}, rm_{j_m} = i \quad (25)$$

$$alwaysIn(u'_{j_m}, a_{j_m}, 0, 0), \qquad\qquad\qquad \forall j_m \in J_{M'} \quad (26)$$

The precedence between pairs of activities is directly stated with constraints involving the associated intervals of a. The first constraint below guarantees the chronological order between each of such pairs, and the second one that the presence of the latter interval depends on whether the former is also present:

$$end(a_{j_1}) + pd_{j_1 j_2} \leq start(a_{j_2}), \qquad \forall j_1 \in J_W, j_2 \in J_W, pc_{j_1 j_2} = 1 \quad (27)$$

$$presence(a_{j_2}) \rightarrow presence(a_{j_1}), \qquad \forall j_1 \in J_W, j_2 \in J_W, pc_{j_1 j_2} = 1 \quad (28)$$

The constraints to force the entire development of a well or its absence are defined between one of the first activities of each well and each of the remaining activities. Let $J_F \subseteq J_W$ be the set of the first development activities, i.e., those which are not preceded by other activities of the same well. Since it is possible that a well k has more than one activity in JF, let $J_{F1} \subseteq J_F$ be a set containing exactly one of such activities of each well. Each pair of activities from a well such that one belongs to J_{F1} and the other to $J_{NF1} = J_W \setminus J_{F1}$ are then both present or absent with the following constraint:

$$presence(a_{j_1}) = presence(a_{j_2}), \qquad \forall j_1 \in J_{F1}, j_2 \in J_{NF1}, loc_{j_1} = loc_{j_2} \quad (29)$$

In order to limit the concurrency on wells and harbors, the cumulative functions of vector x are upper limited according to the type of each location:

$$x_k \leq 1, \qquad \forall k \in K_W \quad (30)$$

$$x_k \leq s_k, \qquad \forall k \in K_H \quad (31)$$

Similarly, the upper and lower limits of inventory on each pipelay vessel are imposed with constraints upon b and BH, respectively:

$$b_i \leq ic_i, \qquad \forall i \in I_V \quad (32)$$

$$bh_{ik} \geq 0, \qquad \forall i \in I_V, k \in K_H \quad (33)$$

In the case of loading activities, the inventory increase due to each interval is limited by its length. Thus, the following constraint limits the increase of inventory associated with each interval from definition (16):

$$heightAtEnd(bh_{ik}, m_{ij}) \leq ic_i * \frac{length(m_{ij})}{mal_i},$$

$$\forall i \in I_V, j \in J_H, k \in K_W,$$

$$c_{ij} = 1, loc_j = k \quad (34)$$

The function $heighAtEnd(f, v)$ represents the variation of cumulative function f caused only by $stepAtEnd$ functions involving interval v.

The inventory is also constrained to be empty before a new load at a harbor:

$$alwaysIn(bh_{ik}, m_{ij}, 0, 0), \quad \forall i \in I_V, j \in J_H, k \in K_H, c_{ij} = 1, loc_j = k \quad (35)$$

With constraint (35), the shipment of each pipe p is always assigned to the last loading activity performed by a resource at the harbor of the pipe before the development activity of connection ca_p. In order to facilitate the satisfaction of (35), the implementation of (16) contains a variable negative step at the start of each load. Such step is aimed to perform small corrections in decisions regarding the weight loaded on previous activities, thus avoiding an excessive use of retraction during the search. Since that was only necessary for performance improvement on current solvers, we do not regard it as part of the model. For an explanation about the relation between search and the propagation of resource constraints, the interested reader is referred to the work of Laborie [13].

In order to reduce the search space associated with the number of loading activities, the bijection f can be used to define a symmetry breaking constraint. Without loss of generality, we can state that a well connection j_w is present if the associated loading activity j_h is also present:

$$presence(a_{j_h}) \rightarrow presence(a_{j_w}), \qquad \forall j_h \in J_H, j_w \in J_W, p \in P,$$
$$f(p) = j_h, ca_p = j_w \quad (36)$$

In order to avoid the assignment of a pipe to a loading activity scheduled before its release date, it was necessary to define an additional vector $\mathbf{z}$. Each cumulative function z_i represents the maximum date of release of a pipe that is carried by resource i along time. One may observe that such cumulative function is actually representing a qualitative property, and that qualitative properties are usually modeled by means of state functions. The rationale for such design choice is that the syntax of cumulative functions is more appropriate in this case, since it enables to constrain that a property only changes at specific moments. Given that the variety of pipes that can be loaded is only altered when loading activities are performed, the change of the maximum date of release of pipes carried by a resource can only occur at the start of loading activities.

Each function z_i is composed in (37) of steps at the start of each loading activity performed by resource i. Its value is constrained by (38) to be always equal to the date of release of the pipe p associated by bijection f with the loading activity j that is performed, i.e., $z_i = rp_p$ for $f(p) = j$ when performing m_{ij}. Thus, the value of each function is always between 0 and **mard**, which is defined as the maximum date of release among all pipes. It is worth observing that the step due to each load is set to vary from $-mard$ to $+rp_p$ in order to account for the state of the resource right before the start of each load:

$$z_i = \sum_{j \in J_H, p \in P : c_{ij} = 1 \land f(p) = j} stepAtStart(m_{ij}, -mard, +rp_p),$$
$$\forall i \in I_V \quad (37)$$

$$alwaysIn(z_i, m_{ij}, rp_p, rp_p), \qquad \forall i \in I_V, j \in J_H, p \in P,$$
$$c_{ij} = 1, f(p) = j \quad (38)$$

Hence, it is possible to constrain each development activity of connection ca_p of a pipe p to be scheduled only when pipe p could have been previously loaded:

$$alwaysIn(z_i, m_{ij}, rp_p, mard), \qquad \forall i \in I_V, j \in J_W, p \in P,$$
$$c_{ij} = 1, ca_p = j \quad (39)$$

It is worth of notice that constraint (39) forces that the load of a pipe p can only be made by an activity j associated by f with a pipe p' such that $rp_{p'} \geq rp_p$. It may appear that such constraint is more restrictive than necessary. However, theorem 1 guarantees that any solution can be represented with the model, and thus that (39) also serves as a symmetry breaking constraint for that reason.

Lemma 1. *Given an ORSP solution S that does not comply with constraint (39), there is always a solution S' equivalent to S with less violations to (39).*

Proof. The following operation can be applied to a solution S to obtain an equivalent solution S' such that the number of pipes of latest release date assigned to loading activities forbidden by (39) is reduced in at least one unit. Suppose that S is a solution that violates such constraint, and that p_1 is the pipe with the latest release date which is assigned to a loading activity j_2 such that $f(j_2) = p_2$ and $rp_{p_2} < rp_{p_1}$. Since there is one loading activity associated with each pipe from the same harbor hp_{p_1} with release date rp_{p_1} onwards and one of such pipes, p_1, is not assigned to neither of them, at least one of such activities, j_1, is either absent or assigned to load pipes with earlier release dates. Thus, there is an equivalent solution S', which differs from S by switching the assignment and schedule of activities j_1 and j_2, and therefore it has less pipes with the same release date as p_1 assigned to a loading activity forbidden by (39) than S has.

Theorem 1. *Constraint (39) does affect the correctness of the ORSP model.*

Proof. After a finite number of applications of the operation above, an equivalent solution satisfying (39) can be achieved from any ORSP solution.

Objective Function. The objective function represents the expected short-term production that would be accumulated from day 0 to day H with the schedule. It is depicted as the summation of the production rate triggered by the finish of each activity in the schedule multiplied by the time left until H:

$$\text{maximize} \qquad \sum_{j \in J_W} MAX(H - end(a_j), 0) * pr_j \qquad (40)$$

Search Phases. The use of search phases is aimed to order the sets of variables of a problem in order to reduce the search effort. As a matter of fact, much of the work of ordering variables and values for assignment still remains to the solver in such a case. However, an outline of search phases according to the modeler point of view can help leveraging domain-specific knowledge to solve the problem.

In the ORSP model, we observed that it was better to assign first the intervals of a related to resource maintenance activities, starting with those demanding full unavailability. The rationale for such strategy is that such intervals are those with mandatory presence. Therefore, it is very likely that a delay in their assignments would cause more conflicts during the search. Besides, we relied on the premise that full unavailability periods are less sensitive to the assignment of activities to each resource, since they do not allow any concurrence.

4.2   Testing the ORSP Model

The experimental evaluation conducted to test the model was based on data from a past scenario of the company. It comprises the activities to develop 171 wells using 73 resources. Such data was used to generate a number of instances by splitting the set of wells, and thus also splitting the corresponding sets of activities and pipes. For confidentiality and to facilitate comparisons, the value of the best solution found for each instance was used to define 100 in an arbitrary scale. In what follows, we will describe the instances and the experiments performed with them, present the results of the tests and discuss those results.

Experiment. The model was tested using a set of instances first described in [21]. Instance O contains the entire set of activities of the past scenario used, which is partitioned approximately into halves for instances $H1$ and $H2$ as well as into quarters for instances $Q1$ to $Q4$. The same set of resources is considered in all cases. It is worth observing that those instances contain data regarding the details that were only introduced in the current work. Nevertheless, such data were ignored on the experiments of previous approaches. Table 1 summarizes how many activities, wells, pipes, rigs and vessels each instance has.

Table 1. Main characteristics of the tested instances

Instance	$Q1$	$Q2$	$Q3$	$Q4$	$H1$	$H2$	O
Activities	116	118	116	115	231	234	465
Wells	46	37	45	43	82	89	171
Pipes	17	17	13	19	32	34	66
Rigs				64			
Vessels				9			

The model was implemented using the OPL language and run using IBM Cplex Studio 12.2 with the CP Optimizer solver [11]. For each instance, it was run four times with different random seeds. The only modification to the standard solving parameters in such runs was that the cumulative function inference level was set as extended. The upper bound of each instance was set by running the MILP model M_F described in [22] on the Cplex solver with an empty schedule as starting solution. To achieve a tighter estimation, an additional constraint was defined to limit the start of each connection activity j, which is represented in M_F by variable S_j, according to the date of release of the pipes it must unload:

$$S_j \geq rp_p, \qquad\qquad\qquad \forall j \in J_W, p \in P, ca_p = j \quad (41)$$

The time limit of each run of either model was set as one hour in accordance to the end user expectation. The computer used had 4 Dual-Core AMD Opteron 8220 processors, 16 Gb of RAM and a Linux operating system.

Results. The results of the runs are summarized in table 2. For each instance, it comprises the average (μ) and the standard deviation (σ) of the time to find the first solution, the production of such solution and the production of the last solution found according to a scale where the best solution found is 100. The upper bound corresponds to the optimal solution of the relaxed model M_F for instances $Q1$ to $Q4$ and $H1$, and to the upper limit reached at the solver halt for instances $H2$ and O. Since greater performance variations were observed only for instance O, the progress of each run to solve that instance is presented in figure 1. Such progress is depicted in terms of the best solution found along time.

Table 2. Summary of test results for each instance

Instance	First solution found				Last solution		Upper bound	
	Time (s)		Value		Value		Value	Worst-case
	μ	σ	μ	σ	μ	σ		opt. gap
Q1	0.44	0.02	99.1	0.3	100.0	0.0	100.1	0.1%
Q2	0.42	0.01	99.0	0.5	100.0	0.0	100.4	0.4%
Q3	0.60	0.02	99.4	0.2	100.0	0.0	100.1	0.1%
Q4	0.45	0.01	98.7	0.7	100.0	0.0	100.2	0.2%
H1	1.53	0.07	94.9	1.0	99.8	0.2	100.2	0.2%
H2	1.58	0.29	91.7	1.3	99.8	0.1	100.8	0.8%
O	4.57	0.24	83.2	1.5	98.9	1.5	107.2	7.2%

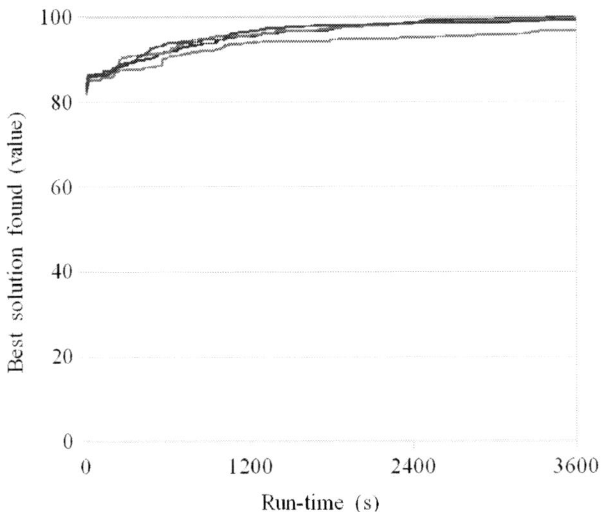

Fig. 1. Best solution found along time for independent runs on instance O

Discussion. The results indicate that our approach was able to deliver good solutions within the time limit set by the end user. A worst-case optimality gap of less than 1% was achieved for all but the largest instance. For the scheduling of the whole set of activities, the gap was 7.2%. It is worth observing that about two thirds of the gap between the first solution and the upper limit vanished after one hour of search. In addition, figure 1 shows that there is a similar trend of improvement for most of the runs. Altogether, such observations evidence the system reliability to handle scenarios like those expected by the company.

Some scalability factors were observed as well. With a fixed time limit, the major impact noticed in solution quality is due to the number of activities that each instance has. The size of the instances also influenced the time and the quality of the first solution found. Among instances of similar size, the tightest gaps were achieved for the instances with a relatively small number of pipes, such as $Q3$ and $H1$. Therefore, the numbers of activities and pipes can be used for a preliminary assessment of instance hardness.

5   Benefits from CP

The greatest advantage sensed with the use of CP in this project was the easiness of prototyping. We were able to develop most of the project with only two professionals in full time dedication, who could deliver new and fully tested versions on a monthly basis. It was crucial that new models were available for evaluation as soon as possible for at least two reasons. First, many of the refinements asked each time by the clients of the project were attempts to describe their process. Second, it was often hard to gather an expressive amount of end users together to discuss the next steps of the development. In this context, the use of concise and easily maintainable models made it possible to work with a continuous deployment of new versions of the system. Hence, in spite of the number of rounds required to achieve a consensus of the problem definition, the short development cycle was essential to the conclusion and success of the project.

The end users showed a great interest in the results and in the technology beneath the system. That was caused by the positive prospect presented at the beginning of the project, which would facilitate a lot their daily work. Many of the questions that they raised referred to their expectation of what a good solution would be, and thus provided guidance to our work. In some of those occasions, they also wanted to understand how the solving process works. Since all of them were engineers and some had an IT background, the outline of the CP framework was easily assimilated. Nevertheless, there was a certain reluctance in accepting the use of a declarative paradigm if, when compared to alternatives like MP, CP was much more focused on feasibility than on optimality. Such opposition could be summarized by the idea that one should only give up the procedural control of a process if the incurred result was guaranteed to be the best possible one. However, the ability presented by the system to deal with large and overconstrained scheduling scenarios due to CP propagation mechanisms came to be considered important as well. In addition, some experiments like those reported in [23] helped to illustrate that it would not be worth to always assume certain assumptions regarding how the problem used to be solved manually. Those assumptions would require imposing unnecessary constraints, which were shown to harm considerably the long-term results on larger instances. With time, we were able to evaluate the solutions of the models through their rationale, and they were able to suggest modifications to the model at a more technical level.

6   Conclusion

This work approached the scheduling of an specialized fleet of oil rigs and pipelay vessels to develop offshore oil wells with the introduction of real-world constraints that have never been considered before. Two direct benefits have been observed by automating decisions related to the use of such resources. First, it reduces the burden over the professionals involved with the control of expensive and highly required machinery. If circumstances change, the time to perform a quick reschedule has passed from days to minutes. Second, a better return in terms of oil production can be pursued by generating and comparing many schedules. Experimental results indicate that a solution with worst-case optimality gap of 7.2% was achieved for a past scenario of the company. In addition, the anticipation of the development of a single well for an oil company usually would be enough to cover the expenses of an entire project like the present one. Hence, this article represents a detailed account on how to approach a type of problem that may interest many capital-intensive industries.

The difficulty to solve some optimization problems is often due to the lack of tools capable of leveraging the specificities of each application domain. We noticed that there is much to be gained by considering the use of Constraint Programming (CP) in such cases, for which reason the use of the technique is being considered in other projects within the company. As a consequence, this work has also focused on the rationale for using the technique as well as on the factors that led to the conclusion that other alternatives were not so appropriate. In a nutshell, CP allows tackling hard problems of large scale without giving up of a systematic approach, thus keeping the chance of finding an optimal solution. That is especially true for scheduling problems, which are quite common in many companies and are usually difficult to handle in such a way.

Acknowledgments. The authors gratefully acknowledge Petrobras for authorizing the publication of the information here present. In addition, the opinions and concepts presented are the sole responsibility of the authors.

We also thank M.I. Harris, S. Harris, D. Ferber, A. Ciré and the anonymous reviewers for their valuable suggestions to improve the quality of this paper.

References

1. Accioly, R., Marcellino, F.J.M., Kobayashi, H.: Uma Aplicação da Programação por Restrições no Escalonamento de Atividades em Poços de Petróleo. In: Proceedings of the 34th Brazilian Symposium on Operations Research, Rio de Janeiro, Brazil (2002)

2. Aloise, D.J., Aloise, D., Rocha, C.T.M., Ribeiro, C.C., Filho, J.C.R., Moura, L.S.S.: Scheduling Workover Rigs for Onshore Oil Production. Discrete Applied Mathematics 154, 695–702 (2006)
3. Baptiste, P., Laborie, P., Le Pape, C., Nuijten, W.: Constraint-Based Scheduling and Planning. In: Rossi, F., van Beek, P., Walsh, T. (eds.) Handbook of Constraint Programming. Elsevier, Amsterdam (2006)
4. Barták, R., Salido, M.: Constraint satisfaction for planning and scheduling problems. Constraints Special Issue 16(3) (2011)
5. BBC. Petrobras raises \$70bn in record share offering (2010), http://www.bbc.co.uk/news/business-11400277 (accessed: March 13, 2012)
6. British Petroleum: BP Statistical Review of World Energy (June 2011), http://bp.com/statisticalreview (accessed: March 13, 2012)
7. Fick, J.: Petrobras Pumps First Crude from Massive Tupi Field Offshore Brazil. Rigzone (2009), http://www.rigzone.com/news/article.asp?a_id=75679 (accessed: March 13, 2012)
8. Glinz, I., Berumen, L.: Optimization Model for an Oil Well Drilling Program: Mexico Case. Oil and Gas Business 1 (2009)
9. Hasle, G., Haut, R., Johansen, B., Ølberg, T.: Well Activity Scheduling - An Application of Constraint Reasoning. In: Artificial Intelligence in the Petroleum Industry: Symbolic and Computational Applications II, pp. 209–228. Technip, Paris (1996)
10. Hooker, J.N.: Testing Heuristics: We Have It All Wrong. Journal of Heuristics 1, 33–42 (1995)
11. IBM: ILOG CPLEX Optimization Studio 12.2 documentation for ODM Enterprise (2010)
12. Iyer, R., Grossmann, I.E., Vasantharajan, S., Cullick, A.S.: Optimal Planning and Scheduling of Offshore Oil Field Infrastructure Investment and Operations. Industrial & Engineering Chemistry Research 37, 1380–1397 (1998)
13. Laborie, P.: Algorithms for propagating resource constraints in AI planning and scheduling: Existing approaches and new results. Artificial Intelligence 143, 151–188 (2003)
14. Laborie, P., Rogerie, J.: Reasoning with Conditional Time-Intervals. In: Proceedings of the 21st International Florida Artificial Intelligence Research Society Conference, Coconut Grove, USA (2008)
15. Laborie, P., Rogerie, J., Shaw, P., Vilím, P.: Reasoning with Conditional Time-Intervals, Part II: An Algebraical Model for Resources. In: Proceedings of the 22nd International Florida Artificial Intelligence Research Society Conference, Sanibel Island, USA (2009)
16. Lustig, I.J., Puget, J.-F.: Program Does Not Equal Program: Constraint Programming and Its Relationship to Mathematical Programming. Interfaces 31, 29–53 (2001)
17. Moura, A.V., Pereira, R.A., de Souza, C.C.: Scheduling Activities at Oil Wells with Resource Displacement. International Transactions in Operational Research 25, 659–683 (2008)
18. do Nascimento, J.M.: Ferramentas Computacionais Híbridas para a Otimização da Produção de Petróleo em Águas Profundas. Master thesis. Universidade Estadual de Campinas (Unicamp), Brazil (2002)
19. Pereira, R.A., Moura, A.V., de Souza, C.C.: Comparative Experiments with GRASP and Constraint Programming for the Oil Well Drilling Problem. In: Nikoletseas, S.E. (ed.) WEA 2005. LNCS, vol. 3503, pp. 328–340. Springer, Heidelberg (2005)

20. Petrobras: Plano de Negócios 2011-2015 (2011),
    `http://www.petrobras.com.br/pt/quem-somos/estrategia-corporativa/`
    `downloads/pdf/plano-negocios.pdf` (accessed: March 13, 2012)
21. Serra, T., Nishioka, G., Marcellino, F.J.M.: A Constraint-Based Scheduling of Off-shore Well Development Activities with Inventory Management. In: Proceedings of the 43rd Brazilian Symposium on Operations Research, Ubatuba, Brazil (2011)
22. Serra, T., Nishioka, G., Marcellino, F.J.M.: On Estimating the Return of Resource Aquisitions through Scheduling: An Evaluation of Continuous-Time MILP Models to Approach the Development of Offshore Oil Wells. In: Proceedings of the 6th Scheduling and Planning Applications woRKshop (SPARK), ICAPS, Atibaia, Brazil (2012)
23. Serra, T., Nishioka, G., Marcellino, F.J.M.: Integrated Project Selection and Resource Scheduling of Offshore Oil Well Developments: An Evaluation of CP Models and Heuristic Assumptions. In: Proceedings of the Constraint Satisfaction Techniques for Planning and Scheduling Problems Workshop (COPLAS), ICAPS, Atibaia, Brazil (2012)
24. Yergin, D.: The Prize: The Epic Quest for Oil, Money & Power. Free Press, New York (2008)

Computational Protein Design as a Cost Function Network Optimization Problem

David Allouche[1,*], Seydou Traoré[2,*], Isabelle André[2], Simon de Givry[1], George Katsirelos[1], Sophie Barbe[2,**], and Thomas Schiex[1,**]

[1] UBIA, UR 875, INRA, F-31320 Castanet Tolosan, France
[2] LISBP, INSA, UMR INRA 792/CNRS 5504, F-31400 Toulouse, France
`Thomas.Schiex@toulouse.inra.fr`, `Sophie.Barbe@insa-toulouse.fr`

Abstract. Proteins are chains of simple molecules called amino acids. The three-dimensional shape of a protein and its amino acid composition define its biological function. Over millions of years, living organisms have evolved and produced a large catalog of proteins. By exploring the space of possible amino-acid sequences, protein engineering aims at similarly designing tailored proteins with specific desirable properties. In Computational Protein Design (CPD), the challenge of identifying a protein that performs a given task is defined as the combinatorial optimization problem of a complex energy function over amino acid sequences.

In this paper, we introduce the CPD problem and some of the main approaches that have been used to solve it. We then show how this problem directly reduces to Cost Function Network (CFN) and 0/1LP optimization problems. We construct different real CPD instances to evaluate CFN and 0/1LP algorithms as implemented in the `toulbar2` and `cplex` solvers. We observe that CFN algorithms bring important speedups compared to the CPD platform `osprey` but also to `cplex`.

1 Introduction

A protein is a sequence of basic building blocks called amino acids. Proteins are involved in nearly all structural, catalytic, sensory, and regulatory functions of living systems [11]. Performance of these functions generally requires the assembly of proteins into well-defined three-dimensional structures specified by their amino acid sequence. Over millions of years, natural evolutionary processes have shaped and created proteins with novel structures and functions by means of sequence variations, including mutations, recombinations and duplications. Protein engineering techniques coupled with high-throughput automated procedures offer today the possibility to mimic the evolutionary process on a greatly accelerated time-scale, and thus increase the odds to identify the proteins of interest for technological uses [29]. This holds great interest for medicine, biotechnology, synthetic biology and nanotechnologies [27,32,15].

---

* These authors contributed equally to this work.
** Corresponding authors.

M. Milano (Ed.): CP 2012, LNCS 7514, pp. 840–849, 2012.

With a choice among 20 naturally occuring amino acids at every position, the size of the combinatorial sequence space is however clearly out of reach of current experimental methods, even for small proteins. Computational protein design (CPD) methods therefore try to intelligently guide this process by producing a collection of proteins, intended to be rich in functional proteins and whose size is small enough to be experimentally evaluated. The challenge of choosing a sequence of amino acids to perform a given task is formulated as an optimization problem, solvable computationally. It is often described as the inverse problem of protein folding [28]: the three-dimensional structure is known and we have to find amino acid sequences that folds into it. It can also be considered as a highly combinatorial variant of side-chain positioning [35] because of possible amino acid changes.

Different computational methods have been proposed over the years to solve this problem and several success stories have demonstrated the outstanding potential of CPD methods to engineer proteins with improved or novel properties. CPD has been successfully applied to increase protein thermostability and solubility; to alter specificity towards some other molecules; and to design various binding sites and construct *de novo* enzymes (see for example [18]).

Despite these significant advances, CPD methods still have to mature in order to better guide and accelerate the construction of tailored proteins. In particular, more efficient computational optimization techniques are needed to explore the vast protein sequence-conformation combinatorial space.

In this paper, we model CPD problems as either binary Cost Function Network (CFN) or 0/1LP problems. We compare the performance of the CFN solver `toulbar2` and the 0/1LP solver `cplex` against that of well-established CPD approaches on various protein design problems. On the various problems considered, the direct application of `toulbar2`, a Depth First Branch and Bound algorithm maintaining soft local consistencies, resulted in an improvement of several orders of magnitude compared to dedicated CPD methods and also outperformed `cplex`. These preliminary results can probably be further improved both by tuning our solver to the specific nature of the problem considered and by incorporating dedicated CPD preprocessing methods.

2   The Computational Protein Design Approach

In CPD, we are given an existing protein corresponding to a native sequence of amino acids folded into a 3D structure, which has previously been determined experimentally. The task consists in modifying a given property of the protein (such as stability or functional efficiency) through the mutation of a specific subset of amino acid residues in the sequence, i.e. by affecting their identity and their 3D orientation (rotamers). The resulting designed protein retains the overall folding of the original protein since we consider the protein *backbone* as fixed and only alter the amino acid *side chains* (Fig. 1). The stability and functional efficiency of a protein is correlated to its energy [1]. Therefore, we aim at finding the conformation possessing the minimum total energy, called *GMEC*

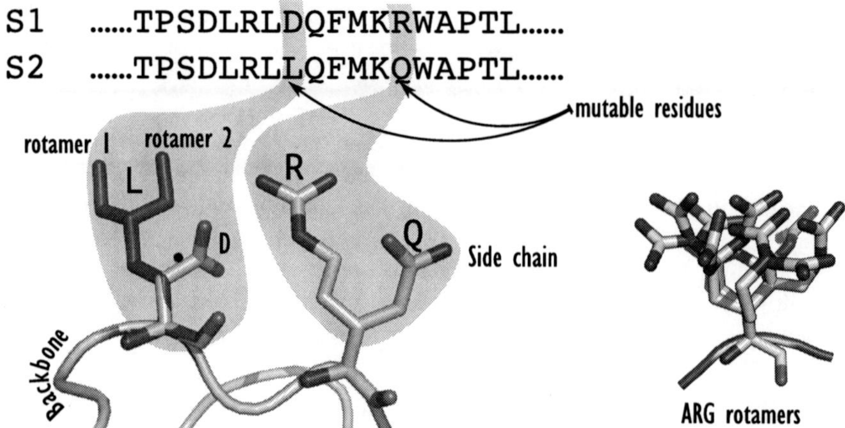

Fig. 1. A local view of combinatorial sequence exploration considering a common backbone. Changes can be caused by amino acid identity substitutions (*for example D/L or R/Q*) or by amino acid side-chain reorientations (rotamers) for a given amino acid. A typical rotamer library for one amino acid is shown on the right (ARG=Arginine).

(Global Minimum Energy Conformation). The energy of a conformation can be directly computed from the amino acid sequence and rotamers by introducing substitutions within the native structure.

Rotamers. The distribution of accessible conformations available to each amino acid side chain is approximated using a set of discrete conformations defined by the value of their inner dihedral angles. These conformations, or *rotamers*, are derived from the most frequent conformations in the experimental repository of known protein structures PDB (Protein Data Bank, `www.wwpdb.org`).

Energy function. Typical energy function approximations [3] use the assumption that the amino acid identity substitutions and rotamers do not modify the folding of the protein. They include non-bonded terms such as van der Waals and electrostatics, often in conjunction with empirical contributions describing hydrogen bond. The surrounding solvent effect is generally treated implicitly as a continuum. In addition, statistical terms may be added in order to approximate the effect of mutations on the unfolded state or the contribution of conformational entropy.

These energy functions can be reformulated in such a way that the terms are locally decomposable. Then, the energy of a given protein defined by a choice of one specific amino acid with an associated conformation (rotamer) for each residue, can be written as:

$$E = E_c + \sum_i E(i_r) + \sum_i \sum_{j>i} E(i_r, j_s) \tag{1}$$

where E is the potential energy of the protein, E_c is a constant energy contribution capturing interactions between fixed parts of the model, $E(i_r)$ is the self energy of rotamer r at position i capturing internal interactions or with fixed regions, and $E(i_r, j_s)$ is the pairwise interaction energy between rotamer r at position i and rotamer s at position j [9]. All terms are measured in $kcal/mol$ and can be pre-computed and cached.

3   Existing Approaches for the CPD

The protein design problem as defined above, with a rigid backbone, a discrete set of rotamers, and pairwise energy functions has been proved to be NP-hard [31]. Hence, a variety of meta-heuristics have been applied to it, including Monte Carlo simulated annealing [21], genetic algorithms [33], and other algorithms [10]. The main weakness of these approaches is that they may remain stuck in local minima and miss the GMEC without notice.

However, there are several reasons motivating the exact solving of the problem. First, because they know when an optimum is reached, exact methods may stop before metaheuristics. Voigt et al. [36] reported that the accuracy of meta-heuristics also degrades as problem size increases. More importantly, the use of exact search algorithms becomes crucial in the usual experimental design cycle that goes through CPD modeling, solving, protein synthesis and experimental evaluation: when unexpected experimental results are obtained, the only possible culprit lies in the CPD model and not in the algorithm.

Current exact methods for CPD mainly rely on the dead-end-elimination (DEE) theorem [9,8] and the A^* algorithm [24,13]. From a constraint satisfaction perspective, the DEE theorem can be seen as an extension of neighborhood substitutability [7,20,2]. DEE is used as a pre-processing technique and removes rotamers that are locally dominated by other rotamers, until a fixpoint is reached. The rotamer r at position i is removed if there exists another rotamer u at the same position such that [9]:

$$E(i_r) - E(i_u) + \sum_{j \neq i} \min_s E(i_r, j_s) - \sum_{j \neq i} \max_s E(i_u, j_s) > 0$$

That is, r is removed if for any conformation with this r, we get a conformation with lower energy if we substitute u for r.

Extensions to higher orders have been considered [14,30,25,12]. These DEE criteria preserve the optimum but may remove suboptimal solutions.

This DEE preprocessing is usually followed by an A^* search method. After DEE pruning, the A^* algorithm allows to expand a sequence-conformation tree, so that sequence-conformations are extracted and sorted on the basis of their energy values. At depth d of the tree, the lower bound used by $A*$ [13] is exactly the PFC-DAC lower bound [37,23] used in WCSP and later obsoleted by soft arc consistencies [34,22,5]:

$$\underbrace{\sum_{i=1}^{d} E(i_r) + \sum_{j=i+1}^{d} E(i_r, j_s)}_{\text{Assigned}} + \sum_{j=d+1}^{n} \Big[\underbrace{\min_s \big(E(j_s) + \sum_{i=1}^{d} E(i_r, j_s) }_{\text{Forward checking}} + \underbrace{\sum_{k=j+1}^{n} \min_u E(j_s, k_u) \big) \Big]}_{\text{DAC counts}}$$

If the DEE algorithm does not significantly reduce the search space, the A^* search tree is too memory demanding and the problem cannot be solved. Therefore, to circumvent these limitations and increase the ability of CPD to tackle problems with larger sequence-conformation space, novel alternative methods are needed. Here, we show that state-of-the-art methods for solving Cost Function Networks offer an attractive alternative to this combined DEE/A^* approach, to solve highly complex case studies of protein design.

4   Cost Function Network Model

A Cost Function Network (CFN) is a pair (X, W) where $X = \{1, \ldots, n\}$ is a set of n variables and W a set of cost functions. Each variable $i \in X$ has a finite domain D_i of values than can be assigned to it. A value $a \in D_i$ is denoted i_a. For a set of variables $S \subseteq X$, D_S denotes the Cartesian product of the domain of the variables in S. For a given tuple of values t, $t[S]$ denotes the projection of t over S. A cost function $w_S \in W$, with scope $S \subseteq X$, is a function $w_S : D_S \mapsto [0, k]$ where k is a maximum integer cost used for forbidden assignments. The Weighted Constraint Satisfaction Problem (WCSP) is to find a complete assignment t minimizing the combined cost function $\sum_{w_S \in W} w_S(t[S])$. This optimization problem has an associated NP-complete decision problem.

Modeling the CPD problem as a CFN is straightforward. The set of variables X has one variable i per residue i. The domain of each variable is the set of *(amino acid,conformation)* pairs in the rotamer library used. The energy function can be represented by 0-ary, unary and binary cost functions respectively capturing the constant energy term E_c, the unary energy terms $E(i_r)$ and the binary energy terms $E(i_r, j_s)$. There is just one discrepancy between the original formulation and the CFN model: energies are represented as arbitrary floating point numbers while CFN use positive integer costs. This can simply be fixed by first subtracting the minimum energy to all energies and then by multiplying energies by a large integer constant M.

5   Integer Linear Programming Model

The resulting CFN can also be represented as a 0/1 linear programming problem using the encoding proposed in [20]. For every value i_r, there is a boolean variable $d_{i,r}$ which is equal to 1 iff $i = r$. Additional constraints enforce that exactly one value is selected for each variable. For every pair of values of different variables (i_r, j_s) involved in a binary energy term, there is a boolean variable $p_{i,r,j,s}$ which is equal to 1 iff the pair (i_r, j_s) is used. Constraints enforce that a pair is used iff the corresponding values are used. Then, finding a GMEC reduces to the following ILP:

$$\min \sum_{i,r} E(i_r).d_{i,r} + \sum_{i,r,j,s} E(i_r, j_s).p_{i,r,j,s}$$

$$\text{s.t.} \sum_r d_{i,r} = 1 \qquad\qquad (\forall i)$$
$$\sum_s p_{i,r,j,s} = d_{i,r} \qquad\qquad (\forall i, r, j)$$

This model is also the ILP model IP1 proposed in [19] for side-chain positioning. The continuous relaxation of this 0/1 linear programming model is known do be the dual of the LP problem encoded by Optimal Soft Arc Consistency [6,5]. When the upper bound k is infinite, OSAC is known to be stronger than any other soft "arc level" arc consistency and especially stronger than the default Existential Directional Arc Consistency (EDAC) [22] used in `toulbar2`. However, as soon as the upper bound k decreases to a finite value, soft local consistencies may prune values and EDAC becomes incomparable with OSAC.

6   Experimental Results

We used a set of 12 protein design cases to evaluate the performance of `toulbar2`, `cplex` and compare them with the DEE/A* approach implemented in `osprey` (open source dedicated Java CPD software). This set comprises 9 protein structures derived from the PDB which were chosen for the high resolution of their 3D-structures and their distribution of sizes and types. Diverse sizes of sequence-conformation combinatorial spaces were considered, varying by the number of mutable residues, the number of alternative amino acid types at each position and the number of conformations for each amino acid (Table 1). The *Penultimate* rotamer library was used [26].

Preparation of CPD instances. Missing heavy atoms in crystal structures and hydrogen atoms were added with the *tleap* module of the AMBER9 software package [4]. Each molecular system was then minimized in implicit solvent (Generalized Born model [17]) using the *Sander* program and the all-atom *ff99* force field of AMBER9. All E_c, $E(i_r)$, and $E(i_r, j_s)$ energies of rotamers (see Equation 1) were pre-computed using `osprey`. The energy function consisted of the Amber electrostatic, van der Waals, and dihedral terms. These calculations were performed on an Altix ICE 8200 supercomputer with 2,816 Intel Nehalem EX 2.8 GHz cores. We used 32 cores and 128GB of RAM. The sequential CPU time needed to compute the set of all energy cost functions is given in Table 1. Although these computation times can be very large, they are also highly parallelizable. For n residues to optimize with d possible (amino acid,conformation) pairs, there are n unary and $\frac{n.(n-1)}{2}$ binary cost functions which can be computed independently.

DEE/A optimization.* To solve the different protein design cases, we used `osprey` version 1.0 (`cs.duke.edu/donaldlab/osprey.php`) which first filters rotamers i_r such that $E(i_r) > 30kcal/mol$ and pairs (i_r, j_s) such that $E(i_r, j_s) > 100kcal/mol$ (*pruningE* and *stericE* parameters). This step is followed by extensive DEE pre-processing (*algOption* $= 3$, includes simple Goldstein, Magic bullet pairs, 1 and 2-split positions, Bounds and pairs pruning) and A^* search. Only the GMEC conformation is generated by A^* (*initEw*=0). Computations were performed on a single core of an AMD Operon 6176 at 2.3 GHz, 4 GB of RAM, and a 100-hour time-out. There were no memory-out errors.

CFN and ILP optimization. The same problems (before DEE preprocessing and using $M = 10^8$) have been tackled by `cplex` version 12.2 (parameters EPAGAP, EPGAP and EPINT set to zero to avoid premature stop) and `toulbar2` version 0.9.5 (`mulcyber.toulouse.inra.fr/projects/toulbar2/`) using binary branching with an initial limited discrepancy search phase [16] with a maximum discrepancy of 2 (options `-d: -l=2`, and other default options including EDAC and no initial upper bound) and domains sorted with increasing unary costs $E(i_r)$. These computations were performed on a single core of an Intel Xeon E5430 core at 2.66 GHz with 64GB of RAM with a 100-hour time-out.

With the exception of one instance (1CM1), `cplex` significantly outperforms `osprey`. On the other hand, `toulbar2` is always faster than both `cplex` and `osprey` by at least one order of magnitude and often many more, even accounting for the performance discrepancy arising from the difference in the hardware we used. We have also verified that the minimum energy reported by all 3 solvers is identical.

Table 1. For each instance: protein (PDB id.), amino acid sequence length, number of mutable residues, maximum number of (amino acid, conformation) pairs, sequential time for computing $E(\cdot)$ energy functions, and CPU-time for solving using `osprey`, `cplex`, and `toulbar2`. A '-' indicates that the 100-hour limit has been reached.

System name	Size	n	d	$E(\cdot)$	osprey	cplex	toulbar2
Thioredoxin (2TRX)	108	11	44	304 min.	27.1 sec.	2.6 sec.	0.1 sec.
Protein G (1PGB)	56	11	45	76 min.	49.3 sec.	14.7 sec.	0.1 sec.
Protein L (1HZ5)	64	12	45	114 min.	1,450 sec.	17.7 sec.	0.1 sec.
Ubiquitin (1UBI)	76	13	45	270 min.	-	405.0 sec.	0.6 sec.
Protein G (1PGB)	56	11	148	1,096 min.	-	2,245 min.	13.9 sec.
Protein L (1HZ5)	64	12	148	831 min.	-	1,750 min.	14.6 sec.
Ubiquitin (1UBI)	76	13	148	1,967 min.	-	-	378 min.
Plastocyanin (2PCY)	99	18	44	484 min.	-	89.5 sec.	0.5 sec.
Haloalkane Dehalogenase (2DHC)	310	14	148	45,310 min.	-	-	77.4 sec.
Calmodulin (1CM1)	161	17	148	11,326 min.	121.9 sec.	1,707 sec.	2.0 sec.
Peptidyl-prolyl cis-trans Isomerase (1PIN)	153	28	148	40,491 min.	-	-	-
Cold-Shock (1C9O)	132	55	148	84,089 min.	-	-	-

6.1   Explaining the Differences

The ILP solver CPLEX is a totally closed-source black box. More generally, solvers are complex systems involving various mechanisms. The effect of their interactions during solving is hard to predict. Therefore, explaining the differences in efficiency observed between the different approaches is not really obvious.

If we consider `osprey` first, it uses an obsolete lower bound instead of the more recent incremental and stronger lower bounds offered by soft local consistencies such as EDAC [22]. This, together with the associated informed value ordering provided by these local consistencies, may explain why `toulbar2` outperforms `osprey`. Similarly, the LP relaxation lower bound used in ILP is known (by duality) to be the same as the Optimal Soft AC lower bound (when no upper bounding occurs, i.e. when $k = +\infty$). Since OSAC dominates all other local consistencies at the arc level, this provides an explanation for the efficiency of `cplex` compared to `osprey`. Finally, the problem is deeply non linear. It can be concisely formulated as a CFN but the ILP formulation is much more verbose. This probably contributes, together with the upper bounding (provided by node consistency) and value ordering heuristics of `toulbar2`, to the efficiency of `toulbar2` compared to `cplex`.

7   Conclusion

The simplest formal optimization problem underlying CPD looks for a Global Minimum Energy Conformation (GMEC) over a rigid backbone and altered side-chains (identity and conformation). It can easily be reduced to a binary Cost Function Network, with a very dense graph and relatively large domains or to 0/1LP with a large number of variables.

On a variety of real instances, we have shown that state-of-the-art CFN algorithms but also 0/1LP algorithms give important speedups compared to usual CPD algorithms combining Dead End Elimination with $A*$ as implemented in the `osprey` package. CFN algorithms are the most efficient by far and have the advantage of requiring reasonable space.

Although existing CFN algorithms still need to be extended and adapted to tackle such problems, the rigid backbone method reported herein may contribute to the development of more sophisticated flexible methods.

Acknowledgements. This work has been partly funded by the "Agence nationale de la Recherche", reference ANR-10-BLA-0214. We would like to thank Damien Leroux for his help in the generation of `cplex` encodings using Python. We thank the Computing Center of Region Midi-Pyrénées (CALMIP, Toulouse, France) and the GenoToul Bioinformatics Platform of INRA-Toulouse for providing computing resources and support.

References

1. Anfinsen, C.: Principles that govern the folding of protein chains. Science 181(4096), 223–253 (1973)
2. Bistarelli, S., Faltings, B., Neagu, N.: Interchangeability in Soft CSPs. In: O'Sullivan, B. (ed.) Constraint Solving and CLP. LNCS (LNAI), vol. 2627, pp. 31–46. Springer, Heidelberg (2003)
3. Boas, F., Harbury, P.: Potential energy functions for protein design. Current Opinion in Structural Biology 17(2), 199–204 (2007)

4. Case, D., Darden, T., Cheatham III, T., Simmerling, C., Wang, J., Duke, R., Luo, R., Merz, K., Pearlman, D., Crowley, M., Walker, R.C., Zhang, W., Wang, B., Hayik, S., Roitberg, A., Seabra, G., Wong, K.F., Paesani, F., Wu, X., Brozell, S., Tsui, V., Gohlke, H., Yang, L., Tan, C., Mongan, J., Hornak, V., Cui, G., Beroza, P., Mathews, D.H., Schafmeister, C., Ross, W.S., Kollman, P.A.: Amber 9. University of California, San Francisco (2006)
5. Cooper, M.C., de Givry, S., Sanchez, M., Schiex, T., Zytnicki, M., Werner, T.: Soft arc consistency revisited. Artificial Intelligence 174, 449–478 (2010)
6. Cooper, M.C., de Givry, S., Schiex, T.: Optimal soft arc consistency. In: Proc. of IJCAI 2007, Hyderabad, India, pp. 68–73 (January 2007)
7. Cooper, M.C.: Fundamental properties of neighbourhood substitution in constraint satisfaction problems. Artificial Intelligence 90(1-2), 1–24 (1997)
8. Dahiyat, B., Mayo, S.: Protein design automation. Protein Science 5(5), 895–903 (1996)
9. Desmet, J., Maeyer, M., Hazes, B., Lasters, I.: The dead-end elimination theorem and its use in protein side-chain positioning. Nature 356(6369), 539–542 (1992)
10. Desmet, J., Spriet, J., Lasters, I.: Fast and accurate side-chain topology and energy refinement (FASTER) as a new method for protein structure optimization. Proteins: Structure, Function, and Bioinformatics 48(1), 31–43 (2002)
11. Fersht, A.: Structure and mechanism in protein science: a guide to enzyme catalysis and protein folding. WH Freemean and Co., New York (1999)
12. Georgiev, I., Lilien, R., Donald, B.: Improved pruning algorithms and divide-and-conquer strategies for dead-end elimination, with application to protein design. Bioinformatics 22(14), e174–e183 (2006)
13. Georgiev, I., Lilien, R., Donald, B.: The minimized dead-end elimination criterion and its application to protein redesign in a hybrid scoring and search algorithm for computing partition functions over molecular ensembles. Journal of Computational Chemistry 29(10), 1527–1542 (2008)
14. Goldstein, R.: Efficient rotamer elimination applied to protein side-chains and related spin glasses. Biophysical Journal 66(5), 1335–1340 (1994)
15. Grunwald, I., Rischka, K., Kast, S., Scheibel, T., Bargel, H.: Mimicking biopolymers on a molecular scale: nano (bio) technology based on engineered proteins. Philosophical Transactions of the Royal Society A: Mathematical, Physical and Engineering Sciences 367(1894), 1727–1747 (2009)
16. Harvey, W.D., Ginsberg, M.L.: Limited discrepency search. In: Proc. of the 14th IJCAI, Montréal, Canada (1995)
17. Hawkins, G., Cramer, C., Truhlar, D.: Parametrized models of aqueous free energies of solvation based on pairwise descreening of solute atomic charges from a dielectric medium. The Journal of Physical Chemistry 100(51), 19824–19839 (1996)
18. Khare, S., Kipnis, Y., Takeuchi, R., Ashani, Y., Goldsmith, M., Song, Y., Gallaher, J., Silman, I., Leader, H., Sussman, J., et al.: Computational redesign of a mononuclear zinc metalloenzyme for organophosphate hydrolysis. Nature Chemical Biology 8(3), 294–300 (2012)
19. Kingsford, C., Chazelle, B., Singh, M.: Solving and analyzing side-chain positioning problems using linear and integer programming. Bioinformatics 21(7), 1028–1039 (2005)
20. Koster, A., van Hoesel, S., Kolen, A.: Solving frequency assignment problems via tree-decomposition. Tech. Rep. RM/99/011, Universiteit Maastricht, Maastricht, The Netherlands (1999)
21. Kuhlman, B., Baker, D.: Native protein sequences are close to optimal for their structures. Proceedings of the National Academy of Sciences 97(19), 10383 (2000)

22. Larrosa, J., de Givry, S., Heras, F., Zytnicki, M.: Existential arc consistency: getting closer to full arc consistency in weighted CSPs. In: Proc. of the 19th IJCAI, Edinburgh, Scotland, pp. 84–89 (August 2005)
23. Larrosa, J., Meseguer, P., Schiex, T., Verfaillie, G.: Reversible DAC and other improvements for solving max-CSP. In: Proc. of AAAI 1998, Madison, WI (July 1998)
24. Leach, A., Lemon, A., et al.: Exploring the conformational space of protein side chains using dead-end elimination and the A* algorithm. Proteins Structure Function and Genetics 33(2), 227–239 (1998)
25. Looger, L., Hellinga, H.: Generalized dead-end elimination algorithms make large-scale protein side-chain structure prediction tractable: implications for protein design and structural genomics1. Journal of Molecular Biology 307(1), 429–445 (2001)
26. Lovell, S., Word, J., Richardson, J., Richardson, D.: The penultimate rotamer library. Proteins: Structure, Function, and Bioinformatics 40(3), 389–408 (2000)
27. Nestl, B., Nebel, B., Hauer, B.: Recent progress in industrial biocatalysis. Current Opinion in Chemical Biology 15(2), 187–193 (2011)
28. Pabo, C.: Molecular technology: designing proteins and peptides. Nature 301, 200 (1983)
29. Peisajovich, S., Tawfik, D.: Protein engineers turned evolutionists. Nature Methods 4(12), 991–994 (2007)
30. Pierce, N., Spriet, J., Desmet, J., Mayo, S.: Conformational splitting: A more powerful criterion for dead-end elimination. Journal of Computational Chemistry 21(11), 999–1009 (2000)
31. Pierce, N., Winfree, E.: Protein design is NP-hard. Protein Engineering 15(10), 779–782 (2002)
32. Pleiss, J.: Protein design in metabolic engineering and synthetic biology. Current Opinion in Biotechnology 22(5), 611–617 (2011)
33. Raha, K., Wollacott, A., Italia, M., Desjarlais, J.: Prediction of amino acid sequence from structure. Protein Science 9(6), 1106–1119 (2000)
34. Schiex, T.: Arc Consistency for Soft Constraints. In: Dechter, R. (ed.) CP 2000. LNCS, vol. 1894, pp. 411–424. Springer, Heidelberg (2000)
35. Swain, M.T., Kemp, G.J.L.: A CLP Approach to the Protein Side-Chain Placement Problem. In: Walsh, T. (ed.) CP 2001. LNCS, vol. 2239, pp. 479–493. Springer, Heidelberg (2001)
36. Voigt, C., Gordon, D., Mayo, S.: Trading accuracy for speed: a quantitative comparison of search algorithms in protein sequence design. Journal of Molecular Biology 299(3), 789–803 (2000)
37. Wallace, R.J.: Directed Arc Consistency Preprocessing. In: Meyer, M. (ed.) Constraint Processing. LNCS, vol. 923, pp. 121–137. Springer, Heidelberg (1995)

A Filtering Technique for Fragment Assembly-Based Proteins Loop Modeling with Constraints

Federico Campeotto[1,2], Alessandro Dal Palù[3], Agostino Dovier[2],
Ferdinando Fioretto[1], and Enrico Pontelli[1]

[1] Dept. Computer Science, New Mexico State University
[2] Depts. Math. & Computer Science, University of Udine
[3] Dept. Mathematics, University of Parma

Abstract. Methods to predict the structure of a protein often rely on the knowledge of macro sub-structures and their exact or approximated relative positions in space. The parts connecting these sub-structures are called *loops* and, in general, they are characterized by a high degree of freedom. The modeling of loops is a critical problem in predicting protein conformations that are biologically realistic. This paper introduces a class of constraints that models a general multi-body system; we present a proof of NP-completeness and provide filtering techniques, inspired by inverse kinematics, that can drastically reduce the search space of potential conformations. The paper shows the application of the constraint in solving the protein loop modeling problem, based on fragments assembly.

1 Introduction

Proteins are macro-molecules of fundamental importance in the way they regulate vital functions in all biological processes. In general, there is a direct correspondence between a protein function and its 3D structure—as the structure guides the interactions among molecules. Thus, proteins structure analysis is essential for biomedical investigations, e.g., drug design and protein engineering.

The natural approach of investigating protein conformations through simulations of physical movements of atoms and molecules is, unfortunately, beyond the current computational capabilities [23,3,26]. This has originated a variety of alternative approaches, many based on *comparative modeling*—i.e., small structures from related protein family members are used as templates to model the global structure of the protein of interest [24,19,34,28,25]. In these methods, named *fragments assembly*, a protein structure is assembled by using small protein subunits as templates that present similarities (*homologous affinity*) w.r.t. the target sequence. The literature has also demonstrated the strength of *Constraint Programming (CP)* techniques in investigating the problem of protein structure prediction—where constraints are used to model the structural variability of a protein [1,2,8,9].

In this paper, we model the problem of assembling rigid fragments as a global constraint. We abstract the problem as a general multi-body system, where each composing body is constrained by means of geometric properties and it is related to other bodies through joint relationships. This model leads to the

M. Milano (Ed.): CP 2012, LNCS 7514, pp. 850–866, 2012.
© Springer-Verlag Berlin Heidelberg 2012

Joined-Multibody (JM) constraint, whose satisfaction we prove to be NP-complete. Realistic protein models require the assembly of hundreds of different body versions, making the problem intractable. We study an efficient approximated propagator, called *JM filtering* (JMf), that allows us to efficiently compute classes of solutions, partitioned by structural similarity and controlled tolerance for error.

We implement and test the constraint and its propagator in the FIASCO framework [10]. We demonstrate our approach in addressing the *loop modeling* problem, which is a special case of the JM constraint. The goal is to connect two bodies that have fixed positions in space and that are connected by a sequence of highly variable bodies. The loop is characterized by a high degree of freedom and resembles the inverse-kinematic problem found in robotics, with some spatial constraints. We demonstrate the strength of the filtering algorithm in significantly reducing the search space and in aiding the selection of representative solutions. We compare our results, based on popular benchmark suites, to other programs specialized on loop modeling, with very encouraging results.

2   Related Work and Background

Loop modeling can be described as a special case of the protein structure prediction problem, where CP has been extensively employed. CP has been used to provide approximated solutions for *ab-initio* lattice-based modeling of protein structures, using local search and large neighboring search [33,15]; exact resolution of the problem on lattice spaces using CP, along with with clever symmetry breaking techniques, has also been investigated [1]. These approaches solve a constraint optimization problem based on a simple energy function (HP). A more precise energy function has been used in [8,11], where information on secondary structures (helices, sheets) are taken into consideration.

Due to the approximation errors introduced by lattice discretization, these approaches do not scale to medium-size proteins. Off-lattice models, based on the idea of fragment assembly, and implemented using Constraint Logic Programming over Finite Domains, have been presented in [9,10] and applied not only to structure prediction but also to other structural analysis problems—e.g., the tool developed in [9] has been used to to generate sets of feasible conformations for studies of protein flexibility [13]. The use of CP to analyze NMR data and the related problem of protein docking has been studied in [2].

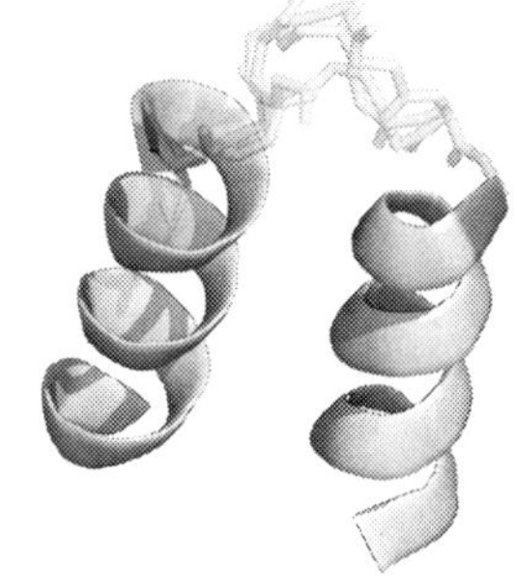

Fig. 1. Helices with a loop

Even when protein structure prediction is realized using homologous templates, the final conformation may present aperiodic structures (*loops*) connecting the known protein segments on the outer region of the protein, where the presence of the solvent lessens the restrictions on the possible movements of the structure. These protein regions are in general not conserved during evolution, and therefore templates provide very limited statistical structural information. The length of a *protein loop* is typically in the range of 2 to 20 amino acids; nevertheless, the flexibility of loops produces very large, physically consistent, conformation search spaces. Figure 1 depicts a possible scenario where two macro-structures (two helices) are connected by a loop—the loop anchors are colored in orange. The loop constraint is satisfied by the loops connecting the two anchor points. Modeling a protein loop often imposes constraints in the way of connecting two protein segments. Restrictions on the mutual positions and orientations (dihedral angles) of the loop anchors are often present. Such restrictions are defined as the *loop closure* constraints.

A procedure for protein loop modeling (e.g., [22]) typically consists of 3 phases: *sampling, filtering*, and *ranking*. In sampling, a set of possible loop conformations is proposed. *Ab initio* methods (e.g., [31,17,21,35,14,16,36]) and methods based on templates extracted from structural databases (e.g., [7]) have been explored. These conformations are checked w.r.t. the loop constraints and the geometries from the rest of the structure, and the loops that are detected as physically infeasible, e.g., causing steric clashes, are discarded by a filtering procedure. Finally, a ranking step—e.g., based on statistical potential energy (e.g., DOPE [32], DFIRE [37], or [18])—is used to select the best loop candidate(s).

Loop sampling plays an important role: it should produce structurally diverse loop conformations, in order to maximize the probability of finding one close to the native conformation. Sampling is commonly implemented as a two-step approach. First, a possible loop candidate is generated, without taking into account geometric or steric feasibility restrictions—this step usually employs dihedral angles sampled from structural databases [16]. Afterwards, the initial structure is altered into a structure that satisfies the loop closure constraints. Popular methods include the *Cyclic Coordinate Descent (CCD)* [6], the *Self-Organizing (SOS)* algorithm [30], and *Wriggling* [5]. Multi-method approaches have also been proposed—e.g., [29] proposes a loop sampling method which combines fragment assembly and analytical loop closure, based on a set of torsion angles satisfying the imposed constraints.

3 The Joined-Multibody Constraint

A *rigid block* B is composed of an ordered list of at least three (distinct) 3D points, denoted by $\mathsf{points}(B)$. The *anchors* and *end-effectors* of a rigid block B, denoted by $\mathsf{start}(B)$ and $\mathsf{end}(B)$, are the two lists containing the first three and the last three points of $\mathsf{points}(B)$. With $B(i)$ we denote the i-th point of the rigid block B. For two ordered lists of points $\boldsymbol{p}$ and $\boldsymbol{q}$, we write $\boldsymbol{p} \frown \boldsymbol{q}$ if they can be perfectly overlapped by a rigid translation/rotation (i.e., a *roto-translation*).

Definition 1 (Multi-body). *A* multi-body *is a sequence* $S_1, \ldots, S_n$ *of nonempty sets of rigid blocks. A sequence of rigid blocks* $B_1, \ldots, B_n$, *is called a* rigid body *if, for all* $i = 1, \ldots, n - 1$, $\mathsf{end}(B_i) \frown \mathsf{start}(B_{i+1})$. *A compatible multi-body is a multi-body where for all pairs of rigid blocks* $B, B' \in \bigcup_{i=1}^{n} S_i$ *and for all* $\boldsymbol{p}, \boldsymbol{q} \in \{\mathsf{start}(B), \mathsf{start}(B'), \mathsf{end}(B), \mathsf{end}(B')\}$ *it holds that* $\boldsymbol{p} \frown \boldsymbol{q}$.

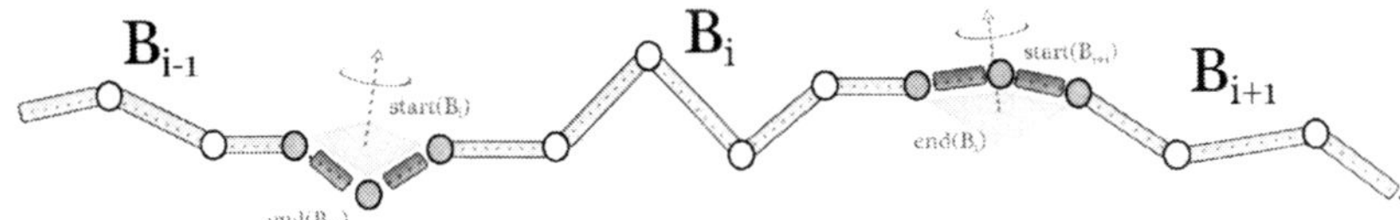

Fig. 2. A schematic representation of a rigid body

A rigid body can be seen as one instance of a multi-body that guarantees the partial overlapping of each two consecutive blocks. The overlapped points end(B_i) and start(B_{i+1}) constitute the i-th *joint* of the rigid body. The number of rigid bodies "encoded" by a single multi-body is bounded by $\Pi_{i=1}^{n}|S_i|$.

Figure 2 provides a schematic representation of a rigid body. The joints connecting two adjacent rigid blocks are marked by orange rectangles and grey circles. The points in points(B) of each rigid block are represented by circles. Each rigid block extends from the first point of a joint to the last point of the successive joint.

A rigid body is defined by the overlap of joints, and relies on a chain of relative roto-translations of its blocks. Each points(B_i) is therefore positioned according to the (homogeneous) coordinate system associated to a rigid block B_{i-1}. Note that once the reference system for B_1 is defined, the whole rigid body is completely positioned.[1] The relative positions of two consecutive rigid blocks B_{i-1} and B_i of a rigid body ($2 \leq i \leq n$) can be defined by a transformation matrix $T_i \in \mathbb{R}^{4 \times 4}$. Each matrix depends on the standard Denavit-Hartenberg parameters [20] obtained from the start and end of the blocks (c.f., [27] for details). We denote the product $T_1 \cdot T_2 \cdot \ldots \cdot T_i \cdot (x, y, z, 1)^T$ by $\nabla_i(x, y, z)$.

Let us focus on the matrix T_1. The block B_1 can be rigidly moved in a desired position and orientation on the basis of additional spatial constraints (e.g., the sets $\mathcal{A}_1, \mathcal{A}_2, \mathcal{A}_3, \mathcal{E}_1, \mathcal{E}_2, \mathcal{E}_3$ in the Def. 2). T_1 is a matrix that allows a roto-translation of B_1 in a position fulfilling the additional constraints.

For $i = 1, \ldots, n$, the coordinate system conversion (x', y', z'), for a point $(x, y, z) \in$ points(B_i) into the coordinate system of B_1, is obtained by:

$$(x', y', z', 1)^T = T_1 \cdot T_2 \cdot \ldots \cdot T_i \cdot (x, y, z, 1)^T = \nabla_i(x, y, z) \qquad (1)$$

Homogeneous transformations are such that the last value of a tuple is always 1. Note that a modification of the matrix T_1 is a sufficient step to place the whole rigid body into a different start position.

Definition 2 (JM-constraint). *The* joined-multibody (JM) constraint *is described by a tuple:* $J = \langle S, V, \mathcal{A}, \mathcal{E}, \delta \rangle$, *where:*

- $S = S_1, \ldots, S_n$ *is a multi-body. Let* $\mathcal{B} = \{B_1, \ldots, B_k\}$ *be the set of all rigid blocks in* S, *i.e.,* $\mathcal{B} = \bigcup_{i=1}^{n} S_i$.
- $V = V_1, \ldots, V_n$ *is a list of finite-domain variables. For* $i = 1, \ldots, n$, *the variable* V_i *is associated to a domain* dom(V_i) = $\{j : B_j \in S_i\}$.
- $\mathcal{A} = \mathcal{A}_1, \mathcal{A}_2, \mathcal{A}_3$, *and* $\mathcal{E} = \mathcal{E}_1, \ldots, \mathcal{E}_{3n}$ *are lists of sets of 3D points such that:*

---

[1] With the exception of the case where the all points of a joint are collinear.

- $\circ$ $\mathcal{A}_1 \times \mathcal{A}_2 \times \mathcal{A}_3$ *is the set of admissible points for* start(B), *with* $B \in S_1$;
- $\circ$ $\mathcal{E}_{3i-2} \times \mathcal{E}_{3i-1} \times \mathcal{E}_{3i}$ *is the set of admissible points for* end(B), *with* $B \in S_i$, $i = 1, \dots, n$;

- $\bullet$ δ *is a constant, used to express a minimal distance constraint between different points. Let us assume that for all $B \in \mathcal{B}$ and for all $a, b \in$ points(B), if $a \neq b$ then $\|a - b\| \geq \delta$ (where $\| \cdot \|$ is the Euclidean norm).*

A solution for the JM constraint J is an assignment $\sigma : V \longrightarrow \{1, \dots, |\mathcal{B}|\}$ s.t. there exist matrixes $T_1, \dots, T_n$ (used in ∇) with the following properties:

<u>*Domain:*</u> *For all $i = 1, \dots, n$, $\sigma(V_i) \in$ dom(V_i).*

<u>*Joint:*</u> *For all $i = 1, \dots, n-1$, let $(a^1, a^2, a^3) =$ end$(B_{\sigma(V_i)})$ and $(b^1, b^2, b^3) =$ start$(B_{\sigma(V_{i+1})})$, then it holds that (for $j = 1, 2, 3$):*

$$\nabla_i(a_x^j, a_y^j, a_z^j) = \nabla_{i+1}(b_x^j, b_y^j, b_z^j)$$

<u>*Spatial Domain:*</u> *Let $(a^1, a^2, a^3) =$ start$(B_{\sigma(V_1)})$, then $T_1 \cdot a^j \in \mathcal{A}_j \times \{1\}$. For all $i = 1, \dots, n$, let $(e^1, e^2, e^3) =$ end$(B_{\sigma(V_i)})$ then*

$$\nabla_i(e_x^j, e_y^j, e_z^j) \in \mathcal{E}_{3(i-1)+j} \times \{1\}$$

where $1 \leq j \leq 3$ and $T_2, \dots, T_i$ (in ∇_i) are the matrices that overlap $B_{\sigma(V_{i-1})}$ and $B_{\sigma(V_i)}$ (the product $\times\{1\}$ is due since we use homogeneous coordinates).

<u>*Minimal Distance:*</u> *For all $j, \ell = 1, \dots, n$, $j < \ell$, and for all points $a \in$ points$(B_{\sigma(V_j)})$ and $b \in$ points$(B_{\sigma(V_\ell)})$, it holds that:*

$$\|\nabla_j(a_x, a_y, a_z) - \nabla_\ell(b_x, b_y, b_z)\| \geq \delta$$

A JM constraint is said to be compatible *if $S_1, \dots, S_n$ is a compatible multi-body.*

If there are no joints with the three points aligned, $T_2, \dots, T_n$ depend deterministically from T_1 and σ. Compatible JM constraints are interesting for our target application. Nevertheless, the additional restriction does not simplify constraint solving, as we discuss below.

Complexity Analysis. The problem of determining *consistency* of JM constraints (i.e., the existence of a solution) is NP-complete. To prove this fact, we start from the NP-completeness of the consistency problem of the constraint *Self-Avoiding-Walk (SAW-constraint)* in a discrete lattice, proved in [12]. In particular, we will use the 3D cubic lattice for this problem. Let $X = X_1, \dots, X_n$ be a list of variables. Each variable has a finite domain dom$(X_i) \subseteq \mathbb{Z}^3$. $\sigma : X \longrightarrow \mathbb{Z}^3$ is a solution of the SAW constraint if:

- $\bullet$ For all $i = 1, \dots, n$: $\sigma(X_i) \in$ dom(X_i),
- $\bullet$ For all $i = 1, \dots, n-1$: $\|\sigma(X_i) - \sigma(X_{i+1})\| = 1$,
- $\bullet$ For all $i, j = 1, \dots, n$, $i < j$, it holds that $\|\sigma(X_i) - \sigma(X_j)\| \geq 1$.

As emerges from the proof in [12], the problem of determine the consistency of a SAW constraint is NP complete even if the domains of dom(X_1) and dom(X_2) are singleton sets. Without loss of generality, we can concentrate on SAW problems where dom$(X_1) = \{(0, 0, 0)\}$ and dom$(X_2) = \{(0, 1, 0)\}$—the other cases can be reduced to this one using a roto-translation.

Theorem 1. *The consistency problem for the JM constraint is NP-complete.*

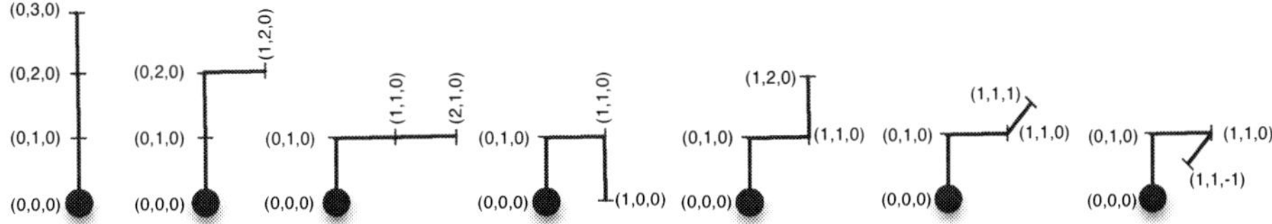

Fig. 3. Rigid blocks for SAW (from left to right, block 1, ..., 7)

Proof (sketch). The proof of membership in NP is trivial; given a tentative solution, it is easy to test it in polynomial time—the most complex task is building the rotation matrixes.

To prove completeness, let us reduce SAW, with the further hypothesis on $\mathsf{dom}(X_1)$ and $\mathsf{dom}(X_2)$ to JM. Let us consider an instance $A = \langle \boldsymbol{X}, \mathsf{dom}(\boldsymbol{X}) \rangle$ of SAW with n $(n > 3)$ variables. We define a equi-satisfiable instance $B = \langle \boldsymbol{S}, \boldsymbol{V}, \boldsymbol{\mathcal{A}}, \boldsymbol{\mathcal{E}}, \delta \rangle$ of JM as follows. Let us choose $\boldsymbol{V} = V_1, \ldots, V_{n-3}$. We select the sets S_i of rigid blocks of the multi-body to be all identical, and consisting of all the (non overlapping) fragments of three contiguous unitary segments of length 1, starting from $(0, 0, 0)$ and with bends of 0 or 90 degrees. A filtering using symmetries is made and the blocks are indicated in Fig. 3. For all $i = 1, \ldots, n-3$ we assign the following sets of 3D points to the end-effectors:

$$\mathcal{E}_{3i-2} = \mathsf{dom}(X_{i+1}) \cap \mathbb{Z}^3, \quad \mathcal{E}_{3i-1} = \mathsf{dom}(X_{i+2}) \cap \mathbb{Z}^3, \quad \mathcal{E}_{3i} = \mathsf{dom}(X_{i+3}) \cap \mathbb{Z}^3$$

Moreover, let $\mathcal{A}_1 = \{(0,0,0)\}, \mathcal{A}_2 = \{(0,1,0)\}, \mathcal{A}_3 = \{(0,2,0), (1,1,0)\}$. Observe that all these sets are subsets of $\mathbb{Z}^3$ and therefore points of the same lattice of the SAW problem. Assigning $\delta = 1$, the reduction is complete. It is immediate to check that SAWs in the 3D lattice and the solutions of the JM constraint defined via reduction are essentially the same 3D polygonal chain.   □

We have a proof for the NP-completeness of the compatible JM constraint (in particular, it does not make use of joints made by collinear points), by reduction from the Hamiltonian path problem in special planar graphs. The proof is omitted due to lack of space. The interested reader can find it at `http://www.cs.nmsu.edu/fiasco`.

4   Filtering Algorithm for the Joined-Multibody Constraint

Since checking the satisfiability (and hyper-arc consistency) of the JM constraint is NP-complete, we studied an approximated polynomial time filtering algorithm. When dealing with multi-bodies, the computation of the end-effectors' spatial domains provides limited filtering information, since it identifies a large volume.

We designed an algorithm (JMf, Algorithm 1) that is inspired by bound-consistency on the 3D positions of end-effectors. The algorithm uses an equivalence (clustering) relation over these bounds, in order to retain precise information about classes of domain variable assignments that produce similar spatial results. This allows a compact handling of the combinatorics of the multi-body, while a controlled error threshold allows us to select the precision of the filtering. The equivalence relation captures those rigid bodies that are geometrically similar and thus compacts small differences among them; relevant gains in computation time can be derived when some errors are tolerated.

Algorithm 1. The JMf algorithm.

Require: $\boldsymbol{S}, \boldsymbol{V}, \boldsymbol{\mathcal{A}}, \boldsymbol{\mathcal{E}}, \mathcal{G}, \delta, \sim$

1: $n \leftarrow |\boldsymbol{V}|$

2: $\mathcal{R}_1 \leftarrow \left\{ B \in S_1 \,\middle|\, \exists T_1 \left(\begin{array}{l} T_1 \cdot \mathsf{start}(B) \in \mathcal{A}_1 \times \mathcal{A}_2 \times \mathcal{A}_3 \ \wedge \\ T_1 \cdot \mathsf{end}(B) \in \mathcal{E}_1 \times \mathcal{E}_2 \times \mathcal{E}_3 \quad \wedge \\ \forall p \in \mathsf{points}(B). \forall q \in \mathcal{G}. \|(T_1 \cdot p) - q\| \geq \delta) \end{array} \right) \right\}$

3: $\mathcal{P}_1 \leftarrow \{ T_1 \cdot \mathsf{end}(B) \mid B \in \mathcal{R}_1, T_1 \text{ as in the line above } \}$

4: $\mathsf{dom}(V_1) \leftarrow \{ \mathsf{label}(B) \mid B \in \mathcal{R}_1 \}$

5: **for all** $i = 2, \ldots, n$ **do**

6:    $\mathcal{P}_i = \emptyset; \mathcal{R}_i = \emptyset;$

7:    **for all** $E \in \mathcal{P}_{i-1}$ **do**

8:      $\mathcal{R}_i \leftarrow \mathcal{R}_i \cup \left\{ B \in S_i \,\middle|\, \begin{array}{l} T = M(E, \mathsf{start}(B)) \ \wedge \ T \neq \mathtt{fail} \ \wedge \\ T \cdot \mathsf{end}(B) \in \mathcal{E}_{3i-2} \times \mathcal{E}_{3i-1} \times \mathcal{E}_{3i} \quad \wedge \\ \forall p \in \mathsf{points}(B). \forall q \in \mathcal{G}. \|(T \cdot p) - q\| \geq \delta \end{array} \right\}$

9:      $\mathcal{P}_i \leftarrow \{ M(E, \mathsf{start}(B)) \cdot \mathsf{end}(B) \mid B \in \mathcal{R}_i \}$

10:   **end for**

11:   **compute** $\mathcal{P}_i / \sim$ **and filter** $\mathcal{R}_i$ **accordingly**

12:   $\mathsf{dom}(V_i) \leftarrow \{ \mathsf{label}(B) \mid B \in \mathcal{R}_i \}$

13: **end for**

The JMf algorithm receives as input a JM-constraint $\langle \boldsymbol{S}, \boldsymbol{V}, \boldsymbol{\mathcal{A}}, \boldsymbol{\mathcal{E}}, \delta \rangle$, along with a set $\mathcal{G}$ of points that are not available for the placement of bodies (e.g., points of other parts of the protein we are studying, points of the loop non-deterministically assigned by previous calls of the algorithm itself) and an equivalence relation $\sim$ on the space of triples of 3D points. The algorithm makes use of a function M (line 8); this function takes as input two lists $\boldsymbol{a}$ and $\boldsymbol{b}$ of 3D points, and computes the homogeneous transformation to overlap $\boldsymbol{b}$ on $\boldsymbol{a}$. A call to this function will fail if $\boldsymbol{a} \not\sim \boldsymbol{b}$. For simplicity, the fourth component (always 1) of the homogeneous transformation is not explicitly reported in the algorithm.

For $i = 1, \ldots, n$, the algorithm computes the sets $\mathcal{R}_i$ and $\mathcal{P}_i$, that will respectively contain the blocks from S_i that can still lead to a solution, and the corresponding allowed 3D positions of their end-effectors. These two sets are strongly related: a data structure linking each block $B \in S_i$ with the list of its corresponding possible end-effectors (and vice versa) is used at the implementation level. For each block $B \in \mathcal{B}$, we denote with $\mathsf{label}(B)$ a unique label identifying it; $\mathsf{dom}(V_i)$ is therefore the set of the blocks' labels that can be used for the variable V_i. The sets $\mathcal{R}_i$ and $\mathcal{P}_i$ are used to determine the domain $\mathsf{dom}(V_i)$ of the variable V_i (lines 4 and 12). In computing/updating $\mathcal{R}_i$ and $\mathcal{P}_i$, only blocks that have end-effectors contained in the bounds $\mathcal{E}_{3i-2}, \mathcal{E}_{3i-1}, \mathcal{E}_{3i}$ are kept. Fragments that would cause points to collapse—i.e., due to a distance smaller than

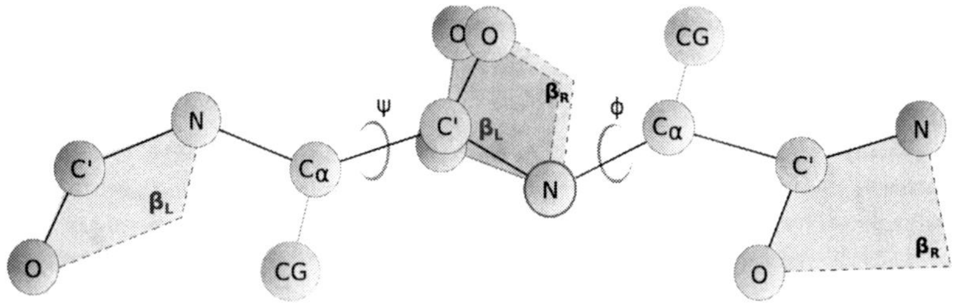

Fig. 4. Fragments are assembled by overlapping the plane β_R, described by the right-most C', O, N atoms of the first fragment (left), with the plane β_L, described by the leftmost C', O, N atoms of the second fragment (right), on the common nitrogen atom

δ from previously placed points—are filtered out (lines 2 and 8). Moreover, the spatial positions of the points of the first block are validated against $\mathcal{A}$ (line 2).

The algorithm performs $|V| - 1$ iterations (lines 5–13). First $\mathcal{R}_i$ and $\mathcal{P}_i$ are computed on the basis of the sets of end-effectors of the previous level $\mathcal{P}_{i-1}$ and the starting point of a selected block B, filtering out those that are not overlapping and those that lead to wrong portions of space (lines 8–9). Then, the $\sim$-based clustering filtering is applied (line 11). During this step, the set of triples of 3D points $\mathcal{P}_i$ is clustered using $\sim$. A representative of each equivalence class is chosen (within $\mathcal{P}_i$). The corresponding block in $\mathcal{R}_i$ is marked. All non marked blocks are filtered out from $\mathcal{R}_i$. Let us also note that the filtering based on clustering is not performed for the initial step $\mathcal{P}_1$, as typically this is already captured by the restrictions imposed by $\mathcal{A}$.

In our tests, the initial domains $\mathcal{A}_1, \mathcal{A}_2, \mathcal{A}_3$ are singleton sets (we start from a rigid block with known end-effectors). Moreover, w.l.o.g., all initial fragments matching with those three points are rotated in the same reference in our database, allowing to use a unique T_1 for all blocks (lines 2 and 3). We experimentally verified that avoiding clustering in the first stage allows to sensibly reduce approximation errors. The filtering algorithm is similar to a directional arc-consistency, where the global constraint is viewed as a conjunction of binary constraints between adjacent blocks. Its peculiarity, however, is the use of representative clustered triples to compactly store the domains.

5   Loop Modeling by the Joined-Multibody Constraint

In this section, we use the joined-multibody constraint to model the protein loop problem. We have implemented the proposed encoding and the constraint solving procedure, based on filtering Algorithm 1, within *FIASCO* (Fragment-based Interactive Assembly for protein Structure prediction with COnstraints) [4].

FIASCO is a C++ tool that provides a flexible environment that allows us to easily manipulate constraints targeted at protein modeling. These constraints do not only concern geometric and energetic aspects but, in general, they allow one to model particular portions of the target protein using arbitrarily long homolog structures. A protein is modeled through the sequence of its amino acids. The *backbone* of each amino acid is represented by all its atoms: $N, C\alpha, C', O$. A single point CG is used for representing the *side chain*, namely the set of atoms characteristic of each amino acid (Fig. 4). *fragment assembly* is used to restrict the allowed spatial positions of consecutive backbone atoms. The final protein conformation is built by combining fragments, treated as basic assembly units. A fragment with $h \geq 1$ amino acids is the concatenation of $4h + 3$ atoms, represented by the regular expression $C' O (N C\alpha C' O)^h N$. The assembling of two fragments is performed by overlapping the planes β_R and β_L, determined by the atoms C', O, N ending the first fragment and starting the second fragment (Fig. 4). The overlapping is made in the N atom in order to best preserve the torsion angle ϕ characteristic of the first amino acid of the second fragment. Each backbone atom is represented by a `Point` variable, whose initial domain can be described by a 3D interval $[L, U]$, that identifies the lower and the upper bounds of a 3D box enclosing the set of possible positions for the atom.[2] A second set of variables (`Fragment`) is used to maintain the sets of possible fragments that can be used to model different segments of the protein. The respective domains are the sets of elements that link the specific protein region modeled by a `Fragment` variable to the related `Point` variables for the atoms in such region. Constraints are introduced to link the domains of `Fragment` variables with those of related `Point` variables.

Loop Modeling. Let us now build on the core constraints of FIASCO and on the JM constraint to address the loop modeling problem. The starting point is a given protein with two known (large) blocks. The model will account for them in the definitions of sets $\mathcal{E}$ (they also allow us to build the set $\mathcal{G}$—see Algorithm 1). The start of the first block and the end of the last block, namely a sequence $C'ON$ (initial anchor) of coordinates $\boldsymbol{a} = (a^1, a^2, a^3)$, and a sequence $C'ON$ (final anchor) of coordinates $\boldsymbol{e} = (e^1, e^2, e^3)$ are known, as well as the sequence $x_1, \ldots, x_n$ of amino acids connecting these two points. Each amino acid can have different 3D forms, depending on the angles Ψ, Φ and the position of CG; a statistically pre-computed set of these forms is loaded from a repository (e.g. `www.pdb.org`) and a weight depending on its frequency is assigned to each fragment. Each fragment is identified by an integer label. Loop modeling can be realized using the joined-multibody constraint $J = \langle \boldsymbol{S}, \boldsymbol{V}, \boldsymbol{\mathcal{A}}, \boldsymbol{\mathcal{E}}, \delta \rangle$ where:

- For $i = 1, \ldots, n$ the set S_i contains all the fragments associated with the amino acid x_i.
- For $i = 1, \ldots, n$, $\mathrm{dom}(V_i)$ is the set of the labels of the fragments in S_i.
- The constant δ (now $\delta = 1.5\text{Å}$) asserts a minimum distance between atoms.
- For the spatial domains, for $j = 1, \ldots, 3$, we set $\mathcal{A}_j = \{a^j\}$ and $\mathcal{E}_{3(n-1)+j}$ is the 3D interval $[e^j - (d_j, d_j, d_j), e^j + (d_j, d_j, d_j)]$, where the values d_j are derived from the covalent radii bond distances $\epsilon_N, \epsilon_O, \epsilon_C$ of the specific types of atoms (specifically, $d_1 = \epsilon_N, d_2 = d_1 + \epsilon_O, d_3 = d_2 + \epsilon_C$). We use this slack for the last 3 points of the loop in order to cushion the error produced during the clustering step, still obtaining solutions that are geometrically eligible.

---

[2] In the current implementation of FIASCO initial domains can be just boxes.

For $\mathcal{E}_1, \ldots, \mathcal{E}_{3(n-1)}$ we allow a "sufficiently large" box of spatial points. Precisely, each $\mathcal{E}_i$ is the box obtained by enlarging by $4n\text{Å}$ in all the directions, the box identified by the two points $[\boldsymbol{a}, \boldsymbol{e}]$ (or $[\boldsymbol{e}, \boldsymbol{a}]$) — 4Å is a rough upper bound to the distance between two consecutive C' in a protein.[3]

Let us observe that the fragments considered lead to a *compatible* multi-body (Def. 1)—thanks to the use of a full-atom description of fragments. A different level of description of the fragments (e.g., the $C\alpha$–$C\alpha$ modeling used in [9] for the fragment-assembly approach to the complete protein folding) would not lead us to a compatible multi-body. Moreover, known larger rigid blocks can be easily inserted in the modeling in an explicit way in some S_i. More loops can be also modeled simultaneously in this way.

Clustering. The JMf algorithm is parametric w.r.t. the clustering relation and the function selecting the representative; they both express the degree of approximation of the rigid bodies to be built. The proposed clustering relation for loop modeling takes into account two factors: *(a)* The positions of the end-effectors in the 3D space and *(b)* The orientation of the planes formed by the fragment's anchor β_L and end-effector β_R (Fig. 4). This combination of clusterings allows to capture local geometrical similarities, since both spatial and rotational features are taken into account.

The spatial clustering *(a)* used is the following. Given a set of fragments, the three end points $C'ON$ (end effectors) of each cluster are considered, and the centroid of the triangle $C'ON$ is computed. We use three parameters: $k_{min}, k_{max} \in \mathbb{N}$, $k_{min} \le k_{max}$, and $r \in \mathbb{R}$, $r \ge 0$. We start by selecting a set of k_{min} fragments, pairwise distant at least $2r$. These fragments are selected as representatives of an equivalence class for other fragments that fall within a sphere of radius r centered in the centroid of the representative. This clustering ensures a rather even initial distribution of clusters, however some fragments may not fall within the k_{min} clusters. We allow to create up to $k_{max} - k_{min}$ new clusters, each of them covering a sphere of radius r. Remaining fragments are then assigned to the closest cluster. Other techniques can be employed, the one used allows a fast implementation and acceptable results.

The orientation clustering *(b)* partitions the fragments according to their relative orientation of planes β_R, β_L (and it can be pre-computed, being independent on the roto-translation the fragment will be subject to). Let us observe that here we consider all four points of a fragment. All fragments are characterized by the normal to the plane β_R, assuming that each fragment is already joined to the previous one. The clustering algorithm guarantees that each cluster contains fragments that pair-wise have normals with a deviation of at most a threshold β degrees. This algorithm produces a variable number of partitions depending on β.

---

[3] $\mathcal{E}_i$ should be intersected with the complement of $\mathcal{G}$ (the region of space occupied by the two known rigid blocks). This kind of domain operation is not yet supported by FIASCO and therefore the control of this part is handled by Algorithm 1.

The final cluster is the intersection of the two partitioning algorithms. This defines an equivalence relation $\sim$ depending on k_{min}, k_{max}, r, and β. The representative selection function selects the fragment for each partition according to some preferences (e.g., most frequent fragment, closest to the center, etc.).

Note that for $r = 0$, $\beta = 0$, and k_{max} unbounded, no clustering is performed and this would cause the combinatorial explosion of every possible end-effector on the whole problem. The spatial error introduced depends on r and β. With $\beta = 0$, the error introduced at each step can be bound by $2r$ for each dimension. At each iteration the errors are linearly increased, since a new fragment is placed with an initial error gathered from previous iterations, thus resulting in a $2nr$ bound for the last end-effector. Clearly this bound is very coarse, and on average the experimental results show better performances. Similar considerations can be argued for rotational errors, however the intersection of the two clusterings, provide, in general, a much tighter bound.

Implementation Details. A pre-processing step clusters "a-priori" the variables' domains (by means of a first call of the JMf algorithm with $\mathcal{G} = \emptyset$). The rotational clustering is made while the fragment databases is computed. This speeds-up the clustering algorithm during the search, since at that time the spatial clusters can only be intersected with the ones already computed.

As soon as the domain for the variables related to the initial anchor of a JM constraint is instantiated, the corresponding constraint is woken up. The algorithm JMf is invoked with the parameters described above. If there are no empty domains after this stage, the search proceeds by selecting the leftmost variable and assigning it a fragment (block) in a leftmost order. All domains are pre-sorted from the most likely to the least likely for each variable (the previous stage of filtering preserves the ordering).

6   Experimental Results

We report on the experimental results obtained from a prototype implementation of the JM constraint, along with its Joined-Multibody filtering algorithm, in the FIASCO system. The system, protein lists, and some examples are available at `http://www.cs.nmsu.edu/fiasco`. Experiments are conducted on a Linux Intel Core i7 860, 2.5 GHz, memory 8 GB, machine. The proposed method has been applied to a data set of 10 loop targets for each of the lengths 4, 8, and 12 residues. The targets are chosen from a set of non-redundant X-ray crystallography structures [6].

We first analyze the performances of JMf filtering by examining the fraction of the search space explored during solution search. Next, we compare the qualities of the loop conformations generated, by measuring the root mean square deviation (RMSD) of the proposed loop with respect to the native conformation. RMSD captures the overall similarity in space of corresponding atoms.

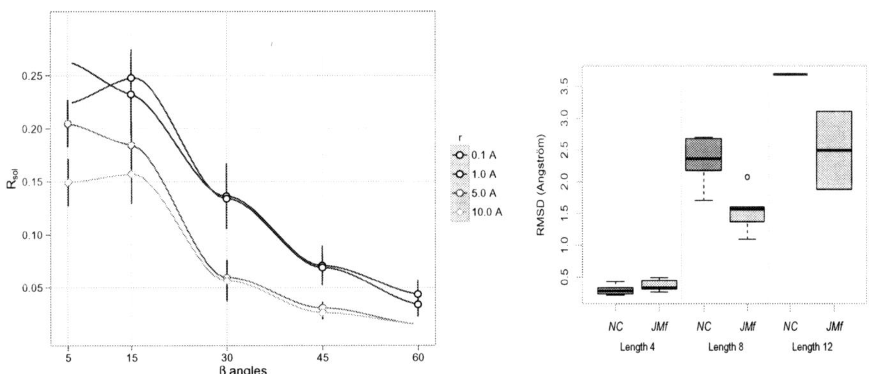

Fig. 5. Ratio of the solutions (left) and RMSD comparison (right)

6.1   Filtered Search Space and Performances

For each of the 10 selected proteins with a loop of length $n = 4$ we compare two
CSPs. The first one enables the JM constraint and the second one disables it
(simple combinatorial fragment assembly). For both problems we have exhaustively generated (without timeout) all the solutions by means of a constraint-
based search using a classical propagate-labeling schema

The number of fragments in each variable domain is 60—this is an adequate
sampling to describe a reasonable amino-acid flexibility. This increases the likelihood to generate a loop structure that is similar to the native one. A loop of
length n generates an exponential search space of size roughly 60^n ($\sim 1.3 \cdot 10^7$ in
this example). The selected variable is the leftmost one. Fragments are selected
in descending frequency order. We have used different values for r and β, while
k_{min} and k_{max} are set to 20 and 100, respectively. In Fig. 5 (left) we show,
for different combinations of clustering parameters, the ratio (R_{sol}) between the
number of solutions to the CSPs with JM constraint and without it. The size of
the prop-labeling tree and running times decrease with a similar trend.

The adopted parameters for the β angles are reported in the x-axes, while the
r values for the clustering distances are plotted in different colors (10Å is the
lightest color). The white dots represent the average values of all the trials and
the vertical bars illustrate the standard error of the mean: $\frac{\sigma}{\sqrt{N}}$, where σ is the
standard deviation and N is the number of samples. It can be observed that the
number of the solutions generated by the JM constraint decreases as the β and
r values increase, as one can expect.

When the clustering is too inaccurate, for some loop targets no solutions are
found by the JM constraint (e.g., with $\beta = 60$ and $r = 10.0$ for 6 loops out of
10). In Section 6.2 we show that the reduced number of solutions does not affect
the approximate completeness of the search.

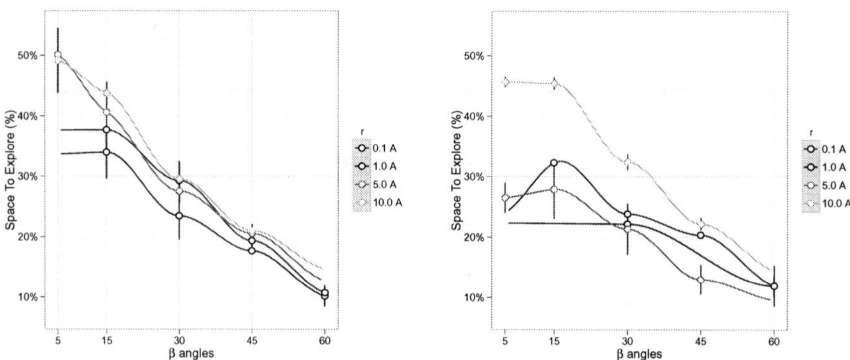

Fig. 6. Ratio of the search space explored using JMf for 8-loops (left) and 12-loops

Let us analyze the case with longer loops (10 proteins of lengths 8 and 10 of lengths 12) where the set of all possible solutions without using JMf cannot be computed in reasonable time. In order to estimate the filtering capability, we build an approximation based on the following algorithm. For $i = 1, \ldots, n$:

- Let us assume we have non-deterministically assigned the variable V_i.
- The variables $V_{i+1}, \ldots, V_n$ have the domains $\mathsf{dom}(V_j) = d_j$, for $j = i+1, \ldots, n$.
- We apply the JM filtering algorithm obtaining the new values $d'_{i+1}, \ldots, d'_n$
- An approximation to the portion of the search space removed at this stage is computed as: $\prod_{j=i+1}^{n} d_j - \prod_{j=i+1}^{n} d'_j$ (2).

To compute the search space ξ removed by the propagation of the JM constraint we sum the values (2) for every node of the search tree visited within the timeout. Let $\varrho = 1 + \sum_{i=1}^{n} \prod_{j=1}^{i} \mathsf{dom}(V_j)$ be the size of the search tree in absence of constraint propagation. In Figure 6 we report the behavior of $\frac{\varrho - \xi}{\varrho}$ for targets loops of length 8 and 12, depending on r and β. Let us observe that ϱ is a rough upper bound of the nodes that really need to be visited (several fragments can immediately lead to a failure due to spatial constraints) and that allotting more time for the computations allows further pruning, thus increasing ξ.

We have set two timeouts: one involving solely the exploration of the prop-labeling tree, fixed to 600 seconds, and another over the total computation (search tree exploration and JM propagation), fixed to 2 hours. In every trial carried out we abort the search due to timeout in exploring the prop-labeling tree. The space filtered by the JM constraint rises according to the increasing values of r and β. In the settings proposed, fixing the values of r and β to 10.0 and 60 respectively, the propagation of the JM constraint produces a filtering over the search space that allows the rest of the search to be carried out merely on about the 10 % of the original prop-labeling tree. However, as stated above, this is just an under-estimation of the pruning capability.

6.2   Loop Conformations Quality

This section provides some insight about the quality of the solutions that are retrieved when using the JM constraint. In particular, we show that the quality in terms of RMSD is not significantly degraded by the large filtering performed by the propagator. As in the previous subsection, we use as reference for the comparisons the results from the search with no JM constraint (named NC). The experiments were carried on with $k_{min} = 20$ and $k_{max} = 100$, while r and β are set respectively to 1.0 and 15 for loops of length 4, and 2.5 and 60 for loops of length 8 and 12. In our analysis, such parameters guarantee a good compromise between filtering power and accuracy of the results.

In Figure 5 (right), the bottom and top point of each vertical line show the RMSD of the best and worst prediction, respectively, within the group of targets analyzed. The results are biased by the fragment database in use: we excluded from it the fragments that belong to the deposited protein targets. Therefore it is not possible to reconstruct the original target loop and none of the searched are expected to reach a RMSD equal to 0. The bottom and top horizontal lines on each box shows the RMSD of the 25th and 75th percentile prediction, respectively, while the line through the middle shows the median. We observe no substantial difference in the distributions related to short loop predictions (length 4), and an improvement for targets of greater size due to time-out. Such results experimentally show the strength of our method: JM filtering algorithm removes successfully redundant conformations; moreover, it quickly direct the search space exploration through predictions that are biologically meaningful.

In Figure 7, we analyze the impact of the k_{max} on computational times (left) and precision (right) of the filtering of the JM constraint. The tests are performed over the protein loops of length 4 adopting as cluster parameters, $r = 1.0$, $\beta = 30$, $k_{min} = 20$. We ignore minimum and maximum values from each data set to smooth out fluctuations and highlight average trends. Each dot in the plot represents the outcome of a trial and the grey area denote the standard error of the mean associated to a particular value of k_{max}. The RMSD values tend to decrease as the number of clusters increase, and it stabilizes when $k_{max} \geq 500$ clusters with a good average value of 0.4Å. As one might expect, instead, the filtering time increases as k_{max} increases.

6.3   Comparison with Other State-of-the-art Loop Samplers

We also compare our method to three other state-of-the-art loop samplers: the Cyclic Coordinate Descent (CCD) algorithm [6], the self-organizing algorithm (SOS) [30], and the FALCm method [29]. Note that our solution does not include specific heuristics and additional information that are used in the other programs. Moreover, our database is not tuned for loop prediction: it is built from any fragment that may appear on a protein (e.g. including helices, β-strands).

Table 1 shows the average RMSD for the benchmarks of length 4, 8 and 12 as computed by the four programs. We report the results as given in Table 2 of [6] for the CCD, Table 1 of [30] for SOS, and Table II of [29] for FALCm method,

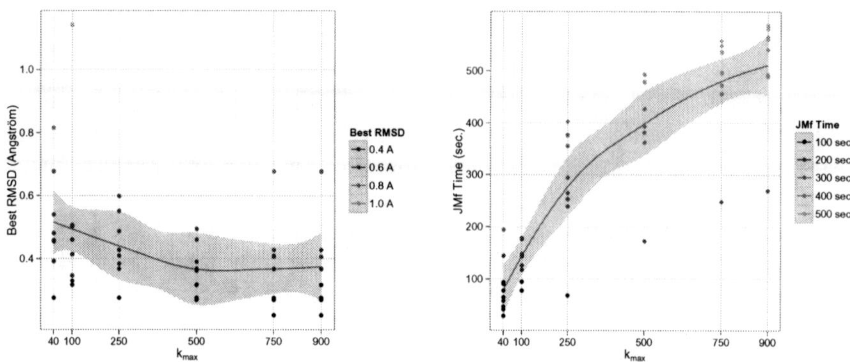

Fig. 7. Comparison of the best RMSD values (left) and JMf computation times (right) at varying of the k_{max} clustering parameter

Table 1. Comparison of loop sampling methods

Loop Length	CCD	SOS	FALCm	JMf
4	0.56	0.20	0.22	0.30
8	1.59	1.19	0.72	1.31
12	3.05	2.25	1.81	1.97

and the RMSD's obtained adopting the best settings for JMf. We do not compare the computational time as they are not provided in the above references. It can be noted that our results are in line with those produced by the other systems, even if a general fragment database has been used in our system.

7   Conclusions

In this paper, we presented a novel constraint (joined-multibody) to model rigid bodies connected by joints, with constrained degrees of freedom in the 3D space. We presented a proof of NP-completeness of the joined-multibody constraint, a filtering algorithm that exploits the geometrical features of the rigid bodies and showed its application in sampling protein loop conformations. Feasibility of the method is shown by performing loop reconstruction tests on a set of loop targets, with lengths ranging from 4 to 12 amino acids. The search space of the protein loop conformations generated is reduced with a controlled loss of quality.

The propagator has been presented as a filtering algorithm based on a directional growth of the rigid body. As future work, we plan to develop a "bi-directional" search that starts from both end anchors. A preliminary study shows that the error propagation, due to the clustering relation, can be bounded with greater accuracy. On the theoretical side, we are interested in proving the NP-hardness in the case of JM constraint based on compatible multi-bodies.

Following the loop conformation direction, we also plan to tune the fragment database and to integrate our filtering algorithm method with other refinements strategies to eliminate infeasible physical loop conformations, (e.g. *DFIRE* potential [37]), to increase the quality of the loop predictions.

As future work, we also plan to apply our filtering method to other related applications: protein-protein interaction, protein flexibility and docking studies. Those systems can be modeled by a set of joined-multibodyconstraints and it appears promising the possibility to explore a large set of conformations by representatives enumeration only.

Acknowledgments. The research has been partially supported by a grant from the Army High Performance Computing Research Center, NSF grants CBET-0754525 and IIS-0812267, and PRIN 2008 (20089M932N).

References

1. Backofen, R., Will, S.: A Constraint-Based Approach to Fast and Exact Structure Prediction in 3-Dimensional Protein Models. Constraints 11(1), 5–30 (2006)
2. Barahona, P., Krippahl, L.: Constraint programming in structural bioinformatics. Constraints 13(1-2), 3–20 (2008)
3. Ben-David, M., Noivirt-Brik, O., Paz, A., Prilusky, J., Sussman, J.L., Levy, Y.: Assessment of CASP8 structure predictions for template free targets. Proteins 77, 50–65 (2009)
4. Best, M., Bhattarai, K., Campeotto, F., Dal Pal'u, A., Dang, H., Dovier, A., Fioretto, F., Fogolari, F., Le, T., Pontelli, E.: Introducing FIASCO: Fragment-based Interactive Assembly for protein Structure prediction with COnstraints. In: Proc. of Workshop on Constraint Based Methods for Bioinformatics (2011), `http://www.dmi.unipg.it/WCB11/wcb11proc.pdf`
5. Cahill, S., Cahill, M., Cahill, K.: On the kinematics of protein folding. Journal of Computational Chemistry 24(11), 1364–1370 (2003)
6. Canutescu, A., Dunbrack, R.: Cyclic coordinate descent: a robotics algorithm for protein loop closure. Protein Sci. 12, 963–972 (2003)
7. Choi, Y., Deane, C.M.: FREAD revisited: Accurate loop structure prediction using a database search algorithm. Proteins 78(6), 1431–1440 (2010)
8. Dal Palù, A., Dovier, A., Fogolari, F.: Constraint logic programming approach to protein structure prediction. BMC Bioinformatics 5, 186 (2004)
9. Dal Palù, A., Dovier, A., Fogolari, F., Pontelli, E.: CLP-based protein fragment assembly. TPLP 10(4-6), 709–724 (2010)
10. Dal Palù, A., Dovier, A., Fogolari, F., Pontelli, E.: Exploring Protein Fragment Assembly Using CLP. In: Walsh, T. (ed.) IJCAI, pp. 2590–2595. IJCAI/AAAI (2011)
11. Dal Palù, A., Dovier, A., Pontelli, E.: A constraint solver for discrete lattices, its parallelization, and application to protein structure prediction. Softw., Pract. Exper. 37(13), 1405–1449 (2007)
12. Dal Palù, A., Dovier, A., Pontelli, E.: Computing approximate solutions of the protein structure determination problem using global constraints on discrete crystal lattices. IJDMB 4(1), 1–20 (2010)
13. Dal Palù, A., Spyrakis, F., Cozzini, P.: A new approach for investigating protein flexibility based on Constraint Logic Programming: The first application in the case of the estrogen receptor. European Journal of Medicinal Chemistry 49, 127–140 (2012)
14. Deane, C., Blundell, T.: CODA. A combined algorithm for predicting the structurally variable regions of protein models. Protein Sci. 10, 599–612 (2001)

15. Dotú, I., Cebrián, M., Van Hentenryck, P., Clote, P.: On Lattice Protein Structure Prediction Revisited. IEEE/ACM Trans. Comput. Biology Bioinform. 8(6), 1620–1632 (2011)

16. Felts, A., Gallicchio, E., Chekmarev, D., Paris, K., Friesner, R., Levy, R.: Prediction of protein loop conformations using AGBNP implicit solvent model and torsion angle sampling. J. Chem. Theory Comput. 4, 855–868 (2008)

17. Fiser, A., Do, R., Sali, A.: Modeling of loops in protein structures. Protein Sci. 9, 1753–1773 (2000)

18. Fogolari, F., Pieri, L., Dovier, A., Bortolussi, L., Giugliarelli, G., Corazza, A., Esposito, G., Viglino, P.: Scoring predictive models using a reduced representation of proteins: model and energy definition. BMC Structural Biology 7(15), 1–17 (2007)

19. Fujitsuka, Y., Chikenji, G., Takada, S.: SimFold energy function for de novo protein structure prediction: consensus with Rosetta. Proteins 62, 381–398 (2006)

20. Hartenberg, R., Denavit, J.: A kinematic notation for lower pair mechanisms based on matrices. Journal of Applied Mechanics 77, 215–221 (1995)

21. Jacobson, M., Pincus, D., Rapp, C., Day, T., Honig, B., Shaw, D., Friesner, R.: A hierarchical approach to all-atom protein loop prediction. Proteins 55, 351–367 (2004)

22. Jamroz, M., Kolinski, A.: Modeling of loops in proteins: a multi-method approach. BMC Struct. Biol. 10(5) (2010)

23. Jauch, R., Yeo, H., Kolatkar, P.R., Clarke, N.D.: Assessment of CASP7 structure predictions for template free targets. Proteins 69, 57–67 (2007)

24. Jones, D.: Predicting novel protein folds by using FRAGFOLD. Proteins 45, 127–132 (2006)

25. Karplus, K., Karchin, R., Draper, J., Casper, J., Mandel-Gutfreund, Y., Diekhans, M., Source, R.H.: Combining local structure, fold-recognition, and new fold methods for protein structure prediction. Proteins 53(6), 491–497 (2003)

26. Kinch, L., Yong Shi, S., Cong, Q., Cheng, H., Liao, Y., Grishin, N.V.: CASP9 assessment of free modeling target predictions. Proteins 79, 59–73 (2011)

27. LaValle, S.: Planning Algorithms. Cambridge University Press (2006)

28. Lee, J., Kim, S., Joo, K., Kim, I., Lee, J.: Prediction of protein tertiary structure using profesy, a novel method based on fragment assembly and conformational space annealing. Proteins 56(4), 704–714 (2004)

29. Lee, J., Lee, D., Park, H., Coutsias, E., Seok, C.: Protein Loop Modeling by Using Fragment Assembly and Analytical Loop Closure. Proteins 78(16), 3428–3436 (2010)

30. Liu, P., Zhu, F., Rassokhin, D., Agrafiotis, D.: A self-organizing algorithm for modeling protein loops. PLOS Comput. Biol. 5(8) (2009)

31. Rapp, C.S., Friesner, R.A.: Prediction of loop geometries using a generalized born model of solvation effects. Proteins 35, 173–183 (1999)

32. Shen, M., Sali, A.: Statistical potential for assessment and prediction of protein structures. Protein Sci. 15, 2507–2524 (2006)

33. Shmygelska, A., Hoos, H.: An ant colony optimisation algorithm for the 2D and 3D hydrophobic polar protein folding problem. BMC Bioinformatics 6 (2005)

34. Simons, K., Kooperberg, C., Huang, E., Baker, D.: Assembly of protein tertiary structures from fragments with similar local sequences using simulated annealing and bayesian scoring functions. J. Mol. Biol. 268, 209–225 (1997)

35. Spassov, V., Flook, P., Yan, L.: LOOPER: a molecular mechanics-based algorithm for protein loop prediction. Protein Eng. 21, 91–100 (2008)

36. Xiang, Z., Soto, C., Honig, B.: Evaluating conformal energies: the colony energy and its application to the problem of loop prediction. PNAS 99, 7432–7437 (2002)

37. Zhou, H., Zhou, Y.: Distance-scaled, finite ideal-gas reference state improves structure-derived potentials of mean force for structure selection and stability prediction. Protein Sci. 11, 2714–2726 (2002)

A Branch and Prune Algorithm for the Computation of Generalized Aspects of Parallel Robots

Stéphane Caro[1], Damien Chablat[1], Alexandre Goldsztejn[2], Daisuke Ishii[3],
and Christophe Jermann[4]

[1] IRCCyN, Nantes, France
`first.last@irccyn.ec-nantes.fr`
[2] CNRS, LINA (UMR-6241), Nantes, France
`Alexandre.Goldsztejn@gmail.com`
[3] National Institute of Informatics, JSPS, Tokyo, Japan
`dsksh@acm.org`
[4] University of Nantes, LINA (UMR-6241), Nantes, France
`Christophe.Jermann@univ-nantes.fr`

Abstract. Parallel robots enjoy enhanced mechanical characteristics that have to
be contrasted with a more complicated design. In particular, they often have parallel singularities at some poses, and the robot may become uncontrollable, and
could even be damaged, in such configurations. The computation of singularity
free sets of reachable poses, called generalized aspects, is therefore a key issue
in their design. A new methodology based on numerical constraint programming
is proposed to compute a certified enclosure of such generalized aspects which
fully automatically applies to arbitrary robot kinematic model.

Keywords: Numerical constraints, parallel robots, singularities, aspects.

1 Introduction

Mechanical manipulators, commonly called robots, are widely used in the industry to
automatize various tasks. Robots are a mechanical assembly of rigid links connected
by mobile joints. Some joints are actuated and they allow commanding the robot operating link, called its end-effector (or platform). One key characteristic of a robot is its
reachable workspace, informally defined as the set of poses its end-effector can reach.
Indeed, its size defines the scope of operational trajectories the robot will be able to perform. Robots comply with either a serial or a parallel (or possibly a hybrid) assembly,
whether their links are connected in series or in parallel. Parallel robots [14,16] present
several advantages with respect to serial ones: They are naturally stiffer, leading to more
accurate motions with larger loads, and allow high speed motions. These advantages
are contrasted by a more complicated design as the computation and analysis of their
workspace present several difficulties. First, one pose of the robot's end-effector may
be reached by several different sets of actuated joint commands (which correspond to
different working modes), and conversely one set of input commands may lead to several poses of its end-effector (which correspond to different assembly modes). Second,
parallel robots generally have parallel singularities, i.e., specific configurations where
they become uncontrollable and can even be damaged.

M. Milano (Ed.): CP 2012, LNCS 7514, pp. 867–882, 2012.

The kinematics of a parallel robot is modeled by a system of equations that relates the pose of its end-effector (which includes its position and possibly its orientation) to its commands. Hence computing the pose knowing the commands, or conversely, requires solving a system of equations, called respectively the direct and inverse kinematic problems. Usually, the number of pose coordinates (also known as *degrees of freedom* (DOF)), the number of commands and the number of equations are the same. Therefore, the relation between the pose and the command is generically a local bijection. However, in some non-generic configurations, the pose and the command are not anymore related by a local bijection. This may affect the robot behavior, e.g., destroying it if some command is enforced with no corresponding pose. These non-generic cases are called robot singularities and they can be of two kinds [1]: Serial or parallel. One central issue in designing parallel robots is to compute connected sets of singularity free poses, so that the robot can safely operate inside those sets. Such a set is called a generalized aspect in [6] when it is maximal with respect to inclusion.

A key issue when computing the aspects is the certification of the results: Avoiding singularities is mandatory, and the connectivity between robot configurations must be certified. This ensures that the robot can actually move safely from one configuration to another. Very few frameworks provide such certifications, among which algebraic computations and interval analysis. The cylindrical algebraic decomposition was used in [5], with the usual restrictions of algebraic methods and with a connectivity analysis limited to robots with two DOF. Interval analysis was used in [17] for robots having a single solution to their inverse kinematic problem. Though limited, this method can still tackle important classes of robots like the Stewart platform. A quad-tree with certification of non-singularity was built in [4] for some planar robots with two DOF. This method extends to higher dimensional robots, but it requires the a priori separation of working modes by adhoc inequalities, and is not certified with respect to connectivity.

In this paper we propose a branch and prune algorithm incorporating the certification of the solutions and of their connectivity. This allows a fully automated computation of the generalized aspect from the model of arbitrary parallel robots, including robots with multiple solutions to their direct and inverse kinematic problems, without requiring any a priori study to separate their working modes. The algorithm is applicable to robots of any dimension, although the complexity of the computations currently restricts its application to robots with three degrees of freedom.

A motivating example is presented in Section 2 followed by some preliminaries about numerical constraint programming and robotics in Section 3. The proposed algorithm for certified aspect computation is presented in Section 4. Finally, experiments on planar robots with two and three degrees of freedom are presented in Section 5.

Notations. Boldface letters denote vectors. Thus $\mathbf{f}(\mathbf{x}) = 0$ denotes a system of equations $\mathbf{f}$ on a vector of variables $\mathbf{x}$: $f_1(x_1, \ldots, x_n) = 0, \ldots, f_k(x_1, \ldots, x_n) = 0$. The Jacobian matrix of $\mathbf{f}(\mathbf{x})$ with respect to variables $\mathbf{x}' \subseteq \mathbf{x}$ is denoted $\mathbf{F}_{\mathbf{x}'}(\mathbf{x})$, and $\det \mathbf{F}_{\mathbf{x}'}(\mathbf{x})$ denotes its determinant. Interval variables are denoted using bracketed symbols, e.g. $[x] = [\underline{x}, \overline{x}] := \{x \in \mathbb{R} \mid \underline{x} \le x \le \overline{x}\}$. Hence, $[\mathbf{x}]$ is an interval vector (or box) and $[A] = ([a_{ij}])$ is an interval matrix. $\mathbb{IR}$ denotes the set of intervals and $\mathbb{IR}^n$ the set of n-dimensional boxes. For an interval $[x]$, we denote $\mathrm{wid}[x] := \overline{x} - \underline{x}$ its width, $\mathrm{int}[x] := \{x \in \mathbb{R} \mid \underline{x} < x < \overline{x}\}$ its interior, and $\mathrm{mid}[x] := (\underline{x} + \overline{x})/2$ its midpoint. These notations are extended to interval vectors.

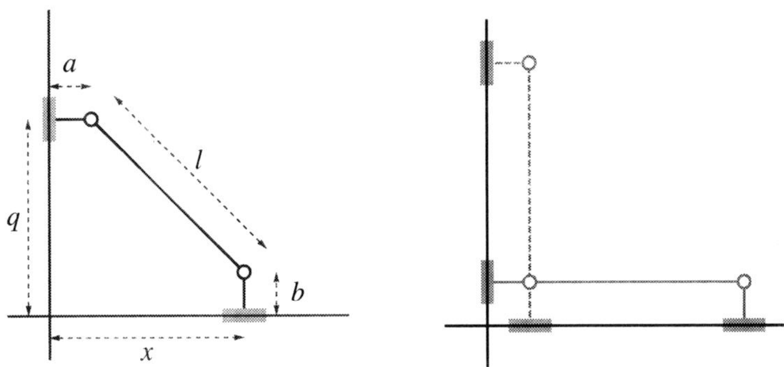

Fig. 1. The P̲RRP in a nonsingular pose (left) and in singular poses (right)

2 Motivating Example

Description. Consider the simple P̲RRP[1] planar robot depicted in Figure 1 which involves two prismatic joints (gray rectangles) sliding along two perpendicular axes. These prismatic joints are connected through three rigid bars (black lines) linked by two revolute joints (circles) which allow free rotations between the rigid bars. The positions of the prismatic joints are respectively denoted by x and q, the end-effector position x being on the horizontal axis and the command q corresponding to the height on the vertical axis. The left-hand side diagram of Figure 1 shows one nonsingular configuration of the robot (note that there is another symmetric pose x associated to the same command q where x is negative). From this configuration, when q changes vertically x changes horizontally, and both are related by a local bijection, hence this configuration is non-singular. The right-hand side diagram shows two singular configurations. In the green, plain line, pose (where the robot's main rigid bar is horizontal), increasing or decreasing the command q entails a decrease of x, hence a locally non-bijective correspondence between these values. In the red, dashed line, pose (where the robot's main rigid bar is vertical), increasing or decreasing the command q would entail a vertical motion of the end-effector which is impossible due to the robot architecture, hence a potential damage to the robot structure. The green configuration is a serial singularity, which restricts the robot mobility without damaging it; the red configuration is a parallel singularity, which is generally dangerous for the robot structure.

Kinematic Model. The kinematic model of this robot is easily derived: The coordinates of the revolute joints are respectively (a, q) and (x, b), where a and b are architecture parameters corresponding to the lengths of the two horizontal and vertical small rigid bars. Then the main oblique rigid bar enforces the distance between these two points to be equal to its length l, a third architecture parameter. Hence, the kinematic model is

$$(x - a)^2 + (q - b)^2 = l^2. \tag{1}$$

---

[1] In robotics, manipulators are typically named according to the sequence of joints they are made of, e.g., P for prismatic and R for revolute, actuated ones being underlined.

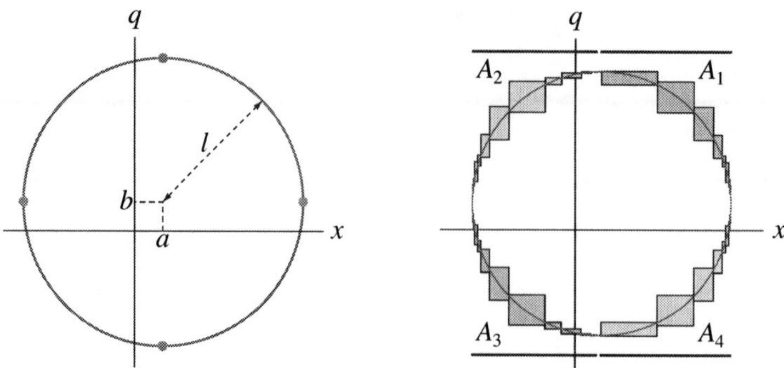

Fig. 2. Left: The P̲RRP kinematic model solutions set. Right: The computed paving.

The solution set of this model, the circle of center (a, b) and radius l, is depicted on the left hand side diagram in Figure 2. The direct kinematic problem consists in computing x knowing q, leading in the case of this robot to two solutions $a \pm \sqrt{l^2 - (q - b)^2}$ if $q \in [b - l, b + l]$, no solution otherwise. Similarly, the inverse kinematic problem consists in computing q knowing x, leading to two solutions $b \pm \sqrt{l^2 - (x - a)^2}$ provided that $x \in [a - l, a + l]$, no solution otherwise. This simple robot is typical of parallel robots, which can have several solutions to their direct and inverse kinematic problems. It is also typical regarding its singularities: It has two serial singularities where the solution set has a vertical tangent (green, leftmost and rightmost, points on the left hand side diagram in Figure 2), and two parallel singularities where the solution set has a horizontal tangent (red, topmost and bottommost, points on the left hand side diagram in Figure 2). These four singularities split the solution set into four singularity free regions, called generalized aspects. Accordingly, we can determine the singularity free reachable workspace of the robot by projecting each aspect onto the x component (see the thick lines above and under the paving on the right-hand side in Figure 2).

Certified Enclosure of Generalized Aspects. This paper aims at using numerical constraint programming in order to compute some certified enclosures of the different aspects. The standard branch and prune algorithm is adapted in such a way that solving the robot kinematic model together with non-singularity constraints leads to the enclosure depicted on the right-hand side of Figure 2. The solution boxes verify:

1. In each box, for each pose x there exists a unique command q such that (x, q) satisfies the robot kinematic model;
2. Each box does not contain any singularity;
3. Each two neighbor boxes share a common solution.

The first two properties ensure that each box is crossed by a single aspect, and that this aspect covers the whole box projection on the x subspace. The third certificate allows connecting neighbor boxes proving that they belong to the same aspect. Therefore, the connected components A_1, A_2, A_3, A_4 of the computed boxes shown in Figure 2 allow separating the four aspects, and provide, by projection, certified inner approximations of the singularity-free reachable workspace of the robot.

3   Preliminaries

3.1   Numerical Constraint Programming

Numerical constraint solving inherits principles and methods from discrete constraint solving [18] and interval analysis [19]. Indeed, as their variable domains are continuous subsets of $\mathbb{R}$, it is impossible to enumerate the possible assignments and numeric constraint solvers thus resorts to interval computations. As a result, we use an *interval extension* $[f] : \mathbb{IR}^n \to \mathbb{IR}$ of each function $f : \mathbb{R}^n \to \mathbb{R}$ involved in a constraint, such that $\forall [\mathbf{a}] \in \mathbb{IR}^n$, $\forall \mathbf{a} \in [\mathbf{a}]$, $f(\mathbf{a}) \in [f]([\mathbf{a}])$.

Numerical Constraint Satisfaction Problems. A *numerical constraint satisfaction problem* (NCSP) is defined as a triple $\langle \mathbf{v}, [\mathbf{v}], c \rangle$ that consists of

- a vector of *variables* $\mathbf{v} = (v_1, \ldots, v_n)$,
- an *initial domain*, in the form of a box, represented as $[\mathbf{v}] \in \mathbb{IR}^n$, and
- a *constraint* $c(\mathbf{v}) := (\mathbf{f}(\mathbf{v}) = 0 \wedge \mathbf{g}(\mathbf{v}) \geq 0)$, $\mathbf{f} : \mathbb{R}^n \to \mathbb{R}^e$ and $\mathbf{g} : \mathbb{R}^n \to \mathbb{R}^i$,
  i.e., a conjunction of equations and inequalities.

A *solution* of an NCSP is an assignment of its variables $\mathbf{v} \in [\mathbf{v}]$ that satisfies its constraints. The *solution set* Σ of an NCSP is the region within its initial domain that satisfies its constraints, i.e., $\Sigma([\mathbf{v}]) := \{ \mathbf{v} \in [\mathbf{v}] \mid c(\mathbf{v}) \}$.

The Branch and Prune Algorithm. The *branch and prune algorithm* [22] is the standard complete solving method for NCSPs. It takes a problem as an input and outputs two sets of boxes, called respectively the undecided ($\mathcal{U}$) and solution ($\mathcal{S}$) boxes. It interleaves a refutation phase, called *prune*, that eliminates inconsistent assignments within a box, and an exploration phase, called *branch*, that divides a box into several sub-boxes to be searched recursively, until a prescribed precision ϵ is reached. Algorithm 1 shows a generic description of this scheme. It involves four subroutines: Extract (extraction of the next box to be processed), Prune_c (reduction of the domains based on refutation of assignments that cannot satisfy a subset of constraint c), Prove_c (certification that a box contains a solution of the constraint c), and Branch (division of the processed box into sub-boxes to be further processed). Each of them has to be instantiated depending on the problem to be solved. The procedure Prune_c obviously depends on the type of constraint in the problem (e.g. inequalities only), as well as on other characteristics of the problem. The procedures Extract and Branch allow defining the search strategy (e.g. breadth-first, depth-first, etc.) which may be tuned differently with regards to the problem. The procedure Prove_c actually defines the aim of the branch and prune: Being a solution can take different meaning depending on the considered problem and the question asked.[2] For instance, if the question is to find the real solutions of a well-constrained system of equations, then it will generally implement a solution existence (and often uniqueness) theorem, e.g., Miranda, Brouwer or interval Newton [21], that guarantees that the considered box contains a (unique) real solution; on the other hand,

---

[2] For discrete CSPs, Prove_c usually checks the given assignment satisfies the constraint.

Algorithm 1. Branch and prune

Input: NCSP $\langle \mathbf{v}, ([\mathbf{v}]), c \rangle$, precision $\epsilon > 0$
Output: pair of sets of boxes $(\mathcal{U}, \mathcal{S})$
 1: $\mathcal{L} \leftarrow \{[\mathbf{v}]\}$, $\mathcal{S} \leftarrow \emptyset$ and $\mathcal{U} \leftarrow \emptyset$
 2: **while** $\mathcal{L} \neq \emptyset$ **do**
 3:    $[\mathbf{v}] \leftarrow \mathrm{Extract}(\mathcal{L})$
 4:    $[\mathbf{v}] \leftarrow \mathrm{Prune}_c([\mathbf{v}])$
 5:    **if** $[\mathbf{v}] \neq \emptyset$ **then**
 6:       **if** $\mathrm{Prove}_c([\mathbf{v}])$ **then**
 7:          $\mathcal{S} \leftarrow \mathcal{S} \cup \{[\mathbf{v}]\}$
 8:       **else if** $\mathrm{wid}[\mathbf{v}] > \epsilon$ **then**
 9:          $\mathcal{L} \leftarrow \mathcal{L} \cup \mathrm{Branch}([\mathbf{v}])$
10:       **else**
11:          $\mathcal{U} \leftarrow \mathcal{U} \cup \{[\mathbf{v}]\}$
12:       **end if**
13:    **end if**
14: **end while**
15: **return** $(\mathcal{U}, \mathcal{S})$

---

if the question is to compute the solution set of a conjunction of inequality constraints, then it will usually implement a solution universality test that guarantees that every real assignment in the considered box is a solution of the NCSP.

3.2   Parallel Robots, Singularities and Generalized Aspects

As illustrated in Section 2, the *kinematic model* of a parallel robot can be expressed as a system of equations relating its end-effector pose $\mathbf{x}$ and its commands $\mathbf{q}$:

$$\mathbf{f}(\mathbf{x}, \mathbf{q}) = 0. \tag{2}$$

The subspace restricted to the pose parameters $\mathbf{x}$ (resp. command parameters $\mathbf{q}$) is known as the *workspace* (resp. *joint-space*). The projection $\Sigma_{\mathbf{x}}$ (resp. $\Sigma_{\mathbf{q}}$) of the solution set of equation 2 is called the robot *reachable workspace* (resp. *reachable joint-space*). The solution set Σ itself is called the *kinematic manifold* and lies in what is known as the (pose-command) *product space*. In this paper, we restrict to the most typical architectures which satisfy $\dim \mathbf{x} = \dim \mathbf{q} = \dim \mathbf{f} = n$, i.e., neither over- nor under-actuated manipulators. Then, by the implicit function theorem, this system of equations defines a local bijection between $\mathbf{x}$ and $\mathbf{q}$ provided the Jacobian matrices $\mathbf{F}_{\mathbf{x}}(\mathbf{x}, \mathbf{q})$ and $\mathbf{F}_{\mathbf{q}}(\mathbf{x}, \mathbf{q})$ are non-singular. The configurations $(\mathbf{x}, \mathbf{q})$ that do not satisfy these regularity conditions are called singularities, respectively serial or parallel whether $\mathbf{F}_{\mathbf{q}}(\mathbf{x}, \mathbf{q})$ or $\mathbf{F}_{\mathbf{x}}(\mathbf{x}, \mathbf{q})$ is singular. These singularity conditions correspond to the horizontal and vertical tangents of the kinematic manifold described in Section 2.

A key issue in robotics is to be able to control a robot avoiding singularities (in particular reaching a parallel singularity can dramatically damage a robot). This leads to the definition of *generalized aspects* [6] as maximal connected sets of nonsingular configurations $(\mathbf{x}, \mathbf{q})$ that can all be reached without crossing any singularity. More formally, the set of reachable nonsingular configurations of the robot is

$$\{(\mathbf{x}, \mathbf{q}) \in \mathbb{R}^n \times \mathbb{R}^n \mid \mathbf{f}(\mathbf{x}, \mathbf{q}) = 0, \det \mathbf{F_x}(\mathbf{x}, \mathbf{q}) \neq 0, \det \mathbf{F_q}(\mathbf{x}, \mathbf{q}) \neq 0\}. \qquad (3)$$

This corresponds e.g. to the left hand side diagram in Figure 2 where the four singularities (green and red points) are removed. As illustrated by this diagram, the set (3) is generally made of different maximal connected components. The generalized aspects of the robots are defined to be these maximal connected singularity free components.

For a given generalized aspect $\mathbb{A}$, its projection $\mathbb{A_x}$ is a maximal singularity-free region in the robot reachable workspace. Knowing these regions allows roboticians to safely plan robot motions: Any two poses in $\mathbb{A_x}$ are connected by at least one singularity-free path. In addition, the study of aspects provides important information about robot characteristics, e.g., if $(\mathbf{x}, \mathbf{q})$ and $(\mathbf{x}, \mathbf{q'})$ exist in an aspect $\mathbb{A}$ and $\mathbf{q} \neq \mathbf{q'}$, i.e., two different commands yield the same pose, then the robot is said to be *cuspidal* [20]. Cuspidal robots can change working mode without crossing singularities, yielding an extra flexibility in their usage. Finally, the computation of aspects allows roboticians to make informed choices when designing a robot for a given task.

4   Description of the Method

The proposed method for the generalized aspect computation relies on solving the following NCSP whose solutions are the nonsingular configurations of the robot:

$$\left\langle (\mathbf{x}, \mathbf{q}) , ([\mathbf{x}], [\mathbf{q}]) , \mathbf{f}(\mathbf{x}, \mathbf{q}) = 0 \wedge \det \mathbf{F_x}(\mathbf{x}, \mathbf{q}) \neq 0 \wedge \det \mathbf{F_q}(\mathbf{x}, \mathbf{q}) \neq 0 \right\rangle. \qquad (4)$$

Let $\Sigma([\mathbf{x}], [\mathbf{q}])$ be the solution set of this NCSP. Our method computes a set of boxes partly covering this solution set, grouped into connected subsets that represent approximations of the aspects of the considered robot. The computed boxes have to satisfy the specific properties stated in Subsection 4.1. The corresponding branch and prune instantiation is described in Subsection 4.2. The connections between the output boxes have to be certified as described in Subsection 4.3.

4.1   From the NCSP Model to the Generalized Aspects Computation

We aim at computing a (finite) set of boxes $\mathcal{S} \subseteq \mathbb{IR}^n \times \mathbb{IR}^n$ together with (undirected) links $\mathcal{N} \subseteq \mathcal{S}^2$, satisfying the following three properties:

(P1) $\forall([\mathbf{x}], [\mathbf{q}]) \in \mathcal{S}, \; \forall \mathbf{x} \in [\mathbf{x}], \; \exists$ a unique $\mathbf{q} \in [\mathbf{q}], \; \mathbf{f}(\mathbf{x}, \mathbf{q}) = 0$;
(P2) $\forall([\mathbf{x}], [\mathbf{q}]) \in \mathcal{S}, \; \forall \mathbf{x} \in [\mathbf{x}], \; \forall \mathbf{q} \in [\mathbf{q}], \det \mathbf{F_x}(\mathbf{x}, \mathbf{q}) \neq 0 \wedge \det \mathbf{F_q}(\mathbf{x}, \mathbf{q}) \neq 0$;
(P3) $\forall(([\mathbf{x}], [\mathbf{q}]), ([\mathbf{x'}], [\mathbf{q'}])) \in \mathcal{N}, \; \exists(\mathbf{x}, \mathbf{q}) \in ([\mathbf{x}], [\mathbf{q}]) \cap ([\mathbf{x'}], [\mathbf{q'}]), \mathbf{f}(\mathbf{x}, \mathbf{q}) = 0.$

Property (P1) allows defining in each $([\mathbf{x}], [\mathbf{q}]) \in \mathcal{S}$ a function $\kappa_{([\mathbf{x}],[\mathbf{q}])} : [\mathbf{x}] \to [\mathbf{q}]$ that associates the unique command $\mathbf{q} = \kappa_{([\mathbf{x}],[\mathbf{q}])}(\mathbf{x})$ with a given position $\mathbf{x} \in [\mathbf{x}]$ (i.e., the solution of the inverse kinematic problem locally defined inside $([\mathbf{x}], [\mathbf{q}])$). Property (P2) allows applying the Implicit Function Theorem to prove that $\kappa_{([\mathbf{x}],[\mathbf{q}])}$ is differentiable (and hence continuous) inside $[\mathbf{x}]$. Therefore, for a given box $([\mathbf{x}], [\mathbf{q}]) \in \mathcal{S}$, the solution set restricted to this box

$$\Sigma([\mathbf{x}], [\mathbf{q}]) = \left\{ \left(\mathbf{x}, \kappa_{([\mathbf{x}],[\mathbf{q}])}(\mathbf{x}) \right) : \mathbf{x} \in [\mathbf{x}] \right\} \tag{5}$$

is proved to be connected and singularity free, and is thus a subset of one generalized aspect. These properties are satisfied by the motivating example output shown on the right-hand side in Figure 2. Remark that given a box $([\mathbf{x}], [\mathbf{q}]) \in \mathcal{S}$ and a position $\mathbf{x} \in [\mathbf{x}]$, the corresponding command $\kappa_{([\mathbf{x}],[\mathbf{q}])}(\mathbf{x})$ is easily computed using Newton iterations applied to the system $\mathbf{f}(\mathbf{x}, \cdot) = 0$ with initial iterate $\tilde{\mathbf{q}} \in [\mathbf{q}]$ (e.g. $\tilde{\mathbf{q}} = \mathrm{mid}[\mathbf{q}]$).

Property (P3) basically entails that $\Sigma([\mathbf{x}], [\mathbf{q}])$ and $\Sigma([\mathbf{x}'], [\mathbf{q}'])$ are connected, and are thus subsets of the same aspect. Finally, assuming $\mathcal{S}_k \subseteq \mathcal{S}$ to be a connected component of the undirected graph $(\mathcal{S}, \mathcal{N})$, the solution set

$$\bigcup_{([\mathbf{x}],[\mathbf{q}]) \in \mathcal{S}_k} \Sigma([\mathbf{x}], [\mathbf{q}]) \tag{6}$$

is proved to belong to one generalized aspect. The next two subsections explain how to instantiate the branch and prune algorithm in order to achieve these three properties.

4.2   Instantiaton of the Branch and Prune Algorithm

Pruning. In our context, implementing the Prune_c function as a standard AC3-like fixed-point propagation of contracting operators that enforce local consistencies, like the Hull [2,15] or the Box consistencies [2,10], is sufficient. Indeed, this allows an inexpensive refutation of non-solution boxes. Moreover, stronger consistency can be achieved at no additional cost thanks to the solution test described below: the interval-Newton-based operator applied for certifying that a box covers an aspect can also refute non-solution boxes and allows pruning with respect to the whole constraint.

Search Strategy. The standard search strategy for NCSPs applies appropriately in our context. Because boxes are output as soon as they are certified or they have reached a prescribed precision, using a DFS approach to the Extract function is adequate and avoids the risk of filling up the memory, unlike a BFS or LFS[3] approach. The Branch function typically selects a variable in a round-robin manner (i.e., all domains are selected cyclically) and splits the corresponding interval at its midpoint (i.e., a domain is split into two halves).

Solution Test. The Prove_c procedure of Algorithm 1 has to return true only when Property (P1) and Property (P2) are verified. The former is related to proving the existence of solution which is performed using a parametric Newton operator as described in the following paragraph. The latter requires checking the regularity of some interval matrices as described in the next paragraph.

---

[3] Largest-first search.

Existence proof. The standard way to prove that a box $([\mathbf{x}], [\mathbf{q}])$ satisfies Property (P1) is to use a parametric interval Newton existence test [8,9,11]. Using the Hansen-Sengupta [21] version of the interval Newton, the following sequence is computed

$$[\mathbf{q}^0] := [\mathbf{q}], \ldots, [\mathbf{q}^{k+1}] := [H]([\mathbf{q}^k]) \cap [\mathbf{q}^k] \tag{7}$$

where $[H]$ is the Hansen-Sengupta operator applied to the system $\mathbf{f}([\mathbf{x}], \cdot) = 0$. As soon as $\emptyset \neq [\mathbf{q}^{k+1}] \subseteq \mathrm{int}[\mathbf{q}^k]$ is verified, the box $([\mathbf{x}], [\mathbf{q}^{k+1}])$ is proved to satisfy Property (P1), and hence so does $([\mathbf{x}], [\mathbf{q}])$ since the former is included in the latter. However, because Algorithm 1 has to bisect the domain $[\mathbf{q}]$ for insuring convergence by separating the different commands associated to the same pose,[4] this test fails in practice in most situations. This issue was overcome in [11], in the restricted context of constraints of the form $\mathbf{x} = \mathbf{f}(\mathbf{q})$, by computing $[\mathbf{q}^{k+1}] := [H]([\mathbf{q}^k])$ in (7), i.e., removing the intersection with $[\mathbf{q}^k]$, in order to allow inflating and shifting $[\mathbf{q}^{k-1}]$ if necessary.[5] As a result, the Hansen-Sengupta acts as a rigorous local search routine allowing the sequence to converge towards the aimed solution set. An inflation factor τ also has to be applied before the Hansen-Sengupta operator so as to ease the strict inclusion test after each iteration. Hence, the computation of $[\mathbf{q}^{k+1}]$ is as follows:

$$[\tilde{\mathbf{q}}^k] := \mathrm{mid}[\mathbf{q}^k] + \tau([\mathbf{q}^k] - \mathrm{mid}[\mathbf{q}^k]) \quad \text{and} \quad [\mathbf{q}^{k+1}] := [H]([\tilde{\mathbf{q}}^k]). \tag{8}$$

Then the condition $\emptyset \neq [\mathbf{q}^{k+1}] \subseteq \mathrm{int}[\tilde{\mathbf{q}}^k]$ also implies Property (P1) and is likely to succeed as soon as $([\mathbf{x}], [\mathbf{q}])$ is small enough and close enough to some nonsingular solution, which eventually happens thanks to the bisection process. A typical value for the inflation factor is $\tau = 1.01$; It would have to be more accurately tuned for badly conditioned problems, but it is not the case of usual robots.

Regularity test. In order to satisfy the regularity constraints, the interval evaluation of each Jacobian $\mathbf{F_x}$ and $\mathbf{F_q}$ over the box $([\mathbf{x}], [\mathbf{q}])$ has to be regular. Testing the regularity of interval matrices is NP-hard, so sufficient conditions are usually used instead. Here, we use the strong regularity of a square interval matrix $[A]$, which consists in checking that $C[A]$ is strongly diagonally dominant, where C is usually chosen as an approximate inverse of the midpoint of $[A]$ (see [21]).

4.3 Connected Component Computation

In order to distinguish boxes in $\mathcal{S}$ belonging to one specific aspect from the rest of the paving, we use transitively the relation between linked boxes defined by Property (P3), i.e., we have to compute connected components with respect to the links in $\mathcal{N}$. This is done in three steps:

Step 1. Compute *neighborhood relations* between boxes, i.e., determine when two boxes share at least one common point;

Step 2. Certify aspect connectivity in neighbor boxes, i.e., check Property (P3);

Step 3. Compute connected components with respect to the certified links.

---

[4] In [8], only problems where parameters have one solution were tackled, hence allowing successfully using the parametric existence test (7).

[5] This was already used in [9] in a completely different context related to sensitivity analysis, and in a recently submitted work of the authors dedicated to the projection of a manifold.

Step 1. Computing Neighborhood Relations. Two boxes $([\mathbf{x}], [\mathbf{q}])$ and $([\mathbf{x}'], [\mathbf{q}'])$ are *neighbors* if and only if they share at least one common point, i.e., $([\mathbf{x}], [\mathbf{q}]) \cap ([\mathbf{x}'], [\mathbf{q}']) \neq \emptyset$. The neighborhood relations between boxes $\mathcal{N}$ are obtained during the branch and prune computation: After the current box has been pruned (line 4 of Alg. 1), its neighbors are updated accordingly (it may have lost some neighbors); also, the boxes produced when splitting the current box (line 9 of Alg. 1) inherit from (some of) the neighbors of the current box, and are neighbors to one another. One delicate point in managing neighborhood comes from the fact that pose or command parameters are often angles whose domains are restricted to a single period, e.g., $[-\pi, \pi]$; the periodicity of these parameters has to be taken into account: Boxes are neighbors when they share a common point modulo 2π on their periodic dimensions.

Step 2. Certifying Connectivity between Neighbors. Once the branch and prune algorithm has produced the paving $\mathcal{S}$ and its neighboring information $\mathcal{N}$, a post-process is applied to filter from $\mathcal{N}$ the links that do not guarantee the two neighbor boxes cover the same aspect: it may happen that two neighbor boxes share no common point satisfying the kinematic relation $\mathbf{f} = 0$, e.g., if they each cover a portion of two disjoint, but close, aspects. Asserting neighborhood Property (P3) requires again a certification procedure: For any neighbor boxes $\big(([\mathbf{x}], [\mathbf{q}]), ([\mathbf{x}'], [\mathbf{q}'])\big) \in \mathcal{N}$, we verify

$$\exists \mathbf{q} \in ([\mathbf{q}] \cap [\mathbf{q}']), \mathbf{f}(\mathrm{mid}([\mathbf{x}] \cap [\mathbf{x}']), \mathbf{q}) = 0. \tag{9}$$

Indeed, for connectivity to be certified, it is sufficient to prove that the intersection of neighbor boxes share at least one point from the same aspect. Because neighbor boxes $([\mathbf{x}], [\mathbf{q}])$ and $([\mathbf{x}'], [\mathbf{q}'])$ are in $\mathcal{S}$, they satisfy Property (P1) and Property (P2), i.e., $\forall \mathbf{x} \in ([\mathbf{x}] \cap [\mathbf{x}']), \exists$ a unique $\mathbf{q} \in [\mathbf{q}], \mathbf{f}(\mathbf{x}, \mathbf{q}) = 0$ and $\exists$ a unique $\mathbf{q}' \in [\mathbf{q}'], \mathbf{f}(\mathbf{x}, \mathbf{q}') = 0$. We need to check these unique values $\mathbf{q}$ and $\mathbf{q}'$ are actually the same for one value $\mathbf{x}$ inside $[\mathbf{x}] \cap [\mathbf{x}']$, e.g., $\mathbf{x} = \mathrm{mid}([\mathbf{x}] \cap [\mathbf{x}'])$. Using the certification procedure described in Section 4.2 allows proving Equation (9). Each link is certified this way. If the certification fails for a given link, it is removed from $\mathcal{N}$.

Step 3. Computing Connected Components. Given the set $\mathcal{N}$ of certified connections between certified boxes in $\mathcal{S}$, a standard connected component computation algorithm (e.g., [13]) is applied in order to obtain a partition of $\mathcal{S}$ into $\mathcal{S}_k$, each $\mathcal{S}_k$ covering a single aspect of the considered robot.

5   Experiments

We present experiments on four planar robots with respectively 2 and 3 degrees of freedom, yielding respectively a 2-/3-manifold in a 4/6 dimensional product space.

Robot Models. Robot RPRPR (resp. RRRRR) has two arms, each connecting an anchor point (A, B) to its end-effector (P), each composed of a revolute joint, a prismatic (resp. revolute) joint and again a revolute joint in sequence. The end-effector P lies at the shared extremal revolute joint and is described as a 2D point $(x_1, x_2) \in [-20, 20]^2$.

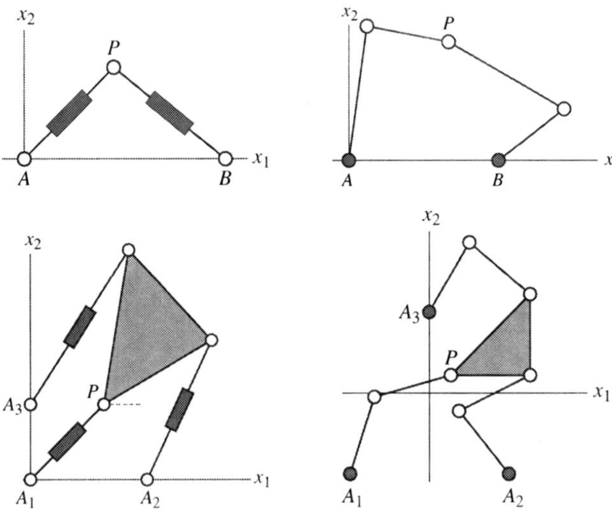

Fig. 3. R$\underline{P}$R$\underline{P}$R (top-left), $\underline{R}$RRRR (top-right), 3-R$\underline{P}$R (bottom-left) and 3-$\underline{R}$RR (bottom-right)

The prismatic (resp. initial revolute) joint in each arm is actuated, allowing to vary the arms lengths (resp. angles). The arm lengths (resp. angles) are considered to be the command $(q_1, q_2) \in [2, 6] \times [4, 9]$ (resp. $[-\pi, \pi]^2$) of the robot. Using the architecture parameters defined in [3] (resp. [6]), their kinematic equations are:

$$x_1^2 + x_2^2 - q_1^2 = 0,$$
$$(x_1 - 9)^2 + x_2^2 - q_2^2 = 0,$$

and, respectively

$$(x_1 - 8\cos q_1)^2 + (x_2 - 8\sin q_1)^2 - 25 = 0,$$
$$(x_1 - 9 - 5\cos q_2)^2 + (x_2 - 5\sin q_2)^2 - 64 = 0.$$

Robot 3-R$\underline{P}$R (resp. 3-$\underline{R}$RR, with the restriction that it has only a fixed orientation, i.e., $x_3 = 0$, its free orientation variant being too complex for the current implementation of the method) has three arms, each connecting an anchor point (A_1, A_2, A_3) to its end-effector (P), each composed of a revolute joint, a prismatic (resp. revolute) joint and again a revolute joint in sequence. The end-effector is a triangular platform whose vertices are attached to the extremal revolute joints of the arms. The position parameters (x_1, x_2, x_3) represent the coordinates $(x_1, x_2) \in [-50, 50]^2$ of one vertex of the platform, and the angle $x_3 \in [-\pi, \pi]$ between its basis and the horizontal axis. The prismatic (resp. initial revolute) joint in each arm is actuated, allowing to vary the arm lengths (resp. angles). The arm lengths (resp. angles) are considered to be the command $(q_1, q_2, q_3) \in [10, 32]^3$ (resp $[-\pi, \pi]^3$) of the robot. Using the architecture parameters defined in [7] (resp. [3]), their kinematic equations are:

$$x_1^2 + x_2^2 - q_1^2 = 0,$$
$$(x_1 + 17\cos x_3 - 15.9)^2 + (x_2 + 17\sin x_3)^2 - q_2^2 = 0,$$
$$(x_1 + 20.8\cos(x_3 + 0.8822))^2 + (x_2 + 20.8\sin(x_3 + 0.8822) - 10)^2 - q_3^2 = 0,$$

and, respectively

$$(x_1 - 10 - 10\cos q_1)^2 + (x_2 - 10 - 10\sin q_1)^2 - 100 = 0,$$
$$(x_1 + 10\cos x_3 - 10 - 10\cos q_2)^2 +$$
$$(x_2 + 10\sin x_3 - 10 - 10\sin q_2)^2 - 100 = 0,$$
$$(x_1 + 10\sqrt{2}\cos(x_3 + \pi/4) - 10\cos q_3)^2 +$$
$$(x_2 + 10\sqrt{2}\sin(x_3 + \pi/4) - 10 - 10\sin q_3)^2 - 100 = 0.$$

Implementation. We have implemented the proposed method described in Section 4 using the Realpaver library [12] in C++, specializing the classes for the different routines in the branch and prune algorithm. Given an NCSP that models a robot and a prescribed precision ϵ, the implementation outputs certified boxes grouped by certified connected components as explained in Section 4. Hence we can count not only the number of output boxes but also the number of output certified connected components. The experiments were run using a 3.4GHz Intel Xeon processor with 16GB of RAM.

Results of the Method. Table 1 provides some figures on our computations. Its columns represent the different robots we consider. Line "# aspects" provides the theoretically established number of aspects of each robot provided in [3,6,7]; This value is unknown for the 3-$\underline{R}$RR robot. Line "precision" gives the prescribed precision ϵ used in the computation. Lines "# boxes" and "# CC" give respectively the number of boxes and the number of connected components returned by our method. Line "time" gives the overall computational time in seconds of the method, including the post-processes.

Table 1. Experimental results

	$\underline{P}$RRP	R$\underline{P}$R$\underline{P}$R	R$\underline{R}$RRR	3-R$\underline{P}$R	3-$\underline{R}$RR
# aspects	4	2	10	2	unknown
precision	0.1	0.1	0.1	0.3	0.008
# boxes	38	2 176	69 612	13 564 854	11 870 068
# CC	4	4	1 767	44 220	56 269
# CC$_{filtered}$	4	2	10	2	25
time (in sec.)	0.003	0.36	38	12 700	10 700

Note that despite the quite coarse precisions we have used, the number of output boxes can be very large, due to the dimension of the search space we are paving. The number of connected components is much smaller, but still it is not of the same order as the theoretically known number of aspects, implying numerous disjoint connected components does in fact belong to the same aspect. This is explained by the numerical instability of the kinematic equations of the robots in the vicinity of the aspect boundaries, which are singularities of the robot. Indeed, in these regions, the numerical certification process cannot operate homogeneously, resulting in disconnected subsets of certified boxes, separated either by non-certified boxes or by non-certified links.

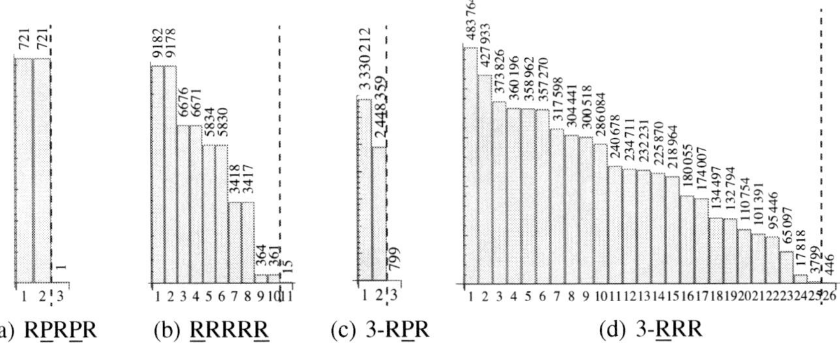

Fig. 4. Number of boxes in each connected component. Each bar corresponding to a connected component shows the number of contained boxes (ordered largest first).

Fig. 5. Computed 3D workspace of of 3-RPR (after filtering). First figure shows the undecided boxes that cover the surface of the workspace. Second and third figures show the certified connected components corresponding to the two aspects.

Under this assumption, each aspect should consist of, on the one hand, one large connected component comprising wide boxes covering the regular inner part of the aspect, and smaller and smaller boxes close to its boundary; And, on the other hand, numerous small connected components comprising tiny boxes gathered on its boundary. As a result, the number of boxes in the "boundary" components should be many times smaller than that of the regular component in an aspect. It should thus be possible to filter out these spurious tiny "boundary" components, based on the number of boxes they contain, as they have no practical use in robotics.

In order to distinguish the relevant components from the spurious ones, we use such a filtering post-process on the output of our method: Output connected components are ordered by decreasing number of constituting boxes; The largest ratio, in number of constituting boxes, between two consecutive components in this order is computed, and used as a separation between relevant and spurious components. Applying this heuristic post-process, the number of obtained connected components, reported at Line "# $CC_{filtered}$" in Table 1, reaches the theoretically known number of aspects in the case of the robots we considered. Figure 4 illustrates the number of boxes of the connected components retained after filtering (the dashed lines represent the computed heuristic

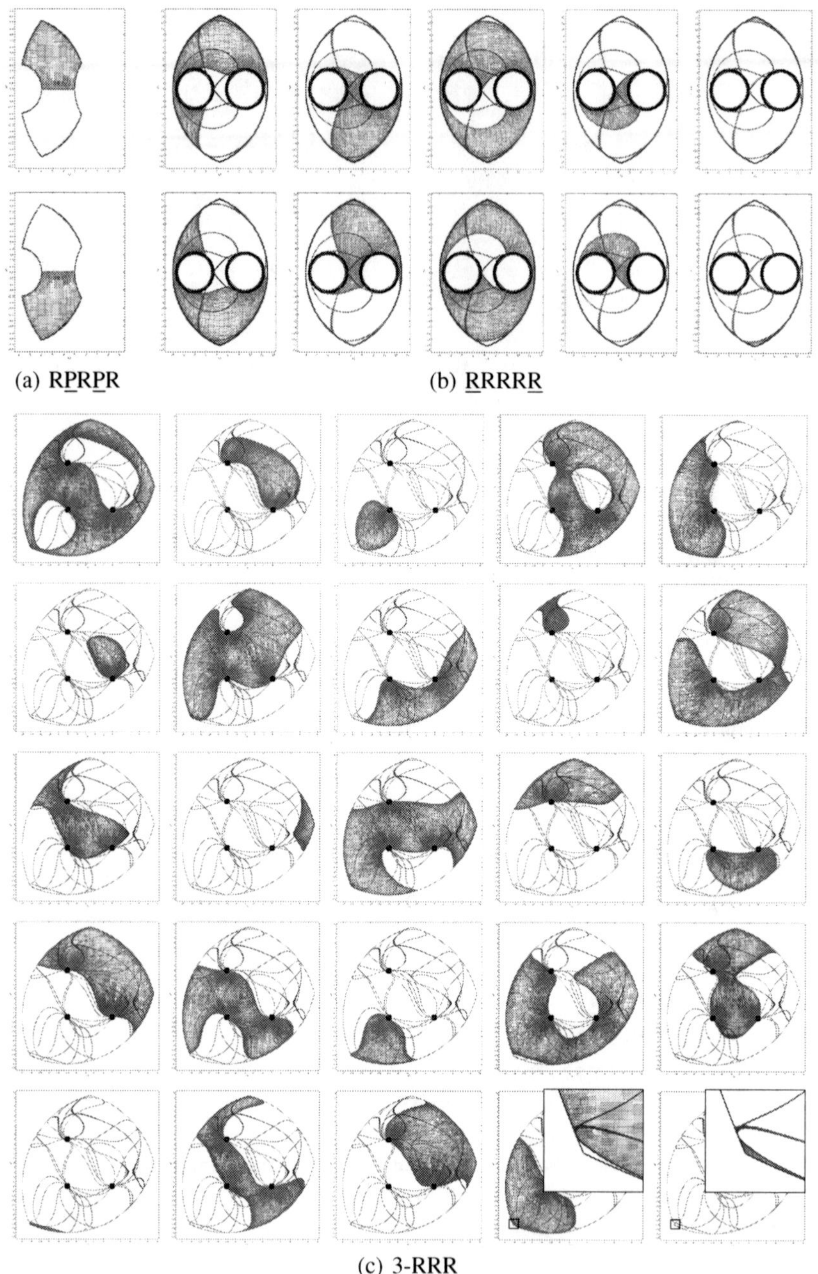

(a) R$\underline{\text{P}}$R$\underline{\text{P}}$R

(b) $\underline{\text{R}}$RRR$\underline{\text{R}}$

(c) 3-$\underline{\text{R}}$RR

Fig. 6. Projections into the 2D workspace of the computed aspects (after filtering). Green boxes are certified; red and black boxes are undecided (i.e., do not satisfy Properties (P1) and (P2), respectively).

thresholds). This seems to indicate that our assumption is correct for the considered robots, i.e., that the major part of each aspect is indeed covered with a single large, regular, connected component.

The retained connected components projected onto the x subspace are depicted in Figures 5 and 6.[6] They graphically correspond to the aspects of the robots for which they are theoretically known (e.g., see [3,4,6,7]). Note that the red boxes, that enclose the singularity curves, seem to cross the aspects due to the projection, while they of course do not cross in the product space.

Because the computation requires an exponentially growing time and space with respect to the prescribed precision ϵ, we need to tweak it for an efficient and reliable aspect determination. For the first three robots, the precision $\epsilon = 0.1$ gave precise enough results to determine the correct number of aspects after filtering out the spurious components. For 3-RPR, we needed the coarser precision $\epsilon = 0.3$ to compute without causing over-consumption of the memory. In the computation of 3-RRR, the threshold between the regular and the spurious components was not as clear as for the other robots, even though we improved the precision up to $\epsilon = 0.008$. Enumerating the obtained components from the largest ones, we assume that this robot (with fixed-orientation) has 25 aspects, the following components being likely to be a part of another component, i.e., spurious. This remains to be formally demonstrated.

6   Conclusion

The computation of aspects, i.e., singularity free connected sets of configurations, is a critical task in the design and analysis of parallel robots. The proposed algorithm uses numerical constraint programming to fully certify this computation. It is worth noting that this is the first algorithm that automatically handles such a large class of kinematic models with fully certifying the configuration existence, non-singularity and connectivity: The only restriction of the algorithm is its computational complexity, which is obviously exponential with respect to the number of degrees of freedom of the robot. The presented experiments have reported the sharp approximations of aspects for some realistic models: The correct number of aspects was computed for well-known planar robots with two and three degrees of freedom. The more challenging 3-RRR, whose number of aspects is still an open question, remains out of reach because of the complexity of the computation, though we have obtained some promising results fixing the orientation of its moving-platform. Tuning the propagation and search strategies of the algorithms should allow fully analyzing it in the future. Finally, although experiments have shown that the proposed method computes approximations of all aspects of well-known robots, it cannot be used for actually rigorously counting the aspects, a challenge we will address in the future on the basis of this method.

Acknowledgments. This work was partially funded by the French agency ANR project SIROPA (PSIROB06_174445) and JSPS (KAKENHI 23-3810). The machine used for the experiments was supported by Prof. Kazunori Ueda.

---

[6] These figures are also available at http://www.dsksh.com/aspects/

References

1. Amine, S., Tale-Masouleh, M., Caro, S., Wenger, P., Gosselin, C.: Singularity conditions of 3T1R parallel manipulators with identical limb structures. ASME Journal of Mechanisms and Robotics 4(1), 011011-1–011011-11 (2012)
2. Benhamou, F., Goualard, F., Granvilliers, L., Puget, J.F.: Revising hull and box consistency. In: Proc. of International Conference on Logic Programming, pp. 230–244 (1999)
3. Chablat, D.: Domaines d'unicité et parcourabilité pour les manipulateurs pleinement parallèles. Ph.D. thesis, École Centrale de Nantes (1998)
4. Chablat, D.: Joint space and workspace analysis of a two-dof closed-chain manipulator. In: Proc. of ROMANSY 18 Robot Design, Dynamics and Control, pp. 81–90. Springer (2010)
5. Chablat, D., Moroz, G., Wenger, P.: Uniqueness domains and non singular assembly mode changing trajectories. In: International Conference on Robotics and Automation, pp. 3946–3951 (2011)
6. Chablat, D., Wenger, P.: Working Modes and Aspects in Fully Parallel Manipulators. In: International Conference on Robotics and Automation, vol. 3, pp. 1964–1969 (1998)
7. Coste, M.: A simple proof that generic 3-RPR manipulators have two aspects. Tech. rep., Institut de Recherche Mathématique de Rennes (IRMAR) (2010)
8. Goldsztejn, A.: A Branch and Prune Algorithm for the Approximation of Non-Linear AE-Solution Sets. In: Proc. of ACM SAC 2006, pp. 1650–1654 (2006)
9. Goldsztejn, A.: Sensitivity analysis using a fixed point interval iteration. Tech. Rep. hal-00339377, CNRS-HAL (2008)
10. Goldsztejn, A., Goualard, F.: Box Consistency through Adaptive Shaving. In: Proc. of ACM SAC 2010, pp. 2049–2054 (2010)
11. Goldsztejn, A., Jaulin, L.: Inner approximation of the range of vector-valued functions. Reliable Computing 14, 1–23 (2010)
12. Granvilliers, L., Benhamou, F.: Algorithm 852: Realpaver: an interval solver using constraint satisfaction techniques. ACM TOMS 32(1), 138–156 (2006)
13. Hopcroft, J., Tarjan, R.: Algorithm 447: efficient algorithms for graph manipulation. Commun. ACM 16(6), 372–378 (1973)
14. Kong, X., Gosselin, C.: Type Synthesis of Parallel Mechanisms. Springer (2007)
15. Lhomme, O.: Consistency Techniques for Numeric CSPs. In: Proc. of IJCAI 1993, pp. 232–238 (1993)
16. Merlet, J.P.: Parallel robots. Kluwer, Dordrecht (2000)
17. Merlet, J.P.: A formal-numerical approach for robust in-workspace singularity detection. IEEE Trans. on Robotics 23(3), 393–402 (2007)
18. Montanari, U.: Networks of constraints: Fundamentals properties and applications to picture processing. Information Science 7(2), 95–132 (1974)
19. Moore, R.: Interval Analysis. Prentice-Hall (1966)
20. Moroz, G., Rouiller, F., Chablat, D., Wenger, P.: On the determination of cusp points of 3-RPR parallel manipulators. Mechanism and Machine Theory 45(11), 1555–1567 (2010)
21. Neumaier, A.: Interval Methods for Systems of Equations. Cambridge University Press (1990)
22. Van Hentenryck, P., Mcallester, D., Kapur, D.: Solving polynomial systems using a branch and prune approach. SIAM Journal on Numerical Analysis 34, 797–827 (1997)

The Semigroups of Order 10

Andreas Distler[1], Chris Jefferson[2], Tom Kelsey[2], and Lars Kotthoff[2]

[1] Centro de Álgebra da Universidade de Lisboa,
1649-003 Lisboa, Portugal
[2] School of Computer Science,
University of St. Andrews, KY16 9SX, UK

Abstract. The number of finite semigroups increases rapidly with the number of elements. Since existing counting formulae do not give the complete number of semigroups of given order up to equivalence, the remainder can only be found by careful search. We describe the use of mathematical results combined with distributed Constraint Satisfaction to show that the number of non-equivalent semigroups of order 10 is 12,418,001,077,381,302,684. This solves a previously open problem in Mathematics, and has directly led to improvements in Constraint Satisfaction technology.

Keywords: Constraint Satisfaction, Mathematics, semigroup, Minion, symmetry breaking, distributed search.

1 Introduction

An important area of investigation is the determination of the number of solutions of a given finite algebraic problem. It is often the case that we are interested in the number of classes of solutions under some type of equivalence relation, since this gives the number of structural types rather than distinct objects. In certain cases, these numbers can be found by deriving counting formulae. It may also be possible, on an *ad hoc* basis, to derive enumerative constructions of larger objects from smaller ones. In both these cases, no systematic computer search is required – the numbers are calculated from mathematical proofs using the structures of the underlying problem.

There is no guarantee, of course, that the use of formally-proven formulae will work for all problems. It may be that no suitable formulae is available. In this event, the only method left is to carefully search for solutions, ensuring that none is missed and none is counted more than once. Examples include the search for all distinct transitive graphs on n vertices [23,24], all binary self-dual codes of length 32 [2], all ordered trees with k leaves [32], and all non-equivalent semigroups up to order 9 and monoids up to order 10 [6,7,9,17,26,28,30]. Large-scale studies often involve a combination of enumeration by formula and computer search. The tautomer enumeration problem [18] from Cheminfomatics is an illustrative example. Commercial and academic software packages used to solve this type of problem typically use a suite of transformation rules that allow enumeration

M. Milano (Ed.): CP 2012, LNCS 7514, pp. 883–899, 2012.

without search, combined with generate-and-test searches for structures not predicted by the rulesets [25].

Whenever computer search is used to solve for types of solutions rather than absolute number of solutions, some method must be employed that ensures that exactly one canonical representative from each equivalence class is returned. This involves breaking the symmetries that allow objects from the same class to be interchanged, and the design and implementation of such methods is an important field of study in its own right [29, chapter 10].

A detailed exposition of search, symmetry-breaking, enumeration, and solution generation is given in [21]. The Constraint Satisfaction methods used in our search are described in [29], details of the specific Constraint Satisfaction solver used – Minion – are in [13], and the computational algebra package – GAP – used to identify and break symmetries is described at [11]. Basics of semigroup theory can be found in [15].

Table 1. Semigroup T of order 10

*	0	1	2	3	4	5	6	7	8	9
0	0	0	0	0	4	4	0	0	4	4
1	0	1	0	0	4	4	0	0	4	4
2	2	2	2	2	5	5	2	2	5	5
3	2	2	2	3	5	5	2	2	5	5
4	0	0	0	0	4	4	4	4	0	0
5	2	2	2	2	5	5	5	5	2	2
6	0	0	2	2	4	5	6	7	8	9
7	0	0	2	2	4	5	7	6	9	8
8	2	2	0	0	5	4	8	9	7	6
9	2	2	0	0	5	4	9	8	6	7

A semigroup $T = (S, *)$ consists of a set of elements S and a binary operation $* : S \times S \to S$ that is *associative*, satisfying $(x * y) * z = x * (y * z)$ for each x, y, z in S. We call two semigroups $(S, *)$ and $(S', \circ)$ *isomorphic* – respectively *anti-isomorphic* – if there exists a bijection $\sigma : S \to S'$ such that $\sigma(x * y) = \sigma(x) \circ \sigma(y)$ – respectively $\sigma(x * y) = \sigma(y) \circ \sigma(x)$ – for all $x, y \in S$; in this case σ is an *isomorphism* respectively *anti-isomorphism*. An element x of T is an *idempotent* if $x * x = x$. The semigroup T is *nilpotent* if there exists an $r \in \mathbf{N}$ such that $|\{s_1 * s_2 * \cdots * s_r \mid s_1, s_2, \ldots, s_r \in S\}| = 1$, in other words if all products of r elements take the same value. If r is the least value for which this is true, then T is *nilpotent of degree r*. Table 1 is an illustrative example: $T = (\{0, \ldots, 9\}, *)$. By inspection, T is 7-idempotent, i.e. has exactly seven idempotents. T is not nilpotent of degree 3 since, for example, $4 * 5 * 6 = 4 \neq 5 = 2 * 3 * 4$. Given a permutation π of the elements of $\{0, \ldots, 9\}$, it is easy to check that a semigroup isomorphic to T is obtained by permuting the rows, the columns, and finally the values according to π, and that additionally transposing yields the table of an anti-isomorphic semigroup.

The problem addressed in this paper is finding all ways of filling in a blank table such that multiplication is associative up to symmetric equivalence, i.e. up to isomorphism or anti-isomorphism.

In this paper we report that the hitherto unknown number of semigroups of order 10, up to equivalence, is 12,418,001,077,381,302,684. The size of the search space for this problem is 10^{100} with $2 \times 10!$ symmetries to be broken, making generate-and-test an intractable solution method. A recent advance in the theory of finite semigroups has led to a formula [8] that gives the number of 'almost all' semigroups of given order. Despite this, 718,981,858,383,872 non-equivalent solutions had to be found by a combination of *ad hoc* constructive enumeration proofs and Constraint Satisfaction search, which took 130 CPU years distributed across two local clusters and the Amazon cloud [1].

Our investigations have directly led to improvements in the Minion Constraint Satisfaction Problem solver. Watched constraints were introduced, and a more efficient lexicographic ordering constraint was implemented. These gave orders of magnitude improvements to our search, and have been included in subsequent releases of Minion.

We stress the interdisciplinary nature of our work: the result cannot – to our knowledge – be obtained by mathematical proof alone, nor can current AI search technologies hope to return the exact number of solutions with realistic resources.

In Section 2 we give detailed descriptions of our models and methods, and of the family of Constraint Satisfaction Problems (CSPs) that were solved to provide the main result. Section 3 contains the results divided into the sub-cases used to overcome computational bottlenecks. In Section 4 we discuss the improvements in CSP solving that we were able to identify and implement, together with a brief analysis of how our solver can often backtrack very early in search, and how distribution of the CSPs across multiple compute nodes is likely to generalise. We give some concluding remarks in Section 5.

2   Methods

We first describe a single CSP for our combinatorial problem (Sec. 2.1), that incorporates *a priori* knowledge regarding the number of semigroups of a certain type. The next stage is the replacement of this CSP by families of CSPs (Sec. 2.2), some of which are both computationally and mathematically difficult, and are solved by distributed search (Sec. 2.3). More computationally tractable families are solved using a single processor, and families having exploitable structure are solved mainly by constructive enumeration (Secs. 2.4, 2.5, and 2.6).

2.1   The Single Constraint Satisfaction Model

We model semigroups of order 10 as a CSP: a set of variables V each with a discrete and finite domain D of potential values, and a set of constraints C that either forbid or require certain instantiations of variables by domain values.

A full instantiation of the variables in V by values from D is a solution whenever no constraint is violated. We make extensive use of the *element* constraint on variables N, M_i and P having natural number domains

$$N = \langle M_0, \ldots, M_{n-1} \rangle [P]$$

which requires that N is the Pth element of the list $\langle M_0, \ldots, M_{n-1} \rangle$ in any solution. Propagators for this constraint are implemented in many CSP solvers, including Minion.

CSP 1. *Let* $V_1 = \{T_{a,b} \mid 0 \leq a, b \leq 9\}$ *be variables representing the entries in a* 10×10 *multiplication table* T, *and* $V_2 = \{A_{a,b,c} \mid 0 \leq a, b, c \leq 9\}$ *be variables representing each of the products of three elements. Our basic CSP has* $V = V_1 \cup V_2$ *as variables, each with domain* $D = \{0, \ldots, 9\}$. *For each triple* (a, b, c) *of values from* D, *we post the pair of constraints*

$$\langle T_{a,0}, \ldots, T_{a,9} \rangle [T_{b,c}] = A_{a,b,c} = \langle T_{0,c}, \ldots, T_{9,c} \rangle [T_{a,b}] \qquad (1)$$

which enforce associativity.

Finding all solutions of CSP 1 would give the number of distinct semigroups of order 10, i.e. all non-identical associative 10×10 multiplication tables. Our task, however, is to search for the number of classes of solutions up to symmetric equivalence, i.e. up to isomorphism or anti-isomorphism. The symmetry group is the set of permutations of $\{0, \ldots, 9\}$ combined with possible transpositions of the tables, which we denote as $S_{10} \times C_2$ using standard group theoretic notation. Let $g = (\pi, \phi) \in S_{10} \times C_2$ be a symmetry and T be a multiplication table. T^g is then the table obtained by

1. permuting the rows and columns of T according to π;
2. permuting the values in T according to π;
3. either
   (a) doing nothing if ϕ is the identity element ϕ_1 of C_2;
   (b) transposing the table if ϕ is the non-identity ϕ_2 element of C_2.

Two multiplication tables T_1 and T_2 are isomorphic if $T_1 = T_2^{(\pi,\phi_1)}$ for some $\pi \in S_{10}$; T_1 and T_2 are anti-isomorphic if $T_1 = T_2^{(\pi,\phi_2)}$ for some $\pi \in S_{10}$. Since $S_{10} \times C_2$ is a group, the set of all multiplication tables can be partitioned into subsets of symmetric equivalents: those tables that are isomorphic or anti-isomorphic to each other form an equivalence class.

Our problem is to find the number of equivalence classes, either by formal enumeration proofs or by solving a variant of CSP 1 that returns exactly one canonical representative from each class. Our general search methodology is to post "lex-leader" symmetry-breaking constraints before search. This is a well-known technique for dealing with symmetries in CSPs [4], made harder to implement in our case because our symmetries involve both variables and values, and made harder to deploy since we need to post $2 \times 10! = 7,257,600$ symmetry-breaking constraints.

To explain our realisation of "lex-leader" we first introduce of another way to describe a solution of CSP 1. A *literal* of a CSP $L = (V, D, C)$ is an element in the Cartesian product $V \times D$. Literals are denoted in the form $(x = k)$ with $x \in V$ and $k \in D$. An instantiation f corresponds to the set of literals $I_f = \{(x = f(x)) \mid f$ is defined on $x\}$, which uniquely defines f (but not every set of literals defines an instantiation). In particular, literals are mapped to literals under the isomorphic and anti-isomorphic transformations described above.

Given a fixed ordering $(\chi_1, \chi_2, \ldots, \chi_{|V_1||D|})$ of all literals in $V_1 \times D$, an instantiation can then be represented as a bit vector of length $|V_1||D| = 100 \times 10$. The bit in the i-th position is 1 if χ_i is contained in the instantiation and otherwise the bit is 0. The bit vector for the instantiation I_f corresponding to the ordering of the literals $(\chi_1, \chi_2, \ldots, \chi_{|V_1||D|})$ is denoted by $(\chi_1, \chi_2, \ldots, \chi_{|V_1||D|})_{I_f}$.

There is one solution in each set of symmetric equivalents for which the corresponding bit vector is lexicographic maximal, which we take to be the property identifying the canonical solution. We denote the standard lexicographic order on vectors by $\geq_{\text{lex}}$.

CSP 2. *Let (V, D, C) be as defined in CSP 1. We extend C by adding for all symmetries $g \in S_{10} \times C_2$ the constraint*

$$(\chi_1, \chi_2, \ldots, \chi_{|V_1||D|}) \geq_{lex} (\chi_1^g, \chi_2^g, \ldots, \chi_{|V_1||D|}^g). \tag{2}$$

The solutions of CSP 2 are canonical representatives of associative multiplication tables, as required.

It is not hard to show that all finite semigroups have at least one idempotent. Our symmetry-breaking constraints ensure that all solution tables will have the value 0 at $T_{0,0}$. Since we ensure that $0 * 0 * 0 = 0$, exactly those solutions that are nilpotent of degree at most 3 will have $a * b * c = 0$ for all values of a, b and c. A formula has recently been derived that gives the number of such solutions without search [8], so we add constraints that rule out these solutions from our CSP. We already have variables that represent each product of three elements, namely V_2.

CSP 3. *Let (V, D, C) be as defined in CSP 2. Form a vector containing the variables $\langle V_2 \rangle = \langle A_{a,b,c} \mid 0 \leq a, b, c \leq 9 \rangle$, and another vector $\langle Z \rangle$ consisting of 10^3 zeros. We post the additional constraint*

$$\langle V_2 \rangle \neq \langle Z \rangle \tag{3}$$

that guarantees that at least one triple product is non-zero in every solution.

Adding the number of solutions of CSP 3 to the formula value for semigroups of nilpotency degree at most 3 will give our full result. It should be noted that no enumeration formula for the solutions of CSP 3 exists: some form of organised search is required to solve the complete problem.

2.2 Case Splits

CSP 3 has too many symmetries to be solved as a single entity. In practice, we apply a well-established technique that has been used in the enumeration of semigroups of orders 6 and 8 [28,30] to subdivide the problem (CSP 4). For each subproblem we prescribe the entries on the diagonal of the multiplication table. There exist 10^{10} distinct diagonals, but only a small number appear in a canonical solution, so that far fewer subproblems are created. A detailed description of the efficient derivation of a set of diagonals containing all canonical ones is given in [5]. This approach heavily influences the symmetry-breaking, as any lex-leader constraint (2) is automatically fulfilled if it corresponds to a symmetry that does not fix the prescribed diagonal. We therefore significantly reduce the overall search space, and also the number of symmetry-breaking constraints needed for most subproblems.

CSP 4. *Let (V, D, C) be as defined in CSP 3, and let $E = \{E_a \mid 0 \leq a \leq 9\}$ be a canonical diagonal. We post the constraints*

$$T_{a,a} = E_a \text{ for } 0 \leq a \leq 9 \tag{4}$$

which prescribe the entries on the diagonal of any solution.

To further reduce the number of instances, we combine certain easy subproblems and prescribe only the number of idempotents, as done in [17]. For more difficult subproblems we do further easily-derived case splits which will be mentioned in Section 3. The choice of method used for a particular subproblem was based on our experience from searching for the semigroups of order 9.

2.3 Distributed CSP Search

Our method of solving constraint problems in a distributed way does not require support for distributed architectures in the constraint solver. Instead, we partition and distribute the problem specification itself, with the different partitions of the search space solved independently. The advantage of this approach over techniques that require communication between the compute nodes is that its implementation and deployment are much simpler.

We partition by splitting on the values in the domain of a variable during search [20]. This is done as follows: first the solution process is stopped, then we compute *restart nogoods* as described in [22] and encode them as constraints that can be added to the original problem. When added, the new constraints enable the solver to restart search from where it was interrupted. In addition, we add constraints that split the remaining search space. We partition the domain for the variable currently under consideration into n pieces of roughly equal size. We then create n new models and to each in turn add the constraints ruling out the previously done search and $n - 1$ partitions of that domain. As an example, consider the case $n = 2$. If the variable under consideration is x and its domain is $\{1, 2, 3, 4\}$, we generate 2 new models. One of them has the constraint $x \leq 2$

added and the other one $x \geq 3$. Thus, solving the first model will try the values 1 and 2 for x, whereas the second model will try 3 and 4.

It is impossible to predict reliably the size of the search space for each of the splits, and the time needed to search it. This directly affects the effectiveness of the splitting – if the search space is distributed unevenly across the splits, some of the compute nodes will be idle while the others do most of the work. We address this problem by repeatedly splitting the search space during search. In this way we create new units of work whenever a worker becomes idle by simply asking one of the busy workers to stop and generate split models. The search is then resumed from where it was stopped and the remaining search space is explored in parallel by the two workers. Note that there is a runtime overhead involved with stopping and resuming search because the constraints which enable resumption must be propagated and the solver needs to explore a small number of search nodes to get to the point where it was stopped before. There is also a memory overhead because the additional constraints need to be stored.

In practice, we run Minion for a specified amount of time, then stop, split and resume instead of splitting at the beginning and when workers become idle. Initially the utilisation of the workers is suboptimal because not enough work units have been generated. However, after some time the number of work units exceeds the number of workers and the utilisation reaches 100%, and the initial phase of under-utilised resources is negligible compared to the total computation time.

We used the distributed computing system Condor [31] to handle distribution of the work units to the worker nodes. Every time new models are generated, they are submitted to the system which queues them and allocates a worker as soon as one becomes available.

2.4   Construction of Certain 2-Idempotent Semigroups

If $T = (\{1, \ldots, 9\}, *)$ is a 1-idempotent semigroup, 1 being the idempotent, we define four multiplications on $\{0, \ldots, 9\}$:

$$i *_{\mathrm{I}} j = \begin{cases} i * j & \text{if } i, j \in T \\ 0 & \text{otherwise} \end{cases} \qquad i *_{\mathrm{II}} j = \begin{cases} i * j & \text{if } i, j \in T \\ i & \text{if } j = 0 \\ j & \text{if } i = 0 \end{cases}$$

$$i *_{\mathrm{III}} j = \begin{cases} i * j & \text{if } i, j \in T \\ i * 1 & \text{if } i \in T, j = 0 \\ 1 * j & \text{if } j \in T, i = 0 \\ 0 & \text{if } i = j = 0 \end{cases} \qquad i *_{\mathrm{IV}} j = \begin{cases} i * j & \text{if } i, j \in T \\ 0 & \text{if } i = 0 \\ 1 & \text{if } i \neq 0, j = 0 \end{cases}$$

Denote $T_{\mathrm{I}} = (\{0, \ldots, 9\}, *_{\mathrm{I}}), \ldots, T_{\mathrm{IV}} = (\{0, \ldots, 9\}, *_{\mathrm{IV}})$. Then it can be readily verified that $T_{\mathrm{I}}, T_{\mathrm{II}}$, and T_{III} are 2-idempotent semigroups, and so is T_{IV} if the element 1 is a left-zero in T, i.e. $1 * i = 1$ for all $i \in T$. These four semigroups are pairwise non-equivalent (except that $T_{\mathrm{II}} = T_{\mathrm{III}}$ if T is a group). Moreover, for two non-equivalent semigroups on $\{1, \ldots, 9\}$ the first three constructions lead to non-equivalent semigroups; and two non-isomorphic semigroups on $\{1, \ldots, 9\}$ lead to non-equivalent semigroups under the fourth construction.

We use the above constructions to reduce the search effort for certain 2-idempotent diagonals, with 0 and 1 being the idempotents. It can then be shown that $T_{0,1}, T_{1,0} \in \{0, 1\}$, and that the only solutions not arising from one of the constructions have $T_{0,1} = T_{1,0} = 1$. Hence we fix these variables of the CSP and in addition add constraints to rule out solutions that are of the form T_{II} or T_{III} for some semigroup T on $\{1, \ldots, 9\}$.

2.5 Construction from Nilpotent Semigroups of Order 9

For a 1-idempotent semigroup $T = (\{0, \ldots, 8\}, *)$ with 0 being the idempotent we define a multiplication $\circ$ on $\{0, \ldots, 9\}$ as follows:

$$i \circ j = \begin{cases} i * j & \text{if } i, j \in T \\ 0 & \text{if } i = j = 9 \\ 9 & \text{otherwise} \end{cases}$$

If 0 is a zero element in T then $\circ$ is associative and we define the semigroup $T_\circ = (\{0, \ldots, 9\}, \circ)$. Two semigroups constructed in this manner are equivalent if and only if the two semigroups on $\{0, \ldots, 8\}$ are equivalent.

To rule out all semigroups that are equivalent to a constructed one from the CSP search for a given 1-idempotent diagonal, we cannot assume that 9 is the distinguished element. Every element whose diagonal entry equals the idempotent element 0, and does not appear on the diagonal itself, is a potential candidate. For each such element k we forbid that the entries in the k-th row and k-th column except the diagonal entry are all equal to k. That is we post the constraint

$$\langle K \rangle \neq \langle T_{0,k}, \ldots, T_{k-1,k}, T_{k+1,k}, \ldots, T_{9,k}, T_{k,0}, \ldots, T_{k,k-1}, T_{k,k+1}, \ldots, T_{k,9} \rangle, \quad (5)$$

where $\langle K \rangle$ is the vector of length 18 containing k in each position. It remains to justify that we do not miss any solution by posting these constraints. If T is a multiplication table for which equality holds in (5) for some k, then either T is equivalent to a semigroup as constructed above or there exists $T_{i,j} = k$ with $i, j \neq k$. This gives

$$k * (i * j) = k * k = 0 \text{ and } (k * i) * j = k * j = k$$

showing that such a table T is not associative.

2.6 Construction from Nilpotent Semigroups of Degree at Most 3

Let $T = (\{1, \ldots, k\}, *)$ be a nilpotent semigroup of degree at most 3 for some $2 \leq k \leq 9$ with 1 being its idempotent. The structure of T yields a natural partition of $\{1,\ldots,k\}$ into 3 sets: the zero element 1 by itself, the non-zero products, and the remaining elements (for details see [5, Section 2.1]). We denote the latter set by A^- and the set of non-zero products by B^+. The superscripts indicate a sign function, or multiplicative parity, that we introduce on $\{0,\ldots,k\}$ with $\mathrm{sign}(0) = -$ and $\mathrm{sign}(1) = +$. We define a multiplication $*_\pm$ on $\{0,\ldots,k\}$ as follows:

$$i *_\pm j = \begin{cases} i * j & \text{if } i,j \in T \text{ and } \mathrm{sign}(i) = \mathrm{sign}(j) \\ 1 & \text{if } i = 0, j \in A^- \text{ or } i \in A^-, j = 0 \text{ or } i = j = 0 \\ 0 & \text{if } \mathrm{sign}(i) \neq \mathrm{sign}(j) \end{cases}$$

Then $T_\pm = (\{0,\ldots,k\}, *_\pm)$ is a semigroup as all products of three elements equal either 0 or 1 depending only on the parity of the elements in the product. In a second step we define a multiplication on the set $\{0,\ldots,9\}$ based on $*_\pm$.

$$i \circ_\pm j = \begin{cases} i *_\pm j & \text{if } i,j \in T_\pm \\ 1 *_\pm j & \text{if } i \in T_\pm, j \geq k+1 \\ i *_\pm 1 & \text{if } i \geq k+1, j \in T_\pm \\ 1 & \text{if } i,j \geq k+1 \end{cases}$$

Two semigroups of order 10 constructed in this way are equivalent if and only if they arise from equivalent semigroups. None of the semigroups are nilpotent as they do not contain a zero and none are equivalent to those constructed in Section 2.5.

In a similar manner to the construction in Section 2.5, given a 1-idempotent diagonal, we have to add constraints for every potential candidate for the distinguished element from the above definition of $*_\pm$. The constraints are similar to (5) but are required for all vectors $\langle K \rangle$ whose entries are the idempotent and the candidate k as in the construction for some semigroup T. Note that this does not result in one constraint for every nilpotent semigroup of degree at most 3, but in one constraint for each of a few specific partitions into sets A^- and B^+ allowed by the diagonal.

In addition we have to search for semigroups not equivalent to one from the construction, in which the idempotent and a candidate for the distinguished element behave as in a construction. This reduces the domains for many table entries to a singleton, and requires one additional constraint that the vector of table entries that would be fixed by the construction does not equal the corresponding vector.

3   Results

For 4 to 7 and for 9 idempotents we added constraints that fixed the idempotents to CSP 3, and were able to solve as single instances on a single computer

Table 2. Semigroups up to equivalence for the cases 3–10 idempotents. Minion[†] denotes specialised search using 289 sub-cases based on possible diagonals for 3 idempotents, and 4 sub-cases for 8 idempotents. The full methodology is described in [7]. Minion[‡] denotes specialised search for bands as described in [5]. The computations took around 920 hours on a machine with 2.66GHz Intel X-5430 processor and 16GB RAM, giving roughly 67,000 solutions per second.

Idempotents	Semigroups	Method
3	219,587,421,825	Minion[†]
4	1,155,033,843	Minion
5	396,258,335	Minion
6	478,396,381	Minion
7	412,921,339	Minion
8	214,294,637	Minion[†]
9	60,036,717	Minion
10	7,033,090	Minion[‡]
Total:	222,311,396,167	

(Table 2). The 3- and 8-idempotent cases were solved by case-split into CSP 4 using the appropriate canonical diagonals. A k-idempotent semigroup of order k is known as a *band*, and there is extensive theory for these objects that can be used to refine search. This has been done in [5] for all bands up to order 10.

The 2-idempotent case is strictly harder: there are $2 \times 9!$ symmetries, and from our experience with semigroups of orders 1–9 we predicted about 4×10^{14} solutions. We therefore derived the constructions given in Section 2.4, which give the majority of such solutions without search. We still had to search for roughly 1.2×10^{14} solutions (Table 3).

Table 3. Semigroups up to equivalence having exactly 2 idempotents. The distributed Minion computation (involving machines with varying architectures) took 73 CPU years, returning an average 50,400 solutions per second. The CPU time taken to search for solutions not given by the constructions described in Section 2.4 was comparatively negligible. Methods T_I etc. denote the constructive multiplication defined in Section 2.4. 'The rest' denotes solutions for those instances that are not ruled out by the constraints that forbid the constructions.

Case	Semigroups	Method
Based on Section 2.4		
– up to equivalence	158,929,640,752,110	T_I, T_{II} and T_{III}
– up to isomorphism	105,945,136,997,613	T_{IV}
– the rest	226,006,150,622	Minion
Not based on Section 2.4	116,179,193,109,431	Distributed Minion
Total:	381,279,977,009,776	

The 1-idempotent case can be split naturally into two cases: solutions that are nilpotent of some degree (Table 4), and solutions that are not nilpotent (Table 5). There is exactly one canonical semigroup of nilpotency degree 2 (the zero semigroup) and one of degree 10 (the monogenic, nilpotent semigroup), and there is an enumeration formula for those of degree 3. The cases for degrees 5 through 9 are small enough to be solved by a single computer. The degree 4 case is again sub-divided into 316 canonical diagonals. For 68 pairs of instances, two distinct diagonals lead to CSPs that have the same search variables (the off-diagonal entries), symmetry group and constraints. In these cases we only solve the first CSP, and double the number of solutions to obtain the true value (Table 4). The final case, 1-idempotent but non-nilpotent semigroups, was approached via the constructions from smaller semigroups described in Sections 2.5 and 2.6, with limited CSP search for the remaining solutions (Table 5).

Table 4. Semigroups up to equivalence having exactly 1 idempotent and being nilpotent of some degree. The distributed Minion computation (involving machines with varying architectures) took 60 CPU years, returning an average 74,000 solutions per second. The CPU times for the 5–9 degree cases were negligible by comparison.

Nilpotency degree	Semigroups	Method
2	1	Zero semigroup
3	12,417,282,095,522,918,811	Formula
4		
– unique	49,304,583,445,962	Distributed Minion
– replicable	91,103,513,956,511	Distributed Minion
– replicas	91,103,513,956,511	No search
5	10,027,051,364	Minion
6	3,395,624	Minion
7	17,553	Minion
8	328	Minion
9	15	Minion
10	1	Monogenic semigroup
Total:	12,417,495,617,164,742,681	

3.1   Improved Solver Performance

As an integral part of the methodology development stage of this investigation, we performed careful profiling of Minion and compared performance against another CSP solver, Gecode [12]. Initial evidence was that Gecode was an order of magnitude faster than Minion on semigroup instances having a large number (over 20,000) of symmetry-breaking constraints, and that the solvers were competitive for other instances. Analysis showed that the speedup was primarily due to the fact that Gecode removes constraints from search once they become entailed, whereas Minion would leave constraints in place throughout search.

We revised Minion to do the same, resulting in two solvers that were broadly comparable in terms of performance.

Through profiling, we found that the revised Minion was spending over 95% of its time in the lexicographic ordering constraint for highly symmetric instances. We identified propagation calls that had a cost but almost always no effect, and as a result were able to design and implement the QuickLex algorithm [16], which looks at only two variables out of the whole constraint at once, leading to massive performance gains for lexicographic ordering constraints. On average, the cost of the QuickLex propagator is 15% of the cost of the hitherto optimal method of Frisch *et al.* [10].

The use of "watched literals" has led to remarkable improvements in SAT solvers, and an implementation for Constraint Satisfaction was proposed in [14]. The careful use of this technique allows efficient maintenance of generalised arc consistency on the element constraints that we use to enforce associativity (Section 2.1).

Our empirical experience for orders smaller than 10 is that the combined use of QuickLex and watched element constraints makes Minion over an order of magnitude faster than Gecode for this class of problems. Minion versions from 0.9 onwards have incorporated these enhancements. We report that solving semigroups appears to be a good stress test for constraint solvers, involving a small number of types of constraints, but large search spaces, large numbers of solutions and large numbers of symmetries.

Table 5. Non-nilpotent semigroups up to equivalence having exactly 1 idempotent. The single CPU Minion calculations took about 110 hours, returning an average 44,700 solutions per second. 'The rest' denotes solutions for those instances that are not ruled out by the constraints that forbid the constructions.

Case	Semigroups	Method
Diagonal does not admit nilpotent of degree 3 solutions	3,673,835,659	Minion
Diagonal admits nilpotent of degree 3 solutions, and has a sub-diagonal that admits nilpotent of degree 3 solutions		
– construction from order 9	52,972,873,141,621	Construction 2.5
– the rest	12,596,375,843	Minion
– construction from orders up to 9	52,968,071,362,553	Construction 2.6
– the rest	712,828,694	Minion
Diagonal admits nilpotent of degree 3 solutions, and **does not** have a sub-diagonal that admits nilpotent of degree 3 solutions	609,690	Minion
Total:	105,957,928,154,060	

4   Discussion

4.1   CSP Search Analysis

Minion is very successful at solving semigroup instances. We found that instances with no solutions were solved almost immediately, and problems with a large number of solutions had just over twice as many search nodes as solutions, approaching the minimum possible.

$*$	0	1	2	3	4	5	6	7	8	9
0	1	7						*0*		
1	7	0								
2			2							
3				2						
4					3					
5						4				
6							3			
7	⊞	*1*						6		
8									6	
9										8

The displayed example shows how Minion performs early backtracks from a partially filled table when no solution exists. In our example, the diagonal entries have been set before search, and we have assigned $T_{0,1} = 7$. By simple inference from the element constraints, three variables are instantly instantiated (shown with values in italics in the figure). The element constraints in Minion can also remove individual values from the domains of variables. In particular, one of these constraints removes domain values for variable $T_{7,0}$ (denoted as ⊞). This constraint gives:

$$\langle T_{7,0}, \ldots, T_{7,9}\rangle[T_{0,1}] = A_{7,0,1} = \langle T_{0,1}, \ldots, T_{9,1}\rangle[T_{7,0}] \tag{6}$$

$$\Rightarrow 7 * 7 = 6 = A_{7,0,1} = \langle 7, 0, T_{2,1}, \ldots, T_{9,1}\rangle[T_{7,0}] \tag{7}$$

$$\Rightarrow T_{7,0} \neq 0 \text{ and } \Rightarrow T_{7,0} \neq 1, \tag{8}$$

contradicting the associativity requirement $7*0 = (0*1)*0 = 0*(1*0) = 0*7 = 0$. The value 7 was arbitrary; Minion will backtrack without further assignment for any value between 3 and 9, and search terminates after only 5 search nodes.

Whilst there are common classes of CSP symmetry that can be broken by posting a polynomially-sized subset of the symmetries – typically pure variable or pure value symmetries [3] – this does not seem to be one of those instances. This is because our symmetries permute both variables and values, and since the underlying group consists of all permutations, we have to post a constraint for each member of the subgroup of $S_{10} \times C_2$ determined by the case splits described in Section 2.2. Moreover, partial symmetry-breaking would lead to more than one solution per equivalence class, and we would then have to implement a potentially expensive maximal image post-process in order to obtain the correct result.

Minion has the further advantage that when symmetry-breaking or case-split constraints are added, the reasoning they generate is automatically combined with the reasoning of the associativity constraints, making it quick and easy to try out new ideas.

4.2   Distributed Search

Using our knowledge of the numbers of smaller semigroups of various types, our *a priori* estimate of the number of solutions of CSP 3 was somewhere between 5×10^{14} and 1×10^{15}. Our development experience with semigroups of order 9 is that, on average, about 70,000 semigroups are found every second using a single compute node. Assuming a search rate of 50,000 solutions per second for order 10 – since the number and length of the constraints increase with increased order – this equates to between 317 and 634 years of wall clock time on a single machine. We therefore developed a distributed strategy.

Our approach works very well for the extremely large problem we are tackling here. The computations took several CPU decades, making negligible the cost of the under-utilisation of nodes in the first few hours of computation. Similarly, we found that the overheads incurred through splitting, restarting and propagating additional constraints were acceptable, because without the distribution across many computers, we would not have been able to solve the problem at all.

There are several advantages to implementing distributed solving in this way. First, by creating regular "snapshots" of the search done, the resilience against failures increases. Every time we stop, split and resume, our modified models are saved. As they contain constraints that rule out the search already done, we can only lose the work done after that point if a worker fails. This means that the maximum amount of work lost in case of a total failure of all workers is the allotted time T_{max} times the number of workers $|w|$. This is especially important for large-scale computations, and our experience shows this to be extremely useful in practice. The modified models can be stored. We exploited this by moving the solving process to a different set of workers, without losing any work. Since our methods require no communication between the individual workers solving the problem, they only need to be able to receive the problem sub-instances, and send either the solution or split models back. We used compute nodes in two local clusters and the Amazon cloud, and, since some parts of the search space required more memory than the machines initially chosen could provide, we were able to seamlessly move those parts of the computation onto more powerful machines.

Our leveraging of existing software to handle the logistics of distribution led to reductions in both development time and systematic errors. For large problems such as these, our experience is that the number of queued jobs will usually exceed the number of workers, ensuring good resource utilisation.

4.3   Validation

For most of the case-splits, we have run the solver exactly once. It is not in-conceivable, therefore, that a miscalculation has occurred. For orders up to 8, the number of solutions is small enough that we can solve CSP 2 with diagonal case splits, using neither enumeration formulae nor constructions. For order 9 we can solve CSP 3, as a family of CSPs using no constructions. We have performed these calculations multiple times on various architectures, using different choices for search heuristics and different implementations of constraint propagators. The expected totals are returned every time. For orders 7, 8 and 9 we have re-calculated using exactly the same case-splits described in this paper. Again, the computed totals match those in the literature for order 7 [17], 8 [30] and 9 [7]. These checks increase our confidence that (a) there are no systematic errors in our splitting of the problem into smaller instances, and (b) our code for identifying symmetries and solving CSPs is correct.

5   Conclusions

Counting semigroups up to equivalence is not easy, being in some sense near the worst point of the combinatoric tradeoff between counting by reasoning and counting by search. At one end of the scale, the number of all distinct 10×10 multiplication tables is trivial to derive: 100 entries each having one of 10 values gives 10^{100} solutions. This triviality is due to the complete lack of structure. As an example from the other end, finite groups are relatively easy to search for due to their higher level of structure, and there are far fewer of them. Semigroups occupy an intermediate zone, having a small amount of exploitable structure and a large number of solutions.

For many algebraic structures we break symmetries in order to deal with the combinatorial explosion in the number of trivially distinct objects. However for semigroups, the breaking of symmetries – which grow factorially with increasing order – still leaves a super-exponential growth in the number of non-equivalent solutions. The formula for finite semigroups that are nilpotent of degree three gives the vast majority of solutions, but, again, the remainder to be found by search grows super-exponentially as order increases.

It is relatively unusual for Constraint Satisfaction modelling and technology to produce new results in Mathematics. The only examples that we are aware of are new instances of graceful graphs [27] and the monoids of order 10 [7]. Finding the number of semigroups of order 10 has involved advances in both Constraint Satisfaction and abstract algebra. The mathematical constructions described in this paper rule out more than half the search needed and without the enumeration formula for nilpotent of degree 3 solutions the problem is effectively intractable using any known approach. Moreover, both the Constraint Satisfaction technology and the Mathematics are vital – semigroups 10 cannot be solved by researchers from either discipline alone.

Acknowledgments. Parts of the computational resources for this project were provided by an Amazon Web Services research grant. This work was developed within the project PTDC/MAT/101993/2008 of Centro de Álgebra da Universidade de Lisboa, financed by FCT and FEDER. TWK is supported by UK EPSRC grant EP/H004092/1. LK is supported by a SICSA studentship and an EPSRC fellowship.

References

1. Amazon Elastic Compute Cloud, Amazon EC2 (2008),
   http://aws.amazon.com/ec2/
2. Bilous, R.T., Van Rees, G.H.J.: An enumeration of binary self-dual codes of length
   32. Des. Codes Cryptography 26(1-3), 61–86 (2002),
   http://dx.doi.org/10.1023/A:1016544907275
3. Cohen, D., Jeavons, P., Jefferson, C., Petrie, K.E., Smith, B.M.: Symmetry Definitions for Constraint Satisfaction Problems. In: van Beek, P. (ed.) CP 2005. LNCS,
   vol. 3709, pp. 17–31. Springer, Heidelberg (2005)
4. Crawford, J.M., Ginsberg, M.L., Luks, E.M., Roy, A.: Symmetry-breaking predicates for search problems. In: Aiello, L.C., Doyle, J., Shapiro, S. (eds.) KR 1996:
   Principles of Knowledge Representation and Reasoning, pp. 148–159. Morgan
   Kaufmann, San Francisco (1996)
5. Distler, A.: Classification and Enumeration of Finite Semigroups. Shaker Verlag,
   Aachen (2010), also PhD thesis, University of St Andrews (2010),
   http://hdl.handle.net/10023/945
6. Distler, A., Kelsey, T.: The Monoids of Order Eight and Nine. In: Autex-
   ier, S., Campbell, J., Rubio, J., Sorge, V., Suzuki, M., Wiedijk, F. (eds.)
   AISC/Calculemus/MKM 2008. LNCS (LNAI), vol. 5144, pp. 61–76. Springer, Heidelberg (2008)
7. Distler, A., Kelsey, T.: The monoids of orders eight, nine & ten. Ann. Math. Artif.
   Intell. 56(1), 3–21 (2009)
8. Distler, A., Mitchell, J.D.: The number of nilpotent semigroups of degree 3. Electron. J. Combin. 19(2), Research Paper 51 (2012)
9. Forsythe, G.E.: SWAC computes 126 distinct semigroups of order 4. Proc. Amer.
   Math. Soc. 6, 443–447 (1955)
10. Frisch, A.M., Hnich, B., Kiziltan, Z., Miguel, I., Walsh, T.: Propagation algorithms
    for lexicographic ordering constraints. Artificial Intelligence 170, 834 (2006)
11. The GAP Group, GAP – Groups, Algorithms, and Programming, Version 4.4.12
    (2008), http://www.gap-system.org
12. Gecode: Generic constraint development environment, http://www.gecode.org/
13. Gent, I.P., Jefferson, C., Miguel, I.: Minion: A fast scalable constraint solver. In:
    Brewka, G., Coradeschi, S., Perini, A., Traverso, P. (eds.) The European Conference
    on Artificial Intelligence 2006 (ECAI 2006), pp. 98–102. IOS Press (2006)
14. Gent, I.P., Jefferson, C., Miguel, I.: Watched Literals for Constraint Propagation in
    Minion. In: Benhamou, F. (ed.) CP 2006. LNCS, vol. 4204, pp. 182–197. Springer,
    Heidelberg (2006)
15. Howie, J.M.: Fundamentals of semigroup theory, London Mathematical Society
    Monographs. New Series, vol. 12. The Clarendon Press, Oxford University Press,
    New York (1995), Oxford Science Publications
16. Jefferson, C.: Quicklex - a case study in implementing constraints with dynamic
    triggers. In: Proceedings of the ERCIM Workshop on Constraint Solving and Con-
    straint Logic Programming, CSCLP 2011 (2011)
17. Jürgensen, H., Wick, P.: Die Halbgruppen der Ordnungen ≤ 7. Semigroup Fo-
    rum 14(1), 69–79 (1977)
18. Katritzky, A., Hall, C., El-Gendy, B., Draghici, B.: Tautomerism in drug discovery.
    Journal of Computer-Aided Molecular Design 24, 475–484 (2010), http://
    dx.doi.org/10.1007/s10822-010-9359-z, doi:10.1007/s10822-010-9359-z

19. Klee Jr., V.L.: The November meeting in Los Angeles. Bull. Amer. Math. Soc. 62(1), 13–23 (1956), http://dx.doi.org/10.1090/S0002-9904-1956-09973-2
20. Kotthoff, L., Moore, N.C.: Distributed solving through model splitting. In: 3rd Workshop on Techniques for Implementing Constraint Programming Systems (TRICS), pp. 26–34 (2010)
21. Kreher, D., Stinson, D.: Combinatorial Algorithms: Generation, Enumeration, and Search. CRC Press (1998)
22. Lecoutre, C., Sais, L., Tabary, S., Vidal, V.: Nogood recording from restarts. In: Proceedings of the 20th International Joint Conference on Artifical Intelligence, pp. 131–136 (2007)
23. McKay, B.D.: Transitive graphs with fewer than twenty vertices. Math. Comp. 33(147), 1101–1121 (1979), contains microfiche supplement
24. McKay, B.D., Royle, G.F.: The transitive graphs with at most 26 vertices. Ars Combin. 30, 161–176 (1990)
25. Milletti, F., Storchi, L., Sforna, G., Cross, S., Cruciani, G.: Tautomer enumeration and stability prediction for virtual screening on large chemical databases. Journal of Chemical Information and Modeling 49(1), 68–75 (2009), http://pubs.acs.org/doi/abs/10.1021/ci800340j
26. Motzkin, T.S., Selfridge, J.L.: Semigroups of order five. Presented in [19] (1955)
27. Petrie, K.E., Smith, B.M.: Symmetry Breaking in Graceful Graphs. In: Rossi, F. (ed.) CP 2003. LNCS, vol. 2833, pp. 930–934. Springer, Heidelberg (2003)
28. Plemmons, R.J.: There are 15973 semigroups of order 6. Math. Algorithms 2, 2–17 (1967)
29. Rossi, F., van Beek, P., Walsh, T.: Handbook of Constraint Programming (Foundations of Artificial Intelligence). Elsevier Science Inc., New York (2006)
30. Satoh, S., Yama, K., Tokizawa, M.: Semigroups of order 8. Semigroup Forum 49(1), 7–29 (1994)
31. Thain, D., Tannenbaum, T., Livny, M.: Distributed computing in practice: The Condor experience. Concurrency – Practice and Experience 17(2-4), 323–356 (2005)
32. Yamanaka, K., Otachi, Y., Nakano, S.-I.: Efficient Enumeration of Ordered Trees with k Leaves (Extended Abstract). In: Das, S., Uehara, R. (eds.) WALCOM 2009. LNCS, vol. 5431, pp. 141–150. Springer, Heidelberg (2009), http://dx.doi.org/10.1007/978-3-642-00202-1_13

Exploring Chemistry Using SMT

Rolf Fagerberg[1], Christoph Flamm[2], Daniel Merkle[1], and Philipp Peters[1]

[1] Department of Mathematics and Computer Science,
University of Southern Denmark
{daniel,phpeters,rolf}@imada.sdu.dk
[2] Institute for Theoretical Chemistry, University of Vienna, Austria
xtof@tbi.univie.ac.at

Abstract. How to synthesize molecules is a fundamental and well studied problem in chemistry. However, computer aided methods are still under-utilized in chemical synthesis planning. Given a specific chemistry (a set of chemical reactions), and a specified overall chemical mechanism, a number of exploratory questions are of interest to a chemist. Examples include: what products are obtainable, how to find a minimal number of reactions to synthesize a certain chemical compound, and how to map a specific chemistry to a mechanism. We present a Constraint Programming based approach to these problems and employ the expressive power of Satisfiability Modulo Theory (SMT) solvers. We show results for an analysis of the *Pentose Phosphate Pathway* and the *Biosynthesis of 3-Hydroxypropanoate*. The main novelty of the paper lies in the usage of SMT for expressing search problems in chemistry, and in the generality of its resulting computer aided method for synthesis planning.

1 Introduction

The rigorous study of the properties of naturally occurring molecules requires their chemical synthesis from simpler precursor compounds. Therefore total synthesis of natural products is one of the fundamental challenges of organic chemistry. Chemical synthesis involves multistep synthetic sequences of elementary reactions. An elementary reaction transforms a set of chemical compounds (*reactants*) in a single step into a new set of chemical compounds (*products*) which are structurally different from the reactants. The step by step sequence of elementary reactions accompanying overall chemical change is denoted as *reaction mechanism.*

Finding a suitable sequence of elementary reactions leading from simple building blocks to a target molecule is in organic chemistry commonly referred to as the *synthesis planning problem.* Synthesis planning is a combinatorial complex problem and several heuristic approaches have been suggested [20] to attack this problem. Among synthetic chemists the *retrosynthetic analysis* [5] is one of the most popular approaches. This strategy systematically simplifies the target molecule by repeated bond disconnections in retrosynthetic direction, leading to progressively smaller precursors until recognized starting material emerges. Heuristic criteria are used to rank competing routes. Several computer programs are available implementing this approach (for a recent review see [4]).

M. Milano (Ed.): CP 2012, LNCS 7514, pp. 900–915, 2012.
© Springer-Verlag Berlin Heidelberg 2012

With the advent of Synthetic Biology and Systems Chemistry the need for rational design of molecular systems with pre-defined structural and dynamical properties has been shifted into the focus of research. Over the last century mathematical prototype models for a great variety of chemical and biological systems with interesting nonlinear dynamic behaviour such as oscillation have been collected and mathematically analysed. The translation of the mathematical formalism back into real world chemical or biological entities, i.e., finding an instantiation of such abstract mechanisms in real chemical molecules (required for the rational design of de-novo molecular systems), is still an unsolved problem. The problem can be rephrased for chemical reaction systems in the following way: finding a set of compatible molecules that react according to a reaction mechanism which was translated from an abstract mathematical prototype model. Of course the solution of this inverse problem is usually not unique, and crucially depends on information that can be provided in a declarative manner via constraints. An example of this is the chosen chemistry, i.e. molecules and reactions that can be employed for solving the underlying problem.

Note that the declarative approach allows for many different levels of modeling, with varying degrees of realism. In this paper, we propose a post-processing step applied to the (possibly intentionally underspecified) declarative solution, to extract real-world chemical solutions.

This paper introduces to our knowledge for the first time a Satisfiability Modulo Theories (SMT) based approach to the problem of chemical synthesis. In addition to the standard product-oriented methods usually employed, our approach covers a significantly wider collection of chemical questions. Satisfiability solvers are used predominantly on computer science related problems, but they have also been used to tackle chemically or biologically relevant topics. In [13] "Synthesizing Biological Theories" were introduced in order to construct high-level theories and models of biological systems. In [1] constraint logic programming was used for chemical process synthesis in order to design so-called Heat Exchanger Networks (which is very different from compound synthesis as it will be discussed in this paper).

This paper is organized as follows. In Section 2, we describe how we model chemistry. In Section 3, we describe the chemical search space, and the constraints on it which the user may impose. In Section 4, we focus on the SMT-formulation, and in Section 5, we explain our post-processing. Finally, we in Section 6 present tests of our approach on several instances from real-world chemistry.

2   Modeling Chemistry

2.1   The Reaction Mechanism

When modeling a chemical synthesis, it is necessary to define the underlying mechanism, i.e., the molecules involved and the chemical reactions specifying how they are transformed. For our purposes, a sufficient modeling of an *elementary reaction* is a change of at most 2 reactant molecules into at most 2 product molecules. Almost all real-world elementary chemical reactions can be viewed this way, if necessary by splitting up reactions with a larger number of participating molecules. Such elementary reactions fall into four sub-categories with different numbers of participating molecules: isomerization (1-to-1), merging (2-to-1), splitting (1-to-2), and transfer (2-to-2).

A reaction mechanism is a combination of elementary reactions. It defines how many and which molecules react in each of these single reactions. Chemists usually denote the elementary reactions of the reaction mechanism as follows:

$$m(A + B \to C + D),$$

where A, B, C, D denote the molecules and m denote the multiplicity of the reaction in the mechanism, i.e. how many times the reaction happens.

In more formal terms, a *reaction mechanism* is a directed multi-hypergraph $G(V, E)$. Each vertex $v \in V$ represents a molecule. The directed hyperedges E represents the elementary reactions in the mechanism: each hyperedge $e \in E$ is a pair (e^-, e^+) of multisets $e^-, e^+ \subseteq V$ of molecules, denoting the reactants and products of the chemical reaction [2], and coefficient m_e represents the multiplicity of the hyperedge. Thus, the reaction $m(A + B \to C + D)$ is represented by the hyperedge $(\{A, B\}, \{C, D\})$ with multiplicity m.

The *balance* $\mathrm{bal}(v)$ of molecule $v \in V$ in a reaction mechanism is defined as an integer number indicating its net production or consumption over the entire synthesis:

$$\mathrm{bal}(v) = \sum_{e \in E} m_e(\mathbf{1}_{e^+}(v) - \mathbf{1}_{e^-}(v)), \tag{1}$$

where $\mathbf{1}_\alpha$ is the multiplicity function on the multiset α. If $\mathrm{bal}(v) < 0$, v is a reactant of the overall synthesis. If $\mathrm{bal}(v) > 0$, $v \in V$ is an end product. If $\mathrm{bal}(v) = 0$, either molecule v does not take part in the synthesis, or is produced and consumed in equal amount during the synthesis.

A related concept is the *overall reaction* of a reaction mechanism. It is defined by summing up the two sides of all reactions (including multiplicities) in the mechanism, cancelling out equal amounts of identical molecules appearing on both sides. Thus, the left hand side of the overall reaction is given by the molecules with negative balance, and the right hand side by the molecules with positive balance.

2.2   The Molecules

In the large field of organic chemistry, the properties of carbon based molecules are studied. Most properties of such molecules are determined by *functional groups* attached to a backbone of carbon atoms. Functional groups are reactive subparts of molecules and define the characteristic physical and chemical properties of families of organic compounds [15]. In Fig. 2 and Fig. 6 some examples are given. Fig. 1 illustrates how a chemical reaction changes the occurrences of functional groups by transferring atoms from the first to the second molecule, and opening the ring of the second molecule. The removed groups are marked

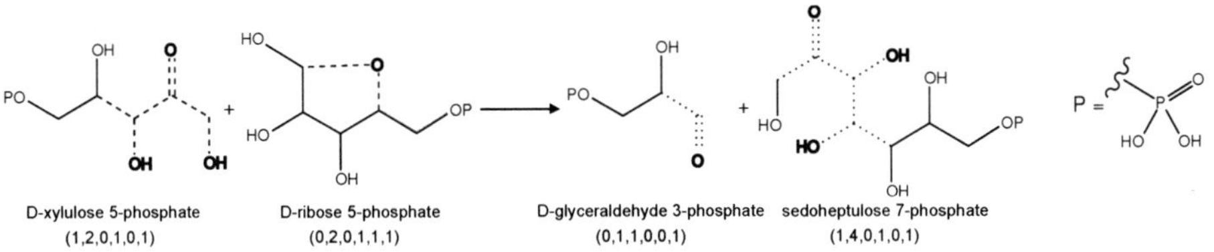

Fig. 1. The chemical reaction of D-xylulose 5-phosphate and D-ribose 5-phosphate to D-glyceraldehyde 3-phosphate and sedoheptulose 7-phosphate. The dashed-marked functional groups on the left side are removed during the reaction, the dotted groups on the right side are created. Non-participating groups are left black. The vectors of functional groups correlate with Fig. 2. For clarity, the phosphate group is substituted by "P".

dashed, the appearing groups are drawn dotted. Note that the number of untouched functional groups (black) in the products does not change. Line segment ends and junctions without annotations signify carbon atoms. In the example each of the reactants has one phosphate group, hence each product has a phosphate group.

In this paper, we model each molecule by a *vector of functional groups*. Position i in molecule A's vector provides the number of occurrences of the functional group x_i in A. The same vector is used for all molecules, i.e., there is one global set of functional groups. This set of functional groups is determined by the user, based on the chemistry, i.e., on what functional groups are deemed relevant to model for the molecules and reactions considered. Once this choice has been made, the functional groups are in our modeling simply positions in a vector. The length of that vector is the number of functional groups modeled, and the mapping between positions in vector and functional groups is arbitrary, but fixed.

This vector representation neglects the spatial structure of a molecule, i.e., only the number of occurrences of a functional group is noted, not its position(s) in the molecule. A chemical reaction is deemed feasible if its participating functional groups are present, irrespectively of whether these appear in the positions necessary for the reaction to take place. This implies that there may not be a real-world chemically valid equivalent to our vector-based reaction mechanism. Note that precise modeling of the chemical implications of the spatial structure of molecules is a hard problem in any formalism, with SMT being no exception. However, in Section 5 we provide a post-processing method which will allow us to filter our set of vector-based reaction mechanism, and retain only chemically viable solutions.

2.3   The Elementary Reactions

An elementary reaction can be defined formally in many ways in artificial chemistry [7], e.g. in a topological way or as a graph rewrite rule [3]. We here model

Fig. 2. The utilized functional groups in the order corresponding to their position in the molecule's vectors for modeling the reaction in Fig. 1 and the mechanism in Sec. 6.1

an elementary reaction by its change of the number of occurrences of the functional groups of the reactants, i.e., its change of their vector representations. We call such a specified change of vectors a *rule*. In addition, a rule specifies preconditions which must hold for the reactants vector representations. Trivially, a specific functional group has to appear in a reactant when a chemical reaction reduces the number of this functional group. However, in chemistry, it may also be necessary that a specific other functional group appears in a reactant for the reaction to take place, and we allow this to be specified, too. This is for each reactant in a reaction expressed by a vector, whose entries state the minimum number of each functional group which is required to be present for the reaction to take place. A concrete example of a rule appears at the start of Section 4.3.

3   Exploring Chemistry

Our overall goal is a system allowing chemists to explore questions of the following general format: can a subset of a given base set of reactions fit together to form a reaction mechanism fulfilling some given constraints? In Section 2.3, we defined rules (as preconditions and change vectors on a predefined set of functional groups), which is how the user will specify the base set of reactions.

In this section, we describe how reaction mechanisms are modeled in our system, how the system attempts to map rules to this, and the multiple ways the user can specify constraints on the reaction mechanism.

We note that our basic modeling of a reaction mechanism is very general. Our philosophy is to not limit beforehand what types of problems the chemist can tackle, while at the same time supplying a large collection of possibilities for expressing constraints on the reaction mechanism. These possibilities the user can utilize to adjust and narrow the search space in each concrete chemical setting, based on preferences, available knowledge, goal of the investigation, and new information learned during the exploration. Two concrete investigative approaches are described in Section 3.3.

3.1   Search Space

Rule Mapping. As defined above, a reaction mechanism is a multi-hypergraph with a vertex set V of molecules and an edge set E of reactions. In the search phase, each molecule in V is considered a vector of integer variables (namely, a counter for each functional group in the modeling), and the task of the system is to find values for these variables compatible with a subset of the rules supplied by the user. This means that for each edge, there must be a rule assigned to it for which the values of the variables in the nodes of the edges fulfil the constraints (change in vectors, preconditions of reactants) of that rule. We call such an assignment of rules to edges a *rule mapping*.

Thus, finding a solution to the specified chemical search problem means finding a rule mapping, and a set of values for the variables in the nodes, compatible with each other, as well as with any further constraints on the reaction mechanism specified via the methods described in Section 3.2. The task of the SMT-solver in our system is to find such a solution.

In the remainder of the paper, we will when needed distinguish between a reaction mechanism with nodes considered as variables (as described above) and a reaction mechanism constrained by requiring nodes to be specific molecules (either in the vector representation or real molecules) by using the terms *abstract reaction mechanism* and *concrete reaction mechanism*, respectively.

In our most general setup, we consider a mechanism of n 2-to-2 reactions (hence with $4n$ vector-valued variables representing the molecules), and m rules. No constraints to the structure of the mechanism are made, and each rule can be mapped onto every reaction (including e.g. a rule with one reactant and one product being mapped to a 2-to-2 reaction, in which case the two unused variables in the reaction mechanism are implicitly defined as null vectors).

Note that a rule mapping can be done in different ways: Consider a specific 2-to-2 rule, which is mapped to a reaction $A + B \rightarrow C + D$. The rules will be defined by two precondition (p_1 and p_2) and two change vectors (δ_1 and δ_2) for the two reactants. When mapping the rule to the reaction, four possibilities exist depending on whether p_1 and δ_1 is taken as the $A \rightarrow C$, $A \rightarrow D$, $B \rightarrow C$, or $B \rightarrow D$ change (with p_2 and δ_2 in all cases giving the change for the remaining part of the reaction).

Equivalence Relation. In our most general setup, the system is free to identify different node variables when looking for a solution. Then part of the output will be an equivalence relation $id : V \times V$ which is used in order to define the identity of two variables in a reaction mechanism. An equivalence relation id on molecules implies:

$$\forall v, w \in V : (v, w) \in id \Rightarrow \forall x_p \in \text{functional groups} : v(x_p) = w(x_p), \quad (2)$$

where $v(\cdot)$ denotes the number of occurrences of x_p in v. I.e., defining two variables to refer to the same molecule implies that occurrences of subgroups are identical.

Multiplicities. *Multiplicities of single reactions* in the mechanism denote how often a reaction takes place. The overall consumption and production of molecules

$$m_1 \; (A + B \rightarrow C + D) \qquad\qquad m_1 \; (A + B \rightarrow C + D)$$
$$\Rightarrow$$
$$m_2 \; (E + F \rightarrow G + H) \qquad\qquad m_2 \; (C + D \rightarrow H)$$

Fig. 3. A general reaction mechanism (left side) without constrained equivalence classes. A predefined equivalence relation $\{(C, E), (D, F)\}$ leads to the reaction mechanism on the right side; G is supposed to be an empty molecule.

is defined by the sum of all produced occurrences minus the sum of all consumed occurrences, including the multiplicities of reactions. To provide the largest generality, in our model the multiplicities do not have to be specified. They are also part of the solution to be found.

3.2   Constraining the Search Space

To answer different chemical questions and to incorporate previous chemical knowledge, our system allows for the specification by the user of a number of additional constraints, which we now present.

Mechanism Specification. By having a predefined equivalence relation id, a generic reaction mechanism can be constrained by specifying predefined equivalence classes of identical molecules. Figure 3 gives an example: On the left side a short generic reaction mechanism of two reactions is shown. Using the predefined equivalence relation $\{(C, E), (D, F)\}$ leads to the reaction mechanism on the right side. Molecules which do not take part (in the example molecule G) in a reaction are left out.

The pre-definition of equivalence classes implies that rules must now map exactly onto the reactions in terms of number of participating compounds. E.g., in contrast to the case where the equivalence relation is not predefined, a rule mapping of a 2-to-1 rule to a 2-to-2 reaction is not allowed. From a chemical perspective, such predefined identities imply an already known reaction mechanism and is an instance of the *Inverse Reaction Mechanism Problem* (see below).

The number n of reactions is always specified in our setup. If the user wants to e.g. search for a minimal n for which solutions exist, several runs with differing value of n can be done.

Balance and Overall Reaction. Another constraint for the reaction mechanism can be set via the balance of molecules. As mentioned in Section 2, for each molecule a specified balance $bal(v), v \in V$ for the whole mechanism holds. Especially in the product-oriented approach it makes sense to specify a desired amount of the product by a positive balance, and an amount of reactants by a negative balance. If it is assumed that there are no side products in a synthesis produced, $bal(v) = 0$ can be set for all other molecules. (However, any so-called food and waste molecules, which provide energy, like Adenosine triphosphate

(ATP), or can be consumed and produced in infinite amounts, like H_2O, should have their balance unconstrained.)

Multiplicities. In the most generic approach, the reaction mechanism and the multiplicities m_e of reactions are not specified in advance. But if it makes chemical sense to restrict the multiplicities, this of course can be specified. This could be done e.g. to prevent the solver to add chemically implausible high frequencies of single reactions to the solution. Note that balances and multiplicities are linked by Eq. (1). Only solutions fulfilling this will be produced.

Molecule Size. Also the number of functional groups, or the number of a specific functional group in a molecule can be restricted. This can be employed in order to find a solution using a minimum necessary (or at least a small) number of functional groups for a desired reaction mechanism. Like a constraint on the multiplicities, a restriction on the molecule size may be used to prevent solutions with (chemically) unrealistic numbers.

3.3   Product-Oriented Exploration vs. IRMP

This generic approach introduced above is used in order to pose and answer different chemical questions. We distinguish two major lines of questions:

In a *product-oriented exploration* of chemistry, the properties of a desired product are known (in terms of functional groups or even as specific molecules). This knowledge serves as constraints to the molecule's vectors defining the number of occurrences of functional groups. This corresponds to a classical question of how to synthesize a specific compound based on a given set of chemical reactions. Based on existing chemical knowledge, a suggestion for the abstract reaction mechanism (including the equivalence relation for its molecules) for the synthesis may actually be known, or, more likely, it may be unknown.

A new approach for synthesis planning, and even more importantly, for understanding chemical reaction patterns, is what we define in this paper as the *Inverse Reaction Mechanism Problem (IRMP)*: In the *IRMP* it is assumed that an underlying reaction mechanism of a synthesis is known. Then, it is investigated if for the same abstract mechanism a *different* set of elementary chemical reactions (rules) can be mapped to it (potentially generating different molecules). In our model this corresponds to finding rule mappings and multiplicities, (but not equivalence classes of identical molecules, as this is assumed to be known), based on set of elementary chemical reactions different from the ones originally participating in the reaction mechanism.

4   The SMT-Implementation

We have implemented the approach delineated above using SMT. In this section, we present central parts of this implementation. The language used is

SMT-LIB [18] and the SMT-Solver is Microsoft's Z3 [6]. Our implementation creates an SMT program, based on input specification files that define preconditions/changes for the rules, the predefined balances, and the equivalence constraints. The auto-generated program is then handed over to the SMT solver.

4.1   Declarations

A subset of the most important data types and functions will be defined here. For concreteness, we as example use the second reaction from Fig. 4, representing the chemical reaction from Fig. 1. The capital letters `A,C,D,E` constitute the type `MOL` representing molecules, the lowercase letters `a,b,c,d,e,f` constitute the type `SUB` of functional groups, and the mechanism's reactions `REACT` are numbered from 1 to n.

```
(declare-datatypes () ((MOL   A C D E)))
(declare-datatypes () ((SUB   a b c d e f)))
(declare-datatypes () ((REACT   react1 react2 ... reactn)))
```

The function `NrOfGroups` provides a non-negative number of occurrences of each functional group for each molecule. Due to simplicity of summing up balances later, *stoichiometric coefficients* `STOI` for the molecules in each reaction are defined. In the notation of chemistry this means that `STOI` for a reactant is the negatively signed value of the multiplicity of the reaction, whereas it is the same but positively signed value for each product. The equivalence relation is implemented as a Boolean matrix `ID` and provides the identity of molecules. `PRODUCT` and `REACTANT` functions provide Boolean values and define if a molecule should be a product or reactant in the whole reaction mechanism.

```
(declare-fun NrOfGroups (MOL SUB) Int)
(declare-fun STOI (REACT MOL) Int)
(declare-fun ID (MOL MOL) Bool)
(declare-fun REACTANT (MOL) Bool)
(declare-fun PRODUCT (MOL) Bool)
```

4.2   Equivalence Relation

Additional to the properties of the equivalence relation (reflexivity, symmetry, transitivity) for the equivalence relation *id*, an implication has been implemented. Two molecules being in the same class implies the same number of occurrences for all functional groups (cmp. Eqn. 2).

```
(assert (forall ((mol MOL)(mol2 MOL))
     (=>  (=(ID mol mol2) true)
     (forall ((sub SUB))
        (= (NrOrGroups mol sub) (NrOrGroups mol2 sub)) )))))
```

4.3 Mapping

In the following, an example of a rule mapping for a 2-to-2 reaction $A + C \rightarrow D + E$ will be presented. Picking the second reaction from Fig. 4, we will restrict the rule mapping to the case where molecule A is changed into molecule D (implying molecule C will change to E). Assume that the rule mapping is defined by the change vector (-1,-1,1,-1,0,0) and the precondition vector (1,1,0,1,0,0) for A, and change vector (1,2,0,0,-1,0) and precondition vector (0,0,0,0,1,0) for C. This leads to the following implementation:

```
(assert (and
    ; stoichiometry constraints:
    ; (the stoi. coeff. of A needs to be the negative of D, etc.)
    (< (STOI react1 A) 0)
    (= (STOI react1 A) (STOI react1 C))
    (= (STOI react1 A) (- (STOI react1 D)))
    (= (STOI react1 D) (STOI react1 E))
    (or  (and
        ;preconditions
          (>= (NrOrGroups A a) 1)   (>= (NrOrGroups A b) 1)
          (>= (NrOrGroups A d) 1)   (>= (NrOrGroups C e) 1)
        ;changes made to A, which results in D
          (= (NrOrGroups D a) (- (NrOrGroups A a) 1))
          (= (NrOrGroups D b) (- (NrOrGroups A b) 1))
          (= (NrOrGroups D c) (+ (NrOrGroups A c) 1))
          (= (NrOrGroups D d) (- (NrOrGroups A d) 1))
          (= (NrOrGroups D e) (NrOrGroups A e))
          (= (NrOrGroups D f) (NrOrGroups A f))
        ;changes made to C, which results in E
          (= (NrOrGroups E a) (+ (NrOrGroups C a) 1))
          (= (NrOrGroups E b) (+ (NrOrGroups C b) 2))
          (= (NrOrGroups E c) (NrOrGroups C c))
          (= (NrOrGroups E d) (NrOrGroups C d))
          (= (NrOrGroups E e) (- (NrOrGroups C e) 1))
        (= (NrOrGroups E f) (NrOrGroups C f))
    ))))
```

4.4 Balance

For all molecules, a balance for the whole mechanism can be defined. In the following example the balance of a product molecule (i.e. `PRODUCT(mol)` is `true`) shall be greater than zero or equal to a specific positive amount.

$$\forall \texttt{mol} \in \texttt{MOL} : \texttt{PRODUCT(mol)} \Rightarrow \sum_{r=1}^{n} \texttt{STOI(r,mol)} > 0$$

The SMT code implementing this constraint is:

```
; balance for product should be > 0, symmetric for reactant
(assert(forall(mol MOL)
      (=> (= (PRODUCT mol) true)
      (> (+(STOI react1 mol)(STOI react2 mol)
        . . .
      (STOI reactn mol)) 0)))))
```

The negative amount for the reactant can be constrained similarly, as well as the balance of value 0 for all other non-product or non-reactant molecules.

5   Post-processing

The solution output by the SMT-solver contains a rule mapping and a set of vector values, and is thus expressed in the vector representation of molecules. As noted earlier, this representation neglects the spatial structure of molecules, implying that false positives can occur in the sense that some found solutions may not have corresponding real-world chemical reactions.

In this section, we describe an automated post-processing method which allows us to filter our set of SMT-solutions, and retain only chemically viable solutions consisting of existing real-world chemical reactions.

The method is based on the existence of large chemical databases of reactions, such as KEGG [17]. The KEGG database is a biochemical database containing biochemical pathways and most of the known metabolic pathways. After converting an entry for a reaction in the KEGG database to a form searchable by the Graph Grammar Library [8,9], we in an automated way search for the appearances of the functional groups. From this, we generate the vector representations of its participating molecules, and deduce the rule version of that reaction. The IDs of the participating real-world molecules are stored with the rule. We then apply a straightforward search algorithm for finding a conversion of the SMT-generated reaction mechanism from vector representation to a form where nodes contain the IDs of real-world molecules and where all edges represent a real-world reaction from the database. This is done by first for each hyperedge e of the SMT-generated reaction mechanism finding the set K_e of KEGG reactions whose vector representation is compatible with that of e. Then for some fixed order $e_1, e_2, \ldots$ of the edges doing a backtracking DFS-type search for an assignment of KEGG reactions to edges for which the implied molecule IDs agree for all nodes. In details, the search starts by assigning the first reaction in K_{e_1} to e_1, recording the implied molecule IDs for the nodes of e_1, and then advancing to the next edge. If for an edge e_i no reaction in K_{e_i} can be found which is compatible with the molecule IDs implied by the currently assigned reactions for e_1 to e_{i-1}, the search backtracks, and tries the next edge of $K_{e_{i-1}}$.

The results of this post-processing may for each SMT-solution provide potentially many reaction mechanisms with real-world chemical reactions and real molecules, or it may find that none can be given based on the database in question. The post-processing may then be repeated with any further solutions from the SMT-solver.

6   Results

In this section, we will present results of our SMT-based exploring approach on chemistry, namely the well studied and well understood *Pentose Phosphate Pathway (PPP)* [12] and the industrial important *biosynthesis of 3-Hydroxypropanoate (3HP)*. As SMT-Solver, *Microsoft's Z3 SMT-Solver* was used on an Intel Core2 Duo CPU T7500 @ 2.20GHz with a memory size of 2 GB.

6.1   The Pentose Phosphate Pathway

The *PPP* can be found in most organisms, including mammals, plants and bacteria such as *E. coli*. It generates the co-enzyme *NADPH*, which takes part in many anabolic reactions as reducing agent, and a sugar with six carbon atoms (here: fructose 6-phosphate) [16]. Its products are used for the synthesis of nucleotides and amino acids. The PPP is also an alternative to the glycolysis which converts glucose into pyruvate and releases highly energetic molecules ATP (adenosine triphosphate). Our model of the PPP takes as input 6 sugar molecules with 5 carbon atoms (*pentoses*, here ribulose 5-phosphate) and releases 5 molecules of fructose 6-phosphate which has 6 carbon atoms.

Product-Oriented Exploration. Starting with a set of reactions, the goal is to identify an abstract reaction mechanism to create a certain amount of fructose 6-phosphate.

Our instance consists of 7 2-to-2 reactions, where not all molecules have to appear in the latter abstract reaction mechanism. The multiplicities of reactions and the equivalence relation *id* are not restricted, and will be part of the solution. The properties and amounts of the reactant and the product are known. These two variables are predefined in the instance, and in addition their balances, too. The latter is done by the constraints: $bal(reactant) = -6$ and $bal(product) = 5$. Additionally, we specify a water and a phosphate molecule, whose balances are not restricted. All other molecules appearing in the mechanism should have balance zero.

For solving this instance, a set of molecules and a set of rules over these molecules are assumed to be given. These sets can be derived for example from a database request (as from KEGG). In our case we chose molecules and rules from the natural appearing PPP and added three additional sugar-molecule rules (giving 10 rules and 11 molecules in total). For this instance, we chose to let functional groups correspond to complete molecules. This implies that all rules remove exactly one "functional group" and add exactly one "functional group", namely the complete molecules.

A valid mapping of these rules to the reaction set instantiates a possible abstract reaction mechanism to synthesize fructose 6-phosphate in 7 reactions.

The auto-generated SMT-Program for this example was solved by Z3 in 128 minutes. Fig. 4 shows a solution of this instance in which the Pentose Phosphate Pathway occurs as it can be found in nature. If the equivalence classes of molecules are specified in advance, the solution is found in less than 10 seconds.

$$
\begin{array}{rl}
2\times & A \to C \\
2\times & A + C \to D + E \\
2\times & D + E \to G + H \\
2\times & A + G \to D + H \\
1\times & D \to F \\
1\times & D + F \to I \\
1\times & B + I \to H + J \\
\end{array}
$$

$$\sum : 6 \cdot \mathbf{A} + \mathbf{B} \to 5 \cdot \mathbf{H} + J$$

$$
\begin{aligned}
\mathbf{A} &= \mathbf{C_5 H_{11} O_8 P} \quad B = H_2 O \quad\quad C = C_5 H_{11} O_8 P \\
D &= C_3 H_7 O_6 P \quad E = C_7 H_{15} O_{10} P \quad F = C_3 H_7 O_6 P \\
G &= C_4 H_9 O_7 P \quad \mathbf{H} = \mathbf{C_6 H_{13} O_9 P} \quad I = C_6 H_{14} O_{12} P_2 \\
J &= H_3 P O_4
\end{aligned}
$$

Fig. 4. The abstract PPP reaction mechanism as it occurs in nature, found by our SMT approach. The letters in the mechanism represents the molecules given to the right. Depicted in bold are ribulose 5-phosphate and fructose 6-phosphate.

Inverse Reaction Mechanism Problem. The solution from the product-oriented approach provides an abstract reaction mechanism, i.e., the equivalence classes of identities of the molecules and the multiplicities are now fixed. By giving this, the *IRMP* is instantiated and a concrete reaction mechanism can be sought after. This means that molecules from the abstract mechanism are now seen as vectors of functional groups and one abstract mechanism can serve as template for several concrete mechanisms. The result will be highly dependent on the given set of rules; in this example we focus on finding the PPP. For testing reasons, the chemistry is chosen in a simple way, it consists of 8 rules from "sugar chemistry", based on 6 functional groups. These are shown in Fig. 2 where they are ordered as in the molecule's vectors. Note that the groups are not overlapping, they only share carbon atoms. As an example we illustrated a rule at the start of Section 4.3, where a *transketolase* (cf. Fig. 1 and second line of Fig. 4) was modeled, including change and precondition vectors. This 2-to-2 reaction transfers a fragment (a keto group) from one molecule to another.

The SMT-solver provided a solution in less than 2 seconds. The solution could by the way be seen to be minimal in the total number of occurring functional groups (using a smaller overall number of functional groups lead to unsatisfiability for the given set of rules), so artificially large molecules seemed to be avoided. Fig, 5 shows the concrete reaction mechanism with the vectors defining the number of functional groups in the molecules. Note that the second to last reaction was modeled here just as a 1-to-1 reaction, because water (B) is always available, also phosphate (J) serves as waste molecule and can be produced in an infinite amount. Based on the SMT solution, the post-processing step generates a known real-world synthesis from the PPP.

6.2   Biosynthesis of 3-Hydroxypropanoate

3-Hydroxypropanoate (3HP) is a high-value organic molecule and is used in numerous reactions. Usually it is organically synthesized, but biosynthetic pathways to this product are in high demand [19]. To produce 3HP from pyruvate

$$
\begin{array}{rlcl}
2\times & (1\,2\,0\,1\,0\,1) & \to & (0\,2\,0\,1\,1\,1) \\
2\times & (1\,2\,0\,1\,0\,1) + (0\,2\,0\,1\,1\,1) & \to & (0\,1\,1\,0\,0\,1) + (1\,4\,0\,1\,0\,1) \\
2\times & (0\,1\,1\,0\,0\,1) + (1\,4\,0\,1\,0\,1) & \to & (0\,2\,1\,0\,0\,1) + (0\,3\,0\,1\,1\,1) \\
2\times & (1\,2\,0\,1\,0\,1) + (0\,2\,1\,0\,0\,1) & \to & (0\,1\,1\,0\,0\,1) + (0\,3\,0\,1\,1\,1) \\
1\times & (0\,1\,1\,0\,0\,1) & \to & (1\,0\,0\,1\,0\,1) \\
1\times & (0\,1\,1\,0\,0\,1) + (1\,0\,0\,1\,0\,1) & \to & (0\,3\,0\,0\,1\,2) \\
1\times & B + (0\,3\,0\,0\,1\,2) & \to & (0\,3\,0\,1\,1\,1) + J
\end{array}
$$

Fig. 5. concrete PPP reaction mechanism with vectors of occurrences of functional groups (cmp. Fig 2) of molecules as found by the SMT-Solver. Post-processing with molecules occurring in this solution of PPP leads to the PPP as it occurs in nature.

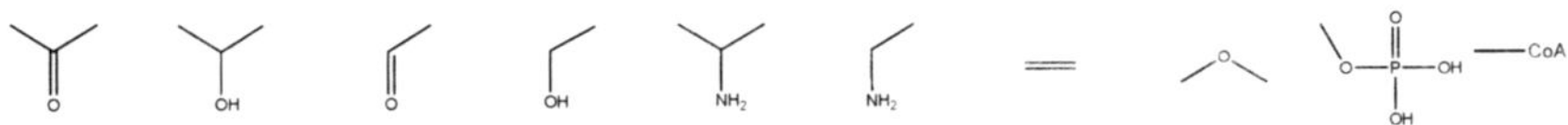

Fig. 6. The utilized functional groups in the order corresponding to their position in the molecule's vectors for the modeling of the biosynthesis of 3HP

(an end product of the glycolysis), biosynthetic pathways have been assembled [11,14] and additionally one pathway has already been implemented in an industrial setup [21].

The biosynthesis of 3HP was investigated here as an instance of the *IRMP*, i.e. an abstract reaction mechanism and the chemistry (a set of rules) are given, and the solution of this instance identifies possible pathways from pyruvate to the desired product 3HP. The vectors of the reactant and the product were predefined. A reaction mechanism of length n, consisting of only 1-to-1 reactions was used, as shown in Fig. 7. The multiplicities of reactions were set to 1. The equivalence classes of molecules were specified, and by doing so, a cascading mechanism $A \to B \to C \to \ldots$ was created. In total, 19 chemical 1-to-1 rules with 10 functional groups were used in order to define the chemistry. These rules were defined by chemical expertise (details omitted in this paper) as well as derived by a recent database-supported approach [10].

The concrete reaction mechanisms provided by the SMT-Solver were post-processed using the KEGG database. Due to space limitation, Fig. 7 shows only 3 of the 27 found pathways from the post-processing. The functional groups marked dashed disappear in the subsequent reaction, the bold-marked functional groups are pre-conditional for the reaction to take place. All pathways generated by [10] could be found.

Additionally, by post-processing a concrete mechanism of length 2, a solution for the *IRMP* could be found that does not produce 3HP. I.e., a pathway was found that employs exactly the same reaction pattern as the synthesis of 3HP but is based on a different set of molecules. The alternative two-step-pathway syntheses 2-phospho-D-glycerate from 3-phosphohydroxypyruvate, using KEGG-notation can be stated as $C03232 \xrightarrow{R01513} C00197 \xrightarrow{R01518} C00631$.

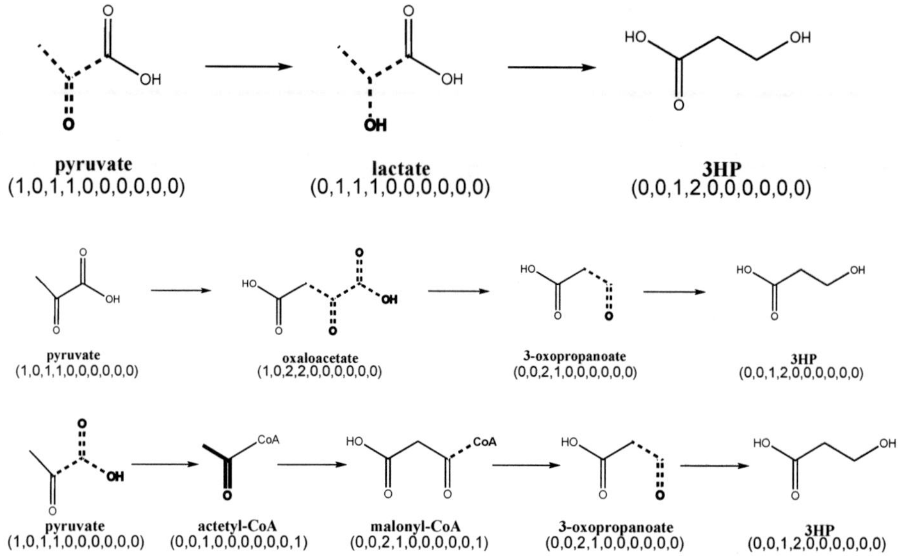

Fig. 7. Three example pathways for the biosynthesis of 3HP from pyruvate; the vectors underneath are using the functional groups and their order from Fig. 6; reacting functional groups are drawn dashed; functional groups which are necessary but remain unchanged are depicted bold

7   Conclusions

We introduced the combination of two rather different fields of research, namely Satisfiability Modulo Theories (SMT) and theoretical and real-world chemistry. Defining chemical questions like the synthesis of a specific compound or the search for pathway patterns formally as instances for SMT solvers allows to answer a large set of chemically highly relevant but so far unasked questions. To underline this we introduced and solved the Inverse Reaction Mechanism Problem (IRMP), which can be used to identify reaction mechanism patterns via SMT. Solutions to the IRMP might have significant impact on chemical compound fabrication and can help to understand patterns in chemical reaction mechanisms. We have shown the applicability of the new approaches on two real-world chemical setups, namely the analysis of the Pentose Phosphate Pathway and the biosynthesis of 3-Hydroxypropanoate.

References

1. Abbass, H., Wiggins, G., Lakshmanan, R., Morton, B.: Heat exchanger network retrofit by constraint logic programming. Computers in Chemical Engineering (1999)
2. Andersen, J., Flamm, C., Merkle, D., Stadler, P.: Maximizing output and recognizing autocatalysis in chemical reaction networks is NP-complete. Journal of Systems Chemistry 3(1) (2012)

3. Benkö, G., Centler, F., Dittrich, P., Flamm, C., Stadler, B., Stadler, P.: A topological approach to chemical organizations. Artificial Life 15(1), 71–88 (2009)
4. Cook, A., Johnson, A.P., Law, J., Mirzazadeh, M., Ravitz, O., Simon, A.: Computer-aided synthesis design: 40 years on. WIREs Comput. Mol. Sci. 2, 79–107 (2012)
5. Corey, E.J.: General methods for the construction of complex molecules. Pure Appl. Chem. 14, 19–38 (1967)
6. de Moura, L., Bjørner, N.: Z3: An Efficient SMT Solver. In: Ramakrishnan, C.R., Rehof, J. (eds.) TACAS 2008. LNCS, vol. 4963, pp. 337–340. Springer, Heidelberg (2008)
7. Dittrich, P., Ziegler, J., Banzhaf, W.: Artificial chemistries - a review. Artificial Life 7(3), 225–275 (2001)
8. Flamm, C., Ullrich, A., Ekker, H., Mann, M., Hogerl, D., Rohrschneider, M., Sauer, S., Scheuermann, G., Klemm, K., Hofacker, I., Stadler, P.F.: Evolution of metabolic networks: A computational frame-work. Journal of Systems Chemistry 1(4) (2010)
9. Graph grammar library, `http://www.tbi.univie.ac.at/~xtof/software/GGL/`
10. Henry, C., Broadbelt, L., Hatzimanikatis, V.: Discovery and analysis of novel metabolic pathways for the biosynthesis of industrial chemicals: 3-hydroxypropanoate. Biotechnology and Bioengineering 106(3), 462–473 (2010)
11. Jiang, X., Meng, X., Xian, M.: Biosynthetic pathways for 3-hydroxypropionic acid production. Applied Microbiology and Biotechnology 82(6), 995–1003 (2009)
12. Kruger, N., von Schaewen, A.: The oxidative pentose phosphate pathway: structure and organisation. Current Opinion in Plant Biology 6(3), 236–246 (2003)
13. Kugler, H., Plock, C., Roberts, A.: Synthesizing Biological Theories. In: Gopalakrishnan, G., Qadeer, S. (eds.) CAV 2011. LNCS, vol. 6806, pp. 579–584. Springer, Heidelberg (2011)
14. Maris, A., Konings, W., Dijken, J., Pronk, J.: Microbial export of lactic and 3-hydroxypropanoic acid: implications for industrial fermentation processes. Metabolic Engineering 6(4), 245–255 (2004)
15. McNaught, A., Wilkinson, A.: IUPAC compendium of chemical terminology, vol 2. Blackwell Scientific Publications (1997)
16. Nelson, D., Cox, M.: Lehninger Biochemie. Springer (2005)
17. Ogata, H., Goto, S., Sato, K., Fujibuchi, W., Bono, H., Kanehisa, M.: Kegg: Kyoto encyclopedia of genes and genomes. Nucleic Acids Research 27(1), 29 (1999)
18. Ranise, S., Tinelli, C.: The SMT-LIB standard: Version 1.2. Department of Computer Science, The University of Iowa, Tech. Rep. (2006)
19. Suthers, P., Cameron, D., et al.: Production of 3-hydroxypropionic acid in recombinant organisms. US Patent 6,852,517 (February 8, 2005)
20. Todd, M.H.: Computer-aided organic synthesis. Chem. Soc. Rev. 34, 247–266 (2005)
21. Willke, T., Vorlop, K.: Industrial bioconversion of renewable resources as an alternative to conventional chemistry. Applied Microbiology and Biotechnology 66(2), 131–142 (2004)

A Pseudo-Boolean Set Covering Machine

Pascal Germain, Sébastien Giguère, Jean-Francis Roy, Brice Zirakiza,
François Laviolette, and Claude-Guy Quimper

Département d'informatique et de génie logiciel, Université Laval, Québec, Canada
{sebastien.giguere.8,jean-francis.roy.1,brice.zirakiza.1}@ulaval.ca,
{pascal.germain,francois.laviolette,claude-guy.quimper}@ift.ulaval.ca

Abstract. The Set Covering Machine (SCM) is a machine learning algorithm that constructs a conjunction of Boolean functions. This algorithm is motivated by the minimization of a theoretical bound. However, finding the optimal conjunction according to this bound is a combinatorial problem. The SCM approximates the solution using a greedy approach. Even though SCM seems very efficient in practice, it is unknown how it compares to the optimal solution. To answer this question, we present a novel pseudo-Boolean optimization model that encodes the minimization problem. It is the first time a Constraint Programming approach addresses the combinatorial problem related to this machine learning algorithm. Using that model and recent pseudo-Boolean solvers, we empirically show that the greedy approach is surprisingly close to the optimal.

1   Introduction

Machine learning [2] studies algorithms that "learn" to perform a task by observing examples. In the classification framework, a learning algorithm is executed on a training set which contains examples. Each example is characterized by a description and a label. A learning algorithm's goal is to generalize the information contained in the training set to build a classifier, i.e. a function that takes as input an example description, and outputs a label prediction. A good learning algorithm produces classifiers of low risk, meaning a low probability of misclassifying a new example that was not used in the learning process.

Among all machine learning theories, *Sample Compression* [4] studies classifiers that can be expressed by a subset of the training set. This theory allows to compute bounds on a classifier's risk based on two main quantities: the size of the *compression set* (the number of training examples needed to describe the classifier) and the *empirical risk* (the proportion of misclassified training examples). This suggests that a classifier should realize a tradeoff between its complexity, quantified here by the compression set size, and its accuracy on the training set.

Based on this approach, the *Set Covering Machine* (SCM) is a learning algorithm motivated by a sample compression risk bound [8]. However, instead of finding the optimal value of the bound, the SCM algorithm is a greedy approach that aims to quickly find a good solution near the optimal bound's value.

In this paper, we address the following question: "How far to the optimal is the solution returned by the SCM algorithm?". To answer this question, one needs

M. Milano (Ed.): CP 2012, LNCS 7514, pp. 916–924, 2012.

to design a learning algorithm that directly minimizes the sample compression bound that inspired the SCM. This task is not a trivial one : unlike many popular machine learning algorithms that rely on the minimization of a convex function (as the famous Support Vector Machine [3]), this optimization problem is based on a combinatorial function. Although Hussain et al. [5] suggested a (convex) linear program version of the SCM, it remains a heuristic inspired by the bound. The present paper describes how to use Constraint Programming techniques to directly minimize the sample compression bound. More precisely, we design a pseudo-Boolean program that encodes the proper optimization problem, and finally show that the SCM is surprisingly accurate.

2   Problem Description

The Binary Classification Problem in Machine Learning. An *example* is a pair $(\mathbf{x}, y)$, where $\mathbf{x}$ is a *description* and y is a *label*. In this paper, we consider binary classification, where the description is a vector of n real-valued attributes (i.e. $\mathbf{x} \in \mathbb{R}^n$) and the label is a Boolean value (i.e. $y \in \{0, 1\}$). We say that a 0-labeled example is a *negative example* and a 1-labeled is a *positive example*.

A *dataset* contains several examples coming from the observation of the same phenomenon. We denote S the *training set* of m examples used to "learn" this phenomenon. As the examples are considered to be independently and identically distributed (iid) following a probability distribution D on $\mathbb{R}^n \times \{0, 1\}$, we have:

$$S \stackrel{\text{def}}{=} \{(\mathbf{x}_1, y_1), (\mathbf{x}_2, y_2), \ldots, (\mathbf{x}_m, y_m)\} \sim D^m \, .$$

A *classifier* receives as input the description of an example and predicts a label. Thus, a classifier is a function $h : \mathbb{R}^n \to \{0, 1\}$. The *risk* $R(h)$ of a classifier is the probability of misclassifying an example generated by the distribution D, and the *empirical risk* $R_S(h)$ of a classifier is the ratio of errors on its training set.

$$R(h) \stackrel{\text{def}}{=} \mathop{\mathbf{E}}_{(\mathbf{x},y) \sim D} I(h(\mathbf{x}) \neq y) \quad \text{and} \quad R_S(h) \stackrel{\text{def}}{=} \frac{1}{m} \sum_{(\mathbf{x},y) \in S} I(h(\mathbf{x}) \neq y) \, ,$$

where I is the indicator function: $I(a) = 1$ if a is true and $I(a) = 0$ otherwise.

A *learning algorithm* receives as input a training set and outputs a classifier. The challenge of a these algorithms is to generalize the information of the training set to produce a classifier of low risk. Since the data generating distribution D is unknown, a common practice to estimate the risk is to calculate the error ratio on a *testing set* containing examples that were not used in the training process.

Overview of the Sample Compression Theory. The sample compression theory, first expressed by Floyd et al. [4], focuses on classifiers that can be expressed by a subset of the training set.

Consider a classifier obtained by executing a learning algorithm on the training set S containing m examples. The *compression set* S_i refers to examples of the training set that are needed to characterize the classifier.

$$S_{\mathbf{i}} \overset{\text{def}}{=} \{(\mathbf{x}_{i_1}, y_{i_1}), (\mathbf{x}_{i_2}, y_{i_2}), \ldots, (\mathbf{x}_{i_n}, y_{i_n})\} \subseteq S \quad \text{with } 1 \leq i_1 < i_2 < \ldots < i_n \leq m.$$

We sometimes use a *message string* μ that contains additional information[1]. The term *compressed classifier* refers to the classifier obtained solely with the compression set $S_{\mathbf{i}}$ and message string μ. Sample compression provides theoretical guarantees on a compressed classifier by upper-bounding its risk. Typically, those bounds suggest that a learning algorithm should favour classifiers of low empirical risk (accuracy) and that are expressed by a few training examples (sparsity). One can advocate for sparse classifiers because they are easy to understand by a human being.

The Set Covering Machine. Suggested by Marchand and Shawe-Taylor [8], the *Set Covering Machine* (SCM) is a learning algorithm directly motivated by the sample compression theory. It builds a conjunction or a disjunction of binary functions that rely on training set data. We focus here on the most studied case where each binary function is a *ball* $g_{i,j}$ characterized by two training examples, a center $(\mathbf{x}_i, y_i) \in S$ and a border $(\mathbf{x}_j, y_j) \in S$.

$$g_{i,j}(\mathbf{x}) \overset{\text{def}}{=} \begin{cases} y_i & \text{if } \|\mathbf{x}_i - \mathbf{x}\| < \|\mathbf{x}_i - \mathbf{x}_j\| \\ \neg y_i & \text{otherwise,} \end{cases} \tag{1}$$

where $\| \cdot \|$ is the Euclidean norm. For simplicity, we omit the case $\|\mathbf{x}_i - \mathbf{x}\| = \|\mathbf{x}_i - \mathbf{x}_j\|$ and consider that a ball correctly classifies its center ($g_{i,j}(\mathbf{x}_i) = y_i$) and its border ($g_{i,j}(\mathbf{x}_j) = y_j$).

We denote $\mathcal{H}_S$ the set of all possible balls on a particular dataset S, and $\mathcal{B}$ the set of balls selected by the SCM algorithm among $\mathcal{H}_S$. Thus, the classification function related to a conjunction of balls is expressed by:

$$h_{\mathcal{B}}(\mathbf{x}) \overset{\text{def}}{=} \bigwedge_{g \in \mathcal{B}} g(\mathbf{x}). \tag{2}$$

As the disjunction case is very similar to the conjunction case, we simplify the following discussion by dealing only with the latter[2]. Figure 1 illustrates an example of a classifier obtained by a conjunction of two balls.

The goal of the SCM algorithm is to choose balls among $\mathcal{H}_S$ to form the conjunction $h_{\mathcal{B}}$. By specializing the sample-compressed classifier's risk bound to the conjunction of balls, Marchand and Sokolova [9] proposed to minimize the risk bound given by Theorem 1 below. Note that the compression set $S_{\mathbf{i}}$ contains the examples needed to construct the balls of $h_{\mathcal{B}}$. Also, the message string μ identifies which examples of $S_{\mathbf{i}}$ are centers, and points out the border example associated with each center. In Theorem 1, the variables n_p and n_b encode the length of the message string μ.

---

[1] See [8] for further details about the *message* concept in sample compression theory.

[2] The disjunction case equations can be recovered by applying De Morgan's law.

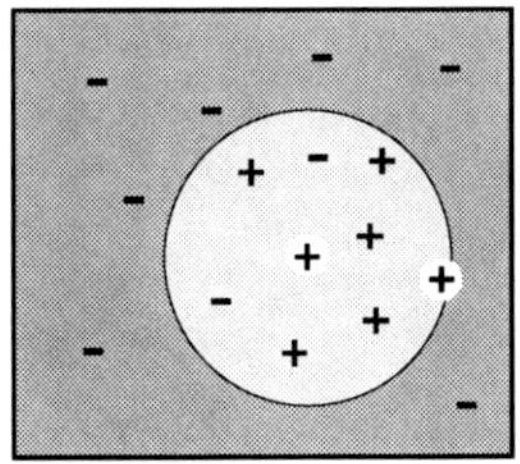 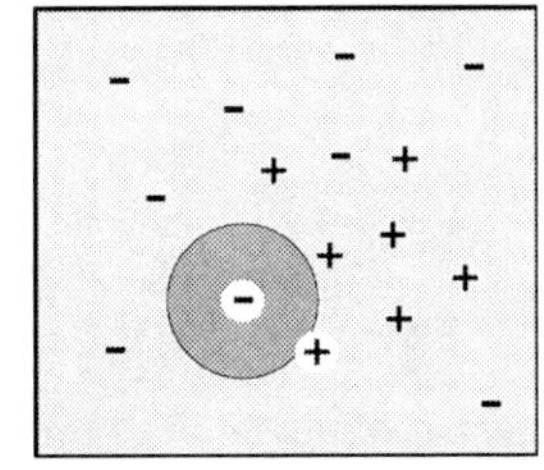 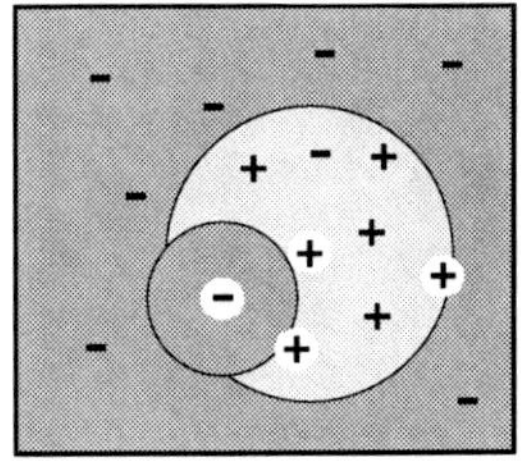

Fig. 1. On a 2-dimensional dataset of 16 examples, from left to right: a positive ball, a negative ball, and the conjunction of both balls. Examples in the light blue region and the red region will be respectively classified positive and negative.

Theorem 1. *(Marchand and Sokolova [9]) For any data-generating distribution D for which we observe a dataset S of m examples, and for each $\delta \in (0,1]$:*

$$\Pr_{S \sim D^m} \left(\forall \mathcal{B} \subseteq \mathcal{H}_S \; : \; R(h_\mathcal{B}) \leq \varepsilon(\mathcal{B}) \right) \geq 1 - \delta,$$

where:

$$\varepsilon(\mathcal{B}) \stackrel{\text{def}}{=} 1 - \exp\left(\frac{-1}{m-(|S_\mathbf{i}|+k)} \ln\left[\binom{m}{|S_\mathbf{i}|+k} \cdot \binom{|S_\mathbf{i}|+k}{k} \cdot \binom{n_p}{n_b} \cdot \frac{1}{\zeta(n_b)\zeta(|S_\mathbf{i}|)\zeta(k)\delta} \right] \right), \quad (3)$$

and where k is the number of errors that $h_\mathcal{B}$ does on training set S, n_p is the number of positive examples in compression set $S_\mathbf{i}$, n_b is the number of different examples used as a border, and $\zeta(a) \stackrel{\text{def}}{=} \frac{6}{\pi^2}(a + 1)^{-2}$.

This theorem suggests to minimize the expression of $\varepsilon(\mathcal{B})$ in order to find a good balls conjunction. For fixed values of m and δ, this expression tends to decrease with decreasing values of $|S_\mathbf{i}|$ and k, whereas $n_b \leq n_p \leq |S_\mathbf{i}|$. Moreover, even if the expression of $\varepsilon(\mathcal{B})$ contains many terms, we notice that the quantity $\left[\binom{n_p}{n_b} \cdot \frac{1}{\zeta(n_b)\zeta(|S_\mathbf{i}|)\zeta(k)\delta} \right]$ is very small. If we neglect this term, it is easy to see that minimizing $\epsilon(\mathcal{B})$ boils down to find the minimum of Equation (4), which is the sum of the compression set size and the number of empirical errors.

$$\mathcal{F}(\mathcal{B}) \stackrel{\text{def}}{=} |S_\mathbf{i}| + k. \quad (4)$$

This consideration leads us to the SCM algorithm. We say that a ball belonging to a conjunction *covers an example* whenever it classifies it negatively. Note that a balls conjunction $h_\mathcal{B}$ negatively classifies an example $\mathbf{x}$ if and only if at least one ball of $\mathcal{B}$ covers $\mathbf{x}$. This implies that if one wants to add a new ball to an existing balls conjunction, he can only change the classification outcome on uncovered examples. A good strategy for choosing a ball to add to a conjunction is then to cover as few positive examples as possible to avoid misclassifying them. This observation underlies the heuristic of the SCM algorithm.

Given a training set S, the SCM algorithm (see Algorithm 1) is a greedy procedure for selecting a small subset $\mathcal{B}$ of all possible balls[3] so that a high

---

[3] More precisely, the heuristic function (Line 6 of Algorithm 1) makes it possible to consider only balls whose borders are defined by positive examples (see [8]).

Algorithm 1. SCM *(dataset S, penalties $\{p_1, \ldots, p_n\}$, selection function f)*

1: Consider all possible balls: $\mathcal{H}_S \leftarrow \{g_{i,j} \mid (\mathbf{x}_i, \cdot) \in S, (\mathbf{x}_j, 1) \in S, \mathbf{x}_i \neq \mathbf{x}_j\}$.

2: Initialize: $\mathcal{B}^* \leftarrow \emptyset$.

3: **for** $p \in \{p_1, p_2, \ldots, p_n\}$ **do**

4:      Initialize: $\mathcal{N} \leftarrow \{\mathbf{x} \mid (\mathbf{x}, 0) \in S\}$, $\mathcal{P} \leftarrow \{\mathbf{x} \mid (\mathbf{x}, 1) \in S\}$ and $\mathcal{B} \leftarrow \emptyset$.

5:      **while** $\mathcal{N} \neq \emptyset$ **do**

6:          Choose the best ball according to the following heuristic:
$$g \leftarrow \operatorname*{argmax}_{g \in \mathcal{H}_S} \left\{ \, \big| \{\mathbf{x} \in \mathcal{N} \mid g(\mathbf{x}) = 0\} \big| \, - \, p \cdot \big| \{\mathbf{x} \in \mathcal{P} \mid g(\mathbf{x}) = 0\} \big| \, \right\}.$$

7:          Add this ball to current conjunction: $\mathcal{B} \leftarrow \mathcal{B} \cup \{g\}$.

8:          Clean covered examples: $\mathcal{N} \leftarrow \{\mathbf{x} \in \mathcal{N} \mid g(\mathbf{x}) = 1\}$, $\mathcal{P} \leftarrow \{\mathbf{x} \in \mathcal{P} \mid g(\mathbf{x}) = 1\}$.

9:          Retain the best conjunction : **if** $f(\mathcal{B}) < f(\mathcal{B}^*)$ **then** $\mathcal{B}^* \leftarrow \mathcal{B}$.

10:    **end while**

11: **end for**

12: **return** $\mathcal{B}^*$

---

number of negative examples of S are covered by at least one ball belonging to $\mathcal{B}$. At each step of the algorithm, the tradeoff between the number of covered negative examples and the number of covered positive examples is due to a heuristic (Line 6 of Algorithm 1) that depends on a penalty parameter $p \in [0, \infty)$. We initialize the algorithm with a selection of penalty values, allowing it to create a variety of balls conjunctions. The algorithm returns the best conjunction according to a *model selection function* of our choice.

Several model selection functions can be used along with the SCM algorithm. The function ε given by Equation (3) leads to excellent empirical results. In other words, by running the algorithm with a variety of penalty parameters, selecting from all generated balls conjunctions the one with the lowest bound value allows to obtain a low risk classifier. This method as been shown by Marchand and Shawe-Taylor [8] to be as good as cross-validation[4]. It is exceptional for a risk bound to have such property.

As we explain, the bound relies mainly on the sum $|S_\mathbf{i}| + k$, and our extensive experiments with the SCM confirms that the simple model selection function $\mathcal{F}$ given by Equation (4) gives equally good results. We are then interested to know if the SCM algorithm provides a good approximation of this function.

To answer this question, next section presents a pseudo-Boolean optimization model that directly finds the set $\mathcal{B}$ that minimizes the function $\mathcal{F}$.

3   A Pseudo-Boolean Optimization Model

A pseudo-Boolean problem consists of linear inequality constraints with integer coefficients over binary variables. One can also have a linear objective function.

To solve our machine learning problem with a pseudo-Boolean solver, the principal challenge is to translate the original problem into this particular form.

---

[4] Cross-validation is a widely used method for estimating reliability of a model, but substantially increases computational needs (see section 1.3 of [2]).

The main strategy to achieve this relies on the following observation:

Observation. As the classification function $h_\mathcal{B}$ is a conjunction (see Equation (2)), we observe that $h_\mathcal{B}$ misclassifies a positive example iff a negative ball covers it. Similarly, $h_\mathcal{B}$ misclassifies a negative example iff no ball covers it.

Equivalence Rules. Let's first state two general rules that will be useful to express the problem with pseudo-Boolean constraints. For any positive integer $n \in \mathbb{N}^*$ and Boolean values $\alpha_1, \ldots, \alpha_n, \beta \in \{0, 1\}$, the conjunction and disjunction of the Boolean values α_i can be encoded with these linear inequalities:

$$\alpha_1 \wedge \ldots \wedge \alpha_n = \beta \iff n - 1 \geq \alpha_1 + \ldots + \alpha_n - n \cdot \beta \geq 0, \tag{5}$$

$$\alpha_1 \vee \ldots \vee \alpha_n = \beta \iff 0 \geq \alpha_1 + \ldots + \alpha_n - n \cdot \beta \geq 1 - n. \tag{6}$$

Program Variables. Let $P \overset{\text{def}}{=} \{i \mid (\mathbf{x}_i, 1) \in S\}$ and $N \overset{\text{def}}{=} \{i \mid (\mathbf{x}_i, 0) \in S\}$ be two disjoint sets, containing indices of positive and negative examples respectively. We define m sets B_i, each containing the indices of the borders that can be associated to center $\mathbf{x}_i$, and m sets C_j, each containing the indices of the centers that can be associated to border $\mathbf{x}_j$. As Marchand and Shawe-Taylor [8], we only consider balls with positive borders. Thus, for $i, j \in \{1, \ldots, m\}$, we have:

$$B_i \overset{\text{def}}{=} \{j \mid j \in P, j \neq i\} \quad \text{and} \quad C_j \overset{\text{def}}{=} \{i \mid i \in P \cup N, j \in B_i\}.$$

In other words, B_k is the set of example indices that can be the border of a ball centered on x_k. Similarly, C_k is the set of example indices that can be the center of a ball whose border is x_k. Necessarily, we have $j \in B_k \iff k \in C_j$.

Given the above definitions of B_i and C_j, the solver have to determine the value of Boolean variables s_i, r_i and $b_{i,j}$ described below:

For every $i \in \{1, \ldots, m\}$:
- s_i is equal to 1 iff the example $\mathbf{x}_i$ belongs to the compression set.
- r_i is equal to 1 iff the $h_\mathcal{B}$ misclassifies the example $\mathbf{x}_i$.
- For every $j \in B_i$, $b_{i,j}$ is equal to 1 iff the example $\mathbf{x}_i$ is the center of a ball and $\mathbf{x}_j$ if the border of that same ball.

Objective Function. The function to optimize (see Equation (4)) becomes:

$$\min \sum_{i=1}^{m} (r_i + s_i). \tag{7}$$

Program Constraints. If an example $\mathbf{x}_i$ is the center of a ball, we want exactly one example $\mathbf{x}_j$ to be its border. Also, if $\mathbf{x}_i$ is not the center of any ball, we don't want any example $\mathbf{x}_j$ to be its border. Those two conditions are encoded by:

$$\sum_{j \in B_i} b_{i,j} \leq 1 \quad \text{for } i \in \{1, \ldots, m\}. \tag{8}$$

An example belongs to the compression set iff it is a center or a border. We then have $s_k = \left[\bigvee_{i \in C_k} b_{i,k} \right] \vee \left[\bigvee_{j \in B_k} b_{k,j} \right]$. Equivalence rule (6) gives:

$$1 - |B_k \cup C_k| \leq -|B_k \cup C_k| \cdot s_k + \sum_{i \in C_k} b_{i,k} + \sum_{j \in B_k} b_{k,j} \leq 0 \quad \text{for } k \in \{1, \ldots, m\}. \tag{9}$$

We denote by $D_{i,j}$ the distance between examples $\mathbf{x}_i$ and $\mathbf{x}_j$. Therefore, D is a square matrix of size $m \times m$. For each example index $k \in \{1, \ldots, m\}$, let E_k be the set of all balls that cover (i.e. negatively classify) the example $\mathbf{x}_k$:

$$E_k \stackrel{\text{def}}{=} \{b_{i,j} \mid i \in P, j \in B_i, D_{i,j} < D_{i,k}\} \cup \{b_{i,j} \mid i \in N, j \in B_i, D_{i,j} > D_{i,k}\} \ .$$

First, suppose that $\mathbf{x}_k$ is a positive example (thus, $k \in P$). Then, recall that the conjunction misclassifies the example $\mathbf{x}_k$ iff a ball covers it (see "observation" above). Therefore, $r_k = \bigvee_{b_{i,j} \in E_k} b_{i,j}$. Using Equivalence Rule (6), we obtain:

$$1 - |E_k| \ \leq \ -|E_k| \cdot r_k + \sum_{b_{i,j} \in E_k} b_{i,j} \ \leq \ 0 \quad \text{for } k \in P \ . \tag{10}$$

Now, suppose that $\mathbf{x}_k$ is a negative example (thus, $k \in N$). Then, recall that the conjunction misclassifies $\mathbf{x}_k$ iff no ball covers it (see "observation" above). We have $r_k = \bigwedge_{b_{i,j} \in E_k} \neg b_{i,j}$. By using Equivalence Rule (5) and $\alpha = \neg\beta \Leftrightarrow \alpha = 1 - \beta$ (where $\alpha, \beta \in \{0, 1\}$), we obtain the following constraints:

$$0 \ \leq \ -|E_k| \cdot r_k + \sum_{b_{i,j} \in E_k} (1 - b_{i,j}) \ \leq \ |E_k| - 1$$

$$\Leftrightarrow 1 \ \leq \ |E_k| \cdot r_k + \sum_{b_{i,j} \in E_k} b_{i,j} \ \leq \ |E_k| \quad \text{for } k \in N \ . \tag{11}$$

4   Empirical Results on Natural Data

The optimization problem of minimizing Equation (7) under Constraints (8, 9, 10, 11) gives a new learning algorithm that we call *PB-SCM*. To evaluate this new algorithm, we solve several learning problems using three well-known pseudo-Boolean solvers, PWBO [6], SCIP [1] and BSOLO [7], and compare the obtained results to the SCM (the greedy approach described by Algorithm 1).

We use the same seven datasets than [8] and [9], which are common benchmark datasets in the machine learning community. For each dataset, we repeat the following experimental procedure four times with training set sizes $m = |S|$ of 25, 50, 75 and 100 examples. First, we randomly split the dataset examples in a training set S of m examples and a testing set T containing all remaining examples[5]. Then, we execute the four learning algorithms (SCM algorithm and PB-SCM with three different solvers) on the same training set S, and compute the risk on the testing set T.

To obtain SCM results, the algorithm is executed with a set of 41 penalty values $\{10^{a/20} \mid a = 0, 1, \ldots, 40\}$ and the model selection function $\mathcal{F}$ given by Equation (4). The PB-SCM problem is solved with the three different solvers. For each solver, we fix the time limit to 3600 seconds and keep the solver's default values for other parameters. When a solver fails to converge in 3600 seconds, we consider the best solution so far. Using the solution of the SCM to provide an initial upper bound to the pseudo-Boolean solvers provided no speed-up.

---

[5] Training sets are smalls because of the extensive computational power needed by pseudo-Boolean solvers.

Table 1. Empirical results comparing the objective value $\mathcal{F}$ obtained by SCM and PB-SCM algorithms, the test risk of obtained classifiers and required running time ("T/O" means that the pseudo-Boolean solver reaches the time limit)

Dataset		SCM			PB-SCM (pwbo)			PB-SCM (scip)			PB-SCM (bsolo)		
name	size	$\mathcal{F}$	risk	time	$\mathcal{F}$	risk	time	$\mathcal{F}$	risk	time	$\mathcal{F}$	risk	time
breastw	25	**2**	0.046	0.04	**2**	0.081	0.03	**2**	0.064	0.71	**2**	0.046	0.05
	50	**2**	0.047	0.07	**2**	0.046	0.06	**2**	0.049	3.7	**2**	0.047	0.64
	75	**2**	0.044	0.12	**2**	0.041	0.16	**2**	0.044	7.4	**2**	0.044	3.7
	100	**2**	0.046	0.16	**2**	0.046	0.43	**2**	0.05	38	**2**	0.046	20
bupa	25	8	0.403	0.31	**7**	0.45	0.31	**7**	0.45	4.1	**7**	0.419	0.64
	50	14	0.431	1.32	**12**	0.495	589	**12**	0.495	47	**12**	0.464	989
	75	21	0.404	4.1	21	0.463	T/O	**19**	0.467	1763	24	0.419	T/O
	100	**27**	0.355	11	32	0.494	T/O	30	0.396	T/O	34	0.367	T/O
credit	25	**4**	0.202	0.11	**4**	0.202	0.08	**4**	0.202	2	**4**	0.202	0.22
	50	6	0.239	0.25	**5**	0.257	9.3	**5**	0.209	21	**5**	0.257	30.1
	75	9	0.216	0.61	**8**	0.266	1920	**8**	0.263	138	**8**	0.268	1862
	100	12	0.233	1.3	11	0.237	T/O	**10**	0.242	798	18	0.302	T/O
glass	25	**5**	0.333	0.11	**5**	0.261	0.03	**5**	0.297	12	**5**	0.261	0.2
	50	9	0.265	0.49	**8**	0.265	10.3	**8**	0.265	35	**8**	0.265	28
	75	16	0.307	1.5	**15**	0.273	T/O	**15**	0.227	736	**15**	0.227	T/O
	100	18	0.222	2.9	**17**	0.222	T/O	**17**	0.206	T/O	22	0.19	T/O
haberman	25	**5**	0.305	0.17	**5**	0.305	0.03	**5**	0.305	3.6	**5**	0.312	0.18
	50	**10**	0.246	0.94	**10**	0.332	34	**10**	0.332	30	**10**	0.246	65
	75	15	0.237	2.5	**14**	0.324	T/O	**14**	0.324	436	16	0.279	T/O
	100	21	0.278	4.5	**20**	0.289	T/O	**20**	0.33	T/O	23	0.289	T/O
pima	25	**8**	0.408	0.33	**8**	0.381	0.36	**8**	0.385	4	**8**	0.381	0.94
	50	15	0.312	0.9	**13**	0.306	2204	**13**	0.311	37	**13**	0.306	1985
	75	20	0.375	3.8	20	0.342	T/O	**19**	0.339	2641	24	0.336	T/O
	100	25	0.326	7.4	26	0.316	T/O	**23**	0.338	T/O	30	0.379	T/O
USvotes	25	**3**	0.112	0.07	**3**	0.11	0.011	**3**	0.107	0.21	**3**	0.12	0.08
	50	5	0.14	0.17	4	0.114	0.141	4	0.127	2.4	4	0.127	1.1
	75	5	0.119	0.28	3	0.131	0.183	3	0.131	54	3	0.131	33
	100	6	0.084	0.35	4	0.146	1.21	4	0.107	100	4	0.137	80

Table 1 shows the obtained results. Of course, except for T/O situations, the minimal value of the heuristic $\mathcal{F}$ is always obtained by solving the PB-SCM problem. However, it is surprising that the SCM often reaches the same minimum value. Moreover, the SCM sometimes (quickly) finds a best value of $\mathcal{F}$ when the pseudo-Boolean programs time out, and there is no clear amelioration of the testing risk when PB-SCM finds a slightly better solution than SCM. We conclude that the greedy strategy of SCM is particularly effective.

5   Conclusion

We have presented a pseudo-Boolean model that encodes the core idea behind the combinatorial problem related to the Set Covering Machine. Extensive experiments have been done using three different pseudo-Boolean solvers. For the first time, empirical results show the effectiveness of the greedy approach of Marchand and Shawe-Taylor [8] at building SCM of both small compression set and empirical risk. This is a very surprising result given the simplicity and the low complexity of the greedy algorithm.

References

1. Achterberg, T.: SCIP-a framework to integrate constraint and mixed integer programming. Konrad-Zuse-Zentrum für Informationstechnik (2004)
2. Bishop, C.M.: Pattern Recognition and Machine Learning (Information Science and Statistics). Springer-Verlag New York, Inc., Secaucus (2006)
3. Cortes, C., Vapnik, V.: Support-vector networks. Machine Learning 20(3), 273–297 (1995)
4. Floyd, S., Warmuth, M.: Sample compression, learnability, and the vapnik-chervonenkis dimension. Machine Learning 21, 269–304 (1995)
5. Hussain, Z., Szedmak, S., Shawe-Taylor, J.: The linear programming set covering machine (2004)
6. Lynce, R.: Parallel search for boolean optimization (2011)
7. Manquinho, V., Marques-Silva, J.: On using cutting planes in pseudo-boolean optimization. Journal on Satisfiability, Boolean Modeling and Computation 2, 209–219 (2006)
8. Marchand, M., Shawe-Taylor, J.: The set covering machine. Journal of Machine Learning Research 3, 723–746 (2002)
9. Marchand, M., Sokolova, M.: Learning with decision lists of data-dependent features. J. Mach. Learn. Res. 6, 427–451 (2005)

Finding a Nash Equilibrium by Asynchronous Backtracking[*]

Alon Grubshtein and Amnon Meisels

Dept. of Computer Science,
Ben Gurion University of the Negev,
P.O.B 653, Be'er Sheva, 84105, Israel
{alongrub,am}@cs.bgu.ac.il

Abstract. Graphical Games are a succinct representation of multi agent interactions in which each participant interacts with a limited number of other agents. The model resembles Distributed Constraint Optimization Problems (DCOPs) including agents, variables, and values (strategies). However, unlike distributed constraints, local interactions of Graphical Games take the form of small strategic games and the agents are expected to seek a Nash Equilibrium rather than a cooperative minimal cost joint assignment.

The present paper models graphical games as a Distributed Constraint Satisfaction Problem with unique k-ary constraints in which each agent is only aware of its part in the constraint. A proof that a satisfying solution to the resulting problem is an ϵ-Nash equilibrium is provided and an Asynchronous Backtracking algorithm is proposed for solving this distributed problem. The algorithm's completeness is proved and its performance is evaluated.

1 Introduction

In a typical Multi Agent setting, agents interact with one another to achieve some goal. This goal may be a globally defined objective such as the minimization of total cost, or a collection of personal goals such as maximizing the utility of each agent. In the latter case, Game Theory predicts that the outcome should be a stable solution – an equilibrium – from which every agent will not care to deviate. This notion of stability is a fundamental concept in game theory literature and has been at the focus of work in the field of Multi Agent Systems.

Graphical Games are a succinct representation of normal form games played over a graph [8]. It is defined by a graph of agents that explicitly defines relations: if an edge e_{ij} exists than agents a_i and a_j interact with one another and therefore affect each other gains. The model exploits the locality of interactions among agents and enables one to specify payoffs in terms of neighbors rather than in terms of the entire population of agents. Two distributed algorithms inspired by Bayesian Networks were initially proposed to find an approximate equilibrium of graphical games, NashTree and NashProp

---

[*] The research was supported by the Lynn and William Frankel Center for Computer Sciences at Ben-Gurion University and by the Paul Ivanier Center for Robotics Research and Production Management.

M. Milano (Ed.): CP 2012, LNCS 7514, pp. 925–940, 2012.

[8,13]. Both rely on an a discretization scheme for the approximation and it is proved in [13] that NashProp can be used to find an ϵ-Nash equilibrium on general graph interactions.

The Graphical Games model is closely related to Distributed Constraint Optimization Problems (DCOPs) [11]. DCOPs are defined by agents with private variables, a finite domain of values for each variable, and constraints mapping each joint assignment to a non negative cost. The values that a DCOP agent assigns to its variables are similar to the choice of strategy made by a Graphical Game agent and the local interactions of the game resembles a constraint. However, despite these similarities the two models are inherently different. Agents connected by a DCOP constraint share a single cost value to their joint assignments and have full knowledge of the constraint's costs structure. In Graphical Games the agents have a personal valuation to each outcome. Furthermore, the standard solution of a DCOP is a joint assignment that minimizes the sum of costs and not necessarily a stable point.

The present paper proposes a model for graphical games as a distributed constraints problem and a new asynchronous algorithm for finding stable points. The model uses constraints which are partially known to each participant to represent private valuation of outcomes [3,7]. First, Asymmetric DCOPs (ADCOPs) [7], an extension of standard DCOPs, provide a natural representation to game-like interactions. It specifies different costs to each agent in a constraint, capturing personal preferences and utilities of different agents. Next, the Partially Known Constraint (PKC) model [3], focusing on asymmetric constraint satisfaction is used to align a satisfying solution with an equilibrium.

By casting the ADCOP representation of a Graphical Game to an asymmetric satisfaction problem one can apply constraint reasoning techniques to find an equilibrium of a multi agent problem. Following NashTree and NashProp [8,13] the present paper presents a new Asynchronous Nash back Tracking (ANT) algorithm for finding ϵ-Nash equilibria. This algorithm is inspired by the well known Asynchronous Back Tracking algorithm (ABT) [18,2] and its asymmetric single phase variant [3] (ABT-1-ph). A proof that a satisfying solution to the revised problem is an ϵ-Nash equilibrium is provided.

Similar to other ABT variants, agents in the ANT algorithm exchange messages specifying their current assignment, Nogood messages and requests for additional communication links (termination messages are not required). However, the ANT algorithm searches through a high arity, asymmetric, distributed constraint satisfaction problem. ANT is proven to find a globally satisfying solution to the graphical game problem – effectively an ϵ-Nash equilibrium.

One former attempt to apply constraint reasoning techniques to game theoretic equilibrium search gave up on the inherently distributed nature of graphical games. A centralized constraint solver is described in [17] where the authors generate very large domains and complex constraints to represent game like structures. The approach of [17] assumes that agents are willing to reveal private information to a third party which carries out the computation. Although this approach is commonly taken when considering the equilibrium search problem (e.g. [15,4,16,6]) it is not suitable to a distributed multi agents interaction.

There were former studies of limited versions of graphical games. Two methods for finding stable points of graphical games examine the relation between local minima and pure strategy equilibria [1,10] (which is not guaranteed to exist). However, both works examine games played on symmetric DCOPs which are a subset of a special class of games known as potential games and do not extend to general form games. Another structured interaction is also specified in [5] which attempt to find the "best" pure strategy equilibrium but is extremely limited to tree based interactions. In contrast to these works the ANT algorithm (always) finds an ϵ-Nash equilibrium on general form graphs and general form games.

The remainder of the paper is organized as follows. Section 2 provides a detailed description of ADCOPs, graphical normal form games and equilibrium solutions. Next comes a description of the asymmetric satisfaction problem generated for finding an ϵ-Nash equilibrium in Section 3. Section 4 presents the Asynchronous Nash back Tracking algorithm (ANT) and its formal properties proof. Section 5 is devoted to experimental evaluation of ANT but also provides a full description of NashProp's backtracking (second) phase. This is, to the best of our knowledge, the first full account of the distributed backtracking phase of NashProp. Finally, Section 6 concludes the present work and discusses future directions.

2 Preliminaries

2.1 Asymmetric Distributed Constraints

Asymmetric Distributed Constraint Optimiaztion Problems (ADCOPs) [7] define a constraint model in which a cost is specified to each agent in a constraint instead of a single cost to all constrained agents. Unlike DCOPs, an assignment change which decreases the cost incurred on one agent in an ADCOP is not guaranteed to decrease the cost incurred on other agents in the constraint. Roughly speaking one can say that the costs of asymmetric constraints are "on the agents" (the vertices in the constraint network) rather than on the "constraints" (edges of the constraint network).

Formally, an ADCOP is a tuple $\langle \mathcal{A}, \mathcal{X}, \mathcal{D}, \mathcal{R} \rangle$. Where $\mathcal{A}$ is a finite set of agents $a_1, a_2, ..., a_n$ and $\mathcal{X}$ is a finite set of variables $x_1, x_2, ..., x_m$. Each variable is held by exactly one agent but an agent may hold more than one variable. $\mathcal{D}$ is a set of domains $d_1, d_2, ..., d_m$, specifying the possible assignment each variable may take. Finally, $\mathcal{R}$ is a set of asymmetric relations (constraints).

A *value assignment* or simply an *assignment* is a pair $\langle x_i, v \rangle$ including a variable x_i, and its value $v \in d_i$. Following common practice we assume each agent holds exactly one variable and use the two terms interchangeably. A *partial assignment* (PA) is a set of value assignments, in which each variable appears at most once.

A constraint $C \in \mathcal{R}$ of an Asymmetric DCOP is defined over a subset of the variables $\mathcal{X}(C)$, and maps the joint values assigned to the variables to a vector of non-negative costs (where the jth entry corresponds to the jth variable in $\mathcal{X}(C)$):

$$C : D_{i_1} \times D_{i_2} \times \cdots D_{i_k} \to \mathbb{R}_+^k$$

We say that a constraint is *binary* if it refers to a partial assignment with exactly two variables, i.e., $k = 2$. A *binary ADCOP* is an ADCOP in which all constraints are binary.

A *complete assignment* is a partial assignment that includes all the variables in $\mathcal{X}$ and unless stated otherwise, an optimal *solution* is a complete assignment of aggregated minimal cost. In maximization problems, each constraint has utilities instead of costs and a solution is a complete assignment of maximal aggregated utility.

For consistency, we also define the satisfaction variant of this problem (denoted AD-CSP). An ADCSP constraint takes the following form:

$$C : D_{i_1} \times D_{i_2} \times \cdots D_{i_k} \to \{0, 1\}^k$$

In this case, constraint C is said to be *satisfied* by a PA p if p includes value assignments to all variables of C and if $C(p)$ is a tuple of all 1's, i.e. all involved agents agree on its satisfiability. The solution of an ADCSP is a complete assignment which satisfies all constraints $C \in \mathcal{R}$.

2.2 Normal Form Games

A normal form game is a model for multiple interacting parties or agents. These parties plan their interaction and may act only once (simultaneously). The joint action of all participants results in an outcome state which is, in general, evaluated differently by the agents.

More formally, a normal form game includes a finite set of n players (agents) $\mathcal{A}$, a non empty set $\mathcal{S}$ of actions or strategies and a preference ordering over outcomes $\mathcal{M}$. Specifically, for each agent a_i, $u_i(x) \in \mathcal{M}$ maps any joint strategy $x = \times_{a_i \in \mathcal{A}} s_i$ to a utility (or cost) value. If $u_i(o_1) > u_i(o_2)$ then agent a_i prefers outcome o_1 to o_2. In the remainder of this paper we follow common practice and denote with x_{-i} the joint action of all agents except for a_i.

The most commonly used solution concept for strategic interactions is that of a Nash Equilibrium [12] defined as a joint assignment x^* to all agents such that:

$$\forall a_i \in \mathcal{A} \quad : \quad u_i(x^*_{-i}, x^*_i) \geq u_i(x^*_{-i}, x_i)$$

If agents can take non deterministic actions, or state a probability distribution over deterministic strategies as their action than Nash's classic theorem states that a mixed strategy Nash Equilibrium always exists (cf. [12,14]).

2.3 Graphical Games

Graphical Games were first described by Kearns et. al in [8]. A graphical game is defined by a pair $(\mathcal{G}, \mathcal{M})$ where $\mathcal{G}$ is an n players interaction graph and $\mathcal{M}$ is the set of *local* game matrices mapping joint assignments to agents utilities. The space required for representing local game matrices is significantly smaller than the standard representation: instead of $n |s|^n$ values required for the representation of each table $u_i \mathcal{M}$, a Graphical Game requires only $n |s|^{d+1}$ values (where d is the maximal degree of an agent). That is, in such games, only neighboring agents affect each other's utility.

We denote a_i's set of neighbors by $N(i)$ and adhering to our previous notation use $x_{N(i)}$ to represent the joint action of a_i's neighbors. It is easy to verify that a Graphical Game defined by a Graph G and local games description $\mathcal{M}$ is very similar to an ADCOP (ADCOPs constraints can provide a slightly more compact representation).

A discretization scheme for computing an approximate equilibrium is presented in [8]. The approximate equilibrium – ϵ - Nash equilibrium – is a mixed strategy profile p^* such that

$$\forall A_i \in \mathcal{A} \;\; : \;\; u_i(p^*_{-i}, p^*_i) + \epsilon \geq u_i(p^*_{-i}, p_i)$$

The scheme presented in [8] constrain each agent's value assignment to a discretized mixed strategy which is a multiple of some value τ which depends on the agents' degree and the desired approximation bound ϵ. Two search algorithms for finding ϵ - Nash equilibrium in graphical games were proposed in [8,13].

3   Nash ADCSP

A simple transformation is defined from a general ADCOP to a Nash Asymmetric Distributed Constraint *Satisfaction* Problem (Nash-ADCSP). The resulting problem can be solved by a distributed constraint satisfaction algorithm capable of handling asymmetric, non-binary constraints. This section is concluded with a proof that a satisfying solution to the Nash-ADCSP is an ϵ-Nash equilibrium.

Given an ADCOP representation of a general multi agent problem, a Nash-ADCSP with the same set of agents and variables is constructed. Using the discretization scheme described in [8] the domains of each agent are revised to represent distributions over values. That is, each agent's domain is a discretized mixed strategy that is a multiple of some τ which is defined according to the desired accuracy level ϵ (cf. [8]).

The new problem includes $n = |\mathcal{A}|$ constraints, each associated with exactly one agent. The arity of constraint C_i associated with agent a_i equals a_i's degree + 1. The set of satisfying assignments (satisfying *all* agents in the constraint) include all joint assignments of a_i and its neighbors $N(i)$ such that a_i's action yields a maximal gain to itself. That is, if $\langle v, x_{N(i)} \rangle \in d_i \times \prod_{k \in N(i)} d_k$ is a joint assignment:

$$C_i(v, x_{N(i)}) = \begin{cases} \langle 1, 1, ..., 1 \rangle & \text{if } u_i(v, x_{N(i)}) \geq u_i(v', x_{N(i)}) \;\; \forall v' \in d_i \\ \langle 1, ..., 0, ..., 1 \rangle & \text{otherwise (the zero is associated with } a_i) \end{cases}$$

As an example, consider the three agent binary interaction presented in the left hand side of Figure 1. In this example there are two separate interactions: between a_1 and a_2 and between a_2 and a_3. Following game theory conventions, the agents' asymmetric constraints stemming from these interactions are described by two cost bi-matrices. That is, each entry in the table correspond to a joint action by the row and column player, where the row player's evaluation of the joint action is given by the left hand cost value and the column player's cost is given by the right hand cost value.

This problem is then cast to a Nash-ADCSP problem (right hand side of Figure 1) with three new asymmetric constraints C_1, C_2 and C_3. Constraints C_1 and C_3 are the binary constraints associated with agents a_1 and a_3 and can only be satisfied when

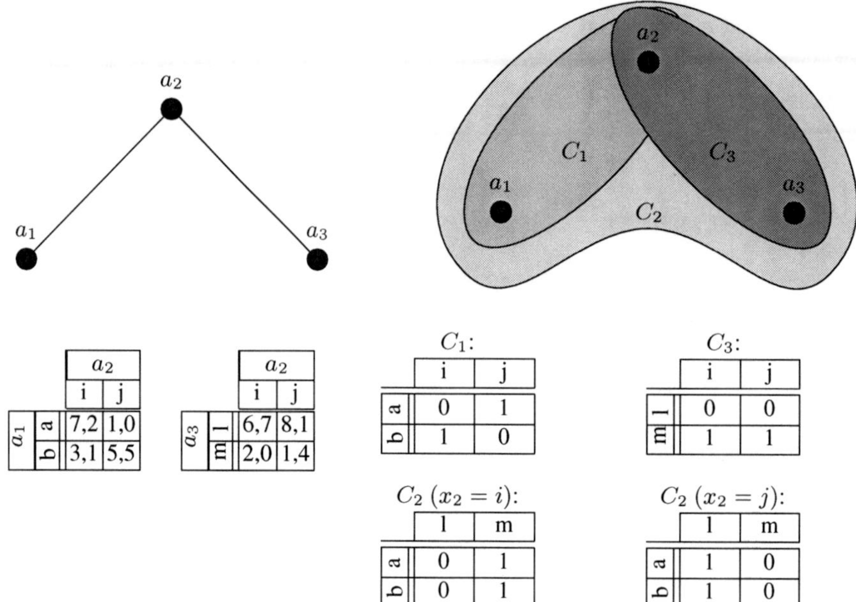

Fig. 1. (left) A simple ADCOP with three agents and two asymmetric constraints. Each agent has two values in its domain $d_1 = \{a, b\}$, $d_2 = \{i, j\}$ and $d_3 = \{l, m\}$. An entry in the constraint represents the cost of each joint action (i.e. the cost of $\langle x_1 = a, x_2 = i \rangle$ is 7 for a_1 and 2 for a_2. (right) The resulting ADCSP with 3 asymmetric satisfaction constraints (C_1, C_2 and C_3) of which C_3 is a trinary constraint. A value of 1 in constraint C_i means that the action corresponding to the entry is a_i's best response.

a_1 and a_3 best respond to a_2's action. Constraint C_2 is a trinary constraint which is satisfied whenever a_2 best responds the joint action of a_1 and a_3. For example, a_2's best response (minimal cost) to the joint action $\langle x_1 = a, x_3 = m \rangle$ is the assignment $\langle x_2 = i \rangle$ which takes the value of 1 (consistent) in C_2.

In this example the assignment $\langle x_1 = b, x_2 = i, x_3 = m \rangle$ is a pure strategy Nash equilibrium (costs are 3, 0 and 2 respectively).It is easy to verify that this joint assignment also satisfies all constraints of the the new Nash-ADCSP problem.

Nash-ADCSP constraints have two noteworthy properties:

1. Constraint C_i's satisfiability state is determined by agent a_i only. Any other agent in the constraint always evaluates its state as "satisfied" and this in turn implies that any agent $a_j \in N(i)$ is ignorant of C_i.
2. Given the joint assignment of all its neighbors, an agent a_i can always satisfy its constraint.

The relation between the satisfaction problem and an equilibrium is supplied by the following proposition:

Proposition 1. *Let p be a complete assignment to a Nash ADCSP with discretized domains corresponding to an approximation value ϵ. Then p is a consistent assignment iff it is an ϵ-Nash equilibrium of the original problem.*

Proof. Necessary: If p is consistent with all constraints then it is easy to see that by definition each agent best responds to the joint action of all its neighbors. Hence p is an equilibrium on the τ grid of the discretization scheme and an ϵ-equilibrium of the original problem.

Sufficient: Let p be an ϵ-Nash equilibrium of the original problem where each of the values taken by the agents exists on the discrete grid (i.e. $p \in \prod_{A_i \in \mathcal{A}} D_i$). Each agent in an ϵ equilibrium must "best-respond" its neighbors joint actions up to some ϵ. Therefore, each constraint C_i associated with A_i is mapped to a tuple of all ones (otherwise the agent can swap its assignment to an alternate one in which its gain is higher with respect to the joint assignment of its neighbors and p is not an equilibrium) and all constraints are satisfied. $\square$

The above proposition also implies that a solution to the constructed Nash ADCSP must always exist (i.e. the problem is always satisfiable).

4   Asynchronous Nash Backtracking

Asynchronous Nash BackTracking (ANT) is a new variant of ABT [18,2,3,11] searching through the space of the Nash-ADCSP problem. ANT extends the original ABT and its PKC variant, ABT-1ph, to support asymmetric non-binary satisfaction. It relies on the agents' total order to cover the space of all joint assignments when attempting to satisfy a constraint. That is, if an agent a_i's assignment is not a best response to the joint assignment of $N(i)$, the lowest priority agent (with respect to the total ordering) in C_i will change its assignment. If this assignment change remains inconsistent, then a Nogood is generated and sent to the lowest priority agent from its list of higher priority neighbors.

ANT's pseudo code is presented below. It adopts the same notation as that described in [2,3] and handles Nogoods similarly. For details and in depth discussion, the reader is addressed to the description provided in [2,3].

Each ANT agent maintains a local copy of the value assignments of which it is aware (its agentView), a store of Nogoods (ngStore) kept as justification for the removal of values from the current domain and a boolean variable isLastAgentInConstraint taking the value TRUE if a_i is the lowest priority agent in C_i. The list of lower priority neighbors is denoted by Γ^+.

Similar to other ABT variants, three message types are used **ok?**, **ngd** and **adl**. **ok?** messages notify other agents of a new value assignment, **ngd** messages are used to request the removal of values and provide an explanation for this request and **adl** messages are used to request an additional link of information be generated between two agents.

An ANT agent begins by assigning itself a value and notifying its neighbors of this value assignment. The agents then react to messages received from others until

---

Asynchronous Nash backTracking - pseudocode

---

procedure ANT
    $myValue \leftarrow chooseValue()$;
    send(**ok?**, neighbors, $myValue$);
    $end \leftarrow false$;
    while $\neg end$ **do**
        $msg \leftarrow getMsg()$;
        switch $(msg.type)$
            case ok? : ProcessInfo(msg);
            case ngd : ResolveConflict(msg);
            case adl : SetLink(msg);
procedure PROCESSINFO(msg)
    updateAgentView($msg.sender \leftarrow msg.assignment$);
    if $\neg$isConsistent($agentView, myValue$) **then**
        if $\neg$ isLastAgentInConstraint **then**
            $ng \leftarrow$ NoGood($agentView, cons\_neighbors$);
            add($ng.addLhs, self \leftarrow myValue$);
            send(**ngd**, ng.rhs, ng);
        else
            checkAgentView();
procedure RESOLVECONFLICT(msg)
    $ng \leftarrow msg.NoGood$;
    if coherent($ng, \Gamma^{+} \cup self$) **then**
        addLinks(ng);
        ngStore.add(ng);
        $myValue \leftarrow$ NULL;
        checkAgentView();
    else if coherent($ng, self$) **then**
        send(**ok?**, msg.sender, $myValue$);
procedure SETLINK(msg)
    $neighbors$.add($msg.sender$);
    send(**ok?**, msg.sender, $myValue$);

---

quiescence is reached. When agent a_i receives an **ok?** message it invokes the Process-Info procedure which first updates its agentView, ngStore and current domain. If the agentView and a_i's assignment are inconsistent with C_i, the agent attempts to find an alternative assignment if it is the lowest priority agent in C_i or generate a new Nogood to the lowest priority agent otherwise. It should be noted that ANT agents consider a constraint C_i to be inconsistent only if all agents in C_i reported their values and these variables result in an unsatisfying solution (as described in 2.1).

If a_i receives a **ngd** message, it first verifies that this is a coherent message with respect to a_i and its lower priority neighbors. If it is, and additional links are needed these are requested. Next, the Nogood is stored and the current assignment is removed until a consistent revised assignment is found (checkAgentView). If the message is not coherent but a_i's current assignment is the same as that assumed in the **ngd** message, an **ok?** message is sent to the originator of the message.

When an **adl** message is received the agents simply add the originator to the list of neighbors and notify it of their current assignment.

Asynchronous Nash backTracking - pseudocode, continued

```
procedure UPDATEAGENTVIEW(assignment)
    add(assignment, agentView);
    for all ng ∈ ngStore do
        if ¬ coherent(ng.lhs, agentView) then
            ngStore.remove(ng);
    currentDomain.makeConsistentWith(ngStore);

procedure CHECKAGENTVIEW
    if ¬isConsistent(agentView, myValue) then
        if (myValue ← chooseValue())==NULL then
            backTrack();
        else
            send(ok?, neighbors , myValue);

procedure BACKTRACK
    resolvedNG ← ngStore.solve();
    send(ngd, resolvedNG.rhs, resolvedNG);
    agentView.unassign(resolvedNG.rhs);
    updateAgentView(resolvedNG.rhs ← NULL);
    checkAgentView();

procedure CHOOSEVALUE
    for all val ∈ currentDomain do
        remove(currentDomain, val);
        if isConsistent(agentView, val) then
            return val;
        else
            ng ← agentView ∩ constrainedNeighbors'
            if ¬ isLastAgentInConstraint then
                add(ng.lhs, self ← val);
                add(ng.rhs, last agent in constraint);
                send(ok?, ng.rhs, val);
                send(ngd, ng.rhs, ng);
                return val;
            else if
                thenadd(ng.rhs, self ← val);
                ngStore.add(ng);
    return NULL;
```

ANT's basic auxiliary functions, updateAgentView, checkAgentView and backtrack are in parts simpler than that of ABT-1ph due to the Nash-ADCSP structure which defines a single constraint to each agent. However, its value choosing mechanism should be addressed with care to ensure completeness of the algorithm. The chooseValue procedure iterates over all values in an agent a_i's current domain. A candidate value is first removed from the current domain and its consistency is checked against the agentView.

If it is not consistent, a new nogood is generated with the values of all agents in C_i. If a_i is the lowest priority agent in C_i this Nogood is stored and a new candidate value is examined. However, if a_i is not the lowest priority agent in C_i the Nogood is updated to include a_i as well, and then sent to the lowest priority agent a_j in C_i *after* an ok? message with the new value is sent to a_j. This ensures that when a_j processes the Nogood, a_i's value is coherent. Finally, it should be noted that the latter case returns a seemingly inconsistent value to a_i. This step is required to ensure all possible joint values are examined, and stems from the fact that constraints are asymmetric. That is, in an asymmetric constraint agents with priority lower than a_i may change their assignment in such a way that will change the consistency state of C_i.

The following proposition provides for ANT's formal proporties:

Proposition 2. *The ANT algorithm always finds an ϵ equilibrium. That is, it is sound, complete and terminates.*

Proof. ANT reports a solution whenever quiescence is reached. In this case, all constraints are satisfied (otherwise, at least one nogood will exist and quiescence will not be achieved). Therefore the reported solution is a globally satisfying solution and by proposition 1 it is also an ϵ Nash Equilibrium.

ANT follows a similar search to that described in ABT-1ph [3]. Its main difference is that it ensures non binary constraints are indeed satisfied by a PA through its nogood handling mechanism and its chooseValue function.

An agent A_i receiving a nogood from an agent A_j of higher priority must be the lowest priority agent in C_j. This nogood also includes all value assignments of the rest of the agents involved in C_j. A_i can therefore select a consistent value (consistent with its personal constraint C_i) from its current domain and propose this value assignment to all agents of C_j. Specifically, this new value assignment will be received by A_j via an ok? message and its impact on the consistency state of C_j with respect to the new PA will be examined. If the new PA is still inconsistent, a new nogood from A_j to A_i will be generated and A_i will seek an alternative value. When A_i's domain is exhausted it generates a new nogood by a similar mechanism to that of ABT-1ph and sends it backward to a higher priority agent.

When A_i receives a nogood from a lower priority agent A_k then it must be due to an inconsistent PA to some constraint C_x (possibly C_i) to which A_k and all other lower priority agents have failed to find a satisfying joint assignment which includes A_i's current value assignment. A_i will therefore attempt to pick an alternate value from its current domain which will not necessarily satisfy its personal constraint C_i(see chooseValue). Despite being in a possible conflict state, the asymmetric constraint may eventually change its state due to a change of assignments by lower priority agents and this step ensures that no value combination is overlooked.

The result of this process is that no agent will change its value to satisfy a constraint unless all value combination of lower priorities agents of the same constraints have been exhausted and no PA which may be extended to a full solution is overlooked.

Finally, due to the exhaustive search of consistent PA to constraints and the nogood mechanism (see [3,2]) which guarantees that discarded parts of the search space accumulate ANT must, at one point, terminate. $\square$

5   Experimental Evaluation

5.1   The NashProp Algorithm

An immediate way of evaluating the performance of the ANT algorithm is to compare it to the only other distributed algorithm for infering an ϵ-Nash equilibrium in graphical games – the NashProp algorithm [13]. This turns out to be a non trivial task, because the NashProp algorithm has been presented only partially in [13].

The algorithm is described in two phases: a table passing phase and an assignment passing one. The first phase is a form of an Arc Consistency procedure, in which every agent a_i exchanges a binary valued matrix T with each of its neighbors a_j. An entry in these tables corresponds to the joint action $\langle x_i = i, x_j = j \rangle$ (or vice versa) and takes the value of 1 if and only if a_i can find a joint action of its neighbors $x_{N(i)} = \langle x_{j1}, ..., x_{jk} \rangle$ such that:

1. The entry corresponding to $\langle x_i, x_{jl} \rangle$ takes the value of 1 $\forall l \ 1 \leq l \leq k$.
2. The assignment $x_i = i$ is a best response to $x_{N(i)}$.

This phase is proven to converge and by a simple aggregation of tables one can hope to prune parts of the agent's domain. It should be noted that this phase is essentially a form of pre processing and will also be used by the ANT algorithm in our evaluation.

Details of the second phase of the NashProp algorithm were omitted from [13]. In fact, NashProp's evaluation is based on experiments in which no backtracking was needed (Kearns et. al specify that backtracking was required only on 3% of the problems in their evaluation). The brief description specifies a simple synchronized procedure in which the initializing agent picks an assignment for itself and all its neighbors and then passes it on to one of the already assigned agents (if possible). The recipient then attempts to extend it to an equilibrium assignment. Once a complete assignment is found an additional pass is made, verifying that this joint assignment is indeed an equilibrium. The authors state that "The difficulty, of course, is that the inductive step of the assignment-passing phase may fail... (in which case) we have reached a failure point and must backtrack". Unfortunately, details on this backtracking are not specified and required a reconstruction which is sound, complete and terminates.

The following pseudo code provides an outline of NashProp's assignment passing phase. The algorithm proceeds in synchronous steps in which only one agent acts. The first agent initializes the search by generating a cpa token – Current Partial Assignment – to be passed between the agents. It then adds a joint assignment of itself and all its neighbors (we use $N(i)_{-cpa}$ to denote all neighbors not on the cpa) to the cpa and passes it to the next agent in the resulting order. An agent receiving a cpa first checks the consistency of its assignment. That is, the agent verifies that its assigned value is a best response to its neighbors assignment. If not all neighbors have values assigned to them then the agent attempts to assign new values which will be consistent with its current assignment. Otherwise, if its assigned value does not correspond to a best response action, the agent reassigns a value to itself and any of its lower priority neighbors. If no consistent assignment can be made, a backtrack message is generated and passed to the previous agent on the cpa. If a joint consistent assignment is found then either the agent passes the cpa to the next agent in the ordering or a backcheck message is generated and passed to the previous agent.

Upon receiving a backtrack message, the agent reassigns its own value and any value of its lower priority neighbors. If the joint assignment is consistent the cpa is moved forward and the search resumes. If it is not, a backtrack message is passed to the previous agent. Finally, once a full assignment is reached agents pass the cpa with this assignment backward to higher priority agents. If an agent encounters an inconsistent assignment for itself, it generates a backtrack message and passes it *to the last agent* in the ordering. When the initializing agent receives a consistent backcheck message then the search is terminated and a solution is reported.

5.2 Evaluation

The performance of both NashProp, ANT and ANT with AC (labeled ANT+AC) is measured in terms of Non Concurrent Constraint Checks (NCCCs) and network load, measured in terms of the total number of messages passed between agents. The algorithms implementation uses the AgentZero framework [9], which provides an asynchronous execution environment for multiple agents solving distributed constraint problems. The source code of all experiment is available at: `http://www.cs.bgu.ac.il/~alongrub/files/code/ANT`.

To refrain from exceedingly large domains required for finding accurate ϵ equilibrium, the problems used for evaluating both algorithms were comprised of random interactions in which a pure strategy Nash equilibrium was guaranteed to exist. This allowed for controlled domain size with no changes to the code.

The evaluation included two setups in which each agent was connected to 3 others. The cost values of every agent in every constraint were uniformly taken from the range 0 to 9, and were then updated to assure that at least one pure strategy Nash equilibrium existed. Each data point was averaged over 10 instances and a time out mechanism limited the execution of a single run to 5 minutes.

The first setup included 7 agents and varied the domain sizes of all agents in the range 2 .. 10. The number of NCCCs as a function of domain sizes is presented in Figure 2. Both variants of ANT provide a dramatic speedup over NashProp – roughly three order of magnitudes less NCCCs. The results also demonstrate that in this setting, ANT+AC is less effective than ANT. The number of Non Concurrent Steps (NC-Steps), approximating performance when communication time is significantly higher than an agent's computation time, was slightly lower for ANT+AC than ANT[1]. This implies that the AC phase failed to significantly prune the agents' domains and can explain the NCCC overhead of ANT+AC.

The second setup, measuring performance as the number of agents increases, shows a similar improvement over NashProp. In this setup, the domain size remained constant at 3, and the number of agents was varied in the range 5 .. 15. Figure 3 presents the number of NCCCs in this setup. ANT+AC is faster than ANT by roughly one order order of magnitude and four orders of magnitude faster than NashProp as the number of agents increase.

---

[1] Not presented herein due to lack of space.

NashProp, assignment passing phase

```
procedure NASHPROPII
    end ← false;
    if is initializing agent then
        cpa ← new CPA();
        cpa.assignValid(self ∪ N(i)₋cpa);
        send(CPA, cpa.next(), cpa);
    while ¬end do
        msg ← getMsg();
        switch (msg.type)
            case CPA : ProcessCPA(msg.cpa);
            case BT : BackTrack(msg.cpa);
            case BC : BackCheck(msg.cpa);
            case STP : end ← true;
procedure PROCESSCPA(cpa)
    if ¬isConsistent(cpa) then
        cpa.assignValid(self ∪ N(i)₋cpa);   // removes all lower priority assignments!
    else
        cpa.assignValid(N(i)₋cpa);
    if ¬isConsistent(cpa) then
        currentDomain← full_domain;
        send(BT, cpa.prev(), cpa);
    else if isLast then
        send(BC, cpa.prev(), cpa);
    else
        send(CPA, cpa.next(), cpa);
procedure BACKTRACK(cpa)
    cpa.assignValid(self ∪ N(i)₋cpa);   // removes all lower priority assignments!
    if ¬isConsistent(cpa) then
        send(BT, cpa.prev(), cpa);
    else
        send(CPA, cpa.next(), cpa);
procedure BACKCHECK(cpa)
    if ¬isConsistent(cpa) then
        send(BT, lastAgent, cpa);
    else if is initializing agent then
        send(STP, all_agents, null);
    else
        send(BC, cpa.prev(), cpa);
```

The algorithms' network load in the second experiment is presented in Figure 4. Interestingly, the number of messages generated by NashProp is proportional to the number of NCCCs. That is, for every message sent, there were roughly 14 constraint checks (all constraint checks are non concurrent due to NashProp's synchronous nature). This ratio remains constant throughout the entire experiment, indicating that NashProp's performance is highly affected by the agents' local environment – the number of adjacent agents and/or the size of their domains.

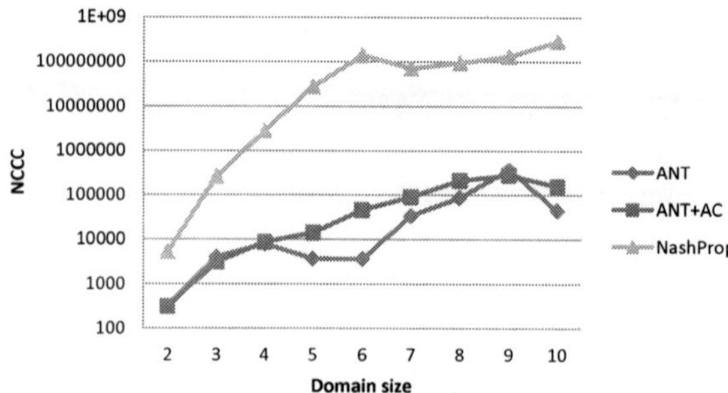

Fig. 2. Number of NCCCs as a function of the domain size

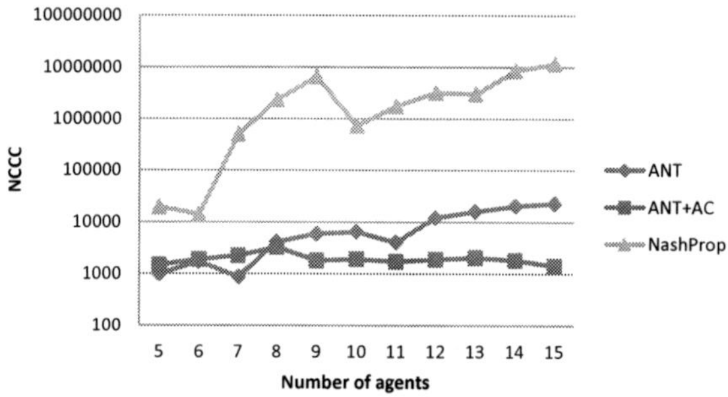

Fig. 3. Number of NCCCs as a function of the number of agents

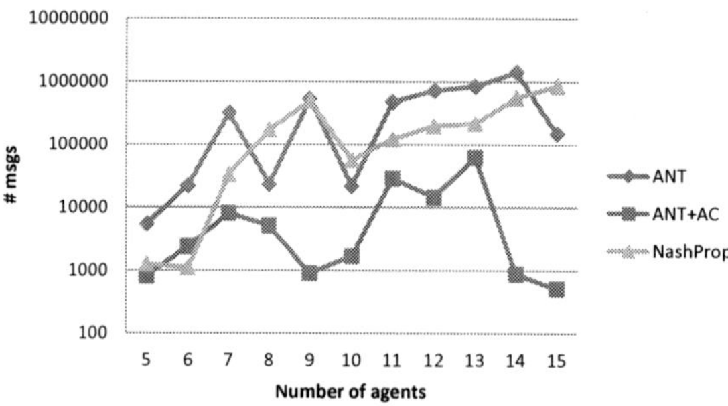

Fig. 4. Number of messages as a function of the number of agents

In sharp contrast to NashProp's constant ratio of NCCCs and network load, the number of messages generated by ANT and ANT+AC does not seem to be aligned with the number of NCCCs. One possible explanation for this discrepancy stems from the combination of a high arity problem and asynchronous execution. If, for some reason, the communication links are not steady and some agents are more active than others, it is possible that a subset of the agents in a constraint exchange messages with each other until the last agent involved in the constraint reacts to the current state. The messages the agents exchange are correct with respect to their agent view and Nogood store, but progress towards the global satisfaction goal can only be made after the final agent in the constraint reacts. As a result, the ANT algorithm can generate redundant messages which increase its communication cost. Nonetheless, ANT+AC almost always requires less messages than NashProp.

6   Conclusions

The present paper explores the similarities between distributed constraint reasoning and graphical games – a well established means for representing multi agent problems. A general form graphical problem is first represented as an Asymmetric Distributed Constraint Optimization Problem (ADCOP) which is capable of capturing the agents' preferences over outcomes. Then, the ADCOP is transformed to a Nash-ADCP with unique, asymmetric, high arity constraints. A satisfying solution to this Nash-ADCP is proven to be an ϵ-equilibrium of the graphical game.

The constraint based formulation of the graphical problem enables one to apply constraint reasoning techniques to solve the underlying problem. Asynchronous Nash back Tracking (ANT), a variant of ABT and ABT-1ph [18,2,3] is presented and a proof that it is always capable of finding ϵ-Nash equilibria is provided.

The performance of ANT and a combination of ANT and the AC mechanism described in [13] is compared to NashProp, the only other distributed algorithm for finding ϵ-Nash equilibrium on general form graphs [13]. The paper also presents the first (to the best of our knowledge) detailed outline of NashProp's second phase. The results of our evaluation indicate a three to four orders of magnitude speedup in terms of run-time as measured by NCCCs and a number of messages which is generally not greater than that of NashProp in favor of the ANT variants.

The connection between distributed constraint reasoning and graphical games induces multiple directions for future work. The special constraint structure of a Nash-ADCSP hinders the performance of some algorithms such as those applying forward checking. However, this structure can hopefully be utilized to find even more efficient distributed algorithms to graphical games. The structure of non random graphical games should also be investigated. Kearns et. al report that the AC phase of Nash-Prop greatly reduced the agents' domains. This phenomenon was not observed on the random instances examined in the present study, and it is our belief that this stems from the fact that in [13], the evaluation only considered a binary action space. Finally, the relation between non cooperative agents, privacy loss and the agents ability to rationally manipulate an algorithm is another topic we intend to research in the near future.

Acknowledgment. The authors would like to thank Benny Lutati for his support throughout the experimental evaluation.

References

1. Apt, K.R., Rossi, F., Venable, K.B.: A comparison of the notions of optimality in soft constraints and graphical games. CoRR, abs/0810.2861 (2008)
2. Bessière, C., Maestre, A., Brito, I., Meseguer, P.: Asynchronous backtracking without adding links: a new member in the ABT family. Artif. Intell. 161(1-2), 7–24 (2005)
3. Brito, I., Meisels, A., Meseguer, P., Zivan, R.: Distributed constraint satisfaction with partially known constraints. Constraints 14(2), 199–234 (2009)
4. Ceppi, S., Gatti, N., Patrini, G., Rocco, M.: Local search techniques for computing equilibria in two-player general-sum strategic-form games. In: 9th Intl. Conf. on Autonomous Agents and Multiagent Systems, AAMAS 2010, pp. 1469–1470 (2010)
5. Chapman, A.C., Farinelli, A., de Cote, E.M., Rogers, A., Jennings, N.R.: A distributed algorithm for optimising over pure strategy nash equilibria. In: Twenty-Fourth AAAI Conference on Artificial Intelligence, AAAI 2010, Atlanta, Georgia, USA (July 2010)
6. Ganzfried, S., Sandholm, T.: Computing equilibria by incorporating qualitative models. In: 9th Intl. Conf. on Autonomous Agents and Multiagent Systems, AAMAS 2010, pp. 183–190 (2010)
7. Grubshtein, A., Zivan, R., Meisels, A., Grinshpoun, T.: Local search for distributed asymmetric optimization. In: 9th Intl. Conf. on Autonomous Agents and Multi-Agent Systems, AAMAS 2010, Toronto, Canada, pp. 1015–1022 (May 2010)
8. Kearns, M.J., Littman, M.L., Singh, S.P.: Graphical models for game theory. In: UAI 2001: 17th Conference in Uncertainty in Artificial Intelligence, San Francisco, CA, USA, pp. 253–260 (2001)
9. Lutati, B., Gontmakher, I., Lando, M.: AgentZero – a framework for executing, analyzing and developing DCR algorithms, http://code.google.com/p/azapi-test/
10. Maheswaran, R.T., Pearce, J.P., Tambe, M.: Distributed algorithms for DCOP: A graphical-game-based approach. In: Proceedings of Parallel and Distributed Computing Systems, PDCS 2004, pp. 432–439 (September 2004)
11. Meisels, A.: Distributed Search by Constrained Agents: Algorithms, Performance, Communication. Springer (2007)
12. Nisan, N., Roughgarden, T., Tardos, É., Vazirani, V.V.: Algorithmic Game Theory. Cambridge University Press (2007)
13. Ortiz, L.E., Kearns, M.J.: Nash propagation for loopy graphical games. In: NIPS 2002: Advances in Neural Information Processing Systems 15, Vancouver, British Columbia, Canada, pp. 793–800 (2002)
14. Osborne, M.J., Rubinstein, A.: A Course in Game Theory. The MIT Press (1994)
15. Porter, R., Nudelman, E., Shoham, Y.: Simple search methods for finding a nash equilibrium. Games and Economic Behavior 63(2), 642–662 (2008)
16. Sandholm, T., Gilpin, A., Conitzer, V.: Mixed-integer programming methods for finding nash equilibria. In: 20th National Conference on Artificial Intelligence, AAAI 2005, vol. 2, pp. 495–501. AAAI Press (2005)
17. Soni, V., Singh, S.P., Wellman, M.P.: Constraint satisfaction algorithms for graphical games. In: 6th International Joint Conference on Autonomous Agents and Multiagent Systems, AAMAS 2007, Honolulu, Hawaii, USA, p. 67 (May 2007)
18. Yokoo, M., Durfee, E.H., Ishida, T., Kuwabara, K.: Distributed constraint satisfaction problem: Formalization and algorithms. IEEE Trans. on Data and Kn. Eng. 10, 673–685 (1998)

Reasoning over Biological Networks
Using Maximum Satisfiability

João Guerra and Inês Lynce

INESC-ID/Instituto Superior Técnico,
Technical University of Lisbon, Portugal
{jguerra,ines}@sat.inesc-id.pt

Abstract. Systems biology is with no doubt one of the most compelling fields for a computer scientist. Modelling such systems is per se a major challenge, but the ultimate goal is to reason over those systems. We focus on modelling and reasoning over biological networks using Maximum Satisfiability (MaxSAT). Biological networks are represented by an influence graph whose vertices represent the genes of the network and edges represent interactions between genes. Given an influence graph and an experimental profile, the first step is to check for mutual consistency. In case of inconsistency, a repair is suggested. In addition, what is common to all solutions/optimal repairs is also provided. This information, named prediction, is of particular interest from a user's point of view. Answer Set Programming (ASP) has been successfully applied to biological networks in the recent past. In this work, we give empirical evidence that MaxSAT is by far more adequate for solving this problem. Moreover, we show how concepts well studied in the fields of MaxSAT and CP, such as backbones and unsatisfiable subformulas, can be naturally related to this practical problem.

1 Introduction

The field of systems biology has seen a tremendous boost due to the advances in molecular biology, which are responsible for the availability of large sets of comprehensive data. The existence of large-scale data sets is a key motivation for developing reliable algorithmic solutions to solve different problems in the field. Understanding those problems comprises reasoning about them.

We address the problem of reasoning over biological networks, in particular gene regulatory networks (GRNs), using influence graphs and the Sign Consistency Model (SCM) to represent GRNs. Reasoning is performed using Boolean Satisfiability (SAT) and Maximum Satisfiability (MaxSAT). These formalisms seem to be particularly well suited to this problem given the Boolean domains of the variables of the problem. SAT is used when reasoning can be formulated as a decision problem, whereas MaxSAT is used when reasoning can be formulated as an optimization problem.

This paper has three main contributions. First, we propose a SAT encoding for modelling biological networks and checking their consistency. This encoding has the advantage of making trivial the task of computing a prediction in case

M. Milano (Ed.): CP 2012, LNCS 7514, pp. 941–956, 2012.

of satisfiability, simply by using available tools for identifying backbones. Second, a MaxSAT encoding is used to repair inconsistent biological networks. Third, we propose an iterative solution, which invokes a MaxSAT solver, to compute predictions under inconsistency. Experimental results show that our contributions outperform existing solutions and as a result solve (many) practical instances which could not be solved by the existing alternatives. All the software implemented in this work is available online[1].

The paper is organized as follows. The next section introduces preliminaries, namely influence graphs, SCM, SAT, MaxSAT and related work. Section 3 describes how SAT and MaxSAT can be applied to reasoning over biological networks, including consistency checking, repairing and predicting. Section 4 is devoted to the experimental evaluation of the solutions proposed. Finally, the paper concludes and suggests future research directions.

2   Preliminaries

A gene regulatory network (GRN) is a kind of biological network in which we are concerned with the interactions between genes. To be precise, genes do not interact with each others directly, but rather through regulatory proteins (and other molecules). Each gene is influenced by the concentration levels of the proteins in its cellular context. Nonetheless, proteins are usually abstracted away and we speak of interactions between genes.

GRN models can be classified as either static or dynamic and qualitative or quantitative. Dynamic models describe the change of gene expression levels over time, whilst static models measure the variation of the gene expression levels between two steady states.

For many biological processes there is no detailed quantitative information available, e.g. accurate experimental data on chemical reactions kinetics is rarely available. This led to the creation of simpler models, the qualitative models. Qualitative models only consider, for example, the sign of the difference between the gene expression levels of two conditions. Despite being a simplification, these models are useful when there is a lack of information about the biological processes and still allow modelling the behaviour of a biological system correctly. Qualitative formalisms have also been successfully applied to other areas besides molecular biology (e.g. see [4]).

Our approach relies on a static qualitative model for GRNs and on the use of SAT and MaxSAT to reason over it. To describe the model, we borrow the notation introduced in [16,12].

For a survey of different models for GRNs refer to the relevant literature (e.g. see [9]).

2.1   Influence Graphs and Sign Consistency Model

Influence graphs are a common representation for a wide range of qualitative dynamical systems, notably biological systems [32]. These kind of graphs offer

---

[1] http://sat.inesc-id.pt/~jguerra/rbnms/

a logical representation of the interactions between the elements of a dynamic system.

An influence graph is a directed graph $\mathcal{G} = (\mathcal{V}, \mathcal{E}, \sigma)$, where $\mathcal{V}$ is a set of vertices representing the genes of a GRN, $\mathcal{E}$ is a set of edges representing the interactions between the genes of the GRN and $\sigma : \mathcal{E} \to \{+, -\}$ is a (partial) labelling of the edges. An edge from a vertex u to a vertex v, with $u, v \in \mathcal{V}$, is denoted as $u \to v$. Biologically, an interaction with label $+$ $(-)$ represents the activation (inhibition) of gene transcription or protein activation (inactivation).

To impose constraints between GRNs and experimental profiles we use the Sign Consistency Model (SCM) [29], which is based on influence graphs. This static qualitative model is particularly well suited for dealing with incomplete and noisy data [16,12]. In the SCM, experimental profiles only contain qualitative information about the observed variation of the gene expression levels.

Given an influence graph $\mathcal{G} = (\mathcal{V}, \mathcal{E}, \sigma)$, an experimental profile $\mu : \mathcal{V} \to \{+, -\}$ is a (partial) labelling of the vertices of the graph. Additionally, each vertex of the graph is classified as either input or non-input. The labelling $\mu(v)$ of a non-input vertex $v \in \mathcal{V}$ is *consistent iff* there is at least one edge that explains its sign, i.e. one edge $u \to v \in \mathcal{E}$ such that $\mu(v) = \mu(u) \cdot \sigma(u \to v)$, where $\cdot$ corresponds to the multiplication of signs in the multiplication of signed numbers. For example, if $\mu(v) = +$ then either $\mu(u) = +$ and $\sigma(u \to v) = +$ or $\mu(u) = -$ and $\sigma(u \to v) = -$. Biologically, label $+$ $(-)$ means that there was an increase (decrease) in the gene expression levels. Note that the definition of consistency does not apply to input vertices.

An influence graph (model) $\mathcal{G} = (\mathcal{V}, \mathcal{E}, \sigma)$ and an experimental profile (data) μ are *mutually consistent iff* there are total labellings $\sigma' : \mathcal{E} \to \{+, -\}$ and $\mu' : \mathcal{V} \to \{+, -\}$, which are total extensions of σ and μ, respectively, such that $\mu'(v)$ is consistent for every non-input vertex $v \in \mathcal{V}$. When considering an influence graph without an experimental profile, i.e. when μ is undefined for every vertex of the graph, we talk about *self-consistency* of the graph [18].

Example 1. Figure 1 illustrates an influence graph (left) and an experimental profile for that influence graph (right). The graph has three vertices, a, b and c and five edges, $a \to b$, $a \to c$, $b \to a$, $b \to c$ and $c \to b$. All vertices are non-input vertices. Lighter (green) edges ending with $\to$ have label $+$, whereas darker (red) edges ending with $\dashv$ have label $-$. Likewise, in the experimental profile, lighter (green) vertices have label $+$, darker (red) vertices have label $-$ and white vertices have no label.

In section 3 we will reason over this example.

2.2   Maximum Satisfiability

The Boolean Satisfiability (SAT) problem is the problem of deciding whether there is an assignment to the variables of a Boolean formula that satisfies it.

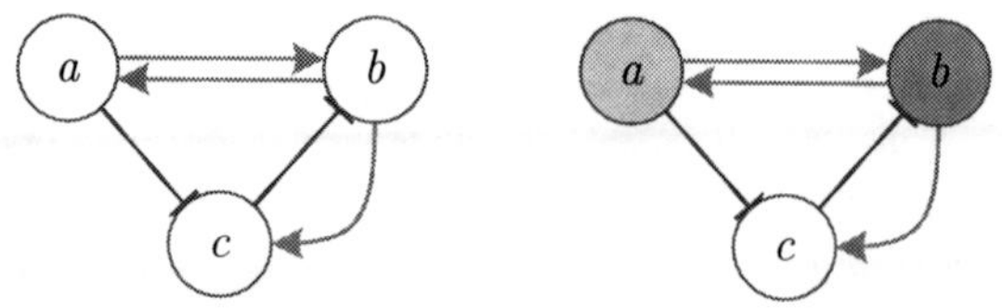

Fig. 1. An influence graph (left) along with an experimental profile (right)

Without lack of generality, one may assume that the formula is in conjunctive normal form (CNF). A CNF formula is a conjunction ($\wedge$) of clauses, where a clause is a disjunction ($\vee$) of literals and a literal is a Boolean variable (x) or its negation ($\neg x$). A CNF formula is satisfied *iff* all of its clauses are satisfied. A clause is satisfied if at least one of its literals is satisfied. A literal x ($\neg x$) is satisfied *iff* the corresponding Boolean variable is assigned the value *true* (*false*).

The Maximum Satisfiability (MaxSAT) problem is closely related to SAT. The goal in MaxSAT is to find an assignment that maximizes (minimizes) the number of satisfied (unsatisfied) clauses.

There are a few interesting variants of the MaxSAT problem. One of them, relevant to the application being of interest, is the Partial MaxSAT problem. In Partial MaxSAT, some clauses are classified as hard, whereas the remaining ones are classified as soft. The goal is to find an assignment that satisfies all hard clauses and maximizes the number of satisfied soft clauses. Hard clauses are usually represented within square brackets, whereas soft clauses are represented within parentheses.

Example 2. Consider the following Partial MaxSAT formula $\mathcal{F} = [x_1 \vee x_2 \vee x_3] \wedge [\neg x_1 \vee x_3] \wedge (\neg x_2) \vee (\neg x_3)$. The two optimal solutions are $\{x_1, \neg x_2, x_3\}$ and $\{\neg x_1, x_2, \neg x_3\}$. In any of these solutions, only one of the two soft clauses is satisfied.

2.3   Related Work

The same problem of modelling and reasoning over biological networks has been tackled in the past using Answer Set Programming (ASP) [16,12,15]. First, the authors analyse whether a biological network is consistent. If the network is consistent that means that there is a solution corresponding to total extensions of σ and μ. Moreover, a prediction corresponding to the intersection of all solutions is computed. If the network is inconsistent then minimal explanations for inconsistency are provided. As an alternative, an optimal repair is given. In addition, the user is also given a prediction, which now summarizes what is common to all optimal repairs. In ASP, the computation of predictions is achieved through cautious reasoning [14].

Reasoning over biological networks with ASP can find similarities with concepts well known in SAT and CP.

Minimal explanations for inconsistency are often called minimal unsatisfiable cores (MUCs) in CP [20] and minimal unsatisfiable subformulas (MUSes) in SAT [24]. Repairing with MaxSAT may be related with the identification of MUSes in SAT. A MaxSAT solution does not satisfy exactly one clause from each MUS of the corresponding SAT formula. The number of unsatisfied clauses may be less than the number of MUSes when some of the MUSes have a non-empty intersection.

The identification of assignments common to all solutions corresponds to the definition of backbone [28,23]. Backbones find applications not only in decision problems, but also in optimization problems [30]. In the context of MaxSAT, backbones have inspired the development of new search strategies [33,27] and the same occurred in other domains (e.g. see [17,21]). Recent work in SAT has focused on implementing efficient algorithms for identifying backbones in practical settings [26,34]. The solution developed in this paper follows one of the solutions proposed for post-silicon fault localization [34].

More sophisticated models exist for GRNs and other biological networks using CP solutions. An example is the framework developed by Corblin et al. [8,7,6]. The authors use a more complex formalism for modelling GRNs that allows multivalued variables and uses transition rules, amongst other particularities. For a survey of CP solutions to solve related biological problems refer to the relevant literature (e.g. see [31,3,1]).

3 Reasoning with Satisfiability

Our goal is to provide SAT and MaxSAT solutions for reasoning over biological networks. We begin by describing how to encode a GRN into SAT using the model introduced in the previous section. This encoding allows to validate an influence graph against an experimental profile. In case the graph and profile are mutually inconsistent, the identification of repairs to restore consistency is of interest. Building on the SAT encoding, we then introduce a MaxSAT encoding that allows to repair the graph and/or profile in order to restore consistency. In addition, we also provide information about what is common to all total labellings, in the case of consistency, or to all optimal repairs, in the case of inconsistency. This information is called prediction.

3.1 Checking Consistency

An influence graph $\mathcal{G} = (\mathcal{V}, \mathcal{E}, \sigma)$ and respective experimental profile μ can be encoded into SAT as follows. For the sake of clarity, the constraints will not be presented in CNF. Translating such constrains to CNF should be a standard task though.

Let us first introduce three types of Boolean variables. For each vertex $v \in \mathcal{V}$, there is a Boolean variable inp_v such that inp_v is assigned the value *true* if v is an input vertex and *false* otherwise. For each vertex $v \in \mathcal{V}$, there is a Boolean variable $lvtx_v$ (label vertex) such that $lvtx_v$ is assigned the value *true/false* if the corresponding label $\mu(v)$ is $+/-$. Likewise, for each edge $u \to v \in \mathcal{E}$, there is Boolean variable $ledg_{uv}$ (label edge) such that $ledg_{uv}$ is assigned the value *true/false* if the corresponding label $\sigma(u \to v)$ is $+/-$.

An additional type of (auxiliary) variables is needed to represent the value of $\mu \cdot \sigma$, which denotes the influence between vertices. For each edge $u \to v \in \mathcal{E}$, create a Boolean variable $infl_{uv}$ such that $infl_{uv}$ is assigned the value *true/false* if $\mu(u) \cdot \sigma(u \to v)$ is $+/-$.

Let us now introduce the constraints, starting with the ones corresponding to unit clauses. For each vertex $v \in \mathcal{V}$, introduce a unit clause (inp_v) if v is an input vertex. Otherwise, introduce a unit clause $(\neg inp_v)$. Given labellings μ/σ, introduce one unit clause for each vertex/edge that has a label. (Remember that μ and σ may be partial labellings and therefore not all vertices/edges may have a corresponding label.) For each vertex $v \in \mathcal{V}$ with $\mu(v) = +/-$ introduce a unit clause $(lvtx_v)/(\neg lvtx_v)$. For each edge $u \to v \in \mathcal{E}$ with $\sigma(u \to v) = +/-$ introduce a unit clause $(ledg_{uv})/(\neg ledg_{uv})$.

Example 3. Consider again the example in Figure 1. Encoding the influence graph into SAT produces the following unit clauses: $(\neg inp_a)$, $(\neg inp_b)$, $(\neg inp_c)$, $(ledg_{ab})$, $(\neg ledg_{ac})$, $(ledg_{ba})$, $(ledg_{bc})$, $(\neg ledg_{cb})$. Moreover, encoding the experimental profile into SAT produces the following unit clauses: $(lvtx_a)$, $(\neg lvtx_b)$.

In order to define the value of variables $infl$ a few additional constraints are needed. The value of these variables is given by $\mu \cdot \sigma$. For each edge $u \to v$, the following constraints are added:

$$
\begin{aligned}
infl_{uv} &\longrightarrow (lvtx_u \wedge ledg_{uv}) \vee (\neg lvtx_u \wedge \neg ledg_{uv}) \\
\neg infl_{uv} &\longrightarrow (lvtx_u \wedge \neg ledg_{uv}) \vee (\neg lvtx_u \wedge ledg_{uv})
\end{aligned}
\tag{1}
$$

Finally, consistency must be ensured. An influence graph and an experimental profile are mutually consistent if total extensions for μ and σ can be found such that all non-input vertices are consistent. The consistency of a vertex v is ensured by making use of variables $infl_{uv}$ where u may be any vertex to which v is adjacent. For each vertex v, the following constraints are added:

$$
\begin{aligned}
inp_v \vee \left(lvtx_v \longrightarrow \bigvee_u infl_{uv} \right) \\
inp_v \vee \left(\neg lvtx_v \longrightarrow \bigvee_u \neg infl_{uv} \right)
\end{aligned}
\tag{2}
$$

Example 4. Equations 1 and 2 applied to vertex c of Figure 1 produce the following constraints:

$$
\begin{aligned}
infl_{ac} &\longrightarrow (lvtx_a \wedge ledg_{ac}) \vee (\neg lvtx_a \wedge \neg ledg_{ac}), \\
\neg infl_{ac} &\longrightarrow (lvtx_a \wedge \neg ledg_{ac}) \vee (\neg lvtx_a \wedge ledg_{ac}), \\
infl_{bc} &\longrightarrow (lvtx_b \wedge ledg_{bc}) \vee (\neg lvtx_b \wedge \neg ledg_{bc}), \\
\neg infl_{bc} &\longrightarrow (lvtx_b \wedge \neg ledg_{bc}) \vee (\neg lvtx_b \wedge ledg_{bc}),
\end{aligned}
$$

$$inp_c \vee (lvtx_c \longrightarrow (infl_{ac} \vee infl_{bc})),$$
$$inp_c \vee (\neg lvtx_c \longrightarrow (\neg infl_{ac} \vee \neg infl_{bc})).$$

A SAT call to the complete CNF encoding of the influence graph reveals that the graph by itself is self-consistent. If we add the encoding of the experimental profile, i.e. $(lvtx_a) \wedge (\neg lvtx_b)$, another SAT call reveals that the influence graph and the experimental profile are mutually inconsistent. Observe that vertex a has only one incoming edge $b \to a$ with label $+$. Given that vertex b has label $-$, vertex a cannot have label $+$.

3.2   Repairing

When an influence graph is inconsistent, whether by itself or mutually with an experimental profile, one may consider repairing the graph and/or the profile in order to restore consistency. The motivation is that some of the observations may be unreliable. To this end, we allow three types of repair operations (and combinations thereof): e, flip edges labels; i, make non-input vertices become input vertices; and v, flip vertices labels. The goal is to identify *cardinality-minimal* repairs.

Restoring consistency is an optimization problem that can be encoded into Partial MaxSAT as follows. The SAT encoding described in the previous section is still valid when allowing repairs. However, the unit clauses referring to non-input vertices, vertices labels and edges labels can be unsatisfied if needed, depending on the type of repair. Hence, these clauses can be made *soft* in the Partial MaxSAT encoding. All other clauses, i.e. clauses encoding constraints (1) and (2), are hard clauses. For repairs of type e, unit clauses referring to edges labels are soft clauses. For repairs of type i, unit clauses referring to input vertices are soft clauses. For repairs of type v, unit clauses referring to vertices labels are soft clauses. All other unit clauses are hard clauses. The MaxSAT solution will satisfy all hard clauses and the largest number of soft clauses. Note that all soft clauses are unit clauses. The size of the repair corresponds to the number of unsatisfied clauses. The actual repairs can be trivially obtained from the unsatisfied clauses.

This encoding can be easily adapted to consider other repairs. The user could manually indicate which repairs would be reasonable to perform. For example, in some cases it can make sense to restrict repairs to a subset of vertices and respective edges.

Example 5. As discussed in Example 4, the influence graph and experimental profile illustrated in Figure 1 are mutually inconsistent. To identify repairs of type e, the following clauses are declared as soft clauses: $(ledg_{ab})$, $(\neg ledg_{ac})$, $(ledg_{ba})$, $(ledg_{bc})$, $(\neg ledg_{cb})$. The solution can be any one of the four cardinality-minimal repairs, all of them with size two: $\{\neg ledg_{ab}, \neg ledg_{ba}\}$, $\{\neg ledg_{ba}, ledg_{ac}\}$, $\{\neg ledg_{ba}, \neg ledg_{bc}\}$ and $\{\neg ledg_{ba}, ledg_{cb}\}$. Had been allowed only repairs of type v, the clauses to be made soft would be: $(lvtx_a)$, $(\neg lvtx_b)$. In this case, two different optimal repairs with size one could be obtained: $\{\neg lvtx_a\}$ and $\{lvtx_b\}$.

3.3 Predicting

Analysing all possible solutions to a given problem instance can become a puzzling task when the number of solutions is too large. In this context, it is certainly useful to know what is common to all solutions. Note that this concept applies both to decision problems and optimization problems as long as (optimal) solutions to a given problem instance can be found.

The intersection of all solutions to a given problem instance is called prediction. This concept can be applied either to consistent problem instances when checking consistency or to inconsistent problem instances when repairing. In SAT and MaxSAT, the assignments which are common to all (optimal) solutions correspond to the backbones of a given formula. However, when predicting, only a subset of the variables is relevant. Still, approaches used for identifying backbones in SAT and MaxSAT can be adapted to compute predictions.

Predicting under consistency can be applied to the SAT encoding described in Section 3.1 using a tool designed to identify backbone variables in SAT formulas. A prediction can be obtained after filtering irrelevant variables from the set of variables returned by the tool.

For the case of prediction under inconsistency, one has to consider the MaxSAT encoding described in Section 3.2. The set of repairs common to all optimal solutions can be obtained from the unit soft clauses that were not satisfied in all optimal solutions. (Observe that each unsatisfied unit soft clause corresponds to a repair.) Next we describe how a MaxSAT solver can be instrumented to efficiently compute predictions.

A naïve approach consists in enumerating all solutions and computing their intersection [26]. This approach requires n calls to a MaxSAT solver, being n the number of optimal solutions. (In practice, this number can be reduced taking into account that only a subset of the variables is relevant.) After each call, a blocking clause corresponding to the negation of the computed solution is added to the formula, in order to prevent the same solution from being found again in a future call. Moreover, after each iteration the actual intersection of the solutions computed so far is updated.

One key optimization to the naïve approach can make a significant difference. Instead of adding one blocking clause for each solution found, there is only one blocking clause, which corresponds to the prediction computed so far. This implies that not all solutions have to be computed. Only a solution that reduces the size of the current prediction can be computed at each iteration.

Algorithm 1 has been implemented to compute predictions under inconsistency using the optimization described above. Similarly to the naïve approach, this algorithm consists in retrieving different optimal solutions and reducing the prediction at each iteration. (Again, we can take advantage of the fact that only a subset of the variables is relevant.) But in practice less iterations are expected to be required.

Given a partial MaxSAT formula $\mathcal{F}$, the algorithm is initialized with an *optimum* value, obtained from calling $\text{MaxSAT}(\mathcal{F})$ and the initial *prediction*, obtained with function GET-REPAIRS. Each call to the MaxSAT solver returns

---

Algorithm 1. Predicting under Inconsistency

   Input   : Partial MaxSAT Formula $\mathcal{F}$
   Output : Predicted Repairs of $\mathcal{F}$, *prediction*

1 $(out, opt, sol) \leftarrow \textsc{MaxSAT}(\mathcal{F})$ // compute initial solution
2 $optimum \leftarrow opt$
3 $prediction \leftarrow \textsc{Get-Repairs}(sol)$

4 **while** $|prediction| \neq 0$ **do**
5      $(out, opt, sol) \leftarrow \textsc{MaxSAT}(\mathcal{F} \cup [\neg prediction])$ // block current prediction
6   **if** out == UNSAT **or** $opt > optimum$ **then**
7   **break**

8      $prediction \leftarrow prediction \cap \textsc{Get-Repairs}(sol)$ // update prediction

9 **return** *prediction*

---

a 3-tuple (out, opt, sol), where out corresponds to the outcome (either SAT or UNSAT); opt corresponds to the optimal value, i.e. the number of *unsatisfied* clauses; and sol corresponds to the variable assignments which result in the optimum value.

The initial prediction corresponds to the set of repairs obtained from the first MaxSAT solution. At each iteration, the MaxSAT solver is called with the initial formula and the blocking clause corresponding to the current prediction, which is added as a hard clause. If the new solution is still optimal, then the corresponding repairs are intersected with the current prediction; otherwise the final prediction has been found. Note that either the current prediction is reduced at each step or the algorithm terminates. In this later case, the final prediction is returned.

Although the worst-case scenario of Algorithm 1 is as bad as the naïve approach, it performs generally well for this domain, as we will see later in Section 4.2.

Example 6. In Example 5 were listed four optimal repairs of type e for the influence graph and the experimental profile in Figure 1. The prediction corresponds to the intersection of all the solutions, i.e. the set $\{\neg ledg_{ba}\}$. A possible run of Algorithm 1 could be to first find repair $\{\neg ledg_{ab}, \neg ledg_{ba}\}$, thus next calling the MaxSAT solver with the original formula and the clause $[ledg_{ab} \lor ledg_{ba}]$. Suppose that afterwards repair $\{\neg ledg_{ba}, ledg_{ac}\}$ is found. So the next step is to call the MaxSAT solver with the original formula and the clause $[ledg_{ba}]$. This call returns UNSAT and so the algorithm terminates. Only two calls to the MaxSAT solver were needed, instead of the four calls of a naïve algorithm.

4 Experimental Evaluation

The experimental evaluation is driven by the goal of comparing the performance of ASP solvers with the performance of SAT and MaxSAT solvers. With this goal in mind, our evaluation is based on the same instances that were used to evaluate the performance of ASP solvers in the past. These instances, as well as the experiments performed, are detailed in the literature [16,12]. In one of the experiments [16], ASP is used to solve randomly generated instances. The target in this case is to check the consistency of each one of the problem instances. In the other experiment [12], real instances are used and the target is to repair unsatisfiable problem instances. In addition, prediction is applied only to the instances for which a repair can be provided.

The computations were performed using the ASP solver *clasp* 2.0.6 together with grounder *gringo* 3.0.4 [13], an improved version of the MiniSat 2.2.0 SAT solver [10] available from *github*, the backbone identification tool for SAT minibones [26] and the MaxSAT solver MSUnCore 2.5 [25]. The experiments were run on several Intel Xeon 5160 machines (dual-cores with 3.00 GHz of clock speed, 4 MB of cache, 1333 MHz of FSB speed and 4 GB of RAM each), running 64-bit versions of Linux 2.6.33.3-85.fc13. All tools were configured to not take advantage of any sort of parallelism.

All times are shown in seconds and correspond to the average execution times taken by each tool for solving a set of problem instances. For each aborted instance, the timeout value of 600 seconds is added to the total sum. The time needed to translate each instance from raw data to ASP, SAT or MaxSAT format is not taken into account, as it is negligible and similar for any of these three formats.

4.1   Checking Consistency and Predicting under Consistency

The first experiment evaluates consistency checking and prediction under consistency for the randomly generated instances. These instances have between 500 and 4000 vertices. The degree of a vertex is on average 2.5, following what is assumed as standard in biological networks [22]. In total, there are 400 instances, 50 for each one of the eight different graph sizes (starting with 500 vertices and up to 4000, with an increment of 500 vertices).

Table 1 shows the average run times for each graph size. We distinguish between consistent and inconsistent instances, denoted as *sat* and *unsat*, respectively. The first columns relate to consistency checking. In these columns, ASP refers to running *gringo* together with *clasp* using the VSIDS heuristic (the most efficient heuristic according to [16]) and SAT refers to MiniSat.

The remaining columns relate to prediction under consistency. Prediction is only applied to satisfiable instances. As before, ASP refers to running *gringo* together with *clasp* using the VSIDS heuristic but now using the cautious reasoning mode (`--cautious`), which makes *clasp* compute the intersection between all answer sets. SAT now refers to running `minibones`, which is used to compute the backbones of SAT formulas.

The results are clear. Checking consistency is trivial for both ASP and SAT solvers. As expected, larger instances require more time but still the time required is not significant. The results obtained with the ASP solver are in conformity with the results available in the literature [16]. Prediction for satisfiable instances is also trivial. We believe that such information is very important from

Table 1. Times for consistency checking and prediction under consistency

		Consistency		Prediction	
		ASP	SAT	ASP	SAT
500	sat	0.11	0.01	0.15	0.05
	unsat	0.10	0.00		
1000	sat	0.26	0.02	0.42	0.18
	unsat	0.23	0.01		
1500	sat	0.42	0.03	0.79	0.39
	unsat	0.37	0.01		
2000	sat	0.58	0.03	1.26	0.69
	unsat	0.51	0.01		
2500	sat	0.75	0.04	1.88	1.08
	unsat	0.66	0.01		
3000	sat	0.91	0.06	2.79	1.57
	unsat	0.79	0.02		
3500	sat	1.08	0.07	3.97	2.14
	unsat	0.95	0.02		
4000	sat	1.24	0.05	5.37	2.82
	unsat	1.10	0.02		

a user's point of view. Whereas analysing dozens or hundreds of solutions is infeasible in practice, knowing which assignments *must* belong to any solution is certainly of interest.

Checking consistency and predicting under consistency is expected to scale well for larger instances using either approach. Nevertheless, at the light of the results obtained for real instances (see next section), extending this evaluation method to larger instances did not seem to be the best way to follow.

4.2   Repairing and Predicting under Inconsistency

This second experiment was conducted using as test set the transcriptional regulatory network of *Escherichia coli* K-12 [11] along with two experimental profiles: the Exponential-Stationary Growth Shift study [5] and the Heatshock experiment [2].

The goal of this experiment is to evaluate the feasibility of (optimally) repairing inconsistent problem instances, as well as of predicting under inconsistency, i.e. computing which assignments are common to all optimal repairs. We used the transcriptional regulatory network of *Escherichia coli* K-12, which contains 5140 interactions between 1915 genes. Each one of the two profiles, Exponential-Stationary Growth Shift and Heatshock, has slightly over 850 gene expression level variations, which correspond to vertex labellings. For a better assessment of the scalability of the approaches used, several data samples were generated by randomly selecting 5%, 10%, 20% and 50% of the whole data for each experimental profile. For each percentage were generated 50 inconsistent instances. The whole test set is made of 400 instances. (In previous work [12] were used percentages 3%, 6%, 9%, 12% and 15% instead, but the better performance of the most recent versions of *gringo* and *clasp* required the generation of more difficult instances.)

Table 2. Times for repair and prediction under inconsistency

Exponential-Stationary Growth Shift							
	e	i	v	ei	ev	iv	eiv
Repair (ASP)							
5%	0.67	0.35	0.27	0.67	0.59	0.41	0.69
10%	0.64	0.35	0.27	0.82	0.75	0.38	1.35
20%	0.94	0.36	0.28	18.28	7.05	0.67	77.98 **(3)**
50%	2.89	0.35	0.29	587.48 **(48)**	572.02 **(46)**	481.16 **(35)**	600.00 **(50)**
Prediction (ASP)							
5%	0.65	0.33	0.26	0.68	0.58	0.39	0.67
10%	0.61	0.32	0.26	0.75	0.67	0.36	0.96
20%	0.90	0.32	0.26	1.80	3.28	0.51	13.86
50%	1.92	0.32	0.27	41.71	320.30	302.43 **(4)**	–
Repair (MaxSAT)							
5%	0.22	0.22	0.21	0.17	0.17	0.17	0.17
10%	0.24	0.24	0.23	0.20	0.20	0.20	0.20
20%	0.37	0.35	0.34	0.29	0.29	0.29	0.29
50%	0.74	0.73	0.72	0.60	0.59	0.59	0.59
Prediction (MaxSAT)							
5%	0.41	0.41	0.39	0.44	0.45	0.45	0.45
10%	0.60	0.58	0.54	0.78	0.76	0.76	0.79
20%	1.41	1.18	0.99	2.11	1.95	1.99	1.97
50%	4.55	3.06	2.26	9.32	7.71	7.83	7.58

Heatshock							
	e	i	v	ei	ev	iv	eiv
Repair (ASP)							
5%	0.69	0.35	0.27	0.66	0.60	0.33	0.67
10%	0.69	0.35	0.27	0.99	0.83	0.34	1.50
20%	1.27	0.34	0.27	112.94 **(8)**	10.61	0.39	42.02
50%	279.43 **(21)**	0.35	0.28	572.03 **(47)**	504.76 **(37)**	202.10 **(12)**	600.00 **(50)**
Prediction (ASP)							
5%	0.67	0.33	0.26	0.66	0.59	0.30	0.66
10%	0.68	0.33	0.27	0.88	0.70	0.30	1.39
20%	0.91	0.31	0.27	17.86	1.07	0.33	9.88
50%	43.76	0.31	0.26	28.01	276.07 **(3)**	126.58 **(1)**	–
Repair (MaxSAT)							
5%	0.21	0.21	0.20	0.17	0.16	0.16	0.16
10%	0.25	0.24	0.24	0.21	0.19	0.19	0.19
20%	0.39	0.38	0.30	0.31	0.25	0.25	0.25
50%	1.02	0.88	0.61	0.86	0.50	0.51	0.50
Prediction (MaxSAT)							
5%	0.43	0.42	0.39	0.48	0.40	0.41	0.40
10%	0.76	0.63	0.56	1.06	0.71	0.71	0.68
20%	1.98	1.46	1.01	3.44	1.68	1.58	1.61
50%	15.14	4.96	2.43	31.71	7.88	6.79	6.58

Table 3. Times for repair and prediction under inconsistency (100% instances)

	Exponential-Stationary Growth Shift				Heatshock			
	Repair		Prediction		Repair		Prediction	
	ASP	MaxSAT	ASP	MaxSAT	ASP	MaxSAT	ASP	MaxSAT
e	4.97	1.46	3.56	13.49	600.00	2.88	–	99.97
i	0.33	1.41	0.30	5.97	0.30	2.77	0.27	16.15
v	0.31	1.43	0.29	4.38	0.28	1.22	0.28	7.73
e i	600.00	1.17	–	28.70	600.00	2.19	–	222.43
e v	600.00	1.14	–	22.01	600.00	0.94	–	26.50
i v	600.00	1.16	–	18.55	600.00	0.94	–	21.45
e i v	600.00	1.14	–	17.25	600.00	0.96	–	16.40

Table 2 shows the average run times for the ASP and MaxSAT approaches. Timeouts, shown within parentheses, represent that the imposed time limit of 600 seconds was exceeded before finding a solution. Observe that there were no timeouts for the MaxSAT runs. In the experiment, we allowed the following repair operations and combinations thereof (previously introduced in Section 3.2): e, flip edges signs; i, make non-input vertices become input vertices; and v, flip vertices signs. This results in 7 types of repairs, thus $400 \cdot 7 = 2800$ instances. Observe that the possibility of making a vertex become an input (i repair operation) makes that vertex trivially consistent.

The repairing phase determines which instances will be used in the prediction phase. It would make no sense to apply prediction to instances for which not even one optimal repair was provided within the time limit. Hence, prediction is applied only to the instances for which repairing was successful.

The results presented in Table 2 were obtained using *gringo* together with *clasp* using flag `--opt-heuristic=1` for better performance. To compute the predictions, *clasp* applies cautious reasoning to all optimal solutions (`--cautious`, `--opt-all=<opt-value>`, with `<opt-value>` being the optimal repair value). To repair an instance using MaxSAT we used MSUNCORE. For prediction, we used Algorithm 1, described in Section 3.3. Note that the operation corresponding to the first line of the algorithm was already computed during the repair phase and therefore the computation is not repeated.

The results are again clear. ASP aborts 357 out of 2800 instances in the repair phase, plus 8 instances in the prediction phase, whereas MaxSAT solves all the instances in a few seconds. MaxSAT is far more adequate than ASP to repair inconsistent instances. Many instances could not be repaired within the time limit of 600 seconds with ASP. As a result, prediction could not be applied to these instances. This is true for the two experimental profiles. In contrast, the MaxSAT solver is able not only to repair all instances, but also to do it in a few seconds. Similar results are obtained for prediction. The number of calls to the MaxSAT solver range from 1 to 52 (on average from 1.08 to 28.30). There seems to be no clear relation between the number of times the MaxSAT solver is called and the prediction size.

Despite the clear trend in favour of the MaxSAT approach, the next step would be to evaluate more difficult problem instances. The most difficult instances would be the ones where the complete experimental profiles are used. This results in 7 instances for each profile, one for each combination of repair operations. Detailed results are given in Table 3. From these 14 instances, the ASP solver was able to repair 3 instances while the MaxSAT solver was able to repair *all* the instances, taking on average less than 3 seconds. Prediction was successfully applied to all the instances but one for which the ASP solver was able to provide a repair. MaxSAT was able to predict all the instances.

Even though the MaxSAT solver is able to provide a prediction taking on average 20 seconds, two outliers exist. Both outliers refer to the Heathstock profile, one to the e repair operation and the other to the ei combination. The first requires 99.97 seconds and the second requires 222.43 seconds. Compared to the remaining instances, these are the ones requiring more calls to the MaxSAT solver (35 and 94, respectively). However, these instances are not the ones with a larger prediction. Actually, the hardest instance has prediction size 0, which means that there are at least two disjoint optimal repairs. If two disjoint optimal repairs had been identified at the first two iterations, then no more iterations would have been needed. This fact suggests instrumenting the MaxSAT solver to find diverse solutions [19], which will be investigated in the future.

Evaluating the accuracy of prediction is out of the scope of this evaluation. Accuracy results have already been provided for the ASP approach [12]. The accuracy of prediction is quite high, being always above 90%. Given that both approaches are equivalent, it makes no sense to repeat such evaluation.

5   Conclusions and Future Work

We have studied how SAT and MaxSAT can be applied to reasoning over biological networks. The use of SAT and MaxSAT is certainly adequate, given that the domains of the variables of the actual problem are Boolean. SAT and MaxSAT encodings have been shown to be more competitive than other approaches used in the past, namely Answer Set Programming (ASP).

As future work we will consider other optimization criteria for repairs. For example, subset-minimal repairs, as already suggested by Gebser et al. [12]. Finding subset-minimal repairs comprises proving that removing any repair from the proposed solution no longer allows to achieve consistency. Additional types of repairs could be considered as well, such as adding edges to the influence graph [12]. As already mentioned, the proposed MaxSAT encoding can easily accommodate other types of repairs.

Another direction for the future is the evaluation of different algorithmic approaches to compute predictions under inconsistency. Ideas coming from existing algorithms for backbone identification in SAT [26] can be discussed as a starting point.

Acknowledgements. This work is partially supported by the Portuguese Foundation for Science and Technology (FCT) under the research project AS-PEN (PTDC/EIA-CCO/110921/2009) and by INESC-ID multiannual funding from the PIDDAC program funds. We thank the reviewers for their insightful comments.

References

1. Workshops on Constraint Based Methods for Bioinformatics (WCB) (2005-2012)
2. Allen, T., Herrgård, M., Liu, M., Qiu, Y., Glasner, J., Blattner, F., Palsson, B.: Genome-scale analysis of the uses of the escherichia coli genome: model-driven analysis of heterogeneous data sets. Journal of Bacteriology 185(21), 6392–6399 (2003)
3. Barahona, P., Krippahl, L., Perriquet, O.: Bioinformatics: a challenge to constraint programming. In: Hybrid Optimization, vol. 45, pp. 463–487. Springer (2011)
4. Bobrow, D.: Qualitative reasoning about physical systems: an introduction. Artificial Intelligence 24(1-3), 1–5 (1984)
5. Bradley, M., Beach, M., de Koning, A., Pratt, T., Osuna, R.: Effects of fis on escherichia coli gene expression during different growth stages. Microbiology 153(9), 2922–2940 (2007)
6. Corblin, F., Bordeaux, L., Fanchon, E., Hamadi, Y., Trilling, L.: Connections and integration with SAT solvers: a survey and a case study in computational biology. In: Hybrid Optimization, vol. 45, pp. 425–461. Springer (2011)
7. Corblin, F., Fanchon, E., Trilling, L.: Applications of a formal approach to decipher discrete genetic networks. BMC Bioinformatics 11, 385 (2010)
8. Corblin, F., Tripodi, S., Fanchon, E., Ropers, D., Trilling, L.: A declarative constraint-based method for analyzing discrete genetic regulatory networks. Biosystems 98(2), 91–104 (2009)
9. de Jong, H.: Modeling and simulation of genetic regulatory systems: a literature review. Journal of Computational Biology 9(1), 67–103 (2002)
10. Eén, N., Sörensson, N.: An Extensible SAT-solver. In: Giunchiglia, E., Tacchella, A. (eds.) SAT 2003. LNCS, vol. 2919, pp. 502–518. Springer, Heidelberg (2004)
11. Gama-Castro, S., Jiménez-Jacinto, V., Peralta-Gil, M., Santos-Zavaleta, A., Peñaloza-Spinola, M., Contreras-Moreira, B., Segura-Salazar, J., Muñiz-Rascado, L., Martínez-Flores, I., Salgado, H., Bonavides-Martínez, C., Abreu-Goodger, C., Rodríguez-Penagos, C., Miranda-Ríos, J., Morett, E., Merino, E., Huerta, A., Treviño-Quintanilla, L., Collado-Vides, J.: RegulonDB (version 6.0): gene regulation model of escherichia coli k-12 beyond transcription, active (experimental) annotated promoters and textpresso navigation. Nucleic Acids Research 36(Database Issue), 120–124 (2008)
12. Gebser, M., Guziolowski, C., Ivanchev, M., Schaub, T., Siegel, A., Thiele, S., Veber, P.: Repair and prediction (under inconsistency) in large biological networks with answer set programming. In: International Conference on Principles of Knowledge Representation and Reasoning, pp. 497–507 (2010)
13. Gebser, M., Kaufmann, B., Kaminski, R., Ostrowski, M., Schaub, T., Schneider, M.: Potassco: the Potsdam answer set solving collection. AI Communications 24(2), 107–124 (2011)
14. Gebser, M., Kaufmann, B., Schaub, T.: The Conflict-Driven Answer Set Solver clasp: Progress Report. In: Erdem, E., Lin, F., Schaub, T. (eds.) LPNMR 2009. LNCS, vol. 5753, pp. 509–514. Springer, Heidelberg (2009)

15. Gebser, M., König, A., Schaub, T., Thiele, S., Veber, P.: The BioASP library: ASP solutions for systems biology. In: IEEE International Conference on Tools with Artificial Intelligence, pp. 383–389 (2010)
16. Gebser, M., Schaub, T., Thiele, S., Veber, P.: Detecting inconsistencies in large biological networks with answer set programming. Theory and Practice of Logic Programing 11(2-3), 323–360 (2011)
17. Gregory, P., Fox, M., Long, D.: A New Empirical Study of Weak Backdoors. In: Stuckey, P.J. (ed.) CP 2008. LNCS, vol. 5202, pp. 618–623. Springer, Heidelberg (2008)
18. Guziolowski, C., Veber, P., Le Borgne, M., Radulescu, R., Siegel, A.: Checking consistency between expression data and large scale regulatory networks: a case study. Journal of Biological Physics and Chemistry 7(2), 37–43 (2007)
19. Hebrard, E., Hnich, B., O'Sullivan, B., Walsh, T.: Finding diverse and similar solutions in constraint programming. In: AAAI Conference on Artificial Intelligence, pp. 372–377 (2005)
20. Hemery, F., Lecoutre, C., Sais, L., Boussemart, F.: Extracting MUCs from constraint networks. In: European Conference on Artificial Intelligence, pp. 113–117 (2006)
21. Hsu, E.I., Muise, C.J., Christopher Beck, J., McIlraith, S.A.: Probabilistically Estimating Backbones and Variable Bias: Experimental Overview. In: Stuckey, P.J. (ed.) CP 2008. LNCS, vol. 5202, pp. 613–617. Springer, Heidelberg (2008)
22. Jeong, H., Tombor, B., Albert, R., Oltvai, Z., Barabási, A.-L.: The large-scale organization of metabolic networks. Nature 407(6804), 651–654 (2000)
23. Kilby, P., Slaney, J., Thiébaux, S., Walsh, T.: Backbones and backdoors in satisfiability. In: AAAI Conference on Artificial Intelligence, pp. 1368–1373 (2005)
24. Marques-Silva, J.: Minimal unsatisfiability: models, algorithms and applications. In: IEEE International Symposium on Multiple-Valued Logic, pp. 9–14 (2010)
25. Marques-Silva, J., Manquinho, V.: Towards More Effective Unsatisfiability-Based Maximum Satisfiability Algorithms. In: Kleine Büning, H., Zhao, X. (eds.) SAT 2008. LNCS, vol. 4996, pp. 225–230. Springer, Heidelberg (2008)
26. Marques-Silva, J., Mikoláš, J., Lynce, I.: On computing backbones of propositional theories. In: European Conference on Artificial Intelligence, pp. 15–20 (2010)
27. Menaï, M.: A two-phase backbone-based search heuristic for partial MAX-SAT – an initial investigation. In: Industrial and Engineering Applications of Artificial Intelligence and Expert Systems, pp. 681–684 (2005)
28. Monasson, R., Zecchina, R., Kirkpatrick, S., Selman, B., Troyansky, L.: Determining computational complexity from characteristic 'phase transitions'. Nature 400(6740), 133–137 (1999)
29. Siegel, A., Radulescu, O., Le Borgne, M., Veber, P., Ouy, J., Lagarrigue, S.: Qualitative analysis of the relation between DNA microarray data and behavioral models of regulation networks. Biosystems 84(2), 153–174 (2006)
30. Slaney, J., Walsh, T.: Backbones in optimization and approximation. In: International Joint Conference on Artificial Intelligence, pp. 254–259 (2001)
31. Soliman, S.: Constraint programming for the dynamical analysis of biochemical systems – a survey. Technical Report Deliverable 1.6, ANR CALAMAR, ANR-08-SYSC-003 (2011)
32. Soulé, C.: Mathematical approaches to differentiation and gene regulation. Comptes Rendus Biologies 329(1), 13–20 (2006)
33. Zhang, W., Rangan, A., Looks, M.: Backbone guided local search for maximum satisfiability. In: International Joint Conference on Artificial Intelligence, pp. 1179–1184 (2003)
34. Zhu, C., Weissenbacher, G., Sethi, D., Malik, S.: SAT-based techniques for determining backbones for post-silicon fault localisation. In: IEEE International High Level Design Validation and Test Workshop, pp. 84–91 (2011)

Properties of Energy-Price Forecasts
for Scheduling

Georgiana Ifrim, Barry O'Sullivan, and Helmut Simonis

Cork Constraint Computation Centre
Department of Computer Science, University College Cork, Ireland
{g.ifrim,b.osullivan,h.simonis}@4c.ucc.ie

Abstract. Wholesale electricity markets are becoming ubiquitous, offering consumers access to competitively-priced energy. The cost of energy is often correlated with its environmental impact; for example, environmentally sustainable forms of energy might benefit from subsidies, while the use of high-carbon sources might be discouraged through taxes or levies. Reacting to real-time electricity price fluctuations can lead to high cost savings, in particular for large energy consumers such as data centres or manufacturing plants. In this paper we focus on the challenge of day-ahead energy price prediction, using the Irish Single Electricity Market Operator (SEMO) as a case-study. We present techniques that significantly out-perform SEMO's own prediction. We evaluate the energy savings that are possible in a production scheduling context, but show that better prediction does not necessarily yield energy-cost savings. We explore this issue further and characterize, and evaluate, important properties that an energy price predictor must have in order to give rise to significant scheduling-cost savings in practice.

1 Introduction

Short-term forecasting of electricity prices is relevant for a range of applications, from operations scheduling to designing bidding strategies. For example, an increasing amount of work has recently focused on the impact of energy-aware schedules for operating data centres and production lines [7,8,18,22,27], where tasks are scheduled based on technical requirements and energy-saving objectives. Furthermore, [22] shows that simple scheduling strategies that can react to the spot-electricity price can save existing systems millions of dollars every year in electricity costs. Since energy costs constitute the largest proportion of a data centre's or production line's running costs [14], designing energy-price-saving schedules becomes critical. A number of recent papers [5,17,19,26] focus on reducing both power usage and power cost by taking real-time energy price into consideration. The scheduling problems addressed differ, e.g., [5] focuses on scheduling in data centres, where tasks can be executed at various time slots.

Prior work has shown that electricity spot prices are one of the most challenging types of time series in terms of simulation and forecasting [4,16,21,29].

M. Milano (Ed.): CP 2012, LNCS 7514, pp. 957–972, 2012.

A large number of techniques have been employed to predict energy prices; see [4,29] for a review. Time series models, Neural Networks (NN) and Support Vector Machines (SVM) were proven to have some degree of success depending on the actual design and volatility of the markets. For example, [6] has studied NN models for forecasting prices in the Pennsylvania, New Jersey, Maryland as well as the Spanish market. NN and SVM were employed in [3,12,23] for forecasting prices in the New England, Ontario and the Australian markets. SVM models were generally shown to perform similarly to the NN models, while often being more scalable and more accurate than competitors. An interesting aspect of price forecasting is that the user may not neccesarily require a good numeric estimate of the actual price, but rather need a good estimation of the price class, e.g., is the price higher or lower than a pre-defined user-threshold. This problem was recently addressed in [29] which trained SVM-based price classification models for the Ontario and Alberta markets.

In this paper, we analyze the Irish electricity market, from market design and publicly available data, to several machine learning modelling strategies and their impact on day-ahead energy price forecasting, as judged by classical error measures (e.g., Mean Squared Error). Furthermore, we investigate the effect of using the proposed forecasting models for designing energy-price-saving schedules. To the best of our knowledge, this is the first analysis of the impact of various properties of day-ahead electricity price forecasts on price-aware scheduling.

Under EU initiatives Ireland has an obligation to supply at least 20% of its primary energy consumption from renewable sources by 2020 [10]. The Irish government has set the following targets in 2007 for its energy usage: no oil in electricity generation by 2020, 15% of electricity from renewable resources by 2010, and 33-40% by 2020. Wind is the most abundant renewable energy source available in Ireland [10,13]. However, introducing such renewable energy sources introduces volatility into the market, making energy price prediction and cost-efficient planning considerably more challenging.

We build on prior literature for short-term (numeric) price forecasting and price classification, and investigate SVM models for these tasks. We propose two SVM models that reduce the numeric price forecasting error of the market operator by 24-28% (Mean Squared Error). Furthermore, we investigate the usage of these models for price classification and for designing price-aware operation schedules. We evaluate the savings that are possible in a production scheduling context, but show that better numeric prediction does not necessarily yield energy-cost savings. We explore this issue further and characterize, and evaluate, important properties that an energy price predictor must have in order to give rise to significant scheduling-cost savings in practice. Therefore, this paper brings together the disciplines of machine learning and combinatorial optimisation to study the most appropriate types of energy price prediction models to use in a scheduling context. In addition, this paper considers this problem in a real-life setting: the Irish energy market.

2   The Irish Electricity Market

The Irish electricity market is an auction-based market, with spot prices being computed every half-hour of a trading day. The methodology for calculating the price in the Irish all-island market is as follows: every half-hour of the trading day, the Single Electricity Market Operator (SEMO)[1] calculates the System Marginal Price (SMP). The SMP has two components: the *Shadow Price* representing the marginal cost per 1MW of power necessary to meet demand in a particular half-hour trading period, within an unconstrained schedule, i.e., no power transmission congestions; and the *Uplift* component, added to the Shadow Price in order to ensure the generators recover their total costs, i.e., start-up and no-load costs [16].

One day ahead of the trade-day the generators have to submit technical and commercial offer data [24]: incremental price-quantity bids summarizing how much supply for what price does a generator offer to provide every half-hour, and the technical specifications of the generator such as start-up costs, maximum capacity, minimum on/off times. Only price-making generator units, that are not under test, are represented individually within the Market Scheduling and Pricing (MSP) software. Non-price making units are scheduled either based on submitted nominations or forecast data in the case of wind units. Once the load met by generation from non-price making units has been removed, then price making generator units are scheduled in merit order according to their bids to meet the remaining load. The SMP is bounded by a Market Price Cap (€1000/MWh) and a Market Price Floor (€-100/MWh), which are set by the Regulatory Authorities.

Two runs of the Market Scheduling and Pricing Software are particularly relevant each trading day. The Ex-Ante (EA) run is carried out one day prior to the trade date which is being scheduled and as such uses entirely forecast wind and load data. A schedule of half-hourly forecasted SMP, shadow price, load and wind generation is produced by the market operator (SEMO) for the coming trade-day. The Ex-Post Initial (EP2) run is carried out four days after the trade date which is being scheduled, and as such is able to utilize full sets of actual wind and load data with no forecast values. The system marginal prices produced in the EP2 run are used for weekly invoicing and the SMP determined in the EP2 run for a given half hour trading period is the price applicable to both generators and suppliers active in such a trading period.

3   Price Forecasting Models

We present an approach to building price forecasting models for the Irish electricity market: the factors influencing the price, data collection and appropriate forecasting models. We show that we significantly improve the forecast of the market operator, thus providing a more reliable price prediction for the next trade-day.

---

[1] http://www.sem-o.com

3.1   Data Collection and Analysis

We begin our study of the Irish electricity market by analyzing the price and load (i.e., demand) profile from January 2009 to June 2011. Figure 1 shows the actual half-hourly price (top frame) and demand (bottom frame). We notice that the load profile is fairly similar over time, showing clear periodicity, with higher load in the cold months. The price is much more volatile, with high variations during both cold and warm months. We also notice that the price volatility and magnitude increased considerably from 2009 to 2011. Table 1 shows some statistics about the price in this period. We notice the increasing median and average price, as well as the increased price volatility (captured by the standard deviation) over time. We believe this could be explained by increasing fuel prices and the ramp-up of wind-generated power, as well as other factors, such as a higher percentage of unscheduled generator outages in 2011.

Table 1. Statistics of the Irish SMP for 2009 to mid-2011

Year	Min	Median	Mean	Stdev	Max
2009	4.12	38.47	43.53	24.48	580.53
2010	-88.12	46.40	53.85	35.49	766.35
2011	0	54.45	63.18	35.79	649.48

Although there seems to be some correlation between demand and price, the price spikes seem to be highly influenced by a combination of additional factors. For example, the half-hourly demand is typically covered by the available wind-generated power, with the remaining load being covered using the generator bids sorted by price merit, with generators using more expensive fuel having higher price bids. If the forecasted load, wind-power and expected supply quantities are unreliable, due to the poor quality of the forecast or unexpected outages, this will affect the market operator scheduling of generators, leading to price volatility. Figure 2 offers a closer look at the price versus load pattern in 2011, for the first week of the year and the week with the maximum price up to mid-2011.

SEMO provides a web interface for public access to the historical SMP, Shadow Price and load details back to January 2008. In November 2009, SEMO started providing day-ahead half-hourly forecasts for load, SMP, Shadow Price and available wind-supply. Considering the changing price profile starting from 2009 to mid-2011 and the later availability of the forecasts (beginning 2010), we decided to use data starting January-2010 to June-2011 for training and evaluating price forecasting models. More concretely, we use year 2010 for training, three months of 2011 (January, March and May) for validation (i.e., calibrating model parameters), and another three months of 2011 (February, April and June) for testing. The choice of training, validation and test is made in order to respect the time dependency in which we train on historical data of the past and forecast prices into the future, as well as testing on months from different seasons to avoid the bias of forecasting prices for summer or winter months exclusively (since prices

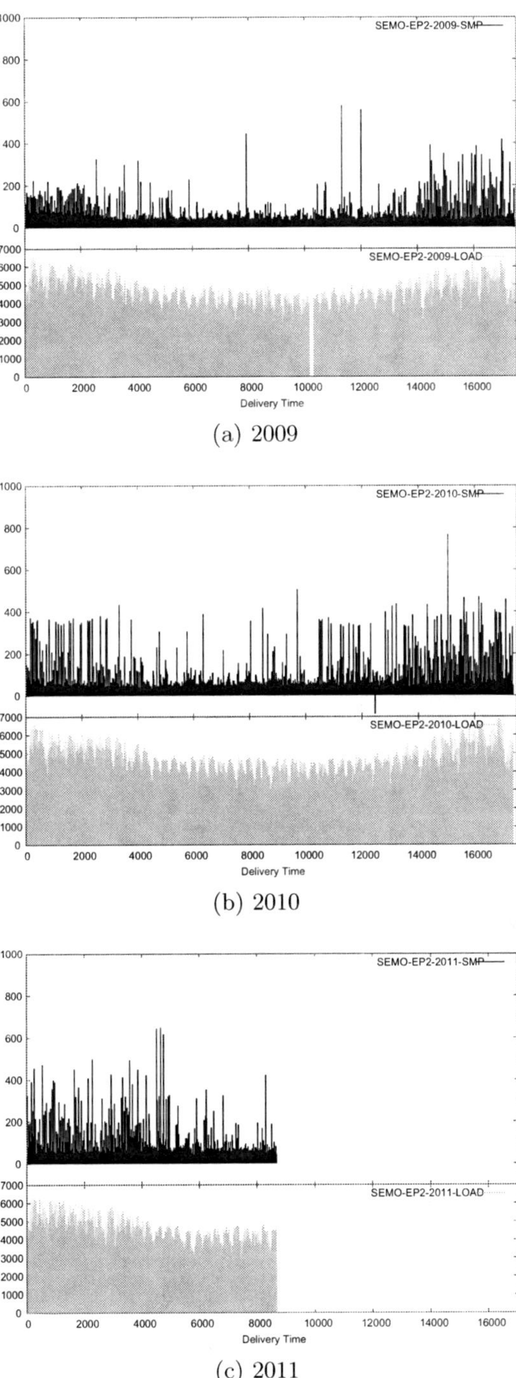

(a) 2009

(b) 2010

(c) 2011

Fig. 1. Half-hourly price (top-black) and demand (bottom-gray) from January-2009 to June-2011. The X axis represents the delivery time (every half-hour of a trading day). The Y axis for the top-black plot represents the SMP in €/MWh and for the bottom-gray plot, the Load in MWh.

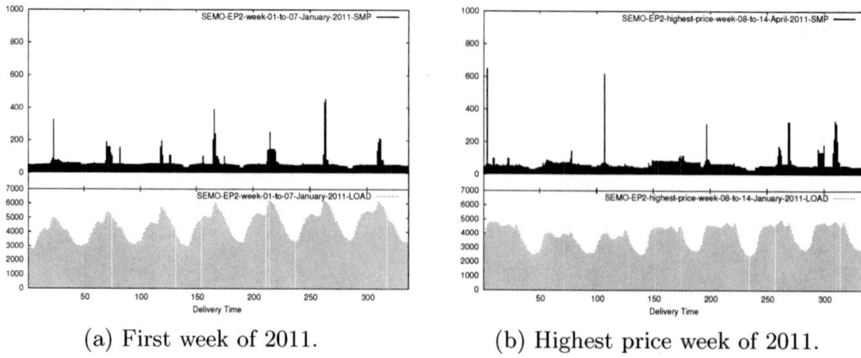

(a) First week of 2011.   (b) Highest price week of 2011.

Fig. 2. Half-hourly price (top-black) and demand (bottom-gray) for two weeks in 2011

in the winter tend to be more volatile than prices in the summer). We also paid attention to the detail that in the Irish electricity market, the actual values for SMP, load, etc., are made available only four days after the tradeday, which means that for prediction we can only use historical data with a gap of four days back into the past from the current day. All evaluations of our models and comparisons to the SEMO price forecasts are done on the three test months that are not otherwise used in any way during training or validation.

We began our analysis with simple models, using only the historical SMP for predicting the SMP of the next tradeday. We then gradually introduced information about the Shadow Price, load and wind-generation and studied the effect of each of these new variables on the prediction quality. Additionally, we investigated the impact of weather forecasts and calendar information (weekend, bank or school holidays) on the quality of the models. In order to estimate the expected supply, we have extracted information on the daily generator bids and planned generator outages available from SEMO and Eirgrid [24,11]. Information about demand and supply is important since price peaks are typically an effect of the mismatch between high load and low supply. Our data integrity checks revealed missing days/hours in the original SEMO data. We have filled in the missing half-hours by taking the data of the closest half-hour. The data collected was available in different granularity (e.g., wind-supply obtained from Eirgrid was sampled every 15 mins) and units (Eirgrid wind-supply was in MW vs SEMO in MWh); we aggregated it to half-hourly granularity and converted the data to same units (MWh). Since we rely on SEMO forecasts for building our models, we estimated the SEMO forecast quality for each of the variables involved: load, wind, shadow price, and SMP. Our evaluation on the training set showed that in terms of forecast quality, the load forecast is most reliable, followed by shadow price, SMP and wind. In our models we use local forecast-quality-estimates as additional features. All the data collected is available online in csv files.[2]

---

[2] http://4c.ucc.ie/~gifrim/Irish-electricity-market

3.2 Machine Learning Models

We have investigated a range of regression models (e.g., linear regression, linear SVM, various kernel-SVM), as well as analysed a variety of features and feature combinations, and present here the two best approaches.

Model 1: Predicting the SMP Using Historical and Forecasted SMP, Shadow Price, Load and Supply. This approach follows the classical line of price prediction in international electricity markets where the main idea is to use historical data, e.g., past prices, load, and supply, for training a price model for the next trade-day. Since machine-learning models were shown to outperform other techniques such as time series models [3,23,29], we focus on non-linear regression techniques for building price forecasting models (e.g., Support Vector Machines).

From the time series data, we extract regression vectors as follows. For each half-hour of a tradeday, we take the actual SMP as a prediction target and use historical data for the same half-hour in the past as features. For example, if the SMP on 1st of January 2010, 7 AM, is € 31.04/MWh, we take this as a learning target and the SMP at 7AM of D past days as features (in this case the most recent historical data is from 27 December 2009, due to the 4 days gap). The number of historical days D is a parameter of the model and is calibrated using the validation dataset.

Since we also have access to day-ahead forecasts of SMP, shadow price, load, wind and other-supply, we study those as additional features. We have additionally investigated weather and calendar information as features, but these have not increased the quality of the model. This may happen since calendar and weather data is already factored into the load and wind-supply forecast, thus it does not add new information to the model. We compute estimates of the weekly and daily available supply from the information on outages and generator bids publicly available from [24,11]. From Eirgrid, we use information on planned outages to estimate the weekly maximum available supply based on the maximum capacity of the available generating plants. From the day-ahead generator bids, we extract features on daily available supply. For example, we set thresholds on the maximum bid price (e.g., € 40) in order to obtain estimates of expected *cheap* supply. The maximum price thresholds of bids are set at € 40, € 50 and € 60, based on the bids and empirical SMP distribution on the validation set (Figure 3). Once the data required for preparing features is processed, we scale all features and use an SVM with an `rbf` kernel for learning. We use the LIBSVM package [9] widely accepted as state-of-the-art for SVM implementations.

Model 2: Predicting the SMP Using the Local Average-SMP and a Learned Difference-from-Average-Model. This approach moves away from the more traditional Model 1, presented above, and builds on the following observation: the actual historical SMP is a good indicator for the *average* electricity price at a given half-hour, but does not capture the particular behaviour of a

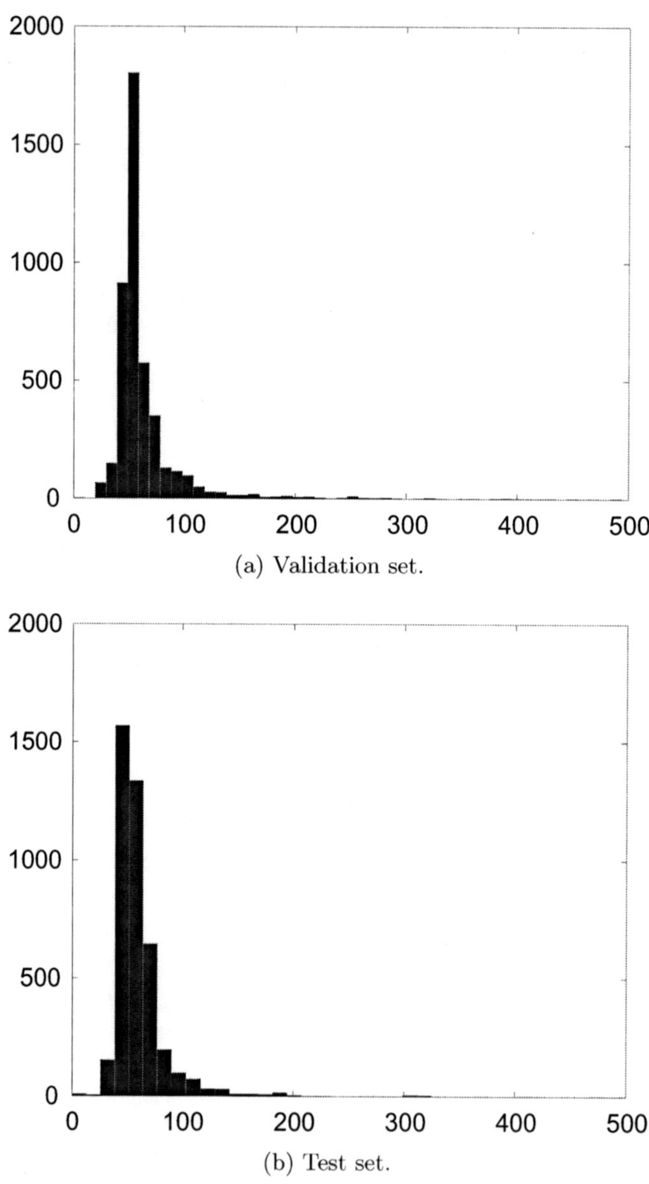

(a) Validation set.

(b) Test set.

Fig. 3. Empirical SMP distribution in our datasets. The X axis represents the SMP truncated at € 500. The Y axis represents the frequency bins for the SMP values.

given day in terms of the magnitude of the SMP peaks and valleys. It may be that due to the particular features of the next tradeday (e.g., strong wind, lower load, enough cheap supply), the SMP stays more or less flat, without exceptional peaks or valleys, thus using the local average SMP itself as an estimate is not sufficient for good prediction quality. Nevertheless, we can estimate the characteristics of the next tradeday using the publicly available forecasts. Hereby, we propose computing the final SMP as a sum of a locally computed average-SMP (e.g., over the last $D = 7$ days) and a learned SMP-difference from the average, estimated from the training set, capturing whether the SMP is going up or down with respect to the average. For example, for forecasting the SMP on 1st of January 2010, 7AM (equal to € 31.04), we use the local average-SMP (equal to € 29.57) over the most recent seven days, respecting the four day gap explained above, as a first component. The second component is the learned-difference between the actual SMP and the average-SMP. As learning features we use the difference between the forecasted tradeday characteristics (load, wind-supply, shadow price) from their local averages. Intuitively, lower than average load and higher than average wind, should trigger a decrease in price, thus a negative difference of SMP from the average. A regression model can estimate the SMP-difference from the differences of its features. We finally compute the SMP as the sum of the local average and the predicted difference for the SMP.

Since this model relies heavily on the quality of the forecasts, for each forecasted variable used in the model, e.g., load, wind, smp, shadow price, we have an additional feature capturing the local quality of that forecast (i.e., we measure over the past week the MSE between true and forecasted value for load, wind, etc.). There is a large body of research literature on using uncertain data for learning [1,2,28]. At the moment we use this simple approach for dealing with forecast uncertainty, and we plan to investigate more advanced approaches in our future work. Our experiments show that even this simple approach of integrating variable uncertainty leads to considerable model improvement.

Evaluation. We use the Mean Squared Error (MSE) as a primary means for evaluating the quality of price forecasts. This is a classical measure of both bias and variance of the models [20]. It typically penalizes gross over or underestimates of the actual values. Additionally we show the Mean Absolute Error (MAE) and the Skill-Score. The Skill-Score indicates the fractional improvement in the MSE over a reference model [20]. We use SEMO (the market operator's forecast) as a reference model, and show the Skill-Score for our forecasts. The parameters of our models are optimized for minimizing the MSE, and can be calibrated for any quality measure the user finds fit.

Table 2 shows the evaluation of our price forecasts (FM1 and FM2), using the market operator's forecast (SEMO) as a strong[3] baseline. The two forecasting models we proposed show between 24-28% improvement over the MSE of the SEMO price forecast.

---

[3] For building price forecasts, SEMO uses more data than what is publicly available.

Table 2. Half-hourly price forecasts: error rates for SEMO and our forecasts FM1 and FM2 with respect to the true price

Method	MAE	MSE	Skill-Score
SEMO	12.64	**1086.25**	NA
FM1	11.14	**821.01**	0.24
FM2	11.21	**781.72**	0.28

Paired t-tests on the MSE and MAE (at 0.95 confidence level) show our price forecasts are statistically significantly better than those of SEMO. Table 3 gives details of the confidence interval of the MSE for all three forecasts.

Table 3. T-tests details for the three price forecasts. Each row is a baseline. Each column (SEMO, FM1, FM2) is compared against it. The upper (U) and lower (L) limits of the 95% confidence intervals are shown.

Baseline Price		SEMO	FM1	FM2
Actual	L	761.8	513.5	486.9
	U	1410.7	1128.4	1076.4
SEMO	L	-	172.4	209.7
	U	-	358.0	399.3
FM1	L	-	-	11.5
	U	-	-	66.9

So far we have focused on developing and analyzing electricity price forecasts with respect to classical error measures (e.g., MSE, MAE). Next, we focus on the effect of these forecasts on price-aware scheduling.

4 Price-Aware Scheduling Model

To test the quality of the price forecasts on a realistic scheduling problem, we adapted a variant of the feedmill scheduling problem from [25]. The schedule is generated from orders on the current day for delivery in the next morning. Tasks i are scheduled on four disjunctive press lines with their allocated machine m_i, duration d_i, power requirement p_i and due date e_i, satisfying an overall power limit l_t at each time point t. We express the problem as a mixed integer programming (MIP) minimization problem following [15], where the main decision variables are 0/1 integers x_{it} indicating whether task i starts at time t, and non-negative, continuous variables pr_t denoting the power use at time t. For this evaluation we choose the MIP formulation over a more conventional constraint programming model, as it allows us to find the optimal solutions for the core problem. The objective function is based on the predicted price v_t, while the evaluation of the quality uses the actual price a_t. This corresponds to a scenario whereby a forecast price is available 24 hours in advance, but the price paid will be the actual market price. As executing the schedule requires significant preparation work, we cannot continuously reschedule based on the current, actual price. Therefore, the energy cost of a schedule is computed as:

$$\text{cost} = \sum_t pr_t^* a_t \tag{1}$$

where pr_t^* is the profile value at time t of the optimal solution to the following MIP problem:

$$\min \sum_t pr_t v_t \tag{2}$$

subject to:

$$\forall_i : \qquad \sum_t x_{it} = 1 \tag{3}$$

$$\forall_t : \sum_i \sum_{t-d_i+1 \le t' \le t} p_i x_{it'} = pr_t \le l_t \tag{4}$$

$$\forall_m \forall_t : \quad \sum_{i|m_i=m} \sum_{t-d_i+1 \le t' \le t} x_{it'} \le 1 \tag{5}$$

$$\forall_i \forall_{t|t+d_i>e_i} : \qquad x_{it} = 0 \tag{6}$$

We generated problem instances randomly, filling each production line to capacity for 24 hours, choosing random durations uniformly between 25 and 100 minutes, the last task generated will be truncated to fit into 24 hours, and power requirements uniformly chosen between 0 and 200 kW. The time resolution was set to 5 minutes, so that optimal solutions could be found within a 10 minute timelimit. For each prediction day, from the same test period used for evaluating the price forecasts, we generated 10 samples, in total 880 runs. For each instance we computed the actual cost based on an optimal solution for the actual price, for the SEMO forecast and for our two forecasts (FM1, FM2). The schedule based on the actual price provides a lower bound, but since the actual price is not known in advance, it is not realizable.

We also experimented with another scenario, where each production line will be busy for only 12 hours. This allows us to avoid the peak price periods completely, which presents a much less challenging problem. The schedule overhead decreases accordingly, and the results are similar to the ones presented here.

Table 4 shows summary results over all sample runs. It provides statistics of the scheduling cost for the different price forecasts. We ran paired t-tests to assess which price forecasts lead to significantly cheaper scheduling. We found that the costs using all the three forecasts are very close to the optimal cost (within

Table 4. Summary Results of Price-Aware Schedule Costs

Price	Min	Median	Mean	Max
Actual	4,383,718	5,934,654	6,093,365	9,805,821
SEMO	4,507,136	6,054,220	6,272,768	10,218,804
FM1	4,499,811	6,058,093	6,266,800	10,070,541
FM2	4,570,552	6,094,818	6,283,261	10,059,264

Table 5. Confidence intervals (95%) for price-aware scheduling costs comparing optimal solutions priced using actual price, and each of the three forecasts (SEMO, FM1, FM2). Baseline for comparison is the method identified on each row.

Price		SEMO	FM1	FM2
Actual	L	$-200,564.9$	$-193,646.7$	$-211,094.4$
	U	$-158,241.3$	$-153,222.5$	$-168,697.4$
SEMO	L	-	$-1,506.1$	$-17,262.6$
	U	-	$13,443.1$	$-3,722.9$
FM1	L	-	-	$-23,968.3$
	U	-	-	$-8,954.2$

5-10%), and that neither FM1 or FM2 forecasts were significantly better than SEMO. In fact, the price forecast with best MSE (FM2, as shown in Table 2) was significantly worse than the other two with respect to the energy cost of the optimal schedule. Table 5 shows the confidence intervals for the average difference between the optimal schedule cost and the costs obtained using the forecasts. This was a somewhat surprising result: the best numeric forecasting model (as judged with respect to classical learning measures, e.g., MSE), was the worst model with respect to scheduling cost. We analyse in the following section what is the correct approach to significantly reduce the energy-costs in a scheduling context.

5 Properties of Energy-Price Forecasts for Scheduling

We analyze here the key properties of price forecasts that positively affect cost-aware scheduling. What seems to matter most for scheduling is that the forecasting model captures the price-trend rather than the exact real value (as measured by MSE). Therefore, forecasting models that capture well the peaks and valleys of the energy price have better behaviour when used for scheduling. To study this hypothesis, we analysed the three previous forecasts in a classification framework, where prices belong to one of two classes: *peak* or *low* price. The class is decided using a threshold inferred from the empirical distribution of the price on the validation set. We set the threshold at the 66th percentile (about €60, see Figure 3). Thus, if a price is above the threshold, it is in the peak class, otherwise it is in the low class.

Analyzed in this context, all three forecasts have similar classification accuracy, about 78%. This could explain the lack of difference with respect to effect on the energy cost of the schedule. To further study this hypothesis, we performed the following experiment. Starting from the SEMO price forecast (with 78% classification accuracy), we artificially obtained better peak-price classifiers by correcting the classification error. There are two types of classification error: false positives (missing lows) and false negatives (missing peaks), and we believe that for scheduling it is more important to reduce the false negatives, than the false positives. To test this, we have first corrected 50% of the false positives (if the classifier predicts *peak*, but the truth is *low*, we replace the SEMO

Table 6. Confidence intervals for scheduling-costs of price forecasts with increasingly better peak-price classification accuracy. FP-82% stands for forecast obtained by correcting the false positive error, to obtain a forecast with 82% classification accuracy. Statistically significant improvements are highlighted in bold.

Price	SEMO-78%	FP-82%	FP-86%
SEMO-78% L	-	**19,610.6**	**29,274.0**
SEMO-78% U	-	**28,795.2**	**39,735.0**
FP-82% L	-	-	**7,388.3**
FP-82% U	-	-	**13,214.8**

Table 7. Confidence intervals for scheduling-costs of price forecasts with increasingly better peak-price classification accuracy. FN-82% stands for forecast obtained by correcting the false negative error, to obtain a forecast with 82% classification accuracy. Statistically significant improvements are highlighted in bold.

Price	SEMO-78%	FN-82%	FN-86%
SEMO-78% L	-	**30,223.1**	**46,582.4**
SEMO-78% U	-	**66,446.1**	**86,151.5**
FN-82% L	-	-	**9,604.2**
FN-82% U	-	-	**26,460.4**

price with the true price), and then 100% of the false positives, to obtain two classifiers with 82% and 86% classification accuracy and their associated price forecasts. Similarly, starting from SEMO's forecast, we have corrected the same number of errors as before, but this time from the false negatives, to obtain two more classifiers with 82% and 86% accuracy. Finally, we have used SEMO versus the four improved forecasts for cost-aware scheduling. The results show that improved classification accuracy leads to reduced scheduling cost, and that the type of classification error matters, with false negatives having more impact on scheduling. Paired t-tests on the scheduling costs obtained with the different forecasts showed that the improvement is statistically significant with very high confidence (higher than 0.99). The schedule-cost improvement over SEMO for the first two artificially improved forecasts is between 0.4-0.5%, and for the second two between 0.8-1.0%. Given that SEMO was already within 5% of the optimum schedule cost, this is a significant improvement. Tables 6 and 7 give detailed results.

In conclusion, we show that developing good regression techniques where quality is evaluated using traditional learning measures is not effective for cost-aware scheduling. We have also tested peak-price classification models trained similarly to [29], but the thresholded regression models had slightly better classification accuracy. We believe one needs to rather focus on various types of *cost-sensitive* peak-price classifiers and their impact on scheduling cost, where ideally learning should inform scheduling decisions and scheduling should inform learning decisions (i.e., the classifications costs have to be motivated by the scheduling application).

6 Conclusion

We have shown that using classical price-prediction features and machine learning techniques, one can obtain better price forecasts with respect to classical error measures (e.g., MSE, MAE). When plugging the improved price forecasts into cost-aware scheduling, we nevertheless do not observe the same benefit on the schedule-cost. This suggests that scheduling requires specific features from the price forecasts. This paper focuses on pruning the large space of learning strategies to identify the most promising models for designing energy-efficient schedules. We have shown that good peak-price classification behaviour is an important model property and that the type of classification error directly affects cost-aware scheduling. This opens new research directions towards designing cost-sensitive price classification forecasts for scheduling.

Our scheduling experiments also give some insights into the usefulness of alternative tariff models. Besides a time-variable tariff based on the actual market price, tariffs based on long- or short-term price prediction have been proposed under the term *time-of-use* tariffs. In this case the customer knows in advance the price to be paid, and the provider carries the risk of a wrong prediction. In our scheduling problem, the value of the optimal solution based on the forecast will then give the final cost of the schedule. This optimal value can be above or below the cost based on the actual prices, but shows much higher variability as compared to the cost using the actual market price. From our preliminary experiments, it therefore seems preferable to use tarrifs based on the actual price. We plan to investigate this further in our future work.

Acknowledgements. This work was funded by Science Foundation Ireland under Grant 10/IN.1/I3032.

References

1. Aggarwal, C., Yu, P.: A survey of uncertain data algorithms and applications. IEEE Transactions on Knowledge and Data Engineering 21(5), 609–623 (2009)
2. Aggarwal, C.C.: On multidimensional sharpening of uncertain data. In: Proceedings of the SIAM International Conference on Data Mining (SDM), pp. 373–384 (2010)
3. Aggarwal, S.K., Saini, L.M., Kumar, A.: Day-ahead price forecasting in ontario electricity market using variable-segmented support vector machine-based model. Electric Power Components and Systems 37(5), 495–516 (2009)
4. Aggarwal, S.K., Saini, L.M., Kumar, A.: Electricity price forecasting in deregulated markets: A review and evaluation. International Journal of Electrical Power and Energy Systems 31(1), 13–22 (2009)
5. Aikema, D., Kiddle, C., Simmonds, R.: Energy-cost-aware scheduling of hpc workloads. In: IEEE International Symposium on a World of Wireless, Mobile and Multimedia Networks, WoWMoM 2011, pp. 1–7 (June 2011)
6. Amjady, N., Keynia, F.: Day-ahead price forecasting of electricity markets by mutual information technique and cascaded neuro-evolutionary algorithm. IEEE Transactions on Power Systems 24(1), 306–318 (2009)

7. Bodík, P., Griffith, R., Sutton, C., Fox, A., Jordan, M., Patterson, D.: Statistical machine learning makes automatic control practical for internet datacenters. In: Proceedings of the 2009 Conference on Hot Topics in Cloud Computing, HotCloud 2009. USENIX Association, Berkeley (2009)

8. Buchbinder, N., Jain, N., Menache, I.: Online Job-Migration for Reducing the Electricity Bill in the Cloud. In: Domingo-Pascual, J., Manzoni, P., Palazzo, S., Pont, A., Scoglio, C. (eds.) NETWORKING 2011, Part I. LNCS, vol. 6640, pp. 172–185. Springer, Heidelberg (2011)

9. Chang, C.-C., Lin, C.-J.: Libsvm: A library for support vector machines. ACM Transactions on Intelligent Systems and Technology 2, 27:1–27:27 (2011)

10. Connolly, D., Lund, H., Mathiesen, B., Leahy, M.: Modelling the existing irish energy-system to identify future energy costs and the maximum wind penetration feasible. Energy 35(5), 2164–2173 (2010)

11. EirGrid (2012), http://www.eirgrid.com/

12. Fan, S., Mao, C., Chen, L.: Next-day electricity-price forecasting using a hybrid network. Generation, Transmission Distribution, IET 1(1), 176–182 (2007)

13. Finn, P., Fitzpatrick, C., Connolly, D., Leahy, M., Relihan, L.: Facilitation of renewable electricity using price based appliance control in ireland's electricity market. Energy 36(5), 2952–2960 (2011)

14. Guenter, B., Jain, N., Williams, C.: Managing cost, performance, and reliability tradeoffs for energy-aware server provisioning. In: Proceedings of the IEEE International Conference on Computer Communications (INFOCOM), pp. 1332–1340 (April 2011)

15. Hooker, J.: Integrated Methods for Optimization. Springer, New York (2007)

16. Jablonska, M., Mayrhofer, A., Gleeson, J.P.: Stochastic simulation of the uplift process for the Irish electricity market. Mathematics-in-Industry Case Studies Journal 2, 86–110 (2010)

17. Kim, T., Poor, H.: Scheduling power consumption with price uncertainty. IEEE Transactions on Smart Grid 2(3), 519–527 (2011)

18. Le, K., Bianchini, R., Martonosi, M., Nguyen, T.: Cost-and energy-aware load distribution across data centers. In: Proceedings of HotPower. Citeseer (2009)

19. Mani, S., Rao, S.: Operating cost aware scheduling model for distributed servers based on global power pricing policies. In: Proceedings of the Fourth Annual ACM Bangalore Conference, COMPUTE 2011, pp. 12:1–12:8. ACM, New York (2011)

20. Pelland, S., Galanis, G., Kallos, G.: Solar and photovoltaic forecasting through post-processing of the global environmental multiscale numerical weather prediction model. Progress in Photovoltaics: Research and Applications (2011)

21. Ptak, P., Jablonska, M., Habimana, D., Kauranne, T.: Reliability of arma and garch models of electricity spot market prices. In: Proceedings of European Symposium on Time Series Prediction (September 2008)

22. Qureshi, A., Weber, R., Balakrishnan, H., Guttag, J., Maggs, B.: Cutting the electric bill for internet-scale systems. In: Proceedings of the ACM SIGCOMM 2009 Conference on Data Communication, SIGCOMM 2009, pp. 123–134. ACM, New York (2009)

23. Saini, L., Aggarwal, S., Kumar, A.: Parameter optimisation using genetic algorithm for support vector machine-based price-forecasting model in national electricity market. Generation, Transmission Distribution, IET 4(1), 36–49 (2010)

24. SEMO (2012), http://www.sem-o.com/

25. Simonis, H.: Models for global constraint applications. Constraints 12(1), 63–92 (2007)

26. Simonis, H., Hadzic, T.: A family of resource constraints for energy cost aware scheduling. In: Third International Workshop on Constraint Reasoning and Optimization for Computational Sustainability, St. Andrews, Scotland, UK (September 2010)
27. Srikantaiah, S., Kansal, A., Zhao, F.: Energy aware consolidation for cloud computing. In: Proceedings of the 2008 Conference on Power Aware Computing and Systems, HotPower 2008, p. 10. USENIX Association, Berkeley (2008)
28. Yang, J.-L., Li, H.-X.: A probabilistic support vector machine for uncertain data. In: IEEE International Conference on Computational Intelligence for Measurement Systems and Applications, CIMSA 2009, pp. 163–168 (May 2009)
29. Zareipour, H., Janjani, A., Leung, H., Motamedi, A., Schellenberg, A.: Classification of Future Electricity Market Prices. IEEE Transactions on Power Systems 26(1), 165–173 (2011)

Aggregating Conditionally Lexicographic Preferences on Multi-issue Domains

Jérôme Lang[1], Jérôme Mengin[2], and Lirong Xia[3]

[1] LAMSADE, Université Paris-Dauphine, France
`lang@lamsade.dauphine.fr`
[2] IRIT, Université de Toulouse, France
`mengin@irit.fr`
[3] SEAS, Harvard University, USA
`lxia@seas.harvard.edu`

Abstract. One approach to voting on several interrelated issues consists in using a language for compact preference representation, from which the voters' preferences are elicited and aggregated. A language usually comes with a domain restriction. We consider a well-known restriction, namely, *conditionally lexicographic preferences*, where both the relative importance between issues and the preference between values of an issue may depend on the values taken by more important issues. The naturally associated language consists in describing conditional importance and conditional preference by trees together with conditional preference tables. In this paper, we study the aggregation of conditionally lexicographic preferences, for several voting rules and several restrictions of the framework. We characterize computational complexity for some popular cases, and show that in many of them, computing the winner reduces in a very natural way to a MAXSAT problem.

1 Introduction

There are many situations where a group of agents have to make a common decision about a set of possibly interrelated issues, variables, or attributes. For example, this is the situation in the following three domains:

- *Multiple referenda*: there is a set of binary issues (such as building a sport centre, building a cultural centre etc.); on each of them, the group has to make a yes/no decision.
- *Committee elections*: there is a set of positions to be filled (such as a president, a vice-president, a secretary).
- *Group product configuration*: the group has to agree on a complex object consisting of several components.

Voting on several interrelated issues has been proven to be a challenging problem from both a social choice viewpoint and a computational viewpoint. If the agents vote separately on each issue, then paradoxes generally arise [6,13]; this rules out this 'decompositional' way of proceeding, except in the restricted case when voters have separable preferences. A second way consists in using a sequential voting protocol: variables are considered one after another, in a predefined order, and the voters know the assignment to the earlier variables before expressing their preferences on later ones (see, e.g., [14,15,2]). This method, however, works (reasonably) well only if we can guarantee that

M. Milano (Ed.): CP 2012, LNCS 7514, pp. 973–987, 2012.

there exists a common order over issues such that every agent can express her preferences unambiguously on the values of each issue at the time he is asked to report them. A third class of methods consists in using a language for compact preference representation, in which the voters' preferences are stored and from which they are aggregated. If the language is expressive enough to allow for expressing any possible preference relation, then the paradoxes are avoided, but at a very high cost, both in elicitation and computation. Therefore, when organizing preference aggregation in multiple interrelated issue, there will always be a choice to be made between (a) being prone to severe paradoxes, (b) imposing a domain restriction or (c) requiring a heavy communication and computation burden.

In this paper, we explore a way along the third class of methods. When eliciting, learning, and reasoning with preferences on combinatorial domains, a domain restriction often considered consists in assuming that preferences are lexicographic. Schmitt et al. [17] address the learning of lexicographic preferences, after recalling that the psychology literature shows evidence that lexicographic preferences are often an accurate model for human decisions [10]. Learning such preferences is considered further in [8,18], and then in [3] who learn more generally *conditionally lexicographic preferences*, where the importance order on issues as well as the local preferences over values of issues can be conditional on the values of more important issues. The aggregation of lexicographic preferences over combinatorial domains has received very little attention (the only exception we know of is [1]). Yet it appears to be – at least in some contexts – a reasonable way of coping with multiple elections. It does imply a domain restriction, and arguably an important one; but, as explained above, domain restrictions seem to be the only way of escaping both strong paradoxes and a huge communication cost, and conditionally lexicographic preference models are not so restrictive, especially compared to the most common domain restriction, namely separability.

The generic problem of aggregating conditionally lexicographic preferences can be stated as follows. The set of alternatives is a combinatorial domain $\mathcal{X}$ composed of a finite set of binary issues.[1] We have a set of voters, each providing a conditionally lexicographic preference over $\mathcal{X}$ under the compact and natural form of a lexicographic preference tree (LP-tree for short) [3], which we will define soon; therefore, a (compactly represented) profile P consists of a collection of LP-trees. Since each LP-tree $\mathcal{L}$ is the compact representation of one linear order $\succ_{\mathcal{L}}$ over $\mathcal{X}$, there is a one-to-one correspondence between P and the (extensively represented) profile P^* consisting of a collection of linear orders over $\mathcal{X}$. Finally, for a given voting rule r, we ask whether there is a simple way to compute the winner, namely $r(P^*)$, where 'simple' means that the winner should be computed directly (and efficiently) from P and in any case we must avoid to produce P^* *in extenso*, which would require exponential space. For many cases where winner determination is computationally hard, we show that these problems can be efficiently converted to MAXSAT problems and thus be solved by sat solvers.

The rest of the paper is organized as follows. Conditionally lexicographic preferences and their compact representation by LP-trees are defined and discussed in Section 2. In Section 3 we state the problem considered in this paper, namely the aggregation of

---

[1] The assumption that variables are binary is made for the sake of simplicity due to the space constraint. Most of our results would easily extend to the non-binary case.

conditionally lexicographic preferences by voting rules. As we will see, some voting rules are better than others in this respect. In the paper we focus on three families of rules. First, in Section 4, k-approval rules: we show that for many values of k, we can give a quite satisfactory answer to our question above, even for our most general models. Note that by 'satisfactory' we do not necessarily mean "computable in polynomial time": for instance, when deciding whether a given alternative is a winner is NP-complete but can be easily translated into a compact maximum (weighted) satisfiability problem, for which efficient algorithms exist, we still consider the answer as (more or less) positive. In Section 5 we then focus on the Borda rule, and show that the answer to our question is satisfactory for some of the simplest LP-tree models, but less so for some general models. We also provide a natural family of scoring rules for which the answer is positive in all cases. Then in Section 6 we consider the existence of a Condorcet winner, and show that for Condorcet-consistent rules, and in particular Copeland and maximin, the answer tends to be negative. Finally, Section 7 is devoted to the specific case of LP-trees with fixed local preferences.

2 Conditionally Lexicographic Preferences and LP-Trees

Let $\mathcal{I} = \{X_1, \ldots, X_p\}$ ($p \geq 2$) be a set of *issues*, where each issue X_i takes a value in a binary *local domain* $D_i = \{0_i, 1_i\}$. The set of alternatives is $\mathcal{X} = D_1 \times \cdots \times D_p$, that is, an alternative is uniquely identified by its values on all issues. Alternatives are denoted by d, e etc. For any $Y \subseteq \mathcal{I}$ we denote $D_Y = \prod_{X_i \in Y} D_i$. Let $L(\mathcal{X})$ denote the set of all linear orders over $\mathcal{X}$.

Lexicographic comparisons order pairs of outcomes (d, e) by looking at the attributes in sequence, according to their importance, until we reach an attribute X such that the value of X in d is different from the value of X in e; d and e are then ordered according to the *local preference* relation over the values of X. For such lexicographic preference relations we need both an *importance* relation, between attributes, and *local preference* relations over the domains of the attributes. Both the importance between attributes and the local preferences may be conditioned by the values of more important attributes. Such lexicographic preference relations can be compactly represented by *Lexicographic Preference trees (LP-trees)* [3], described in the next section.

2.1 Lexicographic Preference Trees

An LP-tree $\mathcal{L}$ is composed of two parts: (1) a tree $\mathcal{T}$ where each node t is labeled by an issue, denoted by $\mathsf{lss}(t)$, such that each issue appears once and only once on each branch; each non-leaf node either has two outgoing edges, labeled by 0 and 1 respectively, or one outgoing edge, labeled by $\{0, 1\}$. (2) A *conditional preference table* $\mathsf{CPT}(t)$ for each node t, which is defined as follows. Let $\mathsf{Anc}(t)$ denote the set of issues labeling the ancestors of t. Let $\mathsf{Inst}(t)$ (respectively, $\mathsf{NonInst}(t)$) denote the set of issues in $\mathsf{Anc}(t)$ that have two (respectively, one) outgoing edge(s). There is a set $\mathsf{Par}(t) \subseteq \mathsf{NonInst}(t)$ such that $\mathsf{CPT}(t)$ is composed of the agent's local preferences over $D_{\mathsf{lss}(t)}$ for all valuations of $\mathsf{Par}(t)$. That is, suppose $\mathsf{lss}(t) = X_i$, then for every valuation u of $\mathsf{Par}(t)$, there is an entry in the CPT which is either $u : 0_i \succ 1_i$ or $u : 1_i \succ 0_i$. For any alternative $d \in \mathcal{X}$, we let the *importance order* of d in $\mathcal{L}$, denoted by $\mathrm{IO}(\mathcal{L}, d)$, to be the order over $\mathcal{I}$ that gives $\boldsymbol{d}$ in $\mathcal{T}$. We use $\triangleright$ to denote an importance order to distinguish it from agents' preferences $\succ$ (over $\mathcal{X}$). If in $\mathcal{T}$, each vertex has no more than one child, then all alternatives have the same importance order $\triangleright$, and we say that $\triangleright$ is the importance order of $\mathcal{L}$.

An LP-tree $\mathcal{L}$ represents a linear order $\succ_{\mathcal{L}}$ over $\mathcal{X}$ as follows. Let $\boldsymbol{d}$ and $\boldsymbol{e}$ be two different alternatives. We start at the root node t_{root} and trace down the tree according to the values of $\boldsymbol{d}$, until we find the first node t^* such that $\boldsymbol{d}$ and $\boldsymbol{e}$ differ on $\mathsf{lss}(t^*)$. That is, w.l.o.g. letting $\mathsf{lss}(t_{\text{root}}) = X_1$, if $d_1 \neq e_1$, then we let $t^* = t_{\text{root}}$; otherwise, we follow the edge d_1 to examine the next node, etc. Once t^* is found, we let $U = \mathsf{Par}(t^*)$ and let d_U denote the sub-vector of $\boldsymbol{d}$ whose components correspond to the nodes in U. In $\mathsf{CPT}(t^*)$, if $d_U : d_{t^*} \succ e_{t^*}$, then $\boldsymbol{d} \succ_{\mathcal{L}} \boldsymbol{e}$. We use $\mathcal{L}$ and $\succ_{\mathcal{L}}$ interchangeably.

Example 1. *Suppose there are three issues. An LP-tree $\mathcal{L}$ is illustrated in Figure 1. Let t be the node at the end of the bottom branch. We have $\mathsf{lss}(t) = X_2$, $\mathsf{Anc}(t) = \{X_1, X_3\}$, $\mathsf{Inst}(t) = \{X_1\}$, $\mathsf{NonInst}(t) = \{X_3\}$, and $\mathsf{Par}(t) = \{X_3\}$. The linear order represented by the LP-tree is $[001 \succ 000 \succ 011 \succ 010 \succ 111 \succ 101 \succ 100 \succ 110]$, where 000 is the abbreviation for $0_1 0_2 0_3$, etc. $\mathrm{IO}(\mathcal{L}, 000) = [X_1 \triangleright X_2 \triangleright X_3]$ and $\mathrm{IO}(\mathcal{L}, 111) = [X_1 \triangleright X_3 \triangleright X_2]$.*

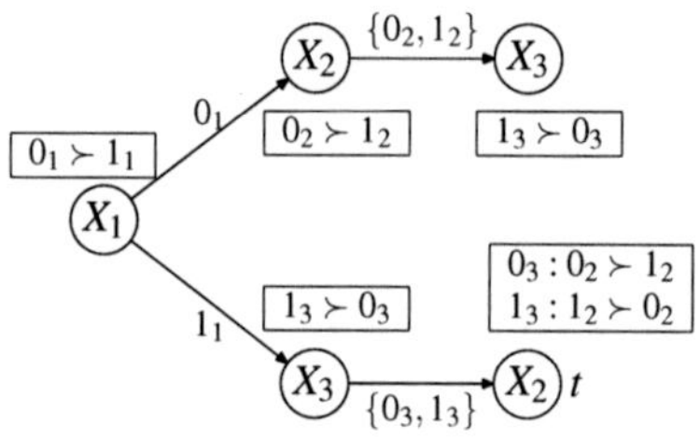

Fig. 1. An LP-tree $\mathcal{L}$

2.2 Classes of Lexicographic Preference Trees

The definition for LP-trees above is for the most general case. [3] also defined some interesting sub-classes of LP-trees by imposing a restriction on the local preference relations and/or on the conditional importance relation.

The local preference relations can be *conditional* (general case, as defined above), but can also be *unconditional* (the preference relation on the value of any issue is independent from the value of all other issues). The most restrictive case is *fixed*, which means that not only are the preferences unconditional, but that they are common to all voters. Formally, UP is the class of LP-trees with *unconditional local preferences*: for every issue X_i there exists a preference relation $\succ_i$ ($1_i \succ_i 0_i$ or $0_i \succ_i 1_i$) and for every node t with $X_i = \mathsf{lss}(t)$, $\mathsf{Par}(t) = \emptyset$, and $\mathsf{CPT}(t) = \{\succ_i\}$. And FP is the class of LP-trees with *fixed local preferences* (FP): without loss of generality, for every node t (with $\mathsf{lss}(t) = X_i$), $\mathsf{CPT}(t) = \{1_i \succ 0_i\}$.

Likewise, the importance relation over issues can be *conditional* (general case), or *unconditional*, of *fixed* when it is common to all voters: (UI) is the set of all linear

LP-trees, i.e., every node has no more than one child. And (FI) is the set of all linear LP-trees with the (unconditional) importance order over issues $[X_1 \rhd \ldots \rhd X_p]$.

We can now combine a restriction on local preferences and a restriction on the importance relation. We thus obtain nine classes of LP-trees, namely, FI-FP, UI-FP, CI-FP, FI-UP, UI-UP, CI-UP, FI-CP, UI-CP, and CI-CP. For instance, UI-CP is defined as the class of all LP-trees with unconditional importance relation and conditional preferences. Note that the FI-FP class is trivial, as it contains a unique LP-tree.

Recall that a LP-tree is composed of a tree and a collection of conditional preference tables. The latter is reminiscent of CP-nets [4]. In fact, it can be viewed as some kind of generalized CP-net whose dependency relations between variables (induced from the importance relation) may be conditional on the values of their parent variables. However, in the case of an unconditional importance relation (UI), then the collection of CP-tables *is* a CP-net, and the LP-tree is a TCP-net [5]. In the general case however, a conditionally lexicographic preferences cannot be represented by a TCP-net.

3   Aggregating LP-Trees by Voting Rules

We now consider n voters. A *(voting) profile* P over a set of alternatives $\mathcal{X}$ is a collection of n linear orders on $\mathcal{X}$. A *voting rule* r maps every profile P to a nonempty subset of $\mathcal{X}$: $r(P)$ is the set of *co-winners* for r and P.

A *scoring function* S is a mapping $L(\mathcal{X})^n \times \mathcal{X} \to \mathbb{R}$. Often, a voting rule r is defined so that $r(P)$ is the set of alternatives maximizing some scoring function S_r. In particular, *positional scoring rules* are defined via a *scoring vector* $v = (v(1), \ldots, v(m))$, where m is the number of alternatives (here, $m = 2^p$): for any vote $V \in L(\mathcal{X})$ and any $c \in \mathcal{X}$, let $S_v(V, c) = v(rank_V(c))$, where $rank_V(c)$ is the rank of c in V; then for any profile $P = (V_1, \ldots, V_n)$, let $S_v(P, c) = \sum_{j=1}^{n} S_v(V_j, c)$. The winner is the alternative maximizing $S_v(P, \cdot)$. In particular, the k-approval rule App_k (with $k \le m$), is defined by the scoring vector $v(1) = \cdots = v(k) = 1$ and $v(k+1) = \cdots = v(m) = 0$, the scoring function being denoted by S_{App}^k; and the *Borda* rule is defined by the scoring vector $(m-1, m-2, \ldots, 0)$, the scoring function being denoted by S_{Borda}.

An alternative α is the *Condorcet winner* for a profile P if for any $\beta \neq \alpha$, a (strict) majority of votes in P prefers α to β. A voting rule is *Condorcet-consistent* if it elects the Condorcet winner whenever one exists. Two prominent Condorcet-consistent rules are *Copeland* and *maximin*. The Copeland winners are the alternatives α that maximize the Copeland score $C(\alpha)$, defined as the number of alternatives β such that a majority of votes in P prefers α to β. The maximin winners are the alternatives α that maximize the maximin score $S_{\text{MM}}(\alpha)$, defined as $S_{\text{MM}}(P, \alpha) = \max\{N_P(\beta, \alpha) : \beta \in \mathcal{X}, \beta \neq \alpha\}$, where $N_P(\beta, \alpha)$ denotes the number of votes in P that rank α ahead of β.

3.1   Voting Restricted to Conditionally Lexicographic Preferences

The key problem addressed in this paper is the following. We know that applying voting rules to profiles consisting of arbitrary preferences on multi-issue domains is computationally difficult. Does it become significantly easier when we restrict to conditionally lexicographic preferences? The question, of course, may depend on the voting rule used.

A *conditionally lexicographic profile* is a collection of n conditionally lexicographic preferences over $\mathcal{X}$. As conditionally lexicographic preferences are compactly represented by LP-trees, we define a *LP-profile P* as a collection of n LP-trees. Given a class $\mathcal{C}$ of LP-trees, let us call $\mathcal{C}$-*profile* a finite collection of LP-trees in $\mathcal{C}$.

Given a LP-profile P and a voting rule r, a naive way of finding the co-winners would consists in determining the n linear orders induced by the LP-trees and then apply r to these linear orders. However, this would be very inefficient, both in space and time. We would like to know how feasible it is to compute the winners directly from the LP-trees. More specifically, we ask the following questions: (a) given a voting rule, how difficult is it to compute the co-winners (or, else, one of the co-winners) for the different classes of LP-trees? (b) for score-based rules, how difficult is it to compute the score of the co-winners? (c) is it possible to have, for some voting rules and classes of LP-trees, a compact representation of the set of co-winners?

Formally, we consider the following decision and function problems.

Definition 1. *Given a class $\mathcal{C}$ of LP-trees and a voting rule r that is the maximizer of scoring function S, in the S-SCORE and EVALUATION problems, we are given a $\mathcal{C}$-profile P and an alternative $\boldsymbol{d}$. In the S-SCORE problems, we are asked to compute whether $S(P, \boldsymbol{d}) > T$ for some given $T \in \mathbb{N}$. In the EVALUATION problem, we are asked to compute whether there exists an alternative $\boldsymbol{d}$ with $S(P, \boldsymbol{d}) > T$ for some given $T \in \mathbb{N}$. In the WINNER problem, we are asked to compute $r(P)$.*

When we say that WINNER for some voting rule w.r.t. some class $\mathcal{C}$ is in P, the set of winners can be compactly represented, and can be computed in polynomial time.

Note that if EVALUATION is NP-hard and the score of an alternative can be computed in polynomial time, then WINNER cannot be in P unless P = NP: if WINNER were in P, then EVALUATION could be solved in polynomial time by computing a winner and its score.

For the voting rules studied in this paper, if not mentioned specifically, EVALUATION is w.r.t. the score functions we present when defining these rules. In this paper, we only show hardness proofs, membership in NP or #P is straightforward.

3.2 Two Specific Cases: Fixed Importance and Fixed Preference

It is worth focusing on the specific case of the class of profiles composed of LP-trees which have a fixed, linear structure: there is an order of importance among issues, which is common to all voters: X_1 is more important than X_2, which is itself more important than X_3, and so on.... Voters of course may have differing local preferences for the value for each issue, and their preferences on each issue may depend on the values of more important issues. A simple, easy to compute, and cheap in terms of communication, rule works as follows [14]: choose a value for X_1 according to the majority rule (possibly with a tie-breaking mechanism if we have an even number of voters); then, choose a value for X_2 using again the majority rule; and so on. The winner is called the *sequential majority winner*. When there is an odd number of voters, the sequential majority winner is the Condorcet winner (cf. Proposition 3 in [14], generalized in [7] to CI-profiles in which all voters have the same importance tree.). This, together with the fact that the sequential majority winner can be computed in polynomial time, shows that the winner of any Condorcet-consistent rule applied to FI profiles can be computed in polynomial time.

The case of fixed preferences is very specific for a simple reason: in this case, the top-ranked alternative is the same for all voters! This makes the winner determination trivial for all reasonable voting rules. However, nontrivial problems arise if we have constraints that limit the set of feasible alternatives. We devote Section 7 to aggregating FP trees.

4 k-Approval

We start by the following lemma. Most proofs are omitted due to the space constraint.

Lemma 1. *Given a positive integer k' such that $1 \leq k' \leq 2^p$ written in binary, and an LP-tree $\mathcal{L}$, the k'-th preferred alternative of $\succ_{\mathcal{L}}$ can be computed in time $O(p)$ by Algorithm 1.*

Algorithm 1. $FindAlternative(\mathcal{L}, k')$

---

1 Let $k^* = (k^*_{p-1}...k^*_0)_2 = 2^p - k'$ and $\mathcal{L}^* = \mathcal{L}$;
2 **for** $i = p - 1$ *down to* $i = 0$ **do**
3   Let X_j be the root issue of $\mathcal{L}^*$ with local preferences $x_j \succ \overline{x_j}$;
4   **if** $k^*_i = 1$ **then**
5   Let $\mathcal{L}^* \leftarrow \mathcal{L}^*(x_j)$ (the subtree of $\mathcal{L}^*$ tracing the path $X_j = x_j$) and let $a_j = x_j$;
6   **end**
7   **else** Let $\mathcal{L}^* \leftarrow \mathcal{L}^*(\overline{x_j})$ and let $a_j = \overline{x_j}$;
8 **end**
9 **return** a.

Similarly, the position of a given alternative d can be computed in time $O(p)$. It follows that the k-approval score of any alternative in a CI-CP profile can be computed in time $O(np)$. However, this does not mean that the winner can be computed easily, because the number of alternatives is exponential in p. For some specific values of k, though, computing the k-approval winner is in P.

Proposition 1. *Let k be a constant independent of p. When the profile is composed of n LP-trees, computing the k-approval co-winners for P can be done in time $O(knp)$.*

Proof: We compute the top k alternatives of each LP-tree in P; we store them in a table together with their k-approval score. As we have at most kn such alternatives, constructing the table takes $O(knp)$. $\qquad\square$

A similar result also holds for computing the $(2^p - k)$-approval co-winners for any constant k.[2]

Theorem 1 (CI-CP). *For CI-CP profiles, WINNER for 2^{p-1}-approval can be computed in time $O(np)$.*

---

[2] However, there is little practical interest in using $2^p - k$ approval for a fixed (small) value of k, since in practice, we will have $kn \ll 2^p$, and almost every alternative will be a co-winner.

Proof: We note that an alternative d is among the first half of alternatives in $\mathcal{L}_j$ *iff* the root issue of $\mathcal{L}_j$ is assigned to the preferred value. We build a table with the following $2p$ entries $\{1_1, 0_1, \ldots, 1_p, 0_p\}$: for every $\mathcal{L}_j$ we add 1 to the score of 1_i (resp. 0_i) if X_i is the root issue of $\mathcal{L}_j$ and the preferred value is 1_i (resp. 0_i). When this is done, for each X_i, we instantiate X_i to 1_i (resp. 0_i) if the score of 1_i is larger than the score of 0_i (resp. vice versa); if the scores are identical, we do not instantiate X_i. We end up with a partial instantiation, whose set of models (satisfying valuations) is exactly the set of co-winners. $\qquad\square$

Applying 2^{p-1}-approval here is both intuitive and cheap in communication (each voter only communicates her most important issue and its preferred value), and the output is computed very easily. On the other hand, it uses a very small part of the LP-trees. We may want to do better and take, say, the most important two issues into account, which comes down to using 2^{k-2}-approval or $(2^{k-1} + 2^{k-2})$-approval. However, this comes with a complexity cost. Let M be a constant independent of p and n and define $N(M, p)$ to be the set of all multiples of 2^{p-M} that are $\leq 2^p$, except 2^{p-1}. For instance, if $M = 3$ then $N(3, p) = \{2^{p-3}, 2^{p-2}, 2^{p-2} + 2^{p-3}, 2^{p-1} + 2^{p-3}, 2^{p-1} + 2^{p-2}, 2^{p-1} + 2^{p-2} + 2^{p-3}\}$.

Theorem 2 (UI-UP). *For any $k \in N(M, p)$, for UI-UP profiles,* EVALUATION *for k-approval is* NP-*hard.*

Proof sketch: When $k = 2^{p-i}$ for some $i \geq 2$, the hardness of EVALUATION is proved by a reduction from the NP-complete problem MIN2SAT [12], where we are given a set Φ of clauses, each of which is the disjunction of two literals, and an integer T'. We are asked whether there exists a valuation that satisfy smaller than T' clauses in Φ. We next show the case $k = 2^{p-2}$ as an example. We note that d is among the first quarter of alternatives in $\mathcal{L}_j$ *iff* the root issue of $\mathcal{L}_j$ is assigned to the preferred value, and the second most important issue in $\mathrm{IO}(\mathcal{L}_j, d)$ is assigned to the preferred value as well. Now, we give a polynomial reduction from MIN2SAT to our problem: given a set Φ of 2-clauses, the negation $\neg C_i$ of each clause $C_i \in \Phi$ is mapped into a UI-UP LP-tree whose top quarter of alternatives satisfies $\neg C_i$ (for instance, $\neg X_3 \wedge X_4$ is mapped into a LP-tree whose two most important issues are X_3 and X_4, and their preferred values are 0_3 and 1_4). The set of co-winners is exactly the set of valuations satisfying a maximal number of clauses $\neg C_i$, or equivalently, satisfying a minimal number of clauses in Φ.

The hardness for any other k in $N(M, p)$ is proved by a reduction from special cases of the MAXSAT problem, which are omitted due to the space constraint. $\qquad\square$

The hardness proofs carry over to more general models, namely $\{\text{UI,CI}\} \times \{\text{UP,CP}\}$. We next present an algorithm that converts winner determination for k-approval to a compact GENERALISED MAXSAT problem ("generalised" here means that the input is a set of formulas, and not necessarily clauses). The idea is, for each LP-tree $\mathcal{L}_j$, we construct a formula φ_j such that an alternative (valuation) is ranked within top k positions *iff* it satisfies φ_j. φ_j is further composed of the disjunction of multiple sub-formulas, each of which encodes a path from the root to a leaf in the tree structure, and the valuations that are ranked among top k positions.

Formally, for each path $\mathbf{u}$, we define a formula $C_\mathbf{u}$ that is the conjunction of literals, where there is an literal X_i (resp., $\neg X_i$) if and only if along the path $\mathbf{u}$, there is an edge marked 1_i (resp., 0_i). For any path with importance order $\mathcal{O}$ (w.l.o.g. $\mathcal{O} = X_1 \rhd X_2 \rhd \cdots \rhd X_p$) and $k = (k_{p-1} \ldots k_0)_2$ in binary, we define a formula $D_{\mathcal{O},k}$. Due to the space constraint, we only present the construction for the CI-UP case, but it can be easily extended to the CI-CP case. For each $i \leq p-1$, let $l_i = X_i$ if $1_i \succ 0_i$, and $l_i = \neg X_i$ if $0_i \succ 1_i$. Let $D_{\mathcal{O},k}$ be the disjunction of the following formulas: for every $i^* \leq p-1$ such that $k_{i^*} = 1$, there is a formula $(\bigwedge_{i > i^* : k_i = 0} l_i) \wedge l_{i^*}$. To summarize, for each LP-tree $\mathcal{L}_j$ in the profile we have a formula φ_j, and we can use a (generalised) MAXSAT solver to find a valuation that maximizes the number of satisfied formulas $\{\varphi_j\}$. Note that there are efficient such solvers; see, e.g., [16] and the *Minimally Unsatisfiable Subset Track* of the 2011 Sat Competition, at http://www.satcompetition.org/2011/#tracks.

Example 2. *Let $\mathcal{L}$ denote the LP-tree in Example 1, except that the preferences for t is unconditionally $0_2 \succ 1_2$. Let $k = 5 = (101)_2$. For the upper path we have the following clause $(\neg X_1) \wedge (\neg X_1 \vee (\neg X_2 \wedge X_3))$. For the lower path we have the following formula $(X_1) \wedge (\neg X_1 \vee (X_3 \wedge \neg X_2))$.*

Theorem 3. *For any $k \leq 2^p - 1$ represented in binary and any profile P of LP-trees, there is a polynomial-size set of formulas Φ such that the set of k-approval co-winners for P is exactly the set of the models of MAXSAT(Φ).*

Therefore, though WINNER for k-approval is hard to compute for some cases, it can be done efficiently in practice by using a generalized MAXSAT solver.

Note that all polynomiality results for k-approval carry on to the *Bucklin* voting rule (that we do not recall): it suffices to apply k-approval dichotomously until we get the value of k for which the score of the winner is more than $\frac{n}{2}$.

Now, we focus on the specific case of fixed importance orders (FI).

Theorem 4 (FI-CP). *Let $k \in N(M,p)$. For FI-CP profiles, WINNER for k-approval can be computed in time $O(2^M \cdot n)$.*

Proof sketch: For simplicity, we only present the algorithm for the case $k = 2^{p-2}$. The other cases are similar. Let $X_1 > X_2 > \ldots$ be the importance order, common to all voters. There are four types of votes: those for which the 2^{p-2} top alternatives are those satisfying $\gamma_1 = X_1 \wedge X_2$ (type 1), those satisfying $\gamma_2 = X_1 \wedge \neg X_2$ (type 2), etc. Let α_i be the number of votes in P of type i ($i = 1, 2, 3, 4$). The 2^{p-2}-approval co-winners are the alternatives that satisfy γ_i such that $\alpha_i = \max\{\alpha_i, i = 1, \ldots, 4\}$. $\square$

5   Borda

We start with a lemma that provides a convenient localized way to compute the Borda score for a given alternative in an LP-tree $\mathcal{L}$. For any $d = (d_1, \ldots, d_p) \in \mathcal{X}$ and any $i \leq p$, we define the following notation, which is an indicator whether the i-th component of d is preferred to its negation in $\mathcal{L}$, given the rest of values in d, denoted by d_{-i}.

$$\Delta_i(\mathcal{L}, d) = \begin{cases} 1 \text{ if in } \mathcal{L}, d_i \succ \overline{d_i} \text{ given } d_{-i} \\ 0 \text{ Otherwise} \end{cases}$$

$\Delta_i(\mathcal{L}, d)$ can be computed in polynomial-time by querying the CPT of X_i along $\mathrm{IO}(\mathcal{L}, d)$. We let $\mathrm{rank}(X_i, \mathcal{L}, d)$ denote the rank of issue X_i in $\mathrm{IO}(\mathcal{L}, d)$.

Lemma 2. *For any LP-tree $\mathcal{L}$ and any alternative d, we have the following calculation:*

$$S_{Borda}(\mathcal{L}, d) = \sum_{i=1}^{p} 2^{p-rank(X_i, \mathcal{L}, d)} \cdot \Delta_i(\mathcal{L}, d)$$

Example 3. *Let $\mathcal{L}$ denote the LP-tree defined in Example 1. We have $S_{Borda}(\mathcal{L}, 011) = 2^2 \cdot 1 + 2^1 \cdot 0 + 2^0 \cdot 1 = 5$ and $S_{Borda}(\mathcal{L}, 101) = 2^2 \cdot 0 + 2^0 \cdot 0 + 2^1 \cdot 1 = 2$.*

Hence, the Borda score of d for profile $P = (\mathcal{L}_1, \ldots, \mathcal{L}_n)$ is $S_{\text{Borda}}(P, d) = \sum_{j=1}^{n} \sum_{i=1}^{p} 2^{p-\text{rank}(X_i, \mathcal{L}_j, d)} \cdot \Delta_i(\mathcal{L}_j, d)$.

Theorem 5 (CI-UP). *For CI-UP profiles,* EVALUATION *is* NP-*hard for Borda.*

Proof sketch: We prove the NP-hardness by a reduction from 3SAT. Given a 3SAT instance, we construct an EVALUATION instance, where there are $q + 2$ issues $\mathcal{I} = \{c, d\} \cup \{X_1, \ldots, X_q\}$. The clauses are encoded in the following LP-trees: for each $j \leq t$, we define an LP-tree $\mathcal{L}_j$ with the following structure. Suppose C_j contains variables $X_{i_1}, X_{i_2}, X_{i_3}$ ($i_1 < i_2 < i_3$), and $d_{i_1}, d_{i_2}, d_{i_3}$ are the valuations of the three variables that satisfy C_j. In the importance order of $\mathcal{L}_j$, the first three issues are $X_{i_1}, X_{i_2}, X_{i_3}$. The fourth issue is c and the fifth issue is d if and only if $X_{i_1} = d_{i_1}$, $X_{i_2} = d_{i_2}$, or $X_{i_2} = d_{i_2}$; otherwise the fourth issue is d and the fifth issue is c. The rest of issues are ranked in the alphabetical order (issues in C are ranked higher than issues in $\mathcal{S}$). Then, we set the threshold appropriately (details omitted due to the space constraint) such that the Borda score of an alternative is higher than the threshold if and only if its d-component is 1, and the its values for $X_1, \ldots, X_p$ satisfy all clauses. $\qquad\square$

Finally, we show that WINNER for Borda can be converted to a weighted generalized MAXSAT problem. We note that $\Delta_i(\mathcal{L}_j, d)$ can be represented compactly by a formula φ_j^i such that a valuation d satisfies φ_j^i iff $\Delta_i(\mathcal{L}_j, d) = 1$. The idea is similar to the logical formula for k-approval, where each path $\mathbf{u}$ corresponds to a clause $C_{\mathbf{u}}$, and there is another clause depicting whether $\Delta_i(\mathcal{L}_j, d) = 1$ in $\mathbf{u}$. For example, let $\mathcal{L}$ denote the LP-tree in Example 1, then $\Delta_2(\mathcal{L}, d)$ can be presented by the disjunction of the clauses for the two paths: $\neg X_1 \wedge \neg X_2$ for the upper path, and $X_1 \wedge ((\neg X_3 \wedge \neg X_2) \vee (X_3 \wedge X_2))$ for the lower path.

Theorem 6. *For any profile P of LP-trees, there is a set of clauses Φ with weights such that the set of Borda co-winners for P is exactly the set of the models of* WEIGHTED MAXSAT(Φ).

Now, we focus on the specific case of unconditional importance orders (UI). When, for each $\mathcal{L}_j$ the importance order is unconditional, $\text{rank}(X_i, \mathcal{L}_j, d)$ does not depend on d: let us denote it $\text{rank}(X_i, \mathcal{L}_j)$. It can be computed in polynomial time by a simple exploration of the tree $\mathcal{L}_j$.

If the preferences are unconditional, then the Borda winner is the alternative d that maximises $\sum_{i=1}^{p} \sum_{j=1}^{n} 2^{p-\text{rank}(X_i, \mathcal{L}_j)} \Delta_i(\mathcal{L}_j, d)$. We can choose in polynomial time the winning value for each issue independently: it is the d_i that maximizes

$$\sum_{j=1}^{n} 2^{p-\text{rank}(X_i, \mathcal{L}_j)} \Delta_i(\mathcal{L}_j, d_i) \quad \text{where } \Delta_i(\mathcal{L}_j, d_i) = \begin{cases} 1 \text{ if in } \mathcal{L}_j, d_i \succ \overline{d_i} \\ 0 \text{ otherwise.} \end{cases}$$

Note that this method still works if the voters have differing importance order – provided they still have unconditional importance.

Theorem 7 (UI-UP). *For UI-UP profiles,* WINNER *for Borda can be computed in polynomial time.*

However, if we allow conditional preferences, computing the Borda winner becomes intractable:

Theorem 8 (FI-CP). *For FI-CP profiles,* EVALUATION *is* NP-*hard for Borda.*

Proof: It is not hard to see that EVALUATION is in NP. We prove the hardness by a reduction from 3SAT. In a 3SAT instance, we are given a formula F over binary variables $X_1, \ldots, X_q$. F is the conjunction of t disjunctive clauses. Let $F = C_1 \wedge \ldots \wedge C_t$ over binary variables. We are asked whether there exists a valuation of the variables under which F is true. Given any 3SAT instance, we construct the following EVALUATION instance.

Issues: There are $q + 1$ issues. For convenience, we use $\mathcal{I} = \{c\} \cup \{X_1, \ldots, X_q\}$ to denote these issues. W.l.o.g. let $\mathcal{O} = [X_1 \rhd X_2 \rhd \cdots \rhd X_q \rhd c]$ denote the fixed importance relation for the LP-trees.

Profile: The profile P is composed of two parts P_1 and P_2, where P_1 encodes the 3SAT instance. For any clause $C_j = l_j^1 \vee l_j^2 \vee l_j^3$, we define the two LP-trees $\mathcal{L}_j$ and $\mathcal{L}_j'$. Suppose $X_{i_1}, X_{i_2}, X_{i_3}$ are the variables that correspond to l_j^1, l_j^2, l_j^3 respectively. In both LP-trees, $\mathsf{Par}(X_{q+1}) = \{X_{i_1}, X_{i_2}, X_{i_3}\}$, and none of the other nodes (issues) has parents. The CPTs are defined as follows.

- $\mathcal{L}_j$: For every $i \leq q$, $0_i \succ 1_i$. For every assignment $(d_{i_1}, d_{i_2}, d_{i_3})$ of $\{X_{i_1}, X_{i_2}, X_{i_3}\}$, the CPT entries for c are $d_{i_1} d_{i_2} d_{i_3} : 1_{q+1} \succ 0_{q+1}$ if and only if C_j is satisfied by $(d_{i_1}, d_{i_2}, d_{i_3})$.
- $\mathcal{L}_j'$: For every $i \leq q$, $1_i \succ 0_i$. The CPT for c is the same as in $\mathcal{L}_j$.

Let $P_1 = \{\mathcal{L}_1, \mathcal{L}_1', \ldots, \mathcal{L}_t, \mathcal{L}_t'\}$. P_2 is composed of t copies of the following two UP LP-trees, which are $\mathcal{L}$, where for every $i \leq q$, $0_i \succ 1_i$. $1_c \succ 0_c$, and $\mathcal{L}'$, where for every issue, $1 \succ 0$. P_2 is used to make sure that we only need to focus on alternatives whose c-component is 1.

Let $P = P_1 \cup P_2$. For any valuation $\boldsymbol{d}_{-c} = (d_1, \ldots, d_q)$, $S_{\text{Borda}}(P, (\boldsymbol{d}_{-c}, 1_c)) = 2t \cdot (2^q - 1) + 2 \cdot K(F, \boldsymbol{d}_{-c})$, where $K(F, \boldsymbol{d}_{-c})$ is the number of clauses in F that are satisfied by $\boldsymbol{d}_{-c}$; and $S_{\text{Borda}}(P, (\boldsymbol{d}_{-c}, 0_c)) \leq 2t \cdot (2^q - 1)$.

It follows that there exists an alternative whose Borda score is more than $T = t \cdot 2^{q+1} - 1$ if and only if the 3SAT instance is a "yes" instance. This completes the proof. $\qquad\square$

6 Condorcet-Consistent Rules

We start by studying the several classes of conditionally lexicographic preferences according to the existence of a Condorcet winner. We recall the following result from [7]:

Lemma 3. *[7] For FI-CP profiles, there always exists a Condorcet winner, and it can be computed in polynomial time.*

Proposition 2. *The existence of a Condorcet winner for our classes of conditionally lexicographic preferences is depicted on the table below, where* yes *(resp.* no*) means that the existence of a Condorcet winner is guaranteed (resp. is not guaranteed) for an odd number of voters.*

	FP	UP	CP
FI	yes	yes	yes
UI	yes	no	no
CI	yes	no	no

Proof: We know from [7] that for FI-CP profiles, there always exists a Condorcet winner, and it can be computed in polynomial time. For CI-FP profiles, since all voters have the same top alternative, the existence of a Condorcet winner is trivial. Finally, here is a UI-UP profile with two variables and three voters, that has no Condorcet winner:
– Voter 1: $[X \rhd Y], x \succ \bar{x}, y \succ \bar{y}$, and the linear order is $[xy \succ x\bar{y} \succ \bar{x}y \succ \bar{x}\bar{y}]$.
– Voter 2: $[Y \rhd X], \bar{x} \succ x, y \succ \bar{y}$, and the linear order is $[\bar{x}y \succ xy \succ \bar{x}\bar{y} \succ x\bar{y}]$.
– Voter 3: $[Y \rhd X], \bar{x} \succ x, \bar{y} \succ y$, and the linear order is $[\bar{x}\bar{y} \succ x\bar{y} \succ \bar{x}y \succ xy]$.   □

Theorem 9 (UI-UP). *For UI-UP profiles, deciding whether a given alternative is the Condorcet winner is* coNP-*hard.*

Proof sketch: We prove the hardness by a reduction from the decision version of MAX HORN-SAT, which is known to be NP-complete [11]. In the decision version of MAX HORN-SAT, we are given a horn formula $F = C_1 \wedge \cdots \wedge C_t$ over variables $\{X_1, \ldots, X_q\}$, where each clause is a *horn clause* (containing no more than one positive literal), and a natural number K. We are asked whether there exists a valuation that satisfy more than K clauses. W.l.o.g. $K \geq t/2$, because there always exists a valuation that satisfies at least half of the clauses. For each horn clause, we define an LP-tree as follows. A clause having the form $C_j = \neg X_1 \vee \cdots \vee \neg X_l \vee X_{l+1}$ corresponds to an LP-tree whose importance order is $[X_1 \rhd \cdots \rhd X_{l+1} \rhd c \rhd \text{Others}]$, and whose local preferences are $0 \succ 1$ for $X_1 \ldots X_l$ and c, and $1 \succ 0$ for other issues. A clause having the form $C_j = \neg X_1 \vee \cdots \vee \neg X_l$ corresponds to an LP-tree whose whose importance order is $[X_1 \rhd \cdots \rhd X_l \rhd c \rhd \text{Others}]$, and whose local preferences are $0 \succ 1$ for issues $X_1 \cdots X_l$, and $1 \succ 0$ for other issues.

We also have $2k - t$ LP-trees with importance order $[c \rhd \text{Others}]$, and the local preferences are $0_c \succ 1_c$, and for other issues, $1 \succ 0$. We can show that $\mathbf{1}$ is the Condorcet winner *iff* the MAX HORN-SAT does not have a solution.   □

Corollary 1. *For UI-UP profiles,* EVALUATION *for maximin is* coNP-*hard.*

7   Fixed Preferences

When the agents' local preferences are fixed (w.l.o.g. $1 \succ 0$), issues can be seen as objects, and every agent has a preference for having an object rather than not, everything else being equal. Obviously, the best outcome for every agent is $\mathbf{1}$, and applying any reasonable voting rule (that is, any voting rule that satisfies *unanimity*) will select this alternative, making winner determination trivial. However, winner determination ceases to be trivial if we have constraints that limit the set of feasible alternatives. For instance, we may have a maximum number of objects that we can take.

Let us start with the only tractability result in this section, with the Borda rule. Recall that, when, for each $\mathcal{L}_j$ the importance order is unconditional, $\mathrm{rank}(X_i, \mathcal{L}_j)$ does not depend on $\boldsymbol{d}$. If, the preferences are fixed, $\Delta_i(\mathcal{L}_j, \boldsymbol{d}) = d_i$, and $S_{\mathrm{Borda}}(P, \boldsymbol{d}) = \sum_{i=1}^{p} d_i \sum_{j=1}^{n} 2^{p-\mathrm{rank}(X_i, \mathcal{L}_j)}$. We have the following theorem, which states that for the UI-FP case, computing the Borda winner is equivalent to computing the winner for a profile composed of importance orders, by applying some positional scoring rule. For any order $\rhd$ over $\mathcal{I}$, let $ext(\rhd)$ denote the UI-FP LP-tree whose importance order is $\rhd$.

Theorem 10 (UI-FP). *Let f_p denote the positional scoring rule over $\mathcal{I}$ with the scoring vector $(2^{p-1}, 2^{p-2}, \ldots, 0)$. For any profile $P_{\mathcal{I}}$ over $\mathcal{I}$, we have $ext(f_p(P_{\mathcal{I}})) = Borda(ext(P_{\mathcal{I}}))$.*

However, when the importance order is conditional, the Borda rule becomes intractable. We prove that using the following problem:

Definition 2. *Let voting rule r be the maximizer of scoring function S. In the K-EVALUATION problem, we are given a profile P that is composed of lexicographic preferences whose local preferences for all issues are $1 \succ 0$, a natural number K, and an integer T. We are asked to compute whether there exists an alternative $\boldsymbol{d}$ that takes 1 on no more than K issues and $S(P, \boldsymbol{d}) > T$.*

Theorem 11 (CI-FP). *For CI-FP profiles, K-EVALUATION is NP-hard for Borda.*

Proof sketch: We prove the NP-hardness by a reduction from restricted X3C where no element in C is covered by more than 3 sets in $\mathcal{S}$, which is NP-complete (problem [SP2] in [9]). Given an X3C instance $\mathcal{S} = \{S_1, \cdots, S_t\}$ over $C = \{c_1, \ldots, c_q\}$, we construct an EVALUATION instance where there are $q + 1 + 1$ issues $\{c\} \cup \mathcal{A} \cup \mathcal{E}$, and the alternative that maximizes the Borda score must take 1 for all issues in $\mathcal{A}$ and $q/3$ issues in $\mathcal{E}$. Then, the Borda score of the alternative is above the quota if and only if the issues chosen in $\mathcal{E}$ consist in a cover of $\mathcal{A}$. $\qquad\square$

Theorem 12 (UI-FP). *Let $k \in N(M, p)$. For UI-FP profiles, K-EVALUATION for k-approval is NP-hard.*

Proof sketch: For simplicity, we only show the proof for the case $k = 2^{p-2}$. The other cases are similar. The hardness is proved by a reduction from X3C, where we are given two sets $\mathcal{A} = \{a_1, \ldots, a_q\}$ and $\mathcal{E} = \{E_1, \ldots, E_t\}$, where for each $E \in \mathcal{S}$, $E \subseteq \mathcal{A}$ and $|E| = 3$. We are asked whether there exist $q/3$ elements in $\mathcal{S}$ such that each element in $\mathcal{A}$ appears in one and exactly one of these elements. There are $t + 1$ issues and q LP-trees. In each LP-tree $\mathcal{L}_j$, suppose $a_j \in E_i$, then we let the importance order be $X_{q+1} \rhd X_i \rhd$ Others. We let $K = q/3+1$ and let $T = q-1$. It shows that the solutions to the K-EVALUATION instance correspond to the solutions to the X3C instance. $\qquad\square$

Theorem 13 (UI-FP). *For UI-FP profiles, Copeland-SCORE is #P-hard.*

Table 1. Summary of computational complexity results

	FP	UP	CP
FI	Trivial	P (Thm. 4)	
UI	NPC (Thm. 12)	NPC (Thm. 2)	
CI			

(a) k-approval, $k \in N(M, p)$.

	FP	UP	CP
FI	Trivial	P (Thm. 7)	NPC (Thm. 8)
UI	P		
CI	NPC (Thm. 11)	NPC (Thm. 5)	

(b) Borda.

	FP	UP	CP
FI	Trivial		Polynomial (Lemma 3)
UI	#P-complete (Thm. 13)		
CI			

(c) Copeland score.

	FP	UP	CP
FI	Trivial		P (Lemma 3)
UI			coNPC (Thm. 9, Coro. 1)
CI			

(d) Maximin and Condorcet winner.

The proof is by polynomial-time counting reduction from #INDEPENDENT SET. Maximin, when the preferences are fixed (to be $1 \succ 0$ for all issues), the maximin score of 1 is 0 and the maximin score of any other alternative is $2^p - 1$. This trivialize the computational problem of winner determination even when with the restriction on the number of issues that take 1 (if $K \neq p$ then all available alternatives are tied). Following Lemma 3, for FI profiles, the winner can be computed in polynomial-time.

8   Summary and Future Work

Our main results are summarized in Table 1. In addition, we can also show that for k-approval (except $k = 2^{p-1}$), Copeland and maximin, there is no observation similar to Theorem 10, and the maximin score of a given alternative is **APX**-hard to approximate.

Our conclusions are partly positive, partly negative. On the one hand, there are voting rules for which the domain restriction to conditionally lexicographic preferences brings significant benefits: this is the case, at least, for k-approval for some values of k. The Borda rule can be applied easily provided that neither the importance relation and the local preference are unconditional, which is a very strong restriction. The hardness of checking whether an alternative is a Condorcet winner suggest that Condorcet-consistent rules appears to be hard to apply as well. However, as we have shown that some of these problems can be reduced to a compact MAXSAT problem. From a practical point of view, it is important to test the performance of MAXSAT solvers on these problems. We believe that continuing studying preference representation and aggregation on combinatorial domains, taking advantages of developments in efficient CSP techniques, is a promising future work direction.

Acknowledgments. We thank all anonymous reviewers of AAAI-12, CP-12, and COMSOC-12 for their helpful comments and suggestions. This work has been partly supported by the project ComSoc (ANR-09-BLAN-0305-01). Lirong Xia is supported by NSF under Grant #1136996 to the Computing Research Association for the CIFellows Project.

References

1. Ahlert, M.: Aggregation of lexicographic orderings. Homo Oeconomicus 25(3/4), 301–317 (2008)
2. Airiau, S., Endriss, U., Grandi, U., Porello, D., Uckelman, J.: Aggregating dependency graphs into voting agendas in multi-issue elections. In: Proceedings of IJCAI 2011, pp. 18–23 (2011)
3. Booth, R., Chevaleyre, Y., Lang, J., Mengin, J., Sombattheera, C.: Learning conditionally lexicographic preference relations. In: Proceeding of ECAI 2010, pp. 269–274 (2010)
4. Boutilier, C., Brafman, R., Domshlak, C., Hoos, H., Poole, D.: CP-nets: A tool for representing and reasoning with conditional ceteris paribus statements. Journal of Artificial Intelligence Research (JAIR) 21, 135–191 (2004)
5. Brafman, R.I., Domshlak, C., Shimony, S.E.: On graphical modeling of preference and importance. Journal of Artificial Intelligence Research (JAIR) 25, 389–424 (2006)
6. Brams, S., Kilgour, D., Zwicker, W.: The paradox of multiple elections. Social Choice and Welfare 15(2), 211–236 (1998)
7. Conitzer, V., Xia, L.: Approximating common voting rules by sequential voting in multi-issue domains. In: Proceedings of KR 2012 (2012)
8. Dombi, J., Imreh, C., Vincze, N.: Learning lexicographic orders. European Journal of Operational Research 183, 748–756 (2007)
9. Garey, M., Johnson, D.: Computers and Intractability. W. H. Freeman and Company (1979)
10. Gigerenzer, G., Goldstein, D.: Reasoning the fast and frugal way: Models of bounded rationality. Psychological Review 103(4), 650–669 (1996)
11. Jaumard, B., Simeone, B.: On the complexity of the maximum satisfiability problem for horn formulas. Information Processing Letters 26(1), 1–4 (1987)
12. Kohli, R., Krishnamurti, R., Mirchandani, P.: The minimum satisfiability problem. SIAM Journal on Discrete Mathematics 7(2), 275–283 (1994)
13. Lacy, D., Niou, E.M.: A problem with referendums. Journal of Theoretical Politics 12(1), 5–31 (2000)
14. Lang, J., Xia, L.: Sequential composition of voting rules in multi-issue domains. Mathematical Social Sciences 57(3), 304–324 (2009)
15. Pozza, G.D., Pini, M.S., Rossi, F., Venable, K.B.: Multi-agent soft constraint aggregation via sequential voting. In: Proceedings of IJCAI 2011, pp. 172–177 (2011)
16. Ryvchin, V., Strichman, O.: Faster Extraction of High-Level Minimal Unsatisfiable Cores. In: Sakallah, K.A., Simon, L. (eds.) SAT 2011. LNCS, vol. 6695, pp. 174–187. Springer, Heidelberg (2011)
17. Schmitt, M., Martignon, L.: On the complexity of learning lexicographic strategies. Journal of Machine Learning Research 7, 55–83 (2006)
18. Yaman, F., Walsh, T., Littman, M., desJardins, M.: Democratic approximation of lexicographic preference models. In: Proceedings of ICML 2008, pp. 1200–1207 (2008)

Constraint Programming for Path Planning with Uncertainty
Solving the Optimal Search Path Problem

Michael Morin[1], Anika-Pascale Papillon[2], Irène Abi-Zeid[3],
François Laviolette[1], and Claude-Guy Quimper[1]

[1] Department of Computer Science and Software Engineering
[2] Department of Mathematics and Statistics
[3] Department of Operations and Decision Systems
Université Laval, Québec, Qc, Canada
{Michael.Morin.3,Anika-Pascale.Papillon.1}@ulaval.ca,
Irene.Abi-Zeid@osd.ulaval.ca,
{Francois.Laviolette,Claude-Guy.Quimper}@ift.ulaval.ca

Abstract. The optimal search path (OSP) problem is a single-sided
detection search problem where the location and the detectability of a
moving object are uncertain. A solution to this $\mathcal{NP}$-hard problem is a
path on a graph that maximizes the probability of finding an object that
moves according to a known motion model. We developed constraint programming models to solve this probabilistic path planning problem for
a single indivisible searcher. These models include a simple but powerful branching heuristic as well as strong filtering constraints. We present
our experimentation and compare our results with existing results in the
literature. The OSP problem is particularly interesting in that it generalizes to various probabilistic search problems such as intruder detection,
malicious code identification, search and rescue, and surveillance.

1 Introduction

The *optimal search path* (OSP) problem we address in this paper is a single-sided
detection search problem where the location and the detectability of a moving
search object are uncertain. The single-sided search assumption means that the
object's movements are independent of the searcher's actions. In other words,
the object does not act, neither to meet nor to escape the searcher. A solution to
this $\mathcal{NP}$-hard problem [1] is a path on a graph that maximizes the probability
of finding an object that moves according to a known motion model. In the
OSP problem, a moving agent must plan its optimal path in order to detect a
mobile search object subject to constraints. This is a path planning problem for a
detection search with uncertainty on the whereabouts of the search object, on the
detection capabilities of the searcher, and on the movement of the search object.
This type of problem arises in many applications related to detection searches
namely, search and rescue [2], military surveillance, malicious code detection [3],

M. Milano (Ed.): CP 2012, LNCS 7514, pp. 988–1003, 2012.

covert messages (violating the security policies of the system) on the Internet [4], and locating a mobile user in a cellular network for optimal paging [5]. In this paper, we introduce *constraint programming* (CP) models that we developed in order to solve the OSP problem. We assume a single indivisible searcher where search effort corresponds to the time available for searching and a probability of detection is associated with each time step. Furthermore, the movement of the searcher is constrained to an accessibility graph.

Most work on the single searcher OSP problem in discrete time and space involved *branch and bound* (BB) algorithms. In [6], Stewart proposed a *depth-first* BB algorithm using a bound that does not guarantee optimality. Eagle [7] considered a Markovian object's motion model and proposed a dynamic programming approach. Eagle and Yee [8] presented an optimal bound for Stewart's BB algorithm. With an object following a Markovian motion model and an exponential *probability of detection* (*pod*) function, their approach produced an optimal bound by relaxing the search effort indivisibility constraint on a set of vertices while maintaining the path constraints. The bound is computed in polynomial time. A review of the BB algorithm procedures and of the OSP problem bounding techniques before 1998 can be found in [9]. Among the recent developments linked to the OSP problem, Lau *et al.* [10] proposed the DMEAN bound which was derived from the MEAN bound found by Martins [11].

The advantage of using CP in the OSP problem context lies in the CP model's expressivity. The model stays close to the formulation of the problem while enabling the use of problem specific constraints, heuristics and bounds. Furthermore, previous results on similar problems (*e.g.*, [12]) show that CP allows to find high quality solutions quickly, an interesting property we explore in this paper.

According to [13], uncertainty in constraint problems may arise in two situations:

- the problem changes over time (*dynamically changing* problems), and
- some problem's data or information are missing or are unclear (*uncertain* problems).

The OSP problem formulation as a constraint program is not uncertain in this sense since it has a complete description. Nonetheless, the location of the search object, its detectability, and its motion are represented by probability distributions. Our CP is not a dynamic formulation since the searcher's detection model, the object's motion model and the prior probability distribution on its location are known *a priori*. More specifically, the OSP problem is a path planning problem with a Markov Decision Process formulation that uses negative information for updating the probabilities in the absence of detection. In the case where the total number of plausible search object's paths is sufficiently low, a situation that rarely occurs in realistic search problems, the problem could potentially be formulated using multiple scenarios and thus be considered a stochastic CP (*e.g.*, [14]) where a scenario would correspond to a possible path of the search object. However, this is not an interesting approach since the Markov OSP problem specialization from search theory enables us to solve the problem without enumerating all the object's plausible paths [15]. Surveys on dealing with uncertainty in constraint problems may be found in [13,16].

Section 2 presents the OSP formalism. Sections 3 and 4 respectively describe the proposed constraint program and the experimentation. The results are discussed in Section 5 and compared to existing results in the literature. We conclude in Section 6.

2 The OSP Problem in Its General Discrete Form

When solving the OSP problem, the goal is to find a path (a search plan), constrained by time, that maximizes the probability of detecting a moving object of unknown location. A continuous search environment may be discretized by a graph[1] $G_A = (\mathcal{V}(G_A), \mathcal{E}(G_A))$ where $\mathcal{V}(G_A)$ is a set of discrete regions. A vertex r is accessible from vertex s if and only if the edge (s, r) belongs to the accessibility graph G_A. The search operation is defined over a given finite set $\mathcal{T} = \{1, \ldots, T\}$ of time steps. Let $y_t \in \mathcal{V}(G_A)$ be the searcher's location at time $t \in \mathcal{T}$. When $y_t = r$, we say that vertex r is searched at time t with an associated probability of detection. A search plan P (*i.e.*, the sequence of vertices searched) is determined by the searcher's path on G_A starting at location $y_0 \in \mathcal{V}(G_A)$:

$$P = [y_0, y_1, \ldots, y_T]. \tag{1}$$

The unknown object's location is characterized by a probability of containment (*poc*) distribution over $\mathcal{V}(G_A)$ that evolves in time, due to the search object's motion and to updates following unsuccessful searches. The poc_1 distribution over $\mathcal{V}(G_A)$ is the *a priori* knowledge on the object's location. A local probability of success (*pos*) is associated with the searcher being located in vertex r at time t. It is the probability of detecting the object in vertex r at time t defined as:

$$pos_t(r) = poc_t(r) \times pod(r), \tag{2}$$

where $pod(r)$ is the probability, conditional to the object's presence in r at time t, of detecting the object in vertex r at time t. This detection model is known *a priori*. For all $t \in \mathcal{T}$, $r \in \mathcal{V}(G_A)$, the detection model is

$$pod(r) \in (0, 1], \qquad \text{if } y_t = r; \tag{3}$$
$$pod(r) = 0, \qquad \text{otherwise.} \tag{4}$$

The OSP formalism assumes that a positive detection of the object stops the search. The probabilities of containment change in time following an assumed Markovian object motion model $\mathbf{M}$ and according to the negative information collected on the object's presence. Thus, for all time $t \in \{2, \ldots, T\}$, we have that

$$poc_t(r) = \sum_{s \in \mathcal{V}(G_A)} \mathbf{M}(s, r)\left[poc_{t-1}(s) - pos_{t-1}(s)\right], \tag{5}$$

---

[1] We restrict ourselves to the case of undirected reflexive graphs (*i.e.*, every vertex has a loop) since they are more natural in search problems. Furthermore, loops enable the searcher and the object to stay at their current location instead of moving on.

where $\mathbf{M}(s, r)$ is the probability of the object moving from vertex s to vertex r within one time step. The optimality criterion for a search plan P is the maximization of the global and cumulative success probability of the operation (COS) over all vertices and time steps defined as:

$$COS(P) = \sum_{t \in \mathcal{T}} \sum_{r \in \mathcal{V}(G_A)} pos_t(r). \tag{6}$$

2.1   An Optimal Search Plan Example

Figure 1 shows an example of an environment with doors and stairs accessibility. Considering the accessibility graph, and assuming $T = 5$, $y_0 = 3$, $poc_1(4) = 1.0$, $pod(y_t) = 0.9$ ($\forall t \in \mathcal{T}$) and a uniform Markovian motion model between accessible vertices, an optimal search plan P^* would then be

$$P^* = [y_0, y_1, \ldots, y_5] = [3, 6, 7, 7, 7, 7]. \tag{7}$$

Using Table 1, we explain why search plan P^* is optimal. Starting from vertex 3,

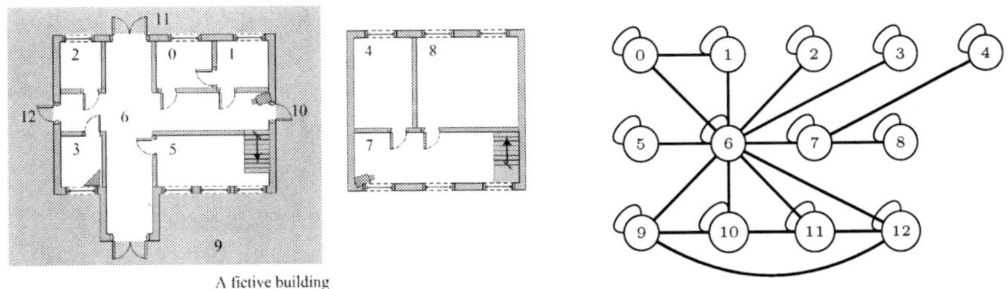

Fig. 1. A fictive building OSP problem environment (left) and its accessibility graph (right)

the searcher first moves to vertex 6 since the probability of containment is high in vertex 4. Then, the only accessible vertex where the probability of containment is nonzero is vertex 7. Therefore, the searcher moves from vertex 6 to vertex 7. Finally, the search plan stabilizes in vertex 7 since it has the highest probability of containment at each subsequent time step. The objective (COS) value of the optimal search plan P^* is equal to 0.889. For a search plan P, the objective value is computed as follows:

- compute the local probability of success in vertex y_1 at time step 1 ($pos_t(y_1)$) using Equation (2);
- for all vertices r, compute the probability of containment at time step 2 ($poc_2(r)$) using Equation (5);
- apply the same process for time steps 2 to T;
- sum all the local success probabilities obtained in time steps 1 to T to compute the objective (COS) value of the search plan P.

992   M. Morin et al.

The objective value at time step t, *i.e.*, COS_t, is

$$COS_t = \sum_{t' \leq t} \sum_{r \in \mathcal{V}(G_A)} pos_{t'}(r). \tag{8}$$

For all search plans P, COS_T is the objective value, *i.e.*, $COS_T(P) = COS(P)$.

Table 1. The probability of containment for each vertex at each time step and the cumulative overall probability of success for each time step for the search plan P^* of the example of Figure 1. The probabilities are rounded to the third decimal.

| | Probability of containment in vertex r at time t ($poc_t(r)$) | | | | | | | | | | | | | |
t \ r	0	1	2	3	4	5	6	7	8	9	10	11	12	$COS_t(P^*)$
1	-	-	-	-	1	-	-	-	-	-	-	-	-	0
2	-	-	-	-	.500	-	-	.050	-	-	-	-	-	.450
3	-	-	-	-	.263	-	.012	.026	.012	-	-	-	-	.686
4	.001	.001	.001	.001	.138	.001	.008	.015	.013	.001	.001	.001	.001	.817
5	.001	.001	.001	.001	.073	.001	.008	.008	.01	.002	.002	.002	.002	.889

3   A Constraint Programming Model for the OSP

We present in this section the CP model and the heuristic we developed to guide the resolution process. We define the following constants:

- $\mathcal{T}$, the set of all time steps;
- $G_A = (\mathcal{V}(G_A), \mathcal{E}(G_A))$, the accessibility graph representing the search environment;
- $y_0 \in \mathcal{V}(G_A)$, the initial searcher's position;
- $poc_1(r)$, the initial probability of containment in vertex r ($\forall r \in \mathcal{V}(G_A)$);
- $pod(r)$, the conditional probability of detecting the object when $y_t = r$ ($\forall t \in \mathcal{T}, \forall r \in \mathcal{V}(G_A)$);
- $\mathbf{M}(s,r)$, the probability of an object's move from vertex s to vertex r in one time step ($\forall s, r \in \mathcal{V}(G_A)$).

Furthermore, we define $poc_t^{\text{Markov}}(r)$, the updated probability of containment in vertex r at time t in the absence of searches as:

$$poc_t^{\text{Markov}}(r) \stackrel{\text{def}}{=} \begin{cases} poc_1(r), & \text{if } t = 1; \\ \sum_{s \in \mathcal{V}(G_A)} \mathbf{M}(s,r) poc_{t-1}^{\text{Markov}}(s), & \text{otherwise.} \end{cases} \tag{9}$$

The Markovian probability of containment poc^{Markov} is an upper bound on the probability of containment in vertex r at time t, *i.e.*, $poc_t(r) \leq poc_t^{\text{Markov}}(r)$. This is due to the fact that an unsuccessful search in vertex r at time t decreases the probability of the object being there at time t (from Equation (5)). Moreover, we observe that the probability of success $pos_t(r)$ in vertex r at time t is bounded by the probability of detection in r ($pod(r)$), *i.e.*, $pos_t(r) \leq pod(r)$. Both of these observations will be used to bound the domains of the probability variables in the CP model.

3.1   The Variables

The model's decision variables used to define the search plan P are:

- $Y_0 = y_0$, the initial searcher's position;
- $Y_t \in \mathcal{V}(G_A)$, the searcher's position at time t ($\forall t \in \mathcal{T}$).

The non-decision variables used to compute the COS criterion value are:

- $POC_1(r) = poc_1(r)$, the probability of containment in vertex r at time 1 ($\forall r \in \mathcal{V}(G_A)$);
- $POC_t(r) \in [0, poc_t^{\mathrm{Markov}}(r)]$, the probability of containment in vertex r at time t ($\forall t \in \mathcal{T}, r \in \mathcal{V}(G_A)$) where $poc_t^{\mathrm{Markov}}(r)$ is defined by Equation (9);
- $POS_t(r) \in [0, pod(r)]$, the probability of success in vertex r at time t ($\forall t \in \mathcal{T}, r \in \mathcal{V}(G_A)$);
- $COS \in [0, 1]$, the COS criterion value, $i.e.$, the sum of all local probabilities of success up to time T.

The domain of Y_t is finite ($\forall t \in \mathcal{T}$). The domains of the probability variables are infinite since these variables are real. Interval-valued domains are used to define these domains, $i.e.$, non-enumerated domains whose values are implicitly given by a lower bound and an upper bound.

3.2   The Constraints

Constraint (10) defines the searcher's path, $i.e.$, the search plan P. It constrains the searcher to move from one vertex to another according to the accessibility graph edges $\mathcal{E}(G_A)$.

$$(Y_{t-1}, Y_t) \in \mathcal{E}(G_A), \qquad\qquad \forall t \in \mathcal{T}. \quad (10)$$

The constraints (11) to (13) compute the probabilities required to evaluate the COS criterion. The first two constraints, (11) and (12) compute the probability of success. Constraint (13) is the probability of containment update equation.

$$Y_t = r \implies POS_t(r) = POC_t(r)pod(r), \qquad \forall t \in \mathcal{T}, \forall r \in \mathcal{V}(G_A). \quad (11)$$
$$Y_t \neq r \implies POS_t(r) = 0.0, \qquad \forall t \in \mathcal{T}, \forall r \in \mathcal{V}(G_A). \quad (12)$$

$$POC_t(r) = \sum_{s \in \mathcal{V}(G_A)} \mathbf{M}(s, r)\left[POC_{t-1}(s) - POS_{t-1}(s)\right], \quad \forall t \in \{2, \ldots, T\},$$

$$\forall r \in \mathcal{V}(G_A). \quad (13)$$

3.3   The Objective Function

We have experimented with two different encodings of the objective function. The first one encodes the objective function as a sum, the second one encodes it as a max. Both encodings are equivalent and lead to the same objective value.

The sum *objective function.* The sum encoding of Equation (14) consists of encoding the objective function as it appears in (6). It is the natural way to represent this function.

$$\max COS, \tag{14}$$

$$COS = \sum_{t \in \mathcal{T}} \sum_{r \in \mathcal{V}(G_A)} POS_t(r). \tag{15}$$

The max *objective function.* The sum constraint does a very poor job of filtering the variables: the upper bound on a sum of variables is given by the sum of the upper bounds of the variable domains. However, since we know that in the summation $\sum_r POS_t(r)$ only one variable is non-null, a tighter upper bound on this sum is given by the maximum domain upper bound. A tighter upper bound on the objective variable generally leads to a faster branch and bound. We therefore have implemented the objective function defined by Equation (6) using the following constraints:

$$\max COS, \tag{16}$$

$$COS = \sum_{t \in \mathcal{T}} \max_{r \in \mathcal{V}(G_A)} POS_t(r). \tag{17}$$

3.4   The Proposed Value Selection Heuristic

In this section we describe the value selection heuristic we developed. The idea of our heuristic is based on a stochastic generalization of a graph based pursuit evasion problem called the *cop and robber* game [17]; the description of its theoretical bases is beyond the scope of this paper. We were also inspired by a domain ordering idea that was successfully used for the *multiple rectangular search areas* problem [12].

Our novel heuristic simplifies the probability system in the OSP problem by ignoring the negative information received by the searcher when s/he fails to detect the object. That is, at each time step $t \in \mathcal{T}$, the heuristic chooses the most promising accessible vertex based on the total probability of detecting the object in the remaining time. Therefore, we call our heuristic the *total detection* (TD) heuristic.

Let $G_A = \langle \mathcal{V}(G_A), \mathcal{E}(G_A) \rangle$ be the accessibility graph where the searcher and the object evolve. Let $t \in \mathcal{T}$ be a time step, and $y, o \in \mathcal{V}(G_A)$ the positions of the searcher and the object. Let $w_t(y, o)$ be the conditional probability that the searcher detects the object in the time period $[t, t+1, \ldots, T]$ given that, at time t, the searcher is in y and the object in o. The function $w_t(y, o)$ is recursively defined as follows:

$$w_t(y, o) \stackrel{\text{def}}{=} \begin{cases} pod(o), & \text{if } o = y \text{ and } t = T, \\ 0, & \text{if } o \neq y \text{ and } t = T, \\ pod(o) + (1 - pod(o))p_t(y, o), & \text{if } o = y \text{ and } t < T, \\ p_t(y, o), & \text{if } o \neq y \text{ and } t < T, \end{cases} \tag{18}$$

where

$$p_t(y, o) = \sum_{o' \in \mathcal{N}(o)} \mathbf{M}(o, o') \max_{y' \in \mathcal{N}(y)} w_{t+1}(y', o'), \tag{19}$$

is the probability of detecting the object in the period $[t + 1, \ldots, T]$. Equations (18) and (19) have the following interpretation:

- If $t = T$, the searcher has a probability $pod(o)$ of detecting the object when the searcher and the object are co-located, *i.e.*, $o = y$; otherwise, the searcher and the object are not co-located and the probability is null.
- If $t < T$ and $o = y$, then the searcher can detect the object at time t with probability $pod(o)$ or fail to detect it at time t with probability $1 - pod(o)$. If the searcher fails to detect the object at time t, s/he may detect it during the period $[t + 1, \ldots, T]$. The probability of detecting the object in the period $[t + 1, \ldots, T]$ is given by $p_t(y, o)$ (Equation (19)) and may be interpreted as follows:

  - in the case where there is only one edge leaving vertex o to vertex o', the searcher chooses the accessible vertex y' that maximizes the conditional probability of detecting the object in the time period $[t + 1, \ldots, T]$, given his/her new position y' and the new object's position o', *i.e.*, $\max_{y' \in \mathcal{N}(y)} w_{t+1}(y', o')$;
  - In the general case where vertex o has many neighbors, $p_t(y, o)$ is the average of all the maximal $w_{t+1}(y', o')$ weigthed by the probability $\mathbf{M}(o, o')$ of moving from o to o'.

  This is reasonable since we do not control the object's movements but we can move the searcher to the vertex that has the highest probability of success.
- Finally, if the search time is not over (*i.e.*, $t < T$) and the object and the searcher are not co-located (*i.e.*, $o \neq y$), the probability of detecting the object at time t is null and the probability of success depends entirely on the probability $p_t(y, o)$ of detecting the object within the period $[t + 1, \ldots, T]$.

A searching strategy $S : \mathcal{T} \times \mathcal{V}(G_A)$ assigns to each time step and plausible searcher's position a set of vertices that are considered to be optimal according to some heuristic. In the TD heuristic case, the strategy sets the new searcher's position to be the accessible vertex that maximizes the probability of detecting the object in the remaining time:

$$S_t(Y_t) \stackrel{\text{def}}{=} \operatorname*{argmax}_{y' \in \text{dom}(Y_t)} \sum_{o \in \mathcal{V}(G_A)} w_t(y', o) poc_t(o), \qquad \forall t \in \mathcal{T}. \tag{20}$$

In order to apply this value selection heuristic, the following *static ordering* of the decision variables is used: $Y_0, \ldots, Y_T$. That is, the solver branches first on Y_0, then on Y_1 and so on. Each time the solver branches on a new path variable Y_t, the strategy $S_t(Y_t)$ is computed in polynomial time.

4   Experimentation

Our experiments were conducted in two phases. In Phase 1, we compared the
two versions of the CP models presented in Section 3 (*i.e.*, CpMax and CpSum).
In Phase 2, we examined the performance of the TD heuristic presented in
Section 3.4 when used as a value selector along with the best CP model retained
from Phase 1. The TD heuristic is compared with the CpMax model using an
increasing domain[2] value selection heuristic. For all experiments, the following
static ordering is used for branching: $Y_0, \ldots, Y_T$.

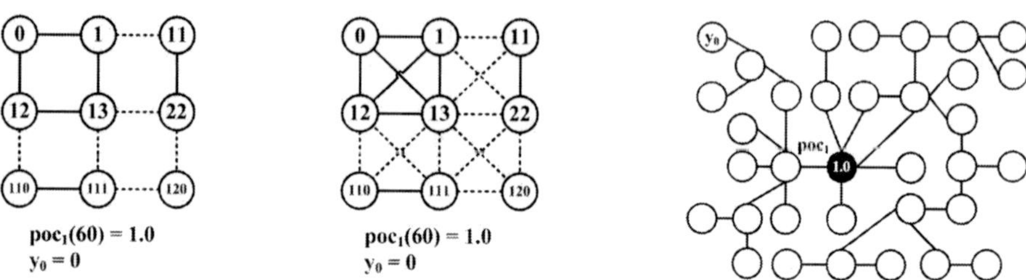

Fig. 2. The 11×11 grid G^+ (left), the 11×11 grid G^* (center), the graph G^L (right)

The graphs used in our benchmark along with the searcher's initial position
y_0 and the initial probability of containment distribution poc_1 are shown on
Figure 2. G^+ is a reflexive 11×11 grid where all adjacent vertices except diagonals
are linked by an edge. G^* is a reflexive 11×11 grid where all adjacent vertices
(diagonals included) are linked by an edge. G^L is a reflexive graph generated
using the Université Laval tunnels map. It is almost a tree. We tried these
graphs with three different probabilities of detection: $pod(r) \in \{0.3, 0.6, 0.9\}$
($\forall r \in \mathcal{V}(G_A)$). The assumed Markovian object's motion model is

$$\mathbf{M}(s,r) = \begin{cases} \frac{1-\rho}{\deg(s)-1}, & \text{if } (s,r) \in \mathcal{E}(G_A), \\ \rho, & \text{if } s = r, \end{cases} \tag{21}$$

where $\deg(s)$ is the degree of s and $\rho \in \{0.3, 0.6, 0.9\}$ is the probability that the
object stays in its current location. The total times allowed for the searches are
$T \in \{9, 11, 13, 15, 17, 19\}$. Usual OSP problem experiments use grids similar to
G^+ (*e.g.*, [8], [10]). Therefore, our G^+ problem instances are comparable with
those used in the literature.

All tests consisted of a single run on an instance $(G_A, T, pod(r)_{r \in \mathcal{V}(G_A)}, \rho)$, as
described above. We allowed a total number of 5,000,000 backtracks and a time
limit of 20 minutes. All implementations are done using Choco Solver 2.1.3 [18],
a solver developed in the Java programming language, and the Java Universal
Network/Graph (JUNG) 2.0.1 framework [19].[3] The probabilities of the OSP

---

[2] When branching on Y_t, the solver selects the integer with the smallest value.
[3] The source code of our experiments is available upon request.

CP model were multiplied by an integer for implementation purposes. Because of numerical errors, our results are accurate to the fourth decimal. All tests were run on an Intel(R) Core(TM) i7-2600 CPU with 4 GB of RAM.

5   Results and Discussion

In this section, we compare the time required to obtain various incumbent solutions (*i.e.*, the best feasible solutions found so far). The time to the last incumbent is the CPU time spent by the solver to obtain the incumbent with the best objective value within a 20 minutes or 5,000,000 backtracks limit.

5.1   Phase 1: Comparing the CP Models

Table 2 compares the results obtained with the CpMax model with the ones obtained with the CpSum model on a 11×11 G^+ grid with $T = 17$, various probability of detection values (pod) and various motion models (ρ). In all cases, the COS value of the last incumbent solution obtained with the CpMax model is higher or equal to the one obtained with the CpSum model. Furthermore, when there is a tie on the COS value, the time required with the CpMax model is lower than the one required with the CpSum model. The tendency of the CpMax model to outperform the CpSum model is present on G^* and on G^L instances with $T = 17$ (not shown). On most instances, the CpMax model requires fewer backtracks than the CpSum model to achieve a higher quality last incumbent solution. We conclude that the use of the constraint "max" leads to a stronger filtering on the variable COS, and thus, computes a tighter bound on the objective function. For this reason, further comparisons involve only the CpMax model.

Table 2. The COS value of the last incumbent solution on a 11×11 G^+ grid with $T = 17$. Bold font is used to highlight the best objective value (higher is better). Ties are broken using the time to last incumbent value.

		CpMax		CpSum	
$pod(r)$	ρ	Time to last incumbent (s)	COS value of the last incumbent	Time to last incumbent (s)	COS value of the last incumbent
0.3	0.3	**1197.15**	**0.0837**	1045.66	0.0831
	0.6	**1198.56**	**0.1276**	990.61	0.1267
	0.9	**1026.02**	**0.3379**	1165.88	0.3379
0.6	0.3	**959.18**	**0.1532**	999.45	0.1532
	0.6	**1168.98**	**0.2202**	1015.64	0.2172
	0.9	**1166.29**	**0.5122**	942.36	0.5014
0.9	0.3	**1161.59**	**0.2162**	1184.86	0.2162
	0.6	**692.16**	**0.3151**	727.57	0.3151
	0.9	**1169.91**	**0.6283**	879.59	0.6252

5.2   Phase 2: Evaluating the TD Value Selection Heuristic

Figures 3 shows the COS value as a function of time (ms) obtained on the G^+, G^* and G^L environments with $pod(y_t) = 0.6$ ($\forall t \in \mathcal{T}$) and a motion model such that the probability ρ that the object stays in its current location equals 0.6. On all instances shown, the benefits of using the TD heuristic as a value selection heuristic are clear as the solver finds incumbent solutions of higher quality in less time when compared to the CpMax model using an increasing domain value

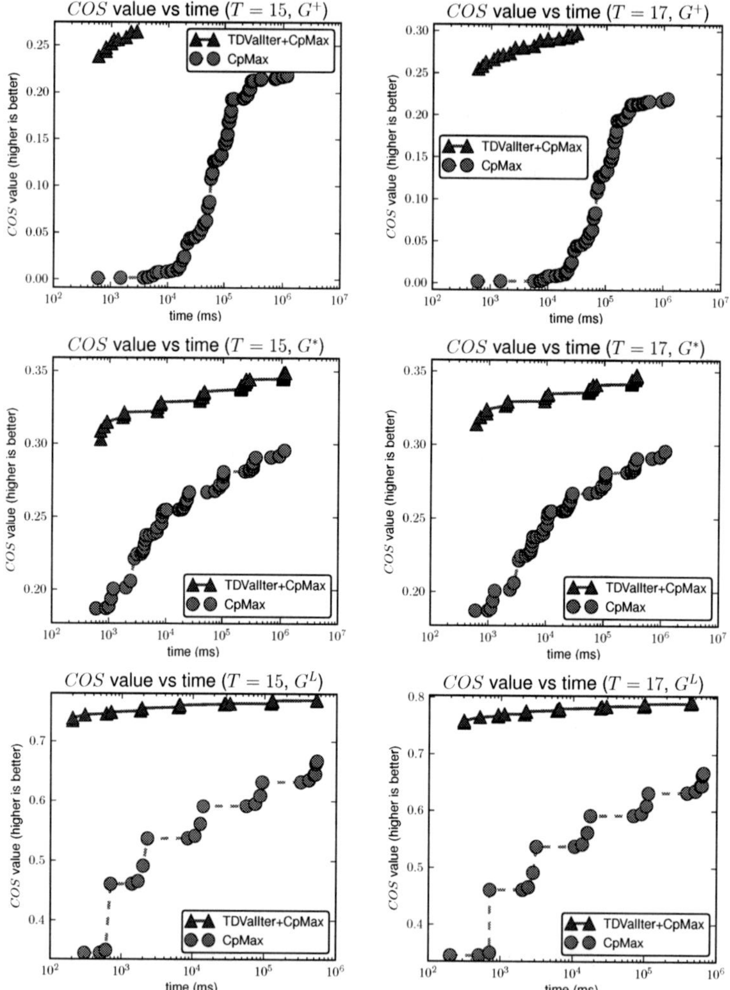

Fig. 3. The COS value as a function of time (ms) (log scale) obtained with the TD-ValSel+CpMax and the CpMax configurations on a 11×11 G^+, on a 11×11 G^* instance and on a G^L instance with $T = 15$ (left column) and $T = 17$ (right column). The $pod(y_t) = 0.6$ ($\forall t \in \mathcal{T}$), and the $\rho = 0.6$.

selection heuristic. In all cases shown, the COS value of the first incumbent solution found with the TDValSel+CpMax configuration, a solution encountered after less than 1 second of solving time, is within 5% of the COS value of the last incumbent solution. On the G^+ instance with $T = 17$, TDValSel+CpMax encountered 21 solutions before settling to an incumbent with $COS = 0.2978$ while the CpMax configuration encountered 98 solutions before settling to an incumbent with $COS = 0.2202$. On the G^* instance with $T = 17$, TDValSel+CpMax encountered 46 solutions before settling to an incumbent with $COS = 0.3478$ while the CPMax configuration encountered 70 solutions before settling to an incumbent with $COS = 0.2959$. Finally, on the G^L instance with $T = 17$, TDValSel+CpMax encountered 29 solutions before settling to an incumbent with $COS = 0.7930$ while the CPMax configuration encountered 23 solutions before settling to an incumbent with $COS = 0.6676$. By looking at the total number of solutions encountered by the two configurations on the three instances, it seems that varying the graph structure leads to very different problem instances. This is partly due to the motion model of the object and to the probability of staying in place ρ: Given an object's position r, the remaining probability mass $1 - \rho$ is distributed among the neighbors of r leading to smaller probability of containment poc values in the neighborhood of r on G^* and G^+ than on most vertices of G^L. For this reason, G^* and G^+ are significantly harder instances to solve than G^L. Furthermore, the G^+ and the G^* accessibility graphs involve more symmetric instances than G^L.

Table 3 compares the results obtained with the TDValSel+CpMax configuration to the ones obtained with the CpMax model using an increasing domain value selection heuristic on a 11×11 G^+ grid with $T = 17$, various probability of detection values (pod), and various motion models (ρ). Again, the TDValSel heuristic is dominant with a time to last incumbent up to 300 times faster for a higher quality solution in terms of COS value. The tendency of the TDValSel+CpMax configuration to outperform the CpMax model using an

Table 3. The COS value of the last incumbent solution on a 11×11 G^+ grid with $T = 17$. Bold font is used to highlight the best objective value (higher is better). Ties are broken using the time to last incumbent value.

		TDValSel+CpMax		CpMax	
$pod(r)$	ρ	Time to last incumbent (s)	COS value of the last incumbent	Time to last incumbent (s)	COS value of the last incumbent
0.3	0.3	**5.31**	**0.1055**	1197.15	0.0837
	0.6	**19.42**	**0.1645**	1198.56	0.1276
	0.9	**3.70**	**0.4418**	1026.02	0.3379
0.6	0.3	**159.42**	**0.1893**	959.18	0.1532
	0.6	**31.02**	**0.2978**	1168.98	0.2202
	0.9	**225.66**	**0.6559**	1166.29	0.5122
0.9	0.3	**54.94**	**0.2595**	1161.59	0.2162
	0.6	**37.73**	**0.4119**	692.16	0.3151
	0.9	**467.34**	**0.8194**	1169.91	0.6283

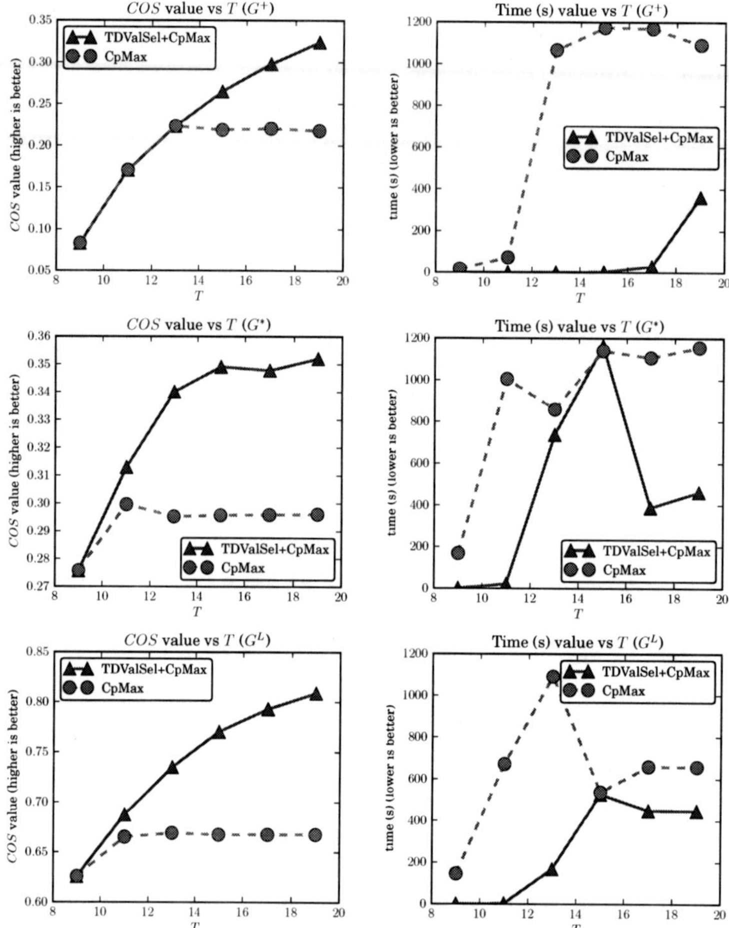

Fig. 4. The COS value (left) and the time to last incumbent (ms) (right) as a function of the total number of time steps (T) with the TDValSel+CpMax and the CpMax configurations on a G^+, a G^* and a G^L instance where $pod(y_t) = 0.6$ ($\forall t \in \mathcal{T}$), and $\rho = 0.6$

increasing value selection heuristic is present on G^* and on G^L instances with $T = 17$ as well. For this reason, the tables for the G^* and the G^L instances are omitted.

Figure 4 compares the COS values obtained on several G^+ 11 × 11 grid with $pod(y_t) = 0.6$ ($\forall t \in \mathcal{T}$), and $\rho = 0.6$ instances of increasing complexity in T. For all values of T and instance types, the COS value of the last incumbent found with the TDValSel+CpMax configuration is higher or equal to the one found with the CpMax model alone. On the G^+ instance, we notice that the solution is found in less than 5 seconds up to $T = 17$. By comparing the first row of figure 4 showing the G^+ instance to the other rows, we notice that time to last incumbent curve of the TDValSel+CpMax is more erratic on the G^* and the G^L

Table 4. The time to last incumbent on a 11×11 G^+ grid with $pod(r) = 0.6$ and $\rho = 0.6$ compared to the time spent by a BB procedure to prove optimality when using various bounds [10]

	Time to incumbent (s)	Time to optimality (s)*			
T	TDValSel+CpMax	DMEAN	MEAN	PROP	FABC
15	2.80	3.14	12.27	8.16	62.64
17	31.02	23.76	71.57	37.20	352.96
*The time values are taken from [10].					
They are used to give a general idea of how our results behave.					

instances. We believe that this may be due to the precision we used to compute the probability variables (a limitation of our solver).

In order to get an idea of the relative performance of our model and value selection heuristic, we compared our results with results published in the literature using a BB algorithm [10]. After communications with the authors of [10] we were able to validate that our solutions for the G^+ instances are optimal up to the fourth decimal. However, the instances are too large for our solver to prove optimality in a reasonable time. Table 4 presents the time to last incumbent on a 11×11 G^+ grid with $pod(r) = 0.6$ and $\rho = 0.6$, and the time spent by a BB procedure to prove the optimality of its last incumbent solution when using various bounds from the literature. The hardware and software configurations used to produce these results differ from ours. Consequently, the goal of this comparison is simply to show the general tendency on the instances for which the optimal value is published rather than proving that our approach outperforms the BB procedure. Recalling that we are not using any problem specific bound on the objective function (except a simple objective function simplification carried out in the CpMax model), our results, comparable to the ones in the literature, highlight the performance of the TD value selection heuristic.

One of the main advantage of using constraint programming is expressivity. The constraint programming model stays close to the natural problem formulation while enabling strong filtering (*e.g.*, the CpMax model) and heuristics (*e.g.*, the TD value selection heuristic). We believe that using a CP model is closer to the natural problem formulation than an IP model for example. In addition, the model can be easily adapted and extended. For instance, the model and the heuristic can be generalized to allow searches from a distance, *i.e.*, the searcher sees a subset of visible vertices including his position.

6   Conclusion

We have presented a CP model to solve the OSP problem. This model includes a very efficient value selection heuristic that branches on vertices leading to a high objective value (*i.e.*, probability of success). We refined the objective function to obtain a tighter bound on the objective value without discarding solutions. Experiments show that our model is competitive with the state-of-the-art in search theory and that constraint programming is a good technique to solve the OSP. Future work includes the development of tight bounds in order to allow the solver to prove the optimality of its incumbent solution. We believe that such a bound could be based on the information already computed for the value selection heuristic presented in this paper.

Acknowledgments. We would like to thank Haye Lau for his help in validating the results of Section 5.2, and the anonymous reviewers for their helpful comments and suggestions.

References

1. Trummel, K., Weisinger, J.: The complexity of the optimal searcher path problem. Operations Research 34(2), 324–327 (1986)
2. Stone, L.: Theory of Optimal Search. Academic Press, New York (2004)
3. Kranakis, E., Krizanc, D., Rajsbaum, S.: Computing with mobile agents in distributed networks. In: Rajasedaran, S., Reif, J. (eds.) Handbook of Parallel Computing: Models, Algorithms, and Applications, pp. 1–26. CRC Press (2007)
4. Chandramouli, R.: Web search steganalysis: Some challenges and approaches. In: Proceedings of IEEE International Symposium on Circuits and Systems, pp. 576–579 (2004)
5. Verkama, M.: Optimal paging - a search theory approach. In: Proceedings of the International Conference on Universal Personal Communications, pp. 956–960 (1996)
6. Stewart, T.: Search for a moving target when the searcher motion is restricted. Computers and Operations Research 6, 129–140 (1979)
7. Eagle, J.: The optimal search for a moving target when the search path is constrained. Operations Research 32(5), 1107–1115 (1984)
8. Eagle, J., Yee, J.: An optimal branch-and-bound procedure for the constrained path, moving target search problem. Naval Research Logistics 38(1), 110–114 (1990)
9. Washburn, A.: Branch and bound methods for a search problem. Naval Research Logistics 45, 243–257 (1998)
10. Lau, H., Huang, S., Dissanayake, G.: Discounted mean bound for the optimal searcher path problem with non-uniform travel times. European Journal of Operational Research 190(2), 383–397 (2008)
11. Martins, G.: A new branch-and-bound procedure for computing optimal search paths. Tech. rep., Naval Postgraduate School (1993)
12. Abi-Zeid, I., Nilo, O., Lamontagne, L.: Resource allocation algorithms for planning search and rescue operations. INFOR 49(1), 15–30 (2011)
13. Brown, K.M., Miguel, I.: Uncertainty and change. In: Rossi, F., Beek, P.V., Walsh, T. (eds.) Handbook of Constraint Programming, pp. 44–58. Springer (2006)
14. Tarim, S., Manandhar, S., Walsh, T.: Stochastic constraint programming: A scenario-based approach. Constraints 11(1), 53–80 (2006)
15. Brown, S.: Optimal search for a moving target in discrete time and space. Operations Research 28(6), 1275–1289 (1980)

16. Verfaillie, G., Jussien, N.: Constraint solving in uncertain and dynamic environments: A survey. Constraints 10, 253–281 (2005)
17. Nowakowski, R., Winkler, P.: Vertex-to-vertex pursuit in a graph. Discrete Mathematics 43(2-3), 235–239 (1983)
18. Laburthe, F., Jussien, N.: Choco Solver Documentation (2012),
    `http://www.emn.fr/z-info/choco-solver/`
19. O'Madadhain, J., Fisher, D., Nelson, T., White, S., Boey, Y.: Jung: Java universal network/graph framework (2010), `http://jung.sourceforge.net`

Feature Term Subsumption Using Constraint Programming with Basic Variable Symmetry*

Santiago Ontañón[1] and Pedro Meseguer[2]

[1] Computer Science Department
Drexel University
Philadelphia, PA, USA 19104
santi@cs.drexel.edu

[2] IIIA-CSIC, Universitat Autònoma de Barcelona
08193 Bellaterra, Spain
pedro@iiia.csic.es

Abstract. Feature Terms are a generalization of first-order terms which have been recently received increased attention for their usefulness in structured machine learning applications. One of the main obstacles for their wide usage is that their basic operation, subsumption, has a very high computational cost. Constraint Programming is a very suitable technique to implement that operation, in some cases providing orders of magnitude speed-ups with respect to the standard subsumption approach. In addition, exploiting a basic variable symmetry –that often appears in Feature Terms databases– causes substantial additional savings. We provide experimental results of the benefits of this approach.

1 Introduction

Structured machine learning (SML) [8] focuses on developing machine learning techniques for rich representations such as feature terms [2, 7, 16], Horn clauses [12], or description logics [6]. SML has received an increased amount of interest in the recent years for several reasons, like allowing to handle complex data in a natural way, or sophisticated forms of inference. In particular SML techniques are of special interest in biomedical applications, where SML techniques can reason directly over the molecular structure of complex chemical and biochemical compounds. One of the major difficulties in SML is that the basic operations required for to design machine learning algorithms for structured representations, have a high computational complexity. Consequently, techniques for efficiently implementing such operations are key for the application of SML techniques in real-life applications with large complex data.

This paper focuses on feature terms, a generalization of first-order terms that has been introduced in theoretical computer science in order to formalize object-oriented capabilities of declarative languages, and that has been recently

---

* Pedro Meseguer is partially supported by the projects TIN2009-13591-C02-02 and Generalitat de Catalunya 2009-SGR-1434.

M. Milano (Ed.): CP 2012, LNCS 7514, pp. 1004–1012, 2012.

received increased attention for their usefulness in SML applications [3, 5, 14, 15, 16]. The most basic operation among feature terms is subsumption, which determines whether a given feature is more general than another, and is the most essential component for defining machine learning algorithms. Inductive machine learning methods work by generating hypotheses (often in the form of rules) that are generalizations of the training instances being provided. The "generality relation" (subsumption) states whether a hypothesis covers a training instance or not, and thus it is one of the most fundamental operations in inductive machine learning. It is well known that subsumption between feature terms has a high computational cost if we allow set-valued features in feature terms [7] (necessary to represent most structured machine learning datasets).

Constraint Programming (CP) is a very suitable technique to implement subsumption. We present the CP modelization of the above operation for set-valued feature terms. In some cases, our CP implementation of feature term subsumption provides speed-ups of orders of magnitude with respect to the standard subsumption approach. In addition, when this CP implementation is enhanced with symmetry breaking constraints that exploit basic variable symmetries in feature terms, we obtain substantial extra gains in performance. Our CP implementation uses JaCoP (an open-source constraint library for Java) [10].

We are aware of a previous use of CP to compute θ-subsumption in ML [13]. However, feature term subsumption is significantly different from θ-subsumption, which is defined as the existence of a variable substitution between logical clauses without considering essential elements of feature terms such as sets or loops [15].

2   Background

Feature Terms. Feature terms [2, 7] are a generalization of first-order terms, introduced in theoretical computer science to formalize object-oriented declarative languages. Feature terms correspond to a different subset of first-order logics than description logics, although with the same expressive power [1].

Feature terms are defined by its *signature*: $\Sigma = \langle \mathcal{S}, \mathcal{F}, \leq, \mathcal{V} \rangle$. $\mathcal{S}$ is a set of sort symbols, including the most general sort ("any"), $\leq$ is an order relation inducing a single inheritance hierarchy in $\mathcal{S}$, where $s \leq s'$ means s is more general than or equal to s', for any $s, s' \in \mathcal{S}$ ("any" is more general than any s which, in turn, is more general than "none"). $\mathcal{F}$ is a set of feature symbols, and $\mathcal{V}$ is a set of variable names. We write a feature term ψ as: $\psi ::= X : s \quad [f_1 \doteq \Psi_1, ..., f_n \doteq \Psi_n]$; where ψ points to the *root* variable X (that we will note as $root(\psi)$) of sort s; $X \in \mathcal{V}$, $s \in \mathcal{S}$, $f_i \in \mathcal{F}$, and Ψ_i might be either another variable $Y \in \mathcal{V}$, or a set of variables $\{X_1, ..., X_m\}$. When Ψ_i is a set $\{X_1, ..., X_m\}$, each element in the set must be different. An example of feature term appears in Figure 1. It is a train (variable X_1) composed of two cars (variables X_2 and X_3). This term has 8 variables, and one set-valued feature (indicated by a dotted line): $cars$ of X_1.

To make a uniform description, constants (such as integers) are treated as variables of a particular sort. For each variable X in a term with a constant value k of sort s, we consider that X is a regular variable of a special sort s_k. For each

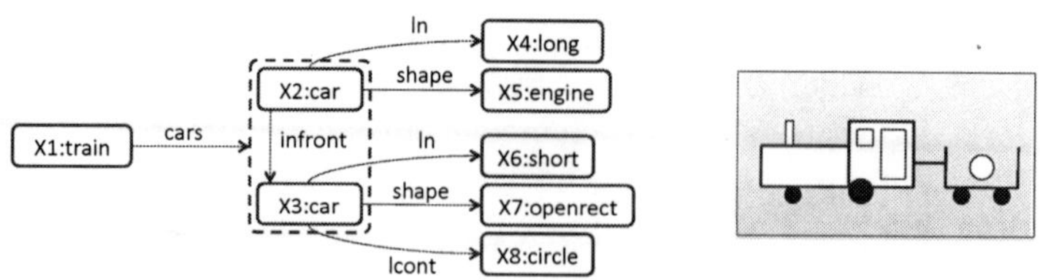

Fig. 1. A simple train represented as a feature term

different constant k, we create a new sort s_k of s. For example, if a variable X had an integer value 5, we would create a new sort s_5 (sub sort of *integer*), and consider that X is a regular variable of sort s_5. Thanks to this representation change, we can forget about constants and just consider all variables in the same way. The set of variables of a term ψ is $vars(\psi)$, the set of features of a variable X is $features(X)$, and $sort(X)$ is its sort.

Feature terms can be represented as directed labelled graphs. Given a variable X, its parents are the nodes connected with X by incoming links, and its children are the nodes connected with X by outgoing links.

Operations on Feature Terms. The basic operation between feature terms is *subsumption*: whether a term is more general than (or equal to) another one.

Definition 1. *(Subsumption) A feature term ψ_1 subsumes another one ψ_2 ($\psi_1 \sqsubseteq \psi_2$)* [1] *when there is a total mapping $m: vars(\psi_1) \to vars(\psi_2)$ such that:*

- *$root(\psi_2) = m(root(\psi_1))$*
- *For each $X \in vars(\psi_1)$*
    - *$sort(X) \leq sort(m(X))$,*
    - *for each $f \in features(X)$, where $X.f = \Psi_1$ and $m(X).f = \Psi_2$:*
        * *$\forall Y \in \Psi_1, \exists Z \in \Psi_2 | m(Y) = Z$,*
        * *$\forall Y, Z \in \Psi_1, Y \neq Z \Rightarrow m(Y) \neq m(Z)$*
        i.e. each variable in Ψ_1 is mapped in Ψ_2, and different variables in Ψ_1 have different mappings.

Subsumption induces a partial order among feature terms, i.e. the pair $\langle \mathcal{L}, \sqsubseteq \rangle$ is a *poset* for a given set of terms $\mathcal{L}$ containing the infimum $\perp$ and the supremum $\top$ with respect to the subsumption order. It is important to note that while subsumption in feature terms is related to θ-subsumption (the mapping m above represents the variable substitution in θ-subsumption), there are two key differences: sorted variables, and semantics of sets (notice that two variables in a set cannot have the same mapping, whereas in θ-subsumption there is no restriction in the variable substitutions found for subsumption).

Since feature terms can be represented as labelled graphs, it is natural to relate the problem of feature terms subsumption to subgraph isomorphism.

---

[1] In description logics notation, subsumption is written in the reverse order since it is seen as "set inclusion" of their interpretations. In machine learning, $A \sqsubseteq B$ means that A is more general than B, while in description logics it has the opposite meaning.

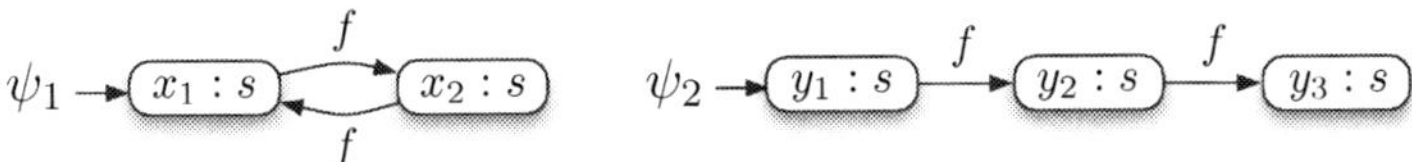

Fig. 2. A bigger feature term subsumes a smaller feature term: $\psi_2 \sqsubseteq \psi_1$

However, subsumption cannot be modeled as subgraph isomorphism because, larger feature terms can subsume smaller feature terms while the corresponding graphs are not isomorphic. See for example the two terms shown in Figure 2, where a term ψ_2 with three variables subsumes a term ψ_1 with two variables (mapping: $m(y_1) = x_1$, $m(y_2) = x_2$, $m(y_3) = x_1$).

Constraint Satisfaction. A Constraint Satisfaction Problem (CSP) involves a finite set of variables, each taking a value in a finite discrete domain. Subsets of variables are related by constraints that specify permitted value tuples. Formally,

Definition 2. *A CSP is a tuple* $(\mathcal{X}, \mathcal{D}, \mathcal{C})$*, where* $\mathcal{X} = \{x_1, \ldots, x_n\}$ *is a set of* n *variables;* $\mathcal{D} = \{D(x_1), \ldots, D(x_n)\}$ *is a collection of finite discrete domains,* $D(x_i)$ *is the set of* x_i*'s possible values;* $\mathcal{C}$ *is a set of constraints. Each constraint* $c \in \mathcal{C}$ *is defined on the ordered set of variables* $var(c)$ *(its scope). Value tuples permitted by* c *are in* $rel(c) \subseteq \prod_{x_j \in var(c)} D(x_j)$*.*

A *solution* is an assignement of values to variables such that all constraints are satisfied. CSP solving is NP-complete.

3   Variable Symmetry in Feature Terms

A *variable symmetry* in feature term ψ is a bijective mapping $\sigma : vars(\psi) \to vars(\psi)$ such that applying σ on term ψ does not modify ψ in any significant way. Often, a basic form of variable symmetry, called *interchangeable variables,* appears in feature terms.[2] Formally,

Definition 3. *Two variables* X *and* Y *of* $vars(\psi)$ *are interchangeable in* ψ *if exchanging* X *and* Y *in* ψ*, the resulting term does not suffer any syntactic change with respect to the original* ψ*.*

Clearly, if X and Y are interchangeable, none of them can be the *root* of ψ. In addition they have to share the same sort, $sort(X) = sort(Y)$. It is easy to see that two variables are interchangeable if and only if they have the same parents and the same children, as proved next.

Proposition 1. *Two variables* X *and* Y *of* $vars(\psi)$ *are interchangeable in* ψ *if and only if they are of the same sort, with the same parents and the same children in* ψ *through the same features.*

---

[2] In CSP terms, this type of symmetry is similar to that between pairs of CSP variables in a graph coloring clique, all variables sharing the same domain.

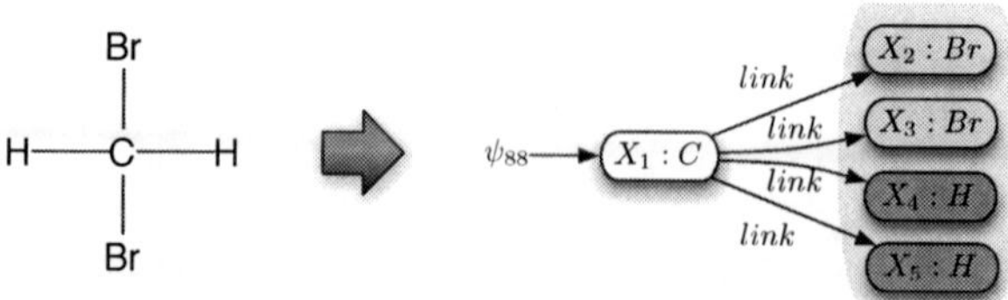

Fig. 3. The chemical structure of methane and its representation as feature term. Variables X_2, X_3, X_4 and X_5 are in the same set. X_2 is interchangeable with X_3, and X_4 is interchangeable with X_5.

Proof. Obviously, if X and Y are of the same sort, with the same parents and children through the same features, exchanging X and Y does not cause any syntactical change in ψ, so they are interchangeable.

If X and Y are not of the same sort, exchanging them causes syntactical changes in ψ. Assuming they share the same sort, if they do not have the same parents or the same children, exchanging X and Y causes syntactical changes in ψ. The same happens when although having the same parents and children, they are connected to them by diferent features. $\square$

Figure 3 shows an example of interchangeable variables in a feature term containing the chemical structure of methane. It happens that Br atoms are all equivalent, to they can be permuted freely without any change in the problem. The same happens with H atoms. Observe that a Br atom is not interchangeable with a H atom, because they are of different sort. As result, variables X_2 and X_3 are interchangeable, and also X_4 with X_5.[3]

4   Subsumption as Constraint Satisfaction

Testing subsumption between feature terms ψ_1 and ψ_2 can be seen as a CSP:

- CSP Variables: for each feature term variable $X \in vars(\psi_1)$ there is a CSP variable x that contains its mapping $m(X)$ in ψ_2. To avoid confusion between the two types of *variables*, feature term variables are written uppercase while CSP variables are written lowercase, the same letter denotes corresponding variables (x is the CSP variable that represents feature term variable X).[4]
- CSP Domains: the domain of each CSP variable is the set $vars(\psi_2)$, except for the CSP variable of $root(\psi_1)$, whose domain is the singleton $\{root(\psi_2)\}$.
- CSP Constraints: three types of constraints are posted
  - Constraints on sorts: for each $X \in vars(\psi_1)$, $sort(X) \leq sort(x)$.

---

[3] Exchanging a Br atom with a H atom would generate another, more elaborated, symmetry than the one we consider. Here we restrict ourselves to the most basic symmetry notion. Nevertheless, exploitation of this kind of symmetry results in very good savings.

[4] For X we use "feature term variable" or "variable". For x we use "CSP variable".

- Constraints on features: for each variable $X \in vars(\psi_1)$ and feature $f \in features(X)$, for each variable $Y \in X.f$ there exists another variable $Z \in x.f$ such that $y = Z$.
- Constraints on difference: If $X.f = \{Y_1, ..., Y_k\}$, where all Y_i's are different by definition, the constraint $all\text{-}different(y_1, ...y_k)$ must be satisfied.

Since ψ_1 and ψ_2 have a finite number of variables, it is direct to see that there is a finite number of CSP variables (exactly $|vars(\psi_1)|$) and all their domains are finite (assuming that the CSP variable x_1 correspond to $root(\psi_1)$, the domain of x_1 will be $\{root(\psi_2)\}$, and the common domain of the other CSP variables is the set $vars(\psi_2)$). If n is the maximum number of variables and m is the maximum number of features, the maximum number of constraints is:

- n unary constraints on sorts (one per CSP variable),
- $O(n^2 m)$ binary constraints on features (number of possible pairs of variables times the maximum number of features),
- $O(nm)$ n-ary constraints on difference (number of variables, each having one $all\text{-}different$ constraint, times the maximum number of features).

Constraints on sorts can be easily tested using the $\leq$ relation amongst sorts; constraints on features and of difference are directly implemented since they just involve the basic tests of equality and difference. Moreover, notice that it is trivial to verify that if the previous constraints are satisfied, the definition of subsumption is satisfied and vice versa. Therefore, the previous CSP problem is equivalent to subsumption in feature terms.

In practice, n varies from a few variables in simple machine learning problems to up to hundreds or thousands for complex biomedical datasets. Most machine learning datasets do not have more than a few different feature labels, and thus m usually stays low. Moreover, in practice, the actual number of constraints is far below its maximum number as computed above.

Consequently, a CP implementation of feature terms subsumption is feasible. We have done it and the results are detailed in Section 5.

4.1   Interchangeable Variables

It is well-known that symmetry explotation can dramatically speed-up CP implementations because it causes substantial search reductions [9]. In this section we explore the simplest variable symmetry in feature terms, variable interchangeability (Definition 3), inside the CP model of feature set subsumption.

Imagine that we want to test $\psi_1 \sqsubseteq \psi_2$ and there are two interchangeable variables X and Y in ψ_1. Interchangeability implies that they have the same parents through the same labels, so they are in the same set. Consequently, $m(X) \neq m(Y)$. Since X and Y are interchangeable, any mapping m satisfying the subsumption conditions for X will also be valid for Y. Therefore, assuming that $m(X) < m(Y)$, there is another mapping m' (symmetric to m) that is equal to m except that permutes the images of X and Y, $m'(X) = m(Y)$ and $m'(Y) = m(X)$. Obviously, $m'(X) > m'(Y)$. Since m and m' are symmetric mappings,

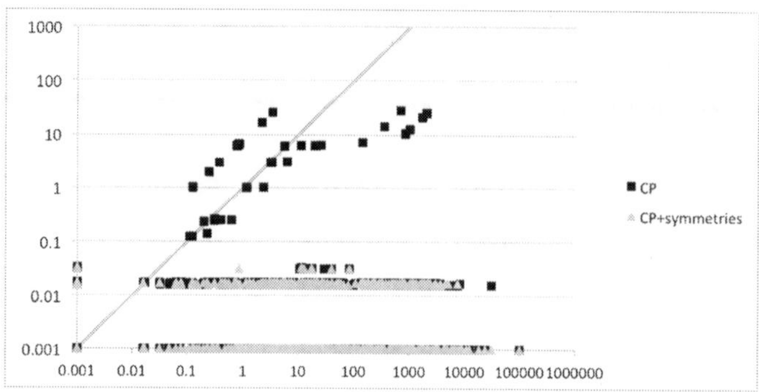

Fig. 4. Time required to compute subsumption in real-world instances. Horizontal axis: time required to compute subsumption by a standard approach; vertical axis: time required to compute subsumption by a CP approach; dots represent subsumption instances. Above the grey line, instances for which the standard approach is faster than the CP implementation, below the grey line the opposite occurs. Square dots correspond to CP implementation without any symmetry breaking, while triangular dots correspond to CP implementation with symmetry breaking constraints.

we choose one of them by adding the symmetry breaking constraint $m(X) < m(Y)$. In consequence, for any pair X, Y of interchangeable variables in ψ_1 we add a symmetry breaking constraint $m(X) < m(Y)$. Often we found subsets $X_1, ..., X_k$ of mutually interchangeable variables in ψ_1, which are also under the constraint $alldifferent(m(X_1), ..., m(X_k))$. Thus, this would add a quadratic number of symmetry breaking constraints:

$$m(X_i) < m(X_j) \quad i : 1..k - 1 \quad j : i + 1..k$$

However, as Puget pointed out in [17], many of these constraints are redundant and it is enough to add a linear number of symmetry breaking constraints:

$$m(X_i) < m(X_{i+1}) \quad i : 1..k - 1$$

to break all symmetries among interchangeable variables.

5   Experimental Results

In order to evaluate our model, we compared the time required to compute subsumption by a standard implementation of subsumption in feature terms [4] with (i) our CP implementation, and with (ii) that CP implementation enhanced with symmetry breaking constraints. We generated 1500 pairs of feature terms using the examples in two relational machine learning data sets as the source of terms: *trains* and *predictive toxicology*. The trains data set was originally introduced by Michalsky as a structured machine learning challenge [11]. Each instance represents a train (different instances have different number of cars, cargos and other properties). Since the size of each instance is different, this dataset cannot be represented using a standard propositional representation, and a relational machine learning representation is required. In the toxicology dataset [5], each instance represents the chemical structure (atoms and their links) of a chemical compound. This is a very complex data set with some instances representing chemical compounds with a large number of atoms. The terms used in our experiments contain between 5 and 138 variables each, and some of them have up to 76 variables belonging to some set.

Figure 4 shows the results of our experiments, where each dot represents one of the 1500 pairs of terms used for our evaluation. The horizontal axis (in a logarithmic scale), shows the time in seconds required by the traditional method, and the vertical axis (also in a logarithmic scale), shows the time required using CP. Square dots correspond to the CP implementation without symmetry breaking, while triangle dots are for CP with symmetry breaking constraints. Dots that lay below the grey line correspond to problems where CP is faster.

We observe that the CP implementation without symmetry breaking is in general faster than the traditional approach because most square dots are below the grey line (in 56 cases out of 1500, the CP implementation is slower than the traditional method). When adding symmetry breaking constraints we observe a dramatic efficiency improvement: almost all triangles are below the grey line (only 34 triangles are above that line) and all instances are solved in less than 0.1 seconds. In some instances there are increments of up to 8 orders of magnitude (observe the triangle in the horizontal axis located at 100000). Adding the time required to perform all the 1500 tests, the traditional method required 669581 seconds, the CP implementation without symmetry breaking required 345.5 seconds, and CP with symmetry breaking lasted 8.3 seconds (4 and 5 orders of magnitude improvement with respect to the traditional method).

Although benefits may vary in other datasets (different from trains [11] and toxicology [5]), these results clearly show the benefits of the CP approach with respect to the traditional method, and the extra benefits we obtain by adding symmetry breaking constraints to the CP implementation. The type of symmetries exploited in our approach is quite simple, but they are very frequent in biomedical data, where feature terms typically represent molecules.

6   Conclusions

A key obstacle when applying relational machine learning and ILP techniques to complex domains is that the basic operations like subsumption have a high computational cost. We presented modeling contributions, including basic variable symmetry exploitation, that allowed us to implement subsumption using CP. As result, this operation is done more efficiently than using traditional methods.

As future work, we would like to improve our CP models to further increase performance. The study of more elaborated symmetries seems to be a promising avenue for research. We would also like to assess the gain in performance of inductive learning algorithms using our CP-based solutions.

References

[1] Aït-Kaci, H.: Description logic vs. order-sorted feature logic. Description Logics (2007)

[2] Aït-Kaci, H., Podelski, A.: Towards a meaning of LIFE. Tech. Rep. 11, Digital Research Laboratory (1992)

[3] Aït-Kaci, H., Sasaki, Y.: An axiomatic approach to feature term generalization. In: Proc. 12th ECML, pp. 1–12 (2001)

[4] Arcos, J.L.: The NOOS representation language. Ph.D. thesis, Universitat Politècnica de Catalunya (1997)

[5] Armengol, E., Plaza, E.: Lazy learning for predictive toxicology based on a chemical ontology. In: Artificial Intelligence Methods and Tools for Systems Biology, vol. 5, pp. 1–18 (2005)

[6] Baader, F., Calvanese, D., McGuinness, D.L., Nardi, D., Patel-Schneider, P.F. (eds.): The Description Logic Handbook: Theory, Implementation, and Applications. Cambridge University Press (2003)

[7] Carpenter, B.: The Logic of Typed Feature Structures. Cambridge Tracts in Theoretical Computer Science, vol. 32. Cambridge University Press (1992)

[8] Dietterich, T., Domingos, P., Getoor, L., Muggleton, S., Tadepalli, P.: Structured machine learning: the next ten years. Machine Learning, 3–23 (2008)

[9] Gent, I., Petrie, K.E., Puget, J.F.: Symmetry in constraint programming. In: Rossi, F., van Beek, P., Walsh, T. (eds.) Handbook of Constraint Programming (2006)

[10] Kuchcinski, K.: Constraint-driven scheduling and resource assignment. ACM Transactions on Design Automaton of Electronic Systems 8, 355–383 (2003)

[11] Larson, J., Michalski, R.S.: Inductive inference of vl decision rules. SIGART Bull. (63), 38–44 (1977)

[12] Lavrač, N., Džeroski, S.: Inductive Logic Programming. Techniques and Applications. Ellis Horwood (1994)

[13] Maloberti, J., Sebag, M.: θ-Subsumption in a Constraint Satisfaction Perspective. In: Rouveirol, C., Sebag, M. (eds.) ILP 2001. LNCS (LNAI), vol. 2157, pp. 164–178. Springer, Heidelberg (2001)

[14] Ontañón, S., Plaza, E.: On Similarity Measures Based on a Refinement Lattice. In: McGinty, L., Wilson, D.C. (eds.) ICCBR 2009. LNCS, vol. 5650, pp. 240–255. Springer, Heidelberg (2009)

[15] Ontañón, S., Plaza, E.: Similarity measuress over refinement graphs. Machine Learning Journal 87, 57–92 (2012)

[16] Plaza, E.: Cases as Terms: A Feature Term Approach to the Structured Representation of Cases. In: Aamodt, A., Veloso, M.M. (eds.) ICCBR 1995. LNCS (LNAI), vol. 1010, pp. 265–276. Springer, Heidelberg (1995)

[17] Puget, J.F.: Breaking symmetries in all-different problems. In: IJCAI, pp. 272–277 (2005)

Author Index

Printed by Printforce, the Netherlands